The Multibody Systems Approach to Vehicle Dynamics

The Multibody Systems Approach to Vehicle Dynamics

Second Edition

Mike Blundell

Damian Harty

Faculty of Engineering and Computing,
Coventry University, Coventry, UK

AMSTERDAM • BOSTON • HEIDELBERG • LONDON
NEW YORK • OXFORD • PARIS • SAN DIEGO
SAN FRANCISCO • SINGAPORE • SYDNEY • TOKYO
Butterworth-Heinemann is an imprint of Elsevier

Butterworth-Heinemann is an imprint of Elsevier
The Boulevard, Langford Lane, Kidlington, Oxford, OX5 1GB, UK
225 Wyman Street, Waltham, MA 02451, USA

First edition 2004

ISBN: 978-0-08-099425-3

British Library Cataloguing in Publication Data
A catalogue record for this book is available from the British Library

Library of Congress Cataloging-in-Publication Data
A catalog record for this book is available from the Library of Congress

For information on all Butterworth-Heinemann publications
visit our website at http://store.elsevier.com/

Typeset by TNQ Books and Journals
www.tnq.co.in

Printed and bound in the UK

Contents

Preface

This book, the second edition, is intended to bridge a gap between the subject of classical vehicle dynamics and the general-purpose computer-based discipline multibody systems (MBS) analysis. Whilst there are several textbooks that focus entirely on the subject, and mathematical foundations, of vehicle dynamics and other more recent texts dealing with MBS there are none yet that link the two subjects in a comprehensive manner.

After 10 years a second edition of this book is indeed timely. Since the first edition there have been notable developments in the understanding and use of active systems, tyre modelling and the use of MBS software.

MBS analysis became established as a tool for engineering designers during the 1980s in a similar manner to the growth in finite element analysis technology during the previous decade. A number of computer programs were developed and marketed to the engineering industry, such as MSC ADAMS™ (*Automatic Dynamic Analysis of Mechanical Systems*), which in this edition still forms the basis for many of the examples provided. During the 1990s MBS became firmly established as part of the vehicle design and development process. It is inevitable that the engineer working on problems involving vehicle ride and handling in a modern automotive environment will be required to interface with the use of MBS to simulate vehicle motion. During the last 10 years several other MBS programmes have become more established, most notably SIMPACK which appropriately receives more coverage in this edition.

The book is aimed at a wide audience including not only undergraduate, postgraduate and research students working in this area, but also practising engineers in industry requiring a reference text dealing with the major relevant areas within the discipline.

The book was originally planned as an individual effort on the part of Mike Blundell drawing on past experience consulting on and researching into the application of MBS to solve a class of problems in the area of vehicle dynamics. From the start it was clear that a major challenge in preparing a book on this subject would be to provide meaningful comment on not only the modelling techniques but also the vast range of simulation outputs and responses that can be generated. Deciding whether a vehicle has good or bad handling characteristics is often a matter of human judgement based on the response or feel of the vehicle, or how easy the vehicle is to drive through certain manoeuvres. To a large extent automotive manufacturers still rely on track measurements and the instincts of experienced test engineers as to whether the design has produced a vehicle with the required handling qualities. To address this problem the book has been co-authored by Damian Harty. At the time of writing the first edition Damian was the Chief Engineer — Dynamics at Prodrive. In the 10 years since the first edition he continued in that role and after a few years working as a Senior Research fellow at Coventry University he moved to his current position with Polaris where he enjoys the additional challenge of modelling vehicles on wide ranging terrain. With experience not only in the area of computer simulation but also the in the practical development and testing of vehicles on the proving ground Damian continues to help in documenting the realistic application of MBS in vehicle development.

Chapter 1 is intended to document the emergence of MBS and provide an overview of its role in vehicle design and development. Previous work by contributors including Olley, Segel, Milliken, Crolla and Sharp is identified providing a historical perspective on the subject during the latter part of the twentieth century.

Chapter 2 is included for completeness and covers the underlying formulations in kinematics and dynamics required for a good understanding of MBS formulations. A three-dimensional vector approach is used to develop the theory, this being the most suitable method for developing the rigid body equations of motion and constraint formulations described later.

Chapter 3 covers the modelling, analysis and postprocessing capabilities of a typical simulation software. There are many commercial programs to choose from including not only MSC ADAMS but also other software packages such as DADS and SIMPACK. The descriptions provided in Chapter 3 are based in the main on MSC ADAMS; the main reason for this choice being that the two authors have between them 25 years of experience working with the software. The fact that the software is also well established in automotive companies and academic institutions worldwide is also a factor. It is not intended in Chapter 3 to provide an MSC ADAMS primer. There is extensive user documentation and training material available in this area from the program vendors MSC Software. The information included in Chapter 3 is therefore limited to that needed to introduce a new reader to the subject and to provide a supporting reference for the vehicle modelling and analysis methodologies described in the following chapters. As discussed, the emergence of SIMPACK and its growing use by the automotive community has led to additional examples to illustrate the modelling approaches with that software.

Existing users of MSC ADAMS will note that the modelling examples provided in Chapter 3 are based on a text-based format of model inputs, known in MSC ADAMS as solver data sets. This was the original method used to develop MSC ADAMS models and has subsequently been replaced by a powerful graphical user interface (GUI) known as ADAMS/View™ that allows model parameterisation, and design optimisation studies. The ADAMS/View environment is also the basis for customised versions of MSC ADAMS such as ADAMS/Car™ that are becoming established in industry and are also discussed in Chapter 3. The use of text-based data sets has been adopted here for a number of reasons. The first of these is that the GUI of a modern simulation program such as MSC ADAMS is subject to extensive and ongoing development. Any attempt to describe such a facility in a textbook such as this would become outdated after a short passage of time. As mentioned, the software developers provide their own user documentation covering this in any case. It is also clear that the text-based formulations translate more readily to book format and are also useful for demonstrating the underlying techniques in planning a model, preparing model schematics and establishing the degrees of freedom in a system model. These techniques are needed to interpret the models and data sets that are described in later chapters and appendices. It is also hoped that by treating the software at this fundamental level the dependence of the book on any one software package is reduced and that the methods and principles will be adaptable for practitioners using alternative software. Examples of the later ADAMS/View command file format are included in Chapters 6 and 8 for completeness.

Chapter 4 addresses the modelling and analysis of the suspension system. An attempt has been made to bridge the gap between the textbook treatment of suspension systems and the MBS approach to building and simulating suspension models. As such a number of case studies have been included to demonstrate the application of the models and their use in the vehicle design process. The chapter concludes with an extensive case study comparing a full set of analytical calculations, using the vector-based methods introduced in Chapter 2, with the output produced from MSC ADAMS. It is intended that this exercise will demonstrate to readers the underlying computations in process when running an MBS simulation.

Chapter 5 addresses the tyre force and moment generating characteristics and the subsequent modelling of these in an MBS simulation. As a major area of importance it deserves to be the largest chapter in the book. Examples are provided of tyre test data and the derived parameters for established tyre models. The chapter concludes with a case study using an MBS virtual tyre test machine to interrogate and compare tyre models and data sets. Since the first edition new tyre models such as the FTire model from Gipser and the TAME Tire model from Michelin have become established and therefore receive a more extended coverage in this edition.

Chapter 6 describes the modelling and assembly of the rest of the vehicle, including the anti-roll bars and steering systems. Near the beginning a range of simplified suspension modelling strategies for the full vehicle is described. This forms the basis for subsequent discussion involving the representation of the road springs and steering system in simple models that do not include a model of the suspension linkages. The chapter includes a consideration of modelling driver inputs to the steering system using several control methodologies and concludes with a case study comparing the performance of several full vehicle modelling strategies for a vehicle handling manoeuvre.

Chapter 7 deals with the simulation output and interpretation of results. An overview of vehicle dynamics for travel on a curved path is included. The classical treatment of understeer/oversteer based on steady state cornering is presented followed by an alternative treatment that considers yaw rate and lateral acceleration gains. The subjective/objective problem is discussed with consideration of steering feel and roll angle as subjective modifiers. The chapter concludes with a consideration of the use of analytical models with a signal-to-noise approach.

Chapter 8 concludes with a review of the use of active systems to modify the dynamics in modern passenger cars. The use of electronic control in systems such as active suspension and variable damping, brake-based systems, active steering systems, active camber systems and active torque distribution is described. A final summary matches the application of these systems with driving styles described as normal, spirited or the execution of emergency manoeuvres.

Appendix A contains a full set of vehicle model schematics and a complete set of vehicle data that can be used to build suspension models and full vehicle models of varying complexity. The data provided in Appendix A were used for many of the case studies presented throughout the book.

Appendix B contains example Fortran Tire subroutines to supplement the description of the tyre modelling process given in Chapter 5. A subroutine is included that uses a general interpolation approach using a cubic spline fit through measured tyre test data. The second subroutine is based on Version 3 of the Magic Formula and has an embedded set of tyre parameters based on the tyre data described in Chapter 5. A final subroutine 'The Harty Model' was developed by Damian at Prodrive and is provided for readers who would like to experiment with a new tyre model that uses a reduced set of model parameters and can represent combined slip in the tyre contact patch.

In conclusion it seems to the authors there still remain two camps for addressing the vehicle dynamics problem. In one is the practical ride and handling expert. The second camp contains theoretical vehicle dynamics experts. This book is aimed at the reader who, like the authors, seeks to live between the two camps and move forward the process of vehicle design, taking full advantage of the widespread availability of convenient digital computing.

There is, however, an enormous difficulty in achieving this end. Lewis Carroll, in *Alice through the Looking Glass*, describes an encounter between Alice and a certain Mr H Dumpty:

> *'When I use a word', Humpty Dumpty said, in rather a scornful tone, 'it means just what I choose it to mean—neither more nor less'.*

> *'The question is', said Alice, 'whether you can make words mean so many different things'.*

There is a similar difficulty between practical and theoretical vehicle dynamicists and even between different individuals of the same persuasion. The same word is used, often without definition, to mean just what the speaker chooses. There is no universal solution to the problem save for a thoughtful and attentive style of discussion and enquiry, taking pains to establish the meanings of even apparently obvious terms such as 'camber' — motorcycles do not have any camber by some definitions (vehicle-body-referenced) and yet to zero the camber forces in a motorcycle tyre is clearly folly. A glossary is included in Appendix C, not as some declaration of correctness but as an illumination for the text. In this edition a new appendix has been added. Appendix D lists some of the test procedures defined by the International Standards Organisation that are used to validate the handling performance of a new vehicle.

Mike Blundell, Damian Harty
April 2014

Acknowledgements

Mike Blundell

In developing my sections of this book I am indebted to my colleagues and students at Coventry University who have provided encouragement and material that I have been able to use. In particular I thank Barry Bolland and Peter Griffiths for their input to Chapter 2 and Bryan Phillips for his help with Chapter 5. I am also grateful to many within the vehicle dynamics community who have made a contribution including Roger Williams, Jim Forbes, Adrian Griffiths, Colin Lucas, John Janevic and Grahame Walter. I am especially grateful to the late David Crolla. He was an inspiration to me as my career took me into the area of vehicle dynamics and my mentor during the preparation of the first edition of this book. I will never forget him. Finally I thank the staff at Elsevier Science for their patience and help throughout the years it has taken to bring both editions of this book to print.

Damian Harty

Mike's gracious invitation to join him and infectious enthusiasm for both the topic and this project has kept me buoyed. At the time of the first edition I acknowledged Robin Sharp, Doug and Bill Milliken for keeping me grounded and rigorous when it is tempting just to play in cars and jump to conclusions. Bill Milliken in particular made a significant contribution to the discipline for an astonishingly long period; Bill passed away in 2012 after a fruitful and remarkable 101-year life that included driving at speed up the hill at Goodwood in 2002 and again in 2007. For those unfamiliar with the clarity of his thinking and the vivacity of a life lived to the full, his autobiography 'Equations of Motion' is an excellent read. Bill's legacy persists with his son Doug continuing to run Milliken Research Associates in Buffalo, NY, USA.

During our work on the first edition of this book the late David Crolla was an ever-present voice of reason keeping this text focused on its *raison d'etre*—the useful fusion of practical and theoretical vehicle dynamics. Professional colleagues who have used banter, barracking and sometimes even rational discussion to help me progress my thinking are too numerous to mention—apart from Duncan Riding, whom I have to single out as being exceptionally encouraging. I hope I show my gratitude in person and on a regular basis to all of them and invite them to kick me if I do not. Someone who must be mentioned is Isaac Newton; his original and definitive brilliance at describing my world amazes me everyday. As Mike, I thank the staff at Elsevier Science for their saintly patience.

Finally, I would just like to say I am very sorry for all the vehicles I have damaged while 'testing' them. I really am.

Nomenclature

a1, a2, a3	Distances for six-mass approximation
a, b	Distance from CM to front and rear axles, respectively
$a_{11}...a_{22}$	Elements of a matrix (generic)
$\{a_I\}_1$	Unit vector at marker I resolved parallel to frame 1 (GRF)
$\{a_J\}_1$	Unit vector at marker J resolved parallel to frame 1 (GRF)
a_x	Longitudinal acceleration (Wenzel model)
a_y	Lateral acceleration (Wenzel model)
b	Longitudinal distance of body mass centre from front axle
c	Damping coefficient
c	Longitudinal distance of body mass centre from rear axle
c	Specific heat capacity of brake rotor
d	Wire diameter
$\left(\frac{dB}{dF_z}\right)$	Variation in scaling factor with load (Harty Model)
$\{d_{IJ}\}_1$	Position vector of marker I relative to J resolved parallel to frame 1 (GRF)
e_1	Path error
f	Natural frequency (Hz)
g	Gravitational acceleration
h	Brake rotor convection coefficient
h	Height of body mass centre above roll axis
i	Square root of -1
k	Path curvature
k	Radius of gyration
k	Stiffness
k	Spring constant in hysteretic model
k	Tyre spring constant
k_1, k_2	Front and rear ride rates, respectively
k_s	Spring stiffness
k_w	Stiffness of equivalent spring at the wheel centre
l	Length of pendulum
m	Mass of a body
$m\{g\}_1$	Weight force vector for a part resolved parallel to frame 1 (GRF)
m_t	Mass of tyre
n	Number of active coils
n	Number of friction surfaces (pads)
p	Brake pressure
q_j	Set of part generalised coordinates
r	Yaw rate
r_1, r_2, r_3	Coupler constraint rotations
$\{r_I\}_1$	Position vector of marker I relative to frame i resolved parallel to frame 1 (GRF)
$\{r_J\}_1$	Position vector of marker J relative to frame j resolved parallel to frame 1 (GRF)

r_u	Unladen radius
r_l	Laden radius
r_w	Wheel radius
s_1, s_2, s_3	Coupler constraint scale factors
t_f	Front track
t_r	Rear track
v_{cog}	Centre of gravity (Wenzel model)
v_x	Longitudinal velocity (Wenzel model)
v_y	Lateral velocity (Wenzel model)
x	Generic variable for describing tanh function
x_i, y_i, z_i	Coordinates of each of the six masses in the six-mass approximation
x_i, y_i, z_i	Components of the ith eigenvector
$x(t)$	Function of time (generic)
x_{CM}, y_{CM}, z_{CM}	Coordinates of body centre of mass
$\{x_I\}_1$	Unit vector along x-axis of marker I resolved parallel to frame 1 (GRF)
$\{y_I\}_1$	Unit vector along y-axis of marker I resolved parallel to frame 1 (GRF)
$\{x_J\}_1$	Unit vector along x-axis of marker J resolved parallel to frame 1 (GRF)
$\{y_J\}_1$	Unit vector along y-axis of marker J resolved parallel to frame 1 (GRF)
y_s	Asymptotic value at large slip (Magic Formula)
z	Auxiliary state variable
z	Heave displacement variable
$\{z_I\}_1$	Unit vector along z-axis of marker I resolved parallel to frame 1 (GRF)
$\{z_J\}_1$	Unit vector along z-axis of marker J resolved parallel to frame 1 (GRF)
A	Area
A	Linear acceleration
A, B, C	Intermediate terms in a cubic equation
A	Scaling for solution form of a differential equation (generic)
A	Step height
A_c	Convective area of brake disc
$[A_{1n}]$	Euler matrix for part n
$\{A_n\}_1$	Acceleration vector for part n resolved parallel to frame 1 (GRF)
A^P	Centripetal acceleration
$\{A^P_{PQ}\}_1$	Centripetal acceleration vector P relative to Q referred to frame 1 (GRF)
$\{A^t_{PQ}\}_1$	Transverse acceleration vector P relative to Q referred to frame 1 (GRF)
$\{A^c_{PQ}\}_1$	Coriolis acceleration vector P relative to Q referred to frame 1 (GRF)
$\{A^s_{PQ}\}_1$	Sliding acceleration vector P relative to Q referred to frame 1 (GRF)
$A_{vehicle}$	Acceleration of vehicle

A_X	Longitudinal curvature factor
A_y	Lateral acceleration
AyG	Lateral acceleration gain
B	Load scaling factor (Harty Model)
B	Stiffness factor (Magic Formula)
[B]	Transformation matrix from frame O_e to O_n
BKid	Bottom kingpin marker
BM	Bump movement
B_T	Brake torque
C	Shape factor (Magic Formula)
[C]	Compliance matrix
C_{D0}	Drag coefficient at zero aerodynamic yaw angle
$C_{D\beta}$	Drag coefficient sensitivity to aerodynamic yaw angle
C_F	Front axle cornering stiffness
C_γ	Camber coefficient
C_{L0}	Coefficient of lift at zero angle of attack
$C_{L\alpha}$	Variation in coefficient of lift with angle of attack
C_{MX}	Overturning moment coefficient
C_r	Rolling resistance moment coefficient
C_R	Rear axle cornering stiffness
C_S	Tyre longitudinal stiffness
C_p	Process capability
CP	Centre of pressure
C_α	Tyre lateral stiffness due to slip angle
$C_{\alpha f}$	Front tyre lateral stiffness due to slip angle
$C_{\alpha r}$	Rear tyre lateral stiffness due to slip angle
C_γ	Tyre lateral stiffness due to camber angle
D	Clipped camber scale constant
D	Mean coil diameter
D	Peak value (Magic Formula)
DZ	Displacement variable (generic)
DM(I,J)	Magnitude of displacement of I marker relative to J marker
DX(I,J)	Displacement in X-direction of I marker relative to J marker parallel to GRF
DY(I,J)	Displacement in Y-direction of I marker relative to J marker parallel to GRF
DZ(I,J)	Displacement in Z-direction of I marker relative to J marker parallel to GRF
E	Camber clip curvature constant
E	Young's modulus of elasticity
E	Curvature factor (Magic Formula)
F	Aerodynamics force
F	Applied force
F	Force generated by hysteretic model
F	Spring force
Fhyst	Amplitude of hysteretic force
Fhyst	Final outcome from sequence of hysteretic calculations
$\{F_{nA}\}_1$	Applied force vector on part n resolved parallel to frame 1 (GRF)

$\{F_{nC}\}_1$	Constraint force vector on part n resolved parallel to frame 1 (GRF)
F_{FRC}	Lateral force reacted by front roll centre
F_{RRC}	Lateral force reacted by rear roll centre
F_x	Frictional force
F_x	Longitudinal tractive or braking tyre force
F_{x_1}	Friction moderated longitudinal load in moderate slip
F_{x_2}	Friction moderated longitudinal load in deep slip
F_y	Lateral tyre force
F_{Y1}	Friction moderated lateral load at moderate slip angles
F_{Y2}	Friction moderated lateral load at deep slip angles
F'_y	Lagged (relaxed) side force
$F_{y\alpha}$	Lateral load due to slip angle
$F_{y\alpha'}$	Friction moderated side force due to slip angle
$F_{y\gamma}$	Lateral load due to camber/inclination angle
$F_{y\gamma'}$	Friction moderated side force due to camber/inclination angle
$\frac{\hat{F}y}{\mu Fz}$	Lateral capacity fraction
F_z	Normal force
F_z	Vertical tyre force
F_z	Time varying tyre load
F_{z0}	Static corner load
F_{zc}	Vertical tyre force due to damping
F_{zk}	Vertical tyre force due to stiffness
$\{F_A\}_1 \{F_B\}_1\ldots$	Applied force vectors at points A, B,... resolved parallel to frame 1 (GRF)
$[F_E]$	Elastic compliance matrix (concept suspension)
F_D	Drag force
FG	Fixed ground marker
G	Shear modulus
GC	Gravitational constant
GO	Ground level offset
GRF	Ground reference frame
$\{H\}_1$	Angular momentum vector for a body
$H(\omega)$	Transfer function
HTC	Half track change
I	Mass moment of inertia
I	Second moment of area
I_2	Pitch inertia of vehicle
I_1, I_2, I_3	Principal mass moments of inertia of a body
I_{wheel}	Mass moment of inertia of road wheel in the rolling direction
$I_{xx}, I_{yy}, I_{zz}, I_{xy}, I_{yz}, I_{xz}$	Components of inertia tensor
ICY	Y-coordinate of instant centre
ICZ	Z-coordinate of instant centre
$[I_n]$	Inertia tensor for a part
J	Polar second moment of area
Jz	Vehicle body yaw inertia (Wenzel model)
K	Drive torque controller constant
K	Spring stiffness

K	Stability factor
K	Understeer gradient
$\mathbf{K_z}$	Tyre radial stiffness
$\mathbf{K_T}$	Torsional stiffness
$\mathbf{K_{Ts}}$	Roll stiffness due to springs
$\mathbf{K_{Tr}}$	Roll stiffness due to anti-roll bar
L	Contact patch length
L	Length
L	Wheelbase
$\{\mathbf{L}\}_1$	Linear momentum vector for a particle or body
$\mathbf{L_{PFZ2}}$	Pneumatic lead scaling factor with load squared
$\mathbf{L_{PFZ}}$	Pneumatic lead scaling factor with load
$\mathbf{L_{PC}}$	Pneumatic lead at reference load
LPRF	Local part reference frame
$\mathbf{L_R}$	Tyre relaxation length
$\mathbf{M_{FRC}}$	Moment reacted by front roll centre
$\{\mathbf{M_{nA}}\}_e$	Applied moment vector on part n resolved parallel to frame e
$\{\mathbf{M_{nC}}\}_e$	Constraint moment vector on part n resolved parallel to frame e
$\mathbf{M_s}$	Equivalent roll moment due to springs
$\mathbf{M_x}$	Tyre overturning moment
$\mathbf{M_{X\gamma\kappa}}$	Overturning moment due to longitudinal forces
$\mathbf{M_y}$	Moment about y-axis
$\mathbf{M_y}$	Tyre rolling resistance moment
$\mathbf{M_z}$	Tyre self aligning moment
$\mathbf{M_{z\alpha}}$	Friction moderated side force due to slip angle
$\mathbf{M_{z\gamma}}$	Friction moderated side force due to camber/inclination angle
$\mathbf{M_{Z\gamma\kappa}}$	Aligning moment due to longitudinal forces
MRF	Marker reference frame
$\mathbf{M_{RRC}}$	Moment reacted by rear roll centre
$\mathbf{N_r}$	Vehicle yaw moment with respect to yaw rate
$[\mathbf{N_t}]$	Norsieck vector
$\mathbf{N_{vy}}$	Vehicle yaw moment with respect to lateral velocity
$\mathbf{O_1}$	Frame 1 (GRF)
$\mathbf{O_e}$	Euler axis frame
$\mathbf{O_i}$	Reference frame for part i
$\mathbf{O_j}$	Reference frame for part j
$\mathbf{O_n}$	Frame for part n
$\mathbf{O_P}$	Lateral offset of contact patch
$\mathbf{P_0}$	Initial tyre pressure at zero load
$\overline{P}$	Average footprint pressure
$\{\mathbf{P_{nr}}\}_1$	Rotational momenta vector for part n resolved parallel to frame 1 (GRF)
$\{\mathbf{P_{nt}}\}_1$	Translational momenta vector for part n resolved parallel to frame 1 (GRF)
Pt	Constant power acceleration
$\mathbf{P_{\Delta z}}$	Change in nominal pressure
$\mathbf{P_{\Delta z}}$	Pressure due to tyre vertical deflection
QG	Position vector of a marker relative to the GRF

QP	Position vector of a marker relative to the LPRF
R	Radius (generic)
R	Radius of turn
R	Fraction of roll moment distributed between front and rear axles
$\mathbf{R_1}$	Unloaded tyre radius
$\mathbf{R_2}$	Tyre carcass radius
$\mathbf{R_d}$	Radius to centre of brake pad
$\mathbf{R_e}$	Effective rolling radius
$\{\mathbf{R_i}\}_1$	Position vector of frame i on part i resolved parallel to frame 1 (GRF)
$\{\mathbf{R_j}\}_1$	Position vector of frame j on part j resolved parallel to frame 1 (GRF)
$\mathbf{R_l}$	Loaded tyre radius
$\{\mathbf{R_n}\}_1$	Position vector for part n resolved parallel to frame 1 (GRF)
$\{\mathbf{R_p}\}_1$	Position vector of tyre contact point P relative to frame 1, referenced to frame 1
$\mathbf{R_u}$	Unloaded tyre radius
$\{\mathbf{R_w}\}_1$	Position vector of wheel centre relative to frame 1, referenced to frame 1
$\{\mathbf{R_{AG}}\}_n$	Position vector of point A relative to mass centre G resolved parallel to frame n
$\{\mathbf{R_{BG}}\}_n$	Position vector of point B relative to mass centre G resolved parallel to frame n
$\mathbf{RC_{front}}$	Front roll centre
$\mathbf{RC_{rear}}$	Rear roll centre
RCY	Y-coordinate of roll centre
RCZ	Z-coordinate of roll centre
$\mathbf{R_Z}$	Reference load (Harty Model)
S	Distance travelled
SA	Spindle axis reference point
$\mathbf{S_{CX}}$	Critical slip ratio
$\mathbf{S_e}$	Error variation
Sh	Horizontal shift (Magic Formula)
Sv	Vertical shift (Magic Formula)
$\mathbf{S_L}$	Longitudinal slip ratio
$\mathbf{S_L}^*$	Critical value of longitudinal slip
SN	Signal-to-noise ratio
$\mathbf{S_T}$	Total variation
$\mathbf{S_\alpha}$	Lateral slip ratio
$\mathbf{S_{L\alpha}}$	Comprehensive slip ratio
$\mathbf{S_\alpha}^*$	Critical slip angle
$\mathbf{S_\kappa}$	Variation due to linear effect
T	Camber clipping threshold fraction
T	Kinetic energy for a part
T	Temperature
T	Torque
$\mathbf{T_B}$	Brake torque
$\mathbf{T_{env}}$	Environmental temperature

T_{PFZ2}	Pneumatic trail scaling factor with load squared
T_{PFZ}	Pneumatic trail scaling factor with load
T_{PC}	Pneumatic trail scaling constant
T_S	Spin up torque
T_0	Initial brake rotor temperature
$\{T_A\}_1 \{T_B\}_1 \ldots$	Applied torque vectors at points A, B,… resolved parallel to frame 1 (GRF)
TK	Top kingpin marker
TR	Suspension trail
$\{U_r\}$	Unit vector normal to road surface at tyre contact point
$\{U_s\}$	Unit vector acting along spin axis of tyre
UCF	Units consistency factor
US	Understeer
V	Forward velocity
V_0	Initial tyre volume at zero load
V_a	Actual forward velocity
V_e	Error variance
V_g	Ground plane velocity
$V_{lowlimit}$	Limiting velocity
$\{V_n\}_1$	Velocity vector for part n resolved parallel to frame 1 (GRF)
$\{V_p\}_1$	Velocity vector of tyre contact point P referenced to frame 1
V_s	Desired simulation velocity
V_x	Sliding velocity
V_{xc}	Longitudinal slip velocity of tyre contact point
Vy	Lateral slip velocity of tyre contact point
Vz	Vertical velocity of tyre contact point
Vref	Reference velocity in hysteretic model
VR(I,J)	Radial line of sight velocity of I marker relative to J marker
VZ	Velocity variable (generic)
$V_{\Delta z}$	Reduced tyre cavity volume
W	Tyre width
WB	Wheelbase marker
WC	Wheel centre marker
WF	Wheel front marker
WR	Wheel recession
XP	Position vector of a point in a marker xz-plane
$\{X_{sae}\}_1$	Unit vector acting at tyre contact point in X_{sae} direction referenced to frame 1
Y_r	Vehicle side force with respect to yaw rate
Y_{vy}	Vehicle side force with respect to lateral velocity
YRG	Yaw rate gain
$\{Y_{sae}\}_1$	Unit vector acting at tyre contact point in Y_{sae} direction referenced to frame 1
$\{Z_{sae}\}_1$	Unit vector acting at tyre contact point in Z_{sae} direction referenced to frame 1
ZP	Position vector of a point on a marker z-axis
α	Angle of attack
α	Tyre slip angle

α_{CY}	Critical slip angle (Harty Model)
$\{\alpha_n\}_1$	Angular acceleration vector for part n resolved parallel to frame 1 (GRF)
α_f	Front axle slip angle
α_r	Rear axle slip angle
β	Aerodynamic yaw angle (or body slip angle surrogate)
β	Side slip angle
$\dot{\beta}$	Rate of change of side slip angle (Beta Dot)
δ	Steer or toe angle
δ_o	Steer angle of outer wheel
δ_i	Steer angle of inner wheel
δ_{mean}	Average steer angle of inner and outer wheels
γ	Camber angle
$\{\gamma_n\}_e$	Set of Euler angles for part n
ζ	Damping ratio
κ	Longitudinal slip (Pacjeka)
κ	Sensitivity of process
θ	2nd Euler angle rotation
θ	Pendulum displacement variable
θ	Pitch displacement variable
θ_1	Orientation of the first principal axis within a plane of symmetry
λ	Eigenvalue (generic)
$\{\lambda\}_1$	Reaction force vector resolved parallel to frame 1 (GRF)
λ_d	Magnitude of reaction force for constraint d
λ_p	Magnitude of reaction force for constraint p
λ_α	Magnitude of reaction force for constraint α
μ	Friction coefficient
μ_o	Tyre to road coefficient of static friction
μ_1	Tyre to road coefficient of sliding friction
η	Signal-to-noise ratio
η	Hysteresis constant/loss factor
ρ	Density
σ	Standard deviation
σ_d	Standard deviation of attribute d
Φ	3rd Euler angle rotation
ψ	1st Euler angle rotation
ψ	Compass heading angle
$\dot{\psi}$	Yaw rate (Wenzel model)
ω	Angular frequency (rads s^{-1})
ω	Yaw rate
ω_d	Damped natural frequency
ω_d	Demanded yaw rate
ω_{err}	Yaw rate error
ω_{fns}	Front axle no-slip yaw rate
$\omega_{friction}$	Yaw rate from limiting friction
ω_{geom}	Yaw rate from geometry
ω_n	Undamped natural frequency
$\{\omega_e\}_1$	Angular velocity vector for part n resolved parallel to frame e

$\{\omega_n\}_1$	Angular velocity vector for part n resolved parallel to frame 1 (GRF)
ω_0	Angular velocity of free rolling wheel
ω_D	Angular velocity of driven wheel
Δ_d	Allowable range for attribute d
Δx	Change in longitudinal position of wheel (concept suspension)
Δy	Change in lateral position (half/track) of wheel (concept suspension)
Δz	Deformation of tyre
ΔV	Change in tyre cavity volume
$\Delta \varepsilon$	Change in steer angle (toe in/out) of wheel (concept suspension)
$\Delta \gamma$	Change in camber angle of wheel (concept suspension)
$\{\Phi_a\}_1$	Vector constraint equation resolved parallel to frame 1 (GRF)
Φ_d	Scalar constraint expression for constraint d
Φ_p	Scalar constraint expression for constraint p
Φ_α	Scalar constraint expression for constraint α
$\dot{\Omega}_{wheel}$	Angular acceleration of road wheel

Introduction

To undertake a numerical process without knowledge of the difference between good and bad numbers is folly.

Damian Harty, 2014

1.1 Overview

In 1969, man travelled to the moon and back, using mathematics invented by Kepler, Newton and Einstein to calculate trajectories hundreds of thousands of miles long and spacecraft with less on-board computing power than a mobile telephone.[1] With today's computing power and the mathematical frameworks handed down to us by Newton and Lagrange, it is scarcely credible that the motor car, itself over 100 years old, can exercise so many minds and still show scope for improvement. Yet we are still repeating errors in the dynamic design of our vehicles that were made in the 1960s. Every car will spin — rotate excessively about a vertical axis until it is no longer pointing where it is going — if driven through an emergency lane change at highway speed without skilled correction from the driver or from computer-controlled stability systems (or both).

Legislation now demands embedded electronic control of brake systems to retain control and stability of vehicles although the driver still retains executive control in terms of choosing speed and path. However, with the exception of braking performance, road vehicle manufacturers are not currently forced by legislation to achieve a measurable standard of vehicle handling and stability. International standards exist that outline procedures for proving ground tests with new vehicles but these are nothing more than recommendations. Vehicle manufacturers make use of many of the tests but in the main will develop and test vehicles using in-company experience and knowledge to define the test programme.

In the absence of legislated standards, vehicle manufacturers are driven by market forces. Journalists report favourably on vehicles they enjoy driving — whether or

---

[1] The Apollo Guidance Computer (AGC) had 2 kB of memory, 32 kB of non-rewritable flash drive and 1 MHz clock speed. A typical smart phone at the time of writing has 1000 kB of memory, 32 million kB of rewritable flash drive and a clock speed of 1000 MHz.

not these are safe in the hands of the general public — and the legal profession seeks every opportunity to blur the distinction between bad driving and poor vehicle design. Matters are further complicated by market pressures driving vehicle designs to be too tall for their width — city cars and sport-utility vehicles have this disadvantage in common.

The growth in media attention and reporting to the public is undoubtedly significant. When the first edition was written, the most well-publicised example of this was the reported rollover of the Mercedes A Class (top left image in Figure 1.1) during testing by the Motoring Press. The test involves a slalom type manoeuvre and became popularly known as the 'Elk Test' or 'Moose Test'. With the arrival of YouTube in 2005, everyone is a journalist. A short search netted three more examples of well-performed tests illustrating other vehicles in disarray. No manufacturer wants videos like these going viral.

It seems to the authors that there are two camps for addressing the vehicle dynamics problem. In the first camp are the practical ride and handling experts. Skilled at the driving task and able to project themselves into the minds of a variety of different possible purchasers of the vehicle, they are able to quickly take an established vehicle design and adjust its character to make it acceptable for the market into which it will be launched. Rarely, though, are experts from this camp called upon to work in advance on the concept or detail of the vehicle design.

The second camp contains theoretical vehicle dynamics experts. They are skilled academics in the mould of Leonard Segel who in 1956 published his 'Theoretical prediction and experimental substantiation of the responses of the automobile to steering control' (Segel, 1956).

FIGURE 1.1

Two-wheel lift abounds. (Top left — Mercedes courtesy of Auto Motor und Sport). There are two very similar tests being used, one is the ISO3888 Lane Change and the other is an ADAC avoidance test.

FIGURE 1.2

What vehicle dynamics looks like to outsiders.

Segel's work, and that of others from this era including the earlier work 'Road manners of the modern motor car' (Olley, 1945), laid the ground for all subsequent 'classical' vehicle dynamic analysis and forms a firm foundation upon which to build. It is rare for an expert from this camp to be part of the downstream vehicle development process.

These two camps, a little like England and the USA, are 'separated by a common language'.[2] They use similar terms very differently (Figure 1.2) and can often have contemptuous relationships in a given organisation.

The 'word cloud' diagram in Figure 1.2 was generated by taking the contents of this chapter and pasting them into www.wordle.net. After removing some trivial common words, this is what remains. The size of the word reflects how often it appears; it seems that conversations about vehicle dynamics are dominated by talk of tyres and yaw. The latter is a rotation of the vehicle about a vertical axis.

1.2 What is vehicle dynamics?

The field known as vehicle dynamics is concerned with two aspects of the behaviour of the machine. The first is isolation and the second is control. Figure 1.3 is the authors' subjective illustration of the intricacy and interconnection of the tasks to be approached; it is not claimed to be authoritative or complete but is rather intended as a thought-starter for the interested reader.

Isolation is about separating the driver from disturbances occurring as a result of the vehicle operation. This, too, breaks into two topics; disturbances the vehicle generates itself (engine vibration and noise, for example) and those imposed upon it by the outside world. The former category is captured by the umbrella term

---

[2]Often attributed to George Bernard Shaw, this quote cannot actually be found in his writings anywhere according to Wikipedia.

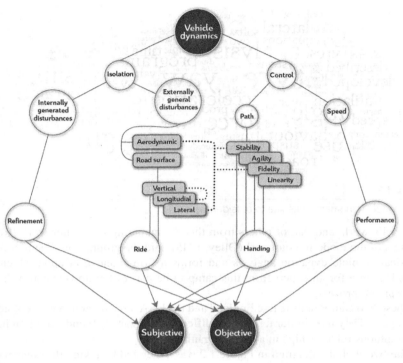

FIGURE 1.3

Vehicle dynamics interactions.

'refinement'. The disturbances in the latter category are primarily road undulations and aerodynamic interaction of the vehicle with its surroundings — crosswinds, wakes of structures and wakes of other vehicles. The behaviour of the vehicle in response to road undulations is referred to as 'ride' and could conceivably be grouped with refinement, though it rarely is in practice.

There is some substantial crossover in aerodynamic behaviour between isolation and control, since control implies the rejection of disturbances ('fidelity') and an absence of their amplification ('stability'). Similarly, one response to road disturbances is a change in the vertical load supported by the tyre; this has a strong influence on the lateral force the tyre is generating at any given instant in time and is thus crucial for both fidelity and stability. It can be seen with some little reflection that one of the difficulties of vehicle dynamics work is not the complexity of the individual effects being considered but rather the complexity of their interactions.

Control is concerned largely with the behaviour of the vehicle in response to driver demands. The driver continuously varies both path curvature and speed, subject to the limits of the vehicle capabilities, in order to follow an arbitrary course.

FIGURE 1.4

Geometric approximations of vehicle behaviour are incorrect.

Speed variation is governed by vehicle mass and tractive power availability at all but the lowest speed, and is easily understood. Within the performance task, issues such as unintended driveline oscillations and tractive force variation with driver demand may interact strongly with the path of the vehicle.

The adjustment of path curvature at a given speed is altogether more interesting. In a passenger car, the driver has a steering wheel, which for clarity will be referred to as a *handwheel*[3] throughout the book. The handwheel is a 'yaw rate' demand — a demand for rotational velocity of the vehicle when viewed from above. The combination of a yaw rate and a forward velocity vector that rotates with the vehicle gives rise to a curved path. There is a maximum path curvature available in normal driving, which is the turning circle, available only at the lowest speeds.

It is generically true that the vehicle does not behave in a 'geometric' manner and its radius of turn cannot usefully be predicted by considering the angle of the front wheels relative to the rear wheels, except below around 30 mph.

The geometric view (Figure 1.4) becomes increasingly inaccurate as speed increases and can lead to an over-estimate of vehicle responses by a factor of up to four at European highway speeds. The lower-than-geometric response of the car is a consequence of pneumatic tyres and modern vehicle engineering practice; it is not necessarily the 'unengineered' behaviour of all vehicle layouts.

In normal circumstances (that is to say in day-to-day road use) the driver moves the handwheel slowly and is well within the limits of the vehicle capability. The vehicle has no difficulty responding to the demanded yaw rate. If the driver increases yaw rate demand slightly then the vehicle will increase its yaw rate by a proportional amount (Figure 1.5). This property is referred to as 'linearity'; the vehicle is described as 'linear'. For the driver, the behaviour of the vehicle is quite instinctive. A discussion of the analysis and interpretation of vehicle non-geometric behaviour, linearity and departure from linearity is given in Chapter 7.

In the linear region, the behaviour of the vehicle can be represented as a connected series of 'steady state' events. Steady state is the condition in which, if the handwheel remains stationary then all the vehicle states — speed, yaw rate, path

---

[3]'Steering Wheel' could mean a roadwheel that is steered, or a wheel held by the driver. In generic discussions including vehicles other than four wheeled passenger cars (motorcycles, tilting tricycles, etc.) 'steering wheel' contains too much ambiguity; therefore 'handwheel' is preferred since it adds precision.

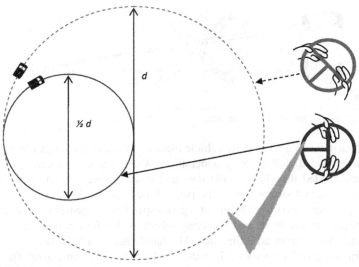

FIGURE 1.5

Linearity: More handwheel input results in proportionally more yaw rate.

curvature and so on — remain constant and is more fully defined in Chapter 7. The steady state condition is easy to represent using an equilibrium analogy, constructed with the help of so-called 'centrifugal force'. It should be noted that this fictitious force is invented solely for convenience of calculation of the analogous equilibrium state or the calculation of forces in an accelerating frame of reference. When a vehicle is travelling on a curved path it is not in equilibrium.

The curved path of the vehicle requires some lateral acceleration. Correctly, the lateral acceleration on a cornering vehicle is a centripetal acceleration — 'centre seeking'. Note that speed is not the same as velocity; travelling in a curved path with a constant speed implies a changing direction and therefore a changing velocity. Even the centripetal acceleration definition causes some problems since everyone

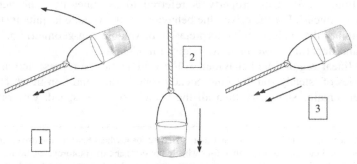

FIGURE 1.6

Thought experiment comparing centripetal acceleration with linear acceleration.

'knows' that they are flung to the outside of a car if unrestrained and so there is much lax talk of centrifugal forces — 'centre fleeing'. To clarify this issue, a brief thought experiment is required. Imagine a bucket of water on a rope being swung around by a subject (Figure 1.6). If the subject looks at the bucket then the water is apparently pressed into the bucket by the mythical 'centrifugal force' (presuming the bucket is being swung fast enough). If the swinging is halted and the bucket simply suspended by the rope then the water is held in the bucket by the downward gravitational field of the earth — the weight of the water pulls it into the bucket. Imagine now a different scenario in which the bucket (on a frictionless plane) is pulled horizontally towards the observer at a constant acceleration in a linear fashion. It's best not to complicate the experiment by worrying about what will happen when the bucket reaches the subject. It is this third scenario and not the second that is useful in constructing the cornering case. If both the first and third cases are imagined in a zero gravity environment, they still work — the water will stay in the bucket. Note that for the third scenario — what we might call the 'inertial' case as against the gravitational case in the second scenario — the acceleration is towards the open end of the bucket. This is also true for the first scenario, in which the bucket is swung; the acceleration is towards the open end of the bucket and is towards the subject — i.e. it is centripetal. That the water stays in the bucket is simply a consequence of the way the bucket applies the centripetal force to the water. Thus the tyres on a car exert a force towards the centre of a turn and the body mass is accelerated by those forces centripetally — in a curved path.

An accelerometer in the car is effectively a load cell that would be between the bucket and the water in the scenarios here and so it measures the centripetal force applied between the calibrated mass within the accelerometer (the water) and its support in the casing (the bucket). The so-called centrifugal force is one half of an action–reaction pair within the system but a free-body diagram of the bucket and rope in all three cases shows tension in the rope as an externally applied force when considering the rope as a separate free body. Only in case 2 is the bucket actually in equilibrium, with the addition of the gravitational force on the bucket and water. Therefore an accelerometer (or an observer) in the vehicle apparently senses a centrifugal force while theoretical vehicle dynamicists talk always of centripetal acceleration. Changing the sign on the inertial force, so that it is now a d'Alembert force, appears to solve the apparent confusion. This can be misleading as we now have the impression that the analysis of the cornering vehicle is a static equilibrium problem. The water is not in equilibrium when travelling in a curved path, and neither is a car.

Centripetal forces accelerate the vehicle towards the centre of the turn. This acceleration, perpendicular to the forward velocity vector, is often referred to as 'lateral' acceleration, since the vehicle broadly points in the direction of the forward velocity vector (see Chapter 7 for a more precise description of the body attitude). It can be seen that the relationship between centripetal acceleration, A^p, yaw rate, ω, forward velocity, V and radius of turn, R is given by:

$$A^p = V^2/R = \omega\,V = \omega^2\,R \qquad (1.1)$$

The absolute limit for lateral acceleration, and hence yaw rate, is set by the friction available between the tyres and the road surface. Competition tyres ('racing slicks') have a coefficient of friction substantially in excess of unity[4] and, together with large aerodynamic downforces, allow a lateral acceleration in the region of 30 m/s^2, with yaw rates correspondingly over 40 deg/s for a speed of 40 m/s (90 mph). For more typical road vehicles, limit lateral accelerations rarely exceed 9 m/s^2, with yaw rates correspondingly down at around 12 deg/s at the same speed. However, for the tyre behaviour to remain substantially linear for a road car, the lateral accelerations must be generally less than about 3 m/s^2, so yaw rates are down to a mere 4 deg/s at the same speed.

While apparently a small fraction of the capability of the vehicle, there is much evidence to suggest that the driving population as a whole rarely exceed the linearity limits of the vehicle at speed and only the most confident exceed them at lower speeds (Lechmer and Perrin, 1993). The 100 Car Naturalistic Driving Study categorises any event greater than 4 m/s^2 as a 'near accident' or 'dangerous occurrence' (Dingus et al., 2004).

When racing or during emergency manoeuvres on the road — typically attempting to avoid an accident — the vehicle becomes strongly 'non-linear'. The handwheel is moved rapidly and the vehicle generally has difficulty in responding accurately to the handwheel. This is the arena called 'Transient Handling' and is correctly the object of many studies during the product design process. In contrast to the steady state condition, all the vehicle states fluctuate rapidly and the expressions above are modified. Steady state and transient behaviour are connected. While good steady state behaviour is connected with good transient behaviour, it is not in itself sufficient (Sharp, 2000).

Transient handling studies concentrate on capturing, analysing and understanding the yaw moments applied to the vehicle and its response to them. Those moments are dominated by the lateral and longitudinal forces from the tyres. For road cars, additional aerodynamic contributions are a small modifier but for racing, the aerodynamic behaviour rises in importance.

The generation of tyre forces is frequently the biggest source of confusion in vehicle dynamics, since both lateral and longitudinal mechanisms are neither obvious nor intuitive. Tyres are dealt with in some depth by Pacejka (2012) in a companion volume in this series and also have some further coverage in Chapter 5.

The tyres generate lateral forces by two mechanisms, 'camber' and 'slip angle'. Camber is the angle at which the tyre is presented to the road when viewed from the front. There exists some confusion when referring to and measuring camber angle; for clarity within this text camber angle is measured with respect to the road[5] unless

---

[4]Some people are troubled by this idea but the authors have never been able to understand why. In any case, the performance of top fuel dragsters should leave no doubt that friction coefficients handsomely above unity exist.

[5]This is referred to as 'inclination angle' in some contexts.

explicitly defined as being relative to the vehicle body. It is the angle with respect to the road that generates a side force. Thus a motorcycle runs a large camber angle when cornering but runs no camber angle with respect to the vehicle body.

Slip angle is the angle at which the moving tyre is presented to the road when viewed in plan. It is important to note that slip angle only exists when the vehicle is in motion. At a standstill (and at speeds under about 10 mph) the lateral stiffness of the tyres generate the forces that constrain the vehicle to its intended path. As speed rises above walking pace, the tyres have a falling static lateral stiffness until above about 5 m/s (about 10 mph) they have effectively none; an applied lateral force, such as a wind load, will move the vehicle sideways from its intended path. It is important to note that the presence of a slip angle does not necessarily imply sliding behaviour at the contact patch.

Slip angle forces are typically more than 20 times camber forces for a particular angle, and are thus the more important aspect for vehicle dynamics. The lateral forces induced by the angles are strongly modified by the vertical loads on the tyres at each moment in time.

The tyres generate longitudinal forces by spinning at a speed different to their 'free-rolling' speed. The free-rolling speed is the speed at which the wheel and tyre would spin if no brake or drive forces are applied to them. The difference in speed is described as 'slip ratio', which is unfortunate since it is confusingly similar to slip angle. It is expressed as a percentage, so for example a tyre turning with a 5% slip ratio will perform 105 revolutions to travel the same distance as a free-rolling tyre performing 100 revolutions. In doing so, it will impart a tractive force to the vehicle. A -5% slip ratio would imply 95 revolutions of the same wheel and the presence of a braking force.

Managing lateral tyre forces by controlling slip and camber angles is the work of the suspension linkage. For the front wheels, the driver has the ability to vary the slip angle using the handwheel. Managing the vertical loads on the tyres is the function of the suspension 'calibration' (springs, dampers and any active devices, if present). Chapter 4 deals with suspension analysis in some detail. Management of longitudinal forces is the role of the vehicle driveline and braking system, including anti-lock braking system or brake intervention systems, dealt with in Chapter 8.

A vehicle travelling in a straight line has a yaw velocity of zero and a centripetal acceleration of zero. When travelling in a steady curve, the centripetal acceleration is not zero and the yaw rate is not zero but both are constant and are related as described in Eqn (1.1). In performing the transition from straight running to a curved path there must be a period of yaw acceleration in order to acquire the yaw velocity that matches the centripetal acceleration. The yaw acceleration is induced and controlled by yaw moments acting on the vehicle yaw inertia.

Transient handling therefore implies the variation of yaw moments applied to the vehicle. Those moments are applied by aerodynamic behaviour and the force-generating qualities of the tyres at a distance from the vehicle's centre of mass. No other mechanisms exist for generating a meaningful yaw moment on the vehicle; while gyroscopic torques associated with camber changes exist, they are small. For road vehicles, the aerodynamic modifications are generally small. Multibody system methods allow the convenient exploration of aspects of the vehicle design that influence those qualities of the tyres. Chapter 6 addresses different methods of modelling those aspects of the vehicle and their relative merits.

The modelling of the tyre forces and moments at the tyre to road contact patch is one of the most complex issues in vehicle handling simulation. The models used are typically not causal[6] but are rather empirical formulations used to represent the tyre force and moment curves typically found through laboratory or road based rig testing of a tyre.

Like all models, tyre models have a spectrum of complexity ranging from simple vertical spring rates with hysteretic damping through to fully causal. Chapter 5 explores this spectrum in some depth. Examples of tyre models used for vehicle handling discussed in this book include:

1. A sophisticated empirical point-follower tyre model known as the 'Magic Formula'. This tyre model has been developed by Pacejka and his associates (Bakker et al., 1986, 1989; Pacejka and Bakker, 1993; Pacejka, 2012) and is known to give usefully accurate representation of measured tyre characteristics. The model uses modified trigonometric functions to represent the shape of curves that plot tyre forces and moments as functions of longitudinal slip or slip angle. For 25 years the work of Pacejka has been well known throughout the vehicle dynamics community. The result of this is a tyre model that is now widely used both by industry and academic institutions and is undergoing continual improvement and development. The complexity of the model does however mean that well over 50 parameters are needed to define a tyre model and that software must be obtained or developed to derive the parameters from measured test data. The latest versions of the model (6.1 at the time of writing) include the effects of inflation pressure and can accommodate motorcycle and passenger car tyres within the same model.

2. An alternative modelling approach is to use a straightforward interpolation model. This was the original tyre modelling method used in MSC ADAMS (Ryan, 1990). This methodology is still used by some companies but has, to a large extent, been superseded by more recent parameter based models. The method is included here as a useful benchmark for the comparison of other tyre models in Chapter 5.

3. Another point-follower tyre model is provided for readers as a source listing in Appendix B. This model (Blundell, 2003) has been developed by Harty and has

---

[6]A Causal model is one in which the causes for every effect produced are explicitly described. An example might be a finite element model of a spring, which infers the behaviour of the whole spring from constituent stress/strain relationships. An empirical model is one assembled entirely from observations but making no attempt to quantify or describe the mechanisms causing the effect. An example might be the linear elastic equation of a spring, which requires and produces no description of the stress or strain state of the metal.

the advantage of requiring only a limited number of input parameters compared to the Pacejka model. The implementation is more complete, however, than the interpolation model and includes representation of the following:

a. Comprehensive slip

b. Load dependency

c. Camber thrust

d. Post limit

It has been found that the Harty model is robust when modelling limit behaviour including for example problems involving low grip or prolonged wheelspin. The same model is suitable for representing motorcycle and passenger car tyre behaviour when different parameters are employed.

4. An elaborate 'semi-causal' model of the tyre structure known as FTire (Gipser, 1999), sold commercially including carcase enveloping and contact patch velocity details. Based on an elegant formulation, conceptually a blend of finite element-style simultaneous solutions of small segments with a modal-component style aggregate behaviour representation, it boasts some impressive abilities to resolve pressure and velocity distributions in the contact patch. It is gaining ground in durability prediction in particular.

5. A further model, which is less elaborate than (4) structurally but incorporates thermal aspects in order to be able to migrate from cold to hot and from fresh to worn during the course of a simulation. Of particular interest in motorsport where the thermal modification of tyres is most marked, the so-called Thermal And MEchanical (TAME Tire) representation (Hague, 2010), complements the FTire approach and is more appropriate for certain applications

These five approaches to tyre modelling represent a comprehensive overview of the breadth of modelling practice now in use and are a substantial expansion over what was common when the first edition was written.

In order to progress from travelling in a straight line to travelling in a curved path, the following sequence of events is suggested:

1. The driver turns the handwheel, applying a slip angle at the front wheels.

2. After a delay associated with the front tyre relaxation lengths (see Chapter 5), side force is applied at the front of the vehicle. Lateral and yaw accelerations exist.

3. The body yaws (rotates in plan), applying a slip angle at the rear wheels.

4. After a delay associated with the rear tyre relaxation lengths, side force is applied at the rear of the vehicle. Lateral acceleration is increased; yaw acceleration is reduced to zero.

In the real world, the driver intervenes and the events run into one another rather than being discrete as suggested here, but it is a useful sequence for discussion purposes. A similar sequence of events describes the return to straight-line travel. Any yaw rate adjustments made by the driver follow similar sequences, too.

During the period of yaw acceleration (stages 2 and 3 above) there exists the need for an excess of lateral forces from the front tyres when compared to the rear in order to deliver the required yaw moment. At the end of this period, that excess must disappear. Side force requirements for the rear tyre are thus increasing while those for the front tyres are steady or decreasing. To understand the significance of this fact, some further understanding of tyre behaviour is necessary.

To a first approximation, camber forces may be neglected from tyre behaviour for vehicles that do not roll (lean) freely. Slip angle is the dominant side force generation mechanism. It is important to note that a tyre will adjust its slip angle to support the required side force, and not the other way around; this is a frequent source of difficulty in comprehending vehicle dynamic behaviour. All tyres display a slip angle at which the maximum side force is generated, sometimes referred to as the 'critical slip angle'. If a force is required which is greater than that which can be generated at the critical slip angle, the tyre will run up to and then beyond the critical slip angle. Beyond the critical slip angle the side force falls off with slip angle, and so an increasing amount of surplus lateral force is available to accelerate the growth of slip angle once the critical slip angle is passed.

Returning to the vehicle, the side force requirement for the rear tyres is increasing while that for the front tyres is steady or decreasing. The rear tyres will be experiencing a growing slip angle, while the fronts experience a steady or reducing one. If at this time the rear tyres exceed their critical slip angle, their ability to remove yaw moment is lost. The only possible way for yaw moment to be removed is by a reduction in the front tyre forces. If the yaw moment persists then yaw acceleration persists. With increasing yaw velocity, the slip angle at the front axle is reduced while that at the rear axle is increased further, further removing the rear tyres' ability to remove yaw moment from the vehicle. If the front tyres are past their critical slip angle, too, the normal stabilisation mechanism is reversed. The result is an accelerating spin that departs rapidly from the driver's control. The modelling and interpretation of such events is dealt with in Chapter 7.

The behaviour of the driver is important to the system performance as a whole. The driver is called on to act as a yaw rate manager, acting on the vehicle controls as part of a closed-loop feedback system to impart the yaw moments required to control the yaw rate of the vehicle. At critical times, the workload of the driver may exceed his or her capability, resulting in a loss of control.

The goal of vehicle dynamics work is to maintain the vehicle behaviour within the bounds that can be comprehended by and controlled by the driver. The increasing use of electronic systems in vehicles drives the application of standards for the safe engineering of such critical standards, the most common of which is the ISO26262 standard for automobiles. So-called 'controllability' is a key factor in this and multibody simulations have an important role in assessing the controllability of many situations, as discussed later in Chapter 8.

1.3 Why analyse?

In any real product-engineering programme, particularly in the ground vehicle industry, there are always time constraints. The need to introduce a new product to retain market share or to preserve competitive advantage drives increasingly tight timetabling for product-engineering tasks. In the western world, and to a growing extent in the developing world, tastes are becoming ever more refined such that the demand for both quality of design and quality of construction is increasing all the while. Unlike a few decades ago, there are few genuinely bad products available.

It seems, therefore, that demands for better products are at odds with demands for compressed engineering timetables. This is true; the resolution of this conflict lies in improving the efficiency of the engineering process. It is here that predictive methods hold out some promise.

Predictive methods notionally allow several good things:

- Improved comprehension and ranking of design variables
- Rapid experimentation with design configurations
- Genuine optimisation of numerical response variables

Therefore the use of predictive methods is crucial for staying 'ahead of the game' in vehicle engineering.

1.4 Classical methods

These methods are taught formally in universities as part of the syllabus. While they can be daunting at first sight, they are elegant and can prove tremendously illuminating in forming a holistic framework for what can easily be a bewildering arena. A quick tour through a classically formulated vehicle model is given in Chapter 7 and an implementation in MS-Excel is available to download with the book.

The best practitioners of the art recommend the use of a body-centred state-space formulation. While full of simplifications, useful insights can be gained by studying a two degree-of-freedom model for typical passenger cars. With a reasonable increase in sophistication but well worth the effort is the elaboration to three degrees of freedom (four states) to include the influence of suspension roll.

Such classical models help the analyst discern 'the wood for the trees' — they easily bring forth, for example, the influence of suspension steer derivatives on straight-line stability. In this they contrast strongly with 'literal' linkage models, in which all the problems of real vehicles (the lack of isolation of single effects) obscures their ranking and comprehension.

Although the task of deriving the equations of motion and arranging the terms for subsequent solution is laborious and may be error-prone, the proponents of the method point quite correctly to the increased comprehension of the problem to which it leads.

1.5 Analytical process

It is clear that the tasks undertaken have expanded to fit the time available. Despite remorselessly increasing computing power, analysis tasks are still taking as long to complete as they always have done. The increased computing power available is an irresistible temptation to add complexity to predictive models (Harty, 1999).

Complex models require more data to define them. This data takes time to acquire. More importantly, it must exist. Early in the design cycle, it is easy to fall into the 'paralysis of analysis'; nothing can be analysed accurately until it is defined to a level of accuracy matching the complexity of the modelling technique. More than the model itself, the process within which it fits must be suited to the tasks at hand. *The model is not the product!*

There is nothing new in the authors' observations; Sharp (1991) comments

> *Models do not possess intrinsic value. They are for solving problems. They should be thought of in relation to the problem or range of problems which they are intended to solve. The ideal model is that with minimum complexity which is capable of solving the problems of concern with an acceptable risk of the solution being "wrong". This acceptable risk is not quantifiable and it must remain a matter of judgement. However, it is clear that diminishing returns are obtained for model elaboration.*

Any method of analysis must be part of a structured process if it is to produce useful results in a timely manner. Interesting results that are too late to influence product design are of little use in modern concurrent[7] engineering practice. Rapid results that are so flawed as to produce poor engineering decisions are also of little use. The use of predictive methods within vehicle design for addressing dynamic issues with the vehicle should follow a pattern not dissimilar to that in Figure 1.7, whatever the problem or the vehicle.

1.5.1 Aspiration

The method properly starts with the recognition of the end goals. In some organisations, confusion surrounds this part of the process, with obfuscation between targets, objectives and goals; the terms are used differently between organisations and frequently with some differences between individuals in the same organisation. Cutting through this confusion requires time and energy but is vital.

1.5.2 Definition

After definition, a clear description of 'success' and 'failure' must exist; without it the rest of the activities are, at best, wasteful dissipation. Aspirations are frequently set in terms of subjective comparisons — 'Ride Comfort better than best in class'.

---

[7]Concurrent = Taking place at the same time. 'Concurrent Engineering' was a fashionable phrase in the recent past and refers to the practice of considering functional, cost and manufacturing issues together rather than the historically derived 'sequential' approach. It was also referred to as 'Simultaneous Engineering' for a while, though the segmented connotations of simultaneous were considered unhelpful and so the 'concurrent' epithet was adopted.

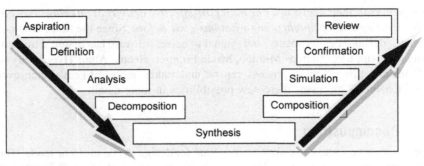

FIGURE 1.7

'V' Process for product design.

To usefully feed these into an analytical process, they must be capable of being quantified — i.e. of having numbers associated with them. Without numbers, it is impossible to address the task using analytical methods and the analytical process should be halted. This is not to say that product development cannot continue but that *to persist with a numerical process in the absence of a knowledge of the difference between good and bad numbers is folly.* Analysts and Development staff must be involved closely with each other and agree on the type of numerical data that defines success, how it is calculated predictively and how it is measured on a real vehicle. Commonly, some form of 'benchmarking' study — a measurement exercise to quantify the current best performers — is associated with this stage. The activity to find the benchmark is a useful shakedown for the proposed measurement processes and is generally a fruitful education for those involved.

1.5.3 Analysis

When success and failure have been defined for the system as a whole, the individual parts of the system must be considered. There is generally more than one way to reach a system solution by combining individual subsystems or elements. During this stage of the process, some decisions must be taken about what combination is preferred. It may be, for example, that in seeking a certain level of vehicle performance there is a choice between increasing power output and reducing weight in order to achieve a given power-to-weight ratio. That choice will be influenced by such simple things as cost (saving weight may be more expensive than simply selecting a larger engine from the corporate library) or by more abstruse notions (the need to be seen to be 'environmentally friendly', perhaps). The task of analysis is to illuminate that choice. The analysis carried out must be sufficiently accurate but not excessively so. 'Simple models smartly used' is the order of the day for analysis work. The analysis may consider many possible combinations in order to recommend a favoured combination. This activity is sometimes referred to by the authors as 'mapping the design space' — producing guidance for those who wish to make design decisions based on wider considerations and who wish to comprehend the consequences of their decisions. *The most cost-effective activity at this stage is accurately recalling and comprehending what has gone before.* Since the first edition of the book, 'design of experiments' and 'signal to noise' software has matured tremendously and with tools such as Minitab, Mode Frontier, Heeds, Altair Hyperstudy or MSC Insight, this mapping process can be undertaken more diligently than ever before. Chapter 7 discusses these new possibilities in more detail.

1.5.4 Decomposition

Once the analytical stage is complete, it is time for design decisions to be made. The whole entity must be decomposed into its constituent parts, each of which has design goals associated with it — cost, performance, weight, etc. It is at this time the first real design decisions are made that shape the product — section properties, geometries, manufacturing process and so on. Those decisions are to be made in the light of the preceding analysis. Many organisations still begin this part of the process too early and as a consequence paint themselves into a corner.

1.5.5 Synthesis

Once the design is decomposed into manageable portions, the task of synthesising (creating) the design begins. During this phase, analytical tools are used to support individual activities and verify the conformance of the proposed design with the intended design goals. An example of this might be the use of kinematic simulation to verify that the suspension geometry characteristics are those required. Discerning the requirement itself is the function of the earlier decomposition phase.

1.5.6 Composition

The reassembly of the separate portions of the design, each of which by now has a high level of confidence at reaching its individual design goals.

1.5.7 Simulation

Before production commences in volume, confidence is needed that the design will be appropriate to go to market. It is often referred to as 'prototyping' and can be real or virtual. The distinguishing feature of this stage is that it is very high fidelity, as distinct from the analysis stage that was no more accurate than necessary.

Prototype vehicles, produced from non-representative tools and/or processes, are physical simulations instead of mathematical ones. The increasing use of 'virtual' prototyping obviates these physical prototypes except for those where an understanding of the man/machine interaction is necessary. One of several arenas where this remains true is the dynamics task.

Predictive models that have been a long time in preparation can be used to assess, virtually and in some detail, the behaviour of the whole design. Models prepared during the Synthesis activity are taken and re-used. It is in this arena that great strides have been made in terms of processing power, model re-use and inter-package integration over the last decade. Unfortunately, in the minds of some, these super elaborate models are all that is useful and anything less is simply worthless, passé and old fashioned. These are valuable models and have a crucial part to play in the process, but without a well-shaped concept design they are unwieldy white elephants. When used unthinkingly they become part of a rather 'Victorian' process in which designs are completed and then inspected for fitness-for-purposes; in the event of them being unfit, an iterative loop is commenced. Far more rapid would have been to have understood the question and to proceed toward the solution in a linear, non-looping fashion.

1.5.8 Confirmation

Sign-off testing is to be carried out on real vehicles that are as representative as possible. This stage should reveal no surprises, as changes at this stage are expensive.

1.5.9 Review

Once the design is successfully signed off, a stage that is frequently omitted is the review. What was done well? What could have been better? What technology do we wish we had then that might be available to us now? Since the most cost-effective analysis activity is to recall accurately and comprehend what has gone before, a well-documented review activity saves time and money in the next vehicle programme.

    The process described is not definitive, nor is it intended to be prescriptive. It should, however, illustrate the difference between 'analysis' and 'simulation' and clearly distinguish them.

1.6 Computational methods

Whether the equations of motion have been derived by hand or delegated to a commercial software package, the primary goal when considering vehicle dynamics is to be able to predict the time-domain solution to those equations. Another important type of solution, the eigensolution, is discussed in Chapter 3.

    Once the equations of motion have been assembled, they are integrated numerically. This is a specialised field in its own right. There are many publications in the field and it is an area rife with difficulties and pitfalls for the unwary. However, in order to successfully use the commercially available software products, some comprehension of the difficulties involved are necessary for users. Chapter 3 deals with some of the more common difficulties with some examples for the reader. By far the most dangerous type of difficulty is the 'plausible but wrong' solution.

Commercial analysts must studiously guard against the 'garbage in, gospel out' mentality that pervades the engineering industry at present.

The equations can be solved in a fairly direct fashion as assembled by the commercial package pre-processor or they can be subject to further symbolic manipulation before numerical solution. So-called 'symbolic' codes offer some tremendous computational efficiency benefits and are being hailed by many as the future of multibody system analysis since they allow real-time computation of reasonably complex models without excessive computing power. The prospect of a real-time multibody system of the vehicle solved on-board in order to generate reference signals for the generation-after-next vehicle control systems seems genuine.

1.7 Computer-based tools

Multibody systems analysis software has become so easy to use that many users lack even a basic awareness of the methods they are using. This chapter charts the background and development to the current generation of multibody systems analysis programs. While the freedom from the purgatory of formulating one's own equations of motion is a blessing, it is partly that purgatory that aids the analyst's final understanding of the problem. Chapter 2, Kinematics and Dynamics of Rigid Bodies, is intended as a reference and also as a 'launch pad' for the enthusiastic readers to be able to teach themselves the process of so-called 'classical' modelling.

Crolla (1995) identifies the main types of computer-based tools, which can be used for vehicle dynamic simulation, and categorises these as:

1. Purpose designed simulation codes
2. Multibody simulation packages, which are numerical
3. Multibody simulation packages, which are algebraic (symbolic)
4. Toolkits such as MATLAB

One of the major conclusions that Crolla draws is that it is still generally the case that the ride and handling performance of a vehicle will be developed and refined mainly through subjective assessments. Most importantly he suggests that in concentrating on sophistication and precision in modelling, practising vehicle dynamicists may have got the balance wrong. This is an important issue that reinforces the main approach in this book, which is to encourage the application of models that lead to positive decisions and inputs to the vehicle design process.

Crolla's paper also provides an interesting historical review that highlights an important meeting at IMechE headquarters in 1956, 'Research in automobile stability and control and tyre performance'. The author states that in the field of vehicle dynamics the papers presented at this meeting are now regarded as seminal and are referred to in the USA as simply 'The IME Papers'.

One of the authors at that meeting, Segel, can be considered to be a pioneer in the field of vehicle dynamics. His paper (Segel, 1956) is one of the first examples where classical mechanics has been applied to an automobile in the study of lateral rigid body motion resulting from steering inputs. The paper describes work carried out on a Buick vehicle for General Motors and is based on transferable experience of aircraft stability gained at the Flight Research Department, Cornell Aeronautical Laboratory. The main thrust of the project was the development of a mathematical vehicle model that included the formulation of lateral tyre forces and the experimental verification using instrumented vehicle tests.

In 1993, almost 40 years after embarking on this early work in vehicle dynamics, Segel again visited the IMechE to present a comprehensive review paper (Segel, 1993), 'An overview of developments in road vehicle dynamics: past, present and future'. In it he provides a historical review that considers the development of vehicle dynamics theory in three distinct phases:

Period 1 — Invention of the car to early 1930s
Period 2 — Early 1930s to 1953
Period 3 — 1953 to the then present (1993)

In describing the start of Period 3 Segel references his early 'IME paper' (Segel, 1956). In terms of preparing a review of work in the area of vehicle dynamics there is an important point made in the paper regarding the rapid expansion in literature that makes any comprehensive summary and critique difficult. This is highlighted by his example of the 1992 FISITA Congress where a total of 70 papers were presented under the general title of 'Total Vehicle Dynamics'. In 2013, searching the SAE website alone using 'vehicle dynamics' (in quotes, so as to preclude matches to only the word vehicle or the word dynamics) produces 2692 papers. There are no fewer than 98 standards produced by the same search.

Following Segel's historical classification of the vehicle dynamics discipline to date, the authors of this text suggest that we have now entered a fourth era that may be characterised by the use of engineering analysis software as something of a 'commodity', bought and sold and often used without a great deal of formal comprehension. In these circumstances there is a need for the software to be absolutely watertight (currently not possible to guarantee) or else for a small number of experts — 'champions' — within organisations to ensure the 'commodity' users are not drifting off the rails, to use a horribly mixed metaphor. This mode of operation is already becoming established within the analysis groups of large automotive companies where analysts make use of customised software programs such as ADAMS/Car, the Simpack Vehicle Wizard or the Dymola Vehicle Dynamics library. These programs have two distinct types of usage. At one level the software is used by an 'expert' with the experience, knowledge and skill and to customise the models generated, the types of simulation to be performed and the format in which selected results will be presented. A larger group of 'standard' users are then able to use the program to carry out suspension or full vehicle simulations assuming little or no knowledge of multibody systems formulations and solution methods. Standard users in many organisations are not full-time analysts but designers or development engineers whose remit includes using analytical tools to inform their opinions according to protocols imposed by the organisation.

1.8 Commercial computer packages

Before the evolution of multibody system programs, engineers analysed the behaviour of mechanisms such as cam-followers and four-bar linkages on the basis of pure kinematic behaviour. Graphical methods were often used to obtain solutions. Chace (1985) summarises the early programs that led to the development of the MSC ADAMS program. One of the first programs (Cooper et al., 1965) was KAM (Kinematic Analysis Method) capable of performing displacement, velocity and acceleration analysis and solving reaction forces for a limited set of linkages and suspension models. Another early program (Knappe, 1965) was COMMEND (Computer-Orientated Mechanical Engineering Design), which was used for planar problems.

By 1969, Chace (1969, 1970) and Korybalski (Chace and Korybalski, 1970) had completed the original version of DAMN (Dynamic Analysis of Mechanical Networks). This was historically the first general program to solve time histories for systems undergoing large displacement dynamic motion. This work lead in 1971 to a new program DRAM (Dynamic Response of Articulated Machinery) that was further enhanced by Angel (Chace and Angel, 1977).

Orlandea published two *American Society of Mechanical Engineers* (ASME) papers (Orlandea et al., 1976a, 1976b). These were a development of the earlier two-dimensional programs to a three-dimensional code but without some of the impact capability contained in DRAM at that time. This program went on to form the core of MSC ADAMS.

General-purpose programs have been developed with a view to commercial gain and as such are able to address a much larger set of problems across a wide range of engineering industries.

A number of other systems based on commercial software have at times been developed specifically for automotive vehicle modelling applications. Several of the larger vehicle manufacturers have at some time integrated MSC ADAMS into their own in-house vehicle design systems. Early examples of these were the AMIGO system at Audi (Hudi, 1988), and MOGESSA at Volkswagen (Terlinden et al., 1987). The WOODS system based on user-defined worksheets was another system at that time in this case developed by German consultants for Ford in the UK (Kaminski, 1990). Ford's Global vehicle modelling activities have since focused on in-house generated linear models and the ADAMS/Chassis™ (formerly known as ADAMS/Pre™) package, a layer over the top of the standard MSC ADAMS pre- and post-processor that is strongly tailored towards productivity and consistency in vehicle analysis.

When the first edition was published there were two leading general-purpose programs, MSC ADAMS and LMS DADS. They dominated the ground vehicle and aerospace markets. LMS DADS has morphed into a product called Virtual Motion and inside MSC ADAMS there has been an explosion of so-called 'Vertical' products, which integrate technology-specific utilities (such as 'wrapping' analysis for caterpillar tracks) with a dual mode software interface geared towards the champion/standard user split described previously.

In the meantime, new products such as Simpack, Dymola and Amesim have emerged as serious commercial general-purpose products (all were somewhat nascent at the time of the first edition) along with open-source codes such as the entire Modelica endeavour (upon which Dymola is built), MBDyn from the university of Milan, SimTK, Metex and so on. What has also emerged is that software that was traditionally for other purposes, such as Matlab Simulink, has begun to extend its reach and coverage such that many useful studies can be carried out entirely within it, without recourse to a 'traditional' multibody solver.

MADYMO is a program recognised as having a multibody foundation with an embedded non-linear finite element capability. This program has been developed by TNO in the Netherlands and complements their established crash test work with dummies. Recent developments in MADYMO have included the development of biofidelic humanoid models to extend the simulation of crash test dummies to 'real-world' pedestrian impact scenarios.

Tremendous advances in computing power were readily foreseeable in 2004 and remain so. A current workstation laptop has 32 GB of RAM (or 32 million Apollo Guidance Computers) serving four separate 'cores' — really separate computers sharing the box and supporting the kind of parallel computing that was spoken of in reverent tones when using Cray-II computers remotely in the early 1990s. Models regarded as awkwardly complex only 10 years ago are now entirely manageable. Tracked vehicles, for example, can be modelled with the track represented link-by-link, including the contact interaction with deformable terrain and all the vehicle wheels; such models are currently close to the limit of what is tolerable in a normal workflow but for the next edition such models may be solving in real time on a mobile phone app.

Blundell (1999a and b, 2000a and b) published a series of four IMechE papers with the aim of summarising typical processes involved with using a general-purpose multibody program to simulate full vehicle handling manoeuvres. The first paper provided an overview of the usage of multibody systems analysis in vehicle dynamics. The second paper described suspension modelling and analysis methodologies. The third paper covered tyre modelling and provided example routines for different tyre models and data. The fourth and final paper brought the series together with a comparative study of full vehicle models, of varying complexity, simulating a double lane change manoeuvre. Results from the simulation models were compared with measured test data from the proving ground. The overall emphasis of the series of papers was to demonstrate the accuracy of simple efficient models based on parameters amenable to design sensitivity study variations rather than blindly modelling the vehicle 'as is'.

Crolla et al. (1994) also define two fundamental types of multiboby systems (MBS) program, the first of which is where the equations are generated in numerical format and are solved directly using numerical integration routines embedded in the package. The second, more recent type of MBS program formulates the equations in symbolic form and often uses an independent solver. The authors also describe toolkits as collections of routines that generate models, formulate and solve equations, and present results.

Other examples of more recently developed general-purpose codes formulate the equations algebraically and use a symbolic approach. Examples of these programs include MESA VERDE (Wittenburg and Wolz, 1985), AUTOSIM (Sayers, 1990), and RASNA Applied Motion Software (Austin and Hollars, 1992). Crolla et al. (1992) provide a summary comparison of the differences between numeric and symbolic code. As stated, MBS programs will usually automatically formulate and solve the equations of motion although in some cases, such as with the work described by Costa (1991) and Holt and Cornish (1992) and Holt (1994), a program SDFAST has been used to formulate the equations of motion in symbolic form and another program ACSL (Automatic Continuous Simulation Language) has been used to generate a solution.

Another customised application developed by the automotive industry is described in Scapaticci et al. (1992). In this paper the authors describe how MSC ADAMS has been integrated into a system known as SARAH (Suspension Analyses Reduced ADAMS Handling). This in-house system for the automotive industry was developed by the Fiat Research Centre Handling Group and used a suspension modelling technique that ignored suspension layout but focused on the final effects of wheel centre trajectory and orientation.

This leads nicely into another category of software available, the single-purpose tool. Unlike the general-purpose tools, which can build models of anything from door latches to spacecraft, some software is written specifically for modelling vehicle dynamics.

At Leeds University a vehicle-specific system was developed under the supervision of Crolla. In this case all the commonly required vehicle dynamics studies have been embodied in their own set of programs (Crolla et al., 1994) known as VDAS (Vehicle Dynamics Analysis Software). Examples of the applications incorporated in this system included: ride/handling, suspensions, natural frequencies, mode shapes, frequency response and steady state handling diagrams. The system included a range of models and further new models could be added using a pre-processor. Single-purpose programs are described as those where the equations of motion have been developed and programmed for a specific model. Model parameters can be changed but the model is fixed unless the program is changed and recompiled. A single-purpose program for passenger cars cannot be used for motorcycles, for example.

A typical example of this type of program would be AUTOSIM described by Sayers (1990), Sharp (1997) and Mousseau et al. (1992) which is intended for vehicle handling and has been developed as a symbolic code in order to produce very fast simulations. Other examples are the Milliken Research Associates VDMS program, which is a single-purpose model for use within the Matlab environment, and IPG Carmaker. For vehicles other than cars, BikeSim and TruckSim are both available.

Generally, single-purpose programs are specifically developed for a given type of simulation but often allow flexibility as to the choice and complexity of the model. An extension of this is where the equations of motion for a fixed vehicle modelling

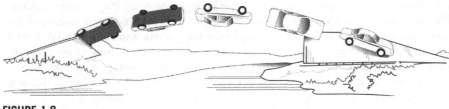

FIGURE 1.8

The Astro Spiral Jump.

approach are programmed and cannot be changed by the user such as the HVOSM (Highway-Vehicle-Object Simulation Model) developed by Raymond McHenry at Calspan in the mid 1960s. The program includes tyre and suspension models and can be used for impact studies in addition to the normal ride and handling simulations. HVOSM deserves a special mention for its contribution to the Astro Spiral Jump (Figure 1.8), a stunt used in the 1974 James Bond film 'The Man with the Golden Gun' that would have been more or less impossible to choreograph using traditional incremental stunt development methods.

The authors (Crolla et al., 1992) indicate that the University of Missouri has also developed a light vehicle dynamics simulation (LVDS) program that runs on a PC and can produce animated outputs. In the mid 1980s, Systems Technology, Inc. developed a program for vehicle dynamics analysis non-linear (VDANL) simulation. This program is based on a 13 degree of freedom, lumped parameter model (Allen et al., 1987) and has been used by researchers at Ohio State University for sensitivity analysis studies (Tandy et al., 1992).

More recently, programs such as ARAS 360 have emerged for vehicle dynamics simulation within the accident reconstruction sphere. ARAS 360 has a 15 degree-of-freedom vehicle model, non-linear tyre behaviour and a rich graphical environment including weather rendering and skid mark production.

The relative ease of computing the vehicle dynamics problem has lead to a number of novel applications for it. In widespread use in aviation, simulators are generating a lot of interest for the study of the human aspect of vehicle control. Real-time capable models that can complete their calculations in the same timescale as the world progresses, or faster, have become relatively common. Several of the commercial packages have a real-time capable implementation although this can depend greatly on the level of complexity selected. In general, single-purpose programs perform better than general-purpose programs in this arena but the relatively low complexity of the computational problem means that even general-purpose programs can be pressed into real-time use if the burden of graphical processing is off-loaded to a dedicated computing resource. The development of compelling motion cueing is still somewhat incomplete with ground vehicle simulators, unlike aircraft simulators that have a different range of sensations to reproduce. Around 2000 a project was carried out on behalf of Prodrive by Harty for Evolution Studios in Cheshire in which a 50 Hz calculation rate produced excellent behaviour including non-linear tyre behaviour for a simulated vehicle on the Sony Playstation platform. The 'accurate' form of the model was not used for the final production release of the games since the phase delays it produced were too realistic (and hence too difficult) for many players to assimilate in a video game environment. A copy of the accurate form of the model is retained by Harty, along with a now-vintage Playstation.

For the Playstation, the tyre model was cut down to the absolute minimum complexity. This model (the 'Harty' model) is described in Chapter 5.

1.9 Benchmarking exercises

A detailed comparison between the various codes is beyond the capability of most companies when selecting an MBS program. In many ways the use of multibody systems has followed on from the earlier use of finite element analysis, the latter being approximately 10 years more mature as applied commercial software. Finite element codes were subject to a rigorous and successful series of benchmarks under the auspices of NAFEMS (National Agency for Finite Elements and Standards) during the 1980s. The published results provided analysts with useful comparisons between major finite element programs such as NASTRAN and ANSYS. The tests performed compared results obtained for a range of analysis methods with various finite elements.

For the vehicle dynamics community, Kortum and Sharp (1991) recognised that with the rapid growth in available multibody systems analysis programs a similar benchmarking exercise was needed. This exercise was organised through the International Association for Vehicle System Dynamics (IAVSD). In this study the various commercially available MBS programs were used to benchmark two problems. The first was to model the Iltis military vehicle and the second a five-link suspension system. A review of the exercise is provided by Sharp (1994) where some of the difficulties involved with such a wide-ranging study are discussed. An example of the problems involved would be the comparison of results. With different investigators using the various programs at widespread locations, a simple problem occurred when the results were sent in plotted form using different size plots and inconsistent axes making direct comparisons between the codes extremely difficult. It was also very difficult to ensure that a consistent modelling approach was used by the various investigators so that the comparison was based strictly on the differences between the programs and not the models used. An example of this with the Iltis vehicle would be modelling a leaf spring for which in many programs there were at the time no standard elements within the main code. Although not entirely successful the exercise was useful in being the only known attempt to provide a comparison between all the main multibody programs at the time. It should also be recognised that in the period since the exercise most of the commercial programs have been extensively developed to add a wide range of capability.

Anderson and Hanna (1989) have carried out an interesting study where they have used two vehicles to make a comparison of three different vehicle simulation methodologies. They have also made use of the Iltis, a vehicle of German design, which at that time was the current small utility vehicle used by the Canadian military. The Iltis was a vehicle that was considered to have performed well and had very different characteristics to the M-151 jeep that was the other vehicle in this study. The authors state that the M-151 vehicle, also used by the Canadian military, had been declared unsafe due to a propensity for rolling over.

Work has been carried out at the University of Bath (Ross-Martin et al., 1992) where the authors have compared MSC ADAMS with their own hydraulic and simulation package. The results for both programs are compared with measured vehicle test data provided in this case by Ford. The Bath model is similar to the Roll Stiffness Model described later in this book but is based on a force roll centre as described by Dixon (1987). This requires the vehicle to actually exist so that the model can use measured inputs obtained through static rig measurements, using equipment of the type described by Whitehead (1995). The roll-centre model described in this book is based on a kinematic roll centre derived using a geometric construction as described in Chapter 4, though there is little to preclude a force-based prediction by modelling the test rig on which the real vehicle is measured.

As a guide to the complexity of the models discussed in Ross-Martin et al. (1992), the Bath model required 91 pieces of information and the MSC ADAMS model although not described in detail needed 380 pieces of information. It is also stated in this paper that the MSC ADAMS model used 150 sets of non-linear data pairs that suggests detailed modelling of all the non-linear properties of individual bushes throughout the vehicle.

Limited studies continue for specific modelling challenges, such as that surrounding mobility for tracked vehicles on deformable terrain. Madsen discusses a comparison between the empirically formulated and widely used Becker soil model and a causally modelled granular soil environment in a package called Chrono::Engine (Madsen et al., 2010). He rather casually notes that a hardware-induced limit of one thousand million (a US billion) contact bodies might preclude modelling grains of sand and call for a slightly coarser approach; again we can be confident that this will probably no longer be an issue for the third edition of this book.

Since the IAVSD benchmarking exercise no new wide-ranging studies have been performed, which is interesting in its own right. It could be interpreted as meaning that most of the software is regarded as 'correct enough' for the applications to which it is put. It is also interesting to note a general shift in terms of the acceptance of predictive modelling such that there is no longer a widely held belief that it does not add value — many organisations have success stories to tell, and software companies carefully collate them to give confidence to new customers.

This leads to an interesting swing of the pendulum; it used to be a popular aphorism that 'nobody believes a simulation except the man who did it, and everyone believes a measurement — except the man who did it'. Currently though, it feels to the authors like some well-presented graphics can give predictive models a credibility out of all proportion to the estimated and uncertain data on which they are based and the often incomplete studies in which they are used.

The basic principle of science — that having formed a hypothesis one should diligently search for proof that it is untrue[8] appears to be cast aside in the rather convenient belief that the computer models are 'probably right' — when nothing of the sort is true. Basic aspects of self-doubt, such as convergence checking and input sensitivity checks, are often overlooked in a misplaced desire for productivity.

---

[8]Also known as 'Black Swan' search; one may produce as many white swans as one likes and while it is consistent with the hypothesis that all swans are white, it is not proof. It takes only one black swan to disprove the hypothesis completely. Black swans are alive and well in Australia.

Kinematics and Dynamics of Rigid Bodies

My God, it's full of stars!

Dave Bowman, 2001: A Space Odyssey

2.1 Introduction

The application of a modern multibody systems (MBS) computer program requires a good understanding of the underlying theory involved in the formulation and solution of the equations of motion. Due to the three-dimensional nature of the problem the theory is best described using vector algebra. In this chapter the starting point will be the basic definition of a vector and an explanation of the notation that will be used throughout this text. The vector theory will be developed to demonstrate, using examples based on suspension systems, the calculation of new geometry and changes in body orientation, such as the steer change in a road wheel during vertical motion relative to the vehicle body. This will be extended to show how velocities and accelerations may be determined throughout a linked three-dimensional system of rigid bodies. The definition of forces and moments will lead through to the definition of the full dynamic formulations typically used in a MBS analysis code.

2.2 Theory of vectors

2.2.1 Position and relative position vectors

Consider the initial definition of the position vector that defines the location of point P in Figure 2.1.

In this case the vector that defines the position of P relative to the reference frame O_1 may be completely described in terms of its components with magnitude Px, Py and Pz. The directions of the components are defined by attaching the appropriate sign to their magnitudes.

$$\{R_P\}_{1/1} = \begin{bmatrix} Px \\ Py \\ Pz \end{bmatrix} \tag{2.1}$$

The Multibody Systems Approach to Vehicle Dynamics.
© 2015 Michael Blundell and Damian Harty. Published by Elsevier Ltd. All rights reserved.

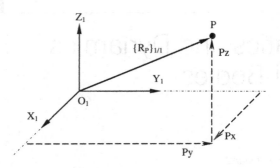

FIGURE 2.1

Position vector.

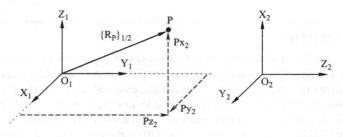

FIGURE 2.2

Resolution of position vector components.

The use of brackets{ } here is a shorthand representation of a column matrix and hence a vector. Note that it does not follow that any quantity that can be expressed as the terms in a column matrix is also a vector.

In writing the vector $\{R_P\}_{1/1}$ the upper suffix indicates that the vector is measured relative to the axes of reference frame O_1. In order to measure a vector it is necessary to determine its magnitude and direction relative to the given axes, in this case O_1. It is then necessary to resolve it into components parallel to the axes of some reference frame that may be different from that used for measurement as shown in Figure 2.2.

In this case we would write $\{R_P\}_{1/2}$ where the lower suffix appended to $\{R_P\}_{1/2}$ indicates the frame O_2 in which the components are resolved. We can also say that in this case the vector is referred to O_2. Note that in most cases the two reference frames are the same and we would abbreviate $\{R_P\}_{1/1}$ to $\{R_P\}_1$.

It is now possible in Figure 2.3 to introduce the concept of a relative position vector $\{R_{PQ}\}_1$. The vector $\{R_{PQ}\}_1$ is the vector from Q to P. It can also be described as the vector that describes the position of P relative to Q.

These vectors obey the triangle law for the addition and subtraction of vectors, which means that

$$\{R_{PQ}\}_1 = \{R_P\}_1 - \{R_Q\}_1$$

or

$$\{R_P\}_1 = \{R_Q\}_1 + \{R_{PQ}\}_1$$

(2.2)

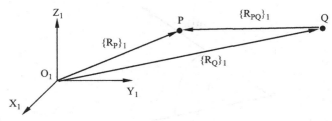

FIGURE 2.3

Relative position vector.

It also follows that we can write

$$\{R_{QP}\}_1 = \{R_Q\}_1 - \{R_P\}_1$$

or
$$\{R_Q\}_1 = \{R_P\}_1 + \{R_{QP}\}_1$$

$$(2.3)$$

Application of Pythagoras' theorem will yield the magnitude $|R_P|$ of the vector $\{R_P\}_1$ as follows

$$|R_P| = \sqrt{Px^2 + Py^2 + Px^2} \qquad (2.4)$$

Similarly the magnitude $|R_{PQ}|$ of the relative position vector $\{R_{PQ}\}_1$ can be obtained using

$$|R_{PQ}| = \sqrt{(Px - Qx)^2 + (Py - Qy)^2 + (Pz - Qz)^2} \qquad (2.5)$$

Consider now the angles θ_X, θ_Y and θ_Z, that the vector $\{R_P\}_1$ makes with each of the X, Y and Z axes of frame O_1 as shown in Figure 2.4. This gives the direction cosines lx, ly and lz of vector $\{R_P\}_1$ where:

$$lx = \cos \theta x = Px / |RP|$$

$$ly = \cos \theta y = Py / |RP| \qquad (2.6)$$

$$lz = \cos \theta z = Pz / |RP|$$

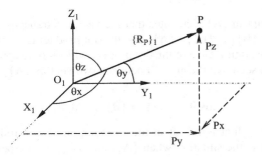

FIGURE 2.4

Direction cosines.

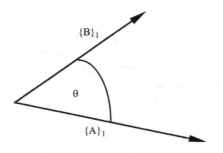

FIGURE 2.5

Vector dot product.

These direction cosines are components of the vector $\{l_P\}_1$ where

$$\{l_P\}_1 = \{R_P\}_1 / |RP| \tag{2.7}$$

It can be seen that $\{l_P\}_1$ has unit magnitude and is therefore a unit vector.

2.2.2 The dot (scalar) product

The dot, or scalar, product $\{A\}_1 \bullet \{B\}_1$ of the vectors $\{A\}_1$ and $\{B\}_1$ yields a scalar C with magnitude equal to the product of the magnitude of each vector and the cosine of the angle between them (Figure 2.5).

Thus

$$\{A\}_1 \bullet \{B\}_1 = |C| = |A|\,|B| \cos \theta \tag{2.8}$$

The calculation of $\{A\}_1 \bullet \{B\}_1$ requires the solution of

$$\{A\}_1 \bullet \{B\}_1 = \{A\}_1^T \{B\}_1 = AxBx + AyBy + AzBz \tag{2.9}$$

where

$$\{B\}_1 = \begin{bmatrix} Bx \\ By \\ Bz \end{bmatrix} \quad \text{and} \quad \{A\}_1^T = [Ax \ \ Ay \ \ Az] \tag{2.10}$$

The ^{T} superscript in $\{A\}_1^T$ indicates that the vector is transposed.

Clearly $\{A\}_1 \bullet \{B\}_1 = \{B\}_1 \bullet \{A\}_1$ and the dot product is a commutative operation. The physical significance of the dot product will become apparent later but at this stage it can be seen that the angle θ between two vectors $\{A\}_1$ and $\{B\}_1$ can be obtained from

$$\cos \theta = \{A\}_1 \bullet \{B\}_1 / |A|\,|B| \tag{2.11}$$

A particular case that is useful in the formulation of constraints representing joints and the like is the situation when $\{A\}_1$ and $\{B\}_1$ are perpendicular making $\cos \theta = 0$.

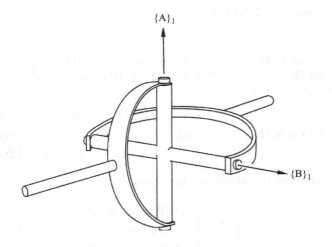

FIGURE 2.6

Application of the dot product to enforce perpendicularity.

As can be seen in Figure 2.6 the equation that enforces the perpendicularity of the two spindles in the universal joint can be obtained from

$$\{A\}_1 \bullet \{B\}_1 = 0 \tag{2.12}$$

2.2.3 The cross (vector) product

The cross, or vector, product of two vectors, $\{A\}_1$ and $\{B\}_1$ is another vector $\{C\}_1$ given by

$$\{C\}_1 = \{A\}_1 \times \{B\}_1 \tag{2.13}$$

The vector $\{C\}_1$ is perpendicular to the plane containing $\{A\}_1$ and $\{B\}_1$ as shown in Figure 2.7.

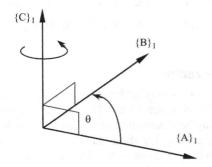

FIGURE 2.7

Vector cross product.

The magnitude of $\{C\}_1$ is defined as

$$|C| = |A|\,|B|\sin\theta \tag{2.14}$$

The direction of $\{C\}_1$ is defined by a positive rotation about $\{C\}_1$ rotating $\{A\}_1$ into line with $\{B\}_1$. The calculation of $\{A\}_1 \times \{B\}_1$ requires $\{A\}_1$ to be arranged in skew-symmetric form as follows

$$\{C\}_1 = \{A\}_1 \times \{B\}_1 = [A]_1\{B\}_1 = \begin{bmatrix} O & -Az & Ay \\ Az & O & -Ax \\ -Ay & Ax & O \end{bmatrix} \begin{bmatrix} Bx \\ By \\ Bz \end{bmatrix} \tag{2.15}$$

Multiplying this out would give the vector $\{C\}_1$

$$\{C\}_1 = \begin{bmatrix} -Az\,By & + & Ay\,Bz \\ Az\,Bx & - & Ax\,Bz \\ -Ay\,Bx & + & Ax\,By \end{bmatrix} \tag{2.16}$$

Exchange of $\{A\}_1$ and $\{B\}_1$ will show that the cross product operation is not commutative and that

$$\{A\}_1 \times \{B\}_1 = -\{B\}_1 \times \{A\}_1 \tag{2.17}$$

2.2.4 The scalar triple product

The scalar triple product D of the vectors $\{A\}_1$, $\{B\}_1$ and $\{C\}_1$ is defined as

$$D = \{\{A\}_1 \times \{B\}_1\} \bullet \{C\}_1 \tag{2.18}$$

2.2.5 The vector triple product

The vector triple product $\{D\}_1$ of the vectors $\{A\}_1$, $\{B\}_1$ and $\{C\}_1$ is defined as

$$\{D\}_1 = \{A\}_1 \times \{\{B\}_1 \times \{C\}_1\} \tag{2.19}$$

2.2.6 Rotation of a vector

In multibody dynamics bodies may undergo motion, which involves rotation about all three axes of a given reference frame. The new components of a vector $\{A\}_1$, shown in Figure 2.8, may be determined as it rotates through an angle α about the X_1-axis, β about the Y_1-axis, and γ about the Z_1-axis of frame O_1.

Consider first the rotation α about O_1X_1. The component Ax is unchanged. The new components Ax', Ay' and Az' can be found by viewing along the X_1-axis as shown in Figure 2.9.

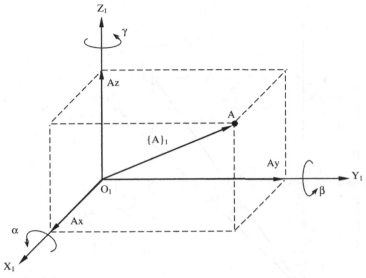

FIGURE 2.8

Rotation of a vector.

By inspection it is found that

$$Ax' = Ax$$

$$Ay' = Ay \cos \alpha - Az \sin \alpha \qquad (2.20)$$

$$Az' = Ay \sin \alpha + Az \cos \alpha$$

In matrix form this can be written

$$\begin{bmatrix} Ax' \\ Ay' \\ Az' \end{bmatrix} = \begin{bmatrix} 1 & 0 & 0 \\ 0 & \cos\alpha & -\sin\alpha \\ 0 & \sin\alpha & \cos\alpha \end{bmatrix} \begin{bmatrix} Ax \\ Ay \\ Az \end{bmatrix} \qquad (2.21)$$

In a similar manner when $\{A\}_1$ is rotated through an angle β about the Y_1-axis the components of a new vector $\{A'\}_1$ can be obtained from

$$\begin{bmatrix} Ax' \\ Ay' \\ Az' \end{bmatrix} = \begin{bmatrix} \cos\beta & 0 & \sin\beta \\ 0 & 1 & 0 \\ -\sin\beta & 0 & \cos\beta \end{bmatrix} \begin{bmatrix} Ax \\ Ay \\ Az \end{bmatrix} \qquad (2.22)$$

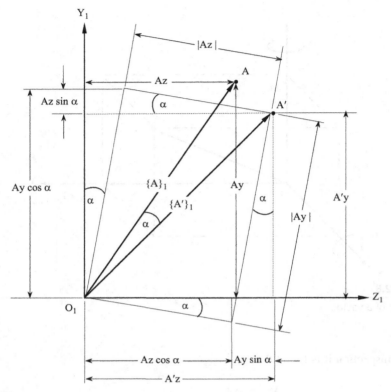

FIGURE 2.9

Rotation of a vector viewed along the X_1-axis.

After a final rotation of $\{A\}_1$ through an angle γ about the Z_1-axis the new components of $\{A'\}_1$ are given by

$$\begin{bmatrix} Ax' \\ Ay' \\ Az' \end{bmatrix} = \begin{bmatrix} \cos\gamma & -\sin\gamma & 0 \\ \sin\gamma & \cos\gamma & 0 \\ 0 & 0 & 1 \end{bmatrix} \begin{bmatrix} Ax \\ Ay \\ Az \end{bmatrix} \qquad (2.23)$$

Applying all three rotations in the sequence α, β and γ would result in the three rotation matrices being multiplied through as follows

$$\begin{bmatrix} Ax' \\ Ay' \\ Az' \end{bmatrix} = \begin{bmatrix} \cos\gamma & -\sin\gamma & 0 \\ \sin\gamma & \cos\gamma & 0 \\ 0 & 0 & 1 \end{bmatrix} \begin{bmatrix} \cos\beta & 0 & \sin\beta \\ 0 & 1 & 0 \\ -\sin\beta & 0 & \cos\beta \end{bmatrix} \begin{bmatrix} 1 & 0 & 0 \\ 0 & \cos\alpha & -\sin\alpha \\ 0 & \sin\alpha & \cos\alpha \end{bmatrix} \begin{bmatrix} Ax \\ Ay \\ Az \end{bmatrix}$$

$$(2.24)$$

$$\begin{bmatrix} Ax' \\ Ay' \\ Az' \end{bmatrix} = \begin{bmatrix} \cos\gamma & -\sin\gamma & 0 \\ \sin\gamma & \cos\gamma & 0 \\ 0 & 0 & 1 \end{bmatrix} \begin{bmatrix} \cos\beta & \sin\beta\sin\alpha & \sin\beta\cos\alpha \\ 0 & \cos\alpha & -\sin\alpha \\ -\sin\beta & \cos\beta\sin\alpha & \cos\beta\cos\alpha \end{bmatrix} \begin{bmatrix} Ax \\ Ay \\ Az \end{bmatrix} \quad (2.25)$$

$$\begin{bmatrix} Ax' \\ Ay' \\ Az' \end{bmatrix} = \begin{bmatrix} \cos\gamma\cos\beta & \cos\gamma\sin\beta\sin\alpha - \sin\gamma\cos\alpha & \cos\gamma\sin\beta\cos\alpha + \sin\gamma\sin\alpha \\ \sin\gamma\cos\beta & \sin\gamma\sin\beta\sin\alpha + \cos\gamma\cos\alpha & \sin\gamma\sin\beta\cos\alpha - \cos\gamma\sin\alpha \\ -\sin\beta & \cos\beta\sin\alpha & \cos\beta\cos\alpha \end{bmatrix} \begin{bmatrix} Ax \\ Ay \\ Az \end{bmatrix}$$

$$(2.26)$$

It should be noted that large rotations such as these are not commutative and therefore the angles α, β and γ cannot be considered to be the components of a vector. The order in which the rotations are applied is important. As can be seen in Figure 2.10, applying equal rotations of 90° but in a different sequence will not result in the same final orientation of the vector.

An understanding that large rotations are not a vector is an important aspect of MBS analysis. Sets of rotations may be required as inputs to define the orientation of a rigid body or a joint. They will also form the output when the relative orientation of one body to another is requested. Note that the convention used here in Figure 2.10 is based on a set of Body (X−Y−Z) rotations. It will be shown later that different conventions may be used, such as the Yaw-Pitch-Roll method based on a set of Body (Z−Y−X) rotations or the Euler angle method used in MSC.ADAMS that is based on a Body (Z−X−Z) combination.

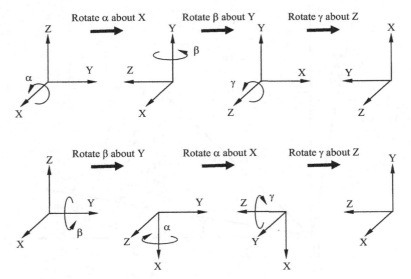

FIGURE 2.10

The effect of rotation sequence on vector orientation.

2.2.7 Vector transformation

In MBS analysis it is often necessary to transform the components of a vector measured parallel to the axis of one reference frame to those measured parallel to a second reference frame. These operations should not be confused with vector rotation. In a transformation it is the magnitude and direction of the components that changes. The direction of the vector is unchanged. Consider the transformation of a vector $\{A\}_1$, or in full definition $\{A\}_{1/1}$, from reference frame O_1 to reference frame O_2. Figure 2.11 represents a view back along the X_1-axis towards the origin O_1. The reference frame O_2 is rotated through an angle α about the X_1-axis of frame O_1.

From Figure 2.11 it can be seen that

$$Ax_2 = Ax_1$$
$$Ay_2 = Ay_1 \cos\alpha + Az_1 \sin\alpha \qquad (2.27)$$
$$Az_2 = -Ay_1 \sin\alpha + Az_1 \cos\alpha$$

In matrix form this can be written

$$\{A\}_{1/2} = \begin{bmatrix} Ax_2 \\ Ay_2 \\ Az_2 \end{bmatrix} = \begin{bmatrix} 1 & 0 & 0 \\ 0 & \cos\alpha & \sin\alpha \\ 0 & -\sin\alpha & \cos\alpha \end{bmatrix} \begin{bmatrix} Ax_1 \\ Ay_1 \\ Az_1 \end{bmatrix} \qquad (2.28)$$

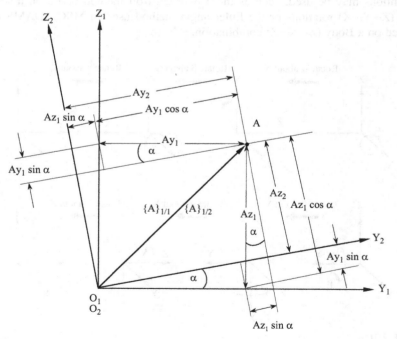

FIGURE 2.11

Transformation of a vector.

This equation may be expressed as

$$\{A\}_{1/2} = [T_1]_2 \ \{A\}_{1/1} \tag{2.29}$$

Consider next the transformation of the vector $\{A\}_{1/1}$ from reference frame O_1 to reference frame O_3. The reference frame O_3 is rotated through an angle β about the Y_1-axis of frame O_1. Following the same procedure as before we get

$$\{A\}_{1/3} = \begin{bmatrix} Ax_3 \\ Ay_3 \\ Az_3 \end{bmatrix} = \begin{bmatrix} \cos\beta & 0 & -\sin\beta \\ 0 & 1 & 0 \\ \sin\beta & 0 & \cos\beta \end{bmatrix} \begin{bmatrix} Ax_1 \\ Ay_1 \\ Az_1 \end{bmatrix} \tag{2.30}$$

$$\{A\}_{1/3} = [T_1]_3 \ \{A\}_{1/1} \tag{2.31}$$

Finally consider the transformation to frame O_4 where O_4 is obtained from a rotation of γ about the Z_1-axis of frame O_1.

$$\{A\}_{1/4} = \begin{bmatrix} Ax_4 \\ Ay_4 \\ Az_4 \end{bmatrix} = \begin{bmatrix} \cos\gamma & \sin\gamma & 0 \\ -\sin\gamma & \cos\gamma & 0 \\ 0 & 0 & 1 \end{bmatrix} \begin{bmatrix} Ax_1 \\ Ay_1 \\ Az_1 \end{bmatrix} \tag{2.32}$$

$$\{A\}_{1/4} = [T_1]_4 \ \{A\}_{1/1} \tag{2.33}$$

The square transformation matrices $[T_1]_2$, $[T_1]_3$ and $[T_1]_4$ are the inverses of the rotation matrices developed in Section 2.2.6. This is to be expected since the method here is the reverse of that shown previously where the vector, rather than the frame, was rotated. It should be noted that a transformation matrix $[T_m]_p$, which transforms a vector from frame m to frame p, has a transpose $[T_m]_p^T$ that is also its inverse $[T_m]_p^{-1}$.

In general terms the transformation of a vector from one frame m to another frame p may be written as

$$\{A\}_{n/p} = [T_m]_p \ \{A\}_{n/m} \tag{2.34}$$

2.2.8 Differentiation of a vector

The differentiation of a vectors $\{A\}_1$ with respect to a scalar variable, such as time t, results in another vector given by

$$\frac{d}{dt} \{A\}_1 = \begin{bmatrix} \dfrac{dAx}{dt} \\ \dfrac{dAy}{dt} \\ \dfrac{dAz}{dt} \end{bmatrix} \tag{2.35}$$

Differentiation with respect to time is often denoted by the Newtonian dot giving

$$\{\dot{A}\}_1 = \begin{bmatrix} \dot{A}_x \\ \dot{A}_y \\ \dot{A}_z \end{bmatrix} \tag{2.36}$$

If the frames used for measurement and reference differ it is necessary to distinguish between, for example, $\dfrac{d}{dt}\{A\}_{m/n}$ and $\{\dot{A}\}_{m/n}$ since

$$\frac{d}{dt}\{A\}_{m/n} \neq \{\dot{A}\}_{m/n} \tag{2.37}$$

In evaluating $\dfrac{d}{dt}\{A\}_{m/n}$, we measure $\{A\}$ in frame m, transform to frame n and then differentiate $\{A\}_{m/n}$ with respect to time. The notation $\{\dot{A}\}_{m/n}$, however, implies that $\{\dot{A}\}_m$ is determined first and that this vector is then transformed to frame n.

Consider the vector $\{A\}_{1/1}$ shown in Figure 2.12. The vector lies in the $X_1 Y_1$ plane of frame O_1 and rotates at a constant speed of ω rad/s about the Z_1-axis. Frame 2 has its Z-axis coincident with Oz_1 and its X-axis is coincident with and rotates with $\{A\}_{1/1}$.

If OX_1 and OX_2 were coincident at time $t = 0$, then where A is the magnitude of $\{A\}_{1/1}$

$$\{A\}_{1/1}^T = [A\ \cos\ \omega t \quad A\ \sin\ \omega t \quad 0] \tag{2.38}$$

Transforming to frame O_2 gives

$$\{A\}_{1/2}^T = [A \qquad 0 \qquad 0] \tag{2.39}$$

Differentiating this with respect to time gives the following since the magnitude A does not vary with time.

$$\frac{d}{dt}\{A\}_{1/2}^T = [0 \qquad 0 \qquad 0] \tag{2.40}$$

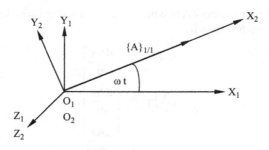

FIGURE 2.12

Vector rotating with constant velocity.

Now

$$\{\dot{A}\}_{1/1}^T = [-\omega A \sin \omega t \quad \omega A \cos \omega t \quad 0] \tag{2.41}$$

and

$$\{\dot{A}\}_{1/2}^T = [0 \quad \omega A \quad 0] \tag{2.42}$$

proving that

$$\frac{d}{dt}\{A\}_{1/2} \neq \{\dot{A}\}_{1/2} \tag{2.43}$$

2.2.9 Integration of a vector

Integration of the vector $\{A\}_1$ with respect to the scalar variable t is given by

$$\int\{A\}_1 \ dt = \begin{bmatrix} \int Ax \ dt \\ \int Ay \ dt \\ \int Az \ dt \end{bmatrix} \tag{2.44}$$

2.2.10 Differentiation of the dot product

The dot product of two vectors $\{A\}_1$ and $\{B\}_1$ is given by

$$\{A\}_1 \bullet \{B\}_1 = \{A\}_1^T \{B\}_1 = [Ax \ Bx + Ay \ By + Az \ Bz] \tag{2.45}$$

Differentiation of $\{A\}_1 \bullet \{B\}_1$ with respect to time t gives

$$\frac{d}{dt}(\{A\}_1 \bullet \{B\}_1) = [Ax \ \dot{B}x + Bx \ \dot{A}x + Ay \ \dot{B}y + By \dot{A}y + Az \ \dot{B}z + Bz \dot{A}z] \tag{2.46}$$

or

$$\frac{d}{dt}(\{A\}_1 \bullet \{B\}_1) = \{A\}_1 \bullet \{\dot{B}\}_1 + \{\dot{A}\}_1 \bullet \{B\}_1 \tag{2.47}$$

The result of this is a scalar. The rule for differentiation of the dot product is similar to the differentiation of the product of two scalars u and v.

$$\frac{d}{dt}(uv) = u\frac{dv}{dt} + v\frac{du}{dt} \tag{2.48}$$

2.2.11 Differentiation of the cross product

The cross product of two vectors $\{A\}_1$ and $\{B\}_1$ is given by

$$\{A\}_1 \times \{B\}_1 = \begin{bmatrix} 0 & -Az & Ay \\ Az & 0 & -Ax \\ -Ay & Ax & 0 \end{bmatrix} \begin{bmatrix} Bx \\ By \\ Bz \end{bmatrix} = \begin{bmatrix} -Az\,By & + & Ay\,Bz \\ Az\,Bx & - & Ay\,Bz \\ -Ay\,Bx & + & Ay\,By \end{bmatrix} \quad (2.49)$$

Differentiation of this vector with respect to time t gives

$$\frac{d}{dt}(\{A\}_1 \times \{B\}_1) = \begin{bmatrix} -Az\,\dot{B}y & - & \dot{A}z\,By & + & Ay\,\dot{B}z & + & \dot{A}y\,Bz \\ Az\,\dot{B}x & + & \dot{A}z\,Bx & - & Ax\,\dot{B}z & - & \dot{A}y\,Bz \\ -Ay\,\dot{B}x & - & \dot{A}y\,Bx & + & Ax\,\dot{B}y & + & \dot{A}y\,By \end{bmatrix} \quad (2.50)$$

or

$$\frac{d}{dt}(\{A\}_1 \times \{B\}_1) = \{A\}_1 \times \{\dot{B}\}_1 + \{\dot{A}\}_1 \times \{B\}_1 \quad (2.51)$$

The result of this operation is a vector. Note again that the rule for the differentiation of the cross product is similar to the rule for the differentiation of the product of two scalars.

2.2.12 Summary

1. A vector is expressed in terms of the magnitudes of its three orthogonal components listed, in natural order, as the elements of a column matrix. The brackets { } are used as a shorthand representation of a vector. Thus the vector $\{A\}_m$ with components of magnitude A_x, A_y and A_z is represented by

$$\{A\}_m = \begin{bmatrix} Ax \\ Ay \\ Az \end{bmatrix} \quad (2.52)$$

2. The suffices m and n appended to the vector $\{A\}_{m/n}$ indicate that its components were measured relative to the axes of frame m but that the vector was then resolved into components parallel to the axes of frame n. We say that the vector is referred or transformed to frame n. If m is equal to n then the vector may be written as $\{A\}_m$.

3. The magnitude of the vector $\{A\}_m$ is represented by $|A|$, or simply by A when the meaning is unambiguous. This magnitude is given by

$$|A| = \sqrt{Ax^2 + Ay^2 + Az^2} \quad (2.53)$$

4. The cosines of the angles that $\{A\}_m$ makes with the X, Y and Z axes respectively of frame m are known as its direction cosines $\{l\}_m$. This vector is derived from $\{A\}_m$ as follows

$$\{l\}_m = \{A\}_m / |A| \qquad (2.54)$$

Since the magnitude of $\{l\}_m$ is unity, it is called a unit vector.

5. If $\{A\}_m$, $\{B\}_m$ and $\{C\}_m$ are vectors of the same dimensions and $\{C\}_m$ is the resultant of $\{A\}_m$ and $\{B\}_m$ then the equation

$$\{C\}_m = \{A\}_m + \{B\}_m \qquad (2.55)$$

expresses the triangle law for the addition or subtraction of vectors. This equation is only valid if all vectors are referred to the same frame.

6. The dot product $\{A\}_m \bullet \{B\}_m$ of the vectors $\{A\}_m$ and $\{B\}_m$ is defined as a scalar whose magnitude is $|A|\,|B|\cos\theta$, θ being the angle between the vectors. The dot product is evaluated in terms of a matrix product as follows

$$\{A\}_m \bullet \{B\}_m = \{A\}_m^T \{B\}_m \qquad (2.56)$$

7. The cross product $\{A\}_m \times \{B\}_m$ of the vectors $\{A\}_m$ and $\{B\}_m$ is defined as a vector $\{C\}_m$ whose magnitude is $|A|\,|B|\sin\theta$, θ being the angle between $\{A\}_m$ and $\{B\}_m$. The vector $\{C\}_m$ is perpendicular to the plane containing the other two and its direction is the direction of advance of a right-handed screw, lying parallel to $\{C\}_m$, when subjected to a rotation which would bring $\{A\}_m$ into alignment with $\{B\}_m$ by the shortest path. The cross product is evaluated in terms of a matrix product as follows

$$\{C\}_m = \{A\}_m \times \{B\}_m = [A]_m \{B\}_m \qquad (2.57)$$

where

$$[A]_m = \begin{bmatrix} 0 & -Az & Ay \\ Az & 0 & -Ax \\ -Ay & Ax & 0 \end{bmatrix} \text{ and } \{B\}_m = \begin{bmatrix} Bx \\ By \\ Bz \end{bmatrix} \qquad (2.58)$$

$[A]_m$ is known as the skew-symmetric form of the vector $\{A\}_m$.

8. The angle θ between the vectors $\{A\}_m$ and $\{B\}_m$ may be determined from

$$\cos\theta = \{A\}_m^T \{B\}_m / |A|\,|B| \qquad (2.59)$$

9. If vectors $\{A\}_m$ and $\{B\}_m$ are parallel then $\{A\}_m$ can be represented using a scalar f as follows

$$\{A\}_m = f\,\{B\}_m \qquad (2.60)$$

10. If vectors $\{A\}_m$ and $\{B\}_m$ are perpendicular then

$$\{A\}_m \bullet \{B\}_m = \{A\}_m^T \{B\}_m = 0 \qquad (2.61)$$

11. The scalar triple product D of the vectors $\{A\}_m$, $\{B\}_m$ and $\{C\}_m$ is a scalar defined by

$$D = \{\{A\}_m \times \{B\}_m\} \bullet \{C\}_m \qquad (2.62)$$

12. The vector triple product $\{D\}_m$ of the vectors $\{A\}_m$, $\{B\}_m$ and $\{C\}_m$ is a vector $\{D\}_m$ defined by

$$\{D\}_m = \{A\}_m \times \{\{B\}_m \times \{C\}_m\} \qquad (2.63)$$

13. If the vector $\{A\}_m^T = [Ax \ Ay \ Az]$ is rotated through angle $+\alpha$ about the x-axis of frame m then the new vector $\{A'\}_m$ is given by

$$\{A'\}_m = \begin{bmatrix} 1 & 0 & 0 \\ 0 & \cos\alpha & -\sin\alpha \\ 0 & \sin\alpha & \cos\alpha \end{bmatrix} \begin{bmatrix} Ax \\ Ay \\ Az \end{bmatrix} \qquad (2.64)$$

For rotation about $O_m Y_m$ and $O_m Z_m$, the square matrix above is replaced by those given in Eqns (2.22) and (2.23)

14. If the X-axes of frames m and p coincide and the Y-axis of frame p is rotated by $+\alpha$, relative to the corresponding axis of frame m, then the vector $\{A\}_{n/m}$ is transformed from frame m to frame p according to the relationship

$$\{A\}_{n/p} = [T_m]_p \{A\}_{n/m} \qquad (2.65)$$

where

$$\{T\}_{m/p} = \begin{bmatrix} 1 & 0 & 0 \\ 0 & \cos\alpha & \sin\alpha \\ 0 & -\sin\alpha & \cos\alpha \end{bmatrix} \qquad (2.66)$$

The other two transformations are given by Eqns (2.30) and (2.32).

15. The transformations of 14 above may be combined as indicated.

$$[T_m]_q = [T_p]_q [T_n]_p [T_m]_n \qquad (2.67)$$

This chain of matrices is to be read from right to left.

16. Differentiation of the vector $\{A\}_m^T = [Ax \; Ay \; Az]$ with respect to the scalar variable t is defined by

$$\frac{d}{dt}\{A\}_m = \begin{bmatrix} \dfrac{dAx}{dt} \\[2mm] \dfrac{dAy}{dt} \\[2mm] \dfrac{dAz}{dt} \end{bmatrix} \qquad (2.68)$$

The result of this operation is a vector.

17. In general,

$$\frac{d}{dt}\{A\}_{m/n} \neq \{\dot{A}\}_{m/n} \qquad (2.69)$$

where the dot denotes differentiation with respect to time.

18. Integration of the vector $\{A\}_m^T = [Ax \; Ay \; Az]$ with respect to the scalar variable t produces another vector defined by

$$\int\{A\}_m \; dt = \begin{bmatrix} \int Ax \; dt \\[1mm] \int Ay \; dt \\[1mm] \int Az \; dt \end{bmatrix} \qquad (2.70)$$

19. Differentiation of the dot product $\{A\}_m \bullet \{B\}_m$ with respect to time t is defined by

$$\frac{d}{dt}(\{A\}_m \bullet \{B\}_m) = \{A\}_m \bullet \{\dot{B}\}_m + \{\dot{A}\}_m \bullet \{B\}_m \qquad (2.71)$$

where

$$\{\dot{A}\}_m = \frac{d}{dt}\{A\}_m \quad \text{and} \quad \{\dot{B}\}_m = \frac{d}{dt}\{B\}_m$$

20. Differentiation of the cross product $\{A\}_m \times \{B\}_m$ with respect to time t follows the same rule as that for the dot product. Hence,

$$\frac{d}{dt}(\{A\}_m \times \{B\}_m) = \{A\}_m \times \{\dot{B}\}_m + \{\dot{A}\}_m \times \{B\}_m \qquad (2.72)$$

2.3 Geometry analysis
2.3.1 Three point method

In order to establish the position of any point in space, vector theory can be used to work from three points for which the coordinates are already established. Consider the following example shown in Figure 2.13.

In this example the positions of A, B and C are taken to be known as are the lengths AD, BD and CD. The position of D is unknown and must be solved. In terms of vectors this can be expressed using the following known inputs:

$$\{R_A\}_1^T = \begin{bmatrix} Ax & Ay & Az \end{bmatrix}$$

$$\{R_B\}_1^T = \begin{bmatrix} Bx & By & Bz \end{bmatrix}$$

$$\{R_C\}_1^T = \begin{bmatrix} Cx & Cy & Cz \end{bmatrix}$$

$$|R_{DA}|$$

$$|R_{DB}|$$

$$|R_{DC}|$$

In order to solve the three unknowns Dx, Dy and Dz, which are the components of the position vector $\{R_D\}_1$, it is necessary to set up three equations as follows:

$$|R_{DA}|^2 = (Dx - Ax)^2 + (Dy - Ay)^2 + (Dz - Az)^2 \qquad (2.73)$$

$$|R_{DB}|^2 = (Dx - Bx)^2 + (Dy - By)^2 + (Dz - Bz)^2 \qquad (2.74)$$

$$|R_{DC}|^2 = (Dx - Cx)^2 + (Dy - Cy)^2 + (Dz - Cz)^2 \qquad (2.75)$$

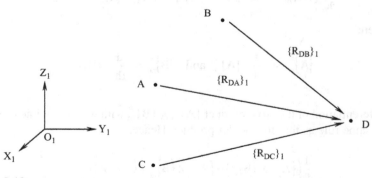

FIGURE 2.13

Use of position vectors for geometry analysis.

Multiplying out these three equations leads to:

$$|R_{DA}|^2 = (Dx^2 - 2Dx\,Ax + Ax^2) + (Dy^2 - 2Dy\,Ay + Ay^2) + (Dz^2 - 2Dz\,Az + Az^2)$$
$$(2.76)$$

$$|R_{DB}|^2 = (Dx^2 - 2Dx\,Bx + Bx^2) + (Dy^2 - 2Dy\,By + By^2) + (Dz^2 - 2Dz\,Bz + Bz^2)$$
$$(2.77)$$

$$|R_{DC}|^2 = (Dx^2 - 2Dx\,Cx + Cx^2) + (Dy^2 - 2Dy\,Cy + Cy^2) + (Dz^2 - 2Dz\,Cz + Cz^2)$$
$$(2.78)$$

At this stage we need to introduce some numerical data to demonstrate how a solution can be obtained. This will also demonstrate how cumbersome the algebra will become and the need to utilise computer software to solve these problems. As an example we can take the following coordinates for points A, B and C:

$$\{R_A\}_1 = [\ 103\ \ 350\ \ 142\ ]\ mm$$
$$\{R_B\}_1 = [-127\ \ 350\ \ 128\ ]\ mm$$
$$\{R_C\}_1 = [\ -15\ \ 500\ \ 540\ ]\ mm$$

The lengths of the three rigid links can be taken as:

$$|R_{DA}| = 172.064\ mm$$
$$|R_{DB}| = 183.527\ mm$$
$$|R_{DC}| = 401.103\ mm$$

Substituting these numerical values into the equations above we get

$$172.064^2 = (Dx - 103)^2 + (Dy - 350)^2 + (Dz - 142)^2$$
$$183.527^2 = (Dx + 127)^2 + (Dy - 350)^2 + (Dz - 128)^2$$
$$401.103^2 = (Dx + 15)^2 + (Dy - 500)^2 + (Dz - 540)^2$$

Multiplying out gives

$$29606.02 = (Dx^2 - 206\,Dx + 10609) + (Dy^2 - 700\,Dy + 122500)$$
$$+ (Dz^2 - 284\,Dz + 20164)$$

$$33682.16 = (Dx^2 + 254\,Dx + 16129) + (Dy^2 - 700\,Dy + 122500)$$
$$+ (Dz^2 - 256\,Dz + 16384)$$

$$160883.617 = (Dx^2 + 30\,Dx + 225) + (Dy^2 - 1000\,Dy + 250000)$$
$$+ (Dz^2 - 1080\,Dz + 291600)$$

Combining the terms in these three equations leads to

$$-123667 \ = (Dx^2 - 206\,Dx) + (Dy^2 - 700\,Dy) + (Dz^2 - 284\,Dz) \qquad (2.79)$$

$$-121331 \ = (Dx^2 + 254\,Dx) + (Dy^2 - 700\,Dy) + (Dz^2 - 256\,Dz) \qquad (2.80)$$

$$-380942 \ = (Dx^2 + 30\,Dx) + (Dy^2 - 1000\,Dy) + (Dz^2 - 1080\,Dz) \qquad (2.81)$$

Subtracting Eqn (2.79) from Eqn (2.80) gives

$$2336 = 460\,Dx \ + \ 28\,Dz$$

$$Dz = 83.43 \ - \ 16.43\,Dx \qquad (2.82)$$

Subtracting Eqn (2.81) from Eqn (2.80) gives

$$259611 = 224\,Dx \ + \ 300\,Dy \ + \ 824\,Dz \qquad (2.83)$$

Substituting the expression for Dz given in Eqn (2.82) into Eqn (2.83) gives:

$$259611 = 224\,Dx \ + \ 300\,Dy \ + \ 68746.32 \ - \ 13538.32\,Dx$$

$$190864.68 = -13314.32\,Dx \ + \ 300\,Dy$$

This gives

$$Dy = 636.22 \ + \ 44.38\,Dx$$

Substituting the Dy and Dz back into Eqn (2.79) gives

$$-123667 = (Dx^2 - 206\,Dx) + (636.22^2 + 1969.58\,Dx^2 - 445354 - 31066\,Dx) +$$

$$(83.43^2 + 269.9\,Dx^2 - 23694.12 + 4666.12\,Dx) - 66355.33 = 2240.52\,Dx^2 \qquad (2.84)$$

$$+27123.5\,Dx$$

Rearranging Eqn (2.84) gives

$$2240.48\,Dx^2 + 27123.5\ Dx \ + 66355.33 = 0 \qquad (2.85)$$

This is now in the familiar form of a quadratic equation $ax^2 + bx + c = 0$ for which the solution is given by

$$x = \frac{-b \pm \sqrt{b^2 - 4ac}}{2a}$$

Therefore the solution to Eqn (2.85) is obtained from

$$Dx = \frac{26605.88 \pm \sqrt{26605.88^2 - 4 \times 2240.48 \times 66355.33}}{2 \times 2240.48}$$

This gives two solutions

$$Dx = -3.403mm \text{ or } -8.703mm$$

The fact that there are two possible solutions for Dx illustrates the nonlinearity of this geometric analysis. In this case inspection of the two solutions does not immediately identify which one should be eliminated. Trying each value in the equation for Dy gives

$$Dy = 636.22 + 44.38 \times (-3.403) = 485.19mm$$

or

$$Dy = 636.22 + 44.38 \times (-8.703) = 249.98mm$$

Similarly trying each value of Dx in the equation for Dz gives

$$Dz = 83.43 - 16.43 \times (-3.403) = 139.34mm$$

or

$$Dz = 83.43 - 16.43 \times (-8.703) = 226.42mm$$

Using a computer program, written in BASIC, to check these answers gives the following two solutions:

$$Dx = -3.403mm \text{ or } -8.703mm$$

$$Dy = 485.194mm \text{ or } 249.986mm$$

$$Dz = 139.338mm \text{ or } 226.413mm$$

In this case the first solution can be identified as the correct one, by inspection only, on the evidence of the Dy coordinate.

2.3.2 Vehicle suspension geometry analysis

The following example, based on a McPherson Strut type of suspension, illustrates the steps that could be followed to determine the geometry for the suspension system shown in Figure 2.14.

The purpose of the analysis is to devise a sequence of calculations that would allow the calculation of the positions of the movable points in the suspension, $\{R_C\}_1$, $\{R_E\}_1$, $\{R_G\}_1$, $\{R_I\}_1$, $\{R_J\}_1$ and $\{R_H\}_1$, once the suspension is displaced from the coordinates given in Figure 2.14. In this example, the two points I and J are used to define the axis of rotation of the wheel.

The following sequence, shown in Figure 2.15, can be used to locate the moveable points.

In order to impart motion to the suspension, resulting from for example hitting a bump, we can shorten the strut DG by a given amount. This will result in links DE and DC also shortening. Points C, E and G will move to new positions that can be

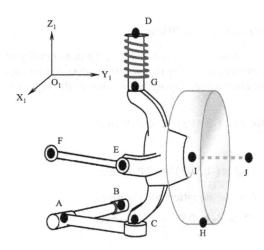

Point	X(mm)	Y(mm)	Z(mm)
A	125	350	-80
B	-115	350	-80
C	5	625	-90
D	-15	530	545
E	155	537	250
F	25	45	270
G	-3	545	125
H	0	670	-270
I	0	65	0
J	0	730	0

FIGURE 2.14

Suspension geometry data.

designated C′, E′ and G′. It should be noted that for the analysis using vectors here C′ and E′ are not the actual final positions but are only used to find the magnitudes of $|R_{DC'}|$ and $|R_{DE'}|$ so that the analysis may progress using the sequence shown in Figure 2.15. Having calculated the new positions of all the movable nodes the movement of the tyre contact patch, in this case taken to be point H, could be used to establish, for example, the lateral movement or half track change. The change in orientation of the wheel will also be of interest. The new positions I′ and J′ can be compared with the undisplaced positions I and J to determine the change in steer angle as shown in Figure 2.16.

The bump steer can be determined by finding the angle δ between the projection of IJ and I′J′ onto the global X_1Y_1 plane. The projection is achieved by setting the z coordinates of all four vectors to zero and then rearranging the vector dot product as shown in Eqn (2.86).

$$\cos \delta = \{R_{IJ}\}_1 \bullet \{R_{I'J'}\} / |R_{IJ}| |R_{I'J'}| \qquad (2.86)$$

2.4 Velocity analysis

Consider the rigid body, Body 2, shown in Figure 2.17. In this case we are initially only interested in motion in the X_1Y_1 plane. The body moves and rotates through an angle $\delta\gamma$, measured in radians, about the Z_1-axis.

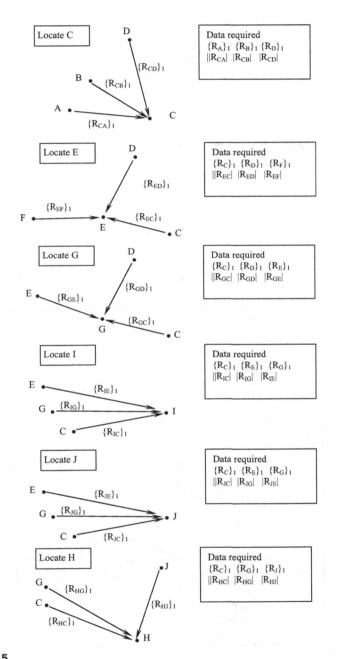

FIGURE 2.15

Calculation sequence to solve suspension geometry.

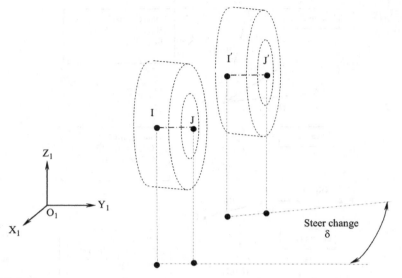

FIGURE 2.16

Using vectors to determine bump steer.

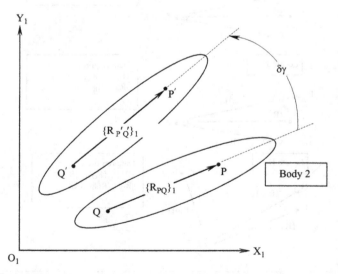

FIGURE 2.17

Motion of a vector attached to a rigid body.

The vector $\{R_{PQ}\}_1$ moves with the body to a new position $\{R_{P'Q'}\}_1$. The new vector $\{R_{P'Q'}\}_1$ is defined by the transformation

$$\{R_{P'Q'}\}_1 = [A] \{R_{PQ}\}_1 \tag{2.87}$$

where $[A]$ is the rotation matrix that rotates $\{R_{PQ}\}_1$ onto $\{R_{P'Q'}\}_1$. Expanding this gives

$$\begin{bmatrix} P'Q'x \\ P'Q'y \\ P'Q'z \end{bmatrix} = \begin{bmatrix} \cos\delta\gamma & -\sin\delta\gamma & 0 \\ \sin\delta\gamma & \cos\delta\gamma & 0 \\ 0 & 0 & 1 \end{bmatrix} \begin{bmatrix} PQx \\ PQy \\ PQz \end{bmatrix} \tag{2.88}$$

Assuming that $\delta\gamma$ is small so that we can take $\cos\delta\gamma = 1$ and $\sin\delta\gamma = \delta\gamma$ (rads) leads to

$$\begin{bmatrix} P'Q'x \\ P'Q'y \\ P'Q'z \end{bmatrix} = \begin{bmatrix} 1 & -\delta\gamma & 0 \\ \delta\gamma & 1 & 0 \\ 0 & 0 & 1 \end{bmatrix} \begin{bmatrix} PQx \\ PQy \\ PQz \end{bmatrix} \tag{2.89}$$

The change in relative position $\delta\{R_{PQ}\}_1$ is given by

$$\delta\{R_{PQ}\}_1 = \{R_{P'Q'}\}_1 - \{R_{PQ}\}_1 \tag{2.90}$$

or

$$\begin{bmatrix} \delta PQx \\ \delta PQy \\ \delta PQz \end{bmatrix} = \begin{bmatrix} 1 & -\delta\gamma & 0 \\ \delta\gamma & 1 & 0 \\ 0 & 0 & 1 \end{bmatrix} \begin{bmatrix} PQx \\ PQy \\ PQz \end{bmatrix} - \begin{bmatrix} 1 & 0 & 0 \\ 0 & 1 & 0 \\ 0 & 0 & 1 \end{bmatrix} \begin{bmatrix} PQx \\ PQy \\ PQz \end{bmatrix} \tag{2.91}$$

This gives

$$\begin{bmatrix} \delta PQx \\ \delta PQy \\ \delta PQz \end{bmatrix} = \begin{bmatrix} 0 & -\delta\gamma & 0 \\ \delta\gamma & 0 & 0 \\ 0 & 0 & 0 \end{bmatrix} \begin{bmatrix} PQx \\ PQy \\ PQz \end{bmatrix} \tag{2.92}$$

If this change takes place in time δt then

$$\frac{\delta}{\delta t} \begin{bmatrix} PQx \\ PQy \\ PQz \end{bmatrix} = \begin{bmatrix} 0 & -\dfrac{\delta\gamma}{\delta t} & 0 \\ \dfrac{\delta\gamma}{\delta t} & 0 & 0 \\ 0 & 0 & 0 \end{bmatrix} \begin{bmatrix} PQx \\ PQy \\ PQz \end{bmatrix} \tag{2.93}$$

In the limit δt approaches zero and we can write

$$\frac{d}{dt}\begin{bmatrix} PQx \\ PQy \\ PQz \end{bmatrix} = \begin{bmatrix} 0 & -\dfrac{d\gamma}{dt} & 0 \\ \dfrac{d\gamma}{dt} & 0 & 0 \\ 0 & 0 & 0 \end{bmatrix}\begin{bmatrix} PQx \\ PQy \\ PQz \end{bmatrix} \tag{2.94}$$

which can be written

$$\begin{bmatrix} V_{PQx} \\ V_{PQy} \\ V_{PQx} \end{bmatrix} = \begin{bmatrix} 0 & -\omega_z & 0 \\ \omega_z & 0 & 0 \\ 0 & 0 & 0 \end{bmatrix}\begin{bmatrix} PQx \\ PQy \\ PQz \end{bmatrix} \tag{2.95}$$

Note that generally rotations cannot be represented as vector quantities unless they are very small, as in finite element programs. Hence angular velocities obtained by differencing rotations over very small time intervals are in fact vector quantities.

If the rigid link also undergoes small rotations $\delta\alpha$ about the X-axis and $\delta\beta$ about the Y-axis then the full expression is

$$\begin{bmatrix} V_{PQx} \\ V_{PQy} \\ V_{PQz} \end{bmatrix} = \begin{bmatrix} 0 & -\omega_z & \omega_y \\ \omega_z & 0 & -\omega_x \\ -\omega_y & \omega_x & 0 \end{bmatrix}\begin{bmatrix} PQx \\ PQy \\ PQz \end{bmatrix} \tag{2.96}$$

Note that this matrix is the skew-symmetric form of the angular velocity vector $[\omega x \ \omega y \ \omega z]^T$. In general terms we write

$$\{V_{PQ}\}_1 = \{\omega_2\}_1 \ \textbf{X} \ \{R_{PQ}\}_1 \tag{2.97}$$

The direction of the relative velocity vector $\{V_{PQ}\}_1$ is perpendicular to the line of the relative position vector $\{R_{PQ}\}_1$ as shown in Figure 2.18. The two points P and Q are fixed in the same rigid body, Body 2. As such there can be no component of motion along the line PQ and any relative motion must therefore be perpendicular to this line.

This relationship can be expressed mathematically in two ways. The first of these uses the vector dot product to enforce perpendicularity as shown in Eqn (2.98).

$$\{V_{PQ}\}_1 \ \bullet \ \{R_{PQ}\}_1 = 0 \tag{2.98}$$

The second method uses the vector cross product as shown in Eqn (2.99).

$$\{V_{PQ}\}_1 = \{\omega_2\}_1 \ \textbf{X} \ \{R_{PQ}\}_1 \tag{2.99}$$

More often than not we use the cross product as this will yield the angular velocity vector $\{\omega_2\}_1$. This may be required for a later acceleration analysis. In

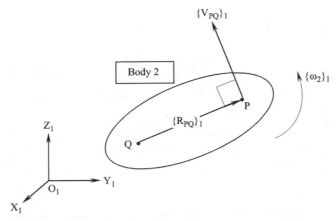

FIGURE 2.18

Relative velocity vector.

developing the equations to solve the velocities in a system of interconnected rigid bodies the triangle law of vector addition can also be used, as shown in Eqn (2.100).

$$\{V_P\}_1 = \{V_Q\}_1 + \{V_{PQ}\}_1 \tag{2.100}$$

2.5 Acceleration analysis

Given that acceleration is the time rate of change of velocity we can develop the equations required for an acceleration analysis using the vectors shown in Figure 2.19.

It is possible to develop equations that would yield the relative acceleration vector $\{A_{PQ}\}_1$ using the angular velocity vector $\{\omega_2\}_1$ and the angular acceleration vector $\{\alpha_2\}_1$ as follows

$$\{A_{PQ}\}_1 = \frac{d}{dt}\{V_{PQ}\}_1 \tag{2.101}$$

$$\{A_{PQ}\}_1 = \frac{d}{dt}\{\{\omega_2\}_1 \times \{R_{PQ}\}_1\} \tag{2.102}$$

$$\{A_{PQ}\}_1 = \{\omega_2\}_1 \times \frac{d}{dt}\{R_{PQ}\}_1 + \frac{d}{dt}\{\omega_2\}_1 \times \{R_{PQ}\}_1 \tag{2.103}$$

Since it is known that

$$\frac{d}{dt}\{R_{PQ}\}_1 = \{V_{PQ}\}_1 \tag{2.104}$$

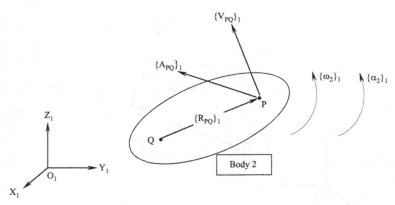

FIGURE 2.19

Relative acceleration vector.

$$\frac{d}{dt}\{\omega_2\}_1 = \{\alpha_2\}_1 \tag{2.105}$$

We can therefore write

$$\{A_{PQ}\}_1 = \{\omega_2\}_1 \times \{V_{PQ}\}_1 + \{\alpha_2\}_1 \times \{R_{PQ}\}_1 \tag{2.106}$$

Since it is also known that

$$\{V_{PQ}\}_1 = \{\omega_2\}_1 \times \{R_{PQ}\}_1 \tag{2.107}$$

This leads to the expression

$$\{A_{PQ}\}_1 = \{\omega_2\}_1 \times \{\{\omega_2\}_1 \times \{R_{PQ}\}_1\} + \{\alpha_2\}_1 \times \{R_{PQ}\}_1 \tag{2.108}$$

The acceleration vector $\{A_{PQ}\}_1$ can be considered to have a centripetal component $\{A^P{}_{PQ}\}_1$ and a transverse component $\{A^t{}_{PQ}\}_1$. This is illustrated in Figure 2.20 where one of the arms from a double wishbone suspension system is shown.

In this case the centripetal component of acceleration is given by

$$\{A^P{}_{PQ}\}_1 = \{\omega_2\}_1 \times \{\{\omega_2\}_1 \times \{R_{PQ}\}_1\} \tag{2.109}$$

Note that as the suspension arm is constrained to rotate about the axis NQ, ignoring at this stage any possible deflection due to compliance in the suspension bushes, the vectors $\{\omega_2\}_1$ for the angular velocity of Body 2 and $\{\alpha_2\}_1$ for the angular acceleration would act along the axis of rotation through NQ. The components of these vectors would adopt signs consistent with producing a positive rotation about this axis as shown in Figure 2.20.

When setting up the equations to solve a velocity or acceleration analysis it may be desirable to reduce the number of unknowns based on the knowledge that a particular body is constrained to rotate about a known axis as shown here. The velocity vector $\{\omega_2\}_1$ could, for example, be represented as follows

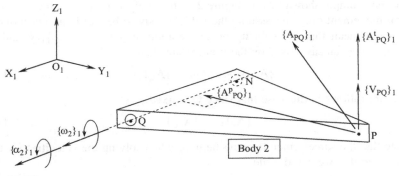

FIGURE 2.20

Centripetal and transverse components of acceleration vectors.

$$\{\omega_2\}_1 = f\omega_2\,\{R_{QN}\}_1 \tag{2.110}$$

In this case, since $\{\omega_2\}_1$ is parallel to the relative position vector $\{R_{QN}\}_1$ a scale factor $f\omega_2$ can be introduced. This would reduce the problem from the three unknown components, ωx_2, ωy_2 and ωz_2 of the vector $\{\omega_2\}_1$ to a single unknown $f\omega_2$. A similar approach could be used for an acceleration analysis with, for example

$$\{\alpha_2\}_1 = f\alpha_2\,\{R_{QN}\}_1 \tag{2.111}$$

It can also be seen from Figure 2.20 that the centripetal acceleration acts towards, and is perpendicular to, the axis of rotation of the body. This relationship can be proved having found the centripetal acceleration by use of the dot product with

$$\{A^P_{PQ}\}_1 \bullet \{R_{QN}\}_1 = 0 \tag{2.112}$$

The transverse component of acceleration is given by:

$$\{A^t_{PQ}\}_1 = \{\alpha_2\}_1 \times \{R_{PQ}\}_1 \tag{2.113}$$

Note that the transverse component of acceleration is also perpendicular to in this case the vector $\{R_{PQ}\}_1$ as defined by the dot product with

$$\{A^t_{PQ}\}_1 \bullet \{R_{PQ}\}_1 = 0 \tag{2.114}$$

Note that although the vector $\{A^t_{PQ}\}_1$ is shown to be acting in the same direction as the vector $\{V_{PQ}\}_1$ in Figure 2.20, this may not necessarily be the case. A reversal of $\{A^t_{PQ}\}_1$ would correspond to a reversal of $\{\alpha_2\}_1$. This would indicate that point P is moving in a certain direction but in fact decelerating.

The resultant acceleration vector $\{A_{PQ}\}_1$ is found to give the expression shown in Eqn (2.115) using the triangle law to add the centripetal and transverse components as follows

$$\{A_{PQ}\}_1 = \{A^P_{PQ}\}_1 + \{A^t_{PQ}\}_1 \tag{2.115}$$

For the example shown here in Figure 2.20 the analysis may often focus on suspension movement only and assume the vehicle body to be fixed and not moving. This would mean that the velocity or acceleration at point Q, $\{A_Q\}_1$ would be zero. Since we can say, based on the triangle law, that

$$\{A_{PQ}\}_1 = \{A_P\}_1 - \{A_Q\}_1 \tag{2.116}$$

it therefore follows in this case that since Q is fixed

$$\{A_P\}_1 = \{A_{PQ}\}_1 \tag{2.117}$$

Note that the same principle could be used when solving the velocities for this problem and that we could write

$$\{V_P\}_1 = \{V_{PQ}\}_1 \tag{2.118}$$

Finally, it should be noted for the particular example, shown here in Figure 2.20, we could work from either point Q or point N when solving for the velocities or accelerations and obtain the same answers for the velocity or acceleration at point P.

Combining the relative acceleration already obtained for points on a rigid body translating and rotating in space with sliding motion will introduce two more components of relative acceleration. This is best explained by considering the situation shown in Figure 2.21 where point P is located on Body 3 and point Q is located on Body 2. In this case Body 3 is constrained to move and rotate with Body 2 but has an additional relative sliding degree of freedom that allows it to move, relative to Body

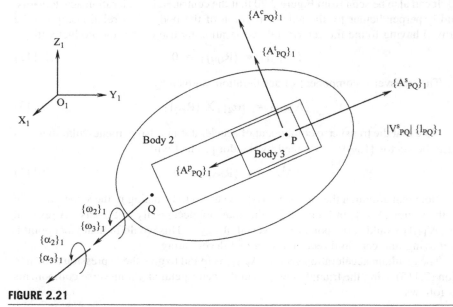

FIGURE 2.21

Relative acceleration with sliding motion.

2, along a slot with an axis aligned with the two points P and Q. To simplify the understanding, the two bodies are assumed to rotate, as shown, about an axis passing through point Q. Hence the relative acceleration vectors can be assumed to be acting in the directions shown, as point P moves away from point Q. The angular velocity and acceleration vectors for Body 2 and Body 3 will be the same and either may be used in the subsequent formulations.

The first of the two new components is easy to comprehend and is associated with the additional sliding motion. Since the direction of motion is known to be constrained to act along the line PQ it is possible to define the sliding acceleration, $\{A^s_{PQ}\}_1$, using the magnitude of the sliding acceleration $|A^s_{PQ}|$ factored with the unit vector $\{l_{PQ}\}_1$, acting along the line from Q to P, as follows

$$\{A^s_{PQ}\}_1 = |A^s_{PQ}| \, \{l_{PQ}\}_1 \tag{2.119}$$

The second of the two new components, $\{A^c_{PQ}\}_1$, is known as the Coriolis acceleration and requires more detailed explanation. As a starting point we can assume that both bodies are rotating as shown in Figure 2.21.

In deriving the Coriolis term, consider first the formulation for the velocity vector $\{V_{PQ}\}_1$. In addition to the component of velocity associated with rigid body motion that would act in a perpendicular direction to the line PQ there is an additional component of sliding velocity. This sliding component can be defined using the magnitude of the sliding velocity $|V^s_{PQ}|$ factored with the unit vector $\{l_{PQ}\}_1$, acting along the line from Q to P. The formulation of $\{V_{PQ}\}_1$ now becomes

$$\{V_{PQ}\}_1 = \{\omega_2\}_1 \times \{R_{PQ}\}_1 + |V^s_{PQ}| \, \{l_{PQ}\}_1 \tag{2.120}$$

Differentiating this with respect to time yields

$$\{A_{PQ}\}_1 = \frac{d}{dt} \{V_{PQ}\}_1 = \frac{d}{dt} \{\{\omega_2\}_1 \times \{R_{PQ}\}_1 + |V^s_{PQ}| \, \{l_{PQ}\}_1\} \tag{2.121}$$

$$\{A_{PQ}\}_1 = \{\omega_2\}_1 \times \{V_{PQ}\}_1 + \{\alpha_2\}_1 \times \{R_{PQ}\}_1 + |V^s_{PQ}| \frac{d}{dt} \{l_{PQ}\}_1 + |A^s_{PQ}| \, \{l_{PQ}\}_1 \tag{2.122}$$

Since

$$\frac{d}{dt} \{R_{PQ}\}_1 = \{V_{PQ}\}_1 = \{\omega_2\}_1 \times \{R_{PQ}\}_1 \tag{2.123}$$

it therefore follows that

$$\frac{d}{dt} \{l_{PQ}\}_1 = \{\omega_2\}_1 \times \{l_{PQ}\}_1 \tag{2.124}$$

and

$$|V^s_{PQ}| \, \{l_{PQ}\}_1 = \{V^s_{PQ}\}_1 \tag{2.125}$$

and

$$|A^s_{PQ}| \{l_{PQ}\}_1 = \{A^S_{PQ}\}_1 \qquad (2.126)$$

Combining these we therefore obtain

$$\{A_{PQ}\}_1 = \{\omega_2\}_1 \times \{V_{PQ}\}_1 + \{\alpha_2\}_1 \times \{R_{PQ}\}_1 + \{\omega_2\}_1 \times \{V^S_{PQ}\}_1 + \{A^S_{PQ}\}_1 \qquad (2.127)$$

Substituting

$$\{V_{PQ}\}_1 = \{\omega_2\}_1 \times \{R_{PQ}\}_1 + \{V^S_{PQ}\}_1 \qquad (2.128)$$

gives the following

$$\{A_{PQ}\}_1 = \{\omega_2\}_1 \times \{\{\omega_2\}_1 \times \{R_{PQ}\}_1 + \{V^S_{PQ}\}_1\} + \{\alpha_2\}_1 \times \{R_{PQ}\}_1 + \{\omega_2\}_1 \times \{V^S_{PQ}\}_1 + \{A^S_{PQ}\}_1 \qquad (2.129)$$

It therefore follows that

$$\{A_{PQ}\}_1 = \{\omega_2\}_1 \times \{\{\omega_2\}_1 \times \{R_{PQ}\}_1\} + \{\alpha_2\}_1 \times \{R_{PQ}\}_1 + 2\{\omega_2\}_1 \times \{V^S_{PQ}\}_1 + \{A^S_{PQ}\}_1 \qquad (2.130)$$

In summary, we can now identify all four components of acceleration associated with the combined rotation and sliding motion as the centripetal acceleration $\{A^P_{PQ}\}_1$, the transverse acceleration $\{A^t_{PQ}\}_1$, the Coriolis acceleration $\{A^c_{PQ}\}_1$ and the sliding acceleration $\{A^s_{PQ}\}_1$, where:

$$\{A^P_{PQ}\}_1 = \{\omega_2\}_1 \times \{\{\omega_2\}_1 \times \{R_{PQ}\}_1\} \qquad (2.131)$$

$$\{A^t_{PQ}\}_1 = \{\alpha_2\}_1 \times \{R_{PQ}\}_1 \qquad (2.132)$$

$$\{A^c_{PQ}\}_1 = 2\{\omega_2\}_1 \times \{V^S_{PQ}\}_1 \qquad (2.133)$$

$$\{A^s_{PQ}\}_1 = |A^s_{PQ}| \{l_{PQ}\}_1 \qquad (2.134)$$

2.6 Static force and moment definition

Before progressing to the development of the equations of motion associated with large displacement rigid body dynamic motion it is necessary to examine the use of vectors for static analysis in MBS. In this case we define static analysis as the study of forces acting on a body or series of bodies where motion takes place in the absence of accelerations. That is, a system that is either at rest or moving in a straight line with constant velocity.

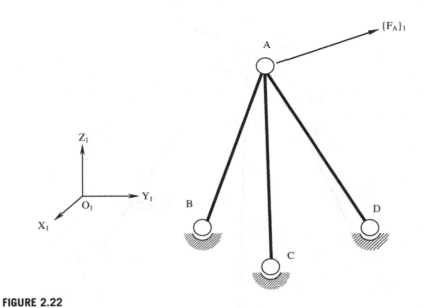

FIGURE 2.22

Tripod structure.

In order to demonstrate the use of vectors for representation of forces consider the following example, shown in Figure 2.22, which involves a tripod structure comprising three links with ball or spherical joints at each end.

Before attempting any analysis to determine the distribution of forces it is necessary to prepare a free body diagram and label the bodies and forces in an appropriate manner as shown in Figure 2.23.

The notation used here to describe the forces depends on whether the force is an applied external force (action-only force) or an internal force resulting from the interconnection of bodies (action–reaction force). In this example we have an applied force at point A. In order to fully define a force we must be able to specify the point of application, the line of sight and the sense of the force. In this case we can use the notation $\{F_A\}_1$ to define the force where the subscript A defines the point of application and the components of the vector F_{Ax}, F_{Ay} and F_{Az} would define both the line of sight and the sense of the force. Where the force is the result of an interaction we can use, for example, the following $\{F_{A52}\}_1$ that specifies that the force is acting at A on Body 5 due to its interaction with Body 2. Note that for this example we are assuming that A is a point that could be, for example, taken to be at the centre of a bush or spherical joint. It should also be noted that in assigning identification numbers to the bodies we are taking Body 1 to be a fixed and unmoving body that may also be referred to as a ground body. In this example the ground body is not in one place as such but can be considered to be located at the positions B, C and D where fixed anchorages are provided.

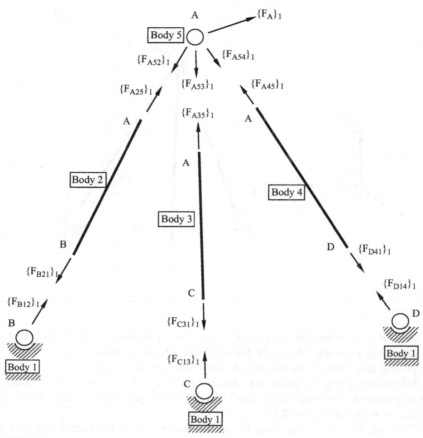

FIGURE 2.23

Free body diagram of tripod structure.

For the action–reaction forces shown here, Newton's Third Law would apply so that for the interaction of Body 5 and Body 2 we can say that $\{F_{A52}\}_1$ and $\{F_{A25}\}_1$ are equal and opposite or

$$\{F_{A52}\}_1 = -\{F_{A25}\}_1 \qquad (2.135)$$

In this example we are looking at linkages that are pin jointed, or have spherical joints at each end. An example of a similar linkage would be a tie rod in a suspension system. As both ends of the linkage are pin jointed, the force by definition must act along the linkage. In this case we could use this information to reduce the number of unknowns by using a scale factor as shown in Eqns (2.136)–(2.138).

$$\{F_{A25}\}_1 = f_2 \{R_{AB}\}_1 \qquad (2.136)$$

$$\{F_{A35}\}_1 = f_3 \{R_{AC}\}_1 \tag{2.137}$$

$$\{F_{A45}\}_1 = f_4 \{R_{AD}\}_1 \tag{2.138}$$

In this case the solution would be assisted as the three unknowns F_{A25x}, F_{A25y} and F_{A25z}, associated with Eqn (2.136), for example, would be reduced to a single unknown f_2. To maintain rigour the choice of position vector, for example, $\{R_{AB}\}_1$ rather than $\{R_{BA}\}_1$, has also been used so that comparing Eqn (2.136) with the forces as drawn in the free body diagram shown in Figure 2.23 would be consistent with the scale factors having positive values. Should the solution yield negative scale factors this would simply involve a reversal of the sense of the force from that initially assumed in the free body diagram.

In this case, since the forces act along the line of the linkages, we equate forces throughout the structure as follows

$$\{F_{A52}\}_1 = -\{F_{A25}\}_1 = \{F_{B21}\}_1 = -\{F_{B12}\}_1 \tag{2.139}$$

$$\{F_{A53}\}_1 = -\{F_{A35}\}_1 = \{F_{C31}\}_1 = -\{F_{C13}\}_1 \tag{2.140}$$

$$\{F_{A54}\}_1 = -\{F_{A45}\}_1 = \{F_{D41}\}_1 = -\{F_{D14}\}_1 \tag{2.141}$$

Clearly by solving the unknown forces acting on the pin, Body 5, which connects to all three linkages, the complete force distribution in this system will be known. Setting up the equation of equilibrium for Body 5 we get

$$\sum \{F_5\}_1 = \{0\}_1 \tag{2.142}$$

$$\{F_A\}_1 + \{F_{A52}\}_1 + \{F_{A53}\}_1 + \{F_{A54}\}_1 = \{0\}_1 \tag{2.143}$$

Using the information developed in Eqns (2.136)–(2.141) we can now write

$$\{F_A\}_1 + f_2\{R_{BA}\}_1 + f_3\{R_{CA}\}_1 + f_4\{R_{DA}\}_1 = \{0\}_1 \tag{2.144}$$

The direction of the position vectors as defined in Eqn (2.144) should be carefully noted. These have been selected to maintain the correct positive sign convention throughout the equation. Expanding Eqn (2.144) would lead to

$$\begin{bmatrix} F_{Ax} \\ F_{Ay} \\ F_{Az} \end{bmatrix} + f_2 \begin{bmatrix} BA_x \\ BA_y \\ BA_z \end{bmatrix} + f_3 \begin{bmatrix} CA_x \\ CA_y \\ CA_z \end{bmatrix} + f_4 \begin{bmatrix} DA_x \\ DA_y \\ DA_z \end{bmatrix} = \begin{bmatrix} 0 \\ 0 \\ 0 \end{bmatrix} \tag{2.145}$$

For a given applied force $\{F_A\}_1$, and taking the geometry of the points A, B, C and D to be known, Eqn (2.145) yields the three equations required to solve the three unknowns f_2, f_3 and f_4 and hence solve the force distribution in this example.

The notation that has been used to define forces in the example shown can be used where the forces represent reactions generated at a point in a multibody system.

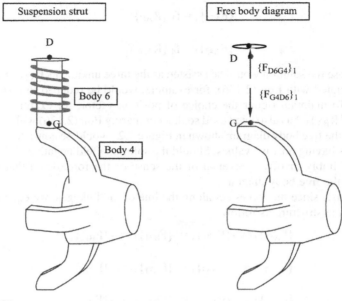

FIGURE 2.24

Vector notation for line of sight forces.

The notation needs to be altered for forces acting along the line of sight of a force element, such as a spring or damper, connecting two bodies as shown in Figure 2.24.

The vector representation of a moment is not as straightforward to interpret as the vector representation of a force. The moment $\{M_P\}_1$ shown acting about point P in Figure 2.25 is represented by a vector that is orientated along an axis about which the moment acts. The length of the vector represents the magnitude of the moment and the direction of the vector is that which is consistent with a positive rotation about the axis as shown. The components of the vector, M_{Px}, M_{Py} and M_{Pz} are resolved parallel to a reference frame, in this case O_1. The double-headed arrows used in Figure 2.25 are intended to distinguish the moment vector from that of a force.

Considering next the equilibrium of the body shown in Figure 2.26 it can be seen that a force acting at point P will produce a reaction moment at point Q. In order to simplify the diagram only the components of the applied force and the resulting moment are shown here. Since we are looking only at the derivation of the moment at Q the reaction force at Q is also omitted.

The moment generated at point Q, which is the point where Body 2 is fixed to the non-moving ground part Body 1, is designated as $\{M_{Q21}\}_1$. Applying the same principle as used for a force we would read this as the moment acting at point Q, on Body 2 due to its connection to Body 1, with components resolved parallel to the axes of reference frame O_1. For convenience and to assist with the interpretation of the derivation that follows the components of the force $\{F_P\}_1$, the moment $\{M_{Q21}\}_1$ and the

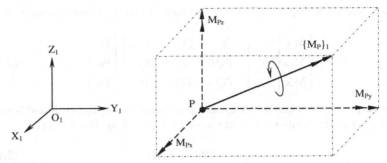

FIGURE 2.25

Vector representation of a moment.

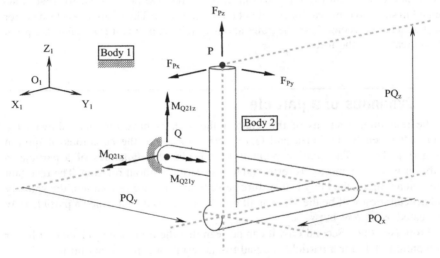

FIGURE 2.26

Reaction moment between two bodies.

relative position vector $\{R_{PQ}\}_1$ have all been set up in Figure 2.26 to have components that are positive.

Using the traditional approach the three equations of moment equilibrium, $\sum M_{Qx} = 0$, $\sum M_{Qy} = 0$ and $\sum M_{Qz} = 0$ could be transformed to produce the following three equations:

$$M_{Q21x} = -PQ_z \times F_{Py} + PQ_y \times F_{Pz} \tag{2.146}$$

$$M_{Q21y} = PQ_z \times F_{Px} - PQ_x \times F_{Pz} \tag{2.147}$$

$$M_{Q21z} = PQ_x \times F_{Py} - PQ_y \times F_{Px} \tag{2.148}$$

Taking the above Eqns (2.146)–(2.148) and arranging in matrix form gives

$$
\begin{bmatrix} M_{Qx} \\ M_{Qy} \\ M_{Qz} \end{bmatrix} = \begin{bmatrix} 0 & -PQ_z & PQ_y \\ PQ_z & 0 & -PQ_x \\ -PQ_y & PQ_x & 0 \end{bmatrix} \begin{bmatrix} F_{Px} \\ F_{Py} \\ F_{Pz} \end{bmatrix}
\tag{2.149}
$$

From Eqn (2.149) it can be seen that the moment $\{M_Q\}_1$ can be found from the cross product of the relative position vector $\{R_{PQ}\}_1$ and the force $\{F_P\}_1$.

$$
\{M_Q\}_1 = \{R_{PQ}\}_1 \times \{F_P\}_1
\tag{2.150}
$$

It should be noted that in using the vector cross product to compute the moment of a force about a point that the order of the operation is critical. The relative position vector is crossed with the force so that it is the relative position vector that is arranged in skew-symmetric form and not the force vector. The relative position vector must also be the vector from the point about which the moment is taken to the point of application of the force.

2.7 Dynamics of a particle

In the remaining sections of this chapter the authors have broadly followed the approach given by D'Souza and Garg (1984) to derive the equations of motion for a rigid body. The starting point is to consider the dynamics of a particle, a body for which the motion is restricted to translation without rotation. The resultant moment acting on the body is therefore zero. In the absence of rotation, the velocity and acceleration will be the same at all points on the body and hence a particle may be treated as a point mass.

From Newton's Second Law it can be seen that the time rate of change of linear momentum $\{L\}_1$ for a particle is equal to the resultant force acting on it.

$$
\sum \{F\}_1 = \frac{d}{dt}\{L\}_1 = \frac{d}{dt}\left(m\{V\}_1\right)
\tag{2.151}
$$

The resultant force is represented by the vector $\sum\{F\}_1$, m is the mass and $\{V\}_1$ is the velocity vector measured relative to an inertial reference frame O_1. The linear momentum of the body is $m\{V\}_1$. The components of the vectors in Eqn (2.151) are all resolved parallel to the axes of O_1. Taking the mass to be constant, Eqn (2.151) may be written as

$$
\sum \{F\}_1 = m \frac{d}{dt}\{V\}_1 = m\{A\}_1
\tag{2.152}
$$

The accelerations in Eqn (2.152) are also measured relative to the inertial reference frame O_1. Since the velocities and accelerations used here are measured relative to a nonmoving frame we refer to them as absolute. Expanding the vector equation given in (2.152) gives

$$\sum \begin{bmatrix} F_x \\ F_y \\ F_z \end{bmatrix} = m \begin{bmatrix} A_x \\ A_y \\ A_z \end{bmatrix} \tag{2.153}$$

2.8 Linear momentum of a rigid body

As a body translates and rotates in space it will have linear momentum $\{L\}_1$ associated with translation and angular momentum $\{H\}_1$ associated with rotation. For the rigid body, Body 2, shown in Figure 2.27 the mass centre is located at G_2 by the vector $\{R_{G2}\}_1$ relative to the reference frame O_2.

A small element of material with a volume δV is located at P relative to O_2 by the position vector $\{R_P\}_1$. Assuming Body 2 to be of uniform density ρ, we can say that the element of material has a mass δm given by

$$\delta m = \rho \, \delta V \tag{2.154}$$

and that as the element becomes infinitesimal the mass of the body m_2 is given by

$$m_2 = \rho \int_{vol} dV = \int_{vol} dm \tag{2.155}$$

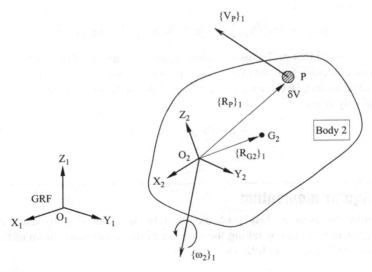

FIGURE 2.27

Linear momentum of a rigid body. GRF, ground reference frame.

The position of the centre of mass $\{R_{G2}\}_1$ is then found by integrating the elemental first moments of mass about frame O_2 and dividing by the total mass m_2.

$$\{R_{G2}\}_1 = \frac{1}{m_2} \int_{vol} \{R_P\}_1 \, dm \tag{2.156}$$

The linear momentum $\{L_2\}_1$ of the body is the linear momenta of the elements of mass that comprise the body. If the position of an element of mass is given by $\{R_P\}_1$ then the velocity $\{V_P\}_1$ of dm is given using the triangle law of vector addition by

$$\{V_P\}_1 = \{V_{O2}\}_1 + \{V_{PO2}\}_1$$
$$\tag{2.157}$$
$$\{V_P\}_1 = \{V_{O2}\}_1 + \{\omega_2\}_1 \times \{R_P\}_1$$

The linear momentum $\{L\}_1$ of the body is therefore found by integrating the mass particles factored by their velocity vectors $\{V_P\}_1$ over the volume of the body

$$\{L_2\}_1 = \int_{vol} \{V_P\}_1 \, dm \tag{2.158}$$

Using the expression for $\{V_P\}_1$ given in Eqn (2.157) leads to

$$\{L_2\}_1 = \{V_{O2}\}_1 \int_{vol} dm + \{\omega_2\}_1 \times \int_{vol} \{R_P\}_1 dm \tag{2.159}$$

Using the expressions given in Eqns (2.155) and (2.156) we get an expression for the linear momentum of Body 2 in terms of the overall mass m_2 and the velocity at the mass centre $\{V_{G2}\}_1$.

$$\{L_2\}_1 = m_2 \{\{V_{O2}\}_1 + \{\omega_2\}_1 \times \{R_{G2}\}_1\} = m_2 \{V_{G2}\}_1 \tag{2.160}$$

It can be noted that problems are often set up in multibody dynamics so that the body frame coincides with the mass centre. In this case for Body 2, O_2 and G_2 would be coincident so that $\{RG_2\}_1$ would be zero and the linear momentum would be obtained directly from

$$\{L_2\}_1 = m_2 \{V_{G2}\}_1 = m_2 \{V_{O2}\}_1 \tag{2.161}$$

2.9 Angular momentum

The angular momentum $\{H_P\}_1$ of the particle of material with mass dm in Figure 2.27 can be found by taking the moment of the linear momentum of the particle about the frame O_2 as follows

$$\{H_P\}_1 = \{R_P\}_1 \times \{\{V_{O2}\}_1 + \{\omega_2\}_1 \times \{R_P\}_1\} \, dm \tag{2.162}$$

Integrating this over the volume of the body leads to the angular momentum $\{H_2\}_1$ of the rigid body about the frame O_2 as follows

$$
\begin{aligned}
\{H_2\}_1 &= \int_{vol} \{R_P\}_1 \times \left\{ \{V\}_{O2} + \{\omega_2\}_1 \times \{R_P\}_1 \right\}.dm \\
&= -\{V\}_{O2} \times \int_{vol} \{R_P\}_1.dm + \int_{vol} \{R_P\}_1 \times \left\{ \{\omega_2\}_1 \times \{R_P\}_1 \right\}.dm
\end{aligned}
\tag{2.163}
$$

Note that in the second line of Eqn (2.163), the vector $\{V_{O2}\}_1$ comes out of the integral and hence the order of the cross product is reversed necessitating the reverse of sign for $\{V_{O2}\}_1$. If for Body 2, frame O_2 is positioned either at the mass centre, so that $\int_{vol} \{R_P\}_1 \ dm = \{0\}_1$, or at a point of attachment to the nonmoving ground, where $\{V_{O2}\}_1 = \{0\}_1$, then Eqn (2.163) reduces to the more convenient form, which we will assume from now on.

$$
\{H_2\}_1 = \int_{vol} \{R_P\}_1 \times \left\{ \{\omega_2\}_1 \times \{R_P\}_1 \right\} dm
\tag{2.164}
$$

If we now take the general case for any body (ignore body subscripts) and expand the vectors into their full form we get

$$
\{H\}_1 = \begin{bmatrix} H_x \\ H_y \\ H_z \end{bmatrix}, \quad \{R\}_1 = \begin{bmatrix} x \\ y \\ z \end{bmatrix} \quad \text{and} \quad \{\omega\}_1 = \begin{bmatrix} \omega_x \\ \omega_y \\ \omega_z \end{bmatrix}
$$

Applying the vector cross product, making use of the skew-symmetric form of a vector in the normal manner, leads to

$$
\begin{aligned}
\{R\}_1 \times \left\{ \{\omega\}_1 \times \{R\}_1 \right\} &=
\begin{bmatrix} 0 & -z & y \\ z & 0 & -x \\ -y & x & 0 \end{bmatrix}
\begin{bmatrix} 0 & -\omega_z & \omega_y \\ \omega_z & 0 & -\omega_x \\ -\omega_y & \omega_x & 0 \end{bmatrix}
\begin{bmatrix} x \\ y \\ z \end{bmatrix} \\
&= \begin{bmatrix} 0 & -z & y \\ z & 0 & -x \\ -y & x & 0 \end{bmatrix}
\begin{bmatrix} -\omega_z y + \omega_y z \\ -\omega_z x - \omega_x z \\ -\omega_y x + \omega_x y \end{bmatrix} \\
&= \begin{bmatrix} y(\omega_x y - \omega_y x) - z(\omega_z x - \omega_x z) \\ z(\omega_y z - \omega_z y) - x(\omega_x y - \omega_y x) \\ x(\omega_z x - \omega_x z) - y(\omega_y z - \omega_z y) \end{bmatrix}
\end{aligned}
\tag{2.165}
$$

Substituting Eqn (2.165) into Eqn (2.164) gives the general expression for the angular momentum $\{H\}_1$ of a body, where for simplicity now the integral sign is taken to indicate integration over the volume of the body.

$$
\{H\}_1 = \int \{R\}_1 \times \left\{ \{\omega\}_1 \times \{R\}_1 \right\} dm
\tag{2.166}
$$

$$H_x = \omega_x \int (y^2 + z^2)\, dm - \omega_y \int xy\, dm - \omega_z \int xz\, dm \qquad (2.167)$$

$$H_y = -\omega_x \int xy\, dm + \omega_y \int (x^2 + z^2)\, dm - \omega_z \int yz\, dm \qquad (2.168)$$

$$H_z = -\omega_x \int xz\, dm - \omega_y \int yz\, dm + \omega_z \int (x^2 + y^2)\, dm \qquad (2.169)$$

It is now possible to substitute into Eqns (2.167)–(2.169) the following general terms for the moments of inertia I_{xx}, I_{yy} and I_{zz} of the rigid body.

$$I_{xx} = \int (y^2 + z^2)\, dm \qquad (2.170)$$

$$I_{yy} = \int (x^2 + z^2)\, dm \qquad (2.171)$$

$$I_{zz} = \int (x^2 + y^2)\, dm \qquad (2.172)$$

In addition we can introduce the products of inertia I_{xy}, I_{yz} and I_{xz}.

$$I_{xy} = I_{yx} = -\int xy\, dm \qquad (2.173)$$

$$I_{yz} = I_{zy} = -\int yz\, dm \qquad (2.174)$$

$$I_{xz} = I_{zx} = -\int xz\, dm \qquad (2.175)$$

This allows Eqns (2.167)–(2.169) to be arranged in matrix form as follows

$$\begin{bmatrix} H_x \\ H_y \\ H_z \end{bmatrix} = \begin{bmatrix} I_{xx} & I_{xy} & I_{xz} \\ I_{xy} & I_{yy} & I_{yz} \\ I_{xz} & I_{yz} & I_{zz} \end{bmatrix} \begin{bmatrix} \omega_x \\ \omega_y \\ \omega_z \end{bmatrix} \qquad (2.176)$$

If we return to our earlier consideration of Body 2 shown in Figure 2.27, the matrix equation in Eqn (2.176) would lead to

$$\{H_2\}_{1/1} = [I_2]_{2/1} \, \{\omega_2\}_{1/1} \qquad (2.177)$$

In writing the vectors $\{H_2\}_{1/1}$ $\{\omega_2\}_{1/1}$ we revert to the full definition of a vector used here where the upper suffix indicates that the vector is measured relative to the axes of reference frame O_1 and the lower suffix indicates that the components of the vector are resolved parallel to the axes of frame O_1. The matrix $[I_2]_{2/1}$ is the moment of inertia matrix for Body 2 about its mass centre G_2 located at frame O_2. The use of the upper and lower suffixes here indicates that the moments of inertia have been measured relative to frame O_2 but transformed to frame O_1. This is necessary so that the vector operation in Eqn (2.171) is consistent. This is only possible if the vectors and matrix are referred to the same frame, which in this case is O_1.

Note that in this form Eqn (2.177) is not practical since the orientation of frame O_2 relative to frame O_1 will change as the body rotates requiring the recomputation of $[I_2]_{2/1}$ at each time step. The matrix $[I_2]_{2/2}$, or in simpler form $[I_2]_2$, is constant since it is measured relative to and referred to a frame that is fixed in Body 2 and hence only needs to be determined once for an undeformable body. When considering the equation for the angular momentum of a body it is preferable therefore to consider all quantities to be referred to a frame fixed in the body, in this case frame O_2.

$$\{H_2\}_{1/2} = [I_2]_{2/2}\,\{\omega_2\}_{1/2} \tag{2.178}$$

Before progressing to develop the equations used to describe the dynamics of rigid bodies translating and rotating in three-dimensional space, the definition of the moments of inertia introduced here requires further consideration.

2.10 Moments of inertia

From our previous consideration of the angular momentum of a rigid body, we see that there are three moments of inertia and three products of inertia the values of which must be specified to analyse the rotational motion of the body. Before considering the three-dimensional situation it is useful to start with the two-dimensional inertial properties associated with plane motion.

For the Body 2 shown in Figure 2.28 we assume that a constraint has been applied that allows the body to move only in the X_1Y_1 plane of frame O_1.

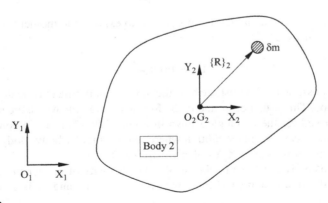

FIGURE 2.28

Moment of inertia for plane motion.

As such the body has three degrees of freedom, which are translation in the X_1 direction, translation in the Y_1 direction and rotation about the Z_1-axis. From the expressions given earlier for the moments of inertia it can be seen that these are in fact the second moments of the mass distribution about the chosen frame fixed in the body.

For the body shown in Figure 2.28, we are only interested in rotation about the Z_1-axis and as such only require the I_{zz} moment of inertia. To indicate that this is for Body 2 we will refer to this as I_{2zz}. Considering the particle of mass δm located by a general position vector $\{R\}_2$ at point P, we see that we are not only measuring the vector with respect to frame O_2 but we are also referring the vector to frame O_2.

Ignoring the z coordinate as this is for plane motion in $X_1 Y_1$ we can take the general case and say that the position of the element is given by $\{R\}_2^T = [x \ y \ 0]$. The moment of inertia I_{2zz} is found by summing the second moments of the elements of mass over the volume of the body.

$$\begin{aligned} I_{2zz} &= \sum |R|_2^2 \ \delta m \\ &= \sum (x^2 + y^2) \ \delta m \end{aligned} \tag{2.179}$$

In the limit as the volume of the element becomes infinitesimal, we can obtain an expression for the moment of inertia I_{2zz} as follows

$$I_{2zz} = \int (x^2 + y^2) \, dm \tag{2.180}$$

The moment of inertia I_{2zz} in Eqn (2.180) is therefore the integral of the mass elements, each multiplied by the square of the radial distance from the z-axis. This distance is referred to as the radius of gyration, in this case k_{2zz} where this is given by

$$k_{2zz} = \sqrt{\frac{I_{2zz}}{m_2}} \tag{2.181}$$

From this we can use the radius of gyration to express the moment of inertia in more general terms as

$$I_{2zz} = m_2 \, k_{2zz}^2 \tag{2.182}$$

It is now possible to demonstrate how the moment of inertia may be derived for a standard shape. This is demonstrated in the following example where the moment of inertia is derived for the rectangle shown in Figure 2.29. This again considers the two-dimensional case for plane motion where the rectangular body, Body 2, is constrained to only move in the $X_1 Y_1$ plane of frame O_1.

Taking this body to have a thickness of t and a density of ρ, we can say that the mass δm of the small element of mass with dimensions δx and δy is given by

$$\delta m = \rho \, t \, \delta x \, \delta y \tag{2.183}$$

The moment of inertia I_{2zz} is again found by summing the second moments of the elements of mass over the volume of the body.

$$I_{2zz} = \Sigma (x^2 + y^2) \, \delta m$$
$$= \rho t \, \Sigma (x^2 + y^2) \, \delta x \, \delta y \qquad (2.184)$$

This leads to the following equation as the volume of the element becomes infinitesimal.

$$I_{2zz} = \rho t \int_{-d/2}^{d/2} \int_{-b/2}^{b/2} (x^2 + y^2) \, dx \, dy \qquad (2.185)$$

Solving this double integral gives

$$I_{2zz} = \rho t \int_{-d/2}^{d/2} \left[\frac{x^3}{3} + y^2 x \right]_{-b/2}^{b/2} dy$$

$$I_{2zz} = \rho t \int_{-d/2}^{d/2} \left(\frac{b^3}{24} + \frac{y^2 b}{2} \right) - \left(-\frac{b^3}{24} - \frac{y^2 b}{2} \right) dy$$

$$I_{2zz} = \rho t \int_{-d/2}^{d/2} \left(\frac{b^3}{12} + y^2 b \right) dy$$

$$\qquad (2.186)$$

$$I_{2zz} = \rho t \left[\frac{b^3 y}{12} + \frac{y^3 b}{3} \right]_{-d/2}^{d/2}$$

$$I_{2zz} = \rho t \left(\frac{b^3 d}{24} + \frac{d^3 b}{24} \right) - \left(-\frac{b^3 d}{24} + \frac{d^3 b}{2} \right)$$

$$I_{2zz} = \rho t \left(\frac{b^3 d + d^3 b}{12} \right)$$

Since the mass of the rectangle body m_2 is given by $m_2 = \rho t b d$, we can write

$$I_{2zz} = m_2 \left(\frac{b^2 + d^2}{12} \right) \qquad (2.187)$$

Another example of a standard shape is the ring shown in Figure 2.30. This again considers the two-dimensional case for plane motion, Body 2 is constrained to only move in the $X_1 Y_1$ plane of frame O_1.

Taking this body to have a thickness of t and a density of ρ, we can say that the mass δm of the small elemental ring of mass, at a radius $R = |\, R\, |_2$, with radial width δr is given by

$$\delta m = \rho \, t \, 2 \, \pi \, R \, \delta r \qquad (2.188)$$

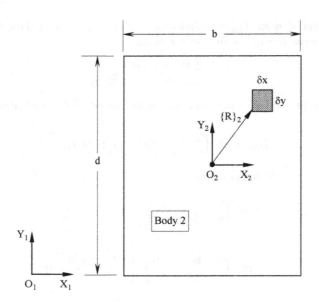

FIGURE 2.29

Moment of inertia for a rectangle.

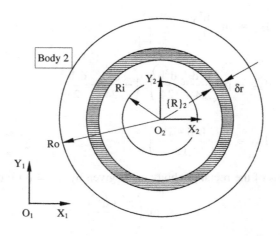

FIGURE 2.30

Moment of inertia for a ring.

The moment of inertia I_{2zz} is again found by summing the second moments of the elements of mass over the volume of the body.

$$\begin{aligned} I_{2zz} &= \sum R^2 \, \delta m \\ &= 2\pi\rho t \sum R \cdot R^2 \, \delta r \end{aligned} \tag{2.189}$$

This leads to the following equation as the volume of the element becomes infinitesimal.

$$I_{2zz} = 2\pi\rho t \int_{Ri}^{Ro} R^3 \, dr \tag{2.190}$$

Solving this integral gives

$$I_{2zz} = 2\pi\rho t \left[\frac{R^4}{4} \right]_{Ri}^{Ro}$$

$$I_{2zz} = \frac{\pi t \rho}{2} \left(Ro^4 - Ri^4 \right)$$

Since the mass of the ring m_2 is given by $m_2 = \rho t \pi (Ro^2 - Ri^2)$, we can write

$$I_{2zz} = \frac{m_2}{2} \frac{\left(Ro^4 - Ri^4 \right)}{\left(Ro^2 - Ri^2 \right)} \tag{2.191}$$

$$I_{2zz} = \frac{m_2}{2} \left(Ro^2 + Ri^2 \right)$$

2.11 Parallel axes theorem

If a rigid body comprises rigidly attached combinations of regular shapes, such as those just described, the overall inertial properties of the body may be found using the parallel axes theorem. Returning to the three-dimensional situation we can consider the two parallel axes systems O_2 and O_3 both fixed in Body 2 as shown in Figure 2.31.

It is now possible to show that there is an inertia matrix for Body 2 associated with frame O_2, which would be written $[I_2]_{2/2}$. In a similar manner it is possible to determine the terms in a moment of inertia matrix $[I_2]_{3/2}$, where the use of the upper suffix here indicates that the moments of inertia have been measured relative to the origin of frame O_3 and the lower suffix indicates that the terms in the matrix are transformed to frame O_2. Since O_2 and O_3 are parallel the matrix $[I_2]_{3/3}$ would be identical to $[I_2]_{3/2}$.

The positions of the frames O_2 and O_3 relative to G_2 the mass centre of Body 2 can be given by

$$\{R_{O2G2}\}_2 = \begin{bmatrix} x_2 \\ y_2 \\ z_2 \end{bmatrix} \quad , \quad \{R_{O3G2}\}_2 = \begin{bmatrix} x_3 \\ y_3 \\ z_3 \end{bmatrix} \tag{2.192}$$

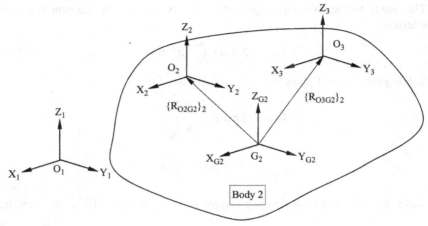

FIGURE 2.31

Parallel axes theorem.

where according to the triangle law of vector addition if a, b and c are the components of the relative position vector $\{R_{O3O2}\}_2$ we can write

$$\{R_{O3O2}\}_2 = \begin{bmatrix} a \\ b \\ c \end{bmatrix} \quad , \quad \{R_{O3G2}\}_2 = \begin{bmatrix} x_2+a \\ y_2+b \\ z_2+c \end{bmatrix} \qquad (2.193)$$

On this basis it is possible to relate a moment of inertia, for example $I_2x_3x_3$ for frame O_3 to $I_2x_2x_2$ for frame O_2.

$$
\begin{aligned}
I_2x_3x_3 &= \int \left(y_3^2 + z_3^2 \right) dm \\
&= \int \left[(y_2 + b)^2 + (z_2 + c)^2 \right] dm \\
&= \int \left[\left(y_2^2 + 2y_2\,b + b^2 \right) + \left(z_2^2 + 2z_2\,c + c^2 \right) \right] dm \qquad (2.194) \\
&= I_2x_2x_2 + 2b \int y_2\,dm + 2c \int z_2\,dm + \left(b^2 + c^2 \right) m_2 \\
&= I_2x_2x_2 + 2m_2 \left(by_2 + cz_2 \right) + m_2 \left(b^2 + c^2 \right)
\end{aligned}
$$

If we take the situation where O_2 is coincident with G_2 the mass centre of Body 2, such that x_2, y_2 and z_2 are zero, then Eqn (2.194) can be simplified to:

$$I_2x_3x_3 \;=\; I_2x_2x_2 + m_2 \left(b^2 + c^2 \right) \qquad (2.195)$$

In a similar manner it is possible to relate a product of inertia, for example $I_2y_3z_3$ for frame O_3 to $I_2y_2z_2$ for frame O_2.

$$
\begin{aligned}
I_2y_3z_3 &= -\int y_3\, z_3\ dm \\[4pt]
&= -\int (y_2 + b)(z_2 + c)\ dm \\[4pt]
&= -\int y_2\, z_2 + cy_2 + bz_2 + bc\ dm \\[4pt]
&= I_2y_2z_2 - m_2(cy_2 + bz_2) - m_2bc
\end{aligned}
\tag{2.196}
$$

Taking again O_2 to lie at the mass centre G_2 we can simplify Eqn (2.196) to

$$
I_2y_3z_3 \;=\; I_2y_2z_2 - m_2\,bc
\tag{2.197}
$$

On the basis of the derivation of the relationships in Eqns (2.195) and (2.197), we can find in a similar manner the full relationship between $[I_2]_{3/2}$ and $[I_2]_{2/2}$ to be

$$
[I_2]_{3/2} \;=\; [I_2]_{2/2} + m_2
\begin{bmatrix}
b^2 + c^2 & -ab & -ac \\
-ab & c^2 + a^2 & -bc \\
-ac & -bc & a^2 + b^2
\end{bmatrix}
\tag{2.198}
$$

A practical application of the parallel axes theorem given in Eqn (2.198) is provided using the simplified representation of a tie rod as shown in Figure 2.32. The body can be considered as an assembly of three components with centres of mass at G_1, G_2 and G_3. The mass centre of the entire body is located at G. The components have masses m_1, m_2 and m_3 and moments of inertia about the local z-axis at each mass centre I_{G1zz}, I_{G2zz} and I_{G3zz}.

Applying the parallel axes theorem would, in this case, give a moment of inertia I_{Gzz} for the body using

$$
I_{Gzz} = I_{G1zz} + m_1\, a_1^{\,2} \;+\; I_{G2zz} + m_2\, a_2^{\,2} \;+\; I_{G3zz} + m_3\, a_3^{\,2}
\tag{2.199}
$$

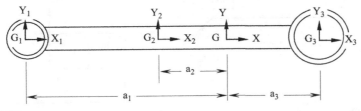

FIGURE 2.32

Application of parallel axes theorem.

2.12 Principal axes

The principal axes of any rigid body are those for which the products of inertia are all zero resulting in an inertia matrix of the form

$$[I] = \begin{bmatrix} I_1 & 0 & 0 \\ 0 & I_2 & 0 \\ 0 & 0 & I_3 \end{bmatrix} \tag{2.200}$$

where in this case I_1, I_2 and I_3 are the principal moments of inertia. The three planes formed by the principal axes are referred to as the principal planes, as shown in Figure 2.33. In this example the geometry chosen is a solid cylinder to demonstrate the concept.

For the cylinder the principal axes are represented by the frame O_2 positioned at the mass centre of the body. In this case each of the principal planes is a plane of symmetry for the body. As can be seen for each element of mass with positive co-ordinates there are other elements of mass, reflected in each of the principal planes, with negative coordinates. The result of this is that the products of inertia are all zero.

Returning to the consideration of the angular momentum of a body given in Eqn (2.178), this can now be written as shown in Eqn (2.201) where I_1, I_2 and I_3 are the principal moments of inertia for Body 2 taken about the origin of frame O_2. The moment of inertia matrix is referred to the principal axes, again frame O_2 and the products of inertia are zero.

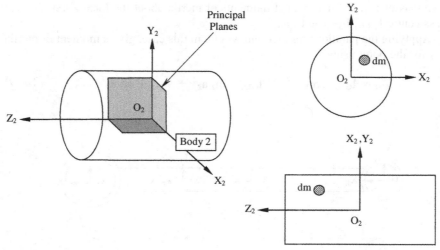

FIGURE 2.33

Principal axes for a body.

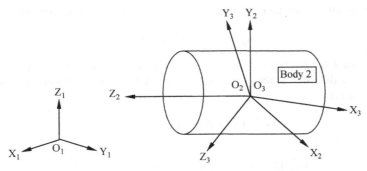

FIGURE 2.34

Transformation to principal axes.

$$\{H_2\}_{1/2} = [I_2]_{2/2}\,\{\omega_2\}_{1/2}$$

$$
\begin{bmatrix} H_{2x} \\ H_{2y} \\ H_{2z} \end{bmatrix}
=
\begin{bmatrix} I_1 & 0 & 0 \\ 0 & I_2 & 0 \\ 0 & 0 & I_3 \end{bmatrix}
\begin{bmatrix} \omega_{2x} \\ \omega_{2y} \\ \omega_{2z} \end{bmatrix}
=
\begin{bmatrix} I_1 & \omega_{2x} \\ I_2 & \omega_{2y} \\ I_3 & \omega_{2z} \end{bmatrix}
\tag{2.201}
$$

The principal axes and the principal moments of inertia may be obtained by considering the two frames O_3 and O_2 both located at the mass centre in Body 2, as shown in Figure 2.34. The axes associated with frame O_2 are again taken to be the principal axes of the body. The angular momentum $\{H_2\}_{1/2}$ referred to O_2 can be transformed from the angular momentum $\{H_2\}_{1/3}$ referred to O_3 using the rotational transformation matrix $[T_3]_2$ as described in Section 2.2.7.

$$\{H_2\}_{1/2} = [T_3]_2\,\{H_2\}_{1/3} \tag{2.202}$$

Since $\{H_2\}_{1/2} = [I_2]_{2/2}\,\{\omega_2\}_{1/2}$ and $\{H_2\}_{1/3} = [I_2]_{2/3}\,\{\omega_2\}_{1/3}$ it follows from Eqn (2.202) that

$$[I_2]_{2/2}\,\{\omega_2\}_{1/2} = [T_3]_2\,[I_2]_{2/3}\,\{\omega_2\}_{1/3} \tag{2.203}$$

If we premultiply a transformation matrix $[T]$ by its transpose $[T]^T$ a property of these matrices is that we produce an identity matrix.

$$[T]^T\,[T] = \begin{bmatrix} 1 & 0 & 0 \\ 0 & 1 & 0 \\ 0 & 0 & 1 \end{bmatrix} \tag{2.204}$$

This allows us to further develop Eqn (2.203) to give

$$[I_2]_{2/2}\,\{\omega_2\}_{1/2} = [T_3]_2\,[I_2]_{2/3}\,[T_3]_2^{\,T}\,[T_3]_2\,\{\omega_2\}_{1/3} \tag{2.205}$$

Since we also know that $\{\omega_2\}_{1/2} = [T_3]_2 \{\omega_2\}_{1/3}$ we can therefore write

$$[I_2]_{2/2} = [T_3]_2 \, [I_2]_{2/3} \, [T_3]_2^T \tag{2.206}$$

In this case we require a transformation matrix $[T_3]_2$, which is orthogonal and will lead to a diagonal matrix $[I_2]_{2/2}$. If we now take Eqn (2.206) and premultiply by $[T_3]_2^T$ we get

$$[T_3]_2^T \, [I_2]_{2/2} = [I_2]_{2/3} \, [T_3]_2^T$$

$$
\begin{bmatrix} T_{11} & T_{21} & T_{31} \\ T_{12} & T_{22} & T_{32} \\ T_{13} & T_{23} & T_{33} \end{bmatrix}
\begin{bmatrix} I_1 & 0 & 0 \\ 0 & I_2 & 0 \\ 0 & 0 & I_3 \end{bmatrix}
=
\begin{bmatrix} I_{xx} & I_{xy} & I_{xz} \\ I_{xy} & I_{yy} & I_{yz} \\ I_{xz} & I_{yz} & I_{zz} \end{bmatrix}
\begin{bmatrix} T_{11} & T_{21} & T_{31} \\ T_{12} & T_{22} & T_{32} \\ T_{13} & T_{23} & T_{33} \end{bmatrix}
$$

$$
\begin{bmatrix} T_{11}I_1 & T_{21}I_2 & T_{31}I_3 \\ T_{12}I_1 & T_{22}I_2 & T_{32}I_3 \\ T_{13}I_1 & T_{23}I_2 & T_{33}I_3 \end{bmatrix}
=
\begin{bmatrix}
I_{xx}T_{11}+I_{xy}T_{12}+I_{xz}T_{13} & I_{xx}T_{21}+I_{xy}T_{22}+I_{xz}T_{23} & I_{xx}T_{31}+I_{xy}T_{32}+I_{xz}T_{33} \\
I_{xy}T_{11}+I_{yy}T_{12}+I_{xz}T_{13} & I_{xy}T_{21}+I_{yy}T_{22}+I_{yz}T_{23} & I_{xy}T_{31}+I_{yy}T_{32}+I_{yz}T_{33} \\
I_{xz}T_{11}+I_{yz}T_{12}+I_{zz}T_{13} & I_{xz}T_{21}+I_{yz}T_{22}+I_{zz}T_{23} & I_{xz}T_{31}+I_{yz}T_{32}+I_{zz}T_{33}
\end{bmatrix}
\tag{2.207}
$$

Equating the first columns from the matrices on either side of Eqn (2.207) leads to the eigenvalue equation in (2.208)

$$
I_1 \begin{bmatrix} T_{11} \\ T_{12} \\ T_{13} \end{bmatrix}
=
\begin{bmatrix} I_{xx} & I_{xy} & I_{xz} \\ I_{xy} & I_{yy} & I_{yz} \\ I_{xz} & I_{yz} & I_{zz} \end{bmatrix}
\begin{bmatrix} T_{11} \\ T_{12} \\ T_{13} \end{bmatrix}
\tag{2.208}
$$

This equation can be rearranged to give

$$
\begin{bmatrix} I_{xx}-I_1 & I_{xy} & I_{xz} \\ I_{xy} & I_{yy}-I_1 & I_{yz} \\ I_{xz} & I_{yz} & I_{zz}-I_1 \end{bmatrix}
\begin{bmatrix} T_{11} \\ T_{12} \\ T_{13} \end{bmatrix}
=
\begin{bmatrix} 0 \\ 0 \\ 0 \end{bmatrix}
\tag{2.209}
$$

For a nontrivial solution to Eqn (2.209), we require the determinant of the square matrix containing I_1 to be zero leading to the characteristic equation

$$
\begin{vmatrix} I_{xx}-I_1 & I_{xy} & I_{xz} \\ I_{xy} & I_{yy}-I_1 & I_{yz} \\ I_{xz} & I_{yz} & I_{zz}-I_1 \end{vmatrix} = 0
\tag{2.210}
$$

The process used to find the determinant is documented in standard texts dealing with the mathematical manipulation of matrices, but in general a three by three square matrix may be summarised as follows

$$\begin{vmatrix} A & B & C \\ D & E & F \\ G & H & I \end{vmatrix} = A(EI - HF) - B(DI - GF) + C(DH - GE) \qquad (2.211)$$

The solution of Eqn (2.210) leads to a cubic equation in I_1 with three positive real roots, these being the three principal moments of inertia I_1, I_2 and I_3. If each of these is substituted in turn into equations that equate all three columns on either side of Eqn (2.207) we get

$$I_1 \begin{bmatrix} T_{11} \\ T_{12} \\ T_{13} \end{bmatrix} = \begin{bmatrix} I_{xx} & I_{xy} & I_{xz} \\ I_{xy} & I_{yy} & I_{yz} \\ I_{xz} & I_{yz} & I_{zz} \end{bmatrix} \begin{bmatrix} T_{11} \\ T_{12} \\ T_{13} \end{bmatrix} \qquad (2.212)$$

$$I_2 \begin{bmatrix} T_{21} \\ T_{22} \\ T_{23} \end{bmatrix} = \begin{bmatrix} I_{xx} & I_{xy} & I_{xz} \\ I_{xy} & I_{yy} & I_{yz} \\ I_{xz} & I_{yz} & I_{zz} \end{bmatrix} \begin{bmatrix} T_{21} \\ T_{22} \\ T_{23} \end{bmatrix} \qquad (2.213)$$

$$I_3 \begin{bmatrix} T_{31} \\ T_{32} \\ T_{33} \end{bmatrix} = \begin{bmatrix} I_{xx} & I_{xy} & I_{xz} \\ I_{xy} & I_{yy} & I_{yz} \\ I_{xz} & I_{yz} & I_{zz} \end{bmatrix} \begin{bmatrix} T_{31} \\ T_{32} \\ T_{33} \end{bmatrix} \qquad (2.214)$$

The solution of Eqns (2.212)–(2.214) thus yields all the terms in $[T_3]_2$, the transformation matrix from frame O_3 to O_2. In summary, I_1, I_2 and I_3 are the eigenvalues of the inertia matrix $[I_2]_{2/3}$ and are also the principal moments of inertia for Body 2 these being the diagonal terms in the matrix $[I_2]_{2/2}$. The three column matrices in Eqns (2.212)–(2.214)

$$\begin{bmatrix} T_{11} \\ T_{12} \\ T_{13} \end{bmatrix} \qquad \begin{bmatrix} T_{21} \\ T_{22} \\ T_{33} \end{bmatrix} \qquad \begin{bmatrix} T_{31} \\ T_{32} \\ T_{23} \end{bmatrix}$$

are the eigenvectors of $[I_2]_{2/3}$. If each vector is now normalised so that the length of the vector is unity, we get the direction cosines between each of the axes of O_2, the principal axes of Body 2, and O_3.

We can now consider a practical application of this with regard to vehicle dynamics where the body of a vehicle will generally be the largest and most significant mass in the model. For the vehicle body, Body 2, shown in Figure 2.35 we can take frame O_3 to be positioned at the mass centre and orientated so that the x-axis is along the centre line and pointing to the rear of the vehicle and the z-axis is vertical. The X_3Z_3 plane is thus a plane of symmetry. It should be noted that in reality this assumption involves some approximation due to the asymmetry of the masses that

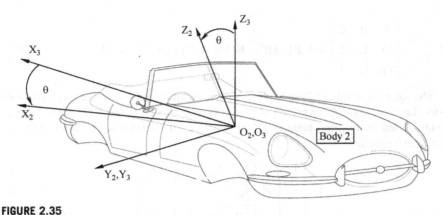

FIGURE 2.35

Vehicle body coordinate system.

may be lumped with the vehicle body, such as the engine, battery, exhaust system and fuel tank. The frame O_2, shown in Figure 2.35 is also positioned at the mass centre and has its Y_2-axis coincident with the Y_3-axis of frame O_3. The frame O_2 represents the principal axes of the vehicle body and is obtained by a transformation from O_3 represented by the rotation through an angle θ about the Y_2- and Y_3-axes.

In determining the products of inertia for this body it can be seen that, for every element of mass with a positive y coordinate there exists an equivalent element with a negative y coordinate. As a result we get

$$I_{xy} = I_{yx} = -\int xy\,dm = 0$$

$$I_{yz} = I_{zy} = -\int yz\,dm = 0$$

The inertia matrix for Body 2 $[I_2]_{2/3}$ measured from frame O_2 and referred to O_3 is therefore

$$[I_2]_{2/3} = \begin{bmatrix} Ixx & 0 & Ixz \\ 0 & I_2 & 0 \\ Ixz & 0 & Izz \end{bmatrix} \tag{2.215}$$

From this it can be seen that the y-axis is a principal axis and is normal to the plane of symmetry. The principal moment of inertia I_2 is therefore equal to I_{yy}. From Section 2.2.7 we can see that the matrix $[T_3]_2$ that transforms from frame O_3 to O_2 is given by

$$[T_3]_2 = \begin{bmatrix} \cos\theta & 0 & -\sin\theta \\ 0 & 1 & 0 \\ \sin\theta & 0 & \cos\theta \end{bmatrix} \tag{2.216}$$

From Eqn (2.206), we can see that using the transformation matrix $[T_3]_2$ given for this particular case will lead to

$$[I_2]_{2/2} = [T_3]_2 \, [I_2]_{2/3} \, [T_3]_2^T$$

$$
\begin{bmatrix} I_1 & 0 & 0 \\ 0 & I_2 & 0 \\ 0 & 0 & I_3 \end{bmatrix} =
\begin{bmatrix} \cos\theta & 0 & -\sin\theta \\ 0 & 1 & 0 \\ \sin\theta & 0 & \cos\theta \end{bmatrix}
\begin{bmatrix} I_{xx} & 0 & I_{xz} \\ 0 & I_2 & 0 \\ I_{xz} & 0 & I_{zz} \end{bmatrix}
\begin{bmatrix} \cos\theta & 0 & \sin\theta \\ 0 & 1 & 0 \\ -\sin\theta & 0 & \cos\theta \end{bmatrix}
$$

$$
\begin{bmatrix} I_1 & 0 & 0 \\ 0 & I_2 & 0 \\ 0 & 0 & I_3 \end{bmatrix} =
\begin{bmatrix} \cos\theta & 0 & -\sin\theta \\ 0 & 1 & 0 \\ \sin\theta & 0 & \cos\theta \end{bmatrix}
\begin{bmatrix} Ixx\cos\theta - Ixz\sin\theta & 0 & Ixx\sin\theta + Ixz\cos\theta \\ 0 & I_2 & 0 \\ Ixz\cos\theta - Izz\sin\theta & 0 & Ixz\sin\theta + Izz\cos\theta \end{bmatrix}
$$

$$
\begin{bmatrix} I_1 & 0 & 0 \\ 0 & I_2 & 0 \\ 0 & 0 & I_3 \end{bmatrix} =
$$

$$
\begin{bmatrix} Ixx\cos^2\theta - 2Ixz\sin\theta\cos\theta + Izz\sin^2\theta & 0 & Ixx\sin\theta\cos\theta + Ixz\cos^2\theta - Ixz\sin^2\theta - Izz\sin\theta\cos\theta \\ 0 & I_2 & 0 \\ Ixx\sin\theta\cos\theta + Ixz\cos^2\theta - Ixz\sin^2\theta - Izz\sin\theta\cos\theta & 0 & Ixx\sin^2\theta + 2Ixz\sin\theta\cos\theta + Izz\cos^2\theta \end{bmatrix}
$$

$$\tag{2.217}$$

Multiplying out the matrix equation in (2.217) leads to the following expressions for the principal moments of inertia I_1 and I_3.

$$I_1 = I_{xx}\cos^2\theta - 2\,I_{xz}\sin\theta\cos\theta + I_{zz}\sin^2\theta \tag{2.218}$$

$$I_3 = I_{xx}\sin\theta\cos\theta + I_{xz}\cos^2\theta + I_{xz}\sin^2\theta - I_{zz}\sin\theta\cos\theta \tag{2.219}$$

Equating now the zero elements on the left-hand side of Eqn (2.217) with the terms in either row1column3 or column1row3 gives

$$0 = I_{xx}\sin\theta\cos\theta - I_{xz}\sin^2\theta + I_{xz}\cos^2\theta - I_{zz}\sin\theta\cos\theta \tag{2.220}$$

This can be rearranged to give

$$0 = I_{xz}(\cos^2\theta - \sin^2\theta) + (I_{xx} - I_{zz})\sin\theta\cos\theta \tag{2.221}$$

From trigonometric addition formulae we can make use of

$$\cos2\theta = \cos^2\theta - \sin^2\theta \qquad \text{and} \qquad \sin2\theta = 2\sin\theta\cos\theta$$

which leads to

$$0 = I_{xz}\cos2\theta + \frac{1}{2}(I_{xx} - I_{zz})\sin2\theta \qquad (2.222)$$

Rearranging Eqn (2.222) leads to an expression from which we can determine a value for θ using the known values for I_{xx}, I_{zz} and I_{xz}.

$$\tan2\theta = \frac{I_{xz}}{\frac{1}{2}(I_{zz} - I_{xx})} \qquad (2.223)$$

Using the value obtained for θ in Eqn (2.223) it is now possible to substitute this back into Eqns (2.218) and (2.219) and obtain values for the two unknown principal moments of inertia I_1 and I_3.

2.13 Equations of motion

If we consider the rigid body, Body 2, shown in Figure 2.36 we can formulate six equations of motion corresponding with the six degrees of freedom resulting from unconstrained motion.

From our earlier consideration of linear momentum given in Eqn (2.161), we can write

$$\sum\{F_2\}_1 = \frac{d}{dt}\{L_2\}_1 = m_2\frac{d}{dt}\{V_{G2}\}_1 \qquad (2.224)$$

Expressing this in the familiar form of Newton's Second Law we get

$$\sum\{F_2\}_1 = m_2\{A_{G2}\}_1 \qquad (2.225)$$

The vector equation given in Eqn (2.225) will thus yield the three equations associated with the translational motion of the body. It may be noted that for these

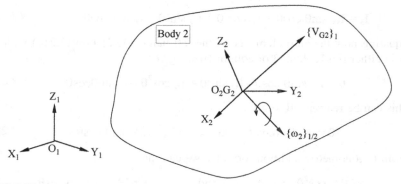

FIGURE 2.36

Rigid body motion.

equations the vectors in Eqn (2.225) may be conveniently referred to the fixed reference frame O_1.

In the same way that a resultant force acting on the body produces a change in linear momentum, a resultant moment will produce a change in angular momentum. If we consider the expression for angular momentum given in Eqn (2.178) we can obtain the equations of motion associated with rotational motion. For the rotational equations it is convenient to refer the vectors to the reference frame O_2 fixed in and rotating with Body 2.

$$\sum\{M_{G2}\}_{\frac{1}{2}} = \frac{d}{dt}\{H_2\}_{\frac{1}{2}} = \frac{d}{dt}[I_2]\{\omega_2\}_{\frac{1}{2}} \tag{2.226}$$

It can also be shown that

$$\frac{d}{dt}\{H_2\}_{\frac{1}{1}} = \frac{d}{dt}\{H_2\}_{\frac{1}{2}} + \{\omega_2\}_{\frac{1}{2}} \times \{H_2\}_{\frac{1}{2}} \tag{2.227}$$

$$\{M_{G2}\}_{\frac{1}{2}} = [I_2]_{\frac{2}{2}}\{\alpha_2\}_{\frac{1}{2}} + \{\omega_2\}_{\frac{1}{2}} \times \{H_2\}_{\frac{1}{2}} \tag{2.228}$$

Expanding Eqn (2.228) gives

$$\begin{bmatrix} M_x \\ M_y \\ M_z \end{bmatrix} = \begin{bmatrix} I_{xx} & I_{xy} & I_{xz} \\ I_{xy} & I_{yy} & I_{yz} \\ I_{xz} & I_{yz} & I_{zz} \end{bmatrix} \begin{bmatrix} \alpha_{2x} \\ \alpha_{2y} \\ \alpha_{2z} \end{bmatrix} + \begin{bmatrix} 0 & -\omega_{2z} & \omega_{2y} \\ \omega_{2z} & 0 & -\omega_{2x} \\ -\omega_{2y} & \omega_{2x} & 0 \end{bmatrix} \begin{bmatrix} H_{2x} \\ H_{2y} \\ H_{2z} \end{bmatrix} \tag{2.229}$$

Substituting in now terms for H_{2x}, H_{2y} and H_{2z} leads to the equations given in Eqns (2.230)–(2.232). For convenience we can drop the subscript for Body 2.

$$M_x = I_{xx}\alpha_x + I_{xy}(\alpha_y - \omega_x\omega_z) + I_{xz}(\alpha_z + \omega_x\omega_y) + (I_{zz} - I_{yy})\omega_y\omega_z + I_{yz}(\omega_y^2 - \omega_z^2) \tag{2.230}$$

$$M_y = I_{xy}(\alpha_x - \omega_y\omega_z) + I_{yy}\alpha_y + I_{yz}(\alpha_z - \omega_x\omega_y) + (I_{xx} - I_{zz})\omega_x\omega_z + I_{xz}(\omega_z^2 - \omega_x^2) \tag{2.231}$$

$$M_z = I_{xz}(\alpha_x - \omega_y\omega_z) + I_{yz}(\alpha_y + \omega_x\omega_z) + I_{zz}\alpha_z + (I_{yy} - I_{xx})\omega_x\omega_y + I_{xy}(\omega_x^2 - \omega_y^2) \tag{2.232}$$

In summary, the rotational equations of motion for Body 2 may be written in vector form as

$$\sum\{M_{G2}\}_{\frac{1}{2}} = [I_2]_{\frac{2}{2}}\{\alpha_2\}_{\frac{1}{2}} + [\omega_2]_{\frac{1}{2}}[I_2]_{\frac{2}{2}}\{\omega_2\}_{\frac{1}{2}} \tag{2.233}$$

Hence we can see that in setting up the equations of motion for any rigid body, the translational equations for all bodies in a system may conveniently be referred to a single-fixed inertial frame O_1. The rotational equations, however, are better referred to a body-centred frame, in this case O_2. A considerable simplification in these equations will result if frame O_2 is selected such that its axes are the principal axes of the body ($I_1 = I_{xx}, I_2 = I_{yy}, I_3 = I_{zz}$) and the products of inertia are zero. The equations that result are known as Euler's equations of motion:

$$M_x = I_1 \alpha_x + (I_3 - I_2) \omega_y \omega_z \qquad (2.234)$$

$$M_y = I_2 \alpha_y + (I_1 - I_3) \omega_x \omega_z \qquad (2.235)$$

$$M_z = I_3 \alpha_z + (I_2 - I_1) \omega_x \omega_y \qquad (2.236)$$

The equations given in Eqns (2.234)–(2.236) become even simpler when the motion of a body is constrained so that rotation takes place in one plane only. If, for example, rotation about the x- and y-axes are prevented then Eqn (2.236) reduces to the more familiar form associated with two-dimensional motion.

$$M_z = I_3 \alpha_z \qquad (2.237)$$

The following example also demonstrates how gyroscopic effects associated with three-dimensional motion may be identified. If we consider the swing arm suspension system shown in Figure 2.37 we can take the suspension arm Body 2 to be constrained by a revolute joint to rotate with a constant angular velocity of 10 rad/s about the axis of the joint as shown. Whilst this motion is in progress the road wheel Body 3 is also rotating with a constant angular velocity of 100 rad/s about the axis of the revolute joint representing the wheel bearing.

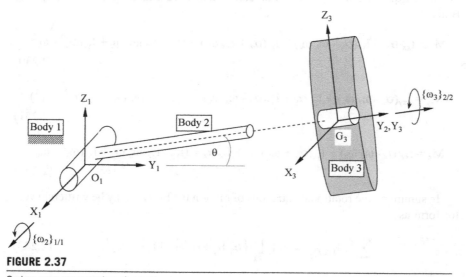

FIGURE 2.37

Swing arm suspension dynamics.

The following data may be used to represent the mass properties of the road wheel

$$m_3 = 16 \text{ kg}$$
$$I_{31} = I_{3xx} = 0.5 \text{ kgm}^2$$
$$I_{32} = I_{3yy} = 1.0 \text{ kgm}^2$$
$$I_{33} = I_{3zz} = 0.5 \text{ kgm}^2$$

In order to determine the reaction torque on the wheel bearing when the axle is still in a horizontal position, $\theta = 0$, we can use the following to give us the required angular velocity vector $\{\omega_3\}_{1/3}$.

$$\{\omega_2\}_{1/1} = \begin{bmatrix} 10 \\ 0 \\ 0 \end{bmatrix} \text{ rad/s} \qquad \{\omega_3\}_{2/2} = \begin{bmatrix} 0 \\ 100 \\ 0 \end{bmatrix} \text{ rad/s} \qquad \{\omega_3\}_{1/1} = \begin{bmatrix} 10 \\ 100 \\ 0 \end{bmatrix} \text{ rad/s}$$

In the absence of angular acceleration, Eqn (2.233) can be adapted to give for this problem

$$\sum \{M_3\}_{1/3} = [\omega_3]_{1/3} [I_3]_{1/3} \{\omega_3\}_{1/3} \qquad (2.238)$$

Expanding this gives

$$\sum \begin{bmatrix} M_{3x} \\ M_{3y} \\ M_{3z} \end{bmatrix}_{1/3} = \begin{bmatrix} 0 & -\omega_{3z} & \omega_{3y} \\ \omega_{3z} & 0 & -\omega_{3x} \\ -\omega_{3y} & \omega_{3x} & 0 \end{bmatrix}_{1/3} \begin{bmatrix} I_{3xx} & 0 & 0 \\ 0 & I_{3yy} & 0 \\ 0 & 0 & I_{3zz} \end{bmatrix}_{3/3} \begin{bmatrix} \omega_{3x} \\ \omega_{3y} \\ \omega_{3z} \end{bmatrix}_{1/3} \qquad (2.239)$$

Substituting in the numerical data for this problem gives

$$\sum \begin{bmatrix} M_{3x} \\ M_{3y} \\ M_{3z} \end{bmatrix}_{1/3} = \begin{bmatrix} 0 & 0 & 100 \\ 0 & 0 & -10 \\ -100 & 10 & 0 \end{bmatrix} \begin{bmatrix} 0.5 & 0 & 0 \\ 0 & 1.0 & 0 \\ 0 & 0 & 0.5 \end{bmatrix} \begin{bmatrix} 10 \\ 100 \\ 0 \end{bmatrix} \text{ Nm}$$

$$\sum \begin{bmatrix} M_{3x} \\ M_{3y} \\ M_{3z} \end{bmatrix}_{1/3} = \begin{bmatrix} 0 & 0 & 100 \\ 0 & 0 & -10 \\ -100 & 10 & 0 \end{bmatrix} \begin{bmatrix} 5.0 \\ 100 \\ 0 \end{bmatrix} \text{ Nm} \qquad (2.240)$$

$$\sum \begin{bmatrix} M_{3x} \\ M_{3y} \\ M_{3z} \end{bmatrix}_{1/3} = \begin{bmatrix} 0 \\ 0 \\ 500 \end{bmatrix} \text{ Nm}$$

As can be seen from the result in Eqn (2.240), the reaction torque on the wheel bearing is about the z-axis and is due to gyroscopic effects as the wheel spins about the y-axis and rotates about the x-axis.

It should be noted that in addition to the derivation of the equations of motion based on the direct application of Newton's Laws, variational methods, including for example Lagrange's equations, provide an elegant alternative and are often employed in MBS formulations. Many texts on classical dynamics, such as (D'Souza and Garg, 1984), include a thorough treatment of these methods.

Variational methods are attractive for a number of reasons. Equations are formulated using kinetic energy and work resulting in scalar rather than vector terms. Solutions can also be more efficient since constraint forces that do not perform work can be omitted. Variational methods also make use of generalised rather than physical coordinates reducing the number of equations required.

The theory and methods described in this chapter form a basis for the MBS formulations covered in the next chapter. The vector notation used here will be used to describe the part equations and the constraint equations required to represent joints constraining relative motion between interconnected bodies. In Chapter 4 the vector-based methods described here will be used to carry out a range of analyses from first principles on a double-wishbone suspension system and to compare the calculated results with those found using MSC.ADAMS.

Multibody Systems Simulation Software

We can only see a short distance ahead, but we can see plenty there that needs to be done.

Alan Turing

3.1 Overview

There exists a range of commercial computer packages that can be used to solve problems in multibody systems analysis. In addition to the commercial packages that may be licensed there are also programs developed by academic institutions that may be available, albeit without the level of development and support that would be expected when buying the software from an established program developer. The first version of this book centred on MSC ADAMS but the progression of other software providers and their diversity of approach has broadened the coverage in this chapter handsomely. The purpose of this chapter is in principle to equip the reader to understand the capabilities of any multibody systems (MBS) analysis programs used in vehicle dynamics.

General-purpose MBS programs are able to address a large set of problems across a wide range of engineering industries and are not restricted to the applications in vehicle dynamics discussed here. MBS software within the automotive industry is used to simulate the performance of anything that moves.

From door latches to gearbox synchromesh, models are increasingly being used for large amplitude nonlinear vibration problems, even in areas that were unthinkable 20 years ago, such as timing chain dynamics. Many of the general-purpose programs have developed toolkits that allow a system to be exercised and validated in isolation before being included in a larger system. Previously, a separate subsystem test rig and full vehicle model would have been prepared, leading to transcription errors when moving between the two and requiring the upkeep of two models when design changes needed to be tracked. The analyst will often wish to validate the performance of a suspension model over a range of displacements and loads before the assembly of a full vehicle model that may be used for ride, handling and durability studies. A detailed model may include representations of the body, sub-frames, suspension arms, struts, anti-roll bars, steering system, engine, drivetrain and tyres. Some of the elements may be structurally compliant and others

```
UNITS/FORCE = NEWTON, MASS = KILOGRAM, LENGTH = MILLIMETER, TIME = SECOND
PART/1, GROUND
MARKER/1, PART = 1, QP = 0, 0, -100
MARKER/2, PART = 1, FLOATING
PART/2, MASS = 11.64, CM = 22, IP = 5.099E+005, 3.851E+005, 6.002E+005
, VX = -40230
MARKER/3, PART = 2, QP = 681.85, 0, 802.28, REULER = 90D, 30.3D, 270D
GRAPHICS/57, CYLINDER, CM = 6, LENGTH = -150, RADIUS = 18
VARIABLE/26, FUNCTION = IF((VARVAL(25)+15.0D):
SFORCE/18, TRANSLATIONAL, I = 38, J = 26, FUNCTION = VARVAL(26) * 8.96
DIFF/4, IC = 0, IMPLICIT, FUNCTION = DIF(4)-varval(29)
TFSISO/100, X = 100, U = 101, Y = 102, NUMERATOR = 986.96, DENOMINATOR = 986.96
, 44.429, 1
```

FIGURE 3.1

A fragment of a model in the language expected by the solver, which is technically human readable but somewhat difficult to work with.

may have elaborate behaviour captured in empirical models embedded within dedicated subroutines of the solver. Examples of the latter include elastomer elements, dampers, tyres and human beings.

The main analysis code consists of a number of integrated programs that perform three-dimensional kinematic, static, quasi-static or dynamic analysis of mechanical systems. These programs may be thought of as the core solver. In addition there are a number of auxiliary programs, which can be supplied to link with the core solver. These programs typically capture the embedded empirical models and are usually kept separate to the core solver to allow the software vendor to sell a modular system and reduce costs for customers who have no need of certain items. Frequently the architecture of the solver is open enough that users may from time to time develop their own embodiment of empirical models and link them to the core solver; the Harty tyre model is one such example of this.

Prior to submitting an analysis the model needs to be described to the solver in a way that it understands and most solvers have some form of proprietary information format, as shown in Figure 3.1, in which they expect to read the description. It is typically true that this format can be a little difficult to prepare and also to read once it has been prepared, and so for several decades now additional programs have been available to allow users to describe a mechanism or system in a broadly 'human-friendly' way[1] before allowing a machine translation to occur into the solver language.

These programs can also generally handle libraries of commonly used blocks and allow their reuse and recombination to analyse hitherto unexamined systems. In this manner it saves a large amount of labour in model preparation. Such a program used to be known as a *preprocessor* and is still often a completely separate executable from the main solver. When working at the edge of the available computational budget, it can be advantageous to work with the preprocessor to define the model

---

[1]In the last 40 years tremendous strides have been made in human–computer interfacing and there is no reason to believe it will stop where we are now.

and then shut it down before starting the solver. Modern preprocessors are typically graphically rich and extremely interactive, which has the effect of making them surprisingly resource-hungry.

Once a model has been defined the core solver will assemble the equations of motion and solve them automatically. It is also possible to include differential equations, transfer functions and other items directly in the solution, which allows the modelling of a variety of control systems, thermal effects and many others.

The preprocessor can link or interface with CAD systems, other analysis software (e.g. finite element, control system or hydraulic modelling), software used for advanced visualisation or software for downstream processing as part of a larger experiment in progress. The combined use of these systems can lead to the development of what may be referred to as 'virtual prototypes': numerical models that can simulate the tests and conditions to which a real prototype would be subject during the development of a new engineering product. The extent of these interfaces is diagrammatically represented in Figure 3.2.

Note that although the software will be continually developed after publication, there is good reason to suspect[2] that this architecture will continue for some time, since it is both computationally and organisationally convenient.

The first step in any simulation is to prepare a data set that will define the system being modelled. This will include a description of the rigid parts, connecting joints, motion generators, forces and compliances. With many modern packages the user is spared from actually looking at anything other than a graphical representation of the data set. Nevertheless in most codes the data set is user friendly in that the data statements are easily understood with few restrictions on format (Figure 3.3); this is a great step compared to the fixed-length formatting often used in early programs, where a 2 (integer 2) would cause a failure if a 2. (real 2.0) was expected and vice versa.

For some applications that may not be commercially worth supporting or that are more advanced, users can prepare their own user-written subroutines in languages, such as Fortran, ANSI 'C', Python, .NET or any of the other myriad of supported languages available. They can be linked with the main code to be invoked at solution time, often in the form of a dynamically linked library. It is worth mentioning that Fortran persists despite its age simply because of the well integrated and very complete function library and the existence of legacy code; however, it could be wise for new practitioners to learn their art in C.

For each rigid body in the system it is necessary to define the mass, centre of mass location, and mass moments of inertia. Each body will possess a set of

---

[2]In his rather brilliant book 'Thinking, Fast and Slow' (2012), Daniel Kahneman observes that if something has already persisted for a period of time, it is reasonable to think that it will continue to persist. The printed page has been around for about five centuries but recording media such as the IBM printed card, magnetic data tape, floppy disks and compact flash cards have come and gone very quickly. So '3 Box' analysis packages have another couple of decades in them, probably.

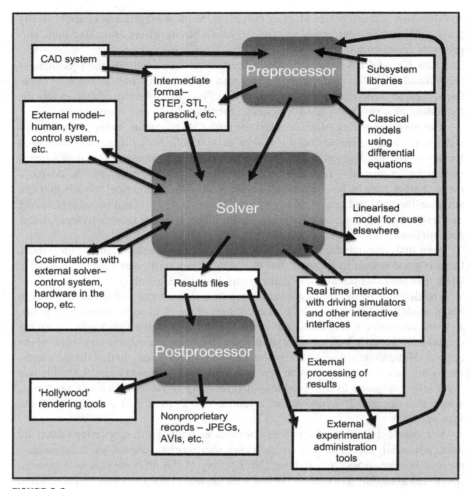

FIGURE 3.2

Interaction of a general-purpose program with other information.

```
marker create marker_name=m_fr_shadow_to_ground &
   location =    &
      (fr_wheel_rad), &
      0.0, &
      (ground_height) &
   orientation=0, 0, 0

constraint create joint planar &
   joint_name=j_ground_to_fr_shadow &
   i_marker=m_ground_to_fr_shadow &
   j_marker=m_fr_shadow_to_ground
```

FIGURE 3.3

A more readable format for text files has emerged.

coordinates that can be defined in global or local coordinate systems and are considered to move with the part during the simulation. These points are used to define centre of mass locations, joint locations and orientations, force locations and directions. The relative motion between different parts in the system can usually be constrained using joints, joint primitives, couplers, gears and user-defined constraints.

There is an important conceptual difference between some of the packages, which is worth describing at this point.

Some packages, such as MSC ADAMS, bring each body into being with six degrees of freedom (DOF) — three translational and three rotational. Once the DOF have been created, they are connected to other DOF using mathematical relationships — such as the relative displacement of two points, each on a different body, being equal to zero (a spherical joint). This is set up as a constraint equation to restrict the DOF, and has to be explicitly solved at each step of the solution.

Other packages, such as Simpack, bring each body into being with only the DOF required to generate the mechanism. In this way there is no iterative solution of constrained DOF to arrive at a solution satisfying the constraint equations — instead those equations are never generated. This is much more like the classical approach, where we might consider a single DOF system and work with only one equation; nobody solves six equations to solutions of zero if they are considering a single DOF model with pencil and paper. The reduced DOF approach has clear benefits at run-time but does carry some overhead in terms of the software assembling the equations of motion.

The two styles can be somewhat confusing if one is unaware of these differences. Unfortunately, language being a rather limited and imprecise thing, the two software packages described typically use the same words but mean different things.

In Simpack, a 'joint' is declared with a body to bring about the DOF of interest. Thus for a falling stone — always the multibody problem of choice in much the same way as 'Hello World' is the task of choice when learning a new coding language — a body is declared with a prismatic joint. This results in only 1 DOF and thus the minimum amount of computation.

To make a pendulum — step two on any journey into new software — Simpack again has only 1 DOF by declaring a 'joint' with only rotation about the pivot point.

In MSC ADAMS, a falling stone can be modelled simply as a single body with 6 DOF and no joints, which eases the model preparation significantly. However, for the pendulum model MSC ADAMS requires five equations to constrain out the five unwanted motions in the solution, which are solved for zero deflection and thus have forces available as part of the solution.

In Simpack, to make a double pendulum the two bodies are 'chained' together, each with a single DOF, to make a 2 DOF model. Each model has a 'joint' which is really a declaration of the desired DOF. For MSC ADAMS there are now 10 equations of motion being solved for a solution of zero, and two we care about. By the time a four-bar linkage is modelled — rather confusingly, with three links and the ground — there are 17 equations being solved to zero.

Using the Simpack-style method of defining bodies chained to one another, it is obvious this approach fails when the chain makes a loop. To close the loop, Simpack makes use of ADAMS-style constraint equations, computing the forces required to solve for a relative motion of zero in the constrained DOF.

Thus in ADAMS, all 'joints' are mathematical constraints but in Simpack, 'joints' are actually DOF and 'constraints' are mathematical constraints.

The next step in building the model would typically be the definition of external forces and internal force elements. External forces can be constant, time histories or functionally dependent on any state variable. These forces can also be defined to be translational or rotational. They can act in the global system or can act in the local system of the body so that they effectively 'follow' the part during the simulation.

Users can also set up internal force elements acting between two parts to represent springs, dampers, cables or rubber mounts. Internal force elements can act along the line of sight between the points the force element connects on the two parts, or can act in some arbitrarily defined direction. In the latter case, some additional effects can be introduced, that the user may not have had in mind; it is up to the user to be in control of what they want to model. These force elements are often referred to as action—reaction forces as they always produce equal and opposite forces on the two parts connected by the force element. The elements can also be defined to act in only tension or compression and may be linear or nonlinear. Additional forces can be defined as action-only forces, in which the universe exerts a force on the model with no internal reaction. Aerodynamic forces might be modelled using an action-only force.

It is also useful if the MBS analysis program allows the definition of elaborate mathematical equations within the data set. This enables the user to formulate an expression involving user-defined constants, system constants, system variables, arithmetic IF's, FORTRAN 90 or ANSI 'C' library functions, standard mathematical functions or 'off-the-shelf' functions supplied with the main code to represent events, such as impacts. The access to system variables can be a powerful modelling tool. The user can effectively access any displacement, velocity, acceleration or other force in the system when defining the force equation. Forces can also be defined as a function of time to vary or switch on and off as the simulation progresses. Caution is needed to ensure formulations are continuous in the time domain to avoid problems during the numerical solution of the resulting equations. Recent versions of many software packages also include a general contact force model between geometries associated with the rigid bodies.

Enforced displacement input can be defined at certain joints to be either constant or time dependent. When a motion is defined at a joint it may be translational or rotational. The motion effectively provides another constraint so that the DOF at that joint is lost to the motion. Motion expressions can be defined using all the functions available as for force definitions, except that the only system variable that should be accessed is time. While it is possible to connect imposed motions to variables other than time, such use is normally deprecated within the documentation for the software since it can lead to very poor numerical conditioning that can be difficult for a novice user to recognise. An example of this might be the steering inputs to a vehicle model, for which the preferred closed loop embodiment would calculate a demanded position in response to vehicle states and apply forces to attempt to deliver the demanded position. Both an open loop embodiment (playing in a recorded steer history) and a closed loop embodiment are discussed in Chapter 6.

An MBS analysis program will often provide a number of elements with the capability to model flexibility of bodies and elastic connections between parts. These may include features for modelling beam elements, rubber bushings or mounts, plus a general stiffness and damping field element. At various positions in a model rigid parts can be elastically connected together in preference to using a rigid constraint element, such as a joint or joint primitive. Vehicle suspension bushes can be represented by a set of six action—reaction forces, which will hold the two parts together. In the simplest form, the equations of force are linear and uncoupled. The user is only required to provide the six diagonal coefficients of stiffness and damping. For more complicated cases a general-purpose statement can be used to provide a linear representation of a flexible body or connection. Further elaboration is difficult with the standard items in the program; simple nonlinearity is usually easy to achieve by making forces reference data stored as splines, but frequency-dependent characteristics, such as displayed by many elastomers, need a more specialised representation that may only be possible using dedicated subroutines, such as those described in Chavan et al. (2010).

Using recent advances in software techniques to allow the combination of component mode synthesis representations for stiff, small amplitude linear dynamic behaviour of structural elements, one or more of the major structural parts of the system may be represented in modal form to study the influence of its flexibility on the behaviour of the system as a whole. A disadvantage of this method of working is the opportunity to consume large amounts of computing resources solving these models, if care is not taken to ensure the flexibility is germane to the task at hand. Where a full representation of the flexibility of the structure is unnecessary, a simpler representation is possible using joints as 'hinges' and an associated stiffness at keypoints in the structure. The preference for one approach or the other is largely governed by the time and data extant. This level of abstraction requires a high degree of understanding of the structural behaviour of elements of the system and can easily lead to poorly conditioned numerical problems if carelessly performed, raising solution times drastically. Worse still, it can lead to 'plausible-but-wrong' answers, particularly if mass properties are poorly distributed.

Using component mode synthesis, a complete set of modal components can be used with a full vehicle comprehensive model. This approach confuses accuracy with usefulness in a manner that is becoming increasingly common. The use of such models works against volatility of design, and such models cannot be effectively used with an emerging design but belong to a new generation of mathematical prototypes for use in a later vehicle programme. The notion that too much complexity is a bad thing is discussed in later chapters of this book.

For full vehicle applications it is important to obtain a usefully accurate model for the tyres and the associated forces generated at the tyre—road surface contact patch. For each tyre on the vehicle model, the program will calculate the three orthogonal forces and three orthogonal torques acting at the wheel centre as a result of the conditions at the tyre—road surface contact patch. In order to perform these calculations it is necessary to continuously update the tyre model regarding the position, velocity and orientation of the wheel centre marker and any changes in the topography of the road surface. Once this information has been received the tyre model must then calculate the set of forces acting at the contact patch. Once these forces have been calculated they can be resolved back to the wheel centre. The MBS analysis program will then integrate through time to find the new position and orientation of the vehicle and then repeat the process.

It should be noted that commercial software is undergoing continual development and as such the description provided here is limited to the software features required to carry out the simulations described in this text. Elements, such as springs, dampers, bushes and bump stops are described in this chapter as these are considered fundamental components of an MBS modelling system.

3.2 Modelling features
3.2.1 Planning the model

Before progressing to the methods used to describe the typical elements of a multibody system model, it is necessary to outline some of the planning that goes into the development of the model. The first step should be to sketch out a system schematic, which would typically illustrate items, such as the parts, joints, imparted motions and applied forces. The model may well include other elements that include, for example, springs, dampers and beams. The drawing of a schematic is an important first step as it will help the user to not only plan the data that will need to be collated but more importantly to estimate the DOF in the system and develop an understanding of how the mechanism will work.

The use of a modern graphical user interface (GUI) should not discourage this process, which has been used to help understand mechanisms for centuries (Figure 3.4).

Many users of modern GUIs are tempted to believe that this step is no longer necessary. The authors suggest this is not the case and will continue not to be the case until a convenient graphical engine for exploring models is incorporated into the interface.[3] The primary issue is that many entities in a multibody model are colocated — i.e. at the same place (Figure 3.5).

---

[3]See, for example, Maneesh Agrawala's work on interactive exploded diagrams, first published in 2004 but not as yet incorporated in any multibody software known to the authors (Wilmot et al., 2004). A 2009 YouTube clip reveals an easy and instinctive way to explore system models were it to be incorporated (http://www.youtube.com/watch?v=NL2QFLiM_mY).

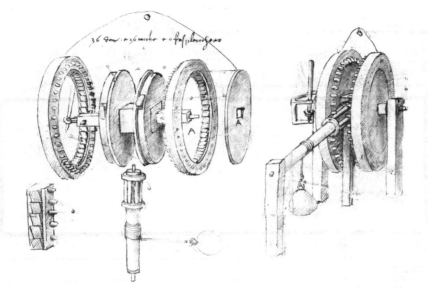

FIGURE 3.4

Good enough for Leonardo da Vinci, a mechanism sketch can help clarify the form the model must have.

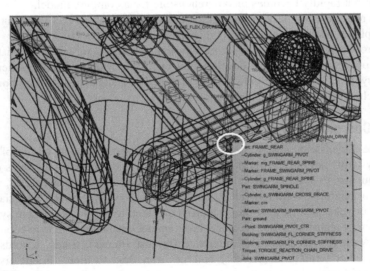

FIGURE 3.5

Illustrating the Graphical User Interface Problem with a Typical Multibody Model. The top left corner of the selection box (circled) often offers the choice of more than 10 items at that point in space.

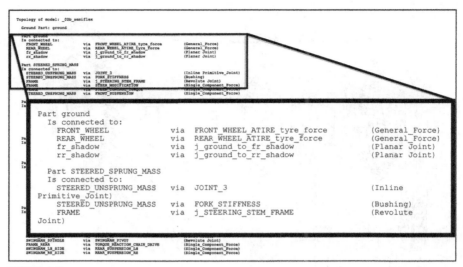

FIGURE 3.6

A text-based topology map.

Many software packages offer some form of topology map for the model, some in a text-based form (Figure 3.6) and some graphical. Some of these are more helpful than others. The MSC textual topology map, for example, is quite useful if the model is simple but rapidly becomes incomprehensible for a complex model.

Both MSC and Simpack provide graphical topology maps (Figure 3.7) but again for complex models this has varying degrees of utility.

Figure 3.8 provides an example of a system schematic for a double wishbone suspension system. Users may develop their own style when drawing a schematic, but the symbols shown in Figure 3.9 are provided as a suggested starting point for the sketching of the various elements of a model.

3.2.2 Coordinate systems

The three-dimensional description of a multibody system requires the use of coordinate systems, not only to set up the configuration and physical properties of the model, but also to describe the calculated outputs, such as the displacements, velocities and accelerations. Although in general coordinate systems can be Cartesian, cylindrical or spherical, general-purpose programs tend to employ Cartesian systems for reasons of algorithmic convenience. Program documentation refers to coordinate systems variously as frames of reference (or just frames), triads, 'coords' and markers; other terms will probably emerge from time to time. 'Frame' is used here to avoid confusion between *multibody system* and *coordinate system*.

The concepts are common to every attempt to describe a multibody system in software and not restricted to any one package. There are three types of right-handed Cartesian frames that will be used in this text:

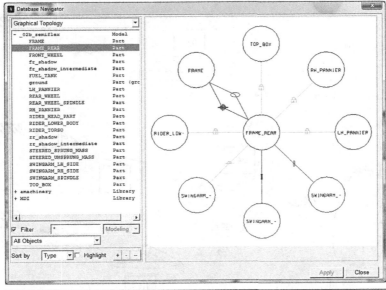

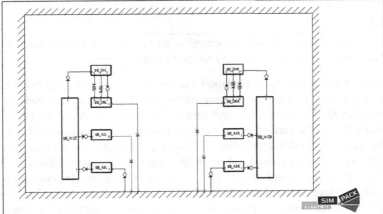

FIGURE 3.7

A Graphical Topology Map as Implemented in Two Different Software Packages. Although somewhat helpful, it still does not assist in assimilating the model in the manner of a really useful mechanism sketch.

3.2.2.1 Ground reference frame

This is by definition the single inertial frame that is considered to be fixed or at 'absolute' rest. Any point defined to be stationary with respect to this frame has zero velocity and acceleration in an absolute sense. The ground reference frame (GRF) is taken to be fixed on a body or part known as the ground part, the physical

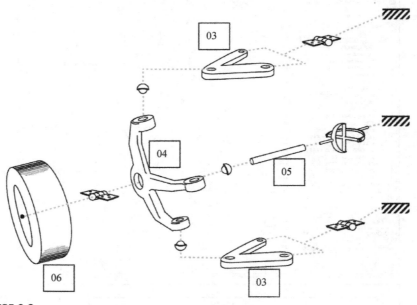

FIGURE 3.8

Double wishbone suspension system schematic of the kind that it would be helpful if commercial software could produce easily for model audit purposes.

significance of which may vary from model to model. For a single suspension model the ground part may be taken to encompass the points on the vehicle body or subframe to which the suspension linkages attach. For a full vehicle model the ground part would relate to the surface of the road used to formulate the contact forces and moments in the tyre model. In addition to providing the single inertial reference, the ground reference can be considered to be the origin of the entire model. As such the absolute coordinates and orientations of all other reference frames and points in the model are measured relative to the GRF. When modelling a complete vehicle, it can be convenient to use the same origin as the master geometric model held in a CAD system. Note that different organisations have different conventions for this master model origin, with some having positive X in the direction of vehicle travel, some with it backward, some having the origin ahead of the vehicle and under the ground, some with the origin at flywheel centre and so on. There is no one universal convention despite the formulation of suggestions for such things by the American Society of Automotive Engineers, for example. Throughout this text the ground part is taken to be the first part in the model and the GRF to be the first frame O_1. Note that some programs do not explicitly declare a ground part but the existence of a GRF is an implicit acknowledgement that it exists. Practitioners may also describe this frame as a 'global' frame. The exact terminology may vary with different multibody system programs but the notion is identical between them.

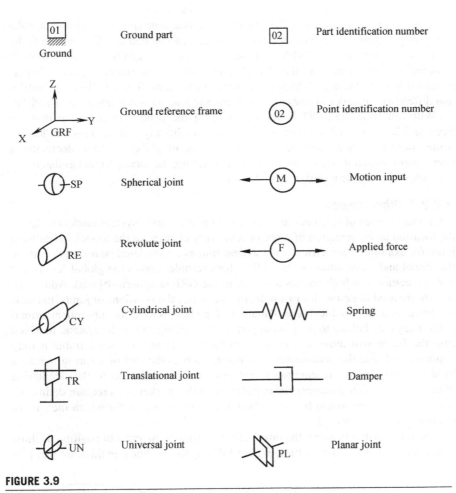

FIGURE 3.9

Suggested symbols for elements of system schematic.

3.2.2.2 The local part reference frame

Each body or part in the system can be considered to have a local part reference frame (LPRF) that moves and changes orientation with the part. The position and orientation of the local part reference system at time zero is defined relative to the GRF. The use of a local axis system on the part may be desirable to facilitate the definition of points on the body by perhaps exploiting the symmetry of the body where the axes of symmetry, at the model definition stage, are not parallel to the axes of the GRF. The explicit declaration of a LPRF is optional and if omitted the LPRF can be considered to be coincident with and parallel to the GRF at the model definition stage. As with the ground part, the fact that an LPRF is not declared does not mean it is not in use behind the scenes.

Some practitioners like to declare the LPRF at some convenient point other than the global origin, perhaps at some distinctive feature like one end of a link or else at the component centre of mass. With increased use of product data management tools, it is increasingly likely that a local part origin established as convenient for geometric purposes will become the default formulation for an LPRF in the future. Different industries and different companies in any case have different habits for the definition of LPRF.

With Simpack, the LPRF is defined by the 'joint' used to define the part — see Section 3.2.1 — so that the user is not presented with any choice. This can lead to some confusion for practitioners used to the use of global position declarations everywhere, which is one of the advantages of setting the local part and global reference frames coincident during model build.

3.2.2.3 Other frames

Additional frames of reference are freely used in multibody system models to define the location and orientation of more or less every element in the model. To define a joint, for example, geometrically coincident frames are declared on different parts in the model and an equation declared that, for example, makes the global X-, Y- and Z-displacements of both frames identical in the GRF (a spherical joint). Additional frames are used to define, for example, mass centres, the positions of joints, the ends of springs and the graphical representation of the bodies used for subsequent animations. They may belong to the ground part or any moving part in the system, in which case the frame will move and rotate with the part. In some cases a frame is only required to define the coordinates of a point, such as the end of a spring, where a local definition of the orientation is not important and defaults to the orientation of the GRF. In other instances the orientation of the marker does require definition. An example of this would be the definition of revolute joints for which the axis of rotation must be specified.

The relationship between the three reference frames, in terms of position, is illustrated in Figure 3.10. The position of the LPRF O_n for any body, in this case part n, is

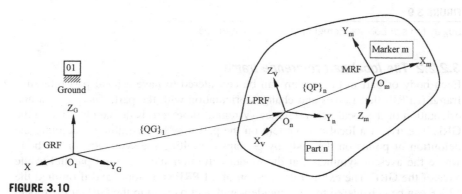

FIGURE 3.10

Relative Position Definition of the GRF, LPRF and Another Frame. GRF, ground reference frame, LPRF, local part reference frame.

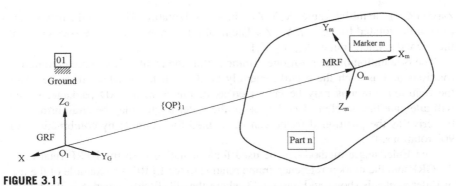

FIGURE 3.11

Relative Position Definition of the Additional Frame when the LPRF is Coincident with the GRF. GRF, ground reference frame, LPRF, local part reference frame.

defined using, in MSC ADAMS terminology, a position vector $\{QG\}_1$. The position of any markers belonging to part n, for example marker m with marker reference frame O_m, are defined relative to O_n, using a relative position vector $\{QP\}_n$. Note that the x, y and z components of $\{QP\}_n$ are resolved parallel to O_n.

As mentioned earlier, the definition of the LPRF is often taken to be coincident with and parallel to the GRF when setting up the model — either explicitly declared as such or by omitting a declaration, depending on the software in use. The position of the marker reference frame O_m is then defined relative to the GRF by the position vector $\{QP\}_1$ as illustrated in Figure 3.11. Note that the x, y and z components of $\{QP\}_1$ are now resolved parallel to the GRF O_1.

There are a number of different methods by which the orientation of one reference frame to another may be established when defining a model. Two commonly used methods are presented here and will be similar to methods used in alternative MBS programs. The first of these methods is a body-fixed 3–1–3 sequence of rotations as shown in Figure 3.12. This Euler angle method involves the definition of three sequential rotations Ψ, θ, and ϕ. The first rotation Ψ acts about the

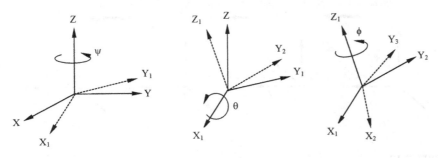

FIGURE 3.12

Orientation of a frame by the euler angle method.

z-axis (Z) of the initial frame (X, Y, Z). The second rotation θ is about the new x-axis (X_1) of the rotated frame (X_1, Y_1, Z). The final rotation ϕ is about the z-axis (Z_1) of the second rotated frame (X_1, Y_2, Z_1).

The orientation of the positioned frame is thus the result of the three cumulative rotations and as such these will generally be difficult to visualise. In some cases the required rotations may be generated as output from a CAD package, which will alleviate the problem. In other cases the Euler angles may be straightforward to derive as the positioned frame can be defined for example by combinations of 90° rotations.

The Euler angle method can be used for orientating both the LPRF relative to the GRF and the marker reference frame relative to the LPRF. An example of the use of Euler angles is shown in Figure 3.13 where the (Ψ, θ, ϕ) sequence is (90d, 90d, -90d), the letter d being used here to denote the use of degrees rather than radians.

An alternative method of orientating a reference frame is referred to as the X-point-Z-point method and involves defining the coordinates of a point that lies on the Z-axis of the positioned frame and another point that lies in the XZ plane of the positioned frame. This is illustrated in Figure 3.14 where this method is used to orientate the LPRF relative to the GRF. The position of the LPRF, O_n, is defined, as stated earlier, by the vector $\{QG\}_1$. The point Q is coincident with O_n. The position of Z is defined by $\{ZG\}_1$. The distance of Z from G along the Z-axis of O_n is arbitrary. The position of X is defined by $\{XG\}_1$ and may lie anywhere in the XZ plane other than on the Z-axis of O_n.

In order to determine the exact orientation of the positioned frame the vector cross product can be applied to first obtain the new Y-axis. The vector cross product of the new Y-axis and the new Z-axis can then be used to find the new X-axis. It will be seen later that if only either the X-axis or Z-axis is important then it is only necessary to specify either $\{XG\}_1$ or $\{ZG\}_1$.

The X-point-Z-point method can also be used to orientate a marker reference frame relative to a LPRF as illustrated in Figure 3.15. The notation is changed using QP, XP, ZP instead of the QG, XG, ZG used to orientate the LPRF. It should also be

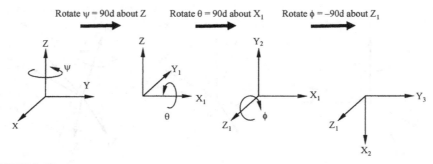

FIGURE 3.13

Example application of the euler angle method to orientate a frame.

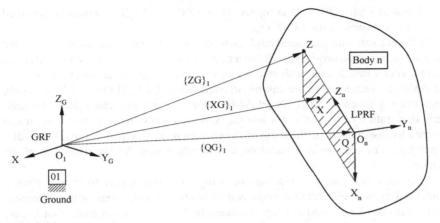

FIGURE 3.14

Orientation of the LPRF Using the X-point-Z-point Method. GRF, ground reference frame, LPRF, local part reference frame.

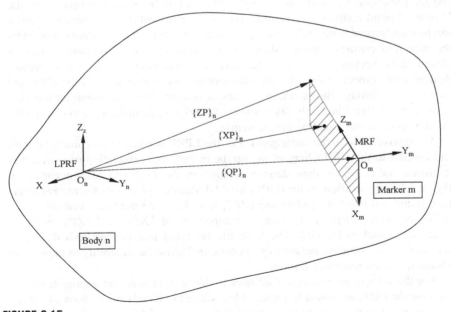

FIGURE 3.15

Orientation of the Marker Reference Frame Using the X-point-Z-point Method. LPRF, local part reference frame.

noted that as with $\{QP\}_n$ the components of $\{XP\}_n$ and $\{ZP\}_n$ would be resolved parallel to the axes of the LPRF O_n.

Different software packages implement this in different ways but in general the idea of starting at an origin, specifying a first vector for an axis and then specifying a second vector that is used with two vector product operations is common. In MSC ADAMS, an additional frame can be oriented using 'ORI_ALONG_AXIS' if only one vector is given or 'ORI_IN_PLANE'. Many current programs allow the arbitrary declaration of any axis, not just the X-point-Z-point logic shown here. It can be seen by inspection that the logic of the operations is identical. Simpack uses dialogue boxes to select which locations are on which axes for the definition of the marker.

This approach has the advantage of being somewhat easier to visualise when compared to the Euler rotation sequence; it is also advantageous in that reference frame axes can be pointed at other locations in the model in a manner that is easy to parameterise and thus have the frames and parts reorient themselves when design variables change the geographical description of the model.

Some packages also allow the definition of direction cosines to orient a reference frame (see Section 3.2.2.1). The direction cosine approach can be visualised as a variation on the X-point-Z-point method, where the length of the vectors specified is unity. Direction cosines are typically given as a 3×3 matrix but some consideration will show that the resulting nine pieces of information are in fact redundant, and the minimum required information collapses into the same information as the X-point-Z-point method. Direction cosines are convenient in that a vector quantity can be transformed from one reference frame to another by directly multiplying it by the direction cosines between them, which form the *transformation matrix* as described in Section 2.2.7. A vector may be transformed between two vectors defined with respect to the GRF by transforming one vector back to the GRF and then forward through the direction cosines; again the direction cosine formulation is convenient algorithmically since the backward transformation matrix is simply the transpose of the direction cosine matrix.

For convenience in the subsequent text, the LPRF is taken to be coincident with and parallel to the GRF when setting up the model. The orientation of the marker reference frame O_m is then defined relative to the GRF. As shown earlier in Figure 3.15, the position vector $\{QP\}_1$ would define the position of the marker reference frame and similarly $\{XP\}_1$ and $\{ZP\}_1$ would now be used to define the orientation. As with $\{QP\}_1$ the x, y and z components of $\{XP\}_1$ and $\{ZP\}_1$ are now resolved parallel to the GRF O_1. It should be noted that the methods described here have been extended and are more general including the capability to implement parameter-based reference frames.

For the vehicle models described later in this text a consistent approach will be used for the GRF, as shown in Figure 3.16, where the X-axis points back along the vehicle, the Y-axis points to the right of the vehicle and the Z-axis is up. The XZ plane will always be taken to be coincident with the centre line of the vehicle so

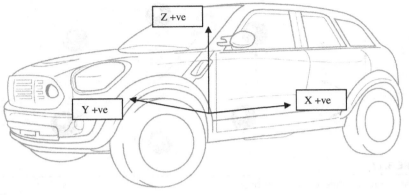

FIGURE 3.16

Ground reference frame for full vehicle models.

(Image courtesy of Prodrive).

as to exploit symmetry when defining, for example, the Y coordinates of left and right suspension systems.

3.2.3 Basic model components

When developing the data set for a model in a MBS analysis program, the following can be considered to be basic model components:

1. Rigid bodies
2. Geometry
3. Constraints equations
4. Forces
5. User-defined algebraic and differential equations

Nonrigid bodies are discussed in principle later in Chapter 6. Constraint equations are meant in the most general sense rather than in any program-specific usage. In the first edition some detailed discussion was given of declaring these items but this information tends to be quite program specific and is thus substantially simplified in this edition.

3.2.4 Parts and frames

The part declaration can be used to define any rigid body or lumped mass. For a vehicle suspension, system components such as the control arms and wheel knuckle would typically be modelled as rigid bodies until quite late in the vehicle development process. For other classes of vehicle, such as motorcycles and commercial vehicles, they tend to be substantially influenced by the compliance of the primary structure and so the parts might be made up of multiple parts joined with some

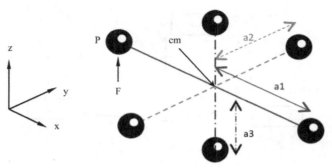

FIGURE 3.17

The six particle concept for a rigid body.

kind of compliance matrix. The type of analysis being performed will dictate the amount of data that must be defined for the body. For a dynamic analysis a full definition will be required to include the mass, centre of mass position, the mass moments of inertia, the orientation of the axes about which the mass moments of inertia are measured and any initial translational and angular velocities to be applied to the body. As mentioned earlier it is also possible to define a LPRF to which any markers belonging to the part can be referenced.

Mass moments of inertia are worth some discussion since, while most people can readily estimate and assimilate mass properties, students and new practitioners are sometimes a little slower to grasp the ideas of inertia.

The equations of motion for a particle are easily understood and are declared by Newton's Second Law of motion. Their derivation is given in Section 2.7 of Chapter 2 where we consider the dynamics of a particle, a body for which the motion is restricted to translation without rotation.

Real bodies have nonzero inertia, which is to say they have a reluctance to rotate in the presence of an applied torque.[4] In order to visualise inertia, a real body may be imagined to be represented by six particles, at distances from the mass centre (cm) of a1, a2 and a3, as shown in Figure 3.17. Each particle has one sixth of the mass of the body.

Considering the leftmost particle, point P, in Figure 3.17, it can be seen that a vertical force, F, applied directly to it will attempt to generate an angular acceleration, α, about the line a_2. The angular acceleration is related to the resulting linear acceleration, A, by the radius, R, about which the rotation occurs.

$$\alpha = A/R \qquad (3.1)$$

---

[4]A torque is taken in this text to be a generalised force acting to rotate a body about an axis. As such it is distinct from a moment, which is taken in this text to mean a force acting at a distance from an axis, attempting to both translate and rotate a body. A couple, a pair of equal and opposite forces symmetrically distributed about an axis and with no net force, is functionally equivalent to a torque. Other authors may use these terms differently.

In this case the radius R is equal to the distance a1. The applied torque, T, is the force multiplied by the radius at which it acts

$$T = F \cdot R \tag{3.2}$$

Since $F = mA$, we can gain some sense of an equivalent angular quantity for the rotational inertia I

$$
\begin{aligned}
I &= T/\alpha \\
&= F \cdot R/(A/R) \\
&= F/A \cdot R^2 \\
&= m \cdot R^2
\end{aligned}
\tag{3.3}
$$

Thus it can be seen that for each of the six particles, their reluctance to rotate is given by m and R^2. Considering each axis of rotation in turn and summing the effects of each mass, it can be seen that

$$I_1 = 2\left(\frac{m}{6}a_2^2 + \frac{m}{6}a_3^2\right) \tag{3.4}$$

$$I_2 = 2\left(\frac{m}{6}a_1^2 + \frac{m}{6}a_3^2\right) \tag{3.5}$$

$$I_3 = 2\left(\frac{m}{6}a_1^2 + \frac{m}{6}a_2^2\right) \tag{3.6}$$

This description is simply a lumped mass variation on the usually given classical description of mass moments of inertia as described in this text in Chapter 2; however, it is the authors' experience that students more readily grasp some kind of intuitive meaning for inertia using the 'six-mass' method compared to the classical method. It has the added advantage that if students can instinctively estimate the mass and physical dimensions of an object, then they are meaningfully able to estimate inertias and 'sense check' supplied data for units conversion errors and the like.

The principal inertias represent some kind of 'equivalent rectangular prism' — an instinctive box (Figure 3.18) we might draw representing the body as an item of uniform density. There is no guarantee that the directions of the principal inertia vectors line up with our chosen axis system, in which case the contributions of the individual masses must be summed in the same consistent reference frame.

The obvious frame in which to undertake the summation is the global one and this is appealingly simple even when the body is not aligned with the global axis (Figure 3.19).

$$I_{xx} = \sum_{i=1}^{6} \frac{m}{6}\left(y_i^2 + z_i^2\right) \tag{3.7}$$

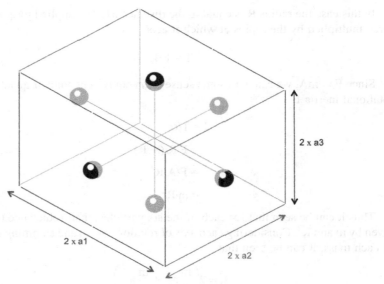

FIGURE 3.18

A graphical representation of what the 'six-mass' method means for a real object.

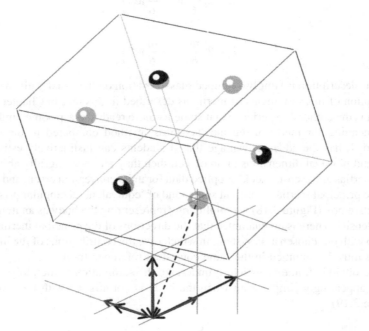

FIGURE 3.19

The contribution of one of the masses in a global reference frame.

$$I_{yy} = \sum_{i=1}^{6} \frac{m}{6} \left(x_i^2 + z_i^2 \right) \qquad (3.8)$$

$$I_{zz} = \sum_{i=1}^{6} \frac{m}{6} \left(x_i^2 + y_i^2 \right) \qquad (3.9)$$

$$I_{xy} = \sum_{i=1}^{6} \frac{m}{6} x_i y_i \qquad (3.10)$$

$$I_{xz} = \sum_{i=1}^{6} \frac{m}{6} x_i z_i \qquad (3.11)$$

$$I_{yz} = \sum_{i=1}^{6} \frac{m}{6} y_i z_i \qquad (3.12)$$

A more logical frame in which to undertake the summation is one which is positioned at the part mass centre but oriented as the global reference frame, as shown in Figure 3.20.

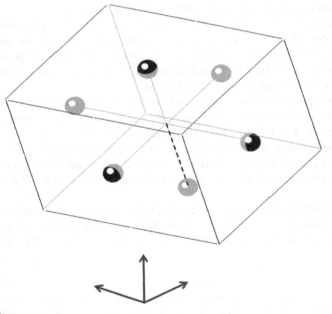

FIGURE 3.20

The contribution of one of the masses in a frame local to the body but oriented as the global reference frame.

$$I_{xx} = \sum_{i=1}^{6} \frac{m}{6}\left((y_i - y_{CM})^2 + (z_i - z_{CM})^2\right) \tag{3.13}$$

$$I_{yy} = \sum_{i=1}^{6} \frac{m}{6}\left((x_i - x_{CM})^2 + (z_i - z_{CM})^2\right) \tag{3.14}$$

$$I_{zz} = \sum_{i=1}^{6} \frac{m}{6}\left((x_i - x_{CM})^2 + (y_i - y_{CM})^2\right) \tag{3.15}$$

$$I_{xy} = \sum_{i=1}^{6} \frac{m}{6}(x_i - x_{CM})^2 (y_i - y_{CM})^2 \tag{3.16}$$

$$I_{xz} = \sum_{i=1}^{6} \frac{m}{6}(x_i - x_{CM})^2 (z_i - z_{CM})^2 \tag{3.17}$$

$$I_{yz} = \sum_{i=1}^{6} \frac{m}{6}(y_i - y_{CM})^2 (z_i - z_{CM})^2 \tag{3.18}$$

There are thus three different possibilities for specifying mass moments of inertia — principals, global and local. This is in contrast with mass, which is simply mass in any reference frame and has a unique position — the mass centre.[5]

The existence of alternative means of specifying gives rise to possibility for errors to trap the unwary. A rigorous and defensible protocol is to always add an additional frame placed at the body mass centre and aligned with the principal axes. Mass moments of inertia can be added directly as their principal components and the opportunity for error is very much reduced.

A second option is to enter a so-called inertia tensor, containing the six moments and products of inertia calculated with respect to the body mass centre.

There are more or less no foreseeable circumstances under which the inertia with respect to some other arbitrary point is relevant. This is a frequent source of difficulty because many CAD systems report it (Figure 3.21), and a frequent habit of students is to retrieve the first piece of data they come to that looks like it might be correct.

In Figure 3.21 the mass centre is reported in Pro-E immediately before the inertia tensor with respect to some other frame. This is an unfortunate consequence of computational convenience (as was seen in Eqns (3.7)–(3.12), the 'global'

---

[5]Note that the mass centre is often referred to colloquially as the 'centre of gravity'. While this makes no sense (gravity is every atom in the universe acting on every other atom in the universe) it is nevertheless widely understood. Speak to your audience in a language they understand, but know that it really is the mass centre.

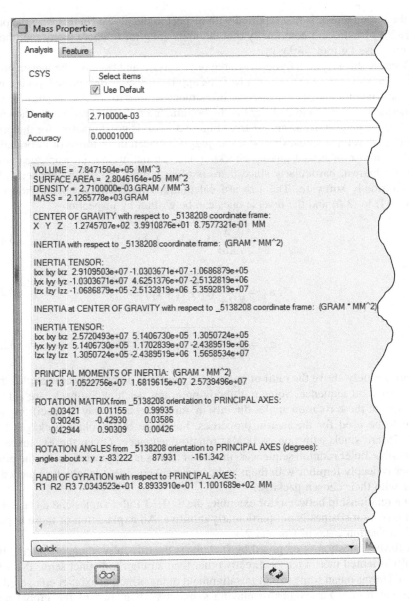

FIGURE 3.21

A typical inertia summary formed from a CAD system.

formulation is substantially simpler and is thus the first one coded by any right-thinking programmer). Further down the printout the quantities of interest are shown. Note the use of centre of gravity for mass centre, as is common. Note also that the inertia tensor is first reported with respect to the global (in this case the part, _5138208) frame, and note finally that the units are g mm² where many multibody programs use kg mm² or kg m².

Because this is such a frequent source of confusion and because the inertia properties of a body determine — along with the applied forces — the applied motion in a very fundamental way, it is worth some investment of effort to develop an intuitive feel for inertia numbers. Note that right at the bottom, the radii of gyration are shown — these are the lengths a1, a2 and a3 from Figure 3.17. It is good practice for the practitioner to develop a spreadsheet that can juggle between these forms of information for the purpose of model quality checking and estimating inertia properties when none are known, particularly since there is often a units transition between CAD and multibody software. The forward calculations have already been shown in Eqns (3.4) to (3.6) and the reverse ones can be written by inspection

$$a_1 = \sqrt{\frac{3}{2m}(I_2 + I_3 - I_1)} \tag{3.19}$$

$$a_2 = \sqrt{\frac{3}{2m}(I_1 + I_3 - I_2)} \tag{3.20}$$

$$a_3 = \sqrt{\frac{3}{2m}(I_1 + I_2 - I_3)} \tag{3.21}$$

Immediately above the radii of gyration in the CAD output are some angles that form a rotation sequence. Some multibody programs, such as Simpack, are able to make use of these rotation angles directly in quite a versatile way in specifying the frame to be used for the inertia properties, but some of the older packages like ADAMS are stuck with one particular rotation sequence. Given the existence of six possible Euler rotation sequences, the use of angles is deprecated unless the practitioner is deeply familiar with them and completely confident in which sequence is in use with their chosen package.

The relationship between, for example, the 3—1—3 Euler angles and a Cartesian set of direction cosines is not particularly intuitive. An explanation of these is provided in Section 2.2.6 of Chapter 2.

A further question often posed by students is the difficulty in understanding the globally oriented inertia tensor directly other than a rather indistinct sense that off-diagonal terms mean some sort of misalignment of the principal axis set compared to the global frame. For bodies displaying symmetry about one of the global axis planes, two of the three off-diagonal terms are zero and the inclination of the principal axis set within the plane of symmetry, accepting that the remaining principal axis is normal to the plane of symmetry, was given in Section 2.12

$$\tan 2\theta = \frac{I_{xz}}{\frac{1}{2}(I_{zz} - I_{xx})} \tag{3.22}$$

Note that the numerical conditioning of these expressions is poor, as the axes get closer to the global axes since the denominators collapse to zero.

For a more general three-dimensional case, the principals are found by finding the eigenvectors of the inertia tensor. Recall that the eigenvector problem is solved by subtracting the eigenvalue, λ, from the matrix of interest and setting the determinant to zero

$$\begin{vmatrix} I_{xx} - \lambda & I_{xy} & I_{xz} \\ I_{xy} & I_{yy} - \lambda & I_{yz} \\ I_{xz} & I_{yz} & I_{zz} - \lambda \end{vmatrix} = 0 \qquad (3.23)$$

This results in a cubic equation in λ

$$\lambda^3 + A\lambda^2 + B\lambda + C = 0 \qquad (3.24)$$

where

$$A = -I_{xx} - I_{yy} - I_{zz} \qquad (3.25)$$

$$B = I_{xx} \cdot I_{yy} + I_{xx} \cdot I_{zz} + I_{yy} \cdot I_{zz} - I_{xy}^2 - I_{xz}^2 - I_{yz}^2 \qquad (3.26)$$

$$C = -I_{xx} \cdot I_{yy} \cdot I_{zz} + I_{xx} \cdot I_{yz}^2 + I_{yy} \cdot I_{xz}^2 + I_{zz} \cdot I_{xy}^2 + 2 \cdot I_{xy} \cdot I_{xz} \cdot I_{yz} \qquad (3.27)$$

This cubic equation produces a graph, such as that shown in Figure 3.22. Note that in Figure 3.22, the curve is typical of a long, thin body with one principal inertia quite low and the other two quite similar to each other. If the body is cylindrical or of square section then the equation has two repeated roots. The roots can be found using a Newton–Raphson iteration or some other solver, such as the linear solver in

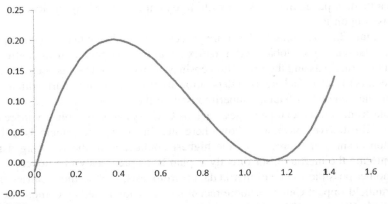

FIGURE 3.22

The roots (zero crossings) for eqn (3.24), shown graphically.

Excel or Open Office; for awkwardly shaped bodies the numerical conditioning of this problem is poor.

$$\lambda_1 = I_1 \tag{3.28}$$

$$\lambda_2 = I_2 \tag{3.29}$$

$$\lambda_3 = I_3 \tag{3.30}$$

Once the eigenvalues have been found, the eigenvectors can be found in the usual way

$$\begin{bmatrix} I_{xx} - I_i & I_{xy} & I_{xz} \\ I_{xy} & I_{yy} - I_i & I_{yz} \\ I_{xz} & I_{yz} & I_{zz} - I_i \end{bmatrix} \begin{bmatrix} x_{i\lambda} \\ y_{i\lambda} \\ z_{i\lambda} \end{bmatrix} = \begin{bmatrix} 0 \\ 0 \\ 0 \end{bmatrix} \tag{3.31}$$

There are an infinite number of possible equivalent eigenvectors, so typically one value, for example $x_{i\lambda}$, is set to unity and then a solution found for the other two. This can be performed using Gaussian elimination or again the linear solver in Excel or Open Office. The resulting vector is typically not of unity length and so is usually normalised to give direction cosines in the familiar form.

Note that in Eqn (3.31) the order of the eigenvalues produced is not necessarily ascending, and in general the roots of an equation do not necessarily appear in any conveniently sorted order. There is not any hard and fast convention concerning the reporting and use of principal inertias, with some practitioners using I_1 as the principal visually closest to the global X-axis, some using I_1 as the numerically lowest and some the numerically highest. The lack of a clear dominance of any one convention suggests that it is unimportant as long as the correct vector is associated with the correct principal inertia. As ever, it is important to verify the behaviour of a particular package with an example before any critical decisions are based upon it.

In Figure 3.23 a sample of the entire process from end-to-end is shown. Input for the calculations is the global inertia matrix about the component mass centre, from which the principals and their direction cosines are calculated as has been described. The vectors a1, a2 and a3 are used to position six masses on a new part, which can be seen in the diagram as baubles superimposed on the component.

Data from commercial packages, such as CAD systems, is typically more exact than the illustrative conversion shown here and thus using the principals and their direction cosines at source gives the highest confidence in the resulting data, or else entering the inertia tensor directly is plan B, as previously mentioned.

Another possible source of inertia data is from tests, such as those carried out by the Cranfield Impact Centre or more recent methods advocated by companies, such as Resonic (Klöpper, 2011). Not all practitioners of such testing are equally competent and when dealing with a new supplier, the ability to vet the measurements for

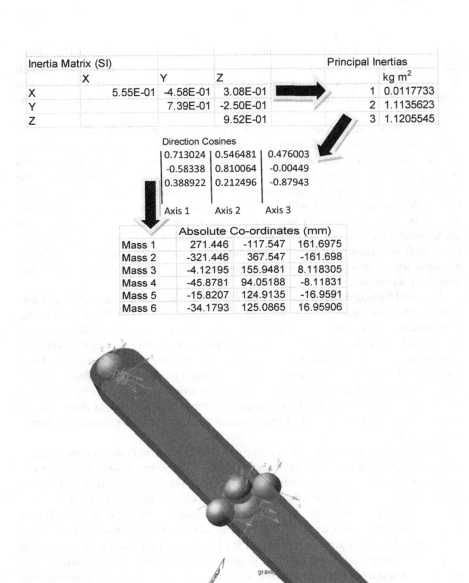

Inertia Matrix (SI)					Principal Inertias	
	X	Y	Z			kg m²
X	5.55E-01	-4.58E-01	3.08E-01		1	0.0117733
Y		7.39E-01	-2.50E-01		2	1.1135623
Z			9.52E-01		3	1.1205545

Direction Cosines

0.713024	0.546481	0.476003
-0.58338	0.810064	-0.00449
0.388922	0.212496	-0.87943
Axis 1	Axis 2	Axis 3

Absolute Co-ordinates (mm)			
Mass 1	271.446	-117.547	161.6975
Mass 2	-321.446	367.547	-161.698
Mass 3	-4.12195	155.9481	8.118305
Mass 4	-45.8781	94.05188	-8.11831
Mass 5	-15.8207	124.9135	-16.9591
Mass 6	-34.1793	125.0865	16.95906

FIGURE 3.23

Sample Six-Mass Calculations for an Arbitrary Body in MSC ADAMS. The link body is shown along with the six baubles positioned according to the spreadsheet output.

physical plausibility is important; only recently the authors were handed an inertia tensor as part of a commercial measurement contract where the basic test — that all the radii of gyration are not negative as calculated in Eqns (3.19)–(3.21) — was failed by a spectacular margin, requiring an adjustment of over 30% just to become physically possible.[6]

As an alternative to entering inertia data at all, many multibody packages offer a conceptual sketching environment allowing for basic solids — blocks, cylinders, extrusions, frusta, tori — and their Boolean combination to give some visual impression of the parts of the model. Although originally something of a 'gaffer-dazzler', the animation of such graphics can give an excellently realistic impression of system motion for communication with a nonexpert audience and also allow debugging when a perfectly valid model has been constructed that is not quite like the model the user *intended* to construct through some oversight or other. These graphical entities, with well-known published solutions for their inertia tensors, can be combined by the software using parallel axis theorem to give an estimated mass and inertia tensor for a component.

It should be noted that many packages presume such entities are solid and also presume they are steel. Anyone who has ever moved a metre of 50 mm radius bar will note that it is a good thing that not many things are actually solid steel and this gives rise to rather excessive properties if used unthinkingly. Since the sketching environment in most multibody packages is nothing like a modern three-dimensional CAD system, a great deal of time can be consumed ineffectively producing 'concept level' graphics. The solution is either to reduce the density until a reasonable mass is obtained and accept the accompanying inertia properties, or to learn to estimate mass and physical dimensions and estimate inertia properties as described.

Finally, when specifying a body, the initial velocity conditions are required. A fully safe approach to initial conditions is to start the entire system from rest and accelerate it with applied forces until it reaches the condition of interest. However, this represents a rather large waste of computing resource and it is much better with vehicle systems, for example, to be able to initialise the model with some known running condition, such as travelling in a straight line at a known speed. There are some risks to this approach if not all the bodies are initialised with compatible speeds since large forces are required at the first iteration step to harmonise velocities and solve the constraint equations. This can give rise to some incomprehensible start-up transients in the model. Making sure all parts have the same translational velocity avoids any difficulties; it is recommended that a body be solved for a

---

[6]Measurements containing such inaccuracies have a negative value to the organisation because they promote confidence where none should exist, which is sure to become apparent later (and more expensively) in the product development process. This is part of a wider issue of unwarranted overconfidence in measurement results, often summed up with 'everybody believes a measurement except the man who did it; nobody believes a calculation except the man who did it'.

few tenths of a second in free-fall with the required initial velocity to uncover any errors in setting initial velocities, which can otherwise be somewhat subtle to uncover. Setting initial velocity as a model parameter and enforcing the same parametric value on every part is good working practice and enables rapid investigation of the influence of speed on the behaviour of interest in any case.

Rotational initial conditions, for example wheel spin speed, can be set a little more approximately since in real systems, such as an aircraft landing, it can be shown that such events can (and indeed should) be tolerated by the system. In fact, the authors' preference is to leave wheel and driveline components nonrotating again as a debug case and ensure the model solves a 'landing spin-up' case effectively as part of the commissioning process. Initial rotational velocities can be parameterised back to initial translational velocities as a final commissioning step.

It is important to note that initial angular velocity conditions are typically resolved using the axes of the centre of mass marker.

3.2.5 Equations of motion for a part

The following sections describe the formulation of the rigid body equations of motion, constraint equations and solution methods used in a general-purpose MBS program, such as MSC ADAMS. The formulations are based on those given by Wielenga (1987) and presented here using the vector methods described in Chapter 2. Although the solvers have been subject to continual development since 1987, the following will serve as an introduction to the typical calculations performed by the software. As a starting point it can be shown that kinematic variables are required to represent the location and orientation of a part with respect to the GRF, as shown in Figure 3.24. Note that here we are using the general form of the position vector $\{R_n\}_1$ to locate the part rather than the equivalent $\{QG\}_1$ used in MSC ADAMS terminology.

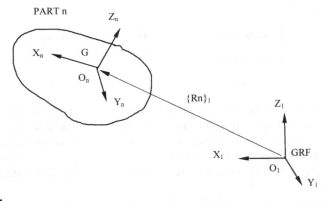

FIGURE 3.24

The Location and Orientation of a Part. GRF, ground reference frame.

The location of any part is specified by a position vector $\{Rn\}_1$ from the GRF to the centre of mass, G, of the part. In this case the part is labelled as the nth part in the system and the GRF is taken to be the first frame O_1. The components of the vector $\{Rn\}_1$ are resolved parallel to the axes of the GRF as indicated by the subscript 1. The velocity is obtained using

$$\{Vn\}_1 = {}^d\!/_{dt}\{Rn\}_1 \qquad (3.32)$$

The orientation of the part reference frame is specified by a set of Euler angles (Ψ, ϕ, θ). Note that the Euler angles are stored in an order that differs from the sequence (Ψ, θ, ϕ) used to change the orientation of a reference frame shown in Figure 3.24.

There are three frames of interest during the transformation. The first is the GRF (X, Y, Z), which is also frame O_1. The second is a frame made up of the axes about which each of the rotations takes place. This is known as the Euler-axis frame (Z, X_1, Z_1) and will be referred to as frame O_e. Note that this is not a reference frame in the true sense as the three axes are not perpendicular to one another. The third frame is the resulting part frame (X_2, Y_3, Z_1). For the nth part in a system this would be the part frame O_n. The matrix $[A_{1n}]$ is the Euler matrix for part n and performs the transformation from the part frame O_n to the GRF O_1.

$$[A_{1n}] = \begin{bmatrix} \cos\psi\,\cos\phi - \sin\psi\,\cos\theta\,\sin\phi & -\cos\psi\,\sin\phi - \sin\psi\,\cos\theta\,\cos\phi & \sin\psi\,\sin\theta \\ \sin\psi\,\cos\phi + \cos\psi\,\cos\theta\,\sin\phi & -\sin\psi\,\sin\phi + \cos\psi\,\cos\theta\,\cos\phi & -\cos\psi\,\sin\theta \\ \sin\theta\,\sin\phi & \sin\theta\,\cos\phi & \cos\theta \end{bmatrix}$$

$$(3.33)$$

Note that the inverse of this matrix $[A_{n1}]$ is simply the transpose and performs the transformation from the GRF to the part frame. Another matrix [B] performs the transformation from the Euler-axis frame O_e (Z, X_1, Z_1) to the part frame O_n (X_2, Y_3, Z_1).

$$[B] = \begin{bmatrix} \sin\theta\,\sin\phi & 0 & \cos\phi \\ \sin\theta\,\cos\phi & 0 & -\sin\phi \\ \cos\theta & 1 & 0 \end{bmatrix} \qquad (3.34)$$

Note that this matrix becomes singular when $\sin\theta = 0$. This corresponds to the situation where Z and Z_1 are parallel and point in the same direction $(\theta = 0)$, or parallel and point in the opposite direction $(\theta = 180°)$. When this occurs an internal adjustment is used to set up a new part frame where the Z_1-axis is rotated through $90°$. Note also that the [B] matrix corresponds with an internal reordering of the Euler angles to (Z, Z_1, X_1).

For large rotations the set of Euler angles for the nth part $\{\gamma n\}_e = [\Psi n\ \phi n\ \theta n]^T$ cannot actually be represented by a vector as indicated here although they can be considered to make up a set of kinematic orientation variables for the nth part. An infinitesimal change in orientation in the part frame O_n can, however, be represented by a vector, which will be denoted $\{\delta\gamma n\}_n$. In a similar manner an infinitesimal change in the Euler angles can be represented by a vector $\{\delta\gamma n\}_e$. The angular velocity vector for the part in the local part frame can also be specified by $\{\omega n\}_n$. MSC ADAMS also requires the components of these vectors in the Euler-axis frame O_e. The angular velocity in the Euler-axis frame is simply the time derivative of the Euler angles.

$$\{\omega n\}_e = {}^{d}/_{dt}\, \{\gamma n\}_e \tag{3.35}$$

The transformation between the part frame and the Euler-axis frame is established using the [B] matrix.

$$\{\delta\gamma n\}_n = [B]\, \{\delta\gamma n\}_e \tag{3.36}$$

$$\{\omega n\}_n = [B]\, \{\omega n\}_e \tag{3.37}$$

In summary, there are now a set of kinematic position and velocity variables for the nth part with components measured in the GRF and also a set of orientation and angular velocity variables measured about the Euler-axis frame

$$\{Rn\}_1 = [\, Rnx \quad Rny \quad Rnz\,]^{T} \tag{3.38}$$

$$\{Vn\}_1 = [\, Vnx \quad Vny \quad Vnz\,]^{T} \tag{3.39}$$

$$\{\gamma n\}_e = [\, \psi n \quad \phi n \quad \theta n\,]^{T} \tag{3.40}$$

$$\{\omega n\}_e = [\, \omega n\psi \quad \omega n\phi \quad \omega n\theta\,]^{T} \tag{3.41}$$

There is also a set of kinematic equations associated with the part, which may be simply stated as

$$\{Vn\}_1 = {}^{d}/_{dt}\{Rn\}_1 \tag{3.42}$$

$$\{\omega n\}_e = {}^{d}/_{dt}\, \{\gamma n\}_e \tag{3.43}$$

The remaining part variables and equations are those obtained by considering the equations of motion for a rigid body. Each part can be considered to have a set of six generalised coordinates given by

$$q_j = [Rnx, Rny, Rnz, \psi n, \theta n, \phi n] \tag{3.44}$$

The translational coordinates are the translation of the centre of mass measured parallel to the axes of the GRF, while the rotational coordinates are provided by the Euler angles for that part. For any part the translational forces are therefore summed in the X, Y and Z directions of the GRF while the summation of moments takes place at the centre of mass and about each of the axes of the Euler-axis frame. Using a form of the Lagrange equations this can be shown as

$$\frac{d}{dt}\left(\frac{\partial T}{\partial \dot{q}_j}\right) - \frac{\partial T}{\partial q_j} - Q_j + \sum_{i=1}^{n}\frac{\partial \Phi_i}{\partial q_j}\lambda_i = 0 \qquad (3.45)$$

The kinetic energy T is expressed in terms of the generalised coordinates q_j and is given by

$$T = \frac{1}{2}\{Vn\}_1^T m\{Vn\}_1 + \frac{1}{2}\{\omega n\}_e^T [B]^T [I_n][B]\{\omega n\}_e \qquad (3.46)$$

The mass properties are specified by m which is the mass of the part and $[I_n]$ which is the mass moment of inertia tensor for the part and given by

$$[I_n] = \begin{bmatrix} I_{xx} & I_{xy} & I_{xz} \\ I_{yx} & I_{yy} & I_{yz} \\ I_{zx} & I_{zy} & I_{zz} \end{bmatrix} \qquad (3.47)$$

The terms Φ and λ represent the reaction force components acting in the direction of the generalised coordinate q_j. The term Q_j represents the sum of the applied force components acting on the part and in the direction of the generalised coordinate q_j. The equation can be simplified by introducing a term for the momenta P_j associated with motion in the q_j direction, and a term C_j to represent the constraints

$$P_j = \frac{\partial T}{\partial \dot{q}_j} \qquad (3.48)$$

$$C_j = \sum_{i=1}^{n}\frac{\partial \Phi_i}{\partial q_j}\lambda_i \qquad (3.49)$$

This results in the equation

$$\dot{P}_j - \frac{\partial T}{\partial q_j} - Q_j + C_j = 0 \qquad (3.50)$$

By way of example, consider first the equations associated with the translational coordinates. The generalised translational momenta $\{Pn_t\}_1$ for the part can be obtained from

$$\{An\}_1 = {}^d/_{dt}\{Vn\}_1 \qquad (3.51)$$

$$\{Pn_t\}_1 = \partial T/\partial\{Vn\}_1 = m\{Vn\}_1 \qquad (3.52)$$

$$ {}^d/_{dt}\{Pn_t\}_1 = m\{An\}_1 \qquad (3.53)$$

This results in $\{An\}_1$, as the acceleration of the centre of mass for that part. It should also be noted that the kinetic energy is dependent on the velocity but not the position of the centre of mass, $\partial T/\partial\{Rn\}_1$ is equal to zero. We can now write the equation associated with translational motion in the familiar form

$$m\{An\}_1 - \sum\{Fn_A\}_1 + \sum\{Fn_C\}_1 = 0 \tag{3.54}$$

where $\{Fn_A\}_1$ and $\{Fn_C\}_1$ are the individual applied and constraint reaction forces acting on the body. The rotational momenta $\{Pn_r\}_e$ for the part can be obtained from

$$\{Pn_r\}_e \;=\; \partial T/\partial\{\omega n\}_e \;=\; [B]^T\,[I_n\,]\,[B]\{\omega n\}_e \tag{3.55}$$

We can now write the equations associated with rotational motion in the form

$$\{Pn_r\}_e \;-\; \partial T/\partial\{\gamma n\}_e \;-\; \sum\{Mn_A\}_e + \sum\{Mn_C\}_e = 0 \tag{3.56}$$

$$\{Pn_r\}_e \;=\; [B]^T\,[I_n\,]\,[B]\{\omega n\}_e \tag{3.57}$$

In this case $\{Mn_A\}_e$ and $\{Mn_C\}_e$ are the individual applied and constraint reaction moments acting about the Euler-axis frame at the centre of mass of the body. Introducing the equation above for the rotational momenta introduces an extra three variables and equations for each part.

The 15 variables for each part are

$$\{Rn\}_1 = [\; Rnx \;\; Rny \;\; Rnz \;]^T \tag{3.58}$$

$$\{Vn\}_1 = [\; Vnx \;\; Vny \;\; Vnz \;]^T \tag{3.59}$$

$$\{\gamma n\}_e = [\; \psi n \;\;\;\; \phi n \;\;\;\; \theta n \;]^T \tag{3.60}$$

$$\{\omega n\}_e = [\; \omega\psi n \;\;\; \omega\phi n \;\;\; \omega\theta n \;]^T \tag{3.61}$$

$$\{Pn_r\}_e = [\; P\psi n \;\; P\phi n \;\; P\theta n \;]^T \tag{3.62}$$

The 15 equations for each part are

$$\{Vn\}_1 \;=\; {}^d\!/_{dt}\{Rn\}_1 \tag{3.63}$$

$$\{\omega n\}_e = {}^d\!/_{dt}\{\gamma n\}_e \tag{3.64}$$

$$\{Pn_r\}_e \;=\; [B]^T\,[I_n\,]\,[B]\{\omega n\}_e \tag{3.65}$$

$$m\{An\}_1 - \sum\{Fn_A\}_1 + \sum\{Fn_C\}_1 = 0 \tag{3.66}$$

$$\{Pn_r\}_e - \partial T/\partial\{\gamma n\}_e - \sum\{Mn_A\}_e + \sum\{Mn_C\}_e = 0 \tag{3.67}$$

3.2.6 Basic constraints

Constraints are used to restrict the motion of parts. There are a number of modelling elements that can be used to do this, and the constraint may restrict the absolute motion of a body relative to the ground or the relative motion between interconnected parts. Constraints can be considered to be of two types:

1. Holonomic constraints are those that are dependent on restricting displacement and result in algebraic equations.
2. Nonholonomic constraints are those where a velocity-dependent motion is enforced and result in differential equations.

There are a wide range of constraint elements available ranging from joint primitives that can constrain combinations of individual DOF between bodies through mechanical type joints and gear elements, to higher pair constraints such as those constraining a point to lie on a curve. The examples shown here are restricted to those that are used to support the examples used in this text.

A typical example of a joint primitive is the use of an inplane type of constraint that restricts the motion of a point on one body to remain in a plane on another body. A typical example of this is given later when the vertical motion of individual suspension models is controlled by using a jack to impart motion to the wheel centre as shown in Figure 3.25.

It is quite common in multibody system programs to refer to joint frames as the 'I' frame and the 'J' frame. The reasons for this are largely traditional and centre around the use of the letters from I onwards being used for implicitly defined integer variables in FORTRAN when taken as the first character of a variable name. In this example the I frame is defined to belong to the wheel knuckle and is located at the wheel centre. The J frame is defined to belong to the jack part and will move vertically with that part according to other constraints that control its motion. The orientation of the J frame must be defined in this case so that the x and y axes define the surface plane of the jack in which the I frame will be constrained to remain. For clarity the jack in Figure 3.25 is shown below the wheel but in actual fact it would be defined to locate the xy plane of J at the wheel centre. The orientation of the I frame is not important, unless used for other additional purposes, but the J frame must be defined as shown. In this case the z direction of the J frame is parallel to the z-axis of the GRF so no further definition is required to change the orientation of the J frame. If this was not the case the ZP method is often the most convenient method to orientate the z-axis of the J frame normal to the surface of the plane.

In this case the I frame on the wheel is identified as 0410 and the J frame on the jack is defined as 0610. This joint primitive would constrain the single translational DOF between the I frame and the J frame in the direction normal to the xy plane of the jack. In other words the I frame has 5 DOF relative to the top surface of the jack defined by the xy plane of the J frame. In MSC ADAMS this would be six equations of motion and one constraint equation, in Simpack it would be simply the five equations of interest. The I frame can translate in the J frame X- and Y-directions and is free to rotate about all axes. There are further types of joint primitive in addition to

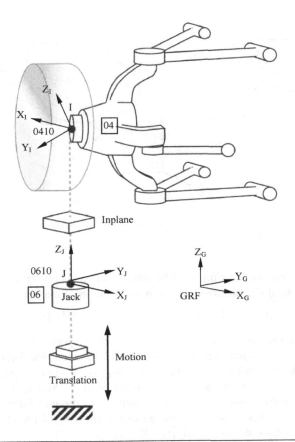

FIGURE 3.25

Application of Inplane joint primitive.

the inplane type introduced here. Others, for example, can constrain a marker on one body to follow a line defined on another body or can constrain the orientation of two markers on different bodies to remain parallel wherever those markers may be in the system model; these concepts are common to most multibody packages.

The various joints and joint primitives can be developed using combinations of four basic constraint elements. For each constraint the resulting forces and moments need to be added into the force and moment balance for a part. Consider first a basic *atpoint* constraint element shown in Figure 3.26, which constrains a point I on one part to remain at the same location in space as a point J on another part, but does not prevent any relative rotation between the two points.

This constraint can be represented by a vector constraint equation working in coordinates parallel to the axes of the GRF.

$$\{\mathbf{\Phi_a}\}_1 = (\{R_i\}_1 + \{r_I\}_1) - (\{R_j\}_1 + \{r_J\}_1) = \{0\} \tag{3.68}$$

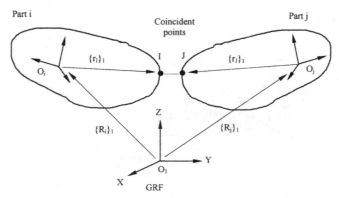

FIGURE 3.26

Atpoint constraint element.

This expression may be simplified by introducing a vector term $\{d_{IJ}\}_1$ to represent the constrained displacement between the I and the J marker.

$$\{d_{IJ}\}_1 = (\{R_i\}_1 + \{r_I\}_1) \ - \ (\{R_j\}_1 + \{r_J\}_1) \tag{3.69}$$

The reaction force on part i can be represented by the vector $\{\lambda\}_1$ with a moment given by $\{r_I\}_1 \ X \ \{\lambda\}_1$. Applying Newton's Third Law the reaction force on part j can be represented by the vector $-\{\lambda\}_1$ with a moment given by $-\{\lambda\}_1 \ X \ \{r_J\}_1$. In order to complete the calculation, the contribution to the term $\sum\{Mn_C\}_e$ in Eqn (3.67), a transformation of the moments into the coordinates of the part Euler-axis frame is required. For part i this would be achieved using $[B_i]^T\{r_I\}_i \ X \ [A_{i1}\ ]\{\lambda\}_1$. For part j this would be achieved using $-[B_j]^T\{r_J\}_i \ X \ [A_{j1}\ ]\{\lambda\}_1$.

The second basic constraint element constrains a point on one part to remain fixed within a plane on another part and is known as the *inplane* constraint or joint primitive, an application of which was shown earlier in Figure 3.25. As such it removes 1 DOF out of the plane as shown in Figure 3.27.

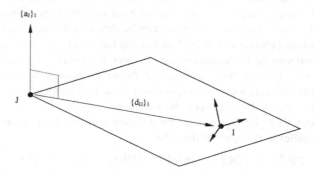

FIGURE 3.27

Inplane constraint element.

The plane is defined by a unit vector $\{a_J\}_1$, fixed in part j and perpendicular to the plane. The I marker belonging to part i is constrained to remain in the plane using the vector dot (scalar) product to enforce perpendicularity.

$$\{d_{IJ}\}_1 \bullet \{a_J\}_1 = 0 \qquad (3.70)$$

Expanding this using the definition given for $\{d_{IJ}\}_1$ in Eqn (3.38) gives the full expression for the constraint Φ_d.

$$\Phi_d = [(\{R_i\}_1 + \{r_I\}_1) - (\{R_j\}_1 + \{r_J\}_1)] \bullet \{a_J\}_1 = 0 \qquad (3.71)$$

This constraint can be represented by a vector constraint equation working in coordinates parallel to the axes of the GRF. The magnitude of the reaction force corresponding to this constraint can be represented by a scalar term λ_d. The reaction force on part i can be represented by the vector $\{a_J\}_1 \lambda_d$ with a moment given by $\{r_I\}_1 \times \{a_J\}_1 \lambda_d$. Applying Newton's Third Law again the reaction force on part j can be represented by the vector $-\{a_J\}_1 \lambda_d$. The moment contribution to part j is given by $-(\{r_J\}_1 + \{d_{IJ}\}_1) \times \{a_J\}_1 \lambda_d$.

Expanding this again using the definition given for $\{d_{IJ}\}_1$ in Eqn (3.69) gives $-(\{R_i\}_1 + \{r_I\}_1 - \{R_j\}_1) \times \{a_J\}_1 \lambda_d$. In order to complete the calculation, the contribution to the term $\sum\{Mn_C\}_e$ in Eqn (3.36), a transformation of the moments into the coordinates of the part Euler-axis frame is needed.

For part i this would be achieved using $[B_i]^T \{r_I\}_i \times [A_{ij}] \{a_J\}_1 \lambda_d$.

For part j this would be achieved using $[B_j]^T \{a_J\}_j \times [A_{j1}](\{R_i\}_1 + [A_{1i}]\{r_I\}_i -\{R_j\}_1)\lambda_d$.

The third basic constraint element constrains a unit vector fixed in one part to remain perpendicular to a unit vector located in another part and is known as the *perpendicular* constraint. The constraint shown in Figure 3.28 is defined using a unit vector $\{a_J\}_1$ located at the J marker in part j and a unit vector $\{a_I\}_1$ located at the I marker belonging to part I.

The vector dot (scalar) product is used to enforce perpendicularity, as shown in Eqn (3.72).

$$\Phi_p = \{a_I\}_1 \bullet \{a_J\}_1 = 0 \qquad (3.72)$$

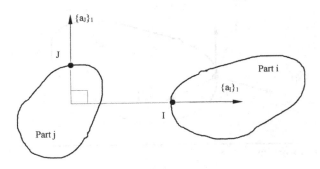

FIGURE 3.28

Perpendicular constraint element.

The constraint can be considered to be enforced by equal and opposite moments acting on part i and part j. The constraint does not contribute any forces to the part equations, but does include the scalar term λ_p in the formulation of the moments. The moment acting on part i is given by $\{a_I\}_1 \times \{a_J\}_1 \lambda_p$. Applying Newton's third law the moment acting on part j is given by $-\{a_I\}_1 \times \{a_J\}_1 \lambda_p$. The moments must be transformed into the coordinates of the part Euler-axis frame.

For part i this would be achieved using $[B_i]^T \{a_I\}_i \times [A_{ij}]\{a_J\}_j \lambda_p$.

For part j this would be achieved using $[B_j]^T \{a_J\}_j \times [A_{ji}]\{a_I\}_i \lambda_p$.

The fourth and final basic constraint element is the *angular* constraint, which prevents the relative rotation of two parts about a common axis. The constraint equation is:

$$\Phi_\alpha = \tan{-1} \left(\{x_i\}_1 \bullet \{y_j\}_1 / \{x_i\}_1 \bullet \{x_j\}_1 \right) = 0 \qquad (3.73)$$

In applying this constraint, it is assumed that other system constraints will maintain the z-axes of the two parts to remain parallel as shown Figure 3.29.

The moment acting on part i is given by $\{z_i\}_1 \lambda_\alpha$ and on part j by $-\{z_j\}_1 \lambda_\alpha$. Transforming into the Euler axis system for each part gives a moment in the coordinate system for part i equal to $[B_i]^T \{z_i\}_i \lambda_\alpha$ and on part j by $- [B_j]^T \{z_j\}_j \lambda_\alpha$.

The equations associated with each of the four basic constraint elements are summarised in Table 3.1.

The force and moment contributions to each part in the generalised coordinates are summarised in Table 3.2 and Table 3.3.

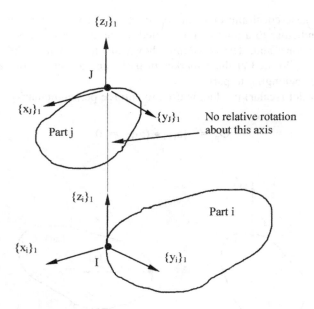

FIGURE 3.29

Angular constraint element.

Table 3.1 Basic Constraint Element Equations

Constraint	Full Equation	Abbreviated Form
Atpoint	$\{\Phi_a\}_1 = (\{R_i\}_1 + \{r_i\}_1) - (\{R_j\}_1 + \{r_J\}_1)$	$\{d_{IJ}\}_1$
Inplane	$\Phi_d = [(\{R_i\}_1 + \{r_i\}_1) - (\{R_j\}_1 + \{r_J\}_1)] \bullet \{a_J\}_1$	$\{d_{IJ}\}_1 \bullet \{a_J\}_1$
Perpendicular	$\Phi_p = \{a_I\}_1 \bullet \{a_J\}_1$	$\{a_I\}_1 \bullet \{a_J\}_1$
Angular	$\Phi_\alpha = \tan^{-1}(\{x_i\}_1 \bullet \{y_j\}_1 / \{x_i\}_1 \bullet \{x_j\}_1)$	α_{IJ}

Table 3.2 Force Contributions for Basic Constraint Elements

Constraint	Part I Force	Part J Force
Atpoint	$\{\lambda\}$	$-\{\lambda\}$
Inplane	$[A_{1j}]\{a_J\}_j\,\lambda_d$	$-[A_{1j}]\{a_J\}_j\,\lambda_d$
Perpendicular	0	0
Angular	0	0

Table 3.3 Moment Contributions for Basic Constraint Elements

Constraint	Part I Moment	Part J Moment
Atpoint	$[B_j]^T\{r_i\}_i \times [A_{i1}]\,\lambda$	$[B_j]^T\{r_J\}_i \times [A_{j1}]\,\lambda$
Inplane	$[B_j]^T\{r_i\}_i \times [A_{ij}]\{a_J\}_j\,\lambda_d$	$[B_j]^T\{a_J\}_j \times [A_{j1}]\,(\{R_i\}_1 + [A_{1j}]\{r_i\}_i - \{R_j\}_1)\,\lambda_d$
Perpendicular	$[B_j]^T\{a_i\}_i \times [A_{ij}]\{a_J\}_j\,\lambda_p$	$[B_j]^T\{a_J\}_j \times [A_{ij}]\{a_i\}_i\,\lambda_p$
Angular	$[B_j]^T\{z_i\}_i\,\lambda_\alpha$	$-[B_j]^T\{z_j\}_j\,\lambda_\alpha$

3.2.7 Standard joints

As stated there are a number of mechanical type joints that may be used to constrain the motion of bodies. Examples of some of the most commonly used joints are shown in Figure 3.30.

Of the joints shown in Figure 3.30 the spherical, revolute, translational, cylindrical and universal will figure most prominently in this text, particularly with regard to the modelling of suspension systems. The concept of an I frame on one part connecting to a J frame on another part is used again for joint elements.

For a spherical joint the I frame and J frame are defined to be coincident at the centre of the joint, but the orientation of the frames is irrelevant. For other joints, such as the revolute, cylindrical and translational, it is necessary not only to position the joint through the coordinates of the I and J frame but also to define the orientation

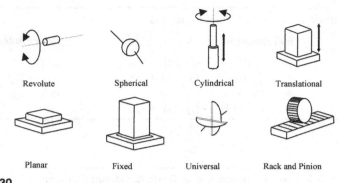

Revolute   Spherical   Cylindrical   Translational

Planar   Fixed   Universal   Rack and Pinion

FIGURE 3.30

Examples of commonly used joint constraints.

of the axis associated with the mechanical characteristics, rotation and/or translation, of the joint. The method used in MSC ADAMS is to use the local z-axis of the markers to define the axis, the most convenient method of doing this often being to define a ZP parameter for each frame. For the universal joint, the axes of the spindles need to be defined perpendicular to one another. For this joint the I and J frame are defined to be coincident with the z-axis of each orientated to suit the axis of the spindle on the side of the joint associated with the body to which the marker belongs.

The vector equations that have been derived earlier for the basic constraints can be combined to generate constraint equations for the standard joints. A spherical joint, for example, fulfils exactly the same function as an atpoint joint primitive. Examples of constraint equations for some commonly used joints are shown in Table 3.4.

It is also possible to define initial conditions associated with a joint, such as a revolute, translational or cylindrical. These are defined to be translational or rotational according to the characteristics of the associated joint. A cylindrical joint,

Table 3.4 Joint Constraints in MSC ADAMS

Joint Type	Constraints			Abbreviated Equation
	Trans'	Rot'	Total	
Spherical	3	0	3	$\{d_{IJ}\}_1 = 0$
Planar	1	2	3	$\{z_I\}_i \bullet \{x_J\}_j = 0, \{z_I\}_i \bullet \{y_J\}_j = 0, \{d_{IJ}\}_1 \bullet \{z_J\}_j = 0$
Universal	3	1	4	$\{d_{IJ}\}_1 = 0, \{z_I\}_i \bullet \{z_J\}_j = 0$
Cylindrical	2	2	4	$\{z_I\}_i \bullet \{x_J\}_j = 0, \{z_I\}_i \bullet \{y_J\}_j = 0, \{d_{IJ}\}_1 \bullet \{x_J\}_j = 0,$ $\{d_{IJ}\}_1 \bullet \{y_J\}_j = 0$
Revolute	3	2	5	$\{d_{IJ}\}_1 = 0, \{z_I\}_i \bullet \{x_J\}_j = 0, \{z_I\}_i \bullet \{y_J\}_j = 0$

for example, may be defined to have an initial translational velocity of zero and a starting displacement of 100mm in the translational direction.

Initial conditions can be enforced at the start of analysis but released once the simulation commences, in other words after time equals zero in some multibody packages. For translational, revolute and cylindrical joints it is also possible to constrain the movement of the joint during the simulation using an applied motion statement.

For the joint referenced it is necessary to define a functional equation, normally only dependent on time, which controls the movement of the I frame relative to the J frame at the associated joint. It can be readily imagined that the functional equation can be extended to encompass more complex formulations using a library of off-the-shelf mathematical functions and expressions of the type associated with an engineering or scientific programming software. Newcomers to MBS analysis often find the concept of a defined motion being a constraint difficult to grasp as the modelling element involves movement. The motion statement here constrains the associated DOF at the joint. The movement defined by the function is enforced and cannot be altered by, for example, changes to the mass properties of the bodies or the introduction of external forces. It should also be noted that where a motion is applied to a joint it would be inconsistent to specify initial conditions for the DOF associated with the motion at the joint.

Another constraint element that will be used in this text is a coupling equation and is used to constrain, or couple, the movement of two or three joints by applying scale factors. One possible application of a coupling equation in this text is to represent the mechanical behaviour of a steering gear and so define the ratio between the rotation of the steering column and the rack. A more detailed discussion of this is given later in Chapter 6.

Note that the ordering and orientation of the frames defining the coupled joints is critical if the correct physical representation of the system is to be obtained.

The coupling equation does not take into account mechanical features, such as play in the joints or backlash, and as such does not model the reaction forces within the real mechanism. The coupling equation also does not consider one joint to drive the other. This will be a function of other forces defined elsewhere in the system model. As with the motion statement a DOF is lost to the system, as the coupling equation has enforced a kinematic relationship between the motion at joints that cannot be changed by external forces.

3.2.8 Degrees of freedom

Having introduced the modelling of rigid bodies and constraints, it is now possible to describe the determination of the degrees of freedom (DOF) in a mechanical system. The concepts are very similar between ADAMS-style multibody models and Simpack-style multibody models; the DOF are brought into being by the declaration of bodies and constrained out of existence by constraint equations. A free-floating rigid body in three-dimensional space will have 6 DOF, as shown in Figure 3.31.

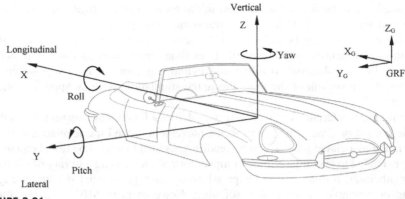

FIGURE 3.31

Degrees of Freedom Associated with an Unconstrained Rigid Body. GRF, ground reference frame.

For the vehicle body shown here and for the handling simulations that will be discussed later the body will have no direct constraint connecting it to the ground part. The only influence of the ground part will be through the forces and moments generated by one or more tyre models. Aerodynamic effects are typically modelled as stand-alone free-body forces and thus do not have any interaction with the ground part. For the axis system shown here, the vehicle will have DOF associated with translational motion in the longitudinal direction X, the lateral direction Y and the vertical direction Z. The rotational motions will involve roll about the x-axis, pitch about the y-axis and yaw about the z-axis. For vehicles such as ships and aircraft the terms surge, sway and heave are terms used to describe the translational motions but these are not commonly used in vehicle dynamics.[7] It should also be noted that for the examples in this text the x-axis is taken to point towards the rear of the vehicle, where in other texts discussing vehicle motion this is often forward to be consistent with the normal direction of travel.

For any multibody system model it is important that the analyst can determine and understand the total DOF in the system. For an ADAMS-style model, this can be achieved by using the Chebychev−Grübler−Kutzbach criterion (sometimes referred to as the 'Gruebler equation' or 'Gruebler count')

$$\text{Total DOF} = 6 \times (\text{Number of parts} -1) - (\text{Number of constraints}) \qquad (3.74)$$

The parts count in Eqn (3.74) is reduced by one to account for the fact that the nonmoving ground is counted as a part in the system. The DOF removed by typical

---

[7]It is noted that the items referred to as 'anti-roll bars' in UK usage are sometimes known as 'anti-sway bars' or just 'sway bars' in US English.

Table 3.5 Degrees of Freedom Removed by Constraint Elements

Constraint Element	Translational Constraints	Rotational Constraints	Coupled Constraints	Total Constraints
Cylindrical joint	2	2	0	4
Fixed joint	3	3	0	6
Planar joint	1	2	0	3
Rack-and-pinion joint[a]	0	0	1	1
Revolute joint	3	2	0	5
Spherical joint	3	0	0	3
Translational joint	2	3	0	5
Universal joint	3	1	0	4
Atpoint joint primitive[b]	3	0	0	3
Inline joint primitive	2	0	0	2
Inplane joint primitive	1	0	0	1
Orientation joint primitive	0	3	0	3
Parallel joint primitive	0	2	0	2
Perpendicular joint primitive	0	1	0	1
Motion (translational)	1	0	0	1
Motion (rotational)	0	1	0	1
Coupler[c]	0	0	1	1

[a] A rack-and-pinion joint is a specific type of coupler.
[b] A Atpoint primitive is functionally identical to a spherical joint.
[c] Couplers may actually couple more than one joint, as in the case of a differential gear represented with three joints and a coupler.

constraint elements are summarised in Table 3.5 and may be used to complete the calculation. The method used in Simpack is demonstrated in Table 3.6.

For a Simpack-Style model, a version of the formula that is conceptually identical is necessary but differs in detail.

$$\text{TOTAL DOF} = (\text{Number of Joint DoF}) - (\text{Number of Constrained DoF}) \quad (3.75)$$

At this stage, it is necessary to introduce the subject of redundant constraints and over constraint checking. As a starting point consider a typical application in a suspension that involves the modelling of the mounts attaching a control arm or wishbone to the body or chassis. This is illustrated in Figure 3.32, where in the real system the suspension arm is attached by two rubber bush mounts. As such the suspension arm is elastically mounted and if modelled in this way loses none of its 6 DOF to the mounts. The geometry and alignment of the bushes might suggest that the rigid body modelling approach might utilise a revolute attachment at each bush location. This in fact would introduce redundant constraints. A single revolute would fix the suspension arm and only allow it to rotate about the axis of the

Table 3.6 Joint DOF and Constrained DOF for Use with Simpack

Joint	Translational DOF	Rotational DOF	Coupled DOF	Total DOF
Free joint	3	3	0	6
Cylindrical joint	1	1	0	2
Fixed joint	0	0	0	0
Planar joint	2	1	0	3
Revolute joint	0	1	0	1
Spherical joint	0	3	0	3
Translational joint	1	0	0	1
Universal joint	0	2	0	2

Constraint	Translational Constraints	Rotational Constraints	Coupled Constraints	Total Constraints
Rack-and-pinion constraint	0	0	1	1
Atpoint constraint	3	0	0	3
Inline constraint	2	0	0	2
Inplane constraint	1	0	0	1
Orientation constraint	0	3	0	3
Parallel constraint	0	2	0	2
Perpendicular constraint	0	1	0	1
Cylindrical constraint	2	2	0	4
Fixed constraint	3	3	0	6
Planar constraint	1	2	0	3
Revolute constraint	3	2	0	5
Spherical constraint	3	0	0	3
Translational constraint	2	3	0	5
Hooke constraint	3	1	0	4
Motion (translational)	1	0	0	1
Motion (rotational)	0	1	0	1
Coupler	0	0	1	1

DOF, degrees of freedom.

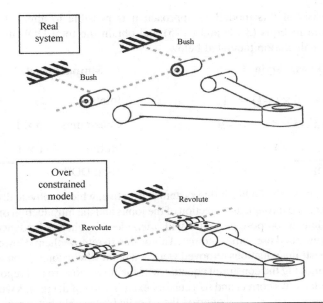

FIGURE 3.32

Redundant constraints in a suspension model.

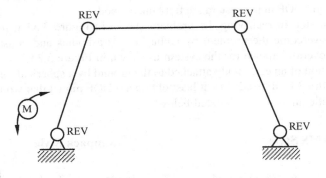

FIGURE 3.33

Overconstrained Four-Bar Linkage Problem. REV, revolutes, M, motion

freedom. Introducing the second revolute constraint only replicates the function of the first and would introduce in this case five redundant constraints.

Taking this a step further, consider the classical four-bar linkage problem shown in Figure 3.33. The mechanism forms a single loop comprising three moving parts and the fixed ground part. Revolutes are used to constrain two of the parts to the ground and a motion is applied at one of these to impart movement to the system. An intuitive modelling attempt might consider all four connections as revolutes as the likelihood is that in the real mechanism they would all appear similar.

On the basis of this modelling approach it is possible to apply the Greubler equation given in Eqns (3.74) and (3.75) and obtain the total DOF in the system based on the calculation tabulated below.

ADAMS Style				**Simpack Style**			
Parts	$6 \times (4\text{-}1) =$	18		Joints	3×1	$=$	3
Revolutes	-5×4	$= -20$		Constraints	-5×1	$=$	-5
Motion	-1×1	$= -1$		Motion	-1×1	$=$	-1
Total DOF		$= -3$		Total DOF		$=$	-3

The total sum of the DOF for this system is negative, which is physically meaningless but has resulted through the selection of the joints and the introduction of redundant constraints. General-purpose programs will often identify when redundant constraints have been applied and use some automated algorithm to remove them. However, it may be that the constraint equations removed were at a point where connection forces were of interest; removing the constraint equation sets the forces to zero and reports them as such. The model looks correct and so a cursory examination of the results would appear plausible. This is the worst kind of error, the so-called 'plausible-but-wrong' error. For this reason, redundant constraints should be approached with a great deal of caution. It is always preferable for the analyst to set the model up in such a way that it is not overconstrained and the DOF in the system are fully understood.

As a first step in rectifying the problem shown in Figure 3.33 it might seem possible to overcome the problem by adding in extra bodies and constraints that result in an overall balance for the system as shown in Figure 3.34.

The addition of an extra body attached to the ground by a spherical joint appears to add the extra 3 DOF needed to at least obtain a 0 DOF model that would be used for a kinematic analysis as tabulated below.

ADAMS Style				**Simpack Style**			
Parts	$6 \times (5\text{-}1) =$	24		Joints	$3 \times 1 +$		
Revolutes	-5×4	$= -20$			1×3	$=$	6
Spherical	-3×1	$= -3$		Revolutes	-5×1	$=$	-5
Motion	-1×1	$= -1$		Motion	-1×1	$=$	-1
Total DOF		$= 0$		Total DOF		$=$	0

Although it appears the problem has been solved there are still redundant constraints in the system. In balancing the DOF in the model, it is necessary that not only is the overall system not overconstrained but also any individual loops within the model. A possible solution in this case is to select the joints shown in Figure 3.35.

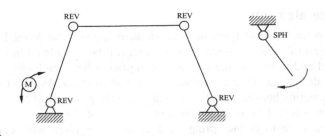

FIGURE 3.34

Overconstrained Loop in a System Model. REV, revolutes, M, motion, SPH, spherical.

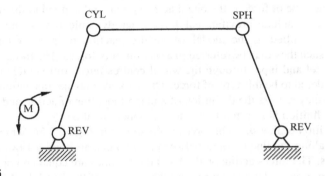

FIGURE 3.35

Zero Degree of Freedom Four-Bar Linkage Model. REV, revolutes, M, motion, SPH, spherical, CYL, cylindrical.

For the joints selected now the result is a 0 DOF model that would be used for a kinematic analysis. The DOF balance is tabulated below.

ADAMS Style			**Simpack Style**		
Parts	$6 \times (4\text{-}1) =$	18	Joints	$1 \times 1 +$	
Revolutes	$-5 \times 2 \quad =$	-10		$2 \times 1 +$	
Spherical	$-3 \times 1 \quad =$	-3		3×1	= 6
Cylindrical	$-4 \times 1 \quad =$	-4	Revolute	-5×1	= -5
Motion	$-1 \times 1 \quad =$	-1	Motion	-1×1	= -1
Total DOF	=	0	Total DOF		= 0

The cylindrical joint is used to prevent an unwanted DOF that would result in the central link spinning about its own axis. In the authors' experience, formulation of linkage models often stumbles on either redundant constraints or the inadvertent inclusion of link-spin DOF.

3.2.9 Force elements

There are two fundamental types of force element that may be defined in a MBS model. The first of these are force elements that can be considered internal to the system model and to involve the effects of compliance between bodies. Examples of these include springs, dampers, rubber bushes and roll bars. As such these forces involve a connection between two bodies and due to the principle of Newton's Third Law are often referred to as action-reaction forces. For the translational class of force elements used to define springs and dampers the force will act along the line between two markers that define the ends of the element and as such this form of definition is referred to as the line-of-sight method.

The second type of force is the one that is external and applied to the model. Examples of these include gravitational forces, aerodynamic forces and any other external force applied to the model where the reaction on another body is not required. As such they may be referred to as action-only forces. The forces generated by a tyre model and input through the wheel centres into a full vehicle model can also be considered to be this type of force. These forces may be translational or rotational and as they require the definition of a magnitude, line of action and sense, the method of definition is referred to as the component method.

The definition of line-of-sight forces is illustrated in Figure 3.36, which shows a force acting along the line of sight between two points, an I and a J frame, on two separate parts. The forces acting on the I and the J frames are equal and opposite. As the line of the force is defined entirely by the location of the I and the J frames, the orientation of these is not relevant when defining the force.

The component method applies to translational action-only forces, where the direction and sense of the force must be defined and to rotational forces, where the axis about which the torque acts is required. In MSC ADAMS, it is the z-axis of the J frame that is used to define the direction and sense of a translational action-only force. In the Simpack interface, a more general method is available for defining the line of action of the force, which can use broadly any frame in the model. The force acts on the I frame as shown in Figure 3.37, and there is no reaction on the J frame.

Rotational forces may also be defined to be action-reaction or action-only. In either case the torque produced is assumed to act on the I frame. For an

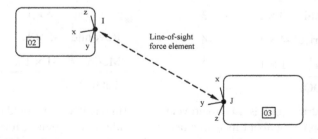

FIGURE 3.36

Line-of-sight force element.

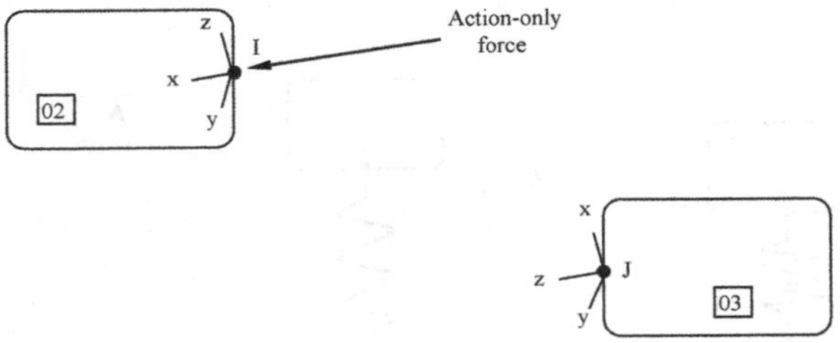

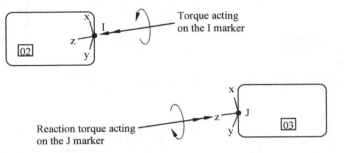

FIGURE 3.37

Action-only force.

FIGURE 3.38

Action-reaction torque.

action-only torque it is again the z-axis of the J frame that is used, in this case to define the axis of rotation. If the torque is action-reaction, the z-axes of the I and J frames must be parallel and point in the same direction, as shown in Figure 3.38.

Considering next the definition of translational spring elements, we can start with a definition that is linear and introduces the use of system variables for the formulation of a force. As can be seen in Figure 3.39, the formulation of the spring force will be dependent on the length of the spring. This is made available through a system variable defined here as DM(I, J), which represents the scalar magnitude of the displacement between the I and the J frames at any point in time during the simulation. The spring force F_S is initially defined here to be linear using

$$F_S = k \, (DM(I,J) - L) \tag{3.76}$$

where k is the spring stiffness and L is the free length of the spring, at zero force.

The equation used in Eqn (3.76) to determine F_S follows the required convention that the scalar value of force produced is positive when the spring is in compression, zero when it is at its free length and negative when it is in tension. This should not be confused with the components of any reaction force recovered at the I or the J frame. The sign of these will be dependent not only on the state of the spring but also the orientation of the spring force line of action and the reference frame in which the components are being resolved. The formulation of the spring force F_S is shown graphically in Figure 3.40.

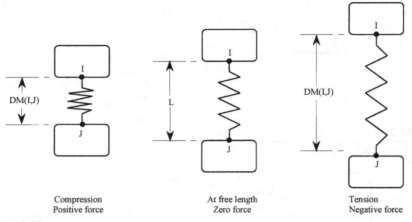

Compression Positive force	At free length Zero force	Tension Negative force

FIGURE 3.39

Spring in compression, at free length and in tension.

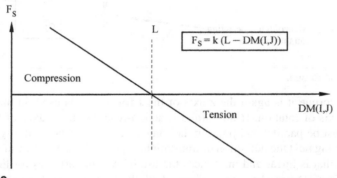

$$F_S = k\,(L - DM(I,J))$$

FIGURE 3.40

Formulation of a linear spring force.

Note that the formulation used here produces a force that is positive in compression and negative in tension. This is opposite to the convention used for stresses in stress analysis and finite element programs.

An alternative form of definition that could be used with most general-purpose programs is a built-in spring element. General-purpose programs often include such built-in elements but they are something of a two-edged sword. There can be a colossal escalation in difficulty between using a built-in element and using a more general form by referring to system variables. This develops something of an institutionalised mentality in many student users that discourages them from doing what needs doing — perhaps raising the deflection to the power 1.1 to simulate a progressively wound spring — by encouraging them take the path of least resistance. It also has the effect of making the spring model seem like something sealed up and inaccessible instead of revealing it as just a force applied to a body as a function of the relative displacement of two markers. This can conceivably (and does, observably, in a teaching environment) prevent the students from seeing the essential brilliance of Newton's work — that acceleration is force over mass and everything flows from that.

In a similar manner to the definition of a spring force the representation of a damper force will involve using the line-of-sight method to formulate an action—reaction force between an I frame on one part and a J marker on another part. We will again start with the linear case where we formulate a damper force F_D using

$$F_D = -c \, . \, VR \, (I,J) \tag{3.77}$$

where

$VR(I, J)$ = radial line-of-sight velocity between I and J marker
c = damping coefficient

Since the force generated in a damper is related to the sliding velocity acting along the axis of the damper we introduce another system variable $VR(I, J)$ that will take a positive sign when the markers are separating, as in suspension rebound, and a negative sign when the markers are approaching, as when a suspension moves upwards relative to the body in bump. It may be thought of as a radial velocity in a spherical coordinate system centred on the I frame and measuring the location of the J frame. The formulation of the damper force F_D in Eqn (3.77) is such that the damper forces are consistent with those of a spring. The force generated is positive in a repelling mode and negative in an attracting mode as illustrated in Figure 3.41.

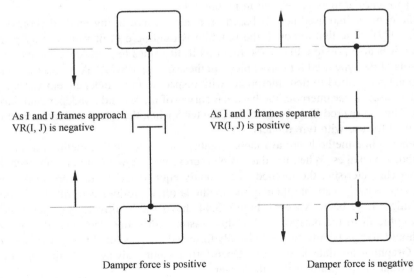

As I and J frames approach $VR(I, J)$ is negative

As I and J frames separate $VR(I, J)$ is positive

Damper force is positive

Damper force is negative

FIGURE 3.41

Sign convention for damper forces and velocities.

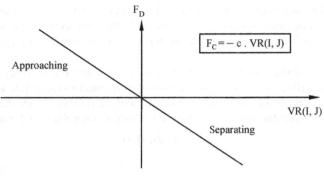

FIGURE 3.42

Formulation of a linear damper force.

The formulation of the damper force F_S is shown graphically in Figure 3.42.

The definitions of spring and damper forces so far have been based on the assumption that the force element can be modelled as linear. This can be extended to consider the modelling of a nonlinear element. The example used will be based on the front and rear dampers for a typical road vehicle.

The nonlinear damper forces are defined in general-purpose programs using xy data sets, where the x values represent the velocity in the damper, VR(I, J), and the y values are the force. During the analysis the force values are extracted using, for example, a cubic spline fit. The damper forces are not only nonlinear but are also asymmetric, having different properties in bump and rebound. Example curves for front and rear dampers are shown in Figure 3.43.

The cubic spline used here is based on a cubic curve fitting method (Forsythe et al., 1977). Note that although the function is used here to fit values to xy pairs of data it is also possible to use the function to fit values to three-dimensional xyz data sets of the type used for carpet plots. In these cases, MSC ADAMS uses a cubic interpolation method to first interpolate with respect to the x independent variables and then uses linear interpolation between curves of the second z independent variables. This is covered in more detail in Chapter 5 when the interpolation method is described for use with tyre models.

Other spline methods are available, notably the Akima spline method, in many multibody packages. When the data to be represented show a smooth and unchallenging characteristic, the methods are broadly equivalent. If the characteristic is more aggressively variable then spline methods often produce a result that is not what might have been expected; Figure 3.44 shows the intuitively expected result and a cubic fit to the datapoints. For this reason with spline data, the package in use should always be interrogated visually to see what it has fitted to the data passed to it. Graphing the data in Excel or Open Office cannot always be relied upon to reveal what the program itself will deliver.

An example of the spline data set used here to represent the nonlinear damper force follows. In this case, the velocities are defined in mm/s on the x-axis and

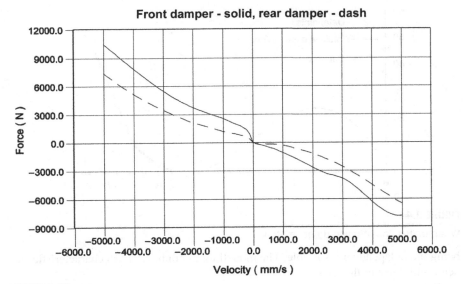

FIGURE 3.43

Nonlinear force characteristics for front and rear dampers.

the values of force (N) are returned on the y-axis. It is important that the data set has sufficient range at both the top and bottom end to encompass the conditions during the simulation. Should the independent variable used on the x-axis reach values outside of the range of the spline data then the program will have to extrapolate values that can lead to unreliable results. Some general-purpose codes have a flag so as not to extrapolate with the last known cubic fit but rather to perform a linear extrapolation off the end of the data set.

```
SPLINE/1
X = -5000, -3150, -2870, -2450, -2205, -1925, -1610, -1260, -910,
-630, -470, -400, -350, -300, -250, -230, -200, -190, -160, -120,
-80, -55, -40, -20, -10, -1, -0.1, 0, 0.3, 3, 30, 40, 60, 80, 100,
200, 250, 400, 490, 770, 1050, 1330, 1820, 2060, 2485, 2590, 2730,
2835, 2940, 3080, 5000
Y = 10, 425, 5800, 5200, 4400, 4000, 3600, 3200, 2800, 2400, 2000,
1800, 1700, 1600, 1500, 1400, 1350, 1310, 1290, 1200, 1000, 700,
400, 210, 80, 40, 4, 0.4, 0, -1, -10, -100, -123, -150, -182, -200,
-260, -300, -400, -500, -800, -1200, -1600, -2400, -2800, -3400,
-3500, -3600, -3700, -3800, -4000, -7840
```

Although the method of modelling a nonlinear force element has been demonstrated here using a damper, the method is equally applicable to other force elements, such as springs and rubber bushes. For road springs it is usually sufficient to model these as linear but should a nonlinear formulation be required, the same approach is used with the magnitude of the displacement in the spring, DM(I, J),

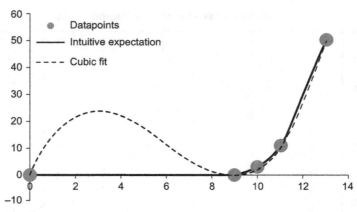

FIGURE 3.44

When maths and intuition do not match.

being the independent x variable. The modelling of nonlinear bush characteristics is dealt with later in this chapter.

Before moving on from the subject of springs and dampers it should be noted that it is also possible to define, in addition to applied torques, rotational springs and dampers. This will be dealt with in Chapter 6, where a rotational spring damper is used to idealise the characteristics of a suspension system in a full vehicle model based on axles that rotate, relative to the vehicle body, about the roll centres of the suspension systems.

Certain types of analysis may also require a suspension model to include the force characteristics of bump and rebound stops. For simulations such as a vehicle traversing off-road terrain, or the road wheel striking a pothole, the bump and rebound stops will need to be modelled. This may be particularly relevant when the vehicle or suspension model is being used to predict the distribution of force as inputs to finite element models.

If we consider the case of a bump stop, as illustrated in Figure 3.45, it can be seen that we face a new modelling problem. The force element needs now to represent the nonlinear problem of a gap, which closes so that the force must not only include the nonlinear characteristics of the rubber on contact but also be able to switch on and off as the gap closes and opens. The approach used to model the bump or rebound stop will depend on its location. The approach used here might need modification for a rebound stop built into a suspension strut.

In this example, the J frame is taken to belong to the vehicle body or chassis and the I frame is located at a point on the suspension that would strike the face of the bump stop. The point at which contact is established can be found here by comparing the z component of the displacement of the I frame from the J frame resolved parallel to the axes of the I frame. This will be described here as DZ (I, J, I). Note that the third frame in brackets has been added to the system variable DZ to indicate that the component of displacement is resolved using this co-ordinate system rather than by default the GRF.

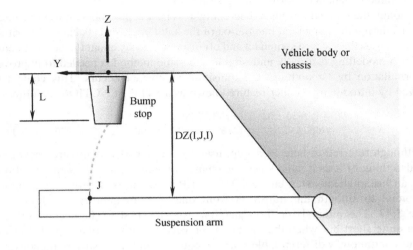

FIGURE 3.45

Modelling a bump stop contact force.

A first attempt at modelling this force could be achieved with the following function statement. In this example 0206 is the I frame, 0307 is the J frame, the length of the bump stop is 50 mm and a linear stiffness for the bump stop of 300 N/mm is used:

```
FUNCTION = IF(DZ(0206,0307,0206)-50: 300 * (50 - DZ(0206,0307,0206)), 0, 0)
```

This example has introduced an arithmetic IF that allows us to conditionally program the value of the FUNCTION and hence the force returned by this statement. The format used here is:

```
IF (expression 1: expression 2, expression 3, expression 4)
```

Expression 1 is evaluated and the value obtained is used to determine which expression is used to evaluate the FUNCTION as follows:

```
IF expression 1 < 0 then the FUNCTION = expression 2
IF expression 1 = 0 then the FUNCTION = expression 3
IF expression 1 > 0 then the FUNCTION = expression 4
```

Note that the arithmetic is in common use in multibody programs and is different from the three-expression IF used inside Excel or Open Office. In this case expression 1 is DZ(0206,0307,0206)-50. Clearly when this is greater than zero the gap is open and so the calculated force from expression 4 is zero. When expression 1 is equal to zero the I marker is just making contact with the face of the bump stop but no deformation has taken place and so the force is still zero. When expression 1 is less than zero then the contact force is generated using

300*(50-DZ(0206,0307,0206)). Note that this has been programmed to ensure that a positive value is generated for the bump stop as it compresses.

Although the method introduced here might work it is possible that it will cause problems during the numerical integration of the solution. This is because the arithmetic IF causes the force to switch on and off instantaneously about the point of contact. Such modelling is usually undesirable and some method is needed to improve the formulation by 'smoothing' the transition as contact occurs. This could be achieved by introducing another feature known as a STEP FUNCTION as follows:

```
VARIABLE/1,FUNCTION = 50 - DZ(0206,0307,0206)
FUNCTION = STEP( VARVAL(1), 0,0, 0.5,1) * (300 * (50 - DZ(0206,0307,0206)))
```

Although referred to here as a step, users of similar MBS software packages may also think of it as a ramp since the change in value returned by the function is not an instantaneous step change. The STEP function used here uses a cubic polynomial to smooth the transition from one state to another, as shown in Figure 3.47. The interval for the transition from the first to the second values can be quite short but when the integrator proceeds very slowly, the STEP form remains continuously differentiable — a key requirement for the successful convergence and completion of the integration step.

This example also introduces the use of VARIABLES that can be used to program equations and substitute the returned value, VARVAL(id), into another FUNCTION. In this case, the line VARIABLE/1... is used to program the deformation of the bump stop. The value of this, VARVAL(1) is used to define the variable on the x-axis that is used to step from one state to another. In this the step function is used to smooth the formulation of the contact force on between 0 and 0.5 mm of bump stop deformation. Note that not every package makes the use of solution variables as straightforward as this.

Additional functions often exist, such as a penalty stiffness modelling function which might be used to switch on a contact force or to extend the description

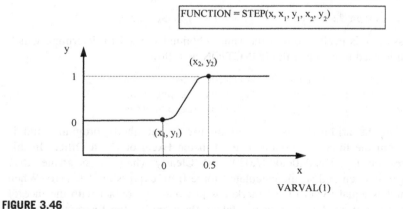

FIGURE 3.46

Step function for a bump stop force.

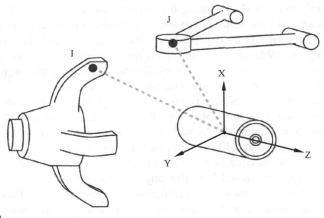

FIGURE 3.47

Modelling of suspension bushes.

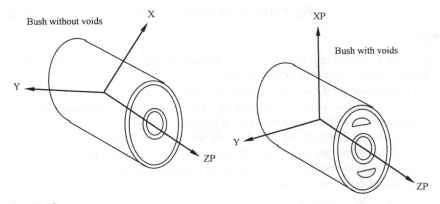

FIGURE 3.48

Orientation of bush axis system.

from the initial linear one used here to a nonlinear model. Another consideration here is that the force formulation takes no account of the possibility that the solution could find a point where the I frame actually moves through the bump stop and past the J frame into the vehicle body. Although this makes no sense physically there is nothing in the formulation to take account of this and should it happen the force would actually reverse direction leading to a probable failure of the solution. Clever programming can take this into account. In addition to improving the description this could include, for example, building a sensor into the model to stop the simulation should this be about to occur, allowing the analyst to investigate the problem further. Alternatively, the I frame could be positioned substantially above its intuitive location and used with a progressively stiffening formulation to give rise to such large forces that the solution is unlikely to pass the point of metal-to-metal contact, but if it momentarily does (perhaps during the iteration associated with an integration step) then the sense of the solution remains and the suspension arm is encouraged to accelerate back towards its real working space. Such an approach also guards against the unpredictable direction in which line-of-sight bump stop forces may act when the two markers are close to each other but undergoing lateral movements, perhaps on suspension elastomer compliances as described next.

The various elastic bushes or mounts used throughout a suspension system to isolate vibration may be represented initially by six linear uncoupled equations based on stiffness and damping. As with a joint, a bush connects two parts using an I frame on one body and a J frame on another body (Figure 3.47). These frames are normally taken to be coincident when setting up the model but it will be seen from the formulation presented here that any initial offset, either translational or rotational, would result in an initial preforce or torque in the bush. This would be in addition to any initial value for these that the user may care to define.

The general form of the equation for the forces and torques generated in the bush is given in Eqn (3.78)

$$\{F_{ij}\}_j = -[k]\ \{d_{ij}\}_j - [c]\ \{v_{ij}\}_j + \{f_{ij}\}_j \qquad (3.78)$$

where

$\{F_{ij}\}_j$ is a column matrix containing the components of the force and torque acting on the I frame from the J frame;
$[k]$ is a square stiffness matrix where all off-diagonal terms are zero;
$\{d_{ij}\}_j$ is a column matrix containing the components of the displacement and rotation of the I frame relative to the J frame;
$[c]$ is a square damping matrix where all off-diagonal terms are zero;
$\{v_{ij}\}_j$ is a column matrix of time derivatives of the terms in the $\{d_{ij}\}$ matrix;
$\{f_{ij}\}_j$ is a column matrix containing the components of the pre-force and pre-torque applied to the I frame.

Expanding Eqn (3.79) leads to the following set of uncoupled equations presented in matrix form as

$$
\begin{bmatrix} F_x \\ F_y \\ F_z \\ T_x \\ T_y \\ T_z \end{bmatrix} = -
\begin{bmatrix}
k_x & 0 & 0 & 0 & 0 & 0 \\
0 & k_y & 0 & 0 & 0 & 0 \\
0 & 0 & k_z & 0 & 0 & 0 \\
0 & 0 & 0 & k_{tx} & 0 & 0 \\
0 & 0 & 0 & 0 & k_{ty} & 0 \\
0 & 0 & 0 & 0 & 0 & k_{tz}
\end{bmatrix}
\begin{bmatrix} d_x \\ d_y \\ d_z \\ r_x \\ r_y \\ r_z \end{bmatrix} -
\begin{bmatrix}
c_x & 0 & 0 & 0 & 0 & 0 \\
0 & c_y & 0 & 0 & 0 & 0 \\
0 & 0 & c_z & 0 & 0 & 0 \\
0 & 0 & 0 & c_{tx} & 0 & 0 \\
0 & 0 & 0 & 0 & c_{ty} & 0 \\
0 & 0 & 0 & 0 & 0 & c_{tz}
\end{bmatrix}
\begin{bmatrix} v_x \\ v_y \\ v_z \\ \omega_x \\ \omega_y \\ \omega_z \end{bmatrix} +
\begin{bmatrix} f_x \\ f_y \\ f_z \\ t_x \\ t_y \\ t_z \end{bmatrix}
$$

$$(3.79)$$

It should be noted that all the terms in Eqns (3.78) and (3.79) are referenced to the J frame and that the equilibrating force and torque acting on the J frame is determined from

$$\{F_{ji}\}_j = -\{F_{ij}\}_j \tag{3.80}$$

$$\{T_{ji}\}_j = -\{T_{ij}\}_j - \{d_{ij}\}_j \times \{F_{ij}\}_j \tag{3.81}$$

The shape and construction of the bush will dictate the manner in which the I frame and J frame are set up and orientated. This is illustrated in Figure 3.48, where it can be seen that for the cylindrical bush where there are no voids the radial stiffness is constant circumferentially. For this bush it is only necessary to ensure the z-axes of the I and J frame are aligned with the axis of the bush. For the bush with voids the radial stiffness will require definition in both the x- and y-directions as shown.

As with springs, it is possible to formulate a linear, viscously damped implementation of a suspension bush using system variables as described.

For a sinusoidal excitation, a linear combination of displacement and velocity will draw a simple Lissajous figure — an oval — with the viscous model. In a real test, a start-up and shutdown transient will be included, as shown in Figure 3.49 where the first and last quarter cycles are described using the expression:

$$DZ = 0.5 \cdot (1 - \cos(2 \cdot \pi \cdot time \cdot frequency)) \tag{3.82}$$

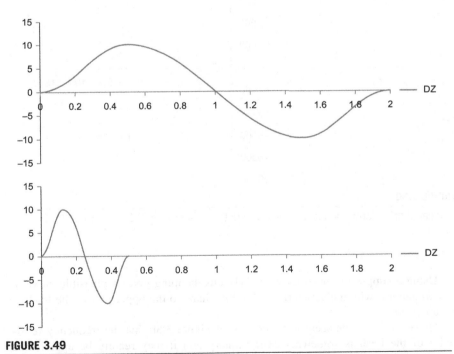

FIGURE 3.49

A comparison of test-style inputs at 0.5 and 2 Hz showing the start-up and shutdown transients in both cases.

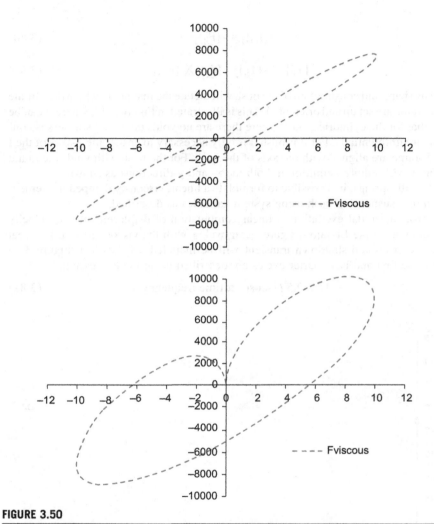

FIGURE 3.50

A linear stiffness and viscous damping description of a bush at 0.5 Hz (top) and 2 Hz (bottom).

Using a simple linear stiffness and viscous damping gives a plausible curve at low frequency, with a characteristic 'top lip' shape to the upper part of the trace — Figure 3.50.

However, it can be seen in the bottom of Figure 3.50 that the frequency sensitivity of the bush is somewhat extraordinary and it may readily be appreciated that either forces are overrepresented at high frequencies or that energy dissipation is underrepresented at low frequencies. A viscous representation is simply not adequate for elastomer elements. They rarely behave in a velocity-dependent manner, and are often better described with a hysteretic formulation.

The essence of hysteretic behaviour is that it is broadly independent of frequency. This is in strong contrast to viscous behaviour which, being proportional to velocity, varies directly with frequency — a doubling of frequency results in a doubling of the energy dissipated.

The key to this is a behaviour not dissimilar to Coulomb friction. Strictly, Coulomb friction is defined thus

$$V < 0, \ F = \mu N \tag{3.83}$$

$$V = 0, \ F = 0 \tag{3.84}$$

$$V > 0, \ F = -\mu N \tag{3.85}$$

However, implementing this in a continuously differentiable manner is not possible directly. A solution used by many practitioners is the TANH function, the hyperbolic tangent

$$\tanh(x) = (1 - e^{-2x}) / (1 + e^{-2x}) \tag{3.86}$$

The derivation of the hyperbolic tangent is somewhat arcane but the hyperbolic tangent looks quite like a Coulomb function except for not having a discontinuity. It is illustrated in Figure 3.51.

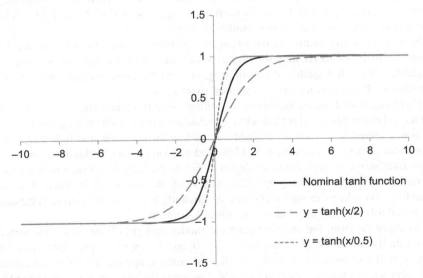

FIGURE 3.51

The tanh function can be used to give a Coulomb-like transition in force with no velocity dependence if velocity is used as the input variable.

Hysteretic damping is often described by the expression:

$$F_{hysteretic} = kx\,(1+i\eta) \qquad (3.87)$$

where

  k is a spring constant;
  x is deflection;
  i is the square root of -1;
  η is a hysteretic damping constant.

Something resembling this can be implemented in a time domain model in the following form:

$$FhystSimple = k\cdot DZ + Fh\cdot tanh(VZ/Vref) \qquad (3.88)$$

where

  k is a spring constant;
  DZ is deflection in the multibody model;
  Fh is the size of the hysteretic force;
  VZ is velocity in the multibody model;
  Vref is the velocity when the full hysteresis effect is present.

It can be seen (Figure 3.52) that with access to a function library (most solvers will support the *tanh* function) and the same two states as discussed previously, quite a different character of behaviour can be produced to the viscous description; even this simplest hysteretic representation is substantially less sensitive to frequency. There is nothing to prevent the elastic term in the equation ($k \cdot DZ$) being represented with a polynomial expression or a nonlinear spline.

Since historically multibody packages were primarily used for low frequency dynamics up to about 15 Hz then the omission of a hysteretic formulation was understandable although arguably disappointing. As can be seen, though, this can be remedied with a relatively simple function expression.

However, with detailed harshness modelling using flexible component representations and expectations of high fidelity simulations to much higher frequency, there are now a selection of toolkits available that allow the sophisticated fitting of elaborate models to suspension bushes. Measured elastomer data is characterised by a distinctive 'scimitar-like' shape, as shown in Figure 3.53, in which the loading characteristic continues in a fairly intuitive way until the maximum is reached; upon unloading, load drops immediately and significantly more steeply before returning to a parallel line some way below the initial loading line.

To move on from the simple hysteretic model and get closer to the measured form rapidly becomes quite arcane in its detail but similar principles can be employed. It will be noted that the described scimitar shape has a level of rotational symmetry when a fully developed loop (i.e. one not including a start-up/shutdown) is used. This means first that the mathematical expressions must discern the difference between loading and unloading:

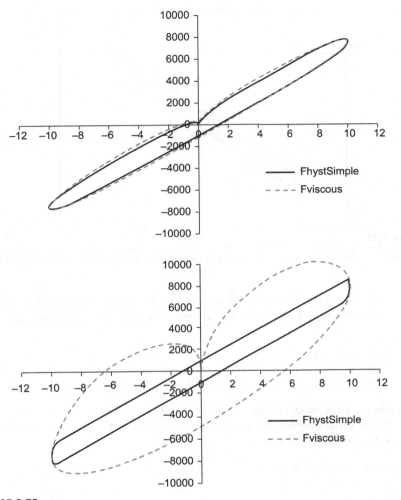

FIGURE 3.52

A simple hysteretic damping response at 0.5 Hz (top) and 2 Hz (bottom). Viscous traces are included as a comparison.

```
SIGN(DZ·VZ) = LOADFLAG
```

Note that in many multibody codes a Fortran-style SIGN function is used in which sign returns 1 if the number is greater than or equal to 0 and −1 if the number is less than 0; the Excel/Star Office-style SIGN function returns 0 if the expression evaluates to 0. Looking for a computationally imprecise real number to actually equal 0 is like looking for a needle in a haystack, but it does mean that the presence of zeroes can trip up test case models in Excel when they would compute in a completely satisfactory manner inside a general-purpose package.

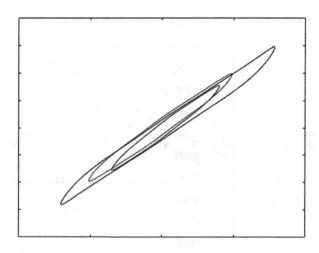

FIGURE 3.53

A Typical 'Scimitar' Shape Hysteresis Loop from an Elastomer. (Karlsson and Persson, 2003.)

The variable LOADFLAG can be used to switch between two different computational schemes to give a rotationally symmetric shape to the resulting trace:

```
LOADFLAG > 0
F = k·DZ + Fh·SIGN(VZ)
LOADFLAG < 0
F = k·DZ + Fh·[2·tanh( VZ/(Vref·ABS(DZ)) -SIGN(VZ) ]
```

In this form the model appears suitable for use as is but detailed investigation shows that when starting from rest a discontinuity appears. To remedy this, a further flag can be calculated and used:

```
SIGN(DZ·AZ) = STARTSTOPFLAG
STARTSTOPFLAG < 0
SCALE = 1
STARTSTOPFLAG > 0
SCALE = tanh(VZ·AccFactor)
```

The hysteretic force model is then given in its entirety by:

```
Fhyst = F·SCALE
```

This model is somewhat more complex than the simple model but does capture the rotationally symmetric aspect of the behaviour as shown in Figure 3.54. Whether or not this is important to the final outcome is for the user to decide; it should be noted that, as the frequency of excitation rises they become very similar to each other, so the more complex model only has value at frequencies under about 10 Hz with the parameters shown.

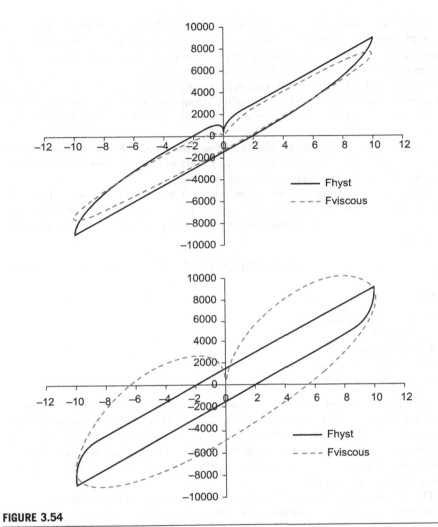

FIGURE 3.54

A rotationally symmetric hysteresis model at 0.5 Hz (left) and 2 Hz (right) with the reference viscous model for comparison. Values for k, Fh, Vref and AccFactor were 750 N/mm, 1500 N, 4 mm/s and 0.1 s/mm, respectively.

Both these models are in essence a spring in parallel with a Coulomb damping term, a variation on the Kelvin–Voigt model. They are just about the simplest representation of elastomer behaviour that captures the essential frequency independence. Other representations for elastomer behaviour are available (Maxwell, Zener and many others), and the interested reader is referred to Karlsson and Persson (2003) for an informed comparison between them.

If a yet more elaborate representation of elastomer behaviour is required, the use of simple function expressions can become somewhat limiting and it becomes more convenient to move into some kind of user subroutine. At this juncture many nonexpert users of the code will wish to pick up an existing library of models and some vendors of general-purpose software will supply a toolkit aimed specifically at this aspect of a system model, crucially including tools to fit measured data to the parameterised modelling scheme — which can be somewhat specialist in its own right.

This discussion of hysteretic elastomer modelling leads nicely on to some more general observations about modelling force generation mechanisms within a multibody model. In essence, there are two approaches.

One is to take some known science and to use it directly; we might call this 'causal' modelling, where the cause of the force is captured directly. An example of this might be the force from a spring, which can be shown to be given by:

$$F_{coil} = (Gd^4 \, DZ) / (8D^3 \, n) \tag{3.89}$$

where:

G = shear modulus;
d = wire diameter;
DZ = deflection;
D = mean coil diameter;
n = number of coils.

A detailed derivation of this expression is beyond the scope of this book but the interested reader is referred to, for example, 'Roark's Formulas for Stress and Strain' (Young et al., 2011) for background in the theory of elasticity. In this example, we can measure the physical dimensions of the spring and look up the standard material properties, and predict the force a spring might produce without having to actually have a spring in our hands.

The second approach is to take some experimental observations and represent them with some form of mathematics that describes the resulting characteristic but makes no attempt to describe the underlying mechanism. We might call this 'empirical'. It is important not to confuse empirical modelling with causal modelling, and it is also important to note that empirical models can often be more accurate since they are not based on an understanding of detailed physical phenomena — which history shows us is an ever-evolving and sometimes strangely shifting — but rather on robust and repeatable experimental observations.

A good example of useful empirical modelling is in the force generated by aerofoils (airfoils in the US). Typically they are represented below stall by some empirical form similar to:

$$F_{aero} = 0.5 \, \rho(C_{L0} + C_{L\alpha}\alpha_A) \, VX^2 \tag{3.90}$$

where:

ρ = local density of air;

C_{L0} = coefficient of lift at zero angle of attack;

$C_{L\alpha}$ = variation in coefficient of lift with angle of attack;

α_A = angle of attack.

A great deal of high quality experimentation was carried out by the National Advisory Committee for Aeronautics (later to morph into NASA) and the results published. With data reduction techniques it is possible to relate a certain description of a wing profile with coefficients C_{L0} and $C_{L\alpha}$ but it should be noted there is no description of aspect ratio, camber or detailed geometry of the aerofoil in these coefficients.

The aerofoil case is a good example, where causal models are remarkably awkward to use but empirical models are very simple. The prospect of modelling the full dynamic behaviour of an aircraft manoeuvring by solving the causal computational fluid dynamics (CFD) problem is heroic in scale, but a comprehensive empirical flight dynamics model can easily be managed on a laptop; such a model is currently provided with every copy of the excellent textbook 'Flight Dynamics' by Stengel (2004).

As a brief aside, there is popular belief that the principle of 'equal transit times' describes the lift generated by an aerofoil. A moment with a pocket calculator shows that the difference in path length on a Boeing 747 wing (about 2%) generates just about enough lift to suspend a European hatchback and nothing like enough to lift 400 tonnes of jumbo off the runway. Some argue that the wing simply deflects the air downwards a bit like a tea tray, but again some time with a pocket calculator shows that it helps but is still nothing like the right amount of lift.[8]

The exact details are somewhat arcane but in principle the special shape of the aerofoil steers a very large amount of air downward and the rate of change of momentum gives a force in the opposite direction, exactly as Newton predicted. The special shape of the aerofoil is necessary to keep the airflow attached and not allow it to separate; if such a separation occurs then a much smaller volume air is induced to change direction and the lift falls away dramatically. This special shape is what makes an aerofoil an aerofoil and why a tea tray does not work as a wing. There appears a great deal of controversy around these relatively simple (and apparently uncontentious) ideas, which the authors find somewhat inexplicable. However, it is not within the remit of this book to address such matters; the student is encouraged to critically consult reputable textbooks, preferably written by people who use the know-how to actually design wings, and not be too credulous of less reliable sources.

Tyres are another subject where causal models are remarkably difficult to formulate, even using sophisticated explicit finite element methods, but a number of

---

[8]Stengel notes that a simple representation such as this actually works quite well for hypersonic flight but grossly underpredicts lift for subsonic flight.

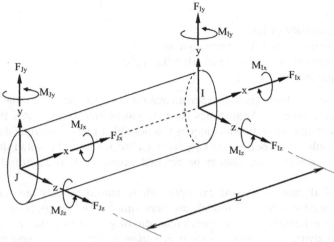

FIGURE 3.55

Massless beam element.

empirical descriptions exist, which are entirely adequate for the subject of researching the whole vehicle behaviour. These are described in detail in Chapter 5.

The most advanced examples of the modelling of force elements extend to the incorporation of finite element type representations of beams and flexible bodies into the MBS model. In modelling a vehicle the most likely use of a beam-type element is going to be in modelling the rollbars or possibly if considered relevant an appropriate suspension member such as a tie rod. So-called 'semi-independent' suspensions, also described as 'twistbeams' also require a beam-type representation, as do leaf springs.

The beam element in general-purpose programs requires the definition of an I frame on one body and a J frame on another body to represent the ends of the beam with length L, as shown in Figure 3.55. The beam element transmits forces and moments between the two markers, has a constant cross section and obeys Timoshenko beam theory.

The beam centroidal axis is defined by the x-axis of the J frame and when the beam is in an undeflected state, the I frame lies on the x-axis of the J frame and has the same orientation. The forces and moments shown in Figure 3.55 are:

Axial Forces F_{Ix} and F_{Jx}
Shear Forces F_{Iy}, F_{Iz}, F_{Jy} and F_{Jz}
Twisting Moments M_{Ix} and M_{Jx}
Bending Moments M_{Iy}, M_{Iz}, M_{Iy} and M_{Jz}

The forces and moments applied to the I frame are related to the displacements and velocities in the beam using the beam theory equations presented in matrix form as

$$
\begin{bmatrix} F_{Ix} \\ F_{Iy} \\ F_{Iz} \\ M_{Ix} \\ M_{Iy} \\ M_{Iz} \end{bmatrix} = - \begin{bmatrix} k_{11} & 0 & 0 & 0 & 0 & 0 \\ 0 & k_{22} & 0 & 0 & 0 & k_{26} \\ 0 & 0 & k_{33} & 0 & k_{35} & 0 \\ 0 & 0 & 0 & k_{44} & 0 & 0 \\ 0 & 0 & k_{35} & 0 & k_{55} & 0 \\ 0 & k_{26} & 0 & 0 & 0 & k_{66} \end{bmatrix} \begin{bmatrix} d_x - L \\ d_y \\ d_z \\ r_x \\ r_y \\ r_z \end{bmatrix} - \begin{bmatrix} c_{11} & c_{21} & c_{31} & c_{41} & c_{51} & c_{61} \\ c_{21} & c_{22} & c_{32} & c_{42} & c_{52} & c_{62} \\ c_{31} & c_{32} & c_{33} & c_{43} & c_{53} & c_{63} \\ c_{41} & c_{42} & c_{42} & c_{44} & c_{54} & c_{64} \\ c_{51} & c_{52} & c_{52} & c_{53} & c_{55} & c_{65} \\ c_{61} & c_{62} & c_{63} & c_{64} & c_{65} & c_{66} \end{bmatrix} \begin{bmatrix} V_x \\ V_y \\ V_z \\ \omega_x \\ \omega_y \\ \omega_z \end{bmatrix}
$$

$$(3.91)$$

The terms dx, dy and dz in Eqn (3.91) are the x-, y- and z-displacements of the I frame relative to the J frame measured in the J reference frame. The terms rx, ry and rz are the relative rotations of the I frame with respect to the J frame measured about the x-axis, y-axis and z-axis of the J frame. It should be noted here that the rotations in the beam are assumed to be small and that large angular deflections are not commutative. In these cases, typically when deflections in the beam approach 10% of the undeformed length, the theory does not correctly define the behaviour of the beam. The terms Vx, Vy, Vz, ωx, ωy and ωz are the velocities in the beam obtained as time derivatives of the translational and rotational displacements.

The stiffness and damping matrices are symmetric. The terms in the stiffness matrix are given by:

$$K_{11} = \frac{EA}{L} \qquad K_{22} = \frac{12EI_{zz}}{L^3(1+P_y)} \qquad K_{26} = \frac{-6EI_{zz}}{L^2(1+P_y)} \qquad K_{33} = \frac{12EI_{yy}}{L^3(1+P_z)}$$

$$K_{35} = \frac{6EI_{yy}}{L^2(1+P_z)} \qquad K_{44} = \frac{GI_{xx}}{L} \qquad K_{55} = \frac{4+P_zEI_{yy}}{L(1+P_z)} \qquad K_{66} = \frac{4+P_yEI_{zz}}{L(1+P_y)}$$

where $P_y = 12\,EI_{zz}A_{SY}/(GAL^2)$ and $P_z = 12\,EI_{yy}A_{SZ}/(GAL^2)$. Young's modulus of elasticity for the beam is given by E and the shear modulus is given by G. The cross-sectional area of the beam is given by A. The terms Iyy and Izz are the second moments of area about the neutral axes of the beam cross section. For a solid circular section with diameter D these would, for example, be given by $I_{yy} = I_{zz} = \frac{\pi D^4}{64}$. Ixx is the second moment of area about the longitudinal axis of the beam. Again for a solid circular section this is given by $I_{xx} = \frac{\pi D^4}{32}$.

The final part of the definition of the terms in the stiffness matrix is the correction factors for shear deflection in the y and z directions for Timoshenko beams. These are given in the y direction by $A_{SY} = \frac{A}{I_{yy}^2}\int_A \left(\frac{Q_y}{l_z}\right)^2 dA$ and in the z direction by

$Asz = \dfrac{A}{I_{zz}^2} \displaystyle\int_A \left(\dfrac{Q_z}{l_y}\right)^2 dA$. The terms Q_y and Q_z are the first moments of area about the beam section y and z-axes respectively. The terms l_y and l_z are the dimensions of the beam cross section in the y and z directions of the cross-section axes.

The structural damping terms c_{11} through to c_{66} in Eqn (3.91) may be input directly or by using a ratio to factor the terms in the stiffness matrix. In a similar manner to the dampers and bushes discussed earlier it is possible to define a beam with nonlinear properties using more advanced elements that allow the definition of general force fields.

As with the bush elements the beam will produce an equilibrating force and moment acting on the J frame using

$$\{F_{ji}\}_j = -\{F_{ij}\}_j \qquad (3.92)$$

$$\{M_{ji}\}_j = -\{M_{ij}\}_j - \{d_{ij}\}_j \times \{F_{ij}\}_j \qquad (3.93)$$

where $\{d_{ij}\}_j$ is the position vector of the I frame relative to the J frame, as measured in the J reference frame.

Beam elements are generally massless and so they are often used in groups with small masses stationed at their interfaces. It can be readily imagined that a large number of beams and masses is functionally indistinguishable from a one-dimensional finite element model; in judging how many beams to use, the practitioner must decide principally on how well the deformed shape of the beam needs to be described; in the case of a twistbeam suspension, for example, the camber compliance mechanism produces a distinctive S-shape when viewed from ahead or behind; a single beam element will not capture this adequately, nor will two. The authors' habits have been to use 8—10 elements along a beam where such a deflection is occurring. If the deflections are substantially more complex than this, the use of full finite element representation of the component could be considered — the practitioner is encouraged to experiment with representative test cases before resting business-critical decisions on matters of modelling convenience.

3.2.10 Summation of forces and moments

Having considered the definition of force elements in terms of model definition we may conclude by considering the formulation of the equations for the forces and moments acting on a body. An applied force or moment can be defined using an equation to specify the magnitude, which may be functionally dependent on displacements, velocities, other applied forces and time. Using the example in Figure 3.56, there is an applied force $\{F_A\}_1$ acting at point A, the weight of the body $m\{g\}_1$ acting at the centre of mass G, a force $\{F_B\}_1$ and a torque $\{T_B\}_1$ due to an element, such as a bush or beam connection to another part. In addition there is an applied torque $\{T_C\}_1$ acting at point C. Note that at this stage all the force and torque vectors are assumed to be resolved parallel to the GRF and that $\{g\}_1$ is the vector of acceleration due to gravity and is again measured in the GRF.

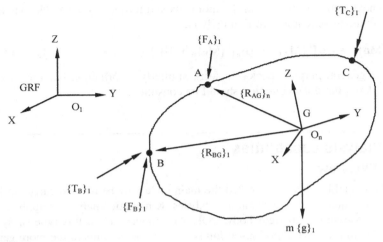

FIGURE 3.56

Applied Forces and Torques on a Body. GRF, ground reference frame.

The summation of applied forces resolved in the GRF as required in Eqn (3.94) is obtained in this example by:

$$\sum \{Fn_A\}_1 = \{F_A\}_1 + \{F_B\}_1 + m\{g\}_1 \tag{3.94}$$

The summation of moments about G is not so straightforward. MSC ADAMS performs the moment calculations about the axes of the Euler-axis frame. It is therefore necessary to use the transformation matrix $[A_{n1}]$ to transform forces and torques to the part frame O_n and to use $[B_n]^T$ to transform from the part frame to the Euler-axis frame.

$$\{F_A\}_n = [A_{n1}] \{F_A\}_1 \tag{3.95}$$

$$\{F_B\}_n = [A_{n1}] \{F_B\}_1 \tag{3.96}$$

$$\{T_B\}_n = [A_{n1}] \{T_B\}_1 \tag{3.97}$$

$$\{T_C\}_n = [A_{n1}] \{T_C\}_1 \tag{3.98}$$

It is now possible to calculate the moments at G due to the forces at A and B working in the part frame.

$$\{M_A\}_n = \{R_{AG}\}_n \times \{F_A\}_n \tag{3.99}$$

$$\{M_B\}_n = \{R_{BG}\}_n \times \{F_B\}_n \tag{3.100}$$

The next step is to transform the moments and torques to the Euler-axis frame and to summate as required in Eqn (3.101).

$$\sum \{Mn\}_e = [B_n]^T \{M_A\}_n + [B_n]^T \{M_B\}_n + [B_n]^T \{T_B\}_n + [B_n]^T \{T_C\}_n \qquad (3.101)$$

Other general-purpose packages work similarly in principle but for package-specific detail the documentation should be consulted.

3.3 Analysis capabilities

3.3.1 Overview

Once the model has been assembled the main code may be used to carry out kinematic, static, quasi-static or dynamic analyses. Kinematic analysis is applicable to systems possessing zero rigid body DOF. Any movement in this type of system will be due to prescribed motions at joints — note this is joints in the more general sense rather than the Simpack-style usage of the word. The program uncouples the equations of motion and force and then solves separately and algebraically for displacements, velocities accelerations and forces.

For static analysis the velocities and accelerations are set to zero and the applied loads are balanced against the reaction forces until an equilibrium position is found. This may involve the system moving through large displacements between the initial definition and the equilibrium position and therefore requires a number of iterations before convergence on the solution closest to the initial configuration. Static analysis is often performed as a precursor to a dynamic analysis. An example would be to perform a static analysis on a full vehicle model before a dynamic handling simulation. This establishes the configuration of the vehicle at 'kerb height' before the vehicle moves forward during the dynamic phase of the simulation. Quasi-static analysis is a series of static equilibrium solutions at selected time steps. Although the system can be considered to be moving the dynamic response is not required. An example would be to perform a quasi-static analysis on a vehicle mounted on a tilting surface. As the surface rotates to different angles with time the static equilibrium of the vehicle can be calculated at selected angles.

Dynamic analysis is performed on systems with one or more DOF. The differential equations representing the system are automatically formulated and then numerically integrated to provide the position, velocities, accelerations and forces at successively later times. Although the user will select output at various points in time the program will often compute solutions at many intermediate points in time. The interval between each of these solution points is known as an integration time step. The size of the integration time step is constantly adjusted using internal logic, although the user may override the system defaults if so desired.

While often recommended as good practice, the authors' experience can be that finding static equilibrium prior to a dynamic analysis might well fail when the model contains errors. The failures are often difficult to diagnose by attempting static equilibrium but often lay themselves open quite clearly if a finely resolved dynamic solution is attempted. For this reason an initial 'drop onto the road' dynamic solution is helpful to debug vehicle models and to assess their plausibility before commencing the manoeuvre of interest. If the model is good it should have no problem solving for this event — including spinning the wheels up from rest if the velocity is nonzero — and should converge to running in a more or less straight line with no driver control input. (A 'more or less' straight line because there are often small force offsets with real measured tyre data at zero slip angle.)

A final possibility for solution of models is sometimes referred to as 'dynamic equilibrium'. This apparently oxymoronic description is essentially a series of dynamic solutions to converged equilibrium positions with no interim reporting.

Before considering the implementation of these analysis methods in more detail we must consider the fundamental methods used to solve linear and nonlinear equations. An additional modal analysis capability may be used to linearise the nonlinear equations of motion about a selected operating point and then find the natural frequencies and mode shapes associated with the system. It is also possible to extract the linearised state-space plant model in a format suitable for input to a control system design package.

3.3.2 Solving linear equations

MBS programs solve both linear and nonlinear equations during the analysis phase. Linear equations can be assembled in matrix form as shown in Eqn (3.102).

$$[A][x] = [b] \qquad (3.102)$$

where

[A] is a square matrix of constants;
[x] is a column matrix of unknowns;
[b] is a column matrix of constants.

The formulations in a MBS program generally lead to a matrix [A], where most of the elements are zero. As such the matrix is referred to as sparse and the ratio of nonzero terms to the total number of matrix elements is referred to as the sparsity of the matrix. In a general-purpose program the sparsity of the matrix is typically in the range of 2–5% (Wielenga, 1987). The computer solvers that are developed in MBS to solve linear equations can be designed to exploit the sparsity of the [A] matrix leading to relatively fast solution times. This is one of the reasons, apart from improvements in computer hardware, that MBS programs can appear to solve quite complex engineering problems in seconds or minutes. These solution speeds can be quite notable when compared with other computer software such as finite elements or CFD programs.

The solution of Eqn (3.102) follows an established approach that initially involves decomposing or factorising the [A] matrix into the product of a lower triangular and an upper triangular matrix.

$$[A] = [L] [U] \qquad\qquad (3.103)$$

For the lower triangular matrix [L], all the elements above the diagonal are set to zero. For the upper triangular matrix [U], all the elements on and below the diagonal are set to zero. In the following step a new set of unknowns [y] is substituted into Eqn (3.102) leading to

$$[L] [U] [x] = [b] \qquad\qquad (3.104)$$

$$[U] [x] = [y] \qquad\qquad (3.105)$$

$$[L] [y] = [b] \qquad\qquad (3.106)$$

The terms in the [L] and [U] matrices may be obtained by progressive operations on the [A] matrix, where the terms in one row are all factored and then added or subtracted to the terms in another row. This process can be demonstrated by considering the expanded [A], [L] and [U] matrices as shown in Eqn (3.103), where to demonstrate the influence of sparsity some of the terms in the [A] matrix are initially set to zero.

$$
\begin{bmatrix}
L_{11} & 0 & 0 & 0 \\
L_{21} & L_{22} & 0 & 0 \\
L_{31} & L_{32} & L_{33} & 0 \\
L_{41} & L_{42} & L_{43} & L_{44}
\end{bmatrix}
\begin{bmatrix}
1 & U_{12} & U_{13} & U_{14} \\
0 & 1 & U_{23} & U_{24} \\
0 & 0 & 1 & U_{34} \\
0 & 0 & 0 & 1
\end{bmatrix}
=
\begin{bmatrix}
A_{11} & A_{12} & 0 & A_{14} \\
A_{21} & 0 & 0 & A_{24} \\
A_{31} & 0 & A_{33} & 0 \\
A_{41} & 0 & A_{43} & A_{44}
\end{bmatrix}
$$

$$(3.107)$$

The fact that the [A] matrix is sparse leads to the [L] and [U] matrices also being sparse with subsequent savings in computer simulation time and memory storage. If we start by multiplying the rows in [L] into the first column of [U] we can begin to obtain the constants in the [L] and [U] matrices from those in the [A] matrix:

$$L_{11} = A_{11}$$

$$L_{21} = A_{21}$$

$$L_{31} = A_{31}$$

$$L_{41} = A_{41}$$

In a similar manner for the second column of [U] we get:

$L_{11} U_{12} = A_{12}$ therefore $U_{12} = A_{12} / A_{11}$

$L_{21} U_{12} + L_{22} = 0$ therefore $L_{22} = -A_{21} A_{12} / A_{11}$

$L_{31} U_{12} + L_{32} = 0$ therefore $L_{32} = -A_{31} A_{12} / A_{11}$

$L_{41} U_{12} + L_{42} = 0$ therefore $L_{42} = -A_{41} A_{12} / A_{11}$

Moving on to the third column of [U] gives:

$L_{11} U_{13} = 0$ therefore $U_{13} = 0$

$L_{21} U_{13} + L_{22} U_{23} = 0$ therefore $U_{23} = 0$

$L_{31} U_{13} + L_{32} U_{23} + L_{33} = A_{33}$ therefore $L_{33} = A_{33}$

$L_{41} U_{13} + L_{42} U_{23} + L_{43} = A_{43}$ therefore $L_{43} = A_{43}$

Finishing with the multiplication of the rows in [L] into the fourth column of [U] gives:

$L_{11} U_{14} = A_{14}$

$L_{21} U_{14} + L_{22} U_{24} = A_{24}$

$L_{31} U_{14} + L_{32} U_{24} + L_{33} U_{34} = 0$

$L_{41} U_{14} + L_{42} U_{24} + L_{43} U_{34} + L_{44} = A_{44}$

therefore

$U_{14} = A_{14} / A_{11}$

$U_{24} = -(A_{24} - A_{21} A_{14} / A_{11}) / (A_{21} A_{12} / A_{11})$

$U_{34} = (-(A_{31} A_{14} / A_{11}) - (A_{31} A_{12} / A_{11}) (A_{24} - A_{21} A_{14} / A_{11}) / (A_{21} A_{12} / A_{11}) / A_{33}$

$L_{44} = A_{44} - A_{41} A_{14} / A_{11} - (A_{41} A_{12} / A_{11}) (A_{24} - A_{21} A_{14} / A_{11}) / (A_{21} A_{12} / A_{11})$

$\quad - A_{43} (-(A_{31} A_{14} / A_{11}) - (A_{31} A_{12} / A_{11}) (A_{24} - A_{21} A_{14} / A_{11}) / (A_{21} A_{12} / A_{11})) / A_{13}$

From the preceding manipulations we can see that, for this example some of the terms in the [L] and [U] matrices come to zero. Using this we can update Eqn (3.107) to give

$$\begin{bmatrix} L_{11} & 0 & 0 & 0 \\ L_{21} & L_{22} & 0 & 0 \\ L_{31} & L_{32} & L_{33} & 0 \\ L_{41} & L_{42} & L_{43} & L_{44} \end{bmatrix} \begin{bmatrix} 1 & U_{12} & 0 & U_{14} \\ 0 & 1 & 0 & U_{24} \\ 0 & 0 & 1 & U_{34} \\ 0 & 0 & 0 & 1 \end{bmatrix} = \begin{bmatrix} A_{11} & A_{12} & 0 & A_{14} \\ A_{21} & 0 & 0 & A_{24} \\ A_{31} & 0 & A_{33} & 0 \\ A_{41} & 0 & A_{43} & A_{44} \end{bmatrix}$$

(3.108)

Consideration of Eqn (3.108) reveals that some elements in the factors [L] and [U] are nonzero where the corresponding elements in [A] are zero. Such elements, for example L_{22}, L_{32}, and L_{42} here, are referred to as 'fills'. Having completed the decomposition process of factorising the [A] matrix into the [L] and [U] matrices the next step, forward−backward substitution, can commence.

The first step is a forward substitution, utilising the now known terms in the [L] matrix to find the terms in [y]. For this example if we expand [L] [y] = [b] as given in Eqn (3.106), we get

$$\begin{bmatrix} L_{11} & 0 & 0 & 0 \\ L_{21} & L_{22} & 0 & 0 \\ L_{31} & L_{32} & L_{33} & 0 \\ L_{41} & L_{42} & L_{43} & L_{44} \end{bmatrix} \begin{bmatrix} y_1 \\ y_2 \\ y_3 \\ y_4 \end{bmatrix} = \begin{bmatrix} b_1 \\ b_2 \\ b_3 \\ b_4 \end{bmatrix} \tag{3.109}$$

therefore

$$y_1 = b_1 / L_{11}$$

$$y_2 = (b_2 - L_{21}y_1) / L_{22}$$

$$y_3 = (b_3 - L_{31}y_1 - L_{32}y_2) / L_{33}$$

$$y_4 = (b_4 - L_{41}y_1 - L_{42}y_2 - L_{43}y_3) / L_{44}$$

The next step is a back substitution, utilising the terms found in the [U] matrix to find the overall solution for the terms in [x]. For this example if we expand [U] [x] = [y] as given in Eqn (3.105), we get

$$\begin{bmatrix} 1 & U_{12} & 0 & U_{14} \\ 0 & 1 & 0 & U_{24} \\ 0 & 0 & 1 & U_{34} \\ 0 & 0 & 0 & 1 \end{bmatrix} \begin{bmatrix} x_1 \\ x_2 \\ x_3 \\ x_4 \end{bmatrix} = \begin{bmatrix} y_1 \\ y_2 \\ y_3 \\ y_4 \end{bmatrix} \tag{3.110}$$

therefore

$$x_4 = y_4$$

$$x_3 = (y_3 - U_{34}x_4)$$

$$x_2 = (y_2 - U_{24}x_4)$$

$$x_1 = (y_1 - U_{12}x_2 - U_{14}x_4)$$

It can be seen that in the solution of Eqn (3.109), it is necessary to divide by the diagonal terms in [L]. These are referred to as 'pivots' and must be nonzero to avoid a singular matrix and failure in solution. There will also be problems if the pivots are so small that they approach a zero value. In these circumstances the condition of the matrices is said to be poor. The answer to this is to rearrange the order of the equations, referred to as pivot selection, until the best set of pivots is available. This process is also called refactorisation. The mathematics has been developed so that the process will also attempt to minimise the number of fills to assist with a faster solution. The sequence of operations can be stored to speed solution as the simulation progresses unless the physical configuration of the system changes to a point where the matrix changes sufficiently to justify the reselection of a set of pivots. It should be noted that, in general, solution of these equations will involve much larger matrices than the four by four examples shown here and the sparsity of the matrix will be more apparent for these larger problems.

3.3.3 Nonlinear equations

In the case of nonlinear equations an iterative approach must be undertaken in order to obtain a solution. A set of nonlinear equations may be described in matrix form using

$$[G][x] = 0 \qquad (3.111)$$

where

[x] is a set of unknown variables;
[G] is a set of implicit functions dependent on [x].

Consider the solution of a single nonlinear equation $G(x) = 0$, where G is an implicit function dependent on x. If we plot a graph of $G(x)$ against x the solution is the value of x where the curve intersects the x-axis, as shown in Figure 3.57.

The solution may be obtained using Newton–Raphson iteration based on the assumption that very close to the solution the curve may be approximated to a straight line where it crosses the x-axis. This is illustrated in Figure 3.57, where

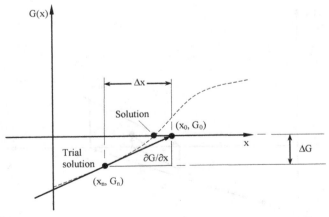

FIGURE 3.57

Application of Newton–Raphson Iteration.

the line is determined by a trial solution, for say the nth iteration, located at (x_n, G_n) and the slope or derivative of the curve $\dfrac{\partial G}{\partial x}$ at this point.

If we take the point (x_o, G_o) to be that at which the straight line crosses the x-axis then the gradient $\partial G/\partial x$ is given by

$$\frac{\partial G}{\partial x} = \frac{G_o - G_n}{x_o - x_n} = \frac{\Delta G}{\Delta x} \tag{3.112}$$

Rearranging this we can write

$$(\partial G/\partial x)\,\Delta x = \Delta G \tag{3.113}$$

We can now extend this to demonstrate how the method can be used to iterate and close in on the solution. This is shown graphically in Figure 3.58.

Extending the single equation $G(x)$ to a set of simultaneous nonlinear equations, we can write

$$\left[\frac{\partial G}{\partial x}\right][\Delta x] = [\Delta G] \tag{3.114}$$

where $\left[\dfrac{\partial G}{\partial x}\right]$ is a matrix of partial derivatives representing the slopes of straight lines used in the solution. This is usually referred to as the Jacobian matrix. $[\Delta x]$ is a column matrix of updates to the values of x for each equation and $[\Delta G]$ is a column matrix of terms representing the current error in each equation.

Once all the terms in $[\Delta G]$ approach zero a solution has been achieved. In practice, since this method involves the programming of a numerical approximation, a small error tolerance is used to determine that the terms in $[\Delta G]$ have converged.

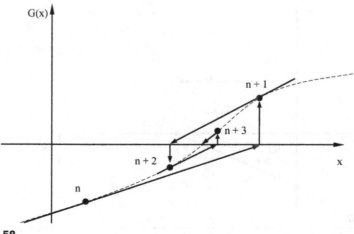

FIGURE 3.58

Convergence of Newton–Raphson Iteration.

For example, if we were to use an error tolerance equal to 10^{-4} then we could say that convergence has been achieved when for each equation in the system.

$$-10^{-4} \leq \Delta G \leq 10^{-4} \qquad (3.115)$$

The iterative process used here can be illustrated using the flow chart shown in Figure 3.59.

The Newton—Raphson method can be modified to speed up the solution process by not performing Step 3 and Step 4 in Figure 3.59. These two steps are the most computationally intensive and a more rapid solution may be achieved by calculating

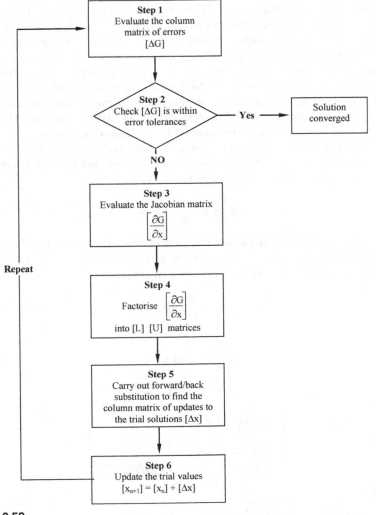

FIGURE 3.59

Flow chart illustrating solution of nonlinear equations.

the Jacobian matrix, for example, on every second or third iteration. The graphical illustration of the process in Figure 3.58 shows how the use of the slope $\dfrac{\partial G}{\partial x}$ from the previous iteration may often be an acceptable modification to the solution process.

3.3.4 Integration methods

The solution of nonlinear equations for dynamic systems with one or more DOF is a fundamental component of the application of MBS analysis to engineering problems. A number of methods have been developed and the process is commonly referred to as integration. The example here is based on the description given by (Wielenga, 1987) of a backwards differentiation formula method. The method described here can be considered to have two phases. The first of these is the use of a polynomial fit through past values of a given equation to predict a value at the next integration time step (solution point). The second step is to use the Newton-Raphson method described in the previous section to correct the prediction and achieve convergence. As such this method may also be described as a predictor-corrector approach. The equations being solved are first order differential equations and are also referred to as state equations. The state equations are implicit and have the general formulation $G(x,\dot{x},t)=0$.

Clearly most dynamic problems such as the unforced vibration of the simple 1 DOF system shown in Figure 3.60 involve acceleration and have an equation of motion that is second order as shown in Eqn (3.116).

$$m\frac{d^2x}{dt^2}+c\frac{dx}{dt}+kx=0 \qquad (3.116)$$

The implementation of this as first order differential equations is simply a matter of introducing a new variable, for example z, for the velocity and writing Eqn (3.116) as two implicit first order differential equations:

$$z-\frac{dx}{dt}=0 \qquad (3.117)$$

$$m\frac{dz}{dt}+cz+kx=0 \qquad (3.118)$$

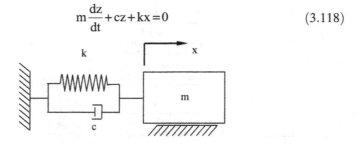

FIGURE 3.60

Simple one degree of freedom mass, spring, damper system.

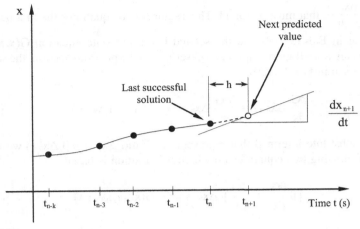

FIGURE 3.61

Use of predictor to fit a polynomial through past values.

The predictor phase of the solution can be explained with the help of the plot of the value of the state variable x as a function of time t as shown in Figure 3.61. The current solution point, or nth integration time step, is shown to be occurring at time t_n. Previous successful solutions have been computed at times t_{n-1}, t_{n-2}, t_{n-3}, ...t_{n-k}. The next value to be predicted, x_{n+1} will occur at time t_{n+1}. The time between each solution point is the integration time step h. Note that this should not be confused with output steps that are generally defined before the analysis and are used to fix the time interval at which results will be calculated for printing and plotting. The integration time step must be at least as small as the output time step to compute solutions but will generally be smaller in order to obtain a solution. In some programs — and notably real-time applications — the integration time steps may be fixed but usually with general-purpose programs they will be variable and will use programmed logic to determine the optimum step size for the problem in hand.

For the solution at the next time step to lie on the polynomial both the value of the state variable x_{n+1} and the derivative $\dfrac{d_{x+1}}{dt}$ must satisfy the polynomial.

The derivative $\dfrac{dx_{n+1}}{dt}$ at the next time step t_{n+1} can be related to the unknown future value x_{n+1} and the past computed values using

$$\frac{dx_{n+1}}{dt} = Pn\ (\ x_{n+1},\ x_n,\ x_{n-1},\ x_{n-2},\ x_{n-3},\ ...x_{n-k}) \tag{3.119}$$

For the solution at the next time step to lie on the polynomial both the value of the state variable x_{n+1} and the derivative must satisfy the polynomial. For successful computation of the next solution at time $= t_{n+1}$ there are therefore two unknowns,

x_{n+1} and $\dfrac{dx_{n+1}}{dt}$, that must be found. This requires two equations: the first being the polynomial in Eqn (3.119) and the second being the state equation $G(x, \dot{x}, t) = 0$. Using the Newton–Raphson approach described in the previous section, the solution takes the form in Eqn (3.120)

$$\frac{\partial G}{\partial \dot{x}} \Delta \dot{x}_{n+1} + \frac{\partial G}{\partial x} \Delta x_{n+1} = -G(\dot{x}_{n+1}, x_{n+1}, t) \qquad (3.120)$$

If we substitute a term β that represents the ratio $\Delta \dot{x}_{n+1} = \beta\, \Delta x_{n+1}$ we end up with the following two equations on which the solution is based

$$[\beta \frac{\partial G}{\partial \dot{x}} + \frac{\partial G}{\partial x}]\, \Delta x_{n+1} = -G(\dot{x}_{n+1}, x_{n+1}, t) \qquad (3.121)$$

$$\Delta \dot{x}_{n+1} = \beta\, \Delta x_{n+1} \qquad (3.122)$$

The integration process can be thought of as having two distinct phases. The first of these is the predictor phase that results in values of x_{n+1} and $\dot{x}_{n+1}$ that satisfy a polynomial fit through past values. The second is the corrector phase that iterates using the process shown in Figure 3.58, until the error is within tolerance and the solution can progress to the next time step. Parameters can be set that control the solution process. Examples of these are the initial, maximum and minimum integration step sizes to be used and the order of the polynomial fit. Parameters used during the corrector phase include the acceptable error tolerance, the maximum number of iterations and the pattern or sequence to be used in updating the Jacobian matrix during the iterations. In general these will default to values programmed into the software but with experience users will find the most suitable settings for the analysis in hand.

The integration step size is variable and if a solution is not achieved at the next prediction point the solver can reduce the integration time step and alter the order of the polynomial to attempt another solution. This typically occurs when there is a sudden change in an equation associated with a physical event, such as an impact or clash of parts. A general rule in modelling is to never program an equation that instantly changes value at a certain time. Problems will be avoided by using, for example, a function that allows a smooth transition from one value to another over a physically small time interval.

Consideration of the predictor corrector approach will indicate that at the start of a transient solution the solver may step forward and backward initially as it establishes a suitable scheme to progress the solution. Experienced users will again be aware of this and avoid programming important events to occur immediately after the start. For example, the simulation of a 5 s lane change manoeuvre may be best accomplished by an initial static analysis followed by the transient simulation for say 1 s of straight line driving before any steering inputs are made, notwithstanding the debugging difficulties mentioned in Section 3.3.1.

Advanced applications may also involve the incorporation of control algorithms based on a discrete time step. A possible solution here is to fix the integration minimum and maximum step size to ensure the solver computes solutions at the fixed time steps of the ABS algorithm. This may of course result in inefficient computation at some stages of the simulation. On the other hand the fixed step scheme must be refined enough to deal with any sudden nonlinear events during the analysis.

Commercial MBS codes often have a range of integrators available that will adopt variations on the solution process discussed here. A commonly used method is referred to as the Gstiff integrator (Gear, 1971). Rather than storing past values of the polynomial a Taylor expansion (3.82) is used to store the polynomial in a form using the current value of the state variable x and the integration step size h. This is known as a Nordsieck vector. At a current time t the Nordsieck vector $[N_t]$ has the form.

$$[N_t]^T = [\; x \quad h\frac{dx}{dt} \quad \frac{h^2}{2}\frac{d^2x}{dt^2} \quad \frac{h^3}{3!}\frac{d^3x}{dt^3} \quad .... \quad \frac{h^k}{k!}\frac{d^kx}{dt^k} \;] \qquad (3.123)$$

The components of $[N_t]$ are added together to predict the next value of x at a time step h forward to a new time t+h. As the simulation progresses the Nordsieck vector is updated by premultiplying by the Pascal triangle matrix to give a new vector $[N_{t+h}]$. An example of this is shown for a polynomial with an order k equal to five in Eqn (3.124).

$$[N_{t+h}] = \begin{bmatrix} 1 & 1 & 1 & 1 & 1 & 1 \\ 0 & 1 & 2 & 3 & 4 & 5 \\ 0 & 0 & 1 & 3 & 4 & 5 \\ 0 & 0 & 0 & 1 & 4 & 5 \\ 0 & 0 & 0 & 0 & 1 & 5 \\ 0 & 0 & 0 & 0 & 0 & 1 \end{bmatrix} [N_t] \qquad (3.124)$$

The predicted values in $[N_{t+h}]$ lie on the polynomial described by $[N_t]$ but are subject to further change as the state equations have not yet been corrected using the process shown in Figure 3.58. During the corrector phase the starting values of x and $\frac{dx}{dt}$ are taken from $[N_{t+h}]$. As the value of x changes during the Newton-Raphson iteration process the components of the Nordsieck vector are updated by $[\Delta N]$ where

$$[\Delta N] = [c] \, [\Delta x] \qquad (3.125)$$

and

$$[c]^T = [c_0 \quad c_1 \quad c_2 \quad c_3 \quad ... \quad c_k] \qquad (3.126)$$

The matrix [c] contains constants with values dependent on the polynomial order k (Orlandea, 1973).

Many other integration schemes are possible and the reader is encouraged to consult the software documentation for details. It is the author's general observation that a well-conditioned model will solve well with the default integrator; if the model will only solve with one particular integrator, this may be a clue that something is poorly conditioned within it. In principle the system should not 'notice' with which integrator it is being solved. However, experimentation may show that some integrators are substantially faster than others for certain classes of problems.

3.4 Eigensolutions

As well as the time domain solution, it is extremely instructional to interrogate a model for its eigensolutions as part of the commissioning process, even if their solution is not part of the immediate analytical requirement. It serves as a rather unimpeachable witness to the actual content of the model (as distinct from the expected or intended content of the model, which is rarely in question).

Eigenvalues have many interesting properties to a mathematician but their primary interest to an engineer is that they allow some familiarity with the character of the time domain solution without actually having to compute it.

If the system has resonances, eigensolutions for the system equations describe them. They are always inherently present even if they are not being excited; the propensity of a passenger car to oscillate on its suspension springs is always there whether or not it is actually doing so.

The equations used to describe the system are differential equations, which have been the subject of study for over 300 years. It is reasonable to investigate the solutions of the differential equations for a form $Ae^{\lambda t}$.

The presumption of a form of solution is common practice in the solution of differential equations but it sometimes mystifies students — it seems as if it is simply not cricket to know the answer before one starts. Be that as it may, if a form for the solution is presumed but turns out to have been inappropriately chosen, it usually emerges in short order. While such a statement hardly constitutes a rigorous mathematical proof, it will have to suffice since unfortunately such a proof is beyond the scope of this book.

In the suggested form of solution, the value λ is also the eigenvalue for the solution being considered. If λ is a positive real number then the response of the system is unbounded and grows with time. If λ is a negative real number then the response of the system is bounded and converges to a constant value. Figure 3.62 illustrates this.

If λ is an imaginary number then the response of the system is oscillatory since

$$e^{-i\omega t} = \cos(\omega t) + i\sin(\omega t) \qquad (3.127)$$

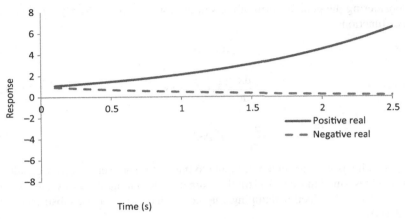

FIGURE 3.62

The expected form of time domain solution for real eigenvalues.

which is oscillatory in time. If λ is a complex number and the real part of λ is positive the response is oscillatory and unbounded. If λ is a complex number and the real part of λ is negative the response is oscillatory and convergent. Figure 3.63 illustrates this.

Systems with multiple DOF have multiple eigenvalues. Eigenvalues can tell us whether the system will converge or not without having to perform repeated calculations to determine the time domain response and are thus appealing in their own right, as well for the audit purposes described earlier. They a logical first step to examining system behaviour.

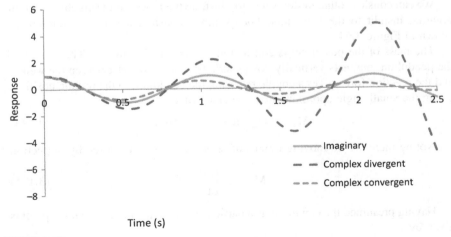

FIGURE 3.63

The expected form of time domain solutions with complex roots.

Considering the general form of the solution, we can immediately write two derivative functions:

$$x(t) = Ae^{\lambda t} \tag{3.128}$$

$$\frac{dx}{dt} = \lambda Ae^{\lambda t} = \lambda x \tag{3.129}$$

$$\frac{d^2x}{dt^2} = \lambda^2 Ae^{\lambda t} = \lambda^2 x \tag{3.130}$$

To find eigensolutions, external inputs to the system are set to zero. Considering the system described in Eqn (3.116), there are no external inputs and so the equation can be used as is. Neglecting damping, the second derivative can be substituted using Eqn (3.130) above

$$-\lambda^2 mx + kx = 0 \tag{3.131}$$

If x is nonzero, which is the case for any nontrivial solutions, this can be seen to lead directly to

$$\lambda^2 = \frac{-k}{m} \tag{3.132}$$

$$\lambda = \sqrt{\frac{-k}{m}} = i\sqrt{\frac{k}{m}} \tag{3.133}$$

This result is intimately familiar to many engineering students — an oscillatory response with a frequency in radians/second given by the magnitude of λ.

We can consider other systems too, to illustrate the power of the eigensolution in granting insight to the behaviour. For example, consider a simple pendulum as shown in Figure 3.64.

The mass of the pendulum is considered concentrated at the tip. The weight of the pendulum, mg, acts vertically downward. The lateral offset between the weight and the reaction force gives a couple restoring the pendulum to the middle position. Using the small angle approximation, we can write:

$$M_y = -mg \sin \theta \cdot 1 \approx -mg\, \theta \cdot 1 \tag{3.134}$$

Noting there are no moving frames of reference, we also can see by inspection

$$M_y = I_y \frac{d^2\theta}{dt^2} \tag{3.135}$$

Having presumed the pendulum is a particle at its tip, its inertia about the pivot is given by

$$I_y = ml^2 \tag{3.136}$$

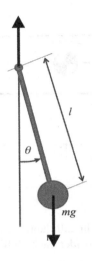

FIGURE 3.64

Pendulum example.

Hence

$$-mg\theta \cdot 1 = \frac{d^2\theta}{dt^2} ml^2 \qquad (3.137)$$

$$-g\theta = \frac{d^2\theta}{dt^2} l \qquad (3.138)$$

As before, we can presume a solution of the form $Ae^{\lambda t}$, which allows us via the same process as previously to arrive at

$$-g\theta = -\lambda^2 \theta l \qquad (3.139)$$

$$\lambda = \sqrt{\frac{-g}{l}} = i\sqrt{\frac{g}{l}} \qquad (3.140)$$

The significance of this is that it also predicts an oscillatory response — but note the oscillation occurs despite the absence of any elastic element in the system. Students who have learned the resonant frequency of a mass on a spring by rote have missed the generality of system behaviour and the possibility for responses other than resonance; it is the authors' experience that this can cause considerable difficulties when trying to understand the behaviour of a vehicle on pneumatic tyres, as described in Chapter 7.

Coupled systems may also be considered using this technique. Consider the vehicle in side view, suspended by elastic elements at each axle, as shown in Figure 3.65.

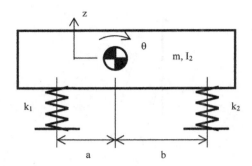

FIGURE 3.65

Vehicle heave and pitch model.

The DOF are z and θ — we might call them 'heave' and 'pitch', respectively. To derive the equations of motion, consider each DOF in turn:

1. Imagine lifting the block a unit displacement in z
   a. Both springs exert a negative heave force
   b. Spring 1 exerts a negative pitch moment
   c. Spring 2 exerts a *positive* pitch moment
2. Imagine rotating the block a unit displacement in θ
   a. Spring 1 exerts a negative heave force
   b. Spring 2 exerts a *positive* heave force
   c. Both springs exert a negative pitch moment

By writing all the terms individually and arranging them together, this allows to write the equations for the system in matrix form:

$$\begin{bmatrix} \dfrac{-(k_1+k_2)}{m} & \dfrac{-ak_1+bk_2}{m} \\ \dfrac{-ak_1+bk_2}{I} & \dfrac{-(a^2k_1+b^2k_2)}{I} \end{bmatrix}\begin{bmatrix} z \\ \theta \end{bmatrix} = \begin{bmatrix} \ddot{z} \\ \ddot{\theta} \end{bmatrix} \tag{3.141}$$

As before, we presume the solution is of the form $Ae^{\lambda t}$ to give:

$$\begin{bmatrix} \dfrac{-(k_1+k_2)}{m} & \dfrac{-ak_1+bk_2}{m} \\ \dfrac{-ak_1+bk_2}{I} & \dfrac{-(a^2k_1+b^2k_2)}{I} \end{bmatrix}\begin{bmatrix} z \\ \theta \end{bmatrix} = \begin{bmatrix} \lambda^2 & 0 \\ 0 & \lambda^2 \end{bmatrix}\begin{bmatrix} z \\ \theta \end{bmatrix} \tag{3.142}$$

$$\begin{bmatrix} \dfrac{-(k_1+k_2)}{m}-\lambda^2 & \dfrac{-ak_1+bk_2}{m} \\ \dfrac{-ak_1+bk_2}{I} & \dfrac{-(a^2k_1+b^2k_2)}{I}-\lambda^2 \end{bmatrix}\begin{bmatrix} z \\ \theta \end{bmatrix} = \begin{bmatrix} 0 \\ 0 \end{bmatrix} \tag{3.143}$$

As before, setting the determinant of the matrix equal to zero will return the eigenvalues:

$$\begin{vmatrix} \dfrac{-(k_1+k_2)}{m} - \lambda^2 & \dfrac{-ak_1+bk_2}{m} \\ \dfrac{-ak_1+bk_2}{I} & \dfrac{-(a^2k_1+b^2k_2)}{I} - \lambda^2 \end{vmatrix} = 0 \qquad (3.144)$$

We can write the individual components of the matrix in a more manageable form:

$$a_{11} = \dfrac{-(k_1+k_2)}{m} \qquad (3.145)$$

$$a_{21} = \dfrac{-ak_1+bk_2}{m} \qquad (3.146)$$

$$a_{12} = \dfrac{-ak_1+bk_2}{I} \qquad (3.147)$$

$$a_{22} = \dfrac{-(a^2k_1+b^2k_2)}{I} \qquad (3.148)$$

$$\begin{vmatrix} a_{11}-\lambda^2 & a_{12} \\ a_{21} & a_{22}-\lambda^2 \end{vmatrix} = 0 \qquad (3.149)$$

This yields a quadratic equation in λ^2, which can be solved in the normal way:

$$(a_{11}-\lambda^2)(a_{22}-\lambda^2) - a_{12} \cdot a_{21} = 0 \qquad (3.150)$$

$$\lambda^4 - \lambda^2(a_{11}+a_{22}) - a_{12} \cdot a_{21} + a_{11} \cdot a_{11} = 0 \qquad (3.151)$$

$$\lambda^2 = \dfrac{-B \pm \sqrt{B^2-4AC}}{2A} \qquad (3.152)$$

$$A = 1 \qquad (3.153)$$

$$B = -(a_{11}+a_{22}) \qquad (3.154)$$

$$C = -a_{12} \cdot a_{21} \qquad (3.155)$$

Even with this relatively simple system, the equations are starting to become unwieldy. The numerical solution of the pitch-heave problem is not unduly onerous, though and can be easily attempted in MS-Excel, Octave or Matlab:

```
m = 1200;    % mass kg
I = 1541;    % pitch inertia kg m^2
a = 1.22;    % CG to front axle m
b = 1.37;    % CG to rear axle m
kf = 43882;  % front axle ride rate N/m
kr = 44583;  % rear axle ride rate N/m

a11 = -(kf + kr)/m
a21 = (-a*kf + b*kr)/m
a12 = (-a*kf + b*kr)/I
a22 = -(a^2*kf + b^2*kr)/I

A = 1;
B = -(a11 + a22);
C = -a12*a21 + a11*a22;
Bsq4AC = B^2-4*A*C
lambda1=((-B + Bsq4AC^0.5)/(2*A))^0.5;
lambda2=((-B - Bsq4AC^0.5)/(2*A))^0.5;

f1 = imag(lambda1)/(2*pi)
f2 = imag(lambda2)/(2*pi)
```

The following results were obtained by the authors using this code and serve as a reference check when implementing it:

```
a11 = -73.721
a21 = 6.2856
a12 = 4.8947
a22 = -96.685
Bsq4AC = 650.42
f1 = 1.3547
f2 = 1.5752
```

Considering the system described by Eqn (3.116) once more, we shall now include the damping, since this is a more general case. For this second order system it is not so obvious how to retrieve the eigenvalues. Recall that for reasons of numerical convenience, the second order system was represented as the combination of two first order systems by introducing the auxiliary system state variable, z. We can write

$$\dot{x} = z \qquad (3.156)$$

$$\dot{z} = \left(\frac{-k}{m}\right)x + \left(\frac{-c}{m}\right)z \qquad (3.157)$$

We may still presume solutions of the form $Ae^{\lambda t}$, as before. This allows us, as previously

$$\dot{\mathbf{x}} = \lambda \mathbf{x} \tag{3.158}$$

$$\dot{\mathbf{z}} = \lambda \mathbf{z} \tag{3.159}$$

Taking these expressions and rearranging things into matrix form allows us to write

$$\begin{bmatrix} \lambda & 0 \\ 0 & \lambda \end{bmatrix} \begin{bmatrix} x_1 \\ x_2 \end{bmatrix} = \begin{bmatrix} 0 & 1 \\ -k/m & -c/m \end{bmatrix} \begin{bmatrix} x_1 \\ x_2 \end{bmatrix} \tag{3.160}$$

Note that nothing new has been added here — this is simply a restatement of the expressions in Eqns (3.156) and (3.157). This can be quite trivially rearranged to give

$$\begin{bmatrix} -\lambda & 1 \\ -k/m & (-c/m) - \lambda \end{bmatrix} \begin{bmatrix} x_1 \\ x_2 \end{bmatrix} = 0 \tag{3.161}$$

The eigenvalues for the system can be found by setting the determinant of the left hand matrix to zero. This process results in the search for the roots of a polynomial — in this case a quadratic — and for this reason, eigenvalues are sometimes referred to as 'roots' or 'roots of the characteristic equation'. For a 2×2 matrix such as this, it can be seen by inspection that

$$\begin{vmatrix} -\lambda & 1 \\ -k/m & (-c/m) - \lambda \end{vmatrix} = 0 \tag{3.162}$$

$$\Rightarrow \lambda = \frac{-c}{2m} \pm \sqrt{\frac{c^2}{4m^2} - \frac{k}{m}} \tag{3.163}$$

Note that the eigenvalues have the same real part with equal and opposite imaginary components — they are known as *complex conjugates*, mirrored about the real axis. Note also that if $c = 0$ then the solution is identical to that of the undamped system.

This approach is very powerful for more complex systems. The matrix immediately to the right of the equal sign in Eqn (3.160) is the Jacobian as mentioned in Section 3.3.3 and can be formed by taking partial derivatives:

$$\begin{bmatrix} 0 & 1 \\ -k/m & -c/m \end{bmatrix} = \begin{bmatrix} \dfrac{\partial}{\partial x}(\dot{x}) & \dfrac{\partial}{\partial z}(\dot{x}) \\ \dfrac{\partial}{\partial x}(\dot{z}) & \dfrac{\partial}{\partial z}(\dot{z}) \end{bmatrix} \tag{3.164}$$

Thus, more generally we can find eigenvalues for systems with or without springs and with coupled, complex behaviour of n system states $x_i = x_1..x_n$ by solving for the determinant of the Jacobian being equal to zero:

$$
\begin{vmatrix}
\dfrac{\partial}{\partial x_1}(\dot{x}_1) - \lambda & \dfrac{\partial}{\partial x_2}(\dot{x}_1) & \cdots & \dfrac{\partial}{\partial x_n}(\dot{x}_1) \\[2mm]
\dfrac{\partial}{\partial x_1}(\dot{x}_2) & \dfrac{\partial}{\partial x_2}(\dot{x}_2) - \lambda & & \dfrac{\partial}{\partial x_n}(\dot{x}_2) \\[2mm]
\cdots & & & \cdots \\[2mm]
\dfrac{\partial}{\partial x_1}(\dot{x}_n) & \dfrac{\partial}{\partial x_2}(\dot{x}_n) & \cdots & \dfrac{\partial}{\partial x_n}(\dot{x}_n) - \lambda
\end{vmatrix} = 0
\qquad (3.165)
$$

With complex systems, the solution of the determinant becomes nontrivial when the matrix gets large. While MS-Excel offers commands such as MDETERM, notionally allowing its built-in solver to hunt for eigenvalues, it typically only solves for a single eigenvalue, and even then will often fail if the matrix becomes poorly conditioned during its numerical hunt for the solution. The impeccable 'Numerical Recipes' (Press et al, 1992) has this to say:

> *You have probably gathered by now that the solution of eigensystems is a fairly complicated business. It is. It is one of the few subjects covered in this book for which we do* not *recommend that you avoid canned routines.*

The reader who is determined to assemble their own solution to the problem is urged to consult *Recipes*, where the routines provided are entirely useful.

3.5 Systems of units

As with all engineering analysis it is important that consistent units are used throughout any calculation or simulation exercise. For static analysis the choice of a system of units can be quite forgiving as long as consistency is observed. For dynamic analysis more care is required. At a basic level if we consider the application of Newton's Second Law to a body n with mass m_n and acceleration $\{A_{Gn}\}_1$ at the mass centre G_n we have with the vector convention used here

$$
\sum \{F_n\}_1 = m_n \{A_{Gn}\}_1
\qquad (3.166)
$$

It is important to note that Eqn (3.166) is only valid for certain systems of units. For a metric system of SI units (force $=$ N, mass $=$ kg, acceleration $=$ m/s^2) Eqn (3.86) can be used as it stands. The system of units used mainly in this text, and popular in Europe, is to use millimetres as the unit of choice for length. Clearly Eqn (3.166) would produce incorrect forces unless the right hand side of Eqn (3.166) was divided by a constant, in this case 1000, to ensure consistency.

Early users of general-purpose programs needed to define such a constant when defining gravitational forces, even for dynamic models operating in a zero gravity

Table 3.7 Units Consistency

Measurement	SI	Metric	Metric	Metric	FPS	FPS	IPS
Length	m	mm	cm	m	ft	ft	in
Velocity	m/s	mm/s	cm/s	m/s	ft/s	ft/s	in/s
Acceleration	m/s^2	mm/s^2	cm/s^2	m/s^2	ft/s^2	ft/s^2	in/s^2
Mass	kg	kg	kg	kg	slug	lbm	lbm
Force	N	N	kgf	kgf	lbf	lbf	lbf
Inertia	kgm^2	$kgmm^2$	$kgcm^2$	kgm^2	$slug\ ft^2$	$lbm\ ft^2$	$lbm\ in^2$
Gravity	9.81 m/s^2	9807 mm/s^2	981 cm/s^2	9.81 m/s^2	32.2 ft/s^2	32.2 ft/s^2	386.1 in/s^2
UCF	1.0	1000	981	9.81	1.0	32.2	386.1

UCF, units consistency factor; FPS, foot–pound–second system; IPS, inch–pound–second system.

environment. As such users were reminded of some of the basic fundamentals of dynamics when using such software. More modern programs require the users to define only a set of units and apply the correction internally. Table 3.7 is provided to show the required corrections to various systems of units. Although for most users of MBS this will be for reference only, advanced users developing subroutines that link with MBS may still be required to implement a constant to ensure consistency. An early term used to define the constant was the 'gravitational constant' although a more recent and applicable definition is the 'units consistency factor' (UCF) where for any system of units the following equation is valid for the given UCF in Table 3.7.

$$\sum \{F_n\}_1 = \frac{m_n \{A_{Gn}\}_1}{UCF} \tag{3.167}$$

3.6 Further comments on pre- and postprocessing

Several general-purpose programs have rather specialist layers over the top of them to facilitate modelling vehicles in particular. In this sense they become a single-purpose general-purpose program, which appears a very complex way of going about things. Nevertheless, they allow the detailed complexity of a general-purpose program and a highly capable numerical engine to be brought to bear on the vehicle analysis problem, which more specialist programs are often unable to achieve — ironically because they lack the generality of the general-purpose programs.

An early customisation of MSC ADAMS for automotive applications was the ADAMS/Vehicle program originally developed as a commercially available product, which has been used by engineers from the Newman/Hass Indy Car racing team (Trungle, 1991). The program allowed a suspension model to be created, carry out an analysis and postprocess the results without specialist knowledge of MSC ADAMS. The program could also be used to automatically generate a full vehicle model, hence the title. The preprocessor included a number of established suspension configurations where the data were input via screen templates using familiar suspension terminology.

Simpack provides a detailed vehicle 'wizard' with a large amount of detail in the models and templates available to allow more or less any conceivable level of complexity to be approached from something significantly less intimidating than a clean sheet of paper. Other packages, such as Dymola, also include vehicle libraries for the same reason.

ADAMS/Car was developed working with a consortium of major vehicle manufacturers including Audi, BMW, Renault and Volvo. The vehicle manufacturers' involvement included developing the specification for the functionality of the software. This included, for example, determining the suspension systems that would be included, the manoeuvres to be simulated and the outputs and their manner of presentation. For this reason, the ADAMS/Car library content and workflow seems to be regarded as something of a benchmark for other software. An example of the ADAMS/Car graphical interface is shown in Figure 3.66.

In practice the different software systems work in different ways but there are two basic categories of user:

1. The Expert User who has access to the fundamental software modelling elements. As such they are able, for example, to create or modify system model templates of suspensions and steering systems. They would also be able to modify or create test procedures to be simulated.
2. The Standard User who is not necessarily a multibody system expert but is able to use the existing templates to enter data and create models using familiar terminology and to produce reliable results according to a robust protocol. They include design, test and development engineers in addition to analysts.

In parallel to ADAMS/Car the ADAMS/Chassis system is also used for vehicle work and offers similar capability. The ADAMS/Chassis program was originally developed in house by Ford in the late 1980s. The program was originally called ADAMS/Pre and as the name suggests the early implementation was a preprocessor that automatically formatted ADAMS data sets. In its current form it has additional capability to run customised simulations and has its own postprocessor. Ford allowed the program to be taken on and developed by another company before the product was acquired by the developers of MSC ADAMS. Due to its origin ADAMS/Chassis appears on the surface unlike any of the other customised MSC ADAMS programs. The program uses a graphical interface based more on data forms to enter the data. An example is shown in Figure 3.67.

Any software requires a high level of programming skill on the part of the Expert User who is going to customise or develop the way models are generated, simulations are run or results are plotted and reported. In addition to vehicle dynamics

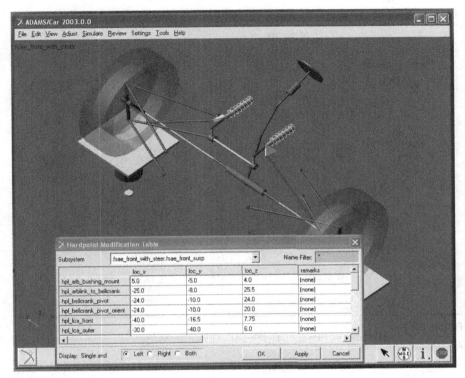

FIGURE 3.66

ADAMS/Car graphical user interface.

(Provided courtesy of MSC.Software.)

knowledge the expert would need a working knowledge of the core solver language, C++ and FORTRAN in order to perform any meaningful customisation. For this reason, such experts are few and far between and tend to be well looked after by organisations that employ them.

All the integrated 'single-purpose general-purpose' programs include a substantial list of preprogrammed vehicle manoeuvres and subsystem test procedures, with both inputs and outputs defined.

It is perhaps in the calculation of outputs such as these that the customised software offers a significant benefit to the analyst but also a risk to the culture of the organisation. With a general-purpose program, every model starts with the proverbial 'blank sheet'. Everything is uniformly difficult, even the easy things. This generates a steep learning curve, which often surprises the authors at the rapidity of its scaling by students — so perhaps it is not as steep as it looks. Once it has been scaled, nothing looks particularly difficult any more and everything becomes possible.

In contrast, with a single-purpose general-purpose program even quite difficult things are very easy *providing someone else has set up the model and procedure.*

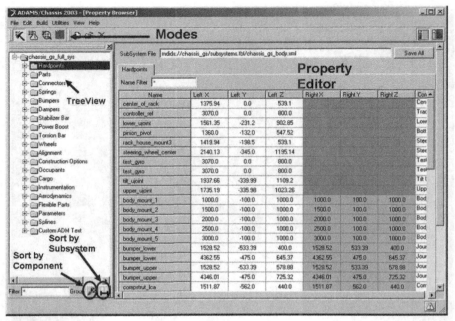

FIGURE 3.67

ADAMS/Chassis user interface.

(Provided courtesy of MSC.Software.)

This leads to a somewhat institutionalised frame of mind in which anything that is not already defined looks impossible and it becomes much more difficult to implement changes in process or procedure; the organisation becomes a slave to what is possible in the currently released version of the software and stops pushing the boundaries. More importantly, things get done that *can be done* instead of that *need to be done*.

It is with this note of caution that the student is encouraged to set their aspiration towards becoming the Expert User, with the labour-saving automation provided by the software but the free-ranging ability to tackle any problem that comes from the clean sheet mentality.

Modelling and Analysis of Suspension Systems

Any suspension will work, if you don't let it

Colin Chapman

In this chapter, the basic role of the suspension system is discussed from a functional, analytical perspective. Several types of suspension are introduced and methods of their analysis are described in a typical commercial multibody systems (MBS) analysis package. Immediately familiar to students is the system shown in Figure 4.1, this being a suspension system referred to in Europe as a double wishbone system. In the USA the practice is to call the same system a short-long arm suspension system. We will be considering the modelling and analysis of suspension systems as separate units, a practice sometimes referred to as quarter vehicle modelling. The discussion in this chapter will be restricted to passive systems and will not address active suspension systems.

It is also intended in this chapter to demonstrate the manner in which the vector based methods described in Chapter 2 can be applied to carry out three-dimensional kinematic, static and dynamic force analyses of a double wishbone suspension system. The model geometry has also been analysed using an equivalent MBS model allowing a comparison of results from theory and MBS and providing readers with an insight into the computational processes involved.

The traditional treatment of suspension systems, dealing usually with a projection in one plane, is used here only where useful to explain standard terms that the student or developing practitioner should be aware of.

The concept of a roll centre is one that has been used extensively by vehicle designers in order to relate suspension layout to vehicle handling performance, particularly when considering understeer or oversteer. The roll centre is introduced in this chapter using a classical treatment together with a description of the typical process used in MBS for computation. Some of the mystery surrounding it is also debunked.

Before progressing to a detailed treatment of the modelling and analysis issues it is important to fully describe the function of a suspension system and the needs that must be addressed. The initial sections of this chapter intend to identify these needs in a systematic manner providing a checklist against which suspension performance may be evaluated during the vehicle design process. The follow-on treatment of modelling and simulation can then be related back to the various needs now described.

FIGURE 4.1

Aston Martin Vanquish double wishbone front suspension system.

4.1 The need for suspension

The term suspension seems at first hearing an odd one when considering the function in a modern vehicle as the vehicle body appears to sit on rather than be suspended from the mechanism. With growing commercial activities, coach bodies hung from so-called 'C-Springs' allowed faster travel over the poor roads of the late eighteenth century and so gave an advantage. Despite the change from hanging to standing, the term 'suspension' has stuck in the English-speaking world.

In its simplest form, a modern road vehicle suspension may be thought of as a linkage to allow the wheel to move relative to the body and some elastic element to support loads while allowing that motion. As suspensions become more complex, the need for well-controlled damping forces and multidirectional compliance emerges. MBS analysis can help quantify an existing design in terms of these parameters or help to synthesise a new design from a set of target parameters.

Most practical vehicles have some form of suspension, particularly when there are four or more wheels. The suspension system addresses two basic needs:

- Reduction of vertical wheel load variations
- Isolation of road inputs from body

However, the introduction of a suspension system introduces some tasks of its own; each additional interface and component brings with it some structural compliance. This may lead to a delay in transmitting loads to the body, with a possible degradation to the vehicle-handling task. It also offers the opportunity (or risk) of modifying the

Table 4.1 Suspension Design Process Activities

	Wheel Load Variation	Body Isolation	Handling Load Control	Compliant Wheel Plane Control	Kinematic Wheel Plane Control	Component Loading Environment
Investigate design strategies	✔	✔	✔	✔	✔	✔
Set design targets	✔	✔	✔	✔	✔	✔
Verify proposed designs	✔	✔	✔	✔	✔	✔

way the wheel is presented to the road. The new components must themselves be capable of surviving the design life when the vehicle is used as intended. Therefore, as a consequence of its existence the suspension generates four more needs:

- Control of transmission of handling loads to body
- Control of wheel plane geometry due to compliant effects
- Control of wheel plane geometry due to kinematic effects
- Comprehension of component load environment

Modern MBS analysis tools offer the opportunity to evaluate each of the six needs listed, either to investigate design strategies for whole systems, set design targets for components or to verify the performance of proposed designs. Any rigorous design process must therefore be able to quantify matters in these 18 different ways as shown in Table 4.1.

It is easy to focus on one small subset of the task above but to do so fails to deliver the full range of benefits possible from MBS analysis software. The software is generally expensive and the skilled personnel to operate it are hard to come by and so such a focus represents a lost opportunity for the organisation.

4.1.1 **Wheel load variation**

Since a vehicle with four wheels is statically indeterminate with respect to the calculation of reaction forces, only with some elastic behaviour ('compliance') in the suspension can a determinate solution be formulated unless some presumption of symmetry is made; otherwise, the 'wobbly table' case exists where there are at least three possible load distributions (Figure 4.2). This leads to abrupt fluctuations in load of a type barely tolerable in restaurants and unconscionable for road vehicle behaviour. For the problem shown we have one system and three possible reaction load solutions. These include being balanced on the two longest legs (left in Figure 4.2) — unlikely in practice but a solution in theory — or being balanced on

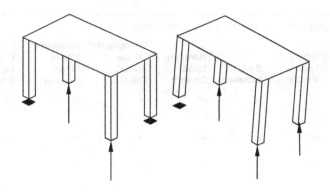

FIGURE 4.2

A classic case of static indeterminacy.

three legs (right in Figure 4.2) in one of two stable states. This system represents a classic case of static indeterminacy.

Consideration of a shopping trolley, a notionally rigid vehicle with four wheels that displays many irksome traits, shows that even on the most well-prepared supermarket floor the absence of suspension causes problems with load distribution between the four wheels due to inconsistent manufacture. The least loaded wheel is prone to shimmy and the uneven wheel-load distribution can emphasise frictional asymmetry in the castors with even the most fastidious grocery-packing practices, giving the familiar 'mind of its own' sensation.

Less infuriating 'rigid' vehicles include some agricultural equipment that uses compliant behaviour in the tyre sidewalls to accommodate uneven terrain and racing karts, which use a significant amount of compliance in the frame structure of the vehicle to modify vertical wheel loads. An example of this is shown in Figure 4.3 where the inclusion of the racing kart frame flexibility clearly influences the yaw rate response for a simulated manoeuvre.

More formally, the problem may be posed as a rigid platform, Body 2, of total weight $m_2\{g\}_1$, with reactions acting at wheel locations A, B, C and D as shown in Figure 4.4.

Consider first the equations that would be needed for equilibrium of forces. Note that for rigour we are continuing with the vector notation established in Chapter 2.

$$\Sigma\{F_2\}_1 = \{0\}_1 \tag{4.1}$$

$$\{F_A\}_1 + \{F_B\}_1 + \{F_C\}_1 + \{F_D\}_1 - m_2\{g\}_1 = \{0\}_1 \tag{4.2}$$

$$\begin{bmatrix} 0 \\ 0 \\ F_{Az} \end{bmatrix} + \begin{bmatrix} 0 \\ 0 \\ F_{Bz} \end{bmatrix} + \begin{bmatrix} 0 \\ 0 \\ F_{Cz} \end{bmatrix} + \begin{bmatrix} 0 \\ 0 \\ F_{Dz} \end{bmatrix} - m_2 \begin{bmatrix} 0 \\ 0 \\ g \end{bmatrix} = \begin{bmatrix} 0 \\ 0 \\ 0 \end{bmatrix} \tag{4.3}$$

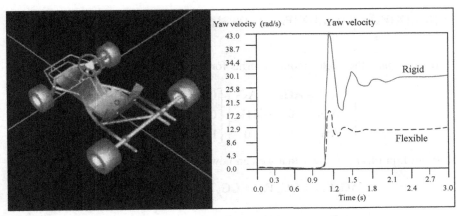

FIGURE 4.3

The influence of racing kart frame flexibility.

Courtesy of MSC software.

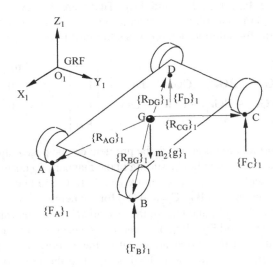

FIGURE 4.4

The wheel load reaction problem. GRF, ground reference frame.

From Eqn (4.3) we now get the expression in Eqn (4.4) that we would normally get quickly by inspection.

$$F_{Az} + F_{Bz} + F_{Cz} + F_{Dz} - m_2 g = 0 \tag{4.4}$$

If we continue now to take moments about the mass centre G of Body 2 we have

$$\sum \{M_G\}_1 = \{0\}_1 \tag{4.5}$$

$$\{R_{AG}\}X\{F_A\}_1 + \{R_{BG}\}_1X\{F_B\}_1 + \{R_{CG}\}_1X\{F_C\}_1 + \{R_{DG}\}_1X\{F_D\}_1 = \{0\}_1$$
(4.6)

If we expand the vector moment terms for the force acting at A only we get

$$\begin{bmatrix} 0 & -AG_z & AG_y \\ AG_z & 0 & -AG_x \\ -AG_y & AG_x & 0 \end{bmatrix} \begin{bmatrix} 0 \\ 0 \\ F_{Az} \end{bmatrix} = \begin{bmatrix} AG_y F_{Az} \\ -AG_x F_{Az} \\ 0 \end{bmatrix}$$
(4.7)

From Eqn (4.7) it is clear that we can now write Eqn (4.6) as

$$AG_y F_{Az} + BG_y F_{Bz} + CG_y F_{Cz} + DG_y F_{Dz} = 0$$
(4.8)

$$-AG_x F_{Az} - BG_x F_{ABz} - CG_x F_{Cz} - DG_x F_{Dz} = 0$$
(4.9)

It can now be seen that we have the classical case of an indeterminate problem where there are insufficient equations available to solve for the unknowns. The four unknowns here are F_{Az}, F_{Bz}, F_{Cz} and F_{Dz}. There are however only the three Eqns (4.4), (4.8) and (4.9) available to effect a solution. Trying to arrange the equations in matrix form for solution demonstrates the futility of the problem.

$$\begin{bmatrix} 1 & 1 & 1 & 1 \\ AG_y & BG_y & CG_y & DG_y \\ -AG_x & -BG_x & -CG_x & -DG_x \\ ? & ? & ? & ? \end{bmatrix} \begin{bmatrix} F_{Az} \\ F_{Bz} \\ F_{Cz} \\ F_{Dz} \end{bmatrix} = \begin{bmatrix} m_2 g \\ 0 \\ 0 \\ ? \end{bmatrix}$$
(4.10)

If a modified formulation is adopted, the problem can be posed more completely. If the rigid platform is presumed to be sprung in the manner of a normal road vehicle with an effective wheel rate, k_A, k_B, k_C and k_D, at A, B, C and D then a new formulation is possible. Writing A_{2z}, B_{2z}, C_{2z} and D_{2z} for the height of corners A, B, C and D on Body 2 and A_{1z}, B_{1z}, C_{1z} and D_{1z} for the ground height corners at A, B, C and D we can then define a preload Fp_A, Fp_B, Fp_C and Fp_D on each spring such that if the z-coordinates of Body 2 at the corner are equal to the z-coordinates of the ground, Body 1, at each corner then the spring load is equal to the preload. This leads to the following equations for the spring force at each corner:

$$F_{Az} = k_A (A_{2z} - A_{1z}) + Fp_A$$
(4.11)

$$F_{Bz} = k_B (B_{2z} - B_{1z}) + Fp_B$$
(4.12)

$$F_{Cz} = k_C (C_{2z} - C_{1z}) + Fp_C$$
(4.13)

$$F_{Dz} = k_D (D_{2z} - D_{1z}) + Fp_D$$
(4.14)

At first sight it would appear this is simply a more elaborate formulation of the previous problem; there are still four unknown quantities (the corner heights of Body 2) and three Eqns (4.4), (4.8) and (4.9). However, consideration of the rigid platform yields a fourth relationship

$$D_{2z} = C_{2z} + (B_{2z} - A_{2z}) \qquad (4.15)$$

This leads to, after some manipulation of Eqn (4.10), substituting in the spring Eqns (4.11)–(4.13) for F_{Az}, F_{Bz}, F_{Cz} and relying on Eqn (4.15) to solve for F_{Dz}, the equations in Eqn (4.16)

$$\begin{bmatrix} k_A - k_D & k_B + k_D & k_C + k_D \\ k_A AG_y - k_D DG_y & k_B BG_y + k_D DG_y & k_C CG_y + k_D DG_y \\ k_A AG_x - k_D DG_x & k_B BG_x + k_D DG_x & k_C CG_x + k_D DG_x \end{bmatrix} \begin{bmatrix} A_{2z} \\ B_{2z} \\ C_{2z} \end{bmatrix} = \begin{bmatrix} \lambda_1 \\ \lambda_2 \\ \lambda_3 \end{bmatrix} \qquad (4.16)$$

where

$$\lambda_1 = m_2 g + k_A A_{1z} - Fp_A + k_B B_{1z} - Fp_B + k_C C_{1z} - Fp_C + k_D D_{1z} - Fp_D \qquad (4.17)$$

$$\lambda_2 = (k_A A_{1z} - Fp_A)AG_y + (k_B B_{1z} - Fp_B)BG_y + (k_C C_{1z} - Fp_C)CG_y + (k_D D_{1z} - Fp_D)DG_y \qquad (4.18)$$

$$\lambda_3 = (k_A A_{1z} - Fp_A)AG_x + (k_B B_{1z} - Fp_B)BG_x + (k_C C_{1z} - Fp_C)CG_x + (k_D D_{1z} - Fp_D)DG_x \qquad (4.19)$$

The equations in Eqn (4.16) may be solved by Gaussian elimination and substitution. It can be seen that preloads (Fp_A, Fp_B, Fp_C and Fp_D), the ground heights (A_{1z}, B_{1z}, C_{1z} and D_{1z}) and the corner stiffness (k_A, k_B, k_C and k_D) and wheel locations are the inputs to the calculations. The unknowns that are found from Eqn (4.16) are the body heights at three corners (A_{2z}, B_{2z} and C_{2z}). These are used to find D_{2z} from Eqn (4.15) leading to the force solution in each corner from Eqns (4.11)–(4.14). Thus, the presence of an elastic suspension for the road wheels allows a solution to the 'wobbly table' problem for vehicles with four or more wheels even when traversing terrain that is not smooth.

4.1.2 Body isolation

The interaction of a single wheel with terrain of varying height is frequently idealised as shown in Figure 4.5. This is a so-called 'quarter vehicle' model and is widely used to illustrate the behaviour of suspension systems. It may be thought of as a stationary system under which a ground profile passes to give a time-varying ground input, z_g. Note that at this point we are assuming the tyre to be rigid. Whether classical or 'literal' MBS models are used, the methods used to comprehend body isolation are the same.

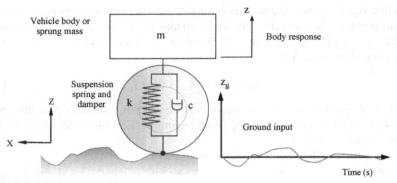

FIGURE 4.5

A classical quarter vehicle ride model.

Classically, the system may be formulated as a single second-order differential equation

$$m\ddot{z} + c(\dot{z} - \dot{z}_g) + k_s(z - z_g) = 0 \tag{4.20}$$

In order to more fully understand the isolation behaviour of the suspension, a fully developed (steady state) harmonic solution of the equation above may be presumed to apply. 'Harmonic' simply means that both input and output may be described with a sine function, which may be conveniently expressed in the form

$$z = A.e^{(i\omega t + \phi)} \tag{4.21}$$

where i is the imaginary square root of -1. Thus for the two derivatives of z it may be written

$$\dot{z} = i\omega A\ e^{(i\omega t + \phi)} = i\omega z \tag{4.22}$$

and

$$\ddot{z} = -\omega^2 A\ e^{(i\omega t + \phi)} = -\omega^2 z \tag{4.23}$$

This is a more problem-specific version of the general Jacobian formulation advanced in Chapter 3. The reason for the problem-specific formulation is to illustrate isolation behaviour specifically. The original Eqn (4.20) can then be written

$$-m\omega^2 z + ic\omega(z - z_g) + k(z - z_g) = 0 \tag{4.24}$$

Rearranging this in the form of a transfer function gives

$$H(\omega) = \frac{output}{input}(\omega) = \frac{z}{z_g} = \frac{(k) + i(c\omega)}{(k - m\omega^2) + i(c\omega)} \tag{4.25}$$

This expression relates the amplitude and phase of ground movements to the amplitude and phase of the body movements and is commonly reproduced in many vibration theory books and courses.

Considering the transfer function, several behaviours may be observed. At frequencies of zero and close to zero, the transfer function is unity. This may reasonably be expected; if the ground is moved very slowly the entire system translates with it, substantially undistorted. The behaviour of the system may be described as 'static' where its position is governed only by the preload in the spring and the mass carried by the spring; dynamic effects are absent.

At one particular frequency where $\omega = \sqrt{k/m}$, the real part of the denominator becomes zero and the transfer function is given by

$$H(\omega) = \frac{(k) + i(c\omega)}{i(c\omega)} = \frac{-ik}{i(c\omega)} + 1 \qquad (4.26)$$

At this frequency, the behaviour of the system is called 'resonant'. Examination of the transfer function shows it is at its maximum value. Since k, c and ω are all positive real numbers, the transfer function shows that ground inputs are amplified at this frequency. If c is zero, i.e. there is no damping present in the system, then the transfer function is infinite. If c is very large, the transfer function has an amplitude of unity. For typical values of c, the amplitude of the transfer function is greater than unity.

At substantially higher frequencies where $\omega \gg \sqrt{k/m}$, the transfer function is dominated by the term $-m\omega^2$ in the denominator and tends towards

$$H(\omega) = \frac{1}{-m\omega^2} \qquad (4.27)$$

At these higher frequencies, the amplitude of the transfer function falls away rapidly.

It may therefore be noted that if it is desired to have the body isolated from the road inputs, the system must operate in the latter region where $\omega \gg \sqrt{k/m}$. In fact this is a general conclusion for any dynamic system; for isolation to occur it must operate above its resonant frequency. Given that the mass of the vehicle body is a function of things beyond the suspension designer's control, it is generally true that the only variables available to the designer are spring and damper calibration.

The preceding analysis would suggest that the softest springs possible, to position the resonant frequency as low as possible, are preferred. In a purely technical sense this is indeed true; however, our vehicles are to have as their primary purpose the transport of people over roads, which constrains our choices further.

Road surfaces are, to a first approximation, a random process passing under the car. They can be described by the expression

$$z_g(\omega) = \frac{K(2\pi V)^{R-1}}{\omega^R} \qquad (4.28)$$

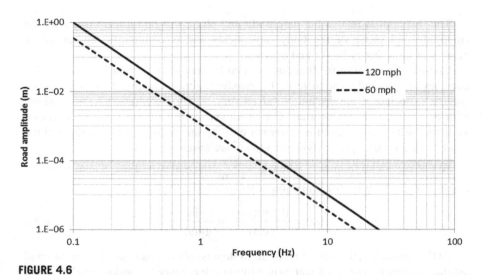

FIGURE 4.6

Road surface frequency content.

where K and R are constants. Figure 4.6 illustrates a road amplitude spectrum with K and R set to 10^{-5} and 2.5 respectively. Values of both K and R vary with the nature of the road being studied. K is a straightforward amplitude measure. For novel applications or for previously unknown markets, some measurements of road profiles is desirable; for the developed Western world the values given are generally appropriate. For military applications, a value of 1 for R may be appropriate. The very smoothest roads are unlikely to have R higher than 4.

It can be seen that the spectral content as seen by the car varies with vehicle speed. Gillespie (1992) gives a 'rational' argument to suggest that spectral content rises with speed squared. However, Hales (1989) gives a different view, reproduced in essence here. The road surfaces is a process that may be approximated by a straight line on a log—log plot and represented by

$$\log[z_g(n)] = \log[K] - R.\log[n] \tag{4.29}$$

implying that

$$z_g(n) = \frac{K}{n^R} \tag{4.30}$$

This is a spatial representation of the road profile, with n defined in cycles per metre. It appears under the car at frequency of V.n Hz, where V is the forward speed of the car, thus

$$V.n = \frac{\omega}{2\pi} \Rightarrow n = \frac{\omega}{2\pi V} \tag{4.31}$$

We can write

$$z_g(n) = \frac{g(n)}{n} = g(n).\frac{2\pi V}{\omega} = z_g(\omega).2\pi V \qquad (4.32)$$

if we acknowledge that $g(n) = g(\omega)$ — i.e. that the process is unchanged whatever we choose to express it as a function of. Hence:

$$z_g(\omega) = \frac{1}{2\pi V}\cdot\frac{K}{n^R} = \frac{1}{2\pi V}\cdot\frac{K}{(\omega/2\pi V)^R} = \frac{K(2\pi V)^{R-1}}{\omega^R} \qquad (4.33)$$

For the purposes of positioning the ride mode, then the frequency domain relationship between ground input, z_g and displacement of the vehicle body, z, is given by

$$\frac{z(\omega)}{z_g(\omega)} = \frac{k^2 + \omega^2 c^2}{(k - \omega^2 m)^2 + \omega^2 c^2} \qquad (4.34)$$

The acceleration environment is the prime concern in ride studies. Assuming harmonic solutions (i.e. made up of sine waves), we may write

$$\ddot{z}(\omega) = \frac{\ddot{z}(\omega)}{\ddot{z}_g(\omega)}\cdot\ddot{z}_g(\omega) = -\omega^2\frac{\ddot{z}(\omega)}{\ddot{z}_g(\omega)}\cdot\ddot{z}_g(\omega) \qquad (4.35)$$

Thus we have all that is needed to calculate a harmonic acceleration response of the car over a typical road profile. If the acceleration response is compared to the curves in ISO2631-1:1997 (1997), then some direct measure of ride comfort can be made.

Figure 4.7 shows a prediction of the same vehicle travelling at 60 and 120 mph on the same road, with threshold figures based on 1 h exposure to the vibration environment. Considering Figure 4.7, it may be supposed that a better riding vehicle could be made by positioning the primary ride resonance at a lower frequency in order to better match the shape of the threshold curves and thus improve the aggregate ride over the frequency range of interest. Notwithstanding the difficulty in general that lower ride frequencies mean larger suspension motions, if we presume these difficulties can be overcome, another difficulty remains.

Figure 4.8 shows the acceleration response for a 0.2 Hz primary ride car, which has a much lower exceedance of the perception threshold. However, superimposed on the graph is a motion sickness threshold for 5% of the population at 1 h exposure. It can be seen that in the region substantially below resonance for the 1 Hz car, the motion sickness threshold exceedance is entirely in the region in which the vehicle does not amplify road inputs; in other words, motion sickness is induced by irregularities in the road and not exacerbated in the vehicle. For the 0.2 Hz car, however, it can be seen that the vehicle resonance on its suspension contributes to a significant exceedance of the threshold motion sickness — the suspension will make people ill. Considering Figure 4.8 in some detail, it would seem that a practical limiting

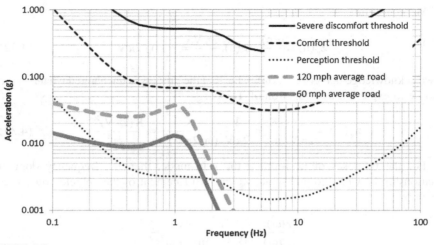

FIGURE 4.7

A prediction of the harmonic response of a vehicle body in comparison with the acceleration curves laid down in ISO2631-1:1997.

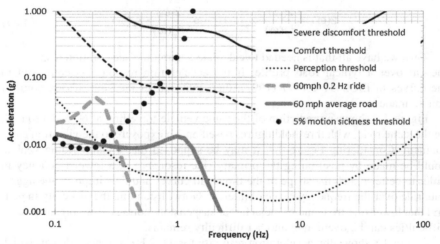

FIGURE 4.8

A prediction of the harmonic response of a vehicle body in comparison with the acceleration curves laid down in ISO2631-1:1997 with the inclusion of motion sickness threshold data, comparing a 0.2 Hz vehicle with a 1 Hz vehicle.

suspension frequency is where the 5% motion sickness line crosses the comfort threshold. This is around 0.65 Hz.

Given that the human frame is primarily engineered for walking, a more typical value for the resonant frequency of the body mass on the road springs is just over

1 Hz, this being the frequency at which we walk for the majority of the time. This is one of several reasons why babies may be readily nursed to sleep in cars; the motion of the vehicle gives an acceleration environment not unlike being held in the arms of a walking adult.

For motorsport or military vehicles, it may be acceptable to use the severe discomfort threshold; in this case the practical lower limit for ride frequencies is around 1.25 Hz.

It should be noted that the threshold values for accelerations drop off significantly above 2 Hz. For this reason, motor vehicles rarely have primary ride modes much above 2.5 Hz, even quite aggressively suspended ones. There is thus a 'window' in which we position the primary ride behaviour of our vehicles to make them compatible with the operators. For autonomous vehicles that never carry humans, these restrictions are relaxed and the suspension primary ride frequency can be chosen on the basis of some other functional aspect.

Our real vehicle systems are rarely so straightforward to address as this, however. Real vehicle suspensions have vertical and longitudinal compliance behaviour. Suspension components rotate as well as translate and the sprung mass has rotational as well as translational freedoms. Modern MBS analysis software allows us to describe the individual components of the system and will automatically calculate the contribution of, for example, sprung mass pitch inertia to the acceleration solution for the body. The software also allows the linearisation of the system about an operating point and the calculation of modes of vibration of the system. For example, the longitudinal compliance typically snubs out under hard braking and the fore-aft isolation suffers; proprietary software will allow the calculation of fore-aft resonant frequencies under cruise and braking conditions and their comparison by an intelligent user will explain (and perhaps generate solutions for) harsh behaviour over small obstacles while braking.

At no stage with added complexity do the basic rules for dynamic systems break down; there remains a subresonant, stiffness-dominated regime, a resonant, damping-dominated regime and a post-resonant, mass-dominated regime. The software user who keeps a grasp of these basic concepts, while allowing the software to undertake the task of assembling the equations of motion and solving them, is productive within an engineering organisation.

4.1.3 Handling load control

The simplest possible representation for a vehicle manoeuvring in the ground plane is shown in Figure 4.9. If we break from the full 3D vector notation the following pair of differential equations describes it fully:

$$\sum M_{2z} = I_{zz}\, \dot{\omega}_{2z} \qquad\qquad (4.36)$$

$$\sum F_{2y} = m_2\, (\dot{V}_{2y} + V_{2x}\, \omega_{2z}) \qquad\qquad (4.37)$$

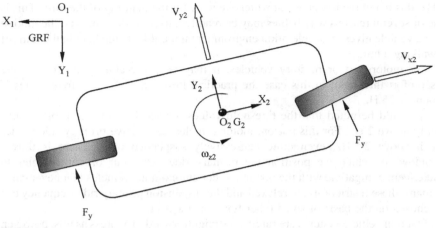

FIGURE 4.9

The simplest possible representation of a vehicle manoeuvring in the ground plane. GRF, ground reference frame.

These equations, their significance and a more formal derivation are well described in the seminal IME paper (Segel, 1956). The formulation states that yaw acceleration is the applied yaw moments divided by the yaw inertia of the vehicle, and that lateral acceleration is the applied lateral forces divided by the mass of the vehicle. The additional term in the lateral force expression reflects a body-centred formulation, which is more convenient when the model is expanded to more than the two degrees of freedom shown.

The two equations are correctly referred to as a two-degree-of-freedom ('2 DoF') model; they are sometimes referred to as a 'bicycle' model but the authors dislike this description since it implies that the description may be suitable for two-wheeled vehicles, which it most certainly is not. Readers interested in two-wheeled vehicles are recommended to study Chapter 10 of 'Tire and Vehicle Dynamics' (Pacejka, 2012).

Even with this simplest possible representation, it can be seen that the vehicle may be thought of as a free-floating 'puck' (as used in ice hockey), to which forces are applied by the tyres in order to manipulate its heading and direction. To many casual observers it appears that the vehicle runs on little 'rails' provided by the tyres and that the function of the tyres is simply to provide a cushion of air beneath the steel wheel rims. This is simply not so, and examination of the behaviour of rally cars in the hands of skilled drivers reveals behaviour which visually resembles that of a hovercraft. All vehicles on pneumatic tyres behave as the rally cars behave, adopting a sideslip angle to negotiate even the slightest curve. Since this angle is typically less than a degree it is not always apparent to the untrained observer; it may however be seen on high speed, steady corners such as motorway interchanges if vehicles are observed attentively.

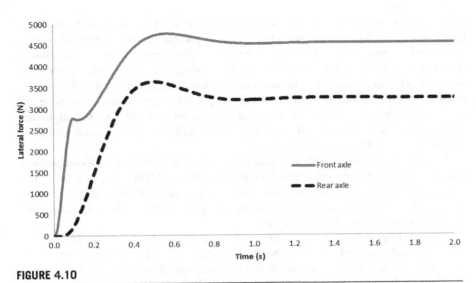

FIGURE 4.10

Side forces to a 0.02 rad input applied between $t = 0$ and $t = 0.1$ with a cosine ramp function — front axle characteristic is common in simulation but rarely observed in practice since such rapid steer inputs are uncommon.

The generation of the forces necessary to initiate the turn, to constrain the vehicle at the correct sideslip angle and to return it to the straight-running condition is the role of the tyres. In order to successfully control the vehicle, however, those loads must be transmitted to the sprung mass. This is a key role for the vehicle suspension system.

Close examination of the vehicle behaviour described by the equations above can demonstrate the slight phasing of the forces necessary to allow the vehicle to accelerate in yaw and be constrained to the desired yaw velocity (Figure 4.10). Errors in that phasing generated by flexibility (compliance) on one axle or the other can lead to an error in the vehicle behaviour as perceived by the driver. This error may take the form of a disconcerting delay in response to the steering if the front axle has more compliance than the rear, or a rather more serious delay in the action of the rear axle in constraining the body slip angle to its required value. In the latter case, particularly for aggressive transient manoeuvres (i.e. accident avoidance) the rear tyre slip angle may exceed its critical value, leading to a divergent behaviour of the vehicle — that is to say a spin.

Since for modern vehicles the isolation of road inputs is a high priority, there is always a desire to introduce some elastomeric elements into the vehicle between the suspension elements and the vehicle body. MBS analysis tools allow the study of handling degradation due to the introduction of such elastomers and allow the ride/refinement compromise to be quantified before excessive experimentation is carried out on the vehicle. Detailed comments on the representation of elastomers are given in Chapter 3. MBS analysis in the right hands allows an understanding of those design parameters that dominate refinement performance and those that dominate handling performance. The understanding so gained allows better conceptual design of suspension systems in order to separate clearly the refinement and handling functionality for elastomeric elements. Modern multilink rear suspensions are a good example of such well-separated systems and are a significant part of the simultaneous improvement in both ride and handling, traditionally areas of mutual exclusivity, that has befallen modern road cars.

Even for a suspension with no elastomers, such as may be used on a competition vehicle, some structural compliance is always present. MBS analysis allows this compliance to be optimised and matched front to rear in order to avoid onerous design constraints using conservative stiffness targets, which would almost certainly incur some sort of weight disadvantage.

A final, important element of the study of handling load transfer is one that allows a single, unified treatment but rarely receives it. The notions of anti-dive/squat and 'roll centre' are rarely discussed together and yet their influence on vehicle handling loads is via the same mechanism; they lend themselves to the consistent rational treatment given here. The so-called 'roll centre' concept has to be one of the most nonintuitive and misunderstood modifiers of vehicle behaviour since empirically observed in racing circles in the 1920s. The understanding developed then is most applicable to beam axles and adds more difficulty than clarity for independent suspensions. The authors prefer the concept of 'sprung' and 'unsprung' loadpaths to the vehicle body.

Figure 4.11 shows a single wheel of an independently suspended vehicle viewed from the side. A braking load is applied at the contact patch, reflecting the fitment of outboard brakes, as is a common practice. Were the brakes inboard or were this a tractive case, the load would be applied at the wheel hub height and the diagrams would be redrawn appropriately. In this case the suspension is of a leading arm type. Since all independent suspensions may be represented as some form of equivalent length virtual swinging arm (although the length and pivot location vary with some types more than others) this remains a useful notion.

Three different orientations for the swinging arm are shown. The first has the swinging arm pivot on a direct line between the contact patch and the mass centre. Considering the wheel and arm together as a single entity and noting the ability of the pivot to support no moments, we may draw the reaction force at the pivot as being on a line between the contact patch and the pivot. The horizontal magnitude is the same as the applied lateral force at the wheel, giving a full solution for the force at the inboard pivot. The reaction on the sprung mass is equal and opposite to the force on the pivot, with a line of action passing directly through the mass centre. This is widely recognised as a 'no-dive' (no-pitch) type of suspension. Although there is no body pitch, this does not mean there is no load transfer between rear and front wheels. We may therefore conclude that the braking load is carried to the vehicle mass centre entirely through the suspension linkage

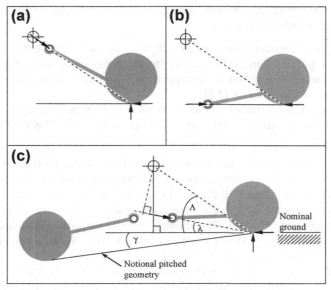

FIGURE 4.11

(a) No-dive suspension — pitch moment transfer solely via an unsprung loadpath.
(b) Pitch moment transfer solely via a sprung loadpath. (c) Typical arrangement; pitch
moment carried by a combination of sprung and unsprung loadpaths.

components and that none is carried in the suspension springs — i.e. via an 'unsprung' loadpath.

The second diagram has the swinging arm pivot at ground level. Using similar
logic as before, the force at the pivot may be drawn as purely lateral, equal and opposite to that at the wheel. This in turn means the horizontal force is applied to the body
at ground level, giving a pitch moment. That pitch moment cannot be reacted until
the suspension has deformed sufficiently to give an equal and opposite moment on
the sprung mass. In this case, the load transfer between rear and front axles is performed entirely by the suspension springing and none is carried in the suspension
linkage components — i.e. via a 'sprung' loadpath.

The third diagram shows a more typical situation with some of the pitch moment
carried by an unsprung loadpath and most carried by a sprung loadpath. Some fraction that is a function of the two angles λ and Λ may be calculated and expressed as
an 'anti-dive' fraction or percentage, or alternatively the anti-pitch angle λ may be
quoted separately. The authors prefer:

$$\text{Anti-dive \%} = 100\,(\lambda+\gamma)/(\Lambda+\gamma) \qquad (4.38)$$

Other texts give differing descriptions and definitions. What matters is not the
definition, although it is important to be certain how the quantities in use are defined

if they are to be compared one with another, but the significance of the sprung and unsprung load transfers themselves:

- Unsprung load transfer occurs via the stiff metallic elements in the system and is thus very rapid. It is limited in speed by the frequency of the wheel hop mode, a mode of vibration in which the unsprung mass oscillates on the tyre stiffness somewhere of the order of 15 Hz.
- Sprung load transfer occurs via the elastic elements of the system and is limited in speed by the frequency of the primary suspension mode. This is of the order of 1.5 Hz.

It may be seen then that unsprung load transfer is some 10 times faster than sprung load transfer. Herein lies the key to understanding some of the most important effects of the so-called 'roll centre'. Figure 4.12 is very similar to Figure 4.11 except that it shows the vehicle from the front instead of the side. Otherwise, the diagrams are identical. Figure 4.12(a) shows a 'no-roll' suspension with load transfer entirely by an unsprung loadpath whereas Figure 4.12(b) shows a suspension that transmits load entirely via a sprung loadpath.

The point frequently but ambiguously referred to as the 'roll centre' is where the line of action of the unsprung loadpath crosses the vehicle centre line. As with the

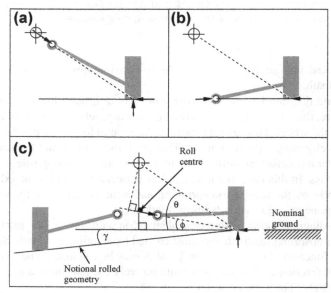

FIGURE 4.12

(a) No-roll suspension — roll moment transfer solely via an unsprung loadpath. (b) Roll moment transfer solely via a sprung loadpath. (c) Typical arrangement; roll moment carried by a combination of sprung and unsprung loadpaths.

anti-pitch behaviour, the absolute height is of less importance than the distribution of loads between sprung and unsprung loadpaths. It is not any kind of centre of motion.

Again as with the anti-pitch behaviour, some fraction that is a function of the two angles ϕ and θ may be calculated and expressed as an 'anti-roll' fraction or percentage, or alternatively the anti-roll angle, ϕ, may be quoted separately. Alternatively, an anti-roll fraction or percentage could be quoted based on the fraction of the 'roll centre height' compared to the mass centre height. The authors prefer to use a ratio of the two angles ϕ and θ to express the anti-roll fraction similarly to before:

$$\text{Anti-roll \%} = 100 \ (\phi+\gamma)/(\theta+\gamma) \qquad (4.39)$$

For lateral handling loads, the same ideas of relative speed between unsprung and sprung loadpaths apply. This has a particular importance when considered in the light of the phasing of front and rear axle forces in order to manipulate the yaw moments on the body. For a vehicle in yaw, the rate of load transfer may thus be set differently at different ends of the vehicle in order to modify the transient behaviour as compared to the steady state behaviour. For example, it is typical for vehicles to run around 20% rear anti-roll and only around 6% front anti-roll. This means that as a manoeuvre develops, load transfer from the outside rear tyre may briefly outpace load transfer from the front tyre. The resulting yaw moment acts to stabilise the vehicle and mitigates sudden, aggressive steer inputs. Vehicles that do not have this type of geometry, notably those with trailing arm rear suspensions, are unable to benefit from these effects.

For both longitudinal and lateral load transfers there is no conceptual reason why either the anti-pitch angle or anti-roll angle may not be negative. Motorcycles, for example, have a negative anti-pitch angle equal to the steer axis rake when they are fitted with conventional telescopic forks. This has the disadvantage of requiring extra performance from the suspension springs since they must carry more than the straightforward load transfer one might instinctively expect. The performance cost of this is discussed in Sections 4.9.2 and 4.9.3.

MBS analysis allows both an understanding of the load transfers in a rig-based environment, such as may be measured on the MIRA Kinematics & Compliance rig (Whitehead, 1995) and also during real driving manoeuvres. In both situations, the ability of an MBS model to retrieve forces in each suspension member in convenient frames of reference while working with quarter, half or full vehicle models is a powerful tool to unscramble some of these less-than-intuitive effects with vehicle designs.

4.1.4 Compliant wheel plane control

Hand-in-glove with an understanding of the load transmission paths and time delays associated with the activity manoeuvring the vehicle in the ground plane comes an understanding of the resulting motion of the wheel plane with respect to the ground. From the treatment of tyres that follows in Chapter 5, and as described briefly in the introductory chapter, it is clear that the angles at which the tyres are presented to the road are of crucial importance in modifying the forces generated by the tyres and hence the resulting motion of the vehicle.

There are some subtle and intricate effects present in real vehicle systems that defy simplistic comprehension and evaluation. For example, the deformation of anti-roll bars re-orients the links with which they are connected to the moving suspension members and may introduce forces that 'steer' the wheel plane with respect to the vehicle. This may not have been considered at the time the suspension was schemed conceptually but yet may modify the behaviour of the vehicle in practice.

MBS analysis allows the construction of rig-like models for such wheel plane compliant behaviour before prototype vehicles exist. It also allows a systematic and well-controlled comparison of behaviour with different levels of compliance in order to establish the influences of the different aspects of wheel plane compliance on vehicle behaviour. Typically such studies carry across several revisions of a model within a market segment and are thus of some strategic benefit.

4.1.5 Kinematic wheel plane control

Suspension arrangements typically consist of some connection of linkages to a device for holding the bearings in which the wheel actually turns. The interaction of those individual elements comprising the linkage means that the wheel plane typically undergoes some sort of translation and rotation as the suspension articulates.

This motion of the wheel plane is traditionally defined with respect to the vehicle body, although for a moving vehicle it is the angles and velocities with respect to the road that are of import. However, using the vehicle body gives a convenient frame of reference and so the descriptions following will do so:

1. Toe change (steer) with suspension articulation gives a lateral force and yaw moment by directly adding to or subtracting from the tyre slip angle and is readily understood. Some care is needed when discussing this characteristic since there are a number of possibilities for definition; steer may be defined as a right hand positive rotation about the vehicle vertical axis or it may be defined as a handed rotation, different on left and right sides of the vehicle ('toe-out') or even as a term relating the behaviour to the yaw moment it induces on the vehicle ('roll understeer').
2. Camber change[1] acts to mitigate the angle with which the tyre is presented to the road due to the roll of the body with respect to road surface. Typical passenger vehicles roll 'out' of turns, with the inside edge of the vehicle platform being further away from the road than the outside. Were the wheels to remain perpendicular to the vehicle platform at all times, they would be presented to the road with a camber/inclination angle equal to the vehicle roll angle. For most independent suspension types, the wheels camber somewhat with respect to the

---

[1] It might be more sensibly named 'inclination change' but the term camber change is nevertheless in popular use.

body such that they are presented to the road at something less than the vehicle roll angle. This is referred to as 'camber compensation'. Were the wheels to remain upright with respect to the road, this would be 100% or 'full' camber compensation. As the wheel is presented to the road progressively more and more upright, the tyre is loaded more and more evenly across its width and lasts longer. This is of primary concern for competition vehicles. Camber/inclination angles also generate forces in their own right and the tendency of a tyre leaned 'out' of the turn is to reduce the cornering forces generated by tyre slip angles; therefore by balancing camber compensations front to rear some influence may be exerted on the overall handling balance of the vehicle. Such treatment is very evident on, for example, the Bugatti Type 35.

3. It is typical for independent suspensions to move the contact patch of the tyre laterally as they articulate (Figure 4.14). Although less intuitively direct than the toe change mechanism, half-track change (lateral displacement) influences the lateral velocity of the tyre contact patch via roll rate. The slip angle of the tyre is affected; since the angle is defined as the arctangent of lateral and longitudinal velocities, an increase in lateral velocity directly increases slip angle. If the track change is plotted against bump travel, with both measured at the contact patch, a direct indication of the anti-roll angle is obtained with no need for knowledge of construction methods for any particular type of suspension.

Some practitioners attempt to calculate combined measures for both suspensions on the same axle or indeed all the suspensions on the vehicle. For beam axles there is some logic in combining the characteristics since the wheels are physically joined but for independent suspensions, calculating some combined metric is of questionable value. For example, attempting to combine anti-roll angles across one axle in a

FIGURE 4.13

Vehicle cornering pose.

Courtesy of Prodrive.

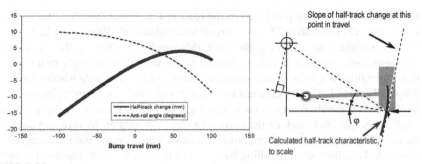

FIGURE 4.14

Relationship between calculated half-track change and anti-roll angle.

purely geometric manner, when their relative importance is determined by wheel loading is clearly nonsensical, as shown in Figure 4.13. If the wheel is in the air, how can it contribute the motion of the vehicle?

Equally, there is often a desire to calculate some of these measures, particularly camber values and anti-roll elements, with respect to the ground. Unfortunately the location and orientation of the ground cannot reliably be determined using only a quarter vehicle model. Nonlinear force–deflection characteristics are typical for the suspension elements and so the frequently used 'symmetric roll' presumption is often flawed. The amount of roll generated for a particular lateral loading varies with suspension calibration and the amount of roll moment carried on a particular axle varies with suspension calibration. Thus the boundary conditions cannot be known for a quarter vehicle model with any useful degree of certainty except for symmetric events. It is strongly suggested that the simplest interpretations of the vehicle kinematic measures be used and comparisons drawn between these simple measures. For comprehending the effects on a full vehicle, a full vehicle model is recommended.

4.1.6 Component loading environment

MBS models of the quarter vehicle type are often used to distribute design loads through the different suspension members with a view to sizing them intelligently in the first instance. A distinction needs to be drawn between design loads and service loads when discussing MBS analyses for this purpose. Design loads are calculated by postulating notional extremes for possible loads that may be induced in reality. An example of this might be the case where a car has been parked between two kerbs, between which it is a snug fit, and the driver then applies maximum effort to the handwheel. The exact philosophy for the selection and implementation of design loads is typically historical; they have been empirically defined and honed over many years of development experience. In practice they often correspond to events described as 'extreme service loads' or 'abuse loads'. They are events that the vehicle must survive but may need some attention immediately following it; in the steering example given, the steering alignment may be distorted by events but the vehicle would probably be required to remain capable of being driven. In industry many companies have standard cases such as the 3G bump case where the static wheel load is factored up to represent dynamic abuse situations such as striking a road hump at speed. The specification of abuse loads can be traced back to a publication in the Automobile Engineer by Garrett (1953) where a range of recommended wheel loads was proposed for the design of vehicles of that period. Design loads are frequently viewed as 'one-off' events in the life of the vehicle.

In contrast to design loads, there is another category of loads to which the vehicle is exposed. These are the service loads. This is the loading environment to which the vehicle is subjected during its durability sign-off testing. Durability sign-off criteria vary widely but they typically consist of a specified number of repetitions of different events at prescribed speeds and loading conditions. They induce a large number of repetitions of events that are probably more commonplace in the life of a vehicle — driving up kerbs, for example. In contrast to the design events, the vehicle is expected to emerge largely undamaged from the durability sign-off procedure — although no expectation of remaining service life is usually associated with durability sign-off. The relationship between durability sign-off criteria and actual usage of the vehicle is another question entirely; durability sign-off procedures themselves are often compiled in the light of historical warranty costs and other such influences.

There is, then, a key difference between design and service events. The design events are slightly fictitious and are usually analysed using a static or quasi-static procedure. While of questionable 'accuracy' they are of tremendous value. They may be calculated with the scarcest data about the vehicle and allow the early intelligent sizing of many different components, once applied to the wheel of a quarter vehicle model and distributed about a conceptual suspension layout. The resulting loading environment for each component may be used by designers to sketch a meaningful first-sight design. This can avoid packaging difficulties induced by ill-informed sketching of components and the subsequent 'stealing' of space needed for section modulus in key components.

Service loads are altogether more detailed. They are typically well-defined events where road profiles and speeds are well known. Once a reasonably representative geometry and suspension setting data set is available, the description of events may be combined with some form of enveloping tyre model in order to predict load-time histories for each component in the suspension. This in turn may be passed to finite element (FE)-based fatigue calculation methods in order to evaluate the number of times a component may be subject to this event before failure. Hypotheses such as the Linear Damage Accumulation Hypothesis (Miner's Rule) may be used to sum up the contribution of different events to the life of the component. Interested readers are referred to the *SAE Fatigue Design Handbook* (Rice, 1997) for more information on component fatigue analysis.

With the calculation of service loads, instead of a single set of peak loads as produced from the design loads there is a vast amount of time-domain data. Typical data rates for service load events are around 200 points per second and service load events last of the order of 10 s. With a repertoire of up to 20 events describing a typical durability sign-off, it may be seen that the calculation of service loads results in an increase in the quantity of results data of the order of five orders of magnitude. Calculation and subsequent processing times rise too, though not quite by the same amount, since linear static FE calculation times are a significant proportion of the overall elapsed time following design load calculations. However, as an overall process time amplification the expectation should be between two and three orders of magnitude for the calculation and use of service loads as compared to design loads. For some simple components, the use of a previously verified and correlated rig test may prove less onerous to the organisation.

4.2 Types of suspension system

There are various suspension systems used on cars and trucks and as described in Chapter 3, single-purpose versions of general purpose programs provide templates with preprogrammed configurations of suspension systems commonly used by automotive manufacturers. Many established textbooks on vehicle dynamics and some that focus on suspensions provide a detailed treatment of the various types of suspension system and their function. The thorough treatment and discussion contained in 'Road Vehicle Suspensions' (Matchinsky, 1998) is particularly commended. The coverage here will be to briefly mention some of the most common systems and to then direct our attention to the MBS modelling and simulation environment. The double wishbone and McPherson strut systems are very common and will provide the basis for the ensuing discussion.

Most of the 'single-purpose general-purpose' program libraries contain a number of different suspension types and are typically open enough to allow the user to add new types (Figure 4.15).

As stated the double wishbone suspension system will form the basis of much of the following discussion in this chapter. For readers new to the subject area and unfamiliar with the system the main components are indicated in Figure 4.16.

Figure 4.16 shows the main components the modelling of which will be dependent on the type of analysis to be performed and the outputs that must be produced. If the model is to be used only for the prediction of suspension characteristics such as camber angle or half-track change (HTC) with bump movement (BM) then an accurate definition of the mass and inertial properties of the rigid bodies will not be required. This information would be required however if the model is to be used for a dynamic analysis predicting the response of the suspension to inputs at the tyre contact patch.

The modelling of the connections between the suspension links will also depend on the type of vehicle and whether the suspension is for the front or rear of the vehicle. On

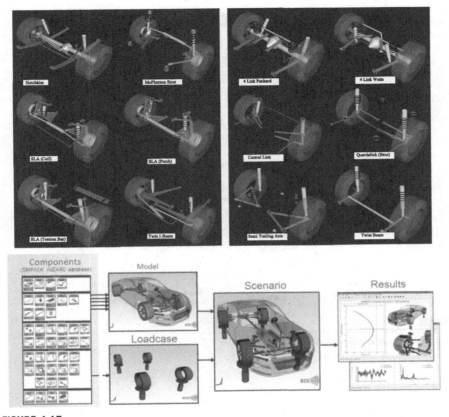

FIGURE 4.15

Graphical representation of suspension configurations in ADAMS/Car and the SIMPACK vehicle wizard.

Provided courtesy of MSC software and Simpack.

the front of the vehicle the connections between the control arms and the wheel knuckle would be modelled using spherical joints to represent the ball joints used here. On the rear these connections are more likely to include the compliance effects of rubber bushes. On a racing car where ride comfort is not an issue the suspension model is likely to be rigidly jointed throughout. Not shown in Figure 4.16 are the bump and rebound stops that would need to be included when considering situations where the wheel loads are severe enough to generate contact with these force elements.

The other type of suspension system that is very common on road vehicles is the McPherson strut system as illustrated in Figure 4.17. The main difference between this system and the double wishbone system is the absence of an upper control arm and the combination of the spring and the damper into a single main strut the body of which is the major component in the system.

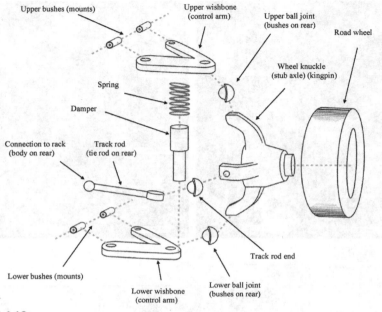

FIGURE 4.16

Double wishbone suspension system.

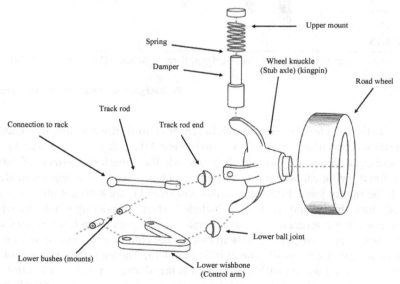

FIGURE 4.17

McPherson Strut suspension system.

4.3 **Quarter vehicle modelling approaches**

One of the earliest documented applications of the MSC ADAMS program by the automotive industry (Orlandea and Chase, 1977) was the use of the software to analyse suspension geometry. This approach is now well established and will be discussed further in the next section of this chapter. The output from this type of analysis is mainly geometric and allows results such as camber angle or roll centre position to be plotted graphically against vertical wheel movement.

The inclusion of bush compliance in the model at this stage will depend on whether the bushes have significant influence on geometric changes in the suspension and road wheel as the wheel moves vertically relative to the vehicle body. With the development of multilink type suspensions, such as the rear suspension on the Mercedes Model W201 (von der Ohe, 1983), it would appear difficult to develop a model of the linkages that did not include the compliance in the bushes. This type of suspension was used as a benchmark during the IAVSD exercise (Kortum and Sharp, 1991) mentioned in Chapter 1 comparing the application of MBS analysis programs in vehicle dynamics.

This modelling issue is best explained by an example using the established double wishbone suspension system. The modelling of the suspension using bushes to connect the upper and lower arms to the vehicle body is shown in Figure 4.18. Vertical motion is imparted to the suspension using a jack upper connected to the jack lower by a translational joint. A translational motion is applied at this joint

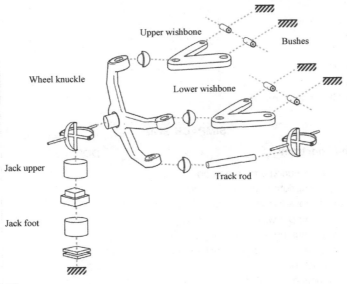

FIGURE 4.18

Double wishbone suspension modelled with bushes.

to move the jack over a range of vertical movement equivalent to moving between the full bump and full rebound positions. Although the jack upper is shown below the upright in Figure 4.18, it is connected to the wheel using a universal joint at the wheel centre (WC); this also illustrates the difficulty in the software being unable to use exploded diagrams referred to previously. Note the orientation of the universal joint is entirely theoretical since no practicably realisable universal joint will work with an angle at or close to 90°. This constraint is necessary to prevent a spin degree of freedom for the jack.

The planar joint constrains the wheel centre or wheel footprint to remain in the plane at the foot of the jack while allowing it to move in the lateral or longitudinal directions while the presence of the universal joint does not constrain the wheel to change orientation. The wheel itself is not explicitly represented since its spin degree of freedom is of no consequence to this model. For the suspension modelled in this manner it is possible to calculate the degrees of freedom for the system as shown in Table 4.2.

Table 4.2 Degree of Freedom (DOF) Calculation for Suspension System with Bushes

ADAMS-Style			
Component	**Number**	**DOF**	**$\sum$ DOF**
Parts	6	6	36
Translationals	1	−5	−5
Universals	2	−4	−8
Sphericals	3	−3	−9
Planars	1	−3	−3
Motions	1	−1	−1

$\sum$ DOF for system = 10

SIMPACK Style	
Joint/Constraint	**DOF**
Universal – steering arm to ground	2
Spherical – upright from steering arm	3
Spherical – lower wishbone from upright	3
Spherical – upper wishbone from upright	3
Universal – jack upper from upright	2
Translational – jack foot from jack upper	1
Planar constraint	−3
Applied motion	−1
Total	10

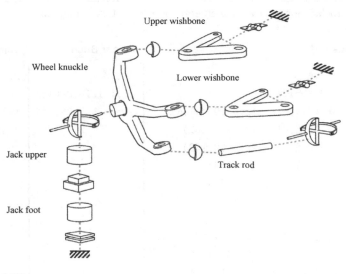

Upper wishbone

Wheel knuckle

Lower wishbone

Jack upper

Track rod

Jack foot

FIGURE 4.19

Double sishbone suspension modelled with joints.

The double wishbone suspension model shown in Figure 4.18 can be simplified to represent the bushes connecting the upper arm and the lower arm to the vehicle body by revolute joints as shown in Figure 4.19.

It can be seen that this system is identical to the previous one, save for the deduction of 10 degrees of freedom, five for each revolute constraint, leaving the system with zero degrees of freedom — a kinematic model — and allows a kinematic analysis to be performed. The fact that at least one of the degrees of freedom constrained in this model is due to a time-dependent motion, input at a joint, means that the model will move and operate as a mechanism rather than 'lock' as a structure.

This model is a slightly different scheme to that presented in the first edition but there are a number of advantages to the slightly more complex layout. When attached at each corner of a full vehicle model, the motion can be removed and replaced with a description of the tyre stiffness and damping to allow an eigensolution of the model to check for expected behaviour before adding a running tyre model. An eigensolution is always recommended to check that the practitioner has modelled what they *think* they have modelled, which with restrictive graphical interfaces is by no means certain. Any worthwhile model audit will contain a stationary eigensolution of the full vehicle less the running tyre model; the addition of compliance is plainly seen to add a large number of high-frequency modes of vibration and thus the penalty in run time is made clear to the user when including compliance. The highest frequency of vibration provides valuable guidance as to the minimum time step that may need to be employed.

Table 4.3 ADAMS Solver-Level Statements for a Joint, Linear Bush and Nonlinear Bush

Nonlinear Bush	Linear Bush	Joint
GFORCE/16,I=1216	BUSH/	JO/
,FLOAT=011600,RM=1216	16,I=1216,J=0116	16,REV,
,FX=CUBSPL(DX(1216,0116,1216),0,161)\	,K=7825,7825,944	I=1216,
,FY=CUBSPL(DY(1216,0116,1216),0,161)\		J=0116
,FZ=CUBSPL(DZ(1216,0116,1216),0,162)\		
,TX=CUBSPL(AX(1216,0116),0,163)\		
,TY=CUBSPL(AY(1216,0116),0,163)\		
,TZ=0.0		
SPLINE/161		
,X=-1.8, -1.5, -1.4, -1.22, -1.123,		
-1.0, -0.75, -0.5, -.25, 0, 0.25, 0.5,		
0.75, 1.0, 1.123, 1.22, 1.4, 1.5, 1.8		
,Y=15350, 10850, 9840, 6716, 5910,		
5059, 3761, 2507, 1253, 0,-1253,		
-2507, -3761, -5059, -5910, -6716,		
-9840, -10850, -15350		
SPLINE/162		
,X=-5, -4, -3, -2.91, -2.75, -2.5,		
-2, -1.5, -1, -0.5, 0, 0.5, 1, 1.5,		
2, 2.5, 2.75, 2.91, 3, 4, 5		
,Y=7925, 3925, 1925, 1790, 1626,		
1450, 1136, 830, 552, 276, 0, -276,		
-552, -830, -1136, -1450, -1626,		
-1790, -1925, -3925, -7925		
SPLINE/163,		
,X=-0.22682, -0.20939, -0.19196,		
-0.17453, -0.1571, -0.13963, -0.10472,		
-0.06981, -03491, 0, 0.03491, 0.06981,		
0.10472, 0.13963, 0.1571, 0.17453,		
0.19196, 0.20939, 0.22682		
,Y=241940, 198364, 160018, 125158, 93387,		
75415, 52951, 35702, 18453, 0, -18453,		
-35702, -52951, -75415, -93387, -125158,		
-160018, -198364, -241940		

With a fixed constraint applied to the vehicle body, the motion can be reinstated and static equilibrium runs performed to retrieve wheel force versus displacement characteristics including bump stops and the influence of bush compliance, which can sometimes be unexpected with, for example, large lateral loads on the lower wishbone when the bump stop is heavily engaged, leading to unexpected steering behaviour.

Remaining with the fixed constraint to the body and removing the applied motion, design loads can be applied to the vehicle corners for load distribution analysis directly, with no need to modify the model further.

The decision to model the bushes will have a significant impact on the collation of data and the effort required to input and check the values. This is illustrated in Table 4.3 where the typical solver-level inputs required to model a connection as a rigid joint, linear bush and nonlinear bush are compared. The fact that modern software presents the user with graphical dialogue boxes to enter the information does not change the fundamental escalation of the required amount of data for the model including compliance.

There are three main types of analysis for which individual suspension models are likely to be used during the design and development of a vehicle. The quantity and type of data will vary from vehicle to vehicle and the type of analysis to be performed. A summary of the typical modelling data is provided for guidance in Table 4.4. Note that for some suspensions, such as a twist beam type, the modelling of structural compliance will also need to be included.

Before leaving the subject of quarter vehicle suspension models it is worth considering the implementation of suspension connections as used in some race cars. The advent of inter-university competitions where students design, build and race a vehicle has become popular on automotive courses in recent years. An example of such a vehicle is a Formula Student car built by students at Coventry University shown in Figure 4.20.

The modelling of part of the suspension system is shown in Figure 4.21. The push-rod connects the suspension arm to the bell crank and uses a spherical joint at one end and a universal joint at the other end to constrain unwanted spin of the pushrod about its own axis. The bell crank is attached to the chassis with a revolute joint, the rotation about which is resisted by the springdamper unit. It can be noted that it is not really necessary to model the spring damper as rigid bodies. The definition of the forces generated will be sufficient to simulate the handling of the full vehicle.

4.4 Determination of suspension system characteristics

The suspension design process discussed at the beginning of this chapter has been summarised in Table 4.1 by six areas in which the suspension performance can be assessed. Ultimately the quality of the design will be judged on the performance of the full vehicle, but an early assessment of the suspension design as an individual unit is essential. In order to quantify the performance of the suspension system a

Table 4.4 Indicative Data Requirements for Individual Suspension Analyses
Kinematic Vertical Rebound to Bump Analysis
Coordinates of suspension linkage connections
Quasi-static Vertical Rebound to Bump Analysis
Coordinates of suspension linkage connections Bush stiffnesses (if this affects the movement) Spring and bump/rebound rubber characteristics
Static or Quasi-static Durability Analysis
Coordinates of suspension linkage connections Bush stiffnesses Spring and bump/rebound rubber characteristics Component flexibility (some suspensions)
Dynamic Durability or Vibration Analysis
Coordinates of suspension linkage connections Mass and inertial properties Bush stiffnesses Bush damping data Spring and bump/rebound rubber characteristics Damper properties Component flexibility (some suspensions)

range of characteristics may be determined through simulation of a single suspension system or quarter vehicle model. During this chapter it will be shown that a single suspension system can be analysed in a number of ways that will provide information to support the six suspension design activities that have been identified.

It should also be noted, however, that while the emphasis in this chapter is to explain the function and modelling of suspension systems using quarter models, so-called 'single-purpose general-purpose' programs such as ADAMS/Car and the SIMPACK Vehicle Wizard extend the modelling as stated to a half vehicle model analysed using a virtual test rig, an example of which is shown in Figure 4.22.

Such a system also allows the additional incorporation of rollbars and a steering system to investigate the cross coupling of left and right suspension systems through rollbar compliance and the effects of steering inputs in isolation or in combination with suspension movement.

A more detailed discussion of suspension analysis methods, such as those used to investigate body isolation issues, will follow later in this chapter but using a virtual rig such as this the following are typical of some of the analyses performed:

1. The wheels may be moved vertically relative to the vehicle body through a defined bump-rebound travel distance. For the half model shown in Figure 4.20

FIGURE 4.20

Coventry University Formula Student Car.

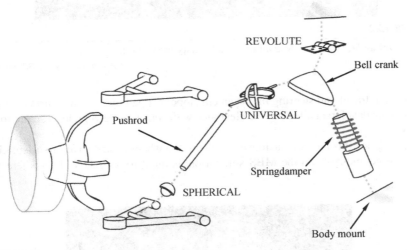

FIGURE 4.21

Modelling of PushRod and Bell Crank mechanism in student race car.

the vertical movement may involve single, opposite or parallel wheel travel representing ride or roll motions for the vehicle. The measured outputs allow the analyst to consider, depending on the model used, aspects of kinematic and compliant wheel plane control.

2. Lateral force and aligning torque may be applied at the tyre contact path allowing measurement for example of the resulting toe angle change and lateral deflection of the wheel (compliant wheel plane control).

In addition to the above basic types of analysis it is also possible to use an MBS suspension model to consider wheel envelopes where under the full range of

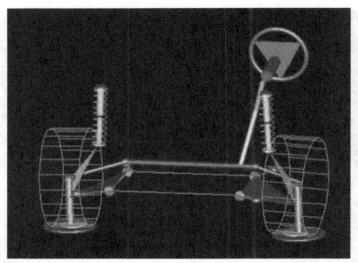

FIGURE 4.22

Use of virtual test rig to analyse a half vehicle suspension model.

Provided courtesy of MSC software.

suspension travel and steering inputs an envelope mapped by the outer surface of the tyre can be developed allowing the clearance with surrounding vehicle structure to be checked.

In practice this has been achieved by using the wheel centre position, orientation and tyre geometry from the MBS simulation as input to a CAD system where the

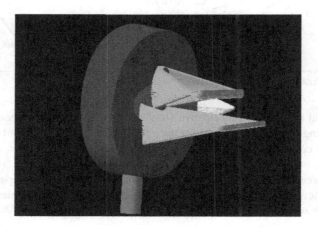

FIGURE 4.23

Superimposed animation frames giving visual indication of wheel envelope.

clearances can be checked. An example of the graphic visualization of a wheel envelope, for vertical wheel travel only, using superimposed animation frames is shown in Figure 4.23. Some CAD packages have an embedded kinematic solver for this purpose and generate the resulting surface within the CAD package directly.

4.5 Suspension calculations
4.5.1 Measured outputs

A glossary of terms providing a formal specification of various suspension characteristics has been provided by the Society of Automotive Engineers (SAE Publication, 1976). In the past, variations in formulations and terminology have been provided by researchers, authors and also practicing engineers following corporate methodologies. The concept of a roll centre has also been subject to a number of definitions (Dixon, 1987).

As discussed in Chapter 3, vehicle-specific programs offer a range of precomputed outputs for suspension characteristics. The user documentation provided with those software systems includes an extensive description of each output and need not be repeated here. For completeness those outputs considered to be most common in their usage and most relevant only to the following discussion in this textbook will be described in this chapter.

As discussed in the previous sections, one of the main uses of an MBS model of a suspension system is to establish during the design process geometric position and orientation as a function of vertical movement between the rebound and bump positions. As the output required does not include dynamic response it is suitable to use a kinematic or quasi-static analysis to simulate the motion. It should be noted that this information could also be obtained using a CAD package or a program developed solely for this purpose. Some CAD packages include an implementation of a spherical kinematic solver, which may come directly from a multibody software provider, and the authors have also seen dedicated Excel and Open Office spreadsheets used with a spherical kinematic formulation to the same end. The fact is that an MBS program used is often associated with the stages of model development described in Chapter 1 that lead through from the individual suspension model to a model of the full vehicle.

A large number of parameters can be measured on an existing suspension system and laboratory rigs such as the Kinematics and Compliance measurement facility (or K&C Rig) described by Whitehead (1995) have been developed specifically for this purpose. The descriptions provided here will be limited to the most commonly calculated outputs, which are:

• Bump motion (spindle rise)
• Wheel recession
• Half track change
• Steer (toe) angle

- Camber angle
- Castor angle
- Steer axis inclination
- Suspension trail
- Ground-level offset
- Wheel rate
- Roll centre height

In each case the parameters are presented as XY plots with the bump motion as the independent variable usually on the x-axis. The calculation of these outputs can be programmed to create a variable that is derived directly from measured system variables or is based on a trigonometric or algebraic derivation. It should be noted that parameters such as the camber angle determined here are measured relative to the vehicle body which is assumed to be fixed and should not be confused with the inclination or presentation angle measured between the tyre and the road surface discussed later in Chapter 5.

The vertical motion is imparted to the wheel using a body to represent a jack with the wheel centre constrained to remain in the plane of the jack using an inplane joint primitive as described in Chapter 3 and shown in Figure 4.24. The motion applied to the translational joint between the jack and the ground will be that needed to move the wheel between the rebound and bump positions. If the model is defined in a position midway between these it is convenient to define the motion as sinusoidal with a cycle of 1 s. This would provide results in the bump position at 0.25 s, the rebound position at 0.75 s and allow for presentation purposes smooth animation of continuous cycles. Running the analysis for 1 s with, say, 80 output steps would ensure that output is calculated in the full bump and rebound positions.

The motion statement applied to the translational joint that would accomplish this for a suspension where the total movement between the rebound and bump is 200 mm would be.

```
FUNCTION = 100*SIN(TIME*2*PI)
```

Note that the TIME variable is in seconds and is converted to radians within the function to represent one cycle over 1 s of simulation time. Some software allows the use of degrees directly within functions with the addition of the letter 'D' after the number but in general computer software works in radians with trigonometric functions.

It should also be noted that if the movement is not symmetric, that is to say that the distance moved in bump is different to that moved in rebound, a more complicated function will be needed for the motion input. In the following example the suspension is required to move 110 mm into the bump position and 90 mm into the rebound position. It is still desirable to have an overall sinusoidal motion for animation purposes and so an Arithmetic If is used in the function to switch the amplitude at the half cycle position of 0.5 s as follows:

```
FUNCTION = IF(TIME-0.5: 110*SIN(TIME*2*PI), 0.0, 90*SIN(TIME*2*PI))
```

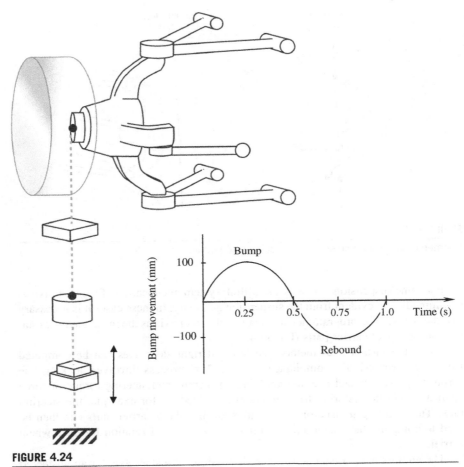

FIGURE 4.24

Input of vertical motion at the wheel centre.

4.5.2 Suspension steer axes

Suspension characteristics, such as castor angle, trail and the steering axis inclination, require an initial computation of the suspension system steer axis. Generally the concept of a steer axis is straightforward when considering, for example,
the double wishbone system described earlier. In such a case it is easy to see that
the wheel will steer about an axis passing through the lower and upper ball
joints.

For a McPherson strut suspension system, the steer axis may be defined to pass
through a point located at the lower ball joint and a point located where the upper
part of the strut is mounted to the vehicle body. Note that this line is not necessarily
parallel to the sliding axis of the upper part of the strut.

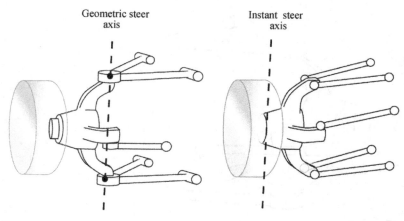

FIGURE 4.25

Geometric and Instant steer axes of a suspension system.

For some suspensions, such as a multilink system, the location of the steer axis is not immediately evident from the suspension geometry. In these cases it is necessary that the software be programmed to calculate the steer axis as the instant axis of rotation of the wheel carrier parts (Figure 4.25).

Using the instant axes method the left and right steer axes can be computed from the suspension's compliance matrix. The process involves locking the spring to prevent wheel rise and applying an incremental steering torque or force against an elastic restoring force temporarily added to, for example, the steering rack. The resulting translation and rotation of the wheel carrier parts can then be used to compute the instant axis, and hence steer axis, of rotation for each wheel carrier.

The authors have often encountered discussions of a steer axis for an unsteered axle but these appear something of a *non sequitur*. It is generally true that a lateral force applied at some point will generate a compound lateral, camber and steer (toe) motion and it is perfectly reasonable to combine the lateral and toe motion to calculate some elastic axis in the same manner as described previously, save for the fact that there is no need to introduce an artificial aligning force. However, some practitioners erroneously advocate arbitrarily joining some suspension hard-points and declaring the resulting line a steering axis; it appears to escape their notice that more than one arbitrary connection of points is generally possible and that in any case the true elastic axis is not generally aligned with suspension hard-points.

Note that the formulations of suspension output that follow are for a quarter vehicle suspension model located on the right side of the vehicle using the general vehicle coordinate system in this text with the x-axis pointing to the rear, the y-axis to the side and the z-axis upwards. Needless to say users must ensure the formulations are consistent with the vehicle coordinate system and the side of the vehicle being considered to ensure the correct sign for the calculated outputs. For each of the suspension characteristics discussed a typical system variable calculation is provided. This will assist users of MBS programs who need to develop their own calculations without access to the automated outputs of more elaborate software.

4.5.3 Bump movement, wheel recession and half track change

As stated earlier it can be the practice to impart vertical motion to a suspension system at either the wheel centre or base of wheel. In the following example the displacements at the wheel centre are used to determine the suspension movement. The displacements at the base of wheel would be corrected for camber, steer and castor angle changes and dependent on the suspension geometry. On the real vehicle the displacements of the tyre contact patch relative to the road wheel would also result due to the effects of tyre distortion. This is discussed later in Chapter 5.

Bump motion is the independent variable and is taken as positive as the wheel moves upwards in the positive z-direction relative to the vehicle body. Similarly wheel recession and half track change are taken as positive as the wheel moves back and outwards in the positive x- and y-directions respectively. The displacements are obtained simply by comparing the movement of a frame at the wheel centre relative to an initially coincident fixed frame on the ground. The displacements are shown in Figure 4.26.

4.5.4 Camber and steer angle

Camber angle, γ, is defined as the angle measured in the front elevation between the wheel plane and the vertical. Camber angle is measured in degrees and taken as positive if the top of the wheel leans outwards relative to the vehicle body as shown in Figure 4.27.

The steer or toe angle, δ, is defined as the angle measured in the top elevation between the longitudinal axis of the vehicle and the line of intersection of the wheel plane and road surface. Steer angle is taken here as positive if the front of the wheel toes in towards the vehicle.

Both camber and steer angle can be calculated using two frames located on the wheel spindle axis. In this case a frame is used at the wheel centre and another on the wheel spindle axis, taken in this example to be outboard of the wheel centre. The calculation of camber and steer angle is converted from radians to degrees by the factor $(180/\pi)$.

4.5.5 Castor angle and suspension trail

Castor angle, ϕ, is defined as the angle measured in the side elevation between the steering (kingpin) axis and the vertical. Castor angle is measured in degrees and taken as positive if the top of the steering axis leans towards the rear as shown in Figure 4.28. Castor is not to be confused with 'caster', which is a type of sugar for throwing (or 'casting') from a device specifically for that purpose (a 'caster').

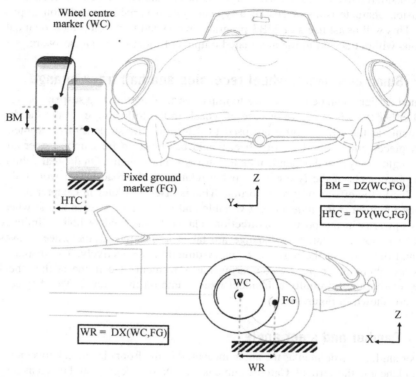

FIGURE 4.26

Bump movement (BM), wheel recession (WR) and half-track change (HTC).

Trail is the longitudinal distance in the x-direction between the wheelbase and the intersection between the steering axis and the ground. Trail generates a measure of stability providing a moment arm for lateral tyre forces that will cause the road wheels to 'centre'. Trail combines with tyre pneumatic trail, discussed in Chapter 5, and contributes to the steering 'feel'.

4.5.6 Steering axis inclination and Ground offset

The steering axis inclination, θ, is defined as the angle measured in the front elevation between the steering axis (sometimes referred to as the 'kingpin' axis from the days when a physical pin was used within a beam axle assembly) and the vertical. The angle is measured in degrees and taken as positive if the top of the steering axis leans inwards as shown in Figure 4.29.

Ground offset is the lateral distance in the y-direction between the wheelbase and the intersection between the steering axis and the ground. Ground offset is

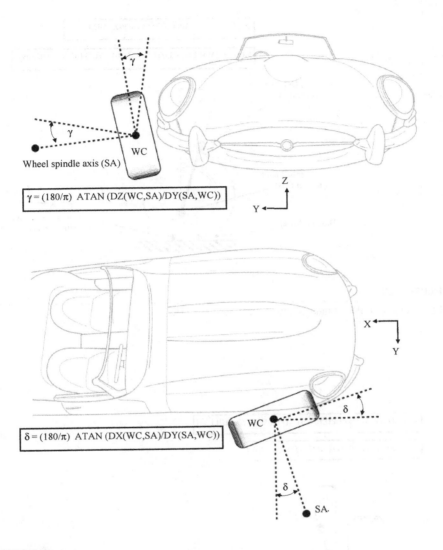

FIGURE 4.27

Calculation of Camber Angle γ and Steer Angle δ.

sometimes also referred to as the scrub radius as the amount of 'scrub' in the tyre as it steers will depend on the Ground offset. Note that US practitioners sometimes use a different definition of scrub radius, which is the hypotenuse of a right triangle drawn using ground offset and trail from the steer axis intercept with the ground to the contact patch centre.

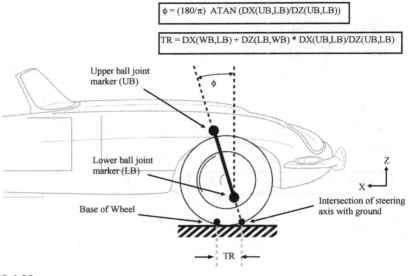

$$\phi = (180/\pi)\ \text{ATAN}\ (\text{DX(UB,LB)/DZ(UB,LB)})$$

$$\text{TR} = \text{DX(WB,LB)} + \text{DZ(LB,WB)} * \text{DX(UB,LB)/DZ(UB,LB)}$$

FIGURE 4.28

Calculation of Castor Angle φ and Suspension Trail (TR).

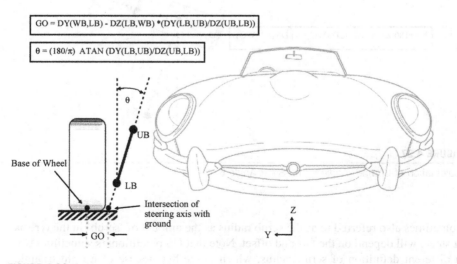

$$\text{GO} = \text{DY(WB,LB)} - \text{DZ(LB,WB)} *(\text{DY(LB,UB)/DZ(UB,LB)})$$

$$\theta = (180/\pi)\ \text{ATAN}\ (\text{DY(LB,UB)/DZ(UB,LB)})$$

FIGURE 4.29

Calculation of Steering Axis Inclination θ and Ground Offset (GO).

4.5.7 Instant centre and roll centre positions

The determination of the instant centre and the roll centre position is more complicated than the previous calculations described here and the following is included to demonstrate an approach that can be used with MBS software to establish these positions.

The following text illustrates the traditional graphical (kinematic) type of construction as described by Gillespie (1992). The approach used to program the computations will require the definition of algebraic equations that calculate the gradients and intersection points of the lines used in the construction. There are two methods that may be used to achieve this in the absence of a dedicated program:

1. Programming in a run-time variable (solution state variable)
2. Preparing, compiling and linking a user-written subroutine

The methods used to formulate the construction will be dependent on the type of suspension system being considered. Contrast this with the generality of the half track change gradient method discussed in Section 4.1.5. Examples are provided here for a double wishbone and a McPherson strut suspension system. For a double wishbone suspension the methods used to determine the instant centre and roll centre position for the front suspension are based on the construction shown in Figure 4.30.

The instant centre is found by intersecting two lines projected along the upper and lower arms. The instant centre is the instantaneous centre of rotation for the complete suspension system. The suspension system can be thought of as an equivalent swing arm that pivots about the instant centre. As such the instant centre is also sometimes referred to as the effective swing arm pivot. The roll centre is found by projecting a line between the base of the wheel and the instant centre. The point at which this line intersects the centre line of the vehicle is taken to be the roll centre. Note earlier caveats on the physical significance of the roll centre.

It should be noted that the two-dimensional representation shown in Figure 4.30 is a simplification of the three-dimensional system and the graphical construction that takes place in a YZ plane passing through the wheel centre WC as shown in Figure 4.31.

Since it cannot be assumed that the axes through the wishbone mount points are parallel to the x-axis the positions of points B and D will need to be obtained by

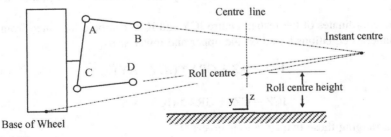

FIGURE 4.30

Instant centre and roll centre positions for a double wishbone suspension.

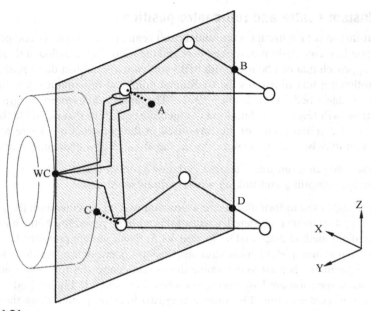

FIGURE 4.31

Position of instant centre construction points on wheel centre YZ plane.

interpolation to the wheel centre YZ plane. Positions A and C are found simply by projecting the upper and lower ball joints on to the same plane.

In order to program this construction algebraically the first step is to set up expressions for the gradients GR1 and GR2, of the upper and lower arms:

$$GR1 = (BZ-AZ) / (BY-AY) \qquad (4.40)$$

$$GR2 = (DZ-CZ) / (DY-CY) \qquad (4.41)$$

where

AY, AZ, BY, BZ, CY, CZ, DY, DZ are the y- and z-coordinates of points A, B, C and D.

The coordinates of the instant centre ICY and ICZ, can be established from two simultaneous equations based on the upper and lower arms:

$$ICZ = AZ + GR1 * (ICY - AY) \qquad (4.42)$$

$$ICZ = CZ + GR2 * (ICY - CY) \qquad (4.43)$$

Rearranging these two equations gives:

$$AZ + GR1 * ICY - GR1 * AY = CZ + GR2 * ICY - GR2 * CY \qquad (4.44)$$

$$ICY * (GR1 - GR2) = GR1 * AY - GR2 * CY + CZ - AZ \qquad (4.45)$$

which allows the instant centre to be located using:

$$ICY = (GR1 * AY - GR2 * CY + CZ - AZ) / (GR1 - GR2) \qquad (4.46)$$

$$ICZ = AZ + GR1 * (ICY - AY) \qquad (4.47)$$

The gradient of the line joining the wheelbase to the instant centre GR3, can be expressed as

$$GR3 = (ICZ-WBZ) / (ICY-WBY) \qquad (4.48)$$

where WBY and WBZ are the y- and z-coordinates of the wheelbase.

This allows the roll centre to be located using:

$$RCY = 0.0 \qquad (4.49)$$

$$RCZ = WBZ + GR3 * (RCY - WBY) \qquad (4.50)$$

The roll centre height (RCH), can now be defined by

$$RCH = RCZ - RZ \qquad (4.51)$$

where RZ is the z-coordinate of the road.

The methods used to determine the instant centre and roll centre position for a McPherson strut suspension are based on the construction shown in Figure 4.32.

The instant centre is found by intersecting lines projected along the transverse arm and perpendicular to the axis of the strut. The roll centre is found by projecting a line between the base of the wheel and the instant centre. The point at which this line intersects the centre line of the vehicle is taken to be the roll centre. All calculations are assumed to take place in the a YZ plane containing the wheel centre.

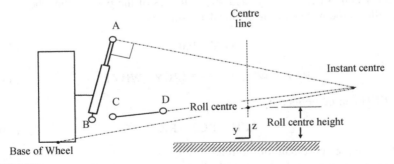

FIGURE 4.32

Instant centre and roll centre positions for a Mcpherson Strut suspension.

The first step is to again set up expressions for the gradients GR1 for the line perpendicular to the strut and GR2 for the line projected along the transverse arm:

$$GR1 = (BY-AY) / (AZ-BZ) \qquad (4.52)$$

$$GR2 = (DZ-CZ) / (DY-CY) \qquad (4.53)$$

where

AY, AZ, BY, BZ, CY, CZ, DY, DZ are the y- and z-coordinates of points A, B, C and D.

The coordinates of the instant centre ICY and ICZ can be established from two simultaneous equations based on the upper and lower arms:

$$ICZ = AZ + GR1 * (ICY - AY) \qquad (4.54)$$

$$ICZ = CZ + GR2 * (ICY - CY) \qquad (4.55)$$

Rearranging these two equations gives:

$$AZ + GR1 * ICY - GR1 * AY = CZ + GR2 * ICY - GR2 * CY \qquad (4.56)$$

$$ICY * (GR1 - GR2) = GR1 * AY - GR2 * CY + CZ - AZ \qquad (4.57)$$

This allows the instant centre to be located using:

$$ICY = (GR1 * AY - GR2 * CY + CZ - AZ) / (GR1 - GR2) \qquad (4.58)$$

$$ICZ = AZ + GR1 * (ICY - AY) \qquad (4.59)$$

The gradient of the line joining the base of the wheel to the instant centre GR3 can be expressed as

$$GR3 = (ICZ-WBZ) / (ICY-WBY) \qquad (4.60)$$

where WBY and WBZ are the y- and z-coordinates of the base of the wheel.

This allows the roll centre to be located using:

$$RCY = 0.0 \qquad (4.61)$$

$$RCZ = WBZ + GR3 * (RCY - WBY) \qquad (4.62)$$

The RCH can be defined by:

$$RCH = RCZ - RZ \qquad (4.63)$$

where RZ is the z-coordinate of the road.

As stated earlier, the calculation of the instant centre and roll centre position can be implemented either by programming a run-time variable directly or by preparing

Table 4.5 Calculation of Roll Centre Height Using a Run-Time Variable in a Solver-Level Set of Commands

```
VAR/14,IC=1,FU=DZ(1414,1411)/(DY(1414,1411)+1E-6)          ! GR1

VAR/15,IC=1,FU=DZ(1216,1213)/(DY(1216,1213)+1E-6)          ! GR2

VAR/16,IC=1,FU=((VARVAL(14)*DY(1411))                      ! ICY
,-(VARVAL(15)*DY(1213))+DZ(1213)
,-DZ(1411))/(VARVAL(14)-VARVAL(15)+1E-6)

VAR/17,FU=DZ(1411)+VARVAL(14)*(VARVAL(16)-DY(1411))        ! ICZ

VAR/18,FU=(VARVAL(17)-DZ(1029))/(VARVAL(16)-               ! GR3
DY(1029)+1E-6)

VAR/19,FU=DZ(1029)+VARVAL(18)*(0.0-DY(1029))              ! RCZ

VAR/20,FU=VARVAL(19)+152.6                                 ! RCH

REQ/1,F2=VARVAL(16)\F3=VARVAL(17)\F4=VARVAL(20)\
,TITLE=NULL:ICY:ICZ:RCH:NULL:NULL:NULL:NULL
```

a user-written subroutine. By way of example these methods are demonstrated for a front suspension system only. Using the run-time variable statement it is possible to program the equations laid out for the double wishbone system as shown in Table 4.5.

Variables such as BZ-AZ are defined using system variables which measure components of displacements between markers, such as DZ(1414,1411). The REQUEST statement REQ/1 demonstrates how to access the information calculated by the VAR statements.

The alternative method of writing a subroutine is demonstrated in Table 4.6 by the listing of a user written REQSUB developed specifically for a double wishbone suspension. The subroutine would be called from the main data set as follows:

```
REQUEST/id,FUNCTION=USER(1,par1,par2,par3,par4,par5,par6,par7,par8,
par9)
```

Where the parameters par1, par2,...par9 are the various items of data outlined in the subroutine.

4.5.8 Calculation of wheel rate

The wheel rate for a suspension system can be thought of as the stiffness of an 'equivalent' spring acting between the WC and the vehicle body as shown in Figure 4.33. This is the definition most useful for developing basic full vehicle MBS models where the wheel will be modelled as rigid with a separate tyre model.

Table 4.6 FORTRAN Subroutine to Calculate Roll Centre Height

```fortran
      SUBROUTINE REQSUB(ID,TIME,PAR,NPAR,
     , IFLAG,RESULT)
C M Blundell    Coventry University    Nov 1994
C Calculation of Roll Centre Height and Instant
C Centre Position -ROVER front suspension.
C
C Definition of Parameters:
C PAR(1)  Subroutine id. Must be 1
C PAR(2)  WC marker
C PAR(3)  WB marker
C PAR(4)  Marker at point A
C PAR(5)  Marker at point B
C PAR(6)  Marker at point C
C PAR(7)  Marker at point D
C PAR(8)  Radius of wheel
C PAR(9)  RZ Height of Road in global Z
C
C Results passed back to MSC.ADAMS as follows:
C Note that the AView does not use
C RESULT(1) or RESULT(5)
C
C RESULT(2) Roll Centre Height above ground
C RESULT(3) Roll Centre Z coordinate
C RESULT(6) ICY coordinate
C RESULT(7) ICZ coordinate
C
      IMPLICIT DOUBLE PRECISION (A-H,O-Z)
      DIMENSION PAR(*), RESULT(8)
      LOGICAL IFLAG
      DIMENSION DATA(6)
      LOGICAL ERRFLG
C
      IDWC=PAR(2)
      IDWB=PAR(3)
      IDA=PAR(4)
      IDB=PAR(5)
      IDC=PAR(6)
      IDD=PAR(7)
      RADIUS=PAR(7)
      RZ =PAR(8)
      CALL INFO ('DISP',IDWC,0,0,DATA,ERRFLG)
      CALL ERRMES(ERRFLG,'WC ID',ID,'STOP')
      WCX=DATA(1)
      WCY=DATA(2)
      WCZ=DATA(3)
      CALL INFO('DISP',IDWB,0,0,DATA,ERRFLG)
      CALL ERRMES(ERRFLG,'WB ID',ID,'STOP')
      WBY=DATA(2)
      WBZ=DATA(3)
      CALL INFO ('DISP',IDA,0,0,DATA,ERRFLG)
      CALL ERRMES(ERRFLG,'IDA',ID,'STOP')
      AY=DATA(2)
      AZ=DATA(3)
      CALL INFO ('DISP',IDB,0,0,DATA,ERRFLG)
      CALL ERRMES(ERRFLG,'IDB',ID,'STOP')
      BY=DATA(2)
      BZ=DATA(3)
      CALL INFO ('DISP',IDC,0,0,DATA,ERRFLG)
      CALL ERRMES(ERRFLG,'IDC',ID,'STOP')
      CY=DATA(2)
      CZ=DATA(3)
      CALL INFO ('DISP',IDD,0,0,DATA,ERRFLG)
      CALL ERRMES(ERRFLG,'IDD',ID,'STOP')
      DY=DATA(2)
      DZ=DATA(3)
      GR1=(BZ-AZ)/(BY-AY)
      GR2=(DZ-CZ)/(DY-CY)
      RICY=((GR1*AY)-(GR2*CY)+CZ-AZ))
     ./(GR1-GR2)
      RICZ=AZ+GR1*(RICY-AY)
      RCY=0.0
      GR3=(RICZ-WBZ)/(RICY-WBY)
      RCZ=WBZ+GR3*(RCY-WBY)
      RCH=RCZ-RZ
      RESULT(2)=RCH
      RESULT(3)=RCZ
      RESULT(6)=RICY
      RESULT(7)=RICZ
      RETURN
      END
```

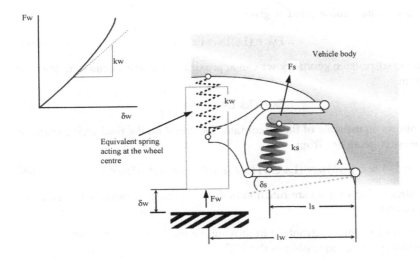

FIGURE 4.33

Equivalent spring acting at the wheel centre.

This differs slightly from other definitions sometimes used for wheel (or suspension rate) where the force displacement curve is measured at the centre of the tyre contact patch. In a quarter vehicle MBS model this would simply involve moving the point of jack contact with the wheel from the WC to the tyre contact patch.

The wheel rate should also not be confused with the term ride rate, which is associated with the force displacement relationship between the vehicle body, or sprung mass, and the ground. To derive ride rate with a quarter vehicle model it would be necessary to model an additional spring, representing the stiffness of the tyre, acting between the wheel centre and the jack with contact at the centre of the tyre contact patch. Note that the previously described quarter model implementation facilitates this readily.

The suspension outputs discussed until this point have been based on the suspension geometry and as such have not required the inclusion of the road spring in the model. By including the road spring and plotting the force against the displacement in the jack translational joint, the wheel rate may be obtained from the slope of the curve at the origin.

An estimate of the wheel rate may also be made as follows. Treating the road spring as linear gives the basic force displacement relationship

$$Fs = ks.\delta s \qquad (4.64)$$

For the equivalent spring we also have

$$Fw = kw.\delta w \qquad (4.65)$$

Taking moments about point A gives

$$Fw = (Ls/Lw) \, Fs \qquad (4.66)$$

From the suspension geometry we can approximate the displacement in the road spring from

$$\delta s = (Ls/Lw) \, \delta w \qquad (4.67)$$

This allows an estimate of the wheel rate, kw, based on the road spring stiffness and suspension geometry from

$$kw = Fw/\delta w = (Ls/Lw) \, Fs \, / \, (Lw/Ls) \, \delta s = (Ls/Lw)^2 \, ks \qquad (4.68)$$

The introduction of a square function in the ratio can be considered a combination of two effects:

1. The extra mechanical advantage in moving the road spring to the WC.
2. The extra spring compression at the WC.

4.6 The compliance matrix approach

The use of a compliance matrix, in programs such as the Milliken Research Vehicle Dynamic Modelling System (VDMS), is a method not commonly described in standard texts on vehicle dynamics but is well suited to an automated computer MBS analysis particularly when the influence of compliance requires consideration. The suspension compliance matrix relates incremental movements of the suspension to incremental forces applied at the wheel centres. The suspension compliance matrix is computed at each solution position as the suspension moves through its range of travel. Characteristics such as suspension ride rate and aligning torque camber compliance are computed based on the compliance matrix.

The compliance matrix for a suspension system, [C], is defined as the partial derivatives of displacements with respect to applied forces

$$[C] = [\partial \Delta / \partial F] \qquad (4.69)$$

If a system is assumed to be linear, the compliance matrix can be used to predict the system movement due to force inputs

$$\{\Delta\} = [C] \, \{F\} \qquad (4.70)$$

Expanding Eqn (4.62) leads to a 12×12 matrix relating the motion of the left and right wheel centres to unit forces and torques applied to the wheel centres. From this perspective, matrix element cij is the displacement of system degree of freedom i due to a unit force at degree of freedom j where the degrees of freedom are the three displacements Δx, Δy and Δz and the three rotations Ax, Ay and Az at each of the left and right WCs.

$$
\begin{bmatrix}
\Delta x_{LW} \\
\Delta y_{LW} \\
\Delta z_{LW} \\
Ax_{LW} \\
Ay_{LW} \\
Az_{LW} \\
\Delta x_{RW} \\
\Delta y_{RW} \\
\Delta z_{RW} \\
Ax_{RW} \\
Ay_{RW} \\
Az_{RW}
\end{bmatrix}
=
\begin{bmatrix}
C_{1,1} & C_{1,2} & C_{1,3} & C_{1,4} & C_{1,5} & C_{1,6} & C_{1,7} & C_{1,8} & C_{1,9} & C_{1,10} & C_{1,11} & C_{1,12} \\
C_{2,1} & C_{2,2} & C_{2,3} & C_{2,4} & C_{2,5} & C_{2,6} & C_{2,7} & C_{2,8} & C_{2,9} & C_{2,10} & C_{2,11} & C_{2,12} \\
C_{3,1} & C_{3,2} & C_{3,3} & C_{3,4} & C_{3,5} & C_{3,6} & C_{3,7} & C_{3,8} & C_{3,9} & C_{3,10} & C_{3,11} & C_{3,12} \\
C_{4,1} & C_{4,2} & C_{4,3} & C_{4,4} & C_{4,5} & C_{4,6} & C_{4,7} & C_{4,8} & C_{4,9} & C_{4,10} & C_{4,11} & C_{4,12} \\
C_{5,1} & C_{5,2} & C_{5,3} & C_{5,4} & C_{5,5} & C_{5,6} & C_{5,7} & C_{5,8} & C_{5,9} & C_{5,10} & C_{5,11} & C_{5,12} \\
C_{6,1} & C_{6,2} & C_{6,3} & C_{6,4} & C_{6,5} & C_{6,6} & C_{6,7} & C_{6,8} & C_{6,9} & C_{6,10} & C_{6,11} & C_{6,12} \\
C_{7,1} & C_{7,2} & C_{7,3} & C_{7,4} & C_{7,5} & C_{7,6} & C_{7,7} & C_{7,8} & C_{7,9} & C_{7,10} & C_{7,11} & C_{7,12} \\
C_{8,1} & C_{8,2} & C_{8,3} & C_{8,4} & C_{8,5} & C_{8,6} & C_{8,7} & C_{8,8} & C_{8,9} & C_{8,10} & C_{8,11} & C_{8,12} \\
C_{9,1} & C_{9,2} & C_{9,3} & C_{9,4} & C_{9,5} & C_{9,6} & C_{9,7} & C_{9,8} & C_{9,9} & C_{9,10} & C_{9,11} & C_{9,12} \\
C_{10,1} & C_{10,2} & C_{10,3} & C_{10,4} & C_{10,5} & C_{10,6} & C_{10,7} & C_{10,8} & C_{10,9} & C_{10,10} & C_{10,11} & C_{10,12} \\
C_{11,1} & C_{11,2} & C_{11,3} & C_{11,4} & C_{11,5} & C_{11,6} & C_{11,7} & C_{11,8} & C_{11,9} & C_{11,10} & C_{11,11} & C_{11,12} \\
C_{12,1} & C_{12,2} & C_{12,3} & C_{12,4} & C_{12,5} & C_{12,6} & C_{12,7} & C_{12,8} & C_{12,9} & C_{12,10} & C_{12,11} & C_{12,12}
\end{bmatrix}
\begin{bmatrix}
Fx_{LW} \\
Fy_{LW} \\
Fz_{LW} \\
Tx_{LW} \\
Ty_{LW} \\
Tz_{LW} \\
Fx_{RW} \\
Fy_{RW} \\
Fz_{RW} \\
Tx_{RW} \\
Ty_{RW} \\
Tz_{RW}
\end{bmatrix}
$$

$$(4.71)$$

From Eqn (4.71) it can be seen that the coefficients on the leading diagonal of matrix [C] directly relate the displacement or rotation to the associated force or torque applied at that degree of freedom. For example, in the absence of any other forces or torques, the vertical motion of the left wheel centre due to a unit vertical force applied at the left wheel centre is given by $\Delta z_{LW} = C_{3,3} \, Fz_{LW}$. Figure 4.34 illustrates this for another example where the vertical motion of the left wheel centre due to the application only of a unit vertical force applied at the right wheel centre

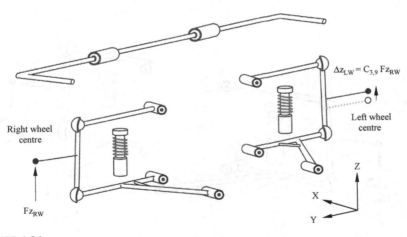

FIGURE 4.34

Application of compliance matrix to suspension system vehicle half model.

given by $\Delta z_{LW} = C_{3,9}\, Fz_{RW}$. From Figure 4.34 it can be seen that for an independent suspension without a rollbar $C_{3,9}$ would be zero in the absence of any mechanical coupling between the left and right suspension systems. The other elements of the compliance matrix are defined similarly.

As stated, the compliance matrix approach is well suited to investigate the effects of suspension movement due to compliance. By way of further example consider the definition used in ADAMS/Car for the calculation of aligning torque — steer and camber compliance.

The aligning torque steer compliance is the change in steer angle due to unit aligning torques applied through the wheel centres. Similarly the aligning torque camber compliance is the change in camber angle due to unit aligning torques acting through the wheel centres.

Figure 4.35 illustrates the determination of steer angle resulting at the right wheel due to unit aligning torques acting through both the left and right wheel centres. Note that the usual symbol for steer angle is δ. For the matrix approach used here however this is given by Az_{RW}. In this system a positive steer angle results when the wheel turns to the left, which in Figure 4.35 is consistent with a positive rotation Az_{RW} about the z-axis for the right wheel. In this case for the right wheel the steer angle would be given by $Az_{RW} = C_{12,6}\, Tz_{LW} + C_{12,12}\, Tz_{RW}$.

Similarly Figure 4.36 illustrates the determination of camber angle, Ax_{RW}, resulting at the right wheel due to unit aligning torques acting through both the left and right wheel centres.

In this system a positive camber angle results when the top of the wheel tilts away from the body, which in Figure 4.36 would actually be a negative rotation Ax_{RW}

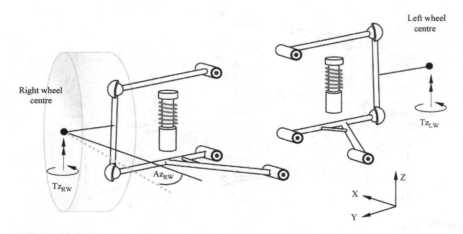

FIGURE 4.35

Steer angle at right wheel due to aligning torques at left and right wheels.

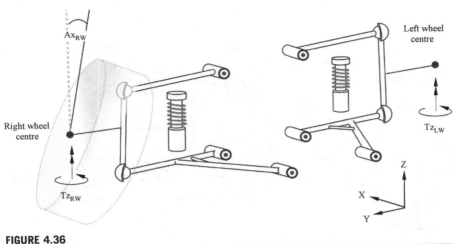

FIGURE 4.36

Camber angle at right wheel due to aligning torques at left and right wheels.

about the x-axis for the right wheel. In this case for the right wheel the camber angle would be given by $Ax_{RW} = C_{10,12} \, Tz_{RW} - C_{10,6} \, Tz_{LW}$.

The sign convention used to define positive steer and camber angles always requires careful consideration particularly when considering the definitions given here using a compliance matrix approach to measure movement of the road wheels relative to the vehicle body.

4.7 Case study 1 — suspension kinematics

The following case study is provided to illustrate the application of the method described in the previous sections to calculate the suspension characteristics as the suspension moves between the bump and rebound positions. Examples of the plotted outputs described here are shown in Figures 4.40–4.45. These plots were from a study based on the front suspension of a passenger car, considering the suspension connections to be joints, linear or nonlinear bushes. The assembly of parts used to make up the front suspension system is shown schematically in Figure 4.37. Example data sets for this model are provided in Appendix A together with more detailed system schematics.

The modelling of the suspension system using bushes is shown in Figure 4.38. The upper link is attached to the body using a connection that is rigid enough to be modelled as a revolute joint. Bushes are used to model the connection of the lower arm and the tie bar to the vehicle body.

Bushes are also used to model the connections at the top and bottom of the damper unit. Where the tie bar is bolted to the lower arm a fix joint has

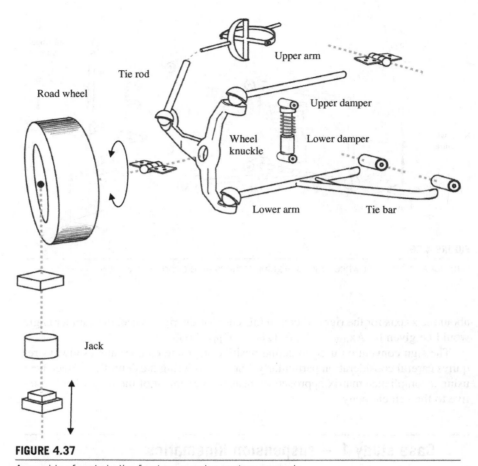

FIGURE 4.37

Assembly of parts in the front suspension system example.

been used to rigidly connect the two parts together. This joint removes all six relative degrees of freedom between the two parts, creating in effect a single lower control arm.

The modelling issue raised here is that rotation will take place about an axis through these two bushes where the bushes are not aligned with this axis. As rotation takes place the bushes must distort in order to accommodate this. The modelling of these connections as nonlinear, linear or as a rigid joint was therefore investigated to establish the effects on suspension geometry changes during vertical movement. For the suspension modelled in this manner it is possible to calculate the degrees of freedom for the system as follows:

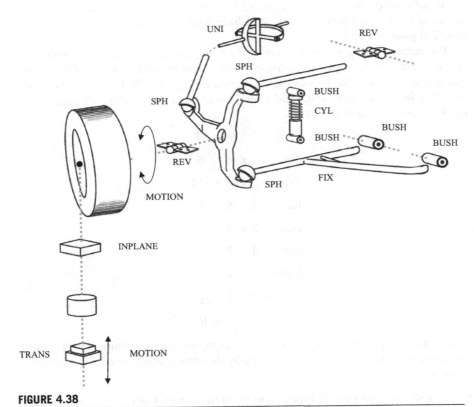

FIGURE 4.38

Modelling the front suspension example using bushes. SPH, Spherical Joint, REV, Revolute Joint; UNI, Universal Joint; CYL, Cylindrical Joint; FIX, Fix Joint; TRANS, Translational Joint.

Parts	9 x 6	= 54
Fix	1 x -6	= -6
Trans	1 x -5	= -5
Rev	2 x -5	= -10
Uni	1 x -4	= -4
Cyl	1 x -4	= -4
Sphs	3 x -3	= -9
Inplane	1 x -1	= -1
Motion	2 x -1	= -2
	Σ_{DOF}	= 13

In order to produce a zero degree of freedom model for this suspension the bushes at the top and bottom of the strut have been replaced by a universal and a spherical joint.

The bushes that were used to connect the lower arm and the tie rod assembly to the vehicle body were replaced in this study by a revolute joint. The axis of this joint was aligned between the two bushes as shown in Figure 4.39. For the suspension modelled in this manner using rigid joints it is possible to calculate the degrees of freedom for the system as follows:

$$
\begin{aligned}
\text{Parts} \quad & 9 \times 6 \;\; = 54 \\
\text{Fix} \quad & 1 \times \text{-}6 \;\; = \text{-}6 \\
\text{Trans} \quad & 2 \times \text{-}5 = \text{-}10 \\
\text{Rev} \quad & 3 \times \text{-}5 = \text{-}15 \\
\text{Uni} \quad & 2 \times \text{-}4 \;\; = \text{-}8 \\
\text{Sphs} \quad & 4 \times \text{-}3 = \text{-}12 \\
\text{Inplane} \; & 1 \times \text{-}1 \;\; = \text{-}1 \\
\underline{\text{Motion} \; 2 \times \text{-}1} \; & \underline{\;\; = \text{-}2} \\
\Sigma_{\text{DOF}} \; & = 0
\end{aligned}
$$

For this suspension it was possible to compare the simulation results with measured suspension rig test data provided by the vehicle manufacturer for the variation of:

- Camber angle (deg) with Bump motion (mm) (Figure 4.40)
- Caster angle (deg) with Bump motion (mm) (Figure 4.41)
- Steer angle (deg) with Bump motion (mm) (Figure 4.42)
- Roll centre height (mm) with Bump motion (mm) (Figure 4.43)
- Half track change (mm) with Bump motion (mm) (Figure 4.44)
- Vertical force (N) with Bump motion (mm) (Figure 4.45)

Examination of the results shown here indicates that despite the alignment of the bushes on the lower arm assembly the calculated suspension characteristics agree well for models using rigid joints, linear bushes or nonlinear bushes. It is noticeable with the front suspension that the plots begin to deviate when approaching the full bump or full rebound positions. This is due to contact with the bump stop or rebound stop generating forces that are then reacted back through the suspension to the bushes. The reaction forces at the bushes lead to distortions that produce the changes in suspension geometry as shown in the plots. This effect is not present in the models using rigid joints that have zero degrees of freedom. Geometry changes are entirely dependent on the position and orientation of the joints.

Considering the merits of each modelling approach it appears from the curves plotted that for the range of vertical movement expected of a handling model there

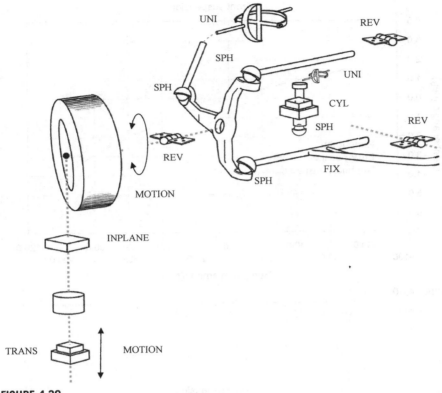

FIGURE 4.39

Modelling the front suspension example using rigid joints. SPH, Spherical Joint, REV, Revolute Joint; UNI, Universal Joint; CYL, Cylinder; FIX, Fix Joint; TRANS, Translational Joint.

is little difference between models using rigid joints, linear bushes or nonlinear bushes. The use of the nonlinear model will significantly increase the effort required to model the vehicle. This is evident from Table 4.4 which compares the data inputs required to model the connection of the front suspension lower arm to the vehicle body.

4.8 Durability studies (component loading)

4.8.1 Overview

Early in the life of computer-aided engineering (CAE) tools, MBS programs were often used to determine the loads acting on suspension components and the body pickup points as inputs to subsequent FE analysis of the components or vehicle structure. Current technology allows a rather more synchronous process to be adopted, notionally allowing embedded FE models to recover stress states and even fatigue damage estimates during a multibody solution run. However, in order for

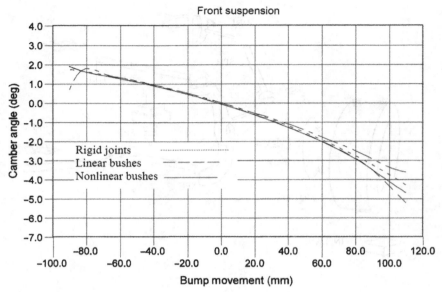

FIGURE 4.40

Front suspension — camber angle with bump movement.

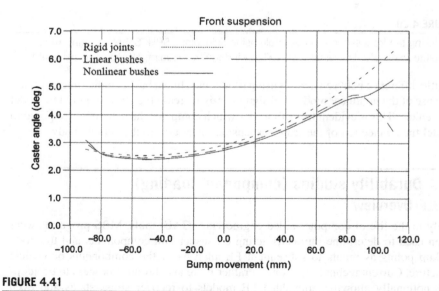

FIGURE 4.41

Front suspension — castor angle with bump movement.

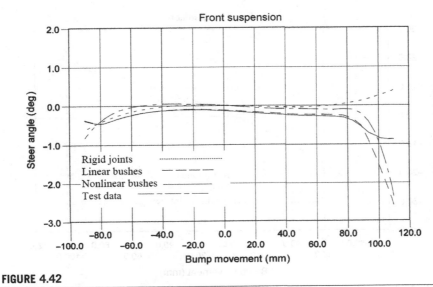

FIGURE 4.42

Front suspension — steer angle with bump movement.

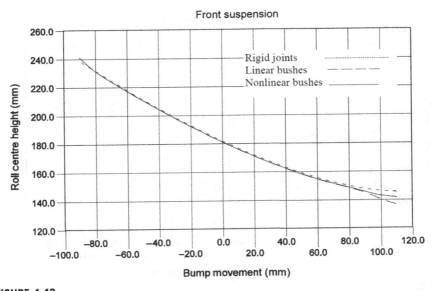

FIGURE 4.43

Front suspension — roll centre height with bump movement.

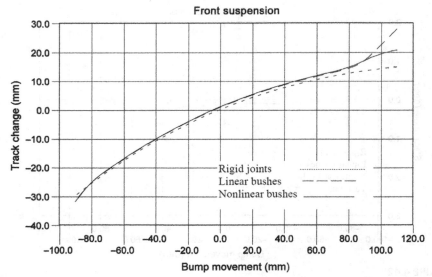

FIGURE 4.44

Front suspension — half-track change with bump movement.

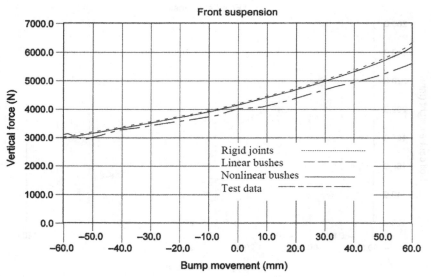

FIGURE 4.45

Front suspension — vertical force with bump movement.

this to be possible, viable models — and therefore designs — of all the components of interest need to exist. There remains a place for what is sometimes referred to as 'first-sight' analysis to usefully estimate applied loads in order to size components of mechanisms, particularly when engineering a class of vehicle that is new to the organisation.

Such simulations intend to reflect the outlying (most extreme) events during the series of tests that a vehicle manufacturer would perform on the proving ground to test the durability of the vehicle. During the test the vehicle will be exposed to a variety of loads in various directions, the entire sequence of which is expected to leave the vehicle undamaged. The durability test is formulated to reflect the loads the vehicle will see in service and so these durability test loads might be called 'service loads'. A substantial topic in its own right, the process of ensuring a durability test matches the actual loads imposed on the vehicle in service is a specialist discipline and not within the scope of this chapter.

Different manufacturers will implement their own procedures but typically these will involve establishing wheel centre loads resulting for example from accelerating, braking, cornering and driving on rough surfaces such as that shown in Figure 4.46. The limited set of loads that notionally represent the outlying values from the service load history might be called 'design loads'.

These loads are distinct from loads to which the vehicle might foreseeably be exposed to but from which it is not expected to emerge fully serviceable. Product liability legislation (and ethical vehicle engineering) requires that readily foreseeable events are managed in a safe way, even if the vehicle is damaged in the process. These events might be termed 'abuse' loads. In a mainstream automotive context, these include large, hard-edged potholes struck under braking at gross vehicle weight (GVW), jump events and so on. Note that in other contexts — for example rallying

FIGURE 4.46

Vehicle durability testing on a rough surface — the Mini WRC undergoes early development testing in February 2011. Picture: Harty.

or Baja racing — jumping may be regarded as a normal service load and not an abuse load.

It may be readily imagined that there is a somewhat blurred line between a large abuse event and a small crash. The large redistribution of loads that occurs when significant plastic deformation occurs is not typically suited to solution with MBS software, even when linear finite models are embedded and so there comes a point — usually determined by the amount of plasticity involved — when the analysis of such events is handed over to an explicitly formulated FE code. Such matters are beyond the scope of this chapter and so the subsequent discussion will confine itself to events where events are substantially elastic (i.e. deformations are recovered once loads are removed).

In rallying in particular, the neat definition of service and abuse loads is not really possible and so the somewhat fatalistic approach must be taken that 'there will always be a bigger rock'; the selection of design loads becomes somewhat arbitrary but over time, guidelines emerge that allow designs which are robust enough for the intended application without becoming overly heavy.

In Figure 4.47, note the total absence of the front left suspension and the complete but bent rear left suspension unit. The front suspension suffered a catastrophic overload when colliding with a large bank after the test driver, Markko Martin, lost control of the car; the subsequent rotation of the car gave a slightly less large impact at the rear, which was big enough to activate the mechanical 'fuse' — the rear toe link — without doing other damage to the suspension sub-frame. The author (Harty) observed all these events from the co-driver's seat and is grateful (thanks, Jonathan Culwick!) for the roll cage optimisation work that went into the design. Nine hours after this picture was taken, the vehicle ran again to continue testing.

The loads that are applied to the suspension may be considered to act at the tyre contact patch or at the wheel centre depending on the type of loadcase. The loads shown in Table 4.7 are typical of those that might be used for a static analysis on a vehicle of the type for which data are provided in this textbook. In this example, the loads are defined in the x-, y- and z-directions for a coordinate frame located at the centre of the tyre contact patch as shown in Figure 4.48.

For the loads shown in Table 4.7 it is possible to calculate values for cases such as cornering and braking using traditional vehicle dynamics and the principles of weight transfer. For cases involving impacts with kerbs and bumps it may be necessary to obtain instrumented road load measurements on the proving ground, or else arbitrary figures may be used. A particularly severe case involves braking while driving through a pothole. To simulate this sort of case the input loads at the contact patch may be set to produce forces, say acting along a tie rod, that are consistent with measured strains on the actual component during the proving ground tests.

The purpose of the MBS model of the suspension in this case is to obtain the distribution of the load through the suspension. This is illustrated in Figure 4.48 where it is indicated that for a given set of loads at the tyre contact patch it is possible to predict the forces and moments that would for example act through the bushes mounting the suspension arm to the body of the vehicle.

FIGURE 4.47

This is not a service load event, nor even a design load event — this is most definitely abuse.

These forces and moments can then be used as boundary conditions for FE models of vehicle structure or of suspension components. A typical FE model of a suspension arm is shown in Figure 4.49.

Once generated the suspension model may be used in two ways. One of these is to apply the load and carry out a static analysis. This may result in the suspension system moving through relatively large displacement to obtain a static equilibrium position for the given load. The static reactions in joints, bushes and spring seats can then be extracted.

The use of equivalent static loads to represent real dynamic effects has been used by the automotive industry for some time, as previously noted in Section 4.1.6. The disadvantage of the static analysis is that any velocity-dependent forces will not be transmitted through the damper without the inclusion of some static equivalent. The author's experience has been to include some static nominal force at the damper in order to exercise damper mounts in some representative manner; this is of less importance when the spring and damper act in a collinear way.

Table 4.7 Example Suspension Loadcases

Loadcase	Fx (N)	Fy (N)	Fz (N)
3G bump			11180
2G rebound			−7460
0.75G cornering (outer wheel)		4290	5880
0.75G cornering (inner wheel)		−1180	1620
1G braking	5530		5530
0.35G reverse braking	−2150		3330
Kerb impact		9270	4120
Pothole braking	15900		12360

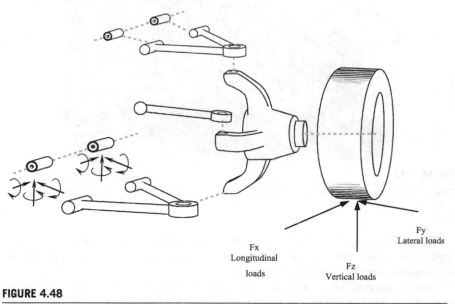

Fx
Longitudinal
loads

Fz
Vertical loads

Fy
Lateral loads

FIGURE 4.48

Application of road loads at the tyre contact patch.

Table 4.7 can easily be generated using a spreadsheet from basic vehicle parameters such as mass, mass centre height, wheelbase and track. With a little planning, the spreadsheet can be used to write out a solution command file to drive the software to perform the required commands automatically. Some 'single-purpose general-purpose' software packages (see Chapter 1) allow such factors to be entered automatically. This means the loading tables can be generated as soon as the basic proportions of the vehicle are known and allows immediate comparison with a known vehicle. This is important if, for example, some judgement is required on

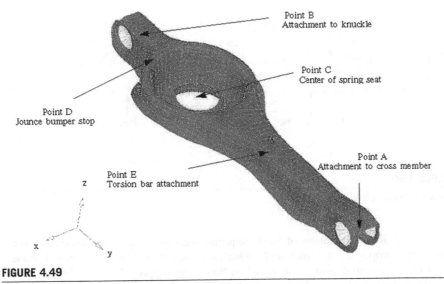

FIGURE 4.49

Finite element model of suspension arm.

Provided courtesy of Jaguar Cars Ltd.

the reuse of a suspension system from one platform to another; the judgement can be made instantly without waiting for weeks to know the results of the FE analysis, or previously known linear results can be scaled by load factors to give an immediate judgement. This kind of rapid feedback is deeply important in the early phases of a vehicle design when strategic choices are being made.

A second method, the extension of the static model to carry out a dynamic simulation, is possible. The model used may be a quarter suspension model or even a simulation of the complete vehicle. An example of this would be the simulation of a sports utility type vehicle in off-road conditions. An early example of this (Rai and Soloman, 1982) was the use of MSC ADAMS to carry out dynamic simulation of suspension abuse tests. A reaction force time history output generated at bush and joint positions may be used in downstream fatigue analysis for individual components. In order to be really useful, the enveloping behaviour of the tyre over small obstacles needs to be well captured — see Chapter 5 for discussion on this matter. As a pragmatic halfway house, load cells built into special purpose wheels can be used to capture the load history as seen by a wheel and this history 'played in' to a dynamic model in order to distribute the loads dynamically through the system and obtain load histories for each component. This is typically much more successful than the fully predictive approach but needs a vehicle of broadly representative proportions and a usefully close rim and tyre configuration to work. Some cautions remain, in particular the damping behaviour of the elastomers, about which more is written in Chapter 3.

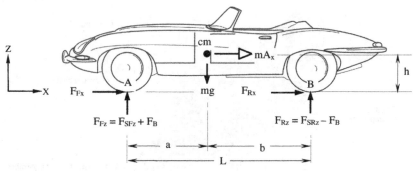

FIGURE 4.50

Forward braking free body diagram.

The dynamic simulation of loads requires usefully accurate estimates of mass and inertia properties of components, which may not exist early in the design phase. Thus the static simulation — the so-called 'first sight' approach — remains useful in order to get started from a clean sheet. The concept of useful accuracy is discussed at some length in Chapter 1.

The input loads used to represent braking and cornering may be obtained using basic vehicle data and weight transfer analysis. As an example consider the free body diagram shown in Figure 4.50 where the wheel loads are obtained for a vehicle braking case. Note that in this example we are ignoring the effects of rolling resistance in the tyre, discussed later in Chapter 5, and the vehicle is on a flat road with no incline.

The vertical forces acting on the front and rear tyres when the vehicle is at rest can be found by the simple application of static equilibrium. These forces, F_{SFz} for the front wheel and F_{SRz} for the rear wheel, are given by

$$F_{SFz} = \frac{m\,g\,b}{2\,L} \qquad\qquad F_{SRz} = \frac{m\,g\,a}{2\,L} \qquad (4.72)$$

It can be noted that the division of the loads by 2 in Eqn (4.72) is simply to reflect that we are dealing with a symmetric case so that half of the mass is supported by the wheels on each side of the vehicle. In this analysis the vehicle will brake with a deceleration A_x as shown in Figure 4.50. During braking weight transfer will result in an increase in load on the front tyres by an additional load F_B and a corresponding reduction in load F_B on the rear tyres. This can be obtained by taking moments about either wheel, for mA_x only and not mg, to give

$$F_B = \frac{m\,A_x\,h}{2\,L} \qquad (4.73)$$

Note that in determining the longitudinal braking forces F_{Fx} and F_{Rx} we have a case of indeterminacy with four unknown forces and only three equations of static equilibrium. The solution is found using another relationship to represent the relationship between the braking and vertical loads. At this stage we will assume that the braking system has been designed to proportion the braking effort so that the coefficient of friction μ is the same at the front and rear tyres. The generation of longitudinal braking force in the tyre is generally not this straightforward and will be covered in the next chapter. We can now combine the static and dynamic forces acting vertically on the wheels to give the full set of forces:

$$F_{Fz} = F_{SFz} + F_B = \frac{m\,g\,b}{2\,L} + \frac{m\,A_x\,h}{2\,L} \tag{4.74}$$

$$F_{Rz} = F_{RFz} - F_B = \frac{m\,g\,a}{2\,L} - \frac{m\,A_x\,h}{2\,L} \tag{4.75}$$

$$F_{Fx} + F_{Rx} = m\,A_x \tag{4.76}$$

$$\frac{F_{Fx}}{F_{Fz}} = \frac{F_{Rx}}{F_{Rz}} = \mu \tag{4.77}$$

Note that the inboard reaction of drive torques means that loads are applied at the wheel centre rather than at ground height for tractive cases involving traditional inboard torque generation and shaft outputs to the wheels. For in-wheel motors the situation is the same as the brake case discussed here.

4.8.2 Case study 2 — static durability loadcase

In order to demonstrate the application of road input loads to the suspension model, a case study is presented here based on the same front suspension system described in Section 4.7 for case study 1. The loading to be applied is for the pothole braking case outlined in Table 4.7. Due to the severity of the loading, the suspension model used here is one that includes the full nonlinear definition of all the bushes, the bump stop (spring aid) and a rebound stop. The model also includes a definition of the dampers and the damping terms in the bushes. These will be required later for an analysis that demonstrates the dynamic input of a road load but are not used for the initial phase where the load is applied quasi-statically.

A schematic for the model is shown in Figure 4.51 where it can be seen that the jack that was used earlier to move the suspension between the rebound and bump positions has been replaced by applied forces acting on a marker located at the bottom of the road wheel. An alternative approach can be taken where longitudinal force and a pitch moment equivalent to the longitudinal force multiplied by the wheel radius is applied at the wheel hub. This simplifies the loading model by removing the wheel part and the motion, saving the solution of six equations of motion to their constrained values; the choice of working practice remains with the practitioner, who must as always take responsibility for the formulation of his/her own model.

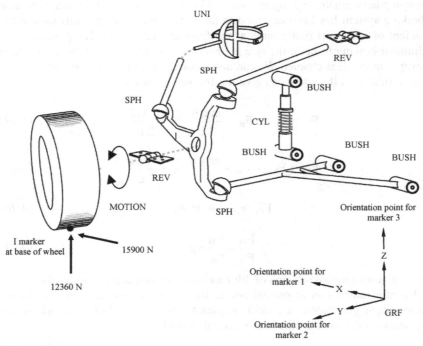

FIGURE 4.51

Application of Pothole Braking loads to suspension model. GRF, Ground Reference Frame; SPH, Spherical Joint; REV, Revolute Joint; UNI, Universal Joint; CYL, Cylindrical Joint.

The loads are applied as action-only single forces acting on the I marker at the contact patch. In this example the wheel is treated as a single rigid body and any compliance in the tyre is ignored. The I frame is in this case located at an undeformed radius directly below the wheel centre. Some practitioners can become obsessed with calculating the actual deformed radius of the tyre under these conditions but to complicate the calculations with second-order effects is to miss the approximate nature of the loading conditions, which are typically expressed to one figure ('3G bump') or the nearest quarter g ('0.75G cornering'); the pursuit of accuracy under such conditions is inappropriate. The motion statement associated with the road wheel revolute joint has a function set to zero to effectively lock the rotation of the wheel. The loads are applied parallel to the axes of the ground reference frame (GRF) at the start and remain parallel to the GRF during the simulation. They do not rotate with the wheel as the suspension deforms under the loading. Different software codes have different ways of achieving this but the nonrotation of the applied forces is important. Again, some practitioners can become obsessed with attempting to compensate for the change in vehicle platform attitude during such a manoeuvre but this again misses the approximate nature of the design conditions.

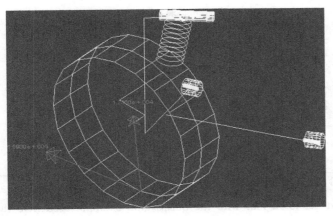

FIGURE 4.52

MSC ADAMS graphics of suspension at maximum Pothole Braking caseload. Graphics are deliberately shown in a 'minimalist' style to emphasise the ability to calculate component loads before the existence of CAD geometry.

Three separate forces are used to define the pothole braking case in this example. It is typically true that solutions starting a long way from equilibrium can have difficulty converging and so it can be desirable to start with the self-weight force, which typically resembles the condition under which the suspension model is defined, and to ramp on the desired additional load in several static equilibrium steps. Each step is in fact a true static equilibrium; the 'time' component of the solution simply serves as a lookup variable to select the level of load application.

```
X Force: FUNCTION = 15900 * TIME
Y Force: FUNCTION = 0
Z Force: FUNCTION = 3727 + 8633 * TIME
```

From this it can be seen that for the pothole braking case an additional longitudinal load in the x-direction and an additional vertical load in the z-direction are applied in comparison to the design definition condition. Note also that different organisations have different definitions for the design condition and so to presume that the design condition is the same as the kerb condition may well be incorrect. The lateral load in the y-direction is set to zero. The functions are set for this example so that for the initial static analysis a vertical load of 3727 N is applied with the additional components due to pothole braking being added over 1 s. It can be seen from this that the functions used to define the forces can be quickly changed to correspond with each of the loadcases given in Table 4.7. The graphics showing the suspension deformed under full load are shown in Figure 4.52 with additional graphics showing the force components at the contact patch. An XY plot showing the development of force magnitude in the spring is shown in Figure 4.53. Examination of the numerical values associated with the components of this force at full load, after 1 s, would provide the inputs for any subsequent FE models.

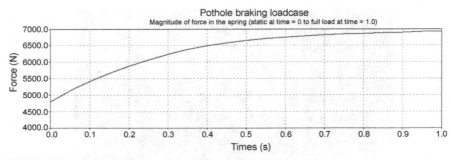

FIGURE 4.53

MSC ADAMS plot of spring load for Pothole braking case.

It should be noted that with the quasi-static example used here there is no velocity-dependent load transmission through the damper. To increase the validity of the results it would be necessary to estimate an equivalent load and apply this as an additional static force. An alternative would be to develop the analysis of the suspension to apply the force as a function of time and carry out a dynamic simulation as described next.

4.8.3 Case study 3 — dynamic durability loadcase

In this case study we extend the model of the single suspension system to include an additional part representing the corner of the vehicle body, quarter model, to which the suspension linkages attach as shown in Figure 4.54.

The vehicle body is attached to the ground part by a translational joint that allows the body to move vertically in response to the loads transmitted through the suspension system. A jack part is reintroduced to reproduce vertical motion inputs representative of road conditions. An additional complexity in the model is the introduction of stiffness and damping terms in a force element that represents the behaviour of the tyre. The force element acts between the centre of the wheel and a point on the jack coincident with the centre of the tyre contact patch. In this case this is assumed to be directly below the wheel centre. It is important that the force element for the tyre acts only in compression and allows the tyre to lift off the top of the jack if the input is severe enough.

The tyre is taken to have linear radial stiffness of 160 N/mm and a damping coefficient of 0.5 Ns/mm. The undeformed radius of the tyre, equivalent to the free length of the tyre spring, is taken here to be 318 mm. A step function can also be used to zero the tyre force if contact with the jack is lost. The full definition can be accomplished using the following function statement that is taken to act between the frame at the wheel centre and the frame on the jack part.

```
FUNCTION=STEP(318—DZ(wheel_frame,jack_frame), 0, 0, 0.1, VARVAL(1) )
   VARIABLE/1,  FU=160*(318-DZ(wheel_frame,jack_frame))-  0.5*VZ(wheel_
frame,jack_frame)
```

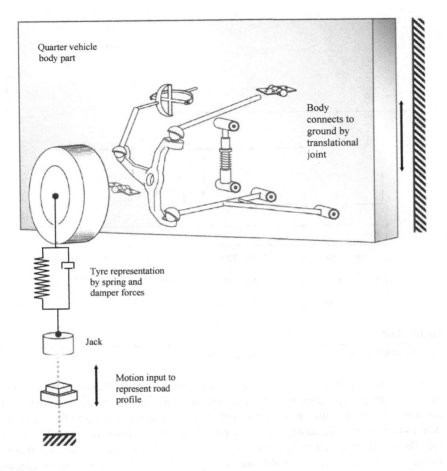

FIGURE 4.54

Quarter vehicle body and suspension model.

The STEP function is common to several multibody programs and consists of a continuously differentiable blend between two states, based on the value of a variable or expression. In this case the expression is 318-DZ(wheel_frame, jack_frame), which may be understood as the distance of the wheel centre above the presumed floor, subtracted from the undeformed radius. When the result of this expression is zero or more, the force resulting has a constant value of zero. When the result of the expression is less than zero it means the wheel is off the floor and the force returned is zero. When the expression is greater than 0.1 mm, a linear stiffness and viscous damping component is enacted. Between 0 and 0.1 mm, the STEP function blends the two functions together in a manner that is sympathetic to the numerical integrators in use.

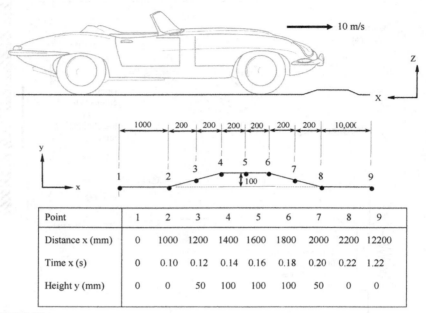

Point	1	2	3	4	5	6	7	8	9
Distance x (mm)	0	1000	1200	1400	1600	1800	2000	2200	12200
Time x (s)	0	0.10	0.12	0.14	0.16	0.18	0.20	0.22	1.22
Height y (mm)	0	0	50	100	100	100	50	0	0

FIGURE 4.55

Road profile for 'sleeping policeman' speed bump.

It can be seen there is a trade-off between the numerical friendliness, which is assured with a large transition region, and absolute rigour of implementation of the transition at any point above zero deformation of the modelled tyre forces. Given that, as George Box famously said, "All models are wrong but some models are useful" (Box and Draper, 1987), an obsessive desire to capture the immediacy of transition is perhaps overlooking the inaccuracy of the linear model when inspected closely in that very transition region. Since it does not dominate the overall behaviour of the system (the reader is invited to verify this assertion by varying the transition region and checking whether it changes the outcome of the analysis in any meaningful way) it is apparently a candidate for another home-grown aphorism: "I'd rather be simple-and-wrong than complicated-and-wrong." This is another way of saying that simple-and-useful is more valuable than complicated-and-useful.

The next step is to define the motion imparted to the jack part to represent the input from the road surface. This is illustrated in Figure 4.55 where the profile for a 'sleeping policeman' road obstacle is given. The profile is defined as a set of xy pairs. Note here that the xy values are local to the definition of the obstacle profile and not associated with the X and Y axes of the GRF. The vehicle is assumed to be moving with a forward speed of 10 m/s so that the x values associated with distance can be converted to time.

One method that could be used to input a motion associated with the profile of the road surface would be to use a form of spline. Most multibody packages offer cubic splines and some offer other spline schemes. All of them have in common that they repeatedly fit an analytical function through a subset of all the points provided in order to provide a continuously differentiable and infinitely resolvable version of the discrete table of points.

Caution is needed here, however, for although the data points may be sufficient to capture the profile of the bump there may not be enough to ensure a good spline fit, as noted in Chapter 3.

A more elegant but elaborate method might be to forgo the use of interpolation and use a combination of arithmetic IF and step functions. This method requires care in formatting but may be applied as follows.

```
FUNCTION=IF(TIME — 0.14: STEP(TIME, 0.1, 0, 0.14, 100), 100,STEP
(TIME, 0.16, 100, 0.22, 0))
```

MSC ADAMS graphics showing the suspension deflecting on the jack and the subsequent departure of the tyre from the road surface are shown in Figure 4.56. A plot showing the time histories for the vehicle body and road wheel vertical displacement is shown in Figure 4.57. The force in the bump stop is provided by way of example Figure 4.58. It should be noted that at this stage the analysis only represents vertical force input and not longitudinal force input from the road surface.

This is no different to the case in the real world, where vertical-only inputs do not capture the full set of inputs to which the real vehicle is subject. Just as in the real world, ever-more elaborate rigs can be used with half and full vehicle models, culminating in running the full vehicle over a complete description of the test surface.

It is tempting but somewhat naive to argue for the most detailed possible model at all times. To do so is fall prey to a common error in predictive engineering: the belief that the model is the product. In fact, the purpose of the model is to support design decisions and to allow a monotonic progression of the design from a concept to a finished product, ready to manufacture, with the minimum number of iterative processes and feedback loops along the way. The model is *not* the product.

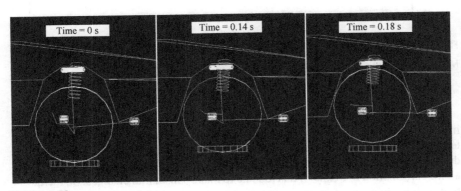

FIGURE 4.56

MSC ADAMS graphics of suspension deflecting on a speed bump.

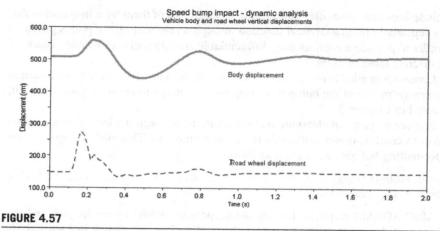

FIGURE 4.57

MSC ADAMS plot vehicle body and road wheel displacements for speed bump strike.

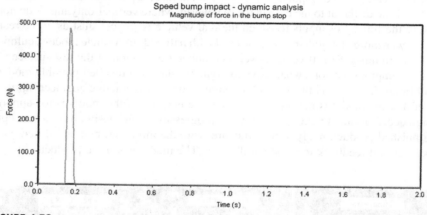

FIGURE 4.58

MSC ADAMS plot of bump stop force time history for speed bump strike.

Some practitioners (Tang et al, 2000) advocate the use of explicit FE solutions instead of the combined multibody/implicit FE approach used here. In durability terms, the use of fully featured explicit FE mesh suffers from most of the same drawbacks that running a real prototype vehicle suffers from: by the time we have all the information to define it, it is too late to make substantial changes. So the question is 'how do we get to a design in which we have confidence that it will not need substantial changes by the time we sign it off?' The answer, it appears to these authors, is a rational and staged approach consisting of a sequence looking something like this:

1. Establish conceptual design based on cost constraints, package constraints and so on.
2. Sketch the conceptual design in a multibody environment with all of the elements represented by sparse line graphics.
3. Perform a functional analysis that distributes static design loads through the various components, as in case study 2.
4. Use the loads to broadly size the components in a manner consistent with their anticipated manufacturing method, sketching proportions in a parametric CAD environment but avoiding final detailing (e.g. paint drain holes) and then solving for linear stress response; the use of linear stress response being consistent with the practices of the organisation (e.g. 'no stresses over 75% yield for design loads' or similar).
5. Detailed progress based on other attributes — ride, refinement and so on — to define details not strongly related to durability.
6. Full dynamic simulation of a model over the expected durability sign-off using all the design data available up to that point to discern dynamic load histories; these being used with detailed material models to search out areas of the design that are not suitably robust using the linear stress sizing criteria. Case study 3 represents the simplest possible approach to this task.
7. Prototyping and real-world sign-off with high confidence.
8. Review sizing in step 4 for future activities; refine if necessary.

Note that if no areas emerge in step 6 then the sizing criteria in step 4 are unnecessarily conservative and the product is probably overweight; if too many emerge then the sizing criteria are not strenuous enough, systematically resulting in an unnecessary redesign for every product.

4.9 Ride studies (body isolation)

The determination of vehicle 'ride' quality is associated with the extent to which the occupants of the vehicle are affected by vehicle motion. Ride is often described as a mainly vertical phenomenon but this can be unhelpful; a better description might be that it is the study of uncommanded vehicle motions, primarily through the interaction of the vehicle with the surface. Automotive companies will often have departments concerned with ride and handling where MBS analysis will be deployed to support design and analysis work. Another area of activity is referred to as noise, vibration and harshness, frequently referred to as 'NVH'. Many authors distinguish between these different phenomena by using arbitrary frequency separations. While superficially helpful, it should be kept in mind that the numbers used represent the centre of some 'transition band' in a continuous spectrum and that the phenomena merge into one another rather like the colours in a rainbow. Figure 4.59 represents this idea graphically.

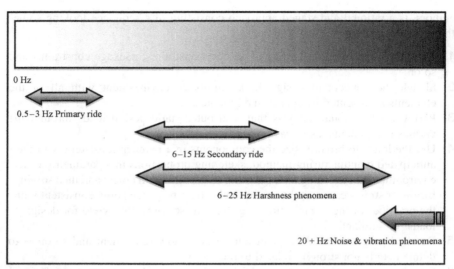

FIGURE 4.59

Ideas of the continuous frequency spectrum and different events on it.

For the purposes of this text, the different dynamic phenomena might be described as follows:

Primary ride: the motion of the whole vehicle on its suspension 'bodily'; excited by transient, unsteady forcing from the road surface.

Secondary ride: the motion of substantial masses within the vehicle such as powertrain-on-mounts or individual suspension assemblies; excited by underlying energy in the road surface in a pseudo-steady state manner and easily recognisable as 'large masses in motion' when *operating shape* information is recorded. Operating shape is a specific type of dynamic measurement in which accelerometers are used with sophisticated postprocessing software to capture the dynamically deformed shape of a system and show how it varies at different frequencies. When a system is in resonance, the operating shape is dominated by the mode shape for the resonance in question; it follows that if the shape does not resemble a mode shape, it is not therefore in resonance but rather being forced statically. See earlier in this chapter for a fuller description.

Harshness: complex motion, sometimes within an individual component, that is touched intermittently or continuously by the operator to give a sense of annoyance (no touch = no harshness); often excited in a broadly steady state manner by powertrain or road surface energy.

Noise & vibration: complex motion, with coupling to the acoustic cavity in the case of noise, detectable by a combination of hearing and touch but occurring at low energy levels; often excited in a broadly steady state manner by powertrain or road surface energy.

Primary and secondary ride vibrations are usually amenable to analysis with the multibody techniques described in this textbook. Some aspects of harshness can be assessed with multibody models when they include intracomponent resonances in the form of flexible representations of components using modal components, a technique described later. Above 20 Hz or so and with complex motions involving distortions of individual components, the analysis of these modes and other acoustic type problems is more in the domain of advanced FE analysis and as such is not covered here.

Vehicles should be thought of as dynamic systems, a mixture of masses, springs and dampers, where vibration is exhibited in response to excitation. The source of excitation may be due to out of balance loads from rotating bodies such as the road wheel or from other sources in the vehicle including the engine and driveline. The other main source of vibration will be associated with the profile of the road surface. At this stage it is easy to envisage that the excitation of vehicle pitch may be in response to a road with an undulating type profile of relatively long wavelength whereas the excitation of a smaller mass such as the road wheel will occur at higher frequencies. This might for example occur whilst driving on a cobbled type of road surface.

4.9.1 Case study 4 — quarter vehicle dynamic performance analysis

A great deal of insight can be gained from the classical study of the quarter vehicle model shown in Figure 4.60. This study can easily be replicated in a commercial multibody program to reproduce the results obtained here, as described later in this section.

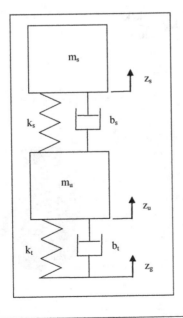

$m_b = 472$ kg
$m_t = 39.1$ kg
$k_s = 36{,}190$ N/m
$k_t = 160{,}000$ N/m

FIGURE 4.60

Quarter vehicle model in mathematical form.

A sprung DoF, m_s, is connected to an unsprung DoF, m_u in series and both are driven by a road excitation profile, u. To perform the calculations, the equations of motion can be written largely by inspection. For the sprung mass:

$$\ddot{z}_s = \frac{b_s(\dot{z}_u - \dot{z}_s) + k_s(z_u - z_s)}{m_s} \qquad (4.78)$$

For the unsprung mass there are some more terms since it is acted upon by both the tyre and suspension but the principle is identical:

$$\ddot{z}_u = \frac{b_s(\dot{z}_s - \dot{z}_u) + k_s(z_s - z_u) + b_t(\dot{z}_g - \dot{z}_u) + k_t(\dot{z}_g - z_u)}{m_u} \qquad (4.79)$$

These terms are convenient for direct numerical solution as is, to obtain time-domain results. Note that the formulation of the tyre forces can be subject to the same sort of conditional switching described in case study 3 to allow the wheel to leave the ground. Before solving the model in the time domain, it is good practice to estimate the natural frequencies for the body on the suspension and for the unsprung mass between the road spring and the tyre spring. Figure 4.61 shows a 2 DoF quarter vehicle model with the data used to support the calculations.

The undamped natural frequencies for the body, f_b, and unsprung mass, f_t, can be estimated using the following equations. Note that for the body we determine an equivalent stiffness, k_{eqv}, to represent the combined contribution of the road and tyre springs.

$$k_{eqv} = \frac{k_s\, k_t}{k_s + k_t} \qquad (4.80)$$

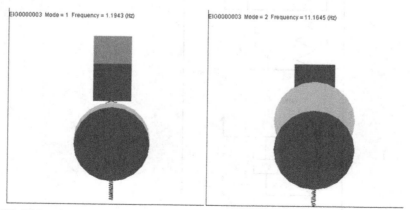

FIGURE 4.61

MSC ADAMS prediction of the modes of vibration of the simplified quarter vehicle model.

$$f_s = \frac{1}{2\pi}\sqrt{\frac{k_{eqv}}{m_s}} \qquad (4.81)$$

$$f_u = \frac{1}{2\pi}\sqrt{\frac{k_s + k_t}{m_t}} \qquad (4.82)$$

Performing the calculations using the data given results in values as follows:

$k_{eqv} = 26639$ N/mm

$f_b = 1.196$ Hz

$f_t = 11.15$ Hz

A multibody model consisting exactly of the model as sketched above can be used to calculate undamped linear modes in a more exact fashion, giving two modes of vibration as might be expected but slightly differing numerical results to the classical calculations.

This particular modal solution uses the ADAMS/Linear™ product, which in turn uses numerical perturbation methods to estimate mass and stiffness matrices and assemble the Jacobian matrix about an operating point before solving for eigenvalues by finding its determinant in the normal fashion.

For purely mechanical systems such as the one modelled, such methods can be relied upon to give good quality results. However for systems where forces are time-dependent and are modelled in specific ways (e.g. as differential equations for modelling turbocharger behaviour as described in Chapter 6, or tyre relaxation length modelling) the results are not necessarily reliable at the time of writing and must be examined on an individual basis before confidence is placed in them. A test solution is recommended to establish confidence in the eigensolution methods for a given software code before critical design decisions are based on it. Note that these types of calculation behaviours are the worst possible kind — the results are plausible but misleading; obvious nonsense can easily be spotted but something which looks reasonable is more difficult to pick up.

All software packages are subject to ongoing modification and development and so functionality of this nature should be evaluated periodically; such evaluations should be part of the software commissioning process within individual organisations, particularly if critical decisions are to be based on software output.

The estimates given above are a simplification of the actual analytical solution to the system. Such a solution is obtained by first dropping the damping terms and external inputs from Eqns (4.69) and (4.70):

$$\ddot{z}_s = \frac{k_s(z_u - z_s)}{m_s} \qquad (4.83)$$

$$\ddot{z}_u = \frac{k_s(z_s - z_u) + k_t(-z_u)}{m_u} \qquad (4.84)$$

These can be arranged more conveniently as:

$$m_s \ddot{z}_s = z_u k_s + z_s(-k_s)$$ (4.85)

$$m_u \ddot{z}_u = z_u(-k_t - k_s) + z_s(k_s)$$ (4.86)

If a solution is assumed of the form

$$x = X e^{\lambda t}$$ (4.87)

then

$$\ddot{x} = \lambda^2 X e^{\lambda t} = \lambda^2 x$$ (4.88)

thus

$$m_u \lambda^2 z_u = z_u(-k_t - k_s) + z_s(k_s)$$ (4.89)

$$m_s \lambda^2 z_s = z_s(k_s) + z_s(-k_s)$$ (4.90)

which may be rearranged into the familiar eigenvalue problem:

$$\begin{bmatrix} -k_t - k_s - m_u \lambda^2 & k_s \\ k_s & -k_s - m_s \lambda^2 \end{bmatrix} \begin{bmatrix} z_u \\ z_s \end{bmatrix} = 0$$ (4.91)

in which the determinant of the matrix can be used to find the eigensolution when set to zero:

$$(-k_t - k_s - m_u \lambda^2)(-k_s - m_s \lambda^2) - k_s^2 = 0$$ (4.92)

$$(m_s m_u)\lambda^4 + (k_s m_u + (k_t + k_s)m_s)\lambda^2 + k_t k_s + k_s^2 - k_s^2 = 0$$ (4.93)

which may be recognised as a quadratic in λ^2 and solved in the normal manner:

$$\lambda^2 = \frac{-b \pm \sqrt{b^2 - 4ac}}{2a}$$ (4.94)

$$a = m_s m_u$$ (4.95)

$$b = k_s m_u + (k_t + k_s)m_s$$ (4.96)

$$c = k_t k_s + k_s^2 - k_s^2 = k_t k_s$$ (4.97)

The calculated roots using this method are

λ^2:

-56.3075
-4920.87

λ:

7.50383i
70.1489i

frequencies:

1.19427 Hz ("Primary Ride")
11.1645 Hz ("Wheel Hop")

which can be seen to agree exactly with the MSC ADAMS model. However, the differences between the exact method and the approximate method are small — less than 0.15%. Thus the approximate method is a 'good enough' check for this system. This is generically true for quarter vehicle models, where the second mode of vibration is typically an order of magnitude higher than the first. However, for particularly stiff suspensions or compliant tyres as may be used on circuit cars, the suitability of the approximate method breaks down and therefore it should be used with some care.

Before leaving the simplified model, it is worth examining its behaviour in the time domain in the context of real vehicle behaviour. A step-up ramp can be implemented by (Figure 4.62):

$$\dot{z}_g(t) = \frac{A}{2}\left(1 - \cos(\pi.t/t_e)\pi\right), t < t_e \qquad (4.98)$$

$$\dot{z}_g(t) = A, t \geq t_e \qquad (4.99)$$

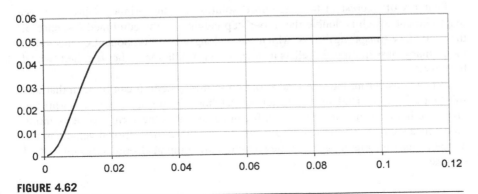

FIGURE 4.62

Cosine step up, a continuously differentiable input function.

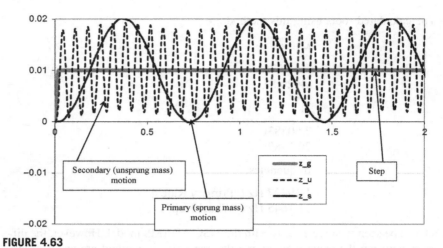

FIGURE 4.63

Undamped solution to a step input for illustration.

where A is the height of the step and t_e is the time at which the vehicle reaches the end of the step, given by step length/velocity. This formulation can be used for both step up and step down; a large, e.g. 1 m, step down is a jump landing simulation.

The step up is strictly a cosine ramp-up as described by Eqn (4.98), since the enveloping behaviour of the tyre means a genuine mathematical step that is never really seen by a real suspension. The length of the ramp is set at 0.5 m. The speed of the vehicle is set at 25 m/s (90 kph) so that the obstacle is encountered in 0.02 s, significantly quicker than the wheel hop dynamics and therefore it may be expected to excite both wheel hop and primary ride. Initially the obstacle height is set at a mere 10 mm.

For illustration, the system was run with zero damping levels to show the base behaviour of the different parts. Figure 4.63 shows how the step input excites both resonances very clearly.

Features of interest in the undamped solution are the primary (sprung) mass displacement which is double the initial step height and the combined response of the secondary (unsprung) mass, which has a 'long wave' component connected to the primary mass motion as well as its own, faster oscillation − the so-called 'wheel hop' mode.

When considering the tyre load variation in this system, it can be seen that it is substantially dominated by the faster, wheel hop vibration. This is unsurprising given the high spring rate of the tyre. It can be seen that the spring force variation (and hence the body accelerations) are very low and thus this notional vehicle might not be as uncomfortable as the somewhat dramatic graphical results in Figure 4.64 suggest.

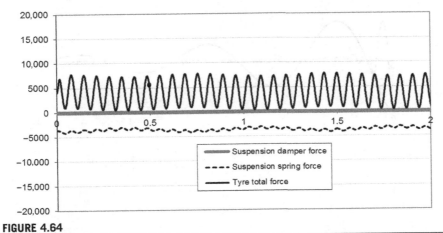

FIGURE 4.64

Forces generated in the undamped solution.

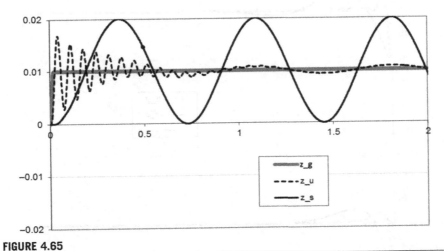

FIGURE 4.65

The addition of tyre damping.

Adding a typical level of tyre damping (estimated at 0.4 Ns/mm) controls the secondary motion, albeit only lightly, but not the primary motion as shown in Figure 4.65.

The road step input is adjusted to 0.1 m height, a typical kerb height. This is an aggressive event at 25 m/s and promotes separation of the tyre from the road. Three results are presented for initial discussion of the suspension damping calculations in Figure 4.66.

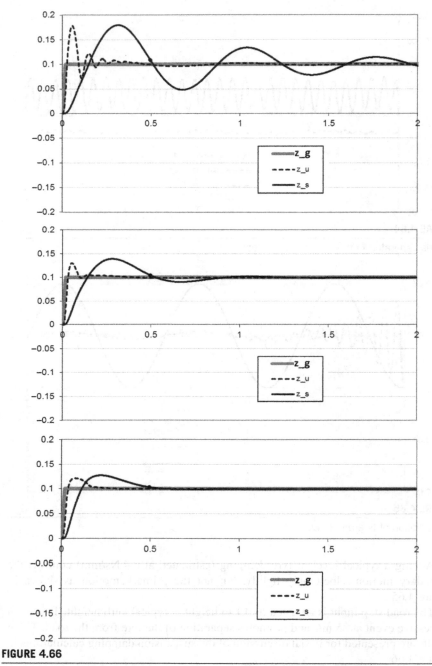

FIGURE 4.66

Suspension damping influence on system performance.

The first damped results, Figure 4.66 (top), are calculated with a suspension damping level of 1 Ns/mm, giving an overall damping ratio of 0.2. The continued oscillation of the primary mass can be clearly seen, although the secondary vibration is clearly reduced compared to the tyre damping alone. Despite this, there are repeated land-and-bounce events experienced by the secondary mass.

The results shown in Figure 4.66 (middle) are calculated with a suspension damping level of 3 Ns/mm, a damping ratio around 0.7 and quite a sensible value for a performance-biased road vehicle. While there is some overshoot, it can be seen that the secondary mass quickly returns to the surface.

The results shown in Figure 4.66 (bottom) are with a suspension damping level of 5 Ns/mm, damping ratio of around 1.1 and 'too hard' by most measures. While the control of the primary mass is better, it seems clear that the 'time in flight' for the secondary mass is longer than the middle solution.

It can thus be seen that the damper calibration has a substantial influence on the behaviour of the system. However, it is less immediately obvious what constitutes 'good' behaviour.

An approach can be taken that looks at several aspects of system performance and attempts to calculate scores for each of them. The desire is to obtain good performance in all areas, with an overall rating being possible if weightings are assigned to the different scores. With enough discipline, it is possible to formulate a genuine multidimensional optimisation problem using these measures.

Quarter vehicle model dynamic performance can be broadly split into:

- Active safety: the ability to stop and steer in emergency situations
- Ride: the ability of the vehicle to absorb disturbances

'Active' safety in this context means the ability to avoid a collision through manoeuvring the vehicle. It is distinct from 'passive' safety, which is the ability of the vehicle to protect the occupants in the event of a collision and includes so-called crumple zones and airbag technology. It does not specifically imply the presence of electronic or other control systems. Other performance measures can be added at will — the response to out-of-travel events, jump landings and so on. Not all items are uniformly important in every application. Active Safety and Ride are considered in detail below and a method for determining a key performance indicator (KPI) is illustrated. It should be noted that these indicators are entirely arbitrary and are a way of using numerical analysis in some meaningful way to influence the product. Indicators should be invented and discarded at will, as long as care is taken to document them carefully and to only compare like with like.

For the KPIs there are some preferable qualities:

- Each should be a 'more is better' measure to allow easy assimilation
- Each should be on a reasonable scale, e.g. 1—10
- Each should have some sort of meaning associated with significant points on the scale, e.g. 0, 10.

To achieve these qualities, KPIs are not always in common engineering units, although they are always traceable to more normal engineering quantities.

Formulating performance metrics in this way means they can be judged easily by nonexperts (since most people have a good idea of the relative difference between 6/10 and 8/10 but many people struggle to assimilate a reduction of $0.5\ m/s^2$ ride acceleration). By expending a little effort on the indicators in their own right, the somewhat arbitrary application of cost functions in an optimisation can be avoided since they are all set to unity if the third requirement is met in some sensible way.

4.9.2 An active safety KPI

The ability to stop and steer the vehicle is completely dominated by the ability of the tyres to generate forces in the ground plane, sometimes referred to as 'grip'. As discussed in Chapter 5, variation in vertical load has the effect of degrading available forces in the ground plane.

Root mean square (RMS) load variation is therefore a useful response variable and can be nondimensionalised by dividing it by the static load. If the static load is first divided by $\sqrt{2}$ then a KPI of less than zero can be seen to induce momentary 'flight' of the wheel under sinusoidal excitation. For smooth road use this is not typical but it is much more common than might be intuitively expected for even moderately degraded surfaces without necessarily running out of suspension travel; some reflection shows that if steps down are more than the corner load divided by the tyre stiffness then the load will indeed get to zero; for a typical modern run-flat tyre this is of the order of 10 mm. It should be noted that the degradation of lateral force does not require a period of zero load — only a variation in vertical load. Finally, the value is subtracted from 1 to give a more-is-better measure and then multiplied by 10

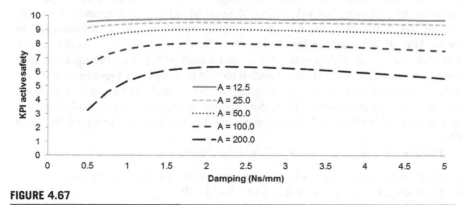

FIGURE 4.67

Variation of Active Safety Key Performance Indicator (KPI) ('Grip') with damping levels over varying steps up — calculated with a 2 degree-of-freedom model.

to give a maximum value of 10 (no load variation). For a rough road, negative values of the KPI are unsurprising.

$$KPI = \left(1 - \frac{\sqrt{\dfrac{\int\limits_{0}^{t}(F_z - F_{z0})^2\, dt}{t}}}{\dfrac{F_{z0}}{\sqrt{2}}}\right) \cdot 10 \qquad (4.100)$$

If this definition is used with a linear quarter vehicle model then a clear optimum emerges in terms of damping level for the modelled step up, as can be seen in Figure 4.67.

Note that Figure 4.67 represents the results of multiple solutions to the model — 100 different solutions, each consisting of 10,000 integration steps[3]. In this case, the picture is worth a million calculations. Several lessons emerge from this exercise:

- Increasing the amplitude of the surface increases the load variation in a manner that cannot be compensated by damper level
- A clear optimum exists that is remarkably similar in level whatever the amplitude of the surface input to the model
- Too much damping degrades the KPI significantly less than too little
- The smaller the input, the less influence damping has on the KPI

These are interesting conclusions for this surface but it is unclear from these results how general these findings might be; it might be argued that the nature of the grip index is very dependent on the surface chosen.

However, this does not appear true. Results were computed as part of a separate exercise using the quarter vehicle corner model for a step up, step down and 'random' surface input. They are shown in Figure 4.68. While the levels of KPI grip are clearly different between the surfaces, they all point to a similar optimum (2–3 Ns/mm in this case), a strong penalty for insufficient damping and only a small penalty for excessive damping. Therefore the KPI is seen as a fairly robust measure that is not unduly influenced by the inputs chosen. The parameters for this exercise were not the same as the reference model provided, suggesting the conclusions drawn are not overly sensitive to the surface profile chosen, nor are they particular artefacts of a specific data set.

Further investigations were made by Harty (2010) to understand the influence of road profile scaling and unsprung mass.

---

[3]The number of steps is high because rather dumb rectangular integration is in use in the Microsoft Excel spreadsheet used to generate this example. This is not a reflection on Microsoft but was convenient to code for the author.

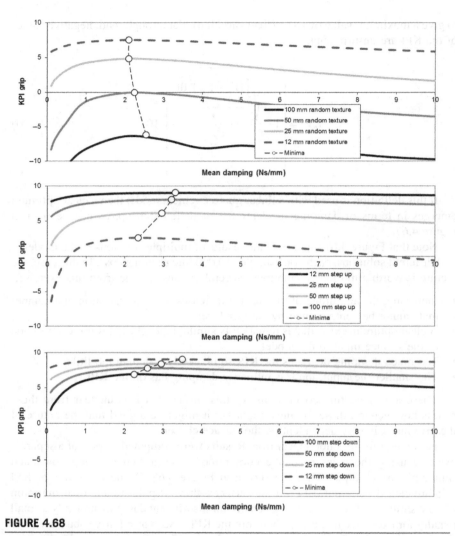

FIGURE 4.68

Key Performance Indicator (KPI) Grip variation for three different surface profiles.

It can be seen in Figure 4.69 that the same type of optimum level appears and that the optimum is very insensitive to unsprung mass. It is also apparent that surface roughness significantly degrades the active safety KPI regardless of damper setting or unsprung mass — as was previously observed. These analyses were carried out using scaled versions of a synthesised textured road profile, one representing a fairly smooth road with low-level inputs, typical of many made roads in the Western world, the second representing a surface with large inputs comparable to the

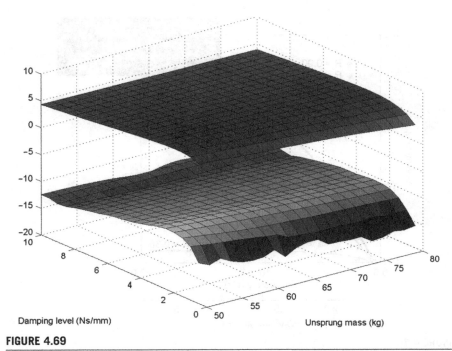

FIGURE 4.69

The Variation of Key Performance Indicator Grip with unsprung mass and damping level for two different road surface roughness levels. The lower surface represents a rougher road.

durability surface known as Pavé or Belgian Block. The rough surface drives the KPI negative, as was previously declared possible, well past the point where the wheel just unloads in some notional sinusoidal cycle.

Without precise tyre characteristics — specifically relaxation length variation with load, which is not typically published or present in empirical data sets — it is impossible to be certain about the effect of load variation on lateral performance, although a useful estimate may be made if lateral acceleration levels and vehicle and wheel motions are logged on a smooth and a rough surface. Care must be taken to avoid damaging the vehicle on particularly rough surfaces at maximum lateral acceleration. The authors' experience testing the handling behaviour of gravel rally cars on the Pavé surface at MIRA suggests that for this level of degradation, the maximum lateral acceleration capability of the vehicle is roughly halved, which helps calibrate the scales in engineering units. This level of degradation is broadly consistent with published data such as that shown in Figure 4.70.

Note the previous observation that the conclusions drawn from such metrics do not seem unduly sensitive to the exact surface profile used. This leads us to the

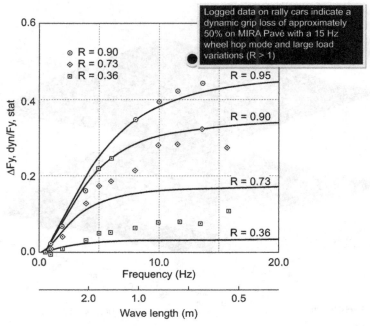

FIGURE 4.70

Measured data showing the loss of lateral force over non-smooth surfaces with an indication of rally car behaviour on the MIRA Pavé surface (Pacejka and Takahashi, 1992).

fortunate position where we can optimise what seems like a useful surrogate for grip *without* detailed knowledge of tyre characteristics apart from knowledge of the vertical dynamics.

4.9.3 Ride KPIs

The vehicle ride environment is dominated by vertical acceleration. Expert assessors often break down ride behaviour into 'primary' and 'secondary' ride. Primary ride is the description given for the vehicle body motions on the suspension and is usually in the window below 3 Hz. Secondary ride is the description given to a number of higher frequency phenomena in the vehicle, such as powertrain shake and wheel vibration on the tyre stiffness, around 6 and 12 Hz, respectively. These ideas were illustrated in Figure 4.59.

The primary ride KPI for this study is RMS vertical acceleration in the 0–3 Hz band for the 2 DoF corner model running over a known road profile. However, since more acceleration is bad, the RMS acceleration in metres per square second is subtracted from 5 m/s²; this reflects the ISO 2631 severe discomfort threshold in the 1–2 Hz region. Finally the value is multiplied by 2 to scale the KPI.

$$\text{KPI} = \left(5 - \sqrt{\dfrac{\displaystyle\int_0^t \left(\text{filt}(\ddot{z}_s) \right)^2 dt}{t}} \right) \cdot 2 \tag{4.101}$$

For a completely immobile ride the RMS acceleration would be zero and therefore the KPI is 10. For an RMS value of 5 m/s^2 (approximately 0.5 g), the ride would be at the severe discomfort level for 1-h exposure and the KPI would be zero. When considering the ISO 2631 curves it may be seen that the progression through the different comfort levels from 'perception' to 'comfort' and onto 'severe discomfort' is substantially linear and therefore the KPI is a useful indicator.

The secondary ride KPI is very similar but just in a different frequency band — 3 to 20 Hz — and using different constants. A value of 3 m/s^2 reflects the ISO 2631 severe discomfort threshold in the 12—15 Hz region.

$$\text{KPI} = \left(3 - \sqrt{\dfrac{\displaystyle\int_0^t \left(\text{filt}(\ddot{z}_s) \right)^2 dt}{t}} \right) \cdot \dfrac{10}{3} \tag{4.102}$$

Again using the simplified model, these KPIs can be investigated to illustrate certain aspects of ride behaviour that may not be obvious. Ride KPIs were investigated as part of the same exercise as that used to generate Figure 4.69 (Harty, 2010).

Smooth road primary ride gets very high KPI scores, close to 10, suggesting that the primary ride environment is largely indistinguishable from a completely quiescent vehicle, which matches well with experience of motorway travel where primary events are confined to specifically notable features in the road such as bridge joints or repair patches (Figure 4.71).

On a rough road the KPI scores fall to around 3, reflecting the fact that surfaces such as the MIRA Pavé are somewhat uncomfortable on prolonged exposure.

Particularly light levels of damping, such as might be present in a vehicle with faulty dampers, produce a larger change of response. The degradation in primary ride with unsprung mass is most marked on a rough road and with light levels of damping.

In general the degradation in rough road primary ride due to 30 kg additional unsprung mass is similar to a slightly miscalibrated damper. (The process capability for a typical automotive damper delivers dampers to around ±15% of the nominal characteristic; the miscalibration required to reproduce the unsprung mass effect is about 50%, see Figure 4.72).

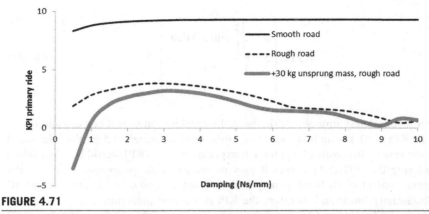

FIGURE 4.71

Variation in Primary Ride Key Performance Indicator (KPI) between rough and smooth roads.

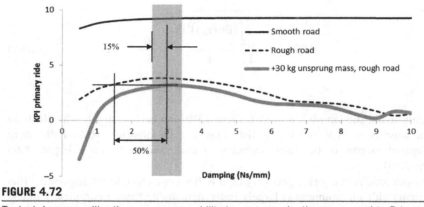

FIGURE 4.72

Typical damper calibration process capability in mass production compared to Primary Ride Key Performance Indicators (KPIs).

The influence of spring calibration on primary ride is also significant (Figure 4.73). In general softer springs produce a better ride with no optimum. Were RMS acceleration the only consideration, then the springs on every vehicle would be driven as soft as possible; motion sickness considerations generally prevent primary ride frequencies being set below about 0.6 Hz as described in Section 4.1.2.

Secondary ride KPI values are illustrated in Figure 4.74 and show a very different trend; immediately noticeable is that for typical levels of damping (around 3 Ns/mm) even the smooth road KPI is less than 4, suggesting that secondary ride is very noticeable and verging on uncomfortable for prolonged exposure even on a smooth road. This may seem a strange assertion but the sceptical reader is invited to spend some time sitting uncushioned in the load area of a panel van to test its

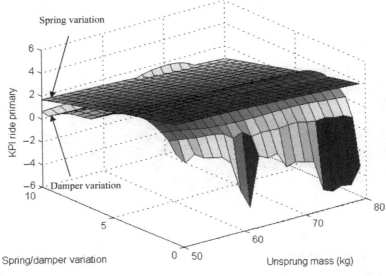

FIGURE 4.73

Relative influence of springs, dampers and unsprung mass on primary ride Key
Performance Indicator (KPI).

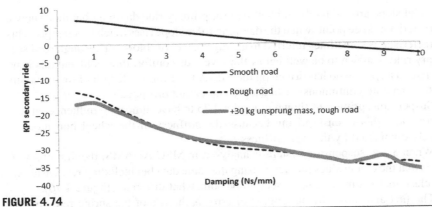

FIGURE 4.74

Key performance indicator (KPI) results show that secondary ride starts from a lower base
and is more degraded by suspension damping on smooth roads than rough.

veracity. Modern automotive seats are primarily aimed at isolating the secondary
ride issue but even so, secondary ride — a restless jiggle — is noticeable on all but
the very best of modern cars even on the motorway.

It is also apparent that any level of suspension damping degrades secondary ride in
a very pronounced manner. For this reason, road vehicle damper curves tend to have

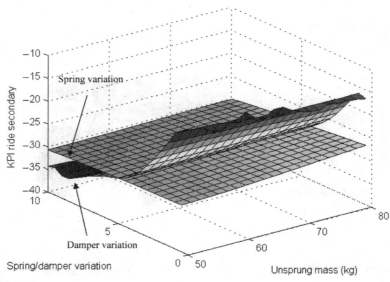

FIGURE 4.75

Relative influence of spring and damper calibration on secondary Ride Key Performance Indicator (KPI).

an initial slope around the level identified for primary ride damping but then show a characteristic 'knee point' when the damping force slope rises much less steeply. This characteristic is sometimes referred to as 'digressive' or 'blow-off'. Rough road secondary ride is shown to be well below the severe discomfort threshold and this is no surprise to anyone who has driven on the MIRA Pavé for any length of time; staying on it for an hour continuously is a deeply unpleasant prospect.

Suspension springs are shown in Figure 4.75 to have almost no influence on secondary ride; this is unsurprising because the stiffness in the wheel hop mode is entirely dominated by the tyre stiffness.

When a real suspension system is analysed in MSC ADAMS, the first observation is that the eigenvalues predicted using the same data but including real wishbone and elastomer geometry can be seen to be somewhat different (Figure 4.76).

The first and most obvious source of error is the use of the spring rate directly from the detailed model in the simplified 2 DoF models. In the literal model, as in the real vehicle, the motion of the wheel does not directly correspond to the motion of the spring. Using the model, the so-called 'motion ratio' can be examined between spring and wheel. It can be seen (Figure 4.77) that the spring changes length by 1.43 mm for every 1 mm of wheel vertical motion. Some practitioners quote this as a motion ratio of 1.43:1 and some as 0.699:1. It is very important that the definition of motion ratio as quoted is understood before the number is used since it is not always true that the spring moves less than the wheel.

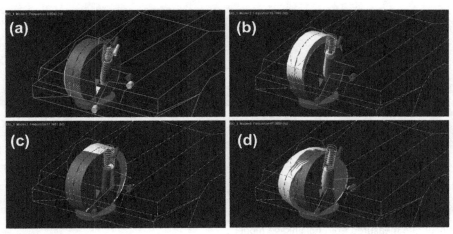

FIGURE 4.76

MSC ADAMS Prediction of the modes of vibration for the full linkage quarter vehicle model with elastomers and individual component mass & inertia data. Primary ride mode (a) 0.95 Hz. Wheel hop mode (b) 10.78 Hz. Fore-Aft compliance mode (c) 17.35 Hz. Unsprung mass lateral mode (d) 47.31 Hz.

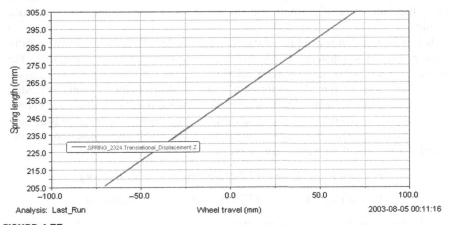

FIGURE 4.77

Spring motion with respect to wheel motion from the model.

Using the motion ratio, the wheel rate due to the spring can be seen to be $31.96/1.43^2 = 15.63$ N/mm. Reassessing the classical calculations, the estimated ride frequency is now 0.874 Hz, with a ride rate of 14.24 N/mm. The estimated frequency is now less than that calculated using the full model (which was 0.954 Hz), suggesting there is some additional stiffness or reduced mass in the mode of

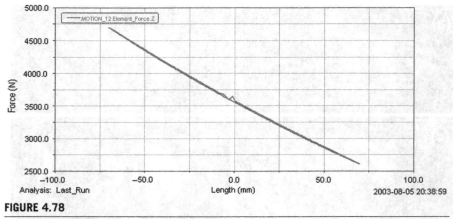

FIGURE 4.78

Wheel rate measured from full linkage model.

vibration. Examining the model again, the so-called 'ride rate' (k_{eqv}) can be taken directly (Figure 4.78). In the case of this particular suspension geometry, the ride rate is 14.97 N/mm. The difference between the two ride rates may be attributed to additional rates arising from the suspension bushes; these are commonly referred to as 'parasitic' rates. Although that term implies something undesirable about them, they are generally small and do not degrade the suspension behaviour unduly.

It may be noted, however, that the difference between the two results is not fully accounted for by the difference in wheel rate. For this particular example, using the analytical solution for the two-mass solution we may calculate 'effective' masses and stiffnesses for each mode of vibration. In the results from the full-linkage model, these effective masses and stiffnesses are reported in the results file. They are known as 'modal' mass and stiffness values and the interested reader is referred to Brüel and Kjær (2003) for further information on the theory of modal decomposition. For the model in question, these values are given as:

Primary Ride Mode : 17.2232 Nmm^{-1} 479.355 kg

Wheel Hop Mode :   177.019 Nmm^{-1} 38.552 kg

It is clear then that there is some other stiffness influence on the ride rate within the model that is not readily apparent to the user. Similarly, the mass is larger than the mass associated with the body alone. This acknowledges the fact that several components are in motion with different velocities when the modes of vibration are excited and thus are storing kinetic energy.

For an appreciation of these and other differences between the simplified and detailed model, the kinetic and strain energy tables calculated by the software can be examined in some detail. Table 4.8 shows the output from the calculations directly. It can be seen that the kinetic energy for the primary ride mode is contained almost

Table 4.8 ADAMS/Linear Output Tables

```
*************************
      Mode number          =        1
      Damping ratio        =   2.20914E-11
      Undamped natural freq.=  9.54000E-01 Cycles per second
      Generalized stiffness =  1.72232E+01 User units
      Generalized mass      =  4.79355E+02 User units
      Kinetic energy        =  2.18135E-01
```

	Percentage distribution of Kinetic energy								
	X	Y	Z	RXX	RYY	RZZ	RXY	RXZ	RYZ
PART/20									
PART/1			98.50						
PART/10		0.06	0.04	0.02					
PART/11		0.03	0.02						
PART/12			0.22	0.05					
PART/13			0.30						
PART/14			0.34	0.03	0.01		0.01		
PART/15			0.04	0.01					
PART/16			0.08						
PART/17		0.01	0.20	0.01					

	Percentage distribution of Strain energy						
	Total	X	Y	Z	RXX	RYY	RZZ
BUSH/16	0.01				0.01		
BUSH/17	0.01				0.01		
BUSH/19	0.14				0.14		
BUSH/21	16.32	0.06	2.47	13.78			
SFOR/1029	5.35	0.01	0.55	4.79			
SFOR/2728							
SFOR/2526							
SFOR/3233							
SPRI/2324	37.71	0.19	0.25	37.27			
GFOR/16	20.16	2.38	17.27	0.02	0.49		
GFOR/17	2.56	0.22	1.79	0.13	0.42		

entirely in the body mass. However, it will be noted that 472/0.985 = 479 kg; thus the kinetic energy table shows the contributions of the other components in the model to the modal mass. The largest of the other contributors to the modal mass are the suspension arms and damper. For the strain energy results, it can be seen that the compliance in the lower arm bushes and the damper lower bush contribute significantly to strain energy storage in the primary ride mode of vibration.

The preceding results were all generated using an undamped eigensolution. However, for vehicle ride work, the influence of the dampers is critical. Moreover, the dampers are typically highly nonlinear devices. Therefore, a further treatment of the existing MSC ADAMS model is required once the modal 'positioning' (i.e. the undamped frequencies for primary ride, wheel hop and fore-aft compliance) is established.

In terms of the modelling approach the only modification here is a change in the motion applied to the jack where the function now represents a sinusoidal input with fixed amplitude but with a frequency that increases as a function of time as illustrated in

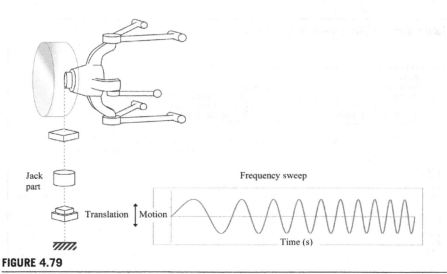

FIGURE 4.79

Input of frequency sweep via jack motion.

Figure 4.79. The motion input is referred to as a frequency sweep, sometimes described as a 'chirp' — for reasons that are obvious if the resulting signal is audible.

The following motion statement is an example of a suitable input function where the amplitude of the road input is fixed at 10 mm and the frequency is increased using the following function from 0 to 20 Hz after 80 s.

```
FUNCTION=10.0*SIN(TIME/8*TIME*360D)
```

Using this simulation we can produce a time history plot showing the change in vertical acceleration of the sprung and unsprung masses of the quarter vehicle model as the simulation progresses. Examination of the response shown in Figure 4.80 reveals that excitation of a system resonance (a 'mode of vibration') occurs at two points during the simulation. The first of these corresponds with the natural frequency of the body and the second with the natural frequency of the unsprung mass.

Another interpretation of the results obtained here is to perform a fast Fourier transform so that the results can be plotted in the frequency rather than the time domain as shown in the lower half of Figure 4.80. The simulation was allowed to run for 81.91 s with an output sample rate of 100 Hz, giving 8192 points including the zero time point. A single buffer transform was performed on the entire record for each of the signals. Although flawed, this method is adequate for identifying frequency peaks. However, for quantifying amplitude content the underlying presumption that the signal repeats itself after the end of the observation buffer is clearly in error; therefore the magnitude results of this exercise should not be used further.

From Figure 4.80 it can be seen that the damped natural frequency of the body occurs at around 0.94 Hz and that the natural frequency of the unsprung mass is about 9.5 Hz. Comparing these values with the previous values, it can be seen

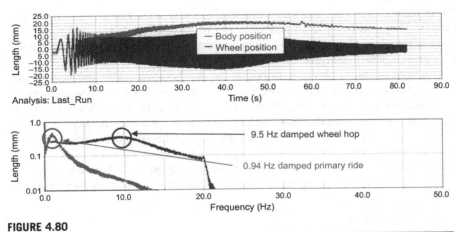

FIGURE 4.80

Quarter vehicle body vertical acceleration time history. (For colour version of this figure, the reader is referred to the online version of this book.)

they are systematically low. This is to be expected since the addition of a damping ratio ζ reduces the damped natural frequency, ω_d, when compared with the undamped natural frequency, ω_n

$$\omega_d = \omega_n \sqrt{1-\zeta^2} \qquad (4.103)$$

The greatest reduction in frequency comes with the heavily damped wheel hop mode. Observing the change between undamped and damped frequency allows an estimate of the damping ratio to be made. For the two modes the damping ratios can be estimated as 0.07 and 0.48 for primary ride and wheel hop, respectively. The damping of the primary ride is low. However, if the exercise were to be repeated at different amplitudes of excitation, the level of damping in the primary ride is certain to vary since the damper characteristics are highly nonlinear. For this reason, the time-domain method and subsequent processing are the preferred methods for evaluating ride behaviour once simplified undamped positioning calculations have been carried out.

4.10 Case study 5 — suspension vector analysis comparison with MBS

4.10.1 Problem definition

The following study is intended to demonstrate the application of the vector theory outlined in Chapter 2 to a range of suspension analyses. Before the advent of computer programs to analyse the motion of suspension linkages, vehicle designers resorted to graphical methods or simplified calculations often using two-dimensional representations to study the suspension in a fore-aft or transverse plane.

These methods are still taught and included in many texts addressing vehicle dynamics. They can develop an understanding of suspension design, and the effect on total vehicle performance, that can be lost using the automated methods in modern computer aided analysis.

The aim of this textbook is to bridge the gap between traditional vehicle dynamics theories and the MBS approach. The following calculations are typical of the processes carried out using MBS software. Whilst the methods used here do not represent exactly the internal machinations of, for example, the MSC ADAMS software, they do give an indication of the computational process involved. The example chosen here is based on a typical double wishbone suspension system. The answers obtained are compared with those running an MSC ADAMS model using the same data. The calculations will include a series of analyses including:

- Geometry analysis
- Velocity analysis
- Acceleration analysis
- Static force analysis
- Dynamic force analysis

The geometric data required to define this problem is defined in Figure 4.81. Note that in this example the x-axis is orientated towards the front of the vehicle rather than pointing to the rear as used generally throughout this chapter. This is therefore the front left suspension system on the vehicle.

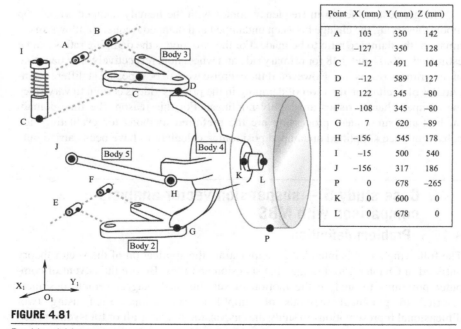

Point	X (mm)	Y (mm)	Z (mm)
A	103	350	142
B	−127	350	128
C	−12	491	104
D	−12	589	127
E	122	345	−80
F	−108	345	−80
G	7	620	−89
H	−156	545	178
I	−15	500	540
J	−156	317	186
P	0	678	−265
K	0	600	0
L	0	678	0

FIGURE 4.81

Double wishbone suspension example geometry data.

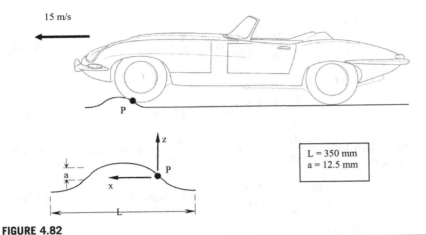

FIGURE 4.82

Road input definition for velocity analysis.

4.10.2 Velocity analysis

The starting point for this analysis is to establish a boundary condition, or road input, at the tyre contact patch, P, as the vehicle negotiates the road hump shown in Figure 4.82 with a forward speed of 15 m/s. The analysis is simplified by ignoring the compliance in the tyre and the profile of the road hump is taken as a sine function.

The local x-z axis taken to reference the geometry of the road hump is located at a point where the vertical velocity, V_{Pz}, of the contact point P reaches a maximum with a corresponding vertical acceleration, A_{Pz} equal to zero. The profile of the road hump can be defined using

$$z = a\sin\left(\frac{2\pi x}{L}\right) \tag{4.104}$$

Working in millimetres and taking the wavelength L as 350 mm and the amplitude a as 12.5 mm, for a total bump height of 25 mm, gives

$$z = 12.5\sin\left(\frac{\pi x}{175}\right) \tag{4.105}$$

We are after the vertical velocity of point P, which can be expressed as:

$$\frac{dz}{dt} = \frac{dz}{dx}\frac{dx}{dt} \tag{4.106}$$

where

$$\frac{dx}{dt} = 15000\,\text{mm/s}$$

and

$$\frac{dz}{dx} = \frac{12.5\,\pi}{175}\cos\left(\frac{\pi x}{175}\right) \tag{4.107}$$

giving

$$\frac{dz}{dt} = \frac{12.5\,\pi}{175}\cos\left(\frac{\pi x}{175}\right)15000 \tag{4.108}$$

The maximum value of $\frac{dz}{dt}$ occurs when $\cos\left(\frac{\pi x}{175}\right) = 1$ and occurs at values of $x = 0, 350, 700 \ldots$ giving

$$\left(\frac{dz}{dt}\right)_{max} = \frac{12.5\,\pi}{175}\,15000 = 3366\,\text{mm/s} \tag{4.109}$$

The acceleration $\frac{d^2z}{dt^2}$ is given by

$$\frac{d^2z}{dt^2} = -\frac{12.5\,\pi^2}{175^2}\sin\left(\frac{\pi x}{175}\right)15000 \tag{4.110}$$

and has a value of zero at $x = 0, 350, 700 \ldots$

This provides inputs for the following velocity and acceleration analyses of

$$V_{Pz} = 3366\ \text{mm/s}$$

$$A_{Pz} = \quad 0\quad \text{mm/s}^2$$

The approach taken here is to initially ignore the spring damper assembly between points C and I shown in Figure 4.82. Solving for the rest of the suspension system will deliver the velocity $\{V_C\}_1$ of point C, thus providing a boundary condition allowing a separate analysis of the spring damper to follow.

Before proceeding with the velocity analysis it is necessary to identify the unknowns that define the problem and the same number of equations as unknowns leading to a solution. The angular velocities of the rigid bodies representing suspension components can be used to find the translational velocities at points within the system. An example of this is shown for the upper wishbone in Figure 4.83.

Referring back to the earlier treatment in Chapter 2 we can remind ourselves that as the suspension arm is constrained to rotate about the axis AB, ignoring at this stage any possible deflection due to compliance in the suspension bushes, the vector $\{\omega_3\}_1$ for the angular velocity of Body 3 will act along the axis of rotation through AB. The components of this vector would adopt signs consistent with producing a positive rotation about this axis as shown in Figure 4.83.

When setting up the equations to solve a velocity analysis it will be desirable to reduce the number of unknowns based on the knowledge that a particular body is constrained to rotate about a known axis as shown here. The velocity vector, $\{\omega_3\}_1$, could for example be represented as follows.

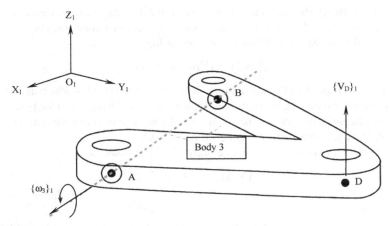

FIGURE 4.83

Angular and translational velocity vectors for the upper wishbone.

$$\{\omega_3\}_1 = f\omega_3 \{R_{AB}\}_1 \qquad (4.111)$$

Since $\{\omega_3\}_1$ is parallel to the relative position vector $\{R_{AB}\}_1$ a scale factor $f\omega_3$ can be introduced. This reduces the problem from the three unknown components, ωx_3, ωy_3 and ωz_3 of the vector $\{\omega_3\}_1$ to a single unknown $f\omega_3$.

Once the angular velocities of Body 3 have been found it follows that the translational velocity of, for example, point D can be found from

$$\{V_{DA}\}_1 = \{\omega_3\}_1 \times \{R_{DA}\}_1 \qquad (4.112)$$

It also follows that since point A is considered fixed with a velocity $\{V_A\}_1$ equal to zero that the absolute velocity $\{V_D\}_1$ of point D can be found from a consideration of the triangle law of vector addition giving

$$\{V_D\}_1 = \{V_{DA}\}_1 \qquad (4.113)$$

A consideration of the complete problem indicates that the translational velocities throughout the suspension system can be found if the angular velocities of all the rigid bodies 2, 3, 4 and 5 are known. Clearly the same approach can be taken with the lower wishbone, Body 2, as with the upper wishbone using a single scale factor $f\omega_2$ to replace the three unknown components, ωx_2, ωy_2 and ωz_2 of the vector $\{\omega_2\}_1$. Finally a consideration of the boundaries of this problem reveals that while points A, B, E, F and J are fixed, the longitudinal velocity, V_{Px} and the lateral velocity, V_{Py} at the contact point P remain as unknowns.

Thus this analysis can proceed if we can develop 10 equations to solve the 10 unknowns:

$$f\omega_2, f\omega_3, \omega x_4, \omega y_4, \omega z_4, \omega x_5, \omega y_5, \omega z_5, V_{Px}, V_{Py}$$

Working through the problem it will be seen that a strategy can be developed using the triangle law of vector addition to generate sets of equations. As a starting point we will develop a set of three equations using

$$\{V_{DG}\}_1 = \{V_D\}_1 - \{V_G\}_1 \qquad (4.114)$$

In this form Eqn (4.114) does not introduce any of the 10 unknowns listed above. It will therefore be necessary to initially define $\{V_{DG}\}_1$, $\{V_D\}_1$ and $\{V_G\}_1$ in terms of the angular velocity vectors that contain unknowns requiring solution. The first step in the analysis can therefore proceed as follows.
Determine an expression for the velocity $\{V_G\}_1$ at point G

$$\{V_G\}_1 = \{V_{GE}\}_1 = \{\omega_2\}_1 \times \{R_{GE}\}_1 \qquad (4.115)$$

$$\{\omega_2\}_1 = f\omega_2\,\{R_{EF}\}_1 = f\omega_2 \begin{bmatrix} 230 \\ 0 \\ 0 \end{bmatrix} \text{ rad/s} \qquad (4.116)$$

$$\begin{bmatrix} V_{Gx} \\ V_{Gy} \\ V_{Gz} \end{bmatrix} = f_{\omega2} \begin{bmatrix} 0 & 0 & 0 \\ 0 & 0 & -230 \\ 0 & 230 & 0 \end{bmatrix} \begin{bmatrix} -115 \\ 275 \\ -9 \end{bmatrix} = \begin{bmatrix} 0 \\ 2070\,f_{\omega2} \\ 63250\,f_{\omega2} \end{bmatrix} \text{mm/s} \qquad (4.117)$$

Determine an expression for the velocity $\{V_D\}_1$ at point D

$$\{V_D\}_1 = \{V_{DA}\}_1 = \{\omega_3\}_1 \times \{R_{DA}\}_1 \qquad (4.118)$$

$$\{\omega_3\}_1 = f\omega_3\,\{R_{AB}\}_1 = f\omega_3 \begin{bmatrix} 230 \\ 0 \\ 14 \end{bmatrix} \text{ rad/s} \qquad (4.119)$$

$$\begin{bmatrix} V_{Dx} \\ V_{Dy} \\ V_{Dz} \end{bmatrix} = f_{\omega3} \begin{bmatrix} 0 & -14 & 0 \\ 14 & 0 & -230 \\ 0 & 230 & 0 \end{bmatrix} \begin{bmatrix} -115 \\ 239 \\ -15 \end{bmatrix} = \begin{bmatrix} -3346f_{\omega3} \\ 1840\,f_{\omega3} \\ 54970\,f_{\omega3} \end{bmatrix} \text{mm/s} \qquad (4.120)$$

Determine an expression for the relative velocity $\{V_{DG}\}_1$ of point D relative to point G

$$\{V_{DG}\}_1 = \{\omega_4\}_1 \times \{R_{DG}\}_1 \qquad (4.121)$$

$$\begin{bmatrix} V_{DGx} \\ V_{DGy} \\ V_{DGz} \end{bmatrix} = \begin{bmatrix} 0 & -\omega_{4z} & \omega_{4y} \\ \omega_{4z} & 0 & -\omega_{4x} \\ -\omega_{4y} & \omega_{4x} & 0 \end{bmatrix} \begin{bmatrix} -19 \\ -31 \\ 216 \end{bmatrix} = \begin{bmatrix} 31\omega_{4z}+216\,\omega_{4y} \\ -19\omega_{4z}-216\,\omega_{4x} \\ 19\omega_{4y}-31\omega_{4x} \end{bmatrix} \text{mm/s} \qquad (4.122)$$

We can now apply the triangle law of vector addition to equate the expression for $\{V_{DG}\}_1$ in Eqn (4.122) with $\{V_D\}_1$ in Eqn (4.120) and $\{V_G\}_1$ in Eqn (4.117).

$$\{V_{DG}\}_1 = \{V_D\}_1 - \{V_G\}_1 \qquad (4.123)$$

$$\begin{bmatrix} 31\omega_{4z}+216\omega_{4y} \\ -19\omega_{4z}-216\omega_{4x} \\ 19\omega_{4y}-31\omega_{4x} \end{bmatrix} = \begin{bmatrix} -3346f_{\omega3} \\ 1840f_{\omega3} \\ 54970f_{\omega3} \end{bmatrix} - \begin{bmatrix} 0 \\ 2070f_{\omega2} \\ 63250f_{\omega2} \end{bmatrix} \text{mm/s} \qquad (4.124)$$

Rearranging Eqn (4.124) yields the first three equations required to solve the analysis.

Equation 1 $\qquad\qquad 3346\,f_{\omega3} + 216\,\omega_{4y} + 31\,\omega_{4z} = 0 \qquad (4.125)$

Equation 2 $\qquad\qquad 2070\,f_{\omega2} - 1840\,f_{\omega3} - 216\,\omega_{4x} - 19\,\omega_{4z} = 0 \qquad (4.126)$

Equation 3 $\qquad\qquad 63250\,f_{\omega2} - 54970\,f_{\omega3} - 31\,\omega_{4x} + 19\,\omega_{4y} = 0 \qquad (4.127)$

We can now proceed to set up the next set of three equations working from point H to point D and using the triangle law of vector addition.

$$\{V_{DH}\}_1 = \{V_D\}_1 - \{V_H\}_1 \qquad (4.128)$$

Determine an expression for the velocity $\{V_H\}_1$ at point H

$$\{V_H\}_1 = \{V_{HJ}\}_1 = \{\omega_5\}_1 \times \{R_{HJ}\}_1 \qquad (4.129)$$

$$\begin{bmatrix} V_{Hx} \\ V_{Hy} \\ V_{Hz} \end{bmatrix} = \begin{bmatrix} 0 & -\omega_{5z} & \omega_{5y} \\ \omega_{5z} & 0 & -\omega_{5x} \\ -\omega_{5y} & \omega_{5x} & 0 \end{bmatrix} \begin{bmatrix} 0 \\ 228 \\ -8 \end{bmatrix} = \begin{bmatrix} -228\omega_{5z}-8\omega_{5y} \\ 8\omega_{5x} \\ 228\omega_{5x} \end{bmatrix} \text{mm/s} \qquad (4.130)$$

We already have an expression for $\{V_D\}_1$ in Eqn (4.120) and can determine an expression for the relative velocity $\{V_{DH}\}_1$ of point D relative to point H using

$$\{V_{DH}\}_1 = \{\omega_4\}_1 \times \{R_{DH}\}_1 \qquad (4.131)$$

$$\begin{bmatrix} V_{DHx} \\ V_{DHy} \\ V_{DHz} \end{bmatrix} = \begin{bmatrix} 0 & -\omega_{4z} & \omega_{4y} \\ \omega_{4z} & 0 & -\omega_{4x} \\ -\omega_{4y} & \omega_{4x} & 0 \end{bmatrix} \begin{bmatrix} 144 \\ 44 \\ -51 \end{bmatrix} = \begin{bmatrix} -44\omega_{4z}-51\omega_{4y} \\ 144\omega_{4z}+51\omega_{4x} \\ -144\omega_{4y}+44\omega_{4x} \end{bmatrix} \text{mm/s} \qquad (4.132)$$

We can now apply the triangle law of vector addition to equate the expression for $\{V_{DH}\}_1$ in Eqn (4.132) with $\{V_D\}_1$ in Eqn (4.120) and $\{V_H\}_1$ in Eqn (4.130).

$$\{V_{DH}\}_1 = \{V_D\}_1 - \{V_H\}_1 \qquad (4.133)$$

$$\begin{bmatrix} -44\,\omega_{4z} - 51\,\omega_{4y} \\ 144\,\omega_{4z} + 51\,\omega_{4x} \\ -144\,\omega_{4y} + 44\,\omega_{4x} \end{bmatrix} = \begin{bmatrix} -3346\,f_{\omega 3} \\ 1840\,f_{\omega 3} \\ 54970\,f_{\omega 3} \end{bmatrix} - \begin{bmatrix} -228\,\omega_{5z} - 8\,\omega_{5y} \\ 8\,\omega_{5x} \\ 228\,\omega_{5x} \end{bmatrix} \text{mm/s} \qquad (4.134)$$

Rearranging Eqn (4.134) yields the next three equations required to solve the analysis.

Equation 4 $\qquad 3346\,f_{\omega 3} - 51\,\omega_{4y} - 44\,\omega_{4z} - 8\,\omega_{5y} - 228\,\omega_{5z} = 0 \qquad (4.135)$

Equation 5 $\qquad -1840\,f_{\omega 3} + 51\,\omega_{4x} + 144\,\omega_{4z} + 8\,\omega_{5x} = 0 \qquad (4.136)$

Equation 6 $\qquad -54970\,f_{\omega 3} + 44\,\omega_{4x} - 144\,\omega_{4y} + 228\,\omega_{5x} = 0 \qquad (4.137)$

We can now proceed to set up the next set of three equations working from point G to point P and using the triangle law of vector addition.

$$\{V_{PG}\}_1 = \{V_P\}_1 - \{V_G\}_1 \qquad (4.138)$$

Determine an expression for the relative velocity $\{V_{PG}\}_1$ of point P relative to point G

$$\{V_{PG}\}_1 = \{\omega_4\}_1 \times \{R_{PG}\}_1 \qquad (4.139)$$

$$\begin{bmatrix} V_{PGx} \\ V_{PGy} \\ V_{PGz} \end{bmatrix} = \begin{bmatrix} 0 & -\omega_{4z} & \omega_{4y} \\ \omega_{4z} & 0 & -\omega_{4x} \\ -\omega_{4y} & \omega_{4x} & 0 \end{bmatrix} \begin{bmatrix} -7 \\ 58 \\ -176 \end{bmatrix} = \begin{bmatrix} -58\,\omega_{4z} - 176\,\omega_{4y} \\ -7\,\omega_{4z} + 176\,\omega_{4x} \\ 7\,\omega_{4y} + 58\,\omega_{4x} \end{bmatrix} \text{mm/s} \qquad (4.140)$$

We already have an expression for $\{V_G\}_1$ in Eqn (4.117) and we can define the vector $\{V_P\}_1$ in terms of the known vertical velocity component, V_{Pz} and the unknown components, V_{Px} and V_{Py}.

$$\{V_P\}_1 = \begin{bmatrix} V_{Px} \\ V_{Py} \\ 3366 \end{bmatrix} \text{mm/s} \qquad (4.141)$$

We can now apply the triangle law of vector addition to equate the expression for $\{V_{PG}\}_1$ in Eqn (4.140) with $\{V_P\}_1$ in Eqn (4.141) and $\{V_G\}_1$ in Eqn (4.117).

$$\{V_{PG}\}_1 = \{V_P\}_1 - \{V_G\}_1 \qquad (4.142)$$

$$\begin{bmatrix} -58\,\omega_{4z} - 176\,\omega_{4y} \\ -7\,\omega_{4z} + 176\,\omega_{4x} \\ 7\,\omega_{4y} + 58\,\omega_{4x} \end{bmatrix} = \begin{bmatrix} V_{PX} \\ V_{Py} \\ 3366 \end{bmatrix} - \begin{bmatrix} 0 \\ 2070\,f_{\omega 2} \\ 63250\,f_{\omega 2} \end{bmatrix} \text{mm/s} \qquad (4.143)$$

Rearranging Eqn (4.143) yields the next set of three equations required to solve the analysis.

Equation 7 $\qquad$ $176\,\omega_{4y} + 58\,\omega_{4z} + V_{Px} = 0$ $\qquad$ (4.144)

Equation 8 $\qquad$ $-2070\,f_{\omega 2} - 176\,\omega_{4x} + 7\,\omega_{4z} + V_{Py} = 0$ $\qquad$ (4.145)

Equation 9 $\qquad$ $-63250\,f_{\omega 2} - 58\,\omega_{4x} - 7\,\omega_{4y} = -3366$ $\qquad$ (4.146)

This leaves us with nine equations and ten unknowns. The last equation is obtained by constraining the rotation of the tie rod (Body 5) to prevent spin about its own axis. This degree of freedom has no bearing on the overall kinematics of the suspension linkage. If there is to be no spin component of angular velocity parallel to the axis of the tie rod, spin, then the angular velocity vector $\{\omega_5\}_1$ must be perpendicular to the line HJ as shown in Figure 4.84. Note that this is equivalent to a common practice in MBS modelling where a universal or Hookes joint is used at one end of the link with a spherical joint at the other end. The universal joint allows the tie rod to articulate in a way that does not constrain overall suspension movement but constrains the spin freedom that would exist if a spherical joint were used at each end.

Using the vector dot product to enforce perpendicularity, as described in Chapter 2, yields the tenth and final equation required to solve this part of the problem.

$$\{\omega_5\}_1 \bullet \{R_{HJ}\}_1 \;=\; \begin{bmatrix} \omega_{5x} & \omega_{5y} & \omega_{5z} \end{bmatrix} \begin{bmatrix} 0 \\ 228 \\ -8 \end{bmatrix} = 0 \text{ mm/s} \qquad (4.147)$$

Equation 10 $\qquad$ $228\,\omega_{5y} - 8\,\omega_{5z} = 0$ $\qquad$ (4.148)

The 10 equations can now be set up in matrix form ready for solution. The solution of a 10×10 matrix will require access to a mathematical or spreadsheet program that offers the capability to invert the matrix.

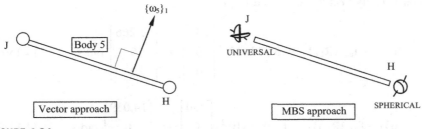

FIGURE 4.84

Constraining the spin degree of freedom of the tie rod. MBS, multibody systems.

$$
\begin{bmatrix}
0 & 3346 & 0 & 216 & 31 & 0 & 0 & 0 & 0 & 0 \\
2070 & -1840 & -216 & 0 & -19 & 0 & 0 & 0 & 0 & 0 \\
63250 & -54970 & -31 & 19 & 0 & 0 & 0 & 0 & 0 & 0 \\
0 & 3346 & 0 & -51 & -44 & 0 & -8 & -228 & 0 & 0 \\
0 & -1840 & 51 & 0 & 144 & 8 & 0 & 0 & 0 & 0 \\
0 & -54970 & 44 & -144 & 0 & 228 & 0 & 0 & 0 & 0 \\
0 & 0 & 0 & 176 & 58 & 0 & 0 & 0 & 1 & 0 \\
-2070 & 0 & -176 & 0 & 7 & 0 & 0 & 0 & 0 & 1 \\
-63250 & 0 & -58 & -7 & 0 & 0 & 0 & 0 & 0 & 0 \\
0 & 0 & 0 & 0 & 0 & 0 & 228 & -8 & 0 & 0
\end{bmatrix}
\begin{bmatrix}
f_{\omega 2} \\ f_{\omega 3} \\ \omega_{4x} \\ \omega_{4y} \\ \omega_{4z} \\ \omega_{5x} \\ \omega_{5y} \\ \omega_{5z} \\ V_{Px} \\ V_{Py}
\end{bmatrix}
=
\begin{bmatrix}
0 \\ 0 \\ 0 \\ 0 \\ 0 \\ 0 \\ 0 \\ 0 \\ -3366 \\ 0
\end{bmatrix}
\tag{4.149}
$$

Solving Eqn (4.149) yields the following answers for the 10 unknowns:

$$f_{\omega 2} = 5.333 \times 10^{-2} \ \text{rad/mm.s}$$

$$f_{\omega 3} = 6.104 \times 10^{-2} \ \text{rad/mm.s}$$

$$\omega_{4x} = -8.774 \times 10^{-3} \text{rad/s}$$

$$\omega_{4y} = -0.945 \, \text{rad/s}$$

$$\omega_{4z} = -1.446 \times 10^{-3} \text{rad/s}$$

$$\omega_{5x} = 14.121 \, \text{rad/s}$$

$$\omega_{5y} = 3.881 \times 10^{-2} \text{rad/s}$$

$$\omega_{5z} = 1.106 \, \text{rad/s}$$

$$V_{Px} = 166.468 \, \text{mm/s}$$

$$V_{Py} = 108.859 \, \text{mm/s}$$

It is now possible to use the two scale factors found, $f_{\omega 2}$ and $f_{\omega 3}$, to calculate the angular velocity vectors $\{\omega_2\}_1$ and $\{\omega_3\}_1$.

$$\{\omega_2\}_1 = f_{\omega 2} \{R_{EF}\}_1 = 5.333 \times 10^{-2} \begin{bmatrix} 230 \\ 0 \\ 0 \end{bmatrix} = \begin{bmatrix} 12.266 \\ 0 \\ 0 \end{bmatrix} \text{rad/s} \tag{4.150}$$

$$\{\omega_3\}_1 = f_{\omega 3} \{R_{AB}\}_1 = 6.1041 \times 10^{-2} \begin{bmatrix} 230 \\ 0 \\ 14 \end{bmatrix} = \begin{bmatrix} 14.039 \\ 0 \\ 0.855 \end{bmatrix} \text{rad/s} \tag{4.151}$$

In summary the angular velocity vectors for the rigid bodies are as follows:

$$\{\omega_2\}_1^T = \begin{bmatrix} 12.266 & 0 & 0 \end{bmatrix} \text{rad/s}$$

$$\{\omega_3\}_1^T = \begin{bmatrix} 14.039 & 0 & 0.855 \end{bmatrix} \text{rad/s}$$

$$\{\omega_4\}_1^T = \begin{bmatrix} -8.774 \times 10^{-3} & -0.945 & -1.446 \times 10^{-3} \end{bmatrix} \text{rad/s}$$

$$\{\omega_5\}_1^T = \begin{bmatrix} 14.121 & 3.881 \times 10^{-2} & 1.106 \end{bmatrix} \text{rad/s}$$

We can now proceed to calculate the translational velocities at all the moving points, C, D, G, H and P, within this part of the model.

$$\{V_C\}_1 = \{V_{CA}\}_1 = \{\omega_3\}_1 \times \{R_{CA}\}_1 \tag{4.152}$$

$$\begin{bmatrix} V_{Cx} \\ V_{Cy} \\ V_{Cz} \end{bmatrix} = \begin{bmatrix} 0 & -0.855 & 0 \\ 0.855 & 0 & -14.039 \\ 0 & 14.039 & 0 \end{bmatrix} \begin{bmatrix} -115 \\ 141 \\ -38 \end{bmatrix} = \begin{bmatrix} -120.555 \\ 435.157 \\ 1979.499 \end{bmatrix} \text{mm/s} \tag{4.153}$$

$$\{V_D\}_1 = \{V_{DA}\}_1 = \{\omega_3\}_1 \times \{R_{DA}\}_1 \tag{4.154}$$

$$\begin{bmatrix} V_{Dx} \\ V_{Dy} \\ V_{Dz} \end{bmatrix} = \begin{bmatrix} 0 & -0.855 & 0 \\ 0.855 & 0 & -14.039 \\ 0 & 14.039 & 0 \end{bmatrix} \begin{bmatrix} -115 \\ 239 \\ -15 \end{bmatrix} = \begin{bmatrix} -204.345 \\ 112.260 \\ 3355.321 \end{bmatrix} \text{mm/s} \tag{4.155}$$

$$\{V_G\}_1 = \{V_{GE}\}_1 = \{\omega_2\}_1 \times \{R_{GE}\}_1 \tag{4.156}$$

$$\begin{bmatrix} V_{Gx} \\ V_{Gy} \\ V_{Gz} \end{bmatrix} = \begin{bmatrix} 0 & 0 & 0 \\ 0 & 0 & -12.266 \\ 0 & 12.266 & 0 \end{bmatrix} \begin{bmatrix} -115 \\ 275 \\ -9 \end{bmatrix} = \begin{bmatrix} 0 \\ 110.394 \\ 3373.150 \end{bmatrix} \text{mm/s} \tag{4.157}$$

$$\{V_H\}_1 = \{V_{HJ}\}_1 = \{\omega_5\}_1 \times \{R_{HJ}\}_1 \tag{4.158}$$

$$\begin{bmatrix} V_{Hx} \\ V_{Hy} \\ V_{Hz} \end{bmatrix} = \begin{bmatrix} 0 & -1.1062 & 3.881 \times 10^{-2} \\ 1.1062 & 0 & -14.121 \\ -3.881 \times 10^{-2} & 14.121 & 0 \end{bmatrix} \begin{bmatrix} 0 \\ 229 \\ -9 \end{bmatrix} = \begin{bmatrix} -252.524 \\ 112.968 \\ 3219.588 \end{bmatrix} \text{mm/s}$$

$$\tag{4.159}$$

The velocity vector $\{V_P\}_1$ is already available. In summary the velocity vectors for the moving points are as follows:

$$\{V_C\}_1^T = [-120.555 \quad 435.157 \quad 1979.499] \text{ mm/s}$$

$$\{V_D\}_1^T = [-204.345 \quad 112.260 \quad 3355.321] \text{ mm/s}$$

$$\{V_G\}_1^T = [0.0 \quad 110.394 \quad 3373.150] \text{ mm/s}$$

$$\{V_H\}_1^T = [-252.524 \quad 112.968 \quad 3219.588] \text{ mm/s}$$

$$\{V_P\}_1^T = [166.468 \quad 108.859 \quad 3366.0] \text{ mm/s}$$

Having found the velocity $\{V_C\}_1$ at the bottom of the springdamper unit point C we can now proceed to carry out a separate analysis of the unit to find the sliding component of velocity acting along the axis CI. For this phase of the analysis we introduce two new bodies, Body 6 and Body 7, to represent the upper and lower parts of the damper as shown in Figure 4.85.

This phase of the analysis can be facilitated by modelling three coincident points, C_3 on Body 3, C_6 on Body 6 and C_7 on Body 7, all located at point C. While points C_3 and C_7 can be considered physically located at point C, point C_6 is a virtual extension of the upper damper as shown in Figure 4.85. The sliding velocity in the damper can then be determined from the relative velocity $\{V_{C6C7}\}_1$ of points C_6 and C_7. Since C_6 and C_7 are coincident, the relative velocity vector can only act along the direction of sliding, axis CI, allowing us to adopt a scale factor and reduce the number of unknowns.

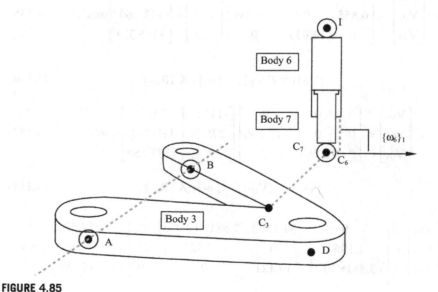

FIGURE 4.85

Modelling the damper unit for a velocity analysis.

$$\{V_{C6C7}\}_1 = f_{vs}\,\{R_{CI}\}_1 \qquad (4.160)$$

Since I is a fixed point and applying the triangle law of vector addition gives

$$\{V_{C6}\}_1 = \{V_{C6I}\}_1 = \{V_{C7}\}_1 + \{V_{C6C7}\}_1 \qquad (4.161)$$

Note also that since points C_3 and C_7 move together and are physically located at point C we already have the velocity for this boundary condition from the preceding velocity analysis of the double wishbone linkage.

$$\{V_{C7}\}_1^T = \{V_{C3}\}_1^T = \{V_C\}_1^T = \begin{bmatrix} -120.555 & 435.157 & 1979.499 \end{bmatrix} \text{mm/s} \qquad (4.162)$$

Combining these last three equations to substitute Eqns (4.160) and (4.162) into Eqn (4.161) gives

$$\{V_{C6}\}_1 = \{V_{C6I}\}_1 = \begin{bmatrix} -120.555 \\ 435.157 \\ 1979.499 \end{bmatrix} + f_{vs} \begin{bmatrix} 3 \\ -9 \\ -436 \end{bmatrix} \text{mm/s} \qquad (4.163)$$

As the suspension moves and the strut component rotates it is also clear that as the only degree of freedom between Body 6 and Body 7 is relative sliding motion then $\{\omega_6\}_1 = \{\omega_7\}_1$.

The velocity vector $\{V_{C6I}\}_1$ can also be defined using

$$\{V_{C6}\}_1 = \{V_{C6I}\}_1 = \{\omega_6\}_1 \times \{R_{CI}\}_1 \qquad (4.164)$$

$$\begin{bmatrix} V_{C6Ix} \\ V_{C6Iy} \\ V_{C6Iz} \end{bmatrix} = \begin{bmatrix} 0 & -\omega_{6z} & \omega_{6y} \\ \omega_{6z} & 0 & -\omega_{6x} \\ -\omega_{6y} & \omega_{6x} & 0 \end{bmatrix} \begin{bmatrix} 3 \\ -9 \\ -436 \end{bmatrix} = \begin{bmatrix} 9\,\omega_{6z} - 436\,\omega_{6y} \\ 3\,\omega_{6z} + 436\,\omega_{6x} \\ -3\,\omega_{6y} - 9\,\omega_{6x} \end{bmatrix} \text{mm/s} \qquad (4.165)$$

Equating the expressions for $\{V_{C6I}\}_1$ in Eqns (4.162) and (4.164) gives

$$\begin{bmatrix} 9\,\omega_{6z} - 436\,\omega_{6y} \\ 3\,\omega_{6z} + 436\,\omega_{6x} \\ -3\,\omega_{6y} - 9\,\omega_{6x} \end{bmatrix} = \begin{bmatrix} -120.555 \\ 435.157 \\ 1979.499 \end{bmatrix} + f_{vs} \begin{bmatrix} 3 \\ -9 \\ -436 \end{bmatrix} \text{mm/s} \qquad (4.166)$$

Rearranging Eqn (4.166) yields three equations that can be used to solve this part of the analysis.

Equation 1 $\qquad\qquad 3\,f_{vs} + 436\,\omega_{6y} - 9\,\omega_{6z} = 120.555 \qquad (4.167)$

Equation 2 $\qquad\qquad -9\,f_{vs} - 436\,\omega_{6x} - 3\,\omega_{6z} = -435.157 \qquad (4.168)$

Equation 3 $\qquad\qquad -436\,f_{vs} + 9\,\omega_{6x} + 3\,\omega_{6y} = -1979.499 \qquad (4.169)$

This leaves us with four unknowns, ω_{6x}, ω_{6y}, ω_{6z} and f_{vs} but only three equations. We can use the same approach here as used with the tie rod in the preceding analysis. Since the spin degree of freedom of Body 6 about the axis CI has no bearing on the overall solution we can again use the vector dot product to enforce perpendicularity of $\{\omega_6\}_1$ to $\{R_{CI}\}_1$ as shown in Figure 4.85. This will yield the fourth equation as follows.

$$\{\omega_6\}_1 \bullet \{R_{CI}\}_1 = 0 \tag{4.170}$$

$$\begin{bmatrix} \omega_{6x} & \omega_{6y} & \omega_{6z} \end{bmatrix} \begin{bmatrix} 3 \\ -9 \\ -436 \end{bmatrix} = 0 \text{ mm/s} \tag{4.171}$$

Equation 4 $\qquad\qquad 3\,\omega_{6x}\ -9\,\omega_{6y} - 436\,\omega_{6z} = 0 \tag{4.172}$

The four equations can now be set up in matrix form ready for solution. The solution of a four by four matrix will require a lengthy calculation or access as before to a program that offers the capability to invert the matrix.

$$\begin{bmatrix} 3 & 0 & 436 & -9 \\ -9 & -436 & 0 & -3 \\ -436 & 9 & 3 & 0 \\ 0 & 3 & -9 & -436 \end{bmatrix} \begin{bmatrix} f_{vs} \\ \omega_{6x} \\ \omega_{6y} \\ \omega_{67} \end{bmatrix} = \begin{bmatrix} 120.555 \\ -435.157 \\ -1979.499 \\ 0 \end{bmatrix} \tag{4.173}$$

Solving Eqn (4.173) yields the following answers for the four unknowns:

$$f_{vs} = 4.561 \text{ s}^{-1}$$

$$\omega_{6x} = 0.904 \text{ rad/s}$$

$$\omega_{6y} = 0.245 \text{ rad/s}$$

$$\omega_{6z} = 1.159 \text{ rad/s}$$

This gives us the last two angular velocity vectors for the upper and lower damper bodies:

$$\{\omega_6\}_1^T = \begin{bmatrix} 0.904 & 0.245 & 1.159 \end{bmatrix} \text{ rad/s}$$

$$\{\omega_7\}_1^T = \begin{bmatrix} 0.904 & 0.245 & 1.159 \end{bmatrix} \text{ rad/s}$$

From Eqn (4.150) we now have

$$\{V_{C6C7}\}_1 = f_{vs}\{R_{CI}\}_1 = 4.561 \begin{bmatrix} 3 \\ -9 \\ -436 \end{bmatrix} = \begin{bmatrix} 13.682 \\ -41.045 \\ -1988.378 \end{bmatrix} \text{mm/s} \tag{4.174}$$

Table 4.9 Comparison of Angular Velocity Vectors Computed by Theory and MSC ADAMS

	Angular Velocity Vectors					
	Theory			MSC ADAMS		
Body	ω_x (rad/s)	ω_y (rad/s)	ω_z (rad/s)	ω_x (rad/s)	ω_y (rad/s)	ω_z (rad/s)
2	12.266	0.0	0.0	12.266	0.0	0.0
3	14.039	0.0	0.855	14.040	0.0	0.855
4	-8.774×10^{-3}	-0.945	-1.446×10^{-3}	-8.774×10^{-3}	-0.945	-1.446×10^{-3}
5	14.121	3.881×10^{-2}	1.106	14.121	5.394×10^{-2}	1.106
6	0.904	0.245	1.159×10^{-3}	0.904	0.245	1.163×10^{-3}
7	0.904	0.245	1.159×10^{-3}	0.904	0.245	1.163×10^{-3}

Since the velocity vector $\{V_{C6C7}\}_1$ acts along the axis of the strut CI. the magnitude of this vector will be equal to the sliding velocity V_s.

$$V_s = |V_{C6C7}| = 1988.641 \text{ mm/s} \qquad (4.175)$$

At this stage it can be seen that the sliding velocity V_s is realistic in magnitude. Given knowledge of the damper force—velocity relationship it would be possible to determine the damping forces produced and reacted at points I and C in the system.

A comparison of the angular velocities found from the preceding calculations and those found using an equivalent MSC ADAMS model is shown in Table 4.9.

A comparison of the translational velocities found at points within the suspension system from the preceding calculations and those found using an equivalent MSC ADAMS model is shown in Table 4.10.

Table 4.10 Comparison of Translational Velocity Vectors Computed by Theory and MSC ADAMS

	Translational Velocity Vectors					
	Theory			MSC ADAMS		
Point	V_x (mm/s)	V_y (mm/s)	V_z (mm/s)	V_x (mm/s)	V_y (mm/s)	V_z (mm/s)
C	-120.555	435.157	1979.499	-120.495	435.224	1979.570
D	-204.345	112.260	3355.321	-204.244	112.316	3355.440
G	0.0	110.394	3373.150	0.0	110.393	3373.130
H	-252.524	112.968	3219.588	-252.210	112.972	3219.588
P	166.468	108.859	3366.0	166.468	108.859	3366.0
C_6C_7	13.682	-41.045	-1988.378	13.682	-41.046	-1988.440

4.10.3 Acceleration analysis

The approach taken here is as before to initially ignore the damper assembly between points C and I. Solving for the rest of the suspension system will deliver the acceleration $\{A_C\}_1$ of point C, thus providing a boundary condition allowing a separate analysis of the damper to follow.

Before proceeding with the acceleration analysis it is necessary to identify the unknowns that define the problem. The angular accelerations and angular velocities of the rigid bodies representing suspension components can be used to find the translational accelerations at points within the system. An example of this is shown for the upper wishbone in Figure 4.86.

Referring back to the earlier velocity analysis we can remind ourselves that as the suspension arm is constrained to rotate about the axis AB, ignoring at this stage any possible deflection due to compliance in the suspension bushes, the vector $\{\alpha_3\}_1$ for the angular acceleration of Body 3 will act along the axis of rotation through AB. The components of this vector would adopt signs consistent with a positive rotation about this axis as shown in Figure 4.86.

When setting up the equations to solve an acceleration analysis it will be, as before, desirable to reduce the number of unknowns based on the knowledge that a particular body is constrained to rotate about a known axis as shown here. The acceleration vector $\{\alpha_3\}_1$ could for example be represented as follows

$$\{\alpha_3\}_1 = f\alpha_3 \{R_{AB}\}_1 \qquad (4.176)$$

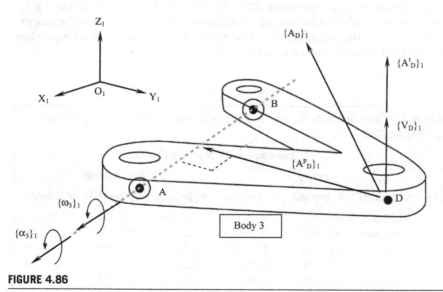

FIGURE 4.86

Angular and translational acceleration and velocity vectors for the upper wishbone.

Since $\{\alpha_3\}_1$ is parallel to the relative position vector $\{R_{AB}\}_1$ a scale factor $f\alpha_3$ can be introduced. This reduces the problem from the three unknown components, αx_3, αy_3 and αz_3 of the vector $\{\alpha_3\}_1$ to a single unknown $f\alpha_3$.

It also follows that since point A is considered fixed with an acceleration velocity $\{A_A\}_1$ equal to zero that the absolute acceleration $\{A_D\}_1$ of point D can be found from a consideration of the triangle law of vector addition giving

$$\{A_D\}_1 = \{A_{DA}\}_1 \tag{4.177}$$

Once the angular accelerations of Body 3 have been found, together with the known angular velocities, it follows now that the translational acceleration $\{A_D\}_1$ of for example point D can be found from

$$\{A_D\}_1 = \{A^P_D\}_1 + \{A^t_D\}_1 \tag{4.178}$$

where the centripetal acceleration $\{A^P_D\}_1$ is given by

$$\{A^P_D\}_1 = \{\omega_3\}_1 \times \{\{\omega_3\}_1 \times \{R_{DA}\}_1\} = \{\omega_3\}_1 \times \{V_D\}_1 \tag{4.179}$$

and the transverse acceleration $\{A^t_D\}_1$ is given by

$$\{A^t_D\}_1 = \{\alpha_3\}_1 \times \{R_{DA}\}_1 \tag{4.180}$$

As with the velocity analysis, clearly the same approach can be taken with the lower wishbone, Body 2, as with the upper wishbone: using a single scale factor $f\alpha_2$ to replace the three unknown components, αx_2, αy_2 and αz_2 of the vector $\{\alpha_2\}_1$. Finally a consideration of the boundaries of this problem again reveals that while points A, B, E, F and J are fixed, the longitudinal acceleration, A_{Px} and the lateral acceleration, A_{Py} at the contact point P remain as unknowns.

Thus this analysis can proceed if we can develop 10 equations to solve the 10 unknowns:

$$f\alpha_2, f\alpha_3, \alpha x_4, \alpha y_4, \alpha z_4, \alpha x_5, \alpha y_5, \alpha z_5, A_{Px}, A_{Py}$$

Working through the problem it will be seen that the same strategy used in the velocity analysis can be developed using the triangle law of vector addition to generate sets of equations. As a starting point we will develop a set of three equations using

$$\{A_{DG}\}_1 = \{A_D\}_1 - \{A_G\}_1 \tag{4.181}$$

In this form Eqn (4.181) does not introduce any of the 10 unknowns listed above. It will therefore be necessary to initially define $\{A_{DG}\}_1$, $\{A_D\}_1$ and $\{A_G\}_1$ in terms of the angular acceleration vectors that contain unknowns requiring solution. The first step in the analysis can therefore proceed as follows.

Determining an expression for the acceleration $\{A_G\}_1$ at point G using values for $\{\omega_2\}_1$ and $\{V_G\}_1$ found from the earlier velocity analysis gives

$$\{A_G\}_1 = \{A_{GE}\}_1 = \{\omega_2\}_1 \times \{V_G\}_1 + \{\alpha_2\}_1 \times \{R_{GE}\}_1 \qquad (4.182)$$

$$\{\alpha_2\}_1 = f\alpha_2 \{R_{EF}\}_1 = f\alpha_2 \begin{bmatrix} 230 \\ 0 \\ 0 \end{bmatrix} \text{ rad/s}^2 \qquad (4.183)$$

$$\begin{bmatrix} A_{Gx} \\ A_{Gy} \\ A_{Gz} \end{bmatrix} = \begin{bmatrix} 0 & 0 & 0 \\ 0 & 0 & -12.266 \\ 0 & 12.266 & 0 \end{bmatrix} \begin{bmatrix} 0.0 \\ 110.394 \\ 3373.150 \end{bmatrix} + f\alpha_2 \begin{bmatrix} 0 & 0 & 0 \\ 0 & 0 & -230 \\ 0 & 230 & 0 \end{bmatrix} \begin{bmatrix} -115 \\ 275 \\ -9 \end{bmatrix} \text{ mm/s}^2 \qquad (4.184)$$

$$\begin{bmatrix} A_{Gx} \\ A_{Gy} \\ A_{Gz} \end{bmatrix} = \begin{bmatrix} 0 \\ -41375.058 \\ 1354.093 \end{bmatrix} + \begin{bmatrix} 0 \\ 2070\, f\alpha_2 \\ 63250\, f\alpha_2 \end{bmatrix} \text{ mm/s}^2 \qquad (4.185)$$

Determining an expression for the acceleration $\{A_D\}_1$ at point D using values for $\{\omega_3\}_1$ and $\{V_D\}_1$ found from the earlier velocity analysis gives

$$\{A_D\}_1 = \{A_{DA}\}_1 = \{\omega_3\}_1 \times \{V_D\}_1 + \{\alpha_3\}_1 \times \{R_{DA}\}_1 \qquad (4.186)$$

$$\{\alpha_3\}_1 = f\alpha_3 \{R_{AB}\}_1 = f\alpha_3 \begin{bmatrix} 230 \\ 0 \\ 14 \end{bmatrix} \text{ rad/s}^2 \qquad (4.187)$$

$$\begin{bmatrix} A_{Dx} \\ A_{Dy} \\ A_{Dz} \end{bmatrix} = \begin{bmatrix} 0 & -0.855 & 0 \\ 0.855 & 0 & -14.039 \\ 0 & 14.039 & 0 \end{bmatrix} \begin{bmatrix} -203.345 \\ 112.260 \\ 3355.321 \end{bmatrix} + f\alpha_3 \begin{bmatrix} 0 & -14 & 0 \\ 14 & 0 & -230 \\ 0 & 230 & 0 \end{bmatrix} \begin{bmatrix} -115 \\ 239 \\ -15 \end{bmatrix} \text{ mm/s}^2 \qquad (4.188)$$

$$\begin{bmatrix} A_{Dx} \\ A_{Dy} \\ A_{Dz} \end{bmatrix} = \begin{bmatrix} -96.030 \\ -47281.651 \\ 1576.804 \end{bmatrix} + \begin{bmatrix} -3346\, f\alpha_3 \\ 1840\, f\alpha_3 \\ 54970\, f\alpha_3 \end{bmatrix} \text{ mm/s}^2 \qquad (4.189)$$

Determining an expression for the relative acceleration $\{A_{DG}\}_1$ of point D relative to point G using values for $\{\omega_4\}_1$ and $\{V_{DG}\}_1$ found from the earlier velocity analysis gives

$$\{A_{DG}\}_1 = \{\omega_4\}_1 \times \{V_{DG}\}_1 + \{\alpha_4\}_1 \times \{R_{DG}\}_1 \qquad (4.190)$$

$$
\begin{bmatrix} A_{DGx} \\ A_{DGy} \\ A_{DGz} \end{bmatrix} = \begin{bmatrix} 0 & 1.446 \times 10^{-3} & -0.945 \\ -1.446 \times 10^{-3} & 0 & 8.774 \times 10^{-3} \\ 0.945 & -8.774 \times 10^{-3} & 0 \end{bmatrix} \begin{bmatrix} -204.345 \\ 1.866 \\ -17.829 \end{bmatrix}
$$
$$
+ \begin{bmatrix} 0 & -\alpha_{4z} & \alpha_{4y} \\ \alpha_{4z} & 0 & -\alpha_{4x} \\ -\alpha_{4y} & \alpha_{4x} & 0 \end{bmatrix} \begin{bmatrix} -19 \\ -31 \\ 216 \end{bmatrix} \, \text{mm/s}^2
$$

(4.191)

$$
\begin{bmatrix} A_{DGx} \\ A_{DGy} \\ A_{DGz} \end{bmatrix} = \begin{bmatrix} 16.851 \\ 0.139 \\ -193.122 \end{bmatrix} + \begin{bmatrix} 31\,\alpha_{4z} + 216\,\alpha_{4y} \\ -19\,\alpha_{4z} - 216\,\alpha_{4x} \\ 19\,\alpha_{4y} - 31\,\alpha_{4x} \end{bmatrix} \text{mm/s}^2
$$

(4.192)

We can now apply the triangle law of vector addition to equate the expression for $\{A_{DG}\}_1$ in Eqn (4.192) with $\{A_D\}_1$ in Eqn (4.189) and $\{A_G\}_1$ in Eqn (4.185).

$$
\{A_{DG}\}_1 = \{A_D\}_1 - \{A_G\}_1
$$

(4.193)

$$
\begin{bmatrix} 16.851 \\ 0.139 \\ -193.122 \end{bmatrix} + \begin{bmatrix} 31\,\alpha_{4z} + 216\,\alpha_{4y} \\ -19\,\alpha_{4z} - 216\,\alpha_{4x} \\ 19\,\alpha_{4y} - 31\,\alpha_{4x} \end{bmatrix} = \begin{bmatrix} -96.030 \\ -47281.651 \\ 1576.804 \end{bmatrix} + \begin{bmatrix} -3346 f_{a3} \\ 1840 f_{a3} \\ 54970 f_{a3} \end{bmatrix}
$$
$$
- \begin{bmatrix} 0 \\ -41375.058 \\ 1354.093 \end{bmatrix} - \begin{bmatrix} 0 \\ 2070 f_{a2} \\ 63250 f_{a2} \end{bmatrix} \text{mm/s}^2
$$

(4.194)

Rearranging Eqn (4.194) yields the first three equations required to solve the analysis.

Equation 1 $3346\,f_{a3} + 216\,\alpha_{4y} + 31\,\alpha_{4z} = -112.881$ (4.195)

Equation 2 $2070\,f_{a2} - 1840\,f_{a3} - 216\,\alpha_{4x} - 19\,\alpha_{4z} = -5906.732$

(4.196)

Equation 3 $63250\,f_{a2} - 54970\,f_{a3} - 31\,\alpha_{4x} + 19\,\alpha_{4y} = 415.833$ (4.197)

We can now proceed to set up the next set of three equations working from point H to point D and using the triangle law of vector addition.

$$
\{A_{DH}\}_1 = \{A_D\}_1 - \{A_H\}_1
$$

(4.198)

Determining an expression for the velocity $\{A_H\}_1$ at point H using values for $\{\omega_5\}_1$ and $\{V_H\}_1$ found from the earlier velocity analysis gives

$$\{A_H\}_1 = \{A_{HJ}\}_1 = \{\omega_5\}_1 \times \{V_H\}_1 + \{\alpha_5\}_1 \times \{R_{HJ}\}_1 \tag{4.199}$$

$$
\begin{bmatrix} A_{Hx} \\ A_{Hy} \\ A_{Hz} \end{bmatrix} =
\begin{bmatrix}
0 & -1.106 & 3.881\times10^{-2} \\
1.106 & 0 & -14.121 \\
-3.881\times10^{-2} & 14.121 & 0
\end{bmatrix}
\begin{bmatrix} -252.524 \\ 112.968 \\ 3219.588 \end{bmatrix}
$$
$$
+ \begin{bmatrix}
0 & -\alpha_{5z} & \alpha_{5y} \\
\alpha_{5z} & 0 & -\alpha_{5x} \\
-\alpha_{5y} & \alpha_{5x} & 0
\end{bmatrix}
\begin{bmatrix} 0 \\ 228 \\ -8 \end{bmatrix} \text{mm/s}^2 \tag{4.200}
$$

$$
\begin{bmatrix} A_{Hx} \\ A_{Hy} \\ A_{Hz} \end{bmatrix} =
\begin{bmatrix} 9.602\times10^{-3} \\ -45743.094 \\ 1605.022 \end{bmatrix}
+ \begin{bmatrix} -228\,\alpha_{5z} - 8\,\alpha_{5y} \\ 8\,\alpha_{5x} \\ 228\,\alpha_{5x} \end{bmatrix} \text{mm/s}^2 \tag{4.201}
$$

We already have an expression for $\{A_D\}_1$ in Eqn (4.189) and can determine an expression for the relative velocity $\{A_{DH}\}_1$ of point D relative to point H using values for $\{\omega_4\}_1$ and $\{V_{DH}\}_1$ found from the earlier velocity analysis.

$$\{A_{DH}\}_1 = \{\omega_4\}_1 \times \{V_{DH}\}_1 + \{\alpha_4\}_1 \times \{R_{DH}\}_1 \tag{4.202}$$

$$
\begin{bmatrix} A_{DHx} \\ A_{DHy} \\ A_{DHz} \end{bmatrix} =
\begin{bmatrix}
0 & 1.446\times10^{-3} & -0.945 \\
-1.446\times10^{-3} & 0 & 8.774\times10^{-3} \\
0.945 & -8.774\times10^{-3} & 0
\end{bmatrix}
\begin{bmatrix} 48.179 \\ -0.708 \\ 135.733 \end{bmatrix}
$$
$$
+ \begin{bmatrix}
0 & -\alpha_{4z} & \alpha_{4y} \\
\alpha_{4z} & 0 & -\alpha_{4x} \\
-\alpha_{4y} & \alpha_{4x} & 0
\end{bmatrix}
\begin{bmatrix} 144 \\ 44 \\ -51 \end{bmatrix} \text{mm/s}^2 \tag{4.203}
$$

$$
\begin{bmatrix} A_{DHx} \\ A_{DHy} \\ A_{DHz} \end{bmatrix} =
\begin{bmatrix} -128.269 \\ 1.121 \\ 45.535 \end{bmatrix}
+ \begin{bmatrix} -44\,\alpha_{4z} - 51\,\alpha_{4y} \\ 144\,\alpha_{4z} + 51\,\alpha_{4x} \\ -144\,\alpha_{4y} + 44\,\alpha_{4x} \end{bmatrix} \text{mm/s}^2 \tag{4.204}
$$

We can now apply the triangle law of vector addition to equate the expression for $\{A_{DH}\}_1$ in Eqn (4.204) with $\{A_D\}_1$ in Eqn (4.189) and $\{A_H\}_1$ in Eqn (4.201).

$$\{A_{DH}\}_1 = \{A_D\}_1 - \{A_H\}_1 \tag{4.205}$$

$$
\begin{bmatrix} -128.269 \\ 1.121 \\ 45.535 \end{bmatrix} + \begin{bmatrix} -44\,\alpha_{4z} - 51\,\alpha_{4y} \\ 144\,\alpha_{4z} + 51\,\alpha_{4x} \\ -144\,\alpha_{4y} + 44\,\alpha_{4x} \end{bmatrix} = \begin{bmatrix} -96.030 \\ -47281.651 \\ 1576.804 \end{bmatrix} + \begin{bmatrix} -3346 f_{a3} \\ 1840 f_{a3} \\ 54970 f_{a3} \end{bmatrix}
$$
$$
- \begin{bmatrix} 9.602 \times 10^{-3} \\ -45743.094 \\ 1605.022 \end{bmatrix} + \begin{bmatrix} -228\,\alpha_{5z} - 8\,\alpha_{5y} \\ 8\,\alpha_{5x} \\ 228\,\alpha_{5x} \end{bmatrix} \text{mm/s}^2
$$

$$(4.206)$$

Rearranging Eqn (4.206) yields the next three equations required to solve the analysis.

Equation 4 $\quad\quad 3346\, f_{a3} - 51\,\alpha_{4y} - 44\,\alpha_{4z} - 8\,\alpha_{5y} - 228\,\alpha_{5z} = 32.229$ (4.207)

Equation 5 $\quad\quad -1840\, f_{a3} + 51\,\alpha_{4x} + 144\,\alpha_{4z} + 8\,\alpha_{5x} = -1539.678$ (4.208)

Equation 6 $\quad\quad -54970\, f_{a3} + 44\,\alpha_{4x} - 144\,\alpha_{4y} + 228\,\alpha_{5x} = -73.753$ (4.209)

We can now proceed to set up the next set of three equations working from point G to point P and using the triangle law of vector addition.

$$\{A_{PG}\}_1 = \{A_P\}_1 - \{A_G\}_1 \tag{4.210}$$

Determining an expression for the relative velocity $\{A_{PG}\}_1$ of point P relative to point G using values for $\{\omega_4\}_1$ and $\{V_{PG}\}_1$ found from the earlier velocity analysis gives

$$\{A_{PG}\}_1 = \{\omega_4\}_1 \times \{V_{PG}\}_1 + \{\alpha_4\}_1 \times \{R_{PG}\}_1 \tag{4.211}$$

$$
\begin{bmatrix} A_{PGx} \\ A_{PGy} \\ A_{PGz} \end{bmatrix} = \begin{bmatrix} 0 & 1.446 \times 10^{-3} & -0.945 \\ -1.446 \times 10^{-3} & 0 & 8.774 \times 10^{-3} \\ 0.945 & -8.774 \times 10^{-3} & 0 \end{bmatrix} \begin{bmatrix} 166.468 \\ -1.535 \\ -7.150 \end{bmatrix}
$$
$$
+ \begin{bmatrix} 0 & -\alpha_{4z} & \alpha_{4y} \\ \alpha_{4z} & 0 & -\alpha_{4x} \\ -\alpha_{4y} & \alpha_{4x} & 0 \end{bmatrix} \begin{bmatrix} -7 \\ 58 \\ -176 \end{bmatrix} \text{mm/s}^2
$$

$$(4.212)$$

$$
\begin{bmatrix} A_{PGx} \\ A_{PGy} \\ A_{PGz} \end{bmatrix} = \begin{bmatrix} 6.755 \\ -0.303 \\ 157.326 \end{bmatrix} + \begin{bmatrix} -58\,\alpha_{4z} - 176\,\alpha_{4y} \\ -7\,\alpha_{4z} + 176\,\alpha_{4x} \\ 7\,\alpha_{4y} + 58\,\alpha_{4x} \end{bmatrix} \text{mm/s}^2 \tag{4.213}
$$

We already have an expression for $\{A_G\}_1$ in Eqn (4.185) and we can define the vector $\{A_P\}_1$ in terms of the known vertical velocity component, A_{Pz} and the unknown components, A_{Px} and A_{Py}.

$$\{A_P\}_1 = \begin{bmatrix} A_{Px} \\ A_{Py} \\ 0.0 \end{bmatrix} \text{ mm/s}^2 \tag{4.214}$$

We can now apply the triangle law of vector addition to equate the expression for $\{A_{PG}\}_1$ in Eqn (4.213) with $\{A_P\}_1$ in Eqn (4.214) and $\{A_G\}_1$ in Eqn (4.185).

$$\{A_{PG}\}_1 = \{A_P\}_1 - \{A_G\}_1 \tag{4.215}$$

$$\begin{bmatrix} 6.755 \\ -0.303 \\ 157.326 \end{bmatrix} + \begin{bmatrix} -58\,\alpha_{4z} - 176\,\alpha_{4y} \\ -7\,\alpha_{4z} + 176\,\alpha_{4x} \\ 7\,\alpha_{4y} + 58\,\alpha_{4x} \end{bmatrix} = \begin{bmatrix} A_{Px} \\ A_{Py} \\ 0.0 \end{bmatrix} - \begin{bmatrix} 0 \\ -41375.058 \\ 1354.093 \end{bmatrix} - \begin{bmatrix} 0 \\ 2070\,f_{a2} \\ 63250\,f_{a2} \end{bmatrix} \text{ mm/s}^2 \tag{4.216}$$

Rearranging Eqn (4.216) yields the next set of three equations required to solve the analysis.

Equation 7 $\qquad\qquad 176\,\alpha_{4y} + 58\,\alpha_{4z} + A_{Px} = 6.755$ (4.217)

Equation 8 $\qquad -2070\,f_{a2} - 176\,\alpha_{4x} + 7\,\alpha_{4z} + A_{Py} = -41375.361$ (4.218)

Equation 9 $\qquad -63250\,f_{a2} - 58\,\alpha_{4x} - 7\,\alpha_{4y} = 1511.419$ (4.219)

As with the velocity analysis this leaves us with nine equations and ten unknowns. The last equation is again obtained by constraining the rotation of the tie rod (Body 5) to prevent spin about its own axis.

$$\{\alpha_5\}_1 \bullet \{R_{HJ}\}_1 = 0 \tag{4.220}$$

$$\begin{bmatrix} \alpha_{5x} & \alpha_{5y} & \alpha_{5z} \end{bmatrix} \begin{bmatrix} 0 \\ 228 \\ -8 \end{bmatrix} = 0 \text{ mm/s} \tag{4.221}$$

Equation 10 $\qquad\qquad 228\,\alpha_{5y} - 8\,\alpha_{5z} = 0$ (4.222)

The 10 equations can now be set up in matrix form ready for solution.

$$
\begin{bmatrix}
0 & 3346 & 0 & 216 & 31 & 0 & 0 & 0 & 0 & 0 \\
2070 & -1840 & -216 & 0 & -19 & 0 & 0 & 0 & 0 & 0 \\
63250 & -54970 & -31 & 19 & 0 & 0 & 0 & 0 & 0 & 0 \\
0 & 3346 & 0 & -51 & -44 & 0 & -8 & -228 & 0 & 0 \\
0 & -1840 & 51 & 0 & 144 & 8 & 0 & 0 & 0 & 0 \\
0 & -54970 & 44 & -144 & 0 & 228 & 0 & 0 & 0 & 0 \\
0 & 0 & 0 & 176 & 58 & 0 & 0 & 0 & 1 & 0 \\
-2070 & 0 & -176 & 0 & 7 & 0 & 0 & 0 & 0 & 1 \\
-63250 & 0 & -58 & -7 & 0 & 0 & 0 & 0 & 0 & 0 \\
0 & 0 & 0 & 0 & 0 & 0 & 228 & -8 & 0 & 0
\end{bmatrix}
\begin{bmatrix}
f_{\alpha 2} \\ f_{\alpha 3} \\ \alpha_{4x} \\ \alpha_{4y} \\ \alpha_{4z} \\ \alpha_{5x} \\ \alpha_{5y} \\ \alpha_{5z} \\ A_{Px} \\ A_{Py}
\end{bmatrix}
=
\begin{bmatrix}
-112.881 \\ -5906.732 \\ 415.833 \\ 32.229 \\ -1539.678 \\ -73.753 \\ 6.755 \\ -41375.361 \\ 1511.419 \\ 0
\end{bmatrix}
$$

(4.223)

Solving Eqn (4.223) yields the following answers for the 10 unknowns:

$$f_{\alpha 2} = -5.126\times10^{-2} \ \text{rad/mm.s}^2$$

$$f_{\alpha 3} = -8.182\times10^{-2} \ \text{rad/mm.s}^2$$

$$\alpha_{4x} = 29.386 \ \text{rad/s}^2$$

$$\alpha_{4y} = 3.737 \ \text{rad/s}^2$$

$$\alpha_{4z} = -20.847 \ \text{rad/s}^2$$

$$\alpha_{5x} = -23.361 \ \text{rad/s}^2$$

$$\alpha_{5y} = 6.466\times10^{-2} \ \text{rad/s}^2$$

$$\alpha_{5z} = 1.843 \ \text{rad/s}^2$$

$$A_{Px} = 558.211 \ \text{mm/s}^2$$

$$A_{Py} = -36163.671 \ \text{mm/s}^2$$

It is now possible to use the two scale factors, $f_{\alpha 2}$ and $f_{\alpha 3}$, to calculate the angular acceleration vectors, $\{\alpha_2\}_1$ and $\{\alpha_3\}_1$.

$$
\{\alpha_2\}_1 = f_{\alpha 2}\{R_{EF}\}_1 = -5.126\times10^{-2}
\begin{bmatrix} 230 \\ 0 \\ 0 \end{bmatrix}
=
\begin{bmatrix} -11.790 \\ 0 \\ 0 \end{bmatrix} \text{rad/s}^2
\qquad (4.224)
$$

$$\{\alpha_3\}_1 = f_{\alpha_3}\{R_{AB}\}_1 = -8.182 \times 10^{-2} \begin{bmatrix} 230 \\ 0 \\ 14 \end{bmatrix} = \begin{bmatrix} -18.819 \\ 0 \\ -1.145 \end{bmatrix} \text{rad/s}^2 \quad (4.225)$$

In summary the angular velocity vectors for the rigid bodies are as follows:

$$\{\alpha_2\}_1^T = \begin{bmatrix} -11.790 & 0 & 0 \end{bmatrix} \text{rad/s}^2$$

$$\{\alpha_3\}_1^T = \begin{bmatrix} -18.819 & 0 & -1.145 \end{bmatrix} \text{rad/s}^2$$

$$\{\alpha_4\}_1^T = \begin{bmatrix} 29.386 & 3.737 & -20.847 \end{bmatrix} \text{rad/s}^2$$

$$\{\alpha_5\}_1^T = \begin{bmatrix} -23.361 & 6.466 \times 10^{-2} & 1.843 \end{bmatrix} \text{rad/s}^2$$

We can now proceed to calculate the translational accelerations at all the moving points, C, D, G, H and P, within this part of the model.

$$\{A_C\}_1 = \{A_{CA}\}_1 = \{\omega_3\}_1 \times \{V_C\}_1 + \{\alpha_3\}_1 \times \{R_{CA}\}_1 \quad (4.226)$$

$$\begin{bmatrix} A_{Cx} \\ A_{Cy} \\ A_{Cz} \end{bmatrix} = \begin{bmatrix} 0 & -0.855 & 0 \\ 0.855 & 0 & -14.039 \\ 0 & 14.039 & 0 \end{bmatrix} \begin{bmatrix} -120.555 \\ 435.157 \\ 1979.499 \end{bmatrix}$$

$$+ \begin{bmatrix} 0 & 1.146 & 0 \\ -1.146 & 0 & 18.819 \\ 0 & -18.819 & 0 \end{bmatrix} \begin{bmatrix} -115 \\ 141 \\ -38 \end{bmatrix} \text{mm/s}^2 \quad (4.227)$$

$$\begin{bmatrix} A_{Cx} \\ A_{Cy} \\ A_{Cz} \end{bmatrix} = \begin{bmatrix} -372.059 \\ -27893.260 \\ 6109.169 \end{bmatrix} + \begin{bmatrix} 161.445 \\ -583.447 \\ -2653.479 \end{bmatrix} = \begin{bmatrix} -210.614 \\ -28476.707 \\ 3455.690 \end{bmatrix} \text{mm/s}^2 \quad (4.228)$$

$$\{A_D\}_1 = \{A_{DA}\}_1 = \{\omega_3\}_1 \times \{V_D\}_1 + \{\alpha_3\}_1 \times \{R_{DA}\}_1 \quad (4.229)$$

$$\begin{bmatrix} A_{Dx} \\ A_{Dy} \\ A_{Dz} \end{bmatrix} = \begin{bmatrix} -96.030 \\ -47281.651 \\ 1576.804 \end{bmatrix} + \begin{bmatrix} 0 & 1.145 & 0 \\ -1.145 & 0 & 18.819 \\ 0 & -18.819 & 0 \end{bmatrix} \begin{bmatrix} -115 \\ 239 \\ -15 \end{bmatrix} = \begin{bmatrix} 177.625 \\ -47432.261 \\ -2920.973 \end{bmatrix} \text{mm/s}^2$$

$$(4.230)$$

$$\{A_G\}_1 = \{A_{GE}\}_1 = \{\omega_2\}_1 \times \{V_G\}_1 + \{\alpha_2\}_1 \times \{R_{GE}\}_1 \quad (4.231)$$

$$\begin{bmatrix} A_{Gx} \\ A_{Gy} \\ A_{Gz} \end{bmatrix} = \begin{bmatrix} 0 \\ -41375.058 \\ 1354.093 \end{bmatrix} + \begin{bmatrix} 0 & 0 & 0 \\ 0 & 0 & 11.791 \\ 0 & -11.791 & 0 \end{bmatrix} \begin{bmatrix} -115 \\ 275 \\ -9 \end{bmatrix} = \begin{bmatrix} 0 \\ -41481.177 \\ -1888.432 \end{bmatrix} mm/s^2$$
$$(4.232)$$

$$\{A_H\}_1 = \{A_{HJ}\}_1 = \{\omega_5\}_1 \times \{V_H\}_1 + \{\alpha_5\}_1 \times \{R_{HJ}\}_1 \qquad (4.233)$$

$$\begin{bmatrix} A_{Hx} \\ A_{Hy} \\ A_{Hz} \end{bmatrix} = \begin{bmatrix} 9.602\times10^{-3} \\ -45743.094 \\ 1605.022 \end{bmatrix} + \begin{bmatrix} 0 & -1.843 & 0.06466 \\ 1.843 & 0 & -23.361 \\ -0.06466 & 23.361 & 0 \end{bmatrix} \begin{bmatrix} 0 \\ 229 \\ -9 \end{bmatrix} = \begin{bmatrix} -422.619 \\ -45953.343 \\ -3744.647 \end{bmatrix} mm/s^2$$
$$(4.234)$$

The acceleration vector $\{A_P\}_1$ is already available from the initial bump analysis and the solution of Eqn (4.223). In summary the acceleration vectors for the moving points are as follows:

$$\{A_C\}_1^T = [-210.614 \quad -28476.707 \quad 3455.690] \, mm/s^2$$

$$\{A_D\}_1^T = [177.625 \quad -47432.261 \, -2920.973] \, mm/s^2$$

$$\{A_G\}_1^T = [0.0 \, -41481.177 \quad -1888.432] \, mm/s^2$$

$$\{A_H\}_1^T = [-422.619 \quad 45953.343 \, -3744.647] \, mm/s^2$$

$$\{A_P\}_1^T = [558.211 \, -36163.671 \quad 0.0] \, mm/s^2$$

Having found the acceleration $\{A_C\}_1$ at the bottom of the damper unit we can now proceed to carry out a separate analysis of the unit to find the components of acceleration acting between bodies 6 and 7.

As with the velocity analysis, this phase of the acceleration analysis can be facilitated by the modelling of three coincident points, C_3 on Body 3, C_6 on Body 6 and C_7 on Body 7, all located at point C. Note that we already have

$$\{A_{C3}\}_1 = \{A_{C7}\}_1 = \{A_C\}_1 = \begin{bmatrix} -210.614 \\ -28476.707 \\ 3455.690 \end{bmatrix} mm/s^2 \qquad (4.235)$$

We can also calculate the acceleration $\{A_{C6}\}_1$ from

$$\{A_{C6}\}_1 = \{A_{C6I}\}_1 = \{\omega_6\}_1 \times \{V_{C6I}\}_1 + \{\alpha_6\}_1 \times \{R_{CI}\}_1 \qquad (4.236)$$

where

$$\{V_{C6I}\}_1 = \{V_{C6}\}_1 = \{V_{C6C7}\}_1 - \{V_{C7}\}_1 \qquad (4.237)$$

$$\{V_{C6I}\}_1 = \begin{bmatrix} 13.628 \\ -41.045 \\ -1988.378 \end{bmatrix} - \begin{bmatrix} -120.555 \\ 435.157 \\ 1979.499 \end{bmatrix} = \begin{bmatrix} 134.183 \\ -476.202 \\ -3967.877 \end{bmatrix} \text{mm/s} \qquad (4.238)$$

therefore $\{A_{C6}\}_1$ is given by

$$\begin{bmatrix} A_{C6x} \\ A_{C6y} \\ A_{C6z} \end{bmatrix} = \begin{bmatrix} 0 & -1.159 & 0.245 \\ 1.159 & 0 & -0.904 \\ -0.245 & 0.904 & 0 \end{bmatrix} \begin{bmatrix} 134.183 \\ -476.202 \\ -3967.877 \end{bmatrix} + \begin{bmatrix} 0 & -\alpha_{6z} & \alpha_{6y} \\ \alpha_{6z} & 0 & -\alpha_{6x} \\ -\alpha_{6y} & \alpha_{6x} & 0 \end{bmatrix} \begin{bmatrix} 3 \\ -9 \\ -436 \end{bmatrix} \text{mm/s}^2$$

$$(4.239)$$

$$\begin{bmatrix} A_{C6x} \\ A_{C6y} \\ A_{C6z} \end{bmatrix} = \begin{bmatrix} -420.212 \\ 3742.479 \\ -463.361 \end{bmatrix} + \begin{bmatrix} 9\,\alpha_{6z} - 436\,\alpha_{6y} \\ 3\,\alpha_{6z} + 436\,\alpha_{6x} \\ -3\,\alpha_{6y} - 9\,\alpha_{6x} \end{bmatrix} \text{mm/s}^2 \qquad (4.240)$$

If we now consider the relative acceleration vector $\{A_{C6C7}\}_1$ we can see that this involves the relative acceleration between points on two bodies where relative rotation and sliding occurs. Referring back to Chapter 2 we can now identify the four components of acceleration associated with the combined rotation and sliding motion as the centripetal acceleration $\{A^P_{C6C7}\}_1$, the transverse acceleration $\{A^t_{C6C7}\}_1$, the Coriolis acceleration $\{A^c_{C6C7}\}_1$ and the sliding acceleration $\{A^s_{C6C7}\}_1$.

$$\{A^P_{C6C7}\}_1 = \{\omega_6\}_1 \times \{\{\omega_6\}_1 \times \{R_{C6C7}\}_1\} \qquad (4.241)$$

$$\{A^t_{C6C7}\}_1 = \{\alpha_6\}_1 \times \{R_{C6C7}\}_1 \qquad (4.242)$$

$$\{A^c_{C6C7}\}_1 = 2\{\omega_6\}_1 \times \{V_s\}_1 \qquad (4.243)$$

$$\{A^s_{C6C7}\}_1 = |A^s_{C6C7}| \{l_{CI}\}_1 \qquad (4.244)$$

Since the C_6 and C_7 are coincident points it follows that $\{A^P_{C6C7}\}_1$ and $\{A^t_{C6C7}\}_1$ are zero. It also follows that the sliding velocity $\{V_s\}_1$ is equal to $\{V_{C6C7}\}_1$. We can also introduce a scale factor, A_s, to simplify the sliding acceleration calculation giving

$$\{A^c_{C6C7}\}_1 = 2\{\omega_6\}_1 \times \{V_{C6C7}\}_1 \qquad (4.245)$$

$$\{A^s_{C6C7}\}_1 = A_s\{R_{CI}\}_1 \qquad (4.246)$$

Combining these components of acceleration gives $\{A_{C6C7}\}_1$ as

$$\{A_{C6C7}\}_1 = 2\{\omega_6\}_1 \times \{V_{C6C7}\}_1 + A_s \{R_{CI}\}_1 \qquad (4.247)$$

$$\begin{bmatrix} A_{C6C7x} \\ A_{C6C7y} \\ A_{C6C7z} \end{bmatrix} = 2 \begin{bmatrix} 0 & -1.159 & 0.245 \\ 1.159 & 0 & -0.904 \\ -0.245 & 0.904 & 0 \end{bmatrix} \begin{bmatrix} 13.682 \\ -41.045 \\ -1988.378 \end{bmatrix} + A_s \begin{bmatrix} 3 \\ -9 \\ -436 \end{bmatrix} \text{mm/s}^2$$

$$(4.248)$$

$$\begin{bmatrix} A_{C6C7x} \\ A_{C6C7y} \\ A_{C6C7z} \end{bmatrix} = \begin{bmatrix} -879.162 \\ 3626.702 \\ -80.457 \end{bmatrix} + A_s \begin{bmatrix} 3 \\ -9 \\ -436 \end{bmatrix} \text{mm/s}^2 \qquad (4.249)$$

Applying the triangle law of vector addition yields

$$\{A_{C6C7}\}_1 = \{A_{C6}\}_1 - \{A_{C7}\}_1 \qquad (4.250)$$

$$\begin{bmatrix} -879.162 \\ 3626.702 \\ -80.457 \end{bmatrix} + A_s \begin{bmatrix} 3 \\ -9 \\ -436 \end{bmatrix} = \begin{bmatrix} -420.212 \\ 3742.479 \\ -463.361 \end{bmatrix} + \begin{bmatrix} 9\alpha_{6z} - 436\alpha_{6y} \\ 3\alpha_{6z} + 436\alpha_{6x} \\ -3\alpha_{6y} - 9\alpha_{6x} \end{bmatrix} - \begin{bmatrix} -210.614 \\ -28476.707 \\ 3455.690 \end{bmatrix} \text{mm/s}^2$$

$$(4.251)$$

Rearranging Eqn (4.251) yields three equations that can be used to solve this part of the analysis.

Equation 1 $\qquad\qquad 3 A_s + 436\,\alpha_{6y} - 9\,\alpha_{6z} = 669.564 \qquad (4.252)$

Equation 2 $\qquad\qquad -9 A_s - 436\,\alpha_{6x} - 3\,\alpha_{6z} = 28592.484 \qquad (4.253)$

Equation 3 $\qquad\qquad -436 A_s + 9\,\alpha_{6x} + 3\,\alpha_{6y} = -3838.594 \qquad (4.254)$

This leaves us with four unknowns, α_{6x}, α_{6y}, α_{6z} and A_s but only three equations. We can use the same approach here as used in the preceding velocity analysis. Since the spin degree of freedom of Body 6 about the axis CI has no bearing on the overall solution we can again use the vector dot product to enforce perpendicularity of $\{\alpha_6\}_1$ to $\{R_{CI}\}_1$. This will yield the fourth equation as follows.

$$\{\alpha_6\}_1 \bullet \{R_{CI}\}_1 = 0 \qquad (4.255)$$

$$\begin{bmatrix} \alpha_{6x} & \alpha_{6y} & \alpha_{6z} \end{bmatrix} \begin{bmatrix} 3 \\ -9 \\ -436 \end{bmatrix} = 0 \text{ mm/s} \qquad (4.256)$$

Equation 4 $\qquad\qquad 3\,\alpha_{6x} - 9\,\alpha_{6y} - 436\,\alpha_{6z} = 0 \qquad (4.257)$

The four equations can now be set up in matrix form ready for solution.

$$\begin{bmatrix} 3 & 0 & 436 & -9 \\ -9 & -436 & 0 & -3 \\ -436 & 9 & 3 & 0 \\ 0 & 3 & -9 & -436 \end{bmatrix} \begin{bmatrix} A_s \\ \alpha_{6x} \\ \alpha_{6y} \\ \alpha_{67} \end{bmatrix} = \begin{bmatrix} 669.564 \\ 28592.484 \\ -3838.594 \\ 0 \end{bmatrix} \qquad (4.258)$$

Solving Eqn (4.258) yields the following answers for the four unknowns:

$$A_s = 7.457 \text{ s}^{-2}$$

$$\alpha_{6x} = -65.730 \text{ rad/s}^2$$

$$\alpha_{6y} = 1.147 \text{ rad/s}^2$$

$$\alpha_{6z} = -0.483 \text{ rad/s}^2$$

This gives us the last two angular acceleration vectors for the upper and lower damper bodies:

$$\{\alpha_6\}_1^T = \begin{bmatrix} -65.730 & 1.147 & -0.483 \end{bmatrix} \text{ rad/s}^2$$

$$\{\alpha_7\}_1^T = \begin{bmatrix} -65.730 & 1.147 & -0.483 \end{bmatrix} \text{ rad/s}^2$$

From Eqn (4.247) we now have

$$\{A_{C6C7}\}_1 = 2\{\omega_6\}_1 X \{V_{C6C7}\}_1 + A_s \{R_{CI}\}_1 \qquad (4.259)$$

$$\{A_{C6C7}\}_1 = \begin{bmatrix} -879.162 \\ 3626.702 \\ -80.457 \end{bmatrix} + 7.457 \begin{bmatrix} 3 \\ -9 \\ -436 \end{bmatrix} = \begin{bmatrix} -856.791 \\ 3559.589 \\ -3331.709 \end{bmatrix} \text{ mm/s}^2 \qquad (4.260)$$

A comparison of the angular accelerations found from the preceding calculations and those found using an equivalent MSC ADAMS model is shown in Table 4.11.

A comparison of the translational accelerations found, at points within the suspension system from the preceding calculations and those found using an equivalent MSC ADAMS model is shown in Table 4.12.

4.10.4 Static analysis

As discussed in this chapter, a starting point for suspension component loading studies is to use equivalent static forces to represent the loads acting through the road wheel, associated with real-world driving conditions. In this example the vector analysis method is used to carry out a static analysis where a vertical load of 10,000 N is applied at the tyre contact patch, this being representative in magnitude of the loads used for a 3G bump case on a typical vehicle of this size.

Table 4.11 Comparison of Angular Acceleration Vectors Computed by Theory and MSC ADAMS

	Angular Acceleration Vectors					
	Theory			MSC ADAMS		
Body	α_x (rad/s^2)	α_y (rad/s^2)	α_z (rad/s^2)	α_x (rad/s^2)	α_y (rad/s^2)	α_z (rad/s^2)
2	−11.790	0.0	0.0	−11.791	0.0	0.0
3	−18.819	0.0	−1.145	−18.823	0.0	−1.146
4	29.386	3.737	−20.847	29.395	3.737	−20.840
5	−23.361	6.466×10^{-2}	1.843	−23.384	0.100	2.053
6	−65.730	1.147	−0.483	−57.204	2.663	−0.449
7	−65.730	1.147	−0.483	−57.204	2.663	−0.449

Table 4.12 Comparison of Translational Acceleration Vectors Computed by Theory and MSC ADAMS

	Translational Acceleration Vectors					
	Theory			MSC ADAMS		
Point	A_x (rad/s^2)	A_y (rad/s^2)	A_z (rad/s^2)	A_x (rad/s^2)	A_y (rad/s^2)	A_z (rad/s^2)
C	−210.614	−28,476.707	3455.690	−210.385	−28,478.60	3456.320
D	177.625	−47,432.261	−2920.973	177.846	−47,433.70	−2921.760
G	0.0	−41,481.177	−1888.432	0.0	−41,480.60	−1888.430
H	−422.619	−45,953.343	−3744.647	−420.215	−45,933.00	−3722.660
P	558.211	−36,163.671	0.0	557.780	−36,161.60	0.0
C_6C_7	−856.791	3559.589	−3331.709	−957.170	3544.09	−2566.940

In this analysis we are ignoring gravity and the self-weight of the suspension components as this contribution tends to be minor compared with overall vehicle loads reacted at the tyre contact patch and diffused into the suspension system. For completeness the effects of self-weight will be included in a follow-on demonstration of a dynamic analysis.

Before attempting any vector analysis to determine the distribution of forces it is necessary to prepare a free body diagram and label the bodies and forces in an appropriate manner as shown in Figure 4.87.

For the action—reaction forces shown acting between the bodies in Figure 4.87 Newton's third law would apply. The interaction, for example, at point D between Body 3 and Body 4 requires $\{F_{D43}\}_1$ and $\{F_{D34}\}_1$ to be equal and opposite equal. Thus instead of including the six unknowns F_{D43x}, F_{D43y}, F_{D43z}, F_{D34x}, F_{D34y} and F_{D34z} we can reduce this to three unknowns F_{D43x}, F_{D43y}, F_{D43z}. In a similar manner,

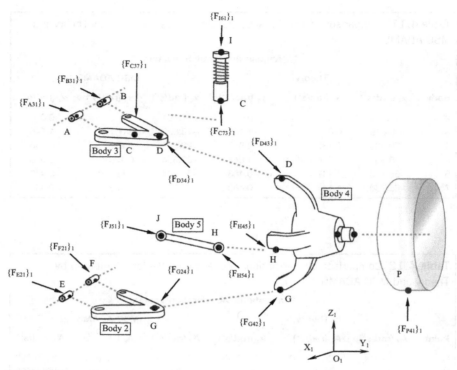

FIGURE 4.87

Free body diagram for double wishbone suspension system static force analysis.

looking at the connections at points G and H we can see for all connections to Body 4 that the following applies:

$$\{F_{D43}\}_1 = -\{F_{D34}\}_1 \tag{4.261}$$

$$\{F_{G42}\}_1 = -\{F_{G24}\}_1 \tag{4.262}$$

$$\{F_{H45}\}_1 = -\{F_{H54}\}_1 \tag{4.263}$$

In this model we are treating the connections and mounts as pin-jointed, or as the equivalent spherical joints in an MBS model. For the track rod, Body 5, both ends of the linkage are pin-jointed and the force by definition must, if we allow ourselves the assumption to ignore gravity for this study, act along the axis HJ. In a similar manner the force acting on Body 7 at the base of the strut at point C must be equal and opposite to the force acting at the top on Body 6 at point I.

$$\{F_{J51}\}_1 = -\{F_{H54}\}_1 \tag{4.264}$$

$$\{F_{C37}\}_1 = -\{F_{C73}\}_1 = \{F_{I61}\}_1 \tag{4.265}$$

The number of unknowns can be reduced even further, by using scale factors to exploit the knowledge that the lines of action of the forces are known.

$$\{F_{H54}\}_1 = f_{S1}\,\{R_{JH}\}_1 \tag{4.266}$$

$$\{F_{C37}\}_1 = f_{S2}\,\{R_{CI}\}_1 \tag{4.267}$$

This results in the following set of 20 unknowns that must be found to solve for static equilibrium:

$$F_{A31x},\ F_{A31y},\ F_{A31z}$$
$$F_{B31x},\ F_{B31y},\ F_{B31z}$$
$$F_{D43x},\ F_{D43y},\ F_{D43z}$$
$$F_{E21x},\ F_{E21y},\ F_{E21z}$$
$$F_{F21x},\ F_{F21y},\ F_{F21z}$$
$$F_{G24x},\ F_{G24y},\ F_{G24z}$$
$$f_{S1},\ f_{S2}$$

The problem can be solved by setting up the equations of equilibrium for Bodies 2, 3 and 4. The use of scale factors to model the forces acting along Body 5 and the strut, Bodies 6 and 7, means that these bodies cannot be used to generate any useful equations to solve the problem. Thus we could generate 18 equations as follows.

For Body 2 summing forces and taking moments about point G gives

$$\sum\{F_2\}_1 = \{0\}_1 \tag{4.268}$$

$$\sum\{M_{G2}\}_1 = \{0\}_1 \tag{4.269}$$

For Body 3 summing forces and taking moments about point D gives

$$\sum\{F_3\}_1 = \{0\}_1 \tag{4.270}$$

$$\sum\{M_{D3}\}_1 = \{0\}_1 \tag{4.271}$$

For Body 4 summing forces and taking moments about point G gives

$$\sum\{F_4\}_1 = \{0\}_1 \tag{4.272}$$

$$\sum\{M_{G4}\}_1 = \{0\}_1 \tag{4.273}$$

This leaves us with the requirement to generate another two equations for solution. The answer comes from a more considered study of the connections or mounts between the upper and lower wishbones and the ground part. Four possible MBS modelling solutions are shown in Figure 4.88.

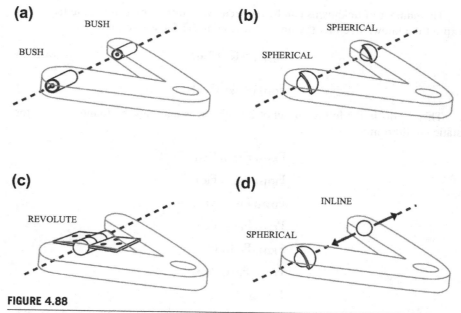

FIGURE 4.88

Wishbone mount modelling strategies. (a) Wishbone mounted by two bushes. (b) Wishbone mounted by two spherical joints. (c) Wishbone mounted by a single revolute joint. (d) Wishbone mounted by a spherical joint and inline joint primitive

In Figure 4.88(a) the wishbone is mounted using two bush force elements. Using this configuration the wishbone is mounted on an elastic foundation and the body has six rigid body degrees of freedom relative to the part on which it is mounted, which for this example is a nonmoving ground part. If the actual wishbone is mounted on the vehicle in this way this would be the MBS modelling solution of choice if as discussed earlier the simulation aimed to produce accurate predictions of the mount reaction forces. The movement of the wishbone relative to the part on which it is mounted is controlled by the compliance in the bushes. This typically would allow relatively little resistance to rotation about an axis through the bushes, while strongly resisting motion in the other five degrees of freedom.

In Figure 4.88(b) the wishbone is constrained by a spherical joint at each bush location. Each spherical joint constrains three degrees of freedom. This is in fact equivalent to our vector-based model shown as a free body diagram in Figure 4.88 where we currently have three constraint reaction forces at each of our mount locations A, B, E and F. The problem with this approach is that the wishbone initially has six degrees of freedom and the two spherical joints remove three, each leaving for the wishbone body a local balance of zero degrees of freedom. This is clearly not valid as, in the absence of friction or other forces, the wishbone is not physically constrained from rotating about an axis through the two spherical joints.

This is a classic MBS modelling problem where we have introduced a redundant constraint or overconstrained the model. It should also be noted that this is the root of our requirement for two more equations for the manual analysis, each equation being related to the local overconstraint of each wishbone. Early versions of MBS programs such as MSC ADAMS were rather unforgiving in these circumstances and any attempt to solve such a model would cause the solver to fail with the appropriate error messages. More modern versions are able to identify and remove redundant constraints allowing a solution to proceed. While this undoubtedly adds to the convenience of model construction it does isolate less-experienced users from the underlying theory and modelling issues we are currently discussing. In any event, if the required outcome is to predict loads at the mount points, the removal of the redundant constraints, although not affecting the kinematics, cannot be relied on to distribute correctly the forces to the mounts.

In Figure 4.88(c) the two wishbone mount connections are represented by a single revolute joint. This is the method suggested earlier as a suitable start for predicting the suspension kinematics but will again not be useful for predicting the mount reaction forces. In this model the single revolute joint will carry the combined translational reaction forces at both mounts with additional moment reactions that would not exist in the real system.

The final representation shown in Figure 4.88(d) allows a model that uses rigid constraint elements and can predict reaction forces at each mount without using the 'as is' approach of including the bush compliances or introducing redundant constraints. This is achieved by modelling one mount with a spherical joint and the other mount with an inline joint primitive, as described earlier in Chapter 3. The inline primitive constrains two degrees of freedom to maintain the mount position on the axis through the two mount locations. This constraint does not prevent translation along the axis through the mounts, this 'thrust' being reacted by the single spherical joint. Thus this selection of rigid constraints provides us with a solution that is not overconstrained. Although the MBS approach would best utilise the model with two bushes to predict the mount reaction forces, the model in Figure 4.88(d) provides us with an understanding of the overconstraint problem and a methodology we can adapt to progress the vector-based analytical solution.

If we return now to the analytical solution and consider the lower wishbone Body 2, we can see in Figure 4.89 that a comparable approach to the use of the MBS inline joint primitive constraint is to ensure that the line of action of one of the mount reaction forces, say $\{F_{F21}\}_1$, is perpendicular to the axis EF through the two wishbone mounts.

Thus we can derive the final two equations needed to progress the analytical solution using the familiar approach with the vector dot product to constrain the reaction force at the mount to be perpendicular to an axis through the mounts at say, point B for the upper wishbone and point F for the lower wishbone.

$$\{F_{F21}\}_1 \bullet \{R_{EF}\}_1 = 0 \qquad (4.274)$$

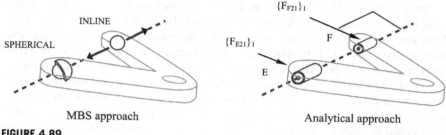

FIGURE 4.89

Comparable multibody systems (MBS) and analytical wishbone mounting models.

$${F_{B31}}_1 \bullet {R_{AB}}_1 = 0 \qquad (4.275)$$

Having established the 20 equations required for solution, it is possible to set up the equations starting with the force equilibrium of Body 2.

$$\Sigma \{F_2\}_1 = \{0\}_1 \qquad (4.276)$$

$$\{F_{E21}\}_1 + \{F_{F21}\}_1 + \{F_{G24}\}_1 = \{0\}_1 \qquad (4.277)$$

$$\begin{bmatrix} F_{E21x} \\ F_{E21y} \\ F_{E21z} \end{bmatrix} + \begin{bmatrix} F_{F21x} \\ F_{F21y} \\ F_{F21z} \end{bmatrix} + \begin{bmatrix} F_{G24x} \\ F_{G24y} \\ F_{G24z} \end{bmatrix} = \begin{bmatrix} 0 \\ 0 \\ 0 \end{bmatrix} N \qquad (4.278)$$

The summation of forces in Eqn (4.278) leads to the first set of three equations:

Equation 1 $\qquad F_{E21x} + F_{F21x} + F_{G24x} = 0 \qquad (4.279)$

Equation 2 $\qquad F_{E21y} + F_{F21y} + F_{G24y} = 0 \qquad (4.280)$

Equation 3 $\qquad F_{E21z} + F_{F21z} + F_{G24z} = 0 \qquad (4.281)$

Taking moments about point G for the forces acting on Body 2 gives

$$\Sigma \{M_{G2}\}_1 = \{0\}_1 \qquad (4.282)$$

$$\{R_{EG}\}_1 \times \{F_{E21}\}_1 + \{R_{FG}\}_1 \times \{F_{F21}\}_1 \qquad = \{0\}_1 \qquad (4.283)$$

$$\begin{bmatrix} 0 & -9 & -275 \\ 9 & 0 & -115 \\ 275 & 115 & 0 \end{bmatrix} \begin{bmatrix} F_{E21x} \\ F_{E21y} \\ F_{E21z} \end{bmatrix} + \begin{bmatrix} 0 & -9 & -275 \\ 9 & 0 & 115 \\ 275 & -115 & 0 \end{bmatrix} \begin{bmatrix} F_{F21x} \\ F_{F21y} \\ F_{F21z} \end{bmatrix} = \begin{bmatrix} 0 \\ 0 \\ 0 \end{bmatrix} Nmm$$

$$(4.284)$$

Multiplying out the matrices in Eqn (4.284) yields the next set of three equations:

Equation 4 $\qquad -9\,F_{E21y} - 275\,F_{E21z} - 9\,F_{F21y} - 275\,F_{F21z} = 0$ $\qquad$ (4.285)

Equation 5 $\qquad 9\,F_{E21x} - 115\,F_{E21z} + 9\,F_{F21x} + 115\,F_{F21z} = 0$ $\qquad$ (4.286)

Equation 6 $\qquad 275\,F_{E21x} + 115\,F_{E21y} + 275\,F_{F21x} - 115\,F_{F21y} = 0$ $\qquad$ (4.287)

Consider next Body 3 and the equations required for force equilibrium.

$$\sum\{F_3\}_1 = \{0\}_1 \qquad (4.288)$$

$$\{F_{A31}\}_1 + \{F_{B31}\}_1 + f_{S2}\{R_{CI}\}_1 + \{F_{D34}\}_1 = \{0\}_1 \qquad (4.289)$$

$$\begin{bmatrix} F_{A31x} \\ F_{A31y} \\ F_{A31z} \end{bmatrix} + \begin{bmatrix} F_{B31x} \\ F_{B31y} \\ F_{B31z} \end{bmatrix} + f_{S2}\begin{bmatrix} 3 \\ -9 \\ -436 \end{bmatrix} + \begin{bmatrix} F_{D34x} \\ F_{D34y} \\ F_{D34z} \end{bmatrix} = \begin{bmatrix} 0 \\ 0 \\ 0 \end{bmatrix} N \qquad (4.290)$$

The summation of forces in Eqn (4.290) leads to the next set of three equations:

Equation 7 $\qquad F_{A31x} + F_{B31x} + 3\,f_{S2} + F_{D34x} = 0$ $\qquad$ (4.291)

Equation 8 $\qquad F_{A31y} + F_{B31y} - 9\,f_{S2} + F_{D34y} = 0$ $\qquad$ (4.292)

Equation 9 $\qquad F_{A31z} + F_{B31z} - 436\,f_{S2} + F_{D34z} = 0$ $\qquad$ (4.293)

Taking moments about point D for the forces acting on Body 3 gives

$$\sum\{M_{D3}\}_1 = \{0\}_1 \qquad (4.294)$$

$$\{R_{AD}\}_1 \times \{F_{A31}\}_1 + \{R_{BD}\}_1 \times \{F_{B31}\}_1 + \{R_{CD}\}_1 \times f_{S2}\{R_{CI}\}_1 = \{0\}_1 \qquad (4.295)$$

$$\begin{bmatrix} 0 & -15 & -239 \\ 15 & 0 & -115 \\ 239 & 115 & 0 \end{bmatrix}\begin{bmatrix} F_{A31x} \\ F_{A31y} \\ F_{A31z} \end{bmatrix} + \begin{bmatrix} 0 & -1 & -239 \\ 1 & 0 & 115 \\ 239 & -115 & 0 \end{bmatrix}\begin{bmatrix} F_{B31x} \\ F_{B31y} \\ F_{B31z} \end{bmatrix} + f_{S2}\begin{bmatrix} 0 & 23 & -98 \\ -23 & 0 & 0 \\ 98 & 0 & 0 \end{bmatrix}\begin{bmatrix} 3 \\ -9 \\ -436 \end{bmatrix} = \begin{bmatrix} 0 \\ 0 \\ 0 \end{bmatrix} Nmm$$
$$(4.296)$$

Multiplying out the matrices in Eqn (4.296) yields the next set of three equations:

Equation 10 $\qquad -15\,F_{A31y} - 239\,F_{A31z} - 1\,F_{B31y} - 239\,F_{B31z} + 42521\,f_{S2} = 0$
$$(4.297)$$

Equation 11 $\qquad 15\,F_{A31x} - 115\,F_{A31z} + 1\,F_{B31x} + 115\,F_{B31z} - 69\,f_{S2} = 0 \quad = 0$
$$(4.298)$$

Equation 12 $\qquad$ $239\,F_{A31x} + 115\,F_{A31y} + 239\,F_{B31x} - 115\,F_{B31y} + 294\,f_{S2} = 0$

$$(4.299)$$

Consider last Body 4 and the equations required for force equilibrium.

$$\sum \{F_4\}_1 = \{0\}_1 \qquad (4.300)$$

$$\{F_{P41}\}_1 - \{F_{D34}\}_1 - \{F_{G24}\}_1 - f_{S1}\{R_{JH}\}_1 = \{0\}_1 \qquad (4.301)$$

$$\begin{bmatrix} 0 \\ 0 \\ 10000 \end{bmatrix} - \begin{bmatrix} F_{D34x} \\ F_{D34y} \\ F_{D34z} \end{bmatrix} - \begin{bmatrix} F_{G24x} \\ F_{G24y} \\ F_{G24z} \end{bmatrix} - f_{S1}\begin{bmatrix} 0 \\ -228 \\ 8 \end{bmatrix} = \begin{bmatrix} 0 \\ 0 \\ 0 \end{bmatrix} N \qquad (4.302)$$

The summation of forces in Eqn (4.302) leads to the next set of three equations:

Equation 13 $\qquad$ $-F_{D34x} - F_{G24x} = 0 \qquad (4.303)$

Equation 14 $\qquad$ $-F_{D34y} - F_{G24y} + 228\,f_{S1} = 0 \qquad (4.304)$

Equation 14 $\qquad$ $-F_{D34z} - F_{G24z} - 8\,f_{S1} = -10000 \qquad (4.305)$

Taking moments about point G for the forces acting on Body 4 gives

$$\sum \{M_{G4}\}_1 = \{0\}_1 \qquad (4.306)$$

$$\{R_{PG}\}_1 \times \{F_{P41}\}_1 - \{R_{DG}\}_1 \times \{F_{D34}\}_1 - \{R_{HG}\}_1 \times f_{S1}\{R_{JH}\}_1 = \{0\}_1 \qquad (4.307)$$

$$\begin{bmatrix} 0 & 176 & 58 \\ -176 & 0 & 7 \\ -58 & -7 & 0 \end{bmatrix}\begin{bmatrix} 0 \\ 0 \\ 10000 \end{bmatrix} - \begin{bmatrix} 0 & -216 & -31 \\ 216 & 0 & 19 \\ 31 & -19 & 0 \end{bmatrix}\begin{bmatrix} F_{D34x} \\ F_{D34y} \\ F_{D34z} \end{bmatrix} - f_{S1}\begin{bmatrix} 0 & -267 & -75 \\ 267 & 0 & 163 \\ 75 & -163 & 0 \end{bmatrix}\begin{bmatrix} 0 \\ -228 \\ 8 \end{bmatrix} = \begin{bmatrix} 0 \\ 0 \\ 0 \end{bmatrix} Nmm \qquad (4.308)$$

Multiplying out the matrices in Eqn (4.308) yields the next set of three equations:

Equation 16 $\qquad$ $216\,F_{D34y} + 31\,F_{D34z} - 60276\,f_{S1} = -580000 \qquad (4.309)$

Equation 17 $\qquad$ $-216\,F_{D34x} - 19\,F_{D34z} - 1304\,f_{S1} = 0 \quad = -70000 \qquad (4.310)$

Equation 18 $\qquad$ $-31\,F_{D34x} + 19\,F_{D34y} - 37164\,f_{S1} = 0 \qquad (4.311)$

Finally applying the vector dot product to ensure that no thrust for the force $\{F_{F21}\}_1$ acts along the axis EF gives

$$\{F_{F21}\}_1 \bullet \{R_{EF}\}_1 = 0 \tag{4.312}$$

$$\left[F_{F21x}\, F_{F21y}\, F_{F21z}\right]\begin{bmatrix} 230 \\ 0 \\ 0 \end{bmatrix} = 0 \text{ Nmm} \tag{4.313}$$

$$230\, F_{F21x} = 0 \tag{4.314}$$

For this particular suspension system the line EF is parallel to the model x-axis yielding the trivial result F_{F21x} being equal to zero. In this case we can therefore ignore F_{F21x} in the following matrix solution of the system equations.

The axis AB for the upper wishbone is not parallel to a model axis and therefore applying the vector dot product to ensure that $\{F_{B31}\}_1$ is perpendicular to the line AB yields the final equation needed to solve the remaining 19 unknowns.

$$\{F_{B31}\}_1 \bullet \{R_{AB}\}_1 = 0 \tag{4.315}$$

$$\left[F_{B31x}\, F_{B31y}\, F_{B31z}\right]\begin{bmatrix} 230 \\ 0 \\ 14 \end{bmatrix} = 0 \text{ Nmm} \tag{4.316}$$

Equation 19 $\qquad\qquad 230\, F_{B31x} + 14\, F_{B31z} = 0 \tag{4.317}$

The 19 equations can now be set up in matrix form ready for solution.

$$
\begin{bmatrix}
0 & 0 & 0 & 0 & 0 & 0 & 0 & 0 & 0 & 1 & 0 & 0 & 0 & 0 & 1 & 0 & 0 & 0 & 0 \\
0 & 0 & 0 & 0 & 0 & 0 & 0 & 0 & 0 & 0 & 1 & 0 & 1 & 0 & 0 & 1 & 0 & 0 & 0 \\
0 & 0 & 0 & 0 & 0 & 0 & 0 & 0 & 0 & 0 & 0 & 1 & 0 & 1 & 0 & 0 & 1 & 0 & 0 \\
0 & 0 & 0 & 0 & 0 & 0 & 0 & 0 & 0 & -9 & -275 & -9 & -275 & 0 & 0 & 0 & 0 & 0 & 0 \\
0 & 0 & 0 & 0 & 0 & 0 & 0 & 0 & 0 & 9 & 0 & -115 & 0 & 115 & 0 & 0 & 0 & 0 & 0 \\
0 & 0 & 0 & 0 & 0 & 0 & 0 & 0 & 0 & 275 & 115 & 0 & -115 & 0 & 0 & 0 & 0 & 0 & 0 \\
1 & 0 & 0 & 1 & 0 & 0 & 1 & 0 & 0 & 0 & 0 & 0 & 0 & 0 & 0 & 0 & 0 & 0 & 3 \\
0 & 1 & 0 & 0 & 1 & 0 & 0 & 1 & 0 & 0 & 0 & 0 & 0 & 0 & 0 & 0 & 0 & 0 & -9 \\
0 & 0 & 1 & 0 & 0 & 1 & 0 & 0 & 1 & 0 & 0 & 0 & 0 & 0 & 0 & 0 & 0 & 0 & -436 \\
0 & -15 & -239 & 0 & -1 & -239 & 0 & 0 & 0 & 0 & 0 & 0 & 0 & 0 & 0 & 0 & 0 & 0 & 42521 \\
15 & 0 & -115 & 1 & 0 & 115 & 0 & 0 & 0 & 0 & 0 & 0 & 0 & 0 & 0 & 0 & 0 & 0 & -69 \\
239 & 115 & 0 & 239 & -115 & 0 & 0 & 0 & 0 & 0 & 0 & 0 & 0 & 0 & 0 & 0 & 0 & 0 & 294 \\
0 & 0 & 0 & 0 & 0 & 0 & -1 & 0 & 0 & 0 & 0 & 0 & 0 & -1 & 0 & 0 & 0 & 0 & 0 \\
0 & 0 & 0 & 0 & 0 & 0 & 0 & -1 & 0 & 0 & 0 & 0 & 0 & 0 & -1 & 0 & 228 & 0 & 0 \\
0 & 0 & 0 & 0 & 0 & 0 & 0 & 0 & -1 & 0 & 0 & 0 & 0 & 0 & 0 & -1 & -8 & 0 & 0 \\
0 & 0 & 0 & 0 & 0 & 0 & 0 & 216 & 31 & 0 & 0 & 0 & 0 & 0 & 0 & 0 & -60276 & 0 & 0 \\
0 & 0 & 0 & 0 & 0 & 0 & -216 & 0 & -19 & 0 & 0 & 0 & 0 & 0 & 0 & 0 & -1304 & 0 & 0 \\
0 & 0 & 0 & 0 & 0 & 0 & -31 & 19 & 0 & 0 & 0 & 0 & 0 & 0 & 0 & 0 & -37164 & 0 & 0 \\
0 & 0 & 0 & 230 & 0 & 14 & 0 & 0 & 0 & 0 & 0 & 0 & 0 & 0 & 0 & 0 & 0 & 0 & 0
\end{bmatrix}
\begin{bmatrix}
F_{A31x} \\ F_{A31y} \\ F_{A31z} \\ F_{B31x} \\ F_{B31y} \\ F_{B31z} \\ F_{D34x} \\ F_{D34y} \\ F_{D34z} \\ F_{E21x} \\ F_{E21y} \\ F_{E21z} \\ F_{F21y} \\ F_{F21z} \\ F_{G21x} \\ F_{G24y} \\ F_{G24z} \\ fS1 \\ fS2
\end{bmatrix}
=
\begin{bmatrix}
0 \\ 0 \\ 0 \\ 0 \\ 0 \\ 0 \\ 0 \\ 0 \\ 0 \\ 0 \\ 0 \\ 0 \\ 0 \\ 0 \\ -10000 \\ -580000 \\ -70000 \\ 0 \\ 0
\end{bmatrix}
\tag{4.318}
$$

Examination of the square matrix in Eqn (4.318) indicates a large number of zero terms, hence the matrix is referred to as sparse. As discussed in Chapter 3 this is a typical characteristic of the matrices generated in MBS and is one of the reasons that fast and efficient matrix inversion techniques can be deployed. The overall result is that MBS programs appear to solve quite complex engineering problems with a much lower requirement for computational effort than comparable other CAE methods such as nonlinear FE analysis. Solving Eqn (4.318) yields the following answers for the 20 unknowns:

$$F_{A31x} = 645.173 \text{ N} \qquad\qquad F_{E21x} = -557.482 \text{ N}$$

$$F_{A31y} = 2006.948 \text{ N} \qquad\qquad F_{E21y} = -1453.911 \text{ N}$$

$$F_{A31z} = 3412.360 \text{ N} \qquad\qquad F_{E21z} = 47.583 \text{ N}$$

$$F_{B31x} = -204.112 \text{ N} \qquad\qquad F_{F21x} = 0 \text{ N}$$

$$F_{B31y} = 3022.800 \text{ N} \qquad\qquad F_{F21y} = -2787.021 \text{ N}$$

$$F_{B31z} = 3353.266 \text{ N} \qquad\qquad F_{F21z} = 91.212 \text{ N}$$

$$F_{D34x} = -557.482 \text{ N} \qquad\qquad F_{G24x} = 557.482 \text{N}$$

$$F_{D34y} = -4680.486 \text{ N} \qquad\qquad F_{G24y} = 4240.9322 \text{ N}$$

$$F_{D34z} = 10154.217 \text{ N} \qquad\qquad F_{G24z} = -138.979 \text{ N}$$

$$f_{S1} = -1.92786767 \text{ N/mm} \qquad\qquad f_{S2} = 38.8069785 \text{ N/mm}$$

It is now possible to use the two scale factors found, f_{S1} and f_{S2}, to calculate the force vectors $\{F_{H54}\}_1$ and $\{F_{C37}\}_1$.

$$\{F_{H54}\}_1 = f_{S1} \{R_{JH}\}_1 = -1.92786767 \begin{bmatrix} 0 \\ -228 \\ 8 \end{bmatrix} = \begin{bmatrix} 0 \\ 439.554 \\ -15.942 \end{bmatrix} \text{ N} \qquad (4.319)$$

$$\{F_{C37}\}_1 = f_{S2} \{R_{CI}\}_1 = 38.8069785 \begin{bmatrix} 3 \\ -9 \\ -436 \end{bmatrix} = \begin{bmatrix} 116.421 \\ -349.263 \\ -16919.843 \end{bmatrix} \text{ N} \qquad (4.320)$$

In summary the force vectors are as follows:

$$\{F_{A31}\}_1^T = \begin{bmatrix} 645.173 & 2006.948 & 3412.360 \end{bmatrix} \text{ N}$$

$$\{F_{B31}\}_1^T = \begin{bmatrix} -204.112 & 3022.800 & 3353.266 \end{bmatrix} \text{ N}$$

$$\{F_{D34}\}_1^T = \begin{bmatrix} -557.482 & -4680.486 & 10154.217 \end{bmatrix} \text{ N}$$

$$\{F_{E21}\}_1^T = \begin{bmatrix} -557.482 & -1453.911 & 47.583 \end{bmatrix} \text{ N}$$

$$\{F_{F21}\}_1^T = \begin{bmatrix} 0 & -2787.021 & 91.212 \end{bmatrix} \text{ N}$$

$$\{F_{G24}\}_1^T = \begin{bmatrix} 557.482 & 4240.932 & -138.979 \end{bmatrix} \text{ N}$$

$$\{F_{H54}\}_1^T = \begin{bmatrix} 0 & 439.554 & -15.942 \end{bmatrix} \text{ N}$$

$$\{F_{C37}\}_1^T = \begin{bmatrix} 116.421 & -349.263 & -16919.843 \end{bmatrix} \text{ N}$$

Table 4.13 Comparison of Force Vectors Computed by Theory and MSC ADAMS

| Force | Force Vectors | | | | | |
| | Theory | | | MSC ADAMS | | |
	F_x (N)	F_y (N)	F_z (N)	F_x (N)	F_y (N)	F_z (N)
F_{A31}	645.173	2006.948	3412.360	645.173	2006.950	3412.360
F_{B31}	−204.112	3022.800	3353.266	−204.112	3022.800	3353.270
F_{C37}	116.421	−349.263	−16919.843	116.421	−349.263	−16919.800
F_{D34}	−557.482	−4680.486	10154.217	−557.482	−4680.490	10154.200
F_{E21}	−557.482	−1453.911	47.583	−557.482	−1453.910	47.582
F_{F21}	0.0	−2787.021	91.212	0.0	−2787.020	91.212
F_{G24}	557.482	4240.932	−138.979	557.482	4240.930	−138.794
F_{H54}	0.0	439.554	−15.942	0.0	439.554	−15.423

A comparison of the forces found, at points within the suspension system, from the preceding calculations and those found using an equivalent MSC ADAMS model is shown in Table 4.13.

4.10.5 Dynamic analysis

Having carried out a static analysis it is possible to progress to a full dynamic analysis of the system. For the suspension system considered here a theoretical solution could be formulated on the basis of the same set of 20 unknown constraint forces as used in the previous static analysis. Referring back to Chapter 2 however, the reader will realise that the addition of inertial forces and the use of a local body-centred coordinate system for the moment balance will add to the complexity of the solution. For brevity a full theoretical solution will not be performed here but rather the six equations of motion for Body 2 will be set up using, by way of example, the velocities and accelerations found earlier. The process of setting up the equations of motion for the other bodies would follow in a similar manner. Body 2 can be considered in isolation as illustrated with the free body diagram shown in Figure 4.90.

For the dynamic analysis we can take it that the physical properties of the suspension component; mass, mass moments of inertia, centre of mass location and orientation of the body principal axis system are all known.

The coordinate data provided with this example have only provided definitions so far for the locations of points such as those defining suspension mounts and joints connecting the linkages. Mass centre positions have not been provided. For a dynamic analysis the mass centre locations of all moving bodies are required in order to set up the equations of motion. For this example using Body 2 the position of the mass centre, G_2, relative to the inertial reference frame, O_1, is defined by the position vector $\{R_{G2O1}\}_1$ and assumed to be

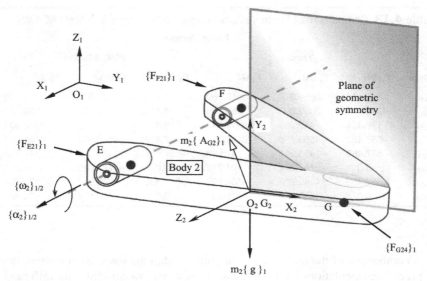

FIGURE 4.90

Free Body diagram for suspension lower wishbone body 2

$$\{R_{G2O1}\}_1^T = \begin{bmatrix} 7 & 500 & -85 \end{bmatrix} \text{mm}$$

The mass of Body 2, m_2, is taken to be 3.5 kg. It should also be noted from Figure 4.90 that the principal axes of Body 2 are located at the mass centre, G_2, and are defined by the reference frame, O_2. The transformation from reference frame O_1 to O_2 is obtained through a set of three Euler angle rotations as shown in Figure 4.91.

The mass moments of inertia for Body 2, measured about the principal axes of the body O_2, are taken to be for this example:

$$I_{21} = I_{2xx} = 1.5 \times 10^3 \text{ kgmm}^2$$

$$I_{22} = I_{2yy} = 38 \times 10^3 \text{ kgmm}^2$$

$$I_{23} = I_{2zz} = 38 \times 10^3 \text{ kgmm}^2$$

The X_2Y_2 plane of O_2 is taken to be a plane of geometric symmetry for the part so that all cross products of inertia are zero. The inertia matrix for Body 2 $[I_2]_{2/2}$ measured from and referred to reference frame O_2 is therefore

$$[I_2]_{2/2} = \begin{bmatrix} 1.5 \times 10^3 & 0 & 0 \\ 0 & 38 \times 10^3 & 0 \\ 0 & 0 & 38 \times 10^3 \end{bmatrix} \text{ kgmm}^2 \qquad (4.321)$$

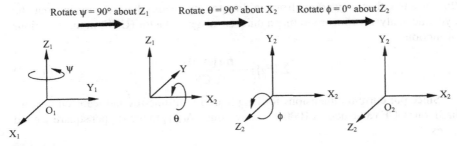

FIGURE 4.91

Definition of body principal axis system using Euler angle rotations.

From the previous velocity and acceleration analysis we also have

$$\{\omega_2\}_1^T = \begin{bmatrix} 12.266 & 0 & 0 \end{bmatrix} \text{ rad/s}$$

$$\{\alpha_2\}_1^T = \begin{bmatrix} -10.642 & 0 & 0 \end{bmatrix} \text{ rad/s}^2$$

Before progressing to set up the equations of motion we need to do one more calculation to find the acceleration $\{A_{G2}\}_1$ of the mass centre for Body 2.

$$\{A_{G2}\}_1 = \{A_{G2E}\}_1 = \{\omega_2\}_1 \times \{V_{G2}\}_1 + \{\alpha_2\}_1 \times \{R_{G2E}\}_1 \qquad (4.322)$$

where

$$\{V_{G2}\}_1 = \{V_{G2E}\}_1 = \{\omega_2\}_1 \times \{R_{G2E}\}_1 \qquad (4.323)$$

$$\begin{bmatrix} V_{G2x} \\ V_{G2y} \\ V_{G2z} \end{bmatrix} = \begin{bmatrix} 0 & 0 & 0 \\ 0 & 0 & -12.266 \\ 0 & 12.266 & 0 \end{bmatrix} \begin{bmatrix} -115 \\ 155 \\ -5 \end{bmatrix} = \begin{bmatrix} 0 \\ 61.33 \\ 1901.23 \end{bmatrix} \text{mm/s} \qquad (4.324)$$

therefore

$$\begin{bmatrix} A_{G2x} \\ A_{G2y} \\ A_{G2z} \end{bmatrix} = \begin{bmatrix} 0 & 0 & 0 \\ 0 & 0 & -12.266 \\ 0 & 12.266 & 0 \end{bmatrix} \begin{bmatrix} 0.0 \\ 61.33 \\ 1901.23 \end{bmatrix} + \begin{bmatrix} 0 & 0 & 0 \\ 0 & 0 & 10.642 \\ 0 & -10.642 & 0 \end{bmatrix} \begin{bmatrix} -115 \\ 155 \\ -5 \end{bmatrix} \text{mm/s}^2$$

$$(4.325)$$

$$\begin{bmatrix} A_{G2x} \\ A_{G2y} \\ A_{G2z} \end{bmatrix} = \begin{bmatrix} 0 \\ -23372.797 \\ -869.336 \end{bmatrix} \text{mm/s}^2 \qquad (4.326)$$

Before progressing further it is important to refer back to Chapter 3 and state that Newton's Second Law is only applicable for a consistent set of units. In effect this means either converting from millimetres to metres before carrying out the dynamic analysis or incorporating a units consistency factor (UCF) in the equations of motion

$$\Sigma\{F_2\}_1 = \frac{m_2\{A_{G2}\}_1}{UCF} \tag{4.327}$$

Since our current dimensions for length are in millimetres and we need to work in SI, our UCF value here is 1000. So converting $\{A_{G2}\}_1$ to metres per square second gives

$$\begin{bmatrix} A_{G2x} \\ A_{G2y} \\ A_{G2z} \end{bmatrix} = \begin{bmatrix} 0 \\ -23.373 \\ -0.869 \end{bmatrix} m/s^2 \tag{4.328}$$

We also need to define a vector $\{g\}_1$ for gravitational acceleration, which for the reference frame O_1 used here would be

$$\begin{bmatrix} g_x \\ g_y \\ g_z \end{bmatrix} = \begin{bmatrix} 0 \\ 0 \\ -9.81 \end{bmatrix} m/s^2 \tag{4.329}$$

For Body 2 summing forces and applying Newton's Second Law gives

$$\Sigma\{F_2\}_1 = m_2\{A_{G2}\}_1 \tag{4.330}$$

$$\{F_{E21}\}_1 + \{F_{F21}\}_1 + \{F_{G24}\}_1 + m_2\{g\}_1 = m_2\{A_{G2}\}_1 \tag{4.331}$$

$$\begin{bmatrix} F_{E21x} \\ F_{E21y} \\ F_{E21z} \end{bmatrix} + \begin{bmatrix} F_{F21x} \\ F_{F21y} \\ F_{F21z} \end{bmatrix} + \begin{bmatrix} F_{G24x} \\ F_{G24y} \\ F_{G24z} \end{bmatrix} + 3.5 \begin{bmatrix} 0 \\ 0 \\ -9.81 \end{bmatrix} = 3.5 \begin{bmatrix} 0 \\ -23.373 \\ -0.869 \end{bmatrix} N \tag{4.332}$$

The summation of forces in Eqn (4.332) leads to the first set of three equations:

Equation 1 $\qquad\qquad F_{E21x} + F_{F21x} + F_{G24x} = 0 \tag{4.333}$

Equation 2 $\qquad\qquad F_{E21y} + F_{F21y} + F_{G24y} = -81.806 \tag{4.334}$

Equation 3 $\qquad\qquad F_{E21z} + F_{F21z} + F_{G24z} = 31.294 \tag{4.335}$

For the rotational equations it is convenient to refer the vectors to the reference frame O_2 fixed in and rotating with Body 2. The rotational equations of motion for Body 2 may be written as Euler's equations of motion in vector form as

$$\Sigma\{M_{G2}\}_{1/2} = [I_2]_2\{\alpha_2\}_{1/2} + [\omega_2]_{1/2}[I_2]_2\{\omega_2\}_{1/2} \tag{4.336}$$

Before progressing this the angular velocity vector $\{\omega_2\}_1$ and angular accelera-
tion vector $\{\alpha\}_1$ need to be transformed from reference frame O_1 to O_2 to give $\{\omega_2\}_{1/2}$ and $\{\alpha\}_{1/2}$. By inspection it can be seen from Figure 4.91 that the transformation is
trivial and that due to the wishbone geometry and constraints ω_{2x} and α_{2x} in frame
O_1 simply become ω_{2z} and α_{2z} when referenced to frame O_2. The process of vector
transformation described in Chapter 2 will however be applied to illustrate the process for more general geometries. In this case we have only two rotations to account
for, the first being $90°$ Ψ about the z-axis, followed by a $90°$ rotation θ about the x-
axis. Thus for the angular velocity vector we have

$$\{\omega_2\}_{1/2} = \begin{bmatrix} \omega_{2x2} \\ \omega_{2y2} \\ \omega_{2z2} \end{bmatrix} = \begin{bmatrix} 1 & 0 & 0 \\ 0 & \cos\theta & \sin\theta \\ 0 & -\sin\theta & \cos\theta \end{bmatrix} \begin{bmatrix} \cos\psi & \sin\psi & 0 \\ -\sin\psi & \cos\psi & 0 \\ 0 & 0 & 1 \end{bmatrix} \begin{bmatrix} \omega_{2x1} \\ \omega_{2y1} \\ \omega_{2z1} \end{bmatrix} \text{rad/s}$$

(4.337)

$$\{\omega_2\}_{1/2} = \begin{bmatrix} \omega_{2x2} \\ \omega_{2y2} \\ \omega_{2z2} \end{bmatrix} = \begin{bmatrix} 1 & 0 & 0 \\ 0 & 0 & 1 \\ 0 & -1 & 0 \end{bmatrix} \begin{bmatrix} 0 & 1 & 0 \\ -1 & 0 & 0 \\ 0 & 0 & 1 \end{bmatrix} \begin{bmatrix} 12.266 \\ 0 \\ 0 \end{bmatrix} = \begin{bmatrix} 0 \\ 0 \\ 12.266 \end{bmatrix} \text{rad/s}$$

(4.338)

The transformation of the angular acceleration vector takes place in a similar
manner so that we have

$$\{\omega_2\}_{1/2}^T = [0 \quad 0 \quad 12.266] \text{ rad/s}$$

$$\{\alpha_2\}_{1/2}^T = [0 \quad 0 \quad -10.642] \text{ rad/s}^2$$

Referring back to Eqn (4.328) and the free body diagram in Figure 4.90 we can
see that it is convenient to sum moments of forces acting on Body 2 about the mass
centre G_2 to eliminate the inertial force $m_2\{A_2\}_1$ acting through the mass centre. In
order to carry out the moment balance we will need to establish new relative position
vectors $\{R_{EG2}\}_{1/2}$, $\{R_{FG2}\}_{1/2}$ and $\{R_{GG2}\}_{1/2}$. We will also need to define the vector
components in metres for consistency. Working first in frame O_1 we have:

$$\{R_{EG2}\}_1^T = [0.115 \quad -0.155 \quad 0.005] \text{ m}$$

$$\{R_{FG2}\}_1^T = [-0.115 \quad -0.155 \quad 0.005] \text{ m}$$

$$\{R_{GG2}\}_1^T = [0.0 \quad 0.120 \quad -0.004] \text{ m}$$

Applying a vector transformation for the vector $\{R_{EG2}\}_1$ from frame O_1 to O_2
gives

$$\{R_{EG2}\}_{1/2} = \begin{bmatrix} 1 & 0 & 0 \\ 0 & 0 & 1 \\ 0 & -1 & 0 \end{bmatrix} \begin{bmatrix} 0 & 1 & 0 \\ -1 & 0 & 0 \\ 0 & 0 & 1 \end{bmatrix} \begin{bmatrix} 0.115 \\ -0.155 \\ 0.005 \end{bmatrix} = \begin{bmatrix} -0.155 \\ 0.005 \\ 0.115 \end{bmatrix} \text{ m} \qquad (4.339)$$

Applying the same vector transformation to $\{R_{FG2}\}_{1/2}$ and $\{R_{GG2}\}_{1/2}$ gives us the three relative position vectors, referenced to the correct frame O_2 and in consistent units, needed for the moment balance:

$$\{R_{EG2}\}_{1/2}^T = 10^{-3}\begin{bmatrix}-155 & 5 & 115\end{bmatrix}\,m$$

$$\{R_{FG2}\}_{1/2}^T = 10^{-3}\begin{bmatrix}-155 & 5 & -115\end{bmatrix}\,m$$

$$\{R_{GG2}\}_{1/2}^T = 10^{-3}\begin{bmatrix}120 & 4 & 0\end{bmatrix}\,m$$

Before writing the rotational equations of motion we can first determine the moment balance of the constraint forces acting at E, F and G.

$$\sum\{M_{G2}\}_{1/2} = \{R_{EG2}\}_{1/2}\times\{F_{E21}\}_{1/2} + \{R_{FG2}\}_{1/2}\times\{F_{F21}\}_{1/2} + \{R_{GG2}\}_{1/2}\times\{F_{G24}\}_{1/2}$$

$$= 10^{-3}\begin{bmatrix}0 & -115 & 5\\115 & 0 & 155\\-5 & -155 & 0\end{bmatrix}\begin{bmatrix}F_{E21x2}\\F_{E21y2}\\F_{E21z2}\end{bmatrix} + 10^{-3}\begin{bmatrix}0 & 115 & 5\\-115 & 0 & 155\\-5 & -155 & 0\end{bmatrix}\begin{bmatrix}F_{F21x2}\\F_{F21y2}\\F_{F21z2}\end{bmatrix} + 10^{-3}\begin{bmatrix}0 & -4 & 0\\4 & 0 & -120\\0 & 120 & 0\end{bmatrix}\begin{bmatrix}F_{G24x2}\\F_{G24y2}\\F_{G24z2}\end{bmatrix}\,Nm$$

$$(4.340)$$

Considering next the rotational inertial terms we have

$$\sum\{M_{G2}\}_{1/2} = [I_2]_{2/2}\{\alpha_2\}_{1/2} + [\omega_2]_{1/2}[I_2]_{2/2}\{\omega_2\}_{1/2} \qquad (4.341)$$

$$\sum\{M_{G2}\}_{1/2} = \begin{bmatrix}I_{2xx} & 0 & 0\\0 & I_{2yy} & 0\\0 & 0 & I_{2zz}\end{bmatrix}\begin{bmatrix}\alpha_{2x2}\\\alpha_{2y2}\\\alpha_{2z2}\end{bmatrix} + \begin{bmatrix}0 & -\omega_{2z2} & \omega_{2y2}\\\omega_{2z2} & 0 & -\omega_{2x2}\\-\omega_{2y2} & \omega_{2x2} & 0\end{bmatrix}\begin{bmatrix}I_{2xx} & 0 & 0\\0 & I_{2yy} & 0\\0 & 0 & I_{2zz}\end{bmatrix}\begin{bmatrix}\omega_{2x2}\\\omega_{2y2}\\\omega_{2z2}\end{bmatrix}\,Nm$$

$$= \begin{bmatrix}1.5\times10^{-3} & 0 & 0\\0 & 38\times10^{-3} & 0\\0 & 0 & 38\times10^{-3}\end{bmatrix}\begin{bmatrix}0\\0\\-10.642\end{bmatrix} + \begin{bmatrix}0 & -12.266 & 0\\12.266 & 0 & 0\\0 & 0 & 0\end{bmatrix}\begin{bmatrix}1.5\times10^{-3} & 0 & 0\\0 & 38\times10^{-3} & 0\\0 & 0 & 38\times10^{-3}\end{bmatrix}\begin{bmatrix}0\\0\\12.266\end{bmatrix}\,Nm$$

$$(4.342)$$

Equating Eqn (4.340) with Eqn (4.342) yields the rotational equations of motion for Body 2:

Equation 4 $\quad(-115F_{E21y2}+5F_{E21z2}+115F_{F21y2}+5F_{F21z2}-4F_{G24y2})\times10^{-3} = 0$

$$(4.343)$$

Equation 5 $\;(115F_{E21x2}+155F_{E21z2}-115F_{F21x2}+155F_{F21z2}+F_{G24x2}-0.120F_{G24z2})\times10^{-3} = 0$

$$(4.344)$$

Equation 6 $\;(-5F_{E21x2}-155F_{E21y2}-5F_{F21x2}-155F_{F21y2}+120F_{G24y2})\times10^{-3} = -404.396\times10^{-3}$

$$(4.345)$$

At this stage the observant reader will note that the inertial terms in Eqn (4.342) have only yielded a numerical value for the moment balance about the principal Z_2 axis. This makes sense as the Z_2 axis has been chosen to be parallel to the fixed axis of body rotation through points E and F. In the absence of components of angular velocity or acceleration about X_2 and Y_2 Eqns (4.343) and (4.344) above simplify to a static moment balance. It should also be noted that when rotation is constrained about a single principal axis, the right-hand part of Eqn (4.341) $[\omega_2]_{\frac{1}{2}} [I_2]_{\frac{1}{2}} \{\omega_2\}_{\frac{1}{2}}$, is entirely zero to indicate a lack of gyroscopic terms in the absence of rotational coupling. Choosing a body-centred axis system for the upper wishbone Body 3, with an axis parallel to an axis through points A and B, would yield a similar formulation. This would not however be the case, for example, if we continued to set up the equations for Body 4 where there is no single fixed axis of rotation.

Before leaving the area of formulating equations of motion for dynamic analysis we should also ensure that the impression is not given that the use of a two-force body type scale factor, as used for the static analysis with the tie rod Body 5, can be employed here. Figure 4.92 shows free body diagrams for both a static and dynamic analysis of the tie rod. For the static analysis it can be seen that, with the assumption that gravity is ignored, the reaction forces at J and H act along the axis of the tie rod allowing a scale factor to be used. For the dynamic analysis it can be seen that the inertial forces do not allow such an assumption and that a set of six equations of motion for Body 5 will be required for the solution.

If at this stage we ignore the mass effects of the damper assembly we can represent the force $\{F_{C37}\}_1$ acting on Body 3 at point C using as before a scalar. Since the line of action of $\{F_{C37}\}_1$ is known to act along the line CI it is possible to define the force using the magnitude of $|F_{C37}|$ factored with the unit vector $\{l_{CI}\}_1$, acting along the line from I to C, as follows:

$$\{F_{C37}\}_1 = F_s\{l_{CI}\}_1 \qquad (4.346)$$

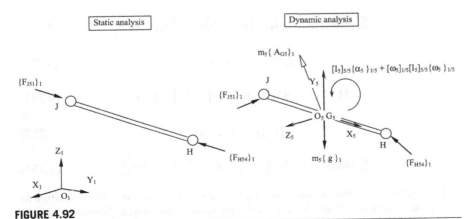

FIGURE 4.92

Free body diagrams for static and dynamic analysis of the tie rod.

where F_s is the magnitude of the force $|F_{C37}|$ with a sign assigned that is positive if the force acts towards point C from I. In this analysis, and under normal driving conditions, F_s will be positive.

A consideration of the complete suspension system indicates that the following set of 25 unknowns must be found to solve for dynamic forces:

$$F_{A31x}, F_{A31y}, F_{A31z}$$

$$F_{B31x}, F_{B31y}, F_{B31z}$$

$$F_{D43x}, F_{D43y}, F_{D43z}$$

$$F_{E21x}, F_{E21y}, F_{E21z}$$

$$F_{F21x}, F_{F21y}, F_{F21z}$$

$$F_{G24x}, F_{G24y}, F_{G24z}$$

$$F_{H54x}, F_{H54y}, F_{H54z}$$

$$F_{J51x}, F_{J51y}, F_{J51z}$$

$$F_S$$

Each moving body, Bodies 2, 3, 4 and 5, yields six equations of motion that can be used to solve the dynamic analysis. Using the same approach as demonstrated with Body 2 the following equations would be generated for all the bodies

$$\sum \{F_2\}_1 = m_2 \{A_{G2}\}_1 \tag{4.347}$$

$$\sum \{M_{G2}\}_{1/2} = \left[I_2\right]_{2} \{\alpha_2\}_{1/2} + \left[\omega_2\right]_{1/2} \left[I_2\right]_{2} \{\omega_2\}_{1/2} \tag{4.348}$$

$$\sum \{F_3\}_1 = m_3 \{A_{G3}\}_1 \tag{4.349}$$

$$\sum \{M_{G3}\}_{1/3} = \left[I_3\right]_{3} \{\alpha_3\}_{1/3} + \left[\omega_3\right]_{1/3} \left[I_3\right]_{3} \{\omega_3\}_{1/3} \tag{4.350}$$

$$\sum \{F_4\}_1 = m_4 \{A_{G4}\}_1 \tag{4.351}$$

$$\sum \{M_{G4}\}_{1/4} = \left[I_4\right]_{4} \{\alpha_4\}_{1/4} + \left[\omega_4\right]_{1/4} \left[I_4\right]_{4} \{\omega_4\}_{1/4} \tag{4.352}$$

$$\sum \{F_5\}_1 = m_5 \{A_{G5}\}_1 \tag{4.353}$$

$$\sum \{M_{G5}\}_{1/5} = \left[I_5\right]_{5} \{\alpha_5\}_{1/5} + \left[\omega_5\right]_{1/5} \left[I_5\right]_{5} \{\omega_5\}_{1/5} \tag{4.354}$$

The equations of motion above yield 24 equations leaving one further equation to be derived to solve the 25 unknowns. The final equation allows us to formulate the scalar F_s with the appropriate magnitude and sign to represent the force acting along the strut. Referring back to the discussion of spring and damper forces in Chapter 3 we are reminded that for a linear formulation based on the spring stiffness, k, free length, L and the damping coefficient, c, of the damper we can formulate the force using

$$\mathbf{F_s} = k \, (\, L - |R_{CI}| \,) - c \times VR_{CI} \tag{4.355}$$

The term $(L - |R_{CI}|)$ represents the deflection of the spring relative to the free length. The term VR_{CI} represents the radial line of sight velocity. This is effectively the magnitude of the velocity vector $\{V_{CI}\}_1$ given a sign so that VR_{CI} is negative when points C and I are approaching each other in bump and is positive when separating in rebound. The result of this is that the component of spring force is positive when the spring is compressed and the damper force component is positive during bump motion.

4.10.6 Geometry analysis

The preceding use of vectors to carry out three-dimensional velocity, acceleration, static force and dynamic force analyses of the double wishbone suspension system should have provided the reader with an insight into the computational work performed by an MBS program during the solution phase. An important aspect of this is that in all the preceding analyses the geometry has been assumed fixed throughout the solution. This is in fact not fully representative of the problem. For example for the static analysis the damper acting between C and I is assumed to be locked so that although the reaction force at C can be determined, the suspension does not move despite a considerable vertical load being applied at the tyre contact patch.

In reality the damper has a sliding degree of freedom that allows the length CI to shorten until the additional compression of the spring produces the force required for the suspension system to be in static equilibrium. This mechanism is the key behind the iterations described in Chapter 3 that take place during a solution step at a given point in time. In effect all the preceding vector analyses can be considered typical of the computations during one of many analysis iterations at a given point in time.

To demonstrate the final phase in this process a vector analysis will now be performed to determine the new position of the movable points throughout the suspension system due to a deflection in the suspension spring unit. In this case we will shorten the line CI by 100 mm taking this to be representative of the movement for this suspension with typical spring and damper properties. We can consider that we are looking here at the suspension moving between the defined or model input position, to the full bump position. During a typical analysis iteration the movement would in fact be far less than this but the following calculations will illustrate the process and complete our treatment of vector analysis in this chapter.

Before proceeding with the analysis, Figure 4.93 is provided to remind us of the suspension configuration, the point labelling system and to illustrate the shortening of the damper unit.

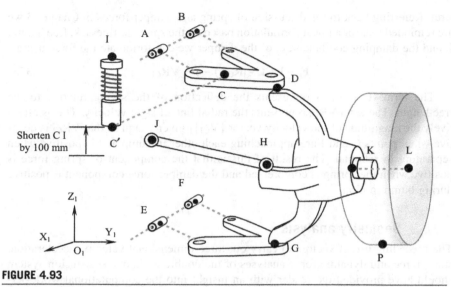

FIGURE 4.93

Shortening of damper unit for double wishbone suspension geometry analysis.

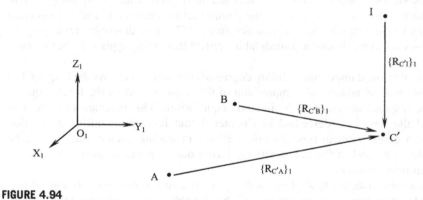

FIGURE 4.94

Locating new point C′ by triangulation.

In order to establish the position of any point that has moved in the suspension system we must work from three points for which the coordinates are already established. To begin the analysis we can consider finding the new position of point C′ working from three points A, B and I that are fixed and cannot move as shown in Figure 4.94.

In this example the positions of A, B and I are known, as are the lengths AC, BC and IC′. The length IC′ takes into account the shortening of the strut by 100 mm but the lengths AC and BC are unchanged. The new position of C′ is unknown and must be solved. In terms of vectors this can be expressed using the following known inputs:

$$\{R_A\}_1^T = \begin{bmatrix} Ax & Ay & Az \end{bmatrix}$$

$$\{R_B\}_1^T = \begin{bmatrix} Bx & By & Bz \end{bmatrix}$$

$$\{R_I\}_1^T = \begin{bmatrix} Ix & Iy & Iz \end{bmatrix}$$

$$|R_{CA}|$$

$$|R_{CB}|$$

$$|R_{C'I}|$$

In order to solve the three unknowns C'x, C'y and C'z, which are the components of the position vector $\{R_{C'}\}_1$, it is necessary to set up three equations as follows:

$$|R_{C'A}|^2 = (Cx - Ax)^2 + (C'y - Ay)^2 + (C'z - Az)^2 \qquad (4.356)$$

$$|R_{C'B}|^2 = (C'x - Bx)^2 + (C'y - By)^2 + (C'z - Bz)^2 \qquad (4.357)$$

$$|R_{C'I}|^2 = (C'x - Ix)^2 + (C'y - Iy)^2 + (C'z - Iz)^2 \qquad (4.358)$$

In Chapter 2 it was demonstrated that the simultaneous solution of Eqns (4.356)–(4.358) results in a quadratic with two solutions, one of which will be correct, for C'x, C'y and C'z. Having demonstrated in Chapter 2 the manipulations required to solve a set of three such equations we will content ourselves here to show the process followed to set up all the equations for this suspension system but use a computer program written in BASIC to solve them.

Figure 4.95 illustrates the process that would be followed to solve the coordinates of all movable points where at each stage the positions of the three reference points must be either fixed or previously found if movable. The length between each of the three reference points and the movable point must also be fixed and known. This will only work if the movable point lies on the same rigid body as each of the reference points.

In addition to locating the movable points just described we will also need to determine the new positions K' and L' of the two points located on the wheel spin axis as shown in Figure 4.96. These two positions will be used with the starting locations K and L to determine the change in steer and camber angle between the two suspension configurations.

Having followed the process outlined here we obtain the new positions shown in Table 4.14. The results obtained using vector theory are compared with those from the equivalent MSC ADAMS model where a motion input has been used to shorten the strut by 100 mm.

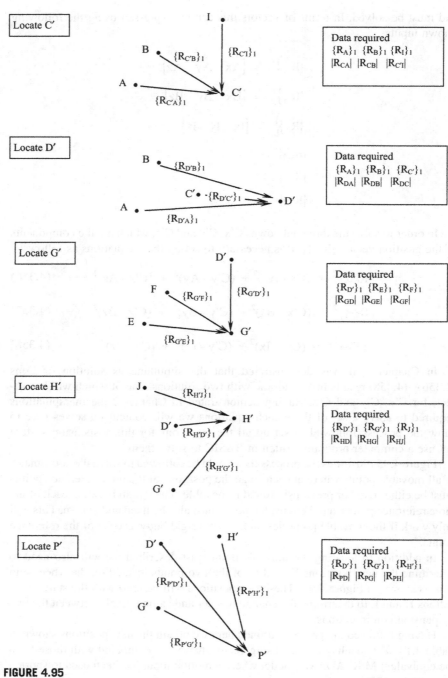

FIGURE 4.95

Calculation sequence to solve double wishbone suspension geometry.

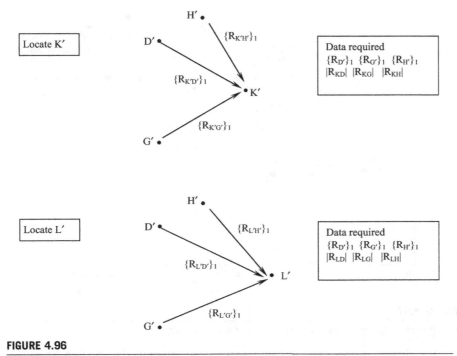

FIGURE 4.96

Location of points K′ and L′ on the wheel spin axis.

Table 4.14 Comparison of Movable Point Locations Computed by Theory and MSC ADAMS

	Suspension Position Vectors					
	Theory			MSC ADAMS		
Point	R_x (mm)	R_y (mm)	R_z (mm)	R_x (mm)	R_y (mm)	R_z (mm)
C′	−18.133	476.250	204.752	−18.133	476.250	204.753
D′	−21.719	534.601	286.699	−21.721	534.604	286.696
G′	7.0	573.627	73.087	7.0	573.629	73.084
H′	−168.982	493.368	330.131	−168.984	493.371	330.127
P′	8.978	638.864	−100.491	8.979	638.867	−100.494
K′	−4.232	550.301	160.835	−4.233	550.305	160.832
L′	−1.781	628.197	164.076	−1.783	628.199	164.073

Having calculated the new positions of all the movable nodes, the movement of the tyre contact patch, in this case taken to be point P, could be used to establish for example the lateral movement. Referring back to Chapter 2 we can also use the methods described there to determine the bump steer as shown in Figure 4.97.

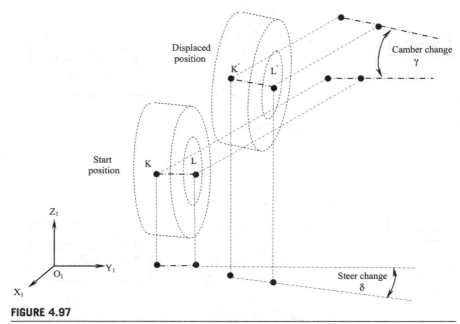

FIGURE 4.97

Using vectors to determine camber and steer angle change.

The change in steer angle or bump steer can be determined by finding the angle δ between the projection of KL and $\text{K}'\text{L}'$ onto the global X_1Y_1 plane. The projection is achieved after setting the z-coordinates of all four position vectors to zero and then applying the vector dot product as shown in Eqn (4.359).

$$\cos \delta = \{R_{KL}\}_1 \bullet \{R_{K'L'}\} / |R_{KL}| \, |R_{K'L'}| \qquad (4.359)$$

The change in camber angle, γ, is obtained in a similar manner where the projection this time takes place in the global Y_1Z_1 plane by setting all the x-coordinates to zero.

$$\cos \gamma = \{R_{KL}\}_1 \bullet \{R_{K'L'}\} / |R_{KL}| \, |R_{K'L'}| \qquad (4.360)$$

Note that in Figure 4.97 the change in steer and camber angles are both shown as positive for this suspension located on the front left-hand side of the vehicle with the X_1-axis pointing forwards. A comparison of the answers found by theory with those from MSC ADAMS is given in Table 4.15

Table 4.15 Comparison of Steer and Camber Angle Change Computed by Theory and MSC ADAMS

	Theory Change in Angle (degrees)	MSC ADAMS Change in Angle (degrees)
Steer	1.802	1.802
Camber	−2.383	−2.382

Tyre Characteristics and Modelling

When people say that tyres are round and black, I tell them they have the black bit right. Mostly.

Jan Prins, Jaguar Land Rover

5.1 Introduction

The handling performance and directional response of a vehicle are greatly influenced by the mechanical force and moment generating characteristics of the tyres. In road vehicle dynamics the manner in which a vehicle accelerates, brakes and corners is controlled by the forces generated over four relatively small tyre contact patches. Even motorsport vehicles predominantly use aerodynamic forces to modify the vertical forces on the tyres in order to harvest the mechanical amplification of those forces by the friction coefficient, which typically exceeds unity in motorsport applications. Applications such as rallying with a lower friction coefficient may use aerodynamic side forces directly but they are both obvious (from the predominance of features approximately parallel with the plane of symmetry of the vehicle) and unusual.

If the tread pattern and the road texture is also considered it is clear that the area of frictional contact is reduced even more significantly. Figure 5.1 shows the deflection of a vehicle's tyres under hard cornering and helps to illustrate the significant requirements on the tyre to produce forces that control the relatively large mass of the vehicle.

It is not intended here to discuss the construction of the tyre carcass, materials or tread pattern. This is addressed by more general texts on vehicle dynamics (Gillespie, 1992) or more focussed books on the subject of tyres (French, 1989), (Moore, 1975). Rather this chapter will start by describing the mechanisms required to generate the vertical tyre forces that support the vehicle, the longitudinal forces required for driving and braking and the lateral forces needed for cornering. The distribution of pressure and stress will also generate local moments acting at the tyre contact patch. A good understanding of these force and moment characteristics is essential before introducing the various mathematical tyre models available and describing the methods used to implement these with multibody systems (MBS) vehicle models.

A spectrum of tyre models exists. Because of the subtlety and deeply nonlinear behaviour of the tyre as a whole, they tend to be isolated in a separate subroutine in all but the simplest multibody model schemes. This chapter will proceed to a discussion of the common elements of tyre modelling and a discussion of the currently

FIGURE 5.1

Examples of tyre deflection under hard cornering.

popular tyre models. When it comes to functional models of tyres it is worth making a distinction between 'a set of expressions that produce tyre-like forces and are formulated in a parametric fashion' — which might perhaps be called the *tyre model architecture* — and 'such expressions, with all parameters present and calibrated to represent a particular tyre' — which might be called a *tyre-specific model*.

Before a computer simulation can be performed, the actual tyre force and moment characteristics must be estimated or obtained from experimental tests. A traditional approach is to test the tyre using a tyre test machine and to measure the resulting force and moment components for various camber angles, slip angles and values of vertical force. There follows the generation of a number of parameters that must be derived from the measured data before the simulation can proceed. The quality of the tyre-specific model will be a compromise between the accuracy of the fit, relevance of the parameters, and the availability of methods to generate the parameters.

As a final introductory comment it should be noted that what is referred to as a 'tyre' model is actually a tyre-road interaction model. Placing the same tyre on a different surface will modify its behaviour substantially, as anyone who has ever driven on ice will have noticed. The most versatile tyre-model architectures allow the tyre-specific model to encounter different surfaces during the course of a single manoeuvre, but require proportionally more data to support this functionality.

5.2 Tyre axis frames and geometry

5.2.1 The SAE J2047 and ISO 8855 tyre axis frames

To assist with the description of the forces and moments generated by a tyre, an axis frame shown in simplified form in Figure 5.2 has been defined by the SAE (1976).

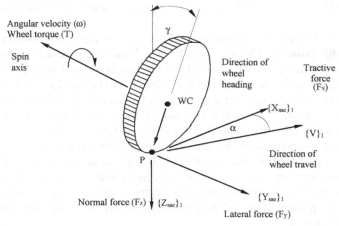

FIGURE 5.2

SAE tyre axis system.

In this frame the X-axis is the intersection of the wheel plane and the road plane with the positive direction taken for the wheel moving forward. The Z-axis is perpendicular to the road plane with a positive direction assumed to be acting downwards. The Y-axis is in the road plane and its direction dictated by the use of a right-handed orthogonal axis frame. The angles α and γ represent the slip angle and camber angle respectively. The SAE frame will be used throughout this text unless stated.

It should be noted that not all practitioners adhere rigidly to this frame in their publications and another frame, the ISO 8855 tyre frame, is gaining favour. This frame,

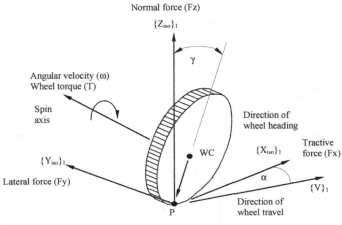

FIGURE 5.3

ISO 8855 tyre axis system.

shown simplified in Figure 5.3 has become a de facto standard for new tyre-model architectures.

The ISO 8855 tyre axis system is one of three axis systems described in the Tyre Data Exchange protocol, TYDEX, with the accompanying note that while the ISO definition presumes a horizontal ground plane — that is to say normal to the gravity vector — TYDEX does not enforce this condition. It will be seen later that distortions in the tyre carcass will cause the contact patch to move away from the rigid wheel plane shown in Figures 5.2 and 5.3. Although the components of force are still assumed to act through the contact point P, the distortions will introduce offsets and additional components of moment also acting about the point P.

While it is preferable to use one of the existing tyre axis systems, from time to time organisational convenience or historical compatibility suggests a different axis system be used. There is no conceptual difficulty in this save for a requirement to avoid errors of presumption, which is avoided with repeated and clear declaration of frames of reference.

5.2.2 Definition of tyre radii

The definition of tyre radii is important for the formulation of slip in the contact patch. In general we consider a tyre to have an unloaded radius, a loaded radius and an effective rolling radius. The unloaded tyre radius, R_u, is straightforward to comprehend and is shown in Figure 5.4. For a rigid disc with radius R_u rolling forward with no sliding (fully geared to the road), during one revolution the disc will move forward a distance $2\pi R_u$.

As can be seen from Figure 5.4, due to tyre deflection the distance moved forward will be less than for the rigid disc and can be related to the effective rolling radius giving

$$R_u > R_e > R_l \tag{5.1}$$

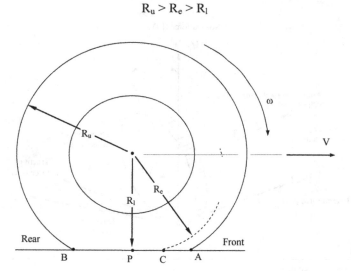

FIGURE 5.4

Definition of tyre radii.

Another definition of effective rolling radius is provided by Moore (1975), this being the distance from the wheel centre to a point C where the distance AC is taken to be one quarter of the total tyre contact patch length AB.

The following are other definitions related to tyre radius provided in SAE J670e (1976):

1. The loaded radius, R_l, is the distance from the centre of the tyre contact patch to the wheel centre measured in the wheel plane.
2. The static loaded radius is the *loaded radius* of a stationary tyre inflated to the normal recommended pressure.
3. The effective rolling radius, R_e, is the ratio of the linear velocity of the wheel centre in the X_{SAE} direction to the angular velocity of the wheel.

While such considerations are easily imagined when the tyre is upright on a smooth surface with a constant load, it may be readily appreciated that discerning the rolling radius for a real tyre with a nonzero inclination angle and on a non-smooth surface is constantly varying and somewhat difficult to discern in anything resembling real time using the above notions.

A more detailed treatment of effective rolling radius is provided by Phillips (2000) based on the representation given in Figure 5.5 that allows the effective rolling radius to be related to the unloaded radius and tyre deflection. For the tyre shown in Figure 5.5, the wheel axle is considered fixed and the road moving such that the relative forward velocity of the wheel is V. If a number of equidistant radial lines are drawn on the tyre the number passing point A in a given time must be the same as the number passing point P in the contact patch.

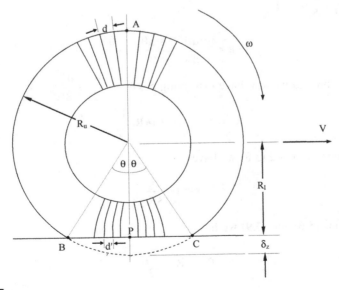

FIGURE 5.5

Deformation of rolling tyre.

If we take

d = the distance between the radial lines at the tyre outer radius near A
d′ = the distance between the radial lines at the contact patch near P

then

$$\frac{\omega R_u}{d} = \frac{V}{d'} \qquad (5.2)$$

therefore

$$R_e = \frac{V}{\omega} = R_u \frac{d'}{d} \qquad (5.3)$$

The tread band is subject to a longitudinal compressive strain within the contact patch ε where

$$\varepsilon = \frac{d - d'}{d} \qquad (5.4)$$

$$d' = d(1 - \varepsilon) \qquad (5.5)$$

therefore

$$R_e = R_u(1 - \varepsilon) \qquad (5.6)$$

Assuming that the strain in the contact line is constant we have (assuming $\sin\theta = \theta - \frac{\theta^3}{3!} + \frac{\theta^5}{5!} \dots \dots)$

$$1 - \varepsilon = \frac{cordBC}{arcBC} = \frac{\sin\theta}{\theta} \approx 1 - \frac{\theta^2}{6} \qquad (5.7)$$

From Figure 5.5 we also have (assuming $\cos\theta = 1 - \frac{\theta^2}{2!} + \frac{\theta^4}{4!} \dots \dots)$

$$\delta_z = R_u(1 - \cos\theta) \approx R_u \frac{\theta^2}{2} \qquad (5.8)$$

From Eqns (5.7) and (5.8) we have

$$1 - \varepsilon = 1 - \frac{\delta_z}{3R_u} \qquad (5.9)$$

From Eqns (5.6) and (5.9) we have

$$R_e = R_u - \frac{\delta_z}{3} \qquad (5.10)$$

If we substitute the loaded radius as $R_l = R_u - \delta_z$ into (5.10) we get

$$R_e = R_1 + \frac{2\delta_z}{3} \qquad (5.11)$$

Despite the complexity of the approach, the final approximation in Eqn (5.11) is pleasingly simple and useful in time-varying applications. With some knowledge of vertical stiffness, characteristics, it can be used to estimate rolling radius with useful accuracy in real time applications.

5.2.3 Tyre asymmetry

Although it is not intended to address the construction of a tyre in this textbook a brief mention is needed on two types of tyre asymmetry that can occur, these being conicity and plysteer. Both types of asymmetry can occur during tyre fabrication and have the effect of introducing small amounts of lateral force and aligning moment when a tyre is running at zero slip angle. It will be seen later in this chapter that these offsets in lateral force or aligning moment are visible when plotted against slip angle and that representation in a simulation will depend on the sophistication of the tyre model used.

In a modern vehicle these effects become more important when considering refinement and the 'on-centre' feel of the vehicle, particularly when driving for long periods at high motorway speeds.

Conicity is an effect that arises due to assuming the tyre to have the shape of a truncated cone as shown in Figure 5.6.

When considering the effect of conicity it must be realised that incorporation in a tyre model must take careful account of the tyre axis system used. In Figure 5.6, for

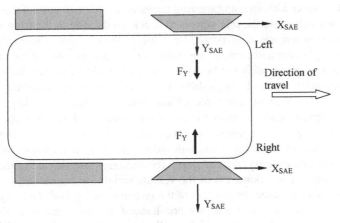

FIGURE 5.6

Generation of tyre lateral forces due to conicity.

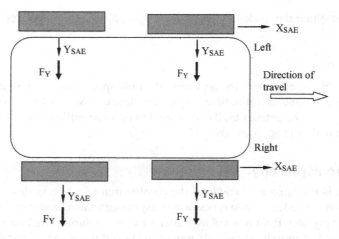

FIGURE 5.7

Generation of tyre lateral forces due to plysteer.

example, the tyres shown with exaggerated conicity produce a force towards the apex of the cone as the vehicle travels on a straight heading. For the tyre on the left side this is a force that is positive when referred to the Y_{SAE} axis. If the same tyre is now switched to the right side of the vehicle, reversing the direction of rotation, the force is still towards the apex but is now negative when referred to the Y_{SAE} axis.

Plysteer is an effect that arises due to a small bias in the positioning of the cords within the tyre belt layers. This is shown in Figure 5.7 where it is evident that the opposite occurs to conicity in that switching a tyre from the left to the right of the vehicle does not reverse the lateral force direction. Thus for a vehicle fitted with tyres all exhibiting the same plysteer there will be a tendency for the vehicle to drift off a straight course without some steering correction. This will correct the course of the vehicle but will cause the rear wheels to 'track' to the side of the front wheels so that the vehicle progresses with a crab like motion, albeit imperceptible to the driver.

In general, plysteer and conicity are 'noise' effects and not something that can be used to meaningfully modify the behaviour of the vehicle. However, knowledge of them is important when distinguishing between real-world measured data and concept-level symmetrical side-force characteristics. It should be kept in mind, though, that typical roads are constructed to have some kind of side slope — called 'cross fall' by highway engineers but popularly and ambiguously referred to as 'camber' by everyone else. Cross fall will induce side forces of comparable magnitude to plysteer and conicity; typical slopes for drainage purposes are of the order of 1%, giving side forces around 1% of the vehicle weight by inspection.

It can be seen that since the features of the tyre influencing both conicity and plysteer are not visible for tyres which are non-handed and non-directional then it is a statistical inevitability that some vehicles will end up with an unfavourable distribution of either plysteer or conicity. When tyres are marked with a rotation arrow or the words 'inside' and 'outside' then these effects are partially controlled. Tyres which use both a rotation arrow and inside/outside markings are completely controlled and are typically uniquely manufactured for each corner of the vehicle; such tyres may expect to show the best control of plysteer and conicity noise effects at the vehicle level.

5.3 The tyre contact patch

5.3.1 Friction

The classical laws of friction as often taught in school can be summarised as:

1. Friction is a property of two contacting surfaces. It does not make sense to discuss friction as if it were a material property.
2. Frictional force is linearly proportional to normal force and can be defined using a coefficient of friction (frictional force/normal force).
3. The coefficient of friction is independent of contact area between the two surfaces.
4. The static coefficient of friction is generally different to (and often greater than) the kinetic (or dynamic or sliding) coefficient of friction.
5. The coefficient of friction is independent of sliding speed.

A detailed treatment of this subject with regard to tyres is given by Moore (1975), where it is shown that the above laws are flawed, or limited in certain conditions such as high tyre pressures. The essential reason for this is the faintly outrageous behaviour of rubber and rubberlike compounds, which were not included in early experiments on friction carried out by, for example, Charles-Augustin de Coulomb, who had materials like glass, wood and leather to experiment with. The concept of a coefficient of friction, which varies with sliding velocity, will however prove useful for describing the tyre models used later in this chapter.

Friction as a force is often poorly understood and it is worth discussing explicitly. Even with the simplified laws of friction as described above, a block of material standing on a surface can be imagined to be subject to frictional forces at the interface of the surface and block. When the surface is level then the block and the surface are pressing on each other with equal and opposite normal forces due to the weight of the block and its reaction. With no other forces acting, there are no frictional forces present, although the *capacity* for them to arise is present. The frictional forces themselves only arise in reaction to an applied external force — an inclination of the surface or an externally applied force parallel to the plane of the surface (Figure 5.8).

The frictional forces calculated by the 'laws' of friction are in fact the maximum capacity for reactive forces to be generated, and when the applied forces are lower than this level then the frictional forces rise and fall to preserve equilibrium. Thus the frictional force in the stationary block is time-varying if the applied force is also time-varying. This concept becomes particularly important for modelling clutches, as discussed in Chapter 8.

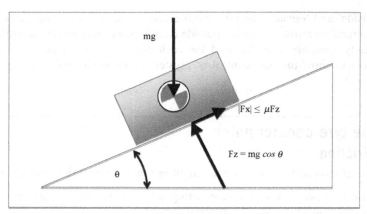

FIGURE 5.8

Frictional forces for a block on an inclined plane.

If the applied force is greater than the maximum capacity for frictional force then the difference between the applied force and the maximum frictional force is a net force available to accelerate the block. This is the simplest possible description of friction, often described as Coulomb friction. It is easy to formulate mathematically

$$\text{For } V_x > 0: \qquad \widehat{F}_x = -\mu\, F_z \qquad (5.12)$$

$$\text{For } V_x = 0: \qquad \widehat{F}_x = 0 \qquad (5.13)$$

$$\text{For } V_x < 0: \qquad \widehat{F}_x = \mu\, F_z \qquad (5.14)$$

However, it can be readily seen that numerically integrating this solution can be challenging owing to the discontinuity present in its formulation. A common work-around is to use the hyperbolic tangent function, *tanh*, as a function of sliding velocity (Figure 5.9).

The *tanh* function has the advantage of being numerically continuous but it does have the disadvantage of having a finite gradient at the origin. This means that, for example, if a block on an inclined plane is modelled then its weight can only be reacted with some nonzero velocity and therefore the block will not stay still. Even compressing the *tanh* formulation using a velocity multiplier cannot make the slope infinite, although it can make it so steep that the block does not move significantly during the simulation period. Students will soon discover, however, that being more and more demanding of the local gradient makes for a very lengthy solution time, requiring as it does a high degree of accuracy.

Sometimes the word 'stiction' is used in discussion of friction but its use is not preferred by the authors since many practitioners either fail to define it or use it inconsistently, or both.

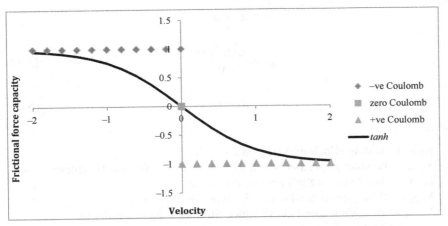

FIGURE 5.9

A dimensionless comparison between Coulomb friction and a numerically continuous implementation with *tanh*.

Other friction models are available, most notably for the authors a comparatively simple formulation that remains continuously differentiable and captures Coulomb, viscous and so-called Stribeck effects, which produce 'law 4' of the friction laws described above. As originally described by Makkar et al. (2005), the model is exceptionally simple, producing a frictional force F as a function of a displacement x and its first derivative

$$F = g_1\left(\tanh(g_2\dot{x}) - \tanh(g_3\dot{x})\right) + g_4\tanh(g_5\dot{x}) + g_6\dot{x} \qquad (5.15)$$

where the terms g_i are parameters: describing the model. The parameters are not particularly usable as is, and can be defined in terms of more readily assimilated parameters

$$R = \log\left(\frac{g_2}{g_3}\right) \qquad (5.16)$$

$$S = -0.099R^6 + 0.7815R^5 - 2.5644R^4 + 4.5815R^3 - 4.9237R^2 + 3.2794R - 1.1697 \qquad (5.17)$$

$$g_1 = \frac{\mu_1 - \mu_0}{10^S} \qquad (5.18)$$

$$g_2 = \frac{\tanh^{-1}(0.99)}{V_{SPk}} \qquad (5.19)$$

$$g_3 = \frac{\tanh^{-1}(0.99)}{V_{Send}} \qquad (5.20)$$

$$g_4 = \mu_0 \tag{5.21}$$

$$g_5 = \frac{\tanh^{-1}(0.99)}{V_{C99}} \tag{5.22}$$

$$g_6 = \mu_2 \tag{5.23}$$

where

μ_0 is the sliding (Coulomb) friction coefficient
V_{C99} is the sliding speed at which 99% of Coulomb friction is achieved
μ_1 is the breakout ('static') friction coefficient
V_{SPk} is sliding speed at which peak friction is achieved
V_{Send} is the sliding speed above which Stribeck effects are absent
μ_2 is the viscous dissipation term

It should be noted that the six-term Coulomb–viscous–Stribeck model described still suffers from the finite gradient problem and cannot hold a block still on an inclined plane. Discussion of yet more elaborate models, such as the Leuven friction model, are beyond the scope of this text but the interested reader is referred to the wider literature for a detailed view (van Geffen, 2009). All these models are essentially empirical, in that they fit some mathematical function to observed behaviour.

For real tyres the friction generated between the tread rubber and the road surface is generated through two mechanisms, these being hysteresis and adhesion.

As already discussed in Chapter 3, rubber displays substantial hysteresis. In order to understand the influence of hysteresis on tyre/road friction, consider a block of rubber subjected to an increasing and then a decreasing load as shown in Figure 5.10. As the rubber is loaded and unloaded it can be seen that, for a given displacement δ, the force F is greater during the loading phase than the unloading phase.

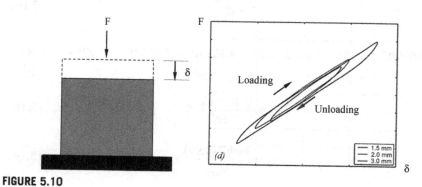

FIGURE 5.10

Hysteresis in rubber.

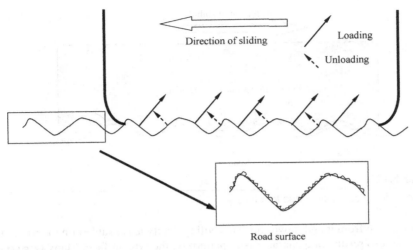

Road surface

FIGURE 5.11

Loading and unloading of a single tread block of tyre rubber in the contact patch.

If we consider the situation where the same block of rubber is sliding over a non-smooth surface, it can be seen from Figure 5.11 that an element of rubber in the contact patch will be subject to continuous compressive loading and unloading.

As the rubber slides over the irregular road surface compressive forces normal to the surface are generated and relieved as the rubber is loaded and unloaded. The resolved vertical sum of all the forces gives the vertical reaction. The resolved horizontal sum of all the forces gives the force due to hysteresis. Due to the hysteresis, the sum of the loading forces is greater than the sum of the unloaded forces, resulting in a force opposing the direction of sliding. It should be emphasised that the scale of Figure 5.11 is frequently misunderstood, being presumed in error as the gaps between aggregate and binder in the road surface; it is rather the texture on the surface of an individual stone that is shown in close-up and is something of a surprisingly intricate 'fractal style' effect. It can be seen that higher hysteresis compounds will deliver more of this effect and that it is greatly sensitive to texture; highly polished surfaces will deliver little of this effect. It is, however, broadly insensitive to surface contamination.

The adhesive component, shown in Figure 5.12, results from momentary intermolecular links generated between the exposed surface atoms of rubber and road material in the contact area (Kummer, 1966). This component of friction 'tops up' the hysteretic force on dry roads but is greatly reduced when the road surface is contaminated with even a thin film of water or powder. 'Slick' tyres, with no tread and increased surface contact area, are used for racing on dry roads to maximise this force.

It is perfectly possible to manufacture a rubber compound in which the affinity for the road surface exceeds the ability of the tread compound to hold itself together, in which case the rubber at the road surface will remain there while the bulk of the

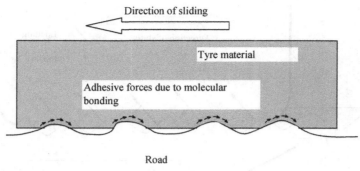

FIGURE 5.12

Frictional force component due to adhesion.

rubber is torn from it and moves on. This will typically leave rubber on the road surface and also results in a characteristic pattern on the tyre surface. Many tyre compounds can generate large forces at the road/tyre interface, which leads to a great deal of energy being applied to the interface; in some cases this raises the temperature sufficiently not only to reduce the shear strength of the rubber compound, but also to locally liquefy a bituminous binder in the road surface, floating it to the top. Thus it is often true that black marks on the road surface are some combination of rubber and bitumen (where such an ingredient is present in the road surface, of course; on concrete this simply cannot be so and the mark is all rubber).

The interested reader is referred to 'The Racing and High Performance Tire' (Haney, 2003) for an excellent empirical discourse on interpreting tyre behaviour in performance-limit conditions by referring to after-the-fact evidence such as tyre surface markings. For a discussion on the liquefaction of road binder and its resulting effects, see (Bullas, 2006) and (Bullas, 2008).

5.3.2 Pressure distribution in the tyre contact patch

In order to understand the manner by which forces and moments are generated in the contact patch of a rolling tyre, an initial appreciation of the stresses acting on an element of tread rubber in the contact patch is required. This pressure distribution interacts with the frictional mechanisms described above in order to produce the aggregate forces and moments within the contact patch. Each element will be subject to a normal pressure p and a shear stress τ acting in the road surface. The element will not slip on the road if $\tau < \mu p$ where μ is the coefficient of friction between the tread rubber and the road surface.

The pressure distribution depends on tyre load and whether the tyre is stationary, rolling, driven or braked. The pressure distribution is not uniform and will vary both along and across the contact patch. In order to understand the mechanics involved with the generation of forces and moments in the contact patch, some simplification of the pressure distribution will be adopted here starting with Figure 5.13 where

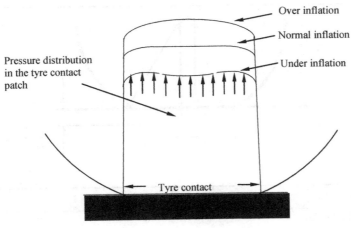

Over inflation

Normal inflation

Under inflation

Pressure distribution
in the tyre contact
patch

Tyre contact

FIGURE 5.13

Pressure distribution in a stationary tyre contact patch.

typical pressure distributions in the tyre contact patch for a stationary tyre and the effects of inflation pressure are considered.

Generally the pressure rises steeply at the front and rear of the contact patch to a value that is approximately equal to the tyre inflation pressure. Overinflation causes an area of higher pressure in the centre of the contact patch while underinflation leads to an area of reduced pressure in the centre of the patch.

When the tyre is rolling it will be shown later that pressure distribution in the contact patch is not symmetric and is greater towards the front of the contact patch.

While it can be asserted with some certainty that the integrated vertical forces throughout the contact patch must sum to the applied load, it is an error to assert that the contact patch dimensions can be simply predicted by, for example, load, width and inflation pressure. It can be generally observed for conventional tyres that vertical stiffness is broadly linear until rim contact occurs (op het Veld, 2006) and (Reimpell and Sponagel, 1988). A moment's consideration suggests that the growth of contact patch area is nonlinear with vertical stiffness and therefore the resulting average contact pressure cannot scale linearly.

In Figure 5.14, it can be seen that the contact patch length can be found using Pythagoras, as given in Eqn (5.24)

$$L = 2\sqrt{r_u^2 - r_l^2} = 2\sqrt{2 r_u^2 \Delta z - \Delta z^2} \tag{5.24}$$

Presuming the width of the contact patch is constant, for a linear stiffness we can say the average contact pressure P must follow a form as follows

$$P = \frac{k \Delta z}{2W \sqrt{2 r_u \Delta z - \Delta z^2}} \tag{5.25}$$

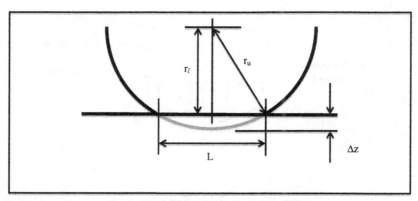

FIGURE 5.14

Contact patch length.

where k is the tyre vertical stiffness and W is the tyre width. Considering the tyre as an isothermal volume, we might estimate the change in nominal pressure P_{Δ_z} from initial pressure P_0 by calculating a change in volume compared to initial volume V_0

$$P_{\Delta z} = P_0 \frac{V_0}{V_{\Delta z}} \qquad (5.26)$$

$$V_0 = \pi \left(r_u^2 - r_w^2\right) W \qquad (5.27)$$

$$V_{\Delta z} = V_0 - \Delta V \qquad (5.28)$$

$$\Delta V = \cos^{-1}\left(\frac{n}{r_u}\right) r_u^2 - \frac{L - n}{2} \qquad (5.29)$$

where r_w is the radius of the wheel rim. Considering some typical values of k = 250 N/mm, W = 205 mm, r_u = 320 mm, P_0 = 2.45 bar and r_w = 228 mm, Figure 5.15 can be plotted.

It can be seen that the average contact pressure never reaches the tyre pressure. The only reasonable conclusion to draw from this is that inflation pressure is a poor surrogate for ground pressure; the tyre is not a balloon in any meaningful sense but rather a structure whose properties are manipulated by the presence of air. The existence of run-flat tyres with a braced sidewall suggests this is so, and recent advances in contact pressure distribution measurements have resulted in a proliferation of data supporting the assertion. Despite this, it is still regarded by some as a matter for debate.

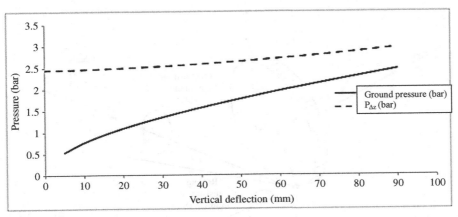

FIGURE 5.15

A comparison between inflation pressure as given in Eqn (5.26) and load divided by contact patch area, referred to as 'ground pressure'.

5.4 Tyre force and moment characteristics
5.4.1 Components of tyre force and stiffness

The local pressures and stresses distributed over the tyre contact patch can be integrated to produce forces and moments referenced to a local coordinate system within the contact patch. Using the SAE tyre axis system the full set of forces and moments are as shown in Figure 5.16.

The following section will explain the mechanical characteristics of each force and moment component. The order in which these components are described will be that which most facilitates an understanding of the mechanisms and dependencies rather than following the local order of the SAE tyre axis system. The tractive force F_x and lateral force F_y depend on the magnitude of the normal force component F_z. Hence the normal force is described first.

It should also be noted that more than one mechanism will be involved in the generation of each component. The tractive force has formulations involving driving, braking and rolling resistance. The lateral force is dependent on both slip and camber angle. It is also not possible to treat components of force and moment in isolation. It is, for example, necessary to provide a single explanation as to how the self aligning moment and lateral force resulting from slip angle arise due to stress distributions within the tyre contact patch.

Finally the reader should also be reminded that vehicle dynamics is traditionally a subject where various terms are used to describe the same thing. For example, vertical force, normal force and tyre load may be used to mean the same thing by various authors. Other examples where confusion may arise include the use of aligning torque, aligning moment or self aligning moment, longitudinal or tractive force and lateral or cornering force.

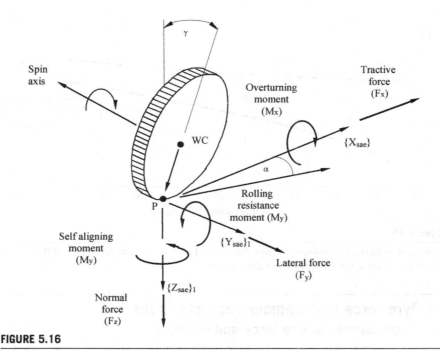

FIGURE 5.16

Tyre forces and moments shown acting in the SAE tyre axis system.

The use of the term stiffness can also add confusion to newcomers to the subject area. A traditional static force/displacement approach is used by Moore (1975) to define longitudinal, lateral and torsional stiffness of a tyre. In each case a non-rolling tyre is mounted on a plate and incrementally loaded as indicated in Figure 5.17 until complete sliding occurs. Plotting graphs of force or moment against displacement or rotation allows the stiffness parameters to be obtained from the slopes at the origin.

We will see later that terms such as cornering stiffness and aligning moment stiffness are associated with a rolling tyre and should not be confused with the lateral and torsional stiffness defined here. The term longitudinal stiffness can be particularly misleading as another definition is commonly used when longitudinal tractive forces due to driving and braking are discussed.

The explanations that follow will initially deal with each force mechanism in isolation, for example lateral forces arising due to slip angle and camber angle are considered separately with no simultaneous longitudinal tractive force. Following this, a more complex treatment involving combinations of the various force components will be addressed.

5.4.2 Normal (vertical) force calculations

The calculation of normal force in the tyre is relatively straightforward compared with the calculation of longitudinal or lateral forces. The normal force will however

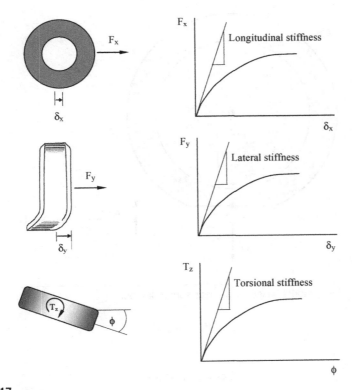

FIGURE 5.17

Measurement of stiffness in a non-rolling tyre.

always be negative when computed using the SAE tyre axis system. This is not particularly elegant when presenting the dependencies of other force components on the normal force. To overcome this, a positive value of this force component is often referred to as vertical force or tyre load. In SAE J670e (1976) vertical load is taken as the negative of normal force.

It is generally sufficient to treat the tyre as a linear spring and damper when computing the vertical force component, notwithstanding the reservations expressed in Chapter 4 about hysteretic damping and ways to model it. The tyre is quite lightly damped and in any case in the running vehicle its motions are dominated by wheel hop, so the use of an equivalent viscous damping term with the equivalence point taken at wheel hop is satisfactory for simple handling models. The calculation of the vertical force F_z acting at point P in the tyre contact patch has a contribution due to stiffness F_{zk} and a contribution due to damping F_{zc}. These forces act in the direction of the $\{Z_{sae}\}_1$ vector shown in Figure 5.18

$$F_z = F_{zk} + F_{zc} \tag{5.30}$$

$$F_{zk} = -k_z \, \delta z \tag{5.31}$$

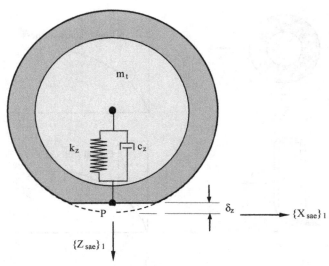

FIGURE 5.18

Vertical tyre force model based on a linear spring damper.

$$F_{zc} = -c_z V_Z \tag{5.32}$$

where

$$c_z = 2.0\, \zeta\, \sqrt{m_t \cdot k_z} \tag{5.33}$$

and

m_t = mass of tyre
k_z = radial tyre stiffness
ζ = radial damping ratio
δ_z = tyre penetration
V_z = rate of change of tyre penetration

A linear model of tyre vertical force may need to be extended to a nonlinear model for applications involving very heavy vehicles or studies where the tyre encounters obstacles in the road or terrain of a similar size to the contact patch or smaller. This could also be applicable for parallel work in the aircraft industry where established tyre models have been formulated to simulate the behaviour of the aircraft on the runway, particularly on landing, and potential problems with wheel shimmy (Smiley, 1957; Smiley and Horne, 1960). Where a nonlinear model of vertical tyre force is required, the most straightforward approach would be to represent the stiffness-based component of the force by a cubic spline interpolation of measured static force—displacement data.

Damping data are often difficult to come by and a simple test used with adequate success by the author (Harty) has been to force fit the results of a drop

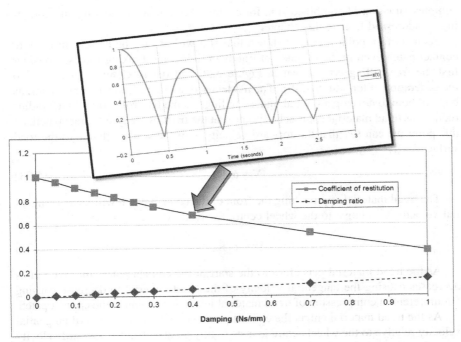

FIGURE 5.19

A simply predicted drop-and-bounce test result with resulting damping.

test to the chosen model. A wheel and tyre assembly, with the tyre at the pressure of interest, is held at some useful distance above the ground, say 1 m. The assembly is dropped — that is to say cleanly released with no initial vertical velocity — and observed. The observation can conveniently be carried out with a video camera suitably mounted so as to largely remove perspective effects. The resulting motion of the wheel — a series of bounces successively decreasing in height — can be captured in a frame-by-frame examination of the video trace where only the highest points are digitised. A simple vertical model, of the kind already described, can be integrated through time and force-fitted characteristics to replicate the observed results. An example is shown in Figure 5.19.

Although somewhat approximate, the influence of tyre damping on the overall solution is relatively small and its measurement in this fashion captures it with adequate resolution in terms of vehicle-level behaviour.

5.4.3 Longitudinal force in a free rolling tyre (rolling resistance)

Under normal driving conditions a tyre is continually subject to a wide range of tractive driving and braking forces. This section discusses the formulation of driving and braking forces under pure slip conditions, i.e. straight-line motion only. The more complex situation of combined slip, for example simultaneous braking and cornering, is addressed later in this chapter.

As a starting point it can be shown that slip will always be present in the tyre contact patch even in the absence of tractive driving and braking forces. Consider first the free rolling tyre shown in Figure 5.20 and the mechanism that leads to the generation of longitudinal slip. The model used in Figure 5.20 has simplifications but will help to develop an initial understanding. As the tyre rolls forward the radius reduces as tread material approaches point A at the front of the tyre contact patch. At this point we can say that the forward velocity V of the wheel relative to the road surface is given by

$$V = \omega \, R_e \qquad (5.34)$$

The tread material approaching the front of the contact patch will have a tangential velocity V^t relative to the wheel centre O given by

$$V^t = \omega \, R_u \qquad (5.35)$$

As the tread material gets close to the start of the contact patch, the tyre radius decreases causing the tangential velocity of the tread material to decrease causing circumferential compression of tread material just before it enters the contact patch.

As the tread material enters the contact patch at point A the rearward tangential velocity relative to the wheel centre is just slightly greater than the forward velocity of the vehicle. This results in initial rearward slip of tread material relative to the road surface between point A and C. At point C it is assumed that the radius has reduced to a value equivalent to the effective rolling radius R_e resulting in the rearward tangential velocity matching the forward vehicle velocity and theoretically producing a point of zero slip in the tyre. Over the central region of the contact patch between C and D the radius reduces to a value below the effective rolling radius reversing the slip in the tyre to the forward direction. At the centre of the patch P the radius reduces to the loaded radius R_l. In theory this point would produce the lowest tangential velocity and the highest forward slip, although experimental observations (Moore, 1975) indicate that the tangential speed does not reduce to this level. Between point D and B the radius recovers to a value greater than the effective rolling radius causing the direction of slip to reverse again to a forward direction.

It is clear that the direction of slip changes several times as tread moves through the contact patch resulting in the distribution of longitudinal shear stress of the type shown at the bottom of Figure 5.20. The shear stress is plotted to be consistent with the SAE reference frame and is not symmetric; the net effect being to produce an overall force, the rolling resistance, acting in the negative X_{SAE} direction.

It should be noted that the two-dimensional model presented is not fully representative as components of lateral slip are also introduced in a free rolling tyre due to deformation of the side walls as shown in Figure 5.21.

As the tyre carcass deforms in the vicinity of the contact patch the deformation of the side walls creates additional inwards movement of the tread material (Moore, 1975).

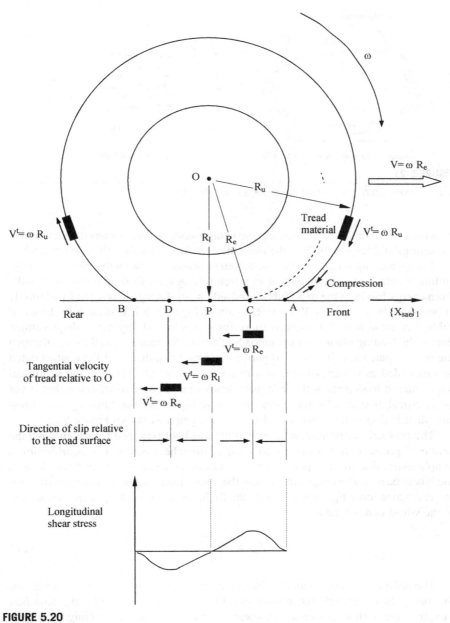

FIGURE 5.20

Generation of slip in a free rolling tyre.

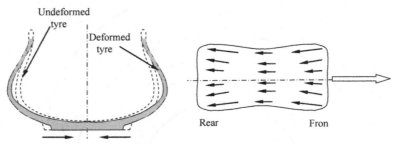

Undeformed tyre

Deformed tyre

Lateral slip movement (Moore, 1975)

Rear   Fron

Squirm through the contact patch

FIGURE 5.21

Lateral distortion of the contact patch for a free rolling tyre.

This causes the contact patch to assume an hourglass shape creating an effect referred to as 'squirm' (Gillespie, 1992) as the tread material moves through the contact patch.

Before moving on to consider the driven or braked tyre we will now consider the rolling resistance forces generated in a free rolling tyre. Rolling resistance results from energy losses in the tread rubber and side walls. Energy loss in the tread rubber is produced by hysteresis. If we refer again to Figure 5.9 it is clear for a block of rubber, or tread material, there is more force required at any given displacement during the loading phase than the unloading phase. As tread material moves through the contact patch it will be loaded until it reaches the midpoint of the contact patch and unloaded as it moves to the rear of the contact patch. This and the additional losses due to hysteresis in the side walls leads to a pressure distribution that is not symmetrical as shown for the stationary tyre in Figure 5.13 and has a greater pressure distribution in the front half of the contact patch as shown Figure 5.22.

The pressure distribution implies that the resultant tyre load F_z acts through the centre of pressure, a distance δx forward of the wheel centre. For equilibrium, a couple exists that must oppose the tyre load and its reaction acting down through the wheel centre. The couple that reacts the wheel load couple results from the rolling resistance force F_{Rx} acting longitudinally in the negative X_{SAE} axis and reacted at the wheel centre where

$$F_{Rx} = \frac{F_z\, \delta x}{R_l} \qquad (5.36)$$

The rolling resistance may also be referenced by a rolling resistance coefficient, this being the rolling resistance force F_{Rx} divided by the tyre load F_z. By definition therefore the rolling resistance moment M_y is $F_z\, \delta x$ and the rolling resistance moment coefficient is δx. Rigorous adherence to the sign convention associated with the tyre reference frame is essential when implementing these formulations in a tyre model. In Figure 5.22, to assist understanding, Fz is represented as the vertical force acting on the tyre rather than the negative normal force computed in the Z_{SAE} direction.

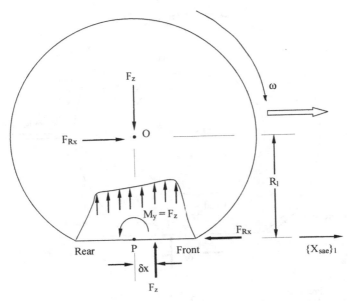

FIGURE 5.22

Generation of rolling resistance in a free rolling tyre.

The rolling resistance force is very small in comparison with other forces acting at the contact patch, a rolling resistance of the order of 1% of vehicle weight being typical for a car tyre. This and the fact that the rolling resistance force may vary by up to 30% of the average value during one revolution (Phillips, 2000) make accurate measurement difficult.

5.4.4 Braking force

During braking the activation of the brake mechanism will apply forces to the rotating wheel that at this stage may be treated as a brake torque T_B acting about the wheel centre and opposing the rolling motion of the wheel. During this process the tread material in the tyre will begin to slide relative to the road, giving rise to slip. As the angular velocity of the wheel reduces, a braking force is generated that tends to move the contact patch rearwards relative to the wheel centre. This effect will introduce circumferential tension in the tread just before entering the contact patch as opposed to the compression noted earlier in this area for a free rolling tyre. As the contact patch distorts rearward during braking, compression will instead be generated in the tread just leaving the patch as shown in Figure 5.23.

During braking, as for free rolling, as tread material approaches the contact patch the radius will reduce with a consequent reduction in the tangential speed of the tread material relative to the wheel centre. For moderate braking, the tread initially entering the contact patch will initially bend rearward under the action of shear stresses for a short distance before the tangential velocity of the tread material slows

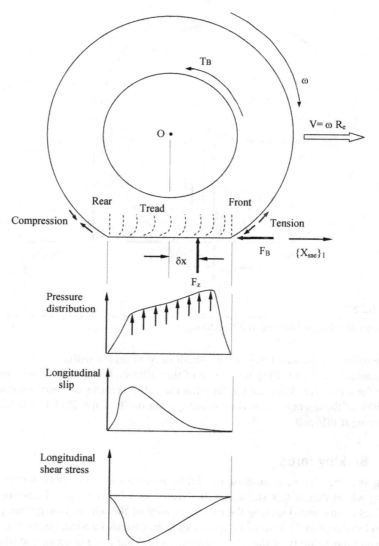

FIGURE 5.23

Generation of force in a braked tyre.

to the forward velocity of the tyre ωR_e and slip begins to progressively develop as the tread material moves back through the contact patch.

As the tread approaches the rear of the tread, the pressure begins to unload, releasing the deformation in the tread. The tangential velocity of the tread material begins to increase and the frictional braking force reduces rapidly to zero at the rear of the patch. These actions produce distributions of pressure, slip and longitudinal shear stress of the type shown at the bottom of Figure 5.23 where the shear stress is plotted as negative in accordance with the resultant braking force resolved in the SAE tyre coordinate system. It should also be noted that when braking, the pressure distribution tends to differ from that in a free rolling tyre as the peak, and hence resultant tyre load, move further forward in the tyre contact patch.

If the resulting braking force is maintained the angular velocity of the tyre will reduce from its free rolling value and will eventually become zero when the wheel is fully locked. A measure of the slip generated can be defined by a slip ratio or percentage slip. In this text the term slip ratio S will be used

$$S = \frac{\omega_0 - \omega_B}{\omega_0} \qquad (5.37)$$

where

   ω_0 is the angular velocity of the free rolling wheel
   ω_B is the angular velocity of the braked wheel

If we consider the case of a braked wheel it can be seen from Eqn (5.18) that for the free rolling state the slip ratio will be zero and that for the fully locked and skidding wheel the slip ratio will be 1.0. Multiplying slip ratio by 100 gives the term percentage slip.

In SAE J670e (1976) the sign is reversed to produce a slip ratio of −1.0 for the fully locked wheel, convenient when plotting braking force that is negative in the SAE system against slip ratio

$$S = \frac{\omega_B - \omega_0}{\omega_0} \qquad (5.38)$$

A further definition that is popular is to substitute $V = \omega_0\, R_e$ into Eqn (5.37) giving

$$S = \frac{V - \omega_B\, R_e}{V} \qquad (5.39)$$

It should be noted that slip ratio has been subject to various definitions by researchers and research groups in tyre companies, several are listed in Milliken and Milliken (1995). Some tyre models use R_l instead of Re when formulating slip ratio, which may require careful consideration when using a general purpose MBS program to model anti-lock braking system (ABS) (Ozdalyan and Blundell, 1998).

Plotting curves of braking force, for convenience shown positive here, against slip ratio for a range of tyre loads will generally produce curves of the type shown in Figure 5.24.

Examination of the curves in Figure 5.24 reveals that at each vertical load the braking force increases rapidly in a linear manner to reach a peak value that,

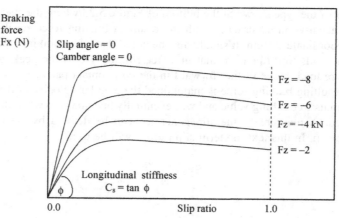

FIGURE 5.24

Braking force versus slip ratio.

depending on tyre design and road conditions, would typically occur at a slip ratio anywhere between 0.15 and 0.3. After this point the braking force will level out or reduce as the wheel approaches the fully locked situation. Examination of curves such as these, sometimes called 'mu-slip' curves, facilitates an understanding of ABS operation where cycling the brake pressure maintains a slip ratio near the peak braking force position for each wheel on the vehicle. This is desirable not only to maximise braking effort but also to maintain a rolling wheel for cornering and directional stability.

An important property of each curve is the slope at the origin, referred to as the longitudinal stiffness, C_s. It can be seen that this is not a constant but increases with load, which is significant when considering the capability of any tyre model to be used in braking simulations. In Figure 5.24 the curves are shown to pass through the origin. In practice a small vertical offset in longitudinal force will be apparent for a free rolling tyre, this being the rolling resistance discussed earlier.

It is important to reiterate that frictional forces are not the property of the tyre alone. The effects of road material and texture, or contamination with water and ice, are also significant. Figure 5.25 demonstrates typical curves of braking force against slip ratio, at a given tyre load for various road conditions (Phillips, 2000). These curves demonstrate that on wet roads peak values of braking force as expected reduce and that for a locked wheel with poor tread a dangerous situation known as hydroplaning or aquaplaning can arise where the tyre runs on a film of water and traction is effectively lost. The curves of braking force can also be categorised by two coefficients of friction associated with the peak braking force and that associated with total sliding at a slip ratio of 1.0. On dry roads it is possible to obtain a coefficient of friction for good tyres substantially in excess of 1.0, on wet roads this could typically reduce to about 0.5 or lower for tyres with poor tread while a road contaminated with ice may only achieve a peak value of 0.1.

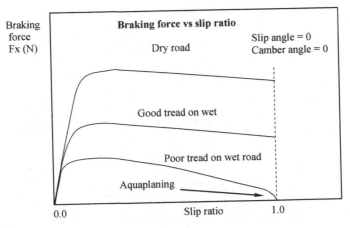

FIGURE 5.25

The effect of road contamination on braking.

It can be seen when examining the curves in Figure 5.25 that the longitudinal stiffness is relatively unaffected by surface contamination. This is particularly dangerous for a road with ice or the poor tyre on a wet road. In these conditions the peak braking force occurs rapidly at low slip ratio causing the vehicle to skid before any possible corrective action from the average driver.

In addition to the above it is also known that an increase in vehicle speed will reduce peak values of braking force and that other parameters such as tyre inflation pressure will have an effect, a more detailed treatment of which is given by Pacejka (2012).

5.4.5 Driving force

During driving the transmission will impart a driving torque T_D to the rotating wheel as shown in Figure 5.26. As the angular velocity of the wheel increases a driving force is generated that tends to move the contact patch forward relative to the wheel centre. This effect will introduce circumferential compression in the tread just before entering the contact patch and tension on leaving as opposed to braking.

As the wheel is driven the tread initially entering the contact patch will initially bend forward under the action of shear stresses for a short distance. As the tread approaches the rear of the tread the pressure begins to unload releasing the deformation in the tread and progressive sliding develops. These actions produce distributions of pressure, slip and longitudinal shear stress of the type shown at the bottom of Figure 5.26. It can also be seen that when driving, the pressure distribution tends to differ from that in a braked tyre as the peak, and hence resultant tyre load, move further to the rear in the tyre contact patch reducing the offset δx.

If the resulting driving force is maintained the angular velocity of the tyre will increase from its free rolling value and will eventually begin to spin. A measure

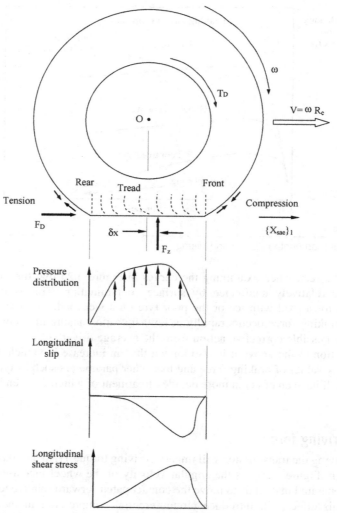

FIGURE 5.26

Generation of force in a driven tyre.

of the slip generated for the driven tyre can be defined by a further modification to the slip ratio S

$$S = \frac{\omega_D - \omega_0}{\omega_0}$$

(5.40)

where

ω_0 is the angular velocity of the free rolling wheel
ω_D is the angular velocity of the driven wheel

For driving a slip ratio of 1.0 is sometimes taken to define the onset of wheel spin. From Eqn (5.40) this will occur when the angular velocity of the driven wheel reaches a value of twice that for free rolling. Unlike braking, the slip ratio in driving can exceed 1.0 as the wheel angular velocity continues to increase. This definition of 'spin' is somewhat arbitrary. For both tractive and braking cases the relationship between longitudinal force and slip ratio is such that the wheel behaviour converges for slip ratios smaller than those at which peak force is produced. However for larger slip ratios the wheel behaviour diverges rapidly. For spin in particular, angular velocity increases very quickly until torque is reduced.

5.4.6 Generation of lateral force and aligning moment

The generation of lateral force and aligning moment in the tyre result from combinations of the same mechanisms and are thus treated together here. As a starting point it is helpful to consider Figure 5.27, which is adapted from the sketches for forces and torques provided by Olley (1945). Figure 5.27 is particularly useful for relating the sign convention for the lateral forces and aligning moments plotted and discussed throughout this chapter.

From Figure 5.27 it can be seen that for a tyre rolling with a slip angle α at zero camber angle the lateral force generated due to the distribution of shear stress in the contact patch acts to the rear of the contact patch centre creating a lever arm known as the pneumatic trail. This mechanism introduces the aligning moment and has a stabilising or 'centring' effect on the road wheel. This is an important aspect of the steering 'feel' that is fed back to the driver through the steering system.

Similarly it can be seen from Figure 5.27 that for a tyre rolling with a camber angle γ at zero slip angle, the lateral force generated is called camber thrust. Due to the conditions in the contact patch the camber thrust acts in front of the contact patch centre creating a mechanism that creates a moment. Although this is referred to here as an aligning moment it has the opposite effect of the aligning moment resulting from slip angle and is sometimes called the camber torque as there is no resultant aligning action on the road wheel. The importance of the camber-induced moment is small for passenger cars but can be large for motorcycles.

5.4.7 The effect of slip angle

In order to understand the mechanisms that lead to the generation of lateral force and aligning moment resulting due to slip angle it is useful to start with Figure 5.28 showing the distribution of pressure p, and the lateral stress in the contact patch. The upper part of the figure provides a side view and the lower part is a top view looking down on to the contact patch. The lateral stress boundary, μp, represents the limit available between the tread rubber and the road surface. If the lateral stress is below this limit no sliding will occur but once the lateral stress reaches this limit the tread rubber will commence sliding.

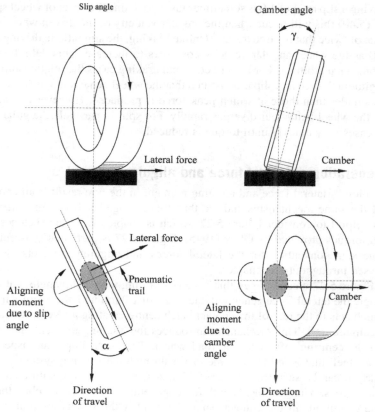

FIGURE 5.27

Forces and moments due to slip and camber angle.

When the tyre rolls at a slip angle α, tread rubber that is put down on the road surface at the front of the contact patch moves back through the patch at the same slip angle, deforming the sidewalls of the tyre, so that the lateral stress in the tread rubber steadily increases as shown. At a certain point in the contact patch the lateral stress reaches the limit boundary after which sliding takes place until the tread rubber leaves the rear of the contact patch and the lateral stress returns to zero.

As the slip angle increases, the rate at which lateral stress is generated as the tread rubber moves back through the contact patch increases so that the point at which slippage commences moves forward in the contact patch. It can also be seen that as the slip angle increases, the area under the lateral stress curve increases. This area is a measure of the resulting lateral force F_y generated by integrating the stress over the contact patch. At low slip angles, when the lateral stress shape is substantially triangular, there is a nearly linear relationship between lateral force and slip angle. In general this linearity only extends to one or two degrees of slip angle.

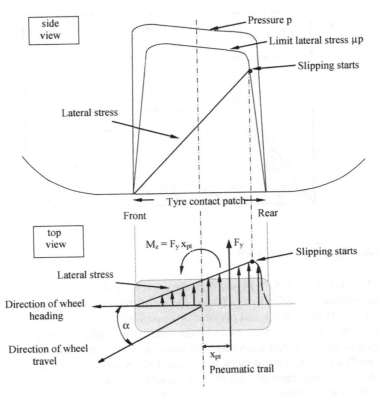

FIGURE 5.28

Generation of lateral force and aligning moment due to slip angle.

As the slip angle increases, the amount of rubber involved in sliding gradually extends from the rear of the tyre contact patch until all the rubber is sliding and the lateral stress follows the boundary limit, μp, distribution.

Since the form of the pressure distribution is a measure of the tyre load $(-F_z)$, it follows that the maximum lateral force $F_{y\ max}$ is found from

$$F_{y\ max} = -\mu F_z \tag{5.41}$$

In practice this maximum is achieved at slip angles around 10 degrees for many road tyres. Motorsport and high performance low profile tyres produce their maximum forces at much lower slip angles, as low as three or four degrees. Figure 5.29 shows a typical plot of lateral force F_y with slip angle α for increasing tyre load with the camber angle set at zero. For the convenience of plotting results in the positive quadrant, negative slip angle is used in this plot. From the plot it can be seen that the cornering stiffness C_α is the gradient of the curve measured at zero slip angle at a given tyre load. As the tyre load increases so does the cornering stiffness, although it will be seen later that at higher tyre loads the magnitude of the cornering

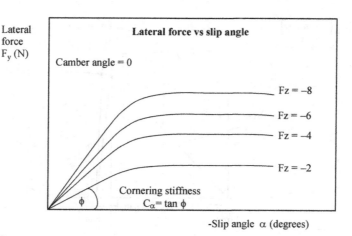

FIGURE 5.29

Plotting lateral force versus slip angle.

stiffness begins to level off. In Figure 5.29 the curves are shown to pass through the origin. In practice a small offset in lateral force will be apparent at zero slip angle due to the effects of conicity and plysteer discussed earlier.

Looking back to Figure 5.28 it can be seen that as the shape of the lateral stress distribution is approximately triangular, the lateral force F_y acts through the centroid of this area that is to the rear of the wheel centre line by a distance referred to as the pneumatic trail. Inspection of Figure 5.28 should indicate that as the slip angle increases the line of action of F_y moves forward reducing the pneumatic trail eventually to zero. The aligning moment M_z is the product of the lateral force and the pneumatic trail and will reduce accordingly, eventually becoming negative usually for lightly loaded tyres at high slip angles. In these situations the extent of sliding occurs to such an extent throughout the contact patch that the lateral stress distribution approaches the shape of the μp curve moving the centroid through which F_y acts forward of the centre. A typical plot of aligning moment with slip angle, for a given tyre load and zero camber angle, is shown in Figure 5.30. From the plot it can be seen that the aligning moment stiffness is the gradient of the curve measured at zero slip angle at a given tyre load.

The aligning moment curve is nearly linear at low slip angles and reaches a maximum value for most tyres at around half the slip angle required for peak lateral force. The gradient of the curve measured at a zero slip angle is the aligning moment stiffness. In Figure 5.30 the curves are shown to pass through the origin. In practice a small offset in aligning moment will be apparent at zero slip angle due to the effects of conicity and plysteer discussed earlier.

5.4.8 The effect of camber angle

The lateral force that arises due to an inclination of the tyre from the vertical is referred to as camber thrust. The SAE definition of positive camber angle is taken

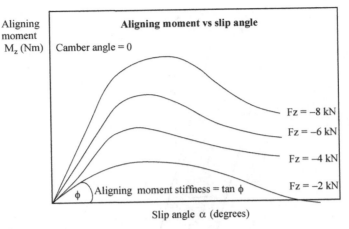

FIGURE 5.30

Plotting aligning moment versus slip angle.

for the top of the tyre leaning outwards relative to the vehicle. The fact that this differs from one side of the vehicle to the other does not lead to consistency when developing a tyre model. For understanding it is useful to remember that the camber thrust will always act in the direction that the tyre is inclined as shown in Figure 5.31. For the SAE system shown here a positive camber angle γ will produce a positive camber thrust for all tyres on the vehicle modelled in that system.

If the tyre is inclined at a camber angle γ, then deflection of the tyre and the associated radial stiffness will produce a resultant force, F_R, acting towards the wheel centre. Resolving this into components will produce the tyre load and the camber thrust.

An alternative explanation provided in Milliken and Milliken (1998) compares a stationary and rolling tyre. For the stationary tyre, experimental observations of tread in the contact patch indicate a curved shape. As the tyre rolls the tread moving through the contact patch is constrained by the road to move along a straight line, the net reaction of these forces being the camber thrust.

Figure 5.32 shows a typical plot of lateral force F_y with camber angle γ for increasing tyre load with the slip angle set to zero. From the plot it can be seen that the camber stiffness C_γ is the gradient of the curve measured at zero camber angle at a given tyre load.

In order to understand why a cambered tyre rolling at zero slip angle produces an aligning moment, it is useful to consider the effect of the shape of the contact patch. Consider the situation shown in Figure 5.33 where the wheel and tyre are rolling at a camber angle γ with the slip angle equal to zero. The lower part of Figure 5.33 is a plan view on the tyre contact patch. The three points A, B and C, shown in Figure 5.33, are initially in line across the centre of the contact patch. If the tyre rolls so that point B moves to B' at the rear of the contact patch and the rubber in the centre line is not subjected to any longitudinal stress.

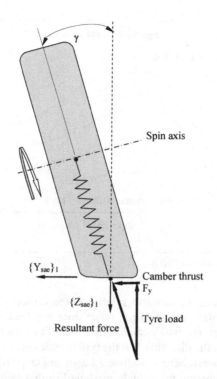

FIGURE 5.31

Generation of lateral force due to camber angle.

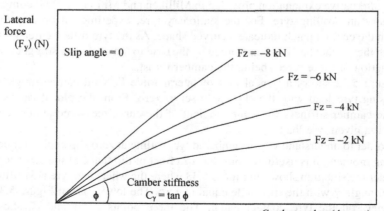

FIGURE 5.32

Plotting lateral versus camber angle.

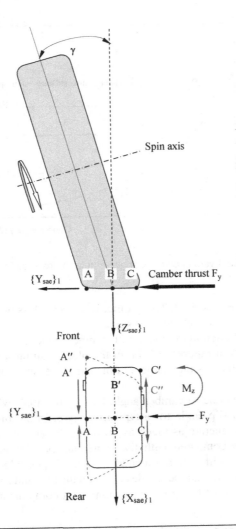

FIGURE 5.33

Generation of self-aligning moment due to camber angle.

Due to the camber the tyre will corner and point A on the inside of the tyre will roll at a smaller radius of bend to point C on the outside of the tyre. If the tyre rubber was not subject to any longitudinal stress these points would move to A″ and C″ respectively to preserve the total circumference on the inner and outer edges. If it is presumed that the stiffness of the tyre restricts this and the points remain in line across the rear of the contact patch (A′, B′ and C′) then a longitudinal compressive stress acts on the inner A side and a tensile stress acts on the outer C side at the front of the contact patch. A similar effect occurs at the rear of the contact patch but the increased pressure in the front of the footprint of a rolling tyre means a net torque into the turn is developed when the effects are summed along the contact patch

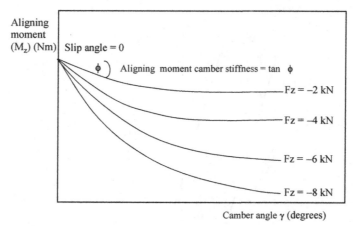

FIGURE 5.34

Plotting aligning moment versus camber angle — note the reversed sign compared to slip angle aligning moment.

length, as shown in Figure 5.33. The geometric aspect is less obvious in motorcycle tyres but the mechanism is essentially similar.

A typical plot of aligning moment with camber angle, for a given tyre load and zero angle, is shown in Figure 5.34. From the plot it can be seen that the aligning moment camber stiffness is the gradient of the curve measured at zero camber angle at a given tyre load.

The lateral forces due to camber angle tend to be small when compared with those resulting from slip angle for a typical car tyre. In the linear range it would not be untypical to generate as much as 20 times the amount of lateral force per degree of slip angle compared with that generated per degree of camber angle. For motorcycle tyres, with a more rounded profile, it is possible for riders to incline the motorcycle to produce camber angles between the tyre and road in excess of 45°, resulting in camber thrust being the most significant component of lateral force.

5.4.9 Combinations of camber and slip angle

The treatment so far has considered the generation of lateral force due to slip angle and camber angle in isolation. For a road car, while slip angle dominates the generation of lateral force, some amount of camber will occur at the same time. The effect of adding camber to slip angle is shown in Figure 5.35 where for a given tyre load the lateral force against slip angle curve is plotted at 0, 5 and 10° of camber angle. It should be noted that the curves here are plotted with assisting camber angle where the wheels are leaning into the turn. A similar reduction in lateral force will occur where the camber angle is reversed and the wheels lean out of the turn.

At zero degrees of slip angle the introduction of camber angle introduces an offset from the origin, this being the camber thrust discussed earlier occurring at

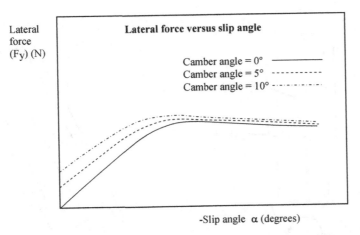

FIGURE 5.35

The effect of combined camber and slip angle on lateral force.

a zero slip angle. The small offsets in lateral force due to conicity and plysteer, discussed in Section 5.2.3, are ignored in Figure 5.35. In the linear range the contributions in lateral force due to slip and camber may be added together but during the transition towards sliding it can be seen that the additive effect of camber will reduce, although the peak value of lateral force is still increased. The maximum increase in peak lateral force will occur at different camber angles for different wheel loads. Thus for a given tyre on a given vehicle it is possible (Milliken and Milliken, 1998) to optimise camber angle for a given combination of slip angle and tyre load.

5.4.10 Overturning moment

Two of the components of moment acting in the tyre contact patch have been discussed. The generation of rolling resistance moment was described while discussing the free rolling tyre in Section 5.4.3. The self-aligning moment arising due to slip or camber angle was discussed in Sections 5.4.7 and 5.4.8. For completeness the final component of moment acting at the tyre contact patch that requires description is the overturning moment that would arise due to deformation in the tyre as shown in Figure 5.36. The forces and moments as computed in the SAE reference frame are formulated to act at P, this being the point where the wheel plane intersects the ground plane at a point longitudinally aligned with the wheel centre.

In Figure 5.36 it can be seen that distortion of the side walls results in a lateral shift of the contact patch that may result from either slip angle or camber angle or a combination of the two. The resulting offset tyre load introduces an additional component of moment M_x. Attention to the sign convention associated with the tyre reference frame is again needed if the moment is to be included in a tyre model. In Figure 5.36, to assist understanding, Fz is represented as the tyre load acting on the tyre rather than the negative normal force computed in the Z_{SAE} direction.

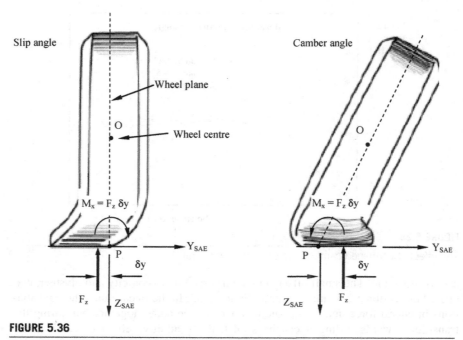

FIGURE 5.36

Generation of overturning moment in the tyre contact patch.

A consideration of the overturning moment is generally more important where relatively large displacements in the tyre occur, as with aircraft tyres (Smiley, 1957; Smiley and Horne, 1960). Overturning effects are also of major importance for motorcycle tyres, particularly in terms of matching the behaviour of front and rear tyres. The lateral offset, δy, also applies to the longitudinal forces and is responsible for the 'stand up under light braking' that all motorcycles display.

5.4.11 Combined traction and cornering (comprehensive slip)

The treatment of longitudinal braking or diving forces and lateral cornering forces has so far dealt with the two components of force in isolation. The simulation of vehicle behaviour involving tyre forces acting in this manner leads to what is termed pure cornering or pure tractive (i.e. driving or braking) behaviour. In reality longitudinal and lateral forces often occur simultaneously during vehicle manoeuvres. A typical situation would be to initiate braking before entering a bend and continue braking into the corner. It is also typical, once the driver feels sufficient confidence, to begin applying throttle, and driving forces, during cornering before exiting the bend. For such situations a tyre model must be able to deal with combined tractive and cornering forces, a situation referred to as comprehensive slip.

The basic law of friction relating frictional force to normal force can be of assistance when considering combinations of longitudinal driving or braking forces with

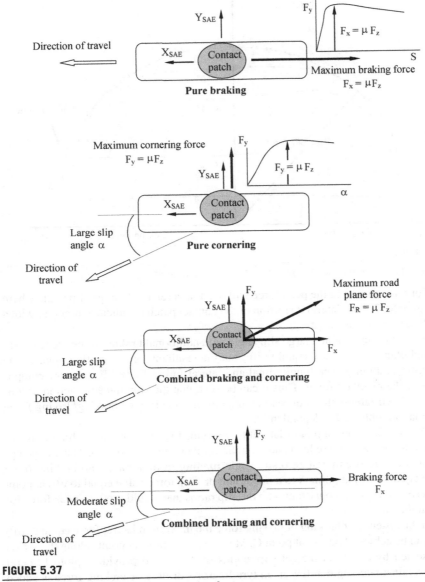

FIGURE 5.37

Pure and combined braking and cornering forces.

lateral cornering forces. The treatment here concentrates on lateral forces due to slip angle with camber angle set to zero. Figure 5.37 initially shows a tyre subject to pure braking or cornering force where in each case the slip in the ground plane is such that the tyre force produced is a peak value, this being μF_z, the peak coefficient of friction multiplied by tyre load.

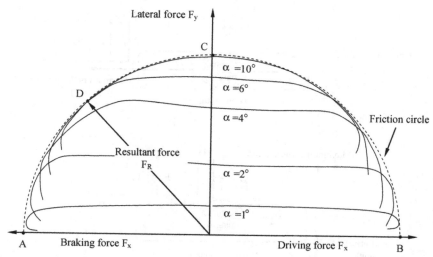

FIGURE 5.38

Plotting lateral force against longitudinal force (friction circle).

For pure cornering the peak force will occur at a relatively large slip angle where in Figure 5.37 some lateral distortion of the contact patch is indicated together with a small amount of pneumatic trail.

For a tyre running at a large slip angle with additional braking force the resultant ground plane force is still equal to μF_z but the resultant force direction opposes the direction of sliding. The longitudinal and lateral forces F_x and F_y are now components of the resultant force. Thus it can be seen that the simultaneous action of longitudinal and lateral slip reduces the amount of cornering or braking/driving force that may be obtained independently.

Figure 5.38 shows a plot of lateral force against longitudinal force for a range of slip angles at a given tyre load and with the camber angle set to zero. The x-axis represents the tyre running at zero slip angle running from a maximum braking force value equal to μF_z at point A to a maximum driving force value equal to μF_z at point B, these points being consistent with the slip ratios that would produce peak force for a straight running tyre.

In the absence of braking or driving force the maximum lateral force equal to μF_z that can be achieved occurs at point C. Measurements of F_y at points along the y-axis intersected by curves at the set slip angles shown would provide a plot of lateral force against slip angle at the given tyre load. As driving or braking force is added the maximum resultant force that can be achieved is defined by points lying on a curve of radius μF_z referred to as the 'Friction Circle' or sometimes the 'Friction Ellipse' as some tyres will have more capability in traction or cornering leading to an elliptical boundary shape.

Figure 5.38 shows, for a typical tyre, the general form of the friction circle diagram for the full range of driving and braking forces. Note that only lateral forces due to positive slip angle are presented and a similar diagram would exist for measurements taken at negative slip angles. Point D represents an example of a position where the tyre is operating at the friction limit for combined braking and cornering, as shown in Figure 5.37 where it is clear that the amount of braking or cornering force that could be produced independently is reduced and that the magnitude of F_R is simply

$$F_R = \sqrt{F_x^2 + F_y^2} \qquad (5.42)$$

It can also be noted that the curves are not symmetric in that lateral forces initially increase slightly as braking force is applied. As discussed earlier the braking force adds circumferential tension to the tyre material entering the contact patch. This stress stiffening effect can be seen to raise the lateral force slightly, while the reversal of longitudinal force to driving leads to a reduction. As the curves approach the friction limit it can be observed that they turn inwards. For a fixed slip angle the longitudinal slip is increased moving along the curves, for either braking or driving, until the point where both lateral force and longitudinal force reduce, hence causing the curves to bend back.

By plotting aligning moment against longitudinal force, it can be seen that the opposite can occur (Phillips, 2000) and that adding braking force reduces aligning moment and adding driving force raises it. Referring back to Figure 5.37 the bottom diagram shows a tyre running at moderate slip angle producing a lateral force F_y along a line of action set back from the centre by the pneumatic trail. The contact patch is shown displaced laterally due to the cornering force so that for simultaneous braking the braking force produces a moment that would subtract from the existing aligning moment due to the product of lateral force and pneumatic trail. From the diagram at the bottom of Figure 5.37 it is also clear that the simultaneous application of a driving force would produce a moment that would add to the aligning moment due to the product of lateral force and pneumatic trail. At higher braking forces the effect may cause the aligning moment to go negative.

The friction circle or ellipse is also a way to monitor the performance of a race car driver using instrumented measurements of lateral and longitudinal accelerations, sometimes called the 'g—g' diagram. Comparing this diagram with known tyre data it is possible to see how well the driver performs keeping the vehicle close to the friction limits of the tyres.

A similar exercise is possible using an MBS model of a vehicle with a road model to represent the circuit. With the MBS model, extraction of longitudinal and lateral tyre force time histories is possible for a simulated lap of the circuit. This in theory allows investigation into the influence of tyre or vehicle model parameter changes and steering inputs on tyre limit behaviour.

One final point is worth mentioning on the subject of comprehensive slip. It is sometimes imagined that a spinning wheel is very helpful for cornering, particularly at the front axle. This is not the case, as can be seen in Figure 5.39. When a vehicle is

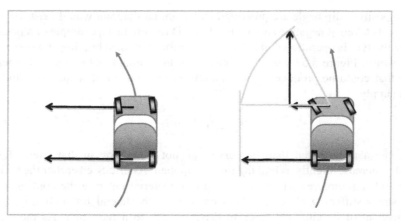

FIGURE 5.39

A comparison of cornering with pure lateral slip compared with pure longitudinal slip at maximum steer angle.

travelling in a curved path using only lateral forces from the tyre, the centripetal acceleration can be seen to be given by the lateral forces on the tyre, acting in a direction perpendicular to the direction of travel. The length of the arrow in the diagram can be taken as an indication of the maximum frictional forces. In contrast, when the steering is turned to the maximum angle of around 30° and the tyre force applied as a pure longitudinal force (with respect to the tyre reference frame) then the lateral force available at the front axle is reduced to about the sine of 30° — about half of what it was previously.

The large imbalance between front and rear causes the car to turn out of the turn and take a substantially wider path than previously. Further, the component along the tangent of the vehicle path causes the vehicle to accelerate, further reducing the available centripetal acceleration; for a 50% reduction in centripetal force and a 10% increase in speed, the radius of turn is increased by around 140% — which is to say it is 2.4 times its original value. The origin of the myth of 'front wheel drive pulling into the corner' is unknown but like many myths it has a persistence beyond reason.

5.4.12 Relaxation length

For cornering it has been shown that the generation of lateral force due to slip angle is of prime importance. In practice the generation of lateral force is not instantaneous but is subject to a delay generally referred to as 'tyre lag'. Without modelling some form of time lag a tyre model will compute the force and pass this to the MBS vehicle model so that the lateral force is applied instantly at that integration time step. It has been shown (Loeb et al., 1990) that the tyre must roll a certain distance, 'relaxation length', for the tyre to deflect sufficiently to generate the lateral force.

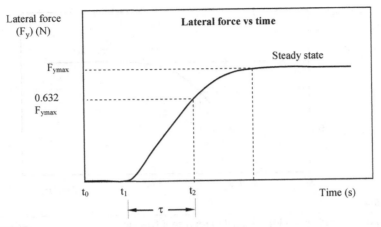

FIGURE 5.40

Development of lateral force following step steering input.

A possible method to measure relaxation length is to set the tyre up at a given slip angle in a test rig with the drum or belt not yet moving. On starting the machine the distance rolled before the forces and moments reach steady state can be recorded hence giving a measure of relaxation length. An improved method for vehicle dynamics (Loeb et al., 1990) is to start the test with the tyre running at low speed in the straight ahead configuration and apply a rapid step input of steer angle to the tyre producing a time history plot, similar to that shown in Figure 5.40, indicating the build-up in lateral force.

The results obtained (Loeb et al., 1990) for the lateral force response appear exponential, indicating a first-order dynamic system where the time constant τ, equal to $t_2 - t_1$ in Figure 5.40, is the time required to achieve 63.2% of the final steady state response.

Incorporation of a lag effect for tyre lateral force within an MBS program requires an understanding of the mathematical integration process used to solve the equations of motion as discussed in Chapter 3 of this book. A typical approach taken is to compute a theoretical value of slip angle, α_1, that includes a lag effect and to input this to the appropriate tyre-model algorithm for lateral force due to slip angle. As a starting point the tyre relaxation length L_R is taken as an input parameter from which, for a forward speed V_x, the time constant τ can be found using

$$\tau = L_R / V_x \qquad (5.43)$$

Thus by this definition the relaxation length L_R is the distance through which the tyre must roll in order to develop 63.2% of the required lateral force. This leads to an initial expression

$$\frac{d\alpha_1}{dt} = \frac{\alpha_c - \alpha_1}{\tau} \qquad (5.44)$$

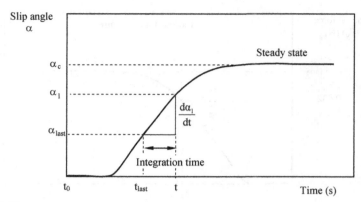

FIGURE 5.41

Build up of slip angle in an MBS model to represent tyre lag.

where:

α_c is the computed value of slip angle (instantaneous) at the current time
α_l is the value of slip angle corrected to account for lag.

An estimate of the term $\dfrac{d\alpha_l}{dt}$ in Eqn (5.44) can be obtained from Eqn (5.45). Additional understanding of the terms can be obtained by reference to Figure 5.41 where for clarity the integration time step, $t - t_{last}$, is shown with exaggerated magnitude

$$\frac{d\alpha_l}{dt} \approx \frac{\alpha_l - \alpha_{last}}{t - t_{last}} \tag{5.45}$$

where

α_{last} is the value of α_l computed at the last successful integration time step
t is the current simulation time
t_{last} is the time for the last successful integration time step

An estimate of the term $\dfrac{\alpha_c - \alpha_l}{\tau}$ in Eqn (5.44) can be obtained from

$$\frac{\alpha_c - \alpha_l}{\tau} = \frac{\alpha_c - \alpha_{last}}{\tau} \tag{5.46}$$

Combining Eqns (5.45) and (5.46) allows the value of α_l required to compute lateral force at the current time step to be obtained from

$$\alpha_l = \left(\frac{\alpha_c - \alpha_l}{\tau} \right) (t - t_{last}) + \alpha_{last} \tag{5.47}$$

Using data for the baseline vehicle used throughout this text it is possible to carry out a calculation to estimate a time delay, tyre lag, for a vehicle travelling at 100 kph. The load on the tyre F_z is taken as 4500 N and the radial stiffness of the tyre k_z is taken as 160 N/mm. From this it is possible to calculate the static tyre deflection δ_z

$$\delta_z = \frac{F_z}{k_z} = 28.1\,\text{mm} \qquad (5.48)$$

Referring back to Eqn (5.10), the effective rolling radius, R_e, can be calculated using the tyre deflection δ_z from Eqn (5.48) and an unloaded tyre radius, R_u, of 318.5 mm from

$$R_e = R_u - \frac{\delta_z}{3} = 309.1\text{mm} \qquad (5.49)$$

Typically a tyre would roll through between 0.5 and 1 revolution (Gillespie, 1992) in order to develop the lateral force following a change in slip angle. If we assume that the tyre must complete 0.5 revolutions then for a speed of 100 kph the tyre lag on this vehicle is 0.035 s.

5.5 Experimental testing

In order to obtain the data needed for the tyre modelling required for simulation, a series of tests may be carried out using tyre test facilities, typical examples being the machines that are illustrated in Figures 5.42 and 5.43. The following is typical of tests performed (Blundell, 2000a) to obtain the tyre data that supports the baseline vehicle used throughout this text. The measurements of forces and moments were taken using the SAE coordinate system for the following configurations:

1. Varying the vertical load in the tyre 200, 400, 600, 800 kg.
2. For each increment of vertical load the camber angle is varied from -10 to $10°$ with measurements taken at $2°$ intervals. During this test the slip angle is fixed at $0°$.
3. For each increment of vertical load the slip angle is varied from -10 to $10°$ with measurements taken at $2°$ intervals. During this test the camber angle is fixed at $0°$.
4. For each increment of vertical load the slip and camber angle are fixed at zero degrees and the tyre is gradually braked from the free rolling state to a fully locked skidding tyre. Measurements were taken at increments in slip ratio of 0.1.

The test programme outlined here can be considered a starting point in the process of obtaining tyre data to support a simulation exercise. In practice obtaining all the data required to describe the full range of tyre behaviour discussed in the preceding sections will be extremely time consuming and expensive. The test programme described here does not, for example, consider effects such as varying the speed of the test machine, changes in tyre pressure or wear, changes in road texture and surface contamination by water or ice. The testing is also steady state and does not consider the transient state during transition from one orientation to another.

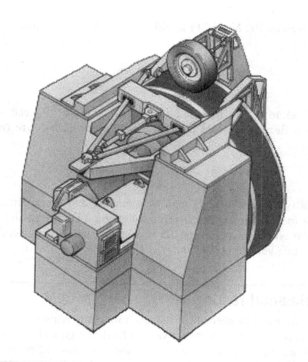

FIGURE 5.42

High Speed Dynamics Machine for tyre testing formerly at Dunlop Tyres Ltd.

Most importantly the tests do not consider the complete range of combinations that can occur in the tyre. The longitudinal force testing described is limited by only considering the generation of braking force. To obtain a complete map of tyre behaviour it would also, for example, be necessary to test not only for variations in slip angle at zero degrees of camber angle but to repeat the slip angle variations at selected camber angles. For comprehensive slip behaviour it would be necessary at each slip angle to brake or drive the tyre from a free rolling state to one that approaches the friction limit, hence deriving the 'friction circle' for the tyre.

Extending a tyre test programme in this way may be necessary to generate a full set of parameters for a sophisticated tyre model but will significantly add to the cost of testing. Obtaining data requires the tyre to be set up at each load, angle or slip ratio and running in steady state conditions before the required forces and moments can be measured. By way of example the basic test programme described here required measurements to be taken for the tyre in 132 configurations. Extending this, using the same pattern of increments and adding driving force, to consider combinations of slip angle with camber or slip ratio would extend the testing to 1452 configurations. In practice this could be reduced by judicious selection of test configurations but it should be noted the tests would still be for a tyre at constant pressure and constant speed on a given test surface. Examples of test results for a wider range of tyres and settings can be obtained by general reference to the tyre-specific

FIGURE 5.43

Flat Bed Tyre Test machine.

(Courtesy of Calspan.)

publications quoted in this chapter and in particular to the textbook by Pacejka (2012).

For the tyre tests described here the following is typical of the series of plots that would be produced in order to assess the force and moment characteristics. The results are presented in the following Figures 5.44–5.53 where a carpet plot format is used for the lateral force and aligning moment results:

1. Lateral force F_y with slip angle α
2. Aligning moment M_z with slip angle α
3. Lateral force F_y with aligning moment M_z (Gough Plot)
4. Cornering stiffness with load
5. Aligning stiffness with load
6. Lateral force F_y with camber angle γ

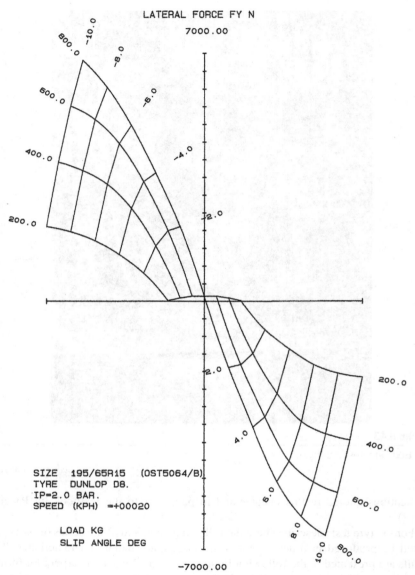

FIGURE 5.44

Lateral force F_y with slip angle α.

(Courtesy of Dunlop Tyres Ltd.)

7. Aligning moment M_z with camber angle γ
8. Camber stiffness with load
9. Aligning camber stiffness with load
10. Braking force with slip ratio

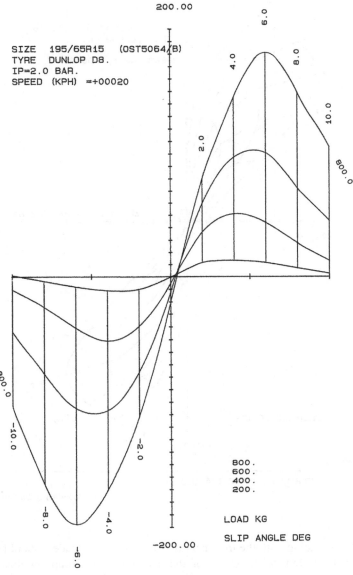

FIGURE 5.45

Aligning moment M_z with slip angle α.

(Courtesy of Dunlop Tyres Ltd.)

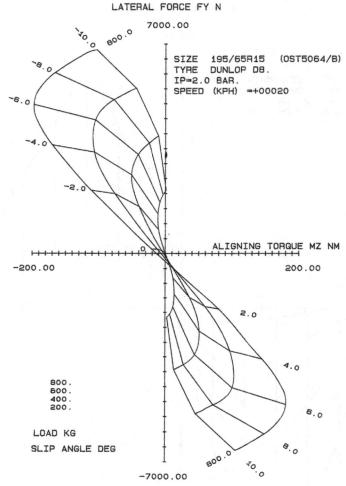

LATERAL FORCE FY N

7000.00

SIZE   195/65R15   (OST5064/B)
TYRE   DUNLOP D8.
IP=2.0 BAR.
SPEED (KPH) =+00020

ALIGNING TORQUE MZ NM

−200.00   200.00

800.
600.
400.
200.

LOAD KG

SLIP ANGLE DEG

−7000.00

FIGURE 5.46

Lateral force F_y with aligning moment M_z (Gough Plot).

(Courtesy of Dunlop Tyres Ltd.)

Before continuing with the treatment of tyre modelling, readers should note the findings (van Oosten et al., 1999) of the TYDEX Workgroup. In this study a comparison of tyre cornering stiffness for a tyre tested on a range of comparable tyre test machines gave differences between minimum and maximum measured values of up to 46%. Given the complexities of the tyre models that are described in the following section the starting point should be a set of measured data that can be used with confidence to form the basis of a tyre model.

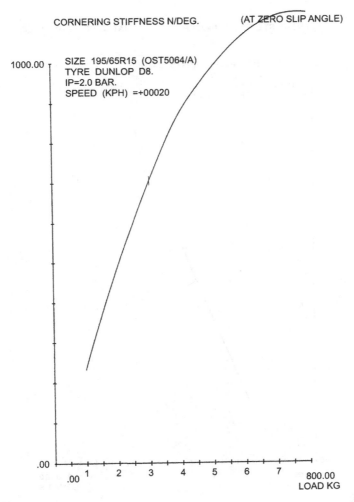

CORNERING STIFFNESS N/DEG.   (AT ZERO SLIP ANGLE)

1000.00

SIZE 195/65R15 (OST5064/A)
TYRE DUNLOP D8.
IP=2.0 BAR.
SPEED (KPH) =+00020

.00

.00   1   2   3   4   5   6   7   800.00
                                         LOAD KG

FIGURE 5.47

Cornering stiffness with load.

(Courtesy of Dunlop Tyres Ltd.)

5.6 Tyre Modelling
5.6.1 Overview

The modelling of the forces and moments at the tyre contact patch has been the subject of extensive research in recent years. A review of some of the most common tyre models was provided by Pacejka and Sharp (1991), where the authors state that it is necessary to compromise between the accuracy and complexity of the model. The

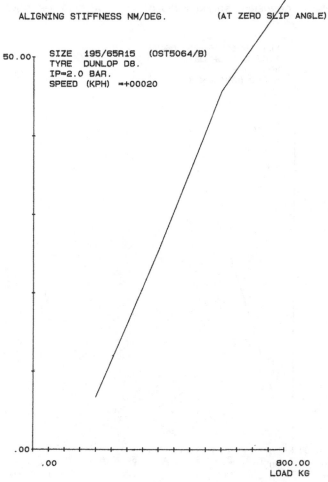

FIGURE 5.48

Aligning stiffness with load.

(Courtesy of Dunlop Tyres Ltd.)

authors also state that the need for accuracy must be considered with reference to various factors including the manufacturing tolerances in tyre production and the effect of wear on the properties of the tyre. This would appear to be a valid point not only from the consideration of computer modelling and simulation but also in terms of track testing where new tyres are used to establish levels of vehicle performance. A more realistic measurement of how a vehicle is going to perform in service may be to consider testing with different levels of wear or incorrect pressure settings.

One of the methods discussed by Pacejka and Sharp (1991) focuses on a multi-spoke model developed by Sharp where the tyre is considered to be a series of radial

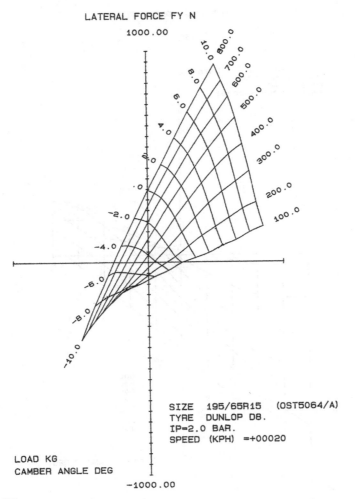

LATERAL FORCE FY N

LOAD KG
CAMBER ANGLE DEG

SIZE 195/65R15 (OST5064/A)
TYRE DUNLOP D8.
IP=2.0 BAR.
SPEED (KPH) =+00020

FIGURE 5.49

Lateral force F_y with camber angle γ.

(Courtesy of Dunlop Tyres Ltd.)

spokes fixed in a single plane and attached to the wheel hub. The spokes can deflect in the radial direction and bend both circumferentially and laterally. Sharp provides more details on the radial-spoke model approach in (Sharp & El-Nashar, 1986), (Sharp, 1990) and (Sharp, 1993). The other method of tyre modelling reviewed is based on the 'Magic Formula' that will be discussed in more detail later in this section. Another review of tyre models is given by Pacejka (1995), where the influence of the tyre is discussed with regard to 'active' control of vehicle motion. The radial-spoke and 'Magic Formula' models are again discussed. More recently, the

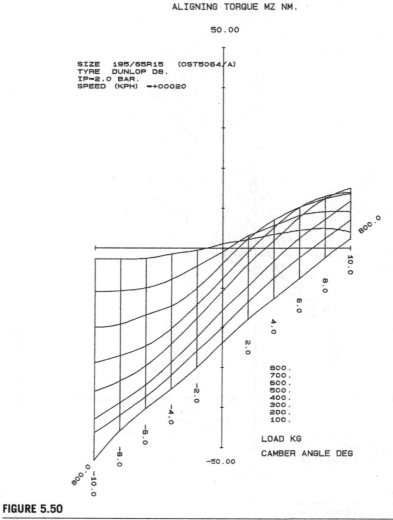

ALIGNING TORQUE MZ NM.

50.00

SIZE 195/65R15 (OST5064/A)
TYRE DUNLOP D8.
IP=2.0 BAR.
SPEED (KPH) =+00020

800.
700.
600.
500.
400.
300.
200.
100.

LOAD KG

CAMBER ANGLE DEG

−50.00

FIGURE 5.50

Aligning moment M_z with camber angle γ.

(Courtesy of Dunlop Tyres Ltd.)

definitive text on tyre modelling is in its third iteration (Pacejka, 2012), covering many types of model.

Before considering tyre models in more detail it should be stated that tyre models are generally developed according to the type of application the vehicle simulation will address. For ride and vibration studies the tyre model is often required to transmit the effects from a road surface where the inputs are small but of high frequency. In the simplest form the tyre may be represented as a simple compression only spring and

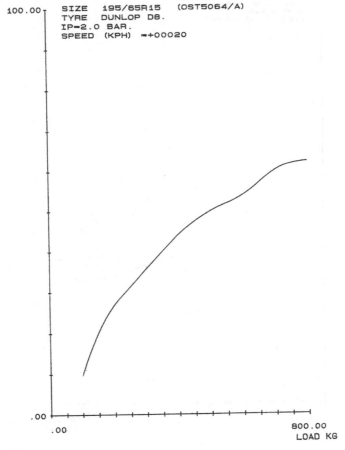

CAMBER STIFFNESS N/DEG.   (AT ZERO CAMBER ANGLE)

100.00   SIZE   195/65R15   (OST5064/A)
          TYRE   DUNLOP D8.
          IP=2.0 BAR.
          SPEED (KPH) =+00020

.00

.00   800.00
                                       LOAD KG

FIGURE 5.51

Camber stiffness with load.

(Courtesy of Dunlop Tyres Ltd.)

damper acting between the wheel centre and the surface of the road. The simulation may in fact recreate the physical testing using a four-poster test rig with varying vertical inputs at each wheel. A concept of the tyre model for this type of simulation is provided in Figure 5.54 where for clarity only the right side of the vehicle is shown.

In suspension loading or durability studies the tyre model must accurately represent the contact forces generated when the tyre strikes obstacles such as potholes and road bumps. In these applications the deformation of the tyre as it contacts the obstacle is of importance and is a factor in developing the model. These sort of tyre models are often developed for agricultural or construction type vehicles

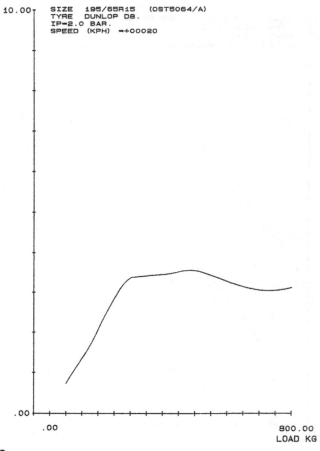

ALIGNING CAMBER STIFFNESS NM/D   (AT ZERO CAMBER ANGLE)

SIZE   195/65R15   (OST5064/A)
TYRE   DUNLOP D8.
IP=2.0 BAR.
SPEED (KPH) =+00020

FIGURE 5.52

Aligning camber stiffness with load.

(Courtesy of Dunlop Tyres Ltd.)

used in an off road environment and dependent on the tyre to a larger extent in isolating the driver from the ground surface inputs. An example of this sort of tyre model is described by Davis (1974) where a radial spring model was developed to envelop irregular features of a rigid terrain. The tyre is considered to be a set of equally spaced radial springs that when in contact with the ground will provide a deformed profile of the tyre as it envelops the obstacle. The deformed shape is used to redefine the rigid terrain with an 'equivalent ground plane'. The concept of an equivalent ground plane model was used in the early ADAMS/Tire™ model for the durability application but has the main limitation that the model is not

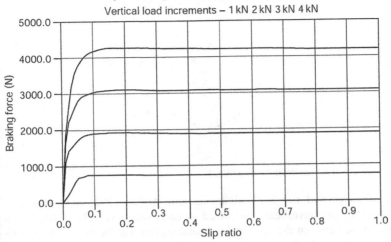

FIGURE 5.53

Braking force with slip ratio.

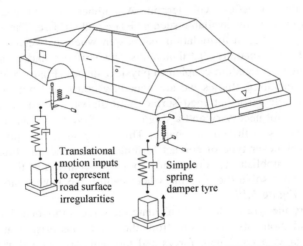

FIGURE 5.54

A simple tyre model for ride and vibration studies.

suitable for very small obstacles that the tyre might completely envelop. This is clarified by Davis (1974) where it is stated that the wave length of surface variations in the path of the tire should be at least three times the length of the tyre to ground contact patch. The other and most basic limitation of this type of model is that the simulation is restricted to straight-line motion and would only consider the vertical and longitudinal forces being generated by the terrain profile. An example of a radial spring tyre model is shown in Figure 5.55.

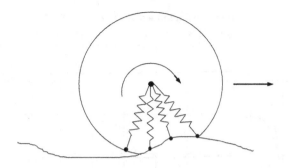

FIGURE 5.55

A radial spring terrain enveloping tyre model.

The work carried out by Kisielewicz and Ando (1992) describes how two different programs have been interfaced to carry out a vehicle simulation where the interaction between the tyre and the road surface has been calculated using an advanced nonlinear finite element analysis program.

More recently, advances in computational budget have allowed the development of a number of tyre models that begin to approach a causal description of the tyre structure and its interaction with terrain. A number of high quality terrain-interaction models are now available such as FTire and CDTire. These models are characterised by an elegant formulation that is somewhere between a coarse finite element model with spatially distributed discrete stiffness and a modal model with scalar degrees of freedom coupled to physical shapes (Figure 5.56).

For vehicle handling studies we are generally concerned with the manoeuvring of the vehicle on a flat road surface and elaborate contact formulations such as these are not necessary and a single point of contact is used to calculate aggregate slip states in the contact patch. This is the so-called 'point follower' formulation. Whichever type of contact formulation is used, the function of the tyre model is to establish the forces and moments occurring at the tyre to road contact patch and resolve these to the wheel centre and hence into the vehicle as indicated in Figure 5.57.

For each tyre the tyre model will calculate the three orthogonal forces and the three orthogonal moments that result from the conditions arising at the tyre to road surface contact patch. These forces and moments are applied at each wheel centre and control the motion of the vehicle. In terms of modelling the vehicle is actually 'floating' along under the action of these forces at each corner. For a handling model the forces and moment at the tyre to road contact patch, which are usually calculated by the tyre model, are:

1. F_x — longitudinal tractive or braking force
2. F_y — lateral cornering force
3. F_z — vertical normal force
4. M_z — aligning moment

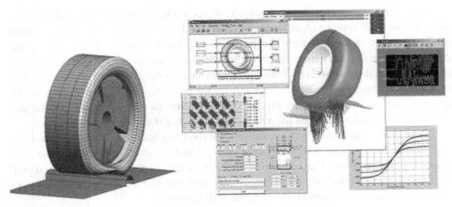

FIGURE 5.56

CD Tire (left) from the Fraunhofer Institute and FTire (right) from Cosin are both examples of semi-causal brush models that have become widely used in the last decade.

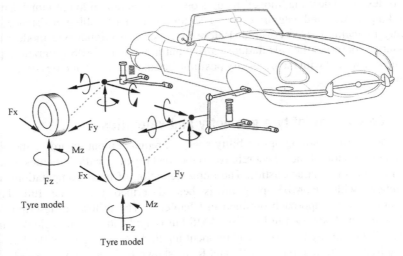

FIGURE 5.57

Interaction between vehicle model and tyre model.

The other two moments that occur at the patch, M_x the overturning moment and M_y the rolling resistance moment, are generally not significant for a handling tyre model for passenger cars. The calculation of forces and moments at the contact patch is the essence of a tyre model and will be discussed in more detail later.

As a simulation progresses and the equations for the vehicle and tyre are solved at each solution point in time there is a flow of information between the vehicle model and the tyre model. The tyre model must continually receive information about the position, orientation and velocity at each wheel centre and also the topography of the road surface in order to calculate the forces and moment at the contact patch. The road surface is often flat but may well have changing frictional characteristics to represent varying surface textures or changes between dry, wet or ice conditions. Inclined or cambered road surfaces can also be modelled if needed. The information from the wheel centre such as the height, camber angle, slip angle, spin velocity and so on are the inputs to the tyre model at each point in time and will dictate the calculation of the new set of forces at the contact patch.

These newly computed tyre conditions are then fed back to the vehicle model at each wheel centre. This will produce a change in the vehicle position at the next solution point in time. The conditions at each wheel centre will change and will be relayed back to the tyre model again. A new set of tyre forces and moments will then be calculated and so the process will continue.

The treatment of tyre models that follows in this section is based on methods that have been developed for vehicle handling simulations. A later section will deal with tyre models for durability analysis. As stated earlier, the computation of vertical force is straightforward based on the equations in Section 5.4.2. For the handling models described here the 'model' focuses on the calculation of longitudinal driving or braking forces and lateral forces. The formulation of rolling resistance and aligning moments is also covered. Before discussing individual tyre models it is necessary to describe the calculations carried out in the main MBS program to provide the tyre model with the necessary position, orientation and velocities of the road wheel.

5.6.2 Calculation of tyre geometry and velocities

A tyre model, for handling or durability analysis, requires input regarding the position and orientation of the wheel relative to the road together with velocities used to determine the slip characteristics. The implementation of these computations as a tyre model with an MBS program is best described using the full three-dimensional vector approach outlined in Chapter 2. The following description is based on the methods used in MSC.ADAMS but is applicable to any vehicle simulation model requiring tyre force and moment input. As a starting point the tyre can be modelled using the input radii R_1 and R_2 as shown in Figure 5.58.

Using the tyre model geometry based on a torus it is possible to determine the geometric outputs that are used in the subsequent force and moment calculations. Consider first the view in Figure 5.59 looking along the wheel plane at the tyre inclined on a flat road surface.

The vector $\{U_s\}$ is a unit vector acting along the spin axis of the tyre. The vector $\{U_r\}$ is a unit vector that is normal to the road surface and passes through the centre of the tyre carcass at C. The contact point P between the tire and the surface of the road is determined as the point at which the vector $\{U_r\}$ intersects the road surface. For the purposes of this document it is assumed the road is flat and only one point of contact occurs.

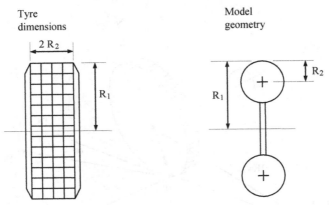

FIGURE 5.58

Tyre model geometry.

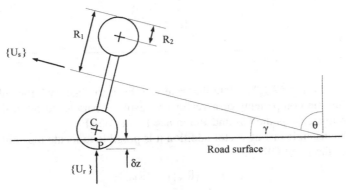

FIGURE 5.59

Inclined tyre geometry.

The camber angle γ between the wheel plane and the surface of the road is calculated using

$$\gamma = \pi/2 - \theta \qquad (5.50)$$

where

$$\theta = \cos^{-1} (\{U_r\} \bullet \{U_s\}) \qquad (5.51)$$

The vertical penetration of the tyre δz at point P is given by

$$\delta z = R_2 - |CP| \qquad (5.52)$$

In order to calculate the tyre forces and moment it is also necessary to determine the velocities occurring in the tyre. In Figure 5.60 the SAE coordinate system is located at the contact point P. This is established by the three unit vectors

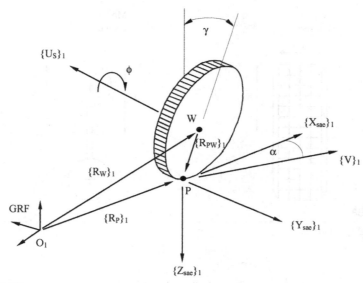

FIGURE 5.60

Tyre geometry and kinematics.

$\{X_{sae}\}_1$, $\{Y_{sae}\}_1$ and $\{Z_{sae}\}_1$. Note that referring back to Chapter 2 the subscript 1 indicates that the components of a vector are resolved parallel to reference frame 1, which in this case is the Ground Reference Frame (GRF).

Using the triangle law of vector addition it is possible to locate the contact point P relative to the fixed GRF O_1

$$\{R_P\}_1 = \{R_W\}_1 + \{R_{PW}\}_1 \qquad (5.53)$$

At this stage it should be said that the vector $\{R_{PW}\}_1$ represents the loaded radius and not the effective rolling radius of the tyre. Should this be significant for the work in hand, such as the modelling and simulation of ABS (Ozdalyan and Blundell, 1998) then further modification of the tyre model may be necessary.

If the angular velocity vector of the wheel is denoted by $\{\omega\}_1$ then the velocity $\{V_P\}_1$ of point P is given by

$$\{V_P\}_1 = \{V_W\}_1 + \{V_{PW}\}_1 \qquad (5.54)$$

where

$$\{V_{PW}\}_1 = \{\omega\}_1 \times \{R_{PW}\}_1 \qquad (5.55)$$

It is now possible to determine the components of $\{V_P\}_1$ that act parallel to the SAE coordinate system superimposed at P. The longitudinal slip velocity V_{XC} of point P is given by

$$V_{XC} = \{V_P\}_1 \bullet \{X_{sae}\}_1 \qquad (5.56)$$

The lateral slip velocity V_Y of point P is given by

$$V_Y = \{V_P\}_1 \bullet \{Y_{sae}\}_1 \tag{5.57}$$

The vertical velocity V_Z at point P, which will be used to calculate the damping force in the tyre, is given by

$$V_Z = \{V_P\}_1 \bullet \{Z_{sae}\}_1 \tag{5.58}$$

Considering the angular velocity vector of the wheel $\{\omega\}_1$ in more detail we can represent the vector as follows. The wheel develops a slip angle α that is measured about $\{Z_{sae}\}_1$, a camber angle γ that is measured about $\{X_{sae}\}_1$ and a spin angle ϕ that is measured about $\{U_S\}_1$. The total angular velocity vector of the wheel is the summation of all three motions and is given by

$$\{\omega\}_1 = \dot{\alpha}\,\{Z_{sae}\}_1 + \dot{\gamma}\,\{X_{sae}\}_1 + \dot{\phi}\,\{U_S\}_1 \tag{5.59}$$

It is possible to consider an angular velocity vector $\{\omega_S\}_1$ that only considers the spinning motion of the wheel and does not contain the contributions due to α and γ. This vector for angular velocity that only considers spin is given by

$$\{\omega_S\}_1 = \dot{\phi}\,\{U_S\}_1 \tag{5.60}$$

Using this it is possible to determine V_C the 'circumferential velocity' component of point P relative to the centre of the wheel W and measured parallel to $\{X_{sae}\}_1$

$$V_C = (\{\omega_S\}_1 \times \{R_{PW}\}_1) \bullet \{X_{sae}\}_1 \tag{5.61}$$

At this stage it is worth considering the usual two-dimensional representation of longitudinal slip for straight-line braking. Referring back to Section 5.4.4 a definition of slip ratio, S, during braking was given by

$$S = \frac{V - \omega_B R_e}{V} \tag{5.62}$$

Based on the velocities, which have been determined for the three-dimensional case it is now possible to calculate a longitudinal slip ratio, S, during braking which is given by

$$S = \frac{V_{XC}}{V_X} \tag{5.63}$$

For this formulation of slip ratio V_{XC} can be considered to be the contact patch velocity relative to the road surface. This is equivalent to $V-\omega_B R_e$ in the two-dimensional model, albeit using the loaded radius in the vector based formulation. The circumferential velocity V_C of P measured relative to the wheel centre can be subtracted from V_{XC} to give V_X the actual longitudinal velocity of P ignoring the rotation effect. This can be thought of as the velocity of an imaginary point in the ground that follows the contact patch and is also equivalent to V in the two-dimensional model. During traction, the longitudinal slip ratio is formulated using

$$S = \frac{V_{XC}}{|V_C|} \tag{5.64}$$

The lateral slip of the contact patch relative to the road is defined by the slip angle α. where

$$\alpha = \arctan \{V_Y / V_X\} \tag{5.65}$$

During braking a lateral slip ratio S_α is computed as

$$S_\alpha = |\tan \alpha| = |V_Y / V_X| \tag{5.66}$$

During braking S_α will have a value of zero when V_Y is zero and can have a maximum value of 1.0, which equates to a slip angle α of 45°. Slip angles in excess of this are not usual for vehicle handling but may occur in other applications where a tyre model is used, for example to simulate aircraft taxiing on a runway.

During traction the formulation becomes:

$$S_\alpha = (1 - S)|\tan \alpha| = |V_Y / V_C| \tag{5.67}$$

5.6.3 Road surface/terrain definition

It is easy to forget when discussing tyres that the resultant forces and moments are a product, as with any friction-based model, of the interaction between the tyre and the road and not a property of the tyre in isolation. Early tyre models treated the road or terrain as simply an infinite flat surface requiring only a vertical datum for definition. This was extended to two-dimensional models that allowed a terrain to be defined as a series of sections. This allowed for durability type analysis, with forward motion only, to encounter potholes, bumps and other road obstacles. A typical current approach is to represent the geometry and frictional characteristics of the road surface or terrain using a finite element approach as shown in Figure 5.61.

The road surface is defined as a system of triangular patches. As with finite elements the outward normal of the element is defined by numbering the nodes for each element using a sequence that is positive when considering a rotation about the outward normal. For each element it is possible to define frictional constants that are factored with the friction parameters associated with a tyre property file. This would allow simulations when the vehicle encounters changing road conditions as with driving from dry to wet conditions or simulation of 'mu-split' conditions where one side of the vehicle is braking in the dry and the other on ice.

The ability to model a road surface as a continuous three-dimensional surface is required not only for durability work but is also needed for handling simulations on

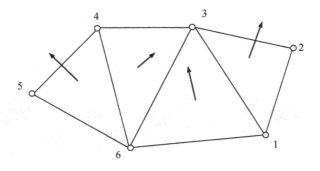

FIGURE 5.61

Definition of road surface using finite element approach.

non-flat road surfaces. As with any finite element model the accuracy of the road surface definition will be a function of the mesh refinement. Current practice amongst some users in industry is to model the road using the more sophisticated mesh generation tools provided with finite element-style software and then run a small translation programme to reformat the data file. The widespread adoption of the so-called 'TeimOrbit®' format has facilitated the separation of road generation, sometimes entirely, from the task of multibody modelling. The format appears widely used, being cited by MSC Software, Altair, Cosin and Intec Dynamics help documentation for roads. Other items are also stored in TeimOrbit files. XML files are also widely used as an easily readable and versatile storage format for finite element-style road grid descriptions. In principle any file format that is topologically similar to a finite element mesh will work for the gridded representation of road surface data, with 'nodes' (points in space) and 'elements' (a list of typically three node numbers that form a triangular facet) — so in principle a stereo lithography (STL) file could be used. However, TeimOrbit and XML appear to dominate applications at the time of writing, for artificial roads.

For durability and detailed ride work, the ready availability of scanned road surface data, (typically on grids around 1 × 1 cm) has led to dissatisfaction with the previous methods, which generate unreasonably large data files. This in turn has led to a standard initially defined and used by the Daimler Group known as 'Curved Rectangular Grid'. It gets around the 'small difference of large numbers' problem of storing detailed road information over a long physical distance by separating macro information (essentially the centreline of the road, including elevation) and micro information (the 'shape of the stones') in two separate data groups. Any given point is synthesised by combining the two pieces of information. A public version was launched in February 2009 in an 'open' form and has been released as a stable protocol with open source Matlab tools available to support it as of July 2013. Sample data from the so-called 'OpenCRG' format is shown in Figure 5.62.

It can be seen that there are a number of options available for defining and using road data files; history suggests it is likely that only a small number will persist but

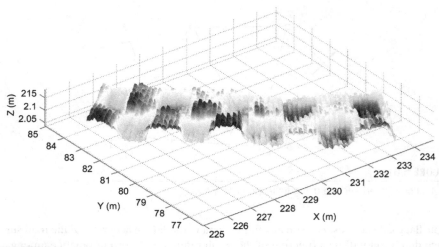

FIGURE 5.62

A sample OpenCRG file rendered using Matlab. Note that the vertical scale is somewhat exaggerated compared to the two horizontal scales.

unfortunately gives few clues as to which ones; for the time being the analyst needs to be adequately familiar with all of them and intimately familiar with the ones supported by the package at hand.

5.6.4 Interpolation methods

Early tyre models for handling used the results of laboratory rig testing directly to generate 'look-up' tables of data that were used directly by the tyre model to interpolate the lateral force and aligning moment at the contact patch. Figure 5.63 illustrates a sample of some results that might typically be obtained from a tyre rig test

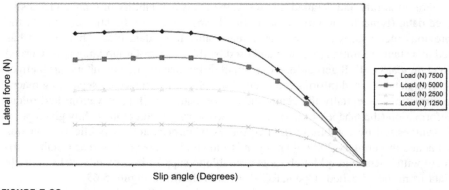

FIGURE 5.63

Interpolation of measured tyre test data as carried out by Microsoft Excel.

where for variations in vertical load F_z the lateral forces F_y are plotted as a function of slip angle at a given camber angle.

For this set of data the independent variables that are set during the test are the camber angle, the vertical force and the slip angle. The measured dependent variable is the lateral force. Using this measured data the tyre model, really a method rather than a model, uses a curve fit to obtain a value for the lateral force for the value of F_z and slip angle determined by the wheel centre position and orientation. If the instantaneous camber angle lies between two sets of measured data at different camber angles set during the test then the tyre model can use linear, quadratic or some other interpolation scheme between the two camber angles. If the instantaneous camber angle is for example 2.4° and measured data is available at 2 and 3°, then the curve fitting as a function of F_z and slip angle is carried out at the two bounding camber angles and the linear interpolation is carried out between these two points. The approach described here for lateral force is applied in exactly the same manner when determining by interpolation, a value for the aligning moment. There are some disadvantages in using an interpolation tyre model:

1. The process of interpolating large quantities of data (easily 4000 plus items at a single tyre pressure) at every integration step in time may not be an efficient simulation approach and is often considered to result in increases in computer solution times for the analysis of any given manoeuvre.
2. This sort of model does not lend itself to any design modification or optimisation involving the tyre. The tyre must already exist and have been tested. In order to investigate the influence of tyre design changes on vehicle handling and stability then the tyre model must be reduced to parameters that can be related to the tyre force and moment characteristics. This has lead to the development of tyre models represented by formulae that will now be discussed.

5.6.5 The 'Magic Formula' tyre model

For smooth road handling work, the tyre model that is now most well established and has generally gained favour is based on the work by Pacejka and as mentioned earlier is referred to as the 'Magic Formula'. The Magic Formula is not a predictive tyre model but is used to empirically represent and interpolate previously measured tyre force and moment curves. The early version (Bakker et al., 1986, 1989) is sometimes referred to as the 'Monte Carlo version' due to the conference location at which this model was presented in the 1989 paper. The tyre models discussed here are based on the formulations described in Bakker et al. (1989) and a later version Pacejka and Bakker, 1993), which is sometimes referred to as Version 3 of the Magic Formula. Mention of Version 4 appears to have disappeared from the literature.

The Magic Formula model is undergoing continual development, which is reflected in a further publication (Pacejka and Besselink, 1997) where the model is not restricted to small values of slip and the wheel may also run backwards. The authors also discuss a relatively simple model for longitudinal and lateral transient responses restricted to relatively low time and path frequencies. The tyre model in this paper also acquired a new name and was referred to as the 'Delft Tyre 97' version, which seems to have reverted over time to MF-Tyre 5.0; 'Delft-Tyre' has become an umbrella term to include not only the base MF-Tyre model but a modified version of it suitable for intermediate frequency events, known as SWIFT (short wavelength intermediate frequency tyre).

MF-Tyre 5.0 switched to a 'normalised' formulation that ostensibly gives improved numerical conditioning, and version 6.1 includes direct support for different tyre pressures, which had been hitherto absent. TNO Automotive have become de facto custodians of the Magic Formula approach and have libraries for it and other tyre models that can be used directly with a number of popular software packages, including Matlab, Simpack, ADAMS, DADS and many others (TNO, 2012). While not apparently claiming a monopoly on the term, it seems that 'MF-Tyre' has become associated specifically with the TNO implementation. At the time of writing, the latest version of the Magic Formula is 6.1.2. Between version 3 and this version, a separate motorcycle version has been introduced ('Pac-MC' or 'MF-MC') and then merged with the basic formulation since version 6.0.

Other authors have developed systems closely based around the Magic Formula approach. The BNPS model (Schuring et al., 1993) is a particular version of the Magic Formula that automates the development of the coefficients working from measured test data. The model name BNPS is in honour of Messrs Bakker, Nyborg and Pacejka who originated the Magic Formula and the S indicates the particular implementation developed by Smithers Scientific Services Inc. MSC Software have a proprietary take on the Magic Formula that they label 'Pac2002' (Kuiper and Van Oosten, 2007). It includes stationary operation by switching to a different model at speeds below walking speed in order to avoid the divide-by-zero problem at rest. Since Magic Formula version 6.0, so-called 'turn-slip' behaviour has been modelled to allow parking loads to be captured.

The general acceptance of the Magic Formula is reinforced by the work carried out at Michelin and described in Bayle et al. (1993). In this paper the authors describe how the Magic Formula has been tested at Michelin and 'industrialised' as a self-contained package for the pure lateral force model. The authors also considered modifications to the Magic Formula to deal with the complicated situation of combined slip. To the authors' knowledge, Pirelli have also had (and may still retain) their own flavour of the Magic Formula model. The US military organisation TARDEC have published Matlab routines for handling the PAC2002 formulation model (Goryca, 2010).

It may thus be seen that talk of 'The' Magic Formula is therefore no longer appropriate and care must be taken to establish which one is in use. The general difficulty of obtaining tyre test data means that tyre data tends to get 'squirreled away' in personal archives and transported from role to role and between organisations; the backwards compatible formulation of most multibody software in use means that original 1987 data sets can still be used in, for example, Simpack, and samples of

1989 ('Pac89') data sets are included in ADAMS/Car at the time of writing. Interchanging coefficient sets between versions may produce that most insidious of errors, the 'plausible-but-wrong' result — although it equally may produce pure garbage. A clear distinction between the different versions available is therefore necessary in the mind of the analyst.

In the original Magic Formula paper, the authors in Bakker et al. (1986) discuss the use of formulae to represent the force and moment curves using established techniques based on polynomials or a Fourier series. The main disadvantage with this approach is that the coefficients used have no engineering significance in terms of the tyre properties and as with interpolation methods the model would not lend itself to design activities. This is also reflected in Sitchen (1983) where the author describes a representation based on polynomials where the curves are divided into five regions but this still has the problem of using coefficients that do not typify the tyre force and moment characteristics.

The generic Magic Formula has been developed using mathematical functions that relate:

1. The lateral force F_y as a function of slip angle α.
2. The aligning moment M_z as a function of slip angle α.
3. The longitudinal force F_x as a function of longitudinal slip κ.

When these curves are obtained from steady state tyre testing and plotted, the general shape of the curves is similar to that indicated in Figure 5.64. It is important to note that the data used to generate the tyre model is obtained from steady state testing. The lateral force F_y and the aligning moment M_z are measured during pure cornering, i.e. cornering without braking, and the longitudinal braking force during pure braking, i.e. braking without cornering.

The basis of this model is that tyre force and moment curves obtained under pure slip conditions and shown in Figure 5.64 look like sine functions that have been modified by introducing an arctangent function to 'stretch' the slip values on the x-axis.

The general form of the model as presented in Bakker et al. (1986) is:

$$Y(X) = D \sin [\, C \arctan\{ Bx - E (Bx - \arctan (Bx))\}] \qquad (5.68)$$

where

$$Y(X) = y(x) + S_v \qquad (5.69)$$

$$x = X + S_h \qquad (5.70)$$

S_h = horizontal shift
S_v = vertical shift

In this case Y is either the side force F_y, the aligning moment M_z or the longitudinal force F_x and X is either the slip angle α or the longitudinal slip, for which

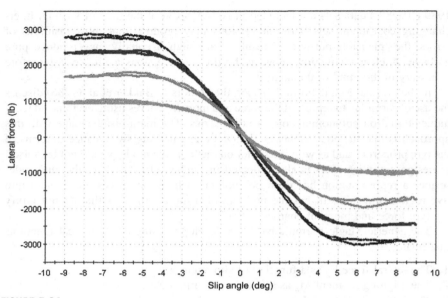

FIGURE 5.64

Typical form of tyre force curves from testing. Note the presence of noise on the data despite high quality data acquisition.

(Courtesy of Calspan.)

Pacejka uses κ. The physical significance of the coefficients in the formula becomes more meaningful when considering Figure 5.65.

For lateral force or aligning moment the offsets S_v and S_h arise due to adding camber or physical features in the tyre such as conicity and plysteer. For the longitudinal braking force this is due to rolling resistance.

Working from the offset XY axis system the main coefficients are:

D — is the peak value.
C — is a shape factor that controls the 'stretching' in the x direction. The value is determined by whether the curve represents lateral force, aligning moment or longitudinal braking force. The very first published implementation of the Magic Formula (Bakker et al., 1986) set C as a constant:

    1.30 — lateral force curve.
    1.65 — longitudinal braking force curve.
    2.40 — aligning moment curve.

As experience emerged with using the formula, C became another one of the constants to be varied during the empirical fit from Version 2 ('Pacejka 89') on.

B — is referred to as a 'stiffness' factor. From Figure 5.65 it can be seen that BCD is the slope at the origin, i.e. the cornering stiffness when plotting lateral force. Obtaining values for D and C leads to a value for B.

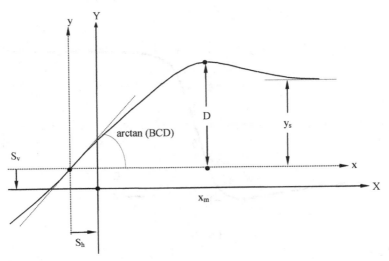

FIGURE 5.65

Coefficients used in the 'Magic Formula' tyre.

E – is a 'curvature' factor that effects the transition in the curve and the position x_m at which the peak value if present occurs. E is calculated using:

$$E = \frac{B_{Xm} - \tan(\pi/2C)}{B_{Xm} - \arctan(B_{Xm})} \tag{5.71}$$

y_s – is the asymptotic value at large slip values and is found using:

$$y_s = D \sin(\pi C/2) \tag{5.72}$$

The curvature factor E can be made dependent on the sign of the slip value plotted on the x-axis.

$$E = E_0 + \Delta E \operatorname{sgn}(x) \tag{5.73}$$

This will allow for the lack of symmetry between the right and left side of the diagram when comparing driving and braking forces or to introduce the effects of camber angle γ. This effect is illustrated in Pacejka and Bakker (1993) by the generation of an asymmetric curve using coefficients $C = 1.6$, $E_0 = 0.5$ and $\Delta E = 0.5$. This is recreated here using the curve shape illustrated in Figure 5.66. Note that the plots have been made non-dimensional by plotting y/D on the y-axis and BCx on the x-axis.

Version 3 of the Magic Formula is discussed in this text rather than the later normalised versions because the algebra can be more readily assimilated. However, the principle of the later normalised versions remains identical but allows the use of

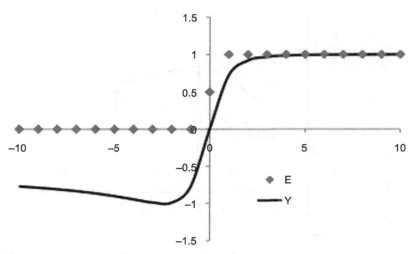

FIGURE 5.66

Generation of an asymmetric curve using different values of E.

convenient scaling factors to manipulate measured data on different surfaces, of which more later.

Version 3 utilises a set of coefficients a_0, a_1, a_2,... as shown in Tables 5.3 and 5.4. In Figure 5.67 it can be seen that at zero camber the cornering stiffness BCD_y reaches a maximum value defined by the coefficient a_3 at a given value of vertical load F_z that equates to the coefficient a_4. This relationship is illustrated in Figure 5.67 where the slope at zero vertical load is taken as $2a_3/a_4$.

The load a_4, sometimes referred to as the 'reference load', is not to be confused with the Rated Load for the tyre. ETRTO, the European Tyre and Rim Technical Organisation, have identified a load beyond which tyres should not be used. This

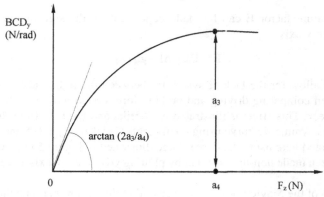

FIGURE 5.67

Cornering stiffness as a function of vertical load at zero camber angle.

is a mean vehicle corner load and reflects some kind of boundary between acceptable durability and unacceptable durability. Rated Load is codified on the tyre sidewall by a so-called 'Load Index', a two- or three-digit code that can be converted into an actual vehicle corner load with the help of the relevant standard. Rated Load generally has some kind of legislative significance, and varies with tyre pressure as well as details of the tyre construction and so on. This is the reason why many vehicles are specified with different tyre pressures for the laden condition. Unfortunately, no clue about the value of the coefficient a_4 is available on the sidewall of the tyre.

The model has been extended to deal with the combined slip situation where braking and cornering occur simultaneously. A detailed account of the combined slip model is given in Pacejka and Bakker (1993) and for more recent implementations in Pacejka (2012). The definitive equation guide from TNO (2010) contains some small but distinct differences from the implementation in the textbook, but has less illuminating commentary. The equations for pure slip only and as developed for the Monte Carlo model (Bakker et al., 1989) are summarised in Table 5.3 and similarly for Version 3 (Pacejka and Bakker, 1993) in Table 5.4. As can be seen a large number of parameters are involved and great care is needed to avoid confusion between each version, something that has become steadily more difficult over time as previously noted.

Apart from implementing the model into a MBS analysis program for vehicle simulation some method is needed to obtain the coefficients from raw test data.

The authors in van Oosten and Bakker (1993) describe their work using measured data and software developed at the TNO Road-Vehicles Research Institute to apply a regression method and obtain the coefficients. The authors in Schuring et al. (1993) have also automated the process for the BNPS version of the model.

The fitting of data for Magic Formula models is often regarded with a certain amount of mystique but there is no particular reason why this should be so, as should already be apparent. Many students are familiar with the 'trendline' function in Microsoft Excel that uses a least-squares method to find the line that is 'least bad' at representing the data. Microsoft Excel also has a solver that will attempt to minimise or maximise the value of a cell by adjusting one or more parameters using the function called 'solver' — note that this is currently unavailable in non-Windows versions of Excel. To use this method, a measured data set at one load from a tyre test machine is subtracted from the output from a calculation of the formula using some seed value coefficients. The resulting value represents an error in the fit with the seed values of coefficients. The solver can be set to minimise the error by varying the parameters. Alternatively, other fitting methods can be used by using a code such as Matlab or Octave. The definitive 'Numerical Recipes' (Press et al, 2011) recommends the Nelder Mead method and provides a 'downhill simplex' implementation of it; both Matlab and Octave have ready-coded versions available.

Some interesting wrinkles lie within such an approach and an important one is illustrated here. A sample curve is generated by setting B, C, D and E to unity. A fictitious data set is invented by adding noise to the data set and then a generic curve-fitting process is let loose on it. A lot of noise can be added and the fitting

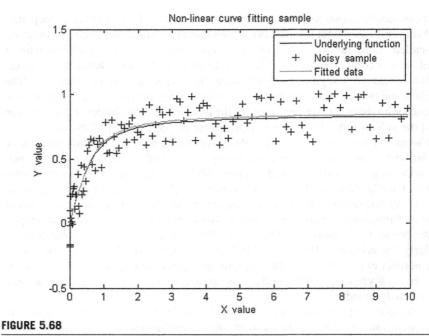

FIGURE 5.68

A generic (Nelder Mead downhill simplex) fit on a fictitious data set with large amounts of noise.

routine generally converges well on the original curve when inspected visually (Figure 5.68).

If the exercise is repeated with different sets of noise — generated using a typical computer-style 'random' function — then the repeated attempts at fitting all converge well and are difficult to distinguish from one another, or the original function, when plotted together, as shown in Figure 5.69.

However, examination of the fitted coefficients themselves gives a different view, as shown in Table 5.1.

1. The fitted parameters vary from run to run, particularly E.
2. The resulting fitted values are nothing like the original parameters, having differences up to 30% in the case of B, C and D and over 160% in the case of E.

In Sharp (1992) a suggested approach is to use an appreciation of the properties of the Magic Formula to fix C based on the values suggested in Pacejka and Bakker (1993) for lateral force, longitudinal force and aligning moment. For each set of load data it is then possible to obtain the peak value D and the position at which this occurs x_m. Using the slope at the origin and the values for C and D it is now possible to determine the stiffness factor B and hence obtain a value for E.

Using exactly this method can result in a disappointing fit but a small variation on it can give good results. Figure 5.70 shows a comparison between the Sharp method

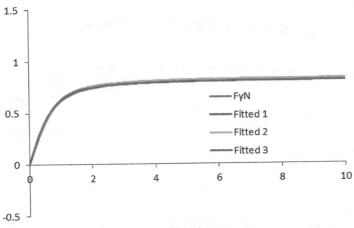

FIGURE 5.69

Results of three different curve fitting exercises with different sets of noise generated to the same parameters.

Table 5.1 Results of Three Different Curve Fitting Exercises with Different Sets of Noise Generated to the Same Parameters

	Actual	Fitted 1	Fitted 2	Fitted 3
B	1	1.2588	1.2113	1.2889
C	1	0.8226	0.8461	0.8480
D	1	0.8540	0.8656	0.8444
E	1	−0.07021	−0.6698	−0.6361

implemented directly and a variation on it where B and D are calculated according to the Sharp protocol while C and E are found by minimising the error between the fitted and measured data using a least squares method.

The comparisons between coefficients are shown in Table 5.2 and are different again. Both C and E need modifying, although this is not unreasonable given the Sharp approach of picking C based on some prior feel for the data; the reference set does not particularly resemble any known tyre characteristic.

It is apparent using the modified Sharp method on this data set that there is a significant interplay between C and E, and also that the overall quality of the fit is not unduly improved by exploring this interplay for levels of C higher than about 2 (it is possible to get an acceptable fit at least up to $C = 5$). In general it is the authors' observation that for sample, artificial data sets such as the modified Sharp method appear quite unpromising but on real tyre data sets they give rather acceptable results. In the examples given in Figure 5.71, below, only E needed finding via the least squares method.

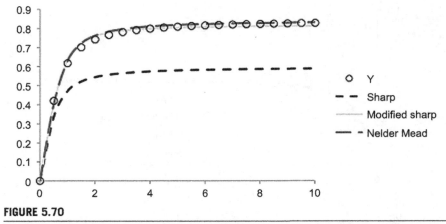

FIGURE 5.70

A comparison of the Sharp, modified Sharp and Nelder Mead fits on a reference data set.

Table 5.2 A Comparison of Fits Using Other Methods

	Actual	Fitted (Sharp)	Fitted (Modified Sharp)
B	1	0.7825	0.7825
C	1	1.3	3
D	1	0.8270	0.3391
E	1	0.8130	−30.49

Having obtained these terms at each load, the various coefficients are determined using curve fitting techniques to express B, C, D and E as functions of load. One issue with this method is that the value of C arbitrarily chosen may not subsequently emerge as the best. Applying the method described to a set of four measurements at zero camber angle gives the results for coefficients B, D and E against load as shown in Figure 5.72.

It can be seen in Figure 5.72 above that the quadratic fit for the parameter E, for example, is not terribly good compared to the measured data. Also apparent in the same figure is that extrapolating significantly beyond the fitted region may be even more unwise than is normal for extrapolation, which is already quite unwise.

A generic issue that occurs when deriving the coefficients for this model is whether those that have a physical significance should be fixed to match the tyre or set to values that give the best curve fit. Given the empirical nature of the model and the stated aim of reproducing measured data, it seems clear to the authors that fit should be favoured over everything else.

Considering the equations given in Tables 5.3 and 5.4, it can be seen that the dependency on vertical load is linear for the Version 3 model with the exception of cornering stiffness, as already discussed. This linear dependency on load

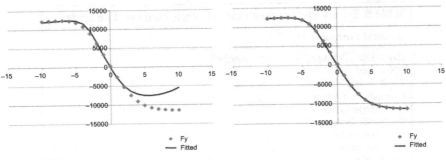

FIGURE 5.71

An improved fit (right) compared to the Sharp method (left) can be made by finding E using a curve fitting method and relinquishing Eqn 5.71.

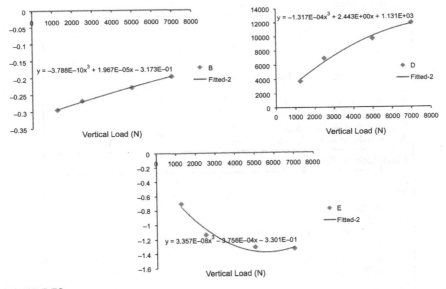

FIGURE 5.72

Fitted values for B, D and E plotted against vertical load in N.

continues into version 6.1. Thus it may be seen that some degradation of fit may be expected at very low loads where E may be poorly served by a linear relationship.

Despite the linear dependency for D and E on load apparently being at odds with the measured data as shown in Figure 5.72, comparisons of output from the Magic Formula with measured test data (Bakker et al., 1986, 1989) indicate good correlation. A study in Makita and Torii (1992) comparing the results of this model with those obtained from vehicle testing under pure slip conditions also indicates the high degree of accuracy that can be obtained using this tyre model.

Table 5.3 Pure Slip Equations for the 'Magic Formula' Tyre Model

General Formula

$y(x) = D\sin[C\arctan\{Bx-E(Bx-\arctan(Bx))\}]$
$Y(X) = y(x) + S_v$
$x = X + S_h$
B = stiffness factor
C = shape factor
D = peak factor
S_h = horizontal shift
S_v = vertical shift
$B = dy/dx_{(x=0)}/CD$
$C = (2/\pi) \arcsin(y_s/D)$
$D = y_{max}$
$E = (Bx_m-\tan(\pi/2C))/(Bx_m - \arctan(Bx_m))$

Lateral Force

$X_y = \alpha$
$Y_y = F_y$
$D_y = \mu_y F_z$
$\mu_y = a_1 F_z + a_2$
$BCD_y = a_3 \sin(2 \arctan(F_z/a_4)) (1 - a_5|\gamma|)$
$C_y = a_0$
$E_y = a_6 F_z + a_7$
$B_y = BCD_y/C_y D_y$
$S_{hy} = a_8 \gamma + a_9 F_z + a_{10}$
$S_{vy} = a_{11} F_z \gamma + a_{12} F_z + a_{13}$

Longitudinal Force

$X_x = \kappa$
$Y_x = F_x$
$D_x = \mu_x F_z$
$\mu_x = b_1 F_z + b_2$
$BCD_x = (b_3 F_z^2 + b_4 F_z) \exp(-b_5 F_z)$
$C_x = b_0$
$E_x = b_6 F_z^2 + b_7 F_z + b_8$
$B_x = BCD_x/C_x D_x$
$S_{hx} = b_9 F_z + b_{10}$
$S_{vy} = 0$

Aligning Moment

$X_z = \alpha$
$Y_z = M_z$
$D_z = c_1 F_z^2 + c_2 F_z$
$BCD_z = (c_3 F_z^2 + c_4 F_z)(1-c_6|\gamma|) \exp(-c_5 F_z)$
$C_z = c_0$
$E_z = (c_7 F_z^2 + c_8 F_z + c_9) (1 - c_{10}|\gamma|)$
$B_z = BCD_z/C_z D_z$
$S_{hz} = c_{11} \gamma + c_{12} F_z + c_{13}$
$S_{vz} = (c_{14} F_z^2 + c_{15} F_z)\gamma + c_{16} F_z + c_{17}$

Monte Carlo Version, also known as 'Pac89' and identifiable by the presence of the coefficient a_0 as distinct from the original version presented in 1986 that had fixed values for C.

Table 5.4 Pure Slip Equations for the 'Magic Formula' Tyre Model

General Formula

$y(x) = D\sin[C\arctan\{Bx - E(Bx - \arctan(Bx))\}]$

$Y(X) = y(x) + S_v$

$x = X + S_h$

B = stiffness factor

C = shape factor

D = peak factor

S_h = horizontal shift

S_v = vertical shift

$B = dy/dx_{(x=0)}/CD$

$C = (2/\pi) \arcsin(y_s/D)$

$D = y_{max}$

$E = (Bx_m - \tan(\pi/2C))/(Bx_m - \arctan(Bx_m))$

Lateral Force

$X_y = \alpha$

$Y_y = F_y$

$D_y = \mu_y F_z$

$\mu_y = (a_1 F_z + a_2)(1 - a_{15}\gamma^2)$

$BCD_y = a_3 \sin(2\arctan(F_z/a_4))(1 - a_5|\gamma|)$

$C_y = a_0$

$E_y = (a_6 F_z + a_7)(1 - (a_{16}\gamma + a_{17})\text{sgn}(\alpha + S_{hy}))$

$B_y = BCD_y/C_y D_y$

$S_{hy} = a_8 F_z + a_9 + a_{10}\gamma$

$S_{vy} = a_{11}F_z + a_{12} + (a_{13}F_z^2 + a_{14}F_z)\gamma$

Longitudinal Force

$X_x = \kappa$

$Y_x = F_x$

$D_x = \mu_x F_z$

$\mu_x = b_1 F_z + b_2$

$BCD_x = (b_3 F_z^2 + b_4 F_z)\exp(-b_5 F_z)$

$C_x = b_0$

$E_x = (b_6 F_z^2 + b_7 F_z + b_8)(1 - b_{13}\text{sgn}(\kappa + S_{hx}))$

$B_x = BCD_x/C_x D_x$

$S_{hx} = b_9 F_z + b_{10}$

$S_{vy} = b_{11}F_z + b_{12}$

Brake force only ($b_{11} = b_{12} = b_{13} = 0$)

Aligning Moment

$X_z = \alpha$

$Y_z = M_z$

$D_z = (c_1 F_z^2 + c_2 F_z)(1 - c_{18}\gamma^2)$

$BCD_z = (c_3 F_z^2 + c_4 F_z)(1 - c_6|\gamma|)\exp(-c_5 F_z)$

$C_z = c_0$

$E_z = (c_7 F_z^2 + c_8 F_z + c_9)(1 - (c_{19}\gamma + c_{20})^*$

$^*\text{sgn}(\alpha + S_{hz}))/(1 - c_{10}|\gamma|)$

$B_z = BCD_z/C_z D_z$

$S_{hz} = c_{11}F_z + c_{12} + c_{13}\gamma$

$S_{vz} = c_{14}F_z + c_{15} + (c_{16}F_z^2 + c_{17}F_z)\gamma$

Version 3, also known as the 'Pac94' version, identifiable by the addition of coefficients a_{15}–a_{17} to improve camber sensitivity.

More detailed texts on tyre modelling (Pacejka, 2012) have additional information about pushing and pulling the Magic Formula around to deliver longitudinal forces and aligning moments. The essential elegance of the Magic Formula is that the same basic expression can be used to calculate all the tyre responses of interest. Tyres operate in conditions other than pure slip and the Magic Formula implementation also allows for the calculation of weighting functions to overlay pure lateral slip functions with mitigation due to longitudinal slip and vice versa. In-depth discussion is beyond the scope of this text but they use another form of the Magic Formula to produce a shape that can reproduce a wide variety and character of weighting functions.

The latest version of the Magic Formula (6.1 at the time of writing) has over 200 parameters and many nested loops of one Magic Formula inside another with complex weighting functions. Even for experienced practitioners this can be more or less impossible to assimilate and one of the longstanding criticisms of the Magic Formula model is that it is very difficult to construct a hypothetical tyre with '+10% cornering stiffness' to see the influence on the vehicle. The more recent embodiments of the Magic Formula, as well as switching to the normalised formulation shown in Table 5.5, have added simple scaling factors such as $\lambda_{Ky\alpha}$ (referred to as LKY inside the property files) to allow just such manipulation. All fitting of the model to measured data is done with these scaling factors set at unity. This allows the intriguing possibility of taking high quality flat-track data, with elaborate fits for all the comprehensive slip character of the tyre and so on, and scaling it according to measured data on a vehicle or tyre test trailer (See Figure 5.73) using high quality load wheel instrumentation to reflect the same tyre on a variety of different surfaces. While still in its infancy, this process shows good promise even with the very noisy data sets that might be acquired on a rolling vehicle, as can be seen in Figure 5.73.

A fundamental limitation of most point followers including the Magic Formula concept is that they are limited to steady state calculations of the tyre forces. Additional layers of calculation can be added to the scheme to permit the representation of transient delays in force build-up. The most commonly described method is the so-called 'relaxation length'. When considering a tyre to which a step change in, say, slip angle is applied, a finite length must be travelled over the ground before the tyre carcass assumes its steady state shape. This length is broadly unchanged with forward speed, and so characterising it as a length and not a time delay is logical. However, it is frequently referred to as a single length as if it were an invariant property for the tyre; there is good evidence that it varies with both slip angle and vertical load. Typical values for relaxation length with passenger car tyres are of the order of 0.3 m at 2000 N, rising to 0.8 m at 7000 N. Practically, they are often measured with a 'plank' style test machine at quite low rolling velocities.

Many texts, Pacejka (2012) and Zegelaar (1998) for example, quote the idea that lateral relaxation length is a function of cornering stiffness divided by static lateral stiffness and that a similar treatment can be applied to longitudinal forces and slip states. In principle a first order differential equation — like charging a

Table 5.5 Pure Lateral Slip Equations for the Magic Formula Model in Later Versions

Load Dependency

Scaled nominal load: $F'_{z0} = \lambda F_{z0} F_{z0}$

Vertical load increment: $df_z = \dfrac{F_z - F'_{z0}}{F'_{z0}}$

Inflation Pressure Dependency

Pressure Increment: $dp_i = \dfrac{P - P_0}{P_0}$

Pure Slip: MF-Tyre

$F_{yp} = D_y \sin\left[C_y \arctan\left\{ B_y \alpha_y - E_y \left(B_y \alpha_y - \arctan\left(B_y a_y\right)\right)\right\}\right] + S_{vy}$

$\alpha_y = \alpha_F + S_{HY}$

$C_y = P_{Cy1} \lambda_{Cy}$

$D_y = \mu_y\, F_z \zeta_2$

$\mu_y = \left(p_{Dy1} + p_{Dy2}\, df_z\right)\left(1 + p_{py3} dpi + p_{py4} dpi^2\right)\left(1 - p_{Dy3}\gamma^2\right)\lambda_{\mu y}$

$E_y = \left(p_{Ey1} + p_{Ey2}\, df_z\right)\left(1 + p_{Ey5}\,\gamma^2 - \left(p_{Ey3} + p_{Ey4}\gamma\right)\operatorname{sgn}\left(\alpha_y\right)\right)\lambda_{Ey}$

$K_{y\alpha} = p_{Ky1} \cdot F'_{z0}\left(1 + p_{py1}\, dpi\right)\cdot \sin\left[p_{Ky4}\arctan\left\{\dfrac{F_z}{\left(p_{Ky2} + p_{Ky5}\gamma^2\right)F'_{z0}\cdot\left(1 + p_{py2} dpi\right)}\right\}\right]\left(1 - p_{Ky3}|\gamma|\right)\lambda_{Ky\alpha}\cdot\zeta_3$

$K_{y\alpha\alpha} = p_{Ky1}\cdot F'_{z0}\left(1 + p_{py1}\, dpi\right)\cdot\sin\left[p_{Ky4}\arctan\left\{\dfrac{F_z}{p_{Ky2}\, F'_{z0}\cdot\left(1 + p_{py2} dpi\right)}\right\}\right]\lambda_{Ky\alpha}$

MF-Tyre 5.2 $K_{y\gamma0} = \left(p_{Hy3}\, K_{y\alpha\alpha} + F_z \left(p_{Vy3} + p_{Vy4}\, df_z\right)\right)\lambda_{Ky\gamma}$

MF-Tyre 6.0 $K_{y\gamma0} = \left(p_{Ky6} + p_{Ky7}\, df_z\right)F_z\, \lambda_{Ky\gamma}$

MF-Tyre 6.1 $K_{y\gamma\gamma} = \left(p_{Ky6} + p_{Ky7}\, df_z\right)F_z\, \lambda_{Ky\gamma}\left(1 + p_{py\gamma} dpi\right)$

$B_y = \dfrac{K_{y\alpha}}{C_y\, D_y}$

$S_{Hy0} = \left(p_{Hy1} + p_{Hy2} df_z\right)\lambda_{Hy}$

Continued

Table 5.5 Pure Lateral Slip Equations for the Magic Formula Model in Later Versions—cont'd

MF-Tyre 5.2 $\quad S_{Hy\gamma} = p_{Hy3} \cdot \gamma \cdot \lambda_{Ky\gamma}$

MF-Tyre 6.0, 6.1 $\quad S_{Hy\gamma} = \dfrac{K_{y\gamma\gamma} \gamma - S_{Vy\gamma}}{K_{y\alpha}} \zeta_0 + \zeta_4 - 1$

$S_{Hy} = S_{Hy0} + S_{Hy\gamma}$

$S_{Vy0} = F_z \left(p_{Vy1} + p_{Vy2} df_z \right) \lambda_{Vy} \cdot \lambda_{\mu y} \cdot \zeta_2$

$S_{Vy\gamma} = F_z \left(p_{Vy3} + p_{Vy4} df_z \right) \gamma \cdot \lambda_{Ky\gamma} \cdot \lambda_{\mu y} \cdot \zeta_2$

$S_{Vy} = S_{Vy0} + S_{Vy\gamma}$

FIGURE 5.73

Loughborough Tyre Test Trailer — CAD model and actual trailer.

(Courtesy of G. Mavros, Loughborough University.)

capacitor — can be used to add a delay to the output of the steady state calculations. It can be applied as either an output filter, limiting rates of change of developed forces, or an input filter, limiting the rate of change of slip state before passing it to the steady state calculations. The disadvantage of applying it as an input filter is that it produces no response to a change in vertical load at a constant slip angle, something that real tyres are known to do (Pacejka and Takahashi, 1992).

An issue with naively delaying the output force with a simple delay function is illustrated in Figure 5.74. A linear cornering stiffness model is calculated on the basis outlined previously, based on vertical force F_z and two coefficients, a_3 and a_4, according to the equation given in Table 5.3. A contact patch longitudinal and lateral velocity are presumed constant to give a slip angle $atan(V_x/V_y)$ and time varying F_z is applied to give a steady state prediction for lateral force, F_y. This prediction is then subject to a simple first order lag equation with τ set at 0.02 s, representing a relaxation length of τV_x. With V_x set to 20 m/s, this gives a relaxation length of around 0.4 m. The resulting lagged F_y is shown in Figure 5.74.

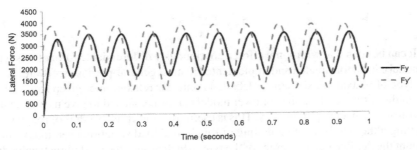

FIGURE 5.74

A naive first order lag applied to tyre force prediction.

It can be seen that while predicting the delayed build up of force in a plausible way, the lag also suggests that the cornering stiffness of tyre dynamically exceeds its steady state value during periods of decreasing load. A moment's consideration suggests that were the tyre operating in its saturated region that the model predicts a transient increase in friction coefficient, which appears unlikely. Measurements of real tyres do not reproduce this behaviour, see Zegelaar (1998).

There are multiple solutions to this but none of them are trivial. One is to detect the impending 'crossover' condition and reduce the value of τ in response to it, as shown in Figure 5.75.

The expressions used to produce Figure 5.75 are:

For $F'_y - F_y > 100N$

$$\tau = 0.02 \qquad (5.74)$$

For $10N < F'_y - F_y > 100N$

$$\tau = 0.0002(F'_y - F_y) \qquad (5.75)$$

For $F'_y - F_y < 10N$

$$\tau = 0.002 \qquad (5.76)$$

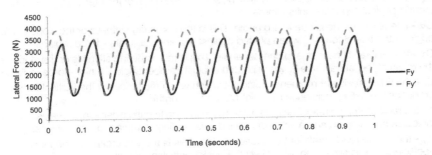

FIGURE 5.75

An asymmetric lag more capable of predicting transient response to a change in load.

$$\dot{F}_y = \left(\frac{F'_y - F_y}{\tau}\right) \tag{5.77}$$

It can be seen that with some effort a continuously differentiable form of this type of model can be formulated and implemented in most general-purpose software. With this type of relaxation, sometimes labelled nonlinear relaxation or given some other title in the software, the point follower model can be persuaded to give useful results up to around wheel hop frequency. This is useful for conceptual studies of rough road handling although detailed modelling of rough road load variation need to take into account the flexible belt dynamics of the tyre, which pushes the modelling bandwidth quite high. An intermediate step is the so-called rigid belt or rigid ring model, which can be used for conceptual studies with high bandwidth systems such as ABS as long as road features are not so complex that multiple points of contact occur.

5.6.6 The Fiala tyre model

The Fiala tyre model (Fiala, 1954) is probably most well known to MSC ADAMS users as it is provided as a standard feature of the program. Although limited in capability this model has the advantage that it only requires 10 input parameters and that these are directly related to the physical properties of the tyre. The input parameters are shown in Table 5.6

The parameters R_1, R_2, k_z, ζ, are all used to formulate the vertical load in the tyre and are required for all tyre models that are used here, including the Pacejka and Interpolation models. As the Fiala model ignores the influence of camber angle, the coefficient, which defines lateral stiffness due to camber angle $C\gamma$, is not

Table 5.6 Fiala Tyre Model Input Parameters

R_1 – The unloaded tyre radius (units - length)

R_2 – The tyre carcass radius (units - length)

k_z – The tyre radial stiffness (units - force/length)

C_s – the longitudinal tyre stiffness. This is the slope at the origin of the braking force F_x when plotted against slip ratio (units - force)

$C\alpha$ – lateral tyre stiffness due to slip angle. This is the cornering stiffness or the slope at the origin of the lateral force F_y when plotted against slip angle α (units - force/radians)

$C\gamma$ – lateral tyre stiffness due to camber angle. This is the cornering stiffness or the slope at the origin of the lateral force F_y when plotted against camber angle γ (units - force/radians)

C_r – the rolling resistant moment coefficient which when multiplied by the vertical force F_z produces the rolling resistance moment M_y (units - length)

ζ – the radial damping ratio. The ratio of the tyre damping to critical damping. A value of zero indicates no damping and a value of one indicates critical damping (dimensionless)

μ_0 – the tyre to road coefficient of 'static' friction. This is the y intercept on the friction coefficient versus slip graph, effectively the peak coefficient of friction

μ_1 – the tyre to road coefficient of 'sliding' friction occurring at 100% slip with pure sliding

used. This means that the generation of longitudinal forces, lateral forces and aligning moments with the Fiala model is controlled using just five parameters (C_s, $C\alpha$, C_r, μ_0 and μ_1).

Tyre states are passed into the model. These are:

- longitudinal slip state (slip ratio) as a decimal fraction with -1 representing a locked wheel, 0 representing a free-rolling wheel and greater than zero representing driving
- lateral slip angle in radians
- presentation (inclination) angle in radians
- vertical deflection of the footprint imposed by the road in mm
- vertical footprint velocity
- a flag denoting spin direction

The effective friction coefficient μ is determined as a function of the comprehensive slip ratio $S_{L\alpha}$ as shown in Figure 5.61. The comprehensive slip ratio $S_{L\alpha}$ is taken to be the resultant of a longitudinal slip coefficient S_x and a lateral slip coefficient $S\alpha$.

$$S_{L\alpha} = \sqrt{S^2 + S_\alpha} \qquad (5.78)$$

The instantaneous value of the tyre to road friction coefficient μ can then be obtained by linear interpolation (Figure 5.76):

$$\mu = \mu_0 - S_{L\alpha} (\mu_0 - \mu_1) \qquad (5.79)$$

Despite the advantage of a simple parameter set the main limitations of the model include:

1. The model cannot in any sense represent combined cornering and braking or cornering and driving.
2. Lateral force and aligning moment resulting from camber angle are not modelled.

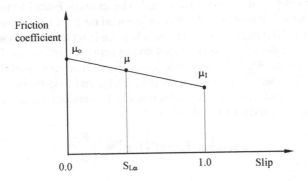

FIGURE 5.76

Linear tyre to road friction model.

3. The variation in cornering stiffness at zero slip angle with tyre load is not considered.
4. The offsets in lateral force or aligning moment at zero slip angle due to conicity and plysteer are not represented.

While points 1, 2 and 4 are acceptable simplifications, the sensitivity of tyres to vertical load plays a primary role in shaping whole vehicle behaviour, as explored in Chapter 6. Because it cannot represent this behaviour, the Fiala model has no practical use in ground vehicle modelling — it is, in fact, that most damaging of calculations, the plausible-but-wrong type.

5.6.7 The Harty tyre model

A desire to have a simple, computationally light tyre model that captured directionally the behaviour of real tyres led to the development of a TIRSUB routine for use with MSC ADAMS in early 1996. The model has progressed over time and has been migrated into a more modern TYR501 subroutine, still in FORTRAN. The source code is reproduced in Appendix B. The starting point for the model was the sample routine for coding the Fiala model supplied with the software and so useful aspects of the Fiala model were re-used.

As with the Fiala model, tyre states are presumed passed into the model:

- longitudinal slip state, S_x (slip ratio) as a decimal fraction with -1 representing a locked wheel, 0 representing a free-rolling wheel and greater than zero representing driving
- lateral slip angle in radians, α
- presentation (inclination) angle in radians, γ
- vertical deflection of the footprint imposed by the road in mm DZ
- vertical footprint velocity, VZ
- a flag denoting spin direction

Vertical load, F_z, is calculated using a linear stiffness and damping ratio and is not described explicitly as it is somewhat trivial. The existing Fiala friction coefficient treatment is retained. It is observed that, over the load range of interest in road vehicle handling, both the slip angle at which peak side force is produced and the slip ratio at which peak longitudinal force is produced do not actually vary much. Thus they are declared as constant. While clearly an approximation, it remains useful.

It is noted that measured tyre data for both lateral and longitudinal force in pure slip displays a characteristic convex form that can be generated using a single additional parameter, a curvature factor, A:

$$|Sx| \leq Scx : Fx = \mu F_z \left(1 - e^{\left(Ax \frac{Sx}{Scx}\right)}\right) \frac{|Sx|}{Sx} \tag{5.80}$$

$$|Sx| > Scx : Fx = \mu F_z \frac{|Sx|}{Sx} \tag{5.81}$$

For the lateral slip angle case, some ability to scale lateral and longitudinal forces independently is desired and so an additional scaling factor, B, is introduced. Load sensitivity is also present in the lateral formulation (but not the longitudinal formulation; this is arguably a simplification too far) by introducing a reference load R_Z and a scaling-factor-to-vertical force sensitivity term, labelled dB_dFz in the software code:

$$|\alpha| \le \alpha_{CY} : F_{y\alpha} = \left(B + \left(|F_z| - R_z \right) \frac{dB}{dF_z} \right) \mu F_z \left(1 - e^{\left(A_Y \frac{\alpha}{\alpha_{CY}} \right)} \right) \frac{|\alpha|}{\alpha} \qquad (5.82)$$

$$|\alpha| > \alpha_{CY} : F_{y\alpha} = \left(B + \left(|F_z| - R_z \right) \frac{dB}{dF_z} \right) \mu F_z \frac{|\alpha|}{\alpha} \qquad (5.83)$$

With inclination angle/presentation angle, the typical taut string idea that force is generated in the plane of the tyre is used with a simple camber coefficient, C_Y. However, a naïve formulation using only the tangent of the angle tends to produce unreasonably high forces as the tyre leans over more, and also does not respect the friction limit. Therefore a formulation is made that sets a fractional threshold, T, beyond which a clipping function is used:

$$|\gamma| \le \frac{\mu T}{C_Y} : F_{y\gamma} = -F_z \tan(C_Y\gamma) \qquad (5.84)$$

$$|\gamma| > \frac{\mu T}{C_Y} : F_{y\gamma} = \frac{|\gamma|}{\gamma} F_z \left[D\left(1 - e^{-E\gamma} \right) - T\mu C_Y \right] \qquad (5.85)$$

$$D = \frac{\left(\frac{1}{(1 - TC_Y)} \right)}{\left(\cos\left(\arctan\left(\frac{T_\mu}{C_Y} \right) \right) \right)^2} \qquad (5.86)$$

$$E = (T - 1)\mu C_Y \qquad (5.87)$$

This apparently complex form is entirely empirical and produces a smooth blend to the friction limit even at extended lean angles, as may be seen in Figure 5.77, where the upper figure was produced with a threshold of 0.7 and the lower figure with a threshold of 0.9.

Lateral forces due to slip and inclination angles are added algebraically to generate a total side force.

This side force and the longitudinal forces together are combined into a comprehensive slip force that is compared to the available friction in a straightforward Pythagorean way. The multiplier B, described earlier, is re-used to make the 'circle of

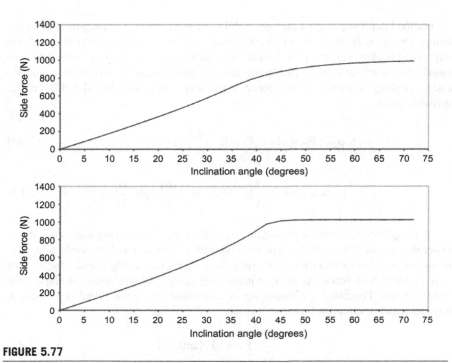

FIGURE 5.77

Harty tyre model showing camber clipping with different clip thresholds, T = 0.7 (upper), T = 0.9 (lower).

friction' into something elliptical; it can be seen that the ellipse can go either way depending on whether B is greater than or less than unity.

$$F_x^2 + \left(F_{y\alpha} + F_{y\gamma}\right)^2 > 1: \tag{5.88}$$

$$F_{x1} = \sqrt{\frac{\mu F_z^2}{1 + \dfrac{\left(F_{y\alpha} + F_{y\gamma}\right)^2}{B^2}}} \, \frac{F_x}{|F_x|} \tag{5.89}$$

$$F_{y1} = \sqrt{\left(1 - \frac{F_{x1}}{\mu F_z}\right)^2 (\mu F_z B)^2} \, \frac{F_y}{|F_y|} \tag{5.90}$$

Although appearing algebraically complex, this is a very simple formulation that simply truncates the combined force vector as it crosses the friction ellipse. The concept is illustrated in normalised form in Figure 5.78.

This formulation behaves well at low levels of slip but becomes confused when the slip levels become high in the tyre, such as the case of a locked wheel or with

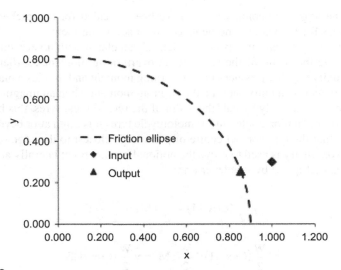

FIGURE 5.78

A normalised representation of the comprehensive slip formulation within the Harty tyre model at low levels of slip.

gross wheelspin. Under these circumstances a different approach is taken, which is to calculate the velocity vector of the contact patch with respect to the ground and to apply a simple Coulomb-style frictional force in the opposite direction. Note that there are none of the numerical difficulties associated with the Coulomb formulation around zero speed because this model is only applied when the contact-patch-to-ground speed is substantial:

$$F_{x2} = \sqrt{\frac{\mu F_z^{\ 2}}{1 + \dfrac{\tan^2 \alpha}{S_x^{\ 2}}}} \frac{F_x}{|F_x|} \qquad (5.91)$$

$$F_{y2} = \sqrt{\left(1 - \frac{F_{x2}}{\mu F_z}\right)^2 (\mu F_z)^2} \frac{F_y}{|F_y|} \qquad (5.92)$$

Note that when using this formulation, slip angle α is expressed in radians and longitudinal slip, S_x is a fraction and not percentage. These two slip quantities are simply blended between a longitudinal slip ratio of 50% and 100% in either direction, which is to say only in the gross wheelspin or near lock condition. The resulting force vectors appear directionally correct. However, it must be said that this final elaboration in the comprehensive slip calculation is really only necessary for simulations where this deep slip condition is of interest, such as perhaps in a video game or perhaps in some A:B comparison animation of, say, behaviour with locked wheels versus behaviour without; in general this deep slip model is unnecessary.

Once the aggregate contact forces have been rescaled for comprehensive slip, computations for the various moments can be made using them.

Overturning moment or capsize moment is calculated with a very simple function involving the width of the tyre and an overturning moment coefficient, C_{MX}; a value of unity for C_{MX} produces no overturning moment and mimics a blade wheel. Values of greater than unity saturate the aligning moment when an imagined contact point has moved laterally by half the width of the tyre. For car tyres, this happens at around 5° of inclination angle but for motorcycle tyres it is often at or beyond 45° of inclination that the tyre contact centre of pressure is closest to the edge of the tyre. Simple trigonometry is used to move the notional contact point laterally and the vertical load is multiplied by this lever arm:

$$|\gamma|\frac{W}{2}(C_{MX}-1) > \frac{W}{2} : M_x = \frac{W}{2}\frac{|\gamma|}{\gamma}F_z \tag{5.93}$$

$$|\gamma|\frac{W}{2}(C_{MX}-1) \le \frac{W}{2} : M_x = \gamma\frac{W}{2}(C_{MX}-1)F_z \tag{5.94}$$

Aligning moment is also comprised of two components, one from slip and one from camber in a similar manner to side force.

A notional contact patch length is calculated geometrically from the deflection passed in to the routine and a notional pneumatic trail used with the slip-angle-induced side force to generate an aligning moment. Note that the side force is that produced by the comprehensive slip truncation F_{y1} or F_{y2} depending on the operating regime. It is observed studying measured tyre data sets that the pneumatic trail decreases in a broadly linear way with increasing slip angle, and so the calculation of pneumatic trail is kept deliberately simple apart from the addition of an empirical, MF-style scaling factor:

$$F_{y\alpha}' = F_{y\alpha}\frac{F_y}{F_{yJ}}, J = 1,2 \tag{5.95}$$

$$L = 2\left(r_u^2 - (r_u - \Delta z)^2\right) \tag{5.96}$$

$$M_{z\alpha} = -F_{y\alpha}'L\left(1 - \frac{\alpha}{\alpha_{CY}}\right)\left(T_{PFZ2}F_z^2 + T_{PFZ}F_z + T_{PC}\right) \tag{5.97}$$

It will be noted that progressing beyond the critical slip angle produces increasingly reversed values of aligning torque, which is unlikely to be true in practice. In most vehicle dynamics simulations (and real driving conditions), slip angles of this magnitude constitute an error condition; in any case reliable tyre measurements in this region are difficult because of the high thermal loads on the tyre — again this formulation is a clear choice of 'simple and wrong' over 'complicated and wrong'.

For camber-induced moments, it is noted that they are in the opposite sense to steer torque moments. A 'pneumatic lead' — conceptually opposite of pneumatic trail — is introduced to calculate camber-induced moments, again scaled with an MF-style factor:

$$F_{y\gamma}' = F_{y\gamma}\frac{F_y}{F_{yJ}}, J = 1,2 \tag{5.98}$$

$$M_{z\gamma} = F_{y\gamma}'\left(L_{PFZ2}F_z^2 + L_{PFZ}F_z + L_{PC}\right) \tag{5.99}$$

Again it is noted that the camber forces are scaled according to the comprehensive slip truncation. Naive but functional calibrations of the model may be obtained by setting all scaling values to zero save for T_{PC} and L_{PC}, which can be set to unity and 25 mm, respectively.

Final aligning and capsize moment contributions come from applied longitudinal forces with the lateral shift of the contact patch centre of pressure under nonzero inclination angles. This is particularly important for motorcycles as it generates an uncommanded steer moment that will steer into the turn and thus reduce roll rate. Its importance is relatively minor for passenger cars although it can improve prediction of overall steering return torque.

$$O_P = 2\gamma\frac{W}{2}(C_{MX} - 1) \tag{5.100}$$

For $O_P \le \dfrac{W}{2}$

$$M_{x\gamma\kappa} = F_Z O_P \tag{5.101}$$

$$M_{z\gamma\kappa} = -F_X O_P \tag{5.102}$$

For $O_P > \dfrac{W}{2}$

$$M_{x\gamma\kappa} = F_Z\frac{W}{2}\frac{O_P}{|O_P|} \tag{5.103}$$

$$M_{z\gamma\kappa} = -F_X\frac{W}{2}\frac{O_P}{|O_P|} \tag{5.104}$$

The rolling resistance moment M_y is identical to the Fiala formulation and is given by:

$$M_y = -C_r F_z \quad \text{(forward motion)} \tag{5.105}$$

$$M_y = C_r F_z \quad \text{(backward motion)} \tag{5.106}$$

5.6.8 Tyre models for durability analysis

The modelling of suspension systems for durability analysis where component loads are needed for follow on stress and fatigue analysis was discussed in Chapter 4. As discussed, the models are developed to simulate tests carried out on the proving ground to represent impacts with road obstacles of the type shown on the left side of Figure 5.79. If an actual vehicle has been taken on the proving ground and instrumented to take readings at the wheel centre, as shown on the right side of Figure 5.79, then the recorded data can be used as input to a suspension or vehicle model and the need for a durability tyre model is negated.

As vehicle design processes move towards more use of virtual prototypes, the need to carry out full dynamic simulations of proving ground procedures requires more sophisticated tyre models to interact with the terrain. In the extreme simulations recreating the testing of off road vehicles involving conditions of the type shown in Figure 5.80 requires a tyre model that can deal with a wide range of terrain. The tyre models developed for such work are based on a physical representation of

Courtesy of Jaguar Cars   *Courtesy of Jaguar Cars*

FIGURE 5.79

Proving ground measurements at the wheel centre for durability analysis.

FIGURE 5.80

Off road testing conditions involving a wide range of terrain.

the tyre carcass geometry and material rather than a mathematical model of the measured force and moment behaviour used for vehicle handling tyre models.

A tyre model that has been developed at TNO on the foundation of the Magic Formula and subsequent Delft-Tyre models has been shown to handle a wide range of road inputs allowing handling simulations to be combined with highly nonlinear road inputs. The SWIFT model is described by the authors van Oosten and Jansen (1999) as a model intended for the development of active chassis control systems and optimising vehicle ride properties with capabilities including:

1. The use of the Magic Formula for slip force calculations
2. A sophisticated contact for short wavelength slip variations
3. An effective method to model road obstacles (durability)
4. A rigid ring model to accommodate tyre belt vibrations to 80 Hz
5. Tyre characteristics that can vary with speed and load

The model has been validated through the extensive tyre test capabilities at TNO and has been shown (van Oosten and Jansen, 1999) to be accurate for durability applications, such as rolling over cleats and enveloping steps in the road surface, when comparing simulations with experimental measurements.

A comprehensive description of this complex tyre model is not possible here. Rather the reader is referred to the companion text in this series (Pacejka, 2012) where a complete chapter is dedicated to describing the formulations within the SWIFT tyre model.

The MSC ADAMS durability tyre model (Vesimaki, 1997) was originally developed to deal with off road applications. An example of this would be the simulation of very large vehicles used by the timber industry in the forests of that author's home country, Finland. The tyre model developed for such an application would be required to deal with a vehicle cornering on a steep uneven slope where the tyres are going to encounter obstacles such as tree stumps. The requirements for such a tyre model are summarised by the author as:

1. to enable handling simulation on an uneven 3D road surface
2. to allow a road/terrain definition based on geometry
3. to accommodate varying friction over the terrain
4. to account for the cross-sectional tyre dimension and geometry

Such a model requires a physical representation of the tyre profile in order to model the boundaries of the tyre carcass as they envelop obstacles. The tyre model input consists of points that model one half of the tyre profile, as shown on the left side of Figure 5.81. The tyre model uses the input geometry to compute interpolated internal points, each of which defines the radius and lateral position of a disc representing a slice of the tyre cross-section.

As with any model based on a physical discretisation, the model refinement or number of cross-sectional elements must be such that the width of the 'slices' is sufficiently small to deal with obstacles that are narrow compared with the overall

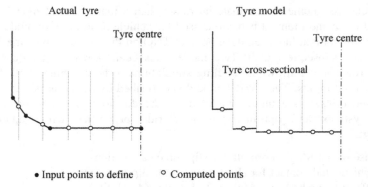

FIGURE 5.81

Discretisation of tyre profile for durability analysis.

width of the tyre. The road model is based on the finite element representation described in Section 5.6.3.

The algorithm developed carries out initial iterations to identify road elements that are subject to potential contact, at the current integration time step, before evaluating the position of each tyre element slice with each of the candidate road elements. An example of this is shown schematically in Figure 5.82 where one tyre cross-sectional element is seen to intersect a step defined by a number of triangular road elements.

For each of the discrete elements used to model the tyre cross-section, the interaction with the road surface elements produces a line projection of the intersection on the tyre element.

From this it is possible to compute the area and hence volume related to the penetration of tyre cross-sectional element by the road, for example by summing the three components shown in Figure 5.83. For a tyre with n cross-sectional elements,

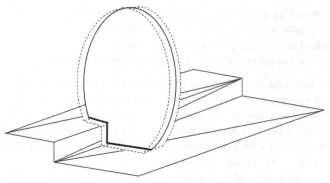

FIGURE 5.82

Intersection of durability tyre model element with road surface element.

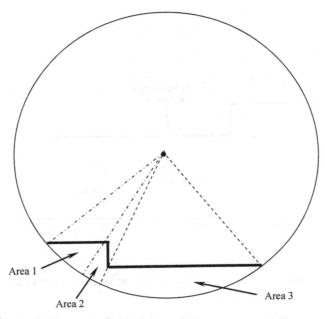

FIGURE 5.83

Penetration of tyre elemental slice by road surface elements.

where each element has m components of penetrated area, the effective penetrated volume, V_{eff}, for the complete tyre is given by

$$V_{eff} = \sum_{i=1}^{n} \sum_{j=1}^{m} A_j w_i \qquad (5.107)$$

where

A$_j$ is the penetrated area of the jth component of area within the ith cross-sectional tyre element

w_i is the width of the ith cross-sectional element of the tyre

The location of the effective contact point, within the deformed volume of tyre, is found by taking a weighted average of the contact point locations associated with each component of area within each cross-sectional element of the tyre. In Figure 5.84 it can be seen that for the component of area shown, the associated contact point is located where the road surface is intersected by a line normal to the road surface and passing through the centre of area. On this basis the x coordinate, X_{ecp}, for the effective contact point of the tyre would be found from

$$X_{ecp} = \sum_{i=1}^{n} \sum_{j=1}^{m} \frac{A_j w_i}{V_{eff}} X_j \qquad (5.108)$$

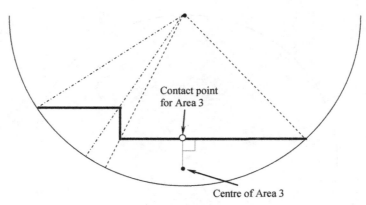

Contact point for Area 3

Centre of Area 3

FIGURE 5.84

Determination of effective contact point.

where

X$_j$ is the x coordinate of the contact point for the jth component of area within ith the cross-sectional tyre element

The y and z coordinates of the effective contact point are found using the same approach used to determine X_{ecp} in Eqn (5.108). Having located an overall effective contact point for the tyre at the given moment of interaction with the terrain it is necessary to determine an effective normal vector to the road surface acting through the contact point. Once again the author (Vesimaki, 1997) uses a weighted average approach. The road normal can therefore be defined by, using again the x component X_{ern} as an example

$$X_{em} = \sum_{i=1}^{n}\sum_{j=1}^{m} \frac{A_j w_i}{V_{eff}} Xn_j \qquad (5.109)$$

where

Xn$_j$ is the x component of the road normal for the jth component of area within ith the cross-sectional tyre element

Having found a volume, contact point and road normal vector for the deformed tyre at the given integration point in time it is necessary to compute a normal force acting on the tyre from the road. This involves an intermediate step where the effective volume of penetration, V_{eff}, is related to effective radial penetration of the tyre. This involves interpolation of a look-up table held within the tyre model that relates, for the defined tyre profile, the tyre penetration to penetrated volume when the tyre is compressed onto a flat surface.

The final computation required by the tyre model is to determine the effective coefficient of friction, μ_{eff}, due to contact with the terrain

$$\mu_{eff} = \sum_{i=1}^{n}\sum_{j=1}^{m} \frac{A_j w_i}{V_{eff}} \mu_j \qquad (5.110)$$

where

μ_j is the coefficient of friction associated with the jth component of area within ith the cross-sectional tyre element

For off road simulation a useful aspect of this approach is that the coefficient of friction can be factored to vary for each road element as described in Section 5.6.3.

As modelling requirements pass beyond handling loads and deeper into durability modelling where the enveloping behaviour of the tyre becomes important, another category of tyre models has emerged. These models take a 'semi-literal' approach to modelling the structure of the tyre.

Another tyre model, FTire, specifically developed for ride and durability simulations, has been developed by COSIN Software in Germany (Gipser, 1999) and made available through an interface in MSC ADAMS.

The model comprises a rigid rim surrounded by elements with elastic interconnections that form a surrounding flexible belt or ring and has been developed to deal with frequencies up to 120 Hz and to encompass obstacles in the longitudinal direction of rolling with wavelengths half the length of the tyre contact patch. In the transverse direction the model can handle inclination of the road surface and also obstacles that vary across the tyre lateral footprint, hence the model is referred to as '2 ½ - dimensional' nonlinear vibration model. The model can also accommodate the effects of stiffening and radial growth associated with high angular spin velocities. The model input parameters comprise tyre geometry and measured physical characteristics, with the optional input of natural frequencies and damping factors associated with the lower vibration modes of an unloaded tyre on a rigid rim. The belt or flexible ring is modelled as 50 to 100 lumped mass elements elastically interconnected and mounted to the rigid rim.

The elements have interconnecting stiffness to account for relative bending, extension, radial and tangential motion in the circumferential and lateral directions. The radial connection between elements on the belt and the rigid rim is a combined spring damper that allows the model to account for centrifugal dynamic stiffening at high angular spin velocities.

Each of the interconnected belt elements has up to 20 mass-less tread blocks each having nonlinear stiffness and damping in the radial, tangential and lateral directions, hence allowing the tread blocks to transmit normal forces from the road directly to the belt. Frictional forces in both the circumferential and lateral directions can be transmitted through the shear forces acting on the mass-less tread elements.

Tread blocks can be given individual heights, allowing the digitization of tread patterns without difficulty.

In this manner the pressure distribution in the contact patch can be discerned to quite high resolution. In conjunction with a good ground plane friction model, the resulting pressure and velocity distributions allow aggregate resultant forces and moments acting on the rigid rim to be found by integrating the forces acting throughout the elastic foundation of the belt.

The 'secret sauce' in the formulation of these types of model is the use of orthogonal shape functions to describe some aspects of the tyre carcass deformation. To explain the significance of this, it is helpful to imagine a cantilever beam subject to a tip load. One possible way of representing this is the classical expression, derived through the process of writing down an expression for shear force, and integrating it successively through bending moment, curvature, slope and finally deflection to calculate the response of the tip to a point load. For a cantilever that does not obviously deform when viewed with the naked eye, careful measurements of deflection will show this representation is usefully accurate. For a cantilever that deforms appreciably and obviously, this approximation becomes increasingly inaccurate due to geometric nonlinearity caused by the changing shape. It has long been noted in Finite Element analysis that the approximation can be improved by breaking the cantilever into multiple elements, each having an elastic stiffness. In general, it is noted that more elements increase the compliance in a convergent way until something resembling the 'real' answer is reached. This is excellent, but leads to a large increase in the number of computations required.

An alternative approach is to use a detailed (classical or finite element) model to calculate a deformed shape for the tip loading scenario and a so-called 'modal stiffness' — which takes into account the stiffness of each individual element and multiplies it by the square of the deformation associated with that mode of deformation *for each individual element*. This is visually represented with the shape function shown in Figure 5.85, where the total shape can be thought of as scaled summations

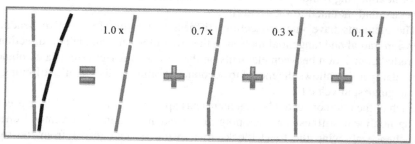

FIGURE 5.85

A visual representation of a fictional shape function for a cantilever modelled with rotational degrees of freedom — the shape function being [1.0 0.7 0.3 0.1].

of the individual deformations. Having calculated the shape functions externally, they are incorporated into the model that now has all the fidelity of the four element model but uses only a single tip deflection, a single so-called modal stiffness and the shape function. If four shape functions are used then the entire deformation is rendered with equal complexity in either the literal co-ordinate space or the modal co-ordinate space, but it is typically true that only a small subset of all possible shape functions are used, thus leading to a tremendous computational advantage while retaining the high fidelity compliance and deformation description of a significantly more complex model. The skill in formulating this model is all in the selection of shape functions to represent as economically as possible all the physical deformed shapes that are likely to be encountered.

Another important aspect of tyre behaviour, particularly in competition vehicles, is the variation of tyre behaviour with temperature. While the carcass stiffness is broadly provided by the metallic elements of the structure and thus is largely insensitive to the temperature variations seen in typical usage (-40 to $+100\,°C$), the frictional behaviour of the contact patch is tremendously influenced over this range. All rubber-like compounds display a somewhat complex behaviour with changes in temperature and a detailed discussion is beyond the scope of this text. However, a simplified understanding can be gained by imagining there is a temperature below which the compound will behave in a much less elastic manner. When subject to intense cold, most objects become extremely brittle but the 'glass transition' displayed in rubber is not a change to an entirely brittle state and is something of a misnomer. The transition temperature is surprisingly close to the lower limit of the operating range, being typically between -50 and $-100\,°C$ for passenger car tyres. Below this limit there is an abrupt change over about a $5\,°C$ temperature window in both stiffness and damping of the compound, with the modulus of elasticity increasing from a typical level around $10\,MPa$ up to around $1000\,MPa$ — approximately the same as polypropylene, from which many lunch-boxes are made.

While the so-called glass transition is never encountered in service, it is empirically observable that compounds with a lower glass transition temperature are 'more rubbery' at a typical operating temperature — which is to say they display more hysteresis, a key mechanism for producing frictional forces as described in Section 5.3.1. It is also generically true that a given compound loses modulus and gains hysteresis as temperature increases.

As temperature increases, the absolute shear strength of the compound reduces — the rubber progresses through a phase somewhat like adhesive putty (which is in fact a rubber compound a long way from its glass transition point at room temperature) before it melts. Thus for use within a tyre there is a point where the increasing hysteresis is of no value because although the surface is in contact with the ground, the shear strength is so low that the contact patch is simply left behind. This means there is an optimum temperature for a given compound to operate at in terms of frictional force, as illustrated in Figure 5.86.

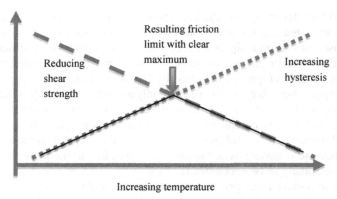

FIGURE 5.86

An illustration of temperature effects on tyre friction.

Since the tread rubber also experiences a 'cooking' (which is to say an irreversible chemical change) effect with temperature, this optimum temperature is unfortunately not constant and varies in use. This fact severely complicates the testing of tyres and also their use in motorsport.

Another rather trivial influence of temperature is to modify the inflation pressure. This can be readily described using Boyle's law, remembering to use absolute temperatures (Kelvin) and not Centigrade. Inflation pressure modifies the pressure distribution in the contact patch and the structural stiffness of the tyre.

It can be seen that for reasons of rubber property changes and for tyre pressure changes that the thermal modelling of tyres is also of interest.

There are two primary thermal inputs to a running tyre. The first is the hysteresis in the whole tyre assembly, which manifests itself globally as a rolling resistance. The rolling resistance torque, typically a few Nm, can be multiplied by the rotational speed to give an input power in Watts. This heat is broadly all applied in the carcase structure by the internal properties of the rubber. A tyre that has been run without excitement for some time at speed is pleasantly warm — between 40 and 50 °C — to a human when the inflation pressure is correct.

The second heat input mechanism is local sliding at the contact patch. A piece of contact patch generating a frictional force and having a certain sliding velocity is generating heat power that is the product of the in-plane force and the sliding velocity. Note that the sliding velocity is not generally the same as the velocity of the vehicle; this is only the case when the wheel is locked. This activity generates a rather large amount of power; for a single wheel carrying 400 kg sliding at 40 m/s on a surface with a friction coefficient of unity, the thermal power at the tyre/road interface is 160 kW, or over 50 English electric kettles. This heat is shared in some proportion between the tyre and the road surface itself.

In motorsport use such gross sliding is uncommon, since it rapidly consumes the tyres, but nevertheless the routine operation of motorsport tyres at 90—100 °C is due almost entirely to their increased use of sliding behaviour in the contact patch. (Some small amount of the temperature increase also comes from direct radiative transfer from brake system components, which operate at average temperatures far higher than road cars.)

Thermal models of tyres track the heat input to the system, its distribution within the tyre and its dissipation via a convective mechanism. The net heat is divided by the thermal mass to give a change in temperature that is integrated over time to give tyre temperature. While in principle point-follower style models allow for bulk modelling of tyre temperature, they typically do not have such calculations included.

Current practice such as the Michelin TAME (Thermal And MEchanical, pronounced to rhyme with *game* and not to rhyme with *gammy*) tyre model (Hague, 2010) tends to split the tyre into two or three chunks thermally — the contact patch tread block, the rest of the carcass and perhaps the air volume, with some kind of empirical heat transfer model between them. The contact patch block may or may not contain a detailed representation of the instantaneous power input on a location-by-location basis; the remainder of the carcass is usually a single thermal mass. Heat transfer coefficients are empirically determined by experiment.

It can be seen that heat transfer is substantially affected by vehicle speed and such effects are generally captured within the model. Models such as FTire can have a thermal layer within them, that can be used with an empirically determined ablation rate to further model tyre wear. Also possible, if data is available to support it, is to predict the change in performance by looking up different frictional properties of the tread compound at different temperatures.

There appear to the authors to be two schools of thought with respect to tyre temperatures. On the one hand, practitioners who regularly use a pyrometer — a sort of thermocouple on a stick — maintain that only the core temperature of the tyre is of interest, and anyone who measures anything else is in error. On the other hand, practitioners who regularly use tyre surface measurements (typically some kind of infrared, contactless, running technology) maintain that after-the-fact measurements of core temperature reveal nothing about which corners put in the most energy and whether or not suspension geometry — most notably camber — is in need of adjustment. As ever with highly polarised discussions, there are aspects of value in both measurements and there are indeed aspects of value in interrogating models for both aspects of the tyre thermal behaviour.

5.7 Implementation with MBS

General purpose MBS software intended for use in vehicle dynamics will often have specialised modules intended for tyre modelling. Implementation of the Magic Formula tyre model and the Interpolation Method can be achieved using commercial modules, or by writing bespoke subroutines and linking these to provide a customised library function, presuming the software allows such linking (as many do). Current versions of programs such as ADAMS/Car or the Simpack Vehicle Wizard make the incorporation of a tyre model appear seamless and can give the user a false sense of comfort with what is in fact an entirely empirical and quite complex piece of modelling. Example tyre model subroutines developed by the authors are provided in Appendix B, some of which form the basis of a tyre modelling, checking and plotting facility (Blundell, 2000a).

An example interface between MSC ADAMS and a tyre model is through a user-written TYR501[1] subroutine. The subroutine receives tyre states from a road/tyre interface module (usually bundled with the overall tyre modelling module but possible to implement as run time variables totally outside any additional modules) and then defines a set of three forces and three torques acting at the tyre-road surface contact patch and formulated in the TYDEX coordinate system. The equations used to formulate these forces and moments have been programmed into the subroutines to represent the various tyre models. Note that in the example provided in Appendix B the source code was migrated from an earlier TIRSUB that used the SAE coordinate frame; just before returning the values to the main solver, the final act is to transform the SAE forces into the TYDEX convention; this is done entirely for reasons of coding convenience and to maintain the availability of legacy support tools — spreadsheets and so on — external to the MBS software. The transformation of the forces and moments from the tyre contact patch to the wheel centre is performed internally by the program.

The TYR501 subroutine is called from within the model data set by a group of statements for each tyre on the vehicle. Appendix B contains a typical group of solver statements used to call the TYR501 model from within MSC ADAMS. It receives the tyre states listed below in the TYDEX ISO co-ordinate frame:

1. Longitudinal Slip Ratio
2. Lateral slip angle (in radians)
3. Camber angle (in radians)
4. Normal deflection of tyre into road surface
5. Normal velocity of penetration of tyre into road surface
6. Longitudinal sliding velocity of contact patch
7. Distance from wheel centre to contact point (loaded radius)
8. Angular velocity about the spin axis of the tyre
9. Longitudinal velocity of tyre tread base
10. Lateral velocity of tyre tread base

5.7.1 Virtual tyre rig model

Because of the complexity of the tyre model and the generic difficulty in assimilating tyre model behaviour as part of a full vehicle model, it is regarded as poor practice not to interrogate tyre models in isolation before they are used in a full

[1]The first edition of the book used a TIRSUB routine but this is no longer supported by the software vendors; TIRSUB examples are provided to illustrate the calculations only and are not usable as is.

vehicle environment. A wide range of failures in data transmission is possible, including but not limited to the use of decaNewtons, kilograms, Newtons or kilo-Newtons for forces in the measured data set, the use of radians or degrees in the measured data set, the use of ISO, SAE, modified ISO or any of the six imaginable co-ordinate systems. There is no substitute for an interrogation of the tyre model and a comparison with common sense — if it has a declared friction coefficient around unity, then frictional forces and vertical load should broadly correspond; a road-use tyre is unlikely to develop peak lateral forces much outside the slip angle range of 5 to 20°; and so on.

A functional model of a generic flat bed tyre test machine has been developed in MSC.ADAMS and forms part of the exemplar process described here. It is clearly desirable to use the tyre data parameters or coefficients to generate the sort of plots produced from a tyre test programme and to inspect these plots before using the data files with an actual full vehicle model. Matlab routines for directly plotting Magic Formula tyre model forces are available, as are built in test rig routines for both ADAMS/Car and the Simpack Vehicle Wizard.

The tyre rig model is also useful where test data has been used to extract mathematical model parameters. The plots obtained from the mathematical model can be compared with test data to ensure the mathematical parameters are usefully accurate, represent the actual tyre and are read by modelling software in the same manner as the parameter-fitting software (these are often quite separate software tools). The process that this involves is shown conceptually in Figure 5.87. Note that this system has also been used to manage the data and model used to develop a low parameter tyre model for use in aircraft ground dynamics (Wood et al., 2012).

Also possible is the use of a dedicated external piece of software for handling, such as Optimum Tire from Optimum G, that can interrogate and render a tire data file with ease as well as handling a large amount of the fitting process described in Section 5.6.5.

Attention must be given to the orientation as well as the location of tyre attachment frames within the model, particularly when suspension adjustments such as static toe and static camber are intended to be made parametrically. Not all tyre models run symmetrically in both forward and backward directions and so it is often good practice to ensure that all tyres are rotating the same way, as shown in Figure 5.88. One possible approach with full vehicle modelling is to set up a global coordinate system or GRF where the x-axis points back along the vehicle, the y-axis points to the right of the vehicle and the z-axis is up. The local z-axis of each tyre part is orientated to point towards the left side of the vehicle so that the wheel spin vector is positive when the vehicle moves forward during normal motion. Note that this is the coordinate system as set up at the wheel centre and should not be confused with the SAE coordinate system that is used at the tyre contact patch in order to describe the forces and moments occurring there. There are many traps for the inattentive; it is clear that the tyre test rig should emulate the vehicle model attachments to be useful.

The model of the tyre test machine presented here contains a tyre part that rolls forward on a flat uniform road surface in the same way that the tyre interacts with a

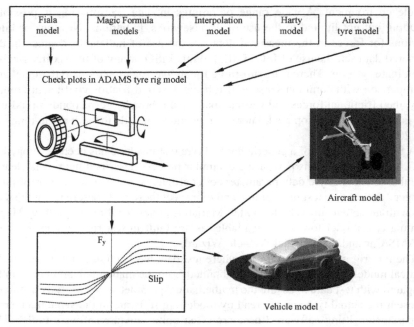

FIGURE 5.87

Overview of the tyre modelling system.

FIGURE 5.88

Orientation of tyre coordinate systems on the full vehicle model.

moving belt in the actual machine. In this model the road is considered fixed as opposed to the machine where the belt represents a moving road surface and the tyre is stationary; modelling a moving belt is surprisingly awkward in a MBS environment. Considering the system schematic of the model shown in Figure 5.89, the tyre part 02 is connected to a carrier part 03 by a revolute joint aligned with the spin axis of the wheel. The carrier part 03 is connected to another carrier part 04 by a

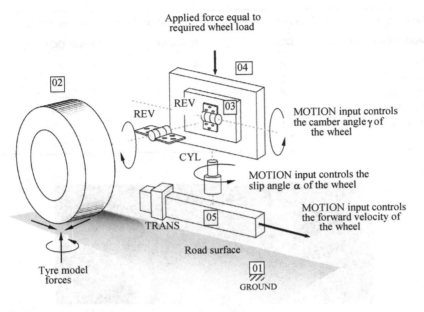

Applied force equal to
required wheel load

MOTION input controls
the camber angle γ of
the wheel

MOTION input controls the
slip angle α of the wheel

MOTION input controls
the forward velocity of
the wheel

Road surface

Tyre model
forces

FIGURE 5.89

Mechanism sketch for a flat bed tyre test machine model.

revolute joint that is aligned with the direction of travel of the vehicle. A motion input applied at this joint is used to set the required camber angle during the simulation of the test process. The carrier part 04 is connected to a sliding carrier part 05 by a cylindrical joint that is aligned in a vertical direction. A rotational motion is applied at this joint that will set the slip angle of the tyre during the tyre test simulation. The cylindrical joint allows the carrier part 04 to slide up or down relative to part 05, which is important as a vertical force is applied downwards on the carrier part 04 at this joint and effectively forces the tyre down on to the surface of the road. The model has been set up to ignore gravitational forces so that this load can be varied and set equal to the required wheel vertical load that would be set during the tyre test process. The sliding carrier part 05 is connected to the ground part 01 by a translational joint aligned with the direction of travel of the wheel. A motion input applied at this joint will control the forward velocity of the tyre during the test.

The joint controlling camber angle can be located at the tyre contact patch rather than at the wheel centre. This will avoid introducing lateral velocity and hence slip angle for the change in camber angle during a dynamic simulation.

The model of the tyre test machine has two rigid body degrees of freedom as demonstrated by the calculation of the DOF balance in Table 5.7. One DOF is associated with the spin motion of the tyre, that is dependent on the longitudinal forces generated and the slip ratio. The other DOF is the height of the wheel centre above the road, that is controlled by the applied force representing the wheel load.

Table 5.7 Degree of Freedom (DOF) Balance Equation for the Tyre Rig Model

Model Component	DOF	Number	Total DOF
Parts	6	4	24
Revolutes	−5	2	−10
Translational	−5	1	−5
Cylindrical	−4	1	−4
Motions	−1	3	−3
			$\Sigma_{DOF} = 2$

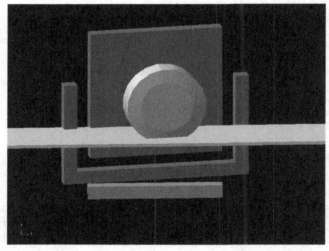

FIGURE 5.90

Computer graphics for the tyre rig model.

The tyre test rig model has been used to read the tyre model data files used in a study (Blundell, 2000a) to plot tyre force and moment graphs. The graphics of the tyre rig model are shown in Figure 5.90.

5.8 Examples of tyre model data

The results obtained from a series of tyre tests (Blundell, 2000a) have been used to set up the data needed for the various modelling approaches described here. In summary the following procedure was followed:

1. For the Interpolation method the measured numerical values were reformatted directly into the SPLINE statements within an MSC.ADAMS data file as shown in Table 5.8. For each spline shown in Table 5.8 the X values correspond to

Table 5.8 Spline Data for Interpolation Model

Lateral Force (n) with Slip Angle (degree) and Load (kg)

SPLINE/100
,X = -10,-8,-6,-4,-2,0,2,4,6,8,10
,Y = 200,2148,2050,1806,1427,867,16,-912,-1508,-1881,-2067,-2151
,Y = 400,3967,3760,3409,2727,1620,75,-1587,-2776,-3482,-3759,-3918
,Y = 600,5447,5099,4436,3385,1962,94,-1893,-3397,-4557,-5049,-5269
,Y = 800,6738,5969,4859,3533,2030,66,-1971,-3662,-5122,-6041,-6500

Aligning Moment (nm) with Slip Angle (degree) and Load (kg)

SPLINE/200
,X = -10,-8,-6,-4,-2,0,2,4,6,8,10
,Y = 200,4.6,-0.1,-6,-11.1,-10.9,-1.3,10.6,11.2,7.9,3.2,-0.3
,Y = 400,-4.8,-19.6,-39,-52.1,-41.9,-6.7,35.8,49.1,38.6,23.4,10.1
,Y = 600,-36.5,-73.1,-102.6,-107.9,-78.7,-14.2,60.6,96.2,93.4,65.8,40.7
,Y = 800,-105.1,-181.1,-206.1,-172.4,-116.0,-23.6,79.9,143.3,172.2,141.5, 98.5

Lateral Force (N) with Camber Angle (degree) and Load (kg)

SPLINE/300
,X = -10,-8,-6,-4,-2,0,2,4,6,8,10
,Y = 100,-123.3,-96.3,-64.6,-39.3,-3,19,46,80.6,108.3,146,173.3
,Y = 200,-142.6,-106.6,-57.3,-14.6,28,78,127,169.6,212.3,255,285.6
,Y = 300,-173.6,-106.6,-44,20.6,87.6,159,223.6,291.3,344.3,393.3,443.6
,Y = 400,-194,-115.6,-31.3,53,141.6,237,319.6,396.3,468.6,526.3,579
,Y = 500,-219.6,-121.6,-17.3,91,199,304,403.3,487,572.6,651.3,717
,Y = 600,-247.6,-128.3,-9.3109.3,234,351,453.3,557.3,651.6,734.6,829.6
,Y = 700,-278,-138.6,-3.6,126.3,254,381,499.3,616,723,827,922.6
,Y = 800,-318.6,-165,-21,128,261.3,404.0,524.3,656,780,895,1012

Aligning Moment (NM) with Camber Angle (degree) and Load (kg)

SPLINE/400
,X = -10,-8,-6,-4,-2,0,2,4,6,8,10
,Y = 100,-5,-5,-4.3,-2.2,-0.9,1.2,2.6,4.2,5.8,7,6.4
,Y = 200,-14.6,-13.7,-12,-9.2,-4.9,-0.9,3.6,6.7,9.6,11,11.7
,Y = 300,-24.1,-22.6,-19.6,-16.7,-11.1,-4.2,2.8,8.1,11.9,15.2,17
,Y = 400,-34.2,-31.8,-28.5,-22.9,-15.8,-8.2,-0.3,6.5,12.2,15.6,17.7
,Y = 500,-41.5,-38,-32.7,-26.5,-18.8,-10.8,-2.5,3.9,10.7,16.5,19.6
,Y = 600,-48.7,-43.6,-38,-31.6,-23.9,-15.9,-8.1,-0.4,6.4,12.1,16.8
,Y = 700,-52.5,-47.5,-40.9,-34.4,-26.6,-19.5,-11.9,-4.7,1.3,7.2,12.6
,Y = 800,-56.9,-51.3,-44.2,-37.9,-30.7,-23.9,-16.7,-10.1,-4,2.4,8.3

either the slip or camber angle and are measured in degrees. The first value in each Y array corresponds to the vertical load measured in kg. The following values in the Y arrays are the measured lateral forces (N) or the aligning moments (Nm) that correspond with the matching slip or camber angles in the X arrays. All the required conversions to the vehicle model units are carried out in the FORTRAN subroutine for the interpolation tyre model listed in Appendix B.

2. The parameters for the Fiala model were obtained by simple measurements from the plots produced during tyre testing. The Fiala model requires a single value of cornering stiffness to be defined although in reality cornering stiffness varies with tyre load. For the purposes of comparing the tyre models the parameters for the Fiala tyre model shown in Table 5.9 have been derived from the test data at the average of the front and rear wheel loads of the vehicle considered in this study. Fiala parameters obtained at front and rear wheel loads are given in Tables 5.10 and 5.11. Using the data for each of these models the tyre rig model described in the previous section was run for vertical loads of 200, 400, 600 and 800 kg. In each case the slip angle was varied between plus and minus $10°$.

3. The coefficients for the Magic Formula model were provided by Dunlop Tyres using in-house software to fit the values. The Magic formula tyre model (Version 3) parameters are shown in Table 5.12. It should be noted that the parameters due to camber effects were not available from this set of tests.

Table 5.9 Fiala Tyre Model Parameters (Average Wheel Load)

$R_1 = 318.5$ mm	$R_2 = 97.5$ mm
$k_z = 150$ N/mm	$C_s = 110000$ N
$C\alpha = 51560$ N/rad	$C\gamma = 2580$ N/rad
$C_r = 0.0$ mm	$\xi = 0.05$
$M_0 = 1.05$	$M_1 = 1.05$

Table 5.10 Fiala Tyre Model Parameters (Front Wheel Load)

$R_1 = 318.5$ mm	$R_2 = 97.5$ mm
$k_z = 150$ N/mm	$C_s = 110000$ N
$C\alpha = 54430$ N/rad	$C\gamma = 2750$ N/rad
$C_r = 0.0$ mm	$\xi = 0.05$
$M_0 = 1.05$	$M_1 = 1.05$

Table 5.11 Fiala Tyre Model Parameters (Rear Wheel Load)

$R_1 = 318.5$ mm	$R_2 = 97.5$ mm
$k_z = 150$ N/mm	$C_s = 110000$ N
$C\alpha = 46980$ N/rad	$C\gamma = 2350$ N/rad
$C_r = 0.0$ mm	$\xi = 0.05$
$M_0 = 1.05$	$M_1 = 1.05$

Table 5.12 Magic Formula Tyre Model (Version 3) Parameters

Lateral Force	Aligning Moment
A0 = 0.103370E+01	C0 = 0.235000E+01
A1 = −0.224482E-05	C1 = 0.266333E-05
A2 = 0.132185E+01	C2 = 0.249270E-02
A3 = 0.604035E+05	C3 = −0.159794E-03
A4 = 0.877727E+04	C4 = −0.254777E-01
A5 = 0.0	C5 = 0.142145E-03
A6 = 0.458114E-04	C6 = 0.00
A7 = 0.468222	C7 = 0.197277E-07
A8 = 0.381896E-06	C8 = −0.359537E-03
A9 = 0.516209E-02	C9 = 0.630223
A10 = 0.00	C10 = 0.00
A11 = −0.366375E-01	C11 = 0.120220E-06
A12 = −0.568859E+02	C12 = 0.275062E-02
A13 = 0.00	C13 = 0.00
A14 = 0.00	C14 = −0.172742E-02
A15 = 0.00	C15 = 0.544249E+01
A16 = 0.00	C16 = 0.00
A17 = 0.379913	C17 = 0.00
	C18 = 0.00
	C19 = 0.00
	C20 = 0.00

5.9 Case study 6 — comparison of vehicle handling tyre models

Using the data derived from the tyre tests and the MSC.ADAMS tyre test rig the following plots have been produced here by way of example.

1. Lateral force for a range of plus and minus 10° of slip angle.
2. Lateral force for a range of plus and minus 2° of slip angle (near zero).

For each plot a set of three curves is presented showing the variation in force or moment for each of the three tyre models. The comparisons are made separately at vertical loads of 200, 400, 600 and 800 kg. Using the Interpolation model as a benchmark the results plotted for the Magic Formula and Fiala models are compared in Figures 5.91 to 5.98.

The plots show that the more detailed Magic Formula will produce a better match than the simpler Fiala model when examining the variation in lateral force with slip angle. Comparing the Magic Formula with the Interpolation data there is little to distinguish between the two sets of curves. The plots also indicate that the Magic Formula can accurately represent offsets in lateral force at zero slip angle due to ply-steer and conicity. Looking at the plots for the Fiala model it can be seen that the

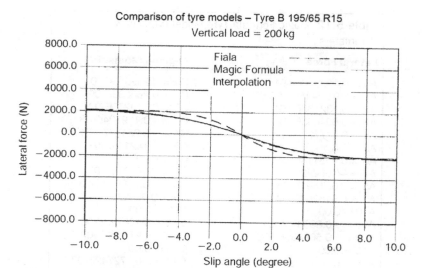

FIGURE 5.91

Comparison of tyre models — lateral force with slip angle (200 kg load).

(This material has been reproduced from the Proceedings of the Institution of Mechanical Engineers, K1 Vol. 214 'The modelling and simulation of vehicle handling. Part3: tyre modelling'. M.V.Blundell, page 19, by permission of the Council of the Institution of Mechanical Engineers.)

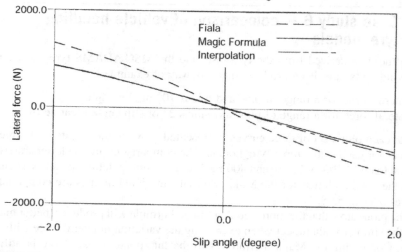

FIGURE 5.92

Comparison of tyre models — lateral force at near zero slip angle (200 kg load).

(This material has been reproduced from the Proceedings of the Institution of Mechanical Engineers, K1 Vol. 214 'The modelling and simulation of vehicle handling. Part3: tyre modelling'. M.V.Blundell, page 19, by permission of the Council of the Institution of Mechanical Engineers.)

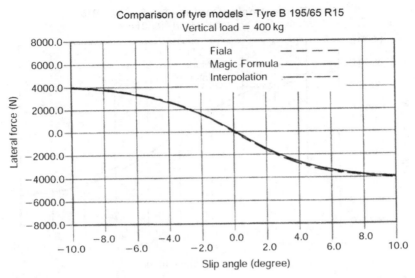

FIGURE 5.93

Comparison of tyre models — lateral force with slip angle (400 kg load).

(This material has been reproduced from the Proceedings of the Institution of Mechanical Engineers, K1 Vol. 214 'The modelling and simulation of vehicle handling. Part3: tyre modelling'. M.V.Blundell, page 20, by permission of the Council of the Institution of Mechanical Engineers.)

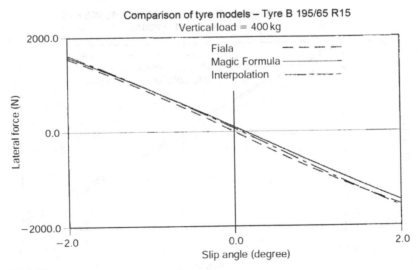

FIGURE 5.94

Comparison of tyre models — lateral force at near zero slip angle (400 kg load).

(This material has been reproduced from the Proceedings of the Institution of Mechanical Engineers, K1 Vol. 214 'The modelling and simulation of vehicle handling. Part3: tyre modelling'. M.V.Blundell, page 20, by permission of the Council of the Institution of Mechanical Engineers.)

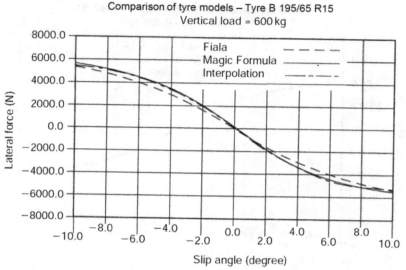

FIGURE 5.95

Comparison of tyre models — lateral force with slip angle (600 kg load).

(This material has been reproduced from the Proceedings of the Institution of Mechanical Engineers, K1 Vol. 214 'The modelling and simulation of vehicle handling. Part3: tyre modelling'. M.V.Blundell, page 21, by permission of the Council of the Institution of Mechanical Engineers.)

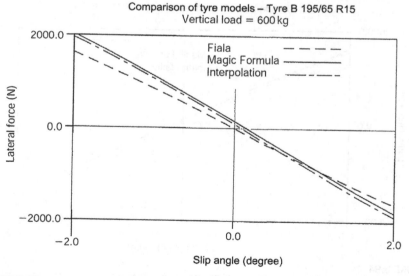

FIGURE 5.96

Comparison of tyre models — lateral force at near zero slip angle (600 kg load).

(This material has been reproduced from the Proceedings of the Institution of Mechanical Engineers, K1 Vol. 214 'The modelling and simulation of vehicle handling. Part3: tyre modelling'. M.V.Blundell, page 19, by permission of the Council of the Institution of Mechanical Engineers.)

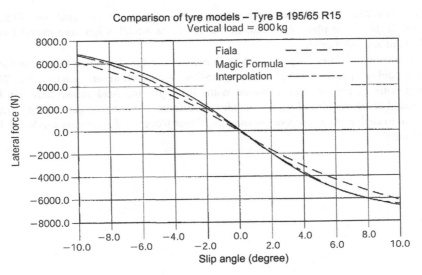

FIGURE 5.97

Comparison of tyre models — lateral force with slip angle (800 kg load).

(This material has been reproduced from the Proceedings of the Institution of Mechanical Engineers, K1 Vol. 214 'The modelling and simulation of vehicle handling. Part3: tyre modelling'. M.V.Blundell, page 22, by permission of the Council of the Institution of Mechanical Engineers.)

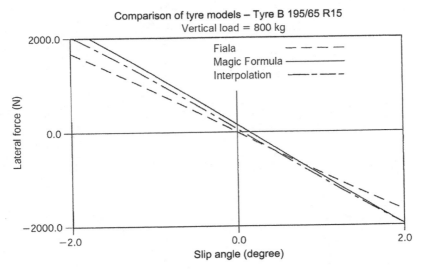

FIGURE 5.98

Comparison of tyre models — lateral force at near zero slip angle (800 kg load).

(This material has been reproduced from the Proceedings of the Institution of Mechanical Engineers, K1 Vol. 214 'The modelling and simulation of vehicle handling. Part3: tyre modelling'. M.V.Blundell, page 22, by permission of the Council of the Institution of Mechanical Engineers.)

model underestimates lateral force where higher slip angles coincide with higher wheel loads. These plots also confirm that the Fiala model is ignoring lateral force offsets at zero slip angle.

The case study here is presented as an example of good practice to interrogate a tyre model and parameters before incorporation in a vehicle simulation exercise. The comparison of lateral force variation with slip angle and load shown here can be extended using the virtual rig to consider the effects of camber angle, braking, traction and combinations thereof on the significant forces and moments predicted by the model.

Modelling and Assembly of the Full Vehicle

6

Everything should be as simple as it can be, says Einstein, but not simpler.
Louis Zukofsky

6.1 Introduction

In this chapter we will address the main systems that must be modelled and assembled to create and simulate the dynamics of the full vehicle system. The term 'full vehicle system' needs to be understood within the context of this textbook. The use of powerful modern multibody systems (MBS) software allows the modelling and simulation of a range of vehicle subsystems representing the chassis, engine, driveline and body areas of the vehicle. This is illustrated in Figure 6.1 where it can be seen that MBS models for each of these areas are integrated together to provide a detailed 'literal' representation of the full vehicle. Note that Figure 6.1 includes the modelling of the driver and road as elements of what is considered to constitute a full vehicle system model.

In this chapter we restrict our discussion of 'full vehicle system' modelling to a level appropriate for the simulation of the vehicle dynamics. As such the modelling of the suspension systems, anti-roll bars, steering system, steering inputs, brake system and drive inputs to the road wheels will all be covered. With regard to steering, the modelling of the driver inputs will also be described with a range of driver models.

Note at this stage we do not consider the active elements of vehicle control. Chapter 8 is dedicated to the modelling of active systems.

For the vehicle dynamics task a starting point involving models of less elaborate construction than that suggested in Figure 6.1 will provide useful insights much earlier in the design process. Provided such models correctly distribute load to each tyre and involve a usefully accurate tyre model, such as the Magic Formula described in Chapter 5, good predictions of the vehicle response for typical proving ground manoeuvres can be obtained.

The modelling of the suspension system was considered in detail in Chapter 4. The treatment that follows in this chapter will discuss a range of options that address the representation of the suspension in the full vehicle as either an assembly of linkages or using simpler 'conceptual' models. It is necessary here to start with the discussion of suspension representation in the full vehicle to set the scene for following sections dealing with the modelling of springs in simple suspension models or the derivation of roll stiffness. A case study provided at the end of this chapter will

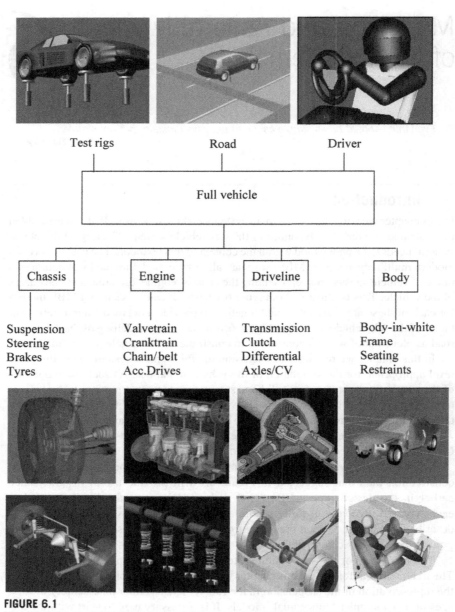

FIGURE 6.1

Integration of subsystems in a full vehicle model.

Provided courtesy of MSC Software.

compare the simulated outputs for a simulated vehicle manoeuvre using a range of suspension modelling strategies that are described in the following section.

6.2 The vehicle body

For the vehicle dynamics task the mass, centre of mass position and mass moments of inertia of the vehicle body require definition within the multibody data set describing the full vehicle. It is important to note that the body mass data may include not only the structural mass of the body-in-white but also the mass of the engine, exhaust system, fuel tank, vehicle interior, driver, passengers and any other payload. A modern CAD system, or the preprocessing capability for example in ADAMS/View, can combine all these components to provide the analyst with a single lumped mass notwithstanding the cautions raised in Section 3.2.4.

Figure 6.2 shows a detailed representation of a full vehicle model. In a model such as this there are a number of methods that might be used to represent the individual components. Using a model that most closely resembles the actual vehicle, components such as the engine might for example be elastically mounted on the vehicle body using bush elements to represent the engine mounts.

The penalty for this approach will be the addition of 6 degrees of freedom (DOF) for each mass treated in this way. Alternatively a fix joint may be used to rigidly attach the mass to the vehicle body. Although this would not add DOF, the model would be less efficient through the introduction of additional equations representing the extra body and the fix joint constraint. The use of fix joint constraints may also

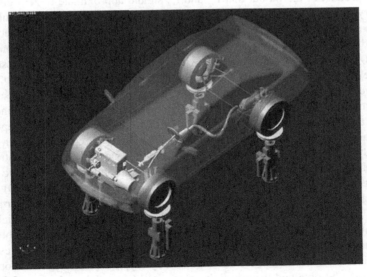

FIGURE 6.2

A detailed multi-body systems vehicle model.

Provided courtesy of MSC Software.

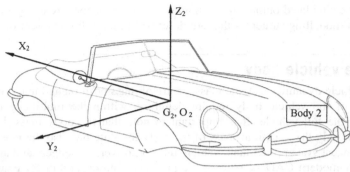

FIGURE 6.3

Vehicle body reference frame.

introduce high reaction moments that would not exist in the model when using elastic mounts distributed about the mass.

An example of a vehicle body referenced frame O_2 located at the mass centre G_2 for Body 2 is shown in Figure 6.3. For this model the XZ plane is located on the centre line of the vehicle with gravity acting parallel to the negative Z_2 direction. Using an approach where the body is a single lumped mass representing the summation of the major components, the mass centre position can be found by taking first moments of mass, and the mass moments of inertia can be obtained using the methods described in Chapter 2. From inspection of Figure 6.3 it can be seen that a value would exist for the I_{xz} cross product of inertia but that I_{xy} and I_{yz} should approximate to zero given the symmetry of the vehicle. In reality there may be some asymmetry that results in a CAD system outputting small values for the I_{xy} and I_{yz} cross products of inertia.

The dynamics of the actual vehicle are greatly influenced by the yaw moment of inertia, I_{zz}, of the complete vehicle, to which the body and associated masses will make the dominant contribution. A parameter often discussed is the ratio k^2/ab, sometimes referred to as the 'dynamic index', where k is the radius of gyration associated with I_{zz} and a and b locate the vehicle mass centre longitudinally relative to the front and rear axles respectively. The significance of this is discussed later in Chapter 7.

The assumption so far has been that the vehicle body is represented as a single rigid body but it is possible to model the torsional stiffness of the vehicle structure if it is felt that this could influence the full vehicle simulations. A simplistic representation of the torsional stiffness of the body may be used (Blundell, 1990) where the vehicle body is modelled as two rigid masses, front and rear half body parts, connected by a revolute joint aligned along the longitudinal axis of the vehicle and located at the mass centre. The relative rotation of the two body masses about the axis of the revolute joint is resisted by a torsional spring with a stiffness corresponding to the torsional stiffness of the vehicle body. Typically, the value of torsional stiffness may be obtained using a finite element (FE) model of the type shown in Figure 6.4. For efficiency, symmetry has been exploited here to model only one half of the vehicle body. This requires the use of anti-symmetry constraints along

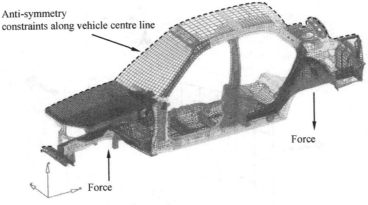

Anti-symmetry
constraints along vehicle centre line

Force

Force

FIGURE 6.4

Finite element model of body-in-white.

the centre line of the FE model, for all nodes on the plane of geometric symmetry, to carry out the asymmetric torsion case.

It is also possible to incorporate a FE representation of the vehicle body within the MBS full vehicle model. An example of this was shown in Figure 4.3 where the flexibility of a racing kart frame was included in the model. Despite the capability of modern engineering software to include this level of detail it will be seen from the case study at the end of this chapter that a single lumped mass is an efficient and accurate representation of a relatively stiff modern vehicle body for the simulation of a vehicle handling manoeuvre for most vehicles, with commercial vehicles and motorcycles being conspicuous exceptions.

6.3 Measured outputs

Before continuing in this chapter to describe the subsystems that describe the full vehicle we need to consider the typical outputs measured on the proving ground and predicted by simulation. The use of instrumented vehicles to investigate handling performance can be traced back to the work of Segal in the early 1950s, which as mentioned in Chapter 1 was the subject of one of the seminal 'IME Papers' (Segel, 1956). Testing was carried out using a 1953 Buick Super, four-door sedan, to investigate steady state behaviour with a fixed steering input at various speeds and also transient response to sudden pulse inputs at the steering wheel. The instrumentation used at that time allowed the measurement of the following:

1. Left front wheel steer
2. Right front wheel steer
3. Steering wheel rotation
4. Lateral acceleration
5. Roll angle

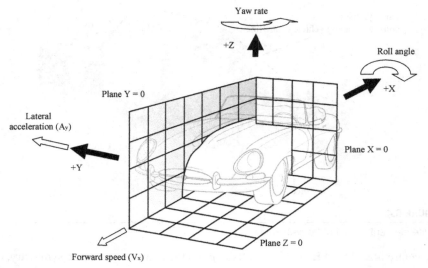

FIGURE 6.5

Typical responses measured in vehicle coordinate frame.

6. Pitch angle
7. Yaw rate
8. Roll rate
9. Forward velocity

Some of these responses are shown in Figure 6.5. The trajectory (path) of the vehicle can also be recorded. With simulation this is straightforward but in the past has been difficult to measure on the test track; testers resorting to measuring a trail of dye left by the vehicle on the test track surface. Modern instrumentation has improved on this somewhat.

Another measure often determined during test or simulation is the body slip angle, β. This being the angle of the vehicle velocity vector measured from a longitudinal axis through the vehicle as shown in Figure 6.6. The components of velocity

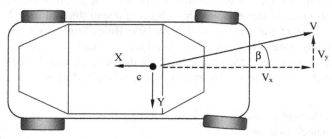

FIGURE 6.6

Body slip angle.

of the vehicle mass centre, V_x and V_y, measured in vehicle body reference frame can be used to readily determine this.

6.4 Suspension system representation

6.4.1 Overview

In Chapter 4 the modelling and analysis of the suspension system was considered in isolation. In this section the representation of the suspension as a component of the full vehicle system model will be considered. As stated the use of powerful MBS analysis programs often results in modelling the suspension systems as installed on the actual vehicle. In the following discussion a vehicle modelled with the suspension represented in this manner is referred to as a *linkage model*.

Before the advent of computer simulation, classical vehicle dynamicists needed to simplify the modelling of the vehicle to a level where the formulation of the equations of motion was manageable and the solution was amenable with the computational tools available at the time. Such an approach encouraged efficiency, with the analyst identifying the modelling issues that were important in representing the problem in hand. The use of modern software need not discourage such an approach. The following sections summarise four vehicle models, one of which is based on modelling the suspension linkages with three other models that use alternative simplified implementations. All four models have been used to simulate a double lane change manoeuvre (Blundell, 2000b) and are compared in the case study at the end of this chapter. The four models described here involve levels of evolving detail and elaboration and can be summarised as follows:

1. A *lumped mass model*, where the suspensions are simplified to act as single lumped masses that can only translate in the vertical direction with respect to the vehicle body.
2. An *equivalent roll stiffness model*, where the body rotates about a single roll axis that is fixed and aligned through the front and rear roll centres.
3. A *swing arm model*, where the suspensions are treated as single swing arms that rotate about a pivot point located at the instant centres for each suspension.
4. A *linkage model*, where the suspension linkages and compliant bush connections are modelled in detail in order to recreate as closely as possible the actual assemblies on the vehicle.

6.4.2 Lumped mass model

For the lumped mass model the suspension components are considered lumped together to form a single mass. The mass is connected to the vehicle body at the wheel centre by a translational joint that only allows vertical sliding motion with no change in the relative camber angle between the road wheels and the body. The camber angle between the road wheels and the road will therefore be directly

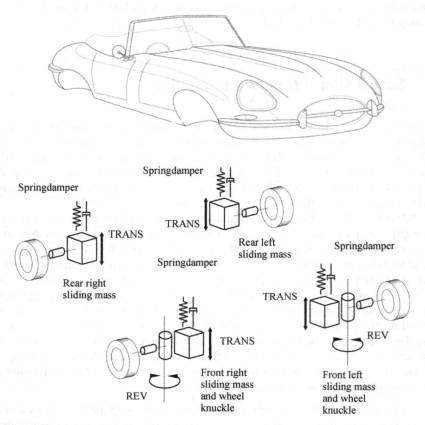

FIGURE 6.7

Lumped mass model approach.
REV, revolute joint; TRANS, translational joint.

related to the roll angle of the vehicle. Spring and damper forces act between the suspensions and the body. Such suspensions have been used on early road vehicles, notably the Lancia Lambda (1908–1927), where it was termed 'sliding pillar'.

The front wheel knuckles are modelled as separate parts connected to the lumped suspension parts by revolute joints. The steering motion required for each manoeuvre is achieved by applying time-dependent rotational motion inputs about these joints. Each road wheel is modelled as a part connected to the suspension by a revolute joint. The lumped mass model is shown schematically in Figure 6.7.

6.4.3 Equivalent roll stiffness model

This model is developed from the lumped mass model by treating the front and rear suspensions as rigid axles connected to the body by revolute joints. The locations of

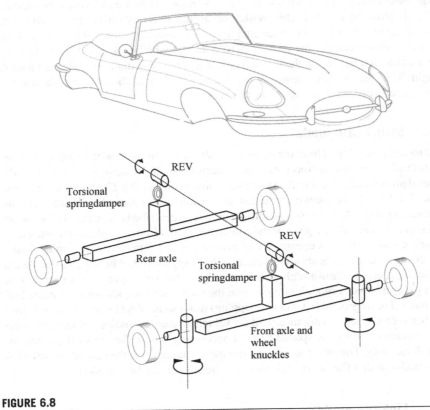

FIGURE 6.8

Equivalent roll stiffness model approach.
REV, revolute joint.

the joints for the two axles are their respective 'roll centres' as described in Chapter 4. A torsional spring is located at the front and rear roll centres to represent the roll stiffness of the vehicle. The determination of the roll stiffness of the front and rear suspensions required an investigation as described in the following section. The equivalent roll stiffness model is shown schematically in Figure 6.8.

Note that this model shows the historical background to much of the current unclear thinking about roll centres and their influence on vehicle behaviour. With beam axles, as were prevalent in the 1920s, this model is a good equivalent for looking at handling behaviour on flat surfaces and ignoring ride inputs. For independent suspensions where the anti-roll geometry remains relatively consistent with respect to the vehicle and where the roll centres are relatively low (i.e. less than around 100 mm for a typical passenger car) — a fairly typical double wishbone setup, for example — then this approximation can be useful despite its systematic inaccuracy. However, drawing general conclusions from such specific circumstances can be dangerous; vehicles that combine a strut suspension at one end (with very mobile anti-roll geometry) and double wishbone at the other (with relatively constant anti-roll geometry) may not be amenable to such simplifications. With this and all other simplified models, the analyst must consider whether or not the conclusions that are drawn reflect upon the simplification adopted or actually reveal some useful insight. The case study presented at the end of this chapter shows a vehicle that behaves acceptably when modelled in this way.

6.4.4 Swing arm model

This model is developed from the equivalent roll stiffness mass model by using revolute joints to allow the suspensions for all four wheels to 'swing' relative to the vehicle body rather than considering using the suspensions linked on an axle. The revolute joints are located at the instant centres of the actual suspension linkage assembly. These positions are found by modelling the suspensions separately as described in Chapter 4. The swing arm model has an advantage over the roll centre model in that it allows the wheels to change camber angle independently of each other and relative to the vehicle body. The swing arm model is shown schematically in Figure 6.9. Although in the sketch the swing arms are shown with an axis parallel to the vehicle axis, this need not be so in general. Also, although in the sketch the swing arms are shown as a 'plausible' mechanical arrangement (i.e. not overlapping) this also need no be so; in general, contact between elements is not modelled for vehicle dynamics studies and in general the instant centres are widely spaced and not necessarily within the physical confines of the vehicle body. The swing arm model has the advantage over the equivalent roll stiffness model in that the heave and pitch ride behaviour can be included.

6.4.5 Linkage model

The model based on linkages as shown in Figure 6.10 is the model that most closely represents the actual vehicle. This sort of vehicle model is the most common approach adopted by MBS software users in the automotive industry, often extending the model definition to include full nonlinear bush characteristics.

A simplification of a model based on linkages is to treat the joints as rigid and generate a kinematic representation of the suspension system. As described in Chapter 4 a double wishbone arrangement is typical of a suspension system that can be modelled in this way and used for handling simulations.

6.4.6 The concept suspension approach

In addition to the four suspension modelling approaches just described, another form of suspension model simplification (Scapaticci et al., 1992) considers an approach where the model contains no elements representing a physical connection between the road wheel and the chassis. Instead the movement of the road wheel with respect to the chassis is described by a functional representation that describes the wheel

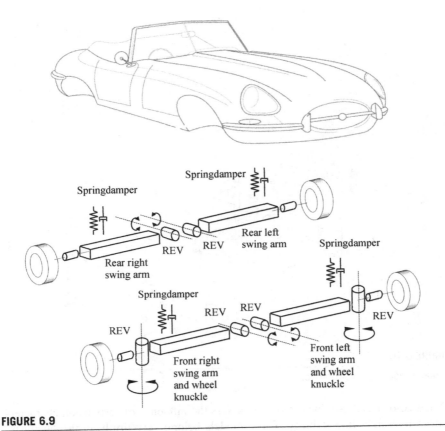

FIGURE 6.9

Swing arm model approach.
REV, revolute joint.

centre trajectory and orientation as it moves vertically between full bump and rebound positions. The authors (Scapaticci et al., 1992) describe this approach as the implementation of synthetic wheel trajectories. Such a method has been adopted within MSC ADAMS where the model is referred to as a 'concept suspension' and is the basis of many dedicated vehicle dynamics modelling software tools such as Milliken Research Associate's VDMS, MSC's CarSim, University of Michigan's ArcSim, Leeds University's VDAS. The way in which such a model is applied is summarised in Figure 6.11. In essence the vehicle model containing the concept suspension can be used to investigate the suspension design parameters that can contribute to the delivery of the desired vehicle handling characteristics without modelling of the suspension linkages. In this way, the analyst can gain a clear understanding of the dominant issues affecting some aspect of vehicle dynamics performance. A case study is given in Section 6.15 describing the use of a reduced

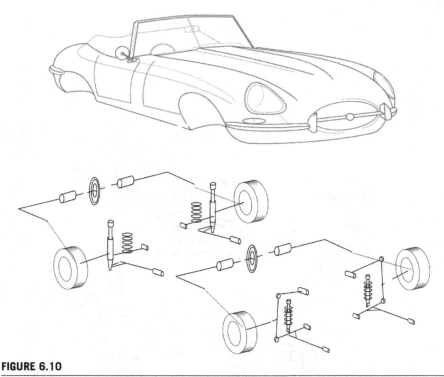

FIGURE 6.10

Linkage model 'as is' approach.

(3 degree-of-freedom) linear model to assess the influence of suspension characteristics on straight line stability. These models belong very firmly in the 'analysis'
segment of the overall process diagram described in Chapter 1, Figure 1.6.

The functional representation of the model is based on components that describe
effects due to kinematics dependent on suspension geometry and also elastic effects
due to compliance within the suspension system. A schematic to support an explanation of the function of this model is provided in Figure 6.12.

If we consider first the kinematic effects due to suspension geometry we can see
that there are two variables that provide input to the model:

- Δz is the change in wheel centre vertical position (wheel travel)
- Δv is the change in steering wheel angle

The magnitude of the wheel travel, Δz, will depend on the deformation of the
surface, the load acting vertically through the tyre resulting from weight transfer
during a simulated manoeuvre and a representation of the suspension stiffness and
damping acting through the wheel centre. The magnitude of the change in steering
wheel angle, Δv, will depend on either an open loop fixed time-dependent rotational
motion input or a closed loop torque input using a controller to feedback vehicle

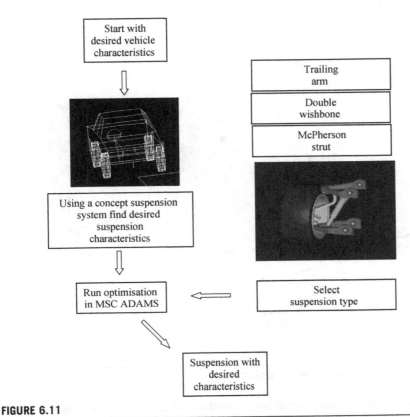

FIGURE 6.11

Application of a concept suspension model.

Provided courtesy of MSC Software.

position variables so as to steer the vehicle to follow a predefined path. The modelling of steering inputs is discussed in more detail later in this chapter. The dependent variables that dictate the position and orientation of the road wheel are:

- Δx is the change in longitudinal position of the wheel
- Δy is the change in lateral position (half-track) of the wheel
- $\Delta \delta$ is the change in steer angle (toe in/out) of the wheel
- $\Delta \gamma$ is the change in camber angle of the wheel

The functional dependencies that dictate how the suspension moves with respect to the input variables can be obtained through experimental rig measurements, if the vehicle exists and is to be used as a basis for the model, or by performing simulation with suspension models as described in Chapter 4. For example the dependence of camber angle, $\Delta \gamma$, on wheel travel can be derived from the curves plotted for case study 1 in Chapter 4.

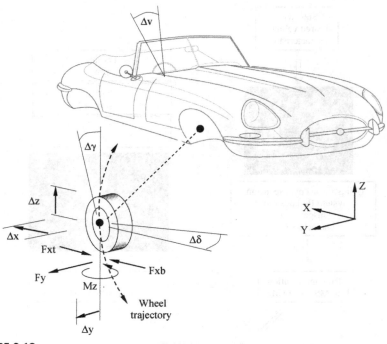

FIGURE 6.12

Concept suspension system model schematic.

The movement of the suspension due to elastic effects is dependent on the forces acting on the wheel. In their paper (Scapaticci et al., 1992) the authors describe the relationship using the equation shown in (Eqn 6.1) where the functional dependencies due to suspension compliance are defined using the matrix, F_E

$$
\begin{bmatrix} \Delta x \\ \Delta y \\ \Delta \delta \\ \Delta \gamma \end{bmatrix} = \begin{bmatrix} & & \\ & F_E & \\ & & \end{bmatrix} \begin{bmatrix} Fxt \\ Fxb \\ Fy \\ Mz \end{bmatrix}
\tag{6.1}
$$

and the inputs are the forces acting on the tyre:

Fxt is the longitudinal tractive force
Fxb is the longitudinal braking force
Fy is the lateral force
Mz is the self-aligning moment

Note that the dimensions of the matrix F_E are such that cross-coupling terms, such as toe change under braking force, can exist. The availability of such data early in the design phase can be difficult but the adoption of such a generalised form allows the user to speculate on such values and thus use the model to set targets for acceptable behaviour.

6.5 Modelling of springs and dampers

6.5.1 Treatment in simple models

The treatment of road springs and dampers in a vehicle where the suspensions have been modelled using linkages is generally straightforward. A road spring is often modelled as linear but the damper will usually require a nonlinear representation as discussed in Chapter 3. It is also common for the bump travel limiter to be engaged early and to have both stiffness and damping elements to its behaviour; both those aspects may be modelled using the methods discussed here. The choice of whether to combine them with the road spring and damper forces is entirely one of modelling convenience; the authors generally find the ease of debugging and auditing the model is worth the carriage of two not strictly necessary additional force generating terms.

For the simplified modelling approach used in the lumped mass and swing arm models the road springs cannot be directly installed in the vehicle model as with the linkage model. Consider the lumped mass model when compared with the linkage model as shown in Figure 6.13.

Clearly there is a mechanical advantage effect in the linkage model that is not present in the lumped mass vehicle model. At a given roll angle for the lumped mass model the displacement and hence the force in the spring will be too large when compared with the corresponding situation in the linkage model.

For the swing arm model the instant centre about which the suspension pivots is often on the other side of the vehicle. In this case the displacement in the spring is approximately the same as at the wheel and a similar problem occurs as with the lumped mass model. For all three simplified models this problem can be overcome as shown in Figure 6.14 by using an 'equivalent' spring that acts at the wheel centre.

As an approximation, ignoring exact suspension geometry, the expression (Eqn 6.2) can be used to represent the stiffness, k_w, of the equivalent spring at the wheel

$$k_w = F_w/\delta_w = (L_s/L_w)\ F_s\ /\ (L_w/L_s)\ \delta_s = (L_s/L_w)^2\ k_s \qquad (6.2)$$

The presence of a square function in the ratio can be considered a combination of both the extra mechanical advantage in moving the definition of spring stiffness to the wheel centre and the extra spring deflection at the wheel centre.

6.5.2 Modelling leaf springs

Although the modelling of leaf springs is now rare on passenger cars they are still fitted extensively on light trucks and goods vehicles where they offer the advantage

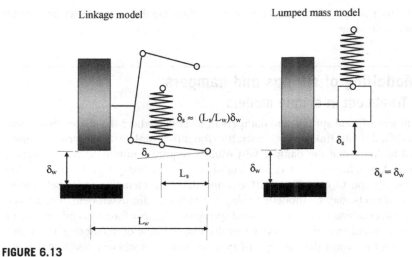

FIGURE 6.13

Road spring in linkage and lumped mass models.

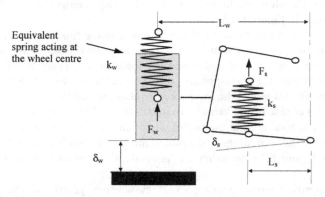

FIGURE 6.14

Equivalent spring acting at the wheel centre.

of providing relatively constant rates of stiffness for large variations in load at the axle. The modelling of leaf springs has always been more of a challenge in an MBS environment when compared with the relative simplicity of modelling a coil spring. Several approaches may be adopted, the most common of which are shown in Figure 6.15.

Early attempts at modelling leaf springs utilised the simple approach based on equivalent springs to represent the vertical and longitudinal force-displacement characteristic of the leaf spring. On the actual vehicle the leaf springs also contribute to the lateral positioning of the axle, with possible additional support from a panhard

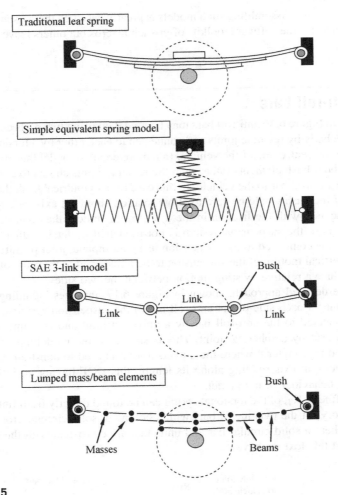

FIGURE 6.15

Leaf spring modelling strategies.

rod. Although not shown in Figure 6.15 lateral springs could also be incorporated to represent this.

The next approach is based on modelling the leaf spring as three bodies (SAE 3-link model) interconnected by bushes or revolute joints with an associated torsional stiffness that provides equivalent force-displacement characteristics as found in the actual leaf spring. The last approach shown in Figure 6.15 uses a detailed 'as is' approach representing each of the leaves as a series of distributed lumped masses interconnected by beam elements with the correct sectional properties for the leaf. This type of model is also complicated by the need to model the interleaf contact forces between the lumped masses with any associated components of sliding friction. Assembling such models is greatly eased by the presence of some kind of macro; some software toolkits offer such macros but others leave the user to devise it.

6.6 Anti-roll bars

As shown in Figure 6.16 anti-roll bars may be modelled using two parts connected to the vehicle body by revolute joints and connected to each other by a torsional spring located on the centre line of the vehicle. In a more detailed model the analyst could include rubber bush elements rather than the revolute joints shown to connect each side of the anti-roll bar to the vehicle. In this case for a cylindrical bush the torsional stiffness of the bush would be zero to allow rotation about the axis, or could have a value associated with the friction in the joint. In this model the connection of the anti-roll bars to the suspension system is not modelled in detail, rather each anti-roll bar part is connected to the suspension using an inplane joint primitive that allows the vertical motion of the suspension to be transferred to the anti-roll bars and hence produce a relative twisting motion between the two sides.

A more detailed approach, shown in Figure 6.17, involves including the drop links to connect each side of the anti-roll bar to the suspension systems. The drop link is connected to the anti-roll bar by a universal joint and is connected to the suspension arm by a spherical joint. This is similar to the modelling of a tie rod as discussed in Chapter 4 where the universal joint is used to constrain the spin of the link about an axis running along its length, this DOF having no influence on the overall behaviour of the model.

The stiffness, K_T, of the torsional spring can be found directly from fundamental torsion theory for the twisting of bars with a hollow or solid circular cross-section. Assuming here a solid circular bar and units that are consistent with the examples that support this text we have

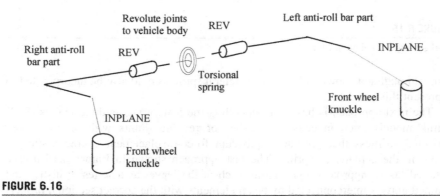

FIGURE 6.16

Modelling the anti-roll bars using joint primitives.
REV, revolute joint.

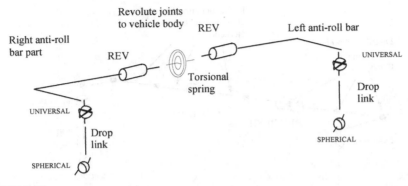

FIGURE 6.17

Modelling the anti-roll bars using drop links.
REV, revolute joint.

$$K_T = \frac{GJ}{L} \qquad (6.3)$$

where

G is the shear modulus of the anti-roll bar material (N/mm^2)
J is the polar second moment of area (mm^4)
L is the length of the anti-roll bar (mm)

Note that the length L used in Eqn (6.3) is the length of the bar subject to twisting. For the configuration shown in Figure 6.17 this is the transverse length of the anti-roll bar across the vehicle and does not include the fore-aft lengths of the system that connect to the drop links. These lengths of the bar provide the lever arms to twist the transverse section of bar and are subject to bending rather than torsion. An externally solved FE model could be used to give an equivalent torsional stiffness for a simplified representation such as this.

Given that bending or flexing of the roll bar may have an influence, the next modelling refinement of the anti-roll bar system uses FE beams, of the type described in Chapter 3, to interconnect a series of rigid bodies with lumped masses distributed along the length of the bar. Such sophistication becomes necessary to investigate anti-roll bar interactions with steer torque, or anti-roll bar lateral 'walking' problems in the vehicle; in general though, such detail is not required for vehicle behaviour modelling.

Lumped masses must be included where the revolute joints connect the anti-roll bar to the vehicle. Again these joints could be modelled with bushes if needed. A final extension of the model shown in Figure 6.18 would be to model the drop links with lumped masses and beams if the flexibility of these components needed to be modelled.

The modelling described so far has been for the modelling of the conventional type of anti-roll bar found on road vehicles. Vehicles with active components in the anti-roll bar system might include actuators in place of the drop links or a coupling device

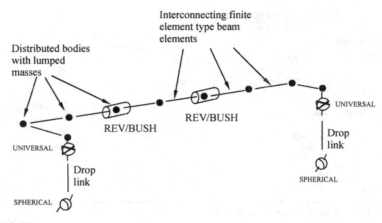

FIGURE 6.18

Modelling the anti-roll bars using interconnected finite element beams.
REV, revolute joint.

connecting the two halves of the system providing variable torsional stiffness at the connection. Space does not permit a description of the modelling of such systems here, but with ever more students becoming involved in motorsport this section will conclude with a description of the type of anti-roll bar model that might be included in a typical student race vehicle. A graphic for the system is shown in Figure 6.19.

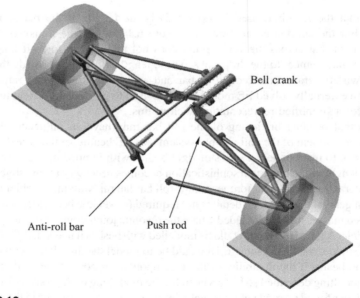

FIGURE 6.19

Graphic of anti-roll bar in typical student race vehicle.

Provided courtesy of MSC Software.

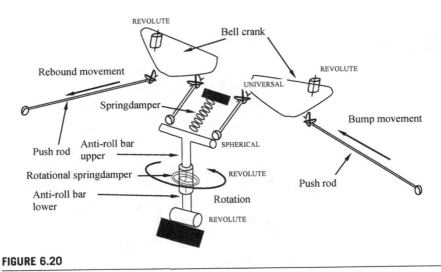

REVOLUTE

Bell crank

Rebound movement

REVOLUTE

UNIVERSAL

Springdamper

Bump movement

Push rod

Anti-roll bar
upper

SPHERICAL

Rotational springdamper

REVOLUTE

Push rod

Anti-roll bar
lower

Rotation

REVOLUTE

FIGURE 6.20

Modelling of anti-roll bar mechanism in student race car.

The modelling of this system is illustrated in the schematic in Figure 6.20 where it can be seen that the anti-roll bar is installed vertically and is connected to the chassis by a revolute joint. The revolute joint allows the anti-roll bar to rock back and forward as the bell cranks rotate during parallel wheel travel but prevents rotation during opposite wheel travel when the body rolls. As the body rolls the torsional stiffness of the anti-roll bar, modelled with the rotational springdamper, resists the pushing motion of one push rod as the suspension moves in bump on one side and the pulling motion as the suspension moves in rebound on the other side. The small springdamper helps to locate the anti roll bar with respect to the vehicle chassis and adds to the heave stiffness and damping. Alternative linkage designs are possible that allow the use of a translational spring element and hence allow independent control of damping in roll compared to damping in heave. Such 'three spring' systems are common in higher formula motorsports events when allowed by the rules.

6.7 Determination of roll stiffness for the equivalent roll stiffness model

In order to develop a full vehicle model based on roll stiffness it is necessary to determine the roll stiffness and damping of the front and rear suspension elements separately. The estimation of roll damping is obtained by assuming an equivalent linear damping and using the positions of the dampers relative to the roll centres to calculate the required coefficients. If a detailed vehicle model is available, the

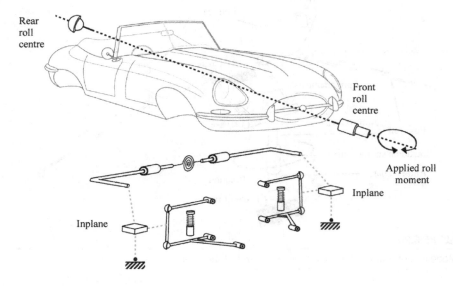

FIGURE 6.21

Determination of front end roll stiffness.

procedure used to find the roll stiffness for the front suspension elements involves the development of a model as shown in Figure 6.21. This model includes the vehicle body, this being constrained to rotate about an axis aligned through the front and rear roll centres. The roll centre positions can be found using the methods described in Chapter 4. The vehicle body is attached to the ground part by a cylindrical joint located at the front roll centre and aligned with the rear roll centre. The rear roll centre is attached to the ground by a spherical joint in order to prevent the vehicle sliding along the roll axis. A motion input is applied at the cylindrical joint to rotate the body through a given angle. By requesting the resulting torque acting about the axis of the joint it is possible to calculate the roll stiffness associated with the front end of the vehicle. The road wheel parts are not included nor are the tyre properties. The tyre compliance is represented separately by a tyre model and should not be included in the determination of roll stiffness. The wheel centres on either side are constrained to remain in a horizontal plane using inplane joint primitives. Although the damper force elements can be retained in the suspension models they have no contribution to this calculation as the roll stiffness is determined using static analysis. The steering system, although not shown in Figure 6.21, may also be included in the model. If the steering system is present, a zero-motion constraint is needed to lock it in the straight-ahead position during the roll simulation.

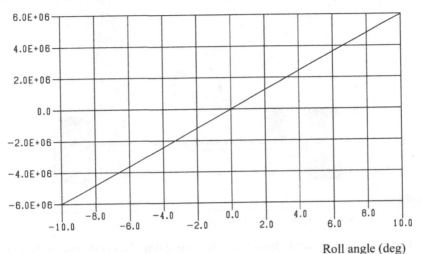

Roll moment (Nmm)

Roll angle (deg)

FIGURE 6.22

Front end roll simulation.

For the rear end of the vehicle the approach is essentially the same as for the front end, with in this case a cylindrical joint located at the rear roll centre and a spherical joint located at the front roll centre.

For both the front and rear models the vehicle body can be rotated through an appropriate angle either side of the vertical. For the example vehicle used in this text the body was rotated 10° each way. The results for the front end model are plotted in Figure 6.22. The gradient at the origin can be used to obtain the value for roll stiffness used in the equivalent roll stiffness model described earlier.

In the absence of an existing vehicle model that can be used for the analysis described in the preceding section, calculations can be performed to estimate the roll stiffness. In reality this will have contributions from the road springs, anti-roll bars and possibly the suspension bushes. Figure 6.23 provides the basis for a calculation of the road spring contribution for the simplified arrangement shown. In this case the inclination of the road springs is ignored and has a separation across the vehicle given by L_s.

As the vehicle rolls through an angle φ the springs on each side are deformed with a displacement, δ_s, given by

$$\delta_s = \phi \, L_s/2 \tag{6.4}$$

The forces generated in the springs F_s produce an equivalent roll moment M_s given by

$$M_s = F_s \, L_s = k_s \, \delta_s \, L_s = k_s \, \phi \, L_s^2 / \, 2 \tag{6.5}$$

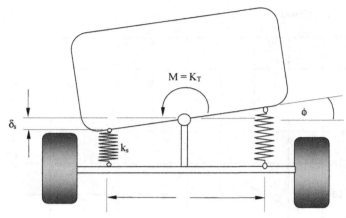

FIGURE 6.23

Calculation of roll stiffness due to road springs.

The roll stiffness contribution due to the road springs K_{Ts} at the end of the vehicle under consideration is given by

$$K_{Ts} = M_s/\phi = k_s \; L_s^2/ 2 \tag{6.6}$$

In a similar manner the contribution to the roll stiffness at one end of the vehicle due to an anti-roll bar can be determined as follows (Figure 6.24).

In this case if the ends of the anti-roll bar are separated by a distance L_r and the vehicle rolls through an angle φ, the relative deflection of one end of the anti-roll bar to the other, δr, is given by

$$\delta_r = a \; \theta = \phi \; L_r \tag{6.7}$$

The angle of twist in the roll bar is given by

$$\theta = \frac{TL_r}{GJ} \tag{6.8}$$

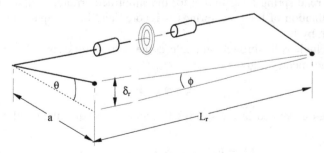

FIGURE 6.24

Calculation of roll stiffness due to the anti-roll bar.

where, as discussed earlier, G is the shear modulus of the anti-roll bar material, J is the polar second moment of area and T is the torque acting about the transverse section of the anti-roll bar. Note that in this analysis we are ignoring the contribution due to bending. The forces acting at the ends of the anti-roll bar, F_r, produce an equivalent roll moment M_r given by

$$M_r = F_r L_r = T L_r /a = \theta \, GJ/a = \phi \, L_r \, GJ/ a^2 \qquad (6.9)$$

The roll stiffness contribution due to the anti-roll bar, K_{Tr}, at the end of the vehicle under consideration is given by

$$K_{Tr} = M_r/\phi = L_r \, GJ/ a^2 \qquad (6.10)$$

The contribution of both the road springs and the anti-roll bar can then be added, ignoring suspension bushes here, to give the roll stiffness, K_T

$$K_T = K_{Ts} + K_{Tr} \qquad (6.11)$$

Note that current practice in vehicles is to have relatively soft springs and fit stiffer anti-roll bars than was the norm some years ago. If vehicles achieve a large proportion of their roll stiffness from anti-roll bars, the subjective phenomenon of 'roll rock' (also known as 'lateral head toss') becomes problematic. A rule of thumb is that such phenomena begin to emerge when the anti-roll bars form more than about one-third of the overall roll stiffness — in other words if K_{Tr} is greater than 0.5 K_{Ts}.

6.8 **Aerodynamic effects**

Some treatment of aerodynamics is generally given in existing textbooks (Milliken and Milliken, 1998; Gillespie, 1992) dealing with vehicle dynamics. Other textbooks are dedicated to the subject (Hucho, 1998). The flow of air over the body of a vehicle produces forces and moments acting on the body resulting from the pressure distribution (form) and friction between the air and surface of the body. The forces and moments are considered using a body-centred reference frame where longitudinal forces (drag), lateral forces, and vertical forces (lift or down thrust) will arise. The aerodynamic moments will be associated with roll, pitch and yaw rotations about the corresponding axes.

Current practice is generally to ignore aerodynamic forces for the simulation of most proving ground manoeuvres but for some applications and classes of vehicles this is clearly not representative of the vehicle dynamics in the real world, for example winged vehicles. Prior to rules limiting downforce, it was often said that for some vehicles of this type the down thrust is so great that this could overcome the weight of a vehicle, allowing it for example to drive upside down through a tunnel, although this has never been demonstrated.

The lack of speed limits on certain autobahns in Germany also means that a vehicle manufacturer selling a high-performance vehicle to that market will need to test the vehicle at speeds well over twice the legal UK limit. The possibility of aerodynamic forces at these high speeds destabilising the vehicle needs to be investigated and where physical testing is to be done, equivalent computer simulation is also desirable. Other effects such as side gusting are also tested for and have been simulated by vehicle dynamicists in the past.

An approach that has been commonly used is to apply forces and moments to the vehicle body using measured results from wind tunnel testing, in look-up tables. As the vehicle speed and the attitude of the body changes during the simulation, the forces and moments are interpolated from the measured data and applied to the vehicle body. A difficulty with such an approach is that the measured results are for steady state in each condition and that transient effects are not included in the simulation. Consideration has been given to the use of a computational fluid dynamics (CFD) programme to calculate aerodynamic forces and moments in parallel with (co-simulation) an MBS programme solving the vehicle equations of motion. The problem at the current time with this approach is the mismatch in the computation time for both methods. MBS models of a complete vehicle can simulate vehicle handling manoeuvres in seconds, or even real time, whereas complex CFD models can involve simulation times running to days. CFD methods can now handle aerodynamic transient effects (e.g. vortex shedding) but the timescale mismatch remains. Thus there is no realistic prospect of the practical use of the interaction of transient aerodynamics effects and vehicle dynamics being modelled in the near future. However, genuine transient aerodynamic effects, such as those involved in so-called 'aeroelastic flutter' — an unsteady aerodynamic flow working in sympathy with a structural resonance — are extremely rare in ground vehicles.

In order to introduce readers to the fundamentals, consider a starting point where it is intended only to formulate an aerodynamic drag force acting on the vehicle body.

The drag force, F_D, can be considered to act along a line of action parallel to the ground and in the vehicle plane of symmetry having the following formulation:

$$F_D = \frac{1}{2} \frac{\rho V^2 C_D A}{GC} \tag{6.12}$$

where

C_D = the aerodynamic drag coefficient
ρ = the density of air
A = the frontal area of the vehicle (projected onto a yz plane)
V = the velocity of the vehicle in the direction of travel
GC = a gravitational constant.

The gravitational constant is included in Eqn (6.12) to remind readers that this is a dynamic force. If the model units are SI then GC is equal to 1. If as commonly used the model units for length are in millimetres then GC is equal to 1000. When formulating the aerodynamic drag force more generally it should be considered that the overall force acts along a line that does not typically pass through the centre of gravity and therefore moments as well as forces are necessary to capture to the correct effects on the vehicle. In the straight line condition generally a pitch moment is all that is needed to reposition the drag vector correctly. It is also generally true that the forces and moments change as the vehicle aerodynamic attitude changes, and one possible way of capturing this is to have an attitude-sensitive formulation for the aerodynamic coefficients. Within MBS models it is difficult to capture effects such as the apparent curvature of the flow when a car is travelling on a tight arc in still air, but tight arcs are typically not possible at anything except the lowest speeds and so the effect is not important in terms of the overall behaviour of the system. At higher speeds, in still air, the body slip angle may be considered a useful surrogate for the aerodynamic yaw angle. In this case, a relatively simple formulation can capture the increase in, for example, drag coefficient with aerodynamic yaw angle:

$$F_D = \frac{1}{2} \rho V^2 \frac{\left(C_{D0} + C_{D\beta} \cdot \beta\right) A}{GC} \tag{6.13}$$

where

C_{D0} = drag coefficient at zero aerodynamic yaw angle
$C_{D\beta}$ = drag coefficient sensitivity to aerodynamic yaw angle
β = aerodynamic yaw angle (or body slip angle surrogate)

This formulation avoids the need for a knowledge of changing aerodynamic frontal area with attitude changes, which can be hard to obtain, and can be calculated from a fairly ordinary set of wind tunnel results. For the position shown in Figure 6.25 it is clear that for anything other than straight line motion it is going to be necessary to model the forces as components in the body-centred axis system. If we consider the vehicle moving only in the xy plane then this is going to require at least the formulation of a longitudinal force, Fx, a lateral force, Fy and a yawing moment, Mz, all in a body-centred axis system, usually located at the mass centre. Wind tunnel testing or CFD analysis is able to yield coefficients for all six possible forces and moments acting on the body, referred back to the mass centre. Note that for passenger vehicles it is typical that the aerodynamic yaw moment is as shown in the diagram, i.e. is such to make the vehicle turn away from the wind. For other

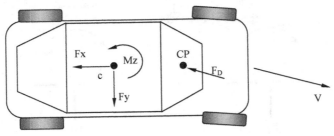

FIGURE 6.25

Application of aerodynamic drag force.

vehicles this may not be true and individual research on the vehicle in question is needed. Note also that over body slip angles of interest, the characteristics are often observed to be usefully linear and so a simple formulation of the type proposed in Eqn (6.13) is sufficient for all the aerodynamic forces and moments; this presumption should be verified when the vehicle has an unusual configuration or operates at extremely large slip angles.

6.9 Modelling of vehicle braking

In Chapter 5 the force and moment generating characteristics of the tyre were discussed and it was shown how the braking force generated at the tyre contact patch depends on the value of the slip ratio which varies from zero for a free rolling wheel to unity for a braked and fully locked wheel. In this section we are not so much concerned with the tyre, given that we would be using a tyre model interfaced with our full vehicle model to represent his behaviour. Rather we now address the modelling of the mechanisms used to apply a braking torque acting about the spin axis of the road wheel that produces the change in slip ratio and subsequent braking force.

Clearly as the vehicle brakes, as shown in Figure 6.26, there is weight transfer from the rear to the front of the vehicle. Given what we know about the tyre behaviour the change in the vertical loads acting through the tyres will influence the braking forces generated. As such the braking model may need to account for real effects such as proportioning the braking pressures to the front and rear wheels or the implementation of ABS. Before any consideration of this we need to address the mechanism to model a braking torque acting on a single road wheel.

If we consider a basic arrangement, the mechanical formulation of a braking torque, based on a known brake pressure, acting on the piston can be derived from Figure 6.27.

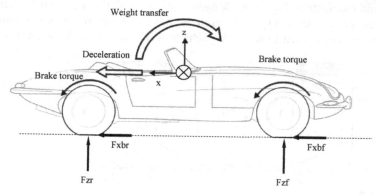

FIGURE 6.26

Braking of a full vehicle.

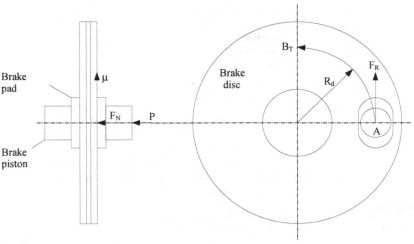

FIGURE 6.27

Braking mechanism.

The braking torque B_T is given by

$$B_T = n \, \mu \, p \, A \, R_d \tag{6.14}$$

where

  $n =$ the number of friction surfaces (pads)
  $\mu =$ the coefficient of friction between the pads and the disc
  $p =$ the brake pressure
  $A =$ the brake piston area
  $R_d =$ the radius to the centre of the pad

Note that depending on the sophistication of the model, the coefficient of friction μ may be constant or defined as a run-time variable as a function of brake rotor temperature. Note also that brake pad/rotor friction is substantially more Newtonian than tyre friction and thus subject to much less complication than the rubber friction previously discussed. Some kind of friction model is still needed to prevent the direct application of a brake torque spinning the wheel backwards, which does not happen in practice. Figure 6.28 shows typical brake friction vs temperature characteristics for different brake rotor materials. Brake rotor temperature, T, can be calculated using the differential equation

$$T = T_0 + \frac{1}{mc} \int_0^t [B_T(t)\omega(t) - hA_c(T(t) - T_{env})].dt \tag{6.15}$$

where

  $T_0 =$ Initial brake rotor temperature (K)
  $\omega(t) =$ Brake rotor spin velocity (rads s^{-1})

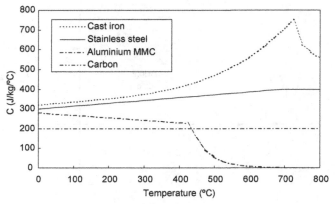

FIGURE 6.28

Specific heat capacity, c, versus temperature, T (Farr, 1999).

t = time (s)
h = Brake rotor convection coefficient (W m^{-2} K^{-1})
A_c = Convective area of brake disc (m^2)
T_{env} = Environmental temperature (K)
m = mass of brake rotor (kg)
c = specific heat capacity of brake rotor (J kg^{-1} K^{-1})
$BT(t)$ = Brake Torque (Nm)

For the most common brake rotor material, cast iron, the specific heat versus temperature characteristic can be approximated in the working range (0 to 730 °C) by the expression

$$c = 320 + 0.15T + 1.164 \times 10^{-9} T^4 \qquad (6.16)$$

Note that in the above expression, temperature T is in centigrade (Celsius) and not Kelvin. The brake torque and temperature models may be used easily within MBS models using a combination of design variables and run-time variables as shown in Table 6.1 where we are using an input format that corresponds to a command language used in MSC ADAMS. Note the need for an explicit iteration since the temperature depends on the heat capacity and the heat capacity depends on the temperature. When modelling such behaviour in a spreadsheet, it is sufficient to refer to the temperature of the preceding time step. Although this is possible within many MBS packages, it can be awkward to implement and can also lead to models with some degree of numerical delicacy. Output from the model during a single braking event is shown in Figure 6.29.

Note also that it is common practice within brake manufacturers to separate the brake energising event from the brake cooling event for initial design calculations, leading to a systematic overestimation of the temperature during fade/recovery testing. This conservative approach is unsurprising given the consequences of brake

Table 6.1 A Brake Rotor Temperature Model Based on Brake Torque

```
!---------------------- Function definitions ----------------------!
!
part create equation differential_equation    &
    differential_equation_name = .model_1.brake_heating_integral   &
    adams_id = 2   &
    comments = "Brake Heat Input Integral"   &
    initial_condition = 0.0   &
    function = "VARVAL(Brake_Torque)*VARVAL(vehicle_velocity)/0.3"   &
    implicit = off   &
    static_hold = off

data_element create variable   &
    adams_id = 102   &
    variable_name = brake_rotor_heat_in   &
    function = "DIF(2)"
!
data_element create variable   &
    variable_name = "rotor_temperature_kelvin_estimate_1"   &
    function = "T_env + VARVAL(brake_rotor_heat_in)/(rotor_mass*350)"
!
data_element create variable   &
    variable_name = "rotor_temperature_estimate_1"   &
    function = "VARVAL(rotor_temperature_kelvin_estimate_1)-273"
!
data_element create variable   &
    variable_name = "rotor_heat_capacity_estimate_2"   &
    function = "320 + 0.15*VARVAL(rotor_temperature_estimate_1)", &
               " + 1.164E-9*VARVAL(rotor_temperature_estimate_1)**4"
!
part create equation differential_equation    &
    differential_equation_name = .model_1.brake_cooling_integral   &
    adams_id = 3   &
    comments = "Brake Heat Rejection Integral"   &
    initial_condition = 0.0   &
    function = "hAc*(VARVAL(rotor_temperature_kelvin_estimate_1)-T_env)"   &
    implicit = off   &
    static_hold = off

data_element create variable   &
    adams_id = 103   &
    variable_name = brake_rotor_heat_out   &
    function = "DIF(3)"
!
data_element create variable   &
    variable_name = "rotor_temperature_kelvin_estimate_2"   &
    function = "T_env +", &
               "( VARVAL(brake_rotor_heat_in) - VARVAL(brake_rotor_heat_out) )/", &
               "(rotor_mass*VARVAL(rotor_heat_capacity_estimate_2)+0.001)"
!
data_element create variable   &
    variable_name = "rotor_temperature_estimate_2"   &
    function = "VARVAL(rotor_temperature_kelvin_estimate_2)-273"
```

system under design. The example given is a relatively simple one, with convection characteristics that are independent of vehicle velocity and no variation of brake friction with brake temperature. Although in practice these simplifications render the results slightly inaccurate, they are useful when used for comparative purposes — for example, if the brake temperature model is used with an ESP algorithm it can rank control strategies in terms of the energy added to individual brake rotors.

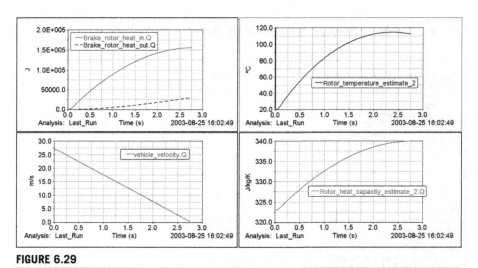

FIGURE 6.29

Output from the brake temperature model shown in Table 6.1 during a 60 mph − 0 stop.

Similar modelling is of course possible for other frictional systems within the vehicle, such as drive or transmission clutches. Typical values of the convection constant, hAc, are around 150 W/K for a front disc brake installation, around 80 W/K for a rear disc brake installation and as low as 20 W/K for a rear drum brake.

A further key factor in modelling brake performance is the distribution of brake torques around the vehicle. While decelerating, the vertical loads on the axles change as described in Section 4.8.1 due to the fact that the mass centre of the vehicle is above the ground.

It may be presumed that for ideal braking, the longitudinal forces should be distributed according to the vertical forces. Using the above expressions, the graphs in Figure 6.30 can be calculated for vertical axle load versus deceleration. Knowing the total force necessary to decelerate the vehicle it is possible to calculate the horizontal forces for 'ideal' (i.e. matched to vertical load distribution) deceleration. Plotting rear force against front force leads to the characteristic curve shown in Figure 6.30. However, in general it is not possible to arrange for such a distribution of force and so the typical installed force distribution is something like that shown by the dashed line in the figure. Note that the ideal distribution of braking force varies with loading condition and so many vehicles have a brake force distribution that varies with vehicle loading condition. For more detailed information on brake system performance and design, Limpert (1999) gives a detailed breakdown of performance characteristics and behaviour, all of which may be incorporated within a MBS model of the vehicle using an approach similar to that shown in Table 6.1 if desired − which is to say the formulation of state variables and their manipulation into quantities of interest, followed by their use to modify applied forces in the model.

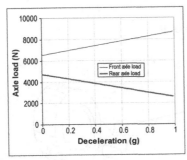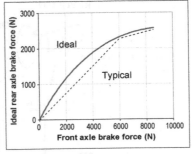

FIGURE 6.30

Force distribution for ideal and typical braking events.

6.10 Modelling traction

For some simulations it is necessary to maintain the vehicle at a constant velocity. Without some form of driving torque the vehicle will 'coast' through the manoeuvre using the momentum available from the velocity defined with the initial conditions for the analysis. Ignoring rolling resistance and aerodynamic drag will reduce losses but the vehicle will still lose momentum during the manoeuvre due to the 'drag' components of tyre-cornering forces generated during the manoeuvre. An example is provided in Figure 6.31 where for a vehicle lane change manoeuvre it can be seen that during the 5 s taken to complete the manoeuvre the vehicle looses about 5 km/h in the absence of any tractive forces at the tyres.

As discussed in Chapter 3 single-purpose general-purpose programs such as ADAMS/Car include a driveline model as part of the full vehicle as a means to impart

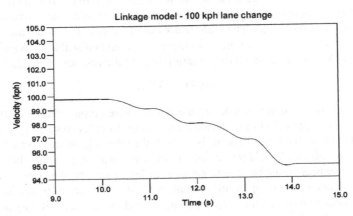

FIGURE 6.31

Loss in velocity as vehicle 'coasts' through the lane change manoeuvre.

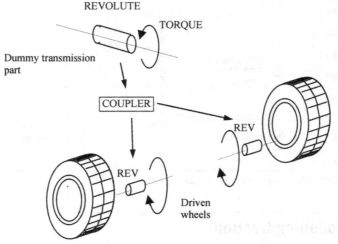

FIGURE 6.32

Simple drive torque model.
REV, revolute joint.

torques to the road wheels and hence generate tractive driving forces at the tyres. Space does not permit a detailed consideration of driveline modelling here but as a start a simple method of imparting torque to the driven wheels is shown in Figure 6.32.

The rotation of the front wheels is coupled to the rotation of the dummy transmission part as shown in Figure 6.32. The coupler introduces the following constraint equation:

$$s_1.r_1 + s_2.r_2 + s_3.r_3 = 0 \qquad (6.17)$$

where s_1, s_2 and s_3 are the scale factors for the three revolute joints and r_1, r_2 and r_3 are the rotations. In this example suffix 1 is for the driven joint and suffixes 2 and 3 are for the front wheel joints. The scale factors used are $s_1 = 1$, $s_2 = 0.5$ and $s_3 = 0.5$ on the basis that 50% of the torque from the driven joint is distributed to each of the wheel joints. This gives a constraint equation linking the rotation of the three joints:

$$r_1 = 0.5\ r_2 + 0.5\ r_3 \qquad (6.18)$$

Note that this equation is not determinant. For a given input rotation r_1, there are two unknowns r_2 and r_3 but only the single equation. In order to solve r_2 and r_3 this equation must be solved simultaneously with all the other equations representing the motion of the vehicle. This is important particularly during cornering where the inner and outer wheels must be able to rotate at different speeds; the constraint equation thus reflects a perfect open differential. A Coulomb style frictional force may also be introduced to reflect the imperfections in real differentials; it may be scaled with applied driveline torque to reflect the function of a Torsen-style ('torque biasing') differential that may or may not have symmetric operation; it may be scaled in response to some externally calculated solution variable to reflect the operation of some electronically controlled proportional clutch across the differential. Alternatively a torque may be applied that is proportional to the speed difference between the two output shafts and thus represents a viscous style element. All these representations may be combined in a single model to represent the components present in the real vehicle.

6.11 Other driveline components

The control of vehicle speed is significantly easier than the control of vehicle path inside a vehicle dynamics model. In the real vehicle, speed is influenced by the engine torque, brakes and aerodynamic drag. As discussed earlier these are relatively simple devices to represent in an MBS model, with the exception of torque converters and turbochargers. Even these latter components can be represented using differential equations of the form:

$$T_{BOOST} = T_2 \cdot \hat{T}_{BOOST} \tag{6.19}$$

$$\frac{d}{dt}(T_2) = \frac{T_1}{k_2} \cdot (t_{boost} - T_2) \tag{6.20}$$

$$\frac{d}{dt}(T_1) = k_1 \cdot (t_{boost} - T_1) \tag{6.21}$$

where $\hat{T}_{BOOST}$ is the maximum possible torque available, t_{boost} is the throttle setting to be applied to the boost torque (which may be different to the throttle setting applied to the normally aspirated torque to model the rapid collapse of boost off-throttle) and $k_{1,2}$ are mapped, state-dependent values to calibrate the behaviour of the engine (i.e. large delays at low engine speed, reducing delays with rising engine speed). An example of the statements required to model the resulting torque is shown in Table 6.2.

In this example the variable throttle runs from −0.3 to 1.0 to simulate overrun torque. The variable `boost_throttle` is a clipped version from 0 to 1.0 since no turbocharger boost is available on overrun. The variable `throttle_derivative` is the first time derivative of throttle. All the other variables (varvals) are retrieved from the relevant curves (splines) plotted in Figure 6.33.

The delays inherent in a torque converter are amenable to such modelling techniques using typical torque converter characteristic data in a similar empirical manner.

Once the physical elements of the system are modelled, the task of modelling the driver behaviour is largely similar to that for path following described later. In order to represent, for example, the effect of a driver using the throttle to maintain a steady velocity through a manoeuvre, a controller can be developed to generate the torque shown in Figure 6.33.

Table 6.2 Example MSC ADAMS Command Statements for an Empricial Mean-State Turbocharger

```
!    -- First First Order Differential Equation --
part create equation differential_equation  &
   differential_equation_name = turbo_lag_equation_1  &
   adams_id = 12 &
   comments = "Lag Equation 1 - Explicit"  &
   initial_condition = 0.0  &
   function = "varval(K1_now) * ( varval(boost_throttle)*100-DIF(12) )" &
   implicit = off  &
data_element create variable &
   variable_name=K2 &
   function="STEP(varval(throttle_derivative),", &
         "-10, 100.0,", &
         " -1, (DIF(12))/varval(K2_divisor_now)", &
         "    )"

!    -- Second First Order Differential Equation --
part create equation differential_equation  &
   differential_equation_name = turbo_lag_equation_2  &
   adams_id = 13 &
   comments = "Lag Equation 2 - Explicit"  &
   function = "varval(K2) * ( varval(boost_throttle)*100-DIF(13) )" &
   implicit = off  &
data_element create variable  &
   variable_name = boost_torque_scaling &
   function="DIF(13)/100"

!    -- Sum both normally aspirated and turbocharged (delayed) component
data_element create variable  &
   variable_name = prop_torque  &
   function = "(", &
            " VARVAL(na_engine_torque)*VARVAL(throttle)*1000", &
"+VARVAL(boosted_engine_torque)*VARVAL(boost_torque_scaling)*1000", &
            ")"
```

A simple but workable solution is to model the driving torque T, with the following formulation:

$$T = K * (Vs - Va) * STEP (Time,0,0,1,1) \tag{6.22}$$

where:

K = a proportional gain that is tuned to stabilize the torque
Vs = the desired velocity for the simulation
Va = the forward velocity of the vehicle, which can be obtained using a system variable

It can be seen that the expression delivers a drive torque proportional to the speed error. The purpose of the STEP FUNCTION is to define a change of state in the expression that is continuous.

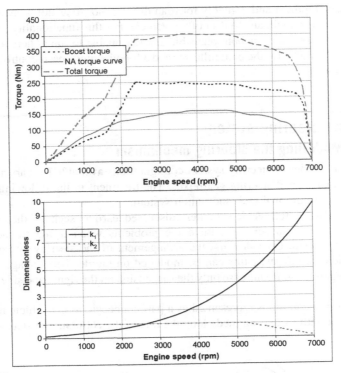

FIGURE 6.33

Empricial mean-state turbocharger model.

The step function can be used to factor a force function by ramping it on over a set time period. In this case the driving torque is being switched on between time $= 0$ and time $= 1$ s. This is important because it is common to perform an initial static analysis of the vehicle at time $= 0$ when Va $= 0$ and the torque must not act.

As can be seen a 'reference' (desired) state is needed, an error term is defined by the difference between the current state and the reference state and finally, responses to that in terms of throttle or brake application to adjust the speed back towards the reference value. There are two possible approaches; the simplest provides a speed 'map' for the track, similar to the curvature map description of it. More elaborately, it is possible to examine the path curvature map locally and decide (through knowledge of the ultimate capabilities of the vehicle, perhaps) whether or not the current speed is excessive, appropriate or insufficient for the local curvature and use brakes or engine appropriately. For the development of vehicles, open-loop throttle or brake inputs may be preferable and are sometimes mandated in defined test manoeuvres, rendering the whole issue of speed control moot.

In many ways the skill of the competition driver lies entirely in this ability to judge speed and adjust it appropriately. It is also a key skill to cultivate for limit handling development and arguably for road driving too, so as not to arrive at hazards too rapidly to maintain control of the vehicle. For this functionality, some form of preview is essential. It is both plausible and reasonable to run a 'here and now at the front axle' model for the path follower and a 'previewing' speed controller within the same model.

6.12 The steering system
6.12.1 Modelling the steering mechanism

There are a number of steering system configurations available for cars and trucks based on linkages and steering gearboxes. The treatment in the following sections is limited to a traditional rack and pinion system.

For the simple full vehicle models discussed earlier, such as that modelled with lumped mass suspensions, there are problems when trying to incorporate the steering system. Consider first the arrangement of the steering system on the actual vehicle and the way this can be modelled on the detailed linkage model as shown in Figure 6.34. In this case only the suspension on the right hand side is shown for clarity.

The steering column is represented as a part connected to the vehicle body by a revolute joint with its axis aligned along the line of the column. The steering inputs

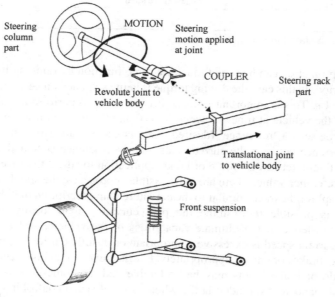

FIGURE 6.34

Modelling the steering system.

required to manoeuvre the vehicle are applied as motion or torque inputs at this joint. The steering rack part is connected to the vehicle body by a translational joint and connected to the tie rod by a universal joint. The translation of the rack is related to the rotation of the steering column by some kind of coupling statement that defines the ratio; such constructs are common to most general purpose software packages.

Attempts to incorporate the steering system into the simple models using lumped masses, swing arms and roll stiffness will be met with a problem when connecting the steering rack to the actual suspension part. This is best explained by considering the situation shown in Figure 6.35.

The geometry of the tie rod, essentially the locations of the two ends, is designed with the suspension linkage layout and will work if implemented in an 'as is' model of the vehicle including all the suspension linkages. Physically connecting the tie rod to the simple suspensions does not work. During an initial static analysis of the full vehicle, to settle at kerb height, the rack moves down with the vehicle body relative to the suspension system. This has a pulling effect, or pushing according to the rack position, on the tie rod that causes the front wheels to steer during the initial static analysis. The solution to this is to establish the relationship between the steering column rotation and the steer change in the front wheels and to model this as a direct ratio using two coupler statements to link the rotation between the steering column and each of the front wheel joints as shown in Figure 6.36. Advanced steer-by-wire research platforms, such as the 'P1' vehicle in use at Stanford University, use this layout in a physical vehicle with a software controller solving the constraint equation by driving electric motors to move the steered road wheels.

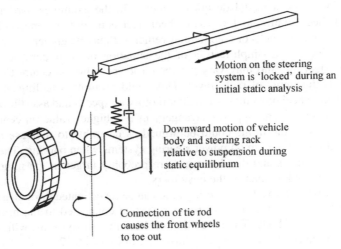

Motion on the steering system is 'locked' during an initial static analysis

Downward motion of vehicle body and steering rack relative to suspension during static equilibrium

Connection of tie rod causes the front wheels to toe out

FIGURE 6.35

Toe change in front wheels at static equilibrium for simple models.

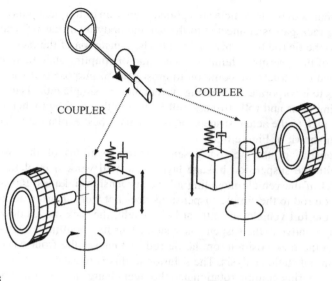

FIGURE 6.36

Coupled steering system model.

6.12.2 Steering ratio

In order to implement the ratios used in the couplers shown in Figure 6.36, linking the rotation of the steering column with the steer change at the road wheels, it is necessary to know the steering ratio. At the start of a vehicle dynamics study the steering ratio can be a model design parameter. In the examples here a ratio of 20° of handwheel rotation to 1° of road wheel steer is used. On some vehicles this may be lower and on trucks or commercial vehicles it may be higher. To treat steering ratio as linear is a simplification of the situation on a modern vehicle. For example the steering ratio may vary between a lower value on centre to a higher value towards the limits of rack travel. This could promote a feeling of stability for smaller handwheel movements at higher motorway speeds and assist lower speed car park manoeuvres. The opposite arrangement — a higher value on centre and a lower level towards the limits could be argued as desirable to increase agility at speed and reduce the burden on power steering systems when parking. The authors suspect the advantages and disadvantages are probably larger in the minds of the steering system vendors than of the customers.

Using the MBS approach the steering ratio can be investigated through a separate study carried out using the front suspension system connected to the ground part instead of the vehicle body. The modelling of these two subsystems, with only the suspension on the right side shown, is illustrated in Figure 6.37.

The approach of using a direct ratio to couple the rotation between the steering column and the steer angle of the road wheels is common practice in simpler

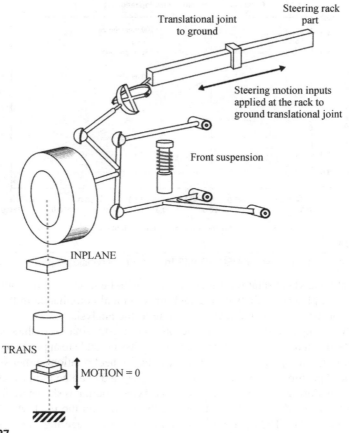

Translational joint to ground

Steering rack part

Steering motion inputs applied at the rack to ground translational joint

Front suspension

INPLANE

TRANS

MOTION = 0

FIGURE 6.37

Front suspension steering ratio test model.

models but may have other limitations in addition to the treatment of the ratio as linear:

1. In the real vehicle and the linkage model the ratio between the column rotation and the steer angle at the road wheels would vary as the vehicle rolls and the road wheels move in bump and rebound.
2. For either wheel the ratio of toe out or toe in as a ratio of left or right handwheel rotation would not be exactly symmetric.

Modelling the suspension with linkages will capture these effects. Although this may influence the modelling of low-speed turning they have little effect for handling manoeuvres with comparatively small steer motions.

With simpler vehicle models, not including suspension linkages, the ratio would need to be functionally dependent on the vertical movement of the suspension and

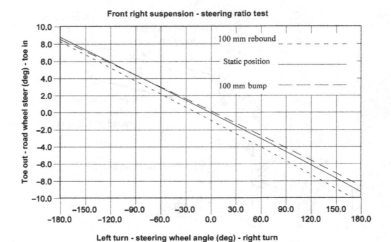

FIGURE 6.38

Results of steering ratio test for MSC ADAMS front right suspension model.

direction of handwheel rotation if the behaviour is to be modelled. It should also be noted that compliance in the steering rack or rotational compliance in the steering column could be incorporated if it adds value to the analysis.

In the following example the geometric ratio between the rotation of the steering column and the travel of the rack is already known, so it is possible to apply a motion input at the rack to ground joint that is equivalent to handwheel rotations either side of the straight ahead position. The jack part shown in Figure 6.37 can be used to set the suspension height during a steering test simulation. Typical output is shown in Figure 6.38 where the steering wheel angle is plotted on the x-axis and the road wheel angle is plotted on the y-axis. The three lines plotted represent the steering ratio test for the suspension in the static (initial model set up here), bump and rebound positions.

Having decided on the suspension modelling strategy and how to manage the relationship between the handwheel rotation and steer change at the road wheels, the steering inputs from the driver and the manoeuvre to be performed need to be considered.

6.12.3 Steering inputs for vehicle handling manoeuvres

The modelling of steering inputs suggests for the first time some representation of the driver as part of the full vehicle system model. Inputs to the handwheel are generally referred to as 'open loop' or 'closed loop'.

An open loop steering input requires a time-dependent rotation to be applied to the part representing a steering column or handwheel in the simulation model. In the absence of these bodies an equivalent translational input can be applied to the joint connecting a rack part to the vehicle body or chassis, assuming a suspension linkage modelling approach has been used.

We can consider an example of an open loop manoeuvre for a steering input where we want to ramp a steering input of 90° on between 1 and 1.5 s of simulation time. The function applied to the steering motion would be:

```
FUNCTION = STEP (TIME, 1, 0, 1.5, 90D)
```

In a similar manner if we wanted to apply a sinusoidal steering input with an amplitude of 30° and a frequency of 0.5 Hz we could use:

```
FUNCTION = 30D * SIN (TIME*180D)
```

For the lane change manoeuvre described earlier the measured steering wheel angles from a test vehicle can be extracted and input as a set of XY pairs that can be interpolated using a cubic spline fit. A time history plot for the steering inputs is shown in Figure 6.39 for lane change manoeuvres at 70 and 100 km/h.

By way of example, the MSC ADAMS statements that apply the steering motion to the steering column to body revolute joint and the spline data are shown in Table 6.3 for a 100 km/h lane change. The x values are points in time and the y values are the steering inputs in degrees. In the absence of measured data it is possible to construct an open loop single or double lane change manoeuvre using a combination of nested arithmetic 'If' functions with embedded step functions with some planning and care over syntax. Note that for a fixed steering input a change in vehicle configuration will produce a change in response so that the vehicle fails to follow a path.

The term "closed loop" is generally associated with some kind of controlled system. Any controlled system can be considered to consist of three elements — the 'plant' (the item to be controlled), the input to the plant and the output from the plant (Figure 6.40).

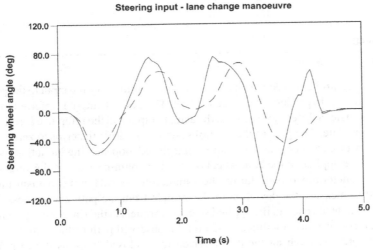

FIGURE 6.39

Steering input for the lane change manoeuvre at 70 km/h (dashed line) and 100 km/h (solid line).

Table 6.3 MSC ADAMS Statements for Lane Change Steering Inputs

```
MOTION/502,JOINT=502,ROT
,FUNC=(PI/180)*CUBSPL(TIME,0,1000)

SPLINE/1000
,X=0,1,2,3,4,5,6,7,8,9
,9.1,9.2,9.3,9.4,9.5,9.6,9.7
,9.8,9.9,10,10.1,10.2,10.3,10.4,10.5,10.6,10.7,10.8,10.9,11
,11.1,11.2,11.25,11.3,11.4,11.5,11.6,11.7,11.8,11.9,12,12.1
,12.2,12.3,12.4,12.5,12.6,12.7,12.8,12.9,13,13.1,13.2,13.3
,13.4,13.5,13.6,13.7,13.75,13.8,13.9,14,14.1,14.2,14.3,14.4,14.5
,14.6,14.7,14.8,14.9,15
,Y=0,0,0,0,0,0,0,0,0,0
,0,0,0,0,0,0,0
,0,0,-5,-17,-40,-55,-57,-52,-43,-30,-5,15,35,55,72,75,70,65,45,10
,-10,-17,-11,-7,15,50,75,67,66,60,50,35,0,-50,-95,-110,-100,-70,
-35,0
,20,20,35,55,20,-6,-3,-2,-1,0,0,0,0,0,0
```

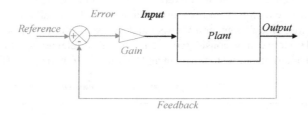

FIGURE 6.40

A generic control system.

When a controller is added to the system, its goal is to allow the input to the plant to be adjusted so as to produce the desired output. The desired output is referred to as the 'reference' state; a difference between the actual output and the reference is referred to as an 'error' state. The goal of the control system is to drive the error to zero.

As a start we can consider an example of a closed loop steering input that requires a torque to be applied to the handwheel or steering column such that the vehicle will follow a predetermined path during the simulation. As such a mechanism must be modelled to measure the deviation of the vehicle from the path and process this in a manner that feeds back to the applied steering torque. As the simulation progresses, the torque is constantly modified based on the observed path of the vehicle and the desired trajectory. Such an input is referred to as closed loop, as the response is observed and fed back to the input thus closing the control loop.

For a closed loop steering manoeuvre a column torque or rack force is applied, which will vary during the simulation so as to maintain the vehicle on a predefined

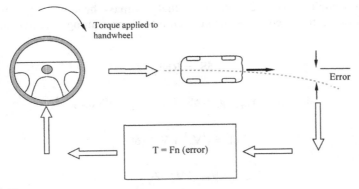

FIGURE 6.41

Principle of a closed loop steering controller.

path. This requires a steering controller to process feedback of the observed deviation from the path (error) and to modify the torque accordingly, as illustrated in Figure 6.41.

Evaluating the path error for the vehicle is somewhat less straightforward than it might appear. Odhams (2006) proposes a method using intrinsic coordinates.

A known trajectory is presumed, consisting of a compass heading angle, Ψ_t, that is defined as a function of trajectory distance, S_t. An absolute compass heading Ψ is calculated using the velocity vector, which may vary from the compass angle of the plane of symmetry of the vehicle θ due to the presence of a body slip angle β. For reasons of habit, East is declared as zero, which corresponds to the X-axis in many schemes:

$$\psi = \arctan\left(\frac{V_y}{V_x}\right) + \beta \tag{6.23}$$

Note that the 2-argument arctangent function ATAN2 may be preferred over the simpler formulation since the trajectory can go in any direction. For trajectories that describe a loop, some care is needed as the calculations wrap through $180°$. Distance travelled is simply formulated as the integral of the velocity vector. Noting that the orientation of the vector varies with time, the actual path on the ground can be calculated as X and Y:

$$S(t) = \int_0^t V(t)\,dt \tag{6.24}$$

$$X(t) = X_0 + \int_0^t V(t)\cos\psi(t)\,dt \tag{6.25}$$

$$Y(t) = Y_0 + \int_0^t V(t)\sin\psi(t)\,dt \tag{6.26}$$

For a vehicle travelling with an actual compass heading angle Ψ at an actual distance travelled of S, from Figure 6.42, presuming small angle theory applies:

$$\delta S_t = R \delta \delta_t \tag{6.27}$$

$$\delta n = \left(R + Y_{err} \right) \delta \psi_t = \delta S_t + Y_{err} \delta \psi_t = \delta S \cos \left(\psi_{err} \right) \tag{6.28}$$

$$\delta S = \sqrt{V_y{}^2 + V_x{}^2} \cdot \delta t \tag{6.29}$$

$$\psi_{err} = \psi - \psi_t \tag{6.30}$$

Rearranging to find δS:

$$\delta S = \frac{\delta S_t + Y_{err} \delta \psi_t}{\cos \left(\psi_{err} \right)} = \frac{\delta S_t + Y_{err} \dfrac{\delta \psi_t}{\delta S_t} \delta S_t}{\cos \left(\psi_{err} \right)} \tag{6.31}$$

Rearranging further:

$$\frac{\delta S}{\delta S_t} = \frac{1 + Y_{err} \dfrac{\delta \psi_t}{\delta S_t}}{\cos \left(\psi_{err} \right)} \tag{6.32}$$

$$\frac{\delta S_t}{\delta S} = \frac{\cos \left(\psi_{err} \right)}{1 + Y_{err} \dfrac{\delta \psi_t}{\delta S_t}} \tag{6.33}$$

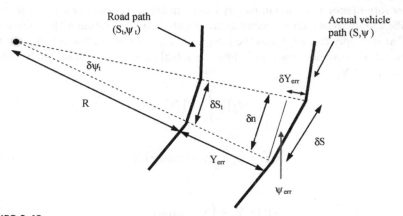

FIGURE 6.42

Intrinsic coordinate calculation proposed.

$$\frac{\delta S_t}{\delta S} = \frac{\cos(\psi - \psi_t)}{1 + Y_{err}\dfrac{\delta \psi_t}{\delta S_t}} \tag{6.34}$$

$$\frac{\delta S_t}{\sqrt{V_y{}^2 + V_x{}^2} \cdot \delta t} = \frac{\cos(\psi - \psi_t)}{1 + Y_{err}\dfrac{\delta \psi_t}{\delta S_t}} \tag{6.35}$$

$$\frac{\delta S_t}{\delta t} = \frac{\cos(\psi - \psi_t)\sqrt{V_y{}^2 + V_x{}^2}}{1 + Y_{err}\dfrac{\delta \psi_t}{\delta S_t}} \tag{6.36}$$

In the limit, as δt tends to zero:

$$\frac{dS_t}{dt} = \frac{\cos(\psi_{err})\sqrt{V_y{}^2 + V_x{}^2}}{1 + Y_{err}\dfrac{d\psi_t}{dS_t}} \tag{6.37}$$

Similarly, it can be written by inspection that

$$\delta Y_{err} = \delta S \sin(\psi_{err}) \tag{6.38}$$

Using the previous relationships:

$$\delta Y_{err} = \delta S \sin(\psi - \psi_t) \tag{6.39}$$

$$\frac{\delta Y_{err}}{\delta S} = \sin(\psi - \psi_t) \tag{6.40}$$

$$\frac{\delta Y_{err}}{\sqrt{V_y{}^2 + V_x{}^2} \cdot \delta t} = \sin(\psi - \psi_t) \tag{6.41}$$

$$\frac{\delta Y_{err}}{\delta t} = \sin(\psi - \psi_t)\sqrt{V_y{}^2 + V_x{}^2} \tag{6.42}$$

In the limit, as δt tends to zero:

$$\frac{dY_{err}}{dt} = \sin(\psi - \psi_t)\sqrt{V_y{}^2 + V_x{}^2} \tag{6.43}$$

In principle, the values for ψ_t, the derivative with respect to distance along the trajectory $\dfrac{d\psi_t}{dS_t}$ must be determined from the model and the differential equations integrated to give current values for actual distance travelled and path error.

When this method is implemented it appears satisfactory but on closer inspection there are some reservations about it. In order to illustrate them, a simple intended path was generated consisting of a straight line followed by a cosine ramp to a constant curvature (constant radius). A vehicle trajectory was generated by presuming a constant value for longitudinal speed and scaling the heading of the vehicle by some factor compared to the intended path. A scaling of unity returns a path error of zero when the calculations are functioning correctly.

Scaling the vehicle path by less than unity results in 'running wide' of the turn. Implementing the calculations as described above with a simple trapezoidal integration returns the results shown in Figure 6.43.

The calculations can be seen to return an error normal to the vehicle when the vehicle is running wide but the results are less obvious when the vehicle turns

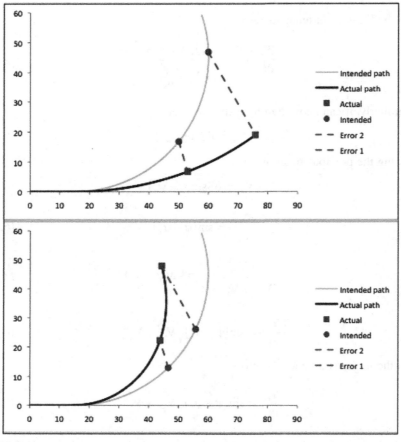

FIGURE 6.43

Odhams calculation for trajectory error.

more tightly than the intended trajectory. Returning an error normal to the vehicle can be seen to risk undefined results (or at least poorly conditioned results) when the vehicle trajectory is substantially perpendicular to the intended path.

A different formulation is possible, that returns the path error measured normal from the path, which gives the results shown in Figure 6.44 for the same two scenarios.

The result in Figure 6.44 is suggested as more intuitively reasonable as it finds the part of the path closest to the current position, but still in the context of a larger journey — which is to say it would not get confused by a figure-of-eight since it looks for a local minimum.

The process is simpler than the Odhams method but has the disadvantage of being iterative. Using similar notation to above, velocity, heading and, distance travelled and

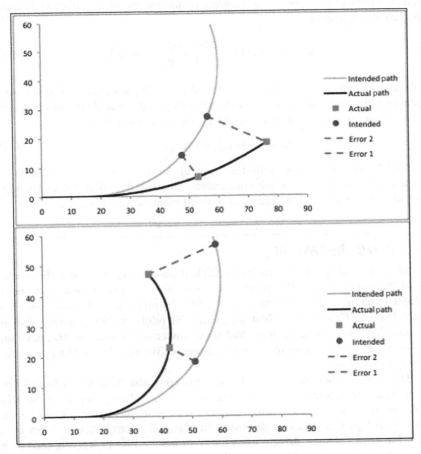

FIGURE 6.44

An alternative path error calculation scheme returns error normal to the intended path.

current location are calculated as before. Using the distance travelled S, the intended location of the vehicle can also be calculated or retrieved as a lookup operation:

$$X_t(S) = X_0 + S\int_0^S \cos(\psi_t(S))\,dS \qquad (6.44)$$

$$Y_t(S) = Y_0 + S\int_0^S \sin(\psi_t(S))\,dS \qquad (6.45)$$

A first estimate of the path error can be calculated simply:

$$e_1 = \sqrt{(X_t(S) - X(S))^2 + (Y_t(S) - Y(S))^2} \qquad (6.46)$$

Also, a relative angle between the error vector and the local trajectory heading, $\psi_t(S)$ can be calculated:

$$\varphi_{err}(S) = \arctan\left(\frac{Y_t(S) - Y(S)}{X_t(S) - X(S)}\right) - \psi_t(S) \qquad (6.47)$$

An assessment can be made of how close this is to 90° and, in the general case an improved estimate S_i can be made for the trajectory length that should be considered in computing the path error based on the previous iteration step:

$$S_i = S_{i-1} - e_{i-1} \cdot \cos(\varphi_{err}(S_{i-1})) \qquad (6.48)$$

This scheme can be iterated until subsequent estimates converge to some convenient amount. Some care is needed in handling the wrapping of trigonometric function and with iteration convergence, as with all numerical approaches.

6.13 Driver behaviour

It becomes inevitable with any form of vehicle dynamics modelling that the interaction of the operator with the vehicle is a source of both input and disturbance. In flight dynamics, the phenomenon of 'PIO' — pilot-induced oscillation — is widely known. This occurs when inexperienced pilots, working purely visually and suffering from some anxiety, find their inputs are somewhat excessive and causing the aircraft to, for example, pitch rhythmically instead of holding a constant altitude (Figure 6.45).

PIO is caused when the operator is unable to recognise the effects of small control inputs and therefore increases those inputs, before realising they were excessive and reversing them through a similar process. It is analogous to the 'excess proportional control gain oscillation' discussed in classical control theory (Leva et al, 2002). For road vehicles, drivers most likely to induce PIO in steering tend to be inexperienced or anxious drivers travelling at a speed with which they are uncomfortable. This type of PIO is not to be confused with the typical experience of drivers

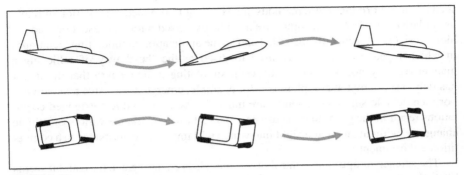

FIGURE 6.45

Pilot-Induced Oscillation (PIO) — not exclusively an aeronautical phenomenon.

of skidding vehicles when the initial skid is corrected but the vehicle subsequently 'fishtails' or simply departs in the opposite direction — this is a 'phasing at resonance' control error; the driver fails to apply a 'feed forward' (open loop, knowledge-based) correction in advance of the vehicle's response to compensate for the delay in vehicle response.

PIO also occurs in tractive (i.e. throttle) control inputs and is the reason even experienced drivers are incapable of travelling at a constant speed on highways; perception of changes in following distance is universally poor. If too little attention is spent on the driving task or if insufficient following distance is left, these PIOs become successively amplified by following drivers until the speed variation results in a 'shunt' accident. Radar-based cruise control systems will alleviate this risk but are no substitute for attentive driving while anything less than the whole vehicle fleet is fitted with it.

6.13.1 Steering controllers

There are a variety of controller models suitable for modelling driver behaviour in existence. Some are very complete while others, such as the two-loop feedback control model used by the authors, are simpler. The analyst must consider the needs of the simulation (and the financial constraints of the company) and choose the most appropriate level of modelling to achieve the task at hand. Driver models in general fall into two categories:

6.13.1.1 Optimum control models

Optimum control models use some form of 'penalty function' — a measure used to assess the quality of control achieved. For example, for a vehicle steering model the appropriate variable might be lateral deviation from the intended path. Optimum control models use repeated simulations of the event and numerical optimisation methods to 'tune' the parameters for a control system to minimise the value(s) of the penalty function(s) over the duration of the event of interest. For learnt events, such as circuit driving, these methods are excellent to produce a prediction of likely driver behaviour. However, some care must be exercised with their use. For road vehicles, drivers are generally unskilled and so the application of modelling techniques in which repeated solutions are used to discover the 'best' way of achieving a manoeuvre may not be appropriate when simulating a manoeuvre that the driver has only one attempt at completing — for example, emergency evasive manoeuvres. For race vehicle simulation, some care must also be exercised lest extended calculations result in the proof that the driver can adapt to a remarkable variety of vehicle changes — without any real insight into which will improve performance in a competition environment.

The science of optimum control driver models is changing at a pace and it seems likely that these will become the architecture for future autonomous driving products, since they can adapt to changing circumstances such as vehicle payload. A very promising thread appears to be appearing with Linear Quadratic Regulators (Sharp, 2005) with a previewing controller. By examining the properties of the vehicle, an optimised set of gains for a given speed and preview distance length can be calculated. The actual control input is the combined influence of all the path errors at multiple preview distances multiplied by their optimised gains. The form of the optimised gains reflects some kind of projection of the 'inverse dynamics' of the car in front of it and is interesting in its own right, as can be seen in Figure 6.46.

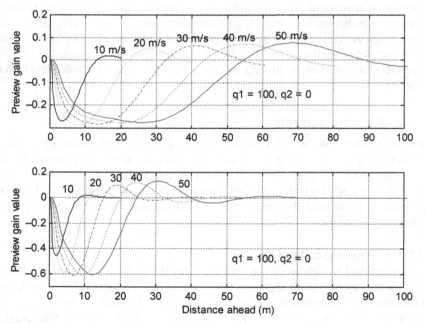

FIGURE 6.46

Preview gains for two different classes of car (Sharp, 2005).

The two vehicles are labelled 'Buick' and 'Ferrari' by Sharp and a number of similarities and differences are immediately apparent between the vehicles. Similar is the fact that a first lobe with a strongly negative character dominates the sequence. Were the controller imagined as a single point previewer, it is quite likely that the distance ahead for the most negative point would reflect a useful preview at different vehicle speeds. The negative sign is a function of the sign convention in use and means that the steering must be turned in opposition to the error in order to reduce it — the familiar 'negative feedback' mechanism.

An obvious difference is that the gain sequence is appreciably longer for the Buick in the top set of gains, requiring over 100 m of preview when travelling at 50 m/s before the gains settle to zero. (It is noted in Sharp's example that the gain sequence does not in fact settle to zero, being truncated before that happens; however, a sound knowledge of vehicle dynamics suggests that nothing particularly unusual is likely to happen to the gain sequence and it will probably converge in some manner similar to the Ferrari in the lower trace.)

The significance of the zero settling of the gain sequence can be imagined in the following way. Imagine a road profile that has an arbitrary discontinuity at some point ahead. Before the discontinuity, the road profile is a straight line described by $y(x) = 0$. After the discontinuity, the road profile is a straight line described by $y(x) = 1$. If the gain sequence is thought of as 'extending' in front of the car, then if the discontinuity is past the zero settling point the sum of all the inputs is zero, and the driver model does nothing. As the nonzero points reach the discontinuity, the driver begins to steer by the same amount as the gain sequence (since the discontinuity is of unit magnitude). It can be seen this will result in a small movement *in opposition to what would be expected to negotiate the discontinuity*. As the gain sequence crosses further, the reversal of the gain sequence will overpower the initial out-of-the-corner input and produce a steer input into the corner. A short while later the steering will reverse again and will undergo yet a further reversal before the final decisive steer input to negotiate the discontinuity. This apparently baffling course of events will be familiar to anyone who has ever enjoyed watching classic rally cars — it is a so-called 'Scandinavian Flick'.

The gain sequence can thus be seen as some kind of measure of how far in advance of any intended path change the driver has to plan, or to view it from the other end, how late the driver can leave things before making a decision. In general, the later a vehicle can adapt to changing circumstances, the better we might declare that elusive thing called 'handling'. What is particularly intriguing about Sharp's approach is that the zero settling point is some kind of invariant property of the vehicle at a given speed; no actual manoeuvre simulations need to be carried out to evaluate whether or not an improvement has been made — a shorter distance to zero settling allows later adaptation; the Ferrari is handsomely better than the Buick.

These types of calculations are not currently implemented in any of the typical MBS general purpose software tools available at the time of writing. However, they are well documented enough in the literature to be implemented with typical commercial tools; there is no 'magic bullet' in the formulation.

6.13.1.2 Moment-by-moment feedback models[1]

Such models are really a subset of the optimum control models described above — the optimum control method repeatedly uses feedback models in order to discern the best state of tune for the controller. When the feedback controllers are used alone, the analyst must set their tuning. Although the absence of 'automated' correlation makes them less appropriate for circuit racing, it also adds clarity in the sense that the parameters, once set, remain constant and so changes in the vehicle behaviour and/or driver inputs can be readily understood.

In general, the driver behaves as the most generic form of loop-closing controller. There are several attractive control technologies represented in the literature and some of their proponents believe they represent a 'one size fits all' solution for the task of applying control to any system. The competing technologies are outlined for comparison.

6.13.1.2.1 Logic controller

A logic controller produces output that has only certain possible values. For example, if a driver model was implemented using logic, the logic might be 'if the vehicle is to the left of the intended path, steer right and vice versa'. With a logic controller, the amount of steer is fixed and so any control of the vehicle would be achieved as a series of jerks, oscillating about the intended path. While probably functional it would be unlikely to represent any normal sort of driver.

6.13.1.2.2 PID controller

PID stands for 'proportional, integral and derivative'. The error state is used in three ways; used directly, a control effort is applied in proportion to (and opposition to) the error — this is the 'P', proportional, element of the control. The fact that the control effort is in opposition to the error is important, since otherwise the control effort would increase the error instead of reducing it. For this reason, such systems are often referred to as 'negative feedback' systems. The error can also be integrated and differentiated, with control forces applied proportional to the integral and the differential — these are the 'I' and the 'D' terms in the controller. One or more of the terms may not be used at all in any particular controller. An analogy for PID controllers can be found in vehicle suspensions. If the ride height is thought of as the desired output, then individual components of the suspension behave as parts of a control system. The springs produce a force proportional to the change in ride height and the dampers produce a force proportional to the derivative of ride height. Real dampers are often nonlinear in performance, and there is nothing to stop nonlinear

---

[1]Feedback models ought to be known as 'instantaneous feedback' models but the word instantaneous has become slightly muddled in recent times. It should be used unambiguously to mean 'existing for a moment in time' but has become sadly confused with 'instant', meaning immediate. Instant feedback would imply the inability to represent transport delays and the like in the controller model, which is incorrect. The use of moment-by-moment is therefore preferred although it introduces confusion with moment in the sense of torque.

gains being used for any of the control terms. The D term has the effect of introducing damping into the control system. An analogy for the I term is a little harder to come by. The best analogy is that of a self-levelling unit fitted to the suspension that applies a restoring force related to the length of time the vehicle has been at the wrong ride height and how wrong the ride height is. (This is an imperfect analogy for many reasons but allows the notion to be understood at least.) In real systems, when the output is *nearly* the same as the reference state it is frequently the case that the control forces become too small to influence the system, either because of mechanical hysteresis or sensor resolution or some similar issue. One important measure of the quality of any control system is the accuracy with which it achieves its goals. Such an offset characterises an inaccurate system; an integral term 'winds up' from a small error until powerful enough to restore the system to the reference state. Thus for classical control, integral terms are important for accuracy. However, since they take some time to act they can introduce delays into the system. In general PID controllers have the advantage that they produce 'continuous' output — that is to say all the derivatives are finite, the output has no steps — which is quite like the behaviour of real people.

6.13.1.2.3 Fuzzy logic

Fuzzy logic was first described in the 1960s but found favour in the 1980s as a fashionable 'new' technology. Notions of 'true' and 'false' govern 'logic' in computer algorithms. Simple control systems assess a set of conditions and make a decision based on whether or not such variables are true or false. Fuzzy logic simply defines 'degrees of truth' by using numbers between 0 and 1 such that the actions taken are some blend of actions that would be taken were something completely true and other actions that would be taken were something completely false. Fuzzy logic is most applicable to control systems where actions taken are dependent on circumstance and where a simple PID controller is unable to produce the correct output in every circumstance. For example, throttle demand in a rear-wheel drive vehicle model might be controlled with a PID controller to balance understeer; however, too much throttle would cause oversteer and some more sophisticated blend of steer and throttle input would be required to retain control under these circumstances. A fuzzy logic model could conceivably manage the transition between the two strategies.

6.13.1.2.4 Neural networks

Where the system of interest is highly nonlinear and a lot of data exists that describes desired outputs of the system for many different combinations of inputs, it is possible to use a neural network to 'learn' the patterns inherently present in the data. A neural network is quite simply a network of devices that are 'neuron-like'. Neurons are the brain's building blocks and are switches with multiple inputs and some threshold to decide when they switch. In general, neural networks are run on transistor devices or in computer simulations. They require a period of 'training' when they learn what settings need to be made for individual neurons in order to produce the required outputs. Once trained, they are extremely rapid in operation since there is very little 'processing' as such, simply a cascade of voltage switching through the transistor network. If the network is implemented as semiconductor transistors then it works at a speed governed only by the latency of the semiconductor medium — extremely fast indeed. Neural networks are extremely useful for controlling highly nonlinear systems for which it is too difficult to code a traditional algorithm. However, the requirement for a large amount of data can make the learning exercise a difficult one. Recent advances in the field reduce the need for precise data sets of input and corresponding outputs; input data and 'desirable outcome' definitions allow neural networks to learn how to produce a desirable outcome by identifying patterns in the incoming data. Such networks are extremely slow in comparison to the more traditional types of network during the learning phase. In general, for driver modelling there is little applicability for neural networks at present due to the lack of fully populated data sets with which to teach them. It is also worth commenting that for any input range that was not encountered during the learning phase, the outputs are unknown and may not prove desirable. This latter feature is not dissimilar to real people; drivers who have never experienced a skid are very unlikely to control it at the first attempt. Also worth mentioning is that if the characteristics of the plant model change once the training period is complete then the controller may not perform well. This is also quite like real drivers; many drivers underapply the brakes in laden vehicles if the loading condition is infrequent.

6.13.1.2.5 System identification

System identification is a useful technique, not dissimilar in concept to neural networking. A large amount of data is passed through one of several algorithms that produce an empirical mathematical formulation that will produce outputs like the real thing when given the same set of inputs. The formulation is more mathematical than neural networking and so the resulting equations are amenable to inspection — although the terms and parameters may lack any immediately obvious significance if the system is highly nonlinear. System identification methods select the level of mathematical complexity required to represent the system of interest (the 'order' of the model) and generate parameters to tune a generic representation to the specific system of interest. As with neural networks, the representation of the system for inputs that are beyond the bounds of the original inputs (used to identify the model) is undefined. System identification is useful as a generic modelling technique and so has been successfully applied to components such as dampers as well as control system and plant modelling. System identification is generally faster to apply than neural network learning but the finished model cannot work as quickly. The same data set availability problems for neural networking also mean system identification is not currently applicable to driver modelling.

6.13.1.2.6 Adaptive controllers

Adaptive control is a generic term to describe the ability of a control system to react to changes in circumstances. In general, people are adaptive in their behaviour and so it would seem at first glance that adaptive control is an appropriate tool for modelling driver behaviour. Optimum control models, described above, generally use some form of adaptive control to optimise the performance of a given controller architecture to the system being controlled and the task at hand. Adaptation is a problem in real-world testing since it obscures differences in performance; equally it can obscure performance changes and so adaptive modelling of driver behaviour is not preferred except for circuit driving. Several techniques come under the headline of adaptive control; the simplest is to change the control parameters in a predetermined fashion according to the operating regime, an operation referred to as 'gain scheduling'. Gain is the term used for any treatment given to an error state before it is fed to an input — thus the PID controller described above has a P-gain, and I-gain and a D-gain. It might be, for example, that under conditions of opposite lock the P-gain is increased since the driver needs to work quickly to retain control, or under conditions of increasing speed the P-gain is reduced since slower inputs are good for stability at higher speeds. A more complex method is to carry a model of the plant on board in the controller and to use it to better inform some form of gain scheduling, perhaps using information that cannot readily be discerned from onboard instrumentation — such as body slip angle. This is referred to as a Model Reference Adaptive Scheme (MRAS). A further variation on the theme is to use the controller to calculate model parameters using system or parameter identification methods (described above). The control system parameters can be modified based on this information — in effect there is an ongoing redesign of the control system using a classical deterministic method, based on the reference state and the plant characteristics according to the latest estimate. This is referred to as a 'self-tuning regulator' and is useful for unpredictably varying systems. Finally, a method known as 'dual control' intentionally disturbs the system in order to learn its characteristics, while simultaneously controlling it towards a reference state. In many ways this is similar to a top-level rally driver stabbing the brakes in order to assess friction levels while disturbing the overall speed of the vehicle as little as possible; the knowledge gained allows the driver to tune their braking behaviour according to recently learnt characteristics. Such behaviour is in marked contrast to circuit drivers, who concentrate on learnt braking points and sometimes have difficulty adapting to changing weather conditions. With the exception of the simplest gain scheduling methods, in general adaptive control techniques are unsuitable for the modelling of driver behaviour as part of any practicable process although they may well have a place in future autonomous systems. Once again the variation in simulation output cannot readily be traced to any particular aspect of the system and hence the success or otherwise of an intended modification is difficult to interpret.

In the light of the preceding description, the authors believe a PID controller, with some form of simple gain scheduling, is most appropriate for the modelling of driver behaviour in order to usefully influence the design of a vehicle in an MBS context. Note that for other outcomes — for example, developing a full authority digital autonomous control system — the preferred method will probably be different. The art of implementing a successful model is in selecting the state variables within the model to use with the controller.

6.13.2 A path following controller model

The first hurdle to be crossed is the availability of suitable state variables and the use of gain terms to apply to them. Typically in an MBS model, many more variables are available than in a real vehicle. Within the model, these variables can be the subject of differential equations in order to have available integral and differential terms. Table 6.4 shows a portion of a command file from MSC ADAMS implementing those terms for yaw rate. While it is a working example, no claim is made that is in any sense optimum.

Such variables can usually be manipulated within the model using the programming syntax provided with the code being used. For general purpose software

---

Table 6.4 A Portion of an MSC ADAMS Command File Showing the Implementation of Differential Equations to Retrieve and Use Integral and Derivative Terms for a State Variable

```
! -- Derivative Term - not generally used --
part create equation differential_equation &
   differential_equation_name = .test.yaw_rate_error_equation_1 &
   adams_id = 3 &
   comments = "Yaw Rate Error Equation - Implicit" &
   initial_condition = 0.0 &
   function = "DIF(3)-varval(yaw_rate_error)" &
   implicit = on &
   static_hold = off
data_element create variable &
   variable_name = yaw_rate_error_derivative &
   function="DIF1(3)"

! -- Integral Term --
part create equation differential_equation &
   differential_equation_name = .test.yaw_rate_error_equation_2 &
   adams_id = 4 &
   comments = "Yaw Rate Error Equation - Explicit" &
   initial_condition = 0.0 &
   function = "varval(yaw_rate_error)" &
   implicit = off &
   static_hold = off
data_element create variable &
   variable_name = yaw_rate_error_integral &
   function="(DIF(4))"

! Steer input torque in response to path error.
force create direct single_component_force &
   single_component_force_name=yaw_rate_handwheel_torque &
   type_of_freedom=rotational &
   action_only = on &
   i_marker_name = .hand_wheel_column.m_wheel_column &
   j_marker_name = .hand_wheel_column.m_wheel_column &
   function="(", &
         "     VARVAL(yaw_rate_error)                * VARVAL(yp_gain) ", &
         "   + VARVAL(yaw_rate_error_integral)    * VARVAL(yi_gain) ", &
         "   + VARVAL(yaw_rate_error_derivative) * VARVAL(yd_gain) ", &
         ")", &
         "* STEP(TIME,0.0,0.0,1.0,1.0)"
```

tools, the format of such calculations can appear a little clumsy but this soon disappears with familiarity. For codes such as MATLAB/Simulink the implementation of control systems is arguably easier since they are written with the prime objective of control system modelling. However, the modelling of the vehicle as a plant is more difficult within these systems and so there is an element of swings and roundabouts if choosing between the codes. General purpose software tools have a history in very accurate simulation of mechanical systems and can be coerced into representing control systems. Codes like MATLAB and MATLAB/Simulink are the reverse; they have a history in very detailed control system simulation and can be coerced into representing mechanical systems. For this reason, a recent development suggests using each code to perform the tasks at which it is best; this is often referred to as 'co-simulation'. Since the first edition of the book, the state of the art has advanced substantially and a recent development has been the development of a software interfacing protocol known as the 'functional mock-up interface' that looks set to ease this difficulty, although its implementation is very incomplete at present. The authors' experiences to date have been universally disappointing for entirely prosaic reasons — the speed of execution remains extremely poor and the robustness of the software suppliers in dealing with different releases of each other's product remains somewhat inconsistent. The effort required to persuade the relevant software to work in an area where it is weak is usually made only once and in any case the additional understanding gained is almost always worthwhile for the analyst involved. Until the performance and robustness of the software improves, the authors do not favour co-simulation except for the most detailed software verification exercises.

The next hurdle to be crossed is the representation of the intended behaviour of the vehicle — the 'reference' states. Competition-developed lap simulation tools use a 'track map' based on distance travelled and path curvature. This representation allows the reference path to be of any form at all and allows for circular or crossing paths (e.g. figures of eight) to be represented without the one-to-many mapping difficulties that would be encountered with any sort of y versus x mapping. Integrating the longitudinal velocity for the vehicle gives a distance-travelled measure that shows itself to be tolerably robust against drifting within simulation models. Using this measure, the path curvature can be surveyed in the vicinity of the model.

Some authors favour the use of a preview distance for controlling the path of the vehicle, with an error based on lateral deviation from the intended path, which has already been discussed.

An alternative method, used by the authors with some success for a variety of extreme manoeuvres, is to focus on the behaviour of the front axle. This model fits with the author's experience (Harty) of driving at or near the handling limit, particularly on surfaces such as snow where large body slip angles highlight the mechanisms used in the driver's mind. High-performance driving coaches (Palmer, 1999) rightly concentrate on the use of a 'model' the driver needs in order to retain control in what would otherwise become stressful circumstances of nonlinear vehicle behaviour and multiple requirements for control — typically vehicle orientation (body slip angle) and velocity (path control). Useful learning occurs on low-grip environments that can be readily transferred across to high grip. In low grip environments, the extreme nonlinearity of response of the vehicle can be explored at low speeds and with low stress levels, allowing the driver to piece together a model to be used within their own heads; it is then a matter of practice to transfer the lessons to a high-grip environment. The same concepts can be used to explore the behaviour of a driver model within an MBS software environment.

The formulation used is described below. All subscripts x and y are in the vehicle reference frame. The ground plane velocity V_g is given from the components V_x and V_y using

$$V_g = \sqrt{V_x^2 + V_y^2} \qquad (6.49)$$

The demanded yaw rate ω_d is found from the forward velocity V_x and path curvature k using

$$\omega_d = V_g\, k \qquad (6.50)$$

The body slip angle β is found from the velocities V_y and V_g using

$$\beta = \arcsin\left(\frac{V_y}{V_g}\right) \qquad (6.51)$$

The centripetal acceleration A^P is given from the components of acceleration A_x and A_y using

$$A^P = A_y \cos(\beta) + A_x \sin(\beta) \qquad (6.52)$$

The front axle no-slip yaw rate ω_{fNS} is found from the centripetal acceleration A^P, the yaw acceleration α_z, the distance, a, from the mass centre to the front axle and the ground plane velocity V_g using

$$\omega_{fNS} = \frac{A^P - \alpha_z\, a}{V_g} \qquad (6.53)$$

The yaw error ω_{err} is then found from the demanded yaw rate ω_d and the front axle no-slip yaw rate ω_{fNS} using

$$\omega_{err} = \omega_d - \omega_{fNS} \qquad (6.54)$$

The implementation of Eqns (6.49)–(6.54) is illustrated, using again an example of the MSC ADAMS command file format, in Table 6.5, representative of real vehicle and driver behaviour (Figure 6.47). The simulated driver and vehicle behaviour for a post-limit turn-in event is compared here to a real vehicle. Note the freewheeling analytical model (of a significantly different vehicle) displays greater body slip angle, while the real vehicle displays greater oversteer.

For a variety of events, this formulation produces good driver/vehicle behaviour.

Table 6.5 Command File Sample for 'Front Axle Control' Driver Model

```
data_element create variable &
    variable_name = ground_plane_velocity &
    function = "(              (", &
        "VX(m_body_CG,base)**2 +",    &
        "VY(m_body_CG,base)**2",    &
              ")**0.5 ) / 1000"

data_element create variable &
    variable_name = demanded_yaw_rate &
    function = "varval(ground_plane_velocity) *", &
        "AKISPL(varval(path_length),0,path_curvature_spline)"

data_element create variable  &
    variable_name = beta &
    function="ASIN(VY(m_body_CG,base,m_body_CG)/", &
        "(varval(ground_plane_velocity)+0.00001))"

data_element create variable  &
    variable_name = centacc &
    function="(VARVAL(latacc)*COS(VARVAL(beta)))+", &
        "(VARVAL(longacc)*SIN(VARVAL(beta)))"

data_element create variable  &
    variable_name = front_axle_no_slip_yaw &
    function="-(", &
            "VARVAL(centacc)   ", &
            "- WDTZ(m_body_CG,base,m_body_CG)", &
            "* DX(m_body_CG,mfr_upright_wheel_centre,m_body_CG)", &
        " )", &
        " /", &
        " ( varval(ground_plane_velocity) + 0.00001 )"

data_element create variable &
    variable_name = yaw_rate_error &
    function = "  varval(demanded_yaw_rate) - varval(front_axle_no_slip_yaw)"
```

6.13.3 Body slip angle control

Skilled drivers, particularly rally drivers, frequently operate at large body slip angles. Colloquially, there is much talk of body slip angles being in excess of 45° but recorded data suggest this is not the case despite appearances. Large body slip angles generally slow progress; although some of the yaw transients are rapid, in general the actual body slip angles are comparatively small (Figure 6.48). In the authors' experience, drivers greatly overestimate body slip angle subjectively.

The steering system on a vehicle has only around 30° of lock and so realistically, control beyond these body slip angles is unlikely without very large amounts of space indeed (Figure 6.49).

Most drivers are acutely sensitive to the rate of change of body slip angle, albeit they do not always respond correctly to it. Instead a 'threshold' behaviour appears common, with drivers neglecting body slip angle until either the angle becomes large or its rate of change becomes large. For road cars, our goals are to have a road car manage its own body slip angle so as not to put pressure on drivers in an area where,

FIGURE 6.47

Driver and vehicle behaviour for a post-limit turn-in event.

Photograph courtesy of Don Palmer, www.donpalmer.co.uk.

in general, skill is lacking. For driver modelling purposes, a separate body slip angle control loop is desirable to catch spins. It need not be terribly sophisticated since if it is invoked then we have to some extent failed in our goal of having the car manage its own body slip angle. Such behaviour is desirable in the real vehicle too and is the goal of active intervention systems such as brake-based stability control systems; however, the robust sensing of body slip angle still proves elusive in a cost-effective manner

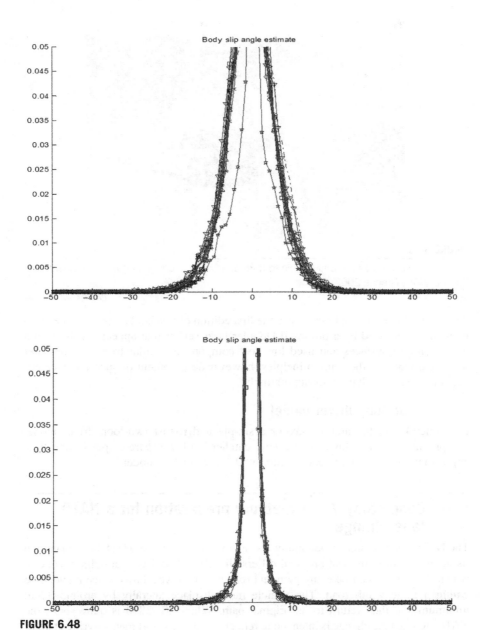

FIGURE 6.48

Probability density for body slip angle estimates — Greece 2002 (top) and Germany 2002 (bottom) for Petter Solberg, Subaru World Rally Team.

FIGURE 6.49

Large body slip angles are unavailable to normal drivers except as part of an accident. Subaru WRC, Greece 2002.

Courtesy of Prodrive.

despite its apparent simplicity. Since the first edition of this book, a frisson of excitement was generated by a proposed Doppler laser sensor that appeared robust on a wide range of surfaces and used low-cost components similar to many computer mice that work on the same principles; however development of such a device appears to have been halted, disappointingly.

6.13.4 Two-loop driver model

For general use, the authors favour a simple and robust two-loop driver model comprising a path follower and a spin catcher, with a separate speed control as appropriate to the task at hand. Figure 6.50 shows such a model.

6.14 Case study 7 — trajectory preparation for a NATO lane change

The NATO lane change is essentially a scaled version of the ISO3888 lane change in its geometry, only the test protocol differs slightly. It consists of a defined zone in which the vehicle can take any path and reflects a real-world avoidance manoeuvre within a finite width road. The vehicle must displace laterally by around 3.5 m minimum and then return to its original path. Where ISO3888 is quite short, the NATO double lane change is intended to reflect maximum effort manoeuvring at highway speeds for larger vehicles and is thus commensurately longer. The length and width of the manoeuvring zone are related to the vehicle proportions, as shown in Figure 6.51.

It can be seen in Figure 6.51 that the course is entirely symmetric. For the NATO lane change the vehicle is driven up to the manoeuvre start and the throttle released;

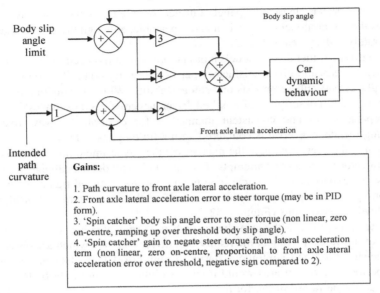

FIGURE 6.50

A two-loop driver model.

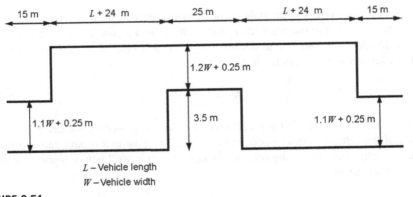

FIGURE 6.51

NATO (AVTP 03-160) lane change test course.

the overrun condition is typically the most difficult in terms of stability. Although the vehicle loses speed as it enters the second of the two manoeuvres, it is typically unsettled dynamically and so the second transition is often the more problematic of the two.

For the subject vehicle, the dimensions used are 6 m length and 2.4 m width.

With real vehicles the performance of a given vehicle through the lane change can be substantially modified with driver skill levels and so when comparing vehicles it is preferable to use a panel of drivers or at least a consistent driver. In the analysis the driver model is consistent and repeatable, although there is no guarantee it accurately reflects real drivers.

In contrast to other driver models it has no preview. It has been used for a number of vehicle classes and gives satisfactory performance. By controlling only the front axle it places particular emphasis on vehicle stability. While it is undoubtedly some way short of a full description of a skilled driver, nevertheless it exercises the vehicle in a representative and consistent manner — for example the yaw overshoot following the steer reversal is countered with a little opposite lock.

When real drivers complete the manoeuvre they experiment noticeably with the exact trajectory over several attempts. This simulation model is unable to learn from its previous experiences but some modification of the intended trajectory is worth mentioning. For simplicity, only the first half of the manoeuvre is considered in the following discussions (Figure 6.52).

It is tempting to draw the lane change path as a series of connected arcs (O'Hara, 2005) but this does not reflect the reality of a driver's inputs, which are observed to be somewhat fluid and continuous during a high-effort lane change, without the dwell periods a series of arcs would imply. An improvement might be to view the path as a 'cosine ramp' of the form:

$$Y = \frac{3.5}{2}\left(1 - \cos\left(\frac{X\pi}{24+L}\right)\right) \tag{6.55}$$

It is presumed that Y is constant before 0 m and after 30 m. It is a trivial matter to differentiate the curve once for 'compass heading' and again for curvature. Curvature is 1/radius and thus some notional centripetal acceleration can be computed with no knowledge of vehicle characteristics:

$$A_y = \frac{d^2Y}{dX^2}V \tag{6.56}$$

With due regard for the lateral acceleration limit imposed by the centre of mass height of the vehicle, it can be seen in Figure 6.53 that using this path geometry will limit the vehicle to something around 40 mph, which is well below reported values for this class of vehicle.

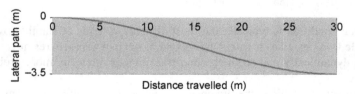

FIGURE 6.52

A cosine ramp trajectory for a lane change.

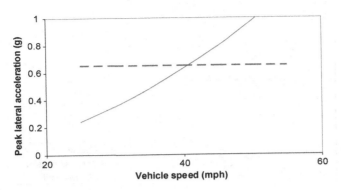

FIGURE 6.53

A comparison of trajectory centripetal acceleration (solid) with the centre of mass height-imposed limit (dashed) for the cosine ramp path.

The intended line across the ground is not continuously differentiable. Some further thought suggests a route across the ground made up of several segments — a turn-in segment represented by a cosine ramp up to maximum curvature, a reversal segment from maximum curvature in one direction to maximum curvature in the other and finally a turn-out segment represented by a third cosine ramp. This function is continuously differentiable and the relative length of the turn-in, reversal and turn-out segments can be tuned.

When considering the task in detail, it is noted that the midsection is wider than the vehicle. By positioning the vehicle fractionally away from the very edge of the entrance gate, by tuning the relative length of the turn-in and turn-out sections and by running out to the very widest part of the path available, the length for the manoeuvre is extended from a nominal 30 to 42.5 m, as shown in Figure 6.54.

If this 'optimised cosine' form is compared with the initial cosine form, a substantial reduction in centripetal acceleration is realised, as shown in Figure 6.55.

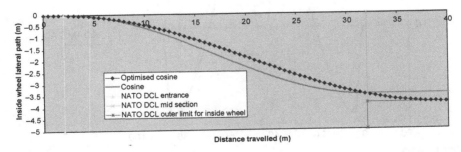

FIGURE 6.54

An improved lane change path that stays within NATO constraints and is more typical of real driver behaviour.

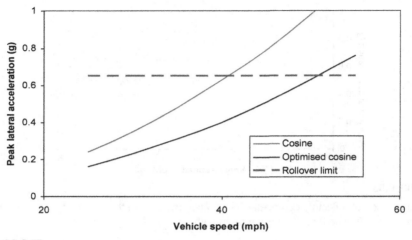

FIGURE 6.55

A comparison between the optimised cosine form and the previous cosine ramp in comparison within the rollover limit.

A comparison can be made between the base vehicle simulation and the trajectory-only values at 30 mph and shows that even at 30 mph the response is being coloured and limited by the dynamics of the vehicle, but that nevertheless there is broad agreement in Figure 6.56.

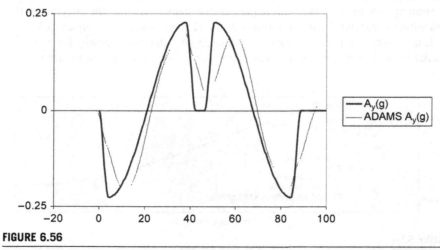

FIGURE 6.56

A comparison between the trajectory-predicted lateral acceleration and the results achieved with a detailed multibody system model at 30 mph.

Thus it can be seen that in order to deliver the best possible performance from a given model, the preparation of the trajectory is of some importance, and in fact the trajectory for a given manoeuvre shows substantial scope for optimisation — this can easily be overlooked in the pursuit of model detail and accuracy.

6.15 Case study 8 — comparison of full vehicle handling models

As mentioned at the start of this chapter the use of modern MBS software provides users with the capability to develop a model of a full vehicle that incorporates all the major vehicle subsystems. Clearly the development of such a model is dependent on the stage of vehicle design and the availability of the data needed to model all the subsystems. For the vehicle dynamics task, however, the automotive engineer will want to carry out simulations before the design has progressed to such an advanced state.

In the following case study the level of vehicle modelling detail required to simulate a 'full vehicle' handling manoeuvre will be explored. The types of manoeuvres performed on the proving ground are discussed in the next chapter but as a start we will consider a 100 km/h double lane change manoeuvre. The test procedure for the double lane change manoeuvre is shown schematically in Figure 6.57.

For the simulations performed in the following case study the measured steering wheel inputs from a test vehicle have been extracted and applied as a time-dependent handwheel rotation (Figure 6.58) as described in Section 6.12.3.

To appreciate the use of computer simulations to represent this manoeuvre, an example of the superimposed animated wireframe graphical outputs for this simulation is given in Figure 6.59.

In this study the influence of suspension modelling on the accuracy of the simulation outputs is initially discussed based on results obtained using the four vehicle

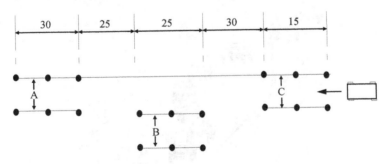

A – 1.3 times vehicle width + 0.25m

B – 1.2 times vehicle width + 0.25m

C – 1.1 times vehicle width + 0.25m

FIGURE 6.57

Lane change test procedure.

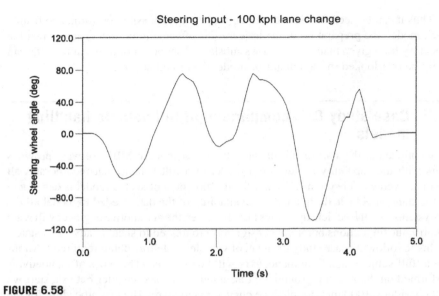

FIGURE 6.58

Steering input for the lane change manoeuvre.

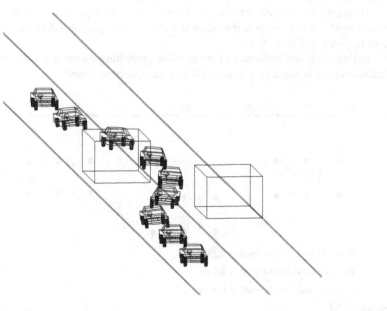

FIGURE 6.59

Superimposed graphical animation of a double lane change manoeuvre.

models described in Section 6.4 and summarised schematically again here in Figure 6.60. The models shown can be thought of as a set of models with evolving levels of elaboration leading to the final linkage model that involves the modelling of the suspension linkages and the bushes.

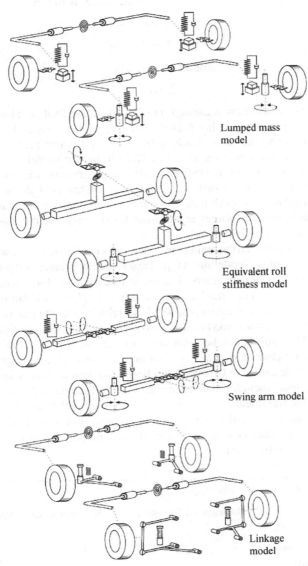

Lumped mass model

Equivalent roll stiffness model

Swing arm model

Linkage model

FIGURE 6.60

Modelling of suspension systems.

For each of the vehicle models described here it is possible to estimate the model size in terms of DOF in the model and the number of equations that MSC ADAMS uses to formulate a solution. The calculation of DOF in a system is based on the Gruebler equation given in Chapter 3. It is therefore possible for any of the vehicle models to calculate DOF in the model. An example is provided here for the equivalent roll stiffness model where DOF can be calculated as follows:

$$
\begin{aligned}
\text{Parts} \quad 9 \times 6 &= 54 \\
\text{Rev} \quad 8 \times -5 &= -40 \\
\underline{\text{Motion } 2 \times -1} &= \underline{-2} \\
\Sigma_{\text{DOF}} &= 12
\end{aligned}
$$

In physical terms it is more meaningful to describe these DOF in relative terms as follows. The vehicle body part has 6 DOF. The two axle parts each have one rotational DOF relative to the body. Each of the four road wheel parts has one spin DOF relative to the axles making a total of 12 DOF for the model.

When a simulation is run in MSC ADAMS the programme will also report the number of equations in the model. As discussed in Chapter 3 the software will formulate 15 equations for each part in the model and additional equations representing the constraints and forces in the model. On this basis the size of all the models is summarised in Table 6.6.

The size of the model and the number of equations is not the only issue when considering efficiency in vehicle modelling. Of perhaps more importance is the engineering significance of the model parameters. The roll stiffness model, for example, may be preferable to the lumped mass model. It is not only a simpler model but is also based on parameters such as roll stiffness that will have relevance to the practicing vehicle dynamicists. The roll stiffness can be measured on an actual vehicle or estimated during vehicle design. This model does, however, incorporate rigid axles eliminating the independent suspension characteristics. Note that in this case study an interpolation tyre model of the type described in Chapter 5 has been used with each vehicle model.

Measured outputs including lateral acceleration, roll angle and yaw rate can be compared with measurements taken from the vehicle during the same manoeuvre on the proving ground to assess the accuracy of the models. By way of example the yaw rate predicted by simulation with all four models is compared with measured track test data in Figures 6.61–6.64.

Table 6.6 Vehicle Model Sizes

Model	Degrees of Freedom	Number of Equations
Linkage	78	961
Lumped mass	14	429
Swing arm	14	429
Roll stiffness	12	265

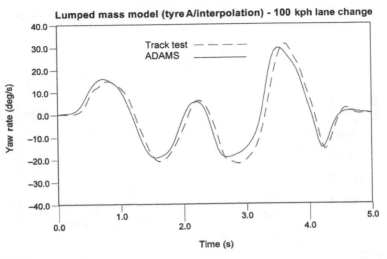

FIGURE 6.61

Yaw rate comparison — lumped mass model and test.

This material has been reproduced from the Proceedings of the Institution of Mechanical Engineers, K2 Vol. 214 'The modelling and simulation of vehicle handling, Part 4: handling simulation', M.V. Blundell, page 80, by permission of the Council of the Institution of Mechanical Engineers.

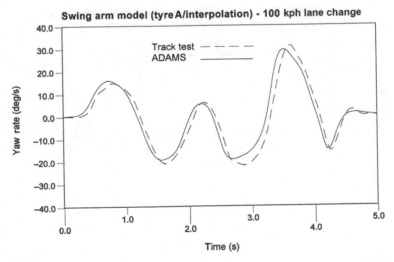

FIGURE 6.62

Yaw rate comparison — swing arm model and test.

This material has been reproduced from the Proceedings of the Institution of Mechanical Engineers, K2 Vol. 214 'The modelling and simulation of vehicle handling, Part 4: handling simulation', M.V. Blundell, page 80, by permission of the Council of the Institution of Mechanical Engineers.

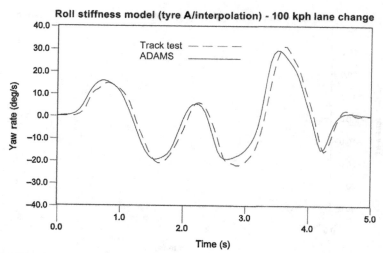

FIGURE 6.63

Yaw rate comparison — roll stiffness model and test.

This material has been reproduced from the Proceedings of the Institution of Mechanical Engineers, K2 Vol. 214 'The modelling and simulation of vehicle handling, Part 4: handling simulation', M.V. Blundell, page 81, by permission of the Council of the Institution of Mechanical Engineers.

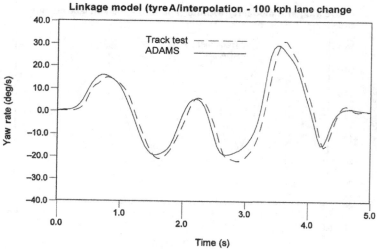

FIGURE 6.64

Yaw rate comparison — linkage model and test.

This material has been reproduced from the Proceedings of the Institution of Mechanical Engineers, K2 Vol. 214 'The modelling and simulation of vehicle handling, Part 4: handling simulation', M.V. Blundell, page 81, by permission of the Council of the Institution of Mechanical Engineers.

Examination of the traces in Figures 6.61–6.64 raises the question as to how an objective assessment of the accuracy of the simulations may be made. Accuracy is not a 'yes/no' quantity, but instead a varying absence of difference exists between predicted (calculated) behaviour and measured behaviour. Such a 'difference' is commonly referred to as an 'error'. This definition neatly sidesteps two other difficulties:

- is the measured data what actually happens in the absence of measurement?
- is the measured data what actually happens during service?

For example, the mass-loading effect of accelerometers may introduce inaccuracies at high frequencies and could mean that the system of interest behaves differently when being measured to when not. The accuracy of controlled measurements in discerning the behaviour of the system when in normal uncontrolled use is another matter entirely. Both topics are far from trivial.

In this case other questions arise such as:

- does the model data accurately represent the vehicle conditions on the day of the test?
- does the tyre test data obtained on a tyre test machine accurately represent the condition of the test surface and tyres used on the day of the test?
- how repeatable are the experimental test results used to make an assessment of model accuracy?
- is there a model data input error common to all the models?

Comparing the performance of the equivalent roll stiffness model with that of the linkage model in Figures 6.63 and 6.64 it is possible to look, for example, at the error measured between the experimental and simulated results for the peaks in the response or to sum the overall error from start to finish. On that basis it may seem desirable to somehow 'score' the models giving say the linkage model 8/10 and the roll stiffness model 7/10. In light of the above questions the validity of such an objective measure is debatable and it is probably more appropriate to simply state:

'For this vehicle, this manoeuvre, the model data, and the available benchmark test data, the equivalent roll stiffness model provides reliable predictions when compared with the linkage model for considerably less investment in model elaboration.'

Clearly it is also possible to use an understanding of the physics of the problem to aid the interpretation of model performance. An important aspect of the predictive models is whether the simplified suspension models correctly distribute load to each tyre and model the tyre position and orientation in a way that will allow a good tyre model to determine forces in the tyre contact patch that impart motion to the vehicle and produce the desired response. Taking this a step further we can see that if we use the equivalent roll stiffness and linkage models as the basis for further comparison it is possible in Figures 6.57 and 6.58 to compare the vertical force in for example the front right and left tyres. The plots indicate the performance of the simple equivalent roll stiffness model in distributing the load during the manoeuvre. The weight transfer across the vehicle is also evident as is the fact that tyre contact with the ground is maintained throughout. It should also be noticed that in determining the load transfer to each wheel the equivalent roll stiffness model does not include DOF that would allow the body to heave or pitch relative to the suspension systems.

In Figures 6.65−6.68 a similar comparison between the two models is made this time considering, for example, the slip and camber angles predicted in the front right tyre.

Although the prediction of slip angle agrees well it can be seen in Figure 6.60 that the equivalent roll stiffness model with a maximum value of about 1.5° underestimates the amount of camber angle produced during the simulation when compared with the linkage model where the camber angle approaches 5°. Clearly the wheels in the effective roll stiffness model do not have a camber DOF relative to the rigid axle parts and the camber angle produced here is purely due to tyre deflection.

As discussed in Chapter 5 it is perhaps fortuitous in this case that for a passenger car of the type used here the lateral tyre force produced due to slip angle is considerably more significant than that arising due to camber between the tyre and road surface. Further investigations can be carried out to establish the significance of a poor camber angle prediction input to the tyre model. In Figure 6.69 the linkage model has been run using an interpolation tyre model where it has been possible to deactivate the generation of lateral force arising from camber angle. In this plot

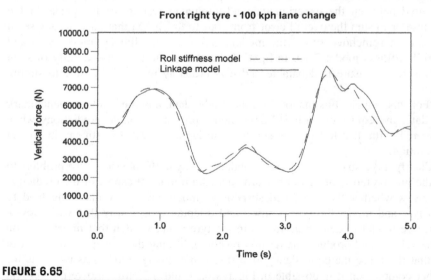

FIGURE 6.65

Vertical tyre force comparison − linkage and roll stiffness models.

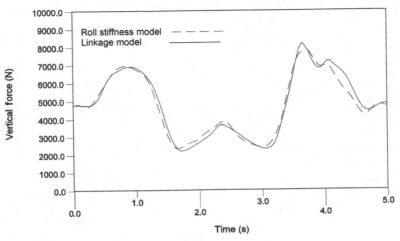

FIGURE 6.66

Vertical tyre force comparison — linkage and roll stiffness models.

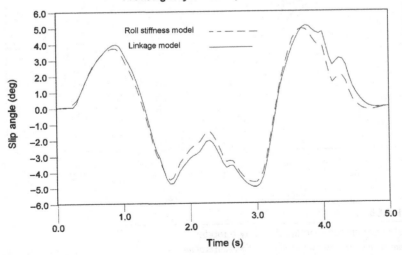

FIGURE 6.67

Slip angle comparison — linkage and roll stiffness models.

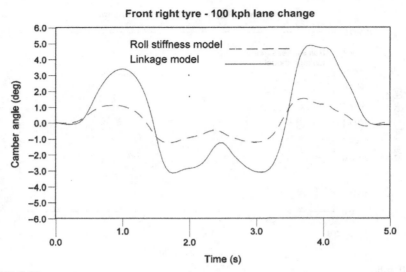

FIGURE 6.68

Camber angle comparison — linkage and roll stiffness models.

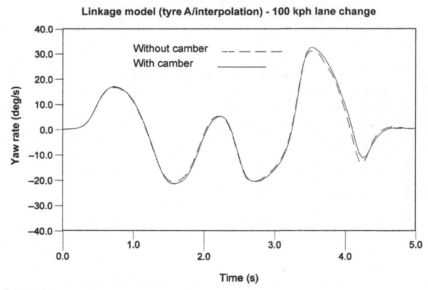

FIGURE 6.69

Yaw rate comparison — interpolation tyre model.

This material has been reproduced from the Proceedings of the Institution of Mechanical Engineers, K2 Vol. 214
'The modelling and simulation of vehicle handling, Part 4: handling simulation', M.V. Blundell, page 83, by
permission of the Council of the Institution of Mechanical Engineers.

it can be seen that the prediction of yaw rate, for example, is not sensitive for this vehicle and this manoeuvre to the modelling of camber thrust.

To conclude this case study it is possible to consider an alternative modelling and simulation environment for the prediction of the full vehicle dynamics. As discussed earlier the incorporation of microprocessor control systems in a vehicle may involve the use of a simulation method that involves:

1. the use of MBS software where the user must invest in the modelling of the control systems.
2. the use of software such as MATLAB/Simulink where the user must invest in the implementation of a vehicle model or,
3. a co-simulation involving parallel operation of the MBS and control simulation software.

In this example the author (Wenzel et al., 2003)[2] has chosen the second of the above options and a vehicle model (Figure 6.70) is developed from first principles and implemented in Simulink. The model developed here is based on the same data used for this case study with 3 DOF: the longitudinal direction x, the lateral direction y and the yaw around the vertical axis z.

The vehicle parameters used in the following model include:

v_x	Longitudinal velocity (m/s)
v_y	Lateral velocity (m/s)
v_{cog}	Centre of gravity velocity (m/s)
a_x	Longitudinal acceleration (m/s^2)
a_y	Lateral acceleration (m/s^2)
Γ	Torque around z-axis (Nm)
δ	Steer angle (rad)
β	Side slip angle (rad)
α_{ij}	Wheel slip angles (rad)
$\dot{\psi}$	Yaw rate (rad/s)
F_{zij}	Vertical forces on each wheel (N)
Ij	Position: $i = $ front(f)/rear(r), $j = $ left(l)/right(r)

Note that steer angle δ and the velocity of the vehicle's centre of gravity v_{cog} are specified as model inputs.

The relationship between the dynamic vehicle parameters can be formulated as differential equations. Most of these can be found in the standard literature. Using

---

[2]Wenzel et al., 2003, describes preliminary work undertaken in a collaborative research project with Jaguar Cars Ltd, Coventry, UK and funded by the Control Theory and Applications Centre, Coventry University, Coventry, UK. It forms the PhD programme for Thomas A Wenzel.

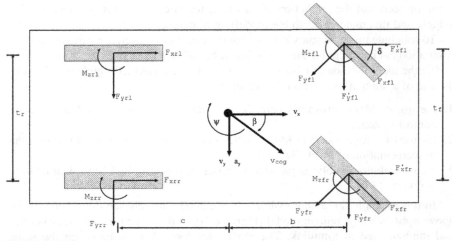

FIGURE 6.70

Three degree of freedom vehicle model (Wenzel et al., 2003).

formulae by Wong (2001) and Will and Żak (1997) the following differential equations for acceleration, torque and yaw rate can be derived:

$$\dot{v}_x = \frac{1}{m}(F_{xfl}\cos\delta - F_{yfl}\sin\delta + F_{xfr}\cos\delta - F_{yfr}\sin\delta + F_{xrl} + F_{xrr}) + v_y\dot{\psi} \qquad (6.57)$$

$$\dot{v}_y = \frac{1}{m}(F_{yfl}\cos\delta + F_{xfl}\sin\delta + F_{yfr}\cos\delta + F_{xfr}\sin\delta + F_{yrl} + F_{yrr}) - v_x\dot{\psi} \qquad (6.58)$$

$$\Gamma = \frac{t_f}{2}F_{xfl}' - \frac{t_f}{2}F_{xfr}' + \frac{t_r}{2}F_{xrl} - \frac{t_r}{2}F_{xrr} + bF_{yfl}' + bF_{yfr}' - cF_{yrl} - cF_{yrr} + M_{zfl} + M_{zfr} + M_{zrl} + M_{zrr} \qquad (6.59)$$

$$\ddot{\psi} = \frac{\Gamma}{J_z} \qquad (6.60)$$

where the additional parameters are defined as:

F_{xij}	Fxij longitudinal forces on tyre ij (N)
F_{yij}	Fyij lateral forces on tyre ij (N)
F_{xij}'	F'xij longitudinal forces on tyre ij in the vehicle's coordinate system (N)
F_{yij}'	F'yij lateral forces on tyre ij in the vehicle's coordinate system (N)
M_{zij}	Mzij self-aligning moment on tyre ij (Nm)
m	m mass of vehicle (kg)
J_z	Jz moment of inertia around vertical axis (Nm^2)
t_f, t_r	tf, tr front and rear track width (m)
b,c	b, c position of centre of gravity between wheels (m)

Other important states are the wheel slip angles α_{ij} and the body slip angle β, defined as follows:

$$\alpha_{fl/r}=\delta-\arctan(\frac{v_y+b\dot{\psi}}{v_x\pm\frac{1}{2}t_f\dot{\psi}}) \tag{6.61}$$

$$\alpha_{rl/r}=\arctan(\frac{-v_y+c\dot{\psi}}{v_x\pm\frac{1}{2}t_r\dot{\psi}}) \tag{6.62}$$

$$\beta=\arctan(\frac{v_y}{v_x}) \tag{6.63}$$

In this model roll and pitch of the vehicle are neglected but weight transfer is included to determine the vertical load at each wheel as defined by Milliken and Milliken (1998):

$$F_{zfl/r}=(\frac{1}{2}mg\pm m\frac{a_yh}{t})\frac{c}{\ell}-ma_x\frac{h}{\ell} \tag{6.64}$$

$$F_{zrl/r}=(\frac{1}{2}mg\pm m\frac{a_yh}{t})\frac{b}{\ell}+ma_x\frac{h}{\ell} \tag{6.65}$$

The additional parameters are the height h of the vehicle's centre of gravity, the wheelbase, λ and the gravitational acceleration, g.

In Eqns (6.66) and (6.67) it has to be considered that $a_x\neq\dot{v}_x$ and $a_y\neq\dot{v}_y$. The yaw motion of the vehicle has to be taken into account (Wong, 2001) giving:

$$a_x=\dot{v}_x-v_y\dot{\psi} \tag{6.66}$$

$$a_y=\dot{v}_y+v_x\dot{\psi} \tag{6.67}$$

In this work the author (Wenzel et al., 2003) has simulated a range of vehicle manoeuvres using both the Magic Formula and Fiala tyre models described in Chapter 5. The example shown here is for the lane change manoeuvre used in this case study with a reduced steer input applied at the wheels as shown in the bottom of Figure 6.71.

Also shown in Figure 6.63 are the results from the Simulink model and a simulation run with the MSC ADAMS linkage model. For this manoeuvre and vehicle data set the Simulink and MSC ADAMS models can be seen to produce similar results.

In completing this case study there are some conclusions that can be drawn. For vehicle handling simulations it has been shown here that simple models such as the equivalent roll stiffness model can provide good levels of accuracy. It is known however, that roll centres will 'migrate' as the vehicle rolls, particularly as the vehicle approaches limit conditions. The plots for case study 1 in Chapter 4 show the vertical movement of the roll centre along the centre line of the vehicle as the suspension moves between bump and rebound. On the complete vehicle, the geometrically constructed roll centre will also move laterally off the centre line as the vehicle rolls, but the exact significance of this is questionable, as shown in Figure 4.13.

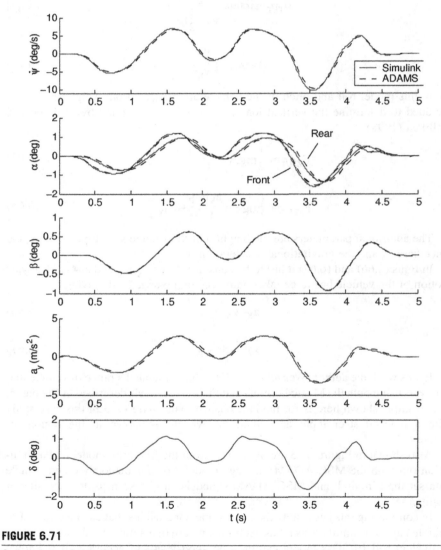

FIGURE 6.71

Comparison of simulink and MSC ADAMS predictions of vehicle response.

Using an MBS approach to develop a simple model may also throw up some surprises for the unsuspecting analyst. The equivalent roll stiffness model, for example, does not include heave and pitch DOF relative to the front and rear axles. During the simulation, however, DOF exist for the body to heave and pitch relative to the ground inertial frame. These DOF must still be solved and in this case are damped only by the inclusion of the tyre model. In the 3 DOF model these motions are ignored and solution is only performed on the DOF that have been modelled. While the main theme in this book is to demonstrate the use of MBS analysis, the Matlab/Simulink model is useful here in providing the basis for additional modelling and simulation of the modern control systems involved in enhancing the stability and dynamics of the vehicle. The effort invested in this modelling approach also provides educational benefits reinforcing fundamental vehicle dynamics theory.

6.16 Summary

Many different possibilities exist for modelling the behaviour of the vehicle driver. That none has reached prominence suggests that none is correct in every occasion. In general, the road car vehicle dynamics task is about delivering faithful behaviour during accident evasion manoeuvres — where most drivers rarely venture. Positioning the vehicle in the linear region is relatively trivial and need not exercise most organisations unduly, but delivering a good response, maintaining yaw damping and keeping the demands on the driver low are of prime importance in the nonlinear accident evasion regime. For this reason, controllers that take time to 'learn' the behaviour of the vehicle are inappropriate — road drivers do not get second attempts. For road vehicles, the closed loop controller based on front axle lateral acceleration gives good results and helps the analyst understand whether or not the vehicle is actually 'better' in the sense of giving an average driver the ability to complete a manoeuvre.

In motorsport applications, however, drivers are skilled and practised and so controllers with some feed-forward capability (to reflect 'learnt' responses), plus closed loop control of body slip angle are appropriate to reflect the high skill level of the driver. Whether or not advanced gain scheduling models, such as the MRAS or Self Tuning Regulator, are in use depends very much on whether or not data exist to support the verification of such a model. The authors' preference is that 'it is better to be simple and wrong than complicated and wrong' — in other words, all other things being equal, the simplest model is the most useful since its shortcomings are more easily understood and judgements based on the results may be tempered accordingly. With elaborate schemes, particularly self-tuning ones, there is a strong desire to believe the complexity is in and of itself a guarantee of success. In truth if a relatively simple and robust model cannot be made to give useful results it is more likely to show a lack of clarity in forming the question than a justification for further complexity.

Simulation Output and Interpretation

Education is not the learning of many facts but the training of the mind to think.
Albert Einstein

7.1 Introduction

Vehicle handling simulations are intended to recreate the manoeuvres and tests that vehicle engineers carry out using prototype vehicles on the test track or proving ground. Some are defined by the International Standards Organization, which outlines recommended tests in order to validate the handling performance of a new vehicle. A list of these tests is given in Appendix D.

Standards also exist for categories of vehicles other than passenger cars and manufacturers in those markets will use them as appropriate. Manufacturers will probably go beyond these minimum procedures in most market segments. The goal of excellence in handling performance is driven not by the need to meet fixed legislation but rather the increasing demands of a competitive marketplace.

The attractions of simulation are summarised in Chapter 1 as

- improved comprehension and ranking of design variables,
- rapid experimentation with design configurations and
- genuine optimisation of numerical response variables.

Also apparent is the ability to consider the behaviour of a vehicle before it exists physically. The cost for prototype vehicles can easily be £250,000 at the time of writing. Prudent use of simulation can ensure that only worthwhile prototype designs are turned into hardware. In many ways, well-considered simulation gives an equivalent to the use of a comprehensively instrumented vehicle in order to understand the sensitivity of the behaviour to the design variables available.

The use of instrumented vehicles to investigate handling performance can be traced back to the work of Segal in the early 1950s, which as mentioned in Chapter 1 was the subject of one of the seminal 'IME Papers' (Segel, 1956). Testing was carried out using a 1953 Buick Super four-door sedan (saloon), to investigate steady state behaviour with a fixed steering input at various speeds and also transient response to sudden pulse inputs at the steering wheel. The instrumentation used at that time allowed the measurement of the following:

1. Left front wheel steer
2. Right front wheel steer
3. Steering wheel rotation
4. Lateral acceleration
5. Roll angle
6. Pitch angle
7. Yaw rate
8. Roll rate
9. Forward velocity.

Some of these responses are shown in Figure 7.1. The trajectory (path) of the vehicle can also be recorded. With simulation this is straightforward but in the past it has been difficult to measure on the test track, testers resorting to measuring a trail of dye left by the vehicle on the test track surface. Modern satellite-based instrumentation has improved on this somewhat, with current systems able to record the trajectory of the front and the rear of the vehicle separately and concurrently at suitably high refresh rates. For each handling manoeuvre, or simulation, it is necessary for vehicle engineers to decide which responses are to be measured during the testing process.

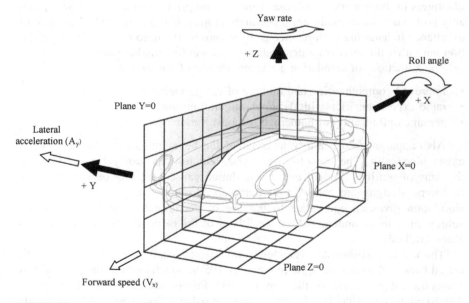

FIGURE 7.1

Typical lateral responses measured in one of several possible vehicle coordinate frames.

All test and analytical activities are directed at understanding and improving the overall dynamic behaviour of the vehicle. As discussed in Chapter 1, the difficulty with the vehicle dynamics field is not the complexity of the effects in play but the level of interaction between them. A further difficulty is the level of change in behaviour required for a vehicle to be usefully 'improved'. Typically, strong impressions of change are made with only modest variations of physical measures. When the difficulty of repeatable testing and the variation of impression with individual testers are both thrown in, the vehicle dynamics process lacks 'capability' — in the sense of quality control — compared to the task it is set. To explain this further, consider the following example.

7.2 Case study 9 — variation in measured data

Several attempts have been made to define single number measures for vehicle dynamic performance. A popular measure in the US is limiting lateral acceleration — frequently referred to as 'grip'. Table 7.1 shows data that might be recorded during a steady state test, using a stopwatch to time a lap of a marked 200 m diameter circle.

If a honest error estimation is made in the recorded data then the results with the stopwatch are probably accurate to ± 0.1 s. Thus the raw data would appear as shown in Figure 7.2, with error bars on the figure as shown. This is a typical set of data for such a test; prolonging the test further might well degrade the tyres on the vehicle and lead to a 'skewed' result favouring the first test configuration with fresher tyres. It could be argued that tests should be repeated on fresh tyres for each different configuration to be examined but while academically sound, this argument neglects the commercial and temporal pressures placed on vehicle testing, particularly in intermediate configurations before matters are finalised and particularly if the tyres themselves are prototype items.

If these measurements are manipulated into lateral acceleration figures using the simple relationship

$$A_y = \frac{\left(\dfrac{200\pi}{t}\right)^2}{100} \tag{7.1}$$

then treated as the results that might be obtained from a production process and manipulated accordingly, the results in Table 7.2 emerge.

Table 7.1 Time to Complete Lap of a Marked 200 m Diameter Circle (s)

	Test 1	Test 2	Test 3	Test 4	Test 5
Configuration A	21.38	21.71	21.57	21.61	21.50
Configuration B	21.60	21.49	21.29	21.39	21.53

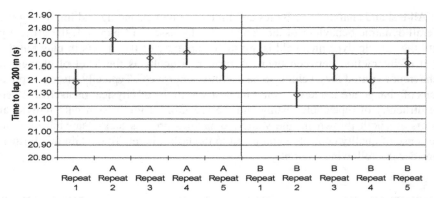

FIGURE 7.2

Raw Data from steady state test with error bars.

Table 7.2 Statistical Summary of Test Data

	Mean	Population Standard Deviation
Configuration A	0.866 g	0.010 g
Configuration B	0.874 g	0.010 g
Difference A − B	0.008 g	

It is clear that the extrapolation to a population from only five samples is somewhat poor practice. However, it is equally clear that the difference between the two configurations is significantly less than the spread of the distribution of the tests. In statistical process control, a process is regarded as 'capable' if six times the standard deviation (so-called '6σ') is less than 75% of the tolerance on the measured attribute. For the measurement processes above, the 'capability' is no less than $6 \times 0.01 \times 4/3 = 0.08$ g − 10 times the difference actually observed in the test. In other words, differences of less than 0.08 g cannot be controlled reliably using such a measurement process. However, such a difference is a significant one between otherwise similar vehicles, representing something around 10% of the total lateral acceleration available.

While the data set shown in the example is fictitious, the statistical character of the measurements is entirely typical, even under well-controlled conditions. When a vehicle is nearly optimised, this problem is typical. The resolution of the process in use − the test facilities and so on − is comparable to or greater than the control required to optimise the vehicle further. For this reason, in both motorsport and production vehicle engineering, a great deal of the final optimisation is based on subjective judgements of a few well-chosen individuals.

7.3 A vehicle dynamics overview

At this point, it is appropriate to develop some basic notions about vehicle dynamics. These definitions will be used later to suggest an interpretation of the subjective/objective relationship; however, it should be clear that by its very nature the absolute quantification of subjective qualities is impossible.

The driver has two primary concerns in controlling the vehicle. These are speed and path. Speed is controlled with engine power and braking systems. The use of separate controls for acceleration and deceleration is logical when the systems are separate but may become less so if vehicle architecture changes significantly; for example, some system with a 'motor in each wheel' that both accelerates and brakes might have a single pedal for control, of the type prototyped by Nomix AB in Sweden. Variation of speed is governed by vehicle mass and tractive/brake power availability at all but the lowest speed, and is easily understood.

7.3.1 Travel on a curved path

The adjustment of path curvature at a given speed is altogether more interesting. In a passenger car the driver has a handwheel, viewed by the authors as a 'yaw rate' demand — a demand for rotational velocity of the vehicle when viewed from above. The combination of a yaw rate and a forward velocity vector, which rotates with the vehicle, gives rise to a curved path. The curved path of the vehicle requires some lateral acceleration. The tyres on a car exert a force towards the centre of a turn and the body mass is accelerated by those forces centripetally — in a curved path. Thus the sum of the lateral forces that the tyres exert on the car is the centripetal force that produces the centripetal acceleration (Figure 7.3).

Note that the authors do not favour the use of the equivalent inertial ('D'Alembert') force since it can be misleading; it gives the impression that the analysis of the cornering vehicle is a static equilibrium problem, which it most certainly is not. The idea of an analogous static equilibrium condition is not in itself problematic but inappropriate 'static' thinking quickly becomes torturous and unwieldy.

For example, a common obsession is to attempt to find a 'centre of rotation in roll'. This is usually performed with some sort of 'point of zero lateral velocity' logic but the reality is that this 'zero velocity' point is with respect to some arbitrary and ill-defined reference frame. Figure 7.4 shows the same vehicle represented in two equally arbitrary reference frames; the first is anchored at the outboard wheel contact patch and the second at the inboard wheel contact patch. Given that most independent suspensions are not symmetric, the 'lateral displacement' for a given roll angle is entirely dependent on the choice of reference frame. Thus the idea of some 'centre of instantaneous motion' is difficult to pin down and even more difficult to ascribe any meaning to. If the reference frame is anchored somewhere on the vehicle body then that point becomes a point of zero translation, and the concept is again seen to be of no value. Nevertheless, within the authors' experience there has been a great deal of effort to track down 'centres of motion' using elaborate static equilibrium analogies, although it has not notably added value to the vehicles involved.

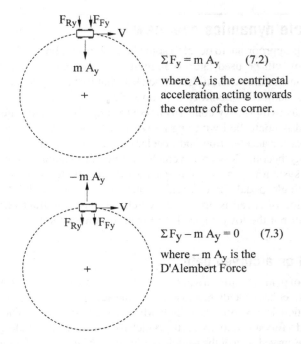

$$\Sigma F_y = m A_y \qquad (7.2)$$

where A_y is the centripetal acceleration acting towards the centre of the corner.

$$\Sigma F_y - m A_y = 0 \qquad (7.3)$$

where $- m A_y$ is the D'Alembert Force

FIGURE 7.3

Representation of inertial force during cornering.

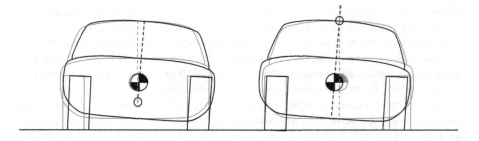

FIGURE 7.4

The difficulty with arbitrary reference frames applied to 'pseudo static' cornering as suggested by the use of D'Alembert forces.

7.3.2 The classical treatment

The classical assessment of the behaviour of a vehicle was developed by the early work of vehicle dynamicists such as Olley, Milliken and Segel and has been documented in several textbooks dealing with the subject. In the following sections the classical treatment will be summarised followed by a consideration based on the transient dynamics of the driven vehicle.

The classical assessment of the behaviour of a vehicle is based on either testing or simulating steady state cornering. 'Steady state' means the vehicle states are unchanging with time — the car is 'settled' in a corner at a constant speed, on a constant radius and so on. Note that steady state is not the same as 'static'. Two traditional evaluation methods exist. The first involves driving the vehicle around a constant radius circle at a range of constant speeds that correspond to a range of increments in lateral acceleration. The second method involves driving the vehicle at a constant speed but with a progressive increase in handwheel angle that reduces the radius of turn with time and consequently increases the lateral acceleration. Which method is in use in a particular organisation is likely to be governed by the test facilities available and previous practice rather than on the merits or otherwise of a one or other test method. In particular, the first method is very practicable to perform and so it will be used for the basis of the following discussion.

7.3.2.1 Low speed behaviour
The starting point for the consideration of steady state cornering behaviour is an assessment of the vehicle cornering at low speed (Figure 7.5).

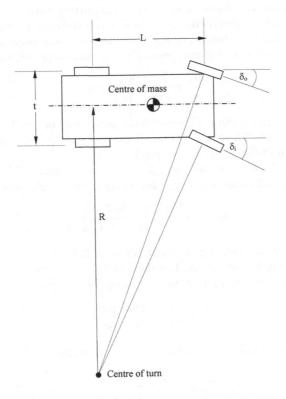

FIGURE 7.5

Cornering at low speed.

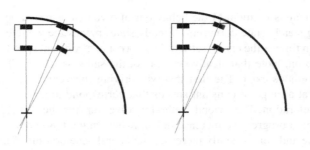

FIGURE 7.6

Different definitions of turning circle.

The minimum radius available in normal driving is the turning circle, determined by turning the steering to its maximum (mechanically limited) extent and following the geometry illustrated in Figure 7.5. Turning circles are usually quoted 'between kerbs' and 'between walls', as illustrated in Figure 7.6.

At higher speeds the limit to path curvature is not the amount the wheels can be turned but rather the maximum centripetal force that can be generated. This is governed by the limiting coefficient of friction (μ) between tyres and road, typically around 0.9. Simplistically, if friction is independent of area (the Newtonian model), the theoretical maximum centripetal acceleration in units of g is identical to the friction coefficient, μ

$$\frac{A_y}{g} = \frac{F_y}{mg} = \frac{mg\mu}{mg} = \mu \tag{7.4}$$

This is for vehicles without additional force pressing them into the road — i.e. not 'winged' racing cars. Therefore, the maximum yaw rate possible (in degrees per second) is shown in Figure 7.7 for a typical vehicle.

The yaw rate from 'geometry', ω_{geom}, as shown in Figure 7.7 is the maximum possible steering and varies with speed according to the simple relationship

$$\omega_{geom} = \frac{V(\delta_i + \delta_o)/2}{L}\left(\frac{180}{\pi}\right) \tag{7.5}$$

where V is the forward velocity in metres per second, δ_i, δ_o are the steer angles in radians as shown in Figure 7.5 and L is the wheelbase in metres.

Yaw rate from limiting friction, $\omega_{friction}$, is equally simple.

$$\omega_{friction} = \left(\frac{180}{\pi}\right)\frac{\mu g}{V} \tag{7.6}$$

So it can be seen that above a limiting speed, given by

$$V_{lowlimit} = \sqrt{\frac{\mu g L}{\left(\frac{\delta_i + \delta_o}{2}\right)\left(\frac{\pi}{180}\right)}} \tag{7.7}$$

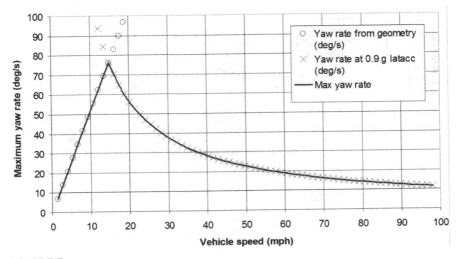

FIGURE 7.7

Maximum possible yaw rate for a typical vehicle.

it is the surface grip that determines the limiting yaw rate and hence path curvature. For typical passenger cars, the region in which geometry dominates steering behaviour is small — up to about 15 mph. This is the speed that may be counted as 'low' and in which Figure 7.5 is a reasonable description of the behaviour of the vehicle. For vehicles such taxis and heavy goods vehicles, the low speed region is of more importance simply because these vehicles make more low speed, minimum radius manoeuvres. However, they rarely perform these manoeuvres at speeds exceeding walking pace and so the yaw rates remain low.

The steer angles required to avoid scrubbing at the inner and outer road wheels, δ_o and δ_i, are

$$\delta_o = \frac{L}{(R+0.5t)}\left(\frac{180}{\pi}\right) \tag{7.8}$$

$$\delta_i = \frac{L}{(R-0.5t)}\left(\frac{180}{\pi}\right) \tag{7.9}$$

The angle of a notional 'average' wheel on the vehicle centreline, δ, is the 'Ackerman' angle — the approximate form is in common usage

$$\delta = \frac{LR}{(R-0.25t^2)}\left(\frac{180}{\pi}\right) \approx \frac{L}{R}\left(\frac{180}{\pi}\right) \tag{7.10}$$

In 1817, Rudolph Ackermann patented geometry similar to this as an improvement over a steered axle as was common on horse-drawn vehicles. That the geometry we today call 'Ackermann' was in fact a modification proposed by the Frenchman

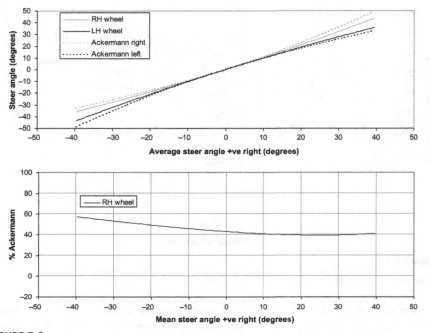

FIGURE 7.8

Steering Geometry in Comparison with Ackermann/Jeantaud Geometry.

Charles Jeantaud in 1878 has been lost in the mists of time. Some measure of how accurately the steering geometry corresponds to the Ackermann/Jeantaud geometry is often quoted although rarely defined. The authors use a description as given in Figure 7.8, which in turn uses the simplified form of Eqn (7.10).

The Ackermann/Jeantaud angles are calculated as given in Eqns (7.8) and (7.9). These are compared with the angles actually achieved by the front wheels. A fraction of the compensation is expressed for the outer wheel using δ_{mean} for the mean of the actual angles:

$$\text{Ackermann Fraction} = \frac{\delta_{o\_actual} - \delta_{mean}}{\left[\left(\dfrac{180}{\pi}\right)\left(\dfrac{L}{\delta_{mean}}\right) + 0.5t\right]\left(\dfrac{180}{\pi}\right) - \delta_{mean}} \qquad (7.11)$$

This is the quantity graphed in Figure 7.8 as '% Ackermann'. Road cars are successfully engineered with widely varying '% Ackerman' values — from 20 to 80%.

Ackermann effects are noticeable at parking speeds. Typically, there is some inclination of the vehicle's steering axis when viewed from the front and the side of the vehicle. These inclinations are known as castor and steer axis inclination or

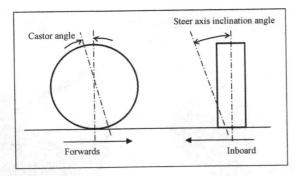

FIGURE 7.9

Steer Angle Geometry Definitions.

kingpin inclination, respectively, and are shown in Figure 7.9. Together with the off-sets at ground level between the geometric centre of the wheel, these inclinations have the effect of moving the inboard wheel down and the outboard wheel up. Figure 7.10 shows the inboard wheel being moved down as it goes onto back lock. Given the constraint of the ground, this has the effect of imposing a roll moment on the vehicle that is reacted by the front and rear suspensions in series (Figure 7.11). The loading of the vehicle wheels becomes asymmetric.

This asymmetry is of the order of 10 N in a typical family saloon and is of little consequence. However, in a more stiffly sprung vehicle if the steering geometry is not constructed with care, the asymmetry can be more than 100 N. If this loading is applied in conjunction with significantly low Ackermann fractions, the result is that the inner wheel is emphasised over the outer wheel.

This emphasis has the consequence of producing an effective side force from the tyre because it is operating at a comparatively large slip angle (Figure 7.12). Normally, side forces from both tyres work in opposition if the wheels are effectively toed in when the Ackermann fraction is less than unity. However, if the forces are not in balance (because of the weight imbalance between inboard and outboard tyres) the additional side force on the inboard tyre has the effect of reducing the steer align-ing torque when the vehicle is in motion. If the Ackermann fraction is particularly low, the vehicle will have a 'wind on to lock' behaviour at car park speeds beyond certain steer angles. Vehicles with stiff suspension, high steer axis angles and wide tyres are more prone to this effect.

There exists some confusion over the significance of Ackermann geometry at speed. For ride and handling work, the significance is sometimes overstated. Consid-ering Figure 7.5, a turn at 50 mph (22 m/s) road speed at 0.4 g lateral acceleration may be calculated as producing a yaw rate of 10.2°/s. For a 2.7 m wheelbase vehicle, this requires a mean steer angle of 1.26°. The radius of turn is 123 m and so the Jeantaud modification gives 1.25018° on the inner wheel and 1.26552° on the outer wheel — an included angle of 0.015°. For a typical cornering stiffness of 1500 N/°, this gives a lateral force variation of 23 N between 0% Ackermann and

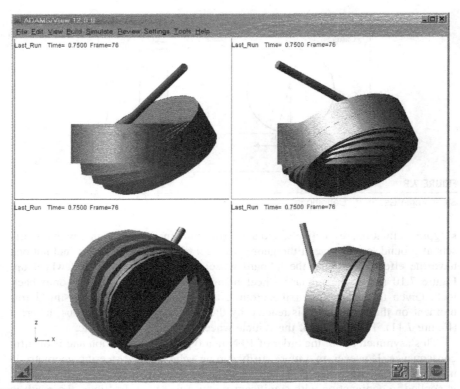

FIGURE 7.10

Steering geometry effects on wheel vertical position: fully constrained body and suspension, model runs from straight ahead to full back lock. Plan view (top left) rear view (top right), left view (bottom left) and three-quarter view (bottom right) of front left wheel with steering axis indicated by cylindrical graphic. Note the steering geometry used is atypical for emphasis.

100% Ackermann. The lateral forces to achieve 0.4 g at 50 mph are over 5900 N for a typical 1500 kg vehicle, so the Ackermann effect amounts for lateral forces of some 0.4% of the total — a small modifier on the vehicle as a whole.

7.3.2.2 Higher speed linear region behaviour

As speeds rise above the low limit calculated in Eqn (7.7), the behaviour of vehicles becomes less instinctive. As illustrated in Figure 1.4, it is generally true that the curvature of path *cannot* readily be estimated using a geometric approximation like Figure 7.5.

For a vehicle manoeuvring in the ground plane, it can be presumed the vehicle is broadly symmetric — note there are some exceptions to this, such as motorcycle combinations, but these are so uncommon as to be obvious.

Ground vehicles tend to be broadly flat on their lower surface and tend to have the majority of their weight distributed around a narrow height band; regardless of

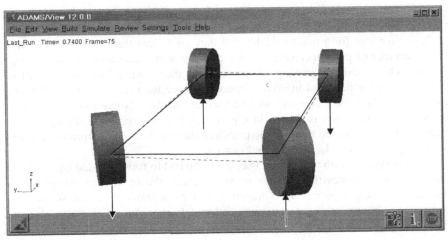

FIGURE 7.11

Platform motion during steering at low speed (black) in comparison with the static platform position (grey dashed). Changes in wheel weight are indicated; front left wheel has an increased reaction force.

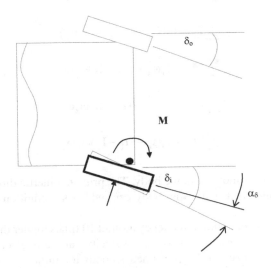

FIGURE 7.12

Effect of inside wheel loading in producing an excess side force due to slip angle α_δ when the vehicle is in motion.

the shape of the superstructure, most of the mass is in a 'stripe' between the wheel hubs and the top of the wheels.

Thus it may be presumed as a simplification that the local x-axis is parallel with the path of the vehicle when running in a straight line. In combination with the symmetry, this gives a rather intuitive inertial axis set for the vehicle. Real vehicles may deviate from these presumptions but they are suitable for overview modelling.

It is often true for ground vehicles that they are longer than they are wide. It is also often true of passenger vehicles that they are wider than they are tall (though modern vehicle design is making this less obvious than it was). Thus it is coincidentally true that the principal inertias of the vehicle are least for the presumed x-axis and most for the presumed z-axis as discussed in Chapter 6 where a convenient axis set for a vehicle model is shown in Figure 6.3. It is conventional (though by no means universally so) to number the principal inertias 1, 2 and 3 from the most to the least and thus $I_1 \approx I_{xx}$, $I_2 \approx I_{yy}$ and $I_3 \approx I_{zz}$.

It is convenient with real bodies (i.e. not particles) to formulate the equations of motion in a body-centred manner to avoid recalculating inertia properties. A detailed treatment is provided in Chapter 2 but for convenience the full set of these equations — known as Euler equations and widely reproduced — is also given in Eqns (7.12)–(7.17) below:

$$\sum F_x = m\left(\dot{v}_x - \omega_z v_y + \omega_y v_z\right) \tag{7.12}$$

$$\sum F_y = m\left(\dot{v}_y - \omega_x v_z + \omega_z v_x\right) \tag{7.13}$$

$$\sum F_z = m\left(\dot{v}_z - \omega_y v_x + \omega_x v_y\right) \tag{7.14}$$

$$\sum M_x = I_1 \dot{\omega}_x - (I_2 - I_3)\omega_y \omega_z \tag{7.15}$$

$$\sum M_y = I_2 \dot{\omega}_y - (I_3 - I_1)\omega_z \omega_x \tag{7.16}$$

$$\sum M_z = I_3 \dot{\omega}_z - (I_1 - I_2)\omega_x \omega_y \tag{7.17}$$

The axes x, y and z move with the body. The principal inertia directions are presumed to be coincident with the body-centred axis definition as previously described.

Lateral motions occur with a frequency around 10 times higher than longitudinal motions, and thus the variation in forward velocity can be neglected in an initial formulation. The surface can be presumed smooth for initial evaluations and so heave, pitch and roll motions can also be neglected. This leaves only lateral velocity, v_y, and yaw velocity, ω_z, as degrees of freedom (DOF):

$$\sum F_y = m\left(\dot{v}_y + \omega_z v_x\right) \tag{7.18}$$

$$\sum M_z = I_3 \dot{\omega}_z \tag{7.19}$$

Despite the large simplification, this representation of the vehicle — sometimes referred to as a 2 DOF model — can give good insight into overall vehicle behaviour.

Lateral forces and yaw moments are generated mainly by the tyres of the vehicle. Aerodynamic forces will be neglected for the initial formulation.

The tyres have been discussed in detail in Chapter 5. For the 2 DOF model they may initially be presumed to generate forces with slip angle — camber is neglected — and may be imagined as a single effective tyre on the centreline of the vehicle at each axle location. Equations 7.18 and 7.19 describe a single track vehicle but it should be noted that the dynamics of real single track vehicles (bicycles and motorcycles) are substantially different to the dynamics of this system. Although the term 'bicycle model' for the single track 2 DOF model is widespread, it is nevertheless very misleading.

Considering Figure 7.13, if the vehicle is viewed in plan, the slip angle at each axle from a lateral velocity v_y is given by

$$\alpha_f(v_y) = \alpha_r(v_y) = \tan^{-1}\left(\frac{v_y}{v_x}\right) \tag{7.20}$$

Similarly, the slip angle at each axle from a yaw velocity ω_z is given by

$$\alpha_f(\omega_z) = \tan^{-1}\left(\frac{\omega_z \cdot a}{v_x}\right) \tag{7.21}$$

$$\alpha_r(\omega_z) = \tan^{-1}\left(\frac{-\omega_z \cdot b}{v_x}\right) \tag{7.22}$$

For small angles, less than around 15°, the value of the tangent function is approximately the same as the angle in radians.

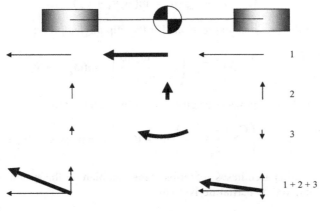

FIGURE 7.13

A graphical representation of the summation of longitudinal, lateral and yaw velocities at the Centre of Mass as seen at the axles.

FIGURE 7.14

Steer angle requires algebraic subtraction.

The aggregate behaviour of the tyres in the presence of both velocities may be presumed to be the linear superposition of the two, since the DOF are mathematically orthogonal Figure 7.13.

The steer angle on the front axle can be seen to need algebraic subtraction with the sign convention chosen (the SAE axis system), in which the steer angle has the same sign as the yaw rate (Figure 7.14).

Assembling all the terms it can be presumed that:

$$\alpha_f = \left(\frac{v_y + \omega_z \cdot a}{v_x}\right) - \delta \qquad (7.23)$$

$$\alpha_r = \left(\frac{v_y - \omega_z \cdot b}{v_x}\right) \qquad (7.24)$$

The behaviour of real tyres can be observed as broadly linear up to about half the available friction budget and can be expressed by a single 'cornering stiffness' coefficient, C_α.

To linearise the model, a constant value will be used initially for the cornering stiffness coefficient. Therefore the lateral equation of motion can now be rendered thus:

$$C_{\alpha f} \cdot \alpha_f + C_{\alpha r} \cdot \alpha_r = m(\omega_z v_x + \dot{v}_y) \qquad (7.25)$$

Substituting the previous expressions for the slip angles gives:

$$C_{\alpha f} \cdot \left(\frac{v_y + \omega_z \cdot a}{v_x} - \delta\right) + C_{\alpha r} \cdot \left(\frac{v_y - \omega_z \cdot b}{v_x}\right) = m(\omega_z v_x + \dot{v}_y) \qquad (7.26)$$

Rearranging the components to group them by system states:

$$\left(\frac{C_{\alpha f} + C_{\alpha r}}{v_x}\right) \cdot v_y + \left(\frac{C_{\alpha f} a - C_{\alpha r} b}{v_x}\right) \cdot \omega_z - \left(C_{\alpha f}\right) \cdot \delta = m(\omega_z v_x + \dot{v}_y) \qquad (7.27)$$

The terms within parentheses are terms in the Jacobian for the system and may be equivalently rendered as partial derivatives:

$$\left(\frac{\partial F_y}{\partial v_y}\right) v_y + \left(\frac{\partial F_y}{\partial \omega_z}\right) \omega_z + \left(\frac{\partial F_y}{\partial \delta}\right) \delta = m(\omega_z v_x + \dot{v}_y) \qquad (7.28)$$

A single character symbolism for the individual Jacobian terms is common in aircraft dynamics where this approach was applied for the seminal IME papers of 1956 and will be familiar to any reader of the definitive 'Race Car Vehicle Dynamics' (Milliken and Milliken, 1995); it simplifies typesetting a great deal. In an affectionate nod to that source, the same symbols are used here:

$$\left(Y_{vy}\right)v_y + \left(Y_\omega\right)\omega_z + \left(Y_\delta\right)\delta = m(\omega_z v_x + \dot{v}_y) \tag{7.29}$$

The yaw equation of motion can be similarly treated:

$$a \cdot C_{\alpha f} \cdot \alpha_f - b \cdot C_{\alpha r} \cdot \alpha_r = I_3 \dot{\omega}_z \tag{7.30}$$

$$a \cdot C_{\alpha f} \cdot \left(\frac{v_y + \omega_z \cdot a}{v_x} - \delta\right) - b \cdot C_{\alpha r} \cdot \left(\frac{v_y - \omega_z \cdot b}{v_x}\right) = I_3 \dot{\omega}_z \tag{7.31}$$

$$\left(\frac{a \cdot C_{\alpha f} - b \cdot C_{\alpha r}}{v_x}\right) \cdot v_y + \left(\frac{a^2 \cdot C_{\alpha f} + b^2 \cdot C_{\alpha r}}{v_x}\right) \cdot \omega_z + \left(-a \cdot C_{\alpha f}\right) \cdot \delta = I_3 \dot{\omega}_z \tag{7.32}$$

$$\left(\frac{\partial M_z}{\partial v_y}\right)v_y + \left(\frac{\partial M_z}{\partial \omega_z}\right)\omega_z + \left(\frac{\partial M_z}{\partial \delta}\right)\delta = I_3 \dot{\omega}_z \tag{7.33}$$

$$\left(N_{vy}\right)v_y + \left(N_\omega\right)\omega_z + \left(N_\delta\right)\delta = I_3 \dot{\omega}_z \tag{7.34}$$

Given initial conditions and a time history for steering input, these equations are suitable for direct integration to find the solution with time. Numerical methods, as previously discussed, are eminently suitable for this. Some practitioners rearrange the equations further but these equations will satisfactorily illustrate the behaviour of the vehicle for the purpose at hand.

If a steer input is added early in the solution over a period of 0.1 s and held constant, and when typical values are used for vehicle parameters, certain features of the solution emerge, which can be seen in Figure 7.15.

The immediately obvious one is that the slip angles on the tyres are negative compared to the steering angle. This arises because the lateral velocity is generally negative when the yaw rate is positive and vice versa. Note that the same sign for both tyre slip angles in the steady state implies the presence of a lateral velocity at the body, which may be thought of as a 'body slip angle' and is often denoted by β — the vehicle crabs sideways as it negotiates the turn (see Figure 7.16).

$$\beta = \tan^{-1}\left(\frac{v_y}{v_x}\right) \tag{7.35}$$

Note that in Figure 7.15 the body slip angle is less than 0.02 rad, or just over $1°$; the angle is not normally visible but is nevertheless always present.

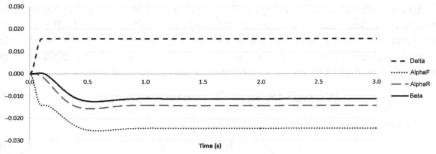

FIGURE 7.15

Response of the two degree-of-freedom model to a sharp steer input.

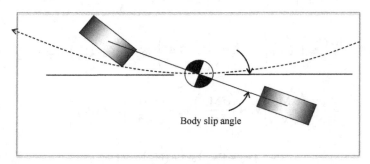

FIGURE 7.16

Body slip angle.

Examination of the equations suggests that a steady state exists in which all time derivatives are zero, and the solutions do indeed display such a form, reaching constant values after a second or so. This may be thought of as a settled condition; some students believe that a 'dynamic steady state' is an oxymoron but this is not so; the steady state implies that all the phase relationships have become constant, as distinct from nothing changing — which would be better described with the word 'equilibrium'.

Various texts reproduce this problem in various forms, which results from Eqns (7.36) and (7.37) with the derivative terms $\dot{v}_y$ and $\dot{\omega}_z$ set to zero. The result is a pair of simultaneous equations with two unknowns:

$$\left(Y_v\right)v_y + \left(Y_\omega - mv_x\right)\omega_z = -\left(Y_\delta\right)\delta \tag{7.36}$$

$$\left(N_v\right)v_y + \left(N_\omega\right)\omega_z = -\left(N_\delta\right)\delta \tag{7.37}$$

The steady state solutions for v_y and ω_z can be calculated directly without recourse to time-domain integration. Rearranging Eqn (7.36) gives:

$$\omega_z = \frac{-Y_\delta\delta - Y_v v_y}{Y_\omega - mv_x} \tag{7.38}$$

Substituting ω_z from Eqn (7.38) into Eqn (7.37) and rearranging yields:

$$N_v v_y + N_\omega \left(\frac{-Y_\delta \delta - Y_v v_y}{Y_\omega - mv_x} \right) = -N_\delta \delta \qquad (7.39)$$

$$N_v v_y - \left(\frac{N_\omega Y_\delta \delta}{Y_\omega - mv_x} \right) - \left(\frac{N_\omega Y_v v_y}{Y_\omega - mv_x} \right) = -N_\delta \delta \qquad (7.40)$$

$$v_y \left(N_v - \frac{N_\omega Y_v v_y}{Y_\omega - mv_x} \right) = -N_\delta \delta + \frac{N_\omega Y_\delta \delta}{Y_\omega - mv_x} \qquad (7.41)$$

$$v_y = \frac{\left(-N_\delta \delta + \dfrac{N_\omega Y_\delta \delta}{Y_\omega - mv_x} \right)}{\left(N_v - \dfrac{N_\omega Y_v v_y}{Y_\omega - mv_x} \right)} \qquad (7.42)$$

Equations (7.42) and (7.38) can be used in that order to calculate the steady state responses. When considering responses, it is helpful to consider the notion of a 'gain' in the control system sense, which is the ratio of response to input. Presuming the small angle approximation once more, the yaw rate gain (YRG) can be expressed as:

$$YRG(v_x) = \frac{\omega_z}{\delta} \qquad (7.43)$$

YRG for a purely geometric vehicle (YRG_{geom}) could be expressed as

$$YRG_{geom} = \frac{\omega}{\delta_{mean}} = \frac{V}{L} \qquad (7.44)$$

This may be recognised as describing the low speed region of Figure 7.17. The geometric yaw rate is achieved by some notional vehicle that may be thought of as running on 'blade' wheels on indestructible ice — no sideslip of the wheels is possible. It is important for vehicle dynamicists to remember that the overwhelming majority of the driving population genuinely believe this is the way in which their vehicles function. For many typical drivers, the belief exists that the tyres are little miniature 'rails' that the vehicle carries around with itself to 'lay tracks' as it goes along.

For the linear 2 DOF model, the calculated YRG characteristic is substantially different from YRG_{geom}. When the characteristic is computed for a range of speeds and compared with the geometric response, a character very different emerges, as shown in Figure 7.17.

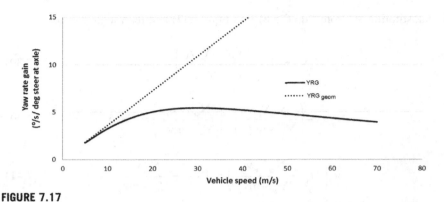

FIGURE 7.17

A comparison between geometric and 2 DoF linear steady state gains. YRG, yaw rate gain; YRG$_{geom}$, geometric yaw rate gain.

This character arises from the yaw moment developed by the vehicle attempting to straighten itself out as it travels in a curved path and also the 'weathercock' influence of the vehicle, again trying to straighten itself out as it crabs somewhat in order to have a slip angle on the rear tyres.

The exact components of the numerator and denominator when calculating the response of the vehicle defy intuitive understanding, but the character in the above graph is typical. Note that the actual YRG is hugely reduced from YRG$_{geom}$ at high speeds.

When YRG is less than YRG$_{geom}$, this is the 'understeer' condition as defined in SAE J670. However, in a real vehicle and with a real operator, there is no sense of the operator being able to understand the actual geometric angle of the front wheels, only whether or not the car is travelling on the intended path. This is true even for open wheel cars where the wheels are directly visible from the driver's seat. Thus the significance of the 'technical understeer' is minimal to the driver; there is a ratio between the angles at the handwheel and the road wheels, and a vehicle at a given speed with more 'technical understeer' but a more direct ratio will be indistinguishable from one with less 'technical understeer' and a less direct ratio.

This all points to the technical definition as being irrelevant from the driver's perspective, since the driver will learn YRG of the vehicle adaptively as the vehicle changes speed. Even skilled development drivers do not report understeer when driving in the linear region — less than about half the available friction budget.

It is important to note that this technical definition of understeer has no connection to journalistic and development drivers' description of it, which is confined exclusively to the nonlinear region and is discussed next.

7.3.2.3 Nonlinear region

The higher speed region, even with a simplistic Newtonian friction model, can be seen to have a nonlinear YRG characteristic even with a completely geometric

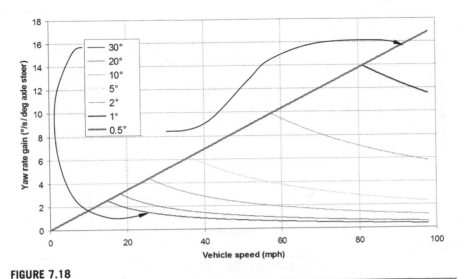

FIGURE 7.18

Yaw rate gains for an idealised vehicle with Newtonian friction.

vehicle description (Figure 7.18) since it is possible to turn the front wheels to an angle that corresponds to a yaw rate greater than that which the vehicle can achieve. The similarity between Figures 7.7 and 7.18 should be apparent to the reader.

As previously noted, when a vehicle yaws less than expected, the term 'understeer' is used — the response of the vehicle is less than ('under') what might have been expected. When a vehicle yaws more than expected, the term 'oversteer' is used — the response of the vehicle exceeds (is 'over') what might have been expected. So far only mechanisms for generating understeer have been discussed. Remaining with the Newtonian friction model to describe the behaviour of the tyres, one further fundamental point is worth establishing. For a vehicle travelling in a circular path, forward speed, V, yaw rate, ω and lateral (centripetal) acceleration, A_y, are related with

$$A_y = \omega V \qquad (7.45)$$

Using the preceding relationship for YRG_{geom}, a geometric lateral acceleration gain (AyG_{geom}) can be deduced for the geometric vehicle in the region where

$$AyG_{geom} = \frac{A_y}{\delta_{mean}} = \frac{\omega_{geom} V}{\delta_{mean}} = V^2 L \qquad (7.46)$$

At high speeds it can be seen that axle steer inputs of much in excess of half a degree cause the available friction to saturate with the geometric formulation. Although it can be seen from Figure 7.17 that the vehicle has substantially reduced gains at speed — and so it might be expected that something of the order 2–3° of steering input might cause the available friction to saturate — the level of steer

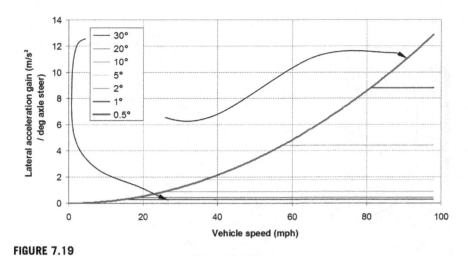

FIGURE 7.19

Lateral acceleration gains for an idealised vehicle with Newtonian friction.

required to saturate is still very low compared to parking manoeuvres. To retain proportional control, the driver must keep inputs below this level. For this reason, quite high reduction ratios are generally used in steering gears to give a reasonable level of input sensitivity at the handwheel. For a European passenger car, a reduction ratio of around 16–18:1 is typical (16° of handwheel giving 1° of axle steer), meaning that at the highest speeds shown in Figure 7.18 and 7.19 handwheel inputs of 10–20° are enough to saturate the vehicle with respect to the available friction in a high-grip environment. For vehicles travelling faster – for example on unrestricted autobahns or competition vehicles – it can be seen that the overall steering ratio is of importance in order not to have the vehicle overly sensitive to driver inputs. There is, however, a trend among vehicle manufacturers to fit numerically lower steering ratios over time – compare a 1966 Ford Cortina at 23.5:1 with its current cousin the Focus at 17.5:1 – to promote a perception of agility. This fashion will require the adoption of more adventurous variable steering ratios (for example, as promoted by Bishop Technologies or as implemented by BMW and ZF in the 2003 five series 'Active Front Steer' system) in order to retain sensitivity at high speeds. In low-grip environments, the handwheel inputs needed to saturate the vehicle at speed are tiny, scaling down with coefficient of friction, μ.

The range of lateral acceleration gains, from 0.6 m/s^2/° at 20 mph to 12.9 m/s^2/° at 100 mph, is quite a wide range to ask the driver to accommodate. When high-speed road systems lack curves of any kind there is no feedback information to allow the driver to adapt and so extremely straight designs of roads are unhelpful in this respect. Thus it is often true that normal drivers get into difficulty when faced with an emergency evasive manoeuvre at high speed. Some component of this is simply because the saturation (in terms of grip) at a comparatively low yaw rate

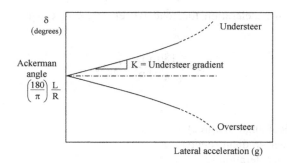

FIGURE 7.20

Determination of understeer gradient.

(Gillespie, 1992.)

of the car surprises the driver. Typically, road infrastructures are constructed so as not to need high lateral accelerations at high speeds. For normal drivers the difficulty is largely masked in day-to-day driving, which contributes to some complacency. In general, vehicles are tuned to have reduced YRG (and hence AyG) at higher speeds to compensate for this problem. The level of this tuning differs widely between different markets (Figure 7.20).

Although the tyres are not linear up until the point of saturation (as discussed at length in Chapter 5), the behaviour of the vehicle is not substantially different to that implied in the preceding discussion, it is merely a slightly distorted and smoothed version of the behaviour implied. To map the transition from linearity to nonlinearity, the constant radius turn test procedure (ISO 4138) may be summarised as:

- Start at slow speed, find Ackerman angle.
- Increment speed in steps to produce increments in lateral acceleration of typically 0.1 g.
- Corner in steady state at each speed and measure steering inputs.
- Produce a graph similar to that shown in Figure 7.12.

Considering the diagram, two regions are apparent. In the 'understeer' region, more steer angle is necessary compared to the Ackermann angle to hold the chosen radius. This may not seem intuitive unless the view is taken that the vehicle steers less than is expected ('under' the Ackermann response) and more steer angle is needed to compensate for it. Similarly, the 'oversteer' region needs less steer angle compared to the Ackermann angle. If the oversteer is large, the steer might need to become negative to trim the vehicle in the steady state. For many, oversteer is marked by the use of steer in the opposite direction to the corner — so-called 'opposite lock'. However, the strict definition only requires that less steer than the Ackermann angle is applied — the transition to opposite lock merely marks a further degree of oversteer but there is nothing especially significant about the sign change. If the steer angle does not vary with lateral acceleration the vehicle is said to be 'neutral steering'.

At low lateral acceleration the road wheel angle, δ, can be expressed using Eqn (7.47).

$$\delta = \left(\frac{180}{\pi}\right)\frac{L}{R} + K A_y \qquad (7.47)$$

where

δ = road wheel angle (degree)
K = understeer gradient (degree/g)
A_y = lateral acceleration (g)
L = wheelbase (m)
R = radius (m).

Note that the use of understeer gradient in degrees per g can be expressed at either the axle or the handwheel if appropriate regard is taken of the steering reduction ratio. For vehicle dynamicists it is easy to declare that the only measure of consequence is the axle steer; however this is to ignore the subjective importance of handwheel angle to the operator of the vehicle. Note also that this parameter K is not to be confused with the more common 'stability factor' K as developed by Milliken and Segel and used later in this chapter.

Olley makes an important distinction between what he calls the primary effects on the car affecting the tyre slip angles and secondary effects affecting handwheel angles and body attitudes, which are acutely sensed by the operator (Milliken and Milliken, 2001). Perhaps the biggest source of difficulty between practical and theoretical vehicle dynamicists is that the large modifiers of the primary vehicle dynamics are generally fixed by the time the practical camp get their hands on a vehicle and so are not considered by them; the secondary modifiers, used to great effect in delivering the required subjective behaviour of the vehicle for its marketplace, are frequently overlooked by the theoretical camp as being 'small modifiers' despite being important to the emotional reaction of the driver to the vehicle. For this entirely prosaic reason it is common that members of each fraternity understand little of what goes on in the other.

7.3.3 The subjective/objective problem

As suggested in the opening chapter, there are two strongly divided camps in the vehicle dynamics field — practitioners and theoreticians. One of the reasons for the difficulties between the two camps is the use of common vocabulary with different meanings. On the theoretical side, for example, there are clear definitions of such basic terms as 'understeer' and 'oversteer'. Subjectively, these terms are used to describe quite different behaviours.

Understeer and oversteer have been defined in Section 7.3.2.3 as yaw rates under and over what might be expected, respectively. The subjective/objective problem has a great deal to do with *the nature of the expectation*.

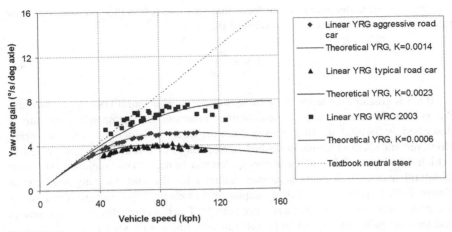

FIGURE 7.21

Yaw rate gain characteristics for three different vehicles. The stability factor, K, is defined later in the text. YRG, yaw rate gain.

Objectively, what is expected is the 'idealised' or 'geometric' yaw rate (Figure 7.19). At the lowest vehicle speeds this corresponds very closely with the performance of the actual vehicle, as noted earlier. As vehicle speeds rise, the previously described effects further reduce YRG even when the tyres are not saturated. Figures 7.21 and 7.22 compare two road cars with a nonwinged competition car.

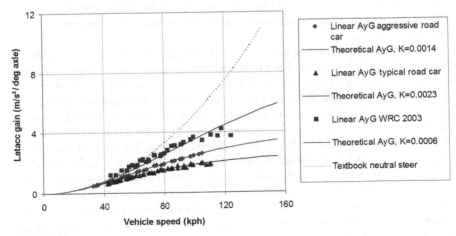

FIGURE 7.22

Lateral acceleration gain characteristics for three different vehicles. AyG, Lateral acceleration gain.

Objectively, understeer is when the actual yaw rate is less than the idealised yaw rate and is often expressed as the ratio of the two

$$US = \frac{\omega_{geom}}{\omega} \qquad (7.48)$$

If US is greater than unity, the vehicle is understeering and if US is less than unity the vehicle is oversteering. A vehicle that produces the geometric yaw rate is described as 'neutral steering' and is often regarded as something of a holy grail. However, for all the reasons described in Section 7.3.2.3 it is rarely engineered in vehicles. In Figure 7.21, all three cars understeer in their linear regions. The range of lateral acceleration gains is greatly reduced for the understeering road car in Figure 7.22, requiring less skilful adaptation to vehicle behaviour as speed varies compared with a textbook neutral steer vehicle. For the competition car the lateral acceleration gain can be seen to be substantially linear with speed.

Subjectively, drivers instinctively learn these base characteristics of vehicles very quickly. Only the most inexperienced novice drivers have difficulty with the steering ratios being higher or lower than expected. Within a few hours of driving experience, errors in steering (as distinct from positional errors caused by the size of the vehicle) are almost entirely absent. When the vehicle is driven at low lateral accelerations (i.e. when there is significant grip in reserve) the vehicle behaves in a substantially linear fashion — i.e. more steering gives more yaw rate in a proportional manner. Under these conditions, vehicles are rarely evaluated as 'understeering' although the vast majority of road cars do in fact understeer at highway speeds. These gains (both yaw and lateral) become the 'datum' condition, which is often rendered as 'neutral' when described subjectively. Differences between vehicles, in particular with YRG, are likely to be ascribed to differences in steering ratio and not differences in fundamental vehicle behaviour.

As lateral accelerations rise towards some significant fraction of the available coefficient of friction (typically above half), the vehicle behaviour becomes nonlinear due to the behaviour of the tyres. Depending on suspension geometry, tyre characteristics, elasto-kinematic behaviour of the suspension and suspension calibration, the yaw rate characteristics in the nonlinear region can vary in either direction — either up or down — from the linear characteristics. Of particular interest is the relationship between AyG and YRG in the nonlinear region. Figure 7.18 graphically suggests three possibilities for the departure from linearity for a vehicle. The situation imagined is that a vehicle has been driven steadily close to the subjective linearity limit and then the vehicle speed is increased in order to increase the lateral acceleration. The increase in vehicle speed is gradual and so details such as driveline layout are not relevant because the drive torque is low. This situation might occur, for example, on a long, constant radius, downhill motorway interchange 'cloverleaf'.

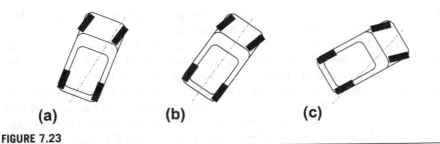

(a)   **(b)**   **(c)**

FIGURE 7.23

Possibilities for departure from linearity.

The three scenarios illustrated in Figure 7.23 can be summarised as:

a. reduced AyG and further reduced YRG
b. reduced AyG and YRG in proportion to each other
c. reduced YRG and further reduced AyG.

In scenario (a), YRG is reduced further than lateral acceleration gain. In order to accommodate the changes in both lateral acceleration and yaw rate, the radius of the path must increase and so the vehicle has a period of adjustment to a new, wider line in the curve. Most drivers notice this and instinctively reduce vehicle speed to restore the desired path over the ground. It is described subjectively as an 'understeer departure' or 'pushing' or perhaps in the USA as 'plowing' (ploughing). If uncompensated, it leads to a vehicle departing the course (road, track, etc.) in an attitude that is basically forwards. This is by far the most common behaviour for road vehicles. It is desirable since, if the vehicle does leave the road, it is least likely to roll over and will correctly present the engineered crash structure between the occupants and any obstacles encountered. For sporty drivers the sensation of the vehicle 'turning out' of the corner as it departs from linearity can become tiresome.

In scenario (b), lateral acceleration gain and YRG change in some connected manner and the vehicle will maintain course although it might need some modification to steering input. Subjectively this vehicle will be described as 'neutral' although objectively it might well be understeering. Excess speed for a curve will lead to the vehicle running wide but with no sense of 'turning out of the curve'. Such a vehicle generally feels benign although the progressive departure can mean it is unnoticed by inattentive drivers. Enthusiastic drivers will not be so frustrated by this behaviour.

In scenario (c), AyG reduces more than YRG. This leads to an 'over rotation' of the vehicle when viewed in plan. Depending on the severity of the mismatch, the change may lead to a spin out of the curve. From inside the vehicle there is a pronounced sense of the rear end of the vehicle departing first but objectively the vehicle may not actually oversteer in the classical sense — it may simply move 'towards neutrality'. This is the nature of rear-wheel-drive vehicles when driven to departure using the throttle. Subjectively, there is a pronounced sense of 'oversteer' —

sometimes described as 'loose' in the USA. Vehicles that preserve YRG as they lose linearity are widely regarded as fun to drive and sporty.

A further difficulty between theoretical and practical dynamicists is that the former group often considers the vehicle on the basis of 'fixed control' and 'free control' where the latter almost always uses 'driver input to complete a set task'. With fixed and free control, the inputs are consistent and the response of the vehicle is used to evaluate it. With driver input, the vehicle response is substantially constant and the vehicle is evaluated on the basis of the required changes in driver input to complete a task. More and more, so-called 'black lakes' — large flat areas of high grip surface — are being added to vehicle testing facilities to allow the evaluation of fixed and free control manoeuvres for experimental correlation purposes. Also gaining in popularity are theoretical 'driver models'. These range from simple path-followers to sophisticated multiloop, multipass adaptive controllers. At present there are many such models and none has gained precedence, suggesting perhaps that none is ideal for the task at hand — understanding and improving vehicle behaviour. Methods for modelling driver behaviour are discussed in Chapter 6. A very real problem with such models is that there is a fine line between evaluating the quality of the driver model and evaluating the change on the vehicle. Did the vehicle performance improve because the modification suited the driver model? Does that behaviour reflect a typical driver in an emergency situation? These criticisms are not unique to modelling and can also be levelled at highly skilled development drivers. Indeed, within motorsport circles this particular difficulty is widely recognised. There are drivers who can drive a given setup in the fastest way that it can be driven but who cannot articulate how the vehicle could be faster. Such drivers are an asset on race days but less so during development testing. For this reason, 'test' or 'development' drivers are frequently employed who have a different set of skills to event drivers.

In summary then, the subjective evaluation of a vehicle depends largely on the nature of its departure from linearity while objective evaluation is an absolute positioning against a geometric datum. From inside the vehicle it is generally difficult to distinguish between a vehicle that is operating at a large body slip angle and one that is truly oversteering. In any case, to control a large body slip angle it is frequently necessary to reduce or reverse steering input ('opposite lock'), changing the measured YRG substantially towards oversteer.

7.3.4 Mechanisms for generating understeer and oversteer

At this stage we must return to the tyre characteristics. Neglecting camber angle, version three of the Magic Formula (Table 5.4) gives cornering stiffness, BCD, by the following expression:

$$BCD = a3\left(\sin\left(2 \cdot \tan^{-1}\left(\frac{F_z}{a4}\right)\right)\right) \tag{7.49}$$

where F_z is the vertical load on the tyre and a3, a4 are empirical coefficients.

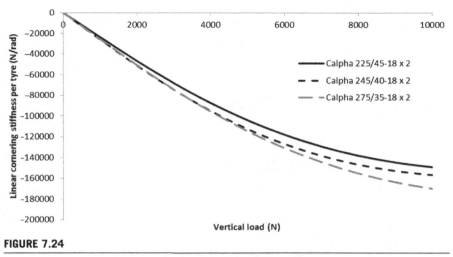

FIGURE 7.24

Influence of vertical load on cornering stiffness — published data.

Using data for three known tyre sizes from the same manufacturer, Figure 7.24 can be plotted.

It can be seen that there is a behaviour that is broadly related to tyre width. Adding data for two additional tyres and plotting both a3 and a4 against tyre width, an empirical approximation can be made (Figure 7.25).

Using the coefficients for the linear fit, which can be seen to be representative but not particularly high quality, the curves in Figure 7.26 can be plotted for notional tyres of different widths.

While clearly not an accurate representation of any particular tyre, these curves represent 'plausible' tyres of the type used on modern vehicles and display the kind of load sensitivity needed to manipulate vehicle behaviour to an extent that is of broadly similar magnitude to that experienced in developing vehicles. In the absence of detailed data, these estimates provide direction; it is suspected they overrepresent the fall-off in performance for smaller tyres and thus underpredict vehicle performance — a conservative error.

Considering one particular tyre, Figure 7.27 shows a typical plot of tyre lateral force with tyre load at a given slip angle. The total lateral force produced at either end of the vehicle is the average of the inner and outer lateral tyre forces. From the figure it can be seen that ΔF_y represents a theoretical loss in tyre force resulting from the averaging and the nonlinearity of the tyre. Tyres with a high load will not produce as much lateral force (in proportion to tyre load) compared with other tyres on the vehicle.

More weight transfer at either end therefore reduces the total lateral force produced by the tyres and causes that end to drift out of the turn. At the front this will produce understeer and at the rear this will produce oversteer. It should be noted that this behaviour is true for all slip angles of interest; at low slip angles, the cornering

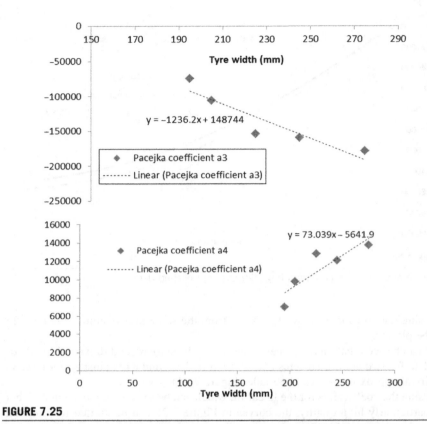

FIGURE 7.25

Fitting a linear regression line to the published data with respect to tyre width.

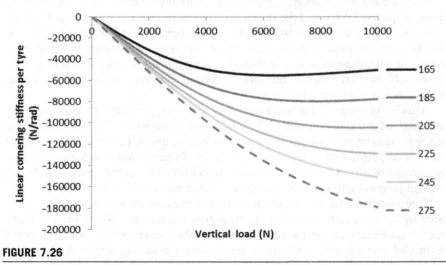

FIGURE 7.26

Reconstructed tyre cornering stiffness versus load characteristics using the fitted equations.

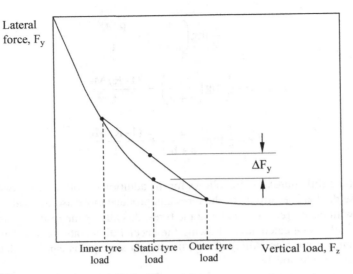

FIGURE 7.27

Loss of cornering force due to nonlinear tyre behaviour.

stiffness of the tyre is reduced nonlinearly with increasing load, promoting a larger slip angle at an axle with a greater roll moment. At saturated slip angles, the peak lateral force is reduced as a proportion of the vertical load, producing a lower coefficient of friction for an axle with a greater roll moment. Thus mechanisms that adjust sublimit understeer and oversteer also adjust departure plough and spin behaviour.

Returning to the 2 DOF model, a common criticism levelled at it is the lack of representation of the effects of load transfer on tyre behaviour. This is easily incorporated into the model by presuming static roll behaviour — i.e. roll moment is proportional to lateral acceleration — and calculating weight transfer on the basis of mass centre height and track width. The frequency of oscillation for the yaw/sideslip mode is something under 0.5 Hz generally. The authors' experience of the roll mode of vibration is that it is well damped and around 2 Hz. Thus there are about two octaves separating two quite well damped modes of vibration. This means the static representation of roll is slightly inaccurate but not unacceptably so for the purpose at hand.

With the tyre characteristics modelled, the lateral acceleration and mass centre height are used together to produce a roll moment:

$$(\dot{v}_y + \omega_z v_x)\, h = M_x \qquad (7.50)$$

This roll moment is distributed arbitrarily between front and rear axles by a fraction, R, sometimes known as the front roll moment distribution (FRMD) and which in turn produces a change in wheel loads via the wheel track distance, t:

$$F_{z\,fr} = \frac{1}{2} mg \left(\frac{b}{a+b} \right) - \frac{RM_x}{t} \qquad (7.51)$$

$$F_{z\,fl} = \frac{1}{2}\,mg\left(\frac{b}{a+b}\right) + \frac{RM_x}{t} \qquad (7.52)$$

$$F_{z\,rr} = \frac{1}{2}\,mg\left(\frac{a}{a+b}\right) - \frac{(1-R)\,M_x}{t} \qquad (7.53)$$

$$F_{z\,rl} = \frac{1}{2}\,mg\left(\frac{a}{a+b}\right) + \frac{(1-R)\,M_x}{t} \qquad (7.54)$$

The static roll approximation allows for the addition of roll moment distribution to exercise the load sensitivity of the tyres and produce more useful results. In combination with the current slip angles at the tyres, this allows the resulting side forces on the vehicle to be calculated, allowing the acceleration state to be calculated for the next integration step. Alternatively, the steady state gains can be calculated as was performed previously.

With the extra static roll DOF, the character of the 2 DOF solution is maintained but the ability to tune the behaviour of the car with roll moment distribution is captured. This model — which perhaps might be called a two-and-a-half DOF ('2½ DOF') model — is all that is needed to examine the behaviour of vehicles in an overview sense for normal driving.

Note that in principle such a model could be further extended to calculate the nonlinear behaviour of the tyres and its validity extended to the entire handling envelope. While the 2½ DOF model is readily soluble in a generic tool such as MS-Excel, as the calculations become more complex it is observed that MS-Excel gets into difficulties and becomes less and less responsive. For this reason a nonlinear model is best implemented in another environment, such as Octave or Matlab/Simulink.

A brief consideration of varying vehicle configurations is warranted with the 2½ DOF model. It is a matter of record that the underlying facts have led to the development of, broadly speaking, three distinct types of vehicles, each of which have their own foibles.

7.3.4.1 50/50 (BMW, Mercedes, Jaguar, MX-5)

The vehicle layout in which the mass is carried midway between the wheels has little or no technical understeer (i.e. a high YRG) and also a large amount of yaw/sideslip damping. That the original layout chosen for horseless carriages also happens to be technically meritorious is simply a matter of serendipity. However, it remains quite robust when subject to development and for this reason it has been the layout of choice for BMW, Mercedes, Jaguar, etc. for sometime.

Practical disadvantages include poor winter mobility when combined with a rear-wheel-drive powertrain and reduced interior package due to the intrusion of transmission and differential into cabin and luggage space. The high YRG generally leads to the addition of technical understeer through roll moment distribution, as shown in Figure 7.28.

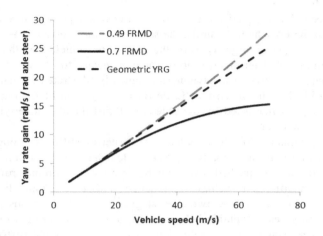

FIGURE 7.28

Yaw Rate Gain manipulation of a 50/50 vehicle with roll moment distribution. FRMD, front roll moment distribution; YRG, yaw rate gain.

For the baseline 50/50 vehicle, the solution does not converge at less than 49% front roll moment distribution (FRMD). The system responds strongly, becoming unbounded and progressing into the nonlinear region, at which time the model is no longer valid — but a disturbing lack of convergence may be expected in a real vehicle as the local slope of the tyre characteristics reduces and the vehicle slides 'over the edge'.

The addition of a strongly front-biased roll moment distribution aids traction performance by reducing the burden on the differential in a rear-wheel-drive powertrain and naturally balances the tendency of the powertrain to add YRG as the lateral capability of the rear tyres reduces with tractive effort.

50/50 cars display a distinctive subjective 'calmness' to their behaviour that some observers describe as competence and others as dullness.

Note also that recent high-power versions of such vehicles tend to display a tyre stagger that is also largely for marketing reasons; homologation documents often support the fitment of the same-size tyres all round if the owner prefers; the author's (Harty) BMW 540 was homologated with an 'M5 style' tyre stagger but was run with 225 tyres all round, also homologated.

7.3.4.2 Tail heavy (VW Beetle, Porsche 911)

Low-powered vehicles like the original 1938 Beetle attained a top speed of around 75 mph. When mass is carried rearward in the vehicle, the equations of motion display technical oversteer, in which YRG exceeds YRG_{geom}. There exists a speed beyond which the car cannot restore itself to a straight ahead condition without intervention from the driver — the yaw damping becomes negative. Early VW Beetles did not exceed this speed and therefore simply harvested the extra traction afforded by the weight over the driven wheels to retain mobility and improve accelerative performance.

When speeds for this platform came up with post-war performance tuning and evolution into the Porsche 911 family, the solution to the problem was empirically determined as the use of larger tyres on the rear of the vehicle than the front. The process of empirical determination took some time and for a period, such cars displayed somewhat difficult behaviour as poignantly illustrated by James Dean's nickname for his Porsche Speedster: 'Little Bastard'. Not until the group of papers surrounding Segel's published by the IMechE (1956) did the underlying system properties become clear.

The axle weights modify the effective tyre cornering stiffness through the vertical load sensitivity mechanism. To recover stable behaviour, the roll moment distribution needs to be pushed even further forward. Extreme roll moment distributions are difficult to achieve in practice, and so it is more likely that a tyre stagger will be adopted. A 'two-size' stagger (using the previously modelled tyre properties), gives acceptable behaviour for a 45% front weight distribution, with a 'four-size' stagger being helpful for a 40% front weight distribution. In both cases, the vehicle tolerates a range of roll stiffness distributions from 50% to 70% to deliver any preferred YRG characteristic. Without the tyre stagger, the vehicle is extraordinarily sensitive to roll moment distribution which would mean, even if it were possible to achieve, that the vehicle would be unduly sensitive to build quality and setup adjustments, rendering it not robust in service, as illustrated in Figure 7.29 where small amounts of roll moment distribution turn the vehicle from stable to unstable.

The high level of roll moment carried on a relatively light axle gives rise to the distinctive wheel-in-the-air cornering stance of Porsche 911s and such like, as shown in Figure 7.30.

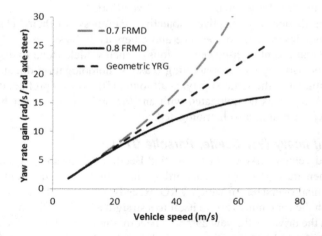

FIGURE 7.29

Tail-heavy vehicles with equal sized tyres are unduly sensitive to roll moment distribution. FRMD, front roll moment distribution; YRG, yaw rate gain.

FIGURE 7.30

High front roll moment distribution gives a distinctive cornering stance.

(Picture courtesy of Pistonheads.com.)

Note that current 911s display rather more tyre stagger than is necessary for the minimum stability criterion, this being driven to some extent by marketing expectations and to some extent by the requirement for acceptable tyre life with increased power outputs.

7.3.4.3 Nose heavy (Issigonis Mini, VW Polo, Ford Focus)

Issigonis recognised that the primary purpose of a car was accommodating passengers and proposed turning the driveline through 90° and pushing it forward in the car. In the equations of motion, this results in a large amount of technical understeer but can lead to very low levels of yaw/sideslip damping. It can be seen from typical eigensolutions that damping is highest at low speed and falls off in a fairly smooth way with increasing speed. Thus the most lightly damped condition is at maximum speed.

While there is no abrupt safe-to-unsafe boundary for yaw/sideslip damping, most organisations will not release a vehicle with a damping ratio of less than about 0.6–0.7. One solution to this would be to fit larger tyres at the front of the vehicle, but since the package is already congested widthwise this is not a preferred solution. Economies of scale with the same tyre all round add to the pressures to find other solutions to replacing the lost yaw damping.

These solutions include roll steer to induce yaw moments out of the turn, compliant steer to the same end and a less front-biased roll moment distribution. Modifying the roll moment distribution to 50% front improves the damping slightly and lifts the control gains from their very low values. The development of nose-heavy vehicles is something of a chase for yaw damping, depending on the required top speed of the vehicle and its mass and inertia characteristics. Again it must be emphasised that this is damping in the yaw/sideslip mode and not suspension damping (Figure 7.31).

The reduced damping and increased yaw/sideslip frequency gives nose-heavy cars a distinctive subjective 'liveliness' to their behaviour that some observers describe as eager and others as nervous.

The requirement to manipulate vehicle characteristics with suspension characteristics means that, unlike the 50/50 platform, nose-heavy cars require a large amount of subjective development to avoid awkward transient sensations. They are also

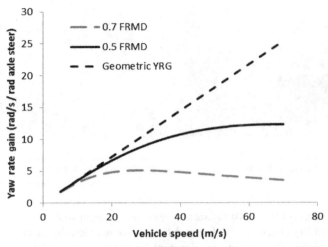

FIGURE 7.31

Influence of roll moment distribution on control gains — nose-heavy vehicle. FRMD, front roll moment distribution; YRG, yaw rate gain.

heavily reliant on the detail of the suspension systems to deliver good behaviour; the last 30 years has seen an explosion of complexity in suspensions, with particular emphasis on the rear suspension.

It can be seen that each of the three categories of vehicle can be tuned for acceptable behaviour without excessive difficulty, and desirable behaviour — stability and adequate yaw damping — looks broadly similar for all the vehicle categories. A fully featured vehicle model must be manipulated in its entirety to represent different vehicles and may not respond well to simple editing of headline wheelbase, mass and inertia properties without accompanying detailed suspension changes. The 2½ DOF linear formulation described here has the advantage of being somewhat simpler and does not obscure the underlying behaviour of the vehicle with such details; the sense that emerges from using such a formulation remains valid despite the simplicity of the model.

It may be readily imagined that most passenger cars cross between two of the identified types with loading condition, and for light commercial vehicles they may transition from nose to tail heavy — that is to say crossing between three of the identified types. While this would seem to be cause for some consternation, it is the author's experience that dynamic considerations are applied to a vehicle with two passengers and an expectation is applied that the vehicle will not be subject to emergencies when fully laden — see for example National Highway Traffic Safety Administration (NHTSA) (2001). Many organisations define a 'performance test weight' that is somewhat less than gross vehicle weight. It would seem that, generally, drivers of laden vehicles are considerably more circumspect.

To understand in more detail the vehicle parameters that have an influence on oversteer, understeer and departure behaviour, we can use the roll stiffness model

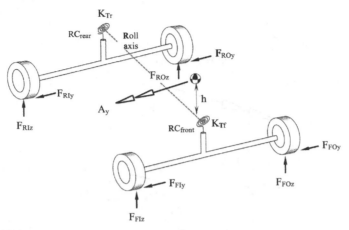

FIGURE 7.32

Free body diagram roll stiffness model during cornering.

described earlier to consider a series of free bodies and the force moment balance on each during steady state cornering. Figure 7.32 shows a version of the roll stiffness model while travelling in a curved path, with a lateral acceleration, A_y.

Consider next the components of force and moment acting on the vehicle body in isolation.

Using the roll stiffness model as the basis for the analysis we are treating the body as a single rigid axis with forces and moments transmitted from the front and rear suspensions (axles) at points representing the front and rear roll centres as shown in Figure 7.33. Consider the forces and moments acting on the vehicle body rigid roll axis. Note that we are ignoring the inclination of the roll axis. A roll moment ($m\, A_y\,.\, h$) acts about the axis and is resisted in the model by the moments M_{FRC} and M_{RRC} resulting from the front and rear roll stiffnesses, K_{Tf} and K_{Tr}.

$$F_{FRCy} + F_{RRCy} - m\, A_y = 0 \tag{7.55}$$

$$M_{FRC} + M_{RRC} - m\, A_y\,.\, h = 0 \tag{7.56}$$

The roll moment causes weight transfer to the inner and outer wheels (Figure 7.34). Taking moments for each of the front and rear axles shown gives:

$$\Delta F_{FzM} = \frac{M_{FRC}}{t_f} = m\, a_y\,.\, h\left(\frac{K_{Tf}}{K_{Tf} + K_{Tr}}\right)\frac{1}{t_f} \tag{7.57}$$

$$\Delta F_{RzM} = \frac{M_{RRC}}{t_r} = m\, a_y\,.\, h\left(\frac{K_{Tr}}{K_{Tf} + K_{Tr}}\right)\frac{1}{t_r} \tag{7.58}$$

It can be seen from Eqns (7.58) and (7.59) that if the front roll stiffness, K_{Tf} is greater than the rear roll stiffness, K_{Tr} there will be more weight transfer at the front

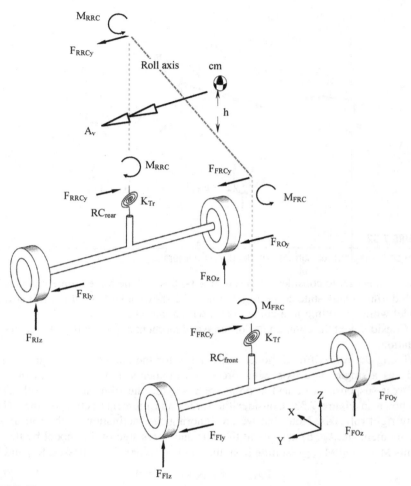

FIGURE 7.33

Forces and moments acting at the roll axis.

(and vice versa). It can also be seen that an increase in track will reduce weight transfer. Consider again a free body diagram of the body roll axis and the components of force acting at the front and rear roll centres (Figure 7.35):

This gives:

$$F_{FRCy} = m a_y \left(\frac{b}{a+b} \right) \tag{7.59}$$

$$F_{RRCy} = m a_y \left(\frac{a}{a+b} \right) \tag{7.60}$$

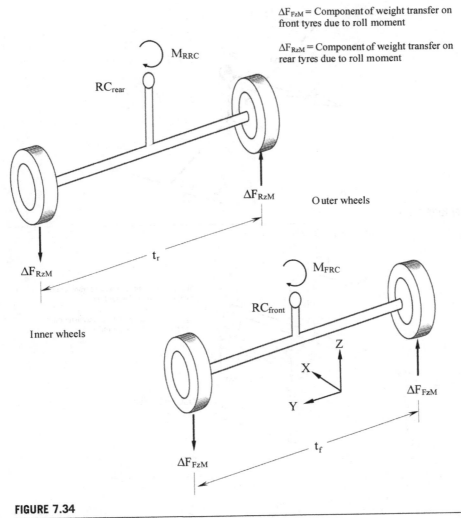

ΔF_{FzM} = Component of weight transfer on front tyres due to roll moment

ΔF_{RzM} = Component of weight transfer on rear tyres due to roll moment

FIGURE 7.34

Components of weight transfer due to roll moment.

From Eqns (7.60) and (7.61) we can see that moving the body centre of mass forward would increase the force, and hence weight transfer, reacted through the front roll centre (and vice versa). We can now proceed to find the additional components, ΔF_{FzL} and ΔF_{RzL}, of weight transfer due to the lateral forces transmitted through the roll centres.

Taking moments again for each of the front and rear axles shown in Figure 7.36 gives:

$$\Delta F_{FzL} = F_{FRCy} \left(\frac{h_f}{t_f} \right) = m a_y . h \left(\frac{b}{a+b} \right) \left(\frac{h_f}{t_f} \right) \tag{7.61}$$

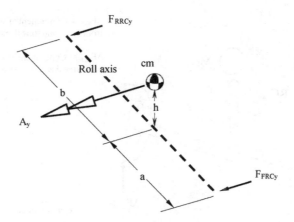

FIGURE 7.35

Forces acting on the body roll axis.

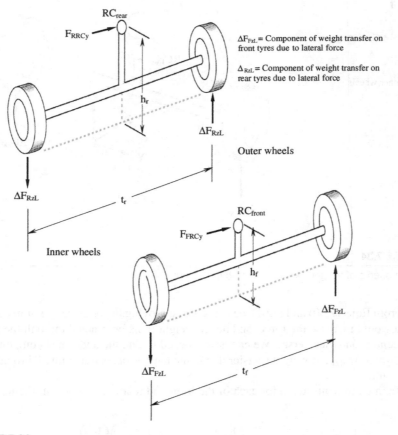

ΔF_{FzL} = Component of weight transfer on front tyres due to lateral force

Δ_{RzL} = Component of weight transfer on rear tyres due to lateral force

FIGURE 7.36

Components of weight transfer due to lateral force.

$$\Delta F_{RzL} = F_{RRCy}\left(\frac{h_r}{tr}\right) = ma_y . h\left(\frac{a}{a+b}\right)\left(\frac{h_r}{t_r}\right) \qquad (7.62)$$

It can be seen from Eqns (7.62) and (7.61) that if the front roll centre height, h_f, is increased there will be more weight transfer at the front (and vice versa).

We can now find the resulting load shown in Figure 7.37 acting on each tyre by adding or subtracting the components of weight transfer to the front and rear static tyre loads (F_{FSz} and F_{RSz}).

This gives:

$$F_{FIz} = F_{FSz} - \Delta F_{FzM} - \Delta F_{FzL} \qquad (7.63)$$

$$F_{FOz} = F_{FSz} + \Delta F_{FzM} + \Delta F_{FzL} \qquad (7.64)$$

$$F_{RIz} = F_{RSz} - \Delta F_{RzM} - \Delta F_{RzL} \qquad (7.65)$$

$$F_{ROz} = F_{RSz} + \Delta F_{RzM} + \Delta F_{RzL} \qquad (7.66)$$

Although substantially simplified, the preceding analysis helps with understanding the essential mechanisms in play when a vehicle is cornering.

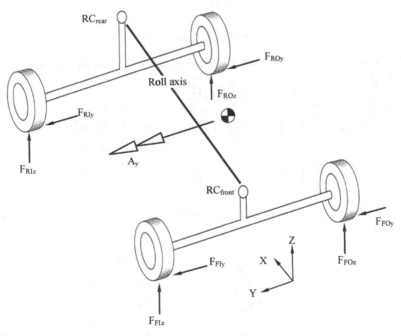

FIGURE 7.37

Resulting forces acting at the inner and outer tyres.

7.4 Transient effects

Vehicle dynamics would be a very simple field if the preceding text described it entirely. The full transient equations developed earlier can be considered again, considering the vehicle travelling in a straight line with no steering input. It is reasonable to presume solutions to the differential equations of the form $Ae^{\lambda t}$, which can represent many different behaviours depending on the value of λ.

λ is the 'eigenvalue' of the equation describing the system. Systems with multiple DOF have multiple eigenvalues. For a system with DOF x_i, the general form of the equations of motion is given by the so-called 'Jacobian' notation, a matrix of partial derivatives:

$$
\begin{bmatrix}
\dfrac{\partial}{\partial x_1}(\dot{x}_1) & \dfrac{\partial}{\partial x_2}(\dot{x}_1) & \cdots & \dfrac{\partial}{\partial x_n}(\dot{x}_1) \\[2mm]
\dfrac{\partial}{\partial x_1}(\dot{x}_2) & \dfrac{\partial}{\partial x_2}(\dot{x}_2) & & \dfrac{\partial}{\partial x_n}(\dot{x}_2) \\[2mm]
\cdots & & & \cdots \\[2mm]
\dfrac{\partial}{\partial x_1}(\dot{x}_n) & \dfrac{\partial}{\partial x_2}(\dot{x}_n) & \cdots & \dfrac{\partial}{\partial x_n}(\dot{x}_n)
\end{bmatrix}
\begin{bmatrix} x_1 \\ x_2 \\ x_3 \\ x_4 \end{bmatrix}
=
\begin{bmatrix} \dot{x}_1 \\ \dot{x}_2 \\ \dot{x}_3 \\ \dot{x}_4 \end{bmatrix}
\tag{7.67}
$$

We have presumed a solution of the form

$$
x(t) = Ae^{\lambda t} \tag{7.68}
$$

$$
\dot{x}(t) = \lambda Ae^{\lambda t} = \lambda x(t) \tag{7.69}
$$

When derivatives are substituted into the equations, it can be seen that

$$
\begin{bmatrix}
\dfrac{\partial}{\partial x_1}(\dot{x}_1) & \dfrac{\partial}{\partial x_2}(\dot{x}_1) & \cdots & \dfrac{\partial}{\partial x_n}(\dot{x}_1) \\[2mm]
\dfrac{\partial}{\partial x_1}(\dot{x}_2) & \dfrac{\partial}{\partial x_2}(\dot{x}_2) & & \dfrac{\partial}{\partial x_n}(\dot{x}_2) \\[2mm]
\cdots & & & \cdots \\[2mm]
\dfrac{\partial}{\partial x_1}(\dot{x}_n) & \dfrac{\partial}{\partial x_2}(\dot{x}_n) & \cdots & \dfrac{\partial}{\partial x_n}(\dot{x}_n)
\end{bmatrix}
\begin{bmatrix} x_1 \\ x_2 \\ x_3 \\ x_4 \end{bmatrix}
=
\begin{bmatrix}
\lambda & 0 & 0 & 0 \\
0 & \lambda & 0 & 0 \\
0 & 0 & \lambda & 0 \\
0 & 0 & 0 & \lambda
\end{bmatrix}
\begin{bmatrix} x_1 \\ x_2 \\ x_3 \\ x_4 \end{bmatrix}
\tag{7.70}
$$

This leads directly to the eigenvalue problem in familiar form as the determinant of the Jacobian:

$$
\begin{vmatrix}
\dfrac{\partial}{\partial x_1}(\dot{x}_1)-\lambda & \dfrac{\partial}{\partial x_2}(\dot{x}_1) & \cdots & \dfrac{\partial}{\partial x_n}(\dot{x}_1) \\[2mm]
\dfrac{\partial}{\partial x_1}(\dot{x}_2) & \dfrac{\partial}{\partial x_2}(\dot{x}_2)-\lambda & & \dfrac{\partial}{\partial x_n}(\dot{x}_2) \\[2mm]
\cdots & & & \cdots \\[2mm]
\dfrac{\partial}{\partial x_1}(\dot{x}_n) & \dfrac{\partial}{\partial x_2}(\dot{x}_n) & \cdots & \dfrac{\partial}{\partial x_n}(\dot{x}_n)-\lambda
\end{vmatrix} = 0
\tag{7.71}
$$

For the 2 DOF linear vehicle system modelled, the solution to the eigenvalue problem is given by

$$\begin{vmatrix} \dfrac{C_{af}+C_{ar}}{mv_{x'}}-\lambda & \dfrac{C_{af}a-C_{ar}b}{mv_{x'}}-v_{x'} \\ \dfrac{a\cdot C_{af}-b\cdot C_{ar}}{I_{zz}v_{x'}} & \dfrac{a^2\cdot C_{af}+b^2\cdot C_{ar}}{I_{zz}v_{x'}}-\lambda \end{vmatrix}=0 \qquad (7.72)$$

Writing the terms in a more convenient form...

$$a_{11}=\frac{C_{af}+C_{ar}}{mv_x} \qquad (7.73)$$

$$a_{21}=\frac{C_{af}a-C_{ar}b}{mv_x}-v_x \qquad (7.74)$$

$$a_{12}=\frac{a\cdot C_{af}-b\cdot C_{ar}}{I_{zz}v_x} \qquad (7.75)$$

$$a_{22}=\frac{a^2\cdot C_{af}+b^2\cdot C_{ar}}{I_{zz}v_x} \qquad (7.76)$$

...allows the solution of the eigenvalue problem with the rather trivial quadratic equation in λ.

$$(a_{11}-\lambda)(a_{22}-\lambda)-a_{21}\cdot a_{12}=0 \qquad (7.77)$$

$$\lambda^2+(-a_{11}-a_{22})\lambda+(a_{11}\cdot a_{22}-a_{21}\cdot a_{12})=0 \qquad (7.78)$$

$$A=1 \qquad (7.79)$$

$$B=(-a_{11}-a_{22}) \qquad (7.80)$$

$$C=(a_{11}\cdot a_{22}-a_{21}\cdot a_{12}) \qquad (7.81)$$

$$\lambda=\frac{-B\pm\sqrt{B^2-4AC}}{2A} \qquad (7.82)$$

It can be seen that there are two types of outcome, the generation of two real roots or the generation of a pair of complex conjugate roots, depending on the magnitude of the terms B and C.

If λ is real and positive then the response of the system is unbounded and grows with time. If λ is real and negative then the response of the system is bounded and converges to a constant value. Both of these outcomes are illustrated in Figure 7.38, which is a repeat of Figure 3.62.

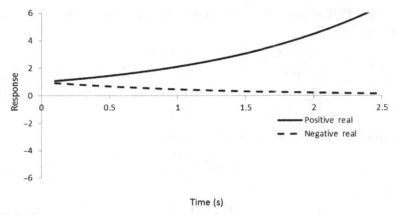

FIGURE 7.38

The influence of the sign of the real part of the roots of the characteristic equation.

For such a response, the time constant of the response is given by the magnitude of the eigenvalue; a numerically large eigenvalue suggests a very rapid response, whether positive or negative.

If λ is imaginary then the response of the system is oscillatory with the frequency given by the eigenvalue since $e^{i\omega t} = \cos(\omega t) + i\sin(\omega t)$ which is oscillatory in time. If λ is complex and the real part of λ is positive the response is oscillatory and unbounded. If λ is complex and the real part of λ is negative the response is oscillatory and convergent. These outcomes are illustrated in Figure 7.39, which is a repeat of Figure 3.63.

For oscillatory responses the period of oscillation is given by the imaginary part of the eigenvalue. Note that the undamped natural frequency is given by the Pythagorean sum of the real and imaginary components, but is of little practical interest.

For an oscillatory response, the damping ratio, ζ, is given by

$$\zeta = \cos\left(\tan^{-1}\left(\frac{\text{imag}\,(\lambda)}{\text{real}\,(\lambda)}\right)\right) \tag{7.83}$$

The eigenvalues for the vehicle system are of particular interest since they change with speed. It can be shown that if the ratio of cornering stiffness front to rear is less than the mass ratio front to rear then the vehicle will always be stable (Milliken and Milliken, 1998).

$$\frac{C_{\alpha f}}{C_{\alpha r}} < \frac{b}{a} \Rightarrow \text{stable} \tag{7.84}$$

If this is not true then there will be some speed above which the eigenvalues become positive real, implying an inability to reject disturbances. It is observed in the literature that vehicles that are always stable display reduced damping in the yaw/sideslip mode of vibration — see for example Milliken and Milliken (1998).

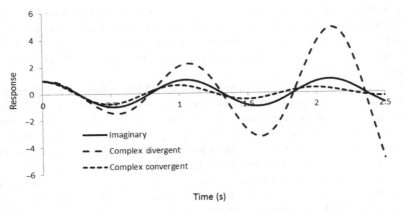

FIGURE 7.39

The influence of imaginary components of the roots of the characteristic equation.

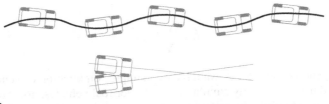

FIGURE 7.40

Yaw/Sideslip mode of vibration for a running car.

The eigenvalue represents a resonant behaviour of the vehicle in the ground plane, often described as a 'yaw/sideslip' mode, illustrated at the top of Figure 7.40. If it is imagined that the car could be shadowed with a helicopter travelling at exactly the same speed, when in resonance the motion of the car as observed from directly above in the helicopter would resemble a pendulum swinging about some 'apparent pendulous centre' — a point ahead of the car when viewed from a platform travelling at the same speed as the vehicle, illustrated at the bottom of Figure 7.40. The figure is an illustrative view and does not capture the phase difference between yaw and side-slip present in the real vehicle motion.

Like a pendulum, a system resonance exists despite the fact that there is no stiffness — a fact that is not always intuitively grasped at first reading. Unlike a pendulum, even the very simple representation of the car has some damping in the mode of vibration; this arises due to the yaw/sideslip coupling terms in the equations of motion. Note that the yaw/sideslip damping ratio has nothing to do with the suspension dampers.

The idea of the body slip angle has already been introduced in Section 7.3.2.2. The need for a body slip angle gives rise to an additional yaw rate, rotating the body to the correct slip angle, for a brief period near the start of any manoeuvre.

There is also a corresponding yaw rate reduction at the end of a manoeuvre required to bring the vehicle back to its trimmed, straight ahead state. These variations in yaw rate are acutely remarked by even the least skilled driver and greatly influence the emotional reaction of the driver to the car — although the driver might not consciously recognise it.

A vehicle that rotates slowly to give the required body slip angle feels sluggish and unresponsive; a vehicle that overshoots and oscillates feels lively and poorly controlled. Subjectively, these two states might be described as 'transient under-steer' and 'transient oversteer', respectively — although the latter might not objectively be oversteer, it is certainly a greater yaw rate than might be expected. Objectively, the quantity of interest is the rate of change of body slip angle, known as $\dot{\beta}$ or 'beta-dot'. The bulk yaw rate of the vehicle is made up of two components: the first is the yaw rate associated with a curved path — it might be thought of as the yaw rate that would be experienced by a stone on a string being swung around. Milliken and others refer to this as the 'no-slip yaw rate' — the yaw rate predicted without any body slip angle. The second is the beta-dot component. By inspection, it can be seen that:

$$\dot{\beta} = \frac{A_y}{V} - \omega \qquad (7.85)$$

Thus in general, beta-dot is available as a simple combination of vehicle states, particularly during multibody simulation work. For real vehicles, noise on acceler-ometer data and the difficulty of knowing the genuine forward speed of the vehicle under conditions of longitudinal tyre slip mean that beta-dot is difficult to discern in real time, although it is amenable to offline processing.

With changes in suspension configuration, elasto-kinematic calibration and damper behaviour, beta-dot can be modified significantly. The overall behaviour of the vehicle may well only be slightly modified and so to concentrate merely on bulk yaw rate would be to underestimate the significance of the change from the driver's perspective.

In the steady state, the body slip angle is constant and so it is of little conse-quence except when something changes. In practice, real vehicles spend very little time in the steady state and so the short-lived events — described as 'transients' — are very important to the driver's reaction to the vehicle because of the acute perception of beta-dot. This perception is not only in terms of absolute levels but also in terms of delays between driver requests and vehicle response. When the handwheel is in motion, the driver is implicitly requesting a change in body slip angle, since body slip angle is linked to lateral acceleration. If the change in body slip angle occurs in a way that is substantially connected to the change in steering, the driver feels reassured. If however the change in body slip angle occurs with perhaps a significant delay or perhaps with a characteristic shape different to the steer rate, the driver forms an impression that the vehicle has 'a mind of its own'. This leads to poor subjective ratings for handling confidence whatever the objective measures might say. Thus the correlation between steer rate and beta-dot is a useful one in measuring improvements in vehicle behaviour using theoretical models and real vehicles alike.

Transients are also important to the objective behaviour of the whole vehicle. Under circumstances of a steering reversal, particularly at or near the grip limit, there is a substantial increase in yaw rate transiently. To understand this, imagine the vehicle travelling in a steady curve with a body slip angle of, say, 5°. Now imagine the steering is reversed to produce a lateral acceleration in the other direction with a corresponding slip angle of −5°. If the adjustment of the body slip angle took place over a 1 second period, this would make a transient yaw rate of 10° per second. Considering Figure 7.6, it can be seen this is a substantial proportion of the total available yaw rate at speeds over about 50 mph. If the steady state lateral acceleration produced a yaw rate of, say, 15°/s (below the friction limit) then the additional 10°/s transient yaw rate would be more than the vehicle is capable of sustaining. For this reason alone, vehicles that converge (i.e. settle to a steady state solution) under normal conditions may well become unstable (spin) under conditions of steer reversal. Therefore any studies of transient vehicle dynamics must at some stage consider responses to steer reversals.

Vehicle mass and particularly inertia properties are a modifier to vehicle behaviour. There is a popular belief that the minimum mass moment of inertia in yaw is the best. However, this is not necessarily so. To understand this, consider a vehicle manoeuvring in the ground plane, which may be described by the classical 2 DOF formulation:

$$C_{af}\alpha_f a - C_{ar}\alpha_r b = \dot{\omega}I_{zz} \tag{7.86}$$

$$C_{af}\alpha_f + C_{ar}\alpha_r = mv\omega + m\dot{v} \tag{7.87}$$

At the moment of turn-in, yaw rate, ω and rear slip angle, α_r are both zero. Assuming a constant acceleration solution for the differential equations (a reasonable assumption for the first few moments of a turn-in event or following a disturbance at the front axle), the response of the builds in the manner given in Eqns (7.90) and (7.91).

$$\omega = \frac{C_{af}\alpha_f a t}{I_{zz}} \tag{7.88}$$

$$v = \frac{C_{af}\alpha_f t}{m} \tag{7.89}$$

Combining these quantities gives an instant centre at a distance c behind the mass centre

$$c = \frac{v}{\omega} = \frac{\dfrac{C_{af}\alpha_f t}{m}}{\dfrac{C_{af}\alpha_f a t}{I_{zz}}} = \frac{I_{zz}}{ma} = \frac{mk^2}{ma} = \frac{k^2}{a} \tag{7.90}$$

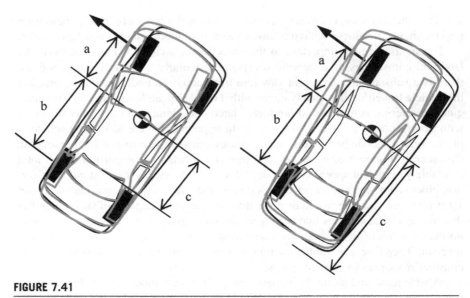

FIGURE 7.41

Dynamic Index influences the centre of rotation of the vehicle in yaw.

Expressing the distance c as a fraction of the distance to the rear axle, b, leads to

$$\frac{c}{b} = \frac{k^2}{ab} \qquad (7.91)$$

This quantity may be recognised as the 'dynamic index' (DI) defined by the SAE and used to good effect by Olley, albeit in pitch rather than yaw. Its importance for vehicle behaviour is that a DI less than unity results in an increase in lateral velocity at the rear axle, and hence an increase in slip angle at the rear tyres (Figure 7.41). If DI is greater than unity then lateral velocity — and hence slip angles — are reduced. In a situation where the vehicle is cornering at the critical slip angle, it can be seen that a DI less than unity will increase the rear slip angle and therefore reduce the cornering force available at the rear axle — promoting the tendency to spin. Low yaw inertia does indeed promote agility and speed of response, but excessively low inertia — as quantified by DI — makes the car difficult even for skilled drivers to manage in yaw. There is thus a trade-off between response and ease of control; for a broad spectrum of vehicles from the Mk 1 Lotus Elan to the BMW M5, this trade-off is a DI of just over 0.9. Motorsport vehicles differ significantly from road cars and the exact figures are jealously guarded. It can be seen with some reflection that a low-mass mid-engined car errs towards a DI that is 'too low' since the wheels have to be at a certain distance from each other to accommodate engine and passengers. The interested reader is invited to compare the mass and wheelbase of a Fiat X/1-9 and an MG-F.

In general, an important goal for vehicle dynamicists is to have the vehicle 'look after itself' from a body slip angle point of view. Most drivers have no conscious knowledge of body slip angle and beta-dot until they become large. The authors estimate that a body slip rate of less than around 3°/s is probably the threshold between subconscious and conscious awareness of body slip rate, but this is necessarily extremely sensitive to context. This means that a body slip angle of 5° or more can develop more or less unnoticed by a typical driver. In general, body slip angles of greater than 10° are difficult for the majority of the driving population to recover control from, and so 5° represents something 'halfway to irrecoverable'. When the body slip rate and/or angle exceeds some threshold, drivers suddenly become aware that something is amiss and so they report that vehicles 'abruptly' skid because the event was well developed before they recognised it. For this reason, an unexpected skid during otherwise normal driving can be surprisingly traumatic and leave people with a large amount of anxiety since they felt 'ambushed' by the vehicle. Drivers become sensitised and acclimatised to variations in body slip rate and angle through familiarity and for this reason, skid pan training for normal drivers is an excellent idea.

7.5 Steering feel as a subjective modifier

A further difficulty for theoretical dynamicists is the question of steering 'feel'. Subjectively, impressions of vehicle behaviour are gathered to a significant extent through the handwheel, whether consciously or subconsciously. A great deal of effort is concentrated in modern road cars on the manner in which torque is transmitted back to the driver up the steering column. Those skilled in the art have no difficulty distinguishing between steering issues and vehicle issues. However, a lack of clarity can lead to confusion if steer effects are not separated from vehicle effects. For example, two otherwise identical vehicles with different steering ratios will be judged quite differently by most drivers. Presuming the underlying behaviour of the vehicle is satisfactory, most drivers will rate a numerical reduction in steering ratio ('quicker' steering) as giving 'better' handling due to the increased YRG of the vehicle as seen by the driver from the handwheel. Yet the vehicles are identical and it would be possible to reproduce the behaviour of one vehicle by steering at a different rate in the other, to different final positions.

Steering feel is correctly given a great deal of importance in road car design since it is the primary means by which the customer comprehends the dynamics of the vehicle. Accurate modelling of steering feel is difficult and requires a great deal of data about friction in individual joints, plus also a good characterisation of the hydraulic or electrical power assistance used in the steering system. Nevertheless, work to understand the relative importance of individual contributions is possible with comparatively inaccurate models so long as good judgement is used and conclusions are correlated with work on real systems.

Changes in steering torque are a primary input for skilled drivers to detect vehicle behaviour. A common description of 'steering feel' involves discussion of tyre aligning torques. When driving normally, the tyres generate forces by distortions in the contact patch (see Chapter 5) that result in a moment attempting to return the tyre to a zero slip angle condition. If the steering system is well designed, it is delivered with very little corruption from vehicle weight and frictional effects directly to the hands of the

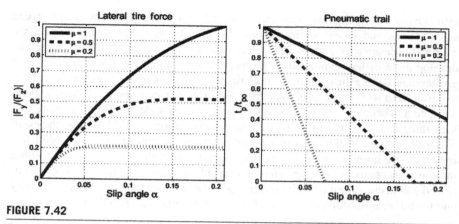

FIGURE 7.42

The use of inferred pneumatic trail for predicting surface friction characteristics even at very low slip angles.

(Hsu, 2006.)

driver. As discussed in Chapter 5, the relationship between side force and aligning moment implies a physical distance known as 'pneumatic trail' — the distance between the geometric centroid of the contact patch and the force centre observed in the contact patch. It is typically observed that pneumatic trail falls away in a fairly straightforward manner as the tyre delivers more and more force. It seems likely that skilled drivers develop an intuitive model in their heads that allows them to read this slope of falling pneumatic trail and thus simultaneously assess the available friction and the front tyre saturation state on a moment-by-moment basis. Hsu (2006) discusses this matter at some length and illustrates it very clearly in Figure 7.42.

The author's (Harty) experience in World Rally Car steering development suggests another mechanism is in play. It is widely recognised in accident reconstruction communities that the skid marks left by a tyre when it is laterally saturated display distinctive patterns described as 'striations'. The formation of such marks, which are a combination of the tyre slip angle, slip ratio, tread block design and rubber compound is described elsewhere in detail (Beauchamp et al., 2009) and will not be repeated here. The key item to observe is that the tyre is typically somewhat deformed during this process and there exists a distinctive pattern of vibration in the carcass of the tyre associated with successive deformation and release of individual tread blocks in the contact patch. In addition, the large lateral forces along with typical suspension geometries cause some energy to bleed into the wheel hop mode of vibration. Taken together, this means there is a distinctive vibration spectrum associated with the tyre operating in a saturated condition. A body of work by practitioners (Giacomin, 2013) suggests that humans are very capable of distinguishing vibration environments from one another even with drastically limited bandwidths and it seems likely that skilled drivers learn to associate the vibration environment presented by the steering system with the stress state of the front tyre.

In addition, if a vehicle starts to spin then the steering system informs the driver within around 0.1 s using the 'castoring' torque generated by operating the entire vehicle at a large slip angle. This mechanism ensures minimum handwheel torque when the wheels are placed so as to recover the skid; in this way the steering system fairly directly signals the current body slip angle to the driver. Skilled drivers are extremely sensitive to these messages, which arrive ahead of the brain's processing of the results of its data from the inner ear and significantly ahead of messages decoded purely from the visual environment.

It can be seen that steering systems that mask any or all of these mechanisms might be less helpful than steering systems that do not. Therefore it is suggested that proposing metrics to capture these aspects of the steering system and examining design variables that influence them will lead to well-behaved steering systems even without extensive prototyping and subjective test activities.

One final aspect of steering feel is worth commenting on, since it is noted in the fore-going discussion that the usefulness of the feedback is predicated on a skilled operator taking time to learn the sensations available and their meaning with respect to the behaviour of the vehicle. When hydraulically assisted power steering was first introduced in the USA, it was clearly an effort-saver in premium (i.e. heavy) vehicles. In the 1970s and 1980s as it started to be applied to more performance-oriented cars in Europe and Japan, there was earnest discussion about whether or not a power-assisted steering system could ever feel 'as good' as a manual system, or whether they were somehow 'impure' and a lot of other such nonsense. It has been interesting to relive most of those discussions with the introduction of electric power-assisted steering (EPAS), which had fairly awful feedback in most of its early incarnations; it has been interesting to see that hydraulic power steering is now held up as the reference and EPAS is somehow viewed as 'not as good', although even at the time of writing the tide of opinion seems to be accepting electric power steering. The important point in noting this is that when assessing steering feedback metrics it is important to compare with some absolute standard of information available to a human about tyre and vehicle states (and their corruption by unintended factors) rather than to concentrate on replicating the characteristics of a currently prevalent system to which people are temporarily habituated.

7.6 Roll as an objective and subjective modifier

So far, no mention has been made of body roll. There are two important effects of roll, they are objective and subjective. In Chapter 4, some discussion of the so-called 'roll centre' was put forward as well as the notion of suspension movement leading to adjustments of the wheel camber and toe angle as the body rolls.

Both of these effects lead to a modification of the way the tyre is presented to the road. These in turn lead to variations in the yaw moment on the car and therefore some real (objective) influence on the behaviour of the vehicle. These are mostly quite straightforward. For example, toe-out on bump on the front suspension will lead to a reduction in the slip angle of the tyre as the vehicle rolls out of the turn, reducing the yaw moment on the vehicle and hence reducing the steady state yaw rate if no adjustment is made by the driver. A slightly more subtle effect is associated with the typical inclination of the so-called roll axis. If the vehicle is imagined to rotate purely about its longitudinal axis (to roll in a pure sense) then the lower front anti-roll geometry will lead to a greater lateral velocity of the front wheels in comparison with the rear. This modification of the lateral velocity will modify the tyre slip angles, with a correspondingly greater increase in front slip angle than rear. As a result, the tyres will produce a yaw moment out of the turn — against the yaw rate but phased with roll rate. Although the motion of the vehicle is not purely roll when entering a turn or reacting to a disturbance, this mechanism may be seen to be one that couples roll and yaw.

Modern road vehicles are comparatively taut in roll, with compliances of 6°/lateral g and less being commonplace on quite ordinary vehicles. This is in comparison with 20 years ago, when roll compliances of 12°/lateral g were quite normal. Circuit competition cars are typically at something under 2°/lateral g. Low levels of roll compliance mean that body roll is no longer the modifier to vehicle dynamics that it once was — its effect is now quite small in the overall scheme of things.

Subjectively, roll has an important effect. Upon entering a turn, the vehicle may be thought of as 'relaxing out' to a final body roll angle. This means the lateral response of the driver's head, high in the vehicle, is reduced during the transient roll-out section of turn-in. Consideration of the expression for beta-dot (Eqn 7.85), shows that subjectively this leads to a momentary overestimation of body slip rate by the driver and hence roll transients often degrade driver confidence by introducing a delay in perceived lateral acceleration. Note that for vehicles typically instrumented, that delay is not captured by an accelerometer mounted on the floor of the passenger compartment and some processing of roll rate is necessary to compute the lateral response of the vehicle at driver's head height. This is also true of MBS models. Although the driver subconsciously compensates for motion of their head on the flexibility of their neck, they do not generally compensate for motion of the platform of the vehicle in the same manner.

The nature of the roll-out event is also important, with many in the vehicle dynamics community believing that roll acceleration profiles and roll jerk (the time derivative of roll acceleration) are important modifiers in the perception of roll as a modifier to platform dynamics. Makers of motion simulators understand and use this effect to 'simulate' angular events, applying an angular jerk and then providing a visual environment to suggest the roll rate is persisting. Although quite reproducible, these small events are difficult to capture with instrumentation on vehicles and so there exists a belief that these phenomena are not amenable to objective quantification. That this is currently true is a consequence of pragmatism rather than any underlying principle — it is very quick to have someone experienced and skilled in the art develop damper tuning for these subjective qualities, compared to instrumenting a vehicle and going through a research programme to define numerical goals at which to aim. In general, a fluid development of initial roll rate and a progressive deceleration to the final roll angle are recognised as necessary to reduce the perception of roll angle. Directional changes towards this end are certainly amenable to predictive analysis with MBS models, and simulation work using the final, released damper calibrations would go a long way towards improving the quality of work on the next vehicle programme. However, institutionally there is little time in modern engineering organisations for such work since it does not immediately contribute to the task at hand. Historically it has been difficult to get good data to define vehicle dampers (primary modifiers for roll transients) although modern system identification techniques mean this is more possible than it once was.

There is some work that suggests that roll—pitch interaction is important for subjective evaluation of roll (Kawagoe et al., 1997) and this certainly seems plausible. Again, such behaviour is amenable to analysis with MBS modelling and allows directional selection of design alternatives if not final tuning on the real vehicle. Essentially, a pitch nose-upward that accompanies roll is often subjectively described as 'the rear of the vehicle rolling more than the front' — a statement that is quite mystifying to objective vehicle dynamicists but common currency among skilled development drivers. It is recognised as undesirable and a small but not excessive amount of nose-down pitch is preferred for road vehicles. The exact amount varies with market segment and changes over time as market tastes change. This is something of a challenge for vehicle dynamicists since often there is a desire to have the vehicle roll onto front bump stops before rear in order to guarantee limit understeer, and hence stability — unfortunately this promotes subjectively undesirable pitch nose-up with large roll transients. Development work with MBS models and real vehicles allows combinations of damper tuning and anti-roll geometry to overcome this difficulty. Once again, a fluid development of initial pitch rate and progressive deceleration to the final pitch angle are recognised as necessary to reduce the perception of pitch angle.

7.7 Frequency response

As mentioned in Section 7.4, a transient demand for yaw rate change (a nonzero steering rate) is also a transient demand for a change in body slip angle. That is to say there is a strong link between expectations of beta-dot and rate of handwheel motion. Delays between the two are also acutely remarked.

The seminal IME papers in 1957 showed the existence of a yaw/sideslip mode of vibration for the vehicle, which has been illustrated with the 2½ DOF model in Section 7.3.4. The mode of vibration may be thought of as analogous to a pendulum but in the ground plane. It is this mode of vibration that rally drivers use when they 'flick' the car from one side to the other before a turn. Like any other vibrating system, the gains of the vehicle vary with frequency. Substantially below the resonant frequency, the behaviour of the vehicle is as already described. Around the resonant frequency, the gain is controlled by the level of damping present and above the resonant frequency the gain is controlled by the mass and inertia of the vehicle.

Also like any other vibrating system, a phase shift builds up between input and output as input frequencies approach resonance. At resonance, response is 90° behind input and beyond resonance, response is 180° behind inputs. For normal drivers this is particularly problematic. When driving normally, the significant frequency content of steer input is very low, typically below the primary ride frequencies at around 1 Hz.

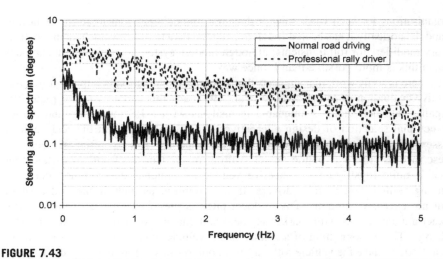

FIGURE 7.43

A frequency-domain comparison between road and competition driving.

A professional rally driver, in contrast, makes rapid and high-frequency steering inputs. Figure 7.43 shows a pair of spectral estimates made from handwheel angle recorded over time. The spectral density measure of degrees per hertz is multiplied by the spectral resolution of the two estimates to allow meaningful overplotting of signals sampled at different rates. A single 4096 point buffer, tapered using the 'Hanning' (cosine) window function has been used for both signals. At the lowest frequency, the professional driver (Petter Solberg during the third special stage of the 2002 Argentina round of the World Rally Championship) is using around three times as much steering input as the road driver, the author (Harty) on a Saturday in Warwickshire. By 0.5 Hz, the normal road driver's steer input is nearing the noise floor and is around one-eighth of its peak. In contrast, the professional driver's input is more or less flat to 1 Hz and tails off slowly, extending all the way to 5 Hz before reaching the same levels as the normal road driver's input at 0.5 Hz. At 0.5 Hz, the professional driver is using between eight and nine times the normal road driver's steer input — and this on a rack ratio that is some 50% faster than the road car on which the data were gathered.

Thus the Milliken assertion (Milliken and Milliken, 1998) that for road use the vehicle can be treated as a series of connected quasi-static events is mostly true — this is the basis of the automotive statics analysis embodied in the MRA Moment Method software (Milliken and Milliken, 1998). However, during emergency manoeuvres, the amplitude and frequency content of driver's steer inputs rise tremendously. Published data suggest that the highest steer rate sustained for 200 ms or more is likely to be around 1000°/s by the population as a whole (Forkenbrock and Elsasser, 1995). Data from the World Rally Championship series confirm this is the highest rate used by those drivers also; the authors' logged data also achieve — but have never exceeded — these levels on occasion.

When such large and high-frequency inputs are used, the response of the vehicle is no longer controlled only by the steady state yaw characteristics but may also be amplified compared to the base level as well having a phase delay imposed as described previously. The dynamic amplification is a result of the yaw/sideslip resonance of the vehicle being excited to produce responses that may be substantially greater as well as delayed by some 150 ms compared to the linear result that the driver might be expecting.

7.8 The problems imposed by ...
7.8.1 Circuit racing

Here the authors draw unashamedly on the material (Milliken and Milliken, 1998) presented in Chapter 1 of the Milliken and Milliken book 'Race Car Vehicle Dynamics'. Circuit racing is about using the acceleration vector of the vehicle to maximum effect at all times, be it in braking, accelerating, turning, or some combination thereof. Circuit racing is a highly rehearsed behaviour with 'braking points' and 'turn-in points' all being prescribed for a given type of vehicle. The driver's task is to apply control in a largely open-loop manner and to add minor trim inputs in a closed-loop manner. The 'Milliken Moment Method' uses, as one of several tools, a diagram that describes the manoeuvring envelope for the vehicle by plotting yaw moment against lateral force (or some variants thereof). Some examples are given of the diagrams for various different possible configurations. Notionally, the 'ideal' circuit car is one that is neutral steering and retains its YRG characteristic right to the limit of adhesion since this maximises the lateral acceleration possible. In truth, most circuit cars exhibit a small amount of 'push' or 'plough' right at the limit but to a first approximation, the yaw moment versus lateral force diagram looks a little like that shown in Figure 7.44. The right-hand vertex aligns with the A_Y axis, indicating a departure that is neither 'push' nor 'loose'.

7.8.2 Rallying

In contrast, rallying is largely unrehearsed, with line-of-sight driving plus prompts from co-drivers. These prompts take the form of memory-joggers: 'left 5, 150 over crest; caution, do not cut'. This is interpreted by the driver as 'left (turn, of severity) 5, (where 1 is gentle and 5 is aggressive based on likely speed before braking commences), 150 (metres) over crest; caution, do not cut (the inside of the corner because there is some hazard not obvious on entry)'. A great deal of skill and rapport is required; both driver and co-driver need to mould each other's expectations on the style of the co-driver's prompts. This information, though, is very incomplete compared to the circuit-racer's knowledge. Rally driving is all about control and adaptation. The driver's inputs have a very small open-loop content and very large and active closed-loop (trim) content. During a typical drift event lasting 2 s, there are five changes of steering angle direction (i.e. from left to right) and handwheel rates as defined in Bartlett (2000)

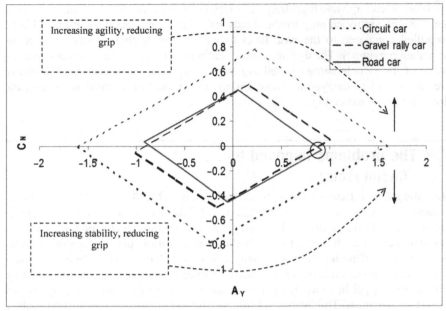

FIGURE 7.44

Force—moment (F—M) Diagrams for circuit racing, rally and road car performance — based on static weight distribution.

exceed 1000°/s regularly despite the use of ultralow ('fast') steering ratios. The yaw moment versus lateral force diagram is typically skewed much more towards control, which often produces stability concerns at the very highest speeds (Figure 7.44). The right-hand vertex is above the A_Y axis, indicating a departure that is 'loose' — that is to say the vehicle will spin unless actively trimmed by the driver.

7.8.3 Accident avoidance

Accident avoidance in road cars is quite different to both the previous scenarios. In terms of the percentage of miles driven it is statistically insignificant; if it were not for the fact that collisions have such serious consequences we would ignore it altogether. This is important because it means the driver is (in general) unpractised and unprepared. The driving task may be thought of as possessing two components — the *command* component, concerned with the speed and path of the vehicle and the *control* component, concerned with maintaining the heading of the vehicle in line with its path. That this split mimics the way the authors model the driver behaviour (Chapter 6) is no coincidence. Under normal circumstances the control part of the driving task is minimal. However, during aggressive manoeuvres it emerges to become an important component of the overall task because of the transient and frequency-domain issues outlines earlier. Typically, the importance of command decisions is also high at these times, leading to either a loss of control or poor command decisions under circumstances where the vehicle could have been physically able to evade the emergent hazards, had the driver been 'more able'. For normal driving tuition, only command driving is taught and control is presumed linear. Progression into the nonlinear region is regarded as a failure and so is inadequately addressed. In the UK, for example, a single 'emergency stop' manoeuvre is the only nonlinear event in DSA test. While notionally it is true that perfect command decisions could maintain the vehicle always in the linear region, most individuals are imperfect and will misjudge matters from time to time — misjudging either their own speed and position or that of others, requiring some emergency avoidance driving of one form or another.

There are two requirements for the vehicle under these circumstances. The first is that it must have sufficient cornering ability to be able to curve the vehicle path in such a manner as to avoid the hazards. The second is that it must be stable — that is to say the control demands on the user are minimised. A strongly desirable attribute is that the vehicle displays behaviour that does not distract the attention of the driver by thinking he 'might' need some control in the near future. In general this is linked with stability but not necessarily in a direct manner. Considering the Force—Moment (F—M) diagram in Figure 7.44, the right-hand vertex (circled) is below the A_Y axis, indicating a departure that is 'push' — that is to say the vehicle loses YRG with increasing lateral acceleration, gaining stability. If stability were the only criterion, then it would be easy to simply ensure an early 'push' departure with the vehicle design. However, increasing expectations for cornering ability to avoid hazards means that excessive sacrifice of grip for stability (a traditional recipe for the North American market) is becoming less and less acceptable.

7.9 The use of analytical models with a signal-to-noise approach

The attraction of using predictive methods is mentioned in Chapter 1 and reiterated in Section 7.1. One key benefit of analytical methods is that they are very repeatable — the same simulation run twice will generally yield the same results. There are specific exceptions to this, where 'random' numbers are included in analyses, but in general it holds as a premise. This removes an important obstacle in vehicle dynamics — the lack of repeatability of measured data as mentioned at the start of the chapter. This is certainly one form of 'noise' but not the form referred to in the title of the section. In general, drivers expect linearity from their vehicles. A departure from linearity may be regarded as 'noise' compared to the 'signal' that represents the driver's inputs.

Most readers will be familiar with the idea of variation in output being some measure of repeatability. For example, if a machine is required to produce a piece of a certain nominal size then a better 'quality' machine will produce pieces closer to that nominal size than a worse 'quality' machine. To quantify the level of match between machine and desired results, we use the notion, introduced in Section 7.1, of 'process capability', C_p

$$C_p = \frac{0.75\Delta_d}{6\sigma_d} \qquad (7.92)$$

where Δ_d is the allowable range for the attribute d and σ_d is the standard deviation of the attribute d as produced by the process. If C_p is greater than unity, the process is 'capable' and if not, it is not. This notion is used to assess production methods for intended tolerances and to focus attention on the least capable processes where resources are limited. As previously noted, the capability for the vehicle dynamics measurement process is questionable, at best — though analytical methods improve the capability of the process at the risk of introducing systematic inaccuracies through modelling errors and the like.

As part of a design process, the idea of a signal-to-noise (SN) ratio becomes much more useful. If we imagine that same manufacturing process, SN ratio is the ability of the process to produce any desired value of d. Thus, for some form of computer numerically-controlled (CNC) machine tool, some dimension d might be anywhere from 10 to 500 mm. A snapshot of the device's ability to produce, say, 276 mm is less useful than an overall knowledge of the relationship between input (desired dimension) and output (dimension produced). There are three types of 'lack of quality' that concern us:

- *Linearity*: the proportionality of output to input. For example, if on small dimensions the stiffness of the workpiece reduced, the machine might systematically make smaller things than required but as the dimension increases this effect might become less significant.
- *Variability*: the consistency of output to input. If the machine is insufficiently stiff then random vibrations induced by the cutting action might lead to random variations in the size of the workpiece.
- *Sensitivity*: the scale of output to input. If some calibration error were present between sensors and tool positioning shafts, it could be that so-called 'millimetres' in one axis were different to 'millimetres' on another axis.

A fourth effect is the ability of our measuring technique to discern the output to the required resolution. If we ask for a 0.1 mm change in size of the product but can only measure to the nearest millimetre, we are unable to discern whether or not we have been successful. This, however, is a matter of good experimental or analytical technique rather than something innate in the process itself and so it is laid aside as a difficulty for the moment.

To capture all three types of 'lack of quality', a single measure is used — the SN ratio. This is simply defined as

$$SN = \eta = 10\log_{10}\left(\frac{\kappa^2}{\sigma^2}\right) \qquad (7.93)$$

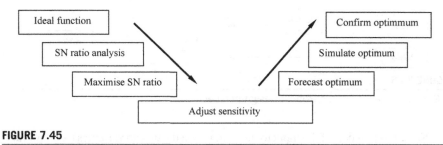

FIGURE 7.45

The Robust Design Process. SN ratio, signal-to-noise ratio.

where κ is the sensitivity of the process (i.e. the amount the output changes for a unit input change) and σ is the standard deviation of the results from the nominal.

To successfully use the idea of SN ratio in vehicle dynamics design, a process is required that looks something like that shown in Figure 7.45. Notice the similarity between this and Figure 1.7 in Chapter 1. This type of process is variously and vaguely known as Robust Design, Parameter Design or the Taguchi Method.

The first and crucial step is to define the so-called 'ideal function'. In vehicle dynamics it is very easy to become mired in talk of detailed effects of springs, dampers, bump rubbers, anti-roll bar bushings and so on and to dive into an intensive programme of investigation that quickly spirals out of control. It is only the experience of the empirical vehicle dynamicists that delivers vehicles that perform under these circumstances. Those same empirical vehicle dynamicists have become rightly suspicious of analytical methods since they can take a long time and run the risk of delivering little if not well directed. This need not be so. The authors suggest that the ideal functions should be generic enough to be applicable to any vehicle, without exception; only the tuning of sensitivities should need matching to the application (Figure 7.45).

If we consider the brake system as an example, it is simple to state an ideal function:

to proportionally transfer the driver's inputs from the pedal to the deceleration of the vehicle up to the available friction level.

Brake systems have several unintended functions as implemented. Squeal, judder and so on are of prime concern to road vehicle manufacturers and the ability to use all the available friction is of prime concern in motorsport applications. The conditions under which this 'ideal function' holds are simple — all conditions under which the vehicle operates. This includes in reverse (when drum brake systems can be startlingly ineffective), in car parks, when cold, when hot, when wet, when new, when worn, without the engine running and so on. Individually, these circumstances lead to a great deal of elaborate testing. However, when using SN ratio methods they can be robustly combined into a relatively small number of tests that will deliver the best performance available from the existing combination of ingredients and design variables (Figure 7.46).

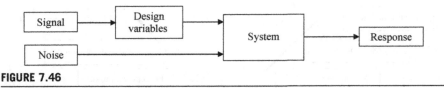

FIGURE 7.46

A Generic system from the Robust Design perspective.

'Signal' is that part of the input to the system, via the design variables ('control factors') under the engineer's control that is intended to produce a response. 'Noise' is the collection of all other factors that influence the response. For the brake system, these are ambient temperature, brake temperature, vehicle loading condition, direction of travel, fluid condition and so on. Design variables available to the engineer will be such things as disc mass, friction material, calliper stiffness, seal material, master cylinder mounting stiffness, pipe stiffness and so on. Some knowledge of the most likely dominant variables is preferred since the number of variables can become unmanageable. Wherever possible, variables should be compounded together for initial studies. If identified as important, further work can be performed to isolate the effects of individual variables. For example, calliper stiffness, master cylinder mounting stiffness and pipe stiffness come together under 'system stiffness'. As long as our tests incorporate some contribution from each of the compounded 'sub-variables', the conclusions will be valid. Similarly, noise inputs can be compounded together to give a 'noise reducing performance the most' and 'noise increasing performance the most'. For example, the former condition might be very cold brakes, fully loaded, worn brake pads. The latter condition might be optimally warm brakes, driver only, new bedded-in brake pads.

Each variable could be varied individually while maintaining the others constant. While this is a valid method, it rapidly leads to a large number of experiments with only a small number of variables. For example, with eight variables the number of experiments required to test two levels of each variable is $2 \times 2 \times 2 \times 2 \times 2 \times 2 \times 2 \times 2 = 256$ experiments. The use of orthogonal arrays allows meaningful results to be produced from a reduced set of experiments. With only eight variables, results can be obtained from only 72 runs. Adhering to so-called 'Taguchi' principles and using dynamic SN ratios as described allows interactions between inputs to be ruled out, allowing eight variables to be handled in just 16 experimental runs. Modern MBS codes usually come with some form of experimental design built into them, although not necessarily the dynamic SN ratio calculations described.

To perform the experiment, several levels of 'signal' are set and the design configuration is set. Results are collected at each signal level, in each design configuration for both noise conditions. The nature of those results should be such that the comparison with the ideal function is a meaningful one. For the brake system, mean deceleration between two speeds is suggested as a suitable response variable. Any modern data logging or MBS analysis software can easily capture and process such data. Signal factors should be chosen to give a reasonable spread of results over the operating envelope. In the case of a brake system, three levels of

	A	B	AxB	C	AxC	BxC	DxE	D	AxD	BxD	CxE	CxD	BxE	AxE	E
1	−	−	+	−	+	+	−	−	+	+	−	+	−	−	+
2	+	−	−	−	−	+	+	−	−	+	+	+	+	−	−
3	−	+	−	−	+	−	+	−	+	−	+	+	−	+	−
4	+	+	+	−	−	−	−	−	−	−	−	+	+	+	+
5	−	−	+	+	−	−	+	−	+	+	−	−	+	+	−
6	+	−	−	+	+	−	−	−	−	+	+	−	−	+	+
7	−	+	−	+	−	+	−	−	+	−	+	−	+	−	+
8	+	+	+	+	+	+	+	−	−	−	−	−	−	−	−
9	−	−	+	−	+	+	−	+	−	−	+	+	+	+	−
10	+	−	−	−	−	+	+	+	+	−	−	−	−	+	+
11	−	+	−	−	+	−	+	+	−	+	−	−	+	−	+
12	+	+	+	−	−	−	−	+	+	+	+	−	−	−	−
13	−	−	+	+	−	−	+	+	−	−	+	+	−	−	+
14	+	−	−	+	+	−	−	+	+	−	−	+	+	−	−
15	−	+	−	+	−	+	−	+	−	+	−	+	−	+	−
16	+	+	+	+	+	+	+	+	+	+	+	+	+	+	+

FIGURE 7.47

An $L_{16}^{(5-1)}$ two-level orthogonal array used for processing five variables in 16 runs. Columns AxB handle interactions between A and B, and so on.

deceleration might be suggested as 0.2, 0.5 and 0.8 g, with the pedal inputs selected to give results around these levels. Note that the signal factors (i.e. pedal inputs) should remain as consistent as possible. The actual array to be used is best selected from an existing library of such arrays; there is no particular need to derive one's own. Figure 7.47 shows a typical such array. Software tools external to the modelling or test environment, such as Minitab, have greatly simplified the experimental design process in recent years. Once the results have been generated — either by an experiment or by simulation — they are processed to calculate SN ratio and sensitivity for each of the conditions in the orthogonal array.

For vehicle dynamic behaviour, the most useful form for the SN ratio calculation is the so-called 'linear' form, which presumes that output is related to input in the following manner

$$y = \kappa(x - \bar{x}) + \bar{y} + e \tag{7.94}$$

where y is the response, x is the signal, and e is the error. $\bar{y}$ and $\bar{x}$ are the averages over the whole experimental data set (Figure 7.48).

If the system is nonlinear (as is typical), the ideal function is defined such that the linear form of this equation can still be applied. As a further example, if acceleration performance is being considered then the ideal function is 'constant power acceleration' such that

$$Pt = \frac{1}{2}mv^2 \tag{7.95}$$

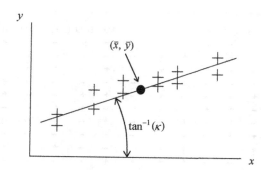

FIGURE 7.48

Linear form of presumed response.

that is to say all of the power of the engine goes into kinetic energy in the vehicle. The ideal function may be expressed as:

$$x = V_0 + \sqrt{\frac{2P(t - t_0)}{m}}, y = V \qquad (7.96)$$

If various speed intervals are used for testing then the constant power formulation allows the expected results, or signal values, to be set according to this formulation. Actual test values can be compared with them using the linear formulation to calculate SN ratios and sensitivity. Changes that improve the sensitivity will be reduced losses in the system (i.e. better efficiency) and changes that improve the SN ratio will be improved robustness. For example lower levels of driveline vibration reducing random losses, reduced aerodynamic drag maintaining constant power acceleration to a higher speed and improved area under the torque curve giving reduced sensitivity to gear ratios will all improve SN ratio.

So how then should this be applied to simulation models to improve the lateral dynamics of vehicles? The key to success lies totally in the definition of ideal functions. We have already discussed the brake system and powertrain. The handwheel remains the only driver input left. Its function is as a yaw rate demand. Previously, the behaviour of typical vehicles and drivers has been discussed. On the basis of this, a suggested ideal function for the handwheel is

to deliver a yaw rate proportional to handwheel angle, related to speed by the stability factor for the vehicle in question, up to the limit of available friction.

Figure 7.21 shows the linear YRG in relation to speed for some different vehicles. Thus, the ideal function for the handwheel could be expressed:

$$x = \frac{V\delta}{L(1 + KV^2)}, y = \omega \qquad (7.97)$$

With n pieces of logged or predicted information, SN ratio, η, is calculated according to the following method

$$\eta = 10 \cdot \log_{10}\left(\frac{S_\kappa - V_e}{\left[\sum_{i=1}^{n}(x_i - \overline{x})^2\right] \cdot V_e}\right) \tag{7.98}$$

where the following intermediate calculations are defined:

Error Variance
$$V_e = \frac{S_e}{n-2} \tag{7.99}$$

Error Variation
$$S_e = S_T - S_\kappa \tag{7.100}$$

Total Variation
$$S_T = \sum_{i=1}^{n}y_i^2 - \frac{\left(\sum_{i=1}^{n}y_i\right)^2}{n} \tag{7.101}$$

Variation due to Linear Effect
$$S_\kappa = \frac{\left(\sum_{i=1}^{n}y_i(x_i - \overline{x})\right)^2}{\sum_{i=1}^{n}(x_i - \overline{x})^2} \tag{7.102}$$

In addition, the sensitivity of the data set (the ratio of input to output) is calculated thus

$$\kappa = \frac{\sum_{i=1}^{n}y_i(x_i - \overline{x})}{\sum_{i=1}^{n}(x_i - \overline{x})^2} \tag{7.103}$$

The similarity to the standard least squares method is apparent. The interested reader is urged to consult Wu & Wu (2000) for further, more detailed description and background to the method. In that text, the symbols used differ slightly. The symbols used here have been chosen to avoid a clash with the familiar vehicle dynamics symbols in use. To use the method, a typical set of predicted or logged values is taken, consisting of yaw rate, $\omega(t)$, handwheel angle, $\delta(t)$ and forward speed $V(t)$. Values for stability factor, K, and wheelbase, L, are presumed known for the vehicle. Using the ideal function, 'expected' values for ω are calculated and taken as the input data series, x. The 'real' — either logged or predicted — values of ω are taken as the output y and the calculations for η and κ performed as described above.

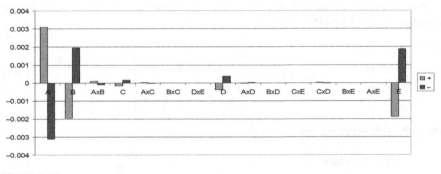

FIGURE 7.49

A typical effects plot produced by processing the results from an orthogonal array.

Once SN ratios have been calculated for each 'state' in the orthogonal array, they are processed to produce an effects plot of the type shown in Figure 7.49. In this case, the array was an $L_{16}^{(5-1)}$ two-level array for processing five design variables and their possible interactions. Variable A is shown to be dominant, with variables B and E also important. The signs of the effects are not all the same; variable A is 'less A gives *more* response' while B and E are 'less B/E gives *less* response'. There are no significant interactions between the variables. Effects plots are produced for κ and η separately. To calculate effects with the array shown in Figure 7.47, results from runs 1,3,5,7,9,11,13 and 15 are averaged to give the '$-$' result for variable A. The remaining columns give the '$+$' result for variable A. For variable B, runs 1,2,5,6,9,10,13 and 14 give the '$-$' result and so on. The combination of design variables is selected to maximise SN ratio to maximise robustness of the system. The following steps in the process are the adjustments to the sensitivity, κ, being made through those variables showing themselves as affecting κ but of less influence on η.

Such an approach to the ideal function for the steering and handling behaviour covers a great many different situations although for motorsport applications, in particular rallying, it may lack one important aspect.

Rally drivers use the nonlinearity of the car to their advantage. Recall the F—M diagram in Figure 7.44. This is derived on the basis of quasi-static calculations. However, professional rally drivers use the dynamic amplification of the yaw/sideslip resonance along with features such as ruts and road surface edges (Figure 7.50) to achieve yaw accelerations significantly greater than that suggested by the Milliken Moment Method, as shown by the comparison in Figure 7.51.

The lightest data in the centre represent the highest speeds achieved during the stage and fit well within the estimated boundaries of the F—M diagram. However, the two other shades show extremely large levels of yaw acceleration at high lateral acceleration in strong contrast to the general form of the Milliken diagram. This is a symptom of the use of the dynamic amplification in the yaw/sideslip mode as used

FIGURE 7.50

Dani Sordo hooks the front left wheel of his 2011 Mini Countryman John Cooper Works WRC off the road and ensures a large yaw moment will be available if he needs it.

(Image: McKlein.)

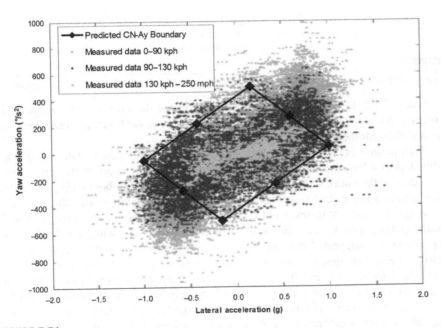

FIGURE 7.51

Yaw acceleration versus lateral acceleration for the Subaru WRC 2002, Petter Solberg, Argentina.

by the rally drivers. The use of an ideal function that relates steer angle to yaw rate will promote an increase in yaw damping, which is good for road cars but precludes the 'flick' style of driving preferred by rally drivers.

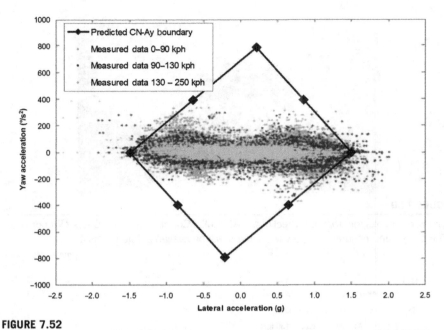

FIGURE 7.52

Yaw acceleration versus lateral acceleration for Subaru WRC 2002, Petter Solberg, Germany.

Figure 7.52, by contrast, shows the behaviour of the rally car on tarmac and suggests that the vehicle stays well within the quasi-static boundary predicted by the Milliken Moment Method.

Examination of the recorded data shows there is little 'steady state' about the rally stage, on loose surfaces and the character of the handwheel angle versus yaw rate trace for the entire stage has a distinctly 'circular' quality about it (Figure 7.53). This indicates that output (yaw) has a phase shift of around $90°$ — an indication that the system is indeed at resonance. In this instance it might be imagined the proposed ideal function is inadequate. However, substantial experience with vehicles for this application has suggested that pursuit of linearity under all possible circumstances delivers behaviour that is easily learnt and does not obstruct good performance. The behaviour of the Mini WRC, admired by a spectrum of drivers from 'gentleman racers' to world-class competitors and finishing only 8 s behind the event winner in its debut tarmac event in France, 2011, suggests that for all the 'technical' shortcomings of the pursuit of linearity, it does actually deliver well-behaved vehicles in a very focused manner and with limited engineering resources.

7.10 Some consequences of using SN ratio

Previously, vehicle dynamicists have sought to address a wide range of phenomena — unintended yaw rate change following lift-off in a turn, delays

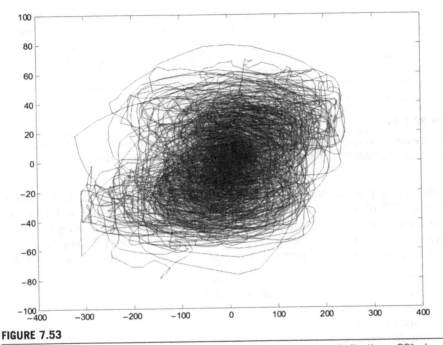

FIGURE 7.53

Yaw versus Steer Angle — note the circular quality of the data traces, indicating a 90° phase difference between the two quantities.

following turn-in, nonlinearity with increasing lateral acceleration and so on. It is eminently possible to attack each of these issues in turn, although it is also likely that solutions to some may become problematic for others.

The use of a single measure — SN ratio — and an ideal function allows experimentation to optimise the performance of the vehicle in its entirety. Questions of balancing one attribute against another no longer become subjective but can be expressed objectively, since an increase in yaw damping will increase SN ratio where it controls yaw overshoots in response to disturbances but reduce SN ratio where it blunts turn-in. This is not to say that the endeavour of optimising vehicle dynamics will become trivial; defining the correct 'handling sign-off' usage over which to record SN ratio remains difficult, instrumentation challenges remain, and the difficulties of accurately simulating vehicle and driver behaviour over a lengthy handling sign-off test are far from trivial.

Also nontrivial is the engineering of areas of vehicle behaviour where the departure from the ideal function might be regarded as 'character' rather than a nonoptimum. An example of such behaviour is the oversteer departure — i.e. the loose behaviour at the limit — of rear-wheel-drive vehicles (Figure 7.54). If smart systems on board disallow such a departure, keeping the vehicle stubbornly linear

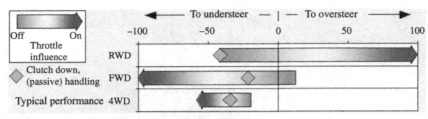

FIGURE 7.54

Vehicle behaviour on throttle — flaws or brand attributes? The illustration is informal, the quantities of understeer and oversteer being some undefined 'Subjective %'.

until the end of the available friction, will it surprise the driver when the friction is exceeded? Will it bore the driver and tempt him or her to buy a different, less perfect brand next time? Or will the retention of YRG as AyG falls off give a uniformly better driving experience, less tainted by some of today's compromises?

Certainly the use of SN ratio can potentially address many issues simultaneously:

- linearity of YRG with lateral acceleration
- delays in response degrading driver confidence
- roll/yaw interaction
- disturbance rejection
- throttle/steer interaction
  - FWD — minimising 'push' under power
  - RWD — reducing oversteer departure tendency under power
  - 4WD — minimising 'push' under power
- brake/steer interaction

Of some importance for the successful application of the method is the balancing of the proportion of different events in the overall SN ratio calculation. For instance, when comparing the influence of crosswind disturbance rejection with the influence of roll/yaw interaction, due account must be taken of how many 'crosswind' events there are for every 'aggressive turn-in' event and so on, lest one become unduly weighted in comparison with others. It is likely that the selection of the sign-off cycle will be what ultimately determines the dynamic brand attributes of the vehicle. This is not so different to the current situation where vehicles are tuned to match the preferences of key individuals at their preferred test facilities. These differences are a source of richness and diversity in vehicle design and are part of the larger commercial process of 'differentiation' to retain or increase market share.

Active Systems

We can do anything — we've been to the moon. The trick is in working out what we want.

Damian Harty, AVEC 10 Keynote Address

8.1 Introduction

Modern passenger vehicles are developed to a very high level in terms of their dynamic behaviour. Indeed, there are many who believe that little more can be done to improve the performance of the road car with passive means. While it may be a little premature to make such a statement, there have been many steps forward that have been made since the start of the vehicle engineering industry:

- addition of suspension damping
- adoption of independent suspension
- adoption of hydraulic suspension damping
- adoption of hydraulic brakes
- progressive stiffening of body structure
- addition of isolating elastomers in suspension
- adoption of radial tyre construction
- decoupling of lateral and longitudinal loadpaths
- optimisation of suspension geometry
- optimisation of elastokinematic behaviour
- adoption of low profile radial tyre construction.

It is probably true that there is less to do than has been done. However, some fundamental difficulties remain with pneumatically tyred vehicles. In particular, the fall-off in yaw damping with speed is problematic. Since the fundamental nature of the tyres is to generate side forces with respect to slip angle, this leads to increased lateral velocities at increased vehicle speeds in order to respond to a given level of disturbance (Figure 8.1). This behaviour leads directly to Fonda's 'tyre-as-damper' analogy (Fonda, 1956). Since lane width in general does not increase with speed, this means the vehicle's ability to reject a given disturbance within the lane is reduced with increasing speed and hence driver workload increases. As well as the

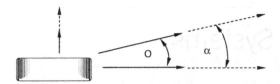

FIGURE 8.1

An increase in forward speed requires an increase in lateral speed to respond to the same disturbance.

straightforward increase in lateral velocity, consideration of the stability derivatives for the classical vehicle dynamics equations given in Milliken and Milliken (1995), and derived again in Chapter 7, shows that four important derivatives have vehicle speed as a denominator:

$$Y_r = \left(\frac{a \cdot C_F - b \cdot C_r}{V_x} \right) \tag{8.1}$$

$$N_r = \left(\frac{a^2 \cdot C_F - b^2 \cdot C_r}{V_x} \right) \tag{8.2}$$

$$Y_{VY} = \frac{C_F + C_R}{V_x} \tag{8.3}$$

$$N_{VY} = \frac{a \cdot C_F - b \cdot C_r}{V_x} \tag{8.4}$$

The dimensions a and b are the longitudinal distances of the vehicle mass centre from the front and rear axles as shown previously in Figure 7.22. The first two derivatives are the side force and yaw moment with respect to yaw velocity. The remaining two are side force and yaw moment with respect to body slip angle, which may be expressed as a function of lateral and forward velocities and hence implicitly contains forward velocity in the denominator. It is thus a fundamental consequence of the tyre behaviour that the vehicle loses restoring force with speed. This is in marked contrast to aircraft (in the subsonic region, at least) where restoring forces increase with speed since aerodynamic forces rise with the square of speed.

Note the contrast between these statements and those at the start of Chapter 7, where the control gains for ground vehicles can be seen to increase significantly with speed. This presents vehicle engineers with some particular problems since control authority is increasing while system stability is reducing — a 'closing gap' scenario in which the safe behaviour of vehicles at Autobahn speeds remains difficult to engineer.

8.2 Active systems

This 'closing gap' means the industry as a whole is considering the use of active systems to complement the passive behaviour of the vehicle. Traditionally, 'active' has meant a system in which energy is added in some substantial amount. In modern usage, this distinction is becoming lost and so the term is being used for any system that performs something other than a passive mechanical reaction. Previously, systems that modified their behaviour without adding energy were referred to as 'adaptive'. In general, systems that combine mechanical actuation with electronic control are referred to as 'mechatronic', although this term appears infrequently in industry literature.

To model any type of mechatronic system requires the introduction of sensors into the vehicle model and the implementation of the control law. Note that sensors in this context are different to software entities that may be also called 'sensors'. One convenient way is to add state variables in a manner similar to that described for driver modelling. One important difference, however, is that the driver model is 'continuous' — that is to say the resulting outputs may be differentiated without discontinuity.

An aspect of many active systems is that they use 'logic controller' principles (see Chapter 6) and thus may branch between conditions in a very short space of time. If carelessly modelled, this may give convergence problems inside the vehicle model and so the use of some smoothing method is required. Note that these difficulties are present when implementing real-life systems and not confined to simulation. In both MSC ADAMS and Simpack, the STEP function described in Chapter 3 allows the transition between one level and another using a half sinusoidal form that is continuously differentiable.

8.2.1 Full authority active suspension and variable damping

Lotus performed a great deal of work in the 1980s looking at 'active suspension', a fast acting system that varied the vertical load on each corner in accordance with control laws attempting to preserve ride height, minimise wheel load variation and minimise body acceleration. This system was undoubtedly very effective, particularly on the attitude-sensitive Formula 1 car. However, expensive components and significant power consumption mean that full authority active systems, which have the authority to substantially raise and lower the vehicle if so commanded, remain uncommon. In the 10 years since the first edition of the book, to the authors' knowledge only the Mercedes 'Active Body Control' system has remained available (Figure 8.2), despite the earnest attempts of marketing departments elsewhere to convince us that there is no difference between variable damping and full authority active suspension.

When compared to the original Lotus system, the Mercedes system is a down-specified version, working only with primary ride motions (up to around 5 Hz) rather than the higher frequency wheel-hop control achieved by the Lotus motorsport and SID research vehicles. In April 1990, Inifiniti, the luxury division of Nissan,

FIGURE 8.2

Mercedes Active Body Control (left) and a passively suspended vehicle (right) in action at high lateral accelerations — around 1° of body roll (a combination of tyre compliance and some programmed roll compliance for 'feel') with the ABC car compared to 4.5° of body roll for the passively suspended car.

(Courtesy of Auto Motor und Sport.)

introduced a similar hydraulic 5 Hz system for the 1991 model year Q45, badged as Q45a when carrying the 'Full Active Suspension' (FAS). This only remained on the market until the 1996 model year, when it was withdrawn the year before the face-lifted model was introduced. Nissan cited low take-up of the FAS option as the reason for its deletion, with internet rumours suggesting only 15% of US customers subscribing to it. Around the same time, Toyota showcased similar technology as part of a well-integrated *tour de force* on the UZZ32 Soarer (Sato et al., 1992) and subsequently on the 1992 Corolla. Neither vehicle sold in large volumes, suggesting the difficulty with full authority active suspension is not the technicalities but the commercial reality of persuading customers that the value is worth the cost.

Recently, Mercedes has introduced the system many engineers dreamed of in the early 1980s by adding a previewing feed-forward aspect to their active suspension, marketing it as 'Magic Body Control' in an attempt to distance it from damper control systems (Figure 8.3). Previewing is a good way to circumvent bandwidth limitations. It is easy to demonstrate conceptually in multibody systems (MBS) models if a deterministic road profile (e.g. a flat road with a cosine step at a certain distance) is employed, by using a location offset in looking up the road profile function. However, in a more general model with an arbitrary road surface — a scanned physical surface, for example — then the preview is more difficult to achieve. Visual terrain mapping of the kind used by the Mercedes system is more or less impossible to implement inside general purpose multibody software. Mercedes' own publicity material suggests the camera previews to about 15 m ahead of the vehicle but does not describe the spatial resolution. If such a preview were desired with an existing road profile then it could conceivably be implemented by conceptually extending the vehicle body component the requisite distance ahead of the vehicle and attaching a large number of small contact blocks to it. The contact properties for the blocks would be set to generate very little force so as not to influence the solution in any

FIGURE 8.3

Mercedes Magic Body Control (top) with road preview compared to an unspecified configuration (bottom) — the improved absolute pitch control is very apparent.

(Courtesy of Mercedes.)

meaningful way. Interrogating the contact elements for their penetration will make a preview of the road profile available to control algorithm.

Partial active systems, working only against body roll, have included Citroën's 'Activa' system fitted to the Xantia (not to be confused with the two concept cars of the same name) from 1995 until 2000, when it was replaced by the C5 model that reverted to the adaptive roll stiffness/damping system known as 'Hydractive II'. Land Rover's 'Active Cornering Enhanchement' is a hydraulically actuated fast roll levelling system, working around 5 Hz and available on Land Rover Discoveries from the 1999 Model year to date, and has made its way upmarket onto Range Rover models in a reversal of the normal trend. BMW's Active Roll Stabilisation system is conceptually similar and available on X5, X6 and X7 series platforms.

Active suspension systems and their cousins, continuously variable damper systems, are easy in concept to implement in MBS models. Action/reaction force pairs are introduced into each vehicle suspension unit in addition to the normal spring, damper and anti-roll bar forces. The magnitude of these forces is controlled by control laws in a similar fashion to that in the real vehicle. For example, a system (Figure 8.4) might use an open-loop roll moment method by applying forces proportional to the output from a lateral accelerometer. A further modification might be the addition of a roll damping term based on relative velocities of left and right suspension units, with a lookup table (spline) for use at different vehicle speeds.

Table 8.1 shows a sample of an MSC ADAMS command file implementing such a system on the front suspension of a vehicle.

Active suspension in principle allows the control of individual tyre reaction forces in a manner decoupled from platform orientation. As well as allowing a level platform during braking and cornering events (preserving suspension travel for disturbance events and improving the angle at which the tyre is presented to the road), it allows the redistribution of roll moment reaction on a moment-by-moment basis. The mechanisms discussed in Chapter 7 for inducing over- and

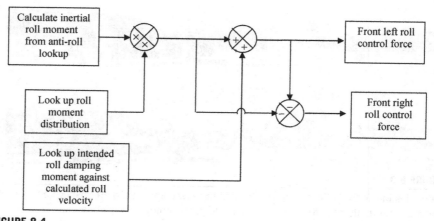

FIGURE 8.4

Simplified active suspension model.

understeer can thus be harnessed and used to improve turn-in behaviour, yaw damping and so on.

Note that in the example, no attempt has been made to filter the output down to any particular bandwidth. Filtering is a lengthy topic in its own right and is best not approached glibly. Many software environments allow the use of filtering techniques within their postprocessing tools. However, for simulating active systems, some kind of filtering is almost always necessary at run-time. Whether to simulate the limited bandwidth of an actuation device or to smooth noisy data, some form of filter must be implemented while the model is running.

Some MBS codes have access to a sophisticated run-time filtering library, such as those set up in the Matlab product environment. Others, like MSC ADAMS and Simpack, lack such run-time tools but do have the ability to enter a generic transfer function command. The filter transfer function itself must be arrived at using some external filter design tool. Particular caution is needed since some software environments use descending order for the terms in the transfer function while some use ascending order, as illustrated in Eqn (8.5), which represents a 1 Hz 2nd order Butterworth filter in the Laplace domain as used in Matlab and MSC ADAMS:

$$g(s) = \underbrace{\frac{39.884}{s^2 + 8.8869s + 39.884}}_{\text{Matlab Description}} = \underbrace{\frac{39.884}{39.884 + 8.8869s + s^2}}_{\text{MSC.ADAMS Description}} \tag{8.5}$$

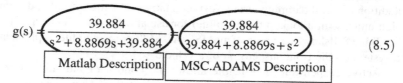

More sophisticated filtering is possible using higher order transfer functions and using the more general state-space modelling methods, which are available in most MBS environments. The interested reader is referred to 'Digital Filtering in the Time and Frequency Domain' (Blinchikoff and Zverev, 2001) for detailed discussion of

Table 8.1 MSC ADAMS Command File Sample for Simplified Active Suspension

```
data_element create variable &
 variable_name = inertial_sprung_roll_moment &
 function = "(VARVAL(latacc)*(mass)*", &
      "(DZ(m_body_CG,base) - ", &
      " (AKISPL(varval(fl_suspension_position),0,front_anti_roll_spline) ", &
      " * (front_track)/2)* varval(fl_suspension_load)/varval(total_load)", &
      " + AKISPL(varval(fr_suspension_position),0,front_anti_roll_spline) ", &
      " * (front_track)/2)* varval(fr_suspension_load)/varval(total_load)", &
      " + AKISPL(varval(rl_suspension_position),0,rear_anti_roll_spline) ", &
      " * (rear_track)/2)* varval(rl_suspension_load)/varval(total_load)", &
      " + AKISPL(varval(rr_suspension_position),0,rear_anti_roll_spline) ", &
      " * (rear_track)/2)* varval(rr_suspension_load)/varval(total_load)", &
      ")))
data_element create variable &
 variable_name = roll_moment_load_front &
 function = "varval(inertial_sprung_roll_moment)*", &
           "AKISPL(varval(latacc),0,roll_moment_distribution_spline)", &
           "/(front_track)"
data_element create variable &
 variable_name = roll_damping_load &
 function = "( ", &
           " (VZ(fl_damper_top,fl_damper_bottom) + ", &
           " VZ(rl_damper_top,rl_damper_bottom)) -", &
           " (VZ(fr_damper_top,fr_damper_bottom) + ", &
           " VZ(rr_damper_top,rr_damper_bottom)) ", &
           ")* ((front_track)+(rear_track))/2* ", &
           "* AKISPL(varval(velocity),0,roll_damping_velocity_spline)
force create direct single_component_force &
      single_component_force_name = front_left_active_force &
      i_marker_name = .fl_damper_top &
      j_marker_name = .fl_damper_bottom &
      function = "(+varval(roll_moment_load_front) + varval(roll_damping_load)", &
                "* STEP(TIME,0.0,0.0,1.0,1.0)"
force create direct single_component_force &
      single_component_force_name = front_right_active_force &
      i_marker_name = .fr_damper_top &
      j_marker_name = .fr_damper_bottom &
      function = "( - varval(roll_moment_load_front) - varval(roll_damping_load)", &
                "* STEP(TIME,0.0,0.0,1.0,1.0)"
```

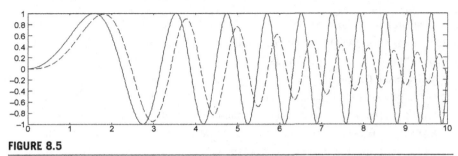

FIGURE 8.5

2-pole Butterworth filter described in Eqn (8.5) applied to signal in Eqn (8.6) (solid) — resulting signal in dashed line.

filtering methods and their repercussions. The most important thing to recognise is that almost any form of real-time/run-time filtering introduces some form of phase delay. Using the example transfer function in Eqn (8.5), an input chirp signal in Matlab defined by Eqn (8.6) produces an output as shown in Figure 8.5.

$$x = \sin(2\pi \cdot \frac{\text{time}^2}{10}) \tag{8.6}$$

Note that the amplitude attenuation is very gentle but that a phase delay has been introduced even at frequencies well below the notional cut-off frequency. Many software packages offer 'zero phase shift' Butterworth filtering by running the filter forwards and backwards on logged data, but a moment's thought should reveal this is unavailable in real-time applications in the absence of time travel technology.

The 2-pole Butterworth filter represents a linear second-order mechanical system and so its intelligent use can be very helpful in avoiding modelling complexity. The interested reader is encouraged to study its formulation and implementation in different forms — pole/zero and Laplace polynomial — as a launch-pad to a deeper understanding of real-time filtering issues and complexities.

Adaptive damping logic can be implemented in a similar fashion to active suspension inside an MBS model, scaling a damper spline or adding a scaled 'damper variation' spline to a 'minimum damping' spline according to a controller demand. The 1988 Lancia Thema 8.32 is believed by the authors to be the first production implementation of an adaptive damping system, with two-state dampers switching rapidly between 'soft' and 'hard' settings according to an algorithm that examined ride accelerations and driver inputs.

Active suspensions and continuously variable dampers seek to address the traditional three-way trade-off in suspension calibration (Figure 8.6). Given enough working space it is possible to improve both ride and handling simultaneously, as evidenced by desert race vehicles.

However, in road cars with suspension travel limited by prosaic packaging constraints, generally ride is achieved at the expense of handling or vice versa. Crolla comments (Crolla, 1995) that good control of the suspension dampers delivers

Body acceleration
(ride)

Tyre load control   Suspension travel
(handling)   (working space)

FIGURE 8.6

Suspension trade-offs.

many of the benefits of active control without the power consumption. Thus it seems to the authors to be the 'thinking man's' solution to improved suspension calibration. It also offers the prospect of software control for calibration of suspension dampers and making inroads into development times. The recent proliferation of adaptive damping systems suggests Crolla's observations were indeed correct.

Adaptive damping systems also offer an intriguing possibility for logistical convenience for companies making use of so-called 'platform engineering', when common hardware is used across a diverse range of products. The Volkswagen group of companies, — making vehicles badged Skoda, Seat, VW and Audi on the same hardware platforms — have a range of dampers that are physically interchangeable. When considered in conjunction with the engine and body style variants this might make something like 20 dampers with unique internals but an interchangeable vehicle interface — which is to say 19 opportunities to fit the wrong damper. A damper that communicates with the vehicle and is instructed by the vehicle to select a different set of parameters could in concept become a single part serving the entire platform family. This certainly brings logistical savings and, given the remorselessly downward trend in electronics costs, there must logically arrive some tipping point when the cost of the adaptive hardware is offset by the savings in logistics costs (which, in contrast, are not changing a great deal).

8.2.2 Brake-based systems

Other systems are less comprehensive but can have large benefits. In 1978, Mercedes introduced the Bosch 'Anti-Blockier System' as an option on the Mercedes S-Class. This system, universally abbreviated to ABS (albeit with some contortions to translate the German expression for 'anti-lock' into English), releases, holds and reapplies brake pressure in individual wheel brakes in order to retain directional stability and steering control with very little sacrifice in braking capability. Although technically it is possible for a skilled driver to outbrake anti-lock braking system (ABS), in truth the level of rehearsal this requires means it is unlikely in a road car in any realistic situation. Highly rehearsed circuit driving is an area where skilled drivers may improve over ABS performance and so circuit cars rarely use ABS.

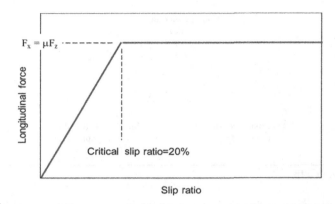

FIGURE 8.7

A simplified slip ratio versus force curve for the following ABS discussions.

Described in some detail in Limpert's work (1999) is the function of a vehicle ABS system. The main variable is the brake pressure. In the work by Ozdalyan (1998) a slip control model was initially developed as a precursor to the implementation of an ABS model. In order to discuss ABS operation, a simplified tyre slip characteristic can be used. Such a characteristic is shown in Figure 8.7. It can be seen to be made up of two linear portions and has a constant friction coefficient, μ.

ABS operation is essentially the interplay between two opposing torques, the frictional torque from the brake rotor and the spin-up torque resulting from friction between the tyre and the ground as illustrated in Figure 8.8. At any speed substantially above zero, the brake friction torque is broadly independent of speed but the tyre friction torque varies significantly with slip ratio (which is in essence a measure

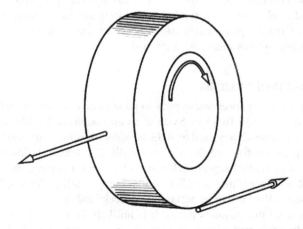

FIGURE 8.8

Anti-lock braking system operation is the balancing of two torques against each other.

of the sliding velocity of the contact patch, as discussed in Chapter 5). Spin-up torque also varies substantially with surface condition, varying by an order of magnitude between dry, smooth road and slightly wet ice.

The angular acceleration of a wheel, $\dot{\Omega}_{wheel}$, can be simply described with Newton's Second Law.

$$\dot{\Omega}_{wheel} = \frac{T_B - T_S}{I_{wheel}} \tag{8.7}$$

Where T_B is brake torque, T_S is spin-up torque and I_{wheel} is the mass moment of inertia in the rolling direction. If this wheel is connected to a single mass representing the vehicle then the linear deceleration of the wheel is given by:

$$A_{vehicle} = \frac{T_S}{R_1 m} \tag{8.8}$$

The above equations can be readily implemented in, say, MS-Excel to investigate their behaviour. Ramping on brake torque to a certain level, the wheel can be made to hold at 19.9% slip, as shown in Figure 8.9. A period of initial deceleration of the wheel can be seen, followed by a settling of the effective linear wheel velocity (its angular velocity multiplied by its radius) to its steady state value of approximately 80% of the vehicle linear velocity during the stop. A more aggressive brake torque ramp to a higher level causes the wheel to rapidly decelerate. The rapid nature of the deceleration can be explained by considering the net torque on the wheel during the non-locked stop in comparison to the locking stop — the saturation of the spin-up torque means the full additional brake torque is applied unopposed to the wheel, and it is very large in comparison to the net torque prior to locking.

Considering Figure 8.9, it can be seen that monitoring wheel angular deceleration gives a clear signature for a wheel entering a lock event. It could be imagined that some software might watch for a value below -400 rad/s and intervene when this is detected.

For real tyres, in Figure 8.10 it can be seen that on initial application of the brakes the brake force rises approximately linearly with slip ratio depending on the wheel load. If the braking is severe the slip ratio increases past the point where the optimum brake force is generated. To prevent the slip ratio increasing further to the point where the wheel is locked, an ABS system will then cycle the brake pressure on and off maintaining peak braking performance and a rolling wheel to assist manoeuvres during the braking event.

In the Ozdalyan (1998) model, the brake pressure is found by integrating the rate of change of brake pressure, this having set values for any initial brake application or subsequent application during the ABS cycle phase. The key ingredient of such a system is the ability to control brake pressure in one of three modes, often described as 'hold, dump and pump'. Hold is fairly self-explanatory, the wheel cylinder pressure is maintained regardless of further demanded increases in pressure from the driver's pedal. 'Dump' is a controlled reduction in pressure, usually at a predetermined rate and 'pump' is a controlled increase in pressure, again usually at a

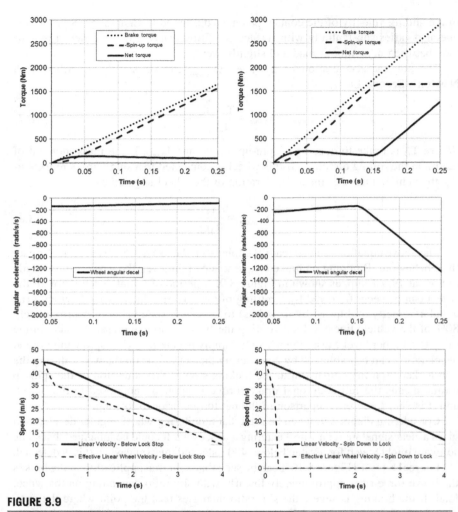

FIGURE 8.9

A 'below lock' stop (left) in contrast to a 'locked wheel' stop (right); note linear velocities are indistinguishable between the two (bottom graphs).

predetermined rate. Implementation of these changing dump, pump and hold states requires care to ensure no discontinuities in the brake pressure formulation.

To explore their influence in an ABS system, we can return to the simplified formulation used earlier and imagine four ascending levels of hardware complexity, as shown in Figure 8.11.

Using the hold valve alone is ineffective for controlling wheel speed, resulting in either a slight slowing of the wheel lock event or a reduction in vehicle deceleration, depending on whether it is triggered early or late (Figure 8.12).

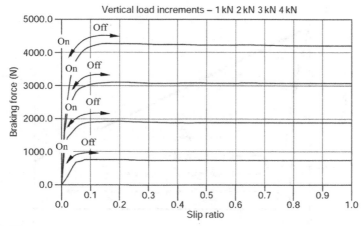

FIGURE 8.10

Principle of a brake slip control model.

Using a dump valve to bleed off excess pressure at a predetermined rate until the wheel spins up again is substantially more successful in principle. This strategy might be called Hold-Dump-Hold and is shown in Figure 8.13. However, although successful at preventing wheel lock, it does not optimise stopping distance and it does not deal well with low-to-high friction transitions.

Closing the dump valve and reopening the hold valve reconnects the master cylinder to the system and allows the reintroduction of pressure at a controlled rate. This is illustrated in Figure 8.14. Note that for the simplified friction model and idealised system shown here the strategy manages to discover the ideal brake

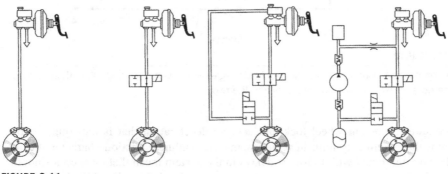

FIGURE 8.11

Ascending levels of brake hardware complexity — from left to right: simple push through, hold valve, dump and hold valves, dump/pump/hold valving. Illustrations based on those in Driving-safety Systems.

(Bosch, 1999.)

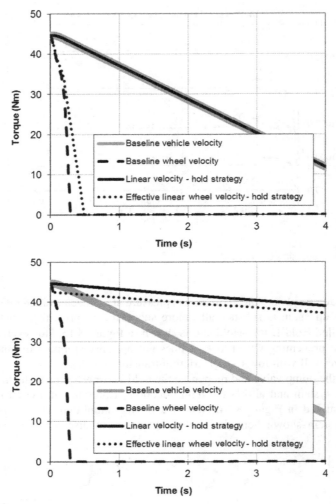

FIGURE 8.12

A comparison of unsuccessful hold-only strategies — too late (top) and too early (bottom); the hold threshold varies by 5% between the two.

pressure to prevent wheel lock and achieve deceleration that is indistinguishable from the baseline system; in real systems the nonlinear friction characteristics of the tyre combined with transport delays in the system mean that even on a smooth, consistent surface the system continues to cycle with the familiar ABS 'pulsing'. It can be readily seen that repeated cycles deplete the fluid from the master cylinder and so a pumped return is often employed. Both the cycling and the pumped return can give haptic sensations that surprise operators.

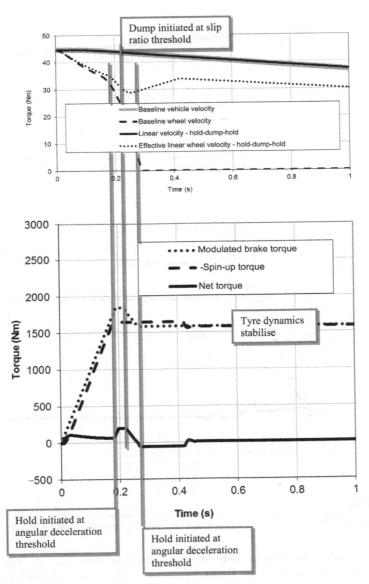

FIGURE 8.13

A hold-dump-hold strategy with the simplified 2 DoF vehicle braking model.

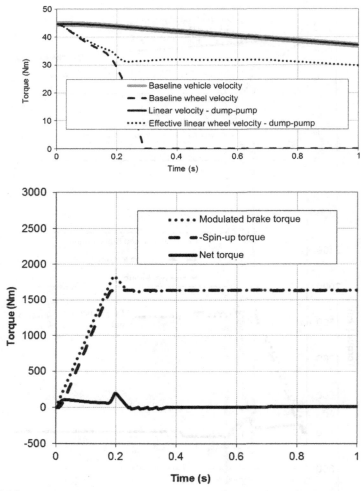

FIGURE 8.14

A hold-dump-hold-pump strategy with the simplified 2 DoF vehicle braking model.

The modelling in MBS of more realistic ABS algorithms (van der Jagt et al., 1989) is more challenging as the forward velocity and hence slip ratio is not directly available for implementation in the model; this of course is also the challenge to implementing the algorithms in real life. The implementation of such a model allows the angular velocity of the wheel to be factored with the rolling radius to produce an output commonly referred to as 'wheel speed' by practitioners in this area. A plot of wheel speed is compared with vehicle speed in Figure 8.15, where the typical oscillatory nature of the predicted wheel speed reflects the cycling of the brake pressure during the activation of the ABS model in this vehicle simulation.

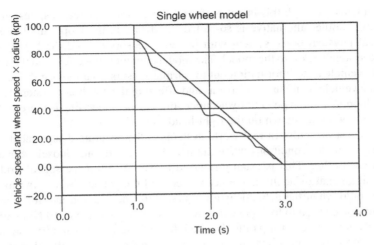

FIGURE 8.15

Plot of vehicle speed and wheel speed during from ABS braking simulation.

In 1995 Mercedes introduced the Bosch 'Electronic Stability Program' (ESP) on the Mercedes S-Class. This system applies the brakes asymmetrically and without the driver's foot applied to the brake pedal, in order to exert restoring yaw moments on the vehicle. While previous ABS strategies had an element of open-loop control to minimise the build-up of yaw moment for asymmetric braking situations, ESP proceeds to the next logical step of closed-loop control of yaw rate. Desired yaw rate is mapped through testing on the vehicle and compared with actual yaw rate. Conceptually this is similar to the driver control model described in Chapter 7. However, a fundamental difference is that the brake actuators are not always controlled as proportional devices. To suppress understeer, brake torque is typically applied at the inside rear wheel in a manner substantially proportional to the magnitude of the understeer. For oversteer, a different strategy is employed. First, a threshold is set below which no intervention occurs and second, the system introduces a defined amount of brake torque for variable lengths of time to trim oversteer. It is also typical that engine torque will be reduced under these conditions. This gives rise to the typical 'lumpy' sensation of the operation of ESP and the belief for many drivers that it is a relatively crude system. In truth, it is extremely elaborate, but the requirement not to burn out the brakes during spirited driving means this 'threshold' control is necessary.

Both ABS and ESP control can be difficult to model for reasons of the discontinuities described. In the real world, finite hydraulic bandwidth imposes its own smoothing on the system but the operation of ABS systems remains subject to a repeated series of 'impact' noises during sharp hydraulic transients. Only the very latest electrohydraulic braking technology succeeds in avoiding this. Cosimulation between multiple modelling codes is the answer suggested by the industry for all these problems although this is not preferred for a variety of reasons, described in Chapter 6. Another alternative is so-called 'hardware-in-the-loop' simulation that uses a real physical brake system with pressure sensors to feed the behaviour of the brake system back into the model. The model computes the effects on the vehicle and feeds vehicle state information out to 'virtual sensors' that replace the real sensors on the vehicle, making its decisions and affecting the real brake system to complete the loop. Although potentially accurate, this type of modelling is more suited to confirmation work as shown on the right-hand side of Figure 7.45.

The biggest difficulty in real-world systems is the accurate discernment of the vehicle speed under conditions when the wheel speeds are only loosely connected to it and each other. This is usually referred to as the 'reference speed' problem. Another significant difficulty is the need to control the vehicle by varying longitudinal slip ratio when the actuator is a torque-control device. On ice, for example, 10 Nm may be enough to lock the wheel whereas on dry pavement, 10 Nm will produce barely any change in slip ratio. For both these reasons, ABS and ESP strategies are more complex than might first be imagined; differences in effectiveness and implementations largely come down to the sophistication of the reference speed calculation and friction estimation.

8.2.3 Active steering systems

In 1986, Nissan launched their High Capacity Active Steering (HICAS) system on the rear axle of the R31 Skyline Coupé. Honda followed with a mechanical system on the Prelude and Mitsubishi were close behind in the Lancer, using an electronically controlled system like the Nissan. Nissan used the notion of a compliant sub-frame with its location controlled by a hydraulic actuator, while Honda and Mitsubishi essentially duplicated the front steering system at the rear of the vehicle, using a central rack. The objectives of the systems were clear and there is a great deal of literature published on the subject (Ro and Kim, 1996). Four-wheel steer seeks to control the body slip angle and rate, and hence has a profound effect on driver impressions of the vehicle. It also has the comparatively trivial effect of tightening the turning circle at low speeds. Honda has subsequently dropped four-wheel steer, declaring that 'advances in tyre technology have rendered it unnecessary'. Mitsubishi kept it until 1999 on their 3000 GTO model, while Nissan continue with a development now called Super-HICAS system on the Skyline.

A difficulty with four-wheel steer is that, while it makes the vehicle feel excellent in the linear region (through enhanced body slip angle control), as the handling limit approaches the flat nature of the tyre side-force-versus-slip-angle curve means that its ability to improve vehicle control disappears. In this sense, it is possibly the worst type of system — enhancing driver confidence without actually improving limit capability. Four-wheel steer is easily simulated in MBS modelling, with an additional part to represent the rear steering rack and forces applied to it according to a control law as with other systems.

BMW announced 'Active Front Steer' in conjunction with ZF for the 2004 model-year 5-Series. Consideration of the behaviour of competition drivers, particularly rally drivers, suggests the potential for this system is high and that it offers a much more continuous control than brake-based systems. Although onerous, the potential failure modes have clearly been overcome. It also offers the chance to overcome the problem of increased control sensitivity at high speeds by reducing the yaw rate gain progressively with speed. It does not suffer any of the problems of rear-wheel steering in terms of limit control.

There are several patents for an 'Active Toe Control' system for application to all four wheels. This is primarily an on-centre modifier for the vehicle and is intended to complement the torque distribution systems they are also known for. Such a system is postulated in Lee et al (1999) and has been prototyped on a research vehicle. The interaction between such a system and a torque distribution system is explored in (He et al., 2003).

Modelling active steer systems is in principle identical to modelling a driver. An external force is applied to the steering system — either the rack or column in response to an algorithm. The algorithm is typically similar to that in use for a brake-based stability system in that it will calculate a target yaw rate and then use the steering system to attempt to achieve that yaw rate.

8.2.4 Active camber systems

Milliken Research Associates produced an active camber Corvette in the mid-1980s with closed-loop control of yaw rate following the success of the 'camber racer', the latter recently run again at Goodwood on modern motorcycle tyres. Mercedes have also produced the F400 Carving that runs an active camber system. Although attractive in function, the increased package requirements in the wheelhouse make active camber something that will probably not be applied to road cars.

8.2.5 Active torque distribution

Nissan's Skyline R32 Coupé had the somewhat bewilderingly named ATTESA-ETS (Advanced Total Traction Engineering System for All-Electronic Torque Split) system in 1989. The Porsche 959 had similar technology in the same year, although without the lengthy acronym. Since then, systems of torque redistribution according to handling (as against traction) priorities have remained elite in production cars. The author's (Harty) work at Prodrive revived interest with several low-cost Active Torque Dynamics (ATD) systems suitable for a wide variety of vehicles in the period 1999–2004, but commercial restrictions prevent discussion of which, if any, achieved production status. In essence, traditional driveline technology systems seek to minimise wheelspin, whereas these systems seek to connect drive torque distribution to vehicle handling via closed-loop feedback. From a modelling point of view, they are simple to implement in that some state variables are declared (typically a target yaw rate and current state, similar to ESP systems) and the actuation forces are implemented accordingly.

There exists a large amount of confusion surrounding torque distribution, which frequently exercises students. It is commonly declared, with some confidence, that 'a locked differential produces a 50/50 torque split'. The origins of this statement are unknown and certainly do not bear scrutiny. Where a traditional open differential of the bevel or epicyclic type does indeed produce a torque split of known proportions, a locked differential — in essence a shaft — produces an equality of speed outputs. How these speeds turn into torques is governed entirely by the wheel to which they are attached — it is a function of vertical load, friction coefficient and slip state. The latter is influenced by vehicle states and, of course, the differential locking condition. These comments apply whichever differential is under consideration.

Figure 8.16 shows a comparison at the rear axle of a vehicle in four different states, combinations of an open and locked differential, applied drive torque and no applied drive torque. The tractive force arrows shown are dimensionally representative of torques in the half shafts, since no other mechanism is applying a force to the tyre.

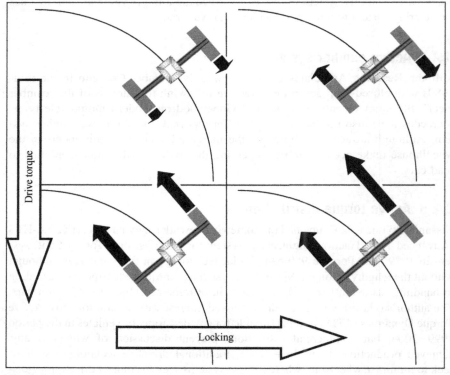

FIGURE 8.16

Locked differentials do not result in equal torque distribution — this is a common myth.

In the upper left quadrant of Figure 8.16, in the presence of an open differential and the absence of driveline torque, the only forces present at the contact patch are the rolling resistance. The outer wheel may be expected to have a slightly higher rolling resistance due to its higher load, associated with weight transfer as discussed in Chapter 7. The individual wheel rotational speeds are related to their individual radius of turn and rolling radius.

In the lower left quadrant of Figure 8.16, in the presence of an open differential and with driveline torque, the forces at the contact patch may be expected to be the sum of the rolling resistance and drive torque. When the drive torque is large and neglecting any frictional torque in the differential, the forces are broadly equal; this is the fundamental purpose of the differential. The individual wheel speeds are the sum of the wheel speeds from the upper left quadrant and the wheel speeds required to deliver the slip ratio, which in turn delivers the required reaction torque to the sideshaft such that it is in equilibrium. Note that a lightly laden inner wheel and a high drive torque may exceed the tyre's ability to generate such a reaction torque, in which case no equilibrium state will be reached and the sideshaft (and hence wheel) rotational speed will continue to increase until torque is interrupted — typically by an attentive driver, an engine speed limiter or a traction control system.

In the upper right quadrant of Figure 8.16, with a locked differential imposing equal rotational speeds across the axle and in the absence of driveline torque, it can be seen that something approximating the average of the speeds from the upper left quadrant is imposed on both wheels. Since this is slower than the free-rolling speed for the outer wheel, it constitutes a negative slip ratio and produces a rearward-acting force at the contact patch. Since it is faster than the free-rolling speed for the inner wheel, it constitutes a positive slip ratio and produces a forward-acting force at the contact patch. Together these produce a yaw moment on the car in opposition to the turn; it is common experience that 'stiff differentials blunt turn-in'. This effect can be used in active systems to introduce an amount of yaw damping to a vehicle. It works with both cross-axle differentials and with the centre differential in an all-wheel-drive vehicle.

In the lower right quadrant of Figure 8.16, with a locked differential imposing equal rotational speeds across the axle and in the presence of driveline torque, positive slip ratios are induced on both wheels. The slip ratios are such that, while satisfying the constraint of equal speeds in both shafts, the total reaction torques sum to the applied shaft torques. The weight difference between the two wheels results in more torque for a given slip ratio, as described in Chapter 5. The geometric effects of the turn tend to slightly increase the slip ratio on the inner wheel but for realistic turn radii these effects are relatively small compared to weight transfer effects. The net result tends to add yaw moment with driveline torque for a cross-axle differential. This effect is readily observable in both front- and rear-wheel drive vehicles when a comparison can be made with and without cross-axle locking.

A locked differential does not result in a 50/50 torque split under most imaginable circumstances. The persistence of the myth that it does is somewhat baffling.

A fully locked drive axle has an unfortunate tendency to spin up both wheels simultaneously, which robs the vehicle of side force on the drive axle by broadly collapsing the cornering stiffness of both rear tyres simultaneously. From the discussions of anti-lock brake control, it can be seen that the runaway phenomenon in the presence of excess drive torque can be very rapid indeed. For rear-wheel-drive vehicles in particular, locked differentials spin the vehicle while open differentials spin a wheel. It can be imagined that spinning the vehicle is a rather more onerous failure mode and it is true that limited slip differentials (which may be regarded as some way along the spectrum to locked) have a reputation for being difficult to drive at the limit. A control system for a proportional limited slip differential that can recognise the vehicle states and reduce the differential locking in response to excess yaw rate is extremely effective, in the author's experience, at limiting vehicle departure behaviour both on split friction and uniform friction surfaces.

Driveline devices that impose a left-right speed difference have also been constructed. The Mitsubishi Active Yaw Control, Super Active Yaw Control and Super All Wheel Control (AYC, S-AYC and S-AWC respectively) use a clutch and additional gear wheels to deliver a limited slip differential in which the 'fully locked' state results in a fixed speed ratio between one sideshaft and the other — typically differing by the order of 10%. The requirement for symmetric control results in a second clutch and a second pair of gear wheels. Honda's Super Handling All Wheel Drive (SH-AWD) and BMW's Dynamic Cornering Performance systems use similar arrangements in principle.

Such a device might be known generically as an 'asymmetric overspeeding differential' and can be represented in a multibody model using the topology shown in Figure 8.17. A three-element coupling equation represents the open differential, an

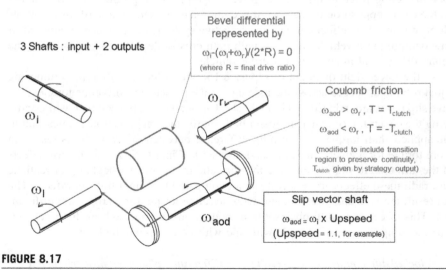

FIGURE 8.17

A modelling scheme for an asymmetric overspeeding differential — right shaft speedup.

additional spin degree of freedom represents the AOD coupling shaft. An additional mass is created, constrained to spin at some multiple of the input speed. In the example shown, a frictional torque using the Coulomb model is then engaged which is scaled by a solution variable T_{clutch}, which in turn is given as the output from some control strategy. When T_{clutch} is high, the upsped AOD shaft is effectively locked to the right-hand shaft. The conventional bevel differential enforces the speed relationship in the model, ensuring that the left-hand shaft is downsped by the same amount to preserve the average speed relationship. Additional details can be added for frictional bias torques and so on, to suit the level of available data. Despite being widely described as Torque Vectoring, this approach is really 'slip vectoring'; the torque distribution is not really enforced. Nevertheless on homogenous surfaces the effect is to deliver a yaw moment that may be used advantageously to manage the vehicle states.

For the rear axle, any use of an asymmetric device tends to reduce the capacity of the axle as a whole and thus it is easier to add yaw moment than to subtract it. The opposite is true for the front wheels. To the author's knowledge there has only been one front axle asymmetric overspeeding device, the Honda Active Torque Transfer System (ATTS) fitted to the Prelude SH in the late 1990s. 'SH' allegedly stood for 'Super Handling'.

A similar principle can be applied between the axles. The Honda SH-AWD system runs an epicyclic overspeed — extremely similar to the electric overdrive unit on 1970s Triumphs in principle — on the propeller shaft to the rear axle. If, say, a 10% overspeed is invoked then the centre differential, when open, simply processes to accommodate the now 10% difference in overall axle ratios. If the centre differential were to be fully locked, the rear axle would have a 10% overspeed compared to the front. This overspeed has an interesting effect on the lateral behaviour of the car; in order to understand it we must return to the tyre characteristics discussed in Chapter 5. To illustrate the effects, a sample tyre longitudinal force curve is presented in Figure 8.18. The curve is normalised to simplify the illustration. A notional 'lateral capacity fraction' can be calculated by presuming a uniform maximum frictional force is available and subtracting the longitudinal force such that the Pythagorean sum of lateral and longitudinal is unity. This sounds complex in words but is simple mathematically:

$$\frac{\widehat{F}_y}{\mu F_z} = \sqrt{1 - \frac{F_x}{\mu F_z}} \tag{8.9}$$

For a slip ratio beyond that at which maximum longitudinal force is generated, the lateral capacity is presumed zero. Normalised longitudinal force and this notional 'lateral capacity fraction' are plotted together in Figure 8.18 against longitudinal slip ratio.

Examining this graph closely, it can be seen for two similarly laden tyres that one with 10% slip ratio has approximately half the lateral capacity of one with 0% slip ratio — a slip ratio difference of 10% between them. If these are the outboard tyres on a vehicle it can be seen that the ability to exert a yaw moment on the vehicle is

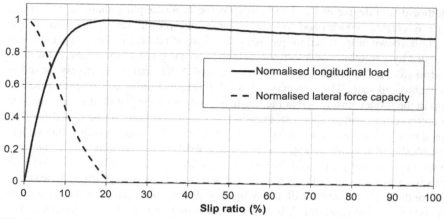

FIGURE 8.18

Normalised Longitudinal Force and Lateral Force Capacity against Slip Ratio.

substantial — this is the effect of slip vectoring and makes a rear-wheel-drive car yaw tail out under power without gross wheelspin. What is interesting is that if in a four-wheel drive driveline, the same slip ratio difference is kept but the slip ratios are now 10% and 20%, then the relative difference in lateral force capacity is of the same order; this means the yaw moment on the vehicle is broadly independent of the mean slip level between 5% and 15% — a substantial torque window. It means that absolute speed control is not vital for interaction with yaw moment, only the inter-axle speed difference. If the mean speed comes up further, to say 15% on one wheel and 25% on the other, lateral force capacity broadly converges and so the consequence of a gross excess driveline torque does not destabilise the vehicle but rather uniformly removes yaw moment as both tyres overspeed.

It is generally true that manipulating side forces via lateral force capacity gives a larger yaw moment than by exerting it directly left—right on the vehicle since generally wheelbase exceeds track. For this reason it is rarely necessary to run such a large speed difference as a left—right device; the first Prodrive prototype slip-vectoring vehicle ran an 11% speed difference but the next and subsequent vehicles dropped to a more typical 5% or so — the values used by other manufacturers.

It can be seen in the Honda-style configuration that the overspeed at the rear can have the effect of reducing lateral force capacity at the rear axle — the Prodrive ATD systems also used this effect — and thus can add yaw moment. When coupled with an epicyclic centre differential having a notional front-biased torque split, a simple proportional clutch across it delivers a front torque bias when open and a rearward torque bias when closed. Using this, a simple linear yaw rate demand model can generate a target yaw rate curve, with the control law for the centre differential responding to insufficient yaw rate by tightening and an excess yaw rate by slackening from its current position. In this manner, the driveline can be set up never to destabilise the car — in other words to decouple the steering from the throttle.

Overall driving impressions of such systems are, on the whole, 'remarkably unre-markable' with the drive experience uncorrupted by drive torque.

One demonstration of the Prodrive system involved wedging the steering wheel with the driver's leg and driving a continuous drift on a 60-m snow circle, managing the line of the car using throttle inputs but leaving the car to look after its yaw rate on the rough, rutted surface of the snow circle by controlling the centre differential. The task of managing the car was so undemanding that the author could fold his arms and eat an apple while doing it. A comparison data set with four-wheel drives but no closed-loop torque vectoring is shown in Figure 8.19; the duration is some 70 s, showing this was no cherry-picked small data segment. At a speed of around 60 mph, this represented of the order of a mile of continuous control. The uncon-trolled car had a sophisticated all-wheel-drive system (including hydraulic centre differential and asymmetric rear axle) and a substantially faster steering ratio than the controlled car, but crucially no loop closure around vehicle yaw rate. It is noted with interest that since 2007 the Mitsubishi S-AWC system has had yaw rate feed-back added compared to the AYC and S-AYC systems, in which it was absent.

Nothing about the learning with geared slip-vectoring systems precludes the same ideas being applied with individually controlled wheel motors. In particular, in-wheel motors such as those produced by Protean Electric Ltd at the time of writing offer high fidelity control and very accurate speed sensing, allowing, for example, a left and right motor to be 'software coupled' for relative speed and managed for mean torque to mimic an asymmetric overspeeding functionality that

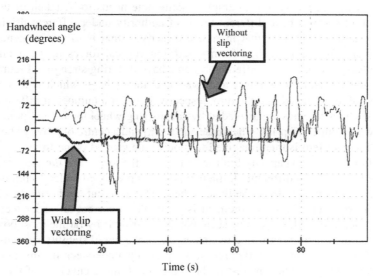

FIGURE 8.19

A comparison between the presence and absence of closed-loop slip vectoring driving on a 200-m snow circle in Arvidsjaur, Sweden — author (Harty) driving both vehicles.

needs no knowledge of friction coefficient. Key to unlocking these benefits is the realisation that what is in play is 'slip vectoring' and not 'torque vectoring'.

8.3 Which active system?

The North American market is currently resisting complexity in its best-selling light truck segment (Pickups and SUVs), since the appeal from the vehicle manufacturer's perspective is the low cost of production, giving margins that are allowing the vehicle manufacturers to remain in business during lean times. Nevertheless, competition from offshore competitors is making the US consumer expectations for vehicle performance and quality increase steadily. In Europe, an already sophisticated consumer is expecting steadily improving dynamic performance from their vehicle, while in Asia the predilection for 'gadgets' on vehicles has seen this market lead the way in terms of satellite navigation and so on. All three major world vehicle markets have their own reason for needing to improve vehicle dynamic performance over and above the level that can be achieved using conventional, passive technology. However, no-one wants to be first to market with a system that initially gives them a cost penalty and runs the risk of delivering benefits that the customer does not notice — and is therefore not prepared to pay for.

This appears to have been a lesson learned by the Japanese manufacturers — notably Toyota with their remarkable Soarer, who have largely withdrawn from the active system war and concentrated significantly on hybrid drivelines. In deciding which active system to employ, some note needs to be taken of the fact that different systems have different levels of authority under different driving regimes. In terms of systems that intervene directly — that is to say by generating ground plane forces, as distinct from modifying vertical forces in the manner of an active suspension — then different systems have differing strengths. Figure 8.20 presents the available 'understeer and oversteer' moments — which is to say moments out of the turn and into the turn, respectively — for a number of different systems. The characteristics are arrived at by considering wheel loads for a point mass vehicle with a finite footprint (wheelbase and track) and a centre of gravity above the ground. Each system is then imagined to modify the ground plane forces up to the friction limit or system capacity limit, whichever is the lower. For the electronically controlled limited slip differential, tractive power is presumed added to remove the presence of braking slip from the inside wheel. The different 'authority envelopes' can be clearly seen. Oversteer moment is available to correct an understeer error state and vice versa. Thus it can be seen that the steering is by far the most powerful actuator on the car for controlling oversteer, but brake-based stability systems are a close second. Steering is ineffective at suppressing understeer in any normally balanced car. Of particular interest is the somewhat limited envelope of the asymmetric overspeeding differential despite its complexity. In particular, in the oversteer case when understeer moment is needed, it is typically reduced in capacity by unloading of the inside rear wheel.

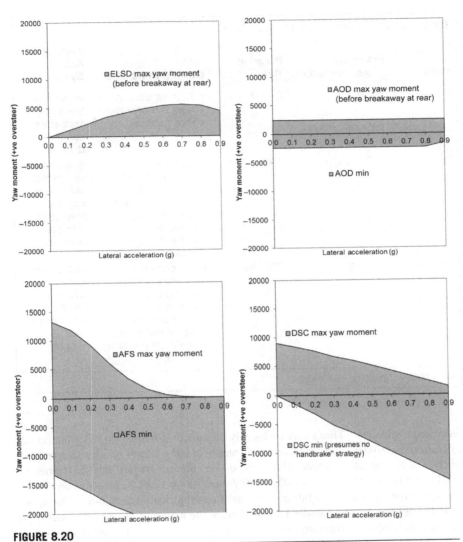

FIGURE 8.20

An illustrative authority envelope comparison between an electronic limited slip differential (ELSD) (top left) and asymmetric overspeeding differential (AOD) (top right), a full authority steering system (AFS) (bottom left) and a brake-based system (bottom right) with no 'handbrake' mode.

Any 'full spectrum' dynamic intervention system is likely to need more than one set of actuators. For this reason, predictive work is needed on a case-by-case basis to sort the useful from the gimmick in terms of vehicle dynamics controls. The author's expectations for the use of active systems on vehicles are given in Table 8.2. Three driving regimes are identified, which may be described as 'normal', 'spirited' and

Table 8.2 Which Active System – and Why?

Name	Lateral Acceleration	Steering Rate	Guiding Principles	Applicable Active System
Normal	0–0.3 g	0–400°/s	• To steer the car, steer the wheels. • Ride matters.	Active rear toe, active front steer, adaptive dampers.
Spirited	0.3–0.6 g	400–700°/s	• Intelligently combine drive and steer forces to deliver control without retardation. • Control body motion.	Active torque distribrution, *active rear toe, active front steer, adaptive dampers*
Emergency	0.6 g+	700+°/s	• Use brakes to reduce kinetic energy of the vehicle. • Minimise wheel load variation for maximum grip and control.	Brake-based system, *adaptive dampers,* active anti-roll bars.

Italics denote duplication – i.e. reuse.

'accident avoidance'. These may be loosely classified as being 0 to 0.3 g lateral acceleration, 0.3 to 0.6 g lateral acceleration and over 0.6 g lateral acceleration respectively. They are also characterised by driver inputs, particularly in terms of handwheel rates — 0 to 400°/s, 400 to 700°/s and 700 to 1200°/s, respectively. These classifications are entirely arbitrary and based on empirical observations. The upper bound is observed to be broadly independent of the nature of the driver, which is to say that motorsport drivers are not observed to be able to put in higher handwheel rates than the rest of us. It's fair to say they have a better idea what handwheel rate to choose than many of us, though.

The results in Table 8.2 are not in any sense definitive but the authors considered views in the light of the available evidence. It is hoped that the readers of this text will be part of the group of engineers discovering just how good these judgements are.

Vehicle Model System Schematics and Data Sets

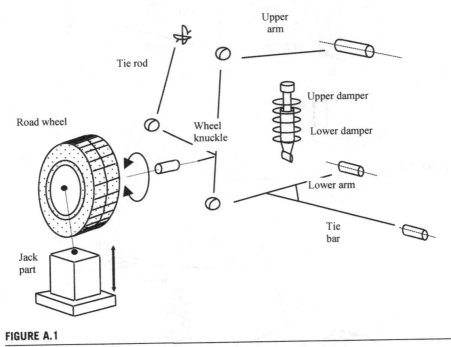

FIGURE A.1

Front suspension components.

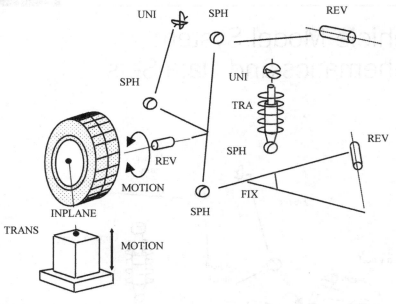

FIGURE A.2

Front suspension with joints.

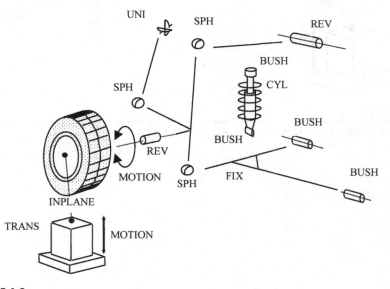

FIGURE A.3

Front suspension with bushes.

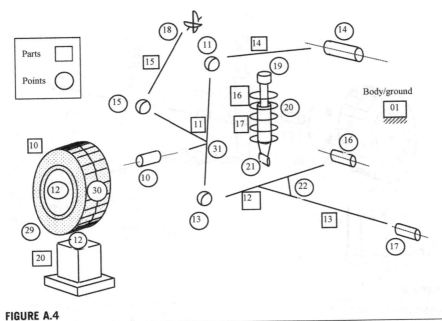

FIGURE A.4

Front suspension numbering convention.

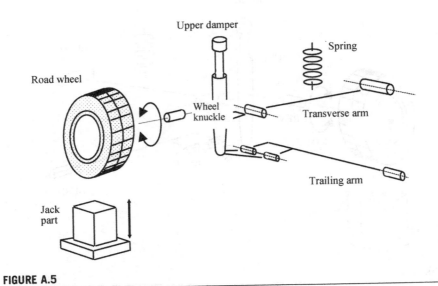

FIGURE A.5

Rear suspension components.

634 Appendix A

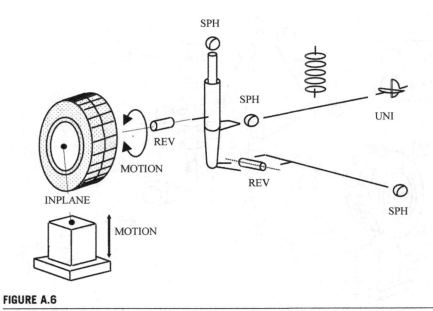

FIGURE A.6

Rear suspension with joints.

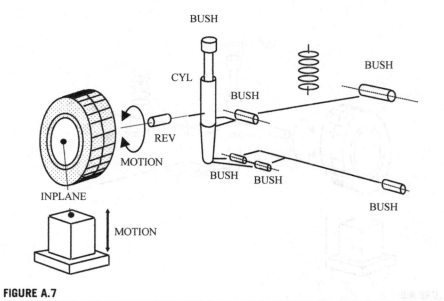

FIGURE A.7

Rear suspension with bushes.

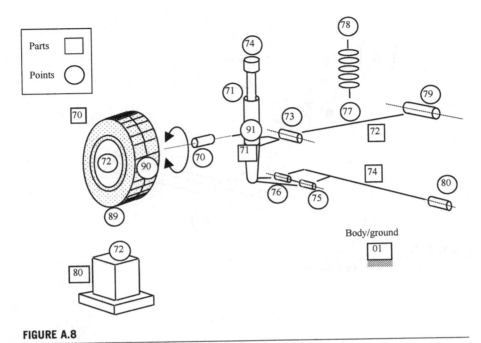

FIGURE A.8

Rear suspension numbering convention.

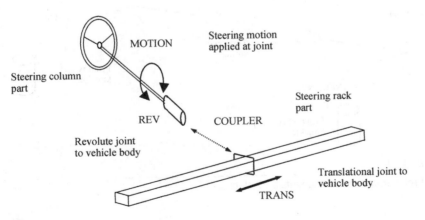

FIGURE A.9

Steering system components and joints.

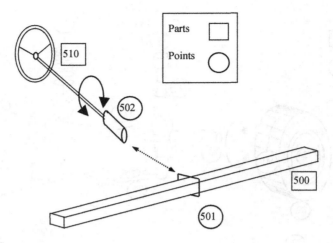

FIGURE A.10

Steering system numbering convention.

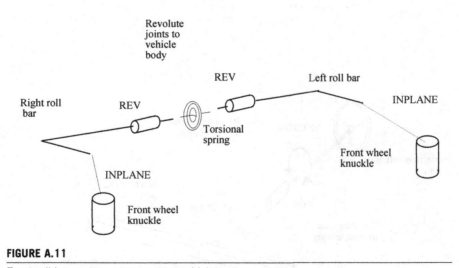

FIGURE A.11

Front roll bar system components and joints.

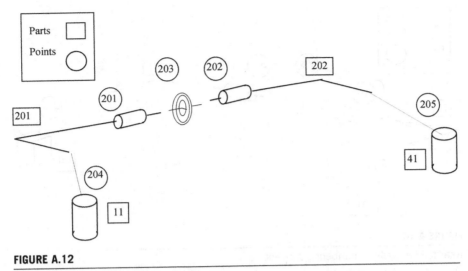

FIGURE A.12

Front roll bar system numbering convention.

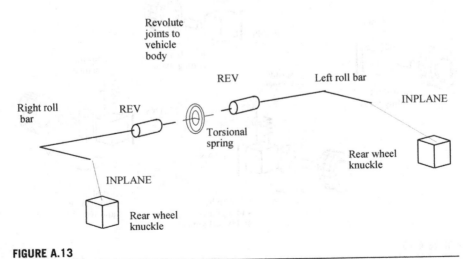

FIGURE A.13

Rear roll bar system components and joints.

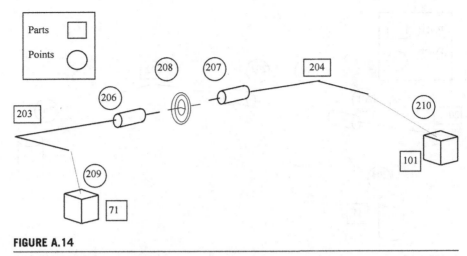

FIGURE A.14

Rear roll bar system numbering convention.

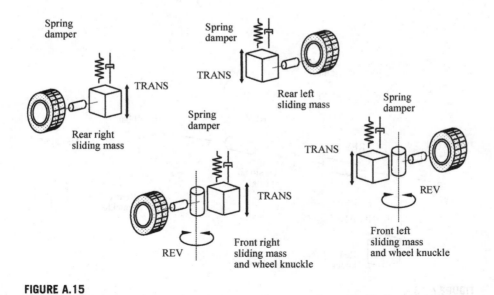

FIGURE A.15

Lumped mass model suspension components and joints.

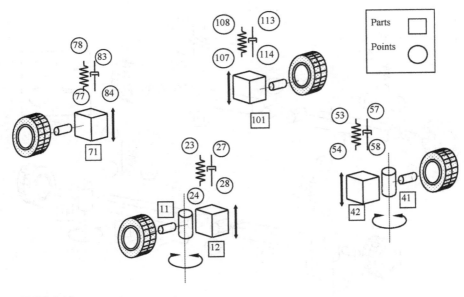

FIGURE A.16

Lumped mass model suspension numbering convention.

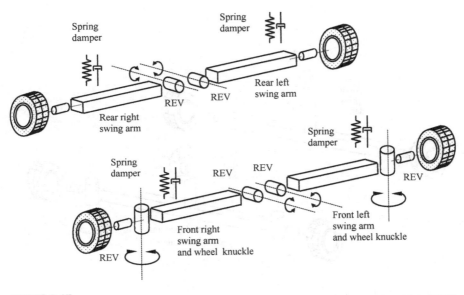

FIGURE A.17

Swing arm model suspension components and joints.

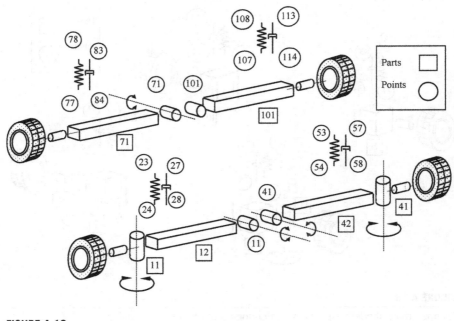

FIGURE A.18

Swing arm model suspension numbering convention.

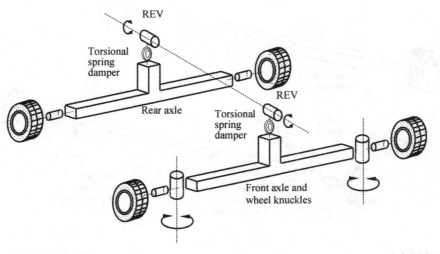

FIGURE A.19

Roll stiffness model suspension components and joints.

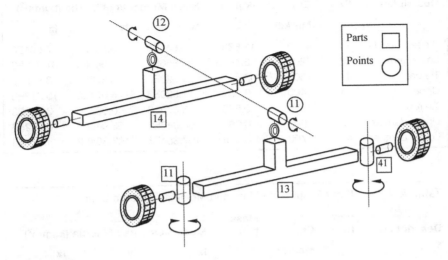

FIGURE A.20

Roll stiffness model suspension numbering convention.

Table A.1 Road Wheel Mass and Moment of Inertia Data

Description	Part ID	CM Marker	Mass (kg)	Mass Moments of Inertia (kgmm²)		
				Ix	Iy	Iz
Front right wheel	10	10	21.204	577.59E3	577.59E3	931.077E3
Front left wheel	40	40[a]	21.204	577.59E3	577.59E3	931.077E3
Rear right wheel	170	70[a]	21.204	577.59E3	577.59E3	931.077E3
Rear left wheel	100	100[a]	21.204	577.59E3	577.59E3	931.077E3

[a] The wheel parts are generated automatically by the TIRE statement. The centre of mass is taken to be at the location of the J marker for each tyre, i.e., locations 10, 40, 70, 100.

Table A.2 Front Right Suspension Mass and Moment of Inertia Data

Description	Part ID	CM Marker	Mass (kg)	Mass Moments of Inertia (kgmm²)		
				Ix	Iy	Iz
Wheel knuckle	11	1100	11.678	65.647E3	120.541E3	77.691E3
Lower arm	12	1200	3.4405	37.856E3	0.348E3	37.933E3
Tie bar	13	1300	2.64	0.023E3	3.876E3	3.876E3
Upper arm	14	1400	2.187	11.520E3	4.071E3	15.218E3
Tie rod	15	1500	0.575	3.876E3	0.023E3	3.876E3
Upper damper	16	1600	0.389	63.948E3	63.948E3	0.822E3
Lower damper	17	1700	6.215	107.364E3	107.364E3	4.972E3

Table A.3 Front Left Suspension Mass and Moment of Inertia Data

Description	Part ID	CM	Mass (kg)	Mass Moments of Inertia (kgmm^2)		
		Marker		Ix	Iy	Iz
Wheel knuckle	41	4100	11.678	65.647E3	120.541E3	77.691E3
Lower arm	42	1200	3.4405	37.856E3	1.348E3	37.933E3
Tie bar	43	1300	2.64	0.023E3	3.876E3	3.876E3
Upper arm	44	1400	2.187	11.520E3	4.071E3	15.218E3
Tie rod	45	1500	0.575	3.876E3	0.023E3	3.876E3
Upper damper	46	1600	0.389	63.948E3	63.948E3	0.822E3
Lower damper	47	1700	6.215	107.364E3	107.364E3	4.972E3

Table A.4 Rear Right Suspension Mass and Moment of Inertia Data

Description	Part ID	CM	Mass (kg)	Mass Moments of Inertia (kgmm^2)		
		Marker		Ix	Iy	Iz
Wheel knuckle	71	7100	12.036	165.994E3	196.457E3	34.224E3
Transverse arm	72	7200	6.424	101.389E3	15.215E3	113.237E3
Trailing arm	74	7400	4.322	12.826E3	190.372E3	199.393E3
Upper damper	73	7300	0.982	1.003E3	1.003E3	0.1E3

Table A.5 Rear Left Suspension Mass and Moment of Inertia Data

Description	Part ID	CM	Mass (kg)	Mass Moments of Inertia (kgmm^2)		
		Marker		Ix	Iy	Iz
Wheel knuckle	101	10,100	12.036	165.994E3	196.457E3	34.224E3
Transverse arm	102	10,200	6.424	101.389E3	15.215E3	113.237E3
Trailing arm	104	10,400	4.322	12.826E3	190.372E3	199.393E3
Upper damper	103	10,300	0.982	1.003E3	1.003E3	0.1E3

Table A.6 Body, Rollbars and Steering Mass and Moment of Inertia Data

Description	Part ID	CM	Mass (kg)	Mass Moments of Inertia (kgmm^2)		
		Marker		Ix	Iy	Iz
Body	200	20,000	1427.3	379.0E6	2235.0E6	2269.0E6
Front right roll bar	201	20,100	1.4	10.0	10.0	20.0
Front left roll bar	202	20,200	1.4	10.0	10.0	20.0
Rear right roll bar	203	20,300	1.4	10.0	10.0	20.0
Rear left roll bar	204	20,400	1.4	10.0	10.0	20.0
Steering rack	500	50,000	0.48	13,357.0	48.0	13,357.0
Steering column	510	51,000	2.2	24.0E3	24.0E3	40,762.0

Table A.7 Lumped Mass Model Mass and Moment of Inertia Data

Description	Part ID	CM	Mass (kg)	Mass Moments of Inertia (kgmm²)		
		Marker		Ix	Iy	Iz
Right wheel knuckle	11	1100	11.678	65.647E3	120.541E3	77.691E3
Left wheel knuckle	41	4100	11.678	65.647E3	120.541E3	77.691E3
Front right mass	12	1200	15.447	86.818E3	159.445E3	102.765E3
Front left mass	42	4200	15.447	86.818E3	159.445E3	102.765E3
Rear right mass	71	7100	23.764	280.209E3	402.062E3	346.854E3
Rear left mass	101	10,100	23.764	280.209E3	402.062E3	346.854E3

Table A.8 Swing Arm Model Mass and Moment of Inertia Data

Description	Part ID	CM	Mass (kg)	Mass Moments of Inertia (kgmm²)		
		Marker		Ix	Iy	Iz
Right wheel Knuckle	11	1100	11.678	65.647E3	120.541E3	77.691E3
Left wheel Knuckle	41	4100	11.678	65.647E3	120.541E3	77.691E3
Front right arm	12	1200	15.447	86.818E3	159.445E3	102.765E3
Front left arm	42	4200	15.447	86.818E3	159.445E3	102.765E3
Rear right arm	71	7100	23.764	280.209E3	402.062E3	346.854E3
Rear left arm	101	10,100	23.764	280.209E3	402.062E3	346.854E3

Table A.9 Roll Stiffness Model Mass and Moment of Inertia Data

Description	Part ID	CM	Mass (kg)	Mass Moments of Inertia (kgmm²)		
		Marker		Ix	Iy	Iz
Right wheel knuckle	11	1100	11.678	65.647E3	120.541E3	77.691E3
Left wheel knuckle	41	4100	11.678	65.647E3	120.541E3	77.691E3
Front axle	13	1300	30.894	173.636E3	318.89E3	205.53E3
Rear axle	14	1400	47.528	560.418E3	804.124E3	693.708E3

Table A.10 Front Right Suspension Geometry

Point Id.	Coordinates (mm)			Euler Angles (deg)			ZP (mm)		
	X	Y	Z	ψ	θ	φ	X	Y	Z
1100	966.1	743.9	165.9						
1200	1006.0	525.0	113.0	12D	−13.5D	0D			
1300	722.0	444.0	115.0						
1400	1064.0	500.0	566.0	28.5D	9D	0D			
1500	1129.0	534.0	189.0						
1600	954.4	509.6	497.3	163.66D	164.61D	0D			
1700	954.4	509.6	497.3	163.66D	164.61D	0D			
10	966.1	743.9	165.9	0D	90D	0D			
11	986.6	639.3	572.1						
13	962.2	703.1	90.7						
14	973.0	417.8	548.4	119.05D	91.03D	0.0D			
15	1106.2	673.9	171.7						
16	1050.0	348.3	137.0	−75.95D	90.87D	−172.78D			
17	474.0	332.0	115.0	−66.15D	90D	0.0D			
18	1145.0	392.0	196.0				1245.0	392.0	196.0[a]
19	945.9	480.4	607.7	163.66D	164.61D	0.0D			
20	954.4	509.6	497.3	163.66D	164.61D	0.0D			
21	987.0	620.0	76.5	−89.72D	98.11D	0.0D			
22	984.2	562.3	113.9						
23	945.9	480.4	607.7						
24	965.5	547.4	354.1						
25	945.9	480.4	607.7						
26	954.4	509.6	497.3						
27	945.9	480.4	607.7						
28	987.0	620.7	76.5						
31	966.1	693.1	165.9						
32	965.3	547.4	354.1						
33	945.9	480.4	607.7	163.66D	164.61D	0D			

[a] The ZP orientation is only applied to the marker which belongs to the rack part and is used for the universal joint connecting the rack to the tie rod.

Table A.11 Front Left Suspension Geometry

Point Id.	Coordinates (mm)			Euler Angles (deg)			ZP (mm)		
	X	Y	Z	ψ	θ	φ	X	Y	Z
4100	966.1	−743.9	165.9	12D	−13.5	0D			
4200	1006.0	−525.0	113.0						
4300	722.0	−444.0	115.0						
4400	1064.0	−500.0	566.0	28.5D	9D	0D			
4500	1129.0	−534.0	189.0						
4600	954.4	−509.6	497.3	−343.66D	164.61D	−180D			
4700	954.4	−509.6	497.3	−343.66D	164.61D	−180D			
40	966.1	−743.9	165.9	0D	90D	0D			
41	986.6	−639.3	572.1						
43	962.2	−703.1	90.7						
44	973.0	−417.8	548.4	−299.05D	91.03D	−180.0D			
45	1106.2	−673.9	171.7						
46	1050.0	−348.3	137.0	−104.05D	90.87D	−7.22D			
47	474.0	−332.0	115.0	−113.85D	90D	−180D			
48	1145.0	−392.0	196.0				1245.0	−392.0	196.0[a]
49	945.9	−480.4	607.7	−343.66D	164.61D	−180.0D			
50	954.4	−509.6	497.3	−343.66D	164.61D	−180D			
51	987.0	−620.0	76.5	89.72D	81.89D	0.0D			
52	984.2	−562.3	113.9						
53	945.9	−480.4	607.7						
54	965.5	−547.4	354.1						
55	945.9	−480.4	607.7						
56	954.4	−509.6	497.3						
57	945.9	−480.4	607.7						
58	987.0	−620.7	76.5						
61	966.1	−693.1	165.9						
62	965.3	−547.4	354.1	−343.66D	164.61D	180D			
63	945.9	−480.4	607.7						

[a] The ZP orientation is only applied to the marker which belongs to the rack part and is used for the universal joint connecting the rack to the tie rod.

Table A.12 Rear Right Suspension Geometry

Point Id.	Coordinates (mm)			Euler Angles (deg)			ZP (mm)		
	X	Y	Z	ψ	θ	φ	X	Y	Z
7100	3732	725	167						
7200	3743	493	112	0D	4.55D	0D			
7300	3701.02	525.66	606.36	163.46D	166.09D	0D			
7400	3630	587	70						
70	3732	725	167	0D	90D	0D			
71	703.6	534.4	569.6	163.46D	166.09D	0D			
72	3732	725	167						
73	3747	706	109.5	88.809D	90D	−4.3376D			
74	3701	525.7	606.4	163.46D	166.09D	−195D			
75	3695.4	684.6	48.1	56.715D	97.334D	0.0D			
76	3803.4	613.7	34.2	56.716D	97.334D	0.0D			
77	3743	496	84.0						
78	3743	475	265.8						
79	3737	225	146	88.809D	90D	−4.3376D			
80	3203	600	144.5	75.0491D	98.3D	0D			
81	3703.5	533	573						
82	3693	499	714						
83	3693.3	499.7	715.9						
84	3729.7	622.2	200						
85	3701	525.7	606.4	163.46D	166.09D	0D			
86	3703.6	534.4	569.6						
91	3732	684.6	167						

Table A.13 Rear Left Suspension Geometry

Point Id.	Coordinates (mm)			Euler Angles (deg)			ZP (mm)		
	X	Y	Z	Ψ	θ	φ	X	Y	Z
10100	3732	−725	167	−180D	4.55D	−180D			
10200	3743	−493	112	−343.46D	166.09D	−180D			
10300	3701.02	−525.66	606.36						
10400	3630	−587	70						
100	3732	−725	167	0D	90D	0D			
101	3703.6	−534.4	569.6	163.46D	166.09D	0D			
102	3732	−725	167						
103	3747	−706	109.5	91.191D	90D	−175.6624D			
104	3701	−525.7	606.4	−343.46D	166.09D	15D			
105	3695.4	−684.6	48.1	123.284D	97.334D	180D			
106	3803.4	−613.7	34.2	123.284D	97.334D	180D			
107	3743	−496	84.0						
108	3743	−475	265.8						
109	3737	−225	146	91.191D	90D	−175.634D			
110	3203	−600	144.5	−255.049D	98.3D	180D			
111	3703.5	−533	573						
112	3693	−499	714						
113	3693.3	−499.7	715.9						
114	3729.7	−622.2	200						
115	3701	−525.7	606.4	−343.46D	166.09D	180D			
116	3703.6	−534.4	569.6						
121	3732	−684.6	167						

Table A.14 Body, Rollbars and Steering Geometry

Point Id.	Coordinates (mm)			Euler Angles (deg)			ZP (mm)		
	X	Y	Z	ψ	θ	φ	X	Y	Z
20000	2150.4	0.0	452.0						
20100	1264	263	87	0D	90D	0D			
201	1264	263	87	0D	90D	0D			
203	1264	0.0 87		0D	90D	0D			
204	966.1	743.9	165.9						
20200	1264	−263	87	0D	90D	0D			
202	1264	−263	87	0D	90D	0D			
203	1264	0.0	87	0D	90D	0D			
205	966.1	−743.9	165.9						
20300	4142	508	268	0D	90D	0D			
206	4142	508	268	0D	90D	0D			
208	4142	0	268	0D	90D	0D			
209	3732	725	167						
20400	4142	−508	268	0D	90D	0D			
207	4142	−508	268	0D	90D	0D			
208	4142	0	268	0D	90D	0D			
210	3732	−725	167						
50000	1145	0	196						
501	1145	0	196	180D	90D	0D			
51000	1964	353	787.6				1698.6	348.6	604.4
502	1964	353	787.6				1698.6	348.6	604.4
5101	2072	353	631						
5102	1961	543	788						
5103	1967	163	788						
5104	1145	338	222						

Table A.15 Lumped Mass Model Geometry

Point Id.	Coordinates (mm)			Euler Angles (deg)			ZP (mm)		
	X	Y	Z	ψ	θ	φ	X	Y	Z
1100	966.1	743.9	165.9						
1200	966.1	500	165.9						
4100	966.1	−743.9	165.9						
4200	966.1	−500	165.9						
7100	3732	500	167						
10100	3732	−500	167						
11	966.1	500	165.9						
41	966.1	−500	165.9						
39	966.1	743.9	165.9						
69	966.1	−743.9	165.9						
99	3732	500	167						
129	3732	−500	167						

Table A.16 Swing Arm Model Geometry

Point Id.	Coordinates (mm)			Euler Angles (deg)			ZP (mm)		
	X	Y	Z	ψ	θ	φ	X	Y	Z
1100	966.1	743.9	165.9						
1200	966.1	500	165.9						
4100	966.1	−743.9	165.9						
4200	966.1	−500	165.9						
7100	3732	500	167						
10,100	3732	−500	167						
39	966.1	743.9	165.9						
69	966.1	−743.9	165.9						
11	966.1	−1361	358.6						
41	966.1	1361	358.6						
71	3732	−576.5	206.2						
101	3732	576.5	206.2						

Table A.17 Roll Stiffness Model Geometry

Point Id.	Coordinates (mm)			Euler Angles (deg)			ZP (mm)		
	X	Y	Z	ψ	θ	φ	X	Y	Z
1100	966.1	743.9	165.9						
4100	966.1	−743.9	165.9						
1300	966.1	0.0	165.9						
1400	3732	0.0	167						
39	966.1	743.9	165.9						
69	966.1	−743.9	165.9						
11	966.1	0.0	27.4						
12	3732	0.0	47.4						

Table A.18 Spring Data

The springs are defined using the SPRINGDAMPER statement.

Front Spring:

Stiffness	$K = 31.96$ N/mm
Free length	$L = 426$ mm

Rear Spring:

Stiffness	$k = 60.8$ N/mm
Free length	$L = 253$ mm

Table A.19 Front Damper Data

The dampers are defined using the SFORCE statement. The dampers are nonlinear.
The data provided shows the variation of force in the damper with velocity.

Velocity (mm/s)   −5000, −3150, −2870, −2450, −2205, −1925, −1610, −1260,
−910, −630, −470, −400, −350, −300, −250, −230, −200, −190,
−160, −120, −80, −55, −40, −20, −10, −1, −0.1, 0, 0.3, 3, 30, 40,
60, 80, 100, 200, 250, 400,490, 770, 1050, 1330, 1820, 2060, 2485,
2590, 2730, 2835, 2940, 3080, 5000

Force (N)   10425, 5800, 5200, 4400, 4000, 3600, 3200, 2800, 2400, 2000, 1800,
1700, 1600, 1500, 1400, 1350, 1310, 1290, 1200, 1000, 700, 400,
210, 80, 40, 4, 0.4,0, −1, −10, −100, −123, −150, −182, −200,
−260, −300, −400, −500, −800, −1200, −1600, −2400, −2800,
−3400, −3500, −3600, −3700, −3800, −4000, −7840

Table A.20 Rear Damper Data

The dampers are defined using the SFORCE statement. The dampers are nonlinear.
The data provided shows the variation of force in the damper with velocity.

Velocity (mm/s)   −5000, −3150, −2800, −2450, −2100, −1750, −1400, −1050, −700,
−560, −500, −450, −400, −350, −300,-250, −200, −150, −100, −50,
−25, −5, −1, −0.1, 0, 0.1, 1, 5, 25,50,100, 200, 300, 400, 500, 700, 1050,
1400, 1750, 2100, 2450, 2800, 3150, 5000

Force(N)   7352,3652,3120,2635,2193,1855,1518,1180,927,843,800,773,722,
686,658,596, 560,488,329,154,77,15.4,3.08,0.308,0,-0.126,-1.26, −6.3,
-31.5,-63,-126,-153.25, −180.5,-208,-235,-253,-380,-675,-970, −1349,
-1788,-2277,-2867,-6567

Table A.21 Roll Bar Data

Front Roll Bar:
Torsional stiffness $Kt = 490E3$ Nmm/rad
Rear roll bar:
Torsional stiffness $Kt = 565E3$ Nmm/rad

Table A.22 Front Suspension Bush Data

The following linear values are used to define the stiffness and damping in the bushes. For each bush data is listed as:

K = kx,ky,kz Stiffness (N/mm)
KT = ktx,kty,ktz Torsional stiffness (Nmm/rad)
C = cx,cy,cz Damping (Ns/mm)
CT = ctx,cty,ctz Torsional Damping (Nmms/rad)

Lower Arm Mount Bush (location 12 and 42)

K = 7825,7825,944
KT = 2.5E6,2.5E6,500
C = 35,35,480
CT = 61000,61000,40

Tie Bar Bush Mount

K = 5723,5723,6686
KT = 543000,543000,500
C = 400,400,300
CT = 18400,18400,4

Upper Damper Mount

K = 14353,14353,10000
KT = 120000,120000,400
C = 400,400,300
CT = 1200,1200,40

Lower Damper Mount

K = 6385,6385,550
KT = 355000,355000,400
C = 640,640,50
CT = 35000,35000,40

Table A.23 Rear Suspension Bush Data

The following linear values are used to define the stiffness and damping in the bushes. For each bush data is listed as:

$K = kx, ky, kz$ Stiffness (N/mm)
$KT = ktx, kty, ktz$ Torsional stiffness (Nmm/rad)
$C = cx, cy, cz$ Damping (Ns/mm)
$CT = ctx, cty, ctz$ Torsional Damping (Nmms/rad)

Rear Trailing Link to Hub Bushes

$K = 10500,10500,870$
$KT = 2.8E5,2.8E5,67500$
$C = 1000,1000,100$
$CT = 25000,25000,40$

Rear Trailing Link to Body Bush

$K = 660,660,175$
$KT = 260300,260300,40000$
$C = 100,100,50$
$CT = 25000,25000,40$

Rear Upper Damper Mount

$K = 540,1300,532$
$KT = 58915,180750,670$
$C = 200,200,70$
$CT = 5800,5800,67$

Rear Lower Arm to Body Mount

$K = 10800,3420,840$
$KT = 790000,380000,400$
$C = 1000,400,100$
$CT = 88000,40000,40$

Rear Lower Arm to Hub Bush

$K = 5540,5540,515$
$KT = 210540,210540,400$
$C = 800,800,50$
$CT = 25000,25000,40$

Fortran Tyre Model Subroutines

B.1 Interpolation tyre model subroutine

```
      SUBROUTINE TIRSUB ( ID, TIME, TO, CPROP, TPROP, MPROP,
     &                    PAR, NPAR, STR, NSTR, DFLAG,
     &                    IFLAG, FSAE, TSAE, FPROP )
C
C    This program is part of the CUTyre system - M Blundell, Feb 1997
C    This version is based on an interpolation approach using measured
C    tyre test data which is include in SPLINE statements. The model is referred
to as the
C    Limited version based on the limited testing where camber and slip
are varied
C    independently.
C
C
C    The coefficients in the model assume the following units:
C    slip angle: degrees
C    camber angle: degrees
C    Fz (load): kg
C    Fy and Fx: N
C    Tz : Nm
C
C    Note this subroutine is developed to not account for offsets
C    twice. The offsets are include for slip interpolation
C    but for camber the offset at zero camber is subtracted.
C
C Inputs:
C
      INTEGER           ID, NPAR, NSTR
      DOUBLE PRECISION  TIME, TO
      DOUBLE PRECISION  CPROP(*), TPROP(*), MPROP(*), PAR(*)
      CHARACTER*80      STR(*)
      LOGICAL           DFLAG, IFLAG, ERRFLG
C
C Outputs:
C
      DOUBLE PRECISION  FSAE(*), TSAE(*), FPROP(*), ARRAY(3)
C
```

```
C Local Variables:
C
      DOUBLE PRECISION  SLIP, ALPHA, DEFL, DEFLD
      DOUBLE PRECISION  R2, CZ, CS, CA, CR, DZ, AMASS, WSPIN
C
      DOUBLE PRECISION  GAMMA,CG,RALPHA,RGAMMA,FZL,TZL,TZLA,TZLG
      DOUBLE PRECISION  CFY,DFY,EFY,SHFY,SVFY,PHIFY,TZLG0,TZLG1
      DOUBLE PRECISION  CTZ,DTZ,ETZ,BTZ,SHTZ,SVTZ,PHITZ
      DOUBLE PRECISION  CFX,DFX,EFX,BFX,SHFX,SVFX,PHIFX
C
      INTEGER           IORD
      DOUBLE PRECISION  ZERO, ONE, SCFACT, DELMAX,FYA,FYG,FYG0,FYG1
      DOUBLE PRECISION  FX, FY, FZ, FX1, FX2, TY, TZ, H, ASTAR, SSTAR
      DOUBLE PRECISION  U, FZDAMP, FZDEFL, WSPNMX, DTOR, RTOD
      LOGICAL           ERFLG
C
      PARAMETER         (ZERO=0.0)
      PARAMETER         (ONE=1.0)
      PARAMETER         (IORD=0)
      PARAMETER         (WSPNMX=5.0D-1)
      PARAMETER         (DTOR=0.017453292)
      PARAMETER         (RTOD=57.29577951)
C
C
C  EXECUTABLE CODE
C
C
C   Extract data from input arrays
C
      SLIP  = CPROP(1)
      DEFL  = CPROP(4)
      DEFLD = CPROP(5)
      WSPIN = CPROP(8)
C
      AMASS = MPROP(1)
C
      R2    = TPROP(2)
      CZ    = TPROP(3)
      CS    = TPROP(4)
      CA    = TPROP(5)
      CR    = TPROP(7)
      DZ    = TPROP(8)
      U     = TPROP(11)
C
      RALPHA = CPROP(2)
      RGAMMA = CPROP(3)
      CG = TPROP(6)
```

```
      ALPHA=RALPHA*RTOD
      GAMMA=RGAMMA*RTOD
C
C    Initialize force values
C
          FX = 0.DO
          FY = 0.DO
          FZ = 0.DO
          TY = 0.DO
          TZ = 0.DO
C
      IF(DEFL .LE. 0.DO) THEN
          GOTO 1000
      ENDIF
C
C    Calculate normal loads due to stiffness (always .LE. zero)
C
      FZDEFL = -DEFL*CZ
C
C    Calculate normal loads due to damping
C
      FZDAMP = - 2.DO*SQRT(AMASS*CZ)*DZ*(DEFLD)
C
C    Calculate total normal force (fz)
C
      FZ     = MIN (0.0DO, (FZDEFL + FZDAMP) )
C
C    Convert to kg and change sign
C
      FZL = -FZ/9.81
C
C    Calculate critical longitudinal slip value
C
      SSTAR = ABS(U*FZ/(2.DO*CS))
C
C    Compute longitudinal force
C
      IF(ABS(SLIP) .LE. ABS(SSTAR)) THEN
        FX = -CS*SLIP
      ELSE
        FX1 = U*ABS(FZ)
        FX2 = (U*FZ)**2/(4.DO*ABS(SLIP)*CS)
        FX = -(FX1-FX2)*SIGN(1.0DO,SLIP)
      ENDIF
C
C    Compute lateral force
C
```

```
       CALL CUBSPL (ALPHA,FZL,100,0,ARRAY,ERRFLG)
       FYA=ARRAY(1)
       CALL CUBSPL (0,FZL,300,0,ARRAY,ERRFLG)
       FYG0=ARRAY(1)
       CALL CUBSPL (GAMMA,FZL,300,0,ARRAY,ERRFLG)
       FYG1=ARRAY(1)
       FYG=FYG1-FYG0
       FY=FYA+FYG
C
C      Compute self aligning moment
C
       CALL CUBSPL (ALPHA,FZL,200,0,ARRAY,ERRFLG)
       TZLA=ARRAY(1)
       CALL CUBSPL (0,FZL,400,0,ARRAY,ERRFLG)
       TZLG0=ARRAY(1)
       CALL CUBSPL (GAMMA,FZL,400,0,ARRAY,ERRFLG)
       TZLG1=ARRAY(1)
       TZLG=TZLG1-TZLG0
       TZL=TZLA+TZLG
C
C      Convert to Nmm
C
       TZ = TZL*1000.0
C
C      Copy the calculated values for FX, FY, FZ, TY & TZ to FSAE
C      and TSAE arrays
C
1000 FSAE(1) = FX
       FSAE(2) = FY
       FSAE(3) = FZ
C
       TSAE(1) = 0.0
       TSAE(2) = 0.0
       TSAE(3) = TZ
C
       FPROP(1) = 0.0
       FPROP(2) = 0.0
C
       RETURN
       END
```

B.2 Magic formula tyre model (version 3) subroutine

```
       SUBROUTINE TIRSUB ( ID, TIME, TO, CPROP, TPROP, MPROP,
      &                    PAR, NPAR, STR, NSTR, DFLAG,
      &                    IFLAG, FSAE, TSAE, FPROP )
C
```

```
C     This program is part of the CUTyre system - M Blundell, Feb 1997
C     This version is based on the Magic Formula tyre model (Version 3).
C     Coefficients are for TYRE B
C
C     The coefficients in the model assume the following units:
C     slip angle: radians
C     camber angle: radians
C     slip ratio %
C     Fz (load): N
C     Fy and Fx: N
C     Tz : Nm
C     Note sign changes between Paceka formulation and SAE convention
C     If camber is not included set A5,A10,A13,A14,A15,A16
C     and C6,C10,C13,C16,C17,C18,C19,C20 to zero
C
C Inputs:
C
      INTEGER           ID, NPAR, NSTR
      DOUBLE PRECISION  TIME, TO
      DOUBLE PRECISION  CPROP(*), TPROP(*), MPROP(*), PAR(*)
      CHARACTER*80      STR(*)
      LOGICAL           DFLAG, IFLAG
C
C Outputs:
C
      DOUBLE PRECISION FSAE(*), TSAE(*), FPROP(*)
C
C Local Variables:
C
      DOUBLE PRECISION  SLIP, ALPHA, DEFL, DEFLD
      DOUBLE PRECISION  R2, CZ, CS, CA, CR, DZ, AMASS, WSPIN
C
C
C
      DOUBLE PRECISION GAMMA,CG,RALPHA,RGAMMA,FXP,FZP,FYP,TZP
      DOUBLE PRECISION A0,A1,A2,A3,A4,A5,A6,A7,A8,A9,A10,A11,A12,A13
      DOUBLE PRECISION A14,A15,A16,A17,SLIPCENT
      DOUBLE PRECISION C0,C1,C2,C3,C4,C5,C6,C7,C8,C9,C10,C11,C12,C13
      DOUBLE PRECISION C14,C15,C16,C17,C18,C19,C20
      DOUBLE PRECISION CFY,DFY,EFY,SHFY,SVFY,PHIFY
      DOUBLE PRECISION CTZ,DTZ,ETZ,BTZ,SHTZ,SVTZ,PHITZ
      DOUBLE PRECISION CFX,DFX,EFX,BFX,SHFX,SVFX,PHIFX,DUMTZ,DUMFY
C
      INTEGER           IORD
      DOUBLE PRECISION  ZERO, ONE, SCFACT, DELMAX
      DOUBLE PRECISION  FX, FY, FZ, FX1, FX2, TY, TZ, H, ASTAR, SSTAR
      DOUBLE PRECISION  U, FZDAMP, FZDEFL, WSPNMX, DTOR, RTOD
      LOGICAL           ERFLG
```

```
C
      PARAMETER          (ZERO=0.0)
      PARAMETER          (ONE=1.0)
      PARAMETER          (IORD=0)
      PARAMETER          (WSPNMX=5.0D-1)
      PARAMETER          (DTOR=0.017453292)
      PARAMETER          (RTOD=57.29577951)
C
C     Define Pacejka Coefficients
C
      A0=.103370E+01
      A1=-.224482E-05
      A2=.132185E+01
      A3=.604035E+05
      A4=.877727E+04
      A5=0.0
      A6=.458114E-04
      A7=.468222
      A8=.381896E-06
      A9=.516209E-02
      A10=0.00
      A11=-.366375E-01
      A12=-.568859E+02
      A13=0.00
      A14=0.00
      A15=0.00
      A16=0.00
      A17=.379913
C
C
      C0=.235000E+01
      C1=.266333E-05
      C2=.249270E-02
      C3=-.159794E-03
      C4=-.254777E-01
      C5=.142145E-03
      C6=0.00
      C7=.197277E-07
      C8=-.359537E-03
      C9=.630223
      C10=0.00
      C11=.120220E-06
      C12=.275062E-02
      C13=0.00
      C14=-.172742E-02
      C15=.544249E+01
      C16=0.00
      C17=0.00
```

```
      C18=0.00
      C19=0.00
      C20=0.00
C
C
C     EXECUTABLE CODE
C
C
C     Extract data from input arrays
C
      SLIP  = CPROP(1)
      DEFL  = CPROP(4)
      DEFLD = CPROP(5)
      WSPIN = CPROP(8)
C
      AMASS = MPROP(1)
C
      R2    = TPROP(2)
      CZ    = TPROP(3)
      CS    = TPROP(4)
      CA    = TPROP(5)
      CR    = TPROP(7)
      DZ    = TPROP(8)
      U     = TPROP(11)
C
C     Convert sign on alpha
C
      RALPHA = CPROP(2)
      RGAMMA = CPROP (3)
      CG = TPROP (6)
      ALPHA=-RALPHA
      GAMMA=RGAMMA
C
C     Initialize force values
C
         FX = 0.D0
         FY = 0.D0
         FZ = 0.D0
         TY = 0.D0
         TZ = 0.D0
C
      IF(DEFL .LE. 0.D0) THEN
        GOTO 1000
      ENDIF
C
C     Calculate normal loads due to stiffness (always .LE. zero)
C
      FZDEFL = -DEFL*CZ
```

```
C
C     Calculate normal loads due to damping
C
      FZDAMP = - 2.D0*SQRT(AMASS*CZ)*DZ*(DEFLD)
C
C     Calculate total normal force (fz)
C
      FZ    = MIN (0.0D0, (FZDEFL + FZDAMP) )
C
C     Convert to kN and change sign
C
      FZP = -FZ
C
C     Compute longitudinal force
C
      IF(ABS(SLIP) .LE. ABS(SSTAR)) THEN
         FX = -CS*SLIP
      ELSE
         FX1 = U*ABS(FZ)
         FX2 = (U*FZ)**2/(4.D0*ABS(SLIP)*CS)
         FX  = -(FX1-FX2)*SIGN(1.0D0,SLIP)
       ENDIF
C
C     Compute lateral force
C
      CFY=A0
      SHFY=A8*FZP+A9+A10*GAMMA
      DFY=(A1*FZP+A2)*(1-A15*GAMMA**2)*FZP
      IF(ALPHA+SHFY.LT.0.0)THEN
        DUMFY=-1.0
      ELSE
        DUMFY=1.0
      ENDIF
      EFY=(A6*FZP+A7)*(1-(A16*GAMMA+A17)*DUMFY)
      BFY=((A3*SIN(2*ATAN(FZP/A4)))*(1-A5*ABS(GAMMA)))/(CFY+DFY)
      SVFY=A11*FZP+A12+(A13*FZP**2+A14*FZP)*GAMMA
      PHIFY=(1-EFY)*(ALPHA+SHFY)+(EFY/BFY)*ATAN(BFY*(ALPHA+SHFY))
      FYP=DFY*SIN(CFY*ATAN(BFY*PHIFY))+SVFY
C
C     Change sign
C
      FY=FYP
C
C     Compute self aligning moment
C
      CTZ=C0
      SHTZ=C11*FZP+C12+C13*GAMMA
      DTZ=(C1*FZP**2+C2*FZP)*(1-C18*GAMMA**2)
```

```
      IF(ALPHA+SHTZ.LT.0.0)THEN
        DUMTZ=-1.0
      ELSE
        DUMTZ=1.0
      ENDIF
      ETZ=(C7*FZP**2+C8*FZP+C9)*(1-(C19*GAMMA+C20)*DUMTZ)
      ETZ=ETZ/(1-C10*ABS(GAMMA))
      BTZ=((C3*FZP**2+C4*FZP)*(1-C6*ABS(GAMMA))*EXP(-C5*FZP))/(CTZ+DTZ)
      SVTZ=C14*FZP+C15+(C16*FZP**2+C17*FZP)*GAMMA
      PHITZ=(1-ETZ)*(ALPHA+SHTZ)+(ETZ/BTZ)*ATAN(BTZ*(ALPHA+SHTZ))
      TZP=DTZ*SIN(CTZ*ATAN(BTZ*PHITZ))+SVTZ
C
C     Convert to Nmm and change sign
C
      TZ = TZP*1000.0
C
C     Copy the calculated values for FX, FY, FZ, TY & TZ to FSAE
C     and TSAE arrays
C
1000  FSAE(1) = FX
      FSAE(2) = FY
      FSAE(3) = FZ
      TSAE(1) = 0.0
      TSAE(2) = 0.0
      TSAE(3) = TZ
      FPROP(1) = 0.0
      FPROP(2) = 0.0
C
      RETURN
      END
```

B.3 The Harty tyre model subroutine
TYR501

```
C MDI TYR501 : Concept Tyre Model
C
C
C A Quick & Dirty Tyre Model which plugs in as the FIALA
C model does, with a "TIRE" statement.
C
C Unlike FIALA, critical slip angle is broadly independent
C of load and initial cornering stiffness is strongly
C load dependent.
C
C These attributes better represent a modern radial tyre
C than does either the FIALA or University of Arizona
C model.
C
```

```
C  The model does handle comprehensive slip.  Lateral force
C  generation is zero at peak longitudinal force slip ratio
C  (typically about 20%) but returns to a value around one
C  tenth of the peak lateral force as the wheel progresses
C  beyond that limit.  This may result in poor post-spin
C  performance.  The force generated with locked wheels is
C  aligned with the wheel plane; this is incorrect.
C
C  Longitudinal force generation is assumed to be symmetric
C  for tractive and braking slip.  This is not generally
C  true beyond the critical slip ratio for real tyres but
C  is reasonable up to that point.  This tyre will over
C  estimate longitudinal forces for tractive slip and
C  slightly underestimate them for braking slip in the
C  post-critical regions.
C
C  -- 29th December 2000 --
C
C  Camber thrust is included as for the motorcycle tire
C  model using "taut string" logic.  Lateral migration of
C  the contact patch is also included, as for the motorcycle
C  tyre model.
C
C  Aligning Torque calculation includes the lateral force
C  due to camber.  This is not quite right as the camber
C  force mechanism has no pneumatic trail associated with
C  it.  Pay attention if using this for motorcycle work;
C  consider reworking it so that TZ does not include the
C  camber force.  The form of the aligning torque is a
C  bit poor and would benefit from some more thought;
C  pneumatic trail collapses linearly with lateral force.
C
C  -- 10th January 2001 --
C
C  Unsuitable Aligning Moment behaviour substantially improved
C  for motorcycle use.
C
C  --
C
C  Relaxation Length is externally imposed as with the
C  Fiala tyre.
C
C  Tyre Data is taken from the tyre parameter file (.tpf)
C  but note that not all the data is used.  The other
C  parameters are passed in via the UPARAMETERS argument
C  on the TIRE statement inside an ADAMS deck.
C
```

```
C  The model is quite empirical and has no basis in any sort
C  of established fact or theory.  It may or may not bear a
C  passing resemblance to "Maltyre", a Malcolm Burgess model
C  implemented at Lotus to the same end.  I don't care, I
C  did it all myself without a grown-up to help with the
C  pointy bits.
C
C  -- 24th April 2001 --
C
C  Banner and zero parameter check added in IFLAG loop.
C
C  -- 7th July 2001 --
C
C  Improved representation of behaviour outside friction
C  circle.  Correct differentiation between lock and
C  wheelspin in terms of force vector.
C
C  -- 6th October 2004 --
C
C  Improved aligning moment form - was significantly too high.
C  Uses passed in paremeter for Pneumatic Trail on-centre.
C  Note that passed-in parameter can be negative, giving
C  pneumatic "lead".
C
C  -- 10th March 2006 --
C
C  Pneumatic lead introduced for camber forces to match
C  measured motorcycle data. Minor error in limit camber
C  clipping corrected
C
C  -- 30th March 2006 --
C
C  Minor error with form of camber clipping  (asymmetric)
C  corrected.
C
C
C  -- 16th February 2009 --
C
C  Damian made the mistake of letting someone else have a go at
C  his model and so I am attempting to migrate it to TYR501
C  since the TIRSUB routine will become defunct at the next release.
C
C  (Teena Gade)
C
C  -- 7th September 2009 --
C
C  Migration to TYR501 completed by DAH after Teena did all the
C  nasty bits getting the right data into the right place.
C
```

```
C  -- 22nd May 2013 --
C
C Sign error in forces carried over from original MSC TYR501
C sample file has led to erratic behaviour of TYR501 until
C pinned down, now fixed. Not that FORCES and TORQUE are the
C variables which actually deliver forces back to the solution.
C VARINF is associated with VPG Tire, a mode of usage I have
C never successfully invoked. VARINF information is of unknown
C provenance and should be used without checking.
C
C Also uncovered some strange behaviour of original TYR501
C that didn't allow it to run backwards - fixed with velocity
C sign check just before FORCES is returned. VARINF not
C corrected.
C
C Resulting bug with rolling resistance fixed, works
C correctly (has it ever done before?)
C
C  --
C
C (c) DAH 24 Oct 1999-2013
C
      SUBROUTINE TYR501( NDEV, ISWTCH, JOBFLG, IDTYRE,
     +              TIME, DIS, TRAMAT, ANGTWC, VEL, OMEGA, OMEGAR,
     +              NDEQVR, DEQVAR, NTYPAR, TYPARR,
     +              NCHTDS, CHTDST, ROAD, IDROAD,
     +              NROPAR, ROPAR, NCHRDS, CHRDST,
     +              FORCES, TORQUE, DEQINI, DEQDER, TYRMOD,
     +              NVARS, VARINF, NWORK, WRKARR,
     +              NIWORK, IWRKAR, IERR )
C
C Inputs:
      INTEGER          NDEV
      INTEGER          ISWTCH
      INTEGER          JOBFLG
      INTEGER          IDTYRE
      DOUBLE PRECISION TIME
      DOUBLE PRECISION DIS(3)
      DOUBLE PRECISION TRAMAT(3,3)
      DOUBLE PRECISION ANGTWC
      DOUBLE PRECISION VEL(3)
      DOUBLE PRECISION OMEGA(3)
      DOUBLE PRECISION OMEGAR
      INTEGER          NDEQVR
      DOUBLE PRECISION DEQVAR(NDEQVR)
      INTEGER          NTYPAR
      DOUBLE PRECISION TYPARR(NTYPAR)
```

```
      INTEGER           NCHTDS
      CHARACTER*256     CHTDST
      INTEGER           IDROAD
      INTEGER           NROPAR
      DOUBLE PRECISION ROPAR(NROPAR)
      INTEGER           NCHRDS
      CHARACTER*256     CHRDST
C Outputs:
      DOUBLE PRECISION FORCES(3)
      DOUBLE PRECISION TORQUE(3)
      DOUBLE PRECISION DEQINI(NDEQVR)
      DOUBLE PRECISION DEQDER(NDEQVR)
      CHARACTER*256     TYRMOD
      INTEGER           NVARS
      DOUBLE PRECISION VARINF(NVARS)
      INTEGER           NWORK
      DOUBLE PRECISION WRKARR(NWORK)
      INTEGER           NIWORK
      INTEGER           IWRKAR(NIWORK)
      INTEGER           IERR
C
C
C
C  Local Variables:
C
C  Locals:
      INTEGER  I
C         DOUBLE PRECISION C_SLIP
      DOUBLE PRECISION C_ALPHA
      DOUBLE PRECISION C_GAMMA
      DOUBLE PRECISION U1
      DOUBLE PRECISION U0
      DOUBLE PRECISION GAIN
      DOUBLE PRECISION R_LEN
      DOUBLE PRECISION URAD(3)
      DOUBLE PRECISION U
      DOUBLE PRECISION F(6)
      DOUBLE PRECISION FCP(3)
      DOUBLE PRECISION TCP(3)
      INTEGER           ARRPTR
      INTEGER           UMODE
      DOUBLE PRECISION RAD(3)
      DOUBLE PRECISION RADIUS
      INTEGER           NROAD
      DOUBLE PRECISION RCP(3)
      DOUBLE PRECISION RNORM(3)
      DOUBLE PRECISION SURFAC
```

```
      DOUBLE PRECISION CN
      DOUBLE PRECISION RDR
C     DOUBLE PRECISION CRR
      DOUBLE PRECISION CPMTX(3,3)
      DOUBLE PRECISION VCPLON
      DOUBLE PRECISION VCPLAT
      DOUBLE PRECISION VCPVRT
      DOUBLE PRECISION VLON
C     DOUBLE PRECISION ALPHA
        DOUBLE PRECISION ALPHA_L
      DOUBLE PRECISION KAPPA
      DOUBLE PRECISION KAPPA_L
      DOUBLE PRECISION GAMMA
      DOUBLE PRECISION FRCRAD
      DOUBLE PRECISION FRCVRT
      DOUBLE PRECISION FRCLON
      DOUBLE PRECISION FRCLAT
      DOUBLE PRECISION TRQALN
      DOUBLE PRECISION FZMAG
C
C     ------------------------------------------------------------------
C  -- Carried across from tirsub --
      DOUBLE PRECISION  FX, FY, FZ, TX, TY, TZ

      DOUBLE PRECISION  SLIP, ALPHA, DEFL, DEFLD
      DOUBLE PRECISION  R1, R2, CZ, CS, C_MX, CR, DZ, AMASS, WSPIN

      DOUBLE PRECISION  ALPHA_C, Ay, By, R_LOAD, dB_dFz, B
      DOUBLE PRECISION  SLIP_C, Ax, SLIP_M, FR_ELLIP, CP_LEN
      DOUBLE PRECISION  LSLIP, USLIP, UNLRAD

      DOUBLE PRECISION  THRSH, CAMB_C, CAMB_INC, A_INT, B_INT, C_INT
      DOUBLE PRECISION  SLIPSQ, TAN_ALPHA_SQ, DIVISOR
      DOUBLE PRECISION  FX1, FY1, FX2, FY2, ABSLIP, PTRAILC, PNOFFSET
      DOUBLE PRECISION  PLEAD, FY_CAMBER, FYWAS
C     ------------------------------------------------------------------
C
C
C  Scaling parameters

C     DOUBLE PRECISION SCLRR
      DOUBLE PRECISION SCLFY
      DOUBLE PRECISION SCLMX
      DOUBLE PRECISION SCLMZ

C  Drift array parameters

      DOUBLE PRECISION PLYFRC
      DOUBLE PRECISION CONFRC
```

```
       DOUBLE PRECISION PLYTRQ
       DOUBLE PRECISION CONTRQ

       LOGICAL          ERRFLG
C
C  Road Declarations:
       INTEGER          MAXDIV
       PARAMETER        (MAXDIV = 10)
       INTEGER          N_T_SHAPE
       DOUBLE PRECISION T_SHAPE(2, MAXDIV)
       DOUBLE PRECISION EFFVOL
       DOUBLE PRECISION EFFPEN
       CHARACTER*256    ERRMSG
       CHARACTER*80     ERRTMP
C
       DOUBLE PRECISION STARTUP
       DOUBLE PRECISION OFF_GRND
C
C  Useful Parameters:
       DOUBLE PRECISION ZERO_VAL
       PARAMETER        (ZERO_VAL = 0.D0)
       DOUBLE PRECISION ONE
       PARAMETER        (ONE = 1.D0)
       DOUBLE PRECISION ZERLIM
       PARAMETER        (ZERLIM = 1.0E-10)
       DOUBLE PRECISION TFULL
       PARAMETER        (TFULL = 0.5)
       DOUBLE PRECISION NO_FRC
       PARAMETER        (NO_FRC = 448)
       DOUBLE PRECISION WSPNMX
       PARAMETER        (WSPNMX = 5.0D-1)
       INTEGER          DYNAMIC
       PARAMETER        (DYNAMIC = 1)
       INTEGER          STATIC
       PARAMETER        (STATIC = 0)
       INTEGER          IORD
       PARAMETER        (IORD=0)
       INTEGER          IMODE
C
       include 'ac_tir_jobflg.inc'
       include 'abg_varptr.inc'
       include 'tyrHarty_501.inc'
C  Functions:
       DOUBLE PRECISION DOT
       EXTERNAL         DOT
C
```

```fortran
      EXTERNAL        ROAD

      LOGICAL         STAFLG
      SAVE            STAFLG

      DATA   STAFLG   /.FALSE./

      IERR = 0
C  Read the tire property file during initialization:
      IF ( JOBFLG .EQ. INIT ) THEN
      CALL USRMES( .TRUE.,
   +  ' ', 0,
   +  'INFO_NOPAD' )
      CALL USRMES( .TRUE.,
   +  ' ', 0,
   +  'INFO_NOPAD' )
      CALL USRMES( .TRUE.,
   +  '**************************************************', 0,
   +  'INFO_NOPAD' )
      CALL USRMES( .TRUE.,
   +  'TYR501 Harty Model: Compiled 12 Aug 2013', IDTYRE,
   +  'INFO_NOPAD' )
      CALL USRMES( .TRUE.,
   +  '**************************************************', 0,
   +  'INFO_NOPAD' )
      CALL USRMES( .TRUE.,
   +  ' ', 0,
   +  'INFO_NOPAD' )
      CALL USRMES( .TRUE.,
   +  ' ', 0,
   +  'INFO_NOPAD' )
         CALL RPF501( NCHTDS, CHTDST, IDTYRE, NTYPAR, TYPARR )
      ENDIF

C  Set DEQINI:
      IF ( JOBFLG .EQ. INQUIRE ) THEN
        DEQINI(1) = 0.0D0
        DEQINI(2) = 0.0D0
      ENDIF

   IF  (JOBFLG .NE. ENDSIM) THEN


C  Decode TYPARR Array:
      UMODE  = NINT( TYPARR( use_mode ) )

      UNLRAD = TYPARR( unloaded_radius )
      TIREW  = TYPARR( width )
      TIREK  = TYPARR( vertical_stiffness )
```

```
      TIREC   = TYPARR( vertical_damping )
      CR      = TYPARR( rolling_resistance )
      CA      = TYPARR( calpha )
      C_GAMMA = TYPARR( cgamma )
      U_MIN   = TYPARR( umin )
      U_MAX   = TYPARR( umax )
      R_LEN   = TYPARR( relaxation_length )
      ALPHA_C = TYPARR( alpha_critical )
      Ay      = TYPARR( curvature_factor_angle )
      By      = TYPARR( scale_factor_lateral )
      R_LOAD  = TYPARR( rated_load )
      dB_dFz  = TYPARR( scale_factor_dim )
      SLIP_C  = TYPARR( slip_ratio_critical )
      Ax      = TYPARR( curvature_factor_ratio )
      PTRAILC = TYPARR( pneum_trailing_scaling )
      PLEAD   = TYPARR( pneumatic_lead_camber )
      THRSH   = TYPARR( limit_camber_onset_fric )
C
      N_T_SHAPE = NINT( TYPARR( n_shape ) )
C
      IF ( JOBFLG .EQ. INIT .OR. JOBFLG .EQ. RESET ) THEN
C     -- Debug only - check we're getting what we think --
C--------------------------------------------------------------------
C        WRITE(*,*) 'UNLRAD ', UNLRAD, ', TIREK ', TIREK
c        WRITE(*,*) ', TIREW ', TIREW
c        WRITE(*,*) 'TIREC', TIREC, ', CR ', CR
c        WRITE(*,*) 'CA ', CA
c        WRITE(*,*) 'C_GAMMA ', C_GAMMA, 'U_min ', U_min
c        WRITE(*,*) 'U_max ', U_max, 'R_LEN ', R_LEN
c        WRITE(*,*) 'ALPHA_C ', ALPHA_C, 'Ay ', Ay
c        WRITE(*,*) 'By ', By, 'R_LOAD ', R_LOAD
c        WRITE(*,*) 'dB_dFz ', dB_dFz
c        WRITE(*,*) 'SLIP_C ', SLIP_C
c        WRITE(*,*) 'Ax ', Ax
C     WRITE(*,*) 'PTRAILC ', PTRAILC, 'PLEAD ', PLEAD, 'THRSH ', THRSH
C--------------------------------------------------------------------
C
C
      ENDIF
C


C====================================================================
C
C  -- All this is standard TYR501 stuff
C     - dynamic or static
C     - soft start to calculations
```

```
C     - road/tyre interaction including profile
C     - states for the tyre model
C
C=============================================================

C  Initialize mode (STATIC or DYNAMIC)
       IMODE = DYNAMIC
       IF ( ISWTCH .EQ. 0 ) IMODE = STATIC

C  Set flag for quasi-statice analyses

       IF ( ISWTCH .EQ. 2 ) STAFLG = .TRUE.

C  Setup Smoothing Function:
C
C  The MDI tire models include a feature for smoothing the
C  tire forces around time=0.0.  So, for example, if there's
C  some initial slip angle at time=0.0, the lateral force
C  builds up slowly instead of acting like a step input.
C  This helps the integrator get started. UMODE comes
C  from the tire property file.

       IF(UMODE .GE. 2 .AND.(.NOT.STAFLG) )THEN
         CALL STEP(TIME,ZERO_VAL,ZERO_VAL,TFULL,ONE,0,
     +             STARTUP,ERRFLG)
       ELSE
         STARTUP = ONE
       ENDIF

C  Setup The Tire Carcase (Cross Section) Shape
C  for use by the durability tire road contact
C  model:

       IF (N_T_SHAPE.EQ.0) THEN
         T_SHAPE(1,1) = 0.D0
         T_SHAPE(2,1) = 0.D0
       ELSE
         ARRPTR = SHAPE
       DO I=1,N_T_SHAPE
         T_SHAPE(1,I)=TYPARR(ARRPTR)
     T_SHAPE(2,I)=TYPARR(ARRPTR+1)
         ARRPTR = ARRPTR + 2
       ENDDO
     ENDIF

C  Offset rolling radius - this is in the original code but
C  I don't know why.
C
C     UNLRAD = UNLRAD + SCLRR
```

```
C  Call ROAD routine
C
C  The road routine calculates the local road normal, the
C  road contact point (contact patch location), the
C  local surface coeffient of friction and the tire's
C  vertical deflection.  The STI passes in the name of
C  the subroutine to be called.  Hence "ROAD" is just a
C  placeholder.

       CALL ROAD(JOBFLG, IDTYRE,
      &          TIME, DIS, TRAMAT,
      &          IDROAD, NROPAR, ROPAR, NCHRDS, CHRDST,
      &          N_T_SHAPE, T_SHAPE, UNLRAD, TIREW,
      &          NROAD, EFFVOL, EFFPEN, RCP,
      &          RNORM, SURFAC, IERR, ERRMSG )



C  Call the TIRE Kinematics Routine (ACTCLC):
C
C  The ACTCLC routine calculates the slip angle (ALPHA),
C  inclination (camber) angle (GAMMA), longitudinal slip
C  (KAPPA), the longitudinal (VCPLON) and lateral (VCPLAT)
C  slip velocities, the longitudinal velocity of wheel
C  center (VLON), the vertical velocity of the wheel center
C  normal to the road (VCPVRT), the unit vector directed
C  from the wheel center to the contact patch (URAD) expressed
c  in global coordinates, and the transformation
C  matrix from SAE contact patch coordinates (CPMTX) to
C  global (ground part) coordinates.
C
C  Calculate the tire kinematics if:
C     The tire is in contact with road (e.g. not flying)
C
C     - and -
C
C     The job is normal execution or differencing for
C     derivatives.
C
      IF(
      .       NROAD .EQ. 1 .AND. IERR .NE. 3
      .       .AND.
      .           (
      .             JOBFLG .EQ. NORMAL .OR.
      .             JOBFLG .EQ. DIFF
      .           )
      .    ) THEN
```

```
            RADIUS = UNLRAD - EFFPEN


        CALL ACTCLC(TRAMAT, VEL, OMEGA, OMEGAR, RADIUS, RNORM,
     &              VLON, VCPLON, VCPLAT, VCPVRT,
     &              ALPHA, GAMMA, KAPPA,
     &              URAD, CPMTX)

c
c  Lag The slip angle to for tire relaxation effects:
c
c  d( Alpha_lagged )/dt =(VLON/Relaxation_Length)*( Alpha - Alpha_lagged )
c
c  If the relaxation length is less than 1e-4 Meters, then don't lag the
c  slips.
c
      IF ( R_LEN .LT. 1D-4  .OR.
   .        IMODE .EQ. STATIC ) THEN

        ALPHA_L  = ALPHA
        KAPPA_L  = KAPPA

      ELSE

        GAIN      = ABS(VLON)/R_LEN
        ALPHA_L   = DEQVAR(1) + DEQINI(1)
        KAPPA_L   = DEQVAR(2) + DEQINI(2)
        DEQDER(1) = GAIN*(ALPHA - ALPHA_L)
        DEQDER(2) = GAIN*(KAPPA - KAPPA_L)

      ENDIF

C=================================================================
C
C  -- End of the Standard TYR501 Stuff --
C
C=================================================================



C  -- Now the tyre modelling proper can start --
C  All forces calculated in SAE reference frame and transformed to TYDEX
format
C  afterwards - ease of continuity with previous model (also true of
reference
C  TYR501 model provided by MSC)
```

```
C  -- SAE Vertical Force like original tirsub calculations --

C     Normal Loads; simple calculations as with sample tirsub.f;
C     Penetrations to hub are not accounted for.

C  -- Calculate normal loads due to stiffness (always .LE. zero) --

      FZDEFL = -EFFPEN*TIREK

C  -- Calculate normal loads due to damping --

      FZDAMP = -VCPVRT*TIREC

C  -- Note the startup modification that was present in the tirsub
model is
C     no longer needed --




C  -- Sum for total normal force --

      FZ = MIN (0.0D0, (FZDEFL + FZDAMP) )




      IF ( IMODE .EQ. DYNAMIC ) THEN

C        Coefficient of friction as function of combined slip:

         U = U_MAX+SQRT(KAPPA_L**2+(TAN(ALPHA_L))**2)*(U_MIN-U_MAX)

C        Modify coefficient of friction based on road surface
C        factor:

         U = U * SURFAC




C Longitudinal Loads

C  -- We're working in percent --

      SLIP=KAPPA*100

      IF(ABS(SLIP) .LE. ABS(SLIP_C)) THEN
```

```
C  -- Exponential Rise (1-e-x) below critical slip ratio --

      FX = (1-EXP(-Ax*ABS(SLIP)/SLIP_C))*U*ABS(FZ)*SIGN(1.0D0,SLIP)

      ELSE

C  -- Linear Decay to Sliding Friction above critical slip ratio --

      FX = ABS(FZ)*(1-EXP(-Ax))*U*SIGN(1.0D0,SLIP)

      ENDIF


C  Lateral force and aligning torque (FY & TZ)

C  -- Scale Factor Diminished with Load FZ --

   B = By+(ABS(FZ)-R_LOAD)*dB_dFz


C  -- We're working in degrees --

   ALPHA_L=ALPHA_L*45/ATAN(1.0)

C  -- Don't let alpha go beyond 80 - the TAN functions go kinda wild --

   IF(ALPHA_L.GT.80.) THEN
      ALPHA_L = 80.0
   ENDIF
   IF(ALPHA_L.LT.-80.) THEN
      ALPHA_L = -80.0
   ENDIF

   IF(ABS(ALPHA_L) .LE. 1.D-10) THEN

      FY = 0.D0
      TZ = 0.D0

   ELSE IF( ABS(ALPHA_L) .LE. ALPHA_C ) THEN

C  -- As for longitudinal forces, Exponential Rise (1-e-x) below
C     critical slip angle --

C  -- This line contains an even number of minus-sign errors --

      FY = (1-EXP(-Ay*ABS(ALPHA_L)/ALPHA_C))
   +       *U*B*FZ*SIGN(1.0D0,ALPHA_L)
```

```
      ELSE

C  -- As for longitudinal forces, Linear Decay to Sliding Friction
C     above critical slip ratio --

C    FY = FZ*U*B*SIGN(1.0D0,SLIP)*(1-(ABS(ALPHA_L)-ALPHA_C)/800)

C  -- Simplified - ADAMS handles transition from static to sliding
C     friction in the calling routine --

      FY = FZ*(1-EXP(-Ay))*U*B*SIGN(1.0D0,ALPHA_L)

    ENDIF



C Aligning Torque based on intermediate FY excluding camber force.

C  -- Contact Patch Length --

   R1=UNLRAD
   R2=TIREW

   CP_LEN = (R1**2 - (R1-ABS(FZ)/TIREK)**2)**0.5 * 2.0

   IF(ABS(ALPHA_L) .GT. 1.D-10) THEN

     IF( ABS(ALPHA_L) .LE. ALPHA_C ) THEN

       TZ = -FY*CP_LEN/6*(1-ABS(ALPHA_L)/ALPHA_C)*PTRAILC

C  -- Divisor is because lever arm is not the entire contact patch
length. --

C  -- Parameter PTRAILC should be set to 1.0 for tyres with recetangular
C  contact patches (i.e. car tyres) and 0.5 for tyres with elliptical
C  contact patches (i.e. motorcycle tyres.) --

       ELSE

         TZ = 0.0

       ENDIF

       ENDIF
```

```
C    -- Add camber force to FY - "Taut String" --

C DAH Sign of Camber Component Changed 13-11-00
C   FY  =  FY - FZ * TAN(GAMMA)

C       CAMBER=GAMMA


C    -- "Clipped" Camber model - improved limit behaviour --
C       DAH 10-01-00

C    -- THRSH represents aggression of departure at limit; high value
C       implies high limit & aggressive departure, lower value implies
C       progression.

C    -- was hard-coded, now user parameter
C    THRSH=0.8

     IF (ABS(GAMMA) .LT. ATAN(THRSH*U/C_GAMMA)) THEN

C       -- Camber term now held separate for aligning moment calculation

        FY_CAMBER = - FZ * TAN(GAMMA) * C_GAMMA



     ELSE

        CAMB_C=ATAN(THRSH*U/C_GAMMA)
        CAMB_INC=ABS(GAMMA)-CAMB_C

        A_INT=(1/(1-THRSH*C_GAMMA))/(COS(ATAN(THRSH*U/C_GAMMA)))**2
        B_INT=-(1-THRSH)*U*C_GAMMA

C       -- Needed when C_GAMMA is not equal to unity --

        C_INT= - FZ * TAN(CAMB_C) * C_GAMMA /
     &            (
     &            SIGN(1.,CAMB_C)*FZ*B_INT*(1-EXP(-A_INT*CAMB_INC)) -
     &            SIGN(1.,CAMB_C)*THRSH*U*FZ*C_GAMMA
     &            )
C    MUX=C_INT


        FY_CAMBER =SIGN(1.,GAMMA)*FZ*B_INT*(1-EXP(-A_INT*CAMB_INC)) -
     &            SIGN(1.,GAMMA)*THRSH*U*FZ*C_GAMMA * C_INT

     ENDIF
```

```
        FY = FY + FY_CAMBER
        FYWAS = FY


C       Mitigate FY depending on "Friction Ellipse"


        FR_ELLIP = (FX/(FZ*U))**2 + (FY/(FZ*U*B))**2



        X_SIGN = SIGN(1.0D0,FX)
        Y_SIGN = SIGN(1.0D0,FY)


        IF ( FR_ELLIP .GT. 1.0 ) THEN

            LSLIP=50.0
            USLIP=100.0
            ABSLIP=ABS(SLIP)


C       -- Friction Ellipse treatment for comprehensive slip below
C       critical slip ratio - revised over previous calculations
C       to preserve ratio of FX, FY but bring them inside the
C       friction ellipse --

        DIVISOR=1 + (FY/FX)**2/B**2

        FX1 = ( (U*FZ)**2 / DIVISOR )**0.5 * X_SIGN



C       -- Alternative term; longitudinal force is preserved at the
C       expense of lateral; seems intuitively more correct but
C       produces apparently poorer results. --
C       FX1 = FX

        FY1 = ( ( 1-(FX1/(FZ*U))**2 ) * (FZ*U*B)**2 )**0.5*Y_SIGN



C       -- Revised formulation for highest slip ratios arrived at
C       by consideration of contact patch velocity.  Gives pleasing
```

```
C      results for wheels locked and wheels spinning cases.  Note
C      conversion of ALPHA from degrees back to radians for
C      this calculation and SLIP back from percent. --

       SLIPSQ = (SLIP/100)**2
       TAN_ALPHA_SQ=(TAN(ALPHA_L*ATAN(1.0)/45))**2

       DIVISOR=( 1 + TAN_ALPHA_SQ/SLIPSQ )

       FX2 = ( (U*FZ)**2 / DIVISOR )**0.5 * X_SIGN
       FY2 = ((1-(FX2/(FZ*U))**2)*(FZ*U)**2)**0.5*Y_SIGN


C      -- Smear between two models using slip ratio --

       CALL STEP(ABSLIP,LSLIP,FX1,USLIP,FX2,IORD,FX,ERRFLG)
       CALL STEP(ABSLIP,LSLIP,FY1,USLIP,FY2,IORD,FY,ERRFLG)
C        FX=FX1
C        FY=FY1


C      -- Mitigate Camber forces too, for subsequent aligning moment
calculations

C      CALL STEP(ABSLIP,LSLIP,FY1,USLIP,FY2,IORD,FY_CAMBER,ERRFLG)

C      Is this right? Doesn't it significantly corrupt aligning torque for
C      a locking wheel at a high slip angle?

       FY_CAMBER=FY_CAMBER*FY/FYWAS


   ENDIF

C C  -- The real MUX and MUY; all others are for debug only --
C C     MUX = (FX/(FZ*U))
C C     MUY = (FY/(FZ*U*B))




C    Rolling resistance moment (TY) as FIALA Tyre:


C    IF ( OMEGAR .GE. 0.0 ) THEN
C
C       TY = -CR * FZ
```

```
C
C     ELSE
C
C        TY =  CR * FZ
C
C     ENDIF

C     No need for loop above - velocity change below takes care of it

         TY =  CR * FZ




C     Compute righting moment due to lateral Contact Patch Shift (TX)
C     Use CA as "shape factor" to add to or subtract righting moment
C     from ADAMS' Toroidal assumption.  CA > 1 = fatter than toroid
C     CA < 1 = more like blade.

C     Lateral Contact Patch Shift clips at tyre extremity
C
C     Add aligning torque based on lateral offset of contact patch and
C     longitudinal forces to give "stand up under braking" behaviour for
C     motorcycles or tramlining for cars.

         PNOFFSET = 2 * GAMMA * R2/2 * (CA - 1)
         IF (ABS(PNOFFSET) .LT. R2/2 ) THEN
            TX = FZ * PNOFFSET
            TZ = TZ - FX * PNOFFSET
         ELSE
            TX = FZ * R2/2 * SIGN(1.,PNOFFSET)
            TZ = TZ - FX * R2/2  * SIGN(1.,PNOFFSET)
         ENDIF

C     Measured data shows evidence of significant "pneumatic lead" on
C     camber force data, aligning moment further modified to reflect
this.
C     Real data shows small dependency on load, some dependency on
camber
C     angle at low cambers; constant lead formulation neglects load
C     dependency and may overestimate torques at small cambers.
However,
C     camber forces are low and so torques are low too.

         TZ = TZ + FY_CAMBER*PLEAD
```

```
          ELSE
c  For static equilibrium zero the forces.

              FX  = ZERO_VAL
              FY  = ZERO_VAL
              TX  = ZERO_VAL
              TY  = ZERO_VAL
              TZ  = ZERO_VAL

          ENDIF



C===========================================================================
C
C  -- After the tyre model giving forces & moments in SAE co-ordinates,
C  the long and arduous business of giving them back to ADAMS. Is this
C  *really* progress? --
C
C  -- Below here, all is standard TYR501 code except for sign mapping in
C  FCP and TCP --
C
C===========================================================================


C  Apply the start-up transient smoothing and force the
C  all other tire forces to zero when the vertical force goes to
C  zero (e.g. when the tire is flying).

          FZMAG = DABS(FZ)

          CALL STEP(FZMAG,ZERO_VAL,ZERO_VAL,NO_FRC,ONE,
      +             0,OFF_GRND,ERRFLG)

          IF( IMODE .EQ. DYNAMIC ) THEN

              FCP(1) = -FX
              FCP(2) = FY
              FCP(3) = FZ

              TCP(1) = TX
              TCP(2) = TY
              TCP(3) = TZ
```

```
        ELSE

          FCP(1) = 0.0
          FCP(2) = 0.0
          FCP(3) = FZ

          TCP(1) = 0.0
          TCP(2) = 0.0
          TCP(3) = 0.0

        ENDIF

C  Transform the contact patch forces and moments to hub
C  coordinates:
C
C    Inputs:
C      FCP tire forces at contact patch in SAE contact patch
c          coordinates.
c
c      TCP Tire moments (torques) at contact patch in SAE
c          contact patch coordinates.
c
c      RAD Tire radius vector express in global (ground)
c          coordinates.
c
c      CPMTX Transformation from SAE contact patch coordinates
c          to ground.
c
c    Outputs:
c
c      F  tire forces at hub in global coordinates
c      T  Tire moments (torques) at hub (wheel center) in
c         global coordinates.
c
c      {F} = [CPMTX]{FCP}
c
c      {T} = [CPMTX]{TCP} + {RAD} X {F}
c
        CALL SVEC( RAD, URAD, RADIUS, 3 )
        CALL XCP2HB(FCP, TCP, RAD, CPMTX, F(1), F(4) )

C  Transformation of forces/torques from global to wheelcarrier
C  axes
        CALL M3T1(FORCES, TRAMAT, F(1) )
        CALL M3T1(TORQUE, TRAMAT, F(4) )

        XVEL=VEL(1)
        IF (XVEL .GE. 0.0) THEN
C          WRITE(*,*) 'Reverse'
          FORCES(1)=-FORCES(1)
```

```
          TORQUE(2)=-TORQUE(2)
C         WRITE(*,*) FORCES
C         WRITE(*,*) TORQUE
      ELSE
C         WRITE(*,*) 'Forwards'
C         WRITE(*,*) FORCES
C         WRITE(*,*) TORQUE
      ENDIF


C ********************************************************
C Assigning output quantities to VARINF array
C ********************************************************

        IF(NVARS .GE. 75) THEN

C Contact Patch Forces/Torques
C   VARINF in ISO coordinates, FCP and TCP in SAE coordinates
C   - Note the VARINF array is for examination inside A/View and
C     does not represent the forces passed back to the solver
C   - Arrays FORCES and TORQUE are the ones which influence
C     the solution

          VARINF(FX_ISO_PTR) =  FCP(1)
          VARINF(FY_ISO_PTR) = -FCP(2)
          VARINF(FZ_ISO_PTR) = -FCP(3)

          VARINF(MX_ISO_PTR) =  TCP(1)
          VARINF(MY_ISO_PTR) = -TCP(2)
          VARINF(MZ_ISO_PTR) = -TCP(3)

C Derivatives of state variables

          CALL COPYD(VARINF(DUDT_PTR), DEQDER(1), 2)

C Slip quantities

C   Kinematic:

          VARINF(SLIPX_PTR) = KAPPA
          VARINF(SLIPI_PTR) = -ALPHA

C   Dynamic:
          VARINF(SLIPX_D_PTR) = KAPPA_L
          VARINF(SLIPI_D_PTR) = -ALPHA_L
```

```
C  Friction coefficients

          VARINF(MUXTYR_PTR) = DABS(FRCLON/FCP(3))
          VARINF(MUYTYR_PTR) = DABS(FRCLAT/FCP(3))

C  Tire characteristics

          VARINF(PT_PTR) = 0.D0
          VARINF(MZR_PTR) = 0.D0
          VARINF(S_PTR) = 0.D0
          VARINF(SIGKPO_PTR) = 0.D0
          VARINF(SIGALO_PTR) = 0.D0
          VARINF(MGYR_PTR) = 0.D0
          VARINF(SVYKAP_PTR) = 0.D0
          VARINF(SVX_PTR) = 0.D0
          VARINF(SVY_PTR) = 0.D0

C  Contact Point

          CALL COPYD(VARINF(RCP1_PTR), RCP, 3)

C  Road Normal

          CALL COPYD(VARINF(RNORM1_PTR), RNORM, 3)

C  Surface Friction

          VARINF(SURFAC1_PTR) = SURFAC
          VARINF(SURFAC2_PTR) = SURFAC
          VARINF(SURFAC3_PTR) = SURFAC

C  Tire kinematics

          VARINF(CAMB_PTR)   = GAMMA
          VARINF(EFFPEN_PTR) = EFFPEN
          VARINF(VCPVRT_PTR) = VCPVRT
          VARINF(RADIUS_PTR) = RADIUS
          VARINF(VCPLON_PTR) = VCPLON
          VARINF(VCPLAT_PTR) = VCPLAT
          VARINF(VLON_PTR)   = VLON
        ELSE
          CALL ZERO(VARINF(1), NVARS)
          IERR = 2
          ERRMSG = 'TYR501: Incorrect Dimension on VARINF Array'
        ENDIF
```

```
C  **********************************************************
C  Use these values if tire is FLYING
C                                   ^^^^^^

      ELSE
         CALL ZERO(VARINF(1), NVARS)
         IF(NVARS .GE. 75) THEN
           CALL COPYD(VARINF(DUDT_PTR), DEQDER(1), 2)
           VARINF(RADIUS_PTR) = UNLRAD
         ELSE
           CALL ZERO(VARINF(1), NVARS)
           IERR = 2
           ERRMSG = 'TYR501: Incorrect Dimension on VARINF Array'
         ENDIF
         ENDIF
      ELSE
         CALL ZERO(VARINF(1), NVARS)
      ENDIF

C  Error Handling

      IF(IERR .EQ. 0) THEN
        TYRMOD = 'TYR501 -> Harty Tyre Model '
      ELSE
        TYRMOD = ERRMSG(1:256)
      ENDIF
C
      RETURN


      END
```

RPF501

```
      SUBROUTINE RPF501( NCHTDS, CHTDST, TIR_ID, NTYPAR, TYPARR )

c  Copyright (C) 2000-1999
c  By Mechanical Dynamics, Inc. Ann Arbor, Michigan
c
c  All Rights Reserved, This code may not be copied or
c  reproduced in any form, in part or in whole, without the
c  explicit written permission of Mechanical Dynamics, Inc.
c

c  DESCRIPTION:
c
c  Reads property file for user tire model based on
c  the Fiala Tire model and initializes the
c  tire parameter array (TYPARR).
```

```
c
c ARGUMENT LIST:
c
c     name    type  storage  use  description
c     ======  ====  =======  ===  ================================
c     NCHTDS  I.S.    -        R   Number of characters in tire
c                                  property file name.
c     CHTDST  C.A.             R   Tire property file name
c     TIR_ID  I.S.    1        R   Tire GFORCE id
c     NTYPAR  I.S.    1        R   Dimension of TYPARR
c     TYPARR  D.A.  NTYPAR     E   Tire parameter array
c
c  *** Legend:  I integer          S scalar   R referenced
c              D double precision  A array    E evaluated
c              C character


C Inputs:
        INTEGER        NCHTDS
        CHARACTER*(*)  CHTDST
        INTEGER        TIR_ID
        INTEGER        NTYPAR

C  Outputs:

        DOUBLE RECISION TYPARR( NTYPAR )

C  Locals:

C  Units conversions:

        CHARACTER*(12)  UNITS(5)
        DOUBLE PRECISION CV2MDL(5)
        DOUBLE PRECISION CV2SI(5)
        DOUBLE PRECISION FCVSI
        DOUBLE PRECISION LCVSI
        DOUBLE PRECISION MCVSI
        DOUBLE PRECISION ACVSI
        DOUBLE PRECISION TCVSI


c  Fiala Property File Map

        INCLUDE 'tyrHarty_501.inc'

C  RTO variables:

        INTEGER          RETURN_VAL
        DOUBLE PRECISION TMPREAL
```

```
C  Shape Array RTO Stuff:

       DOUBLE PRECISION TMP1, TMP2
       INTEGER          N_NODES
       INTEGER          ARRPTR
       CHARACTER*80     FORM
       INTEGER FLEN
       CHARACTER*80     TABLE
       INTEGER TLEN

       LOGICAL          ERRFLG
       CHARACTER*80     MESSAG

c+----------------------------------------------------------------------*
c
c  Open the file:

       CALL RTO_OPEN_FILE_F2C ( CHTDST, NCHTDS, RETURN_VAL )

       ERRFLG = RETURN_VAL .EQ. 0
       MESSAG = 'Harty Tyre 501:  No Error opening tire property file.'
       CALL ERRMES ( ERRFLG, MESSAG, TIR_ID, 'STOP' )

c  Read [UNITS] block from property file:

c     Parameters in the property file may be given in any consistent
c     set of units.  The [UNITS] block identifies those units.
c     During evaluation, however, SI Units are used. So as parameters
c     are read from the property file they are converted
c     to SI units.

c     SI unit system.
c        LENGTH = meter
c        FORCE  = newton
c        ANGLE  = radians
c        MASS   = kg
c        TIME   = second

c     UNITS(1)-> FORCE  UNITS(2)-> MASS  UNITS(3)-> LENGTH
c     UNITS(4)-> TIME   UNITS(5)-> ANGLE

       CALL ATRTOU( TIR_ID, UNITS )
       CALL ACUNFN( UNITS, CV2MDL, CV2SI )

       FCVSI = CV2SI(1)
C   Force Conversion
       MCVSI = CV2SI(2)
C   Mass Conversion
       LCVSI = CV2SI(3)
```

```
C    Length Conversion
        TCVSI = CV2SI(4)
C    Time Conversion
        ACVSI = CV2SI(5)
C    Angle Conversion

C************************ TIRPRP POPULATION ************************

c  Read [MODEL] block:

        CALL RTO_READ_REAL_F2C
      .    (
      .       'MODEL', 5, 'USE_MODE', 8,
      .        TYPARR( USE_MODE ), RETURN_VAL
      .    )

        ERRFLG = RETURN_VAL .EQ. 0
        CALL ERRMES( ERRFLG ,
      .     'Harty Tyre 501:  No Use_mode?'
      .     ,TIR_ID,'STOP')


c  Read [DIMENSION] block:
        CALL RTO_READ_REAL_F2C
      . (
      . 'DIMENSION', 9, 'UNLOADED_RADIUS', 15,
      .    TMPREAL, RETURN_VAL
      . )

        ERRFLG = RETURN_VAL .EQ. 0
        CALL ERRMES( ERRFLG,
      . 'Harty Tyre 501:  No UNLOADED_RADIUS?'
      . ,TIR_ID,'STOP')

        TYPARR( UNLOADED_RADIUS ) = TMPREAL  *  LCVSI

        CALL RTO_READ_REAL_F2C
      . (
      . 'DIMENSION', 9, 'WIDTH', 5,
      .    TMPREAL, RETURN_VAL
      . )

        ERRFLG = RETURN_VAL .EQ. 0
        CALL ERRMES( ERRFLG, 'Harty Tyre 501:  No WIDTH?'
      .                 ,TIR_ID,'STOP')

        TYPARR( WIDTH ) = TMPREAL  *  LCVSI
```

```
c  Read [PARAMETER] block

      CALL RTO_READ_REAL_F2C
    . (
    . 'PARAMETER', 9, 'VERTICAL_STIFFNESS', 18,
    .   TMPREAL, RETURN_VAL
    . )

      ERRFLG = RETURN_VAL .EQ. 0
C-------------------------------------------------------------- ----
      CALL ERRMES( ERRFLG, 'Harty Tyre 501:  No VERTICAL_STIFFNESS?'
    .              ,TIR_ID,'STOP')

      TYPARR( VERTICAL_STIFFNESS ) = TMPREAL * (FCVSI / LCVSI)

      CALL RTO_READ_REAL_F2C
    . (
    . 'PARAMETER', 9, 'VERTICAL_DAMPING', 16,
    .   TMPREAL, RETURN_VAL
    . )

      ERRFLG = RETURN_VAL .EQ. 0
      CALL ERRMES( ERRFLG, 'Harty Tyre 501:  No VERTICAL_DAMPING?'
    .              ,TIR_ID,'STOP')

      TYPARR( VERTICAL_DAMPING ) = TMPREAL * (FCVSI * TCVSI / LCVSI)


      CALL RTO_READ_REAL_F2C
    . (
    . 'PARAMETER', 9, 'ROLLING_RESISTANCE', 18,
    .   TMPREAL, RETURN_VAL
    . )

      ERRFLG = RETURN_VAL .EQ. 0
      CALL ERRMES( ERRFLG,
    . 'Harty Tyre 501:  No ROLLING_RESISTANCE?',
    .   TIR_ID,'STOP')

      TYPARR( ROLLING_RESISTANCE )  = TMPREAL


      CALL RTO_READ_REAL_F2C
    . (
    . 'PARAMETER', 9, 'CMX', 3,
    .   TMPREAL, RETURN_VAL
    . )
```

```
        ERRFLG = RETURN_VAL .EQ. 0
        CALL ERRMES( ERRFLG,
      . 'rpf501: CMX undefined.',
      .   TIR_ID,'STOP')

        TYPARR( CMX ) = TMPREAL


        CALL RTO_READ_REAL_F2C
      . (
      . 'PARAMETER', 9, 'CGAMMA', 6,
      .   TMPREAL, RETURN_VAL
      . )

        ERRFLG = RETURN_VAL .EQ. 0
        CALL ERRMES( ERRFLG,
      . 'Harty Tyre 501:  No CGAMMA?',
      .   TIR_ID,'STOP' )

        TYPARR( CGAMMA ) = TMPREAL * (FCVSI / ACVSI)

        CALL RTO_READ_REAL_F2C
      . (
      . 'PARAMETER', 9, 'UMIN', 4,
      .   TMPREAL, RETURN_VAL
      . )

        ERRFLG = RETURN_VAL .EQ. 0
        CALL ERRMES( ERRFLG,
      . 'Harty Tyre 501:  No UMIN?',
      .   TIR_ID,'STOP')

        TYPARR( UMIN ) = TMPREAL


        CALL RTO_READ_REAL_F2C
      . (
      . 'PARAMETER', 9, 'UMAX', 4,
      .   TMPREAL, RETURN_VAL
      . )

        ERRFLG = RETURN_VAL .EQ. 0
        CALL ERRMES( ERRFLG,
      . 'Harty Tyre 501:  No UMAX?',
      .   TIR_ID,'STOP')

        TYPARR( UMAX ) = TMPREAL
```

```
      CALL RTO_READ_REAL_F2C
    . (
    . 'PARAMETER', 9, 'RELAXATION_LENGTH', 17,
    .   TMPREAL, RETURN_VAL
    . )

      ERRFLG = RETURN_VAL .EQ. 0
      CALL ERRMES( ERRFLG,
    . 'Harty Tyre 501:  No RELAXATION_LENGTH?',
    .   TIR_ID,'STOP')

      TYPARR( RELAXATION_LENGTH ) = ABS(TMPREAL * LCVSI)


      CALL RTO_READ_REAL_F2C
    . (
    . 'PARAMETER', 9, 'ALPHA_CRITICAL', 14,
    .   TMPREAL, RETURN_VAL
    . )

      ERRFLG = RETURN_VAL .EQ. 0
      CALL ERRMES( ERRFLG,
    . 'Harty Tyre 501:  No ALPHA_CRITICAL?',
    .   TIR_ID,'STOP')

      TYPARR( ALPHA_CRITICAL ) = ABS(TMPREAL * ACVSI)


      CALL RTO_READ_REAL_F2C
    . (
    . 'PARAMETER', 9, 'CURVATURE_FACTOR_ANGLE', 22,
    .   TMPREAL, RETURN_VAL
    . )

      ERRFLG = RETURN_VAL .EQ. 0
      CALL ERRMES( ERRFLG,
    . 'Harty Tyre 501:  No CURVATURE_FACTOR_ANGLE?',
    .   TIR_ID,'STOP')

      TYPARR( curvature_factor_angle ) = ABS(TMPREAL)


      CALL RTO_READ_REAL_F2C
    . (
    . 'PARAMETER', 9, 'SCALE_FACTOR_LATERAL', 20,
    .   TMPREAL, RETURN_VAL
    . )
```

```
      ERRFLG = RETURN_VAL .EQ. 0
      CALL ERRMES( ERRFLG,
    . 'Harty Tyre 501:  No SCALE_FACTOR_LATERAL?',
    .  TIR_ID,'STOP')

      TYPARR( SCALE_FACTOR_LATERAL ) = ABS(TMPREAL)


      CALL RTO_READ_REAL_F2C
    . (
    . 'PARAMETER', 9, 'RATED_LOAD', 10,
    .  TMPREAL, RETURN_VAL
    . )

      ERRFLG = RETURN_VAL .EQ. 0
      CALL ERRMES( ERRFLG,
    . 'Harty Tyre 501:  No RATED_LOAD?',
    .  TIR_ID,'STOP')

      TYPARR( rated_load ) = ABS(TMPREAL * MCVSI)


      CALL RTO_READ_REAL_F2C
    . (
    . 'PARAMETER', 9, 'SCALE_FACTOR_DIM', 16,
    .  TMPREAL, RETURN_VAL
    . )

      ERRFLG = RETURN_VAL .EQ. 0
      CALL ERRMES( ERRFLG,
    . 'Harty Tyre 501:  No SCALE_FACTOR_DIM?',
    .  TIR_ID,'STOP')

      TYPARR( scale_factor_dim ) = ABS(TMPREAL)


      CALL RTO_READ_REAL_F2C
    . (
    . 'PARAMETER', 9, 'SLIP_RATIO_CRITICAL', 19,
    .  TMPREAL, RETURN_VAL
    . )

      ERRFLG = RETURN_VAL .EQ. 0
      CALL ERRMES( ERRFLG,
    . 'Harty Tyre 501:  No SLIP_RATIO_CRITICAL?',
    .  TIR_ID,'STOP')
```

```
        TYPARR( slip_ratio_critical ) = ABS(TMPREAL)


        CALL RTO_READ_REAL_F2C
      . (
      . 'PARAMETER', 9, 'CURVATURE_FACTOR_RATIO', 22,
      .    TMPREAL, RETURN_VAL
      . )

        ERRFLG = RETURN_VAL .EQ. 0
        CALL ERRMES( ERRFLG,
      . 'Harty Tyre 501:  No CURVATURE_FACTOR_RATIO?',
      .    TIR_ID,'STOP')

        TYPARR( curvature_factor_ratio ) = ABS(TMPREAL)


        CALL RTO_READ_REAL_F2C
      . (
      . 'PARAMETER', 9, 'PNEUM_TRAILING_SCALING', 22,
      .    TMPREAL, RETURN_VAL
      . )

        ERRFLG = RETURN_VAL .EQ. 0
        CALL ERRMES( ERRFLG,
      . 'Harty Tyre 501:  No PNEUM_TRAILING_SCALING?',
      .    TIR_ID,'STOP')

        TYPARR( pneum_trailing_scaling ) = ABS(TMPREAL)


        CALL RTO_READ_REAL_F2C
      . (
      . 'PARAMETER', 9, 'PNEUMATIC_LEAD_CAMBER', 21,
      .    TMPREAL, RETURN_VAL
      . )

        ERRFLG = RETURN_VAL .EQ. 0
        CALL ERRMES( ERRFLG,
      . 'Harty Tyre 501:  No PNEUMATIC_LEAD_CAMBER?',
      .    TIR_ID,'STOP')

        TYPARR( pneumatic_lead_camber ) = ABS(TMPREAL * LCVSI)
```

```
      CALL RTO_READ_REAL_F2C
    . (
    . 'PARAMETER', 9, 'LIMIT_CAMBER_ONSET_FRIC', 23,
    .    TMPREAL, RETURN_VAL
    . )

      ERRFLG = RETURN_VAL .EQ. 0
      CALL ERRMES( ERRFLG,
    . 'Harty Tyre 501:  No LIMIT_CAMBER_ONSET_FRIC?',
    .   TIR_ID,'STOP')

      TYPARR( limit_camber_onset_fric ) = ABS(TMPREAL * ACVSI)

      n_nodes  = 0
      arrptr   = shape

C   READ [SHAPE] BLOCK IF IT EXISTS:

      CALL RTO_START_TABLE_READ_F2C
    . (
    . 'SHAPE', 5, FORM, FLEN, RETURN_VAL
    . )

      IF ( RETURN_VAL .EQ. 1 ) THEN

800      CONTINUE

         CALL RTO_READ_TABLE_LINE_F2C( TABLE, TLEN, RETURN_VAL )
         if ( return_val . eq. 1.and. tlen .gt. 3 ) then

           call act_line_parse (table, tmp1, tmp2, tlen)

           if ( n_nodes .lt. max_shape .and. tlen .eq. 2) then
               n_nodes  = n_nodes  + 1
               typarr( arrptr ) = tmp1
               typarr( arrptr + 1) = tmp2
               arrptr = arrptr + 2
            else

               if ( n_nodes .gt. max_shape) then
         CALL ERRMES( .true.,
       . 'Harty Tyre 501:  Shape table has more than 10 nodes',
       .   TIR_ID, 'STOP' )
               endif
```

```
                    if (tlen .ne. 2) then
                      CALL ERRMES( .true.,
                    . 'Harty Tyre 501: Error parsing line of SHAPE table',
                    .  TIR_ID, 'STOP' )
                    endif

              endif

          goto 800

        endif

        typarr( n_shape ) = n_nodes

    else

call usrmes( .true.,
    . 'Harty Tyre 501: No shape table. Cylinder will be used'
    . ,tir_id, 'WARN')

    endif

C  Close tire property file:

        CALL RTO_CLOSE_FILE_F2C ( CHTDST, NCHTDS, RETURN_VAL )

        ERRFLG = RETURN_VAL .EQ. 0
        MESSAG = 'exa_fiaini: Error closing tire property file.'
        CALL ERRMES( ERRFLG, MESSAG, TIR_ID, 'STOP' )

        RETURN
        END
```

Sample .TIR file

```
$-------------------------------------------------------------
------MDI_HEADER
[MDI_HEADER]
 FILE_TYPE    = 'tir'
 FILE_VERSION = 2.0
 FILE_FORMAT  = 'ASCII'
(COMMENTS)
{comment_string}
'Tyre      - Dunlop 100/90 19 D401'
'Pressure - Unknown'
'Test Date - Estimated DAH 2004'
```

```
'Harty Tire Model 2013'
'New File Format v2.1'
$------------------------------------------------------------------
------units
[UNITS]
LENGTH = 'mm'
FORCE  = 'newton'
ANGLE  = 'radians'
MASS   = 'kg'
TIME   = 'sec'
$------------------------------------------------------- ---model
[MODEL]
$                use mode   1   2
$                ---------------------------------
$                smoothing      X
$
  PROPERTY_FILE_FORMAT = 'USER'
  FUNCTION_NAME        = 'HTire501_2013::TYR501'
  USE_MODE             = 2.0
$---------------------------------------------- -----------dimension
  [DIMENSION]
  UNLOADED_RADIUS = 341
  WIDTH = 100.0
$--------------------------------------------- ------parameter
  [PARAMETER]
  VERTICAL_STIFFNESS = 146.0
  VERTICAL_DAMPING   = 0.2
  ROLLING_RESISTANCE = 0.02
  CMX   = 1.70
  CGAMMA   = 1.00
  UMIN     = 1.40
  UMAX     = 1.30
  RELAXATION_LENGTH = 100.0
  ALPHA_CRITICAL = 10.0
  CURVATURE_FACTOR_ANGLE = 2.70
  SCALE_FACTOR_LATERAL = 1.6417
  RATED_LOAD = 1662
  SCALE_FACTOR_DIM = -1.0E-4
  SLIP_RATIO_CRITICAL = 20.0
  CURVATURE_FACTOR_RATIO = 5.5
  PNEUM_TRAILING_SCALING = 1.25
  PNEUMATIC_LEAD_CAMBER = 20.0
  LIMIT_CAMBER_ONSET_FRIC = 0.80
 $
```

Sample build file

```
@echo off
rem for Intel Fortran 2013 and ADAMS 2013.1

dir /b *.f > build.lst

call \msc.software\adams_x64\2013_1\common\mdi  cr-us  n  @build.lst
HTire501_2013_for_Adams2013_1.dll

del build.lst
copy HTire501_2013_for_Adams2013_1.dll
"C:\MSC.Software\Adams_x64\2013_1\win64\HTire501_2013.dll"
```

Glossary of Terms

Agility

Ground vehicles are free roaming devices (with the exception of railed vehicles) in which the driver/rider seeks to influence path curvature and speed in order to follow an arbitrary course. The ease, speed and accuracy with which the operator can alter path curvature is agility. For motorcycles, the primary dictator of path curvature is roll angle, and therefore motorcycle agility discussions focus on roll behaviour; the handlebar is a roll acceleration demand. For finite track vehicles, path curvature arises from yaw rate; the handwheel is a yaw rate demand. Measures such as maximum speed through a given slalom crudely discern agility but not in a way that provides information on how to improve the design if more agility is desired.

Anti-aliasing

Anti-aliasing is analogue filtering applied to a signal before it is digitally sampled. Digital data should not be collected without an appropriate anti-aliasing filter. Aliasing is when something happens between subsequent samples that is simply 'missed' by the sampling process.

Consider Figure C.1; a 5 Hz process has been sampled at 4.762 Hz. Nyquist's theory predicts that spectral content is wrapped around according to multiples of half the sampling frequency. For the above example, the wrapping occurs at $5 - (2*4.762/2) = 0.238$ Hz. Once data have been allowed to alias, they cannot be 'unaliased'.

Anti-lift

Anti-lift is a geometric property of the suspension which means the reaction of traction-induced pitch moment is reacted entirely by tension in the mechanical suspension members (100% anti-lift), entirely by the roadspring(s) (0% anti-lift) or some combination of the two.

Anti-lift of less than 0% is described as 'pro-lift'. The concept is applicable only to a driven front wheel/axle and is thus generally inapplicable to motorcycles. Anti-lift is only applicable to a driven front wheel/axle. For a rear wheel/axle, the appropriate measure is anti-squat (q.v.).

For the behaviour of rear suspension geometry under braking this behaviour is sometimes referred to as 'anti-pitch'.

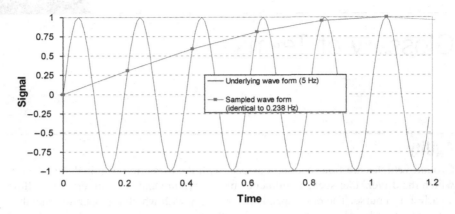

FIGURE C.1

Aliasing, which can only be prevented with an anti-aliasing filter.

Anti-pitch

Anti-pitch is a geometric property of the suspension which means the reaction of brake-induced pitch moment is reacted entirely by compression in the mechanical suspension members (100% anti-pitch), entirely by the roadspring(s) (0% anti-pitch) or some combination of the two. Anti-pitch of less than 0% is described as 'pro-pitch'.

Road motorcycles and comfort-oriented passenger cars have a typical inclination of the front suspension elements that provides pro-pitch behaviour. Rear suspension elements are often inclined to provide some anti-pitch under braking, but the effectiveness of this reduces with rear brake apportioning — little brake effort is apportioned to the rear of motorcycles braking hard and so it is generally ineffective at preventing pitch.

The term 'anti-dive' is not preferred since it has been used as a proprietary description for systems that modify front suspension damper calibration in order to resist brake-induced pitch in motorcycles.

For finite track vehicles, excessive anti-pitch geometry degrades refinement by forcing the wheel into any obstacles it encounters instead of allowing it to retreat from them. Brake-induced pitch behaviour is substantially modified by the use of inboard brakes.

Anti-roll

Anti-roll is a geometric property of the suspension that means the reaction of roll moment is reacted entirely by compression in the mechanical suspension members

(100% anti-roll), entirely by the suspension calibration (0% anti-roll) or some combination of the two. Less than 0% anti-roll is described as 'pro-roll'. This concept is meaningless for motorcycles.

One effect of anti-roll geometry is to speed the transfer of load into the tyre for a given roll moment. By modifying this load transfer differentially from front to rear strong changes in character can be wrought, though they rarely result in fundamental changes in vehicle behaviour. (This is not to say they are of no import; race performance hinges on 'character' since it leads towards or away from a confident driver.)

Different anti-roll geometry from front to rear also acts to provide a yaw/roll coupling mechanism. For typical saloons that coupling is such that roll out of a turn produces a yaw moment out of the turn. This is given by more anti-roll at the rear than at the front axle. Vehicles with less anti-roll at the rear have a distinctive impression of sitting 'down and out' at the rear when driven aggressively; it rarely results in confidence.

Anti-squat

Anti-squat is a geometric property of the suspension where the traction-induced pitch moment is reacted entirely by compression in the mechanical suspension members (100% anti-squat), entirely by the roadspring(s) (0% anti-squat) or some combination of the two. Less than 0% anti-squat is described as 'pro-squat'.

The calculation of anti-squat on a chain-driven motorcycle is nontrivial and must include chain tension reaction components in the free-body diagram for the swingarm. Anti-squat is only applicable to a driven rear wheel/axle. For a front wheel/axle, the appropriate measure is anti-lift (q.v.).

Anti-squat is not a preferred term for the behaviour of front suspension geometry under braking; the preferred term for this behaviour is 'anti-pitch'. The term 'anti-dive' is not preferred since it has been used as a proprietary description for systems that modify front suspension damper calibration in order to resist brake-induced pitch.

Articulated

An articulated vehicle is one in which two significant bodies are present and must pivot with respect to one another in order for the vehicle to follow a curved path. Motorcycles and tractor—trailer combinations are articulated.

Beta dot

See body slip rate.

Body slip angle

Body slip angle is the angle measured in the ground plane between the heading angle and the path of the vehicle. Heading may be thought of as what would be read by an onboard compass calibrated to the frame of reference in use. Path is the vector extending in the direction of travel at the current instant. Note that for all vehicles using pneumatic tyres, some body slip angle is inevitable to allow the rear tyres to generate a slip angle (q.v.) when travelling in a curved path.

Body slip rate

Body slip rate is the first time derivative of body slip angle (q.v.). Body slip rate is an angular velocity associated solely with creating or removing body slip angle. The total vehicle yaw rate (q.v.) is made up of the no-slip yaw rate (q.v.) — that part necessary for travelling in a curved path — plus the body slip rate. Body slip rate is zero in the steady state; the relationship between body slip rate and handwheel rate is crucial in modifying driver confidence. Unfortunately, for vehicles other than road cars on high friction surfaces it is difficult to discern experimentally since it relies on 'the small difference of two large numbers' and is thus prone to sensor noise.

Bump

Bump is a term used specifically to describe a motion of the suspension arrangement in which the wheel travels closer to the body. Its opposite is rebound (q.v.). Bump is also known as jounce.

Camber

Camber is the angle between a vertical line passing through the wheel centre and the lateral projection of that line onto the wheel plane (the projection being in the lateral direction). Strictly, it is absolute relative to the ground. A totally upright wheel has zero camber. A wheel lying flat on the ground has $90°$ camber.

Confusion can occur with finite track vehicles as to whether camber is measured with reference to the body coordinate system or a coordinate system on the ground. To ease this, the angle relative to ground can be referred to as the 'inclination angle' and that relative to the body as camber angle. A clear and unambiguous statement of the reference frame as often as is necessary is recommended to preclude ambiguity.

Castor

Castor is often referred to as rake (q.v.) for motorcycles. For finite track vehicles, it is the angle that the steering axis makes with the vertical when projected onto the plane of symmetry of the vehicle. For motorcycles, care needs to be taken that the angle of the fork legs, if fitted, may not be the same as the angle of the steering axis. In these conditions, 'rake' may be used for the fork legs and 'castor' for the steering axis.

Among the car community, there is frequent surprise at the notion that the steering axis need not pass through the wheel centre but may be offset from it when viewed the side; thus it is possible to have trail (q.v.) but no castor, or vice versa.

Some sugars are formulated to be thrown ('cast') from a device specially made for the purpose (a 'caster'). They are not connected to steering systems in any way.

Centre of percussion

See inertial conjugate.

Centripetal force

Centripetal forces are the unbalanced forces that result in acceleration of a body towards the centre of its path curvature. Without them there is no curvature of path. A body travelling in a curved path is not in equilibrium.

Cepstrum

Cepstrum is a contrived word. It is an anagram of spectrum and is used to describe the Fourier transform of something already expressed in the frequency domain. The technique is used widely in sonar to identify patterns of frequency content such as that generated by rotating machinery. It is slowly creeping into industrial usage and may yet become a tool in the ground vehicle industry.

Coherence

Coherence is an unambiguously defined calculated quantity that describes the consistency of phase relationship between two spectral estimates. Whenever a cross-spectrum or transfer function is calculated, coherence should be calculated and used to assess the credibility of the results. Cross-spectra with a coherence of less than 0.9 should be regarded cautiously and those below about 0.8 should be rejected. Note that these judgements are made on a spectral-line-by-spectral-line basis and it is acceptable for data at 5Hz to be believable while that at 7 Hz is discarded.

Complex numbers

Complex numbers are numbers containing real and imaginary components. The imaginary component of a complex number is a real number multiplied by the square root of -1. 'Complex' thus has a specific meaning and should be avoided to represent 'complicated' or 'elaborate' in general discussion.

Eigenvalues (q.v.) calculated using complex numbers are 'complex eigenvalues'; those calculated without are simply 'real eigenvalues', as calculated in many finite element (q.v.) packages. Undamped eigenvalues for oscillatory systems as calculated in MBS systems are often purely imaginary.

Computational fluid dynamics

Computational fluid dynamics is a tool for addressing fluid dynamic problems. Closed form (classical) fluid dynamics solutions quickly become extremely cumbersome for any but the simplest problems and boundary layer complexity render closed form fluid dynamics forms inapplicable for any object of interest except perhaps for artillery shells and the like.

In a manner similar to finite element analysis (q.v.), computational fluid dynamics uses a large number of finite volumes, known as cells, for which closed form solutions are known. The equations are coupled by imposing conditions of compatibility between adjacent cells and solving the resulting problem numerically. The solution is a time domain integration across the whole problem region, which is extremely time-consuming for even quite simple bodies.

Contact patch (tyre)

The contact patch for a pneumatic tyre is that part of the tyre in contact with the ground.

Couple

A couple is one or more forces acting at a distance from some point of interest such that a moment is exerted at that point.

In rigorous usage it is distinct from a torque, which is sometimes referred to as a 'pure moment' to differentiate it from a couple. Pure moments are unusual; couples are much more common.

This distinction is of little importance in day-to-day use, and unless the distinction between couples and torques is germane to the discussion it is best left unmade; it usually adds more confusion than it avoids. 'Torque' is commonly used for both, and 'pure torque' as a distinction is acceptable.

Damper

A damper is a device that produces a force opposing a motion applied to it. Gyroscopic torques are not dampers although they are velocity-dependent, since they do not act to oppose a motion applied but instead act at 90° to that motion.

The term damper is frequently applied to mean a hydraulic device to the exclusion of other devices, but this is merely habit rather than preference. Dampers are not shock absorbers — they transmit shocks, not absorb them. The spring and damper assembly together could be described as a shock absorber. The term shock absorber, or 'shock' for short, is common in US usage and is often used to mean only the damper; a combined spring and damper unit is sometimes referred to as a 'coilover' in US usage. It is wise to check exactly which components are being discussed as often as necessary.

A dampener is something that adds moisture and has nothing to do with dynamics.

Dynamics

The Greek word 'δύναμις' (pronounced dunn-a-miss) means energy and is where we get our word dynamic from. In the sense frequently used today by mechanical engineers it is used to signify a time-varying exchange of energy between kinetic (motion), strain (stretching) and/or potential (height) states. Similar phenomena exist with electrical circuits and are crucial to the operation of radios and other elaborate equipment.

When this exchange occurs easily the system is said to be at resonance. It happens at characteristic frequencies (speeds of a repeated motion) for any structure. If an event happens at a speed substantially similar to or greater than the speed (frequency) of the slowest resonance, it will be dynamic or fast. If an event happens at less than that speed, it will be static or slow. For a full system, such as a motorcycle, the same is true. The correct understanding of the dynamic behaviour of vehicle systems at the design stage is vital to avoid costly mistakes being carried forward to the prototype stage or, worse, to production.

Dynamic absorber

A dynamic absorber is an additional spring-mass-damper system added to a mechanical system to take a single resonance and modify it by forming two resonances, each of which is more damped than the single resonance in the unmodified system. One of the two resonances is at a lower frequency than the original and the other is at a higher frequency.

Mass-damper systems can also be added to dissipate energy uniformly across a wide frequency range, and these devices are also sometimes referred to as 'dynamic absorbers'. Finally, discrete lumped masses can be firmly attached to change resonant frequencies in order to decouple an excitation from a structural resonance. Such a device is frequently and confusingly referred to as a 'mass damper' although it would be less ambiguous to call it an inertial attenuator.

Eigensolution, eigenvalues, eigenvectors

There is a clear mathematical description of characteristic solutions ('eigen' is German for 'characteristic') of matrix problems, but it is unhelpful for this glossary.

Eigensolutions, which consist of eigenvalues and eigenvectors, are quite simply free vibration solutions — that is to say resonances, or 'natural solutions' for the system described by the matrix problem. If the system is heavily damped (more than around 5% damping ratio) then complex numbers are necessary for useful eigensolutions of the system. If it is lightly damped then only real numbers are necessary.

Eigensolutions are also sometimes referred to as 'modal solutions', 'normal modes' and a variety of other titles. Thus a 'real normal modes' solution is one that solves the eigenvalue problem for a lightly damped system, using only real numbers. Note that it is a property of modes of vibration calculated in this manner, that they are normal (orthogonal) to one another; they produce a dot product of zero if the vectors are multiplied together. The concept of orthogonality requires at least two items for it to be meaningful.

Expected and unexpected response

Expected response is not the subjective vagary it might be supposed. Expected response is defined clearly for dynamics usage as the product of the system inputs and the idealised characteristics of the system. For example, in yaw for a finite track vehicle the expected response is a yaw rate, the value of which is the forward speed multiplied by the steer angle and divided by the wheelbase. This might also be called 'idealised yaw rate'. 'Idealised' does not imply 'ideal' in the sense of most optimum.

The reason for wishing to capture the expected response is to compare it with the actual response. The difference between actual and expected is then logically termed 'unexpected response' and is important in quantifying operator interpretation of the vehicle. One vehicle might display a yaw response that differs from another by only a few percent; for example, a comparison of the unexpected components of response might show that one has double the unexpected response of the other and explain the strong preference of operators for one over the other even though the objective over-all response is very similar.

If the terms 'expected' and 'unexpected' cause consternation then the control theory terms 'reference' (demanded) and 'actual' could be substituted.

Finite element method

Simple engineering structures can be represented using a closed-form equation derived using differential calculus. The beam equation is the most widely used of these forms, being familiar to most engineers. These might be thought of as infinitesimal (infinitely small) element solutions, being derived by considering an infinitesimal slice then summing (integrating) the results.

The derivation of closed form solutions for more complex structures subject to complex loading patterns rapidly becomes cumbersome and impractical. The finite element method uses small but not infinitesimal chunks of the structure, each of which has a closed form solution, and assembles them imposing conditions of force and displacement continuity at the boundaries between elements. The resulting set of simultaneous equations is converted to matrix form and is well suited to being solved using a digital computer.

The finite element method was invented in principle during the latter stages of World War II and immediately thereafter, but did not achieve widespread use until the Apollo programme in the 1960s in America. In the last 30 years it has become more and more popular, with FE tools now available for use with home computers.

Forced response

Forced response is the response of a system under some external excitation. Excitation is usually time-varying; if it is not then the problem is a static one.

If the input excitation has been established for some time then the behaviour of the system will have achieved 'steady state' (q.v.); the response will be stationary (q.v.) in character. Such solutions can be calculated using the 'harmonic forced response' method. A transfer function is calculated based on the free vibration (natural solution) response of the system then multiplied with the frequency spectrum of the input excitation to provide a response spectrum. Refinement (q.v.) problems are frequently addressed using the harmonic forced response method.

If the input excitation has not been established for some time, then a solution method is required that can capture the developing phase relationships in the system. This is a 'transient forced response'. There is typically a computational resource penalty of an order of magnitude when switching to transient solution methods from harmonic ones.

Gain

Gain is often used when discussing control systems to describe the relationship between output(s) and input(s). A system with a higher gain produces larger outputs for a given input than a system with a lower gain. Gains may or may not be linear and often vary with input frequency.

In vehicle dynamics, quantities like yaw rate and lateral acceleration are regarded as the output when compared to handwheel angle as an input. For a motorcycle, roll acceleration is the output and steer torque is the input.

Occasionally gain is used when talking about the rate of increase of a quantity, for example when discussing the development of roll angle during a transient manoeuvre. Such confusing usage is unhelpful when clearer terms like "increase" and "rate" are available.

Gyroscope, gyroscopic torques

Much mystique surrounds gyroscopes and gyroscopic torques. Gyroscopes do not produce forces, only torques (moments). Gyroscopic torques are a logical consequence of Newton's third law in its correct and full form. Sometimes expressed succinctly as 'applied force is equal to rate of change of momentum', if a full 6x6 formulation is pursued then differentiating the product of the inertia tensor and velocity vector yields some off-diagonal terms if the inertia tensor varies with time, as it does with a rotating body that is not spherical.

Handwheel

Handwheel is the preferred term for what is popularly called the steering wheel. The reason for this distinction is that a typical car has three steering wheels — two are road wheels and the other is the handwheel. Using handwheel avoids ambiguity although it may seem a little cumbersome.

Harmonic

Harmonic in dynamic terms means 'consisting of one or more sine waves'.

When the phrase 'assume harmonic solutions' is used, it means 'assume the solution consists of one or more sine waves'. There is no implication of fixed frequency relationships; the term harmonic is sometimes confused in careless usage with 'harmonies' as used in music.

Heave

Heave is one of three motions performed by the whole vehicle on its suspension, referred to collectively as 'primary ride'. It is a motion whereby the whole vehicle rises and falls evenly, with no rotation about any axis. The other two ride motions are pitch and roll. For motorcycles the roll motion is not a primary ride motion but the fundamental degree of freedom for the vehicle. Heave is also known as bounce, or jounce.

In reality, ride motions are never pure heave or pitch but always some combination of the two. This fact is a frequent source of confusion between development staff and analysis staff; the two groups use the terms differently. The problem is even more acute when including roll, with confusion around the notion of the 'roll centre' as a motion centre for the vehicle.

Inertial conjugate (centre of percussion)

Inertial conjugate is the preferred term for the location at which no translation occurs when a free body with finite mass and inertia properties is loaded in a direction that does not pass through its centre of gravity. No translation occurs at this point during load application, whether or not the load is percussive (an impact). This phenomenon gives rise to the behaviour of 'rigid' bats in some sports, particularly baseball and the like, that is described as the 'sweet spot'. This is not to be confused with the use of the same term for strung rackets.

The term 'centre of percussion' is not preferred to describe this location since it implies a percussive loading must be present for the concept to be useful.

Jounce

Jounce is another term for heave. It is of US origin.

Kinematics

Kinematics is the study of motion. In a mechanism it is the study of the motion of the individual motions of the components and how they relate to and constrain each other.

Modes, modal analysis

Modes is shorthand for 'modes of vibration'. See 'eigensolution' for a description. 'Modal' is an adjective meaning 'of or relating to a mode or modes'. Modal analysis is thus a fairly loose term; its preferred use is as part of the expression, Experimental Modal Analysis, or better still Modal Test, in which experimental methods are used to discern the modes of vibration of a structure. Analysis to predictively calculate modal behaviour is distinguished by the expression 'modal solution' (solution being a general word for the submission, retrieval and interpretation of a set of results to a problem using a digital computer).

Multibody system analysis, multibody codes

Mechanical systems can be viewed as a connection of separate items, or bodies, connected by various means. Such a system, comprised of multiple bodies, can be analysed by the application of Newtonian or Lagrangian methods to formulate the equations of motion, which may then be interrogated in a variety of ways — integrated through time, solved for an eigensolution and so on. This is multibody system analysis.

A group of software packages, or codes as they are informally referred to, have emerged that greatly ease the task of formulating and solving the equations of motion. The best known is called MSC ADAMS — Automated Dynamic Analysis of Mechanical Systems — developed by MSC Software.

Until recently, an implicit assumption in this type of analysis has been that the elements comprising the system are rigid, but this limitation is being removed by the elegant integration of multibody methods with structural dynamics methods.

No-slip yaw rate

No-slip yaw rate is the yaw rate required to support a vehicle travelling in a curved path, allowing it to change heading in order to have the correct orientation when it leaves the corner.

If the vehicle did not yaw when travelling in, say, a 90° corner then it would be travelling sideways upon exiting the corner. If it is to be travelling forwards (in a vehicle-centred sense — i.e. as noted by the operator) when it leaves the corner then it must have rotated in plan during the corner.

Consideration of basic physics leads to the observation that no-slip yaw rate is centripetal acceleration divided by forward velocity.

Non-holonomic constraints

'Non-holonomic constraints involve nonintegrable relationships between velocities. In vehicle dynamics they arise typically if wheels are assumed to roll without slip in problems of more than one dimension. Suppose a car is parked in an open, flat, high-friction area and radial line marks are appended to the tyre sidewalls and to the points on the ground nearest to them. The car is then driven slowly, without tyre slip, round the area and eventually returned to the precise location where it started. Although the car body can be repositioned precisely the tyre marks will not, in general, align now'.

Sharp, R.S. Multibody Dynamics Applications in Vehicle Engineering (1998).

NVH

See Refinement.

Objective

Objective is unfortunately ambiguous. In one sense (as a noun) it is similar to 'target' but that is not its preferred usage. The preferred usage is in contrast with 'subjective' (q.v.) and is as an adjective. It refers to measurements or conclusions that are independent of the person who observes. The term arises from basic English sentence construction which is 'subject-verb-object'. If A observes B then B is the object of A's observation; B is always B whether A or C observes.

Objectivity is essential to dynamic activities; without it work cannot be credible, reproducible or professional. These three are prerequisite for any scientific activity.

Operating Shape

When harmonically (q.v.) excited, a system will respond with a characteristic motion that varies over time and for lightly damped systems may often be represented as a scaled sine wave for each point in the system with a fixed phase relationship.

Examining the system at the maxima of these individual sine waves gives a characteristic shape at which the system operates. When operating at resonance, the operating shape strongly resembles the mode shape or eigenvector (q.v.) associated with that resonance.

For heavily damped systems the scaling is usually complex (q.v.), meaning that operating shapes cannot readily be assimilated without animation.

Oversteer, understeer

Oversteer is strictly the condition in which the slip angle of the rear tyres exceeds that of the fronts. Understeer is strictly the reverse.

Before debating this matter, a review of Segel and Milliken's papers from 1956 are in order, where all this was laid down as fact with test work and mathematical development. The definitions arrived at in there are applicable in the linear region only. The persistence of their usage for limit behaviour is an extension beyond their validity. Sometimes, drivers may refer to oversteer and mean a yaw overshoot after turn-in, which is somewhat confusingly a consequence of an understeering car. Equally, a car that has genuine oversteer behaviour will have non-oscillatory roots to its characteristic equation, resulting in excessive yaw damping that manifests itself as a reluctance to turn in and is reported as transient understeer.

Some modern cars have steady state oversteer behaviour on low grip surfaces since it dulls response and makes the vehicle manageable, even though it guarantees the need for driver correction.

Preferred alternative forms of reference are such subjective terms as 'push on' or 'loose rear' (that is loose, not lose) or more technical descriptions, such as 'real roots for the characteristic equation', 'rear slip angle exceeds front' and so on.

The importance of both oversteer and understeer as steady state phenomena in road vehicles in the linear region is greatly exaggerated. Its prime effect in the steady state is to encourage the driver to adjust the handwheel slightly, since it results in a path error due to the difference between demanded [idealised, expected (q.v.)] and no-slip yaw rates (q.v.). In normal road driving this is completely trivial; fitting a 'faster' steering rack will mask the perception of path error related to understeer completely.

For transient handling, the development of body slip angle (q.v.) is of much more importance; the yaw rate associated with changes in body slip angle is acutely sensed by the driver and is used to give warning of an impending spin even though it is only a small fraction of the total yaw rate. If the vehicle manages its body slip rate poorly, then it can give the impression of an impending spin even if no such event was likely. Even at low lateral accelerations, body slip rate is a strong modifier on perceived transient performance.

Path error

Path error is a preferred term for steady state understeer. When the vehicle 'runs wide' from the driver's intended path, the normal driver response is to add more steer angle. If the lateral acceleration is high and the path error is large, then more steer angle may not help reduce path error. This situation is reported by many drivers as 'pushing on' and is common limit handling behaviour for safe road vehicles.

Path error is strictly the difference between idealised [demanded, expected (q.v.)] yaw rate and no-slip yaw rate (q.v.). It is a numerical quantity suitable for discernment from logged data or mathematical models.

Pitch

Pitch is one of three motions performed by the whole vehicle on its suspension, referred to collectively as 'primary ride'. It is a motion whereby the front of the vehicle rises and falls in opposition to the rear of vehicle, rotating about the lateral axis of the vehicle. The other two ride motions are heave and roll. For powered two wheelers the roll motion is not a primary ride motion but the fundamental degree of freedom for the vehicle.

Sometimes compound heave (q.v.) and pitch motions are referred to as 'front end heave' or 'rear end heave' in a verbal effort to describe the operating shape (q.v.).

Predictive methods

Predictive methods is an umbrella term for all forms of mathematical modelling, from statistical to explicit, from numerical to algebraic.

'The only relevant test of the validity of a hypothesis is comparison of prediction with experience' (Milton Friedman)

PTW

PTW is an acronym for 'powered two wheeler' — a generic term encompassing mopeds, scooters, motorcycles, enclosed motorcycles, feet firsts (FFs) and any other thing which has two wheels and is self-powered.

Rake

Rake is the angle between the steering axis and the vertical for an upright motorcycle. Typical values for rake vary from 23° for a sports machine and over 30° for a cruiser. Rake is also known as castor (q.v.), and is occasionally quoted as an angle from the ground plane. Some designs of motorcycles have a different angle for the steering axis and the telescopic fork legs; in this case rake is sometimes used for the fork leg angle and castor reserved for the steering stem angle.

Rate

Rate means, clearly and unambiguously, the first time derivative. It is shorthand for 'rate of change with time'. Using rate for derivatives other than time is not preferred. In particular, using rate as in 'spring rate' is undesirable; 'stiffness' is preferred in this instance, though the authors make this error frequently.

Rebound

Rebound is a term used specifically to describe a motion of the suspension arrangement in which the wheel travels away from the body. Its opposite is bump (q.v.).

Refinement

Refinement is a general term that refers to the ability of a vehicle to isolate its operator and other occupants from external disturbances and from disturbances generated by the vehicle itself — for example engine vibration.

Refinement is frequently referred to as 'NVH'. This stands for 'Noise, Vibration and Harshness'. The distinction between the three phenomena is not well defined. It is suggested that noise refers to audible phenomena from 30 Hz upward, vibration to dynamic phenomena below about 10 Hz and harshness to tactile sensations of vibration of intermediate frequency, for example, steering column shake at idle.

Segment

'Segment' is industry shorthand for 'market segment' — the term used to define the objectives for the vehicle in terms of who will buy how many copies of it.

Shock absorber

See Damper.

Slip, slip angle (of tyres)

Slip is perhaps the poorest word possible to describe the behaviour of a tyre under lateral loading. A rolling tyre will not support a lateral load due to the repeated relaxation of the sidewalls just behind the contact patch. This action causes the tyre to 'walk' sideways under an applied lateral load. Combined with the forward motion, this results in an angle between the wheel plane and the direction of motion. That angle is dubbed the 'slip angle'. The presence of a slip angle does not imply sliding friction at the contact patch, despite the linguistic link between slip and slide. This is a frequent source of confusion for newcomers to the field.

To further add confusion, the behaviour of a tyre under a tractive or braking torque is also referred to as slip, and is often quantified in percentage. A free-rolling wheel is at zero slip, a completely locked wheel is at 100% brake slip. 100% tractive slip is when the wheel is spinning twice as fast as the free-rolling wheel. The presence of tractive/braking slip does not imply sliding friction at the contact patch, though this is the case at high slip percentages (greater than 20%), as it is at high slip angles (greater than about 6°).

Stability

Stability means the response to a disturbance is bounded (non-infinite). In practical terms for a motorcycle this would mean it would not fall over if nudged. This is clearly not the case when the machine is at rest but is generally true of the motorcycle/rider combination at all speeds above zero. For running vehicles, the notion of stability is generally associated with the control of body slip angle (q.v.) and

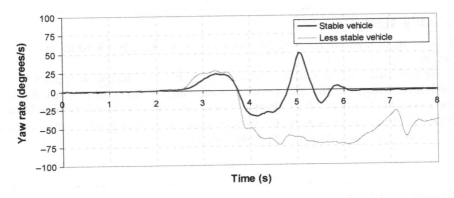

FIGURE C.2

Time Domain Comparison of Stable and Less Stable Vehicle during 120 kph Entry Chicane Manoeuvre.

hence implicitly the yaw rate behaviour of the vehicle. In practice a response that is simply 'finite' is insufficient to guarantee stability in any useful sense; vehicles are not simply 'stable' or 'instable' but have degrees of stability that are often related to the nature of the nonlinearity of the system. For example, the front wheel wobble (q.v.) problem on a motorcycle is unstable on-centre but has a 'bounded instability' in the sense that the increasingly nonlinear behaviour of the tyres dissipates more and more energy as the amplitude of the cycle rises. This has the effect of 'containing' the instability in the system within bounded limits.

Figure C.2 shows a comparison between a vehicle that has a passively stable response and another with a less stable response during a chicane manoeuvre. Both vehicles have comparable mass, wheelbase, inertia properties and tyre fitments. The manoeuvre is completed off-throttle and it can be seen that the less stable vehicle produces a very aggressive yaw acceleration at the steer reversal (3.8 s) and a high yaw rate persists for some time, resulting in a half spin of the vehicle.

It can be seen for a well-controlled manoeuvre like this one, performed under identical conditions with the same driver, it would be easy to formulate some measure of stability based on area under the yaw rate curve. To compare vehicles on an absolute basis requires a well-controlled test that is tester-, venue- and weather-independent — a more difficult task but addressed in part by the ISO manoeuvres.

Stationary

In dynamics, stationary does not mean immobile. It is a property of a time-domain signal and is such that lengthening the observation period does not alter the fundamental characteristics (mean, RMS, Standard Deviation, etc.) of the signal.

Steady state

In general, steady state refers to the condition where phase relationships are fixed and no longer vary with time.

In the sense used in vehicle dynamics work, it is that definition applied to a vehicle cornering. The phase relationship between all motions of the vehicle is fixed; there is neither yaw nor roll acceleration that is anything other than established and harmonic. In the simplest sense the vehicle is travelling steadily on a curved path. Note that steady state does not imply equilibrium.

Steering offset

Steering offset is a fairly general term that has specific meanings when considering the steering system of a finite track vehicle. It is the lateral distance between the steer axis and the wheel plane. It can be seen that if the steer axis is inclined when viewed from the front of the vehicle (as is generally the case) then this offset has different values at different heights above ground.

Wherever it is measured, positive steering offset is when the wheel plane is outboard of the steer axis. Negative steering offset is the reverse.

Steering offset is of interest at ground level since it determines steering system loads (and hence handwheel torque) under differential longitudinal loads. Excessive ground level offset promotes nervousness under braking and sensitivity to ABS operation. However, it also assists the driver in retaining control under split mu braking conditions, with or without ABS. Negative ground level offset makes the handwheel and vehicle insensitive to ABS operation.

The steering offset at wheel centre height, usually called 'hub level offset' dominates the sensitivity of the vehicle to tractive effort imbalance — usually called 'torque steer'. It also strongly colours the amount of vibration passed through to the driver from the road surface under normal driving — so-called 'texture feel'. With a driven axle, steering offset is normally the outcome of drive shaft plunge calculations and not an independent design variable in its own right.

Subjective

Subjective is in contrast with 'objective' (q.v.) and is an adjective. It refers to measurements or conclusions that are dependent on the person who observes.

It would be unfair to suggest that subjectivity has no place, but excessive reliance on unreproducible observations is unhelpful, unproductive and unprofessional.

Symbolic codes (multibody system analysis)

'Conceptually, the simplest approach to solving…(a multi-body system problem) is to give the whole set of differential and algebraic equations to a powerful and appropriate solver… Simulations tend to be painfully slow…' 'Symbolic preparation of the problem for numerical solution implies that the symbol manipulations, which are performed only once ahead of numerical processing, can be devoted to making the (later repeated) numerical operations minimal. It also implies independence of the symbolic equation building and the numerical solution'.

Sharp, R.S. Multibody Dynamics Applications in Vehicle Engineering (1998). Simpack is one of several symbolic codes on the market at the time of writing.

Traction, tractive

Traction for ground vehicles is the process of deploying power via a torque at the road wheels in order to provide a forward force on the vehicle. Such force is sometimes referred to as 'tractive effort'. Multiplying that force by the vehicle velocity gives 'tractive power'.

Trail

Trail is the distance between the point at which the steering axis intercepts the ground plane and the front wheel contact patch centre. Positive trail implies the steering axis intercepts the ground ahead of the front wheel contact patch. Trail can be measured at two different points; the geometric centre of the contact patch and the centre of lateral load of the contact patch. The difference between the two measurements is often referred to as 'pneumatic trail'. The measurement to the geometric centre of the contact patch is then 'mechanical'.

With finite track vehicles, trail is an important contributor to the self-aligning behaviour of the steering system and handwheel efforts.

With motorcycles, negative trail machines existed into the 1920s but the destabilising effects with speed rendered them impractical for high-speed use. Negative trail occasionally appears on mountain bicycles today and does not guarantee instability of the machine/rider system. At low speeds, high roll angles and significant steer angles, trail is substantially modified by the wheel geometry.

Transient (cornering)

In general, transient refers to the condition where phase relationships are not fixed and are varying with time. In the sense used in this work, it is that definition applied to a vehicle cornering. Transiently, the phase relationship between all motions of the vehicle is developing and changing. Rates of change are generally decaying with time and the transient motion becomes the steady state motion if sufficient time is allowed. For the transient aspects of the motion to decay completely, something around 3 s is required for cars and motorcycles. This is rarely achieved in practice and so reality consists of a connected series of transient events.

Understeer

See Oversteer.
  See Expected Response.

Vehicle dynamics

For the purpose of this work, a vehicle is defined as a wheeled device capable of propelling itself over the ground. Dynamics, from the Greek 'dunamis (δύναμις)' meaning power or the capability to achieve something, is the study of time-varying phenomena. Most dynamic phenomena involve the shuffling or transforming of energy between one item in a system and another. The phrase 'vehicle dynamics' is generally used to refer to phenomena of interest to the rider or driver of the vehicle in guiding its path — roll, turn-in, and so on. It includes consideration of elements such as tyres and suspension dampers, and is very much a system level activity.

Vehicle programme

'Vehicle programme' is industry shorthand for 'vehicle design, development and sign-off programme'. It is that defined activity that takes as input a desire to have a vehicle in a certain market segment and produces as output a fully mature design, ready to be mass produced. The start of vehicle programmes is shrouded in obfuscation in order that each manufacturer can claim their vehicle programmes are shorter than others; the end of the vehicle programme is frequently referred to as 'job 1' or 'launch'.

Weave

Weave is one of the lowest fundamental modes of vibration of a motorcycle. It is a combination of roll, yaw and lateral motion that results in the machine following a sinuous path. Modern road machines rarely display weave behaviour poorly damped enough to hinder the rider's control of the machine but even current machines can display a low enough level of weave damping to diminish the rider's confidence, particularly when laden. The highest performance machines typically employ an aluminium beam frame concept and have excellent weave damping, to the point where weave mode cannot readily be discerned with a typical 'rider rock' manoeuvre (a hands-off hip flick) but must instead be found using a swept sine handlebar input.

Wheelbase

Wheelbase is the distance between the front and rear wheel centres of a wheeled vehicle.

Wheel hop

A description of the system resonance that consists of the unsprung mass (wheel, tyre, brake components, hub, bearings and some proportion of suspension members, springs and dampers) as the dominant kinetic energy storage and the tyre as the dominant strain energy storage, with the tyre providing the dominant energy dissipation route. Typical frequencies are usually around 12 Hz on a finite track vehicle, 15 Hz on a road motorcycle.

Wheel trajectory map

A collective name for the characteristics that describe the orientation and location of the wheel plane in relation to the vehicle body. The characteristics are all defined with respect to bump (q.v.) motion. In a finite track vehicle, the wheel trajectory map includes bump steer, bump camber, bump recession and bump track change. The interaction of each of these characteristics with the vehicle is subtle and complicated. For motorcycles it is generally only bump recession, also known as anti-pitch or anti-squat, that is of interest.

Wobble

Wobble is a faintly comical term used to describe a distinctly serious trait of motorcycle behaviour. Originally referred to as 'speedmans's wobble' this was shortened to just wobble by frequent usage. It describes the mode of vibration of a motorcycle in which the steered mass is in motion more-or-less independently of the non-steered mass and rider. It is typically around 8 Hz and generally well damped enough that it does not present a problem. However, when ill-damped it can present a terrifying experience whereby the amplitude of the motion set-off by some small disturbance grows until it is only restrained by the steered mass colliding with some part of the motorcycle at each extreme. It is also referred to as flutter, shimmy or headshake.

Yaw, yaw rate

Yaw is simply rotation when viewed in plan view. Yaw rate is the first time derivative and is often referred to as yaw velocity. Yaw is measured in degrees or radians and yaw rate in degrees/second or radians/second.

Standards for Proving Ground Tests

Vehicle handling simulations are intended to recreate the manoeuvres and tests that vehicle engineers carry out using prototype vehicles on the test track or proving ground. Some are defined by the International Standards Organization (ISO), which outlines recommended tests in order to substantiate the handling performance of a new vehicle:

ISO 3888-1:1999	Passenger cars – Test track for a severe lane-change manoeuvre – Part 1: Double lane-change
ISO 3888-2:2002	Passenger cars – Test track for a severe lane-change manoeuvre – Part 2: Obstacle avoidance
ISO 4138:1996	Passenger cars – Steady-state circular driving behaviour – Open-loop test procedure
ISO 7401:2003	Road vehicles – Lateral transient response test methods – Open-loop test methods
ISO 7975:1996	Passenger cars – Braking in a turn – Open-loop test procedure
ISO/TR 8725:1988	Road vehicles – Transient open-loop response test method with one period of sinusoidal input
ISO/TR 8726:1988	Road vehicles – Transient open-loop response test method with pseudo-random steering input
ISO 9815:2003	Road vehicles – Passenger-car and trailer combinations – Lateral stability test
ISO 9816:1993	Passenger cars – Power-off reactions of a vehicle in a turn – Open-loop test method
ISO 12021-1:1996	Road vehicles – Sensitivity to lateral wind – Part 1: Open-loop test method using wind generator input
ISO 13674-1:2003	Road vehicles – Test method for the quantification of on-centre handling – Part 1: Weave test
ISO 14512:1999	Passenger cars – Straight-ahead braking on surfaces with split coefficient of friction – Open-loop test procedure
ISO 15037-1:1998	Road vehicles – Vehicle dynamics test methods – Part 1: General conditions for passenger cars
ISO 15037-2:2002	Road vehicles – Vehicle dynamics test methods – Part 2: General conditions for heavy vehicles and buses
ISO 17288-1:2002	Passenger cars – Free-steer behaviour – Part 1: Steering-release open-loop test method
ISO/TS 20119:2002	Road vehicles – Test method for the quantification of on-centre handling – Determination of dispersion metrics for straight-line driving

References

Allen, R.W., Rosenthal, T.J., Szostak, H.T., 1987. Steady State and Transient Analysis of Ground Vehicle Handling. SAE Paper 870495, Society of Automotive Engineers, 400 Commonwealth Drive, Warrendale, PA 15096, USA.

Anderson, R.J., Hanna, D.M., 1989. Comparison of three vehicle simulation methodologies. In: Proceedings 11th IAVSD Symposium.

Austin, A., Hollars, M.G., November 1992. Concurrent Design and Analysis of Mechanisms. Kinematics and Dynamics of Multi-body Systems Seminar (S057). Institution of Mechanical Engineers.

Bakker E., Nyborg L., Pacejka, H.B., 1986. Tyre Modelling for Use in Vehicle Dynamics Studies. SAE Paper 870421, Society of Automotive Engineers, 400 Commonwealth Drive, Warrendale, PA 15096, USA.

Bakker, E., Pacejka, H.B., Lidner, L.A., 1989. New Tyre Model with Application in Vehicle Dynamics Studies. SAE Paper 890087, 4th Auto Technologies Conference, Monte Carlo.

Bartlett W., 2000. Driver Abilities in Closed Course Testing. SAE paper 2000-01-0179, Society of Automotive Engineers, 400 Commonwealth Drive, Warrendale, PA 15096, USA.

Bayle, P., Forissier, J.F., Lafon, S., 1993. A new tyre model for vehicle dynamic simulations. Automotive Technology International, pp. 193–198.

Beauchamp, G., Hessel, D., Rose, N.A., Fenton, S.J., Voitel, T., 2009. Determining Vehicle Steering and Braking from Yaw Mark Striations. SAE Paper 2009-01-0092. http://www.kineticorp.com/publications/2009-01-0092-yaw-striations-steering-braking.pdf.

Blinchikoff, Zverev, 2001. Digital Filtering in the Time and Frequency Domain. Noble.

Blundell, M.V., 1990. Full Vehicle Modelling and Simulation of a Rolls Royce. Internal consulting report to Rolls Royce Motor Cars Ltd, Tedas Ltd, Coventry.

Blundell, M.V., November 1991. Full Vehicle Modelling and Simulation Using the ADAMS Software System. Institution of Mechanical Engineers Paper C427/16/170, Autotech '91, Birmingham.

Blundell, M.V., 1999a. The modelling and simulation of vehicle handling part 1: analysis methods. Journal of Multi-body Dynamics, Proceedings of the Institution of Mechanical Engineers 213 (Pt K), 103–118.

Blundell, M.V., 1999b. The modelling and simulation of vehicle handling part 2: vehicle modelling. Journal of Multi-body Dynamics, Proceedings of the Institution of Mechanical Engineers 213 (Pt K), 119–133.

Blundell, M.V., 2000a. The modelling and simulation of vehicle handling part 3: tyre modelling. Journal of Multi-body Dynamics, Proceedings of the Institution of Mechanical Engineers 214 (Pt K), 1–32.

Blundell, M.V., 2000b. The modelling and simulation of vehicle handling part 4: handling simulation. Journal of Multi-body Dynamics, Proceedings of the Institution of Mechanical Engineers 214 (Pt K), 71–94.

Blundell, M.V., 2003. Computer modelling of tyre behaviour for vehicle dynamics. In: The 3rd European Conference for Tire Design and Manufacturing Technology, Hamburg, Germany.

Blundell, M.V., Harty, D., 2006. Intermediate tyre model for vehicle handling simulation. Proceedings of the Institution of Mechanical Engineers Part K: Journal of Multi-body Dynamics 221 (K1), 41–62. http://dx.doi.org/10.1243/14644193JMBD51.

Bosch, R., 1999. Driving-Safety Systems, second ed.

Box, G.E.P., Draper, N.R., 1987. Empirical Model-Building and Response Surfaces. Wiley, ISBN 0471810339, p. 424.

Brüel, Kjær, 2003. Sound & Vibration Measurement Experimental Modal Analysis (Modal Analysis 84).

Bullas, J.C., 2006. Bituplaning A Low Dry Friction Phenomenon of new bituminous road Surfaces (Ph.D. thesis).

Bullas, J.C., 2008. Low dry friction—measurement and imaging. In: SURF 2008 8th International Symposium on Pavement Surface Characteristics of Roads and Airfields, Slovenia.

Chace, M.A., 1969. A Network-Variational Basis for Generalised Computer Representation of Multifreedom, Constrained, Mechanical Systems. Design Automation Conference, Miami, Florida.

Chace, M.A., June 1970. DAMN-A prototype program for the design analysis of mechanical networks. In: 7th Annual Share design Automation Workshop, San Francisco, California.

Chace, M.A., February 1985. Modeling of Dynamic Mechanical Systems. CAD/CAM Robotics and Automation Institute, Tucson, Arizona.

Chace, M.A., Angel, J.C., February 1977. Interactive Simulation of Machinery with Friction and Impact Using DRAM. SAE Paper No. 770050, Society of Automotive Engineers, 400 Commonwealth Drive, Warrendale, PA 15096, USA.

Chace, M.A., Korybalski, K.E., April 1970. Computer Graphics in the Schematic Representation of Nonlinear, Constrained, Multifreedom Mechanical Systems. Computer Graphics 70 Conference, Brunel University.

Chavan, A., Koppenaal, J., Dhakare, Y.A., July 18—22, 2010. General Bushing Model for Vehicle Ride Studies. In: ICSV 17, the 17th International Congress on Sound and Vibration, Cairo. http://www.iiav.org/archives_icsv/2010_icsv17/papers/1162.pdf.

Cooper, D.W., Bitonti, F., Frayne, D.N., Hansen, H.H., May 1965. Kinematic Analysis Method (KAM). SAE Paper No. SP-272, Society of Automotive Engineers, 400 Commonwealth Drive, Warrendale, PA 15096, USA.

Costa, A.N., 1991. Application of Multibody Systems Techniques to Vehicle Modelling. Colloquium Model Building Aids for Dynamic Simulation, IEE Digest No. 1991/196.

Crolla, D.A., 1995. Vehicle Dynamics—Theory into Practice. Automobile Division Chairman's Address, Institution of Mechanical Engineers.

Crolla, D.A., Horton, D., Firth, G., 1992. Applications of Multibody Systems Software to Vehicle Dynamics Problems. Kinematics and Dynamics of Multi-body Systems Seminar (S057). Institution of Mechanical Engineers.

Crolla D.A., Horton, D.N.L., Brooks, P.C., Firth, G.R., Shuttleworth, D.W., Yip, C.N., 1994. A Systematic Approach to Vehicle Design Using VDAS (Vehicle Dynamics Analysis Software). SAE paper 940230, Society of Automotive Engineers, 400 Commonwealth Drive, Warrendale, PA 15096, USA.

Davis, D.C., 1974. A radial-spring terrain-enveloping tire model. Vehicle System Dynamics 3, 55—69.

Dingus, T.A., Klauer, S.G., Neale, V.L., Petersen, A., Lee, S.E., Sudweeks, J., Perez, M.A., Hankey, J., Ramsey, D., Gupta, S., Bucher, C., Doerzaph, Z.R., Jermeland, J., 2004. The 100-Car Naturalistic Driving Study; Phase II- Results of the 100-Car Field Experiment. Contract No. DTNH22-00-C-07007 (Task Order No. 06). Virginia Tech Transportation Institute, Blacksburg, VA.

Dixon, J.C., 1987. The roll-centre concept in vehicle handling dynamics. Proceedings of the Institution of Mechanical Engineers 210 (D1), 69—78.

D'Souza, A.F., Garg, V.K., 1984. Advanced Dynamics: Modelling and Analysis. Prentice Hall, Englewood Cliffs, London, ISBN 0130113123.

Fiala, E., 1954. Seitenkrafte am rollenden Luftreifen. Vdi-zeitschrift 96, 973.

Fonda, A.G., 1956. Tire tests and interpretation of experimental data. In: Proceedings of the Automobile Division, Institution of Mechanical Engineers.

Forkenbrock, G.J., Elsasser, D., June 1995. An Assessment of Human Driver Steering Capability. National Highway Traffic Safety Administration (NHTSA). Report DOT HS 809 875. http://www.nhtsa.gov/DOT/NHTSA/NRD/Multimedia/PDFs/VRTC/ca/capubs/NHTSA_forkenbrock_driversteeringcapabilityrpt.pdf.

Forsythe, G.E., Malcolm, M.A., Moler, C.B., 1977. Computer Methods for Mathematical Computations. Prentice Hall, Inc., Englewood Cliffs, NJ.

French, T., 1989. Tyre Technology. IOP Publishing, ISBN 0-85274-360-2.

Gallrein, A., Baecker, M., 2007. CDTire: a tire model for comfort and durability applications. Vehicle System Dynamics 45 (Suppl.), 69–77.

Garrett, T.K., February 1953. Automobile Dynamic Loads Some Factors Applicable to Design. Automobile Engineer.

Gear, C.W., January 1971. Simultaneous numerical solution of differential-algebraic equations. IEEE Transactions on Circuit Theory CT-18 (1), 89–95.

van Geffen, V., 2009. A Study of Friction Models and Friction Compensation. Report: DCT 2009.118, Dynamics and Control Technology Group, TU Eindhoven. http://www.mate.tue.nl/mate/pdfs/11194.pdf.

Giacomin, J., 2013. Perception enhancement for automotive steering systems: cognitive cues and the natural language of automobile-to person communication. In: 7th Annual Conference Steering Systems, Frankfurt. http://www.steering-conference.com/AgendaSection.aspx?tp_day=26880&tp_session=23161.

Gillespie, T.D., 1992. Fundamentals of Vehicle Dynamics. SAE Publications, Society of Automotive Engineers, 400 Commonwealth Drive, Warrendale, PA 15096, USA.

Gipser, M., November 1999. FTire, a new fast tire model for ride comfort simulations. In: Proceedings of the 14th European ADAMS Users' Conference, Berlin, Germany.

Goryca, J., 2010. Force and Moment Plots from Pacejka 2002 magic formula tire model Coefficients. TARDEC Technical Report 21187RC, US Army Research. http://www.dtic.mil/dtic/tr/fulltext/u2/a535124.pdf.

Hague, O.B., 2010. New handling simulation opportunities of tyre and vehicle interaction by using a "physical tyre". In: IPG Technology Conference.

Hales, F.D., 1989. Vehicle Dynamics Undergraduate Course Notes. Loughborough University.

Haney, P., 2003. The Racing and High Performance Tire. Society of Automotive Engineers.

Harty, D., January 1999. The myth of accuracy. The Journal of the Engineering Integrity Society.

Harty, D., 2010. The Influence of Unsprung Mass on Dynamic Key Performance Indicators. The 10th Advanced Vehicle Control (AVEC10). Loughborough.

He, J., Crolla, D.A., Levesley, M.C., Manning, W.J., September 9–11, 2003. Coordinated Active Rear Steering and Variable Torque Distribution Control for Vehicle Stability Enhancement. In: Proceedings of the Sixteenth International Conference on Systems Engineering (ICSE2003), vol. 1. Coventry, UK, pp. 243–248. ISBN 0-905949-91-9.

Holt, M.J., 1994. Simulation Tools for Vehicle Handling Dynamics. Multi-body System Dynamics Codes for Vehicle Dynamics Applications Seminar (S275). Institution of Mechanical Engineers.

Holt, M.J., Cornish, R.H., November 1992. Simulation Tools for Vehicle Handling Dynamics. Kinematics and Dynamics of Multi-body Systems Seminar (S057). Institution of Mechanical Engineers.

Hsu, Y.H., 2006. Estimation and Control of Lateral Tire Forces Using Steering Torque (Ph.D. thesis). Stanford University. http://ddl.stanford.edu/sites/default/files/Judy_dissertation.pdf.

Hucho, W.R., 1998. Aerodynamics of Road Vehicles: From Fluid Mechanics to Vehicle Engineering, fourth ed. SAE Publications, Society of Automotive Engineers, 400 Commonwealth Drive, Warrendale, PA 15096, USA, ISBN 0768000297.

Hudi, J., October 1988. AMIGO—a modular system for generating ADAMS models. In: Proceedings of the 5th European ADAMS News Conference, Tedas GmbH, Marburg, FR Germany.

van der Jagt, P., Pacejka, H.B., Savkoor, A.R., 1989. Influence of Tyre and Suspension Dynamics on the Braking Performance of an Anti-lock System on Uneven Roads. Institution of Mechanical Engineers. Paper No. C382/047, pp. 453–460.

Kahneman, D., 2012. Thinking Fast and Slow. Penguin.

Kaminski, S., April 1990. WOODS—a worksheet orientated design system. In: Proceedings of the 6th European ADAMS News Conference, Kurhaus Wiesbaden, FR Germany.

Karlsson, F., Persson, A., 2003. Modelling Non-linear Dynamics of Rubber Bushings—Parameter Identification and Validation, Master's Dissertation, Lund University. http://www.lth.se/fileadmin/byggnadsmekanik/publications/tvsm5000/web5119.pdf.

Kawagoe, K., Sume, K., Watanabe, M., 1997. Evaluation and Improvement of Vehicle Roll Behaviour. SAE paper 970093, Society of Automotive Engineers, 400 Commonwealth Drive, Warrendale, PA 15096, USA.

Kisielewicz, L.T., Ando, K., November 1992. Accurate Simulation of Road Tire Vehicle Interactions Using PAM_CRASH™/ADAMS™ Coupled Solution. '92 ISID ADAMS Users Conference—Tokyo.

Klöpper, R., 2011. A New Approach to Measuring Inertia Properties. Automotive Testing Expo, Stuttgart. http://www.testing-expo.com/europe/11txeu_conf/pdfs/day_3/38._Thursday_12.20_Dr_Robert_Kloepper.pdf.

Knappe, L.F., May 1965. A computer-orientated mechanical system, mechanical engineering. ASME 87 (5).

Kortum, W., Sharp, R.S. (Eds.), 1991. Multibody Computer Codes in Vehicle System Dynamics. Supplement to Vehicle System Dynamics, vol. 22.

Kortum, W., Sharp, R.S., de Pater, A.D., 1991. Application of Multibody Computer Codes to Vehicle System Dynamics. Progress Report to the 12th IAVSD Symposium on a Workshop and Resulting Activities, Lyon.

Kuiper, E., Van Oosten, J.J.M., 2007. The PAC2002 Advanced handling tire model. Vehicle System Dynamics 45 (Suppl. 1).

Kummer, H.W., 1966. Unified Theory of Rubber and Tire Friction. Issue 94 of Engineering Research Bulletin. Pennsylvania State University.

Lechmer, D., Perrin, C., 1993. The actual use of dynamic performance of vehicles. Proceedings of the Institute of Mechanical Engineers 207.

Lee, Lee, Ha, Han, 1999. Four wheel independent steering system for vehicle handling improvement by active rear toe control. JSME International Journal 42 (4).

Leva, A., Cox, C., Ruano, A., 2002. Hands-on PID autotuning: a guide to better utilisation. International Federation of Automatic Control. http://www.oeaw.ac.at/ifac/publications/pbriefs/PB_Final_LevaCoxRuano.pdf.

Limpert, R., 1999. Brake Design and Safety. Society of Automotive Engineers, 400 Commonwealth Drive, Warrendale, PA 15096, USA.

Loeb, J.S., Guenther, D.A., Chen, H.H., Ellis, J.R., 1990. Lateral Stiffness, Cornering Stiffness and Relaxation Length of the Pneumatic Tyre. SAE Paper 900129, Society of Automotive Engineers, 400 Commonwealth Drive, Warrendale, PA 15096, USA.

Madsen, J., Heyn, T., Negrut, D., 2010. Methods for Tracked Vehicle System Modeling and Simulation. Technical Report 2010-01. University of Wisconsin. http://sbel.wisc.edu/documents/TR-2010-01.pdf.

Makita, M., Torii, S., September 1992. An analysis of tire cornering characteristics using a magic formula tire model. In: JSAE Paper 923063, Proceedings of the International Symposium on Advanced Vehicle Control 1992 (AVEC 92), Yokohama.

Makkar, Dixon, Sawyer, Hu, 2005. A new continuously differentiable friction model for control systems design. IEEE(R)/ASME.

Matchinsky, W., 1998. Road Vehicle Suspensions. Wiley.

Milliken, W.F., Milliken, D.L., 1998. Race Car Vehicle Dynamics. SAE Publications, Society of Automotive Engineers, 400 Commonwealth Drive, Warrendale, PA 15096, USA.

Milliken, W.F., Milliken, D.L., 2001. Chassis Design—Principles and Analysis. SAE Publications, Society of Automotive Engineers, 400 Commonwealth Drive, Warrendale, PA 15096, USA.

Milliken, W.F., Whitcomb, D.W., 1956. General introduction to a programme of dynamic research. In: Proceedings of Automobile Division, Institution of Mechanical Engineers.

Moore, D.F., 1975. The Friction of Pneumatic Tyres. Elsevier Scientific Publishing Company, ISBN 0-444-41323-5.

Mousseau, C.W., Sayers, M.W., Fagan, D.J., 1992. Symbolic quasi-static and dynamic analyses of complex automobile models. In: Sauvage, G. (Ed.), The Dynamics of Vehicles on Roads and on Tracks. Swets and Zeitlinger, Lisse, pp. 446—459. Proceedings of the Twelfth IAVSD Symposium.

NHTSA, 2001. NCAP Rollover Resistance Final Policy Statement, NHTSA-2001-9663, Section V C.

Odhams, A.M.C., 2006. Identification of driver steering and speed control (Ph.D. thesis). Cambridge University. http://www2.eng.cam.ac.uk/~djc13/vehicledynamics/downloads/Odhams_PhDthesis_Sep06.pdf.

O'Hara, S.R., 2005. Vehicle Path Optimization of Emergency Lane Change Maneuvers for Vehicle Simulation (Master's thesis). University of Maryland. http://drum.lib.umd.edu/bitstream/1903/2823/1/umi-umd-2817.pdf.

von der Ohe M., 1983. Front and rear Suspension of the new Mercedes model W201. SAE Paper No. 831045.

Olley, M., 1945. Road manners of the modern motor car. Proceedings of the Institution of Mechanical Engineers.

van Oosten, J.J.M., Bakker, E., 1993. Determination of magic tyre model parameters, tyre models for vehicle dynamic analysis. In: Pacejka, H.B. (Ed.), Proceedings 1st International Colloquium on Tyre Models for Vehicle Dynamic Analysis. Swets & Zeitlinger, Lisse, pp. 19—29.

van Oosten, J.J.M., Jansen, S.T.H., November 1999. High frequency tyre modelling using SWIFT-tyre. In: Proceedings of the 14th European ADAMS Users' Conference, Berlin, Germany.

van Oosten, J.J.M., Unrau, H.J., Reidal, A., Bakker, A., July—September 1999. Standardization in Tire Modelling and Tire Testing—TYDEX Workgroup, TIME Project. Tire Science and Technology, TSTCA 27 (3), 188—202.

Orlandea, N., 1973. Node-Analogous, Sparsity-Orientated Methods for Simulation of Mechanical Dynamic Systems (Ph.D. thesis). The University of Michigan.

Orlandea, N., Chace, M.A., Calahan, D.A., October 1976a. A Sparsity-Orientated Approach to the Dynamic Analysis and Design of Mechanical Systems—Part I. Paper No. 76-DET-19,

ASME Mechanisms Conference, Montreal. (Also Trans. ASME Journal of Engineering for Industry, 1977.)

Orlandea, N., Chace, M.A., Calahan, D.A., October 1976b. A Sparsity-orientated Approach to the Dynamic Analysis and Design of Mechanical Systems—Part II. Paper No. 76-DET-20, ASME Mechanisms Conference, Montreal. (Also Trans. ASME Journal of Engineering for Industry, 1977.)

Orlandea, N., Chase, M., 1977. Simulation of a Vehicle Suspension with the ADAMS Computer Program. SAE paper 770053, Society of Automotive Engineers, 400 Commonwealth Drive, Warrendale, PA 15096, USA.

Ozdalyan, B., Blundell, M.V., 30 September—2 October 1998. Anti-lock braking system simulation and modelling in ADAMS. In: International Conference on Simulation '98 (IEE), Conference Publication Number 457. University of York, UK, pp. 140—144.

Pacejka, H.B., 2012. Tyre and Vehicle Dynamics, third ed. Butterworth_Heinemann.

Pacejka, H.B., Bakker, E., 1993. The magic formula tyre model, Tyre models for vehicle dynamic analysis. In: Pacejka, H.B. (Ed.), Proceedings 1st International Colloquium on Tyre Models for Vehicle Dynamic Analysis. Swets & Zeitlinger, Lisse, pp. 1—18.

Pacejka, H.B., Besselink, I.J.M., 1997. Magic formula tyre model with transient properties. In: Proceedings Berlin Tyre Colloquium, Vehicle System Dynamics Supplement 27. Swets & Zeitlinger, Lisse, pp. 145—155.

Pacejka, H.B., February 1995. The role of tyre dynamic properties. In: Seminar on Smart Vehicles, Delft, Proccedings: Vehicle System Dynamics, Special Issue.

Pacejka, H.B., Sharp, R.S., 1991. Shear force generation by pneumatic tyres in steady state conditions: a review of modelling aspects. Vehicle System Dynamics 20, 121—176.

Pacejka, H.B., Takahashi, T., 1992. Pure slip characteristics of tyres on flat and on undulated road surfaces. AVEC 92.

Palmer, D., 1999. Course notes—high performance driving training. Driving Developments.

Phillips, B.D.A., 2000. Chassis Engineering. Course Notes, Coventry University.

Press, W.H., Teukolsky, S.A., Vetterling, W.T., Flannery, B.P., 1992. Numerical Recipes in C: The Art of Scientific Computing, second ed. Cambridge University Press.

Rai, N.S., Soloman, A.R., 1982. Computer Simulation of Suspension Abuse Tests Using ADAMS. SAE paper 820079, Society of Automotive Engineers, 400 Commonwealth Drive, Warrendale, PA 15096, USA.

Reimpell, J., Sponagel, P., 1988. Reifen und Räder. Vogel Verlag.

Rice, R.C. (Ed.), 1997. SAE Fatigue Design Handbook. Society of Automotive Engineers, 400 Commonwealth Drive, Warrendale, PA 15096, USA, ISBN: 1-56091-917-5.

Ro, Kim, 1996. Four Wheel Steering for Vehicle Handling Improvement. Institution of Mechanical Engineers.

Ross-Martin, T.J., Darling, J., Woolgar, R., 1992. The Simulation of Vehicle Dynamics Using the Roll Centre Concept. Institution of Mechanical Engineers Paper C389/047, FISITA92.

Ryan, R., 1990. ADAMS - multibody systems analysis Software. In: Schielen, W. (Ed.), Multibody Systems Handbook. Springer-Verlag, Berlin.

Ryan, R., 1993. ADAMS Mechanical Systems Simulation Software. In: Multibody Computer Codes in Vehicle Systems Dynamics, vol. 22. Swets & Zeitlinger B.V, Amsterdam/Lisse, ISBN 90-265-1365-8. Suppl.

SAE Publication, 1976. Vehicle Dynamics Terminology, Handbook Supplement, SAE J670e. Society of Automotive Engineers, 400 Commonwealth Drive, Warrendale, PA 15096, USA.

Sato, S., Inoue, H., Tabata, M., Inagaki, S., 1992. Integrated chassis control System for improved vehicle dynamics. In: Proceedings of AVEC'92, pp. 413—418.

Sayers, M.W., 1990. Automated formulation of efficient vehicle simulation codes by symbolic computation (AUTOSIM). In: Proceedings 11th IAVSD Symposium of Vehicles on Roads and Tracks, Kingston, Ontario, Swets and Zeitlinger, Lisse.

Scapaticci, D., Coeli, P., Minen, D., October 1992. ADAMS implementation of synthetic trajectories of the wheels with respect to the car body for handling manoeuvres simulations. In: Proceedings of the 8th European ADAMS News Conference, Munich.

Schielen, W. (Ed.), 1990. Multibody Systems Handbook. Springer-Verlag, Berlin.

Schuring, D.J., Pelz, W., Pottinger, M.G., 1993. The BNPS model—an automated implementation of the "magic formula" concept. In: IPC-7 Conference and Exposition, Phoenix.

Segel, L., 1956. Theoretical prediction and experimental substantiation of the response of the automobile to steering control. In: Proceedings of Automobile Division, Proceedings of the Institution of Mechanical Engineers, pp. 310−330.

Segel, L., 1993. An overview of developments in road vehicle dynamics: past, present and future. In: Proceedings Institution of Mechanical Engineers Conference on "Vehicle Ride and Handling", London, pp. 1−12.

Sharp, R.S., 1990. On the accurate representation of tyre shear forces by a multi-radial-spoke model. In: Anderson, R.J. (Ed.), Proceedings 11th IAVSD Symposium on the Dynamics of Vehicles on Roads and Tracks. Swets & Zeitlinger, Lisse, pp. 528−541.

Sharp, R.S., November 1991. Computer Codes for Road Vehicle Dynamic Models. Institution of Mechanical Engineers Paper 427/16/064, Autotech '91, Birmingham.

Sharp, R.S., 1992. Magic Formula Method for Representation of Tyre Forces and Moments. Course Notes, Cranfield Institute of Technology.

Sharp, R.S., 1993. Tyre structural mechanics influencing shear force generation: Ideas from a multi-radial-spoke model, Tyre models for vehicle dynamic analysis. In: Pacejka, H.B. (Ed.), Proceedings 1st International Colloquium on Tyre Models for Vehicle Dynamic Analysis. Swets & Zeitlinger, Lisse, pp. 145−155.

Sharp, R.S., 1994. Review of Currently Available Codes—Benchmarking Attempt. Multi-body System Dynamics Codes for Vehicle Dynamics Applications Seminar (S275). Institution of Mechanical Engineers.

Sharp, R.S., 1997. Use of the Symbolic Modelling Code AUTOSIM for Vehicle Dynamics. School of Mechanical Engineering, Cranfield University.

Sharp, R.S., December 1998. Multi-body dynamics applications in vehicle dynamics. Institution of Mechanical Engineers, International Conference on Multi-body Dynamics (1 86058 152 8), C553/7/045/98, pp. 215−228.

Sharp, R.S., 2000. Aspects of the Lateral and Longitudinal Control of Automobiles. Cranfield University.

Sharp, R.S., 2005. Driver steering control and a new perspective on car handling qualities. Journal of Mechanical Engineering Science, Proceedings of the Institution of Mechanical Engineers 219 (C8), 1041−1051.

Sharp, R.S., El-Nashar, M.A., 1986. A generally acceptable digital computer based model for the generation of shear forces by pneumatic tyres. Vehicle System Dynamics 15, 187−209.

Sitchen, A., 1983. Acquisition of Transient Tire Force and Moment Data for Dynamic Vehicle Handling Simulations. SAE paper 831790, Society of Automotive Engineers, 400 Commonwealth Drive, Warrendale, PA 15096, USA.

Smiley, R.F., 1957. Evaluation and Extension of Linearized Theories for Tire Motion and Wheel Shimmy. Technical Report NACA-TM-1299, NASA.

Smiley, R.F., Horne, W.B., 1960. Mechanical Properties of Pneumatic Tires with Special Reference to Modern Aircraft Tires. Technical Report NASA-TR-64, Langley Research Center, NASA.

Stengel, R.F., 2004. Flight Dynamics. Princeton University Press.

Tandy, K., Heydinger, G.T., Christos, J.P., Guenther, D.A., 1992. Improving Vehicle Handling Simulation via Sensitivity Analysis. Institution of Mechanical Engineers Paper C389/396, FISITA92.

Tang, A., Tamini, N., Yand, D., 2000. Virtual proving ground—a CAE tool for automotive durability, ride & handling and NVH applications. In: 6th International LS-DYNA Conference.

Terlinden, M.W., Langer, W., Hache, M., September 1987. MOGESSA—A modular full vehicle simulation system based on ADAMS. In: Proceedings of the 4th European ADAMS News Conference, Tedas GmbH, Marburg, FR Germany.

TNO. MF-Tyre/MF-Swift 6.1.2 Equation Manual, 2010.

TNO. MF-Tyre/MF-Swift 6.1.2 Help Manual, 2012.

Trungle, C.V., 1991. Engineering a winner at Newman/Hass Racing. Dynamic Dimensions 2 (3). Mechanical Dynamics Inc.

op het Veld, I.B.A., 2006. Run Flat Tires versus Conventional Tires — An Experimental Comparison. Technical Report, Eindhoven University of Technology. http://www.mate.tue.nl/mate/pdfs/6865.pdf.

Vesimaki, M., November 1997. 3D contact algorithm for tire-road interaction. In: Proceedings of the 12th European ADAMS Users' Conference, Marburg, Germany.

Wenzel, T.A., Blundell, M.V., Burnham, K.J., Williams, R.A, September 9—11, 2003. Modelling and estimation of automotive vehicle states. In: Proceedings of the Sixteenth International Conference on Systems Engineering (ICSE2003), vol. 2. Coventry, UK, pp. 744—749. ISBN: 0-905949-91-9.

Whitehead, J., November 1995. The Essential Contribution of Rig Measurement to Suspension Design and Development. Institution of Mechanical Engineers Paper C498/7/061/95, Proceedings of Autotech '95, Birmingham.

Wielenga, T., October 1987. Analysis Methods and Model Representation in ADAMS. MDI Technical Paper No. 41, Mechanical Dynamics Inc., 2301 Commonwealth Blvd., Ann Arbor, Michigan, USA.

Will, A.B., Żak, S.H., 1997. Modelling and control of an automated vehicle. In: Lugner, P., Hedrick, J.K. (Eds.), Vehicle System Dynamics, vol. 27. Swets & Zeitlinger, Lisse.

Wilmot, L., Agrawala, M., Salesin, D., 2004. Interactive Image-Based Exploded View Diagrams. Graphics Interface 2004, 203—212.

Wittenburg, J., Wolz, U., 1985. MESA VERDE: a symbolic program for non-linear articulated rigid body dynamics. In: Proceedings of Tenth ASME Design Engineering Division Conference on Mechanical Vibration and Noise, Cincinnati.

Wong, J.Y., 2001. Theory of Ground Vehicles, third ed. John Wiley & Sons, Inc., New York.

Wood, G., Blundell, M., Sharma, S., April 2012. A low parameter tyre model for aircraft ground dynamic simulation. Journal of Material & Design 35, 820—832.

Wu, Y., Wu, A., 2000. Taguchi Methods for Robust Design. ASME Press, New York.

Young, W.C., Budynas, R.G., Sadegh, A.M., 2011. Roark's Formulas for Stress and Strain, eighth ed. McGraw-Hill.

Zegelaar, P.W.A., 1998. The dynamics Response of tyres to Brake Torque Variations and road Unevennesses (Ph.D. thesis). Technical University of Delft. http://www.tno.nl/downloads/DT_PhD_Thesis_Zegelaar1.pdf.

Index

Evolutionary Genetics

From Molecules to Morphology

Richard Lewontin is undoubtedly one of the most distinguished evolutionary biologists of our time. He has contributed to science not only by his own work on evolutionary theory and molecular variation and by his influence on the many young scientists who have worked with him but also by asking us to think about the relationship between the science we do and that world in which we do it. Sciences in general, and the life sciences in particular, need their own critic, and Lewontin has been an untiring critic of science and its relevance to society.

This collection of essays is produced in honor of Lewontin's 65th birthday. The volume is unique, as it has a comprehensive coverage of modern evolutionary genetics from molecules to morphology by a group of star authors, including his students and colleagues. Such a comprehensive treatment of evolutionary genetics has never been attempted before. The volume is set in a historical perspective, but it has an up-to-date coverage of material in the various fields. The areas covered are the mathematical and the molecular foundations of population genetics, molecular variation and evolution, selection and genetic polymorphisms, linkage and breeding system evolution, quantitative genetics and phenotypic evolution, gene flow and population structure, speciation, behavior, and ecology.

The volume brings out the central role of evolutionary genetics in all aspects of its connection to evolutionary biology and is a must for all graduate students and researchers in evolutionary biology.

Rama Singh is a Professor in the Department of Biology at McMaster University. Costas Krimbas is Professor of Philosophy and History of Science at the University of Athens.

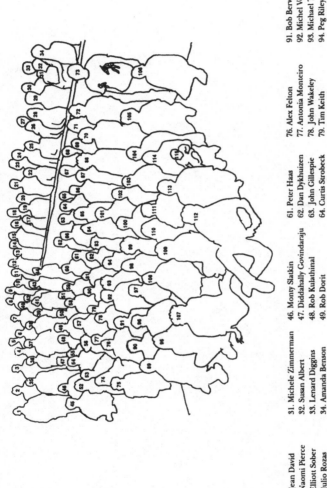

1. Tim Prout
2. Chris Dick
3. Mark Siegal
4. Mike Dietrich
5. Dmitri Petrov
6. Michael Bradie
7. Joe Felsenstein
8. Will Provine
9. Norman Johnson
10. Jeff Dole
11. Marcos Antezana
12. Lloyd Demetrius
13. Christian Biemont
14. Bill Piel
15. Dan Hartl

16. Jean David
17. Naomi Pierce
18. Elliott Sober
19. Julio Rozas
20. Yvonne Parsons
21. David Rand
22. Lisa Lloyd
23. Michael Cummings
24. Dan Weinreich
25. Carmen Segarra
26. Suzanne Kaplan
27. Michael Hammer
28. Rachel Nasca
29. Manyuan Long
30. David Glaser

31. Michele Zimmerman
32. Susan Albert
33. Lenard Diggins
34. Amanda Benson
35. Janet Collett
36. Diane Paul
37. John Beatty
38. Antonio Barbadilla
39. Teresa Sebastiá
40. Donal Hickey
41. Montsé Aguadé
42. Richard Kliman
43. Rasmus Nielsen
44. Kristin Ardlie
45. Toshinori Endo

46. Monty Slatkin
47. Diddahally Govindaraju
48. Rob Kulathinal
49. Rob Dorit
50. Hiroshi Akashi
51. Tomoko Steen
52. Hongping Tian
53. Weber Amaral
54. Ken Weber
55. Richard Morton
56. Tracy McLellan
57. Dave Parker
58. Einar Arnason
59. John Ramshaw
60. Marty Kreitman

61. Peter Haas
62. Dan Dykhuizen
63. John Gillespie
64. Curtis Strobeck
65. Fred Cohan
66. Dick Lewontin
67. Jeff Powell
68. Mark Kirkpatrick
69. Lisa Brooks
70. Maryellen Ruvolo
71. Gisella Caccone
72. Richard Burian
73. Russ Lande
74. Dana Campbell
75. Belinda Chang

76. Alex Felton
77. Antonia Monteiro
78. John Wakeley
79. Tim Keith
80. Walt Eanes
81. Steven Irwin
82. Jerry Coyne
83. William Wimsatt
84. Marcy Uyenoyama
85. Adriana Briscoe
86. Rama Singh
87. Costas Krimbas
88. Laura Landweber
89. Peter Godfrey-Smith
90. Eleutherios Zouros

91. Bob Berwick
92. Michel Veuille
93. Michael Turelli
94. Peg Riley
95. Simon Frost
96. Peter Goss
97. Sally Otto
98. Maria Pompeiano
99. Evan Balaban
100. Carlos Bustamante
101. Chuck Taylor
102. Geoff Morse
103. Jean-Pierre Berlan
104. Doug Futuyma
105. Yaneer Bar-Yam

106. William Brown
107. Cay Craig
108. Cathy Laurie
109. Soojin Yi
110. Anna Haynes
111. Jeff Townsend
112. Andrew Berry
113. Dennis Cullinane
114. Lynn King
115. William Clark King.

Photo taken at the Harvard Museum of Comparative Zoology, September 6, 1998 by Liza Green, Harvard Medical School Media Services.

Evolutionary Genetics

From Molecules to Morphology

Edited by

R. S. SINGH

McMaster University

C. B. KRIMBAS

University of Athens

PUBLISHED BY THE PRESS SYNDICATE OF THE UNIVERSITY OF CAMBRIDGE
The Pitt Building, Trumpington Street, Cambridge, United Kingdom

CAMBRIDGE UNIVERSITY PRESS
The Edinburgh Building, Cambridge CB2 2RU, UK
40 West 20th Street, New York, NY 10011-4211, USA
10 Stamford Road, Oakleigh, VIC 3166, Australia
Ruiz de Alarcón 13, 28014 Madrid, Spain
Dock House, The Waterfront, Cape Town 8001, South Africa

http://www.cambridge.org

First published 2000
Reprinted with corrections 2000

Typeface New Baskerville 9.75/12 pt.   *System* LaTeX 2_ε [TB]

A catalog record for this book is available from the British Library.

Library of Congress Cataloging in Publication Data
Evolutionary genetics : from molecules to morphology /
    edited by R. S. Singh and C. Krimbas.
        p.   cm.
    Includes bibliographical references and index.
    ISBN 0-521-57123-5
    1. Evolutionary genetics.   2. Lewontin, Richard C., 1929–
I. Singh, R. S.   II. Krimbas, Costas B.
QH390.E96   1999
576.8 – dc21   99-14415
                                                     CIP

ISBN 0 521 57123 5 hardback

Transferred to digital printing 2002

Contents

List of Contributors

Aguadé, M., Department of Genetics, University of Barcelona, Barcelona, Spain

Antezana, M., Department of Ecology and Evolution, University of Chicago, Chicago, IL 60637

Árnason, E., Institute of Biology, University of Iceland, 108 Reykjavik, Iceland

Barbadilla, A., Department of Genetics and Microbiology, University Autonoma de Barcelona, Barcelona 08193, Spain

Barker, J. S. F., Department of Animal Science, University of New England, Armidale NSW 2351, Australia

Berry, A., Population Genetics, Harvard University, Cambridge, MA 02138

Carson, H. L., Department of Genetics and Microbiology, University of Hawaii, Honolulu, Hawaii 96822

Charlesworth, B., Institute of Cellular, Animal and Population Biology, University of Edinburgh, Edinburgh EH9 3JN, UK

Christiansen, F. B., Department of Ecology and Genetics, University of Aarhus, DK-8000, Aarhus C, Denmark

Cohan, F. M., Department of Biology, Wesleyan University, Middletown, CT 06457-0170

Coyne, J. A., Department of Ecology and Evolution, University of Chicago, Chicago, IL 60637

Demetrius, L., Department of Organismic and Evolutionary Biology, Harvard University, Cambridge, MA 02138

Eanes, W., Department of Ecology and Evolution, State University of New York, Stony Brook, NY 11794

Ewens, W. J., Department of Biology, University of Pennsylvania, Philadelphia, PA 19104-6018

Feldman, M. W., Department of Biological Sciences, Stanford University, Stanford, CA

Felsenstein, J., Department of Genetics, University of Washington, Seattle, WA 98195-7360

Frankham, R., Key Centre for Biodiversity and Bioresources, School of Biological Sciences, Macquarie University, NSW 2109, Australia

Franklin, I., CSIRO, Division of Animal Production, Prospect, NSW, Australia

Gordon, D. M., Department of Biological Sciences, Stanford University, Stanford, CA 94305-5020

Hedrick, P., Department of Biology, Arizona State University, Tempe, AZ 85287

Hickey, D. A., Department of Biology, University of Ottawa, Ottawa, Ontario, Canada K1N 6N5

Hughes, K. A., Department of Life Sciences, Arizona State University West, Phoenix, AZ 85069

Jain, S. K., Department of Agronomy and Range Science, University of California, Davis, CA 95616-8515

Kim, T. J., Department of Biology, Arizona State University, Tempe, AZ 85287

King, L. M., Department of Biology, University of Miami, Coral Gables, FL 33124-0421

Kreitman, M., Department of Ecology and Evolution, University of Chicago, Chicago, IL 60637

Krimbas, C. B., Department of Philosophy and History of Science, University of Athens, 121 Athens, Greece

Lande, R., Department of Biology, University of Oregon, Eugene, OR 97403-1210

Levin, B. R., Department of Biology, Emory University, Atlanta, GA 30322

Lewontin, R. C., Museum of Comparative Zoology, Harvard University, Cambridge, MA 02138

Masters, J. C., Natal Museum, Private Bag 9070, Pietermaritzburg, 3200, South Africa

Maynard Smith, J., School of Biological Sciences, University of Sussex, Falmer, Brighton BN1 9QG, UK

Niklasson, M., Molslaboratoriet, Ebeltoft, Denmark

Orr, H. A., Department of Biology, The University of Rochester, Rochester, NY 11794

Parker Jr., E. D., Department of Ecology and Genetics, Aarhus University, Aarhus C, Denmark

Powell, J. R., Department of Biology, Yale University, New Haven, CT 06511

Prout, T., Department of Genetics, University of California, Davis, CA 95616

Rand, D. M., Department of Ecology and Evolutionary Biology, Brown University, Providence, RI 02912

Riley, M. A., Department of Biology, Yale University, New Haven, CT 06511

Schaeffer, S. W., Department of Biology, Pennsylvania State University, University Park, PA 16802-5301

Singh, R. S., Department of Biology, McMaster University, Hamilton, Ontario Canada L8S 4K1

Slatkin, M., Department of Integrative Biology, University of California, Berkeley, CA 94720-3140

Uyenoyama, M. K., Department of Zoology, Duke University, Durham, NC 27708-0325

Wallace, B., Department of Biology, Virginia Polytechnic Institute and State University, Blacksburg, VA 24061

Weber, K. E., Department of Biological Sciences, University of Southern Maine, Portland, ME 04193, USA

Zouros, E., Department of Biology, Dalhousie University, Halifax, NS, Canada B3H 4J1 and Department of Biology, University of Crete, Iraklion, Crete, Greece

Preface

Scientists earn their reputation by making special contributions in a variety of ways. Some become known for a discovery that revolutionizes their science. Others are respected as intellectual leaders for significant contributions leading to sustained progress in their field. Still others become known for providing guidance and opportunity to and uniquely inspiring rapport with a large number of graduate students, writers, and research colleagues. A rare few do all the above, and remarkably enough still find time to deal with the broader issues of epistemology, philosophy, history, and sociology of science. Richard Lewontin is one of these rare scientists.

If we are to attach a major discovery or a conceptual breakthrough to Lewontin's name (such as Haldane's cost of natural selection, Fisher's fundamental theorem of natural selection, Wright's shifting-balance theory, or Maynard Smith's game theory applications), then the successful completion of the genetic variation research program of the Chetverikov–Dobzhansky School will be known as the outstanding highlight of Lewontin's career. Dobzhansky and his students and collaborators pursued the twin problems of the amount and the adaptive role of genetic variation for nearly 25 years without a satisfactory solution. All estimates of genetic variation were indirect or inadequate as there was no reductionist research program that could allow the study of genetic variation at the level of the gene. Lewontin's pioneering success in the application of protein electrophoresis to the problem of genetic variation changed the scene radically. The estimation of electrophoretic variation was direct and more useful than anyone had expected. The technique also removed the experimental limitations imposed by genetic incompatibility among species and allowed reliable comparisons of genetic variation among populations and species without any need to make genetic crosses. The impact and the anticipation of the avalanche of future results from the use of electrophoresis were discussed in his well-known book, *The Genetic Basis of Evolutionary Change* (1974). This book sets out the problem of population genetics in a rationally constructed historical context and is required reading for all aspiring population geneticists.

Evolutionary research requires broad interest and versatility in modeling, experimental design, statistics, field biology, and much more. Such breadth

xv

allowed Lewontin to be successful, time and again, in designing new experimental systems or suggesting key concepts to answer old questions or pursue new ones. Lewontin became interested in the uniqueness of the phenotype— and the genotype–environment interactions inspired mainly by the book *Factors of Evolution* by the Russian biologist I. Schmalhausen. His doctoral thesis studied fitness as a function of genotype frequency and density and showed that "viability of a genotype is a function of the other genotypes which coexist with it, the result of any particular combination not being predictable on the basis of the viabilities of the coexisting genotypes when tested in isolation." This was followed by studies of interlocus epistatic interactions in fitnesses and the evolution of naturally occurring inversion polymorphism in *Drosophila*. His mathematical work on linkage disequilibrium provided a new direction for research and results from a series of papers on multilocus fitness effects anticipated discussion on the units of selection. His experimental work on norms of reaction in *Drosophila* was exemplary in exposing the problem of the genetic determination and led to a new appreciation of genotype–environment interaction and phenotypic plasticity. He pointed out the importance of developmental time in fitness, something which is usually forgotten when describing fitness components. His 1972 paper on the "Apportionment of Human Diversity," pointing out that any genetic differences between races has to be compared with genetic variation within population and races, is a landmark in human genetics and evolution. More recently his laboratory has been a major center for studies of DNA sequence variation. Lewontin has provided training and guidance to a large number of graduate students and postdoctoral fellows. The number is well over one hundred! Many more have worked in Lewontin's laboratory but have not necessarily coauthored publications with him.

But what makes Lewontin known more in the wider circle of evolutionary biology and in science in general is his role as a critic of how science is done, on the one hand, and his passionate engagements with the issues of science and society, on the other. He has made important contributions and has influenced research workers in the history and philosophy of science and in the areas of science and society, such as agriculture, social health problems, bioethics, and genetics, and IQ. Dropping Lewontin's name in any group of biologists is sure to spark an animated discussion – not about science but about its relevance and applications to human affairs. His concern about social issues springs directly from his unique perspective of evolutionary biology. Lewontin's research program may be reductionist but he is not. He has encouraged and challenged evolutionary biologists to find the most desirable combination of Platonic and Aristotelian traditions in studying nature. Accordingly the mathematical rigor of early population biology must be extended to accommodate interactive, hierarchical, probabilistic, and historical factors as learned empirically in the field. To him "Context and interaction are of the essence" (Lewontin, 1974, p. 318), whether one is talking about interactions between hierarchical levels, between organisms and the environment, or between causes and effects. A reductionist approach to science does not necessitate a reductionist view of the world. No level of analysis is specially privileged for a general understanding of causality. Genetic and environmental effects are interdependent and the phenotypic variance can not be partitioned into fixed components. Organisms do not fit in preexisting ecological niches but create their own niches. History and contingencies are so important in evolution that looking for adaptive explanations for all organismic traits undermines the role of natural history. These ideas essentially follow from his belief that relationships between organisms and their environments, and likewise, those between groups and hierarchical levels, are governed by forces so weak that the outcomes are neither fixed nor predetermined.

John Maynard Smith has written (this volume, Chap. 30) that "Richard Lewontin has contributed to science not only by his own work on evolutionary theory and molecular variation and by his influence on the many young scientists who have worked with him but also by asking us to think about the relationships between the science we do and the world we do it in." While one may not agree with Lewontin on all issues (he would be surprised if you did!) one thing is sure – Lewontin has been a colorful personality who has made evolutionary biology rigorous and interesting at the same time. We affectionately dedicate this volume to him.

This volume has been long in preparation and we thank the authors for their patience. We are extremely grateful to all our section editors who read all manuscripts and wrote introductory remarks and to many colleagues who provided help as reviewers. At Cambridge University Press, we express our sincere thanks to Robin Smith for his early enthusiasm and help in the preparation of this book, and to Jackie Mahon, Michael Penn, and Wendy Wagner for their help at various stages of its completion. Finally we would like to thank Kathy McIntosh of McMaster University for her enthusiasm and the enormous amount of work that she has put in, from communication with authors to preparation of final manuscripts of this book, and Aaron Thomson for compiling the index.

Rama Singh, Hamilton, Canada
Costas Krimbas, Athens, Greece

With Editorial Assistance from R. Frankham, I. Franklin, P. W. Hedrick, S. K. Jain, M. E. Kreitman, T. Prout, and M. Slatkin.

SECTION A

POPULATION GENETICS: PROBLEMS, FOUNDATIONS, AND HISTORICAL PERSPECTIVES

Introductory Remarks

COSTAS B. KRIMBAS

"[I]f I would not do anything else but to produce these two, Bruce Wallace and more recently Dick Lewontin, at present professor in Rochester, I think my earthly existence would be justified." – Th. Dobzhansky, 1962, p. 485

The first four chapters of this book (Section A) examine the foundations and the basic problems of population genetics, "the automechanics of [neo-Darwinian] evolutionary biology," in Lewontin's words. Their historical dimension is obvious; therefore my intention is to introduce the reader to these chapters by rearranging them in a historical narrative. This does not mean that a time succession of the events will be strictly followed but rather an articulation of these events according to a logical succession, something like "une histoire raisonnée."

The controversy between geneticists on one hand and biometricians/ Darwinists on the other (Provine, 1971) was resolved only when it was shown that the two fields were not only compatible but necessary to each other: mendelism could solve problems that Darwinian selectionism could not answer in a satisfactory way. The first proof of this was provided by mathematical models, and Ewens, in the second chapter of this section, together with a reappraisal of their theoretical constructions, examines the foundational and the historical aspects of the work of the three great pioneers, Ronald Fisher, Sewall Wright, and J. B. S. Haldane. Their work created the new field of population genetics and at the same time provided the foundations of the synthetic theory. Ewens reviews the achievements of these three legendary figures, their significant coincidence of opinions on what mattered and that provided the main fabric of the synthetic theory (the acceptance of the Darwinian and Mendelian marriage) as well as their personal contributions and disagreements (the fundamental theorem of Fisher, the balance shift theory of Wright and the concept of effective size, the tactical approach of Haldane and the theory of load). He also mentions briefly the rebirth of mathematical population genetics by Kimura. Ewens discusses the present status of knowledge and also the unsettled questions, as is the problem of the evolution of many genes and the relevance to the one-gene models; the contribution of R. C. Lewontin to this particular problem is significant.

1

The third chapter, by Bruce Wallace, gives us the latest view of this long sequel that was the genetic load controversy. The experimental population genetics was born from another happy marriage, that of the genetics of the Morgan school with the Russian tradition of naturalists and Darwinists, exemplified in the person of their leader, Sergei S. Chetverikov. Theodosius Dobzhansky's *Genetics and the Origin of Species* (1937) was the neo-Darwinian manifesto of the geneticists and naturalists. From Dobzhansky (not from J. Huxley, neither from E. B. Ford) originated the approach of both the study of natural and laboratory populations as well as the new systematics [it is noteworthy that in Huxley's (1940) *The New Systematics* Dobzhansky was not asked to contribute although he has been the most cited author, with 43 citations in total, the second being Muller with 25 and Darwin, Haldane, Ford, Huxley, Wright, Timoféeff-Ressovsky and others with less, ranging between 10 and 20]. From Dobzhansky, Mayr's (1942) *Systematics and the Origin of Species* was inspired as well as G. G. Simpson's (1944) *Tempo and Mode in Evolution*.

Dobzhansky worked with inversions and with entire chromosomes, the viability of which he studied. The panselectionism of Dobzhansky soon led to the question of Haldane's genetic load. Dobzhansky's preferred student, Bruce Wallace, spent considerable time and energy on that question. This has not been a futile enterprise since the concepts of hard and soft selection (in relation to frequency- and density-dependent selection) were first suggested and discussed (for a review, see Wallace, 1991). Chapter 3 provides the final interesting remarks on this historical controversy, presenting a naturalistic point of view that provides us with a significant insight into the selection process in nature: "Neutrality of phenotypic variation arises as an average of many, non-neutral selective roles played by individual variants during decisive encounters, [... which] cull an excessive number of zygotes."

The last two chapters of this section refer directly to the very significant contribution of R. C. Lewontin to the field of population genetics. Lewontin was able to accomplish the scientific program of Chetverikov, something which Dobzhansky himself did not succeed in doing. The saga is well known and presented here in Chap. 4 by his students, Rama Singh, Walter Eanes, Donal Hickey, Margaret Riley, and Lynn King. It starts with Lewontin's proposal to estimate the genetic variability of populations (the main aim of Chetverikov's program) from the prime products of the gene action, the enzymes and the constitutive proteins, by using electrophoresis. From a sample of several genes Lewontin was able to infer the amount of genic variation. From 1966, when the first report was published, until 1984, more than 1100 independent studies of assessment of genic variation were published in which this suggestion was used. Of course, allozymes may fail to uncover the entire variation, since they may group in a heterogeneous class different variants with the same electrophoretic mobility. Refinements of the method were used in order to distinguish heterogeneity within the established allozyme classes, based on the temperature sensitivity of allozymes, their denaturation by urea, and finally the sequential electrophoresis.

The ultimate method of analysis was DNA sequencing, and this was first performed in Lewontin's lab. This sequencing permitted not only the identification of all alleles but also the comparison of the variation at the first and the second positions with that of the third, that is, the comparison among differences at the protein level with others probably not so important functionally to be distinguished as targets by natural selection. The problem of natural selection is a key one: Is this uncovered variation neutral or a target to selective forces? Is it the variation Chetverikov had in mind or is it a variation without any evolutionary importance? Singh et al. examine all methodologies and attempts to uncover the biological (physiological, ecological, etc.) significance of this variation.

However, Lewontin in Chap. 1 deals in detail with the problems raised by data of past decade, especially those from DNA sequences. He provides us with a general description of the problem with the help of a well-known diagram, which depicts the relation between genotypes and phenotypes, on which selection is acting. Lewontin distinguishes gametic from phenotypic selection, regarding only the latter as true selection, and relegating segregation distortion to another process. Genetic drive may be widespread, as a recent report of Merçot et al. (1995) suggests. Of course, one of the main problems is that most of our data are of static nature; they refer to precise time slices, and thus frequency changes are not recorded. This lack of information creates problems for interpreting the data.

Complete narratives covering all paths, depicted in Lewontin's diagram, that is, cases of selection for which we are confident and that explain the frequency changes of alleles with the help of an ecological narrative, amount to only 5% of all narratives at our disposal (Endler, 1986). In these cases the target is linked to one or a few genes (e.g., see Vouidibio et al. 1989). Because of the importance of selective forces involved, they may constitute a specific minority. Most of the time, changes might be more subtle and many genes may be simultaneously involved; problems raised by sequence data, such as those referring to codon bias, or to unexpected regions of high conservation, may point to a probable presence of selective processes. However, selective forces might be very difficult to quantify, since their rates might be extremely small. The amount and the nature of molecular data may in a sense distance us from the part of the story that belongs to natural history. All these problems are discussed in a very clear and thought-provoking way by Lewontin. He discusses, in the best way, the actual problems population genetics faces now, old questions as well as new ones created by the sequence data at our disposal. An impossibility in solving them might turn the field of evolutionary theory to a consideration of evolution at only the phenotypic level (the streetcar approach; Marrow and Johnstone, 1996). But this will significantly depauperate the evolutionary theory.

REFERENCES

Dobzhansky, Th. 1937. *Genetics and the Origin of Species*. New York: Columbia U. Press.
Dobzhansky, Th. 1962. Oral transcript. New York: Columbia University (transcript).

4   *Costas B. Krimbas*

Endler, J. A. 1986. *Natural Selection in the Wild*. Princeton, NJ: Princeton U. Press.
Huxley, J., ed. 1940. *The New Systematics*. Oxford: Oxford U. Press.
Marrow, P. and Johnstone, R. A. 1996. Riding the evolutionary streetcar: where population genetics and game theory meet. *TREE* 11:445–446.
Mayr, E. 1942. *Systematics and the Origin of Species*. New York: Columbia U. Press.
Merçot, H., Atland, A., Jacques, M., and Montchamp-Moreau, C. 1995. Sex-ratio distortion in *Drosophila simulans:* co-occurrence of a meiotic drive and a suppressor of drive. *J. Evol. Biol.* 8:283–300.
Provine, W. 1971. *The Origin of Theoretical Population Genetics*. Chicago: U. Chicago Press.
Simpson, G. G. 1944. *Tempo and Mode in Evolution*. New York: Columbia U. Press.
Wallace, B. 1991. *Fifty Years of Genetic Load, An Odyssey*. Ithaca, NY: Cornell U. Press.
Vouidibio, J., Capy, P., Defaye, D., Pla, E., Sandrin, J., Csink, A., and David, J. 1989. Short range genetic structure of *Drosophila melanogaster* populations in an Afrotropical urban area and its significance. *Proc. Natl. Acad. Sci. USA* 86:8442–8446.

The Problems of Population Genetics

RICHARD LEWONTIN

The science of population genetics is the auto mechanics of evolutionary biology.

Organic evolution, in the Darwinian scheme, is a consequence of the conversion of variation among members of an ensemble into differences between ensembles in time and space. Classically those ensembles are collections of individual organisms, populations in the usual sense, but the Darwinian scheme can be applied as well to ensembles of organelles within cells or to collections of populations that make up a species. The essential features of the Darwinian scheme that determine both the subject and form of population genetics are that

1. there are processes that produce variation among individuals within a population,
2. there are processes that result in changes in the relative frequencies of the variants within a generation,
3. there is a hereditary process across generations that may result in further change in the relative frequencies of the different variants, but population frequencies are correlated across generational lines so that the frequency distribution of variants in any time interval is some nontrivial function of their distribution in the previous interval of time.

It is the relation between the processes of the generation and the modulation of variation within generations in features 1 and 2 and the processes occurring between generations in feature 3 that both create the science of population genetics and pose its methodological dilemmas and its shape as an inquiry. These relations are most easily understood as the transformations in the state of the population, shown in Fig. 1.1 (from Lewontin, 1974). The population change can be represented both in a genotypic and a phenotypic space, and the complete laws of transformation make use of both these spaces.

R. S. Singh and C. B. Krimbas, eds., *Evolutionary Genetics: From Molecules to Morphology*, vol. 1. © Cambridge University Press 2000. Printed in the United States of America. ISBN 0-521-57123-5. All rights reserved.

6   *Richard Lewontin*

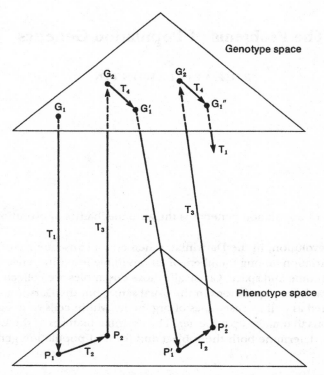

Figure 1.1. Schematic representation of the paths of transformation of population genotype from one generation to the next. *G* and *P* are the spaces of genotypic and phenotypic description. G_1, G_1', G_2, and G_2' are genotypic descriptions at various points in time within successive generations. P_1, P_1', P_2, and P_2' are phenotypic descriptions. T_1, T_2, T_3, and T_4 are laws of transformation. Details are given in the text. (From *Evolutionary Genetics* by R. C. Lewontin. Copyright © 1974 by Columbia University Press. Reprinted with permission of the publisher.)

G_1, G_2, P_1, and P_2 are genotypic and phenotypic descriptions at different points within a generation, while the same quantities with primes are the descriptions of the equivalent states in the next generation. The laws of transformation are

T_1: a set of epigenetic laws that gives the distribution of phenotypes, including fitness, that results from the development of various genotypes in various environments.

T_2: the laws of mating, migration, and natural selection including stochastic elements that transform the phenotypic array of potential reproducing units in a population within a generation.

T_3: the set of reverse epigenetic transformations that allows inferences about the distribution of genotypes G_2 corresponding to the distribution of phenotypes P_2.

T_4: the phenomenology of genetics such as segregation, recombination, mutation, horizontal transfer, etc., including stochastic elements, that allows us to predict the probability distribution of genotypes in the next generation produced from gametogenesis and fertilization, given an array of parental genotypes.

This representation of the structure of population genetic inference points immediately to two major sources of difficulty that have shaped the problems of population from its beginnings and still do so. First there is the problem of development. T_4, the complete apparatus of genetics that carries the population transformation across generations, is operating in the genotypic space to transform genotypic frequencies and cannot be framed in phenotypic terms. But T_2 that specifies the transformation of frequencies within generations is a set of physiological, ecological, and behavioral relations that operates in the phenotypic space. In the absence of a knowledge of the epigenetic transformations, the circuit is cut and nothing can be done. Nor can one depend on developmental genetics to provide the missing laws, because the main direction of developmental biology has never been, and shows no promise of ever being, to understand the environmental contingency of variations in development. Much of the past and the present problems of population genetics can be understood only as an attempt to finesse the unsolved problem of an adequate description of development.

The second problem that appears in Fig. 1.1 is the question of rates. Generally the movement in the phenotypic and the genotypic spaces is extremely small, the number of interacting genetic and phenotypic variables is large, and the number of generations and individuals over which observations can be taken is extremely limited. Thus the possibility of inferring the forces that are operating from the differences between G_1 and G_2 or G'_1 or between P_1 and P_2 is very remote. This has led to the other major preoccupation of population genetics, the possibility of estimating dynamic forces without ever actually measuring them. To the extent that population genetics has succeeded in this effort, it has provided a uniquely powerful methodology to evolutionary reconstruction that is not available from studies of physiology, ecology, and behavior, whose resolving power is necessarily low compared with the size of the actual forces operating in nature.

Finally, there is a problem that is not so much a difficulty within population genetics, but rather lies in its relationship with other branches of biology and with the rest of evolutionary theory. In these latter fields there is a strong value on finding universals, or at least in claiming universals even when there is no compelling evidence for them. So, even in a historical science like evolutionary biology, an overwhelming emphasis has been placed on selective, and even adaptive, explanations for all phenomena, to the exclusion of other possibilities, partly as a consequence of the belief that a science is validated to the degree that it can make universal claims. But population genetics has developed in a different direction. The schematic given in Fig. 1.1 is one of the historical trajectories of a character change as a result of the stochastic realizations of the interaction of a large number of forces that alter gene frequencies. Moreover, the strength of these various forces and the signal-to-noise ratio in their dynamics depend on circumstances that are different for different species and different genes. There are no universal claims, or even great generalizations, that can be made about selection intensities, population sizes, migration rates, or about the temporal history of these forces, which are certainly varying in time, or about the amount of genetic variance likely to be present at any moment, on which these forces act. Thus the population genetic research problem is short on general hypotheses to be tested. There are some very large statistical characterizations that can be achieved by repeated observations on a variety of species and genes, and, at the other extreme, there is the possibility of finding examples of the operation of particular forces like balancing selection or neutral evolution, but neither statistical generalities nor individual cases satisfy the model of what a powerful science is supposed to be.

1.1. Coping with Development

Population genetics has taken two opposite paths to solving the problem of reconciling the phenotypic and the genotypic spaces, paths that essentially finesse the problem by operating in only one of them. The first of these, characteristic of investigation before the emergence of molecular biology but now almost entirely abandoned, is to operate entirely in the phenotypic space. Phenotypic observations were used to make rough inferences about the genotypic space, but without providing a detailed description of the population in terms of genotypic frequencies. The tools and concepts were those of biometrical genetics: selection experiments on phenotypic traits in a variety of environments and a variety of selection schemes, estimates of components of genetic variance and covariance for phenotypic characters, studies of norms of reaction of fitness components, observations on fitness components or morphological traits from segregations, and recombinations involving large parts of the genome.

While lacking in any genetic details, these observations and experiments produced an immense richness of information about heritable phenotypic variation in populations and left unresolved a large number of important problems that are untouched by the more recent observations of molecular population genetics. The concentration on a molecular genetic description of population variation has depauperized the problematic of population genetics by marginalizing studies of the phenotype.

A few among the many discoveries of phenotypic population genetics that have yet to be dealt with at the level of the genotype are

1. Populations sampled from nature can usually be artificially selected for almost any morphological character, with some notable exceptions such as directional

asymmetry, and some physiological characters such as alcohol tolerance, while many physiological characters such as temperature and salt tolerance or speed of development are much less selectable. The relevant evidence on genic polymorphism is either in the opposite direction or almost entirely lacking. Enzyme loci are more genetically variable than genes coding for structural proteins, and artificial selection for adult alcohol tolerance in *Drosophila melanogaster* produces no change in allele frequencies at the *Adh* locus (Cohan and Graf, 1985; Weber, 1986). While homeobox genes and other similar loci active in early development have been extensively studied, nothing is known about the loci responsible for continuous selectable variation in either morphological or physiological phenotype.

2. Little is known about the number of loci segregating and the distribution of gene effects for phenotypic traits with heritable variation. Older gene mapping studies in which sparsely distributed markers were used could localize some gene effects to chromosome arms and occasionally localize a large effect. Modern quantitative trait loci (QTL) mapping studies in which much more densely distributed molecular markers are used have a much greater discriminatory power, but it is a long way from localizing the loci responsible for phenotypic differences between two highly divergent lines to giving a description in allelic frequency terms of the standing heritable variation in a population.

3. Characters with no phenotypic variation may nevertheless have considerable underlying genetic variation that is revealed under conditions of extreme developmental stress. Despite the clear evidence that such canalized characters are common and that the degree of developmental plasticity of a character can also be selected for (Waddington, 1960; Rendel, 1967), including, for example, the amount of fluctuating asymmetry in characters that show no directional asymmetry (Reeve, 1960), nothing is known at the genotypic level. Some glimpse of the possibilities is offered by the recent finding of Gibson and Hogness (1996) that selection for genetic assimilation of the bithorax ether phenocopy is accompanied by a change in the frequency of sequence variants at the *Ubx* locus.

4. Many correlations between morphological characters or between morphology and fitness are a consequence of linkage disequilibrium, but some are not or at least have not been broken by recombination and selection. The entire phenomenon of the resistance to artificial selection by countervailing fitness (usually fertility) effects needs to be investigated at the genotypic level. What is the explanation, for example, of the appearance in the classic selection experiments of Mather and Harrison (1949) of balanced sterile lines at intermediate levels of the selected phenotype so that no selection progress could be made in either direction, despite the presence of genetic variance for the character?

5. There are constraints on the pathway that phenotypic evolution can take under selection. As shown, for example, by Hall (1982) for the *ebg* locus in *Escherichia coli*, the same selected phenotype may have different genotypic bases because of different phenotypically similar mutations, some of which allow for further selectable mutations while others are dead ends for selection because the subsequent one-step mutations do not give appropriate phenotypes.

The mapping of genotypic mutational space onto phenotypic space is a major task for population genetics if an adequate genetic explanation of phenotypic selection and evolution is to be given.

The alternative solution to the problem of development has been to study characters for which there is no morphogenesis, that is, to study the genome directly. This approach is an attempt to collapse the two spaces in Fig. 1.1. It is this ploy that is the contribution of molecular population genetics, beginning, imperfectly, with electrophoretic and immunological studies of proteins and ending with the complete genotypic information contained in DNA sequences. Electrophoretic studies eliminate much of the problem of morphogenesis but there remains a residual ambiguity about genotype because even the most discriminating sequential electrophoresis will not distinguish some amino acid substitutions and there is no information on which or how many amino acid differences separate two electromorphs. There is, of course, no information about silent substitutions. Restriction enzyme studies are completely ambiguous about amino acid substitutions but at least provide information on silent sites. It is only with complete DNA sequences that a complete collapse of the phenotypic into genotypic description is possible.

The euphoria that has accompanied our ability, finally, to give a complete and unambiguous description of the allelic composition of a natural population for some arbitrarily chosen piece of the genome has hindered us from seeing that the problem of epigenesis has not been eliminated. While it is possible now to give a genotypic description of the variation in a population, the scheme of evolutionary transformation in Fig. 1.1 still requires phenotypic information on whole organisms. Formally, what DNA sequencing does is to allow a complete description of G_1 and so we have the illusion that we do not need the morphogenetic rules T_1. But the rules of transformation T_2, the rules of mating, migration, and natural selection, are rules about phenotypes. Gene sequences do not mate, they do not migrate, they do not live or die differentially, except as a consequence of the physiology and the metabolism of their organismic carriers in interaction with the physical and biotic environment. To eliminate totally the problem of morphogenesis from our observations, we must avoid all phenotypic descriptions above the level of genotype, including fitness. A direct attempt to measure the demography and the fitness of genotypes once again introduces the contingencies of development. The epigenetic transformation T_1 is still needed to predict the distribution of phenotypes P_1 that are transformed into the selected phenotypes P_2, and we still require T_3 for carrying the phenotypes back into the genotypic space.

The problem of directly measuring the fitness of genotypes has been considerable, and the results not generally encouraging. The direct estimation of net fitnesses of genotypes in nature and the discrimination of fitness differences from nonrandom mating, segregation biases, migration, and stochastic elements demand the possibility of observing mating pairs and associating their offspring with them, so that fertility as well as viability estimates can be made (see Prout, 1971, and Christiansen and Frydenberg, 1973, for the general methodology of such estimation). The most favorable material would be annual seed plants or an animal with single clutches that can be associated with parental pairs like birds or ovoviviparous fish. The most complete and sophisticated attempt to measure all the components of the life cycle in nature for genotypes has been, in fact, the work of Christiansen et al. (1973) on the esterase enzyme polymorphism in an ovoviviparous fish, *Zoarces*, but no strong fitness differences were found. Nor should this be very surprising. Given the immense amount of genotypic polymorphism present in natural populations, it is unlikely that large fitness variance can be associated with many individual polymorphic loci, and even when there are cases of strongly selected polymorphisms, it is unlikely that an arbitrary polymorphism, chosen for its ease of observation, will be one of these cases.

As substitutes for net fitness measurements, there have been attempts to measure the physiological properties of defined genotypes and then to argue that such properties are manifest as fitness differences in nature. An example of this approach is the work on the phosphoglucose isomerase enzyme polymorphisms in Colias butterflies (see Watt, 1991, for a review), in which it has been demonstrated that activity differences associated with different alleles appear to cause differences in male flight ability, which in turn must influence male mating success. The question that remains open, however, and a deep difficulty of all such work on fitness in nature, is that this is a *ceteris paribus* argument: "All other things being equal," male mating ability is an important component of fitness, but we do not know whether male mating ability is, in fact, an important component of the standing variance in fitness in nature. The problem is similar to the situation of melanism in *Biston betularia*, in which melanic forms on dark backgrounds are certainly less likely to be taken by predators, but it is by no means clear that predation pressure is sufficiently high to represent an important cause of differential mortality. Moreover, the laboratory demonstration of fitness differences associated with allelic variation makes use of deliberately constructed stress conditions that may be irrelevant in nature. The polymorphic alleles at the *Adh* locus in *D. melanogaster* have very different activities on an ethanol substrate and are selected in the presence of toxic levels of ethanol in the medium (Bijlsma-Meeles and Van Delden, 1976). These ethanol levels, however, are above any known in nature, including those in fermentation sheds in wineries (McKenzie and Parsons, 1974; McKenzie and McKechnie, 1978). The classical work of the Dobzhansky school on fitness in *D. pseudoobscura* was almost entirely in terms of larval competitive viability, yet in nature larvae of this species are never found in crowded conditions, and it seems more likely that the differential ability of females to find an egg-laying site and the ability of larvae to pupate before the small food sources dry up are much more important determinants of fitness variance. The problem of directly measuring fitness differences in nature is one that has plagued population genetics for nearly a century, and our ability to identify genotypes at the DNA level has not made it disappear.

1.2. The Problem of Rates

The difficulty of measuring fitness differences in nature is one facet of the more general issue that the individual forces operating on genotypic composition of a population at a locus are small so that the change from one generation to another, even the real stochastic change, is usually well below the observable limit in samples of feasible size. The question then is how we are to reconstruct the dynamic process in Fig. 1.1 when G and G' are not observably different. The answer has been to invert the process of deduction and to appeal to the cumulative effects of weak forces over long periods. From theoretical considerations we can predict the probability distribution of changes in genotypic composition under various models from one generation to the next. If the forces have remained constant for a very long period, then the population genotypic composition will come to a steady state under these forces, and, if we are lucky, that steady state will bear the unique signature of the forces, making an inference about them possible. We can use, for this purpose, not only the genotypic composition of a given population at a given locus, but the variation among loci, among populations, and among closely related species. Such synchronic or static data replace the direct measurement of the forces, and the direct diachronic observation of temporal change as the preferred method of population genetics.

Inference from static data has a long history in population genetics. In its simplest form, an observed intermediate allelic frequency of a polymorphism could be taken as evidence of balancing selection or of an equilibrium between mutation and selection. In the case of sickle-cell hemoglobin, in which the homozygote S/S is lethal, the inferred mutation rate would have to be of the order of 10^{-1} in West Africa but virtually absent elsewhere to explain the facts of the polymorphism. Clearly the postulate of balancing selection would make more sense even in the absence of the geographical correlation with malaria. The use of B/A and D/L ratios in the 1950s and 1960s were attempts to make inferences about the dominance of deleterious fitness effects when these effects could not be directly measured (see Lewontin, 1974, for a review of these and other similar methods). But the wholesale use of static data as the bread and butter of population genetics has depended, first, on the provision of a rich and precise genotypic data set by molecular survey methods and second, on the development of an articulated stochastic theory of gene frequency evolution that can, in principle, predict different static signatures for different mixtures of forces.

The first attempts to match stochastic theory to molecular data involved the use of the Ewens (1972) and Watterson (1977) distributions to test whether the frequency distributions of electrophoretic alleles at multiallelic loci like *Est5-B* and *Adh* in *Drosophila* were compatible with nonselective processes, with balanced polymorphism, or with purifying selection (Keith, 1983; Keith et al., 1985). This method assumed, as do all subsequent forms of the same approach, that

1. the experimental method distinguishes all genotypes,
2. the forces of selection and effective population size have been constant for a long time so that

3. the current allelic frequency distribution is at steady state, and
4. the current allelic frequency distribution is independent of the initial conditions.

These earlier attempts were not continued partly because the method of electrophoresis does not fulfill requirement 1 and partly because the tests of Watterson and Ewens turned out not to have sufficient power for such data sets to distinguish many selective hypotheses from nonselective hypotheses. In the case of *Est-5b* the data were not compatible with any of the alternatives, suggesting that the steady-state assumption was wrong or that a more complex selective hypothesis needed to be invoked. It soon became apparent that electrophoretic data simply did not have the necessary structure to distinguish hypotheses among forces. In the first place, most polymorphic loci are, at the electrophoretic level, biallelic or triallelic, and there is usually insufficient evidence in the frequencies of two or three alleles to test most hypotheses. There are exceptions. For example, the repeatable altitudinal clines in the two common alleles of the *Adh* (Grossman et al., 1969) and similar geographical clines in different continents of *Gpdh* and *Tpi* electromorphs (Oakshott et al., 1984) make a selective rather than historical explanation extremely likely. Second, even for the occasional locus like *Xdh* or *Esterase* in *Drosophila*, which has 20 or more alleles present in a single sample in which the pattern is of one or two common alleles and a large menu of rare alleles, extreme similarity of distribution between different populations can be explained either as repeatable selection or as a consequence of a small amount of migration between the populations. Third, the bulk of loci that are electrophoretically homozygous provide no information at all about the causes of their homozygosity. It is this problem that lies at the crux of the difficulty, a problem that has been solved by a fortunate quirk of nature.

The problem of genetic identity is to distinguish two possible cases for apparent allelic identity. It may be that two gene copies are identical because they stem from a recent common ancestor, so recent that no mutations have appeared since their genealogical splitting. That is, they are identical by descent. The other possibility is that their most recent common ancestor was indeed very ancient and that many mutations have occurred since that ancestor, but these mutations have been selected against, leaving only the allowable gene form. That is, the present genes are identical by state. Nor does this distinction apply at only a completely monomorphic locus. If there are two alleles in intermediate frequency at a polymorphic locus, one might be the ancient form kept uniform by a constant purifying selection, while the other might be a recent mutation that is selectively favored and has spread rapidly since its origin, having reached a polymorphic equilibrium with the older allele or being in the process of replacing it. On the other hand, both alleles may have been around for a very long time by chance because both are acceptable by natural selection while all other mutations have been rejected. The difficulty is that two alleles do not carry with them any information about how long they have been separated, and this is true whether they are identified by electrophoresis or by total amino acid sequencing. If only alleles also had temporal information that allowed us to distinguish identity by recent descent from enforced identity by state. But, of course, they do as a consequence of a lucky quirk of nature: the redundancy of the DNA code.

Alleles that are identical in the amino acid state may nevertheless differ in their nucleotides and if, as a first order of approximation, natural selection does not distinguish among alternative codons for the same amino acid, then gene copies can accumulate silent nucleotide substitutions unimpeded by natural selection. The number of such silent differences is then an index to the time that has passed since their common ancestor and can be used to distinguish among competing theories. It is vital to understand that DNA sequencing as the method of choice for population genetics does not stem from the fact that somehow DNA is a more basic or fundamental state description than amino acid sequences, but from the many–one correspondence between DNA sequences and amino acid sequences. Had the relation turned out to be one–one, there would be no advantage to DNA sequencing except as a cheap way to determine protein sequence.

More broadly, it is the presence, side by side in the same short gene region, of coding sites together with silent sites, with introns, and with untranslated and nontranscribed flanking DNA, all of which are subject to the same temporal processes of mutation and reproduction, but that differ from one another in the physiological consequences of their variation, that allows us to distinguish selective from purely historical similarities and differences. While the rationale for DNA sequencing is often stated to be that silent sites in codons are selectively neutral and so can serve as a selection-free molecular clock against which amino acid substitutions can be calibrated, that neutrality is not necessary (fortunately, since it is not true). All that we require is that different DNA positions have physiological constraints operating at different levels of intensity and in different contextual situations, so that the static variation observed within and between populations bears the distinguishable signatures of different temporal processes.

The most striking use of the temporal information in DNA sequences and the one with the greatest implications for our understanding of dynamical processes was, in fact, the first application of the method by Kreitman (1983) in his study of the standing variation at the *Adh* locus in *D. melanogaster*. Except for the single Fast/Slow (F/S) electrophoretic polymorphism that was deliberately introduced into the sample by the choice of lines, there was not a single amino acid polymorphism in a geographically very diverse sample, yet silent sites and introns were approximately 6% polymorphic (14% in exon 3). Thus the identity of the amino acid sequences could not be the result of recent common ancestry, but must be the outcome of a selection process in which all amino acid replacements, except the single widespread F/S polymorphism, have been removed by selection.[1] This result was made even more striking by that of Schaeffer and Miller (1992), which showed that in a sample of 99 genomes of *D. pseudoobscura*, there is no amino acid polymorphism at all at *Adh*, with the exception of one line with an isoleucine/valine change, despite the fact that there is ~6% silent nucleotide polymorphism. The implications of these results for our understanding of natural selection are powerful. Three-quarters of all random single nucleotide changes in coding regions will cause an amino acid substitution, yet every one of these has been screened out by natural selection in a protein that is, for example, 28% leucine/isoleucine/valine. While selection differences need not be large for any substitution, since population sizes for these species are estimated to be, conservatively, in excess of 10^5, selection has been discriminating enough to weed out every substitution. The problem is how every amino acid substitution can make a physiological difference that is ultimately translated into an average difference in viability and fertility. Nor is this discriminatory power of natural selection confined to a small enzyme like alcohol dehydrogenase. When similar studies were done on the very large and electrophoretically very polymorphic enzymes xanthine dehydrogenase (Riley et al., 1992) and esterase-5b (Veuille and King, 1995) a similar but less extreme result was found. Approximately 90% of all amino acid substitutions in xanthine dehydrogenase and 85% in esterase-5b have been rejected by natural selection.

If these results are indeed general, then a major problem has been posed for cellular physiology and metabolism and for evolutionary biology. Given environmental contingency and the accidents of development, how can the amino acid composition of proteins be translated so exquisitely into differential fertility and viability of individuals? If that translation does occur, how can the amino acid compositions of enzyme proteins change by multiple amino acids during speciation, as they indeed have?

Of course, speciation may be accompanied by drastic reduction in population size so that Ns is temporarily small and random fixations of amino acids differences may occur. But that cannot be the whole story. As pointed out by McDonald and Kreitman (1991), neutral fixations during speciation should reproduce between species the same ratio of silent to replacement substitutions as are polymorphic within species. Yet in some cases, for example glucose-6-phosphate dehydrogenase in *D. melanogaster* and *D. simulans* (Eanes et al., 1993) there is a great excess of amino acid differences compared with the intraspecific polymorphism.

The McDonald–Kreitman approach to detecting selectively driven divergence between species is another opening into an old problem made possible by the redundancy of the DNA code. It is widely recognized by systematists, even those trained to avoid typologies, that many characters that distinguish taxa have no detectable variation within species. The genus *Drosophila* is distinguished from other members of the *Drosophilidae* by the possession, by every individual in every species in the genus, of one proclinate and two reclinate orbital bristles. Given the lack of intrataxon variation, where did the variation for taxonomic differentiation come from and where did it go after the event? Many more or less plausible stories can be told, but there is no available method for demonstrating that natural selection was involved. For nucleotide sequences the possibility of implicating natural selection exists because we can make internal comparisons of polymorphism and species divergence for nucleotide positions of different function.

Regional heterogeneity in the density of polymorphism along a DNA sequence can arise for reasons other than the direct constraints on the sequence itself because of the extremely tight linkage among the nucleotides in short genomic sections. Thus a region of unusually high nucleotide variation can be taken as evidence that random fixation of mutations has been resisted by some form of balancing selection. Kreitman and Hudson (1991) used this approach to infer heterosis for the F/S electrophoretic alleles of *Adh* in *D. melanogaster*, although their data show an excess nucleotide polymorphism even within the Slow allelic class, so that something else besides heterosis of the two electrophoretic forms must also be happening.

In contrast, extremely low variation in a genomic region might be the equilibrium signature of a high functional constraint on the region, but it might also be the historical leftover of a recent rapid replacement of one or a few sites that carried adjacent nucleotides to fixation by hitchhiking. This explanation by Berry et al. (1991) of nucleotide homogeneity of the fourth chromosome must surely be the correct one. Such a rapid fixation, however, might be the result of a selectively advantageous mutation's having been fixed, a so-called selective sweep, but it might also be the consequence of a segregation distortion. Because the segregation distorters (SDs) that have been studied are necessarily those that are in some sort of balanced equilibrium in populations, like the SD in *Drosophila* and the *t locus* in mice, we do not know how often distorters, unopposed by selection, may sweep through populations. Whether or not one can distinguish selective sweeps from distorter sweeps depends on the rapidity with which the two processes occur and on the time since the event. In the absence of any ancillary evidence about replacement speed and relaxation time, it is hard to see how these processes can be separated.

The problem of distinguishing selection from segregation distortion applies as well to the inference of adaptive evolution between species. Both are illustrations of a deep structural problem in making inferences from static variation. Such data are, at the first level, information on the times of divergence from a common ancestral sequence, in mutational units, between genomes in populations. But the distribution of divergence (coalescence) times is a dual representation of the distribution of rates, in mutational units, of fixation of particular sequences through the population. An unusually low fixation rate of some stretch of genome compared with the average for the genome may mean that there are forces holding sequences in equilibrium against random fixation or that the time scale for the production of variation in the region is more compressed. Such a compression can come from a local excess in mutability or from a migration into the population of genomes that were divergent in this genic region.

In a like manner, a deficiency of variation means either that some genome has spread unusually rapidly through the population, for whatever reason, or that the time scale for the production of variation has been extended because of selective constraints. The decision from among these alternatives and the biological interpretation of the causes of the changes in time rates must be made on other grounds not contained in the static data themselves.

The discovery by molecular biology of different functional classes of nucleotides has not only led to the possibility of obtaining strong evidence about old problems, but has revealed a new set of problems for population genetics, problems that arise from considering the standing variation within those nucleotide classes. These are the problems of explaining regional heterogeneity in DNA sequences even within functional types. Thus, at the *dpp* locus in *Drosophila*, a gene that sits at the center of many of the processes of early embryonic development, there is a major heterogeneity in the pattern of nucleotide polymorphism in the large intron. There is a concentration of polymorphism at the two ends of the intron and little variation in the middle of the intron both within and between species (Richter et al., 1997; Newfeld et al., 1997). Some of this conservation of the middle of the intron coincides with multiple short repressor motifs in the region, but much of it remains to be explained on functional grounds. There is also a region of perfect conservation within and between species in the 3′ untranslated region of Intron 3, in the middle of a region of considerable nucleotide divergence and polymorphism. What are the functional sources of such conservations and how frequent are they?

One of the most interesting and general unresolved problems arising from observations of genomic heterogeneity is the question of codon bias. A reasonable explanation of unequal codon usage is that, especially for highly transcribed genes, unequal availability of tRNA species constrains the codon usage to match it for the most efficient translation. But this explanation raises many problems when different species and different genes are compared. For example, the AUA codon is totally avoided for isoleucine in *Adh* in all species of *Drosophila*, yet for other genes in *D. melanogaster*, it is used on average ∼5% of the time. In *D. virilis*, however, it comprises 38% of the isoleucine codons for the *dpp* gene. Moreover, although a clear bias toward G/C ending alternative codons is general for genes in *D. melanogaster*, in *D. virilis* the *dpp* gene is biased toward A/T in eight amino acid groups. Have there been major changes in tRNA availability from one species to another, and, if so, why do different genes in the same species not agree in their codon bias differences from other species? The existence of specific usages of gene and species make simple explanations of the evolution of codon usage doubtful.

The issue of codon usage raises a problem of constraint on DNA sequences that is not ordinarily considered. The secondary structure of message, its stability, and the time course of its translation will be influenced by the nucleotide sequence. We should not suppose that a maximum rate of translation is necessarily optimal for the translation process. The final folding of a protein occurs as a consequence of the formation of folding intermediates during the process of translation, and, presumably, some slowing down and speeding up of chain elongation at appropriate places is necessary for a realization of local free-energy minimizations during folding. These variations in speed will be affected by the particular DNA sequence, partly in its relation to the availability of tRNA, so that the problem is more complicated than that of simply maximizing the use of common tRNA species. Nor should we even suppose that fidelity of translation is an unmitigated good. The translational process, like any molecular mechanism, makes mistakes. There must be a population of variant protein molecules in the cell. Are these variants of some physiological use? After all, if one argues that heterosis comes from the advantage of having two slightly different copies of a molecule in each cell, then why not have more than two different copies and in unequal numbers? But all of these considerations lead us to suppose that the DNA sequence itself, even in coding regions, may be constrained over stretches of various length. It is the common assumption that constraints on amino acid variation are the consequence of requirements on the physiological function of the protein in question. But if we suppose that DNA sequence itself is constrained over even short stretches, then the amino acid sequence will be held constant as a consequence of the constancy of the underlying nucleotide sequence. At the moment we do not know how much of the variation and the conservation of protein sequences is a secondary consequence of requirements on the nucleotide sequence itself. Nor do we know how to find out.

There is yet another possibility opened by data from DNA sequence that returns us to a previous stage in the history of inference. The attempt to make judgements about forces from static data on electrophoretic variation foundered for three reasons: ambiguity about the actual genetic identity of the observed classes, insufficiently rich data sets even for the few multiallelic loci, and a stochastic theory that was derived from the consideration of the moments of stationary distributions rather than the richer structure of gene phylogenies. The first two problems are dealt with directly by the nature of DNA sequence data and by an immense amount of haplotypic diversity that appears even in modest data sets. The third problem has been attacked by developments in stochastic theory that take into account the details of haplotypic phylogenies, although still suffering from the need to make many simplifying assumptions (Hudson, 1983; Hudson et al., 1987; Tajima, 1989). Essentially these developments attempt to compare the observed distribution of DNA haplotypic differences within a sample with those expected from purely neutral evolution. The null hypothesis that produces the expected distribution of differences among haplotypes consists, necessarily, of a long list of assumptions, so what is being tested when the data are fit to the null hypothesis is the conjunction of a large number of neutral assertions. Some of these are background assumptions about population history and linkage that are not of primary interest in the test. Others are relevant to whether natural selection needs to be invoked to explain the observed pattern of differences. For example, the method given by Hudson et al. (1987) involves the differences between two species for two or more genomic regions and the polymorphism within these regions. The background assumptions are that generations are discrete, that mutations at different sites occur independently and with a common mean for sites within a locus, that the species are at a stationary state, that they diverged from an ancestral population whose population size was the average of the stationary sizes of the current species, and that there is complete linkage within gene regions but free recombination between them. The selectively relevant assumptions are that the segregating mutations that are actually segregating within the species are selectively equivalent, that the degree of purifying selection may differ between regions but is the same between species, and that no change in selective regime occurred in the process of speciation. We do not know, in general, how deviations from the background assumptions affect the power of the test compared with the selective assumptions, nor do we know, in general, what the power of the test is to detect one sort of selective deviation as opposed to another. Thus we do not know at present how to interpret observations of significant or nonsignificant test results. In practice, such tests can be subject to an extensive investigation of their operating characteristics. Simonsen et al. (1995) have shown for the simpler test of Tajima (1989) that the power to detect selection is low, a result that is reminiscent of the earlier situation of similar tests on electrophoretic data. The important question is less whether the tests are powerful in any absolute sense, but what their relative power is to detect deviations from the background assumptions as opposed to deviations from the selective assumptions. Tests of low power are only useless. Tests that powerfully detect the wrong deviations are destructive.

1.3. Problems of Generality

Twenty years of electrophoretic surveys and 15 years of DNA sequencing and restriction analysis make it clear that there is a lot of standing genetic variation in populations both at the amino acid level and the DNA level, but there is a great deal of difference among species and gene regions in how much genetic variation is to be found. It is also possible from theoretical considerations to make some predictions about correlations, as for example that regions of the genome with low recombination should turn out to be regions of lower genetic variation, all other things being equal, as indeed has been observed for a couple of chromosomal regions in *D. melanogaster* (Berry et al., 1991; Begun and Aquadro, 1992; Martin-Campos et al., 1992). How well this correlation holds in other organisms with different demographic histories remains to be seen.

The alternative to statistical generalization is the exemplification of particular phenomena without any strong implication about their frequency in nature. So, for example, a convincing case has been made that the *Adh* electrophoretic variation in *D. melanogaster* is a balanced polymorphism (Kreitman and Hudson, 1991), but, unlike the universalist claims about heterosis that were common in population genetics in the 1950s and 1960s, it has never been suggested that *Adh* in *D. melanogaster* provides any general insight into genetic variation in general. In like manner, the strong case made by Eanes et al. (1993) that *G6pdh* diverged selectively in *D. melanogaster* and *D. simulans* is not taken as a general refutation of claims of the neutral evolution of amino acid sequences. But producing an example, or even several, of selective divergence, neutral evolution, balanced polymorphism, a selective sweep-out of variation, or linkage equilibrium or disequilibrium fails to match the universalist model on which so much of science is built. It is, moreover, a self-terminating research program since its objective, to demonstrate that certain evolutionary cases do indeed occur in nature under appropriate circumstances, is soon realized.

There is a third pathway of research in population genetics that is exceedingly difficult of accomplishment partly for reasons of technique and partly for reasons of the intellectual division of labor in biology. That pathway is to bring population genetics into the main methodological tradition of functional biology by attempting to measure the forces operating in populations directly, rather than by statistical inference, and by carrying out functional experiments that lead to a knowledge of the physiological, ecological, and behavioral mediations for evolutionary change. To do so means solving the problem of rates.

First, it is necessary to make cellular and organismic physiological measurements of the effects of nucleotide substitutions, by use of the available apparatus of *in vitro* mutagenesis and recombination followed by a molecular and physiological analysis of the relevant metabolic functions including transcription and translation, as has been attempted for the *Adh* locus in Drosophilae (Laurie-Ahlberg and Stam, 1987; Laurie and Stam, 1994). But the demonstration of some difference in some parameter of cellular activity is not sufficient. It must be possible to show that these physiological differences mediate fitness differences whose values are at least 2 orders of magnitude smaller than what has been previously measured in the laboratory. The problem is not one of sufficient experimental population size because even in Drosophilae populations of effective size, 10^5 can be managed. The major problem is the detection, in sufficiently large samples taken over sufficiently long time periods, of the small changes brought about by selection intensities of the order of 10^{-3} or smaller. The time problem can be solved by the choice of an organism with generation times of the order of a day. The sample size problem can only be solved by labor or by mechanization.

And to what end? If population genetics can only illustrate a few exemplars of processes or find very general correlations, what purpose is served by being able to provide the complete path of mediation from DNA to fitness? The answer lies in the essentially methodological nature of the science. While progress in molecular biology has consisted both of learning how things actually are in nature (the double-helical nature of DNA) and in finding methods for asking such questions (how to sequence DNA), the discoveries, both theoretical and experimental, that have had the most impact on population genetics are not those that revealed a particular fact about the world, but those that have provided methodologies for answering particular questions when desired. That is why theoretical investigations have played such a central role in population genetics as opposed to any other branch of biology.

Population genetics is a science of methodologies that provides the concepts and techniques that can be used to resolve, in favorable circumstances, problems in evolution, human genetics, plant and animal breeding, and ecology. The greatest methodological challenge that it now faces is to connect the observations about the outcome of evolutionary processes to the tradition of experimental functional biology.

NOTE

1. The only alternative would be to suppose a differential correction mechanism by which amino acid substitution mutations are corrected much more efficiently than silent changes. While no such mechanism is known, it cannot be excluded finally without an experimental search for such a possibility.

REFERENCES

Berry, A. J., Ajioka, W., and Kreitman, M. 1991. Lack of polymorphism on the *Drosophila* fourth chromosome resulting from selection. *Genetics* **129**:1111–1117.

Begun, D. J. and Aquadro, C. F. 1992. Levels of naturally occurring DNA polymorphism correlate with recombination rates in *Drosophila melanogaster*. *Nature (London)* **356**:519–520.

Bijlsma-Meeles, E. and Van Delden, W. 1976. Intra- and inter- population selection concerning the alcohol dehydrogenase locus in *Drosophila melanogaster*. *Nature (London)* **247**:369–371.

Christiansen, F. B. and Frydenberg, O. 1973. Selection component analysis of natural polymorphisms using population samples including mother child combinations. *Theor. Popul. Biol.* **4**:425–445.

Christiansen, F. B., Frydenberg, O., and Simonsen, V. 1973. Genetics of *Zoarces* populations. IV. Selection component analysis of an esterase polymorphism using population samples including mother–offspring combinations. *Hereditas* **73**:291–304.

Cohan, F. W. and Graf, J. 1985. Latitudinal cline in *Drosophila melanogaster* for knockdown resistance to ethanol fumes and for rates of response to selection for further resistance. *Evolution* **39**:278–293.

Eanes, W. F., Kirchner, M., and Yoon, J. 1993. Evidence of adaptive evolution of the *G6pd* gene in *Drosophila melanogaster* and *Drosophila simulans* lineages. *Proc. Natl. Acad. Sci. USA* **90**:7475–7479.

Ewens, W. 1972. The sampling theory of selectively neutral alleles. *Theor. Popul. Biol.* **3**:87–112.

Gibson, G. and Hogness, D. S. 1996. Effect of polymorphism in the *Drosophila* regulatory gene *Ultrabithorax* on homeotic stability. *Science* **271**:200–203.

Grossman, A. I., Koreneva, L. G., and Ulitskaya, L. E. 1969. Variation of the alcohol dehydrogenase locus in natural populations of *Drosophila melanogaster*. *Genetika* **6**: 91–96.

Hall, B. G. 1982. Evolution on a petri dish: using the evolved beta-galactosidase system as a model for studying acquisitive evolution in the laboratory. *Evol. Biol.* **15**:85–149.

Hudson, R. R. 1983. Testing the constant-rate neutral allele model with protein sequence data. *Evolution* **37**:203–217.

Hudson, R. R., Kreitman, M., and Aguade, A. 1987. A test of neutral molecular evolution based on nucleotide data. *Genetics* **116**:153–159.

Keith, T. P. 1983. Frequency distribution of Esterase-5 alleles in two populations of *Drosophila pseudoobscura*. *Genetics* **105**:125–155.

Keith, T. P., Brooks, L. D., Lewontin, R. C., Martinez-Cruzado, J. C., and Rigby, D. L. 1985. Nearly identical allelic distributions of Xanthine dehydrogenase in two populations of *Drosophila pseudoobscura. Mol. Biol. Evol.* **2**:206–216.

Kreitman, M. 1983. Nucleotide polymorphism at the alcohol dehydrogenase locus of *Drosophila melanogaster. Nature (London)* **304**:412–417.

Kreitman, M. and Hudson, R. R. 1991. Inferring the evolutionary histories of the *Adh* and *Adh–dup* loci in *Drosophila melanogaster* from patterns of polymorphism and divergence. *Genetics* **127**:565–582.

Laurie-Ahlberg, C. C. and Stam, L. F. 1987. Use of P-element mediated transformation to identify the molecular basis of naturally occurring variants affecting *Adh* expression in *Drosophila melanogaster. Genetics* **115**:129–140.

Laurie, C. C. and Stam, L. F. 1994. The effect of an intronic polymorphism on alcohol dehydrogenase expression in *Drosophila melanogaster. Genetics* **138**:379–385.

Lewontin, R. C. 1974. *The Genetic Basis of Evolutionary Change.* New York: Columbia U. Press.

Martin-Campos, J. M., Comeron, J. M., Miyashita, N., and Aguade, M. 1992. Intraspecific and interspecific variation at *y–ac–sc* region of *Drosophila simulans* and *Drosophila melanogaster. Genetics* **130**:805–816.

Mather, K. and Harrison, B. S. 1949. The manifold effects of selection. *Heredity* **3**:1–52.

McDonald, J. H. and Kreitman, M. 1991. Adaptive protein evolution at the *Adh* locus in *Drosophila. Nature (London)* **351**:652–654.

McKenzie, J. A. and McKechnie, S. W. 1978. Ethanol tolerance and the *Adh* polymorphism in a natural population of *Drosophila melanogaster. Nature (London)* **272**:75–76.

McKenzie, J. A. and Parsons, P. A. 1974. Microdifferentiation in a natural population of *Drosophila melanogaster* to alcohol in the environment. *Genetics* **77**:385–394.

Newfeld, S. J., Padgett, R. W., Richter, B., Finley, S. F., de Cuevas, M., and Gelbart, W. M. 1997. Molecular evolution at the *decapentaplegic* locus in *Drosophila. Genetics* **145**:297–309.

Oakshott, J. G., McKechnie, S. W., and Chambers, G. K. 1984. Population genetics of the metabolically related *Adh, Gpdh* and *Tpi* polymorphisms in *Drosophila melanogaster.* I. Geographic variation in *Gpdh* and *Tpi* allele frequencies in different continents. *Genetica* **63**:21–29.

Prout, T. 1971. The relation between fitness components and population prediction in *Drosophila.* I. The estimation of fitness components. *Genetics* **68**:127–149.

Reeve, E. C. R. 1960. Some genetic tests on asymmetry of sternopleural chaetae number in *Drosophila. Genet. Res.* **1**:151–172.

Rendel, J. M. 1967. *Canalization and Gene Control.* London: Academic, Logos Press.

Richter, B. G., Long, M., Lewontin, R. C., and Nitasaka, E. 1997. Nucleotide variation and conservation at the *dpp* locus, a gene controlling early development in *Drosophila. Genetics* **145**:311–323.

Riley, M. A., Kaplan, S. R., and Veuille, M. 1992. Nucleotide polymorphism at the xanthine dehydrogenase locus in *Drosophila pseudoobscura. Mol. Biol. Evol.* **9**:56–69.

Schaeffer, S. W. and Miller, E. I. 1992. Estimates of gene flow in *Drosophila pseudoobscura* determined from nucleotide sequence analysis of the alcohol dehydrogenase region. *Genetics* **132**:471–480.

Simonsen, K. L., Churchill, G. A., and Aquadro, C. F. 1995. Properties of statistical tests of neutrality for DNA polymorphism data. *Genetics* **141**:413–429.

Tajima, F. 1989. Statistical method for testing the neutral mutation hypothesis by DNA polymorphism. *Genetics* **123**:585–595.

Veuille, M. and King, L. M. 1995. Molecular basis of polymorphism at the esterase-5B locus in *Drosophila pseudoobscura*. *Genetics* 141:255–262.

Waddington, C. H. 1960. Experiments on canalizing selection. *Genet. Res.* 1:140–150.

Watt, W. B. 1991. Biochemistry, physiological ecology, and population genetics – the mechanistic tools of evolutionary biology. *Funct. Ecol.* 5:145–154.

Watterson, G. A. 1977. Heterosis or neutrality? *Genetics* 85:789–814.

Weber, K. E. 1986. The effect of population size on the response to selection. Ph.D. dissertation Cambridge, MA: Harvard University.

CHAPTER TWO

The Mathematical Foundations
of Population Genetics

WARREN EWENS

2.1. Introduction

My allocated task in this chapter is to focus on foundational and historical aspects of the application of mathematical and statistical methods in population genetics theory. I will therefore only briefly discuss the technical aspects of the recent mathematical theory and will say little about specific aspects of the current theory. These are in any event discussed in other chapters in this book. Thus I will take a historical and philosophical approach to the subject of mathematical population genetics, focusing on the question of the extent to which the mathematical aspects of population genetics are central, perhaps even crucial, to the subject. In particular I will discuss the extent to which this was so in the works of the three great masters, Fisher, Haldane, and Wright. I will then outline what I see to be the broad trends of the present theory and the extent to which it relies on mathematical analyses. To do all this, it is necessary to discuss what we mean by a mathematical, as opposed to a more-or-less verbal, analysis in any scientific discipline, and I will start by taking up this matter.

2.2. What Is a Mathematical Analysis?

One can put forward two extreme views on the place of mathematics within the foundations of population genetics. The first is that a mathematical approach has been crucial to an understanding of the operation of the Darwinian paradigm when based on Mendelian genetics, and supporters of this view claim that it can hardly have been a coincidence that the three initial analysts of the Darwinian–Mendelian process followed a largely mathematical approach. The other view is that biological, ecological, and evolutionary processes are so complex that a mathematical approach necessarily glosses over crucial aspects and thus can hardly give useful information on the main themes of the Darwinian

R. S. Singh and C. B. Krimbas, eds., *Evolutionary Genetics: From Molecules to Morphology*, vol. 1. © Cambridge University Press 2000. Printed in the United States of America. ISBN 0-521-57123-5. All rights reserved.

evolutionary theory. A watered-down version of this latter view is that to the extent that a mathematical component to population genetics is useful at all, all the necessary mathematics was done by Fisher, Haldane, and Wright and that present-day work consists largely of tinkering at the edges of the edifice that these three created. Lewontin himself, modestly and no doubt provocatively, has from time to time proposed this latter view. Taking this cue, I will not hesitate to be provocative in this essay and will air my own views more than is perhaps seemly and appropriate.

If the second of the above extreme viewpoints is correct, there is little to be said in this chapter – there may well be very little mathematics in the properly perceived foundations of population genetics. But of course the truth lies in between: The proposal that I shall make is that a mathematical analysis (in the sense that I define below) is important but not central to the foundations of population genetics in the evolutionary field. The exact level of importance at the foundational level will always be controversial, but whatever conclusions are reached at this level, no one will deny that mathematics has a significant place in the subject at a more tactical level, taking up smaller well-defined problems, and insofar as they are relevant to the theme of this chapter, I shall discuss aspects of tactical questions also.

What is a reasonable requirement of a mathematical analysis in evolutionary theory, assuming that we want the analysis to advance our understanding of evolution in a way that nonquantitative reasoning could not do? The answer is quite different from the corresponding requirement in, say, physics, and I start by discussing some of these differences.

First, in physics, we often assume an extreme precision in the models we use and thus we assume that any mathematical model describes the real world extremely accurately, so that even minutely small discrepancies between observation and the predictions of a theory are enough for us to discard that theory. This precision allows an extremely accurate prediction of future phenomena, for example the future positions of the planets. Second, the use of mathematical models allows general principles (for example, least-action principles) to be brought to bear to discuss physical phenomena: this is not only of practical value, but also has a most pleasing aesthetic and philosophical appeal. Finally, the mathematical theory in physics presents a magnificent overall structure – indeed, it is no doubt the finest creation of humanity's genius.

No one would expect a similar story in evolutionary theory. First, it is accepted from the start in biology that the complexity of the subject implies that any mathematical analysis is based on models that are never more than caricatures of reality, so that we are not disturbed by minor deviations of the observations from the predictions of our analysis. We also accept that completely unpredictable events, from mutations on upward, occur all the time. Thus we cannot hope to predict the future progress of biological evolution. On the other hand we might hope to establish broad general principles of evolution as a genetical process, and I will argue that Fisher, in particular, aimed at this high level in much of his work. However, it is certainly inappropriate to aim at an overarching structure such as that created by mathematics in physics. The complexities of biological evolution are too great for us to do this, and we are left, in population genetics theory, with a collection of results that embraces aspects of a grand strategic theory on the one hand and a number of rather unconnected tactical-level conclusions on the other.

The fact that mathematical models in evolutionary theory are caricatures at best introduces the concept of the robustness of the model used to describe the aspect of evolution of interest or, more specifically, of the extent to which conclusions derived from a model are still useful even when the model describes reality only rather imprecisely. This question hardly arises at all, for example, in physics, but does arise (or perhaps I should say should arise) in a subject like mathematical economics, in which models also describe the real world only approximately. I will further discuss the modeling aspect of mathematical population genetics below.

A further important difference between mathematical modeling in physics and in biology is the following. Broadly speaking, the word theoretical has two connotations, namely, mathematical and pertaining to a general theory. In the context of physics, these two interpretations become more or less identified: the general theory of Newtonian physics, for example, from the foundations to the intricate details, is unavoidably mathematical. In quantum theory the situation is even more extreme: here we believe the theoretical implications of the mathematics and note their important practical applications, while at the same time we are completely bewildered by the seeming bizarre physical principles on which they are based. In evolutionary theory, by contrast, there is no identification of the two interpretations of the word theoretical, and we would never believe any prediction of the mathematical analysis of any model of evolutionary behavior if it were clearly bizarre – the caricature-like nature of our mathematical models ensures this. To avoid appearing to imply the general theory connotation when discussing the place of mathematics in evolutionary theory, I will refer below to a mathematical analysis rather than a theoretical result, implying a perhaps quite restricted analysis of a quite specific problem.

What is a mathematical analysis? Many arguments in population genetics use mathematics in quite a minor way, often little more than using algebraic symbols for the quantities discussed. This is often appropriate, but these arguments are essentially purely verbal. By mathematical analysis I mean here more than this, namely, the use one or another technique of mathematics, for example, finding the solution of a differential equation, finding the maximum of some complicated function subject to various constraints, or analyzing the stability behavior of a complex system, to arrive at a qualitative conclusion that cannot be reached nonmathematically. It would be a major attractive feature of the mathematical analysis if it arrived at some important general evolutionary principle, and it would be an interesting consequence if it showed that common-sense nonmathematical conclusions on some point are incorrect.

An example of this definition of a mathematical analysis is as follows. It has been known since Fisher's early days (Fisher, 1930) that the condition under selective forces only, for a stable equilibrium of allelic frequencies with both alleles represented at positive frequencies and that the heterozygote have (viability) fitness exceeding that of each of the two homozygotes. While Fisher found this result mathematically, it can be established by more-or-less nonmathematical methods, and, further, the stable allelic frequencies can be found without resorting to any serious mathematics. Thus one may reasonably claim that the condition that the heterozygote must have the highest fitness is not one for which a mathematical analysis is necessary. Whatever the truth of such a claim might be, no one could make the corresponding claim in a k-allele system for general k. Finding the condition that there be a stable equilibrium point with all k alleles at positive frequency unavoidably involves a mathematical analysis of some intricacy: indeed, even simply stating the required condition (that the matrix of fitness values have one positive eigenvalue and at least one negative eigenvalue) can hardly be done in other than mathematical terms. Similarly, the conditions for maintaining two alleles at each of two linked loci cannot be approached other than through a thoroughgoing mathematical analysis, and, in general, results in the multilocus theory cannot be found other than by mathematical methods. Finally, results in the stochastic theory can only seldom be reached nonmathematically, intuition being notoriously misleading in this area.

2.3. The Three Masters

It is appropriate to begin by discussing the extent to which the three masters, Fisher, Haldane, and Wright, whose work set the entire direction of population genetics theory, saw the mathematical aspects of their respective works as being identified with their respective overall viewpoints about evolution and to comment on the extent to which mathematics supports these viewpoints. In doing so, I must emphasize that any brief account such as this must leave many significant areas untouched or with distorted emphases. It would take an entire book to flesh out the details and to give a balanced account, not least because their views changed over time and the personal interactions among them distorted the presentation of their opinions. Subject always to this reservation, I will take Fisher (1930, 1958), Wright (1968, 1969, 1977, 1978), and Haldane (1932) as representing and summarizing their overall statements about evolution, with Haldane's various later papers providing changes and additions to his earlier views.

There is of course a large degree of identity of view among the three: all three saw evolution very much from the genetical side and thus, inevitably, from the quantitative side. All three also realized that mathematics could not tell the whole story, and each developed an evolutionary world view only partly derived from his mathematical analysis. In my opinion, Fisher, following his avowed aim of putting the Darwinian theory on Mendelian foundations and in establishing the general principles of the new science of population genetics, aimed far more at finding broad general evolutionary principles than did Haldane and Wright. I begin my discussion of the three masters by developing this view, referring for illustration to the 1930 version of Fisher's book.

In a letter to Wright (cited in Provine, 1986) following the publication of this book, Fisher said, "In some ways the first chapter is the most important, and in some the second." The first chapter in the book uses mathematics to show that variation is conserved in a Mendelian hereditary system, as contrasted to its loss in a blending system, thus preserving variation for natural selection to act on. This observation came as a revelation to me when I was a student, overcoming as it does the main obstacle seen by Darwin to his evolutionary theory. No doubt this conclusion was well known in 1930, 22 years after the publication of the Hardy–Weinberg law, but it certainly needed stating, and even today I find many students of evolution completely unaware of the fact that this (rather than the binomial form of the genotype frequencies) is the important implication of Hardy–Weinberg. Further, I believe that only the quantal Mendelian hereditary system will have this crucial variation-preserving property. It would be interesting to develop a genetic anthropic principle from this: if intelligent life is found on other worlds, then the hereditary system will be Mendelian, since only under such a system will intelligent life evolve. Furthermore, as Fisher stated, we could have inferred in 1860 (i.e., after Darwin but before Mendel) by mathematical arguments that our own hereditary system would be Mendelian. A little bit of mathematics has indeed gone a long way.

The second chapter consists of an attempt, the details of which I now believe to have been misinterpreted for many years, to recast in genetical and quantitative terms the main tenets of the Darwinian theory, that evolution needs variation in a population and that the rate of improvement is related to that level of variation, through the "Fundamental Theorem of Natural Selection." This was a new venture, contained entirely within the then-infant subject of population genetics theory, and no doubt perceived by Fisher as the crowning masterpiece of his work. Unfortunately his cryptic mode of writing led to what I believe was a misunderstanding of the content of the theorem (a misunderstanding of which I was as guilty as any), leading in particular to the view that the theorem was erroneous. In what I believe is the correct statement of the theorem (Ewens, 1989), it involves what I call the partial change in mean fitness, does not assume random mating, and is true as a mathematical theorem in an extremely wide range of cases. While its value as an evolutionary principle is not perhaps as high as Fisher would have wished, is nevertheless of great importance and value. From the point of view of this essay, the theorem could not have been established other than by a mathematical analysis, especially in its multilocus version, and certainly could not at all have been found by a purely verbal argument.

Quantitative arguments were also central in Fisher's arguments in the following three chapters of the book. Chapter 3, concerning the evolution of dominance, was important to Fisher first as an example of an evolved phenomenon and second as showing the efficacy of even a small amount of selection in directing evolutionary change. Chapter 4 is the most mathematical in the book, and perhaps the main conclusion he drew from it is that, in the presence of selection, random drift has negligible effect in determining gene frequency changes in large populations. This view, concerning as it does the relative importance of stochastic and selective effects, is discussed at greater length below. Chapter 5 discusses the evolutionary importance of epistatic interaction between two loci and the consequent evolution of tight linkage between them. Of these three chapters, I have long believed that Fisher's mathematics and his conclusions on the evolution of dominance are incorrect, that the argument on the importance of random drift is satisfactory as far as it goes (but that in ignoring the effective population size concept and simply assuming very large population sizes and in not considering such problems as the effect on a neutral locus of an evolutionary process at a linked selected locus, does not go nearly as far as Fisher might have thought), and that the argument on the evolution of linkage is essentially correct but with some serious flaws, which I discuss below in connection with Wright's calculations on the subject.

Much of the remainder of the book is devoted to Fisher's eugenic views in relation to humanity, without general evolutionary principles' being discussed. They have very little serious mathematical content and thus are not directly relevant to this discussion.

Thus mathematical arguments were important to Fisher, perhaps even central, in forming his main evolutionary views. Nevertheless it has to be emphasized that Fisher had a balanced view on the extent to which mathematics can be used in evolutionary arguments. In a second 1930 letter to Wright, he said, "Mathematicians always tend to assume that the hardest mathematics will be the most important . . . . It is certainly not true of my book, where the . . . nonmathematical parts . . . are often of greatest ultimate interest" (Provine, 1986). Certainly many of the results in the most mathematical chapter in his book (Chap. 4) follow up arcane and ultimately irrelevant topics, while the essentially verbal later chapters contain a wealth of interesting arguments and observations.

Wright's main evolutionary arguments focus on a somewhat different level than Fisher's. While he and Fisher agreed completely on the need to base the Darwinian theory on Mendelian principles, much of Wright's work, in particular the centerpiece of his evolutionary work, namely the shifting-balance theory of evolution, appears to attack in the main a perhaps secondary question, namely that of assessing the optimal conditions under which natural selection can act. Clearly mathematics played a major role in his analysis of this matter, but the extent to which the mathematical treatment helps is open to question. I have severe reservations about several points in his mathematical analysis, of which I mention here only two.

First, Wright's focus was on the building up of harmonious multilocus gene complexes, so that the necessary vehicle for his mathematical analysis is the algebra of multilocus evolution. It is well known that the appropriate quantities to consider for this are gametic frequencies, gene frequencies not being sufficient – yet from start to finish, and even after the mid-1950s, when the necessity for using gametic frequencies became very well known, he used gene frequencies in all his analyses. He clearly knew that this involved some approximation and sometimes justified this approach by assuming that the loci involved are loosely linked, but the assumption of loose linkage implies problems with the shifting-balance theory, as noted below. Thus any conclusions that he drew from his work by using a gene frequency analysis are clearly liable to reconsideration, and some have been shown subsequently to be misleading. Indeed, several of his analyses used models that cannot even begin to be useful. For example, he used (Wright, 1952) gene frequencies in an attempt to analyze the equilibrium behavior of a two-locus model in which there is no stable internal equilibrium, as an analysis in which gametic frequencies are used reveals. So here, at any rate, there is no sense in which mathematics is being used constructively in a concrete problem.

This leads to my second, and far stronger, reservation. I have never been convinced by the mathematical analysis of the shifting-balance theory. My concern is not so much with the notoriously changing interpretation of the axes in the adaptive topography diagram, as discussed by Provine (1986) and many others, since this diagram provides no more than a rough image of the evolutionary process. Nor is it with the reliance on random genetic drift in one phase of the process. Rather, my problem is with the reliance on subdivided populations as a central aspect of the theory and the almost complete lack of any sort of serious mathematical analysis of the mechanics of the shifting-balance process as it relates to gene frequency changes within and between populations.

The theory was motivated by the argument that the appropriate focus in evolution is on harmonious gene complexes and that the single-locus theory cannot satisfactorily describe the evolution of such complexes. The aim of the theory was to prescribe a way in which these complexes would best, and perhaps could only, evolve. Under the theory, a small subpopulation is assumed to proceed from one peak in the adaptive surface across a valley to higher selective peak, with random drift playing an essential role in the initial stages of the process, thus creating a subpopulation of organisms with a gene complex more favored than had existed before, and also more favored than those remaining in neighboring subpopulations. This complex is then exported by migration to other subpopulations, leading to its eventual fixation in the entire population. There are, to me, many difficulties with this paradigm, all based in part on what appears to me to be an incomplete mathematical modeling of the process. Here I will mention only two of these, both relating to the first phase of the shifting-balance theory.

The first problem is that no analysis is made of the effects of migration into the subpopulation in question of nonharmonious complexes while the favorable new gene complex becomes fixed, initially through stochastic effects, subsequently by selection. Such migration might perhaps fatally hinder this fixation process. The amount of migration between subpopulations that makes the subpopulational structure more or less irrelevant for evolutionary processes was later analyzed by Moran (1959), who showed that only a small amount of migration is needed for this to occur. Wright's stationary distribution for gene frequencies in the presence of selection and migration is not sufficient for the required dynamic, nonstationary analysis of this phase of the process.

The second problem concerns the breaking apart, through recombination, of the favorable complex being built up during this first phase (especially since, as noted above, Wright's analysis implicitly assumed loose linkage). The crucial role of the nature of the epistatic fitness relations (supermultiplicative versus submultiplicative) in the relative rates of buildup and breakdown of multilocus complexes has only recently been analyzed in the amount of detail required (see Otto et al., 1994), and the conclusions of this analysis have never been explored in the shifting-balance paradigm concurrently with migration between subpopulations.

These are but two of the problems that must be attacked mathematically for a proper development of the shifting-balance theory, and my own (clearly extreme) view is that despite sometimes desperate attempts to support the theory by mathematical methods, no such support, certainly in its entirety, has ever been offered. Indeed, recent mathematical examinations of the theory (see in particular Moore and Tonsor, 1994; Michalakis and Slatkin, 1996; and Phillips, 1996) indicate the very great difficulties involved with sustaining it. Note in particular that Moore and Tonsor show that, for a given fitness configuration favorable to the theory, there is only a very small window of migration rates for which the gene exportation central to the theory can succeed, in which a low migration rate implies only infrequent exportation to other demes of any favored gene complex built up in one deme and a high rate implies rare fixation of the favored complex in the first place. This problem with the theory has been clear from the start, and I see no reason to change the view I offered almost 20 years ago (Ewens, 1979, p. 32): "[Wright's] picture of evolution may well rely on an equipoise of migration rates, fitness differentials and deme sizes of an unrealistically finely-tuned nature."

These criticisms are, of course, strong and no doubt personal ones. However, they sustain one of the themes of this chapter, that a reasonably complete mathematical analysis is necessary before any evolutionary argument, including the shifting-balance theory, can be put forward with any confidence. Of the broad evolutionary concepts discussed in this chapter, the mathematical support for the shifting-balance theory is, to me, the least satisfactory.

Having said all the above, I hasten to add two points. First, it is a truism that evolution concerns complexes of genes rather than individual genes. The point made above is different from this and concerns problems with the particular paradigm advanced by Wright, focusing on subdivided populations, to bring gene-complex evolution about. Second, perhaps as a result of his interest in the shifting-balance theory, Wright recognized the importance of stochastic factors in evolutionary genetics and integrated the stochastic theory into his world view rather more than did Fisher and Haldane. Quite apart from their relevance to the first phase of the shifting-balance paradigm, stochastic processes are central to recent work, in particular with coalescent processes and the neutral theory discussed further below. Here Wright's work has been quite fundamental.

Haldane tended to focus more on providing general support for the Darwinian theory in a wide and important range of tactical, as opposed to strategical, questions. Since the compexities of nature preclude a description of evolution in purely mathematical terms, his focus might well be the most appropriate one. His best-known work, on the other hand, is quite mathematical and is intended as a description of part of the evolutionary process: the series of papers written in the 1920s and summarized in his 1932 book used differential equations to predict gene frequencies in future generations, given values of fitnesses, mutation rates, and so on, in a wide-ranging variety of real-world situations. This analysis was successful, for example, in discussing the increase in frequency of the melanic form of the peppered moth *Amphidasys betularia* during nineteenth-century industrial England. However, it cannot be expected to be equally valuable for very long-term evolutionary processes, since changes in fitness values through altered environments and through the evolutionary process itself, not to mention random and unpredictable phenomena such as mass extinctions from extrinsic nonbiological events, imply that the assumptions made at the start of the process will not hold throughout. Nor was this intended. It is nevertheless interesting that Haldane's gene frequency predictions by using differential equations come closest, in all of mathematical population genetics, to the form of predictive activity, again by using differential equations, occurring so often in physics.

One of Haldane's contributions, however, I believe to be misguided. This is his concept of the evolutionary load, or cost of natural selection. (Note that I feel equally strongly about the so-called segregational load, although I have no problem with the mutational load concept.) I have given the reasons for holding this view, which of course is shared by many, in detail elsewhere (Ewens, 1993), and will not repeat them at length here: From the point of view of this chapter, the deficiencies in the analysis derive from inappropriate mathematical modeling relating to the concept (see also Feller, 1967). I should also point out the apparently little-known fact that Mayr's well-known criticism of beanbag genetics was specifically directed against the evolutionary load concept, as a reading of Mayr (1963) makes quite clear. I might also mention that Haldane's subsequent defense of beanbag genetics, by using the debating trick of avoiding Mayr's comments and inventing and discussing his own issues, not only avoids the thrust of Mayr's argument, but is also, to me – and here no doubt I again differ from the conventional wisdom – a quite unsatisfactory defense of theoretical population genetics and, in particular, of the value of its mathematical analyses. I have also expanded on these views (Ewens, 1993).

Over the years I often sensed a frustration among biologists (including Mayr) with whom I have discussed the question of the evolutionary and the segregational loads, who felt that the concepts were misguided but who could not see their way through the mathematical derivations and so find the errors in the load arguments. I do not for one moment believe that ability in mathematics is necessary in a population geneticist, but I do feel that there is real value in having those in the subject who are well versed in mathematics and do not lose the woods for the trees in a mathematical analysis. In my view, the clarification of the load question is a case in point.

2.4. Technical Aspects

Although my focus is on general issues, I now turn briefly to the purely technical mathematical aspects of the work of the three great pioneers. These were wonderfully innovative and influenced many areas of science other than population genetics.

The main innovations were largely in the stochastic theory. The Wright–Fisher model is a Markov chain (a phrase never, to my knowledge, used by either). It is one of the first Markov chain models used and analyzed in any scientific application and, despite the later introduction of more complex Markov chain models in genetics, still stands as the benchmark model in population genetics, not least through its relation to the often-misunderstood concept of the effective population size. Besides Markov chain models, the pioneers studied the closely associated diffusion processes – indeed, Fisher's introduction of these in 1922 predated formal introduction of the Wright–Fisher model (although his diffusion theory analysis was clearly based on knowledge of properties of that model). Branching processes were also introduced early and used to calculate the survival probability of a new mutant. Their masterly analysis of all three processes forms the basis for essentially all work in the stochastic part of the theory to this very day.

It is inevitable, as in any great pioneering work, that there were several deficiencies in the early stochastic theory that, to present a rounded picture, should be mentioned. In the early 1930s, Kolmogorov presented his general theory of stochastic processes, including both the forward and the backward equations. His work was not mentioned by the pioneers until Wright used it, tangentially, in the 1940s. It is a pity that this is so: Fisher used forward equation methods in a roundabout way for problems immediately soluble by the backward equation, and none of the pioneers evidently realized that Kolmogorov's work implied that the solution they offered to the forward equation in the simplest case (no selection, no mutation, etc.) was incomplete, being but the first term in an infinite series that is not well described by this first term alone.

Two further mathematically based concepts introduced by the pioneers should also be mentioned. The first of these is Wright's concept of the effective population size, which shows that the size that best describes the evolutionary behavior of the population might be far smaller than the actual breeding number. Fisher never paid as much attention to this concept as he should have, possibly associating it with the shifting-balance theory to which he was so antipathetic, and used extremely high population sizes (up to 10^{12}) in his analyses, surely far too large in general. It might be mentioned that, even today, very few who use the effective population size seem to realize its true definition – or, more precisely, its variety of definitions – and in particular its close relationship to the Wright–Fisher model is not widely understood.

The second concept is Fisher's additive genetic variance. It is true that Fisher first introduced this in the context of the correlation between relatives (Fisher 1958), but he was quick to realize its evolutionary importance. It plays a key role in his Fundamental Theorem of Natural Selection, in plant and animal breeding theory, and in much else. The additive genetic (or, perhaps better, genic) variance is one component of the total genetic variance and the idea, first arising in Fisher's genetic work, of extracting meaningful components from a total variance led to the basic concept of the analysis of variance, whose impact in experimental science generally has been so immense.

How much did the mathematically based theories influence workers in evolution such as Mayr, Huxley, Dobzhansky, and Simpson in developing their broad genetically based syntheses? I disagree with Provine's view that "the tension between [Fisher's and Wright's] evolutionary views was a highly creative force in evolutionary biology" and that "[their] differences in viewpoint ... are central ... in evolutionary theory" (Provine, 1986). To me, almost the very opposite is the case. Fisher, Haldane, and Wright agreed on most of the essentials, in particular the enormous importance of the Mendelian hereditary system to the Darwinian theory, and it was the features commonly agreed on among them that, on the whole, entered the evolutionary syntheses and textbooks. So far as their disagreements are concerned, a fully informed discussion, especially of the more mathematical aspects of their theories, was probably well beyond anyone else at the time that they wrote. A full resolution would require more information about genetic parameters than is available even today. Further, several of the differences between Fisher and Wright, argued almost ad nauseam, concerned arcane points that were ultimately found to be unimportant to the broad general theory.

It is also interesting to note how often such mathematical calculations as were included in the textbooks were either not sufficiently seriously discussed or were simply wrong. I mention here three examples from the book most closely allied to the mathematical analysis, namely, that of Dobzhansky (1970). First, the only reference in this book to Fisher's Fundamental Theorem is in a seemingly irrelevant addendum (page 204) in a section on heritability and the additive genetic variance: hardly a serious discussion of a "Fundamental Theorem." Second, Dobzhansky's analysis of the rate of loss of genetic variation was out of date, relying on the incomplete 1930's solution to the forward equation, commented on above, rather than to Kimura's complete solution (Kimura, 1955), which had been available for 15 years when the book was published. Finally, his conclusion (page 220) that "higher vertebrates and man do not possess enough 'load space' to maintain more than a very few balanced polymorphisms" relies on what I believe to be a sadly misleading (segregational) load theory, also commented on above. (Given Dobzhansky's views on the ubiquity of polymorphism in real populations, it is also difficult to believe that this conclusion was anything other than a gesture to the load argument.)

There is one point that the three masters understood thoroughly and on which they agreed completely, but for which, unfortunately, their views did not sufficiently permeate into the common body of knowledge. This concerns the argument, which has persisted in one form or another since Darwin's time, that the amazingly intricate adaptations that we see in living organisms today could not have arisen in the lifetime of our planet through the processes of natural selection, so that our initial biological material must have arrived from extraterrestrial sources. Hoyle and Wickramasinghe (1981) and Denton (1986) are but the latest to promote this sorry speculation, which arises through incorrect mathematical modeling of the genetics of the evolutionary process. (It is important to note that the argument I am making is not that biological material cannot arrive on earth from extraterrestrial sources, but that it is unnecessary to make this assumption to explain evolution on our planet. The center of the Hoyle–Wickramasinghe argument is that there has not been enough time for the observed terrestrial biological evolution to occur since the creation of our planet. It is interesting to note that their models do not even allow sufficient time since the Big Bang for the observed terrestrial biological evolution!) As stated above, we do not expect our mathematical models to reflect reality in a complete way. However, we do expect them to contain the essence of reality and that the broad conclusions reached from them describe in outline, even if not in exact detail, the implications of that reality. In creating quite misleading mathematical models of the evolutionary process, these writers have promoted incorrect conclusions that have been detrimental not only scientifically, but also in the realm of public policy.

2.5. Where Are We Today?

I will discuss first the advances since the time of the masters in the mathematical parts of population genetics theory that follow the general directions of their own work, again focusing on broad outlines and general points. I have elsewhere described the fleshing out of the work of the masters as belonging to the prospective theory, that is, the theory that takes the current state as given and looks to properties of evolution in the future. The huge development in the prospective stochastic theory, with which one must primarily associate the name of Kimura, is one such area. Far more is known about this subject than was known 40 years ago, and a more advanced use of Markov chain and diffusion theory has allowed far more rapid and general calculations than were possible before. We freely use backward and forward Kolmogorov equations and know the complete solutions of these equations, we use multidimensional diffusions to describe multiallelic and multilocus processes, we have a thorough understanding of boundary types and of the behavior of diffusions near boundaries, we have a firm idea of the accuracy of these (continuous) processes in describing real-world discrete processes, and we have developed the theory of conditional diffusion processes (i.e., processes for which we make the condition that a specified boundary is the one eventually reached), a development inspired by the retrospective theory that I will discuss in a moment. We have a complete stochastic theory (see Gillespie, 1991) in which the stochastic element arises not from random sampling within a finite population, but from temporal or spatial variations (or both) in the fitness values themselves. This theory appears often to be more useful in describing the real world than that deriving from the classical fixed-fitness theory. We have the ability to generalize our models beyond those allowing a mathematical analysis to those in which computer simulation, or calculation, plays an essential role in our attempts to understand nature. Although no new issue of principle is involved with this vastly expanded theory, there is no doubt that as to a knowledge about innumerable details, we are very much better off than we were 40 years ago.

In particular, this increased knowledge of the mathematics of the stochastic theory has led to one important practical development, namely, the introduction of different models (together with their different analyses) appropriate for the different forms of genetic data obtained in recent decades. In increasing levels of genetical sophistication we have the charge-state model appropriate for electrophoretic data and some simple models of microsatellite evolution, the infinitely many alleles model that recognizes the effective infinity of possible allele DNA sequences for any gene, and finally (and ultimately) the infinitely many sites model, in which the data used are DNA sequences. In this succession of models we may discern the influence of the form of data used on the theory used. This change of focus is central to retrospective population genetics, which I will discuss in a moment, and is one in which mathematics and statistics play a fundamental role.

Of further recent mathematical work extending that of the masters, one should mention in particular the theory of inclusive fitness, as developed largely by Hamilton (1964) and alternatives to this based on frequency-dependent selection (Cavalli-Sforza and Feldman, 1978). Inclusive fitness was well known to the three masters, all of whom referred to it explicitly in their major works, but the recent mathematical analysis has led to a clearer understanding of the principles involved.

The second major area in which theory has blossomed is in multilocus analysis, following the introduction (by Kimura, 1956, and Lewontin and Kojima, 1960) of the correct vehicle for this analysis, namely, the evolutionary analysis of gametic frequencies. It was noted above that Wright's view of evolution centered around coadapted gene complexes, but that he did not use the appropriate vehicle for analyzing the evolution of these complexes, namely, gametic frequencies. The debate between gene-complex and single-gene approaches to evolution continues unresolved to this day. No one doubts, of course, that gene complexes are the correct entities whose evolution is to be considered. The real question is whether the single-locus work now making up such a large proportion of population genetics theory is sufficiently accurate to answer important questions within an acceptable margin of error. If not, much of the present theory is irrelevant. Of recent writers, Lewontin (1974) appears largely to believe that it is irrelevant and Crow and Kimura (1970) largely to believe that it is not. Clearly a resolution of this question is one of the outstanding unresolved issues of population genetics, especially as far as the value of the mathematical theory is concerned.

It is appropriate to mention here Fisher's Fundamental Theorem of Natural Selection. It is an irony of the generally perceived views of Wright and Fisher, with the former seen as introducing and favoring the gene-complex approach and the latter as favoring the single-gene approach, that the Fundamental Theorem, at least in the version I prefer, considers the evolution of an arbitrary constellation of gene loci (even the entire genome), correctly uses the evolution of gamete rather than gene frequencies, and takes epistatic interactions and recombination fully into account. It is therefore a fully gene-complex theorem. Given Wright's inappropriate use of gene rather than gamete frequencies to analyze gene-complex evolution, their generally perceived positions are here reversed.

Returning to the multilocus theory generally, it is clear that this is of necessity largely mathematical and that many conclusions found from the mathematical analysis could not have been arrived at by nonquantitative methods. I mention here only a few. First, the fact that, even under random mating, the mean-fitness can decrease in multilocus systems came, in the 1960s, as a great shock to many and was interpreted by many as showing that the Fundamental Theorem is not true in multilocus systems. As noted above, I now believe that this interpretation, which I shared at the time, is incorrect and arose as a result of a misunderstanding of what the theorem says. Nevertheless, the decrease in mean fitness phenomenon is a curious matter. Second, the number and the locations of points of equilibrium that are possible in a multilocus setting and the effect of the recombination fraction on these and their stability behavior (as established largely in a series of papers in the 1970s by Karlin and Feldman) could not be arrived at nonquantitatively and formed the beginning of a large-scale attack on multilocus problems. More broadly, it is obvious that the evolutionary effect of recombination can be attacked in only a multilocus setting, as can the converse retrospective question of the effect of evolution on recombination values. Finally, only a thorough analysis of multilocus evolutionary models can shed light on what was described above as a central unresolved question of population genetics, namely, the extent to which the essential features of a multilocus evolutionary process can be described by its constituent single-locus marginal processes.

One central area of multilocus theory is that of modifiers: in discussing the evolution of recombination rates, Fisher (1958, p. 117) appears to be using an interpopulational selection argument quite at variance with his general intrapopulation selection viewpoint. It is interesting to note that this is at variance with his approach to the evolution of dominance, which used modifier loci. Modern work on the evolution of recombination rates, mutation rates, etc., starting with Nei (1968, 1969), also uses modifier loci and multilocus methods, considering the evolution of primary and modifier loci jointly. This obviates the need for interpopulational arguments.

The analysis of the evolution of genetic phenomena brings me to the final area of modern-day population genetics analysis, one in which I believe mathematics plays a key part. This is the retrospective theory, that is, the theory in which we observe the genetic make-up of contemporary populations, and asks how we got here. As just noted, the pioneers asked these questions, but focused on the (extremely important) question of the broad evolution of genetic phenomena. Recent work tends more to be data driven, in particular by taking currently observed gene frequencies as its starting point and asks about the nature of the evolutionary process that produced these frequencies. Since these frequencies are the result of a stochastic process, this inquiry necessarily uses statistical methods and represents a new departure in the application of population genetic methods to the discussion of evolutionary questions.

As a first example, I take the case in which we have a sample of n genes of various allelic types and wish to test the hypothesis of selective neutrality, that is, that the alleles observed are selectively equivalent. Assuming the infinitely many alleles model, this hypothesis may be tested as proposed by Watterson (1978). Leaving aside the details of this testing procedure, we note that this is clearly an activity of the retrospective theory. Because of the use of numerical data, we might expect greater precision in such testing processes than that arising in the quantitative but data-free prospective theory. From this it follows that we may expect a greater role for significant mathematical and statistical analysis in the retrospective theory. There are, however, at least three cautionary notes to sound in this regard. First, any statistical analysis will be based on some assumed evolutionary model, and the simplifications and inaccuracies that almost certainly will be made in choosing this model will imply inaccuracies in the result of the analysis. Second, even if we get the model approximately right, the variances of estimators of important parameters in evolutionary population genetics are often of the order of $1/\log n$ rather than the more classical $1/n$ (where n is the population size) essentially because of the high correlations between observations due to common ancestry. This clearly implies weak precision in estimation and that tests of hypotheses will have little power (unless we can make clever use of data from many loci simultaneously). Finally, there is the danger of being carried away by some of the more elegant aspects of the sampling theory, attributing (with Keats) truth to beauty when a more prosaic assessment is probably necessary.

The best-known retrospective activity in areas close to population genetics concerns the construction of phylogenetic trees from contemporary genetic data. Tree construction often proceeds on a purely algorithmic basis and has no direct connection with much of population genetics theory, as discussed above. Perhaps the most important new area of retrospective research grounded within population genetics theory is that associated with the coalescent, the classic papers on which are due to Kingman (1982a, 1982b). It is hardly an exaggeration to say that coalescent theory has revolutionized the subject of mathematical population genetics, with much of current research being carried out within its framework. The theory will be discussed elsewhere in this volume (see Chaps. 21 and 29), and I mention it here so as to finish this account of the use of mathematics in theory of population genetics on a high and optimistic note. I believe that a retrospective analysis of the history of contemporary populations, whose theoretical aspects will almost certainly be centered around the coalescent, will provide one of the main areas of scientific research over the next few decades and that mathematics will be an essential, indeed the leading, component of this analysis.

2.6. Summary

Population genetics theory arose in the early years of this century as part of the process of placing the Darwinian theory on a Mendelian foundation. From the start (and specifically with its three great pioneers) it has had a significant mathematical component in it, involving mathematical models of evolution as a genetic process. This mathematical approach was a key factor in validating and in part quantifying the Darwinian theory and thus a key factor in the creation of the so-called neo-Darwinian synthesis.

These successes should not, however, lead us to excessive optimism about the value and the use of quantitative models in evolution. The use of mathematics in biological evolution is different from its use in physics, and the predictive value of quantitative models in biological evolution is far less than that of quantitative models in physics. Mathematical modeling has been of use first in establishing broad general principles of biological evolution, and second in a great variety of specific limited questions. It has also been misused, in particular in load theory and in the argument that there has not been time for the observed evolution that we see to have occurred purely on our planet.

The present-day theory uses quite sophisticated mathematics, building on but inevitably going considerably beyond that of the pioneers. This is even more true of the new retrospective theory than of the older prospective theory, and it seems likely that the present focus on retrospective questions will carry the value of mathematics in population genetics theory to new heights.

REFERENCES

Cavalli-Sforza, L. L. and Feldman, M. W. 1978. Darwinian selection and "altruism." *Theor. Popul. Biol.* 14:268–280.

Crow, J. F. and Kimura, M. 1970. *An Introduction to Population Genetics Theory.* New York: Harper & Row.

Denton, M. 1986. *Evolution: A Theory in Crisis.* Bethesda, MD: Adler and Adler.

Dobzhansky, Th. 1970. *Genetics of the Evolutionary Process.* New York: Columbia U. Press.

Ewens, W. J. 1989. An interpretation and proof of the Fundamental Theorem of Natural Selection. *Theor. Popul. Biol.* 36:167–180.

Ewens, W. J. 1993. Beanbag genetics and after. In *Human Population Genetics: A Centennial Tribute to J. B. S. Haldane,* P. P. Majumder, ed., pp. 7–29. New York: Plenum.

Feller, W. 1967. On fitness and the cost of natural selection. *Genet. Res.* **9**:1–15.

Fisher, R. A. 1930. *The Genetical Theory of Natural Selection.* Oxford: Clarendon.

Fisher, R. A. 1958. *The Genetical Theory of Natural Selection.* 2nd revised edition. New York: Dover.

Gillespie, J. H. 1991. *The Causes of Molecular Evolution.* New York: Oxford U. Press.

Haldane, J. B. S. 1932. *The Causes of Evolution.* London: Longmans Green.

Hamilton, W. D. 1964. The genetical evolution of social behavior. I. *Theor. Popul. Biol.* **7**:1–16.

Hoyle, F. and Wickramasinghe, N. C. 1981. *Evolution from Space.* London: Dent.

Kimura, M. 1955. Solution of a process of random genetic drift with a continuous model. *Proc. Natl. Acad. Sci. USA* **41**:144–150.

Kimura, M. 1956. A model of a genetic system which leads to closer linkage by natural selection. *Evolution* **10**:278–287.

Kingman, J. F. C. 1982a. On the genealogy of large population. *J. Appl. Prob. A* **19**:27–43.

Kingman, J. F. C. 1982b. The coalescent. *Stochast. Proc. Appl.* **13**:235–248.

Lewontin, R. C. and Kojima, K. 1960. The evolutionary dynamics of complex polymorphisms. *Evolution* **14**:458–472.

Lewontin, R. C. 1974. *The Genetic Basis of Evolutionary Change.* New York: Columbia U. Press.

Mayr, E. 1963. *Animal Species and Evolution.* Cambridge, MA: Harvard U. Press.

Michalakis, Y. and Slatkin, M. 1996. Interaction of selection and recombination in the fixation of negative-epistatic genes. *Genet. Res.* **67**:257–269.

Moore, F. B.-G. and Tonsor, S. J. 1994. A simulation of Wright's shifting-balancing process: migration and the three phases. *Evolution* **48**:69–80.

Moran, P. A. P. 1959. The theory of some genetical effects of population subdivision. *Aust. J. Biol. Sci.* **12**:109–116.

Nei, M. 1968. Evolutionary change of linkage intensity. *Nature (London)* **218**:1160–1161.

Nei, M. 1969. Linkage modification and sex differences in recombination. *Genetics* **63**:681–699.

Otto, S. P., Feldman, M. W., and Christiansen, F. B. 1994. Some advantages and disadvantages of recombination. In *Frontiers of Mathematical Biology*, S. A. Levin, ed., pp. 198–211. Berlin: Springer-Verlag.

Phillips, P. C. 1996. Waiting for a compensatory mutation: phase zero of the shifting-balance process. *Genet. Res.* **67**:271–283.

Provine, W. 1986. *Sewall Wright and Evolutionary Biology.* Chicago: U. Chicago Press.

Watterson, G. A. 1978. The homozygosity test of neutrality. *Genetics* **88**:405–417.

Wright, S. 1952. The genetics of quantitative variability. In *Quantitative Inheritance*, E. C. R. Reeve and C. H. Waddington, eds., pp. 5–41. London: Her Majesty's Stationery Office.

Wright, S. 1968. *Evolution and the Genetics of Populations. Vol 1. Genetic and Biometric Foundations.* Chicago: U. Chicago Press.

Wright, S. 1969. *Evolution and the Genetics of Populations. Vol 2. The Theory of Gene Frequencies.* Chicago: U. Chicago Press.

Wright, S. 1977. *Evolution and the Genetics of Populations, Vol 3. Experimental Results and Evolutionary Deductions.* Chicago: U. Chicago Press.

Wright, S. 1978. *Evolution and the Genetics of Populations. Vol 4. Variability Within and Among Natural Populations.* Chicago: U. Chicago Press.

CHAPTER THREE

A Natural Historian's View of Heterosis and Related Topics

BRUCE WALLACE

3.1. Introductory Remarks

The publication *Heterosis,* edited by John W. Gowen of the Iowa State University Press, appeared in 1952, more than four decades ago. One of the chapters of that book, "Nature and Origin of Heterosis," was prepared by Th. Dobzhansky. Although in doing so I reveal my ancient roots, I propose here to revisit that largely forgotten age, an age that is receding into oblivion at an increasingly rapid rate as the number of scientific papers published annually grows exponentially. What I propose to do is to join the late M. Kimura (1991) and J. Crow (1992) in reviewing and providing my perception of words and concepts over which we have had and may still have differences of opinion. I suspect, however, that each of us has somewhat altered his views over time, but none of us completely understands the others' current position – nor, unfortunately, do many latecomers to the population-genetics scene attempt such understanding.

3.2. Breaking the Vocabulary

A former colleague at Cornell once told me that in his course on the history of labor movements in the United States, he spent the first two lecture periods destroying his students' vocabulary. He did this because, early in his career, he learned that words that he had used casually throughout a term had entirely different meanings to his students and different meanings to different students. As a result, his carefully constructed lectures were thoroughly scrambled by students who thought they understood, but did not. I shall follow his example.

Among quantitative geneticists, heterosis means only hybrid vigor, a commonly observed consequence of crossing unrelated individuals. Darwin (1877)

This paper is dedicated to my friend Richard Lewontin who at 65 years of age, of necessity, must be a treasured colleague of four decades.

R. S. Singh and C. B. Krimbas, eds., *Evolutionary Genetics: From Molecules to Morphology*, vol. 1. © Cambridge University Press 2000. Printed in the United States of America. ISBN 0-521-57123-5. All rights reserved.

made systematic observations on the differing outcomes of self-fertilization and cross-fertilization among plants of many species and genera. His, for example, seems to be the first record of hybrid vigor in corn (*Zea mays*).

Possible explanations for heterosis commonly cited are overdominance and dominance. The first postulates that the presence of two dissimilar alleles at a single gene locus may result in the observed hybrid vigor. The second recognizes the commonly observed dominance of normal alleles over their deleterious homologues, with each parent providing the dominant alleles that conceal the deleterious recessive alleles contributed by the other.

Related to the concept of overdominance, is the phrase "heterozygosity *per se*" that is seemingly attributed to I. Michael Lerner. At any rate, Lerner (1959) felt compelled to make the following remark at the Cold Spring Harbor symposium:

"What "heterozygosity *per se*" refers to today is obviously to interchangeability of heterotic effects on some phenotypic level of loci with polygenic action. Such interchangeability of homozygous polygenes has never caused a raised eyebrow and there is no reason why it should for heterzygotes."

In this quotation, Lerner uses the word heterotic, a term used by many as an adjective derived from heterosis. Lerner's point, if I understand it correctly, is that among the allelic combinations involving many gene loci (polygenic action) that are responsible for hybrid vigor in some aspect of the phenotype are those that require heterozygosity, not homozygosity, at this or that locus. At which loci heterozygosity is required, in Lerner's view, would depend on the particular one of many possible gene combinations. That, at any rate, is my interpretation of his use of the word interchangeability.

I suspect that our current understanding of the molecular structure of the gene blurs the seemingly sharp distinction between dominance and overdominance as possible explanations for heterosis. The gene that is defined as a unit of physiological action and that is generally inherited as a Mendelizing unit has, at the molecular level, regions of DNA that possess a variety of functions: promoters, transcription initiation sites, enhancers, as well as the better-understood structural gene with its introns and exons. These pieces frequently are physically located at distances that allow genetic recombination, thus leading to a multiplicity of genes under the definition of the gene as a unit of recombination. Here, then, is a problem: Is hybrid vigor that is ascribable to genetic differences at the functional gene level to be regarded as overdominance if the functional unit can be disassembled by recombination? This question is related, I believe, to matters that Kimura (1991) has addressed under the concept of "compensatory neutral mutations."

Overdominance as an explanation for hybrid vigor has lost favor over the years, largely, I suspect, through the demonstration of increasing levels of gene–gene and gene–environment interactions. These interactions would be demonstrated especially well by means of long-term studies during which recombination would continuously reduce the sizes of nonrecombined blocks of genes.

In terms of evolutionary, ecological, or population genetics, I believe that calculating the long-term consequences of overdominance at a single locus is a futile exercise. I do not believe that any one allele when homozygous or that any one (heterozygous) combination of dissimilar alleles can be best suited for all possible environments (or genetic milieux). That belief might stand repetition: No one allele nor any combination of alleles is likely to be best suited for all possible environments or all possible background genotypes. Perhaps I have now managed to disassociate myself from the commonly held view that I support overdominance; I do, but only with the qualifying term marginal, which I shall explain shortly. Perhaps my position resembles Lerner's that was cited earlier: Within any one combination of genes (polygenes) that produces a phenotype that suits an individual to its environment, many loci may be heterozygous for dissimilar alleles.

The preceding remarks have raised the concept of marginal overdominance, a concept that recognizes the variation in effects on relative fitness in different situations that are ascribable to different genotypes (homozygous or heterozygous) at a single gene locus. The concept assumes that the allele with the proper effect tends to be the dominant one of a heterozygous pair (see Table 3.1 and

Table 3.1. *Several Possible Mechanisms by which Heterozygous Individuals may Attain Higher Average Fitnesses than their Corresponding Homozygotes**

	AA	Aa	aa
1	1	1	$1-s_1$
2	$1-s_2$	1	1
3	$1-s_3$	1	1
4	1	1	$1-s_4$
.	.	.	.
.	.	.	.
.	.	.	.
i	$1-s_i$ (or 1)	1	1 (or $1-s_i$)
PRODUCT	$1-s_A$	1	$1-s_a$

The succession of numbers 1, 2, 3, ..., i may represent (a) different ecological niches within a single locality, (b) different environmental conditions prevailing in successive generations, or (c) different developmental processes governed by pleiotropic gene actions. Note that the heterozygotes would possess highest marginal fitness (marginal overdominance) if their average fitness in each situation merely exceeds the average of the fitnesses of the two homozygotes. Note also that the dominance hypothesis of heterosis is a form of marginal overdominance that is achieved by multiplying across gene loci.

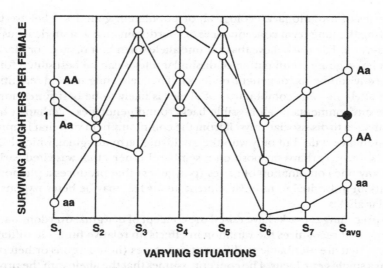

Figure 3.1. A full-face view of the population arena. The population exists under a variety of situations (S_1, $S_2 \cdots$) within each of which carriers of the three genotypes (*AA*, *Aa*, and *aa*) exhibit different relative fitnesses. Norms of reaction consist of the reactions of individuals possessing certain genotypes to all possible situations.

Fig. 3.1). Multiplied through all situations – temporal, spatial, or developmental – the heterozygote emerges with the highest relative fitness (Wallace, 1959, 1981, p. 246). The term marginal overdominance, which Dobzhansky disliked because he regarded marginal as a perjorative term meaning negligible, was coined by Clark Cockerham during a discussion at North Carolina State University during the early 1950s. (Heterosis that results from the concealment of deleterious alleles by their dominant counterparts is, in fact, an example of marginal overdominance in which the multiplication is carried out across gene loci.)

An additional point follows from the admission that even the marginal overdominance of single-locus heterozygotes can be lost if an altered array of challenging situations happens to consistently favor one of the homozygotes for an extended time; in this case, the favored allele (because of its already high frequency in the population) quickly eliminates its less favored homologue (see Fig. 3.2). The once-favored heterozygote, through close linkage (hitchhiking), may also have lifted an otherwise rare, beneficial, recessive allele to a frequency (20% or more) that allows it to quickly approach fixation in the population (Wallace, 1986, 1987a, 1987b).

3.3. Heterosis and Coadaptation

Referring once more to Dobzhansky's views on heterosis, one frequently reads (Lewontin et al., 1981, p. 103; cited by Crow, 1992) of the conflict that he entertained regarding heterosis and coadaptation. Lewontin describes how the

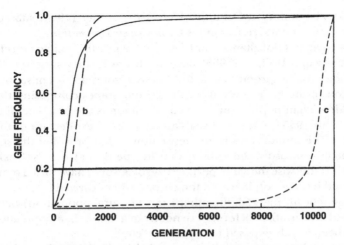

Figure 3.2. A diagram illustrating the expected increases in the frequencies of a, dominant, b, semidominant, and c, recessive mutations, each possessing a 1% selective advantage and each starting from an initial frequence of 0.01. Superimposed on the basic diagram is a baseline drawn at a frequency of 0.20; above this baseline, further increases in frequency for all three types of mutations occur at essentially the same rapid rate. At intervals of 160 generations experimental populations of *Drosophila melanogaster* underwent apparent genetic changeovers involving second-chromosome elements that exhibit selective advantages greater than 25% in heterozygous carriers. Should the advantage of either allele be lost, the favored one would quickly approach fixation. Furthermore, each of these favored elements could easily pull other (hitchhiking) alleles lying within chromosomal segments 25 or more centimorgans in length to frequencies exceeding the base line.

interpopulation F_1 hybrid vigor observed by Brncic, Vetukhiv, and myself seemed to Dobzhansky to negate his views on coadaptation.

Two quotations from Dobzhansky's publications reveal the conflict that existed in his mind. Dobzhansky and Wallace (1953), in a study involving replicated test cultures of both homozygous and heterozygous combinations of wild-type chromosomes of *Drosophila melanogaster* and *D. pseudoobscura*, found that the estimated viabilities of the homozygotes varied greatly from one culture to the next, while those of the heterozygotes varied no more than would be expected by chance. In the case of homozygous test cultures, statistical probabilities were consistently near 0.001 whereas, in the case of heterozygous combinations, the corresponding probabilities were consistently near 0.30. These observations led to a discussion of genetic homeostasis. The discussion included the following paragraph: "It should be emphasized, however, that our data are based on the relative frequencies of two classes of flies (D/+ and +/+) and, consequently, statistically homogeneous data within sets of replications of heterozygous combinations indicate the existence of homeostasis within *both* [i.e., both the genetically marked and the wildtype] classes of flies." This sentence was composed in the presence of I. Michael Lerner, who happened to be visiting

Dobzhansky's laboratory at Columbia University; Lerner (1954) subsequently cited these experimental results in his book *Genetic Homeostasis*.

In a later paper (Dobzhansky and Levene, 1955), the following reference to Dobzhansky and Wallace (1953) occurs: "It was further surmised that the superior homeostatic properties of the heterozygotes are a consequence of coadaptation of the gene contents of the chromosomes composing the gene pool of a Mendelian population. The coadaptation is the outcome of natural selection." This 1955 quotation reveals an erroneous recollection of the 1953 conclusion. The mutant flies in the test cultures that provided the standard against which to measure the viability of wild-type flies of necessity exhibited homeostasis (otherwise the wildtype heterozygotes would have varied more than that expected by chance); however, the chromosomes carried by these mutant flies (one genetically marked, the other from a natural population) had never encountered one another before, were not from a Mendelian population, and had never been jointly exposed to natural selection.

The term coadaptation, one might recall, has itself evolved through time. In his *Origin of Species*, Darwin used it in reference to the functional relationships between parts of an individual's body. Dobzhansky used it in reference to chromosomal inversions in *D. pseudoobscura*. These inversions (Standard, Chriicahua, Arrowhead, and others) almost invariably established balanced polymorphisms in laboratory populations kept at 25°C if they had been obtained from the same geographic locality. They did not always do so if they had been obtained from two different localities. The inversions found within a single locality were said by Dobzhansky to be coadapted, and the coadaptation was said to be the result of natural selection (see review in Wallace, 1981, pp. 514–523).

I extended the term coadaptation to encompass the entire gene pool of a population: The particular alleles representing various gene loci that are found in a population are there because, on average, they interact well with one another in nonadditive ways. Unlike Dobzhansky (apparently), I saw no obvious conflict between the coadaptation of genes within populations and the heterosis of F_1 interpopulation hybrids. After all, many interspecific hybrids such as those obtained by crossing *D. pseudoobscura* and *D. persimilis* are physically robust; following gene recombination and chromosomal segregation in female F_1 hybrids, however, the majority of backcross progeny (F_1 hybrid males are sterile) are physically delicate, even moribund. I must not have expressed myself clearly on this matter for Merrell (1981, page 209), in reference to Wallace (1955), says: "Wallace also stated, 'The higher average viability of interpopulation F_1 hybrids is a measure of the price paid by local populations for an integrated gene pool capable of being transmitted successfully from generation to generation,' a statement whose meaning remains obscure." Perhaps viewing the problem from an interspecific rather than an interpopulation perspective helps in understanding my ancient claim; further, "provides a measure" might have been a better expression than "is a measure."

Coadaptation was a topic that I have discussed twice before the National Poultry Breeders Roundtable (Wallace, 1956, 1989a). The earlier talk was given at a time when coadaptation was still a concept that needed to be demonstrated.

Table 3.2. *The Calculated Relative Fitnesses of Various Genotypes in Three Experimental Populations within which an Abrupt Increase in the Frequency of Lethal Heterozygotes Suggested the Occurrence of a Genetic Changeover*

Population	$+/+$	$+/l_i$	$+/l_f$	l_i/l_i	l_i/l_j	l_i/l_f	l_f/l_f
5	1.00	0.96	1.10	0	0.93	1.35	0
7	1.00	0.97	1.07	0	0.93	1.32	0
19	1.00	0.99	1.05	0	0.98	1.22	0

Nonlethal chromosomes, $+$; ordinary lethals, l_i or l_j; selectively favored lethals, l_f. (see Wallace, 1986, 1987a).

At that time, I argued that for coadaptation to be a reasonable outcome of natural selection, five conditions must be met:

1. Selection must operate on heterozygous individuals.
2. A select few rather than all heterozygotes must be favored.
3. Selective forces must be sufficiently large to bring about changes in populations within short periods of time.
4. Selective forces must vary considerably with genetic background.
5. The magnitudes of selection must be sufficiently large to alter the frequencies of these backgrounds.

Thirty-three years later, I was able to report that supporting data were now available for each of these points. This evidence has been summarized in an article entitled, "Coadaptation revisited" (Wallace, 1991a) and need not be reviewed here (however, see Table 3.2 and Fig. 3.3 for relevant observations).

The discussion of coadaptation and Dobzhansky's views concerning coadaptation can be terminated by a citation of these summarizing comments from Dobzhansky and Spassky (1968): "The viability of the lethal heterozygotes is equal to, or greater than that of quasinormal heterozygotes on the genetic background of the natural population in which the chromosomes tested were found. **It is otherwise on foreign genetic backgrounds**; here the heterozygotes for lethal are equal or inferior in viability to heterozygotes for quasinormal chromosomes (emphasis added)." Although the term was omitted, this discussion clearly refers to coadaptation in its most general sense, the coadaptation of the gene pool.

3.4. Norms of Reaction, Culling through Decisive Encounters, and Diversifying Selection

I shall conclude this account by discussing matters that I view as being interrelated but that are often treated separately as if they are independent phenomena (see Wallace, 1994): norms of reaction and diversifying selection.

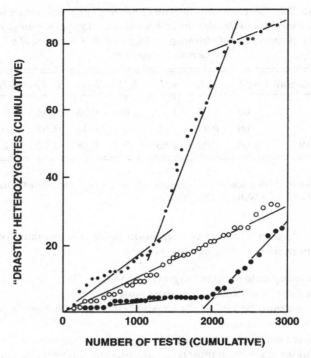

Figure 3.3. Changes in the frequencies with which combinations of randomly chosen second chromosomes from three *Drosophila* populations proved to be lethal (or semilethal) to their heterozygous carriers. Abrupt changes in the slopes of these lines reflect the retention and the accumulation within the populations of lethals that enhanced the overall fitness of their heterozygous carriers. Dots, population 5; open circles, population 6; filled circles, population 7.

Natural populations are not uniform nor do they inhabit uniform environments. Thus, as Dobzhansky (1970, pp. 36–41) and Lewontin (1974; see Suzuki et al., 1986, p. 514) have stressed, a complete understanding of the response of genotypes to selection can be gained only by knowing their norms of reaction, that is, their fitnesses relative to other genotypes under all possible environments and in combination with all possible background genotypes. I have considered this same problem (Wallace, 1989b) from the perspective of two alleles at a single locus whose carriers encounter different situations. If the relative fitnesses of three genotypes (*AA*, *Aa*, and *aa*) are altered or reversed under different situations, the heterozygote may easily exhibit marginal overdominance.

Diagrams illustrating norms of reaction generally depict the challenges as differing only slightly from one another. This need not be so. On encountering a predator, the potential victim may escape by hiding, by fleeing, or by fighting back. To the extent that success in surviving whatever challenge by whatever means may have a genetic basis, natural selection may impose contradictory demands on the genetic composition of a population. Tachida and Mukai (1985) stressed this point with reference to the role of diversifying selection in maintaining genetic variation.

Culling a large number of young zygotes to a number of adult individuals consistent with the carrying capacity of the environment requires surprisingly few decisive encounters – encounters that effectively remove half of the individuals involved. Three such encounters will remove 87.5% of all zygotes, leaving only one-eighth as surviving adults. Six such encounters will leave a single survivor for every 64 starting zygotes. The number of encounters is sufficiently small that in a complex environment, each surviving adult may have met and successfully coped with a unique set of challenges. Each may have succeeded in surviving because of a genotype differing from those of all other survivors. Here, in my opinion, is the basis for much of the genetic variation one finds in natural populations. Dobzhansky (1970, p. 167) discusses diversifying selection from the standpoint of the superior fitness of heterozygous individuals; I do not believe that this restriction is necessary. On the other hand, to the extent that heterozygous individuals tend to share the advantage possessed by the favored homozygotes of each encounter, these heterozygotes might tend to exhibit marginal overdominance.

In recent publications (Reeve et al., 1990; Wallace, 1991b, 1993), I have stressed the overall neutrality of components of fitness. During culling, each surviving individual may owe its continued existence to this component or that one, but among the final population of surviving adults no component need confer an obvious advantage to its carriers. Prize fighters provide an illustration: Winners and losers in championship and near-championship bouts do not differ significantly with respect to any of the 15 physical measurements generally listed in sports-page accounts for each combatant. The demonstration that stabilizing selection operates on these attributes in the case of heavyweights reveals, however, that they are indeed involved in winning or losing. In my opinion, the demonstration that components of fitness – perhaps because each must have an optimal value, perhaps because so many are negatively correlated with others – are themselves neutral with respect to total fitness goes far in explaining the success of Kimura's neutral theory. In a complex environment, the vital role played by a particular component of fitness in a particular decisive encounter cannot be discerned; no component dominates over the others. Neutrality of phenotypic variation arises as an average of many, nonneutral selective roles played by individual variants during decisive encounters. These encounters, few in number, cull an excessive number of zygotes (and aging parents if generations overlap) to K, the number that the environment can sustain at a given moment. If, however, phenotypes can be shown to be selectively neutral even though they comprise components of fitness, their underlying genetic bases must also be selectively neutral.

REFERENCES

Crow, J. F. 1992. Mutation, mean fitness, and genetic load. *Oxford Sur. Evol. Biol.* 9:32–42.

Darwin, C. 1877. *The Effects of Cross and Self Fertilization in the Vegetable Kingdom.* New York: Appleton.

Dobzhansky, Th. 1970. *Genetics of the Evolutionary Process.* New York: Columbia U. Press.

Dobzhansky, Th. and Levene, H. 1955. Genetics of natural populations. XXIV. Developmental homeostasis in natural populations of *Drosophila pseudoobscura. Genetics* 40:797–808.

Dobzhansky, Th. and Spassky, B. 1968. Genetics of natural populations. XL. Heterotic and deleterious effect of recessive lethals in populations of *Drosophila pseudoobscura. Genetics* 59:411–425.

Dobzhansky, Th. and Wallace, B. 1953. The genetics of homeostasis in *Drosophila. Proc. Natl. Acad. Sci. USA* 39:162–171.

Gowen, J. W. 1952. *Heterosis.* Ames, IA: Iowa State U. Press.

Kimura, M. 1991. Recent development of the neutral theory viewed from the Wrightian tradition of theoretical population genetics. *Proc. Natl. Acad. Sci. USA* 88:5969–5973.

Lerner, I. M. 1954. *Genetic Homeostasis.* Edinburgh: Oliver and Boyd.

Lerner, I. M. 1959. Discussion following Hampton Carson's paper on species formation. *Cold Spring Harbor Symp. Quant. Biol.* 24:105.

Lewontin, R. C. 1974. The analysis of variance and the analysis of causes. *Am. J. Hum. Genet.* 26:400–411.

Lewontin, R. C., Moore, J. A., Provine, W. B., and Wallace, B. 1981. *Dobzhansky's Genetics of Natural Populations, I–XLIII.* New York: Columbia U. Press.

Merrell, D. J. 1981. *Ecological Genetics.* Minneapolis, MN: U. Minnesota Press.

Reeve, R., Smith, E., and Wallace, B. 1990. Components of fitness become effectively neutral in equilibrium populations. *Proc. Natl. Acad. Sci. USA* 87:2018–2002.

Suzuki, D. T., Griffiths, A. J. F., Miller, J. H., and Lewontin, R. C. 1986. *An Introduction to Genetic Analysis.* New York: Freeman.

Tachida, H. and Mukai, T. 1985. The genetic structure of natural populations of *Drosophila melanogaster.* XIX. Genotype–environment interaction in viability. *Genetics* 111:43–55.

Wallace, B. 1955. Inter-population hybrids in *Drosophila melanogaster. Evolution* 9:302–316.

Wallace, B. 1956. The role of subvital mutations in the genetics of populations. In *Proceedings of the Fifth Annual National Poultry Breeders Roundtable,* pp. 149–164. Kansas city, MO: Poultry Breeders of America.

Wallace, B. 1959. The role of heterozygosity in *Drosophila* populations. *In Proceedings of the Tenth International Congress on Genetics,* pp. 408–419.

Wallace, B. 1981. *Basic Population Genetics.* New York: Columbia U. Press.

Wallace, B. 1986. Genetic change-over in *Drosophila* populations. *Proc. Natl. Acad. Sci. USA* 83:1374–1378.

Wallace, B. 1987a. Analyses of genetic change-over in *Drosophila* populations. Z. Zool. Syst. Evolutionsforschung 25:40–50.

Wallace, B. 1987b. Selectively favored elements, genetic change-over, and the fate of mutant alleles. *Genet. Iber.* 39:565–574.

Wallace, B. 1989a. Selection for genetic variation in *Drosophila* populations. In *Proceedings of the Thirty-Eighth Annual National Poultry Breeders Roundtable,* pp. 8–27. Kansas City, MO: Poultry Breeders of America.

Wallace, B. 1989b. One selectionist's perspective. *Q. Rev. Biol.* 64:127–145.

Wallace, B. 1991a. Coadaptation revisited. *J. Hered.* **82**:89–95.

Wallace, B. 1991b. The manly art of self-defense: on the neutrality of fitness components. *Q. Rev. Biol.* **66**:455–465.

Wallace, B. 1993. Toward a resolution of the neutralist-selectionist controversy. *Perspect. Biol. Med.* **36**:450–459.

Wallace, B. 1994. Norms of reaction and diversifying selection. *Genetica* **92**:139–146.

CHAPTER FOUR

The Molecular Foundation of Population Genetics

RAMA S. SINGH, WALTER F. EANES, DONAL A. HICKEY,
LYNN M. KING, AND MARGARET A. RILEY

4.1. Introduction

Dobzhansky's dictum that "nothing in biology makes sense except in the light of evolution" (Dobzhansky, 1973) implies that evolutionary theory is the intellectual glue that binds together the several disparate disciplines within the biological sciences. Population genetics, in turn, provides both the experimental and the quantitative foundations upon which evolutionary theory is built. Specifically, it is the job of population genetics to provide a mechanistic explanation for the evolutionary process. Historically, the development of population genetics was first fueled by its relevance to Darwinian evolution (Chetverikov, 1926; Fisher, 1930; Haldane, 1932; Wright, 1931) and later by its importance in evolutionary biology through such fields as systematics and paleontology (Mayr and Provine, 1980). However, it was not until the late 1930s and early 1940s that a population genetics framework became part of the core impetus for both the initiation (Dobzhansky, 1937; Huxley, 1942; Simpson, 1944) and the development of the evolutionary synthesis (Jespen et al., 1949; Stebbins, 1950; Dobzhansky, 1951; Simpson, 1953; Mayr, 1963; Grant, 1963).

In contrast to its rich theoretical foundations, the amount of experimental data available in population genetics, up until 30 years ago, was extremely meager. For instance, even the most extensively studied polymorphic systems that illustrated the level of genetic variation segregating in natural populations, such as shell markings in snails, wing patterns in Lepidoptera, blood group and hemoglobin polymorphisms in humans, or inversion polymorphism in Drosophila (Lewontin, 1974) were only selected, textbook examples that illustrated how evolution could work. As Lewontin (1991) has pointed out, none of these polymorphic systems lent themselves to a quantitative functional evolutionary explanation. Thus our knowledge about the primary ingredient required for formulating a satisfactory theory of evolution, i.e., knowledge about

R. S. Singh and C. B. Krimbas, eds., *Evolutionary Genetics: From Molecules to Morphology*, vol. 1.
© Cambridge University Press 2000. Printed in the United States of America. ISBN 0-521-57123-5. All rights reserved.

the amount and the pattern of genetic variation in natural populations, re-mained meager until the technique of gel electrophoresis became available (Harris, 1966; Hubby and Lewontin, 1966; Lewontin and Hubby, 1966).

In this chapter, we have outlined the manner in which the discovery of al-lozymic variation opened up new opportunities for the investigation of both the quantity and the functional nature of genetic variation and how it prepared the ground for the more recent applications of molecular biology in population ge-netics. Our aim is not to review the vast amount of literature on electrophoretic variation and its biological significance. This has been done in numerous books (e.g., Lewontin, 1974; Nei, 1987; Gillespie, 1991), symposia proceedings (e.g., Ayala, 1976; Karlin and Nevo, 1976; Nei and Koehn, 1983), and many, many review articles (e.g., Ayala, 1975; Hamrick et al., 1979; Brown, 1979; Coyne, 1982; MacIntyre and Collier, 1986; O'Brien and MacIntyre, 1978; Nevo et al., 1984; and Singh, 1989, 1990).

4.2. Gel Electrophoresis – A Methodological Revolution

The problems of studying genetic variation before the arrival of gel electro-phoresis were summarized by Lewontin (1974). The main difficulty was that there was no method that could be easily used to identify different alleles and their frequencies at different loci. Visible markers were clearly unrepresentative of the genome; tests of allelism were too cumbersome and time consuming; inbreeding methods failed to differentiate between the effects of different loci; and inversion polymorphisms at the chromosomal level confounded variation at many loci. Therefore research programs in experimental population genetics that depended on detailed knowledge of allelic variation at different loci had hit a roadblock, and the various indirect methods used for studying genetic variation provided only inadequate guesses.

Protein electrophoresis introduced a new dimension in the field of popu-lation genetics. Before electrophoresis there was no way, short of sequencing proteins, to relate the existing genetic variation within populations directly to genetic divergence between populations and species. Electrophoresis, by by-passing the problem of genetic crosses and species barriers, allowed genetic comparisons to be made between closely related species. It is this feature of electrophoresis, more than anything else, that was responsible for its rapid spread in population biology. The technique became a powerful tool for sorting out relationships between populations and between species (e.g., Hubby and Throckmorton, 1968; Selander et al., 1969; Prakash, 1972; Ayala and Powell, 1972; Richmond, 1972; Zouros, 1973).

We tend to forget, however, that the electrophoretic revolution did not start with the invention of the method itself, which occurred 10 years earlier (Smithies, 1995), nor did it start with the demonstration of electrophoretic variation at many protein-coding loci (Shaw, 1965); the real revolution started only after the power and the usefulness of gel electrophoresis had been shown step by step with a case study and the findings had been interpreted in the context of the problems of population genetics (Hubby and Lewontin, 1966; Lewontin and Hubby, 1966; Harris, 1966).

4.3. Patterns of Variation and Population Genetics' Dilemma

In the preelectrophoretic era, a simple demonstration of substantial genetic variation in a natural population would, in itself, have been proof that variation was significant for evolutionary change! This is because the classical concept of wild-type alleles (Muller, 1950) had no room for genetic variation of any type except for rare, deleterious mutations, and the balance school was undeterred by the arguments of genetic load and its effect on limiting genetic variation under natural selection (Wallace, 1991). The discovery of substantial genetic variation by gel electrophoresis simultaneously solved one problem and revealed a second. The mere presence of substantial genetic variation in natural populations supported the balance school in no uncertain terms but opened the question of its significance for evolutionary change. There was simply too much genetic variation to be explained either by the classical theory or by the balance theory, which led to Lewontin and Hubby's (1966) dilemma.

The old controversy, centering on the quantity of genetic variation, was replaced with a modified new version that now centered on the quality of genetic variation and its evolutionary significance. A new angle on the nature of genetic variation was introduced by the old concept of the isoallele (Stern, 1943) and neutral genetic variation (Kimura and Crow, 1964). As a result, a theory of selectively neutral evolution of protein differences between species (Kimura, 1968; King and Jukes, 1969) and neutral protein polymorphisms within species (Kimura and Ohta, 1971) was born (see Gillespie, 1987, 1991, for reviews). The possibility of neutral variation was a new twist on the old classical theory, and most experimental population geneticists, particularly those working with natural populations (with the exception of a few such as Terumi Mukai; see Mukai et al., 1982), were suspicious. However, even Darwin (1859, p. 81) had previously considered the possibility of neutral variation when he pointed out that "variation neither useful nor injurious would not be affected by natural selection, and would be left a fluctuating element, as perhaps we see in the species called polymorphic!"

4.4. Demonstration of Selection and Approaches

The response to Lewontin and Hubby's dilemma was swift. Several complex selection models were immediately proposed (King, 1967; Milkman, 1967; Sved, et al., 1967). However, as electrophoretic variation data accumulated it became obvious that we needed more than a theoretical possibility to determine if the observed variation was detectable by natural selection. Three distinct approaches were pursued: ecological, functional, and statistical.

By an ecological approach we mean the use of various properties of a population's structure to infer the role of natural selection. In the early phases, this approach became by far the most popular, partly because at that time population structure was assumed, as a matter of faith, to result from the action of natural selection. Some studies, such as the study of coadaptation in *Drosophila pseudoobscura* (Prakash and Lewontin, 1968) and of genetic variation in the phylogenetic relic horseshoe crab (Selander et al., 1970), provided critical ways of evaluating the significance of enzyme variation. Others, such as studies showing the association among protein and spatial and temporal variations (e.g., Vrijenhoek et al., 1992) or among protein heterozygosity and growth rate and life-history traits (e.g., Singh and Zouros, 1978; Allendorf and Leary, 1986; Mitton and Grant, 1984; Zouros and Foltz, 1987) raised hopes of finding more general ecological explanations. In general, the mere presence of electrophoretic variation was taken as evidence of natural selection. There are many studies in which selection was invoked because the amount of genetic variation was lower than expected on the basis of population size, such as that of Drosophila and bacteria (for a review, see Nei and Graur, 1984), or higher than expected on the basis of breeding systems, such as those of in plants. For example, high levels of electrophoretic variation in selfing plants were explained by natural selection on ecotypic differentiation (Hamrick and Allard, 1972; Allard et al., 1972) with full realization that in these systems the whole genome becomes the target of selection. Likewise, geographic variation, e.g., in Drosophila (Singh et al., 1982, and others), was used as evidence of natural selection without sufficient knowledge of gene flow. Indeed, the large number of studies published during the first 10 years after the introduction of electrophoresis that uniformly, and mostly uncritically, invoked the role of natural selection is astonishing. In retrospect, this is understandable; population genetics was going through a paradigm shift, and much new data needed to be collected before the adaptive significance of electrophoretic variation could be fully evaluated.

The statistical approaches took basically two routes. One approach of course was to develop appropriate tests of selective neutrality and apply them to the observations (Ewens, 1972; Watterson, 1977), and the other was to modify and improve the original neutral theory (Kimura, 1968; King and Jukes, 1969; Kimura and Ohta, 1971) to make it in accord with the observed data (reviewed in Gillespie 1987, 1991). The latter approach was pursued by Ohta (reviewed in Ohta, 1992), who introduced the concept of slightly deleterious alleles and proposed the so-called nearly neutral theory of molecular evolution, which was able to explain the observations better than the original theory. Nei and his colleagues, on the other hand, pursued a different approach by introducing fluctuations in population sizes and historical effects to explain the data (see Nei and Graur, 1984; Nei, 1987).

In general, there were very few large, single-locus data sets to which the Watterson test could be applied, and tests involving heterozygosity were meaningless as they were insensitive to small deviations and thus could explain everything! Thus it is a paradox that while individually most field population geneticists continued to defend the role of natural selection, collectively they were being convinced of the position of selective neutrality! So powerful did the neutral theory become that almost any pattern of variation could be explained by invoking appropriate variations in historical population size, mutation rate, and (negative) selection intensity. Alternative theories of natural selection that could account for enzyme polymorphisms were slow to develop and, as a result, had less impact when they did arrive (reviewed in Gillespie, 1987, 1991). By this time the molecular mode of thinking based on the role of deleterious mutations and purifying selection had reached its height.

As the field of electrophoretic studies matured, it was realized that the general lack of evidence for selection may result from several limitations. First, it was possible that electrophoretically detectable alleles, or allelic classes, were not the target of selection and quantitative differences in enzyme activities needed to be considered in selection experiments (Hickey, 1979). Second, the electrophoretic data set may not be appropriate for the statistical tests that were being applied. These tests assumed an infinite allele model of mutation, and it was possible that simple gel electrophoresis may not detect all segregating alleles at a locus; in other words the electrophoretic alleles (or allozymes as they were called) were heterogeneous classes and contained many more hidden true alleles (Singh et al., 1976). Third, laboratory selection experiments involving electrophoretic alleles would be insensitive unless the experiments were designed to consider the physiological nature of the enzymes and the metabolic flux of their pathways under consideration (Koehn, 1969). And finally, protein electrophoretic data were phenotypic and of a static nature, and they did not allow an examination of historical relationships of alleles at polymorphic loci (Kreitman, 1983; Riley et al., 1989). Below we discuss these limitations and their solutions.

4.5. Allozymes and Physiological Adaptation

Lewontin (1974) emphasized the importance of measuring biochemical phenotypes of allozymes as evidence of an adaptive potential. The epistemological issues of detecting and assessing physiological and biochemical differences among molecular genotypes have been a central problem in the study of allozyme variation and its association with fitness variation (see Watt, 1994, for a recent discussion and one view). At face value, the approach of directly evaluating fitness differences among electrophoretic genotypes appears straightforward, and many studies in the 1970s reported significant results. However, numerous problems are associated with this approach (Eanes, 1987). Small selection differentials, lack of statistical powers, insensitivity of most experimental approaches (see Dykhuizen and Hartl, 1983), and assumption of independent loci, leading to overestimation in the variation of total fitness, are just a few examples.

Interest in the so-called functional or measured genotype approach to studying allozyme variation began soon after Lewontin and Hubby's (1966) classic observations. Studies such as Koehn's (1969) report of activity variation at a serum esterase locus in the freshwater sucker *Catostomus clarkii* were early attempts simply to demonstrate that different electrophoretic variants possess different activity levels. From this start, more sophisticated issues, methods, and philosophies have developed, and there is now ample evidence that allozyme variants can possess large functional differences (see Watt, 1994). How these differences translate into fitness differences such that polymorphisms persist in natural populations is yet another issue. While such genotypes may be adaptive solutions to maintain maximum fitness across a range of environments, how this is solved mechanistically is unknown. Furthermore, it is not clear from the standpoint of simple pathway flux whether different genotypes are solutions to maintaining continuous high catalytic efficiency and flux in different environmental contexts, or conversely whether selection is requiring that flux be different in different contexts. In a branching pathway, where metabolic intermediates may be shunted into different pools, differential allocation may be the target of selection and even low-activity genotypes may be favored in different environmental contexts. Resolving these two views is an important issue, which will not be resolved here. Instead, we discuss below several examples of rather exhaustive examinations of features associated with allozyme polymorphisms.

The *G6pd* allozyme polymorphism in *D. melanogaster* is represented by two electrophoretic alleles (termed *A* and *B*) and is the consequence of a single amino acid polymorphism (proline to leucine substitution at residue 382; Eanes et al., 1993). The frequency of the *A* allozyme is less than 5% in sub-Saharan Africa, but increases to ~90% in Europe and is reciprocally clinal with latitude in both North America and Australia (Oakeshott et al., 1983), increasing with more temperate climate. Much of the work has focused on the question of whether *in vitro* measures of such classic kinetic parameters as K_m and k_{cat} actually translate into *in vivo* differences in function for these two variants. The study of *G6PD* protein levels and steady-state kinetics on purified enzyme predicted that at 25 °C the *A* genotype possesses ~28% lower activity than *B*, and this is derived largely from that variant's relatively lower affinity (higher KM) for glucose-6-phosphate (Eanes et al., 1990). A further aim of these studies was to estimate *in vivo* function by means of two independent approaches. One method was to measure flux directly by use of radiolabeled substrates. Cavener and Clegg (1981) compared the relative fluxes between the putative high- and low-activity dilocus allozyme genotypes at the *G6pd* and *6Pgd* loci in *D. melanogaster* by using the method of Wood et al., (1963). They observed statistically significantly greater pentose shunt flux associated with the proposed higher activity *G6pdB–6PgdS* dilocus genotype, but the experiment was not specifically designed to partition the contribution of the genotypes at the *G6pd* and the *6Pgd* loci separately. The analysis was extended to focus on the *G6pd* genotypes specifically (Labate and Eanes, 1992), and one can estimate that the low activity *G6pdA* allele possesses ~32% lower pentose shunt flux than the *G6pdB* allele.

Clear differences were also found in the ability of the *G6pdA* and the *G6pdB* genotypes to suppress the lethality associated with a 6-phosphogluconate (*6PGD*) low-activity mutation (Eanes, 1984; Eanes and Hey, 1986; and Eanes et al., 1990).

This reflects a lower *in vivo* activity associated with the *G6pdA* genotype, resulting in a reduced accumulation of 6-phosphogluconate. Eanes and Hey (1986) extended this to a series of rare *G6PD* variants. Two clusters of different *in vivo* activities were observed, presumably separating rare variants into two sets that were derived from the *A* and the *B* common allozyme alleles. A recent analysis of the amino acid sequences of those 11 alleles is, without exception, consistent with that prediction (Eanes et al., 1996); the major activity difference among rare variants is associated with the Pro/Leu polymorphism. Finally, Eanes et al., (1990) concluded that the *A*-allele-bearing genotypes possess an *in vivo* activity that is 40% lower than that of the *B* genotypes. The results obtained with direct flux measures with radiolabeled glucose are therefore consistent with that magnitude of difference. The failure to find trade-offs in function, in which one *G6PD* genotype becomes catalytically better in one environment, is not seen here. In contrast, it would appear that a reduced pentose shunt function is favored in more temperate parts of the range. These results certainly support the argument that *in vitro* studies are reasonable predictors of variant function, although whenever possible *in vivo* measures will be preferred over *in vitro* measures. The problem is that designing *in vivo* studies for many enzymes and species is a challenging problem.

One of the most extensive analyses of an allozyme polymorphism is that of the LDH-B isozyme in *Fundulus heteroclitus*, carried out by Powers and his associates. This polymorphism varies in frequency dramatically across the latitudinal temperature gradient along the Atlantic coast of the United States (Powers and Place, 1978). Place and Powers (1979) showed that the two variants differ significantly in a number of parameters across temperature and pH, and these favor the two alleles with respect to catalytic efficiency. Genotypes from these alleles differ with respect to hatching time (DiMichele and Powers, 1982a), swimming performance (DiMichele and Powers, 1982b), developmental rate (DiMichele et al., 1986), and differential mortality at high temperatures (DiMichele and Powers, 1991).

Another classic case is that of the leucine aminopeptidase polymorphism in *Mytilus edulis* (see review by Koehn and Hilbish, 1987). This polymorphism shows dramatic local geographic variation, generally in association with recognized climatic boundaries in marine habitats, and, in particular, salinity changes and temperature. Koehn and associates have shown that there are differences in specific activity among the different variants and that the genotypes are associated with cell volume regulation under osmotic stress (Hilbish et al., 1982), an observation consistent with the clinal variation involving gradients in salinity and modulation of the free amino acid pool, which is a well-established response in marine bivalves.

An analogous case has also been reported by Burton and Feldman (1983) for the intertidal pool copepod *Tigriopus californicus*, in which there are distinctive associations of metabolic amino acid pools with allozyme genotypes. Watt (1992, and references therein) carried out extensive analyses of the phosphoglucose isomerase genotypes in Colias butterflies and has reported dramatic associations of male mating success and female fecundity with electrophoretic phenotype.

Finally, the ADH polymorphism in *D. melanogaster* has been studied extensively from a catalytic and regulatory standpoint and remains one of the best-studied loci for which functional differences have been mapped to different parts of the gene itself. In early studies of ADH it was quickly observed that the two variants differed 2–3.5-fold in overall activity, an apparent adaptive response to ethanol selection (see review by Van Delden, 1982). However, the dissection of the causal sources (catalytic or regulatory) of the differences has been a long process. Studies of catalytic function on purified enzyme have determined that at least half of the activity difference is due a catalytic difference in k_{cat} that is attributed to the Thr/Lys difference responsible for the allozymes. On the other hand, the two variants differ by nearly twofold in protein levels, which is not caused by different levels of *Adh* transcript (Laurie and Stam, 1988). The source of this difference in enzyme activity has remained an enigma, with the amino acid polymorphism and a small insertion/deletion polymorphism found upstream vying for attention (Laurie and Stam, 1994). Unfortunately these two sites are in tight linkage disequilibrium.

Numerous questions about adaptation and the contribution of molecular polymorphisms to genetic variation for life-history variation and physiological adaptive variation remain unresolved. The challenge is now to understand the way in which natural selection acts on such polymorphisms in a metabolic context. Are trade-offs important, or is it differential allocation that is important? Are branch point enzymes more receptive targets for adaptive polymorphisms? How old are such adaptations? Are environments too ephemeral for the establishment of old balanced polymorphisms, but are amino acid polymorphisms nevertheless adaptive in the short run? We might predict that future work will focus on integrating a genealogical approach that uses DNA sequence data with functional analyses of genotypes. There will also be an emphasis on inventing new ways to assess proximal flux in a physiological context.

4.6. Allozyme Variation and Gene Expression

It is generally accepted that regulatory gene variation is of the utmost importance in evolution. The adaptive enzyme systems of prokaryotes, as exemplified by the *lac* operon, are an elegant illustration of the direct link between gene regulatory characteristics and ecological adaptation. It is also believed that adaptive mutations affecting regulatory genes are equally, if not more, important in eucaryotic evolution. For instance, it has been suggested that a relatively small number of genetic changes affecting the processes of gene regulation may account for the relatively large morphological and ecological differences between humans and chimpanzees (King and Wilson, 1975). These ideas stimulated a considerable amount of research aimed at characterizing the regulatory systems that control enzyme production in mammalian systems (Berger and Paigen, 1979; Paigen, 1979). Similarly, it has been argued that regulatory gene variation is of central importance in the adaptation of plants to environmental stresses (Matters and Scandalios, 1986).

Experimental population geneticists working on Drosophila were also quick to see the potential for exploiting allozymes to study the evolution of gene regulatory processes (Laurie-Ahlberg et al., 1982). McDonald et al. (1977) found that *D. melanogaster* populations could adapt to a high-ethanol environment where the amount of ADH enzyme produced was increased, and they later showed (McDonald and Ayala, 1978) that some of the genetic elements affecting ADH quantity were not linked to the *Adh* structural gene locus. Genetic mapping showed that some of these regulatory genes were located on the third chromosome. Subsequent studies (Birchler et al., 1990) have shown that the expression of Drosophila ADH is subject to autosomal dosage compensation and that this dosage compensation is mediated through the *Adh* promoter sequences. In addition to the intraspecific studies of *Adh* regulatory gene variation, interspecific comparisons of the tissue distributions of ADH have also provided evidence for adaptive regulatory gene evolution (Dickinson et al., 1984).

The amylase gene–enzyme system in Drosophila provides a second example of the use of the allozymic approach to study regulation in gene regulatory processes. Amylase variants have been studied extensively. The early population genetics studies on this system followed a familiar pattern: selection experiments to test for gene frequency changes in experimental populations (DeJong et al., 1972; Hickey, 1977, 1979), functional characterization of enzyme variants (Doane, 1969), and geographical surveys of allozyme variation in natural populations (Singh et al., 1982). There were two early indications, however, that some of the genetic variation at the amylase locus was due to variation in associated regulatory elements. First, there was the finding by Abraham and Doane (1978), who discovered that the tissue distribution of amylase was under the control of a separate, but tightly linked genetic locus. Second, it was shown that strains isogenic for different allozymes produced different quantities of the enzyme (Hickey, 1981). Later studies confirmed the existence of regulatory variation at the amylase locus and showed that even within a single isogenic strain the levels of enzyme expression could change over several orders of magnitude because of repression by dietary glucose (Hickey and Benkel, 1982; Benkel and Hickey, 1986). Moreover, it was shown that the level of glucose repression was itself genetically variable between strains (Benkel and Hickey, 1986a). Consequently, different Drosophila strains show different norms of reaction in response to dietary composition. These results provide us with some hint of the exquisite regulation of enzyme production at this locus; they also make us aware, in retrospect, of the naiveté of the early selection experiments.

The advent of recombinant DNA techniques opened up new possibilities for investigating the mechanisms of gene regulation at the amylase locus. It was shown that the regulation was not only at the level of enzyme quantity but also at the level of mRNA quantity (Benkel and Hickey, 1987). P-element-mediated transformation by hybrid genes, in which the amylase promoter was combined with the ADH coding region, demonstrated that the promoter could confer glucose-repressible regulation on the ADH reporter sequence (Grunder et al., 1993). Deletion analysis of the promoter region, combined with site-specific mutagenesis, has shown that the glucose repressibility of amylase expression is controlled by multiple short elements located within a 70-bp region just upstream of the TATA box (Magoulas et al., 1993). The fact that glucose regulation in Drosophila appears to perform an analogous function to glucose repression in microbial systems raises the question of possible homology of the two systems of glucose repression. Strong evidence in favor of such homology was provided by the fact that the Drosophila amylase promoter can mediate the glucose repressible expression of firefly luciferase in transgenic yeast (Hickey et al., 1994).

There is ample evidence that allelic variation affecting the levels of amylase expression in *D. melanogaster* is common to other species of *Drosophila*. We have ample evidence that the effect is not confined to this species. Genetic variations that control the levels of amylase gene expression have been reported among other species in the *melanogaster* species subgroup (Payant et al., 1988), *D. pseudoobscura* (Powell and Lichtenfels, 1979), *D. persimilis* (Norman and Prakash, 1980), and *D. ananassae* (Da Lage et al., 1996). Most of these studies have concentrated on variation in the tissue-specific patterns of amylase expression. In addition to the repressive effects of dietary glucose, there is also evidence for a positive effect of dietary starch (Yamazaki and Matsuo, 1984), although the magnitude of this effect is less dramatic than the repression by glucose.

More recently, a number of studies have focused on the evolutionary patterns of regulatory genes themselves, rather than merely inferring their presence through their effects on target enzyme-coding genes, such as ADH and amylase. Examples of such studies are the survey of engrailed expression patterns in a variety of arthropods (Patel et al., 1989) and the many studies on the *HOM/HOX* class of homeobox genes (see Kappen and Ruddle, 1993, for a review). It is somewhat surprising, given the recognition that regulatory loci must play a major role in eukaryotic evolution (Hedrick and McDonald, 1980), that population geneticists have not focused more closely on their patterns of variation within species. Even the well-characterized heat shock response genes in Drosophila have received very little attention from population geneticists. The most obvious explanation that comes to mind is that much of the basic molecular characterization of these regulatory genes still remains to be done. It is only after we have a basic understanding of the structure and the function of regulatory genes that we can begin to look for adaptive allelic variants.

4.7. Hidden Allelic Variation and the Stepwise–Charge-State Mutation Model

The central problem with describing and explaining protein variation by use of electrophoresis is that of the identification of true alleles. As Gillespie (1991) points out, the ability to assess levels of variation is critical to our understanding of the mechanisms that maintain variation. Estimates of population parameters and genetic relationships among species that rely on the resolving power of gel electrophoresis, e.g., estimates of average heterozygosity and genetic differences among species and any population comparisons or conclusions drawn from the shape of allele frequency distributions, are meaningful only if protein electrophoretic classes are true alleles. It has been clearly recognized ever since protein electrophoresis was first applied to the study of genetic variation in natural populations (Harris, 1966; Lewontin and Hubby, 1966) that the technique is not a perfect detector of protein polymorphism and could underestimate the number of alleles in a population and consequently the average heterozygosity of individuals in these populations.

The charge-state model, or stepwise mutation model, was originally proposed by Ohta and Kimura (1973) to accommodate the limited resolving power of protein gel electrophoresis. The model assumes that a mutation can be detected electrophoretically only when it leads to the replacement of an amino acid with an unlike charge. Under standard electrophoretic conditions, nonpolar amino acids have a charge state of 0, the basic amino acids arginine and lysine are positive ($+1$), and the acidic amino acids glutamic and aspartic acid are negative (-1). It is assumed that mutations can change an electrophoretic allele by only one unit or two unit charges, either positive or negative, although a series of mutations can lead back to the original charge state. Thus alleles at a locus belong to discrete classes that are determined by the net electrostatic charge of the protein. Based on the charge-state model, the effective number of alleles under equilibrium conditions is $n_e = (1 + 8N_e u)^{1/2}$, where N_e is the effective population size and u is the neutral mutation rate (Ohta and Kimura, 1973). When $N_e u$ is small, this estimate differs little from the infinite alleles model, but it gives a much smaller estimate of n_e when $4N_e u$ is very large. Brown et al. (1981) review the charge-state model and extensions of it (various contexts of the model are also reviewed by Kimura, 1983). Although the model is considered somewhat simplistic, they conclude that it is "at one end of the spectrum" of types of models about mutational processes that the data generally support.

The lack of a satisfactory fit of electrophoretic data to neither selectionist nor neutralist hypothesis (Lewontin, 1974) led to the questioning of the nature of the data itself. Although it was recognized that the technique would underestimate the amount of genic variation, it was not widely appreciated that observed allele frequency distributions based on undetected heterogenous classes cannot meaningfully be compared with theoretical frequency distributions that require the detection of true alleles (e.g., theoretical frequency distributions predicted by the infinite alleles model of Kimura and Crow, 1964). King and Ohta (1975) addressed this problem, emphasizing that electromorphs are phenotypic classes distinguished by protein gel electrophoresis and may be heterogenous assemblages.

4.8. The Sequential Electrophoresis and the New Data

Rigorous approaches to detect hidden variation within electrophoretic alleles were undertaken by use of sequential gel electrophoresis (Singh et al., 1976;

Coyne, 1976; Ramshaw et al., 1979), a method that uses several running conditions that differ by pH, buffer, and acrylamide gel concentration. A general outcome of surveys of genetic variation in natural populations by this method (see Coyne, 1982, for a review) was that loci characterized as being monomorphic under single conditions tended to remain monomorphic or nearly so when sequential methods were used (e.g., Coyne and Felton, 1977; Beckenback and Prakash, 1977; Kreitman, 1980), whereas highly polymorphic loci revealed many more electrophoretic classes (e.g., Coyne, 1976; Singh et al., 1976; Singh, 1979; Loukas et al., 1981; Keith, 1983; Keith et al., 1985). Most studies of polymorphic loci revealed an allele frequency distribution characterized by one main electrophoretic class flanked by less-frequent alleles (Ohta, 1976; Bulmer, 1971; Maynard Smith, 1972). The result of increased resolution at polymorphic loci resulted in a more skewed allele frequency distribution, with the detection of more rare alleles in excess of neutral theory predictions (Ohta, 1976; Keith, 1983; Keith et al., 1985). Lewontin (1985) reviewed the development and principal studies of sequential studies and observed that patterns of protein electrophoretic variation fell into four general classes: (1) monomorphic loci (or nearly so); (2) polymorphic loci with only two or three electrophoretic classes, with the less frequent alleles at a frequency of 5% or more; (3) polymorphic loci in which one electrophoretic class is in high frequency and many classes are in low frequency; and (4) polymorphic loci in which two electrophoretic classes are in nearly equal frequency and many classes are in low frequency.

The relationship between protein electrophoretic variation and amino acid variation can be determined explicitly by nucleotide sequence analysis, and studies in Drosophila have shown a general pattern: electrophoretically monomorphic loci are also virtually monomorphic at the amino acid level, and polymorphic loci may show considerable amino acid variation within an electrophoretic class (Table 4.1). For example, 99 nucleotide sequences of electrophoretically monomorphic *D. pseudoobscura* (Singh et al., 1976; Coyne, 1976) show only one Ile/Val polymorphism (Schaeffer and Miller, 1993). Similarly, loci that have two common electrophoretic classes in *D. melanogaster* do not show amino acid variation within an electrophoretic class and only a single amino acid difference distinguishes the two classes. The Fast and Slow electrophoretic classes of ADH are distinguished by a single threonine/lysine polymorphism (Kreitman, 1983). The two electrophoretic alleles of superoxide dismutase differ by a single asparagine/lysine polymorphism (Hudson et al., 1994). Sequence analysis of *G6pd* genes shows no amino acid variation except for a single proline/leucine polymorphism that distinguishes two cosmopolitan electrophoretic variants (Eanes et al., 1993). Finally, a single asparagine/lysine replacement differentiates the Fast/Slow electrophoretic variants encoded by *Gpdh* (Takano et al., 1993).

Polymorphic loci are markedly different. Many amino acid sites in the proteins are polymorphic, segregate amino acids that are conservative and nonconservative in charge and are unique to a single sequence (Table 4.1). What is striking is that nucleotide sequence analysis has shown that not only are

Table 4.1. *Summary of Amino Acid Variation Corresponding to Different Classes of Electrophoretic Variation*

Locus	Species*	Nelec[†]	NNT Seq[‖]	Nelec Seq[♯]	Poly. AA Sites** (Total)	Charge Noncon[‡‡]	Unique Poly[§§]	References
Adh	*D. pse*	1	99	1	1 (254)	0	0	Schaeffer and Miller (1993)
Adh	*D. mel*	2	11	2	1 (265)	1	0	Kreitman (1983)
G6pd	*D. sim*	1	12	1	0 (516)	0	0	Eanes et al. (1993)
G6pd	*D. mel*	3[‡]	32	3	2 (516)	0	0	Eanes et al. (1993)
Gpdh	*D. mel*	2	11	2	1 (352)	1	0	Takano et al. (1993)
Sod	*D. mel*	2	41	2	1 (153)	1	0	Hudson et al. (1994)
Amy	*D. mel*	12[§]	18	6	9[††] (494)	5	2	Dalnou et al. (1987); Inomata et al. (1995)
Est-6	*D. mel*	14	13	10	16 (544)	3	10	Cooke et al. (1987); Labate et al. (1989); Cooke and Oakeshott (1989)
Xdh	*D. pse*	20	7	6	28 (1342)	9	24	Keith et al. (1985); Riley et al. (1992)
Est-5	*D. pse*	41	16	14	33 (545)	11	20	Keith (1983); Veuille and King (1995)

* *D. pse, D. pseudoobscura; D. mel, D. melanogaster; D. sim, D. simulans.*
[†] Nelec, number of electrophoretic classes from population surveys.
[‡] Two common and one rare electromorph.
[§] Number of electromorphs using a single electrophoretic condition.
[‖] NNT Seq, number of nucleotide sequences.
[♯] Number of electrophoretic classes sequenced.
** The number of polymorphic amino acid (Poly. AA) sites and the total number (in bracket) of amino acids in the protein. At some sites, *Amy, Est-6, Est-5,* and *Xdh* have more than two amino acids segregating.
[††] Excluding two null alleles.
[‡‡] Nonconservatively charged amino acid polymorphisms.
[§§] Number of unique amino acid polymorphisms.

electrophoretic classes heterogeneous, but that members of a single electrophoretic class can be quite dissimilar in amino acid sequence. At the duplicated amylase (*Amy*) locus in *D. melanogaster*, the encoded proteins differ on average by 3.9 amino acids, identical electromorphs differ on average by 1 amino acid, and 5 nonconservatively charged amino acid polymorphisms completely explain the electrophoretic classes (Inomata et al., 1995). Moreover, alleles from different electromorphs may be more similar in amino acid sequence than alleles of identical electromorphs. For example, one allele of *Amy3* differs from two other *Amy3* alleles by four amino acids, but differs from an *Amy4* allele by only one amino acid. Thus it appears that members of an electrophoretic class may have different ancestors and converge to identical electrophoretic mobility. The amylase data may be atypical, however, because concerted evolution of the duplicated amylase coding sequences could affect the patterns of allelic variation (Hickey et al., 1991). Sequences of *Est-6* in *D. melanogaster* representing 10 electrophoretic classes differ on average by 3.8 amino acids, and two sequences representing a single *EST-6* allele Fast differ by three amino acids (Cooke and Oakeshott, 1989). The large number of electrophoretic classes observed at this locus is explained by nonconservatively charged amino acid polymorphisms and minor mobility differences by conformational variation.

The convergence of dissimilar sequences to identical electrophoretic mobility also characterizes the highly polymorphic *Est-5* locus in *D. pseudoobscura*, where 41 electrophoretic classes, with two in nearly equal frequency and the rest mostly rare, have been detected (Keith, 1983). *Est-5* sequences representing 14 electrophoretic classes show that the encoded proteins differ on average by 7.8 amino acids (Veuille and King, 1995). Members of the two common electrophoretic classes each differ by five conservatively charged amino acids, and no single amino acid polymorphism distinguishes the two classes. One of the common alleles differs from a rare electrophoretic allele by only four amino acids. The sequence analysis shows that similarly charged electrophoretic variants may be quite dissimilar at the amino acid (and nucleotide) level. The tremendous number of electrophoretic variants can be explained by different combinations of amino acid polymorphisms that affect the overall charge on a protein, amino acid polymorphisms unique to a single electrophoretic class, and conformation differences that affect mobility.

For the loci examined so far, nucleotide sequencing shows that electrophoretically monomorphic loci and loci that show two common electrophoretic classes are essentially monomorphic in amino acid sequence within an electrophoretic class, whereas polymorphic loci show amino acid heterogeneity within an electrophoretic class. More importantly, sequencing has shown that electrophoretic classes may be heterogeneous assemblages whose members may be quite dissimilar in sequence and coancestry. Therefore, from these observations, we can ask, what information can be extracted from the patterns of electrophoretic variation in natural populations?

Barbadilla et al., (1996) developed a charge state model that explains two general features of protein electrophoretic variation at polymorphic loci: the intermediate electrophoretic mobility of electromorphs in high frequency and rare alleles. There are two important differences between this model and previous charge-state models. First, it does not assume that allele frequencies are at equilibrium. This implies that any evolutionary process, selective or neutral, may be happening and no specific evolutionary model is assumed. Second, it assumes linkage equilibrium, or effective linkage equilibrium, among segregating sites in a protein. This condition means that, independent of the evolutionary forces acting on a gene, recombination must be relatively important with respect to mutation to generate linkage equilibrium. Given these two conditions, for a moderate number of segregating amino acids in a protein, Barbadilla et al. show that the commonly observed frequency distribution of electrophoretic variants is purely a consequence of statistical relations and carries no information on underlying evolutionary forces. The importance of this outcome is that patterns of electrophoretic variation (frequency distributions) are spurious, that is, they emerge solely from statistical relationships.

Using *Est-5* in *D. pseudoobscura* as an example, Barbadilla et al. show that the basic symmetric frequency distribution of electrophoretic variants is similar between single and sequential gel electrophoresis and that independent segregation of the nonconservative amino acid polymorphisms detected by sequencing can explain the major electrophoretic classes. Using Watterson's test of Ewen's distribution, Keith (1983) tested the fit of this distribution to the expected neutral distribution and found a significant deviation in the direction of an excess of rare alleles. However, this test is based on the infinite allele model, which assumes the identification of true alleles. From the sequence data, which shows heterogeneity within electrophoretic classes, and assuming that all of the polymorphic amino acid sites (under a 10% criterion) are segregating independently, Barbadilla et al. show that the distribution of true alleles no longer shows common alleles or a relative excess of rare alleles. The frequency distribution of electrophoretic variants appears to result from the technique itself, which can resolve both stepwise and nonstepwise variants.

There are several consequences of this charge-state model: (1) A symmetrical frequency distribution of electromorphs is expected for single electrophoresis; (2) all parameters measuring genetic diversity will be underestimated, and the degree of underestimation will be proportional to the amount of variation; (3) tests based on fitting electrophoretic data to the infinite allele model are not informative; and (4) heterogeneity within electrophoretic classes decreases the power of any analysis based on associations, such as fitness, with allozyme classes. Our conclusion is that electrophoresis provides information on only the total amount of allele variation. The distribution of that variation, in terms of testing evolutionary hypothesis, is uninformative.

4.9. The Limitations of Protein Variation Data and the Dawn of Sequence Variation

One can construct *a posteriori* neutralist and selectionist evolutionary examples to account for why different proteins harbor more or less variability. For example, a neutralist explanation for the electrophoretic differences between ADH and *Est-5 Drosophila pseudoobscura* would be that ADH experiences elevated levels of purifying selection and therefore is monomorphic at the protein level, while *Est-5* is more tolerant of amino acid polymorphism, so that the observed electrophoretic polymorphism is essentially neutral. Kimura (1983) describes protein variation at the population level as a transient phase of neutral molecular evolution. Therefore if a population of *D. pseudoobscura* was sampled at a later date, the pattern of electrophoretic variation for *Est-5* would probably be different because many of the neutral alleles would have been lost from the population. However, a selectionist could provide several opposing arguments. For example, it is possible that ADH is monomorphic because an advantageous allele was recently fixed in the population through positive Darwinian selection. *Est-5* variability could be caused by a balanced polymorphism for the two major alleles, or it could be that some other form of frequency-dependent selection acts on the alleles. It is the inability to determine the evolutionary history that produces the electrophoretic identity among the *ADH* alleles or the differences among the *Est-5* alleles that precludes us from distinguishing between these neutralist and selectionist hypotheses. If we could determine identity by kind from identity by descent we would have an increased power in discriminating between alternative hypotheses.

It has been argued that surveys of DNA sequence polymorphism will provide the kind of data required for testing hypotheses regarding the forces operating on genetic variation (Lewontin, 1985; Kreitman, 1988). The power in this approach lies in the detail provided by complete sequence information. Nucleotide positions can be examined that are closely linked but functionally distinct, i.e., synonymous and nonsynonymous positions. By comparing levels of polymorphism and divergence at these functionally distinct positions with levels predicted by the neutral mutation theory (Kimura, 1983), it may be possible to distinguish between neutral versus selective forces operating to produce the observed variation (see, for example, Hudson et al., 1987).

Estimates of neutral parameters from DNA sequences allow the resolution of conflicting hypotheses. This is because (1) only the neutral mutation rate governs the rate of sequence divergence between species, assuming that the number of selection events is not large relative to the neutral fixation rate; (2) natural selection can have transient effects on polymorphism without affecting levels of divergence; and (3) under the neutral theory, for strictly neutral mutations, levels of polymorphism are expected to covary with levels of divergence because both are functions of the neutral mutation rate. It is possible to compare the level and the pattern of divergence between the nonsynonymous and synonymous sites of a coding region with the level and pattern of polymorphism. The neutral prediction is a positive correlation between nucleotide divergence and polymorphism.

Selection decreases the correlation of polymorphism with divergence. For example, Strobeck (1983) shows how the presence of a site under balancing selection changes the frequency of polymorphism at closely linked sites. Neutral mutations tend to accumulate in the region tightly linked to the selected site because they are held in the population with the selected site at a higher frequency than neutral mutations at unlinked sites. This phenomenon is known as the hitchhiking effect (Maynard Smith and Haig, 1974).

A second example is a locus with a recent favorable mutation driven to high frequency in the population under directional selection. This event will produce a decreased level of polymorphism at linked sites, also caused by genetic hitchhiking. The task of experimental population geneticists interested in determining the relative importance of various evolutionary forces in maintaining the standing levels of genetic variation is quite clear. Numerous loci from the same organism and in closely related species must be surveyed to allow the sorts of comparisons described above.

4.10. Examples with DNA Sequence Information

In a population survey of 11 *Adh* genes in *D. melanogaster*, Kreitman (1983) was able to distinguish unambiguously that selection operates consistently to weed out all the amino acid substitutions, with the exception of the single amino acid position involved in the Fast/Slow mobility polymorphism described in protein electrophoretic surveys. The data also show an uneven distribution of silent site polymorphism. The third exon of *Adh*, where the amino acid polymorphism resides, has approximately 3 times the level of silent polymorphism (14.3%) as seen in either the remaining exons (3.9%) or in the introns (1.8%–7.1%). This pattern of silent polymorphism is consistent with a selectionist interpretation regarding the Fast/Slow amino acid polymorphism. As described above, selection of a balanced polymorphism would result in an increased level of silent polymorphism in the sites closely linked, exactly the situation observed at *Adh*.

In an attempt to distinguish the importance of positive selection for allozyme variability in *Xdh*, Riley et al. (1992) sequenced seven *Xdh* alleles in *D. pseudoobscura*. The most striking result of this study is the high level of amino acid variation detected at *Xdh*. 2% of the amino acids are polymorphic and allozymes differ by an average of nine amino acids. The polymorphism data for *Xdh* suggest that the encoded protein is not under the influence of positive selection. Not only do members of the major allozyme class not differ in amino acid sequence, the patterns of amino acid and silent variation also do not differ. Further, when one compares the patterns of polymorphism and divergence at *Xdh*, the positive correlation predicted by the neutral theory is obtained.

These studies illustrate the power of molecular data in addressing population genetics questions. At this point, each new locus has provided a radically different story with respect to the levels of variation segregating and the evolutionary forces at work. It is clear that many surveys of nucleotide polymorphism, in particular surveys of additional loci within the same species, will be required before general patterns of molecular evolution at the population level are revealed.

4.11. Conclusions

The introduction of gel electrophoresis into population genetics initiated a fruitful and continuing cross-fertilization between biochemistry and evolutionary biology. In the early years, gel electrophoresis overcame the limitations of genetic crosses and allowed direct comparison of closely related species. More recently, gene cloning and DNA sequence alignments have allowed evolutionary geneticists to collect data on both microevolutionary and macroevolutionary genetic changes. This does not mean that the previous electrophoretic surveys have been eclipsed by the modern DNA-based technologies. The two approaches are, in fact, complementary and will continue to be so for the foreseeable future. In retrospect, we realize that it was the application of enzyme electrophoresis to population studies that constituted the major paradigm shift whereby state-of-the-art molecular and biochemical techniques were introduced into evolutionary studies. As such, it stands as the crucial link between the earlier morphological phase of evolutionary biology and the current DNA phase.

Even if the initial impetus for the electrophoretic studies of natural populations was merely to quantify the amount of genetic polymorphism, these studies have obviously achieved much more than that. They have provided an invaluable insight into the link between genotype and phenotype and they have allowed us the first glimpses of how adaptations happen at the level of individual genes, both structural and regulatory. They have also provided the database for testing the neutral theory of molecular evolution. The earlier studies are now being complemented by DNA-based studies, many of them on the same gene–enzyme systems. These latter studies yield a much clearer view of the partitioning of genetic variation within and between populations (Riley et al., 1992; Veuille and King, 1995). In addition, they are providing information on nonrandom mutational processes that can affect both the DNA sequences of the genes themselves and the amino acid sequences of their protein products (Foster et al., 1997). Such results have obvious implications for the testing of models of neural evolution in that they represent nonrandom, but nonadaptive, evolutionary changes in proteins.

Perhaps the greatest potential pay-off of the molecular approach to evolutionary studies will come from the discovery of the genetic changes that could have given rise to the major morphological transitions during macroevolution. An example of such a study is the recent report by Swalla and Jeffery (1996) of a gene that is essential for the development of chordate features in ascidian larvae. Thus the marriage of biochemistry and population biology, whose first goal was merely to give a more precise quantification of the genetic basis of microevolution, has turned out to be an important cornerstone in the foundation of much broader studies in evolutionary biology.

REFERENCES

Abraham, I. and Doane, W. W. 1978. Genetic regulation of tissue-specific expression of amylase structural genes in Drosophila. *Proc. Natl. Acad. Sci. USA* **75**:4446–4450.

Allard, R. W., Babbel, G. R., Clegg, M. T., and Kahler, A. L. 1972. Evidence for coadaptation in *Avena barbata. Proc. Natl. Acad. Sci. USA* **69**:3043–3048.

Allendorf, W. F. and Leary, R. F. 1986. Heterozygosity and fitness in natural populations of animals. In *Conservation Biology: The Science of Scarcity and Diversity*, M. Soule, ed. Sunderland, MA: Sinauer.

Ayala, F. J. and Powell, J. R. 1972. Allozymes as diagnostic characters of sibling species of *Drosophila. Proc. Natl. Acad. Sci. USA* **69**:1094–1096.

Ayala, F. J. 1975. Genetic differentiation during the speciation. *Process. Evol. Biol.* **8**: 1–78.

Ayala, F. J. 1976. *Molecular Evolution.* Sunderland, MA: Sinauer.

Barbadilla, A., King, L. M., and Lewontin, R. C. 1996. What does electrophoretic variation tell us about protein variation? *Mol. Biol. Evol.* **13**:427–432.

Beckenback, A. T. and Prakash, S. 1977. Examination of allelic variation at the hexokinase loci of *Drosophila pseudoobscura* and *D. persimilis* by different methods. *Genetics* **87**:743–761.

Benkel, B. F. and Hickey, D. A. 1986. Glucose repression of amylase gene expression in *Drosophila melanogaster. Genetics* **114**:137–144.

Benkel, B. F. and Hickey, D. A. 1986a. The interaction of genetic and environmental factors in the control of amylase gene expression in *Drosophila melanogaster. Genetics* **114**:943–954.

Benkel, B. F. and Hickey, D. A. 1987. A Drosophila gene is subject to glucose repression. *Proc. Natl. Acad. Sci. USA* **84**:1337–1339.

Berger, F. G. and Paigen, K. 1979. *Cis*-active control of mouse beta-galactosidase biosynthesis by a systemic regulatory locus. *Nature (London)* **282**:314-316.

Birchler, J. A., Hiebert, J. C., and Paigen, K. 1990. Analysis of autosomal dosage compensation involving the alcohol dehydrogenase locus in *Drosophila melanogaster. Genetics* **124**:679–686.

Brown, A. H. D. 1979. Enzyme polymorphism in plant populations. *Theor. Popul. Biol.* **15**:1–42.

Brown, A. H. D., Marshall, D. R., and Weir, B. S. 1981. Current status of the charge state model for protein polymorphism. In *Genetics Studies of Drosophila Populations: Proceedings of the Kioloa Conference.* Canberra: Australia National U. Press.

Bulmer, M. G. 1971. Protein polymorphism. *Nature (London)* **234**:410–411.

Burton, R. S. and Feldman, M. W. 1983. Physiological effects if an allozyme polymorphism: glutamate-pyruvate transaminase and response to hyperosmotic stress in the copepod *Trigriopus californicus. Biochem. Genet.* **21**:239–251.

Cavener, D. R. and Clegg, M. T. 1981. Evidence for biochemical and physiological differences between genotypes in *Drosophila melanogaster. Proc. Natl. Acad. Sci. USA* **78**:4444–4447.

Chetverikov, S. S. 1926. On certain aspects of the evolutionary process from the standpoint of modern genetics. Z. Eksp. Biol A **2**:3-54 (in Russian) (English translation, 1961, Proc. Am. Philos. Soc. **105**:167–195).

Cooke, P. H. and Oakeshott, J. G. 1989. Amino acid polymorphisms for esterase-6 in *Drosophila melanogaster. Proc. Natl. Acad. Sci. USA* **86**:1426–1430.

Cooke, P. H., Richmond, R. C., and Oakeshott, J. G. 1987. High resolution electrophoretic variation at the esterase-6 locus in a natural population of *Drosophila melanogaster. Heredity* **59**:259–264.

Coyne, J. A. 1976. Lack of genic similarity between two sibling species of *Drosophila* as revealed by varied techniques. *Genetics* **84**:593–607.

Coyne, J. A. 1982. Gel electrophoresis in cryptic protein variation. In *Isozymes, Vol. 6 of Current Topics in Biological and Medical Research Series,* pp. 1–32. New York: Liss.

Coyne, J. A. and Felton, A. A. 1977. Genic heterogeneity at two alcohol dehydrogenase loci in *Drosophila pseudoobscura* and *Drosophila persimilis. Genetics* **87**:285–304.

Dalnou, O., Cariou, M. L., David, J. R., and Hickey, D. 1987. Amylase gene duplication: an ancestral trait in the *Drosophila melanogaster* species subgroup. *Heredity* **59**:245–251.

Da Lage, J. L., Klarenberg, A., and Cariou, M. L. 1996. Variation in sex-, stage- and tissue-specific expression of the amylase genes in *Drosophila ananassae. Heredity* **76**: 9–18.

Darwin, C. 1859. *On the Origin of Species by Means of Natural Selection, or the Preservation of Favored Races in the Struggle for Life.* London: Murray.

De Jong, G., Hoorn, A. J. W., Thorig, G. E. W., and Scharloo, W. 1972. Frequencies of amylase variants in *Drosophila melanogaster. Nature (London)* **238**:453–454.

Dickinson, W. J., Rowan, R. G., and Brennan, M. D. 1984. Regulatory gene evolution: adaptive differences in expression of alcohol dehydrogenase in *Drosophila melanogaster* and *Drosophila simulans. Heredity* **52**:215–225.

DiMichele, L. and Powers, D. A. 1982a. LDH-B genotype specific hatching times of *Fundulus heteroclitus* embryos. *Nature (London)* **296**:563–564.

DiMichele, L. and Powers, D. A. 1982b. Physiological basis for swimming endurance differences between LDH-B genotypes of *Fundulus heteroclitus. Science* **216**:1014–1016.

DiMichele, L. and Powers, D. A. 1991. Allozyme variation, developmental rate and differential mortality in the model teleost, *Fundulus heteroclitus. Physiol. Zool.* **64**:1426–1443.

DiMichele, L., Powers, D. A., and DiMichele, J. A. 1986. Developmental and physiological consequences of genetic variation at enzyme synthesizing loci in *Fundulus heteroclitus. Am. Zool.* **26**:210–208.

Doane, W. W. 1969. Amylase variants in *Drosophila melanogaster*: linkage studies and characterization of enzyme extracts. *J. Exp. Zool.* **171**:321–342.

Dobzhansky, Th. 1937. *Genetics and the Origin of Species,* 1st ed. New York: Columbia U. Press.

Dobzhansky, Th. 1951. *Genetics and the Origin of Species,* 2nd ed. New York: Columbia U. Press.

Dobzhansky, Th. 1973. Nothing in biology makes sense except in the light of evolution. *Am. Biol. Teacher* **35**:125–129.

Dykhuizen, D. E. and Hartl, D. L. 1983. Functional effects of PGI allozymes in *E. coli. Genetics* **105**:1–18.

Eanes, W. F. 1984. Viability interactions, *in vivo* activity and the G6PD polymorphism in *Drosophila melanogaster. Genetics* **106**:95–107.

Eanes, W. F. and Hey, J. 1986. *In vivo* function of rare G6pd variants from natural populations of *Drosophila melanogaster. Genetics* **113**:679–693.

Eanes, W. F. 1987. Allozymes and fitness: evolution of a problem. *Trends Ecol. Evol.* **2**:44–48.

Eanes, W. F., Katona, L., and Longtine, M. 1990. Comparison of *in vitro* and *in vivo* activities associated with the G6PD allozyme polymorphism in *Drosophila melanogaster. Genetics* **125**:845–853.

Eanes, W. F., Kirchner, M., and Yoon, J. 1993. Evidence for adaptive evolution of the G6PD gene in *Drosophila melanogaster* and *D. simulans. Proc. Natl. Acad. Sci. USA* **90**:7475–7479.

Eanes, W. F., Kirchner, M., Taub, D. R., Yoon, J., and Chen, J. 1996. Amino acid polymorphism and rare electrophoretic variants of G6PD from natural popualtions of *Drosophila melanogaster. Genetics* **143**:401–406.

Ewens, W. 1972. The sampling theory of selectively neutral alleles. *Theor. Popul. Biol.* **3**:87–112.

Fisher, R. A. 1930. *The Genetical Theory of Natural Selection.* Oxford: Oxford U. Press.

Foster, P. G., Jermiin, L. S., and Hickey, D. A. 1997. Nucleotide composition bias affects amino acid content in animal mitochondria. *J. Mol. Evol.* **44**:282–288.

Gillespie, J. H. 1987. Molecular evolution and the neutral allele theory. *Oxford Surv. Evol. Biol.* **4**:10–37.

Gillespie, J. H. 1991. *The Causes of Molecular Evolution.* Oxford: Oxford U. Press.

Grant, V. 1963. *The Origin of Adaptation.* New York: Columbia U. Press.

Grunder, A. A., Loverre-Chyurlia, A., and Hickey, D. A. 1993. Expression of an amylase-alcohol dehydrogenase chimeric gene in transgenic strains of *Drosophila melanogaster. Genome* **36**:954–961.

Haldane, J. B. S. 1932. *The Causes of Evolution.* London: Longmans, Green.

Hamrick, J. L., and Allard, R. W. 1972. Microgeographical variation in allozyme frequencies in *Avena barbata. Proc. Natl. Acad. Sci. USA* **69**:2100–2104.

Hamrick, J. L., Linhart, Y. B., and Mitton, J. B. 1979. Relationships between life history characteristics and electrophoretically detectable genetic variation in plants. *Annu. Rev. Ecol. Syst.* **10**:173–200.

Harris, H. 1966. Enzyme polymorphism in man. *Proc. R. Soc. London Ser. B* **164**:298–310.

Hedrick, P. W. and McDonald, J. F. 1980. Regulatory gene adaptation: an evolutionary model. *Heredity* **45**:83–97.

Hickey, D. A. 1977. Selection for amylase allozymes in *Drosophila melanogaster. Evolution* **31**:800–804.

Hickey, D. A. 1979. Selection on amylase allozymes in *Drosophila melanogaster*: selection experiments using several independently derived pairs of chromosomes. *Evolution* **33**:1128–1137.

Hickey, D. A. 1981. Regulation of amylase activity in *Drosophila melanogaster*: variation in the number of enzyme molecules produced by different amylase genotypes. *Biochem. Genet.* **19**:783–796.

Hickey, D. A. and Benkel, B. F. 1982. Regulation of amylase activity in *Drosophila melanogaster*: effects of dietary carbohydrate. *Biochem. Genet.* **20**:1117–1129.

Hickey, D. A., Bally-Cuif, L., Abukashawa, S., Payant, V., and Benkel, B. F. 1991. Concerted evolution of duplicated protein-coding genes in Drosophila. *Proc. Natl. Acad. Sci. USA* **88**:1611–1615.

Hickey, D. A., Benkel, K. I., Fong, Y., and Benkel, B. F. 1994. A Drosophila gene promoter is subject to glucose repression in yeast cells. *Proc. Natl. Acad. Sci. USA* **91**:11109–11112.

Hilbish, T. J., Deaton, L. E., and Koehn, R. K. 1982. Effect of an allozyme polymorphism on regulation of cell volume. *Nature (London)* **298**:688–689.

Hubby, J. L. and Lewontin, R. C. 1966. A molecular approach to the study of genic heterozygosity in natural populations. I. The number of alleles at different loci in *Drosophila pseudoobscura. Genetics* **54**:577–594.

Hubby, J. L. and Throckmorton, L. H. 1968. Protein differences in Drosophila. IV. A study of sibling species. *Am. Nat.* **102**:193–205.

Hudson, R. R., Bailey, K., Skarecky, D., Kwiatowski, J., and Ayala, F. J. 1994. Evidence for positive selection in the superoxide dismutase (Sod) region of *Drosophila melanogaster. Genetics* **136**:1329–1340.

Hudson, R. R., Kreitman, M., and Aguade, M. 1987. A test of neutral molecular evolution based on nucleotide data. *Genetics* **116**:153–159.

Huxley, J. 1942. *Evolution, the Modern Synthesis.* London: Allen and Unwin.

Inomata, N., Shibata, H., Okuyama, E., and Yamazaki, T. 1995. Evolutionary relationship and sequence variation of α-amylase variants encoded by duplicate genes in the *Amy* locus of *Drosophila melanogaster. Genetics* 141:237–244.

Jespen, G. L., Mayr, E., and Simpson, G. G. 1949. *Genetics, Palaeontology and Evolution.* Princeton, NJ: Princeton U. Press.

Kappen, C., and Ruddle, F. H. 1993. Evolution of a regulatory gene family: HOM/HOX genes. *Curr. Opin. Genet. Dev.* 3:931–938.

Karlin, S. and Nevo, E. 1976. *Population Genetics and Ecology.* New York: Academic.

Keith, T. P. 1983. Frequency distribution of esterase-5 alleles in populations of *Drosophila pseudoobscura. Genetics* 105:135–155.

Keith, T. P., Brooks, L. D., Lewontin, R. C., Martinez-Cruzada, J. C., and Rigby, D. L. 1985. Nearly identical allelic distributions of xanthine dehydrogenase in two populations of *Drosophila pseudoobscura. Mol. Biol. Evol.* 2:206–216.

Kimura, M. 1968. Evolutionary rate at the molecular level. *Nature (London)* 217:624–626.

Kimura, M. 1983. *The Neutral Theory of Molecular Evolution.* New York: Cambridge U. Press.

Kimura, M. and Crow, J. 1964. The number of alleles that can be maintained in a finite population. *Genetics* 49:725–738.

Kimura, M. and Ohta, T. 1971. Protein polymorphism as a phase of molecular evolution. *Nature (London)* 229:467–469.

King, J. L. 1967. Continuously distributed factors affecting fitness. *Genetics* 55:483–492.

King, J. L. and Jukes, T. H. 1969. Non-Darwinian evolution: random fixation of selectively neutral mutations. *Science* 164:788–798.

King, J. L. and Ohta, T. 1975. Polyallelic mutational equilibria. *Genetics* 79:681–691.

King, M.-C. and Wilson, A. C. 1975. Evolution at two levels in humans and chimpanzees. *Science* 188:107–116.

Koehn, R. K. 1969. Esterase heterogeneity: dynamics of a polymorphism. *Science* 163:943–944.

Koehn, R. K. and Hilbish, T. J. 1987. The adaptive importance of genetic variation. *Am. Sci.* 75:134–141.

Kreitman, M. 1980. Assessment of variability within electromorphs of alcohol dehydrogenase in *Drosophila melanogaster. Genetics* 95:457–475.

Kreitman, M. 1983. Nucleotide polymorphism at the alcohol dehydrogenase locus of *Drosophila melanogaster. Nature (London)* 304:412–417.

Kreitman, M. 1988. Molecular population genetics. *Oxford Surv. Evol. Biol.* 4:38–60.

Labate, A., Bortoli, A., Game, A. Y., Cooke, P. H., and Oakeshott, J. G. 1989. The number and distribution of esterase-6 alleles in populations of *Drosophila melanogaster. Heredity* 63:203–208.

Labate, J. and Eanes, W. F. 1992. Direct measurement of *in vivo* flux differences between electrophoretic variants of G6PD in *Drosophila melanogaster. Genetics* 132:783–787.

Laurie, C. C. and Stam, L. F. 1988. Quantitative analysis of RNA produced by Slow and Fast alleles of *Drosophila melanogaster* Adh. *Proc. Natl. Acad. Sci. USA* 85:5161–5165.

Laurie, C. C. and Stam, L. F. 1994. The effect of an intronic polymorphism on alcohol dehydrogenase expression in *Drosophila melanogaster. Genetics* 138:379–385.

Laurie-Ahlberg, C. C., Wilton, A. N., Curtsinger, J. W., and Emigh, T. H. 1982. Naturally occurring enzyme activity variation in *Drosophila melanogaster*. I. Sources of variation for 23 enzymes. *Genetics* **102**:191–206.

Lewontin, R. C. 1974. *The Genetic Basis of Evolutionary Change*. New York: Columbia U. Press.

Lewontin, R. C. 1985. Population genetics. *Annu. Rev. Genet.* **19**:81–202.

Lewontin, R. C. 1991. Twenty-five years ago in genetics: electrophoresis in the development of evolutionary genetics: milestone or millstone? *Genetics* **128**:657–662.

Lewontin, R. C. and Hubby, J. L. 1966. A molecular approach to the study of genetic heterozygosity in natural populations. II. Amount of variation and degree of heterozygosity in natural populations of *Drosophila pseudoobscura*. *Genetics* **54**:595–609.

Loukas, M., Vergini, Y., and Krimbas, C. B. 1981. The genetics of *Drosophila subobscura* populations. XVII. Further genic heterozgeneity within electrophoretic electromorphs by urea denaturation and the effect of the increased genic variability on linkage disequilibrium studies. *Genetics* **97**:429–441.

MacIntyre, R. J., and Collier, G. E. 1986. Protein evolution in the genus Drosophila. In *The Genetics and Biology of Drosophila*, Vol 3., M. Ashburner, H. L. Carson, and J. N. Thompson Jr., eds., pp. 39–146. London: Academic.

Magoulas, C., Bally-Cuif, L., Loverre-Chyurlia, A., Benkel, B., and Hickey, D. 1993. A short 5′ flanking region mediates glucose repression of amylase gene expression in *Drosophila melanogaster*. *Genetics* **134**:507–515.

Matters, G. L. and Scandalios, J. G. 1986. Changes in plant gene expression during stress. *Dev. Genet.* **7**:167–175.

Maynard Smith, J. 1972. Protein polymorphism. *Nature (London) New Biol.* **237**:31.

Maynard Smith, J. and Haigh, J. 1974. The hitch-hiking effect of a favorable gene. *Genet. Res. Cambridge* **23**:23–35.

Mayr, E. 1963. *Animal Species And Evolution*. Cambridge, MA: Harvard U. Press.

Mayr, E. and Provine, W. B. 1980. *The Evolutionary Synthesis. Perspective on the Unification of Biology*. Cambridge, MA: Harvard U. Press.

McDonald, J. F. and Ayala, F. J. 1978. Gene regulation in adaptive evolution. *Can. J. Genet. Cytol.* **20**:159–175.

McDonald, J. F., Chambers, G. K., David, J., and Ayala, F. J. 1977. Adaptive response due to changes in gene regulation: a study with Drosophila. *Proc. Natl. Acad. Sci. USA* **74**:4562–4566.

Milkman, R. D. 1967. Heterosis as a major cause of heterozygosity in nature. *Genetics* **55**:493–495.

Mitton, J. B. and Grant, M. C. 1984. Association among protein heterozygosity, growth rate, and developmental homeostasis. *Annu. Rev. Ecol. Syst.* **15**:479–499.

Muller, H. J. 1950. Our loads of mutations. *Am. J. Hum. Genet.* **2**:111–176.

Mukai, T., Yamaguchi, O., Kusakabe, S., Tachida, H., Matsuda, M., Ichinose, M., and Yoshimaru, H. 1982. Lack of balancing selection for protein polymorphisms. In *Molecular Evolution, Protein Polymorphism and the Neutral Theory*, M. Kimura, ed., pp. 81–120. Tokyo: Japan Scientific Societies Press; Berlin: Springer-Verlag.

Nei, M. 1987. *Molecular Evolutionary Genetics*. New York: Columbia U. Press.

Nei, M. and Graur, D. 1984. Extent of protein polymorphism and neutral mutation theory. *Evol. Biol.* **17**:73–118.

Nei, M. and Koehn, R. K. 1983. *Evolution of Genes and Proteins*. Sunderland, MA: Sinauer.

Nevo, E., Beiles, A., and Ben-Shlomo, R. 1984. The evolutionary significance of genetic diversity: ecological, demographic and life history correlates. *Lect. Notes Biomath.* **53**:13–213.

Norman, R. A. and Prakash, S. 1980. Developmental variation in amylase allozyme activity associated with chromosome inversions in *Drosophila persimilis*. *Genetics* **95**:1001–1011.

Oakeshott, J. G., Chambers, G. K., Gibson, J. B, Eanes, W. F., and Willcocks, D. A. 1983. Geographic variation in G6pd and Pgd allele frequencies in *Drosophila melanogaster*. *Heredity* **50**:67–72.

O'Brien, S. J. and MacIntyre, R. J. 1978. Genetics and biochemistry of enzymes and specific proteins of Drosophila. In *The Genetics and Biology of Drosophila*, M. Ashburner, H. L. Carson, and J. N. Thompson, Jr., eds., pp. 396–552. London: Academic.

Ohta, T. 1976. Role of very slightly deleterious mutations in molecular evolution and polymorphism. *Theor. Popul. Biol.* **10**:254–275.

Ohta, T. 1992. The nearly neutral theory of evolution. *Annu. Rev. Ecol. Syst.* **23**:263–286.

Ohta, T. and Kimura, M. 1973. A model of mutation appropriate to estimate the number of electrophoretically detectable alleles in a finite population. *Genet. Res. Cambridge* **22**:201–204.

Paigen, K. 1979. Acid hydrolases as models of genetic control. *Annu. Rev. Genet.* **13**:417–466.

Patel, N. H., Martin-Blanco, E., Coleman, K. G., Poole, S. J., Ellis, M. C., Kornberg, T. B., and Goodman, C. S. 1989. Expression of engrailed proteins in arthropods, annelids, and chordates. *Cell* **58**:955-968.

Payant, V., Abukashawa, S., Sasseville, M., Benkel, B. F., Hickey, D. A., and David, J. 1988. Evolutionary conservation of the chromosomal configuration and regulation of amylase genes among eight species of the *Drosophila melanogaster* subgroup. *Mol. Biol. Evol.* **5**:560–567.

Place, A. R. and Powers, D. A. 1979. Genetic variation and relative catalytic efficiencies: lactate dehydrogenase-B allozyme. *Proc. Natl. Acad. Sci. USA* **76**:2354–2358.

Powell, J. R. and Lichtenfels, J. M. 1979. Population genetics of Drosophila amylase. I. Genetic control of tissue-specific expression in *D. pseudoobscura*. *Genetics* **92**:603–612.

Powers, D. A. and Place, A. R. 1978. Biochemical genetics of *Fundulus heteroclitus*. I. Temporal and spatial variation in gene frequencies of Ldh-B, Mdh-A, Gpi-B and Pgm-A. *Biochem. Genet.* **16**:593–607.

Prakash, S. 1972. Origin of reproductive isolation in the absence of apparent genic differentiation in a geographic isolate of *Drosophila pseudoobscura*. *Genetics* **72**:143–155.

Prakash, S. and Lewontin, R. C. 1968. A molecular approach to the study of genic heterozygosity. III. Direct evidence of coadaptation in gene arrangements of Drosophila. *Proc. Natl. Acad. Sci. USA* **59**:398–405.

Ramshaw, J. A. M., Coyne, J. A., and Lewontin, R. C. 1979. The sensitivity of gel electrophoresis as a detector of genetic variation. *Genetics* **93**:1019–1037.

Richmond, R. C. 1972. Enzyme variability in the *Drosophila willistoni* group. III. Amounts of variability in the super species *D. paulistorum*. *Genetics* **71**:87–112.

Riley, M. A., Hallas, M. E., and Lewontin, R. C. 1989. Distinguishing the forces controlling genetic variation at the Xdh locus in *Drosophila pseudoobscura*. *Genetics* **123**:359–369.

Riley, M. A., Kaplan, S. R., and Veuille, M. 1992. Nucleotide polymorphism at the xanthine dehydrogenase locus in *Drosophila pseudoobscura*. *Mol. Biol. Evol.* **9**:56–69.

Schaeffer, S. and Miller, E. L. 1993. Estimates of linkage disequilibrium and the recombination parameter determined from segregating nucleotide sites in the alcohol dehydrogenase region of *Drosophila pseudoobscura*. *Genetics* **135**:541–552.

Selander, R. K., Hunt, W. A., and Yang, S. Y. 1969. Protein polymorphism and genetic heterozygosity in two European subspecies of the house mouse. *Evolution* **23**:379–390.

Selander, R. K., Yang, S. Y., Lewontin, R. C., and Johnson, W. E. 1970. Genetic variation in the horseshoe crab (*Limulus polyphemus*), a phylogenetic "relic." *Evolution* **24**:402–414.

Shaw, C. R. 1965. Electrophoretic variation in enzymes. *Science* **149**:936–943.

Simpson, G. G. 1944. *Tempo and Mode in Evolution*. New York: Columbia U. Press.

Simpson, G. G. 1953. *The Major Features of Evolution*. New York: Columbia U. Press.

Singh, R. S. 1979. Genic heterogeneity within electrophoretic "alleles" and the pattern of variation among loci in *Drosophila pseudoobscura*. *Genetics* **93**:997–1018.

Singh, R. S. 1989. Population genetics and evolution of species related to *Drosophila melanogaster*. *Annu. Rev. Genet.* **23**:425–453.

Singh, R. S. 1990. Patterns of species divergence and genetic theories of speciation. In *Population Biology: Ecological and Evolutionary Viewpoints.*, K. Wohrmann and S. K. Jain, eds., pp. 231–265. New York and Berlin: Springer-Verlag.

Singh, R. S., Lewontin, R. C., and Felton, A. A. 1976. Genetic heterozygosity within electrophoretic alleles of xanthine dehydrogenase. *Genetics* **117**:255–271.

Singh, R. S., Hickey, D. A., and David, J. 1982. Genetic differentiation between geographically distant populations of *Drosophila melanogaster*. *Genetics* **101**:135–156.

Singh, S. M. and Zouros, E. 1978. Genetic variation associated with growth rate in the American oyster (*Crassosferea virginica*). *Evolution* **32**:342–353.

Smithies, O. 1995. Early days of gel electrophoresis. *Genetics* **139**:1–4.

Stebbins, G. L. 1950. *Variation and Evolution in Plants*. New York: Columbia U. Press.

Stern, C. 1943. Genic action as studied by means of the effects of different doses and combinations of alleles. *Genetics* **28**:441–475.

Strobeck, C. 1983. Expected linkage disequilibrium for a neutral locus linked to a chromosomal arrangement. *Genetics* **103**:545–555.

Sved, J. A., Reed, T. E., and Bodmer, W. F. 1967. The number of balanced polymorphisms that can be maintained in a natural population. *Genetics* **55**:469–481.

Swalla, B. J. and Jeffery, W. R. 1996. Requirement of the manx gene for expression of chordate features in a tailless ascidian larva. *Science* **274**:1205–1208.

Takano, T. S., Kusakabe, S., and Mukai, T. 1993. DNA polymorphism and the origin of protein polymorphism at the Gpdh locus of *Drosophila melanogaster*. In *Mechanisms of Molecular Evolution*, N. Takahata and A. G. Clark, eds., pp. 179–190. Sunderland, MA: Japan Scientific Societies and Sinauer.

Van Delden, W. 1982. The alcohol dehydrogenase polymorphism in *Drosophila melanogaster*. *Evol. Biol.* **15**:187–222.

Veuille, M. and King, L. M. 1995. Molecular basis of polymorphism at the esterase-5B locus in *Drosophila pseudoobscura*. *Genetics* **141**:255–262.

Vrijenhoek, R. C., Pfeiler, E. and Wetherington, J. D. 1992. Balancing selection in a desert stream-dwelling fish, *Poeciliopsis monacha*. *Evolution* **46**:1642–1657.

Wallace, B. 1991. *Fifty Years of Genetic Load*. Ithaca, NY: Cornell U. Press.

Watt, W. B. 1992. Eggs, enzymes, and evolution – natural genetic variants change. *Proc. Natl. Acad. Sci. USA* **89**:10608–10612.

Watt, W. B. 1994. Allozymes in evolutionary genetics – self-imposed burden or extraordinary tool. *Genetics* **136**:11–16.

Watterson, G. A. 1977. Heterosis or neutrality? *Genetics* **85**:789–814.

Wood, H. G., Katz, J., and Landau, B. R. 1963. Estimation of pathways of carbohydrate metabolism. *Biochem. Z.* **338**:809–847.

Wright, S. 1931. Evolution in Mendelian populations. *Genetics* **16**:97–159.

Yamazaki, T. 1971. Measurement of fitness at the esterase-5 locus in *Drosophila pseudoobscura*. *Genetics* **67**:579–603.

Zouros, E. 1973. Genic differentiation associated with the early stages of speciation in the *mulleri* subgroup of Drosophila. *Evolution* **27**:601–621.

Zouros, E. and Foltz, D.W. 1987. The use of allelic isozyme variation for the study of heterosis. *Isozyme* **15**:1–59.

SECTION B

MOLECULAR VARIATION AND EVOLUTION

Introductory Remarks

MARTIN KREITMAN

The focus on characterizing allelic polymorphism at single gene loci to understand forces governing genetic variation in natural populations can be attributed directly to R. C. Lewontin (Lewontin and Hubby, 1966). Indeed, the paradigmatic shift in focus from quantitative and classical genetic approaches to a strictly molecular one for investigating polymorphism is one of Lewontin's proudest achievements (personal communication). The advent of allozyme electrophoresis did two things for population genetics. First, it resulted in a quantum leap in the precision with which polymorphism could be measured within and among populations. This greater precision in measurement led to an explosion in its use in the estimation of the genetic differences between populations and species, and a number of important estimators, such as Nei's D, were proposed precisely for this purpose. Second, and more importantly, electrophoresis provided the very data – allele frequencies of individual loci – that could be related to the predictions of a broad body of theoretical work. Population genetics is, after all, about gene frequencies.

The advent of DNA sequencing, offers the opportunity to study genetic variation in the greatest possible detail, and I, for one, jumped at this opportunity. Lewontin, initially suspicious of the new technology, quickly converted to the new theology, offering his own rationale for it, "God is in the detail." The ability to distinguish natural selection from genetic drift with the new data would be as direct (and inevitable) as determining the melting temperature of this double-stranded helix.

On the theoretical front, it was Kimura's development of the neutral theory of molecular evolution and, in particular, his formulation of a model of neutral alleles with infinitely many sites (Kimura 1969) that provided evolutionary and molecular biologists alike with a conceptual framework for making sense of DNA-level variation and change. But more than that, the theory led Kimura to a far-reaching insight about the molecular evolutionary process, one which, in my view, has been the single most important factor in propelling the field forward. This insight was summed up most succinctly in his 1983 book on the neutral theory of molecular evolution: "Polymorphism is simply a transient phase of molecular evolution" (Kimura, 1983). But more than merely connecting population genetic variation within species and evolutionary changes between species, his theory does so for the irreducible currency of DNA change, the single nucleotide substitution. With DNA, all degrees of divergence between two sequences could be quantitatively treated under his neutral theory. Stated another way, the number of nucleotide differences between sequences taken from individuals in the same population or in different species could be understood as simply reflecting different timepoints in the evolutionary process. With Kimura's insight, population genetics no longer had to stop at the doorstep of speciation. Rather, the analysis of empirical data and the modeling of molecular evolutionary processes could now scale essentially the complete range of evolutionary time.

From a more practical perspective, it was now possible to fit data representing more than one timepoint in the evolutionary process to the predictions of the neutral theory. Thus, rather than being restricted to asking whether estimates of allele frequencies were compatible with a particular theoretical prediction, one could now ask whether the patterns of nucleotide substitution at more than one timepoint in the process are compatible with a theoretical prediction. This liberation, like the one that preceded it for allozymes, led to a proliferation of statistical tests of neutral evolution. Chapter 7, by Schaeffer and Aguade, provides an up-to-date survey of statistical methods for testing the neutral theory and also a summary of knowledge about the forces governing the patterning of nucleotide variation across the *Drosophila* genome.

As highlighted in Chap. 7, recent empirical findings in *Drosophila* underscore the fact that molecular variation and evolution cannot be understood in terms of forces acting in isolation at individual nucleotide sites, but must instead take into account genetic linkage and the selection that might be acting at all of these sites. Chapter 6, by Berry and Barbadilla, provides compelling arguments for the need to include gene conversion in consideration of the patterning of nucleotide polymorphism on the scale of individual gene loci. A simple model of recombination that includes both classical crossing over and gene conversion is sufficient to make the point that, within a locus, gene conversion is far more likely to shuffle mutations among alleles than cross over. It also leads, perhaps, to the disappointing suspicion that the greater-than-neutral level of nucleotide polymorphism that is expected to build up in linkage disequilibrium with a site under balancing selection may not extend much beyond the tract length of a gene conversion event. In *Drosophila*, regions of such small size may be hard to detect statistically.

On a more encouraging note, Chap. 5, by Kreitman and Antezana, reviews and extends knowledge about the evolution of synonymous substitutions – single nucleotide changes in the coding regions of genes that do not affect the amino acid sequence of a protein. Although it is perfectly plausible to have expected that synonymous substitutions would be selectively neutral, this turns out not to be the case in a number of species, including *Drosophila*. More surprising, perhaps, is the progress that has been made in understanding the nature of the selection's acting on synonymous substitutions. Chapter 5 summarizes that progress and discusses empirical observations about synonymous changes that have not yet been fully accounted for with the current theoretical model of codon bias: the mutation–selection–drift model. Chapter 5 also provides specific applications of and clever extensions to the statistical tests described in Chap. 7. In particular, it points to the remarkable statistical sensitivity of the molecular population genetic approach to detect extremely weak processes in molecular evolution. Lewontin, I am sure, as confident as he was about the power of DNA analysis, would have bet the ranch that this approach would not allow the rejection of a completely neutral model for synonymous substitutions and at the same time would provide an estimate of the product of population size and the strength of selection $N_e s$ for synonymous substitutions in *Drosophila* of ~2! With certain kinds of data and the appropriate characterization of them, the population genetic approach can be a very powerful one.

REFERENCES

Kimura, M. 1969. The number of heterozygous nucleotide sites maintained in a finite population due to steady flux of mutations. *Genetics* **61**:893–903.

Kimura, M. 1983. *The neutral theory of molecular evolution.* Cambridge: Cambridge U. Press.

Lewontin, R. C. and Hubby, J. L. 1966. A molecular approach to the study of genic heterozygosity in natural populations, II. Amount of variation and degree of heterozygosity in natural populations of *Drosophila pseudoobscura*. *Genetics* **54**:595–609.

CHAPTER FIVE

The Population and Evolutionary Genetics
of Codon Bias

MARTIN KREITMAN AND MARCOS ANTEZANA

5.1. Introduction

The claim that the calculus of population genetic theory can be applied to evolutionary and molecular population genetic data to decipher the forces of genetic drift and selection is one that is indelibly associated with R. C. Lewontin. In point of fact, Lewontin's vision has yet to be realized for the object of most of his research career – protein polymorphism – and the advent of DNA sequence-level information has not improved the situation very much (Kreitman and Akashi, 1995). For although it is almost certainly true that major features of protein evolution cannot be explained by genetic drift alone (Gillespie, 1991; Ohta, 1995), little progress has been made in understanding the forces governing protein polymorphism (but see Watt, 1994). Nevertheless, not all has been lost in the flood of DNA sequence data inundating the databases. Rather, in an irony perhaps typical for population genetics, a field notorious for having two opposing theories that have been able to account for each new set of data, synonymous nucleotide substitutions, *a priori* thought to be the least likely to be governed by selection, now provide the most convincing case for it.

Synonymous changes are sufficiently common as polymorphisms and substitutions to allow detailed population genetic and evolutionary analyses. They also offer a unique opportunity to study the joint influences of mutation, finite population size, and selection on the patterning of variation within and between species. In this chapter we investigate how well the prevailing theory of codon bias – the mutation/selection/drift (MSD) theory – accounts for data on codon bias. The broad subject of biased codon usage has focused primarily on three species, *Escherichia coli*, *Saccharomyces cerevisiae*, and *Drosophila melanogaster*. We will pay particular attention here to published work on *Drosophila* and will also present new data and analyses. Information about other species will be introduced only when relevant. This chapter is not intended as a review of

R. S. Singh and C. B. Krimbas, eds., *Evolutionary Genetics: From Molecules to Morphology*, vol. 1.

codon usage and selection, as there are numerous excellent reviews of the subject (Ikemura, 1985; Sharp and Li, 1986; Sharp, 1989; Andersson and Kurland, 1990; Sharp and Matassi, 1994; Sharp et al., 1995). Rather, we focus attention on aspects of data that are not predicted by the MSD theory and provide plausible alternatives or modifications to the theory to account for the data.

There is overwhelming evidence that natural selection governs codon usage in yeast, enterobacteria, and *Drosophila*. Some of this evidence is presented in the following sections. However, a basic set of facts about codon bias provides a necessary background to the subject and motivates the appeal to selection. First, the span of codon usage frequencies in genes with low and high codon biases is far too large to be due to mutational processes alone. In addition, genes with high bias use a common set of preferred, or major, codons, generally one for each codon family. For example, CCC, a member of the fourfold degenerate codon family encoding proline, is the major codon in *D. melanogaster*. It is utilized 54% of the time in highly biased genes of this species but only 25% of the time in genes with low bias (see Table 5.1 and Section 5.6). Second, in yeast and *E. coli*, the major codon is always recognized by the most abundant tRNA for each codon family and is referred to as the major codon preference. Third, the extent of biased codon usage is strongly correlated with gene expression levels in yeast and *E. coli*. This correlation suggests that the strength of selection for a favored codon in each gene is directly related to the number of times the gene's mRNA is translated.

5.2. MSD Model of Codon Bias

Several authors have developed models of weak selection and finite population size to study the joint actions of mutation, selection, and genetic drift on codon usage bias (Sharp and Li, 1986; Li, 1987; Bulmer, 1987, 1991; Akashi, 1995). The importance of genetic drift is apparent in genes with low codon bias, in which the base composition of synonymous sites and the levels of synonymous polymorphism and divergence are similar to noncoding regions of the genome. At the other extreme, polymorphism levels and divergence rates in highly biased genes are sufficiently large to require that natural selection for favored codons must be counterbalanced by genetic drift of mutations to nonmajor codons. Li (1987) showed that even with a substantial amount of genetic drift, weak selection is sufficient to allow for the evolution of extreme codon bias.

Under the theory of weak selection and finite population size, the equilibrium frequency of the major codon is given by

$$P = e^S V / (e^S V + U), \tag{5.1}$$

where $S = 4N_e s$, $V = 4N_e u$, and $U = 4N_e v$ for diploid organisms. The mutation parameters u and v are the forward and the backward mutation rates, respectively, to the major codon, and $1 + s$ is the selective advantage of carrying the major codon relative to all others (Bulmer, 1991). (For simplicity, it is

Table 5.1. Codon Usage in Drosophila

Amino Acid	Codon	mel (24)*	pse (24)*	sub (24)*	mel High†	mel Low†	mel 4th‡	Amino Acid	Codon	mel (24)*	pse (24)*	sub (24)*	mel High†	mel Low†	mel 4th‡
Phe	TTT	81	93	99	166	1506	298	Ser	TCT	32	53	51	205	1028	230
	TTC	243	230	222	1385	1319	136		TCC	146	117	119	1110	1349	181
Leu	TTA	12	18	10	32	897	251		TCA	37	25	23	82	1116	235
	TTG	105	104	106	432	1549	208		TCG	101	108	108	663	1170	138
Leu	CTT	36	41	37	183	1070	233	Pro	CCT	38	42	33	173	972	114
	CTC	102	101	113	598	771	86		CCC	164	168	165	1062	1270	94
	CTA	34	39	24	97	908	135		CCA	69	62	69	342	1671	214
	CTG	365	336	355	2207	1884	131		CCG	64	64	63	388	1194	101
Ile	ATT	149	158	183	528	1850	375	Thr	ACT	59	95	95	275	1200	213
	ATC	263	247	227	1684	1248	136		ACC	216	187	173	1547	1439	142
	ATA	40	53	49	53	1125	251		ACA	33	48	54	191	1450	245
Met	ATG	182	192	194	1037	1739	187		ACG	80	81	93	351	984	91
Val	GTT	83	74	66	395	1350	275	Ala	GCT	150	154	136	697	1613	324
	GTC	164	186	178	924	985	111		GCC	376	352	346	2445	1828	179
	GTA	23	47	29	102	884	177		GCA	55	60	72	221	1546	267
	GTG	280	244	273	1443	1648	149		GCG	81	85	103	358	1011	107

Amino Acid	Codon	mel (24)*	pse (24)*	sub (24)*	mel High†	mel Low†	mel 4th‡	Amino Acid	Codon	mel (24)*	pse (24)*	sub (24)*	mel High†	mel Low†	mel 4th‡
Tyr	TAT	70	75	85	215	1125	188	Cys	TGT	32	37	35	61	558	110
	TAC	177	167	156	1087	1135	141		TGC	127	118	122	487	882	102
End	TAG							Trp	TGG	77	78	78	437	718	64
His	CAT	55	79	84	216	1030	210	Arg	CGT	67	79	95	582	760	112
	CAC	130	110	109	741	1055	166		CGC	114	150	148	1141	976	95
Gln	CAA	54	54	53	274	1821	361		CGA	34	28	22	86	820	104
	CAG	196	205	214	1784	2482	275		CGG	50	41	38	137	577	52
Asn	AAT	93	133	154	346	2232	461	Ser	AGT	43	42	28	149	1319	188
	AAC	232	201	177	1548	1887	343		AGC	97	102	119	667	1561	198
Lys	AAA	67	72	77	306	2267	476	Arg	AGA	32	15	23	50	655	79
	AAG	367	344	333	2926	2739	314		AGG	31	25	20	152	555	36
Asp	GAT	208	203	204	1057	2715	407	Gly	GGT	141	144	150	684	1194	206
	GAC	185	184	183	1437	1710	237		GGC	283	335	350	1434	1497	164
Glu	GAA	92	93	85	554	2543	453		GGA	164	103	82	585	1755	281
	GAG	335	329	324	2963	2755	234		GGG	26	44	31	50	452	78

* 24 genes (Zeng et al., 1998).
† 1000 genes containing at least 190 codons were translated from GenBank. Nc scores were calculated for each gene, assuming 60% A + T content. High: 100 genes with highest Nc scores; Low: 100 genes with lowest Nc scores.
‡ 10 fourth-chromosome genes (see text for detail).

assumed that there is a single major codon for each codon family.) If one further assumes that $u = v$, then

$$P = (1 + e^{-S})^{-1}. \tag{5.2}$$

Thus the expected proportion of sites with the major codon is dependent on only S, the scaled selection coefficient. Equations (5.1) and (5.2) are appropriate only when $U + V \ll 1$, a condition likely to be true in virtually all organisms. For example, in *D. melanogaster*, nucleotide diversity at synonymous sites, an estimator of $4N_e u$, falls in the range $0 < 4N_e u < 0.01$. For some other *Drosophila* species the maximum estimated value of $4N_e u$ is slightly larger, but does not generally exceed 0.03. As pointed out by Bulmer, the model predicts monomorphism at each synonymous site, with a fraction P of the sites having the major codon when $U + V$ are small.

5.3. Codon Usage Evolution among Distantly Related Species

There is a good reason to be interested in comparing codon bias patterns among related species. Under the MSD theory, codon bias levels are expected to be sensitive to S, the product of the population size and the selective advantage of a major codon. Once codon bias has evolved in a species, a change in a major codon preference can come about either through sustained oscillation in population size, so that $S \ll 1$, or through selection for a major codon shift within a codon family. A population size shift should affect all codon families, particularly in genes in which S is not very large. In contrast, major codon switches should affect only particular codon families. A major codon shift, such as one that tracks a change in the relative abundance of the tRNAs of a codon family, is expected to impose a large genetic load. Therefore one might expect major codon shifts to be rare in evolution. On the other hand, population size fluctuations and evolutionary changes in effective population sizes are expected to be common occurrences.

The knowledge that one or even a few codon families have changed their codon preference between two species does not allow distinguishing between a major codon shift hypothesis and population size fluctuation scenario. In fact, the two hypotheses are not mutually exclusive: a reduction in population size so that $S \ll 1$ can ameliorate the genetic load imposed by a major codon shift. An interesting empirical question is whether there is evidence for the systematic reduction or loss of codon bias over micro- and macro-evolutionary time. Such a reduction or loss, if it occurred in all codon families, would be strong evidence in support of the population fluctuation hypothesis.

Here we summarize findings about the conservation and the divergence in codon usage patterns among distantly related taxa and among genes. We know of no convincing case in which closely related species have shifted from one major codon to another (see Maynard Smith and Smith, 1996, for a possible exception), although there are distinct shifts among very distantly related species. Comparisons of codon usage among distantly related prokaryotes and eukaryotes have been made by Sharp (1989). A summary of his findings is as follows. The gram-negative enterobacteria, *E. coli* and *Salmonella typhimurium*, have been estimated to have diverged approximately 140 million years ago, on the basis of 16*S* rRNA sequence divergence (Ochman and Wilson, 1987). Despite extensive sequence divergence, the two species exhibit very similar codon usage patterns (Ikemura, 1985; Sharp and Li, 1986; Sharp, 1989). *Bacillus subtilis*, a member of the gram-positive bacteria, is very distantly related to the gram-negative bacteria and exhibits sharply biased codon usage, but the order of codon preferences is different for 6 of the 18 degenerate amino acid codon families (Sharp, 1989). Similarly, *S. cerevisiae* and *Schizosaccharomyces pombe* are very distantly related species, having been separated for as long as 10^9 years, but have similar, although not identical, codon preferences. Amazingly, the degree of codon bias, while lower overall in *S. pombe*, is claimed to be highly concordant between homologous genes in both species (Sharp, 1989).

We have recently collected the DNA sequences of 24 genes from three Drosophila species: *D. melanogaster, D. pseudoobscura*, and *D. subobscura* (Zeng et al., 1997). All three species are members of clades belonging to the sophophoran radiation of fruit flies that took place approximately 30–65 million years ago (Spicer, 1988; Sharp and Li, 1989). The radiation includes the *melanogaster* group (including *D. melanogaster*), the *obscura* group (including *D. pseudoobscura* and *D. obscura*), and the *willistoni – paulistorum* groups. *D. pseudoobscura* and *D. subobscura* differ on average by 0.29 synonymous nucleotide substitutions per site; *D. melanogaster* differs from the two obscura species by an average of 0.81 synonymous substitutions per site.

In these species, not only is the order of codon usage preference identical for every codon family, even the actual values within each codon family are remarkably similar (Table 5.1). In addition, both the range of codon bias values for the 24 genes and the average codon bias are nearly identical in the three species. Given the extent of the evolutionary divergence among these species, these data show that codon usage can be maintained for relatively long periods of evolution and that the selective mechanism(s) governing codon bias can be conserved.

Exactly the opposite conclusion about the evolutionary stability of codon usage can be reached for *D. willistoni*, however. This species is a member of the *willistoni* species group, the third clade in the sophophoran radiation. Powell and colleagues compared codon usage in three genes, *Adh, Sod*, and *per*, in this species (Anderson et al., 1993; Powell and Gleason, 1996). The authors found a consistent shift in all three genes away from C-ending codons, generally preferred in the *melanogaster* and *obscura* groups, toward U-ending codons. They interpreted this shift as evidence for the evolution to a new pattern of codon preferences; we would argue instead for the simple loss of codon bias. Evidence for this alternative hypothesis is provided by the patterns observed in the four-fold degenerate leucine family, CUN, in which the G-ending codon is strongly preferred in *D. melanogaster* over the C-ending codon (45% versus 16%; Sharp and Lloyd, 1993). For this codon family there is a shift in *D. willistoni* away from G in the third position, but to A rather than to U. These shifts in *D. willistoni*, G to A and C to U, are consistent with a mutational bias that favors transitions over transversions. Thus we can interpret the specific changes in codon usage in *D. willistoni* as a relaxation of selection patterned by mutational tendencies rather than as a shift in codon usage driven by natural selection. More data are required, however, for establishing whether this interpretation of the shift in codon usage in *D. willistoni* is correct.

One interesting question, which we will return to again shortly, is why the pattern of codon bias has remained nearly constant in the *melanogaster* and the *obscura* species, but appears to have shifted so dramatically in *D. willistoni*. Under the MSD hypothesis of codon bias, constancy is expected only if scaled selection parameter S does not decrease below 1.0 for any protracted period of time, especially in the recent past. An indication that this may not be the case is the fact that nucleotide diversity in *D. willistoni*, a measure of extant effective population size, is substantially lower than for other *Drosophila* species, including sibling species in its own group (Antezana, 1993). This is consistent with the idea that the shift in codon bias in this species may be the loss resulting from small population size ($S \ll 1$).

5.4. Codon Usage Evolution among Closely Related Species

Akashi, in two particularly insightful papers, studied patterns of synonymous codon changes between *D. melanogaster* and *D. simulans* (Akashi, 1995, 1996). He first identified the codons within each codon family that are preferentially used in genes with biased codon usage (Akashi, 1995). All twofold codon families have a major codon, always ending in C or G, and fourfold families have either one or, in several cases, two major codons, again always ending in C or G. He then defined two categories of synonymous mutations, those that change an unpreferred to a preferred (i.e., major) codon, termed a preferred change, and those that cause the opposite change, from a preferred to an unpreferred codon, termed an unpreferred change.

Akashi then used the sequence of at least one other member of the *melanogaster* species subgroup as an outgroup (*yakuba, teissieri, erecta,* or *orena*) in order to determine the branch on which the substitutions occurred which show up as fixed synonymous differences between the two species. This allowed him to categorize these changes into unpreferred and preferred substitutions. He could then test two predictions about the expected numbers of the two types of substitution under the MSD model at equilibrium. First, the number of preferred and unpreferred substitutions should be the same on each species branch if one assumes no change in S, the parameter governing this selection process. Second, when the total numbers of synonymous changes on the two branches are compared, the numbers should also be the same, implying an equal rate of evolution in the two species.

In total, Akashi compared eight genes in the two species (*Adh, Adhr, Amy, boss, Mlc1, per, Pgi,* and *Rh3*). In *D. simulans,* the number of preferred and unpreferred changes was identical, 9 versus 9, whereas in *D. melanogaster* a large excess of unpreferred changes compared with preferred changes (45 versus 5) was found. This result was not due to substitutions in just a few genes: a separate analysis of 34 genes that could be compared between the two species showed that codon bias is consistently lower in *D. melanogaster* than in *D. simulans.* Thus, not only did he find strong evidence for accelerated synonymous evolution in the *D. melanogaster* lineage, but he showed that this acceleration could be explained as an increase in the rate of substitution of preferred codons with unpreferred ones.

Akashi suggested that this pattern of change could be explained by a period of relaxation of selection for codon bias in this species and argued that a reduction in population size relative to *D. simulans* (and presumably to the common ancestor) is the most plausible explanation. The rationale for this contention follows directly from the MSD theory of weak selection: The establishment and maintenance of biased codon usage requires only very weak selection, $1 > S > 4$ (Li, 1987; Bulmer, 1991). A sustained decrease in population size in the *D. melanogaster* lineage should result in the drift to fixation of slightly deleterious unpreferred mutations. Note that such mutations are more common in genes with strongly biased codon usage, since most of their third positions are in the preferred state. A loss of codon bias is predicted by the MSD theory for modest reductions of population size, and it should occur in all genes in the genome that have biased codon usage.

This short-term evolutionary change seen in *D. melanogaster* provides an important clue about the changes in codon usage among more distantly related species. For instance, the possible loss of codon bias in *D. willistoni* can be explained as a possible long-term reduction in population size in this lineage. We prefer the loss hypothesis to a gain of codon bias in the other lineages because other more distantly related *Drosophila* species, such as *D. virilis,* exhibit strongly biased codon usages, which are similar to those of *D. melanogaster.* In addition, this explanation is congruent with very low nucleotide polymorphism observed in *D. willistoni,* as mentioned above.

5.5. Relationship between Codon Bias and Synonymous Polymorphism

Further support for the MSD theory of codon bias comes from the analysis of synonymous polymorphism within species. The MSD model makes specific predictions about relative levels of polymorphism and divergence for preferred and unpreferred changes. First, the number of preferred and unpreferred substitutions between species must be equal because otherwise codon bias will have changed in at least one of the species, violating the equilibrium assumption of the model.

Second, polymorphism levels for preferred and unpreferred mutations are not expected to be the same. Consider a gene such as *Adh*, which has strongly biased codon usage in *D. melanogaster*. In such a gene, approximately 90% of the codons are major codons, and they can mutate only to the unpreferred state. This implies that the majority of newly arising mutations must be of the unpreferred kind. These mutations are expected to contribute to polymorphism to a greater extent than to divergence because selection will reduce their fixation to a greater extent than their presence as polymorphisms (Kimura, 1983). At equilibrium, selection reduces the substitution rate of the more abundant unpreferred mutations and accelerates that of the less abundant preferred mutations so that an equal number of the two types of mutation eventually fix. Without any selection operating, preferred and unpreferred mutations are expected to fix exactly in proportion to the frequencies at which they arise.

Akashi tested these predictions with data from five genes of *D. melanogaster* and *D. simulans*. As with the fixed differences between species, he used a sequence from the other species to establish the ancestral state at polymorphic sites. The data for both species showed the expected skew in polymorphism toward unpreferred changes (70 unpreferred versus 5 preferred changes in *D. simulans* and 57 unpreferred versus 2 preferred changes in *D. melanogaster*). The skew of fixed differences in *D. melanogaster* toward unpreferred substitutions was discussed above and indicated a violation of the equilibrium assumption. However, this was not the case for *D. simulans*, in which the fixed differences in this lineage were evenly divided between preferred and unpreferred mutations (8 versus 7, respectively). These data are perhaps the most compelling evidence available in direct support of the selection model of codon bias: Relative to preferred mutations, unpreferred mutations are present as polymorphisms in higher proportion than as fixed differences. Akashi used this skew to estimate the scaled selection parameter to be $S = 2.2$, which indicates very weak selection.

5.6. Relationship between Codon Bias and Recombination Rate in *Drosophila*

Additional evidence for the influence of population size on codon bias in *Drosophila* can be found in the analysis of the relationship between the degree of codon bias and the chromosomal location of a gene. In *D. melanogaster* and many other organisms, the recombination rate per physical unit length of DNA is not constant, but rather varies both along and among chromosomes. Interest in recombination rates in relation to codon bias levels stems from the fact that in *Drosophila* the nucleotide polymorphism level but not the rate of divergence is strongly reduced in regions of low recombination (Begun and Aquadro, 1992). As explained in Chapter 7, in Kaga to mutations being selected outside a region of interest can reduce the effective population size of the region and thus its neutral polymorphism. This is expected under models of selection and genetic hitchhiking (Maynard Smith and Haigh, 1974; Kaplan et al., 1989) and under background selection (Charlesworth et al., 1993; Hudson and Kaplan, 1995; Charlesworth, 1996).

Whatever the mechanism, the efficacy of selection against weakly selected mutants is expected to be reduced in a region of low recombination in proportion to the relative reduction in the effective population size of that region. (See Kliman and Hey, 1993, for an informative discussion of selection mechanisms.) Relative effective population sizes in different regions of the genome can be estimated directly from comparisons of nucleotide polymorphism levels in those regions. If synonymous substitutions are under weak selection, then their dynamics should be influenced by these local differences in effective population size. Thus codon bias is expected to be positively correlated with recombination rate.

Kliman and Hey (1993) tested this prediction with 385 *D. melanogaster* genes. The recombination rate for each gene was estimated from a least-squares polynomial describing the relationship between genetic and cytological map positions of gene loci. Although the authors did not find a general relationship between recombination rate and the degree of codon bias, they observed that genes located in regions of most strongly suppressed recombination – six centromeric genes, eight genes on the tip of the X chromosome, and three genes on the fourth chromosome – have lower codon bias than the remaining genes. In a more recent analysis involving 428 genes, they estimated that 16% of variation in codon bias could be attributable to recombination rate (Kliman and Hey, 1994).

This finding, however, does not imply that codon bias is only weakly dependent on the recombination rate. As seen above, codon bias was entirely absent for the three genes located on the fourth chromosome of *D. melanogaster*, a chromosome that is not known to recombine under normal conditions (Hochman, 1976). We have compiled codon usage data for 1000 *D. melanogaster* genes, each with coding regions of at least 190 amino acids in length. Genes with fewer than 190 codons were excluded from this analysis because the variance in the estimation of codon bias explodes in smaller-sized genes (J. Comeron, personal communication). Genes were chosen without regard to chromosome location and represented a large range of codon biases.

As expected, a plot of codon bias, as measured by the effective number of codons (Nc; Wright, 1990) versus the GC base composition at synonymous positions (Fig. 5.1) reveals that genes with high bias strongly favor G- and C-ending codons, whereas genes with low codon bias tend to favor A- and T-ending codons. Nc scores were renormalized in this plot to 60% AT usage to reflect the general base composition of the *Drosophila* noncoding genome (Moriyama and Hartl, 1993). Genes with the smallest Nc scores are, indeed, slightly AT biased, and this should be interpreted as reflecting mutation pressure rather than selection for A- or T-ending codons.

The plot also shows the extent of codon bias versus the $G + C$ content of synonymous sites for 10 fourth-chromosome genes (*ankyrin* [L35601], *Arf2* [L25062], *AtpsynB* [X71013], *bent* [L35899], *CaMKII* [D13330], ci^D [X54360], *ey* [X79493], *RfaBp* [U62892], *RpS3A* [Y10115], and *zfh2* [M63450]). All of these genes fall into the lowest codon bias classes and have high AT synonymous

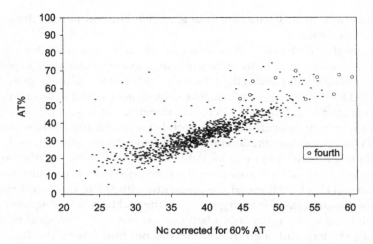

Figure 5.1. The effective number of codons (Nc) versus synonymous A + T percentage for 1010 *D. melanogaster* genes, including 10 fourth-chromosome genes (see text). Lower codon bias is indicated by a larger *Nc* score. The maximum possible *Nc* value is 62.

composition (62%, on average). Some of the products of these genes, particularly *ankaryn* and *RpS3A*, are abundant in every cell and therefore the genes can be assumed to be highly expressed. Given that the level of gene expression and codon bias is positively correlated in *Drosophila* (Shields et al., 1988), it is likely that these genes are subject to codon selection.

As an alternative, it is likely that the efficacy of selection is insufficient to maintain codon bias in genes on the fourth chromosome. Support for this idea comes from a study in *D. melanogaster* and *D. simulans* of nucleotide polymorphism at ci^D, one of the fourth-chromosome genes (Berry et al., 1991). The study showed severely reduced levels of silent polymorphism compared with those of a second chromosome gene (*Adh*) in both species, indicating reduced effective population sizes for the fourth chromosome. With a smaller effective population size, S can become smaller than 1 and genetic drift becomes the stronger force. Interestingly, the effective population size of *D. melanogaster*, which has been estimated from nucleotide polymorphism levels in regions of high recombination to be at least 10^6, suggests that S can be as small as 10^{-6} for weak selection to be operating. Given the scantiness of actual changes in the study by Berry et al., the estimates of nucleotide polymorphism for the fourth chromosome have a large variance. A more accurate estimate could provide the basis for calculating with better confidence an upper limit on the strength of selection maintaining codon bias in these species.

5.7. Selection on Minor Codons in *Drosophila*

It is likely that selection governs the frequency of usage of every codon in *Drosophila* rather than only that of the major codons. That this is true can be gleaned from the examination of codon usage in fourfold degenerate codon families in the 1000 genes mentioned above, as given in Table 5.1. The data indicate that minor codons are used differentially in highly biased genes and that these minor codon preferences are evolutionarily conserved. For example, in the threonine codon family the major codon, ACC, is used 65% of the time in highly biased genes. ACT (12%) and ACA (8%) are the least-preferred codons in these genes, but the usage is significantly different ($X^2 = 15.1$, $P < 0.01$), indicating a preference for the T-ending codon in highly biased genes. However, this preference is reversed in the least-biased genes, in which ACA is significantly more frequent than ACT ($X^2 = 21.6$, $P < 0.001$), indicating a possible mutational bias toward A in the third position. Furthermore, this small preference for ACT over ACA in genes with high codon bias is preserved in *D. melanogaster*, *D. pseudoobscura*, and *D. subobscura*.

These two observations provide compelling evidence for natural selection governing all codon frequencies in highly biased genes, not merely that of the major codons. In fourfold degenerate codon families, selection coefficients may differ for each of the four codons, and in some cases the selection differential between the two least-preferred codons can be greater than that between the major codon and the next most-preferred codon. This can be seen in the glycine codon family, in which the major codon, GGC, is present in highly biased genes at approximately 2.1 times the frequency of the next most-frequent codon, GGT, whereas GGG, the least-used codon, is 11.5 times less frequent than GGA, the second least-used codon. This implies that Akashi's analyses – which lumped all changes from a major codon to a nonmajor codon into a single category, unpreferred changes – can be improved. Therefore the actual strength of selection governing individual codon usages may be weaker or stronger than the estimates given by Akashi (1995).

5.8. MSD Predicts Mostly Strong or Weak Bias

Perhaps the strongest challenge to the MSD theory of codon usage is the fact that so many genes have intermediate levels of codon bias, whereas only a limited range of scaled selection parameter values yields intermediate codon bias values. Under this theory and assuming that $U + V \ll 1$ [this assumption is justified in the explanations for Eqs. (5.1) and (5.2)], the equilibrium frequency of the major codon is dependent only on S, the scaled selection coefficient. As pointed out by Bulmer, this equation predicts monomorphism at each synonymous site, with a fraction $P = 1/(1 + e^{-S})$ of the sites having the major codon.

Now, consider the range of values of the selection parameter S that yields intermediate values of P. For a twofold degenerate codon to have intermediate bias for the major codon, i.e., $0.6 < P < 0.9$, it is required that $0.4 < S < 2.2$. Within this range of values a small change in population size will have a large effect on the frequency of the major codon and hence on the extent of codon bias. For greater or smaller values of S, the frequency of the major codon will be less sensitive to small changes in the effective population size. One might expect therefore that genes with intermediate codon bias will exhibit greater

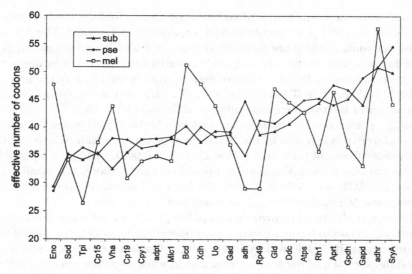

Figure 5.2. Comparison of codon bias, as measured by the effective number of codons, in 24 genes of *D. melanogaster*, *D. pseudoobscura*, and *D. subobscura*. Data are from Zeng et al. (1997).

differences in their extent of codon bias among species than genes with either high or low codon bias. In fact, a comparison of codon bias levels in 24 genes of *D. melanogaster*, *D. subobscura*, and *D. pseudoobscura* lineages (Fig. 5.2) indicates that this is not the case. The data plotted in Fig. 2 show that codon bias has remained relatively constant for all 24 genes in two obscura group species, including those with intermediate codon bias. We are forced to conclude that either population size has remained remarkably constant in these two species' or the MSD theory is not correct. It also shows, surprisingly, that codon bias levels are different for many of the 24 genes in *D. melanogaster*. Nevertheless, the magnitude of the differences in codon bias levels between *D. melanogaster* genes and the *obscura* species' genes is not related to the level of codon bias.

Another indication that the MSD model is not a correct characterization of codon selection comes from a consideration of the relationship between codon bias and recombination rates in *Drosophila*. As described above, Kliman and Hey (1993) found no evidence of a correlation between recombination rate and codon bias for the 345 genes of *D. melanogaster* they examined, which were located throughout the genome but not in regions known to have severely restricted recombination. In contrast, nucleotide polymorphism in this species is positively correlated with recombination rate throughout the genome, even excluding loci in regions of highly restricted recombination – all of which show little or no nucleotide polymorphism (Begun and Aquadro, 1992; Hudson and Kaplan, 1995; Charlesworth, 1996). Since nucleotide polymorphism levels are thought to reflect the effective population size of each gene, one would also have expected to find codon bias to be correlated with recombination rate under the MSD model. That it is not correlated with recombination rate suggests that codon bias levels are not strongly sensitive to differences in effective population size below an order of magnitude, contrary to the MSD prediction.

5.9. Alternatives to the MSD Model

We saw above that many genes in *Drosophila* have intermediate levels of codon bias and that these levels are stable in the *D. subobscura* and *D. pseudoobscura* lineages. In addition, codon bias levels are not noticeably correlated with recombination rates for most of the genome (except for regions of highly restricted recombination). All three observations indicate that codon bias levels may not be highly sensitive to small differences in effective population size, contrary to the MSD model's prediction. This model also cannot explain the observed correlation between codon bias and divergence rate.

On the other hand, certain features of synonymous site evolution are indeed compatible with the MSD theory and must be taken into account. In particular, the amounts of synonymous nucleotide polymorphism and divergence within and between species are similar in magnitude to the levels in adjacent noncoding DNA (Moriyama and Powell, 1996), such as introns, in which genetic drift is thought to be the major evolutionary force. Both observations are compatible with an assumption of weak selection for synonymous mutations, but not with one of strong selection. In support of this conclusion, Akashi (1995) showed that the patterning of unpreferred mutations within and between species can be accounted for by the MSD model with weak selection ($S \cong 2$). Thus we are forced to consider alternative models that preserve weak selection as an active force in codon bias evolution, at least at equilibrium, but that can also accommodate the facts mentioned above.

We consider here two possible alternatives to the MSD model. First, as proposed by Akashi (1996), selection coefficients for synonymous mutations may not be constant, but might instead decrease in magnitude as a gene evolves higher codon bias levels. In essence, there is a diminishing return on the strength of selection's favoring a mutation to a preferred codon as the bias level of a gene increases. Theoretical models containing diminishing returns and biologically plausible justifications for them are discussed by Gillespie (1994). With diminishing returns codon bias will evolve in a gene until it stalls out at approximately $S = 1$, at which point further evolution becomes quasi-neutral. The level of bias for each gene will depend on the specific fitness function describing the relationship between selection and codon bias for that gene. Under such a model, weakly expressed genes with low codon bias are assumed to have a more strongly concave fitness function than highly expressed genes with high codon bias. A key feature of this model is that codon bias levels are buffered from small changes in population size: When population size decreases so that S is less than 1, codon bias will not disappear entirely, as predicted by the MSD theory, but will instead equilibrate to a lower value. A similar buffering against changes in N_e related to recombination rate will also apply. Therefore many of the discrepancies between the MSD theory and the facts can be reconciled with a model that incorporates diminishing returns.

As an alternative to a model in which the selective value of a synonymous substitution changes with the level of codon bias in a gene, we now consider a model in which the selective value of each synonymous substitution is constant, but differs, depending on the individual codon and on its location in the gene. Akashi (1994), for example, showed that functionally constrained regions of a protein have higher codon bias than less constrained regions, a strong indication that the strength of selection varies among the sites. Under this view, small shifts in population size will influence codon bias evolution most strongly only for the proportion of synonymous substitutions satisfying the condition $S \cong 1$, but will only weakly influence synonymous substitutions for which $S \gg 1$. More recently, Maynard Smith and Smith (1996) showed that codon usage is site specific in two enterobacterial genes, *gapA* and *ompA*, and that, furthermore, this site-specific usage is evolutionarily conserved. They argue that selection must be site specific and that the selective value of particular codons must also vary. In this manner, unpreferred substitutions can occur at a certain fraction of sites in a gene, even in species with large population sizes.

5.10. Selection Intensity Must Vary among Codon Families and along a Gene

Codon selection is likely to act through a number of different biophysical and biochemical mechanisms. Several of these mechanisms are outlined below. As a consequence of the variety of selection mechanisms influencing codon usage, codons are likely to differ one from another in their selective values, and these are likely to change, depending on their physical location within a gene.

One widely acknowledged form of selection is for translational efficiency, which is governed by (and possibly leads to) each major codon's matching the most abundant isoaccepting tRNA (Ikemura, 1981, 1982; Bennetzen and Hall, 1982). Bulmer presented a model predicting the coevolution of codon usage and tRNA abundance in which rapid cell division in a clonally reproducing organism favors selection for decreased translation time (Bulmer, 1987). In bacteria and yeast (and possibly *Drosophila*) there is also a strong positive correlation between the extent of codon bias in a gene and its expression level (Grantham et al., 1981; Grosjean and Fiers, 1982; Ikemura, 1985). Codon selection is thought to act through increased protein elongation rates associated with major codons (Sorenen et al., 1989; Xia, 1996) and is expected to be an increasing function of the number of times a gene is translated.

In *Drosophila*, selection for translational efficiency may also be acting to favor wobble codons ending in U, which can pair with isoaccepting tRNAs with anticodons starting with either A or G. There are nine codon families that have both A- and U-ending codons (Leu, Ile, Val, Ser, Pro, Thr, Ala, Arg, and Gly). In *Drosophila* none of these codon families utilize either an A- or a U-ending codon as the major codon. However, in highly biased genes, eight of the nine families (the exception being proline) utilize the U-ending codon significantly more frequently than do the corresponding A-ending codons (see Table 5.1). This means that these genes have a DNA strand bias favoring T over A on the coding stand. In the absence of a strand bias in the mutational process, this excess of T over A on the coding strand must indicate that selection has favored T-ending codons relative to A-ending codons in genes with high codon bias, even though T is never the major codon. Selection in this case may be for translational efficiency because there is no net trend favoring the use of these T-ending codons in low-biased genes, as indicated in Table 5.1.

Translational accuracy is a second form of selection acting on codon usage (Akashi, 1994, and other references therein). As with translational efficiency, selection acts at the level of translation. Translational accuracy is determined by those aspects of translation that favor the incorporation of the correct amino acids into a growing peptide chain. In *E. coli* a 10-fold reduction in the misincorporation rate has been found for major codons in comparison with other members of the same codon family (Precup and Parker, 1987). Akashi reasoned that if the use of a major codon reduced misincorporation of incorrect amino acids, the selection for favored codons will be more intense in those regions of proteins that are more critical for function than in regions that are less critical for function. Translational efficiency, in contrast, does not make any prediction that selection intensity for a major codon will depend on its physical location. By comparing codon bias levels in evolutionarily conserved and functionally important regions of genes with those in less critical portions of genes, Akashi was able to show that codon bias levels are higher in the conserved/functionally important regions in *Drosophila* genes. Translational accuracy may also influence codon bias in *E. coli*, as evidenced by the finding of a positive relationship between codon bias and gene length in this species (Eyre-Walker, 1996).

A third, largely unexplored, form of codon selection is that affecting mRNA tertiary structure. Tertiary structure in RNA is expected to influence both its stability as well as its ability to be translated efficiently. The severe underrepresentation of the codon GGG, a member of the fourfold degenerate glycine family, is a possible case of codon avoidance in *Drosophila* that may be due to this codon's effects on RNA tertiary structure. Runs of guanine are known to favor the formation of stable tetraplex structures that are created by cyclic hydrogen bonding among four guanines (reviewed by Williamson, 1994). G-quartet structures have been identified at the ends of telomeres, in other DNA sequences, as well as in certain RNA sequences, such as the tRNA *supF* gene (Akman et al., 1991) and the *Leptomonas collosoma* spliced leader RNA (LeCuyer and Crothers, 1993).

The cohesiveness of poly-G sequences for other such sequences is likely to be deleterious in many mRNA sequences. Accordingly, we find that in our sample of 1000 genes of *Drosophila*, the codon GGG is the least frequently used of any fourfold codon, regardless of the level of codon bias. For example, in the 100 least-biased genes of *D. melanogaster*, the relative frequency of GGG is 0.09, whereas the least-frequent codons for the other fourfold codon families – CTC

(Leu), GTA (Val), TCT (Ser), CCT (Pro), ACG (Thr), GCG (Ala) and CGG (Arg) – all range between 0.17 and 0.22 frequencies. GGG is, moreover, exceedingly rare in highly biased genes, being present at only 0.02 frequency in the 100 most biased genes. The same is true in *E. coli*, in which it is a rare codon in highly biased genes (Sharp and Li, 1986). The rarity of this codon, even in genes in which codon bias is largely absent, cannot be explained by its codon – anticodon binding properties. The codon CCC, a member of the four-fold degenerate proline family, is, for example, the major codon for this family in highly biased genes, and it is also abundant in low-biased genes. Therefore we suggest that the codon GGG is strongly selected against in *Drosophila* and *E. coli* to avoid its propensity to form stable structures with other ribonucleotides. This negative selection may extend to all codons ending in GG in *E. coli* (Maynard Smith and Smith, 1996) and in *Drosophila* (Antezana and Kreitman, 1999).

5.11. Conclusions

Synonymous changes are perhaps better understood than any other class of mutations in the genome with respect to how the joint action of selection and genetic drift influences their fates. Data from *Drosophila* have provided a first direct estimate of the scaled selection coefficient S, which is in accord with the idea that unpreferred mutations are nearly neutral in *D. simulans*. Consistent with this view is the fact that systematic loss of codon bias in *D. melanogaster* can be ascribed to a sustained reduction in population size in this species but not to switches in the preference rankings of synonymous codons. But, on a longer evolutionary time scale, codon bias can remain remarkably stable, and this suggests the possibility that stronger selection dominates its evolutionary dynamics. Furthermore, the fact that no correlation was found in *Drosophila* between recombination rate and codon bias except in regions of extremely reduced recombination, and the inability of the MSD theory to account for the existence of intermediate levels of codon bias suggest that stronger forms of selection might also be operating. Possibly contradicting this view, however, is the indication that codon bias has been lost in *D. willistoni*. These are observations deserving of further investigation. In particular, if codon bias is sensitive to moderate population size fluctuations, as expected when S is approximately 1, then the kinds of differences in synonymous substitutions seen between *D. melanogaster* and *D. simulans* should also be found between other closely related species. It will also be worth investigating codon bias in many additional *Drosophila* species, and in *D. willistoni* in particular, to investigate the lability of codon bias. In addition, with more extensive population genetic datasets, it will be possible to estimate the values of S for each codon in fourfold degenerate codon families and to avoid lumping the nonmajor codons together. We believe that additional data will stimulate further refinement of the models and will play a catalytic role in the search for a more comprehensive understanding of codon bias evolution.

Now, if only a parallel effort could be made to assess the contributions of drift and selection for amino acid changes in proteins. That, of course, would make Dick a very happy man, but the higher structural dimensionality of proteins compared with codons makes this a much more difficult endeavor.

5.12. Acknowledgments

We thank J. Comeron for commenting on this manuscript and for providing data and Tao Pan for information on mRNA tertiary structure. This work was supported by a grant (GM39355) to MK by the U.S. National Institutes of Health.

REFERENCES

Akashi, H. 1994. Synonymous codon usage in *Drosophila melanogaster*: natural selection and translational accuracy. *Genetics* **136**:927–935.

Akashi, H. 1995. Inferring weak selection from patterns of polymorphism and divergence at "silent" sites in Drosophila DNA. *Genetics* **139**:1067–1076.

Akashi, H. 1996. Molecular evolution between *Drosophila melanogaster* and *D. simulans*: reduced codon bias, faster rates of amino acid substitution, and larger proteins in *D. melanogaster*. *Genetics* **144**:1297–1307.

Akman, S. A., Lingeman, R. G., Doroshow, J. H., and Smith, S. S. 1991. Quadruplex DNA formation in a region of the tRNA gene *SupF* associated with hydrogen peroxide mediated mutations. *Biochemistry* **30**:8648–8653.

Anderson, C. E., Carew, A., and Powell, J. R. 1993. Evolution of the *Adh* locus in the *Drosophila willistoni* group: the loss of an intron and shift in codon usage. *Mol. Biol. Evol.* **10**:605–618.

Andersson, S. G. E. and Kurland, C. G. 1990. Codon preferences in free-living microorganisms. *Microbiol. Rev.* **54**:198–210.

Antezana, M. 1993. Speciation and molecular evolution in the *Drosophila willistoni* group of species: inferences from autosomal and mitochondrial sequencing data. Ph.D. dissertation, U. California, Irvine.

Antezana, M. A. and Kreitman, M. 1999. The nonrandom location of synonymous codons suggests that reading frame-independent forces have patterned codon preferences, *J. Mol. Evol.* **48** (In press).

Begun, D. J. and Aquadro, C. F. 1992. Levels of naturally occurring DNA polymorphism correlate with recombination rate in *Drosophila melanogaster*. *Nature (London)* **356**:519–520.

Bennetzen, J. L. and Hall, B. D. 1982. Codon selection in yeast. *J. Biol. Chem.* **257**:3026–3031.

Berry, A. J., Ajioka, J. W., and Kreitman, M. 1991. Lack of polymorphism on the Drosophila fourth chromosome resulting from selection. *Genetics* **129**:1111–1117.

Bulmer, M. 1987. Coevolution of codon usage and transfer RNA abundance. *Nature (London)* **325**:728–730.

Bulmer, M. 1988. Are codon usage patterns in unicellular organisms determined by selection–mutation balance? *J. Evol. Biol.* **1**:15–26.

Bulmer, M. 1991. The selection–mutation–drift theory of synonymous codon usage. *Genetics* **129**:897–907.

Charlesworth, B. 1996. Background selection and patterns of genetic diversity in *Drosophila melanogaster. Genet. Res. Cambridge* **68**:131–149.

Charlesworth, B., Morgan, M. T., and Charlesworth, D. 1993. The effect of deleterious mutations on neutral molecular variation. *Genetics* **134**:1289–1303.

Comeron, J. M. 1997. Estudi de la variabilitat nucleotidica a Drosophila: Regio *Xdh* a *D subobscura.* Ph.D. thesis, Universitat de Barcelona, Spain.

Eyre-Walker, A. 1996. Synonymous codon bias is related to gene length in *Escherichia coli*: selection for translational accuracy? *Mol. Biol. Evol.* **13**:864–872.

Gillespie, J. H. 1991. *The Causes of Molecular Evolution.* Oxford Series in Ecology and Evolution. New York: Oxford U. Press.

Gillespie, J. H. 1994. Substitutional processes in molecular evolution. III. Deleterious alleles. *Genetics* **138**:943–952.

Grantham, R., Gauthier, C., Gouy, M., Jacobzone, M., and Mercier, R. 1981. Codon catalog usage is a genome strategy modulated for gene expressivity. *Nucleic Acids Res.* **9**:43–79.

Grosjean, H. and Fiers, W. 1982. Preferential codon usage in prokaryotic genes: the optimal codon–anti-codon interaction energy and the selective codon usage in efficiently expressed genes. *Gene* **18**:199–209.

Hochman, B. 1976. The fourth chromosome of *Drosophila melanogaster.* In *The Genetics and Biology of Drosophila,* M. Ashburner and E. Novitski, eds., pp. 903–928. London: Academic.

Hudson, R. R. and Kaplan, N. L. 1995. Deleterious background selection with recombination. *Genetics* **141**:1605–1617.

Ikemura, T. 1981. Correlation between the abundance of *Escherichia coli* transfer RNAs and the occurrence of the respective codons in its protein genes: a proposal for a synonymous codon choice that is optimal for the *E. coli* translation system. *J. Mol. Biol.* **151**:389–409.

Ikemura, T. 1982. Correlation between the abundance of yeast transfer RNAs and the occurrence of the respective codons in protein genes: differences in synonymous codon choice patterns of yeast and *Escherichia coli* with reference to the abundance of isoaccepting transfer RNAs. *J. Mol. Biol.* **158**:573–597.

Ikemura, T. 1985. Codon usage and tRNA content in unicellular and multicellular organisms. *Mol. Biol. Evol.* **2**:13–34.

Kaplan, N. L., Hudson, R. R., and Langley, C. H. 1989. The hitchhiking effect revisited. *Genetics* **123**:887–899.

Kimura, M. 1983. *The Neutral Theory of Molecular Evolution.* Cambridge: Cambridge U. Press.

Kliman, R. M. and Hey, J. 1993. Reduced natural selection associated with low recombination in *Drosophila melanogaster. Mol. Biol. Evol.* **10**:1239–1258.

Kliman, R. M. and Hey, J. 1994. The effects of mutation and natural selection on codon bias in the genes of Drosophila. *Genetics* **137**:1049–1056.

Kreitman, M. and Akashi, H. 1995. Molecular evidence for natural selection. *Annu. Rev. Ecol. Syst.* **26**:403–422.

LeCuyer, K. A. and Crothers, D. M. 1993. The Leptomonas collosoma spliced leader RNA can switch between two alternative structural forms. *Biochemistry* **32**:5301–5311.

Li, W.-H. 1987. Models of nearly neutral mutations with particular implications for nonrandom usage of synonymous codons. *J. Mol. Evol.* **24**:337–345.

Maynard Smith, J. and Haigh, J. 1974. The hitch-hiking effect of a favorable gene. *Genet. Res.* **23**:23–35.

Maynard Smith, J. and Smith, N. H. 1996. Site-specific codon bias in bacteria. *Genetics* **142**:1037–1043.

Moriyama, E. N. and Hartl, D. L. 1993. codon usage bias and base composition of nuclear genes in Drosophila. *Genetics* **134**:847–858.

Moriyama, E. N. and Powell, J. R. 1996. Intraspecific nuclear DNA variation in *Drosophila*. *Mol. Biol. Evol.* **13**:261–277.

Ochman, H. and Wilson, A. C. 1987. Evolution in bacteria: evidence for a universal substitution rate in cellular genomes. *J. Mol. Evol.* **26**:74–86.

Ohta, T. 1995. Synonymous and nonsynonymous substitutions in mammalian genes and the nearly neutral theory. *J. Mol. Evol.* **40**:56–63.

Powell, J. R. and Gleason, J. M. 1996. Codon usage and the origin of *P* elements. *Mol. Biol. Evol.* **13**:278–279.

Precup, J. and Parker, J. 1987. Missense misreading of asparagine codons as a function of codon identity and context. *J. Biol. Chem.* **262**:11351–11356.

Sharp, P. M. 1989. Evolution at "silent" sites in DNA. In *Evolution and Animal Breeding: Reviews in Molecular and Quantitative Approaches in Honour of Alan Robertson*, W. G. Hill and T. C. McKay, eds., pp. 23–31. Wallingford: CAB International.

Sharp, P. M., Averof, M., and Lloyd, A. T. 1995. DNA sequence evolution: the sounds of silence. *Phil. Trans. R. Soc. London Ser. B* **349**:241–247.

Sharp, P. M. and Li, W-H. 1986. An evolutionary perspective on synonymous codon usage in unicellular organisms. *J. Mol. Evol.* **24**:28–38.

Sharp, P. M. and Li, W-H. 1989. On the rate of DNA sequence evolution in Drosophila. *J. Mol. Evol.* **28**:398–402.

Sharp, P. M. and Lloyd, A. T. 1993. Codon usage. In *An Atlas of Drosophila Genes: Sequences and Molecular Features*, G. Maroni, ed., pp. 378–397. New York: Oxford U. Press.

Sharp, P. M. and Matassi, G. 1994. Codon usage and genome evolution. *Curr. Biol.* **4**:851–860.

Shields, D. C., Sharp, P. M., Higgins, D. G., and Wright, F. 1988. Silent sites in Drosophila genes are not neutral: evidence of selection among synonymous codons. *Mol. Biol. Evol.* **5**:704–716.

Sorenen, M. A., Kurland, C. G., and Pedersen, S. 1989. Codon usage determines translation rate in *Escherichia coli*. *J. Mol. Biol.* **207**:365–377.

Spicer, G. S. 1988. Molecular evolution among some Drosophila species groups as indicated by two-dimensional electrophoresis. *J. Mol. Evol.* **27**:250–260.

Watt, W. B. 1994. Allozymes in evolutionary genetics: self-imposed burden or extraordinary tool? *Genetics* **136**:11–16.

Williamson, J. R. 1994. G-quartet structures in telomeric DNA. *Annu. Rev. Biophys. Biomol. Struct.* **23**:703–730.

Wright, F. 1990. The 'effective number of codons' used in a gene. *Gene* **87**:23–39.

Xia, X. 1996. Maximizing transcription efficiency causes codon usage bias. *Genetics* **144**:1309–1320.

Zeng, L.-W., Comeron, J. M., Chen, B., and Kreitman, M. 1998. The molecular clock revisited: the rate of synonymous vs. replacement change in Drosophila. *Genetica* **102/103**:369–382.

CHAPTER SIX

Gene Conversion Is a Major Determinant of Genetic Diversity at the DNA Level

ANDREW BERRY AND ANTONIO BARBADILLA

6.1. Summary

Recombination generates genetic diversity by shuffling preexisting mutations. We analyze here the relative importance in this process of the two kinds of recombination, gene conversion and crossing over. By combining simple theory and data from intensive studies of gene conversion at the *Drosophila melanogaster rosy* locus (e.g., Hilliker et al., 1994), we show that gene conversion, which results in the exchange of short tracts between chromosomes, is the major force in the generation of diversity within loci. Between loci, crossing over, which results in the exchange of large chromosome segments, is more important. We point out that gene conversion has hitherto been neglected by population geneticists, and we introduce techniques for its analysis in DNA polymorphism datasets from natural populations. Contrasting analyses of two such datasets, one from the *rp49* region of *D. subobscura* and the other from *D. melanogaster Adh*, reveal that gene conversion in each case is the major determinant of within-locus genetic diversity. Gene conversion is an important population genetic process and one that will inevitably attract considerably more attention as we improve our understanding of the evolutionary dynamics of the nuclear genome.

6.2. Introduction

This chapter is about how population geneticists apply models of recombination to evolutionary problems. Because we can study recombination only in species for which we have genetic markers, the history of our understanding of recombination reflects the technological history of empirical genetics. As we have developed ever more refined ways to detect genetic variation, so we have come to understand more about the effect of recombination in reconfiguring

R. S. Singh and C. B. Krimbas, eds., *Evolutionary Genetics: From Molecules to Morphology*, vol. 1. © Cambridge University Press 2000. Printed in the United States of America. ISBN 0-521-57123-5. All rights reserved.

that variation. However, our current notion of the role of recombination in evolutionary genetics is both out of date and misleading.

Before the application of protein electrophoresis to evolutionary problems in 1966, empirical population geneticists had typically to rely on visible markers. As a result, we know a great deal about banding patterns in *Cepaea* (Lamotte, 1959), about hindwing spots in *Maniola jurtina* (Ford, 1971), and about the distribution of color forms in any number of organisms (e.g., Kettlewell, 1973). Dobzhansky's studies of *Drosophila* karyotypes took population genetics a step away from phenotypic markers and toward genotypic ones, but these markers, whole segments of chromosomes (inversions), bore little relation to the bean-bag genes of theory. A single-locus two-allele model could not realistically be applied to a polymorphic inversion occupying two thirds of a chromosome arm. However, the phenotypic markers of the ecological geneticists corresponded, at least roughly, to the single loci whose behavior had been so thoroughly explored by theoretical population geneticists. The problems with such visible markers are twofold. (1) They are limited in number: Even under ideal conditions, an insect wing cannot manifest a great many variable traits. Once you have studied the genetics of wing spot number, of spot size, of spot color, and of background color, it is probably time to move on to another species. (2) The mapping between genotype and phenotype is often neither linear nor simple. A visible marker, such as a color pattern, is inevitably the product of a complex pathway, and many factors may therefore intervene in the production of a phenotype from a given genotype. Other genetic factors, modifiers, or, in cases of variable penetrance or expressivity, uncontrolled environmental factors can affect the phenotype.

1966 changed all that. The observable phenotypes, allozymes, were direct translations of genotypes and it was simple enough to demonstrate that a given electromorph was produced by a single locus. Theory and experiment were finally on speaking terms. But the projected population genetic utopia did not emerge. With hindsight (Lewontin, 1991), we appreciate that protein electrophoresis not only failed to resolve many of the evolutionary problems it was supposed to resolve but it may actually have fostered misleading interpretations of evolutionary process. For example, Barbadilla et al. (1996) have shown that many of the distribution patterns of supposed evolutionary significance are attributable to the statistical properties of a number of segregating and recombining charge-altering amino acids.

One area in which protein electrophoresis stymied both theoretical and empirical progress was in our understanding of recombination. Implicit to most studies of allozymes was the notion of the fixity of alleles: New alleles could arise by mutation but, in essence, an allozyme was thought of as an internally homogeneous fixed entity. The Fast and Slow alleles at the *Drosophila melanogaster* alcohol dehydrogenase locus were two such fixed entities. With the advent of protein sequencing, we know that the Fast/Slow (F/S) difference is encoded by a single amino acid difference (Lysine/Threonine, respectively) so a Fast allele could change into a Slow allele only through mutation and vice versa.

Whatever was happening elsewhere in the locus was irrelevant (and invisible) to allozyme studies. Recombination could occur between loci, those fixed entities, resulting eventually in linkage equilibrium among loci, but could not occur within an entity. Recombination, to someone interested in allozymes, was important only as a force over large chunks of chromosomes – It eroded linkage disequilibrium between distant markers. The theoretical offspring of allozyme analyses, Neutral Theory, reflects this position (Kimura, 1983): None of the major models (infinite sites, infinite alleles, or charge state) considered intragenic recombination because of both the theoretical difficulties associated with incorporating recombination and the belief that recombination was negligible on an intragenic scale.

Allozyme studies thus effectively reinforced the existing understanding of recombination. To most geneticists, recombination was a tool in, say, mapping studies. The rate of generation of recombinants was a measure of the distance separating markers on a chromosome. Both classical genetic and allozyme approaches emphasized the importance of recombination over large chromosomal distances.

We argue here that that conceptualization of recombination remains with us to this day and that such a conceptualization tells only half the story: that recombination is crucially important on a very much more local scale than has hitherto been recognized. We highlight a qualitative difference between recombination processes operating on intergenic and intragenic scales, between crossing over and gene conversion. Whereas crossing over results in the reciprocal exchange of whole segments of chromosomes, gene conversion, which is typically nonreciprocal and results in the recombination of short tracts, is much more local in its effect. That effect may be local but, we show below, gene conversion is nevertheless the main force in the generation of diversity within loci: Much of the genetic diversity we find in studies of DNA polymorphism at single loci is the product of gene conversion processes.

It is the brief of population genetics to describe, analyze, and understand patterns of genetic variability in nature. An appreciation of gene conversion is necessary if we are to make progress toward these goals. We first discuss the general importance of an understanding of recombination in population genetics before briefly reviewing previous attempts to address the problem. We then analyze a dataset in two different ways in an empirical attempt to gauge the importance of gene conversion in generating haplotypic diversity.

6.3. Recombination in Population Genetic Theory and Experiment

Multilocus theory inevitably takes into account recombination, but that theory is treating loci in exactly the manner that we have seen allozyme studies to have treated them. They are genetic entities capable of change only through mutation; recombination addresses the relationships among loci, not what is going within loci.

Although it is possible to introduce certain types of recombination into co-alescent models (e.g., Hudson, 1993), the whole approach is basically unable to deal with recombination. The problem for a coalescent theorist confronted with data from a recombining locus is the same as that facing a systematist interested in reconstructing the history of a group of sequences generated by both mutation and recombination processes. Normally systematists can ignore recombination because they are analyzing sequences derived from different species – i.e., sequences that cannot recombine with each other and that can therefore diverge only mutationally. Recombination is problematic because it results in reticulation of the genealogy: Different parts of a molecule have different evolutionary histories. Thus a tree reconstructed with one part of the molecule will differ from one reconstructed with another part of the molecule. We have no *a priori* basis for determining which of the two trees, if either, reflects the history of the species: Recombination has confounded our attempt at phylogenetic reconstruction.

Coalescent theory is thus torpedoed by recombination for the same reason. It is a theory, after all, that is based on the reconstruction of common ancestral states. However, any population genetic theory that cannot incorporate a realistic model of recombination must eventually go extinct: we can mine for only so long those nonrecombining mitochondrial and chloroplast genomes for population genetic gold before the richness of those seams starts to dwindle. *D. melanogaster's* nuclear genome is some 4×10^8 base pairs (bp's) in size while its mitochondrial genome weighs in at approximately 2×10^4 bp's – the nuclear genome is some 20,000 times larger than its mitochondrial counterpart. The future of population genetics must inevitably lie amid the recombining sequences of the nuclear genome. Coalescent theory must come to terms with recombination, or, failing that, an alternative theoretical approach will be necessary.

Currently empirical population geneticists working with recombining sequences cursorily acknowledge the problem. Everyone conscientiously explores patterns of linkage disequilibrium in their data, finds some, and mentions the discovery in passing in the discussion. Nobody is quite sure what, if anything, the patterns they have discovered signify. A large number of nuclear DNA studies include an intraspecific tree generated by any of a number of methods ranging from unweighted pair-group method with arithmetic mean (UPGMA) to parsimony. These trees are inevitably misleading, each contemporary sequence being a hodgepodge of recombined segments, with its own peculiar history. But the trees are nevertheless published, along with a sort of shoulder-shrugging apology, because we all appreciate the importance of understanding something, anything, about the genealogical history of the sample. Their presentation may be well motivated, but these trees are basically useless.

How can we incorporate recombination into phylogenetic reconstruction? Given a number of simplifying assumptions, such as a uniform (Poisson) distribution of recombination-initiation sites, it should be possible to reconstruct, by means of a parsimony-based exhaustive search, past recombination events in cases in which the density of such events is not prohibitively high. Various attempts at this are under way. The addition of a recombination-identifying algorithm to our already unwieldy phylogenetic programs may have to await yet another generation of computer, but this nevertheless represents an important area of potential progress. It is worth mentioning that, because of the limited genetic distances involved when sequences are compared from within a species, the computational problems are not currently too great at this level. The addition of recombination event analysis to standard phylogenetic algorithms may not, therefore, add too great a computational burden. We have here something of a trade-off: Recombination analysis is necessary only when the standard phylogenetic problems are straightforward (i.e., within a species in which the genetic distances between sequences are small); when the standard problems are greater (i.e., between species), recombination is irrelevant. If, however, such a project is to be both useful and successful, we need as complete as possible an understanding of the recombination processes incorporated into the model.

A review of the empirical population genetic literature of the past, say, 10 years leaves one in no doubt about the importance of recombination in current studies of population genetics. Many recent advances are based on hitchhiking, whereby evolutionary events at a given site affect variation at linked sites. Kreitman and Hudson (1991) have, for example, shown that variable sites linked to the *D. melanogaster adh* F/S site, which is apparently under balancing selection, also have ancient coalescences. Thus selection governs variation at a specific site and additionally affects, through hitchhiking, sites that are not directly under selection. Linkage disequilibrium causes the impact of selection to be detectable some distance away from the target site. That distance is determined by the strength of selection and by rates of recombination: If rates are high, then there is relatively little hitchhiking; if low, then there is potential for considerable hitchhiking. If we are to make strong predictions about the importance of hitchhiking in a particular case in which selection is putatively operating, we need to have a thorough understanding of local patterns of recombination. As, in our current population genetic paradigm, the inference of selection is our major goal and the major method for realizing that goal is by studying patterns of hitchhiking, we need hardly reiterate that an understanding of recombination is an essential prerequisite for such a project.

Two of the most important advances, empirical and theoretical, in population genetics in recent years are similarly recombination entrenched. Begun and Aquadro (1992) found that *D. melanogaster* nucleotide heterozygosity levels are positively correlated with recombination rates. Charlesworth et al. (1993) have advanced a model to explain this pattern, background selection, based on hitchhiking in response to purifying selection against deleterious mutations. There is no doubt of the centrality of recombination in modern population genetics. And yet, as shown above, it remains difficult or impossible to analyze, and its phenomenonology and implications are poorly understood.

In order to analyze the recombinational history of a sample, we must have accurate perception of the way in which recombination works on the scales that we are dealing with. Our current notions of recombination are out of date, derived as they are from multilocus theory, classical genetics, and protein electrophoresis studies. Evolutionary genetics now operates on an entirely new scale: DNA sequence within a locus. We know now that linkage disequilibrium among polymorphisms segregating within a locus (e.g., Berry and Kreitman, 1993) can be selectively significant. Are the recombination processes that affect evolution within genes the same as the familiar ones that affect evolution among genes?

6.4. Recombination within the Gene: Conversion versus Crossing Over

To understand intragenic recombination, we have to disentangle the relative importance of the two basic recombination processes, gene conversion and crossing over, each of which produces a characteristic footprint on patterns of nucleotide variation. The amount of recombination generated by each of these two mechanisms depends on two basic parameters: (1) the per-event average length of the region undergoing recombination and (2) the per-generation rate of events. Any comparison between conversion and crossing over inevitably requires an estimation of these two basic parameters for each one. Current population genetic models are content to summarize the effect of recombination in a single parameter, $4Nr$ (Hudson and Kaplan, 1985; Hudson, 1987, 1993), where r is the recombination rate and N is the population size. Although formally r is the weighted average of the contributions of both gene conversion and crossing over, it seems that, in practice, it is taken to embody solely crossing over. Gene conversion is implicitly ignored, and the relative importance of each process is nowhere addressed. This bias, we suggest, is the product of the classical genetic/allozyme traditions that have both focused on intergene recombination. We show below that such a bias is unjustified at the intragenic level of analysis.

When markers, such as separate allozyme or visible loci, are distant with respect to the average tract length transferred in a gene conversion event (this is inevitable in the intergenic case), crossing over is typically much the most important form of recombination (Hilliker et al., 1994; Navarro et al., 1996). Within a gene, however, markers are only a few bp's [or, maximally, kilobase pairs (kbp's)] apart and, for such short distances, the rate of each event becomes the crucial parameter. Thus, on the local, intragenic scale, gene conversion may be a more important form of recombination than crossing over if it occurs at a higher rate than crossing over. To flesh out these intuitive arguments, here we derive an expression for the rate of recombination between two markers as a function of the distance between the markers for both crossing over and gene conversion.

Gene conversion and crossing over are intimately related processes, and we must take into account this relationship when comparing them. During meiosis, three separate parameters govern the composition of a tetrad: the number of potential recombination events (C), each of which can be resolved either as conversions with associated crossing-over (Cx) events or as conversions without associated crossing-over (Co) events (Mortimer and Fogel, 1974; Foss et al., 1993; Navarro et al., 1996). Hence the rate of conversion will always be larger than or equal to that of crossing over. Let $(m + 1)$ be the ratio of gene conversions to crossing overs, where m is the expected number of Co events for each Cx event (Foss et al., 1993; Navarro et al., 1996). For $m > 0$, that is, when a nonzero proportion of C events results in Co events, the conversion rate will exceed the crossing-over rate. If λ is the average number of C (recombination) events in a four-strand chromosome during meiosis, then the expected number of crossing-over events per chromosome, $\lambda\{Cx\}$, will be $\lambda/(m+1)$. Let Lct and L be the average length of a conversion tract and the chromosome length (both in bp's), respectively, where $Lct \ll L$.

We focus first on the case of two consecutive markers, that is, two contiguous bp's. A conversion event can cause recombination between these markers only if one of the two end bp's of the conversion tract captures one of the markers. There are thus only two events that can cause recombination; the probability of recombination, RF (recombination frequency), in this case is therefore $2/L$ for any conversion event. (Note that, because $Lct \ll L$, the border effect of chromosome is negligible). If the probability of simultaneous backconversion (the probability that a site is converted two or more times in simultaneous conversion events) is disregarded, the RF for λ events per chromosome is the sum of RF for each conversion event divided by 4 (given a conversion event, on average only one gamete out of four resulting from a meiosis is converted; Griffiths et al., 1993). Thus RF $(d = 2) = 2\lambda/(4L) = \lambda/(2L)$, where d is the distance between the markers, including the markers themselves. When the two markers (or sites) are separated by one bp (i.e., $d = 3$), then either the end bp of the tract or the last-but-one bp must capture one of the markers to cause recombination. Hence, four different events can cause recombination, so RF $(d = 3) = 4\lambda/(4L) = \lambda/L$.

By following this reasoning (Fig. 6.1), we derive a linear relationship between RF and d when $d = {<}Lct$. The expression is RF$= (d - 1)\lambda/2L$. λ is the slope of the line. For $d = Lct$, RF reaches its maximum value of RF $= (Lct - 1)\lambda/(2L)$. This value remains constant and independent of d for $d > Lct$ (i.e., all along the chromosome). The reason for this independence of d is that conversion events can occur between the markers without including either of them.

Crossing over will give a similar linear relationship between RF and d, but with a slope of $\lambda/(m+1)$. For $m > 0$, the slope is greater under conversion than under crossing over, revealing conversion as the main recombination process under these conditions. However, because the segment exchanged by crossing over is, on average, much longer than a conversion tract, the linear relationship will extend for a considerable fraction of the chromosome length. The relationship between d and RF is shown in Fig. 6.1 for both processes. Three interesting regions can be distinguished: (1) When $2 = {<}d = {<}Lct$, RF resulting

Markers or Sites (M)	Distance (d) between markers (including markers)	Recombination Frequency (RF)
...MM...	2	$\lambda/(2L)$
...M.M...	3	$2\lambda/(2L)$
...M..M...	4	$3\lambda/(2L)$
...M.. ..M...	$d < Lct$	$(d-1)\lambda/(2L)$
...M.. ..M...	$d > Lct$	$(Lct-1)\lambda/(2L)$

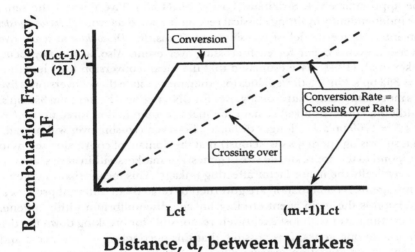

Distance, d, between Markers

Figure 6.1. The relationship between the distance between two markers and their probability of recombining under both conversion and crossing over. The table shows how the relationship is established for conversion. L is the length of the chromosome, Lct is the length of a conversion tract, m is the ratio of gene conversion events to crossing-over events, and λ is the per-generation rate of conversion. See text for more details.

from conversion is $(m + 1)$ times greater than that resulting from crossing over; (2) when $Lct = <d = <(m + 1)Lct$, conversion remains the prevalent process, but crossing over is catching up; (3) when $d > (m + 1)Lct$, crossing over dominates.

Thus we see that the relative importance of crossing over and gene conversion is completely dependent on the distance between the sites considered. The relationship in fact reverses when $d = (m + 1)Lct$: For DNA regions shorter than $(m + 1)Lct$, conversion dominates, but, for longer distances, crossing over is the chief factor. This analysis explains why classical genetic and allozyme studies have concentrated on crossing over: The distance between markers in such studies is well in excess of $(m + 1)Lct$.

Our analysis reveals clearly the importance of both Lct and m in determining the size of the domain in which conversion is predominant. What are their values – how big is that domain? We need empirical estimates of m and Lct to answer this question. Chovnick's long-term studies on the fine structure of the *D. melanogaster rosy* locus have provided the best available estimates of these parameters. Until very recently, much of this work has slipped through molecular evolution's disciplinary literature net (Foss et al., 1993; Betrán et al., 1996; Navarro et al., 1996). If this chapter serves just one purpose, we hope that that will be to draw attention to this body of work and its relevance to evolutionary studies. Hilliker and Chovnick (1981) and Hilliker et al. (1991) estimated m, the ratio of conversions resolved as Cos to those resolved as Cxs, to be approximately 4. Strikingly, Foss et al. (1993) have derived the same value independently by fitting classical recombination data from *D. melanogaster* to an interference model of crossing over. Thus, for *Drosophila* at least, five conversion events occur for each crossing-over event. Also, at the *rosy* locus, Hilliker et al. (1994) have estimated that the mean conversion tract length is $Lct = 352$ bp's. Putting these values together, we can state that conversion is five times more important than crossing over for DNA regions shorter than 352 bp's. In addition, conversion will remain the main recombination force up to $d = 352 \times 5 = 1760$ bp's. For longer distances, however, crossing over will prevail.

It is surprising (perhaps suspicious!) that the domain of conversion seems to correspond to the approximate size of genes: For markers within genes, conversion is typically the major factor affecting linkage. This means that intergenic and intragenic recombination are governed by related but different processes. Conversion is the main agent eroding linkage disequilibrium within a gene, while crossing over is almost exclusively responsible for breaking down the disequilibria among genes. It is possible that this decoupling of intragenic and intergenic recombination processes has evolutionary (even adaptive) repercussions. For example, genes under selective pressure for high variability (e.g., *MHC* loci in humans; Hughes et al., 1993) might undergo an elevated rate of conversion without increasing the rate of crossing over. In this way, the increased intragenic recombination rate is achieved independently of the chromosome-scale recombination rate, thus avoiding the maladaptive effects of a general increase of recombination. In fact, it appears that the ratio of Co to Cx events could indeed be adjusted by natural selection. Carpenter's (1982) genetic dissection of the *D. melanogaster rosy* locus revealed mutations that decrease the number of Cx's without affecting the number of Co's. Likewise, the fact that conversion events occur on an intragenic scale plays into the hands of proponents of the exon shuffling model of the origin of new genes: Perhaps (only perhaps) m and Lct are fine tuned by selection to maximize the rate at which potentially useful new combinations are generated.

6.5. Detecting Recombination from DNA Sequence Data

The effect of recombination on nucleotide variation in samples of DNA sequences is typically summarized by the parameter $4Nr$, which captures the distribution of linkage disequilibria among sites and the variance of the number of segregating sites in the sampled sequences (Hudson, 1987). $4Nr$ can be compared with the neutral mutation parameter $4N\mu$ and, when other parameters are known, permits the estimation of r, the recombination rate, or N, the effective size of the population. For instance, Schaeffer and Miller (1993) estimated that 7–17 recombination events occur for each mutation event at *D. pseudoobscura Adh*, demonstrating that intragenic recombination is the main factor generating haplotypic diversity.

Hudson and Kaplan (1985) first developed an estimator for $4Nr$ based on the four-gamete test. The empirical basis for this estimator is the observation, in a sample of sequences, of all four gametic types for any two polymorphic sites (i.e., AB, Ab, aB, ab). Excluding backmutation and recurrent mutation (the assumptions of the infinite-site model), such a pattern can be explained only by the occurrence of at least one recombination event between the focal sites some time in the history of the sample. The frequency of the observed four gametic types is then used to obtain an underestimate of $4Nr$. This inevitably is not just an underestimate; it is a major underestimate. In most cases, one or both of the products of a recombination event is or are lost through drift – thus only a small proportion of all recombination events will leave a four-gametic-type footprint. Furthermore, correcting for this huge bias is generally difficult because of the complex statistical relationship between the true and the observed number of recombination events (Hudson and Kaplan, 1985).

Hudson (1987) introduced an alternative estimator that is both easier to compute and more appropriate to DNA polymorphism data. It is based on the statistic $S^2\{k\}$, the variance of the number of pairwise differences between sequences in a sample. By Monte Carlo simulation, he showed that the distribution of this estimator has, for large datasets, better statistical properties than the four-gamete test. Hudson (1993) has suggested a third estimator, which seems to outperform the previous ones. Data are simulated from a coalescent process in which a multi-site model with recombination is considered and the sampling properties for different values of $4Nr$ are investigated.

Note that $4Nr$ does not distinguish between the two basic recombination processes discussed above. The parameter and its three estimators fail to take advantage of each process's specific footprint on patterns of nucleotide variation. All these estimators nevertheless are utterly dependent on the omnipresent assumptions of the steady-state neutral mutation–drift model, assumptions that pervade most molecular analyses of DNA sequence and that are seldom justified (Gillespie, 1991). The study of recombination at the DNA level is thus doubly handicapped. First, current approaches fail to take advantage of much of the power inherent to the data, and, second, they rely on untested and inappropriate steady-state assumptions.

In addition to estimating $4Nr$, we can distinguish between the actions of the two recombination processes by identifying conversion events in the sampled sequences. Conversion events are often intuitively inferred from the observation of mosaic haplotypes that comprise segments of putative parental haplotypes (Riley et al., 1989; Hughes et al., 1993; Rozas and Aguadé, 1994). However, there are rigorous statistical methods for their detection. Stephens's method (1985) is based on the clustering pattern of variable sites in a sample of homologous DNA sequences. Two (or more) unique mutations occurring in the same branch of a gene tree will generate two (or more) sites in complete linkage disequilibrium (congruent sites). If the mutation rate is uniform along the DNA sequence, the distance between these two linked sites is randomly distributed. Conversely, if congruent sites are produced by the transfer of a small continuous segment between two sequences (that is, by gene conversion), then a nonrandom clustering is expected for the congruent sites. Therefore, from a significant non-uniform distribution of congruent sites, it is inferred that one or more intragenic recombination event or events has or have occurred among the sample sequences. Stephens's method is especially appropriate for small sample sizes, of the order of two or three sequences. When sample sizes are larger, the probability of finding two or more congruent sites is low, especially when the sequences are reasonably polymorphic. Furthermore, the method treats each set of congruent sites separately, failing to correct for multiple comparisons.

Sawyer (1989) introduced a more powerful and general statistical test for detecting conversion. Consider n homologous DNA sequences with s segregating sites. For any two random sequences, the number of segregating sites distinguishing any two regions that differ depends on whether the difference is the product of mutation or of conversion. Conversion can be inferred by comparison of the square of the sum of the numbers of segregating sites obtained over all pairwise comparisons with the distribution of an artificial data set generated by random permutation of segregating sites. The application of this test to two different *Escherichia coli* loci provided strong evidence for multiple intragenic conversion events.

The approaches of both Stephens and Sawyer simply infer that gene conversion has occurred among the sampled sequences. They ignore the pattern, the number, and the tract lengths of gene conversion events. Ideally, we would like to detect each individual conversion event and from these derive the frequency and the tract length distributions. This, however, is impossible because (1) conversion events do not always leave an unambiguous footprint in the sequence, and (2) the length in base pairs of this footprint only seldom coincides with the true length of the transferred tract. The ideal could be realized only if we are studying sequences that differ at every nucleotide – a situation that does not seem to arise very often in natural populations! The reverse case is utterly uninformative: Exchange occurring between identical sequences or between sequences in linkage equilibrium for all their segregating sites would leave no footprint whatsoever. Typical of population genetic datasets is the intermediate case, in which sequences differ at several sites and exhibit some linkage disequilibrium. Although here again not all conversion events can be detected, a proportion of them can be inferred from the unambiguous mosaic pattern left in some cases. In practice, however, population geneticists have to make do with what we have: These comparative sequence data offer us our only opportunity to assess the role of conversion.

Recently, Betrán et al. (1996) put forward a statistical model that allows the estimation, from the observation of individual gene conversion tracts, of the number and the tract length distribution of conversion events in a set of sequences exhibiting haplotypic differentiation. The reasoning underlying the model is as follows: Consider two sets of differentiated (i.e., exhibiting linkage disequilibrium) sequences (e.g., sequences derived from two different polymorphic inversions). A fraction of the tracts exchanged between each set, or subpopulation, will produce mosaic haplotypes. Note that at least two subpopulation-specific variants must be transferred in order to distinguish a conversion event from mutation. These are the observed tracts, which will constitute the empirical basis of the whole model. We assume here that each observed tract is produced by a single conversion event, i.e., each tract has not subsequently been reconfigured by additional conversion or crossing-over events. This assumption holds for DNA regions with a moderate or low density of conversion events. The length (in bp's) of an observed tract is the distance between the two outermost informative nucleotides; those nucleotides reveal the mosaic origin of the tract. Because informative nucleotides are a subset of all nucleotides, the observed tract length will always be less than or equal to the true tract length. Therefore we must determine the relationship between observed and true tract lengths if we are to estimate the latter from the former. The relationship depends both on the frequency of informative sites in the two subpopulations and on the true tract length distribution. Assuming that the latter follows a geometric distribution (Hilliker et al., 1994), estimators of the parameter of the length distribution and its mean and variance can be developed. Once the true distribution is estimated, the fraction of detected (and undetected) tracts can also be determined. From this, the true number of conversion tracts (observable plus hidden) can be estimated as well.

The results show that, in general, only a small percentage of all conversion events in a sample are detected: Solely those conversion events occurring between very divergent haplotypes or having very long tracts will typically be detected. This explains why gene conversion is recognized when it occurs between very different haplotypes (as in the case of conversion between paralogous sequences; Slightom et al., 1980), but is often overlooked when occurring among homologous sequences. Finally, if each conversion event is assumed to be unique, then conversion becomes theoretically equivalent to point mutation, and $4Nc$ and c (the conversion rate) can also be estimated. This general approach allows us to assess the importance of conversion in generating haplotypic differentiation.

6.6. The Power of Gene Conversion: an Example

The estimators of Betrán et al. have been applied to the sequence data of Rozas and Aguadé (1994). Thirty-four sequences of the *rp49* locus, which is close to an inversion breakpoint, were analyzed for two polymorphic arrangements (*Ost* and *O*3 + 4) in *D. subobscura*. Five different conversion events were identified, with an average observed tract length of 51 bp's. The maximum likelihood estimate of the average true tract length was 122 bp's (2.4 times longer than the observed one). Only approximately one fourth of conversion events between the arrangements was detected, the true number of exchanged conversion events in the sample being approximately 19. Assuming that the frequency of heterokaryotypes has remained approximately constant throughout the history of the sample, the conversion rate per bp per generation is 2.73×10^{-7}, a value more than 100 times greater than the value of the neutral mutation rate per bp per generation estimated for this region (Rozas and Aguadé, 1994).

This approach allows us to infer the relative importance of gene conversion in generating an observed set of haplotypes. Let $d\{xy\}$ be the probability of a site differing between two sequences, one taken at random from each of two subpopulations [$d\{xy\}$ is the average number of nucleotide substitutions per site between subpopulations (Nei, 1987; Eq. 10.29)]. On average, $Lct^*d\{xy\}$ sites will be converted from one subpopulation to the other per conversion event. If, therefore, we find a haplotypic network (a two-dimensional plot relating haplotypes on the basis of the number of differences between them; see Fig. 6.2 for an example) in which sequences differ by approximately (Lct $d\{xy\}$) sites, then, having ruled out parallel mutation as a possible cause, we can assert that conversion is responsible for this pattern. Consider again the data Rozas and Aguadé. They found 12 polymorphic sites segregating within both subpopulations, arrangements *Ost* and *O*3 + 4. With reference to a hypergeometric distribution, they showed that, given the total number of segregating sites, the probability that mutation alone could result in 12 shared polymorphisms was negligible. Analysis of the data of Betrán et al. concluded that 10 out of the 19 estimated conversion events would result in the transfer of one or more different nucleotides, generating polymorphic sites shared between the arrangements. This prediction agrees very well with the observed 12 shared sites, and it is reasonable to conclude that conversion is responsible for the observed pattern. Gene conversion has thus produced 12 new haplotypes in a sample of 34 sequences; it is clearly a major factor in the generation of intragenic sequence diversity. Note that this is an underestimate of the role of gene conversion in generating haplotypic diversity in this study. We examined instances of conversion only between inversion types; it is likely that conversion has additionally played an important role in haplotypic diversification within inversion types.

6.7. The Power of Gene Conversion: Another Example, Another Approach

We introduce another general approach to assessing the importance of gene conversion in generating intragenic diversity by example. Berry and Kreitman (1993) conducted a four-cutter RFLP study of variation in a region approximately 3 Kb in length encompassing the *D. melanogaster* alcohol dehydrogenase (*adh*) locus. They surveyed 1533 chromosomes sampled from 25 populations

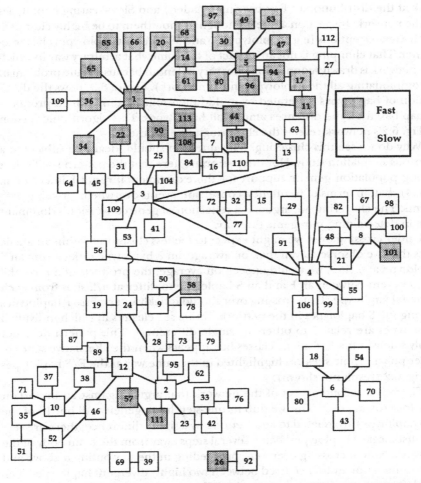

Figure 6.2. *D. melanogaster Adh* haplotype network (from Berry and Kreitman, 1993) showing all haplotypes related to at least one other by a single difference. Lines connect haplotypes differing at a single site. Haplotypes are numbered by their overall abundance rank: number 1 is the most common in the entire sample and number 113 the rarest. Haplotypes are shaded according to their allozyme type (Fast or Slow).

throughout the east coast of the U.S. and scored each one for 44 variable sites, recognizing in the process 113 haplotypes. Figure 6.2 shows a network in which each haplotype is connected to every other haplotype from which it differs by only a single site difference. Most variable sites are segregating at low frequency (i.e., the lower frequency variant occurs on only a few haplotypes), but the F/S allozyme difference, a single nucleotide substitution that results in an amino acid replacement and is located near the middle of the surveyed region, is maintained by selection at intermediate frequencies. It is informative, therefore, to look at the distribution of Fast-bearing (shaded) and Slow-bearing haplotypes in the network. Inspection alone is enough to show them to be highly clumped: With a few exceptions (e.g., haplotypes 75 and 59) the Fast haplotypes cluster together. That clumping is not an artifact of the somewhat arbitrary way in which the network is drawn (connecting lines vary in length because of the problem of accommodating all the haplotypes in the network). Figure 6.3 shows the distribution of haplotypes generated by multidimensional scaling, which produces a map based on the distances among all haplotypes. This algorithmic version of Fig. 6.2's network reveals the clustering to be robust.

Why do we see this clustering? This question should ideally be subject to a rigorous simulation study. However, we argue here that the pattern itself offers strong population genetic support for the experimental result Hilliker et al. (1994) that, in terms of frequency, conversion events overwhelm crossing-over events. The clustering, we argue, is a reflection of gene conversion's dominant role in determining intragenic diversity.

Under crossing over, we might expect to see less clustering within an allelic class than we observe: Given that, on average, an F haplotype differs from an S haplotype at n sites, we would expect, on average, the products of a recombination event between an F and an S haplotype to differ at $n/2$ sites from each parental haplotype. Thus crossing over should quickly intersperse F haplotypes among the S haplotypes in the network. Yet we see clusters in which individual haplotypes are related to others by single differences. This pattern does not apply solely to the F/S allelic classes but appears when the alternative states of other polymorphic sites are highlighted in the same way as the F/S haplotypes in Fig. 6.2 (result not shown).

There is a second aspect of the network that argues against an important role for crossing over. This we dub its one-step homogeneity. All but 10 of the 113 haplotypes are related to another one by a single difference; there are very few instances of haplotypes' being several steps away from the main body of the network. Now, if crossing over were proceeding in the way outlined above and a recombination event occurred between two highly divergent haplotypes (i.e., n is large), then each product would differ by a substantial number of steps from existing haplotypes. Recombination between those products and existing haplotypes would, of course, produce intermediates. However, to produce the observed network's one-step homogeneity would require many recombination events, none of whose products could go extinct because they represent the crucial intermediates. With 44 variable sites (i.e., 2^{44} possible haplotypes under free recombination) and only 113 haplotypes, though, recombination can neither be particularly extensive nor the retention of its products particularly efficient.

The one-steppedness (for want of a better term) of the network is underlined by further analysis (Fig. 6.4). Under almost any evolutionary model, we expect common haplotypes to give rise more frequently than rare ones to novel haplotypes, either through mutation or some form of recombination. Assuming some stability of haplotype frequencies through time, we therefore

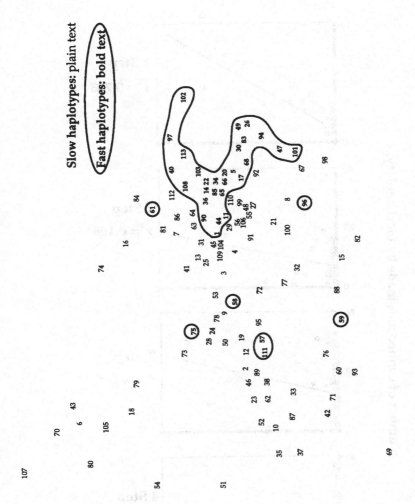

Figure 6.3. Multidimensional scaling (SAS Institute) plot showing haplotype relatedness for the *Adh* haplotypes shown in Fig. 6.2. Haplotypes are numbered in the same way as in Fig. 6.2.

Slow haplotypes: plain text

Fast haplotypes: bold text

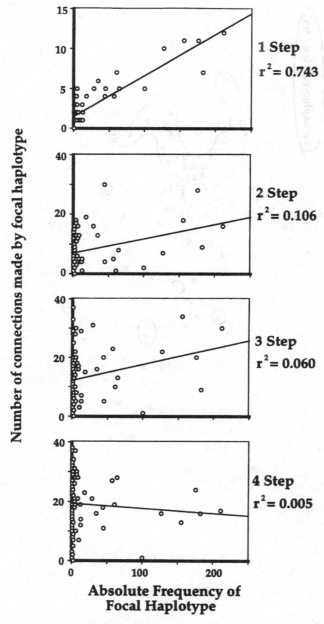

Figure 6.4. The relationship, for all *Adh* haplotypes, between the overall frequency of a focal haplotype and the number of haplotypes from which it differs by one, two, three, or four variable sites. Each connection is counted twice (e.g., the one-step connection between haplotype *x* and haplotype *y* is counted for both *x* and *y*), but this does not affect the relationship. Linear regressions are plotted and their r^2 values given.

expect to see a positive correlation between the frequency of a haplotype H and the rate at which H produces new haplotypes. Figure 6.4 shows the relationship between the frequency of a haplotype in the entire sample and the number of haplotypes it can be connected to by each of one, two, three, and four steps. Because the expected correlation exists only in the case of the one-step-connected plot, Fig. 6.4 suggests strongly that diversification is based on a one-step process (i.e., not by crossing over, which, as we have just seen, results in multistep diversification insofar as the products of crossing-over events often differ by many sites from the parental haplotypes).

Imagine instead that the process in operation were a four-step one in which a haplotype produced a new haplotype differing from itself at four sites. The most common haplotype, H1, would generate more new haplotypes than any other haplotype but all of these would be four steps distant from H1. We would then expect to see a positive correlation in the bottom plot of Fig. 6.4 (for four-step connections). We would also expect NOT to see the observed positive correlation in the top plot of Fig. 6.4 (for one-step connections) because the derived haplotypes are four steps away from the focal parental haplotype, say, H1, and will therefore likely be one-step connected to haplotypes whose frequencies differ from that of H1. They would be one-step connected to, say, H2 and H3. In other words, the relationship, under a four-step process, between haplotype frequency and number of one-step connections is decoupled.

Presumably the clustering we see in the network is the product of the same process that generates its one-step homogeneity; they are effectively flip sides of the same coin. What mechanism can account for this pattern? We present three models to explain the observations:

(1) Mutation is frequent relative to recombination and accordingly drives the diversification of haplotypes. This would account comfortably for the observations. However, we have already seen that Schaeffer and Miller (1993) have estimated for *Adh* in another *Drosophila* species, *D. pseudoobscura*, the recombination rate per gene per generation to be of the order of an order of magnitude greater than the mutation rate per gene per generation. Also, such a model predicts that all the changes between haplotypes are likely to be unique except for the occasional instance of parallel mutation. That 15 of the 20 changes that distinguish among F haplotypes are also segregating among S haplotypes argues strongly against such a mutation model. The amount of parallel mutation required for generating the observed haplotypes without recombination could be formally investigated by generation of a parsimony tree from the dataset. The number of parallel mutation events would then be the number of mutational steps on the tree minus 44, the number of variable sites.

(2) Crossing over typically occurs between similar haplotypes. This may be a function of population structure (i.e., of the available haplotypes with which a given haplotype may recombine) or of some genetic mechanism that permits pairing and exchange between only very similar chromatids. The latter mechanism is discussed by Stephan and Langley (1992), who present a model based on the observation that efficient repair of double-strand breaks in yeast is greatly

reduced when homologous chromosomes are diverged in sequence (Resnick et al., 1989). However, it seems unlikely that the minor levels of divergence we see among *Adh* haplotypes would affect chiasma formation. Recombination between similar haplotypes due to quirks of population structure is conceivable (e.g., two separate populations, one all *Adh*-F and the other all *Adh*-S, with recombination occurring freely within each) but unlikely, as this would also result in divergence of the different allelic classes, which we do not see.

(3) Gene conversion is the principle mode of recombination. As noted above, Hilliker et al. (1994) found the average conversion tract length at the *rosy* locus to be 352 bp's. The average distance between contiguous sites for the 19 variable sites whose low-frequency variant is segregating at a frequency greater than approximately 1% (i.e., in population genetic terms, the informative sites; see Berry and Kreitman, 1993) is 151 bp's. (We have approximated the position of two insertions/deletions whose exact location is unknown to the middle of fragment to which they could be localized. See Berry and Kreitman, 1993, Table 2). A large proportion of all gene conversions will therefore result in the transfer of a single site between haplotypes. Note that the fact that the putatively typical conversion tract would, on average, contain two of these variable sites does not mean that we necessarily expect, on average, two differences to be transferred in a conversion event. The relative frequencies of the variants at each of the sites governs that number. Imagine two contiguous variable sites, 151 bp's apart. If we take these numbers, 151 and 352, to be fixed, even in the case most likely to produce a two-step transfer (i.e., one that has introduced two new variants to a haplotype), there is a higher probability of producing a one-step transfer: If both sites are segregating at 50%, there is a 0.25 chance that both sites will be novel on the haplotype to which they are transferred, yet there is a higher probability (0.5) that only one will differ between the recombining haplotypes, resulting in a one-step transfer. Now imagine that both sites are segregating at frequency 10% (as is typical); the probability that a two-step transfer will occur is only 0.01. Only 7 out of the 44 variable sites scored in this study were segregating at a frequency greater than 10%. Assuming that the estimate of conversion tract length by Hilliker et al. (1994) applies to *Adh*, the spacing of the variable sites in this study thus predicts that the vast majority of conversion events that result in the introduction of novel variant sites into a haplotype will introduce one. This explains why we see only a weak positive correlation between focal haplotype frequency and the number of haplotypes to which it is connected by two steps (Fig. 6.4, second plot).

We see therefore that the pattern accords well with the model advanced in this chapter, that conversion is the major determinant of intragenic diversity. The haplotype network is consistent with the observation of Hilliker et al. (1991) of a considerable excess of conversion events with respect to crossing-over ones. Gene conversion is the only model that can account for both the local clustering and the one-step homogeneity of the network.

6.8. Conclusion

This chapter has brought the important results of the *rosy* locus studies of Hilliker et al. (1991) and Hilliker et al. (1994) into the domain of population genetics. We have demonstrated, both theoretically and empirically, that gene conversion is the most important factor promoting haplotypic diversification on an intragenic scale. In particular, we have shown that recombination processes are effectively decoupled between intragenic and intergenic levels: Crossing over is the major player between genes and conversion the major player within genes. Although our two empirical examples may both reasonably be regarded, in evolutionary terms, as special cases – *D. melanogaster Adh* is subject to balancing selection, and *D. subobscura rp49* is associated with inversions – we are confident that this will prove to be a general result.

Population geneticists' current treatment of recombination is still part of a world view born of classical genetics and allozyme studies. With the production of ever more detailed studies of sequence evolution within loci, the time has come to recognize that it is gene conversion, not crossing over, that shapes the patterns of variation we see within genes. These results will, we hope, serve to promote the incorporation of realistic models of recombination into evolutionary analyses.

6.9. Acknowledgments

We thank E. Betrán, A. Navarro, A. Ruiz, and D. Weinreich for their comments. A. Berry was funded by the Harvard University Society of Fellows and by the Milton Fund of Harvard University. We thank Percy Grainger for inspiration.

REFERENCES

Barbadilla, A., King, L. M., and Lewontin, R. C. 1996. What does electrophoretic variation tell us about protein variation? *Mol. Biol. Evol.* 13:427–432.

Begun, D. J. and Aquadro, C. F. 1992. Levels of naturally occurring DNA polymorphism correlate with recombination rates in *D. melanogaster. Nature (London)* 356:519–520.

Berry, A. and Kreitman, M. 1993. Molecular analysis of an allozyme cline: alcohol dehydrogenase in *Drosophila melanogaster* on the east coast of North America. *Genetics* 134:869–893.

Betrán, E., Rozas, J., Navarro, A., and Barbadilla, A. 1996. The estimation of the number and length distribution of gene conversion tracts from population DNA sequence data. *Genetics* 146:89–99.

Carpenter, A. T. C. 1982. Mismatch repair, gene conversion, and crossing-over in two recombination-defective mutants of *Drosophila melanogaster. Proc. Natl. Acad. Sci. USA* 79:5961–5965.

Charlesworth, B., Morgan, M. T., and Charlesworth, D. 1993. The effect of deleterious mutations on neutral molecular variation. *Genetics* 134:1289–1303.

Ford, E. B. 1971. *Ecological Genetics*, 3rd ed. London: Chapman & Hall.

Foss, E., Lande, E., Stahl, F. W., and Stainberg, C. M. 1993. Chiasma interference as a function of genetic distance. *Genetics* **133**:681–691.

Gillespie, J. 1991. *The Causes of Molecular Evolution.* Oxford: Oxford U. Press .

Griffiths, A. J. F., Miller, J. H., Suzuki, D. T., Lewontin, R. C., and Gelbart, W. G. 1993. *An Introduction to Genetic Analysis.* 5th ed. New York: Freeman.

Hilliker, A. J. and Chovnick, A. 1981. Further observations on intragenic recombination in *Drosophila melanogaster. Genet. Res.* **38**:281–296.

Hilliker, A. J., Clark, S. H., and Chovnick, A. 1991. The effect of DNA sequence polymorphisms on intragenic recombination in the *rosy* locus of *Drosophila melanogaster. Genetics* **129**:779–781.

Hilliker, A. J., Harauz, G., Reaume, A. G., Gray, M., Clark, S. H., and Chovnick, A. 1994. Meiotic gene conversion tract length distribution within the *rosy* locus of *Drosophila melanogaster. Genetics* **137**:1019–1026.

Hudson, R. R. 1987. Estimating the recombination parameter of a finite population model without selection. *Genet. Res.* **50**:245–250.

Hudson, R. R. 1993. The how and why of generating gene genealogies. In *Mechanisms of Molecular Evolution,* N. Takahata and A. G. Clark, eds. Sunderland, MA: Sinauer.

Hudson, R. R. and Kaplan, N. L. 1985. Statistical properties of the number of recombination events in the history of a sample of DNA sequences. *Genetics* **111**:147–164.

Hughes, A. L., Hughes, M. K., and Watkins, D. I. 1993. Contrasting roles of interallelic recombination at the HLA-A and HLA-B loci. *Genetics* **133**:669–680.

Kettlewell, H. B. D. 1973. *The Evolution of Melanism. The Study of Recurring necessity; with Special Reference to Industrial Melanism in the Lepidoptera.* Oxford: Clarendon.

Kimura, M. 1983. *The Neutral Theory of Evolution.* Cambridge: Cambridge U. Press.

Kreitman, M. and Hudson, R. R. 1991. Inferring the evolutionary histories of the *Adh* and *Adh-dup* loci in *Drosophila melanogaster* from patterns of polymorphism and divergence. *Genetics* **127**:565–582.

Lamotte, M. 1959. Polymorphism of natural populations of *Cepaea nemoralis. Cold Spring Harbor Symp. Quant. Biol.* **24**:65–86.

Lewontin, R. C. 1991. 25 years ago in genetics: electrophoresis in the development of evolutionary genetics: milestone or millstone. *Genetics* **128**:657–662.

Mortimer, R. K. and Fogel, S. 1974. Genetical interference and gene conversion. In *Mechanisms in Recombination,* R. F. Grell, ed. New York: Plenum.

Navarro, A., Barbadilla, A., Betrán, E., and Ruiz, A. 1996. Gene flux caused by conversion and by crossing over in inversion systems. *Genetics* **146**:695–709.

Nei, M. 1987. *Molecular Evolutionary Genetics.* New York: Columbia U. Press.

Resnick, M. A., Skaanild, M., and Nilsson-Tillgren, T. 1989. Lack of DNA homology in a pair of divergent chromosomes greatly sensitizes them to loss by DNA damage. *Proc. Natl. Acad. Sci. USA* **86**:2276–2280.

Riley, M. A., Hallas, M. E., and Lewontin, R. C. 1989. Distinguishing the forces controlling genetic variation at the *Xdh* locus of *Drosophila pseudoobscura. Genetics* **123**:359–369.

Rozas, J. and Aguadé, M. 1994. Gene conversion is involved in the transfer of genetic information between naturally occurring inversions of Drosophila. *Proc. Natl. Acad. Sci. USA* **91**:11517–11521.

Sawyer, S. 1989. Statistical tests for detecting gene conversion. *Mol. Biol. Evol.* **6**:526–538.

Schaeffer, S. W. and Miller, E. L. 1993. Estimates of linkage disequilibrium and the recombination parameter determined from segregating nucleotide sites in the alcohol dehydrogenase region of *Drosophila pseudoobscura*. *Genetics* **135**:541–552.
Slightom, J. L., Blechl, A. E., and Smithies, O. 1980. Human fetal Gg- and Ag-globin genes: complete nucleotide sequences suggest that DNA can be exchanged between these duplicated genes. *Cell* **21**:627–638.
Stephan, W. and Langley, C. H. 1992. Evolutionary consequences of DNA mismatch inhibited repair opportunity. *Genetics* **132**:567–574.
Stephens, J. C. 1985. Statistical methods of DNA sequence analysis: detection of intragenic recombination or gene conversion. *Mol. Biol. Evol.* **2**:539–556.

CHAPTER SEVEN

Evidence for Balancing, Directional, and Background Selection in Molecular Evolution

STEPHEN W. SCHAEFFER AND MONTSERRAT AGUADÉ

7.1. Introduction

Two questions have dominated population genetics this century: How much genetic variation exists in natural populations and what evolutionary forces are responsible for the observed diversity? The answers to these questions required innovations in molecular biology and theoretical population genetics for unraveling the forces that generate and maintain polymorphisms. The analysis of nucleotide sequence polymorphisms as they are arranged on chromosomes and the theoretical underpinnings of the neutral theory allow adaptive evolution to be detected in the genomes of plants and animals. This chapter presents a review of the statistical methods used to infer when selection has caused the rapid fixation of a new beneficial mutation and the types of nucleotide data that are appropriate for each analysis. We review the evidence for balancing selection within species, directional selection within and between species, and background selection. Evidence for epistatic selection is also discussed.

7.2. Historical Development

Natural selection, as envisioned by Darwin, can act as a purifying force that removes deleterious genes from populations, but it can also create new phenotypes by increasing frequency of beneficial alleles (Provine, 1971). How often adaptive selection has acted to increase allele frequencies has been a dominant question of population genetics. The challenge of detecting positive Darwinian selection in natural populations was limited by advances in theoretical and molecular genetics.

Population genetics has undergone three developmental phases, the classical, neoclassical, and modern periods (Ewens, 1993). The main architects

R. S. Singh and C. B. Krimbas, eds., *Evolutionary Genetics: From Molecules to Morphology*, vol. 1. © Cambridge University Press 2000. Printed in the United States of America. ISBN 0-521-57123-5. All rights reserved.

of the classical period, Fisher, Wright, and Haldane, used forward recursion equations to unify Mendelian genetics with Darwinian selection on continuous traits (Provine, 1971). Empirical data were limited during the classical period, but two hypotheses emerged about the structure of genetic variation within individuals and the pace of evolution (Dobzhansky, 1955; Lewontin, 1974; Muller, 1950). H. J. Muller suggested that most genetic loci in the genome would be homozygous in populations such that adaptive evolution would be limited by selectable genetic variation (Muller, 1950). Alternatively, Dobzhansky thought that most individuals would be heterozygous for a unique set of alleles such that phenotypic variation was always present for adaptive evolution to act on (Dobzhansky, 1955). These two views of genomic evolution would not be resolved until population genetics entered the neoclassical period (Ewens, 1993), when protein electrophoresis allowed the first glimpses of molecular genetic diversity to be sampled (Lewontin and Hubby, 1966).

Protein electrophoretic data seemed to support Dobzhansky's hypothesis because the levels of polymorphism were higher than those predicted by Muller. Several important theoretical advances arose during the neoclassical period to explain what evolutionary forces modulated the levels of genetic diversity (Ewens, 1993), including the neutral theory (Kimura, 1983) and multilocus theory (Karlin and Feldman, 1970). The neutral theory suggested that protein variation was a transient phase of molecular evolution, in which mutation introduced selectively neutral variants that were either fixed or lost by random genetic drift. Alternatively, natural selection was thought to play an active role in the maintenance of allozymic diversity. Experimental approaches designed to detect selection at allozyme loci attempted to demonstrate physiological and biochemical differences among allozymic genotypes (Koehn, 1978). In hindsight, experiments designed to determine the molecular basis of adaptive evolution for an enzyme polymorphism were not guaranteed to yield interpretable results without evidence that a protein was under positive Darwinian selection.

None of the allozyme data for any species conformed to the predictions of either Muller or Dobzhansky. The observed levels of protein variation were too high if all mutations were deleterious, but too low if all genetic variation was beneficial. In an effort to resolve this discrepancy, Maynard Smith and Haigh (1974) used a deterministic approach to explore how the action of selection at one locus influenced variation at a linked neutral locus. They showed that fixation of a new adaptive mutation by directional selection could reduce heterozygosity at linked neutral sites, thus providing an explanation for why some loci might not be variable. Alternatively, Ohta and Kimura (1975) used a diffusion approach that showed that "... the hitchhiking effect is generally unimportant as a mechanism for reducing heterozygosity. It may be significant only when the recombination fraction is smaller than the selection coefficient." Given that most allozyme loci are not tightly linked, Ohta and Kimura were right to conclude that genetic hitchhiking was an unimportant mechanism. The conclusions of Ohta and Kimura do suggest, however, that genetic hitchhiking is an important mechanism for reducing diversity when genetic markers are extremely tightly linked, as are polymorphic nucleotides.

Detecting selection within a single allozyme locus proved to be a difficult problem because observed patterns of allozyme diversity could be explained by neutral and selective hypotheses. The two-allele polymorphism in the alcohol dehydrogenase (*Adh*) locus of *Drosophila melanogaster* is a good example of this problem. *Adh* encodes two allozymes, *Adh* Fast and *Adh* Slow, that are polymorphic across the worldwide distribution of the species, and allozyme frequencies vary in a latitudinal cline in North America, Europe, and Australia (David, 1982; Oakeshott et al., 1982; Vigue and Johnson, 1973). These data are consistent with neutral forces in which *Adh* Fast originates in higher latitudes and slowly diffuses to lower latitudes by limited migration. Alternatively, selection could explain the gradients in gene frequency in which individuals are free to migrate among populations, but differential selection on genotypes alters the frequency of alleles in various environments (Berry and Kreitman, 1993; Simmons et al., 1989). The magnitude of the migration rate is a critical parameter to estimate if one is to reject neutrality, but one cannot use the allozyme frequencies to estimate gene flow if the amino acid polymorphism is also the target of selection. Nucleotide sequencing overcame this difficulty because rates of migration could now be inferred from linked neutral markers whose fixation rate is independent of selected sites (Birky and Walsh, 1988).

Population genetics has now entered what Ewens defined as its modern period (Ewens, 1993). The development of rapid nucleotide sequencing methods in experimental population genetics and the use of coalescent models in theoretical population genetics have provided powerful tools that can infer when natural selection has acted on a genetic locus. Nucleotide sequences from protein-encoding genes have two useful properties that allow for neutral and selective hypotheses to be discriminated, the redundancy of the genetic code, and the tight linkage of variable sites. The redundancy of the genetic code causes protein-coding sequences to be composed of selected and neutral sites. Selection acting at the protein level does not see synonymous sites because they fail to change the encoded amino acid sequence of a protein. Nonsynonymous sites are the targets of selection because mutations in these nucleotides change the amino acid sequence of an encoded gene product. The neutral model provides the null hypothesis that predicts the frequency spectra for synonymous and nonsynonymous variations. The action of positive Darwinian selection at nonsynonymous sites will decrease or increase the numbers of linked neutral polymorphisms compared with the expectations of a neutral model.

The modern period of theoretical population genetics saw a fundamental shift from progressive models that predict future levels of diversity to backward models that generate expected times back to a common ancestor for an extant set of sequences. This fundamental shift in how population genetics hypotheses are tested has generated a more rigorous approach to population genetics. The modern period was launched by the experimental study of nucleotide diversity at the *Adh* locus of *D. melanogaster* (Kreitman, 1983). Kreitman sequenced 11 alleles of *Adh* collected from around the world, including five *Adh* Slow and six

Adh Fast alleles. His data showed how selection leaves its footprint in the multi-site genotypes generated by the tightly linked synonymous and nonsynonymous nucleotides. One of the striking features of the *Adh* data was that only one of the 43 polymorphic sites occurred in a nonsynonymous site and many of the synonymous polymorphisms were nonrandomly associated with the segregating amino acid replacement. Kreitman's paper stimulated an interaction between theoretical and experimental population geneticists that has helped to develop the methods currently used to detect selection.

7.3. Theories

The statistical methods and theoretical models necessary to analyze and understand nucleotide sequence data have kept pace with or surpassed experimental studies of DNA in populations. The major theoretical development was the coalescent that allows the mutation–random genetic drift process to be followed backward through time so that neutral expectations may be derived for numbers of segregating sites in a given sample (Kingman, 1982a, b). The null hypothesis is that alleles within a neutral genealogy are expected to have an average coalescence time of $4N$ generation, where N is the effective population size. Directional selection causes genealogies of alleles to be more shallow than neutrality, while balancing selection causes alleles to trace their ancestry to greater than $4N$ generations back in time. Critical population parameters and test statistics have been designed by use of a coalescent approach (Hudson, 1983; Hudson et al., 1987; Tajima, 1983, 1989). Coalescent methods allow neutral gene genealogies to be simulated under various population genetic models without an extensive need for computer time or space (Hudson and Kaplan, 1988; Slatkin, 1991; Tavaré, 1984). A reexamination of a genetic hitchhiking model with a coalescent approach has demonstrated that adaptive fixation of a new mutation can have a profound effect on an assemblage of linked neutral polymorphisms typical of a nucleotide sequence (Kaplan et al., 1989).

7.4. Methodological Approaches

7.4.1. Methods to Sample Nucleotide Diversity within Populations

Nucleotide diversity within populations has been examined with two approaches, restriction endonuclease mapping (Aquadro et al., 1992) and nucleotide sequencing (Kreitman, 1983). Restriction endonuclease mapping establishes differences in restriction site locations in a genomic region. Six-base enzymes allow large segments of the genome to be inexpensively surveyed for nucleotide polymorphisms, and the positions of restriction sites can be mapped to within 50 nucleotides. The disadvantage of the six-cutter technique is that variable sites tend to be separated by an average of 4096 bases, which does not generate dense polymorphism maps sufficient to detect genetic hitchhiking. Kreitman and Aguadé (1986a) refined the restriction mapping by introducing the mapping of four-base restriction enzymes so that nucleotide polymorphisms could be resolved to within several nucleotides. The prior availability of a gene sequence allows four-cutter restriction fragments to be predicted from a standard individual and the maps of unknown individuals are inferred from the standard pattern. Kreitman and Aguadé (1986a) estimated that 19% of the total sites of a genetic region could be surveyed with four cutters.

Complete nucleotide sequences provide the best data for inferring the past action of selection at a genetic locus. Before automated Polymerase Chain Reaction (PCR), nucleotide sequences were generated with tedious genetic engineering approaches (Sambrook et al., 1989). The generation of a sequence from a single strain took several months under ideal laboratory conditions, but could take longer when technical problems arose. Automated PCR (Saiki et al., 1988) has significantly reduced the time necessary to determine one nucleotide sequence. The amplification of a genetic locus takes hours rather than weeks, and generally 10 to 20 strains may be sequenced in the time required for generating one sequence with molecular cloning methods. PCR may also be used with four-cutter restriction mapping to estimate levels of diversity rapidly (Bénassi and Veuille, 1995).

7.4.2. Statistical Methods that Detect Selection

The statistical tests of neutrality are designed to detect selection within populations or between species. The Hudson, Kreitman, and Aguadé (HKA) (1987) method uses intraspecific and interspecific nucleotide data to test the null hypothesis that the ratio of polymorphism to divergence is equivalent among loci. The HKA test rejects a neutral model if the ratio of polymorphism to divergence differs significantly among independent loci, provided that one genetic locus evolves according to predictions of the neutral theory. A ratio of polymorphism to divergence that is significantly lower than neutral expectation indicates that directional selection has recently fixed a new adaptive mutation within species. When the observed polymorphism levels exceed neutral expectations, then balancing selection has maintained alleles within populations for many generations. Various corrections have been derived for the HKA test to account for comparisons of loci in which the sample sizes differ (Berry et al., 1991), comparisons between autosomal and X-linked loci (Begun and Aquadro, 1991), and comparisons between cytoplasmic and nuclear markers (Ford et al., 1994). This test may not always detect directional selection events if the adaptive increase in allele frequency is fairly recent (Hudson et al., 1994).

McDonald and Kreitman (1991) proposed a test of neutral protein evolution that asks if the ratio of nonsynonymous to synonymous mutation is equivalent for fixed differences between species and polymorphisms within species. Intraspecific and interspecific data are used to test if the ratio of nonsynonymous to synonymous change is equivalent for divergent and polymorphic sites. Directional selection is detected if the nonsynonymous/synonymous ratio is greater for fixed differences than for polymorphisms, suggesting that too many amino acid replacements have accumulated between species relative to the interspecific number of synonymous changes. This test was criticized because the number of fixed differences between species may underestimate the ratio (Graur and Li, 1991; Whittam and Nei, 1991).

Hughes and Nei (1988) designed a test for the neutral mutation hypothesis that compares the rate of substitution between synonymous and nonsynonymous sites. The neutral theory predicts that the rate of synonymous substitution will be greater than that in nonsynonymous sites, assuming that mutations are introduced at random, and that purifying selection removes deleterious amino acid changes. The target of adaptive selection is most likely nonsynonymous sites, which will cause the rate of nonsynonymous substitutions to be greater than that of synonymous sites.

Tajima (1989) designed a test statistic that uses intrapopulation data to detect departures from an equilibrium neutral model. The test statistic is the difference of two estimates of the neutral mutation parameter $4N\mu$, where $4N$ is the effective population size and μ is the neutral mutation rate. The first estimate of $4N\mu$ is based on the number of segregating sites S (Watterson, 1975), and the second estimate is based on the number of pairwise differences k_{ij} (Nei, 1987). The expected difference between the two estimates of $4N\mu$ should be zero under a neutral model of molecular evolution. Significant negative values of the standardized distance D indicate either purifying selection, directional selection, or population subdivision, and significant positive values of D indicate balancing selection. The Tajima test has limited power to detect selection because the test statistic is sensitive to sample size, S and timing of selection events (Simonsen et al., 1995).

Fu and Li (1993) derived a test of selective neutrality that examines the number of mutations that occur on external versus internal branches of a genealogy. The test uses within- and between-species data to estimate the number of internal and external mutations, η_i and η_e, respectively. An external mutation is a nucleotide site where the frequency of the rare base at a segregating site is $1/n$ within species and is not shared with an outgroup species, while an internal mutation is any other polymorphic nucleotide site. The test statistic D is used to determine if there is an excess or deficiency of external mutations. A significant negative value of D indicates directional selection that is due to an excess of external mutations, while a significant positive value of D denotes balancing selection that is due to a deficiency of external mutations.

7.5. Results

7.5.1. Balancing Selection

Before we get to results that are considered good evidence for the action of natural selection in shaping DNA variation, it is worth stressing that it is difficult in some cases to infer what selective model is responsible for observed polymorphism patterns in a static data set. In fact, the expectations about the amount of variation in neutral sites linked to a balanced polymorphism are different if the polymorphism has been maintained for a long time by balancing selection

(Hudson et al., 1987; Kreitman, 1983) or if it is rather recent and is on its way to equilibrium (Hudson et al., 1994); in this latter case, it is not possible to discern whether balancing or directional selection is operating.

With these difficulties in mind, our examples of data sets supporting the existence of a balanced polymorphism will necessarily be partial and will concentrate on old balanced polymorphisms. Compared with a neutral mutation, a polymorphism maintained by balancing selection will persist much longer in the population, and so will neutral variants tightly linked to the balanced polymorphism. This has two consequences that have been used in assessing the existence of balanced polymorphism through molecular data: (1) an excess of silent nucleotide variation will be the signature of an old balanced polymorphism, and (2) old selected alleles will be highly divergent and often will have been maintained from a time before the formation of species.

The first approach was developed in *Drosophila* by a comparison of levels of polymorphism and divergence in two different regions, the *Adh* region that codes for alcohol dehydrogenase and its 5′ flanking region (Kreitman and Aguadé, 1986b). The original fine restriction-map variation survey of the *Adh* region in natural populations of *D. melanogaster* revealed a high level of silent polymorphism in the coding region compared with that observed in its 5′ flanking region, which was in contrast with similar levels of silent divergence in both regions (Kreitman and Aguadé, 1986b). This excess of polymorphism proved to be inconsistent with the neutral prediction of a direct relationship between polymorphism and divergence in any given region (Hudson et al., 1987) and therefore was consistent with the presence of a balanced polymorphism in the *Adh* region. However, only sequence comparison of these regions in a worldwide sample of 11 alleles of *D. melanogaster* and one allele of *D. simulans* by use of a sliding window approach (Hudson and Kaplan, 1988; Kreitman and Hudson, 1991) indicated that the threonine–lysine amino acid replacement polymorphism distinguishing the *Adh* Fast and the *Adh* Slow electrophoretic alleles (or a closely linked polymorphism) was the polymorphism being maintained by balancing selection (Fig. 7.1). Additionally, the fine restriction-map survey of different populations of *D. melanogaster* along the North American east coast revealed a significant clinal variation at the Fast/Slow polymorphism, but not at other silent polymorphic sites, an assessment that shows that balancing selection acting at the two electrophoretically different alleles is mainly responsible for the cline despite extensive migration rates among populations (Berry and Kreitman, 1993).

The three following data sets illustrate the second approach:

(1) Major histocompatibility complex antigens: Major histocompatibility complex (MHC) loci are generally very highly polymorphic in mammals. When several alleles of MHC class I loci were sequenced in humans, they showed a high level of nucleotide polymorphism ($\pi = 0.04$) compared with that of other loci in the same species (Nei and Hughes, 1991). The intraspecific variation was even higher than the interspecies divergence between man and chimpanzee for other noncoding regions. These highly divergent alleles had a long genealogy

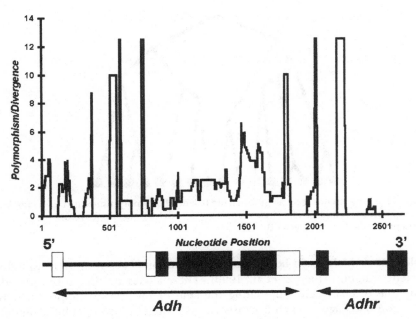

Figure 7.1. Sliding window plot of the polymorphism/divergence ratio for the alcohol dehydrogenase region of *D. melanogaster* (Hudson and Kaplan, 1988). Polymorphism is estimated as the number of pairwise differences in a sample of 11 alleles (Kreitman, 1983), and divergence is estimated from the average number of pairwise differences between *D. simulans* and each *D. melanogaster* allele scaled by the coalescence time $(T + 1)$. The window size is 150 potential synonymous sites. All insertions and deletions observed within or between species were removed from the analysis.

in which some of the genetic distances among alleles were smaller in interspecific than in intraspecific comparisons (Fig. 7.2), such that the time of coalescence for intraspecific alleles was older than the split of the species. Additionally, knowledge of the amino acid residues involved in the recognition of processed foreign antigens (Bjorkman et al., 1987) allowed Hughes and Nei (1988) to detect an excess of nonsynonymous polymorphisms in the antigen recognition site (ARS) as compared with the rest of the molecule. If balancing selection was acting to maintain different protein variants at the ARS, the average number of nonsynonymous differences per site between alleles ($d_N = 0.133$ for HLA-A alleles in humans) should be higher than the average number of synonymous differences per site in that same region ($d_S = 0.035$), but the reverse would hold for the rest of the residues given the action of purifying selection.

(2) Self-incompatibility alleles of plants: Molecular analysis of alleles at the self-incompatibility locus in Solanaceae, both within and between species, has revealed that, as expected from the strong selection acting on this locus, these alleles are highly divergent and unusually old (Clark and Kao, 1991; Ioerger et al., 1990; Richman et al., 1995). Alleles from the different species share some polymorphisms, i.e., sites that are segregating for the same pair of different

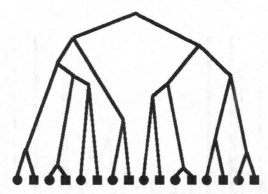

Figure 7.2. A hypothetical genealogy for an old balanced polymorphism with alleles sampled from two species, A and B. The circles represent the alleles from species A and the squares represent the alleles from species B.

nucleotides, which could be attributed to common ancestry and not to independent mutation (Ioerger et al., 1990). These alleles are therefore very old, their time of coalescence predating the split of the species (approximately 40 million years ago), which is consistent with the self-incompatibility model. On the other hand, Clark and Kao (1991) showed that among shared polymorphisms there is an excess of nonsynonymous polymorphisms. Given that only nonsynonymous mutations, which confer some selective advantage, are likely to be maintained for a long time (as opposed to deleterious or slightly deleterious mutations that will be much shorter lived), the excess of nonsynonymous polymorphisms among shared polymorphisms strongly points to these amino acid changes being involved in the function of the S protein through pollen recognition.

(3) Color Vision: Both squirrel monkeys and marmosets harbor a triallelic polymorphism at the only X-linked color vision gene present in New World monkeys (Platyrrhini). The spectral sensitivities of the photopigments coded by these alleles lie in both cases within the spectral range delimited by the human red and green pigments, which, as in Old World monkeys and apes, are coded by two different tandemly arranged genes on the X chromosome. A region encompassing the fourth intron of the three functionally different alleles of squirrel monkeys and marmosets has been sequenced (Shyue et al., 1995). The three alleles of each species are highly divergent; the within-species variation for this gene (0.02–0.04) is of the same order as the between-species variation in another noncoding region in New World monkeys. This high level of variation for the X-linked color vision gene indicates a very long allele genealogy, consistent with balancing selection maintaining the three alleles in the population. In addition, the nucleotide divergence between the squirrel monkey and the marmoset alleles is higher (0.06) than within-species variation, which together with their being differentiated by 14 length variants supports the fact that each triallelic system has arisen and is being maintained independently.

7.5.2. Directional Selection Driving Protein Evolution

Directional selection acting on a favorable mutation will cause an increase in frequency of the selected variant and eventually its fixation in the population. This will result in a loss of intraspecific neutral variation at linked neutral sites if the selective sweep is recent enough and the population is therefore still in the phase of neutral variation recovery, i.e., the adaptive fixation must have occurred within the last $4N$ generations, this being the average time to the common ancestor of all present $2N$ copies under neutrality.

On the other hand, older episodes of directional selection will have no impact either on neutral polymorphism or on neutral divergence. However, those selectively fixed variants that are amino acid replacement substitutions will cause an increase in the number of nonsynonymous differences between two species compared with the number of synonymous differences. Both the McDonald and Kreitman test (1991), which compares the levels of nonsynonymous polymorphism and divergence with the levels of synonymous polymorphism and divergence in a given coding region, and the Hughes and Nei test (1988), which compares the rates of nonsynonymous and synonymous substitutions, have been used to infer the action of directional positive selection in protein evolution.

As shown in Table 7.1, comparison of synonymous and nonsynonymous variation within and between species of *Drosophila* has been extensively used in *Drosophila* since the first application of the McDonald and Kreitman test. In three of these cases, the observed pattern of variation revealed an excess of fixed amino acid differences, which is consistent with adaptive directional selection acting at the protein level. The first two data sets concentrate on nucleotide variation of the coding region of two genes that code for well-studied enzymes,

Table 7.1. *Synonymous and Nonsynonymous Differences within and between Species of Drosophila*[†]

Locus	F/P	
	Nonsynonymous	*Synonymous*
Adh	7*/2	17/42
G6pd	21*/2	26/36
jgw	21*/4	16/27
per	3/10	35/84
yp2	7/13	4/11
boss	8/13	71/106

[†] References to the primary data are in Brookfield and Sharp (1994). F, fixed; P, polymorphic; *, a significant excess of fixed nonsynonymous differences was detected with the McDonald and Kreitman (1991) test.

alcohol dehydrogenase (*Adh*) (McDonald and Kreitman, 1991) and glucose-6-phosphate dehydrogenase (*G6pd*) (Eanes et al., 1993), both harboring two cosmopolitan electrophoretically different variants. The pattern of intraspecific and interspecific variation observed, with an excess of replacement substitutions between species, is consistent with selection causing the fixation of advantageous replacement mutations in the evolution of both *Adh* and *G6pd* during the divergence of the different species.

The third data set that shows an excess of fixed replacement substitutions between species consists of multiple sequences of the *Adh*-derived part of a chimeric gene, *jingwei* (Long and Langley, 1993), of the two species of the melanogaster group where it is present, *D. yakuba* and *D. teissieri*. This chimeric gene, *jgw*, was formed by a 5′ coding region captured from a gene of unknown function called *yande* (Long et al., 1996) and an *Adh*-derived part, originated by retrotranscription of the processed *Adh* transcript. Analysis of the *Adh*-derived part sequences revealed an excess of fixed nonsynonymous differences since the origin of the chimeric gene. The case in which directional selection accelerates protein evolution in this new gene whose function is still unknown can be viewed within the more general framework of duplicated diverging genes.

An additional data set in *Drosophila* reveals the action of directional selection in protein evolution in the process of speciation. The duplicated *Amy* locus has been sequenced in eight species of the *melanogaster* subgroup (Shibata and Yamazaki, 1995). Both copies of the gene are functional (except in *D. sechellia*) and have evolved in a concerted manner, as revealed by the strong within-species similarity. Nucleotide substitutions, either synonymous or nonsynonymous, were classified into equal and not-equal groups: Substitutions are equal when there is a nucleotide difference between species but no nucleotide difference between the two copies of the gene in the same species; otherwise they are considered not equal. Equal substitutions are therefore those observed between species, and they have occurred before the last homogenization (gene conversion) event between the two copies of the gene. If both nonsynonymous and synonymous mutations in a given region were neutral, and given that it can be assumed that they have the same gene genealogy, one would predict a direct relationship between both kinds of substitutions in any partition of the phylogeny, in this case before and after the last homogenization event in each species. However, when the overall number of replacement and synonymous substitutions in the equal and the not-equal groups was compared, an excess of equal replacement substitutions was observed. This is an indication of an acceleration of amino acid replacement fixations in the upper parts of the phylogeny, i.e., those parts closer to the different speciation events. When the number of nucleotide substitutions specific to each species was compared in a similar way, again an overall excess of replacement equal substitutions was observed, mainly attributable to the excess observed in the *D. erecta* lineage. *D. erecta* is a specialist that uses only one genus of plant for feeding and breeding sites. Changes in ecology and breeding habits in the speciation process might have accelerated the fixation of adaptive amino acid mutations in amylase, a digestive enzyme.

In organisms other than *Drosophila*, in which in general data on only inter-specific divergence are collected, a significantly positive ratio of replacement versus synonymous divergence ($K_a/K_s > 1$) is considered good evidence for directional selection acting on advantageous replacement mutations. Abalone are marine mollusks with external fertilization with a requirement for sperm–egg recognition proteins. Lysin is an acrosomal protein of the sperm used in the first step of fertilization; actually it disrupts the egg vitelline envelope by a nonenzymatic mechanism, a reaction that *in vitro* and possibly *in vivo* is species specific. Comparison of the cDNA sequences of the gene coding for the abalone lysin of 20 different species revealed, under the Hughes and Nei test (1988), an excess of replacement versus synonymous substitutions in some comparisons of the most closely related species (Lee et al., 1995; Lee and Vacquier, 1992). This observation, together with a general positive value of the corrected percentage of nucleotide substitutions per site in the first and the second positions of codons versus that in the third positions, makes the authors claim that directional selection is the main force differentiating these proteins between species. Selection pressure might act directly on lysin or indirectly on the egg receptor protein. This might even be an example in which adaptive evolution acts on two interacting molecules.

7.5.3. Directional and Background Selection in Regions with Different Rates of Recombination

As discussed in Section 7.5.2, an advantageous mutation on its way to fixation will drag a particular variant of all polymorphic neutral sites tightly linked to it (the hitchhiking effect). The decrease of neutral linked variation will depend on the strength of selection, the recombination rate (Stephan et al., 1992) (which will or will not allow other variants to recombine into the haplotype being driven to fixation), and the level of population subdivision at the time of the selective sweep. Episodes of directional selection that occurred during the last $4N$ generations will therefore leave a signature of reduced neutral variation in a larger or a smaller region, depending directly on the recombination rate. In fact, the hitchhiking effect model predicts a positive correlation between levels of DNA sequence variation and rates of recombination (see Fig. 1 in Begun and Aquadro, 1992).

The reduction of nucleotide variation in regions of low recombination is well documented in several species of *Drosophila* (Aguadé and Langley, 1994; Aquadro et al., 1994; Kreitman and Wayne, 1994; Stephan, 1994). In *D. melanogaster* nucleotide variation at regions close to the X chromosome telomere [y–ac–sc, erw, $su(s)$, $s(w^a)$] and centromere [$su(f)$], the second-chromosome left-arm telomere [$l(2)gl$] and centromere (cta), the second-chromosome right-arm centromere [Lcp-ψ], the third-chromosome left-arm telomere ($Lsp1$-g) and at the nonrecombining fourth chromosome (ci^D) are much lower than in regions of normal recombination in the same chromosome. The larval cuticle protein pseudogene of *D. melanogaster* (Lcp-ψ) is a dramatic case of reduced nucleotide variation in a region of low recombination as this cuticle pseudogene lacks diversity, despite a lack of functional constraints on the nucleotide sequence (Pritchard and Schaeffer, 1997). Likewise the y–ac, $su(f)$, and ci^D regions of $D.$ $simulans$, and the centromere proximal v and fw regions of $D.$ $ananassae$ show low levels of nucleotide variation. However, no reduction of silent nucleotide divergence has been detected when these regions between $D.$ $melanogaster$ and its sibling species $D.$ $simulans$ were compared (see Fig. 2 in Begun and Aquadro, 1992).

This decoupling of intraspecific and interspecific variations, a deviation of the neutral expectations in the sense of a reduction of neutral within-species variation, was initially attributed to the hitchhiking effect of advantageous mutations. In all cases population bottlenecks could be discarded as the cause of the observed reduction of polymorphism because the same lines showed normal levels of polymorphism for other loci in regions exhibiting no restricted recombination. Additionally, the positive correlation detected between levels of recombination and estimates of nucleotide variation within species and the lack of the corresponding correlation with estimates of between-species variation (Aquadro et al., 1994; Begun and Aquadro, 1992) was considered an indication of directional selection. In fact, advantageous mutations happening at different parts of the $Drosophila$ genome would go to fixation and leave a footprint of reduced neutral variation in a smaller or larger tightly linked region.

These initial observations motivated the development of an alternative model, the background selection model (Charlesworth et al., 1995; Charlesworth et al., 1993) that also predicts a positive correlation between levels of polymorphism and rates of recombination. In the background selection model, the reduction of variation is caused by deleterious mutations that are maintained in the population by a mutation–selection balance and whose effect is a reduction of the effective population size. Although the background selection model could in general account for the observed reduction of variation, the original parameters used (deleterious mutation rates and selection coefficients) could not account for the extreme reduction of variation observed both at the y–ac–sc region at the tip of the X chromosome (Charlesworth, 1994; Charlesworth et al., 1995; Charlesworth et al., 1993) or at the $Lsp1$-γ region at the tip of 3L (Hudson and Kaplan, 1995). It seems, however, that when the deleterious effect of transposable elements and their distribution in the $D.$ $melanogaster$ genome are considered, even these extreme cases can be explained by the background selection model (Charlesworth, 1996). How does this, however, affect loci at the rest of those chromosomes and what is the situation in $D.$ $simulans?$

The hitchhiking and the background selection models make different predictions with respect to the frequency spectrum of individual DNA polymorphisms: The hitchhiking model predicts a frequency spectrum skewed toward an excess of rare polymorphisms, while no such skewness is predicted for the neutral-behaving fraction of variation that remains after deleterious-mutation-carrying alleles have been eliminated by background selection. Although some data sets in *Drosophila* show such a skewness, which might be an indication of local hitchhiking events, there is not a consistent pattern. It is, however, worth mentioning that, because of the reduction of variation observed in such regions, some of the data sets harbor a low number of polymorphisms that renders them inadequate to test for deviations of the frequency spectrum expected under neutrality. The relative contribution of advantageous and deleterious mutation to the observed pattern of reduced variation in regions of low recombination is therefore an open question.

7.5.4. Other Situations

As discussed in Subsection 7.5.1, the excess of neutral variation associated with a balanced polymorphism is expected only when the polymorphism arose a long time ago (more than $4N$ generations) and has been able to accumulate neutral variation in excess of that expected under neutrality. However, if a recently arisen variant has increased in frequency because of balancing selection, then the pattern of neutral variation in tightly linked sites will be similar to that of a recently arisen advantageous variant that is being driven by directional selection and that is on its way to fixation. Using a coalescent approach, Hudson et al. (1994) developed a test to determine whether selection (either balancing or directional) is increasing the frequency of particular haplotypes. The first application of the haplotype test at the *Sod* region, which codes for the enzyme superoxide dismutase, proved that the fast A haplotype has recently increased in frequency because of selection and that the other amino acid variant (*Sod* Slow), which differs from the fast A haplotype only by the replacement change, has also increased in frequency recently either because of selection on the amino acid change or selection on the fast A haplotype.

Finally, if the rate of change of the selective agent that confers advantage to a new variant is high, the advantageous character of the variant can change while its frequency is increasing because of directional selection, and that will result in transient polymorphisms. Wayne et al. (1996) have analyzed variation at the *ref(2)P* region, which confers viral resistance, in a worldwide sample of alleles (both permissive and restrictive) of *D. melanogaster*. They observe an excess of replacement polymorphism at the *N*-terminal part of the protein, where the cause for viral restrictivity has been identified, but no excess of linked neutral variation. The observed pattern of variation could be explained either by a recent balanced polymorphism or by transient selection for new restrictive alleles.

7.5.5. Epistatic Selection

The past action of balancing selection and directional selection can create significant linkage disequilibrium among polymorphic nucleotide sites (Hudson et al., 1994; Kreitman, 1983); however, strong nonrandom associations among nucleotide sites may also result from strong epistatic selection. Epistatic selection results when a random assortment of genetic variations at two loci generates genetic combinations with lower fitness than the parental genotypes. Ultimately, epistatic selection leads to linkage disequilibrium among two or more genetic loci. Linkage disequilibrium may result from low recombination rates, population subdivision, and genetic hitchhiking associated with adaptive selection as well as strong epistatic selection. Epistatic selection is the likely explanation of significant linkage disequilibrium among loci if the region experiences high recombination rates, if populations are not subdivided, and if no recent adaptive selection events are detected (Miyashita et al., 1993; Schaeffer and Miller, 1993). Although the pattern of variation observed at the *white* locus of *D. melanogaster* (Kirby and Stephan, 1995; Kirby and Stephan, 1996; Miyashita and Langley, 1988; Miyashita et al., 1993) and the *Adh* locus of *D. pseudoobscura* (Kirby et al., 1995; Schaeffer and Miller, 1993) were considered good examples of epistatic selection among nucleotide sites, a more complex multi-locus model has been proposed to explain the pattern of variation at the *white* locus (Kirby and Stephan, 1996).

7.6. Conclusions

The effects of genetic hitchhiking associated with adaptive fixations or selection against deleterious alleles are dramatically observed in regions of the genome where recombination rates are low. In general background selection can explain modest reductions in diversity, but directional selection is implicated when variation levels are less than those predicted by a background selection model. Reductions of variation in regions of high recombination are likely to be due to directional selection, but the effects of genetic hitchhiking induced by adaptive fixations are more local. Balancing selection leaves a footprint of excess levels of diversity, either as an elevated number of polymorphisms at linked neutral sites or as an elevated divergence in nonsynonymous sites between alleles.

REFERENCES

Aguadé, M. and Langley, C. H. 1994. Polymorphism and divergence in regions of low recombination in *Drosophila*. In *Non-Neutral Evolution: Theories and Molecular Data*, B. Golding, ed., pp. 67–76. New York: Chapman & Hall.
Aquadro, C. F., Begun, D. J., and Kindahl, E. C. 1994. Selection, recombination, and DNA polymorphism in *Drosophila*. In *Non-Neutral Evolution: Theories and Molecular Data*, B. Golding, ed., pp. 46–56. New York: Chapman & Hall.
Aquadro, C. F., Noon, W. A., and Begun, D. J. 1992. RFLP analysis using heterologous probes. In *Molecular Genetic Analysis of Populations: A Practical Approach*, A. R. Hoelzel, ed., pp. 115–157. New York: IRL Press at Oxford U. Press.
Begun, D. J. and Aquadro, C. F. 1991. Molecular population genetics of the distal portion of the X chromosome in Drosophila: evidence for genetic hitchhiking of the *yellow-achaete* region. *Genetics* **129**:1147–1158.
Begun, D. J. and Aquadro, C. F. 1992. Levels of naturally occurring DNA polymorphism correlate with recombination rates in *D. melanogaster. Nature (London)* **356**:519–520.

Bénassi, V. and Veuille, M. 1995. Comparative population structuring of molecular and allozyme variation of *Drosophila melanogaster Adh* between Europe, West Africa and East Africa. *Genet. Res. Cambridge* **65**:95–103.

Berry, A. and Kreitman, M. 1993. Molecular analysis of an allozyme cline: alcohol dehydrogenase in *Drosophila melanogaster* on the east coast of North America. *Genetics* **134**:869–893.

Berry, A. J., Ajioka, J. W., and Kreitman, M. 1991. Lack of polymorphism on the Drosophila fourth chromosome resulting from selection. *Genetics* **129**:1111–1117.

Birky, C. W. and Walsh, J. B. 1988. Effects of linkage on rates of molecular evolution. *Proc. Natl. Acad. Sci. USA* **85**:6414–6418.

Bjorkman, P. J., Saper, M. A., Samraoui, B., Bennett, W. S. Strominger, J. L., et al. 1987. The foreign antigen binding site and T cell recognition regions of class I histocompatibility antigens. *Nature (London)* **329**:512–518.

Brookfield, J. F. Y. and Sharp, P. M. 1994. Neutralism and selectionism face up to DNA data. *Trends Genet.* **10**:109–111.

Charlesworth, B. 1994. The effect of background selection against deleterious mutations on weakly selected, linked variants. *Genet. Res. Cambridge* **63**:213–227.

Charlesworth, B. 1996. Background selection and patterns of genetic diversity in *Drosophila melanogaster. Genet. Res. Cambridge* **68**:131–149.

Charlesworth, B., Charlesworth, D., and Morgan, M. T. 1995. The pattern of neutral molecular variation under the background selection model. *Genetics* **141**:1619–1632.

Charlesworth, B., Morgan, M. T., and Charlesworth, D. 1993. The effect of deleterious mutations on neutral molecular variation. *Genetics* **134**:1289–1303.

Clark, A. G. and Kao, T.-H. 1991. Excess nonsynonymous substitution at shared polymorphic sites among self-incompatibility alleles of Solanaceae. *Proc. Natl. Acad. Sci. USA* **88**:9823–9827.

David, J. R. 1982. Latitudinal variability of *Drosophila melanogaster*: Allozyme frequencies divergence between European and Afrotropical populations. *Biochem. Genet.* **20**:747–761.

Dobzhansky, Th. 1955. A review of some fundamental concepts and problems of population genetics. *Cold Spring Harbor Symp. Quant. Biol.* **20**:1–15.

Eanes, W. F., Kirchner, M., and Yoon, J. 1993. Evidence for adaptive evolution of the *G6pd* gene in the *Drosophila melanogaster* and *Drosophila simulans* lineages. *Proc. Natl. Acad. Sci. USA* **90**:7475–7479.

Ewens, W. J. 1993. History and development of population genetics. In *Seventeenth International Congress of Genetics*, p. 33.

Ford, M. J., Yoon, C. K., and Aquadro, C. F. 1994. Molecular evolution of the *period* gene in *Drosophila athabasca. Mol. Biol. Evol.* **11**:169–182.

Fu, Y.-X. and Li, W.-H. 1993. Statistical tests of neutrality of mutations. *Genetics* **133**:693–709.

Graur, D. and Li, W.-H. 1991. Neutral mutation hypothesis test. *Nature (London)* **354**:114–115.

Hudson, R. R. 1983. Testing the constant-rate neutral allele model with protein sequence data. *Evolution* **37**:203–217.

Hudson, R. R., Bailey, K., Skarecky, D., Kwiatowski, J., and Ayala, F. J. 1994. Evidence for positive selection in the superoxide dismutase (*Sod*) region of *Drosophila melanogaster. Genetics* **136**:1329–1340.

Hudson, R. R. and Kaplan, N. L. 1988. The coalescent process in models with selection and recombination. *Genetics* **120**:831–840.

Hudson, R. R. and Kaplan, N. L. 1995. Deleterious background selection with recombination. *Genetics* **141**:1605–1617.

Hudson, R. R., Kreitman, M., and Aguadé, M. 1987. A test of neutral molecular evolution based on nucleotide data. *Genetics* **116**:153–159.

Hughes, A. L. and Nei, M. 1988. Pattern of nucleotide substitution at major histocompatibility complex class I loci reveals overdominant selection. *Nature (London)* **335**:167–170.

Ioerger, T. R., Clark, A. G., and Kao, T.-H. 1990. Polymorphism at the self-incompatibility locus in Solanaceae predates speciation. *Proc. Natl. Acad. Sci. USA* **87**:9732–9735.

Kaplan, N. L., Hudson, R. R., and Langley, C. H. 1989. The "hitchhiking effect" revisited. *Genetics* **123**:887–899.

Karlin, S. and Feldman, M. W. 1970. Linkage and selection: two locus symmetric viability model. *Theor. Popul. Biol.* **1**:39–71.

Kimura, M. 1983. *The Neutral Theory of Molecular Evolution.* New York: Cambridge U. Press.

Kingman, J. F. C. 1982a. The coalescent. *Stochast. Process. Appl.* **13**:235–248.

Kingman, J. F. C. 1982b. On the genealogy of large populations. *J. Appl. Prob.* **19A**:27–43.

Kirby, D. A., Muse, S. V., and Stephan, W. 1995. Maintenance of pre-mRNA secondary structure by epistatic selection. *Proc. Natl. Acad. Sci. USA* **92**:9047–9051.

Kirby, D. A. and Stephan, W. 1995. Haplotype test reveals departure from neutrality in a segment of the *white* gene in *Drosophila melanogaster. Genetics* **141**:1483–1490.

Kirby, D. A. and Stephan, W. 1996. Multi-locus selection and the structure of variation at the *white* gene of *Drosophila melanogaster. Genetics* **144**:635–645.

Koehn, R. K. 1978. Physiology and biochemistry of enzyme variation: the interface of ecology and population genetics. In *Ecological Genetics: The Interface,* P. F. Brussard, ed., pp. 51–72. New York: Springer-Verlag.

Kreitman, M. 1983. Nucleotide polymorphism at the alcohol dehydrogenase locus of *Drosophila melanogaster. Nature (London)* **304**:412–417.

Kreitman, M. and Aguadé, M. 1986a. Genetic uniformity in two populations of *Drosophila melanogaster* as revealed by filter hybridization of four-nucleotide-recognizing restriction enzyme digests. *Proc. Natl. Acad. Sci. USA* **83**:3562–3566.

Kreitman, M. and Aguadé, M. 1986b. Excess polymorphism at the *Adh* locus in *Drosophila melanogaster. Genetics* **114**:93–110.

Kreitman, M. and Hudson, R. R. 1991. Inferring the evolutionary histories of the *Adh* and *Adh-Dup* loci in *Drosophila melanogaster* from patterns of polymorphism and divergence. *Genetics* **127**:565–582.

Kreitman, M. and Wayne, M. L. 1994. Organization of genetic variation at the molecular level: lessons from Drosophila. In *Molecular Ecology and Evolution,* B. Schierwater, ed., pp. 157–184. Basel: Birkhauster Verlag.

Lee, Y.-H., Ota, T., and Vacquier, V. D. 1995. Positive selection is a general phenomenon in the evolution of abalone sperm lysin. *Mol. Biol. Evol.* **12**:231–238.

Lee, Y.-H. and Vacquier, V. D. 1992. The divergence of species-specific abalone sperm lysins is promoted by positive Darwinian selection. *Biol. Bull.* **182**:97–104.

Lewontin, R. C. 1974. *The Genetic Basis of Evolutionary Change.* New York: Columbia U. Press.

Lewontin, R. C. and Hubby, J. L. 1966. A molecular approach to the study of genic heterozygosity in natural populations. II. Amount of variation and degree of heterozygosity in natural populations of *Drosophila pseudoobscura*. *Genetics* **54**:595–609.

Long, M., Alvarez, C., Lozovskaya, E. R., Langley, C. H., and Gilbert, W. 1996. Origin of new genes: story of *yande* and its daughter *jingwei*. In *37th Annual Drosophila Research Conference*, p. 10A. Bethesda, MD: The Genetics Society of America.

Long, M. and Langley, C. H. 1993. Natural selection and the origin of *jingwei*, a chimeric processed functional gene in *Drosophila*. *Science* **260**:91–95.

Maynard Smith, J. and Haigh, J. 1974. The hitch-hiking effect of a favorable gene. *Genet. Res. Cambridge* **23**:23–35.

McDonald, J. H. and Kreitman, M. 1991. Adaptive protein evolution at the *Adh* locus in *Drosophila*. *Nature (London)* **351**:652–654.

Miyashita, N. and Langley, C. H. 1988. Molecular and phenotypic variation of the *white* locus region in *Drosophila melangaster*. *Genetics* **120**:199–212.

Miyashita, N. M., Aguadé, M., and Langley, C. H. 1993. Linkage disequilibrium in the *white* locus region of *Drosophila melanogaster*. *Genet. Res. Cambridge* **62**:101–109.

Muller, H. J. 1950. Our load of mutations. *Am. J. Hum. Genet.* **2**:111–176.

Nei, M. 1987. *Molecular Evolutionary Genetics*. New York: Columbia U. Press.

Nei, M. and Hughes, A. L. 1991. Polymorphism and evolution of the major histocompatibility complex loci in mammals. In *Evolution at the Molecular Level*, R. K. Selander, A. G. Clark, and T. S. Whittam, eds., pp. 222–247. Sunderland, MA: Sinauer.

Oakeshott, J. G., Gibson, J. B., Anderson, P. R., Knibb, W. R., Anderson, D. G., et al. 1982. Alcohol dehydrogenase and glycerol-3-phosphate dehydrogenase clines in *Drosophila melanogaster* on different continents. *Evolution* **36**:86–96.

Ohta, T. and Kimura, M. 1975. The effect of selected linked locus on heterozygosity of neutral alleles (the hitch-hiking effect). *Genet. Res. Cambridge* **25**:313–326.

Pritchard, J. K. and Schaeffer, S. W. 1997. Polymorphism and divergence at a Drosophila pseudogene locus. *Genetics* **147**:199–208.

Provine, W. B. 1971. *The Origins of Theoretical Population Genetics*. Chicago: U. Chicago Press.

Richman, A. D., Kao, T.-H., Schaeffer, S. W., and Uyenoyama, M. K. 1995. S-allele sequence diversity in natural populations of *Solanum carolinense* (Horsenettle). *Heredity* **75**:405–415.

Saiki, R. K., Gelfand, D. H., Stoffel, S., Scharf, S. J., Higuchi, R. et al. 1988. Primer-directed enzymatic amplification of DNA with a thermostable DNA polymerase. *Science* **239**:487–491.

Sambrook, J., Fritsch, E. F., and Maniatis, T. 1989. *Molecular Cloning: A Laboratory Manual*. Cold Spring Harbor, NY: Cold Spring Harbor Laboratory.

Schaeffer, S. W. and Miller, E. L. 1993. Estimates of linkage disequilibrium and the recombination parameter determined from segregating nucleotide sites in the alcohol dehydrogenase region of *Drosophila pseudoobscura*. *Genetics* **135**:541–552.

Shibata, H. and Yamazaki, T. 1995. Molecular evolution of the duplicated *Amy* locus in the *Drosophila melanogaster* species subgroup: concerted evolution only in the coding region and an excess of nonsynonymous substitutions in speciation. *Genetics* **141**:223–236.

Shyue, S.-K., Hewett-Emmett, D., Sperling, H. G., Hunt, D. M., Bowmaker, J. K., et al. 1995. Adaptive evolution of color vision genes in higher primates. *Science* **269**:1265–1267.

142   *Stephen W. Schaeffer and Montserrat Aguadé*

Simmons, G. M., Kreitman, M. E., Quattlebaum, W. F., and Miyashita, N. 1989. Molecular analysis of the alleles of alcohol dehydrogenase along a cline in *Drosophila melanogaster*. I. Maine, North Carolina, and Florida. *Evolution* 43:393–409.

Simonsen, K. L., Churchill, G. A., and Aquadro, C. F. 1995. Properties of statistical tests of neutrality for DNA polymorphism data. *Genetics* 141:413–429.

Slatkin, M. 1991. Inbreeding coefficients and coalescence times. *Genet. Res. Cambridge* 58:167–175.

Stephan, W. 1994. Effects of genetic recombination and population subdivision on nucleotide sequence variation in *Drosophila ananassae*. In *Non-Neutral Evolution: Theories and Molecular Data*, B. Golding, ed., pp. 57–76. New York: Chapman & Hall.

Stephan, W., Wiehe, T. H. E., and Lenz, M. W. 1992. The effect of strongly selected substitutions on neutral polymorphism: Analytical results based on diffusion theory. *Theor. Popul. Biol.* 41:237–254.

Tajima, F. 1983. Evolutionary relationship of DNA sequences in finite populations. *Genetics* 105:437–460.

Tajima, F. 1989. Statistical method for testing the neutral mutation hypothesis by DNA polymorphism. *Genetics* 123:585–595.

Tavaré, S. 1984. Line-of-descent and genealogical processes, and their applications in population genetic models. *Theor. Popul. Biol.* 26:119–164.

Vigue, C. L. and Johnson, F. M. 1973. Isozyme variability in species of the genus *Drosophila*. VI. Frequency-property-environment relationships of allelic alcohol dehydrogenases in *D. melanogaster*. *Biochem. Genet.* 9:213–227.

Watterson, G. A. 1975. On the number of segregating sites in genetical models without recombination. *Theor. Popul. Biol.* 7:256–276.

Wayne, M. L., Contamine, D., and Kreitman, M. 1996. Molecular population genetics of *ref(2)P*, a locus which confers viral resistance in *Drosophila*. *Mol. Biol. Evol.* 13:191–199.

Whittam, T. S. and Nei, M. 1991. Neutral mutation hypothesis test. *Nature (London)* 354:115–116.

SELECTION AND GENETIC POLYMORPHISM

Introductory Remarks

PHILIP HEDRICK AND TIMOTHY PROUT

The maintenance of genetic polymorphism by selection was one of the major research interests of Dobzhansky and his students, including Dick Lewontin. In particular, much effort was spent in trying to understand the selective factors important in maintaining inversion polymorphisms in *Drosophila pseudoobscura*. Other classic cases of genetic polymorphism being intensively investigated contemporaneously included those of sickle-cell hemoglobin and blood groups in humans and various ecological genetics studies, such as banding and color patterns in *Cepea* and color and pattern polymorphism in butterflies. Although much was learned about each of these polymorphisms, no generalizations seemed to emerge. In addition to the question of maintenance of polymorphism, directional selection due to human disturbance was found and studied, for example, industrial melanism, heavy metal tolerance, and resistance to insecticides and antibiotics were all found to be adaptive changes due to genes of large effect (e.g., Bishop and Cook, 1981).

While these classic studies stimulated theoretical research and generally allowed insight into the specific selective factors in these cases, for all of them there are important evolutionary questions that appear to be unresolved to this day. For example, how variable are the fitnesses of the various genotypes over time and space and are the fitnesses frequency dependent? Detailed studies over the years of these specific cases, generally without molecular information except for sickle-cell hemoglobin and insecticide resistance, obviously have shown that, even for genes (or inversions) with large and important selective effects, effects of selection by itself are often complicated and difficult to understand completely.

Dick Lewontin was originally interested in heterozygote advantage per se as a mechanism to maintain genetic polymorphism. For example, he gave a general method to determine the polymorphic equilibrium maintained by selection (Lewontin, 1958). Then, in Lewontin and Hubby (1966), he proposed heterozygote advantage as a possible explanation for the maintenance of the extensive allozyme variation he and Hubby documented in *D. pseudoobscura*. However, to maintain variation with heterozygote advantage at a large number of loci, he showed that the fitness of the population would be intolerably low.

Later, Lewontin et al. (1978) suggested that heterozygote advantage by itself would be an unlikely explanation for maintenance of multiple allele polymorphism.

Possible alternative explanations suggested by Lewontin et al. (1978) to maintain molecular variation were frequency-dependent selection and variable selection in different environments. Obviously, the brief note by Levene (1953), written as a potential mathematical explanation of how variable selection in different environments could maintain inversion and other polymorphisms known at the time, initiated much of the interest in this mechanism and the research by Levins (e.g., Levins, 1968) further stimulated it. The studies by Kojima demonstrated that fitness may be a function of the frequency of genotypes and the potential for maintenance of polymorphism by means of frequency-dependent selection (e.g., Kojima, 1971). Today, the relative importance of these selection models in maintaining polymorphism has yet to be determined by experimental studies. Indeed, the extraordinary amount of allozyme polymorphism initially described by Lewontin and Hubby (1966) has yet to be clearly explained.

The interest in single genes with large adaptive effects has somewhat waned with the recent focus on molecular evolution by use of DNA sequence information to understand the relationships among populations, species, alleles, and genes. Information from DNA sequences may determine subtle cumulative selective differences that we could not detect directly in a single or a few generations. However, these subtle differences, in which the selective difference need be only somewhat larger than the reciprocal of the effective population size, may not be the immediate stuff of evolution that is required for responding to large environmental challenges. On the other hand, the recent identification of major genes in quantitative genetics, or quantitative trait loci (QTL), by molecular mapping techniques is focusing on genes that potentially have a large effect on the phenotype. Variants at these genes may be ones that contribute important adaptive morphological or physiological effects to natural populations. Nevertheless, at this point the explanation for the large amount of variation found for these quantitative characters also remains in the realm of theoretical speculation.

In the Chaps. 8–12, it is generally assumed that some proportion of the genes at any given time in evolution are influenced by strong selection pressures. By focusing on understanding the problems and dynamics of these genes of large effects, these chapters will provide insight into understanding the variation in other genes with lesser effects whose examination is experimentally more difficult.

This section contains five contributions, with a range of approaches and topics but focused on genes with strong selective effects. First, Christiansen and Prout (Chap. 8) give a general overview of the possible types of selection that can occur in a population. They divide selection into four major components: viability, fecundity, sexual, and gametic selection. Each of these components may in turn result in different forms or outcomes of selection, some of which maintain polymorphism and others that result in fixation of favored types. Although genetic polymorphisms are often thought to be maintained by selection with complementary opposing effects, Prout (Chap. 9) next presents an essay detailing examples of how difficult it is to maintain a polymorphism for a variety of different models of selection with opposing effects. For example, he shows that in an infinite population, the region of fitnesses that gives a stable polymorphism is often smaller than the region for fixation for a variety of models that include variable environments, meiotic drive, or antagonistic pleiotropy.

The last three chapters discuss situations in which it is assumed that selection at a single gene or group of genes is either large or may be easily detected. First, Arnason and Barker (Chap. 10) discuss models to detect selection with an example from the *Adh* (alcohol dehydrogenase) locus in *D. melanogaster*. In addition to the use of F_{ST} to detect selection, they also discuss a case of strong selection for insecticide resistance in the sheep blowfly, one of the now-classic cases of adaptive evolution. Hedrick and Kim (Chap. 11) present a review of the major histocompatibility complex (MHC) in vertebrates and the evidence that it confers resistance to infectious disease, particularly in humans. Although there is substantial evidence for selective maintenance of MHC variation, including detailed molecular information, the exact details of the selective mechanisms are still difficult to document. Finally, Levin (Chap. 12) gives a historical and personal perspective on antibiotic resistance, initially detailing some of the mechanisms of inheritance and the nature of resistance. Using population genetics approaches, he discusses the maintenance of these resistant genetic elements and the relationship between the incidence of antibiotic use and antibiotic resistance.

REFERENCES

Bishop, J. A. and Cook, L. M. 1981. *Genetic Consequences of Man Made Change.* New York: Academic.

Kojima, K. 1971. Is there a constant fitness for a given genotype? No! *Evolution* **25**: 281–285.

Levene, H. 1953. Genetic equilibrium when more than one ecological niche is available. *Am. Nat.* **87**:131–133.

Levins, R. 1968. *Evolution in Changing Environments.* Princeton, NJ: Princeton U. Press.

Lewontin, R. C. 1958. A general method for investigating the equilibrium of gene frequency in a population. *Genetics* **43**:419–433.

Lewontin, R. C., Ginzburg, L. R., and Tuljapurkar, S. D. 1978. Heterosis as an explanation for large amounts of genic polymorphism. *Genetics* **88**:149–170.

Lewontin, R. C. and Hubby, J. L. 1966. A molecular approach to the study of genic heterozygosity in natural populations. II. Amount of variation and degree of heterozygosity in natural populations of *Drosophila pseudoobscura. Genetics* **54**:595–609.

CHAPTER EIGHT

Aspects of Fitness

FREDDY BUGGE CHRISTIANSEN AND TIMOTHY PROUT

The definition of fitness is one of the perpetual questions in evolutionary biology. It seems like every generation of evolutionary biologists ponders this question, and in the long term rather little progress toward a universally acceptable definition has been made, but this issue has promoted fruitful and recurring discussions on Darwinian natural selection. The concept of fitness emerges from the last remark in the definition of natural selection:

... This preservation of favourable individual differences and variations, and the destruction of those which are injurious, I have called Natural Selection, or the Survival of the Fittest... (Darwin, 1872).

(This remark was added by Darwin in the sixth edition of *The Origin of Species* after suggestion and encouragement by Thomas Huxley, but Darwin later regretted this.) Of course, fitness has entered evolutionary biology as the metaphysical character that natural selection acts on, and this substitution of an object for a process provides more freedom of expression in the exposition of the evolutionary processes. However, the word fitness originates through the equivalence of natural selection and the survival of the fittest, which itself is a tautology. The character fitness therefore cannot be simpler than the process it describes, and so fitness is no simple character. Darwin's definition of fitness is a metaphor in that survival should be seen as a representation of the action of natural selection and not as alluding to literal survival of individuals. Also, fitness is not a character which evolves; rather, it is a character that helps describe the process of evolution of other characters of the individual.

The source of evolutionary change in a character is hereditary variation. Mendelian inheritance separates the object of natural selection, the phenotype,

Research supported in part by grant 94-0163-1 from the Danish Natural Science Research Council to F. B. C. and by pension benefits from the University of California to T. P.

R. S. Singh and C. B. Krimbas, eds., *Evolutionary Genetics: From Molecules to Morphology*, vol. 1. © Cambridge University Press 2000. Printed in the United States of America. ISBN 0-521-57123-5. All rights reserved.

from the record of the change, the genotype. Natural selection on a character, viewed as an evolutionary mechanism, should therefore be specified in terms of the induced effects in terms of genotypic selection, even though the biological description and understanding of the causes of natural selection are centered on the phenotype in its environment. The repression of this neo-Darwinian challenge has been the cause of much controversy in evolutionary biology. The approach of evolutionary ecology is largely to describe the connection between phenotypic variation and natural selection, whereas the approach of population genetics focuses on the process of genetic change caused by natural selection. A full description of the evolutionary process needs reference to both of these aspects.

More than 20 years ago, the observation of genotypic selection induced by natural selection was summarized as follows:

To the present moment no one has succeeded in measuring with any accuracy the net fitness of genotypes for any locus in any species in any environment in nature ... (Lewontin, 1974).

Progress has been made, but the statement is still largely true.

Darwin recognized two basic components of natural selection expressed as variation in the survival of individuals to maturity and as variation in the number of offspring a mature individual leaves; these two components of selection correspond to the birth and death rates in the Malthusian model of population growth. The distinction between variations in birth rate and in death rate reflects the different manifestations of action. The differential elimination of certain types from the population provides an immediate expression of natural selection, whereas individual differences in fecundity or fertility are more subtle aspects. The Malthusian model of population growth is the reference model of population ecology, and so this basic description of natural selection is closely related to the parameters of models of population growth and interaction. This is most vividly expressed in studies of the evolution of life histories, in which the interaction of characters related to individual growth, survival, and reproduction is considered.

The third component of selection that Darwin considered is sexual selection expressed in a sexual organism as individual variation in participation in matings. This component is related to fertility, and in its expression it is even more subtle than fecundity or fertility selection, a characteristic of reproductive components of selection. Darwin singled out sexual selection because of its obvious cause in the active interaction among the individuals in a population, choices among partners of the opposite sex, and competition for the attention of the other sex. The two basic components describe the effects of the environment and of more passive interactions like exploitative competition. Direct and active interactions among the individuals in a population may also influence survival and fecundity, and antagonistic and cooperative interactions, for instance, are known and important in many groups of animals.

The description of the widespread haplont–diplont cycle in the life of eucaryotes added an extra dimension to the description of natural selection. A component of selection, gametic selection, is usually added as an extra reproductive component in higher organisms. Selection in the haploid stage and selection in the diploid stage have different consequences for the resulting change in the genetic composition of the population. This dependence on organizational level extends in higher organisms to fecundity selection, in that diploid offspring are ascribed to both the mother and the father individual in the mated pair. Thus a simple summary of natural selection is that on the haploid level gametic selection acts, and pairs of gametes form zygotes subject to differential survival to maturity, often referred to as zygotic selection. Finally, zygotes form mating pairs in the process of reproduction. The reproductive pairs are formed through sexual selection and mating behavior, and the genetic composition of the offspring population is determined by these pairs, gametic selection, and fecundity selection. This defines the four basic components of selection in an organism with nonoverlapping generations (Bundgaard and Christiansen, 1972; Christiansen and Frydenberg, 1973):

zygotic selection: differential survival of zygotes to maturity
sexual selection: differential recruitment of mature individuals to parents
fecundity selection: variation in fecundity of parental pairs
gametic selection: differential survival of gametes

The strict separation of the components needs reference to the biological description of the organism in question. The reproductive components, in particular, refer to the integrated process of procreation. The number of progeny an individual leaves depends on the individual's being recruited as a parent and on its fecundity as a parent, which in turn is reflected as the number of its gametes that form offspring. Sexual selection and fecundity selection, however, refer to very different biological processes. Sexual selection may emerge as the result of mate choice or competition for mates, even though not all deviations from randomness in mating result in sexual selection. Fecundity selection is the result of variation in the number of progeny, given the matings. Gametic selection is a seemingly simple survival component, but it often occurs within adult individuals, and this makes it a rather complex component. Meiotic drive is usually viewed as gametic selection dependent on the parental origin of the gamete, whereas pollen or sperm competition is viewed as gametic selection possibly dependent on the recipient of the gamete. The four components describe and categorize the various aspects of natural selection, but this does not mean that every action of natural selection can be caught within this framework. Self- incompatibility alleles in plants, for instance, may cause fecundity selection, but their primary action may be viewed as gametic or sexual selection. The former is attributed to the incompatibility type of the pollen (or gametophyte) and the latter is attributed to the genotype in the style (or sporophyte). With a gametophytic self-incompatibility system, the selection in pollen is gametic selection, but the pollen type is the genotype of the donor plant for sporophytic self-incompatibility.

The four components may be described as quantitatively equivalent in an organism with nonoverlapping generations, but time becomes an issue already for annual organisms with an extended breeding period. Then variation in survival until maturity will have a larger effect than survival between successive breeding events, and so zygotic selection, in particular, needs to be extended to a description of survival of individuals between events of reproduction. Most organisms, however, have overlapping generations, and for these the reproductive components need to be specified as age dependent to describe the action of natural selection (Bodmer, 1968; Lewontin, 1974; Christiansen and Frydenberg, 1976). Other complications may occur, but the four basic components of selection are sufficient to illustrate the complexity necessary in the description of natural selection for an adequate determination of the induced genotypic selection.

The basic components of selection show quite different dynamical properties, although zygotic selection and sexual selection in simple models are very similar. As an example we may consider an autosomal locus with two alleles A_1 and A_2 in a population that reproduces by random mating. Three boxes summarize the dynamics when only one selection component acts on this locus, and these are zygotic selection (Box 8.1), fecundity selection (Box 8.2), and gametic selection (Box 8.3). The qualitative differences in the dynamics that are due to the action of the three components emphasize that separation of the description of natural selection into its components is necessary to arrive at an evolutionarily sufficient description. Formally, the simple selection models considered here may be combined into a fecundity selection model, but they cannot be fused into a model of zygotic selection unless the fecundity selection is of a very special nature. Simplification of the description of fecundity selection may be achieved if only females influence fecundity, or if additive or multiplicative interaction of males and females on fecundity exists (Box 8.4).

Boxes 8.1–8.3 correspond to very simple models of constant natural selection. All selection components may, and sexual selection usually does, depend on the composition of the population, in which case they are frequency dependent.

The general problems of analysis of data on the action of natural selection is covered in Manly (1985). Here we focus on the specific problems concerning the description of the various component of selection.

Characters

Certain characters, for instance offspring number, are obviously subject to natural selection, and as such must be an aspect of the compound character fitness. Fecundity selection is described in terms of this character. Other characters, for instance individual size, may be under natural selection in that they may have a direct influence on survival or fecundity. Yet other characters, for instance number of dorsal spots, may show rather haphazard variation, but may be

Box 8.1. Zygotic selection.

Suppose the zygotes of genotypes A_1A_1, A_1A_2, and A_2A_2 in both sexes survive from fusion to maturity according to the fixed values w_{11}, w_{12}, and w_{22}, respectively. Then the frequency p of A_1 increases to fixation when $w_{11} > w_{12}$ and $w_{12} > w_{22}$ (directional selection),

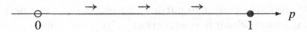

($\circ$, unstable equilibrium; $\bullet$, stable equilibrium) and it decreases to loss when $w_{11} < w_{12}$ and $w_{12} < w_{22}$. For $w_{11} > w_{12}$ and $w_{12} < w_{22}$ (underdominant selection), the frequent allele increases to fixation, and an unstable polymorphic equilibrium exists:

(we have used a symmetric model in which $w_{11} = w_{22}$ in the drawing). For $w_{11} < w_{12}$ and $w_{12} > w_{22}$ (overdominant selection) the fixation equilibria are both unstable and the population will end up polymorphic (again $w_{11} = w_{22}$ in the drawing):

When the survival values of females and males are different, then the same dynamical situations as in the above drawings result, plus two additional possibilities (Selgrade and Ziehe, 1987). First, two polymorphic equilibria, one stable and one unstable, exist, and one fixation equilibrium is stable the other unstable:

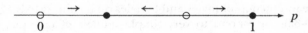

(and the symmetric case with alleles A_1 and A_2 interchanged). Second, three polymorphic equilibria, two stable and one unstable, exist, and the fixation equilibria are unstable:

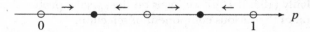

(again we have used a symmetric model in the drawing).

under selection if involved in mating behavior. The amount of acid phosphatase in red blood cells is a character whose variation may or may not have physiological effects that influence individual survival. It is an interesting question whether these characters are subject to natural selection or, to remain in a character framework, whether their variation is correlated to variation in fitness.

Box 8.2. Fecundity selection.

A very rich array of dynamical possibilities exists with fixed fecundities for the nine different pair combinations (Hadeler and Liberman, 1975; Feldman et al., 1983). In addition to those for zygotic selection (Box 8.1) a very simple model (symmetric, in that A_1A_1 and A_2A_2 have the same phenotype) provides the dynamics,

and a slightly more general symmetric model provides for stable limit cycles in the genotypic composition of the population (Hadeler and Liberman, 1975).

Box 8.3. Gametic selection.

Suppose the gametes survive from production to fusion according to the fixed values v_1 for A_1 and v_2 for A_2. Then the frequency of A_1 increases to fixation when $v_1 > v_2$ and decreases to loss when $v_1 < v_2$. Polymorphism is not possible, and only dynamics similar to zygotic directional selection (Box 8.1) occur.

A character (obviously) under natural selection is sometimes referred to as a fitness-related character, which is alright as long as the name is not used in place of evidence that natural selection is indeed working. A clear distinction, however, must be made among the survival component of fitness, say, and characters that influence the survival of the individual. A character influencing survival is subject to natural selection, whereas the survival component of fitness describes the action of natural selection.

A particularly interesting set of characters comprises those characters that define the characteristics of life and reproduction in a species, the life-history characters. The description of natural selection on a character necessarily refers to the life history of the organism, and so evolution of life-history characters is the evolution of the framework of natural selection. Among life-history characters, however, two classes usually emerge: those which in any circumstance should be viewed as components of a fitness description (they are the demographic parameters) and those that place the organism broadly in its biotic and abiotic environmental context. For instance, Stearns (1992) provides a list of principal life-history characters:

1. size at birth
2. growth pattern
3. age of maturity
4. size at maturity
5. number, size, and sex ratio of offspring

Box 8.4. Analysis of fecundity counts.

A simple description of fecundity selection is, by average fecundity counts,

$$
\begin{array}{cccc}
 & A_1A_1 & A_1A_2 & A_2A_2 \\
A_1A_1 & F_{11,11} & F_{11,12} & F_{11,22} \\
A_1A_2 & F_{12,11} & F_{12,12} & F_{12,22} \\
A_2A_2 & F_{22,11} & F_{22,12} & F_{22,22}
\end{array}
$$

Due to the sex symmetry at an autosomal locus, selection induced by these fecundities is equal to that of the fecundities:

$$
\begin{array}{cccc}
 & A_1A_1 & A_1A_2 & A_2A_2 \\
A_1A_1 & F_{11,11} & \frac{1}{2}(F_{11,12}+F_{12,11}) & \frac{1}{2}(F_{11,22}+F_{22,11}) \\
A_1A_2 & \frac{1}{2}(F_{11,12}+F_{12,11}) & F_{12,12} & \frac{1}{2}(F_{12,22}+F_{22,12}) \\
A_2A_2 & \frac{1}{2}(F_{11,22}+F_{22,11}) & \frac{1}{2}(F_{12,22}+F_{22,12}) & F_{22,22}
\end{array}
$$

This observation shows that additive fecundities where $F_{ij,k\ell} = U_{ij} + V_{k\ell}$ are dynamically equivalent to a unisexual fecundity where $F_{ij,k\ell} = U_{ij} + V_{ij}$ and that a unisexual fecundity $F_{ij,k\ell} = W_{ij}$ is dynamically equivalent to the additive model $F_{ij,k\ell} = \frac{1}{2}(W_{ij} + W_{k\ell})$ (Feldman et al., 1983). A multiplicative fecundity model, $F_{ij,k\ell} = u_{ij}v_{k\ell}$, is dynamically equivalent to a zygotic selection model with different selection in the two sexes (Bodmer, 1965). A unisexual or additive fecundity selection model therefore is dynamically equivalent to unisexual zygotic selection. Thus, if the effects on fecundity of the genotypes in the two sexes are additive or multiplicative, then fecundity selection is dynamically equivalent to a zygotic selection model with different selection in the two sexes, and in this sense the action of fecundity selection and zygotic selection may be described by a model of individual fitnesses. On the other hand, when the action of fecundity selection may be described by a model of selection on the individual, then the fecundities are dynamically equivalent to a multiplicative fecundity model.

6. age- and size-specific reproductive investment
7. age- and size-specific mortality schedules
8. length of life

Only some of these life-history characters specifically identify fitness components. The fecundity component is contained in item 5, but size and sex ratio are not fitness components. The survival component is included in item 7 as mortality. The reference to age is related to the temporal description of the components of selection in age-structured populations. Age-specific survival and reproduction of the sexes comprise the demographic description of the life history. The age at maturity and the length of life (items 3 and 8) are characteristics derived from the demographic description. The demography and the characteristics in items 1, 2, and 4 describe in very general terms the kind of organism under study. The size-related features in the list could be under natural selection and therefore have important effects on fitness components. For instance, a very good generalization in insects is that female fecundity is positively correlated with size.

Some of the characters in Stearn's list are important as supposed mediators of interaction between the various life-history characters. Size-related constraints, however, are not the only sources for correlations in traits among the life-history characters, and other currencies of trade-offs are possible. In the examples of trade-offs provided by Stearns (1992):

- current reproduction and survival
- current reproduction and future reproduction
- number, size, and sex of offspring

the first two may as well be a reflection of a higher exposure of reproducing individuals to predators.

Our current interest is the discussion of fitness and natural selection, and we focus on the demographic description of the life history of the organism. The basic demographic parameters of the life history are the age-dependent survival and fecundity, and variation in these parameters causes variation in fitness. Genetic influence on these parameters causes natural selection on the genotypes, and this may cause them to evolve. The interesting question of life-history evolution is how the demographic characteristics will change as a result of natural selection. The effect of the environment on fitness is described by the life-history parameters, and so the main concern of the models is to formulate reasonable correlations or trade-offs among genetics effects on the various parameters to avoid the trivial result that survival and fecundity should increase. The reference models of these deliberations are the continuous-time Euler–Lotka demographic models or the discrete-time Leslie model, and these models are the reference models for selection in age-structured populations (Norton, 1928; Charlesworth, 1980).

These classical demographic models have been extended to cover more realistic situations; for instance, in recent years the treatment of stochastic models has advanced significantly (Tuljapurkar, 1990) and the consideration of stage-structured models has added to the applicability of the population models (Metz and Diekmann, 1986). The models, however, have their origin in the Malthusian model of population growth, which treats both survival and reproduction as individual characteristics, and so they generally equate fitness with population growth rate or the analog corresponding to the demographic description of individuals of a certain genotype, that is, the rate which emerges as the solution to the characteristic Euler equation (Norton, 1928; Fisher, 1930). The only interaction among individuals usually included in these kinds of models is density dependence of individual survival and reproduction, a kind of interaction which is usually quite trivial (Fisher, 1930, 1958; Kostitzin, 1936, 1938). The processes of mating and reproduction of pairs of individuals signify interactions among individuals which are hard to handle in demographic models (Jagers, 1975).

Even though the models have to make simplifying assumptions to produce interesting results, this does not legitimate the assumption that interactions among individuals are unimportant in nature. Therefore investigations of the evolution of life-history characteristics need to acknowledge the complexity in the levels of genetic organization during the life cycle, a complexity which is assumed absent in the classical models of age-structured populations. Thus the Euler rate is a measure of fitness only in the simplified theoretical models. In a natural population, fitness in an age- or stage-structured population is a compound character for which the four basic fitness components have to be described with reference to the age or the stage structure.

The process of natural selection does not necessarily benefit the population or the species, as is frequently implied. A general population mean fitness that increases with time cannot be found, because fecundity selection allows for stable limit cycles in the genotypic composition of the population (Hadeler and Liberman, 1975). Sexual selection can result in bizarre phenomena such as traumatic insemination in Cimicid bugs. In population dynamics, variations in individual female fecundity can drive a population into complex limit cycles or chaos (Prout and McChesney, 1985).

Spurious frequency dependence of estimates of fitness values may result from an incomplete design of observations due to neglect of the complexity of fitness (Prout, 1965). Prout (1971a, 1971b) described selection on a pair of pseudoalleles on chromosome IV in *Drosophila melanogaster* and revealed differences in survival among the genotypes, fecundity selection, and sexual selection. The zygotic selection component acted differently in the two sexes, whereas fecundity depended on only female genotype. Sexual selection worked on males only, but the genotypic composition of the sire population depended on the genotype of the female. In this case the action of natural selection cannot be described by one fitness parameter for individual females and one for individual males. Other observations in experimental populations or *Drosophila* have yielded a similar richness in the action of natural selection (see Hedrick and Murray, 1983), and the effect of incomplete design may easily be illustrated in experimental populations (Bundgaard and Christiansen, 1972). These observations and observations from other organisms show clearly that the description of natural selection in natural and experimental populations cannot be founded on simplifying assumptions about the action of selection.

Natural selection and fitness are common words in the evolutionary literature, and they are frequently used in a very abstract, even ritualistic, way. The field would benefit from a custom to identify the aspects of fitness that a particular trait affects or might affect. The aspects of fitness should be specified in terms of the effect on the probability of survival through a life stage or the effect on the number of offspring (or number of mates). Characters such as shape, size, color, and behavior are not components of the character fitness, but they could have evolved to their present phenotype only through the action of natural selection. The complexity of natural selection on phenotypic variation and the indirect nature of biological inheritance necessitate the description of natural selection and fitness in terms of the basic components, that is, if the aim is to understand the process of evolution.

REFERENCES

Bodmer, W. F. 1965. Differential fertility in population genetic models. *Genetics* **51**: 411–424.

Bodmer, W. F. 1968. Demographic approaches to the measurement of differential selection in human populations. *Proc. Natl. Acad. Sci. USA* **59**:690–699.

Bundgaard, J. and Christiansen, F. B. 1972. Dynamics of polymorphisms. I. Selection components in an experimental population of *Drosophila melanogaster*. *Genetics* **71**: 439–460.

Charlesworth, B. 1980. *Evolution in Age-Structured Populations*. Cambridge: Cambridge U. Press.

Christiansen, F. B. and Frydenberg, O. 1973. Selection component analysis of natural polymorphisms using population samples including mother–offspring combinations. *Theor. Popul. Biol.* **4**:425–445.

Christiansen, F. B. and Frydenberg, O. 1976. Selection component analysis of natural polymorphisms using mother-offspring samples of successive cohorts. In *Population Genetics and Ecology*, S. Karlin and E. Nevo, eds., pp. 277–301. New York: Academic.

Darwin, C. 1872. *The Origin of Species*, 6th ed. London: Murray.

Feldman, M. W., Christiansen, F. B., and Liberman, U. 1983. One some models of fertility selection. *Genetics* **105**:1003–1010.

Fisher, R. A. 1930. *The Genetical Theory of Natural Selection*. Oxford: Clarendon.

Fisher, R. A. 1958. *The Genetical Theory of Natural Selection*, 2nd ed. New York: Dover.

Hadeler, K. P. and Liberman, U. 1975. Selection models with fertility differences. *J. Math. Biol.* **2**:19–32.

Hedrick, P. W. and Murray, E. 1983. Selection and measures of fitness. In *The Genetics and Biology of Drosophila*, M. Ashburner, H. L. Carson, and J. N. Thomson, eds., Vol. 3d, pp. 61–104. London: Academic.

Jagers, P. 1975. *Branching Processes with Biological Applications*. London: Wiley.

Kostitzin, V. A. 1936. Équation différentielles generales du problème de sélection naturelle. *C. R. Acad. Sci. Paris* **203**:156–157.

Kostitzin, V. A. 1938. Sur les coefficients mendélians d'hérédité. *C. R. Acad. Sci. Paris* **206**:883–885.

Lewontin, R. C. 1974. *The Genetic Basis of Evolutionary Change*. New York: Columbia U. Press.

Manly, B. F. J. 1985. *The Statistics of Natural Selection on Animal Populations*. London: Chapman & Hall.

Metz, J. A. J. and Diekmann, O., eds. 1986. *The Dynamics of Physiologically Structured Populations*, Vol. 68 of Lecture Notes in Biomathematics Series. Berlin: Springer-Verlag.

Norton, H. T. J. 1928. Natural selection and Mendelian variation. *Proc. London. Math. Soc. Ser. 2* **28**:1–45.

Prout, T. 1965. The estimation of fitness from genotypic frequencies. *Ecology* **19**:546–551.

Prout, T. 1971a. The relation between fitness components and population prediction in *Drosophila*. I. The estimation of fitness components. *Genetics* **68**:127–149.

Prout, T. 1971b. The relation between fitness components and population prediction in *Drosophila*. II. Population prediction. *Genetics* **68**:151–167.

Prout, T. and McChesney, F. 1985. Competition among immatures affects their adult fertility: population dynamics. *Am. Nat.* **126**:521–558.

Selgrade, J. F. and Ziehe, M. 1987. Convergence to equilibrium in a genetic model with differential viability between the sexes. *J. Math. Biol.* **25**:477–490.

Stearns, S. C. 1992. *The Evolution of Life Histories*. Oxford: Oxford U. Press.

Tuljapurkar, S. 1990. *Population Dynamics in Variable Environments*, Vol. 85 of Lecture Notes in Biomathematics Series. Berlin: Springer-Verlag.

CHAPTER NINE

How Well Does Opposing Selection Maintain Variation?

TIMOTHY PROUT

9.1. Introduction

The extensive genetic variation in natural populations is one of the most important, largely unsolved questions of population genetics.

This chapter, which is about the effects of opposing selection, is prompted by the frequent statements in the literature that opposing selection results in stable polymorphisms, and this is frequently invoked to explain the variation observed. There are four outcomes of opposing forces: (1) one is stronger than the other and wins – directional selection; (2) there is some kind of stalemate – a stable equilibrium; (3) either one might win, depending on which starts first, – unstable equilibrium; and (4) the forces somehow cancel – neutrality, which is of little interest here. However, net neutrality is often implied (see Prout, 1967).

This chapter consists of a series of very simple models involving different modes of opposing selection and an objective evaluation of the possible outcomes of opposing selection. Intrinsic heterozygote advantage is excluded.

This is intended for nontheoreticians. The source of the theory is cited, but only the conditions are shown explicitly. However, the pattern of selection in each case is diagrammed and the parameter space is mapped so that the areas of possible outcomes can be assessed visually. Five very simple, mostly single-locus cases will be considered: (1) a Levene model of spatial variation; (2) a simple case of temporal variation; (3) opposing selection between the sexes; (4) antagonistic pleiotropy, or trade-offs; and (5) meiotic drive.

It will be seen that, except in special cases, a stable equilibrium is the most unlikely outcome, and the message is that more intensive studies of each observed case should be pursued.

R. S. Singh and C. B. Krimbas, eds., *Evolutionary Genetics: From Molecules to Morphology*, vol. 1. © Cambridge University Press 2000. Printed in the United States of America. ISBN 0-521-57123-5. All rights reserved.

9.2. Spatial Variation

Since opposing selection implies two causes of selection in opposite directions, the environmental Levene (1953) model will be reduced to the simple situation of just two localities, as in Hoekstra et al. (1985). This is the simplest form of genotype by environment interaction. Levene's two niches will be in equal abundance, which maximizes the possibility of an equilibrium standoff. Such a situation is not completely unlikely where, for instance, there are two adjoining habitats such as a forest and a field. The application of the Levene model would require that members of some insect species mate at random, but half lay eggs in the forest and half in the field. There are many adjoining habitats where such a situation might occur.

The simplest opposing direction of selection is illustrated by additive fitnesses as follows:

Frequency of Niches	A_1A_1		A_1A_2		A_2A_2
1/2	1	>	$\dfrac{1+W}{2}$	>	W
1/2	V	<	$\dfrac{1+V}{2}$	<	1

The Levene harmonic mean conditions result in the conditions

$$W < \frac{1}{2-V}, \tag{9.1}$$

$$W > \frac{2V-1}{V}. \tag{9.2}$$

Figure 9.1 is a graphical representation of these conditions over a reasonable range of fitnesses. The shaded areas are the regions satisfying the equilibrium conditions, and the unshaded areas indicate directional selection. The areas here and below are referred to as the likelihood of the combination of the fitness sets occurring. First note that if the fitnesses in different pairs of locations range from equality ($V = W = 1$) to the extremes of $V = 2$, $W = 2$ or down to $V = 0$, $W = 0$, then the likelihoods are 0.38 for equilibria and 0.61 for directional selection in the upper right or lower left quadrants. Now consider the result of less intense selection shown in Fig. 9.2, where $0.5 \leq V \leq 1$ (equivalent to $1 \leq V \leq 1.5$). The squares show diminishing areas as V increases by increments of 0.10 to neutrality at $V = 1$. These diminishing areas are given by p on the right. This shows that the likelihood of directional selection is very high, ranging from 0.750 to 0.938. So the likelihood of obtaining an equilibrium becomes very small with the weakening of the intensity of opposing selection. Weaker selection differences may be more likely in nature.

In this additive case, the fitnesses among the three genotypes are constrained by having the heterozygote exactly intermediate, which restricts the possible outcome of opposing selection.

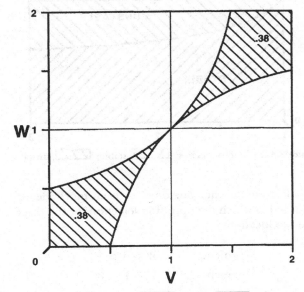

Figure 9.1. Two-niche space: ⟨⟨⟨⟩, stable; ⟨⟩, directional.

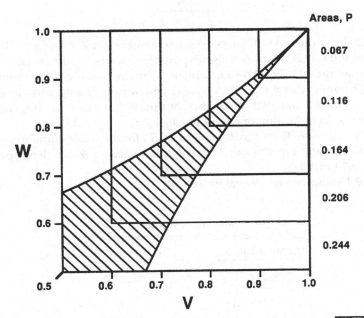

Figure 9.2. Two-niche space: Weak selection, 0.5^2V, W^21. p is areas of squares. ⟨⟨⟨⟩, stable; ⟨⟩, directional.

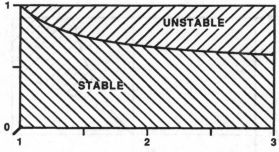

Figure 9.3. Symmetric niches: ⬛, stable; ▨, unstable.

In order to illustrate another outcome, an even more idealized two-niche model is considered in which the opposing selection for the three genotypes is reversed in the two locations:

$$\text{niche 1,} \qquad W > 1 > V;$$
$$\text{niche 2,} \qquad V < 1 < W.$$

The harmonic mean fitness criterion for a stable equilibrium results in

$$W < \frac{V}{2V - 1}. \tag{9.3}$$

Figure 9.3 shows the relevant part of this space where $0 < V < 1$ and $1 < W < 3$. The figure shows that there are only two possible outcomes, either a stable equilibrium or an unstable equilibrium, and, incidentally, because of the symmetry the equilibrium is always 0.50. An unstable equilibrium is a situation in which there is an equilibrium allele frequency, but perturbation in one direction or the other results in fixation or loss of the allele.

When W increases from 1 to 2, the area is 0.77 for the stable equilibrium and 0.23 for the unstable equilibrium. For an extreme value of $W = 3$, the respective areas are 0.70 and 0.30.

Table 9.1 shows two numerical examples.

Table 9.1. *Fitnesses*

Location Pair 1			Location Pair 2		
Stable			*Unstable*		
1.5	1	0.70	1.5	1	0.80
0.70	1	1.5	0.80	1	1.5

This shows that if a population colonized a new area where, for environmental reasons, the inferior type (0.70) now became less inferior (0.80), then this immigrant population would become unstable and fix one or the other alleles. Or if two different locations were colonized in this area, then one might be fixed for one allele and the other fixed for the alternative allele.

This very special case has only two possible outcomes, a stable or an unstable equilibrium. The latter is not possible in the additive case. The purpose is to show that more general models of spatial variation giving unstable equilibria are possible. This will be shown below.

The general purpose of these contrived cases is to show that opposing selection caused by variable spatial environments can have three outcomes: a stable equilibrium, an unstable equilibrium, or directional selection. Also, the additive case shows that when environmental effects are weak, the likelihood of a stable equilibrium is very small and directional selection is much more likely.

There are a very small number of documented cases involving simple genetics and opposing selection. This section on opposing environmental effects concludes with a brief summary of and conclusions from data for such an unusual case. The case involves color and pattern polymorphism in the Leaf Hopper, *Mocydia crocea*, which is highly polymorphic in Eastern Europe. The following is summarized from Prout and Savolainen (1996).

The phenotypes and genotypes are shown in Figure 9.4. The genetics is relatively simple with two loci, one with two alleles with dominance and the other with three alleles and hierarchical dominance. The phenotypes are expressed in the early nymphs so that Müller (1987) was able to determine the viability by placing contrived mixtures on 20 different species of host plants. Figure 9.5 shows schematically the viability of the six phenotypes across a random order of the 20 host plants.

Opposing selection, or genotype by environmental interaction, is clearly demonstrated when pairs of hosts are compared. This is an extension of the above cases of two environments to 20 and an extension of the genetics from one locus with two alleles to two loci and one with two and the other with three alleles. There is no theory available to analyze such genetics. In fact, there are very few cases in which the Levene model (one locus, two alleles) might have been applied but was not applied for practical reasons (e.g., Clark et al., 1963, and Smith, 1993). So a simulation analysis was done. The result was that the six phenotypes were reduced to two, namely, two banded P2- and striped pp on a homozygous background of red YY. (This may be the only data set conforming to the one-locus two-allele Levene model conditions for dominant–recessive polymorphisms. For the record, these data are shown with the conditions in Table 9.2). It is worth noting that by simulating with subsets of hosts, one subset of four hosts could give a two-locus stable equilibrium, showing a real set of environments that can in fact result in a stable equilibrium for a real case with more complex genetics. However, the six phenotypes in this Leaf Hopper case are in fact polymorphic in many localities in Eastern Europe, which clearly shows

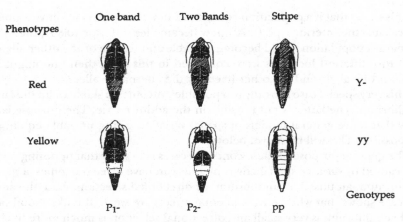

Figure 9.4. Leaf Hopper genotypes and phenotypes.

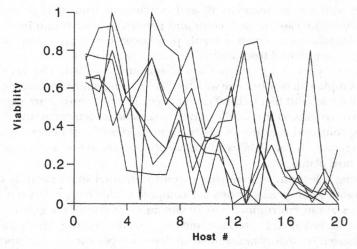

Figure 9.5. Viabilities of the six phenotypes across 20 species of host plants.

that there are additional factors involved. Apart from hidden heterosis, the most likely reason is uneven distribution of hosts and population subdivision with limited migration or metapopulation structure. This possibility is clearly tractable to investigation.

This is the message of this chapter, which will be repeated. When genetic variation is observed and environmental variation is the suspected cause, additional investigation is required for confirming the proposed reason. Population subdivision with limited migration is always a possibility. The robustness of limited migration for maintaining variation does not require theoretical elaboration here.

Table 9.2. *Leaf Hopper Viabilities Raised on 18 Host Plants**
(see Prout and Savolainen, 1996)

Host	Red Two Bands Y−, P_2−	Red Striped Y−, pp	Reciprocal
1	1	0.818	1.224
2	1	1.349	0.741
3	1	0.80	1.250
4	1	0.718	1.393
5	1	23.00	0.043
6	1	0.35	2.857
7	1	0.28	2.967
8	1	0.649	1.541
9	1	0.567	1.764
10	1	5.667	0.176
11	1	1.031	0.969
14	1	1.55	0.645
15	1	0.162	6.173
16	1	0.075	13.333
17	1	1.25	0.800
18	1	0.143	6.993
19	1	2.40	0.417
20	1	0.071	14.085
N = 18	Mean	2.2711	3.2208
	Harmonic mean		0.3105

* Red striped normalized to red two banded. Host plants 12 and 13 were omitted because of zero viabilities. These viabilities conform to the Levene model case for a recessive (Prout, 1968), arithmetic mean >1 and harmonic mean <1.

9.3. Temporal Variation

A very simple case for temporal variation is the Haldane–Jayakar (1963) theory for temporal variation of the fitness of a simple recessive (Crow and Kimura, 1970, p. 281). When there are only two alternative environments resulting in fitnesses W and V, the conditions for a stable equilibrium are

$$2 - V < W < 1/V.$$

This relation is shown in Fig. 9.6 (see Hedrick 1995, p. 211, Fig. 6.4). The region giving a stable equilibrium is small, 0.133 of the total space shown. In

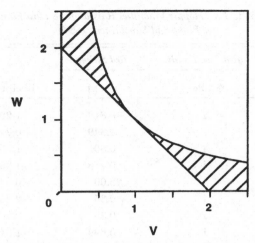

Figure 9.6. Temporal variation: Haldane-Jayakar recessive model with two fitnesses, W and V. Geometric mean <1, arithmetic mean >1. ▨▨▨, stable; ▭▭▭, directional.

fact, temporal variation is more restrictive than spatial variation. For the recessive case with two equal niches, the area for a stable equilibrium is 0.186 for the same total space in Fig. 9.6. This relation is shown in Hedrick (1986), Fig. 6.4. There are, of course, much more general studies of the theory of cycling (e.g., Hoeckstra, 1975).

The point here is to show again that the most likely outcome of opposing selection is directional selection, as shown for the spatial case in Fig. 9.1.

9.4. Opposing Selection between the Sexes

This section concerns simple opposing selection between the two sexes. This subject of sex-dependent selection will further be developed in Section 9.6. Specific biological examples of opposing selection between the sexes are not obvious, but, in general, such selection could help to explain sexual dimorphism as well as polymorphism. This kind of opposing selection has a strong link with opposing selection between environments, as will be noted below, but biologically these kinds of selection are virtually unrelated. Although spatial environmental effects as a means of maintaining variation are of much greater interest, this sex dependence is possible and worth pursuing.

For example, color brightness in the two sexes could have opposite effects in the two sexes. In the males, in spite of increased predation, there is net higher fitness of bright color because of mating success, while in the females predation alone will reduce the fitness of those with brighter color. In a dioecious plant more flowers will increase the fitness in males by producing more pollen while in the females of some species more flowers might reduce the nutrient in the seeds, and if the seeds drop directly to the ground, high density could reduce the seedling survival rate.

The theoretical conditions for a stable equilibrium in the sex-dependent case are the same harmonic mean conditions for the two equally abundant Levene niches (Hartl and Clark, 1989, p. 179). This is not obvious, but not surprising, because the two sexes could be regarded as two equally abundant environments. Also the two-sex case, of course, has none of the kinds of limitations as does the contrived two-niche case.

The parameters for additive effects are

	A_1A_1		A_1A_2		A_2A_2
female	F	$<$	$\dfrac{1+F}{2}$	$<$	1
male	1	$>$	$\dfrac{1+V}{2}$	$>$	V

and the conditions are

$$F < \frac{1}{2-V}, \tag{9.4}$$

$$F > \frac{2V-1}{V}, \tag{9.5}$$

which are identical to the conditions in Section 9.3 with F substituted for W, and the spaces in Figs. 9.1–9.3 apply.

In this section the effects of dominance are developed, all of which apply to the two-niche model. It is important to note that dominance could result from behavior or ecology on what could be an underlying additive expression. For example, in the mating–predation example above, among the three genotypes the heterozygotes might be more similar to the most colorful homozygote in the male's mating success and the predation on females. In the dioecious plant, dominance might be reversed, with the heterozygotes being more similar to the low-pollen and low-seeds homozygote.

Two patterns of dominance are discussed: parallel, or normal, dominance and dominance reversal.

For parallel dominance the parameters are

A_1A_1		A_1A_2		A_2A_2
1	$>$	$1-h_1(1-F)$	$>$	F
V	$<$	$1-(1-h_2)(1-V)$	$<$	1

and the conditions are

$$F > \frac{V(h_1+h_2)-h_2}{Vh_1}, \tag{9.6}$$

$$F < \frac{(1-h_1)}{1-h_1+(1-h_2)(1-V)}. \tag{9.7}$$

The effects of these dominance parameters can be seen by letting $h_1 = h_2 = 1$ and $h_1 = h_2 = 0$, as follows:

	A_1A_1		A_1A_2		A_2A_2
$h_1 = h_2 = 1$	1	>	F	=	F
	V	<	1	=	1

in which case A_2 is dominant, or

	A_1A_1		A_1A_2		A_2A_2
	1	=	1	>	F
$h_1 = h_2 = 0$	V	=	V	<	1

in which case A_1 is dominant.

A surprising result is that if h_1 and h_2 are equal, denoted by h, it can be seen that substituting h into conditions (9.6) and (9.7) results in,

$$F < \frac{1}{2 - V},$$

$$F > \frac{2V - 1}{V}.$$

Dominance disappears and the conditions become conditions (9.4) and (9.5) for the additive case, independent of the true degree of dominance. (Kidwell et al., 1977, p. 180).

Interesting results occur when h_1 and h_2 are in the same direction (parallel) but are of different magnitude. Consider the case in which they are both large, resulting in A_2 dominance. The midpoint between 0.5 and 1 is 0.75, and if $h_1 = 0.75 + 0.15 = 0.90$ and $h_2 = 0.75 - 0.15 = 0.60$, the result is as shown in Fig. 9.7(a). An unstable region appears, and the stable region is reduced compared with Fig. 9.1, which is the result when $h_1 = h_2$. When h_1 and h_2 are small, favoring A_1 dominance, the midpoint between 0 and 0.5 is 0.25. When $h_1 = 0.25 - 0.15 = 0.10$ and $h_2 = 0.25 + 0.15 = 0.40$, the result is as shown in Fig. 9.7(b), showing a large increase in the stable area compared with Fig. 9.7(a) and especially Fig. 9.1.

These results can be understood by considering two limiting cases. First, for the large values, A_2 dominant, the limiting differences are $h_1 = 1$ and $h_2 = 0.50$, and the second, for the small values, A_1 dominant, the differences are $h_1 = 0$ and $h_2 = 0.50$:

		A_1A_1		A_1A_2		A_2A_2
A_1 dominant	females	1	>	F	=	F
$h_1 = 1, h_2 = 0.5$	males	V	<	$\dfrac{1 + V}{2}$	<	1
A_2 dominant	females	1	=	1	>	F
$h_1 = 0, h_2 = 0.50$	males	V	<	$\dfrac{1 + V}{2}$	<	1

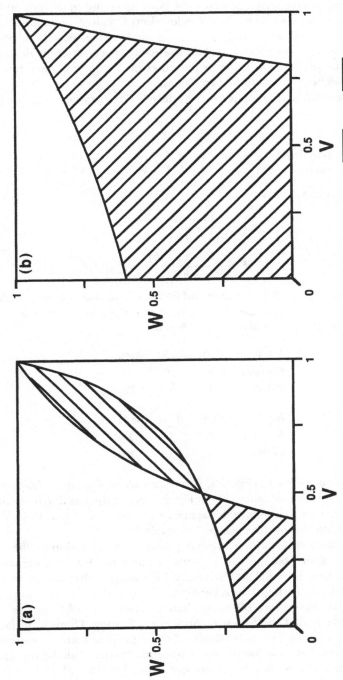

Figure 9.7. Opposing sexes (or niches), parallel dominance. (a) $h_1 = 0.90$, $h_2 = 0.60$; (b) $h_1 = 0.10$, $h_2 = 0.40$. ☐, stable; ▨, unstable; ☐, directional.

The sexes can, of course, be reversed. In both cases the males are additive ($h_2 = 0.50$). It appears that when the deleterious female F is dominant, this produces the unstable space and reduces the stable space. When the favored female allele is dominant, this increases the stable space. There is no *a priori* reason why one would expect one or the other of these dominance reversals between the sexes or between the two niches.

Now the reversal of dominance is considered. The parameters are defined as

	A_1A_1		A_1A_2		A_2A_2
females	F	$<$	$1 - h(1 - F)$	$<$	1
males	1	$<$	$1 - h(1 - V)$	$<$	V

and the stability conditions are

$$F < \frac{(1 - h)}{(1 - hV)}, \tag{9.8}$$

$$F > \frac{[V - (1 - h)]}{hV}. \tag{9.9}$$

h measures the degree of dominance and $0 \le h \le 1$, so that selection is always directional in each sex or niche. The effect of h on dominance reversal can be shown by the limiting cases of $h = 1$ and $h = 0$:

$h = 1$	A_1A_1	A_1A_2	A_2A_2
females	F	F	1
males	1	V	V

$h = 0$	A_1A_1	A_1A_2	A_2A_2
females	F	1	1
males	1	1	V

Two example are shown in Fig. 9.8. The results are analogous to those of the parallel but unequal dominance case in Fig. 9.7. When the less fit types tend to be dominant there is a large unstable space and small stable space, while if the more fit type is dominant this enlarges the stable space.

This dominance reversal case inevitably leads to a discussion of the polymorphism for pattern baldness in humans in which the trait is dominant in males and recessive in females. The results here suggest that dominant bald males leave more offspring than recessive nonbald males, and the dominant nonbald females have more offspring than the recessive bald females. The assumption here is that the fitness component is fertility. There is, in fact, a polymorphism that appears to be stable. If these proposed fitnesses between sexes were reversed, the likelihood of a stable equilibrium would be very small. In fact, if dominance is complete, as it appears to be, then when $h = 0$ the whole space is stable and when $h = 1$ the whole space is unstable. Even if there

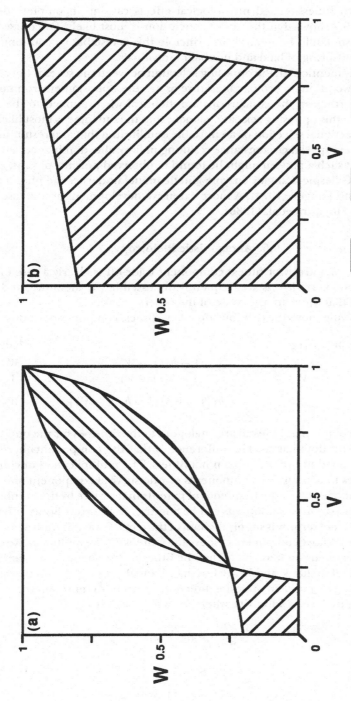

Figure 9.8. Opposing sexes (or niches), dominance reversal. (a) $h = 0.80$; (b) $h = 0.20$. ▨, stable; ▨, unstable; ☐, directional.

are unknown fitness-related physiological effects causing incomplete dominance, Fig. 9.7 shows that the above conclusion is most likely. Of course, the polymorphism could be neutral. In principle this is testable by counting the number of offspring of bald and nonbald individuals.

As already mentioned, the opposing selection between the sexes is not a subject of the evolutionary ecology literature as are the opposing environmental effects. Nevertheless, the conclusions are identical when applied to the two-niche case in this opposing sex case, namely, that the outcome is most likely to result in directional selection or an unstable equilibrium that also results in fixation. The very special circumstance resulting in favorable dominance reversal predicts that such a pattern might be found in observed polymorphisms, especially if the Gillespie–Langley enzyme kinetic model underlies the phenotypes (Gillespie and Langely, 1974). A more extensive development of this case can be found in Hoekstra et al. (1985).

9.5. Antagonistic Pleiotropy

Antagonistic pleiotropy or fitness component trade-off is clearly a case of opposing selection. Trade-offs are frequently discussed in the literature, and it is often stated that these are the cause of the observed variation.

The following shows the relationships of two fitness components for this case:

Component	A_1A_1		A_1A_2		A_2A_2
1	1	>	$1 - h_1(1 - W)$	>	W
2	V	>	$1 - h_2(1 - V)$	>	1
Net	V,		$[1 - h_1(1 - W)][1 - h_2(1 - V)]$,		W

The two component fitnesses are analogous to the fitnesses for the opposing sexes case with dominance. The difference is that the components here are simply multiplied in order to obtain net fitness. The multiplication of components applies to two survival components or one survival component and one fertility component, i.e., the frequency of parents that survive by the number of offspring per parent. Components can be additive, such as fertility at different ages, in which net fertility is simply the sum of the fertilities at different ages. The results of the additive component case are the same as for the multiplicative case, which will be discussed here. Note that the fitness components have been normalized to 1.00 so that V and W are less than 1 and $0 \leq h_1, h_2 \leq 1$, so that selection is always directional in opposite directions in order to represent trade-offs.

The limiting case of additivity, where $h_1 = h_2 = 1/2$, is

$$1 > \frac{1 + W}{2} > W,$$

$$V < \frac{V + 1}{2} < 1,$$

$$\text{Net} \quad V, \frac{(1 + W)(V + 1)}{4}, W.$$

The case of parallel dominance, $h_1 = 1$ and $h_2 = 0$ (or the reverse), is

	A_1A_1		A_1A_2		A_2A_2
	1	>	W	=	W
	V	<	1	=	1
Net	V		W	=	W

In this case of complete parallel dominance therefore, trade-offs will always result in directional selection, the direction determined by the relative magnitudes of V and W (a biological example with be discussed below).

There are two ways dominance can be reversed in the two components. If $h_1 = h_2 = 1$,

$$1 > W = W,$$
$$V = V < 1,$$
$$\text{Net } V > WV < W.$$

Since V and W are less than 1.00, this results in net underdominance or unstable equilibrium.

If $h_1 = h_2 = 0$,

$$1 = 1 > W,$$
$$V < 1 = 1,$$
$$\text{Net } V < 1 > W.$$

This is overdominance that gives a stable equilibrium.

The additive case is discussed first and then the underdominance and the overdominance cases follow.

For the additive case, net overdominance is necessary for a stable equilibrium. The conditions are

$$\frac{(1 + W)(1 + V)}{4} > V \quad \text{and} \quad W.$$

resulting in

$$W > \frac{3V - 1}{1 + V}, \tag{9.10}$$

$$W < \frac{1 + V}{3 - V}. \tag{9.11}$$

Figure 9.9(a) shows the space for $0 \le V, W \le 1$. This figure is analogous to Fig. 9.1 for the two-niche cases. The stable area here is somewhat smaller than the two-niche case with stable equilibrium area 0.24 and a directional selection of 0.76. It is even more plausible in this case that weaker selection is more likely because fitness less than 0.50 suggests that the two components are approaching lethality and sterility. Such extreme trade-offs are uncommon. If

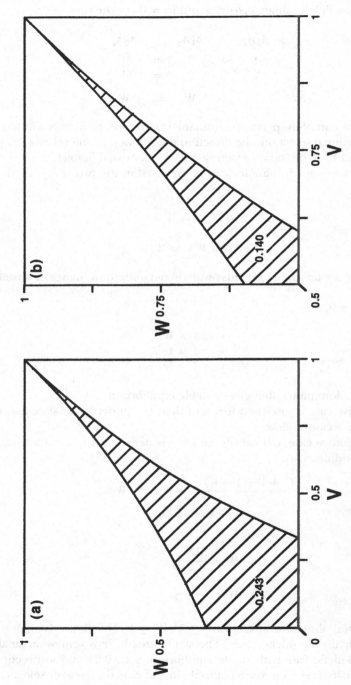

Figure 9.9. Trade-off space with additive phenotypes ($h = 1/2$). (a) 0^2V, W^21; (b) trade-off with additive phenotypes but weaker selection, 0.50^2W, V^21. ▨, stable; ☐, directional.

the fitness components are restricted to $1 \geq V, W \geq 0.50$, Fig. 9.9(b) shows that the equilibrium area is 0.14 and the directional selection 0.96.

Parallel dominance can be formulated with a single parameter h, as follows:

Component

1	1	$>$	$1 - (1-h)(1-W)$	$>$	W
2	V	$<$	$1 - h(1-V)$	$<$	1
Net			$V, [1 - (1-h)(1-W)][1-h(1-V)], W$		

When $h = 1$ the result is the same as the parallel case above when $h_1 = h_2 = 1$ or 0.

The conditions for net overdominance, when $0 < h < 1$, are

$$W > \frac{V - h(1-V)}{1 - h(1-V)}, \qquad (9.12)$$

$$W < \frac{1 - h(1-V)}{2 - V - h(1-V)}. \qquad (9.13)$$

An example is shown in Fig. 9.10, where $h = 0.30$. The area is not reduced very much from that shown in Fig. 9.9. The mean area was calculated over values of h for $0.10 < h < 0.90$, giving a mean area of 0.224 and, for less extreme selection values, $0.5 < W, V < 1$, where the mean area is 0.109. The

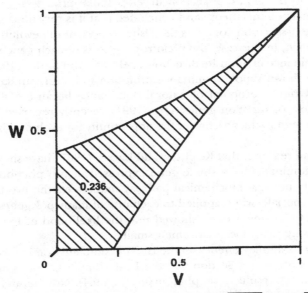

Figure 9.10. Trade-off with parallel dominance of $h = 0.30$. ▨▨▨, stable; ▭▭▭, directional.

extreme values of $0 < h < 0.10$ and $1 > h > 0.99$ were not included because these can give extremely weak equilibria, approaching neutrality.

Situations in which there is dominance reversal are conceivable. One could imagine a not uncommon situation in plants in which seedlings could either grow into small plants with many flowers but small size or grow into larger plants with less flowers. In the first case this gives high fertility (pollen and ovules) but poor survival because of the small size, and in the second case, there is high survival but lower fertility. Then among the three genotypes in some cases the low fertility and survival might be largely expressing in the homozygote, giving overdominance or the opposite giving underdominance.

This discussion of trade-offs so far has considered only selection on single-locus two-allele genotypes, while nearly all of the cases reporting such trade-offs are for quantitative characters. The following is a brief discussion of trade-offs in quantitative characters taken from Curtsinger et al. (1994).

It can be easily shown that a fundamental property of trade-off selection on quantitative characters is that it results in an overall optimum or stabilizing selection (Curtsinger et al., 1994, p. 224, Fig. 6), which, all else being equal, tends toward fixation of alternate polygenic alleles and does not maintain variation without recurrent mutation (Caballero and Keightley, 1994). However, it has been suggested that there might be reversal of dominance at each of the many polygenic loci so that, as shown in Section 9.4 in the table following condition (9.9), each one is overdominant, resulting in the variation's being maintained by stabilizing selection. This dominance reversal is very different from the special ecological situations suggested above. In this case, the dominance reversal would refer to the genes' effects on developmental processes. Curtsinger et al. (1994) analyzed this theory and concluded that it is very unlikely for three reasons. The first reason concerns the Fisher theory of the evolution of dominance in which, in this case, the pleiotropic effects of each gene would have evolved dominance in opposite directions. Feldman and Karlin (1971), among others, have shown simple dominance evolution to be very unlikely. Also, in support of Wright's proposal of natural dominance because of the enzyme kinetics in loss of function genes, Orr (1991) recently reported that dominance is common in the absence of the opportunity for evolution in a haploid alga.

The second reason is that Keightley and Kacser (1987) have shown that the reversal of dominance of a single gene expressed in two phenotypes would require a very unusual biochemical pathway that affects the two phenotypes in this way, especially when applied to each one of many polygenes. The final reason is that Curtsinger et al. showed that the likelihood of the parameter space's involving many loci is extremely small.

Returning to simple genetics, in the case of complete dominance there are only two phenotypes. In Section 9.2, the Leaf Hopper case was discussed in which dominance results in six phenotypes (Fig. 9.4). Not mentioned then is the fact that, apart from variation in survival on the host plants, there is overall nonhost-related variation among the phenotypes in female fecundity, which was, of course, included in the simulations.

Table 9.3.

	$Y-, P_1$	$Y-, P_2$	$Y-, pp$	yyP_1-	$, yyP_2-$	$yp\ pp$
Fecundity, F	0.564	0.653	0.871	0.883	1.00	0.842
Viability, V	1.430	1.015	0.745	0.582	1.00	0.77
Net $F \times V$	0.806	0.663	0.648	0.513	1.00	0.648

If the host variation is ignored and mean viabilities are compared with fecundity, the results show a trade-off relationship among the six phenotypes.

Table 9.3 shows the fecundity and the viabilities both normalized to yyP_2-. It can be seen that there is a trade-off between yyP_2- with the highest fecundity and $Y-P_1-$ with the highest viability. However, yyP_2- has the highest net fitness. The point here is that this system has much more complex genetics than the simple three genotype cases discussed so far, which with dominance has only two phenotypes. In this case, with complete dominance there are six phenotypes with all four single-locus heterozygotes masked by dominance. Nevertheless, just one of the phenotypes has the highest net fitness and, ignoring the host effects, there would be directional selection toward homozygosity for yy P_2P_2.

This therefore is a simpler example of trade-offs that result in an optimum, analogous to stabilizing selection on a quantitative character.

9.6. Sex-Limited Trade-offs

This Leaf Hopper example in which only females showed trade-offs and males were subject to directional selection (ignoring host effects) suggests that trade-offs can be sex limited, which requires a return to the subject of opposing selection between the sexes.

The trade-offs frequently reported refer to fecundity in females as in the Leaf Hopper, which may be due to fundamental differences in genetic effects on reproductive biology of the two sexes. Also there are trade-offs in males that have evolved by sexual selection, gaudy displays (or mating calls) that subject them to predation. There are cases in birds in which such bright display characters are expressed in both sexes, such as the Paradise Kingfisher of Australia and New Guinea, perhaps representing a primitive stage in such evolution before sex-limiting modifiers have evolved.

A simple additive model of this sex-limited trade-off is

Male (or Female)	Female (or Male)

$$1 > \frac{1+F}{2} > F$$

$$V < \frac{1+V}{2} < 1 \qquad\qquad V < \frac{1+V}{2} < 1$$

$$\text{Net}\ \ V, \frac{(1+F)}{2}\frac{(1+V)}{2}, F \qquad V < \frac{1+V}{2} < 1$$

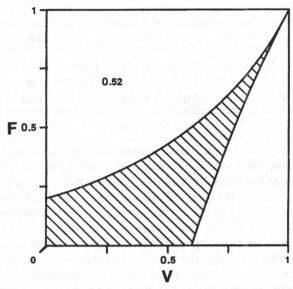

Figure 9.11. Sex-limited trade-off: trade-off in females, directional selection in males. ▨▨▨, stable; ▭, directional.

When the sex-limited selection criteria are applied as above, the conditions are

$$F < \frac{1+V}{5-3V}, \tag{9.14}$$

$$F > \frac{5V-3}{1+V}. \tag{9.15}$$

The result is shown in Fig. 9.11.

Compared with the simple sex-opposing selection of Fig. 9.1, the equilibrium areas of Fig. 9.11 are not symmetrical because of the directional selection on females (or males). The stable area is somewhat less, 0.266 compared with 0.38, and for total direction selection, 0.734 compared with 0.61. When dominance is introduced the results are analogous to the results of the opposing sex or niche case in which favorable dominance results in a very high likelihood of equilibrium compared with that of the favorable recessive.

It might be noted in Fig. 9.11 that the largest of the three spaces is directional selection against V (0.52), which, in the male trade-off case, may have occurred in the past, resulting in drab females and males that attract fewer biologists and others, including predators.

If trade-offs are limited to one sex, such as mating calls that also attract predators, the results are the same as for the sex-independent trade-offs in Fig. 9.9.

The equilibrium conditions are not known for quantitative genetic traits for which there is a net optimum in one sex and directional selection in the other. This is frequently reported.

9.7. Meiotic Drive

The final case of opposing selection to be considered is meiotic drive versus zygotic selection. Meiotic drive has been well documented in mice (t locus, Lewontin, 1968), *Drosophila melanogaster* (SD), and also sex ratio distortion (SR) in *D. pseudoobscura* and some mosquitoes.

This is selection, in which in male heterozygotes postmeiotic young spermatids compete, greatly favoring the one carrying the drive allele d, in which segregation values can be 0.9 to 0.99 compared with the expected Mendelian 0.50. The opposing selection is in the homozygotes, dd, which can be lethal, semilethal, or sterile. Homozygote, lethal dd will result in a stable equilibrium. For example, if the drive is 0.9, the equilibrium for the drive allele is 0.466 (Crow and Kimura, 1970, pp. 311–312). The actual frequencies in mice are lower than this, approximately 0.20, so it is suspected that the heterozygotes may have somewhat lower fitness.

An interesting property of this system is that if the dd has reduced fitness, but is not lethal and if the heterozygote has some reduced fitness, as suggested by the mouse data, then an unstable equilibrium is possible. For example, if the drive is 0.90, the fitnesses are

$$\begin{array}{ccc} DD & Dd & dd \\ 1 & 0.70 & 0.50 \end{array}$$

The unstable point $\hat{p} \cong 0.140$ so that if initial $p > 0.14$ then d runs to fixation, resulting in the fitness of the whole population being reduced to 0.50.

The space for four possible outcomes is shown Fig. 9.12 (see Appendix).

The fitness parameters refer to

$$\begin{array}{ccc} DD & Dd & dd \\ 1 & 1 - hs & 1 - s \end{array}$$

where s is the directional selection coefficient against the drive allele and h measures the degree of dominance. It can be seen that with moderately strong selection against dd and dominance greater than $h = 0.5$, the result is an unstable equilibrium. For the case of $k = 0.90$, most of this space results in directional selection for the drive allele d.

One might speculate that if an organism with strong selection against the homozygote dd, but only mildly deleterious heterozygotes were to colonize an archipelago where the environment caused the heterozygote to be more deleterious, and if the migrants went through a bottleneck to each island, then the islands would become differentiated by being fixed for the drive allele or for the nondrive allele. If crosses between islands were then made, the presence

178 *Timothy Prout*

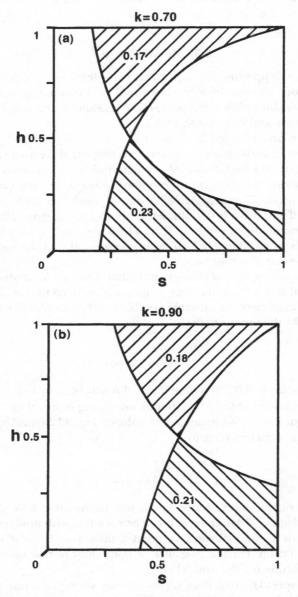

Figure 9.12. Meiotic drive parameter space: *k* is the drive parameter, *s* is the selection coefficient, and *h* is the dominance parameter. ⟨⟨⟨⟩⟩⟩, stable; ⟨⟨⟨⟩⟩⟩, unstable; ⟨⟨⟨⟩⟩⟩, directional.

of meiotic drive would be revealed. Crosses between races have been made that purportedly show drive, but the problem has been that the apparent segregation distortion was not distinguished from simple viability effects of the marker.

This final case results again in a large likelihood for directional selection and a substantial area for an unstable equilibrium. The term likelihood has to be used with caution. It simply represents our ignorance of the factors involved.

In this case of meiotic drive, a great deal is known about the mechanism so that further studies could be pursued that would result in a biological basis for the limits within the space shown here.

9.8. Conclusion

There are numerous discussions and reviews in the literature concerning the robustness of models, simple cases of which have been discussed here. Of special interest are environmental effects such as the Levene model and those which have evolved from it (Felsenstein, 1976; Hedrick, 1986; Hoekstra et al., 1985; Maynard Smith and Hoekstra, 1980). The object of this chapter is to start with opposing selection, with no intrinsic heterozygote advantage, and show the specific outcome of such selection. As already noted, this chapter was promoted by the very common verbal theory in the literature that opposing forces balance, resulting in balancing selection.

These simple one-locus cases show that directional selection is the most likely outcome, which is the most intuitive prediction when one force is stronger than the other. However, diploid Mendelism is involved so stable equilibria can occur. The very robust cases of reversal of favorable dominance require, as noted, very special and probably unusual ecological circumstances, and such relationships should be sought when polymorphisms are observed.

In every case discussed, unstable equilibrium appears. The only such cases that have been investigated are chromosomal races in which underdominant heterokaryotypes have been proposed as the cause, such as White's stasipatric speciation (White, 1978; also King, 1993). However, this has been subject to serious questions (Sites, 1995; Coyne, 1993; Coyne et al., 1997). Another very special case is alternative asymmetry in the snail Partula (Johnson, 1982). The unstable equilibrium that always appears in these cases discussed suggests the occasional occurrence of polytypic sets of isolated populations, which, when differences are observed, are usually attributed to unknown differences among environments. In principle, the unstable equilibrium possibility could be tested by perturbation experiments.

The message of these simple exercises has already been noted several times. When polymorphisms are observed, more intense study should be pursued to discover the reason for the polymorphism. One of the most robust cases for maintaining variation is opposing selection and population subdivision with limited migration, which can be investigated. Or, if migration is ruled out, additional data should be collected in an attempt to demonstrate conformation with the current theory predicting stable equilibrium. If it does not conform, then a new theory should be proposed. For example, in addition to restricted migration there are more hypothetical theories that are analogous to population subdivision and are very robust, such as habitat choice (Hedrick, 1990, and references therein) and sublines that are separated by timing of germination or diapause in a temporally varying environment (Hedrick, 1995).

One of the most important, if not the most important, question of population genetics is the cause of the extensive genetic variation found in a great many population studies. Abundant theory on this subject has accumulated, but empirical confirmation of this theory is very rare.

Appendix

The two curves in Fig. 9.12 are concave, up to the right,

$$h = \frac{2s - 2k + 1}{s(3 - 2k)},$$

and convex, down to the right,

$$h = \frac{2k - 1}{s(2k + 1)}.$$

The two curves intersect at

$$h = \frac{1}{2}, \quad s = \frac{4k - 2}{2k + 1}$$

REFERENCES

Caballero, A. and Keightley, P. D. 1994. A pleiotropc non additive model of variation in quantitative traits. *Genetics* **138**:883–900.

Clark, C. A., Dickson, C. G., and Shepard, P. M. 1963. Larval color pattern in *Papilio domodocus. Evolution* **17**:130–137.

Coyne, J. 1993. Speciation by chromosomes. *Trends Ecol. Evol.* **8**:76–77.

Coyne, J., Barton, N. H., and Turelli, M. 1997. A critique of Wright's shifting balance theory of evolution. *Evolution* **51**:643–671.

Crow, J. F. and Kimura, M. 1970. *An Introduction to Population Genetics Theory*. New York: Harper & Row.

Curtsinger, J., Service, P., and Prout, T. 1994. Antagonistic pleiotropy, reversal of dominance and genetic polymorphism. *Am. Nat.* **144**(2):210–228.

Feldman, M. W. and Karlin, S. 1971. The evolution of dominance: a direct approach through the theory of linkage and selection. *Theor. Popul. Biol.* **2**:482–492.

Felsenstein, J. 1976. The theoretical population genetics of variable selection and migration. *Annu. Rev. Genet.* **10**:253–280.

Gillespie, J. H. and Langley, C. H. 1974. A general model to account for enzyme variation in natural populations. *Genetics* **76**:837–884.

Haldane, J. B. S. and Jayakar, S. D. 1963. Polymorphism due to selection of varying direction. *J. Genetics* **58**:237–242.

Hartl, D. L. and Clark, A. G. 1989. *Principles of Population Genetics.* 2nd ed. Sunderland, MA: Sinauer.

Hedrick, P. W. 1983. *Genetics of populations.* New York: Van Nostrand and Reinhold.

Hedrick, P. W. 1986. Genetic polymorphism in heterogenous environments: a decade late. *Annu. Rev. Ecol. Syst.* **17**:535–566.

Hedrick, P. W. 1990. Genotypic-specific habitat selection: a new model and its application. *Heredity* **65**:145–149.

Hedrick, P. W. 1995. Genetic polymorphism in a temporally varying environment: effects of delayed germination or diapause. *Heredity* **75**:164–170.

Hoekstra, R. 1975. A deterministic model of cyclical selection. *Genet. Res. Cambrigde* **25**:1–15.

Hoekstra, R., Bijlsma, R., and Dolman, A. J. 1985. Polymorphism from environmental heterogeneity: models are only robust if the heterozygote is close in fitness to the favored homozygote in each environment. *Genet. Res. Cambridge* **45**:299–314.

Johnson, M. G. 1982. Polymorphism in direction of coil in *partula suturalis*: behavioral isolation and positive frequency dependent selection. *Heredity* **49**: 307–313.

Keightley, P. D. and Kacser, H. 1987. Dominance, pleiotropy and metabolic structure. *Genetics* **117**:319–329.

Kidwell, J. F., Clegg, M. T., Stewart, F. M., and Prout, T. 1977. Regions of stable equilibria for models of differential selection in the two sexes under random mating. *Genetics* **85**:171–181.

King, M. 1993. *Species Evolution: The Role of Chromosomal Change.* New York: Cambridge U. Press.

Levene, H. 1953. Genetic equilibriums when more than one ecological niche is available. *Am. Nat.* **87**:331–333.

Lewontin, R. C. 1968. The effect of differential viability on the population dynamics of *t* alleles in the house mouse. *Evolution* **22**:262–275.

Maynard Smith, J. and Hoekstra, R. 1980. Polymorphism in a varied environment: how robust are the models? *Genet. Res. Cambridge* **35**:45–57.

Müller, H. J. 1987. Uber die Vitalität der Larvenformen der Jasside *Mocydia crocea* (H.-S.) (Homoptera Auchenorrhycha) und ihre ökologische Bedeutung. *Zool. Jahrb. Syst.* **114**:105–129.

Orr, H. A. 1991. A test of Fisher's theory of dominance. *Proc. Natl. Acad. Sci. USA* **88**:11413–11415.

Prout, T. 1967. Selective Forces in *Papilio glaucus.* *Science* **156**:534.

Prout, T. 1968. Sufficient conditions for multiple niche polymorphism. *Am. Nat.* **102**:493–496.

Prout, T. and Savolainen, O. 1996. Genotype by environment interaction is not sufficient to maintain variation: Levene and the Leaf Hopper. *Am. Nat.* **148**:930–936.

Sites, J. W. 1995. Chromosomal speciation. *Evolution* **49**:218–222.

Smith, T. B. 1993. Disruptive selection and the genetic basis of bill size in the African finch pyrenestes. *Nature* **363**:618–620.

White, M. J. D. 1978. *Modes of Speciation.* San Francisco: Freeman.

CHAPTER TEN

Analysis of Selection in Laboratory and Field Populations

EINAR ÁRNASON AND J. S. F. BARKER

10.1. Introduction

When selection in laboratory populations is analyzed, the focus will commonly be on a particular polymorphic locus, and the aim is to determine the magnitude and the mechanism of selection. That is, a particular locus is studied because of prior information indicating the possibility of fitness differences among genotypes; prior information that usually will have come from studies of natural populations.

The analysis of selection essentially involves three stages:

(1) Detection of selection – has selection affected or is it affecting allele and genotype frequencies, either in terms of changes over time or the maintenance of polymorphism?
(2) Given apparent selection, what are the relative fitnesses of different genotypes?
(3) What are the mechanisms of selection? How do the fitness differences relate to the biology and ecology of the species?

Thus we emphasise, as others have previously (e.g., Clarke, 1975; Koehn, 1978) that analyses of selection in laboratory and field populations are not separate endeavors. Selection at a specific locus may be indicated in nature. Then, fitness differences and possible mechanisms of selection may be demonstrated in the laboratory under particular environmental conditions, followed by evaluation of those fitnesses and postulated mechanisms in nature, which may point to the need for further laboratory analyses, and so on. Evolutionary data obtained in the laboratory or the field are longitudinal in time, and such historical processes must be analyzed as an autocorrelated series of data (Diggle et al., 1994).

Within this program, a complete understanding of selection at the locus in question will require further detailed study, both in the field and in the

R. S. Singh and C. B. Krimbas, eds., *Evolutionary Genetics: From Molecules to Morphology*, vol. 1. © Cambridge University Press 2000. Printed in the United States of America. ISBN 0-521-57123-5. All rights reserved.

laboratory. In the field, critical aspects of the biology and ecology of the species will need to be understood (e.g., Watt, 1992). In the laboratory, apparent functional differences among genotypes will need to be related to the primary variation in DNA sequences and the sequence–structure–function relationships determined for the structural gene itself and for its promoter (e.g., Saad et al., 1994).

Clearly, such an analysis for any one gene represents a daunting task, but it is essential that it be completed for a number of genes in a variety of species if we are to progress from merely making inferences about selection to an understanding of direct cause–effect relationships between DNA sequence variation and fitness differences among genotypes.

10.2. Fitness Estimation

Population genetics by deduction provides various models describing the kinetics of the selective process. These can be models of selection in haploid or diploid phase, with constant or variable fitness, etc. However, the problem is how these models can be turned into estimators of fitness differences. There are two main ways to do this: (1) the estimation of tautological fitnesses, and (2) the estimation of functional fitnesses.

10.3. Tautological Fitnesses

Under tautological fitness estimation, a change is observed in the genetic constitution of a population. Selection is then invoked to explain the change. The model, being a deductive construct, predicts a change in allele frequency, given some fitness differences. By accepting the model, it can be turned around and a change in allele frequency tautologically turned into fitness differences. This, in a nutshell, is the tautological method. However, it has several difficulties.

First, selection is invoked as an explanation for a frequency change. But it is not the only force that changes allele frequencies, for they can change by mutation, migration leading to gene flow, random genetic drift due to the finiteness of populations, and by linkage to other genes undergoing selection. "Selection is a wastebasket category" (Wright, 1955), and invoking selection over these other processes must be justified in some way. On the experimental level, ways must be found to eliminate these other forces as explanations for change. Second, it may be possible that the model is wrong. But it will nevertheless yield unique fitness estimates. In that event, an incorrect mode of selection is inferred even if the hypothesis of selection is supported. Third, the number of parameters specifying a (complex) model may be larger than the number of observed transitions, leading to an insufficient number of degrees of freedom to estimate all parameters.

To sum up, tautological fitnesses represent a composite or net selective effect of all components in the life cycle contributing to fitness differences of genotypes at the monitored loci. By themselves, tautological fitnesses do not identify what aspects of the biology of the experimental organisms are important. Revealing that is the aim of functional fitness estimation.

10.4. Functional Fitnesses

Differences may exist in sexual activity, fecundity, and fertility between genotypes. Likewise, there may exist differences in longevity, developmental rate, and time of maturity, which may, in turn, indirectly influence sexual activity, fecundity, and fertility. Also, differences in viability due to differences in competitive ability or predator avoidance and differential susceptibility to disease may exist that interact with other components. Dependent on the norm of reaction of different genotypes, fitness differences in any of these or other such components may arise through development of phenotypes in normal as well as perturbed environments. As an example, we refer to the application of insecticide and selection for resistance (see below). In turn these overall components (e.g., fertility and viability) depend in complex ways on interactions of various physiological and biochemical processes during development of a phenotype. If fitness effects of genotypes at a single locus (e.g., an allozyme locus) are being studied, the products of that locus are but a component of the physiological and the biochemical interactions.

At a different level, we may consider the net or total fitness of an individual (see deJong, 1994, for further discussion on fitness). An individual's fitness is a composite of the various components of fitness of all its traits: All traits contribute something to a single selective value or fitness. Should we then be concerned with estimating net fitnesses of individuals? Darwin's natural selection is variation, heritability, and differential fitness. Individuals have differential fitnesses, but, being genetically unique, individuals do not show heritability. Only their traits do. We therefore return to studying traits and fitness differences of genotypes contributing to particular traits. Perhaps we should consider net fitness a trait.

Given that various components of fitness of genotypes contributing to a particular trait have been estimated, a question arises as to how one should combine the components to give overall fitness. Components may be separate or independent, allowing addition or multiplication of components, or they may interact in specific ways, thus not allowing simple rules of combining components. There is no reason to think that there are necessary or *a priori* rules for combining components. This is probably true both for traits and for taxa: Equivalent components of fitness of different traits in the same species will not combine in the same manner across traits. Likewise, components of the same trait will not combine in the same way across species. Rules for combining components of fitness are contingent on the evolutionary history of a species. There may be empirical rules, however, but in that instance they have to be discovered. Ideally such overall fitnesses obtained by combining components should be used for population prediction and compared with tautological fitnesses. Such a comparison will reveal if some components were missed or if the rules for combining components are reasonable.

10.5. Analysis of Selection in Laboratory Populations

Bringing a problem of selection into the laboratory allows us to control other forces that change allele frequencies, thus justifying the invocation of the wastebasket category of selection as an explanation of changes. A population cage perturbation is a standard method (Wright and Dobzhansky, 1946). Migration is eliminated, and mutation is rare enough that it can be discounted. That leaves random genetic drift and linkage disequilibrium as the remaining problems to be controlled experimentally. Random drift is indeterminate only in the direction of change but it is determinate in the amount of change. Thus directionality and not rate of change is the criterion for invoking selection and excluding drift as an explanation for change. Furthermore, any random walk can appear directional if viewed myopically over a short time span of a few generations. The perturbation method therefore requires the setup of a number of replicate populations, and directionality is judged as the statistical behavior of the ensemble of replicates. If changes of the ensemble are directional, then selection can be invoked over drift. Selection coefficients can then be estimated tautologically from allele frequency changes.

Having thus eliminated random drift as an explanation, linkage disequilibrium is now a remaining problem for an experimentalist. Selection at linked loci can generate apparent directional or balancing selection at a monitored locus. True balancing selection as well as background selection of linked detrimentals can produce apparent balancing selection. To avoid generating linkage disequilibrium, an attempt is made to randomize genetic background of linked loci either by crossing isogenic lines or by letting random isochromosomal lines represent each allele (Jones and Yamazaki, 1974). Linkage disequilibrium is a problem in experiments even if there is no linkage disequilibrium in nature. The reason is that only a limited number of chromosomes are used in setting up a population cage, and therefore random linkage disequilibria are generated. A reperturbation experiment (Árnason, 1991) addresses the question of whether linkage disequilibria are responsible for changes in a perturbation experiment, thus allowing us to invoke selection instead of linkage as an explanation for allele frequency change. A reperturbation is done by founding a new population by changing frequency away from a putative equilibrium. A second return to the equilibrium would be proof of selection. However, a new bottleneck is necessarily generated during the founding of a new population. But, since the replicates are independent and are reperturbed independently, any such secondarily generated linkage disequilibria will be random in sign and thus any new linkage disequilibria will mimic the effects of drift. If the ensemble of reperturbed populations changes directionally then there is evidence for selection either at the monitored locus or at tightly linked genes that recombination did not break up. The size of the recombinational unit thus discovered is dependent on the time scale of the experiment.

Thus, when a problem is brought into the laboratory, alternative explanations of change are controlled and the wastebasket category of selection can be invoked. A qualitative answer is not sufficient, however, for the outcome of evolution depends on a balance of evolutionary forces. In order to predict evolutionary outcomes, we need quantitative statements about the strength of these forces. The next task therefore is to apply models and estimate fitness differences.

10.6. Modeling and Estimation

Theory plays a dual role in population genetics, as exemplified in the work of the founders, Fisher, Haldane, and Wright. On one hand, there is theory of processes and, on the other, theory that guides the analyses of experimental data. In our view, fitness estimation is in part a theoretical problem of the second kind that has not been sufficiently thought about. We present in this section an example of how one might approach the problem of quantitative fitness estimation.

All selection and fitness analyses, like other statistical analyses, must start with an exploratory and visualizing phase. Tukey (1977) and Cleveland (1993) discuss some of the available tools for this phase. The exploratory phase can also make use of smooth nonparametric curves (Hastie and Tibshirani, 1990, and references therein). Such detective work will identify or suggest patterns in the observations that are useful for formulating specific hypotheses about the selective processes at work. This phase is in dialectical relationship with the confirmatory phase that weighs the evidence available in the observations for or against the hypothesis. The confirmatory analysis consists of four steps: (1) formulation of hypothesis and specification of the model, (2) estimation of parameters, (3) inference by testing hypothesis and setting confidence limits, and (4) diagnostic checking that the model adequately summarizes the data.

Like other statistical models, fitness estimation models contain elements describing both systematic and random effects. In symbols, let $Y_{ij} = \mu_i + \varepsilon_{ij}$, where the systematic effect $\mu_i = \Sigma X_i \beta_i$ can be a regression, an analysis of variance, or an analysis of covariance model, depending on the nature of the predictors or explanatory variables X. A fundamental assumption of the linear model is that the random effects ε_{ij} are independent and identically distributed random variables. Frequently an additional assumption specifies a normal distribution for the random effects, $\varepsilon \sim N(0, \sigma^2)$. The parameter vector β specifying the systematic effect of the model is usually estimated by the method of least squares. The constant σ^2 specifying the random effect can then be estimated from the residual sum of squares over the remaining degrees of freedom. The linear model is described in more detail, for example, by Draper and Smith (1981) and Sokal and Rohlf (1995). Alternatively, the parameters β can be estimated by maximum likelihood (ML) by specifying the probability distribution of the data Y given the parameters β, $f(Y/\beta)$. Maximizing the likelihood, which is interpreted as a function of the parameters given the data $L(\beta/Y) = f(Y/\beta)$, yields estimates of the parameters. An objection is that ML produces biased estimators. In terms of the above model, the ML estimator for the constant σ^2 is the residual sum of squares over the number of observations.

Thus, in not being adjusted for the degrees of freedom used for the estimating the systematic effect, the estimator for σ^2 is biased.

Although useful and widely applied, the linear model has problems as a data analysis tool and this applies in fitness estimation. Distributional assumptions frequently do not hold, and the response often is confined to a limited range of values when the model expectations are not limited to stay within that range. Transformation of observations (arcsine, log, etc.) frequently is used to remedy the problem, but a transformation that cures one problem may create another. Therefore transformations are not a general solution. This is true for variables in fitness estimation, such as proportions and counts that follow binomial or multinomial and Poisson distributions for which the variance is not independent of the mean and the response is constrained to lie in a certain interval (e.g., 0 to 1 for binomial). Also, selection changes the mean and uses up genetic components of variance and therefore assumptions of constant variance may be suspect.

Generalized linear models (GLMs) (McCullagh and Nelder, 1989) unify linear models for independent, discrete, and continuous responses and generalize them to allow different forms of the variance. Also, they transform the model instead of the observations and thus lift the restriction that the model expectations stay within a certain range. First, in a GLM the mean response $\mu_i = E(Y_i)$ is linked by means of a function (g) to a linear predictor η, $g(\mu) = \eta = \Sigma X\beta$. Second, the variance of Y_i is a function of its mean, μ_i: $\text{var}(Y_i) = \phi V(\mu_i)$, where V is a known function of the mean. The scaling factor ϕ may be known for some GLM distributions but may have to be estimated for others. Third, the GLMs belong to the exponential family of distributions for which a general form of the likelihood function can be specified. Examples include logistic regression with a logit link ($g(\mu) = \text{logit } p = \log p/q = \eta$, which expands model expectations from $-\infty$ to ∞) and variance function of $\mu(1 - \mu)$ for binomial data and Poisson regression with a link and variance of μ for count data. The ML estimators of GLMs are obtained by iteratively reweighted least squares of the specified likelihood function.

In studies of fitness (and other biological studies) the variance of Y is often greater than the GLM assumes, a phenomenon known as overdispersion. One way to deal with this is allowing the dispersion parameter $\phi > 1$. Then the variance of the estimators of the regression parameters β specifying the systematic effect is regarded as inflated by a factor of ϕ. When overdispersion is present, ignoring it in an analysis would yield underestimation of standard errors and consequent overinterpretation of confidence limits or significance in hypothesis testing. Overdispersion is observed in fitness estimation studies and adjusted for in different ways (e.g., Prout, 1971a; Williams et al., 1990; Árnason, 1991). The GLM family of models contains an important property that the score function, the derivative of the likelihood function, used for estimation depends on only the mean and the variance of the Y. Therefore a quasi-likelihood (Wedderburn, 1974) can be defined without knowledge of the distribution if only the mean and the variance relationships are

known. A quasi-likelihood, therefore, can be used for estimation even in cases
in which details of the probabilistic mechanisms generating the data are missing as, for example, when data are overdispersed and do not follow a regular GLM family distribution. This is useful in fitness estimation in which
one does not know if and how background effects and epistasis influence
observations.

The GLMs carry over from the classical linear models the fundamental assumption of independent, uncorrelated errors. This assumption frequently is
violated in fitness estimation as in many other biological phenomena in which
the history of a process creates correlated observations. Such data can be repeated measures of a single subject in the form of longitudinal data (Diggle
et al., 1994), observations that are ordered by temporal or spatial position. Se-
rial correlation frequently results. In terms of the model $Y_{ij} = \mu_i + \varepsilon_{ij}$, the ε
are correlated random variables. Consequences of ignoring serial correlation
when it is present are (1) incorrect inferences about the regression parameters β and (2) inefficient estimates. Serial correlations of the random effect
can be modeled, however, and estimated along with other parameters. This is
a strength of longitudinal studies. A merit of doing so is a gain in efficiency
because some of the variation can be eliminated. An analogy can be drawn with
a gain in efficiency of a paired t-test over a two-sample t-test in which variation
among subjects can be eliminated under the right conditions.

Diggle et al. (1994) discuss three extensions to GLMs for longitudinal data:
marginal, random effects, and transition (Markov) models, extensions that
model the serial correlation in different ways. All these models can be useful for fitness estimation in both laboratory and field studies. Endler (1986)
refers to it as genotypic cohort analysis. Suppose we have repeated measures
for replicates in a perturbation study (Árnason, 1991) or replicates in a fitness component study. First, under the marginal model, the mean or marginal
expectation, for example, the average behavior of a group of replicates (e.g.,
Árnason, 1991) is modeled as a function of the explanatory variables. Here
the heterogeneity among replicates, due perhaps to random intercept terms or
random starting conditions, creates correlations between observations (j and
k) within replicates, $\mathrm{corr}(Y_{i,j}, Y_{i,k})$, because they share the same initial state. In
this instance interest may nevertheless focus on average behavior (e.g., directionality in the behavior of the ensemble), and therefore a marginal model may
be appropriate. Second, in random-effect models, the response is a function
of the linear predictors with coefficients that may vary from one replicate to
another. Such variability reflects heterogeneity due to unknown factors (e.g.,
due to genetic background in fitness studies). It can be assumed that the data
of a single subject, for example data from a single replicate in fitness studies,
are independent observations that can be modeled by some GLM, but that the
regression coefficients vary between replicates: that the regression coefficients
follow some distribution that is determined by some unknown variables that
differ among replicates (cf heterogeneity of theoretical expectations that may
be due to unobserved linkage disequilibria or epistatic interactions; Árnason,

1991). Correlations among observations for a single replicate arise because they share the combination of unobserved or latent (Bartholomew, 1987) variables that cause the regression coefficient of that particular replicate. If interest focuses on the individuality of response, i.e., interest is focused on understanding the nature of the unobserved variables (i.e., linkage disequilibria or epistasis), then the random-effect model may be appropriate. Third, under a transition or Markov model, observations $Y_{i1}, \ldots, Y_{in}$ are correlated because past values $Y_{i1}, \ldots, Y_{ij-1}$ directly influence present observations. As one example of a transition model consider $Y_{ij} = \Sigma X_{ij}\beta + \varepsilon_{ij}$, where $\varepsilon_{ij} = \alpha\varepsilon_{ij-1} + Z_{ij}$, where the Z_{ij} are independent random variables. Substituting $\varepsilon_{ij} = Y_{ij} - \Sigma X_{ij}$ into the second equation and rearranging gives the response given previous responses as $Y_{ij} = \Sigma X_{ij}\beta_i + \alpha(Y_{ij-1} - \Sigma X_{ij-1}\beta_i)$. Thus, present observations are modeled as explicit functions of the past. Many population genetics phenomena are Markovian, and models of this nature may be suitable for fitness estimation studies if interest focuses on the effects of random drift due to finite population size (see Diggle et al., 1994, and references therein for further details on these statistical models).

Parameters of longitudinal models such as those described above can be estimated by ML or by restricted maximum likelihood (REML), which often gives similar results. When the two estimators differ, Diggle et al. (1994) recommend REML because it is not biased. In many instances the likelihood is intractable, however, and may involve estimation of several nuisance parameters besides the parameters of interest. In such cases it may be possible to use generalized estimating equations (GEEs) that are multivariate analogs of the quasi-likelihood referred to above. Furthermore, as an alternative to the above parametric modeling, smooth nonparametric models can be fitted in a longitudinal setting (Hastie and Tibshirani, 1990; Hart, 1991; Diggle et al., 1994). We give a brief example below that indicates the flexibility of a nonparametric approach. Various of these modern methods are also being used in fitness studies of quantitative traits (e.g., Brodie et al., 1995).

Lambert and Roeder (1995) discuss overdispersion diagnostics for GLM. Finally, for model diagnostics in longitudinal data a variogram is useful. For a stochastic process it is defined as $\gamma(\mu) = \frac{1}{2}E\{[Y(t) - Y(t - u)]^2\}$ and is thus related to an autocorrelation function. For diagnostics of longitudinal data Diggle et al. (1994) suggest superimposing the fitted variogram on the empirical one to highlight discrepancies in the model effects. If the diagnostics indicate adequacy of the model, the estimation task is completed.

10.6.1. Model Selection in Fitness Estimation

In estimating fitnesses, a selection model is fitted to allele frequency changes and parameters estimated by methods such as least squares (Wright and Dobzhansky, 1946), ML (DuMouchel and Anderson, 1968; Wilson et al., 1982), or generalized least squares (Schaffer et al., 1977; Árnason, 1991). Estimates thus obtained, however, apply only to the extent that the model is correct.

Fitting, for example, a strict viability model when fertility selection is operative will yield spurious fitness estimates. The time of census can also seriously influence the estimates (Prout, 1965, 1969, 1971a, 1971b). For example, spurious frequency dependence could ensue even if fitnesses were, in fact, constant. By fitting a model of constant fitness, a model of differential selection in the sexes, and a model of frequency dependence to a chromosomal polymorphism in laboratory populations, Wright and Dobzhansky (1946) showed that all models fit approximately equally well in the range of frequencies observed in the experiment. Similarly, Árnason (1991) found confluent fits in an allozyme locus perturbation/reperturbation experiment with a constant fitness and a frequency-dependent model. If the large fitness differences observed in chromosomal polymorphisms do not allow discrimination between models, then much less can we hope to be able to discriminate between models for single (allozyme) loci. Selection is likely to be much weaker on such loci, and consequently the statistical power to detect selection is much less.

These difficulties of estimation would seem to suggest that we have to give up hope of estimating the correct mode of selection. However, taking the attitude of giving up model discrimination would seem like throwing out the baby with the bathwater. If there is selection, we want to be able to discriminate between directional and balancing selection because we want to know whether selection maintains variation or only uses it up. Relative fitnesses determine the outcome of selection in subtle ways, and therefore it is not sufficient to have evidence for fitness differences; their magnitude must also be determined. The current fascination with molecular studies of DNA variation have shown the strength of purifying selection. However, with notable exceptions, balancing selection is out of fashion in such studies. We need to reintroduce the possibility of balancing selection. We suggest that perturbation/reperturbation experiments can be designed so as to give at least qualitative answers to questions of whether balancing or directional selection is operating, if indeed selection is operating at the locus of interest. If the ensemble of replicates consistently tracks an internal equilibrium both from above and below, that is evidence for balancing selection. Other patterns of directional change are evidence for directional purifying or positive selection. Given such a qualitative evidence for balancing selection, any of a number of models could then be used to estimate and quantify effective selection coefficients. Such estimates could be useful for predicting rates of evolution by selection even if judgement about balancing or purifying selection were based on the qualitative results of the experiment.

10.6.2. An Example of a Flexible Nonparametric Approach

If we give up estimating fitness by fitting specific selection models, it opens up ways to use more flexible statistical tools for estimating effective fitness differences. As an example, consider some results from a perturbation/reperturbation experiment at the alcohol dehydrogenase locus in *Drosophila melanogaster*

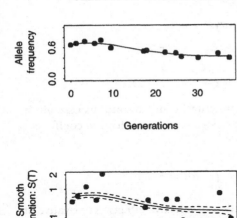

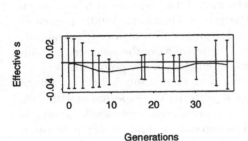

Figure 10.1. An example of model fitting in a single replicate from a perturbation/reperturbation experiment of *Adh* in *D. melanogaster*. Top panel: observed allele frequencies (dots) with a fitted smooth nonparametric local regression (a loess curve; Cleveland, 1993); middle panel: a smooth curve representing a summation of selective effects $S(T)$ estimated with a generalized additive model (Hastie and Tibshirani, 1990) with 95% confidence limits (dashed curves) and partial deviance residuals (dots); bottom panel: effective time-dependent selection coefficients estimated as a derivative of $S(T)$ with 95% error bars.

(Fig. 10.1). Briefly, the results of the experiment are similar to the *Est-5* results of Árnason (1991): perturbation shows directional changes but a reperturbation does not. Therefore the initial selection must be due to linkage disequilibrium that has dissipated. The qualitative result is therefore that effective selection is decaying with time because of recombinational breakup. A biologically realistic way of modeling the situation would be to model a selection coefficient that exponentially decays in time, $s_t = s_0 e^{-\lambda t}$. This model is attractive because of realism but behaves badly by being very sensitive to outliers at the start or at the end of the experiment. For example, the model gives estimates of enormous selection that decays instantly when an outlying observation occurs at the start of a sequence of values. Alternatively it gives estimates of tiny selection that grows exponentially to large values after a long time if an outlier occurs near the end of a sequence of observations. Thus, even if biologically reasonable, such a model is useless for fitness estimation.

Another and perhaps a more flexible modeling approach is the following. Consider semidominant selection for a two-allele single-locus viability model with selection coefficient $2s$ against homozygotes:

Genotypes	FF	FS	S
Frequency	p^2	$2pq$	q^2
Fitness	1	$1-s$	$1-2s$

Under this additive genetic model we can write in the continuous case allele frequency change dp per generation time t as a function of selection coefficient s, $dp/dt = spq$, and integrate to get

$$\ln\left(\frac{p_t}{1-p_t}\frac{1-p_0}{p_0}\right) = \int_0^T s\,dt = S(T)$$

or $logit(p_t) + constant = S(T)$, a summation of selective effects. A derivative of that summation function $S(T)$ is then a time dependent selection coefficient s.

Recent statistical methods offer a way to model this. An extension of classical linear models ($Y = a + b_1 X_1 + b_2 X_2$) by means of a GLM is provided by generalized additive models (GAMs): $Y = a + f_1(X_1) + f_2(X_2)$, which, instead of specifying a parametric functional form, model the response as nonparametric smooth functions of predictors (Hastie and Tibshirani, 1990). Therefore modeling logits as a smooth function of time, $logit(p_t) \sim smooth(generation)$, gives the integral of selection over time. The derivative of the estimated smooth function then gives an effective selection coefficient at any point in time.

An example is presented in Fig. 10.1. The top panel shows the results from a single replicate of the above-mentioned perturbation experiment with *Adh*. Allele frequencies (dots) decrease in time, and a nonparametric local regression (loess curve) shows the trend with an apparent attainment of equilibrium. In this instance the qualitative result of the perturbation/reperturbation suggested modeling an effective selection coefficient decaying in time. The middle panel shows the estimated smooth curve with confidence limits (dashed curves) and partial deviance residuals (dots), which on inspection indicate an adequate fit to the data. The bottom panel gives an effective temporal selection coefficient (the derivative of the smooth function) of $\sim$1.5% at generation 10 decaying to nil at approximately generation 30. The error bars enclose the hypothesis of no selection $H_0 : s = 0$ for all sampling points, so a formal significance is not reached in spite of considerable sampling effort. This data analysis thus completes our example of how one might approach quantitative fitness estimation.

10.7. Analysis of Selection in Field Populations

10.7.1. Detecting Selection in the Field

Endler (1986) discussed 10 methods for detecting selection in natural populations. Some methods give only presumptive evidence and many are not sufficient by themselves, but the usefulness of some has increased significantly in the past 10 years with the development of a range of DNA-based markers, and new methods and tests that use DNA markers and sequences also have been developed (see also Schaeffer and Aguadé, 1999). This is a static or comparative approach in contrast to a dynamic or functional approach that we discuss. We consider here only those methods that are dependent on DNA data or in which DNA markers add to the feasibility or power of the method.

10.7.1.1. F-Statistics

Wright's (1951) F_{ST} statistic is the correlation of genes between individuals within the same population and can be interpreted as a measure of the amount of differentiation among populations, relative to the limiting amount under complete fixation. As the effects of drift and migration are the same for all alleles at all neutral polymorphic loci, all such alleles have the same expected F_{ST}, and there should be no significant heterogeneity of the F_{ST} estimates. Selection at any locus, however, will cause the estimated F_{ST} for that locus to be either greater or less, depending on the nature of the selection. Thus balancing selection that is common to all populations will reduce differentiation (lower F_{ST} estimate), while differential selection pressures in different populations will increase differentiation (higher F_{ST} estimate). Lewontin and Krakauer (1973) derived the sampling distribution of F_{ST} and proposed a test of heterogeneity by a comparison of the observed variance with the theoretical expected variance. Problems with estimating the expected variance (Robertson, 1975) preclude general use of the Lewontin–Krakauer test, but it is still applicable in certain cases (Tsakas and Krimbas, 1976).

However, a modification of the Lewontin–Krakauer test provides a direct and powerful method for detecting selection. Suppose that we have available a set of polymorphic markers that can reasonably be assumed to be neutral to selection. The observed variance of estimates from these markers is then an empirical estimate of the expected variance for neutral loci, which can be used instead of an expected theoretical variance. Nuclear DNA markers such as restriction fragment length polymorphism (RFLP) loci or microsatellites are the obvious ones for use in this way. Such a comparison of variances could be useful for initial exploration of the possibility that selection is affecting at least some of a number of protein-coding loci. But particularly in cases in which only a small number of protein-coding loci are being studied, comparison of variances may well be misleading. For example, if all of these loci were subject to diversifying selection, the variance of their F_{ST} estimates may not differ significantly from that derived from the set of DNA markers. Yet in this case, the mean F_{ST} for the protein-coding loci may be significantly greater than the mean for the DNA markers. That is, both variance and mean should be considered. One particularly striking example of the information to be gained by comparing protein-coding and DNA polymorphisms comes from studies of the American oyster, *Crassostrea virginica* , on the Atlantic and the Gulf of Mexico coasts of the southeastern United States. Karl and Avise (1992) analyzed four anonymous single-copy nuclear genes in nine populations and compared results with previous data for 18 allozyme polymorphisms (Buroker, 1983). Whereas the allozymes had a near uniformity of allele frequencies across all populations, the nuclear genes revealed significant differences in allele frequencies between the Atlantic and the Gulf coasts. The contrast between the two classes of markers was so great that one hardly needed analyses of F_{ST} to make inferences about selection, gene flow, and population structure, but the F_{ST} estimates are indeed significantly different (McDonald, 1994). One might be tempted to conclude that balancing selection is maintaining similar allele frequencies in all populations for all allozyme loci. This may seem rather unlikely. Alternatively, the constancy of allozyme frequencies may reflect high rates of gene flow, while the nuclear genes differ between the Atlantic and the Gulf because of diversifying selection. This too may seem unlikely, but as we will reiterate again in relation to other methods of detecting selection, the nuclear genes used are just markers, and allele frequencies at marker loci that are themselves neutral may be influenced by selection that is acting on closely linked loci. This is a fundamental problem. Thus neither alternative conclusion is completely satisfactory, although a survey of additional DNA markers could help to resolve the issue. Nevertheless selection clearly has influenced at least some of the genetic markers.

Comparisons of F_{ST} estimates are not restricted to the simultaneous testing of a number of loci; the F_{ST} estimated for just one locus may be compared with the mean F_{ST} for a number of (presumed neutral) DNA markers (Berry and Kreitman, 1993). However, whether a single locus or a set of loci is being studied, the most appropriate method of making the test should be considered. McDonald (1994) used simulation to investigate the properties of the two common statistical estimators of F_{ST} (Nei, 1986; Weir and Cockerham, 1984), weighted and unweighted averages of these estimates, numbers of alleles, and sample sizes. He shows that unweighted means of the estimates are less biased, that pooling alleles so that all polymorphisms are treated as two allele may be necessary to avoid statistical artifacts, and that for a given total sample size, two or three population samples can be as efficient at detecting selection as a larger number of smaller samples. Given the power and potential use of this method for detecting selection in natural populations, further study of these questions and the development of a parametric test would be desirable. Although our focus is on single-locus polymorphisms and quantitative genetics is the subject of Section E comparisons of F_{ST} estimators also may be used to detect selection that affects quantitative traits in natural populations (Prout and Barker, 1993; Spitze, 1993; Podolsky and Holtsford, 1995; Long and Singh, 1995).

10.7.1.2. Perturbation of Natural Populations

Endler (1986) considered six types involving genetic or environmental perturbations. Here we discuss only two of these, viz., intentional artificial perturbations of natural populations as a direct method of detecting selection and known abrupt changes in the environment, which, in the context of insecticide resistance, we return to later as a paradigm for the analysis of selection.

If genetic variation at some locus is being actively maintained by natural selection, the action of this selection should be detectable following artificial perturbation of allele frequencies produced either by addition of certain genotypes to the population or by removal of them from it. Following perturbation, if allele frequencies at the perturbed locus merely drift around their new values, variation at the locus must be assumed neutral. On the other hand, if allele frequencies change consistently to preperturbation values, then either selection is effective or there is migration from surrounding populations. Simultaneous perturbation of a number of loci should allow selection and migration to be distinguished (Barker and East, 1980). In that study, three allozyme loci were perturbed and all returned to pre-perturbation frequencies, but migration was excluded because of significant heterogeneity among loci in the migration rates necessary to account for the allele frequency changes following perturbation. Clearly that was fortuitous, and a more efficient design for future experiments would be to simultaneously perturb for one or more loci of interest, and preferably two or more RFLP or microsatellite markers.

Although the results of Barker and East (1980) demonstrate that selection influences all three loci, they do not show that selection is acting directly on any of them nor is there any information on the nature of the selection. However, where evidence already exists that selection is influencing allele frequencies at a particular locus, appropriately designed perturbation experiments can provide strong support for direct selection and information on the nature and the magnitude of the selective forces operating (McKenzie et al., 1994).

10.7.1.3. Comparisons among Age Classes or Life-History Stages

Simple comparisons among these classes or stages can directly demonstrate ongoing selection, but we focus here on a particular type of comparison, namely fitness component analysis. This is not only an efficient method for detecting selection; it also provides for analysis of selection in terms of sexual, gametic, fecundity, and viability components and allows estimation of the magnitude of selection.

The method depends on being able to identify at least one parent of each individual in a progeny group, which has limited its use to experimentally manipulated populations, (e.g., Bundgård and Christiansen, 1972; Clegg et al., 1978) or to species such as the live-bearing fish Zoarces in nature (Østergård and Christiansen, 1981, and references quoted therein).

However, molecular markers and new methods for estimating pairwise relatedness in natural populations (Ritland, 1996) dramatically increase the potential application of this method to many other species. For analysis of a single locus, if females produce a number of offspring, such as a clutch or litter, that can be associated unequivocally with the mother, and the breeding biology of the species is such that these offspring have been fathered by one (but unknown) male, and the genotypes of mother and offspring are determined, then the genotype of the father may be inferred, and a complete fitness component analysis made.

In contrast, imagine a species with a discrete population of potential parents, for which groups of young can be associated with particular females (say, as a clutch), but not all of which are necessarily her offspring, and for which females may mate with one or more males. Further, assume (for simplicity in the first instance) that only two alleles are segregating at the locus of interest and that the genotypes of all individuals are known for this locus and for a number of molecular markers. From the genotypes for the locus being studied, two types of young can be excluded immediately as not offspring of their putative mother (*aa* young with *AA* putative mother, and the reciprocal). For other young–putative mother combinations, DNA-marker-based estimates of pairwise relationships will assess whether a particular young is an offspring of its putative mother and, if not, which other female is likely to be its mother. Once young–mother combinations are defined, pairwise relationships among the offspring of a particular female will determine if they are a group of full-sibs or of full- and half-sibs. Finally, pairwise relationships between each young and potential fathers will identify the true father of each. Depending on the number of DNA markers used (and this needs investigation), some ambiguities will no doubt remain, and thus some data lost. Nevertheless, a detailed selection component analysis would be feasible even for a species such as that in the above example.

10.7.2. Insecticide Resistance and the Analysis of Selection

Insecticide resistance is a major agricultural problem, but it is also an excellent model for the analysis of selection in the field. Each introduction of an insecticide to control a pest species represents an abrupt change in the environment of that species, to which it usually responds (often rapidly) by evolving resistance. That is, there is a direct cause–effect relationship for which the selective agent is known, resistance is usually due to allelic substitution at a single locus, the mechanism of resistance can be identified, and allelic differences characterized at the biochemical and molecular levels (Roush and McKenzie, 1987; Russell et al., 1990).

Despite the diversity of structures among the five major classes of chemical insecticides (Hutson and Roberts, 1985) and the diversity of the some 450 species of insects and mites that developed resistance to insecticides by the 1980s (Georghiou, 1986), the number of resistance mechanisms apparently is limited (reviewed by Soderlund and Bloomquist, 1990). Nevertheless, analysis of selection for insecticide resistance is not trivial, and no one case has yet been analyzed completely from its evolution in the field to a specification of its molecular basis. We consider here one example of the analysis of resistance in the field and briefly comment on some of the molecular mechanisms that have been uncovered.

The evolution of resistance to diazinon (an organophosphate) in the Australian sheep blowfly (*Lucilia cuprina*) is primarily due to allelic substitution at a

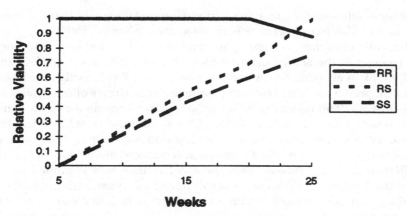

Figure 10.2. Relative larval viabilities of *Rop-1/Rop-1* (*RR*), *Rop-1/+* (*RS*), and +/+ (*SS*) diazinon resistance genotypes of *L. cuprina* at times after treatment of sheep with diazinon (adapted from McKenzie and Whitten, 1982).

single locus, designated *Rop-1* (McKenzie, 1987). The usual genetic model for the evolution of resistance assumes a reversal of fitnesses between absence and presence of the insecticide, but with constant fitnesses in each case. Thus in the absence of the insecticide, susceptible (+/+) individuals are fitter than resistant (+/R and R/R), and the R allele is maintained at low frequency by mutation–selection balance. With insecticide exposure, resistant phenotypes are fitter, but the relative fitnesses of the three genotypes will depend on whether, with respect to fitness, resistance is dominant, additive or recessive. For genotypes at the *Rop-1* locus in *L. cuprina*, however, their relative fitnesses depend on the concentration of diazinon to which they are exposed (McKenzie and Whitten, 1982; McKenzie, 1987) (Fig. 10.2) so that frequency- and density-dependent interactions between genotypes become important (McKenzie, 1993). One of the most notable features of the analysis of selection in this system has been the demonstration of selection for fitness modifiers of diazinon-resistant genotypes and the concurrent study of fitness and developmental homeostasis (McKenzie et al., 1990; McKenzie, 1993).

When an allele of major effect (such as a resistance allele) is selected following some change in the environment, it may disrupt existing developmental constraints that stabilize phenotypic expression (Maynard Smith et al., 1985; Scharloo, 1991). In the face of continuing challenge from the new environment, however, modifiers that increase developmental stability and fitness are expected to be selected (Fisher, 1958). As diazinon was used for a significant period after resistance had evolved, this provided a rare opportunity to confirm these predictions. The introduction of the resistant allele resulted in phenotypes of lowered developmental stability, increased bilateral asymmetry (Palmer and Strobeck, 1986), and lowered relative fitness in the absence of the insecticide (Clarke and McKenzie, 1987; McKenzie and Clarke, 1988), with consequent selection for a modifier that ameliorated these effects and that has increased to high frequency in field populations (McKenzie, 1993).

Metabolic resistance to organophosphate insecticides has been associated with changes in the activity of carboxylesterases in many insect species, and the *Rop-1* locus encodes the carboxylesterase isozyme E3 (Russell et al., 1990). With PCR primers designed from the α-esterase gene cluster of *D. melanogaster* (Russell et al., 1996) regions of six presumptively homologous α-esterase genes were amplified from *L. cuprina* DNA, with one of these (*LcαE7*) recovered from all DNA sources predicted to be sites of expression for E3 (Newcomb et al., 1996a). Subsequently, *LcαE7* has been sequenced and shown to encode E3 (Newcomb et al., 1996b). Thus the *Rop-1* locus is now identified at the molecular level, so that DNA sequence differences between resistant and susceptible genotypes can be determined and related to the functional differences in organophosphate resistance.

Given that meaningful data on selection can be obtained, as exemplified above, and that a single locus is generally implicated, these studies to elucidate the molecular mechanisms are necessary to bring the analysis of selection to its fundamental level. Do fitness differences result from a change in the coding sequence of a structural gene, or a change in gene regulation, or some other mechanism? As noted above, the number of resistance mechanisms at the biochemical level is limited, and they are broadly of two classes – altered target sites in the insect nervous system and altered metabolic systems involved in insecticide degradation. Nevertheless, the molecular basis of resistance is poorly understood, largely because of inadequate genetic knowledge of most pest species for which resistance has been detected, and hence a limited capacity to clone the resistance genes. Thus Wilson (1988) proposed *D. melanogaster* (reviewed by ffrench-Constant et al., 1993a) as a model for insecticide resistance studies, and this approach already has proved particularly fruitful.

More than 60 reported cases of insecticide resistance involve resistance to cyclodiene insecticides, with resistance due to insensitivity of a γ-aminobutyric subtype A (GABAA) receptor in the nervous system (ffrench-Constant et al., 1991). The responsible gene in *D. melanogaster* has been cloned and sequenced and a single base-pair mutation shown to be the only consistent difference between resistant and susceptible strains (ffrench-Constant et al., 1993b). Most significantly, by using the *D. melanogaster* clone of this gene, Thompson et al. (1993) have cloned the homologous cyclodiene resistance gene from the mosquito *Aedes aegypti* and shown that resistant strains have the same single base-pair substitution as that in *D. melanogaster*.

10.8. Conclusions

In his book Lewontin (1974) claimed that "to the present moment no one has succeeded in measuring with any accuracy the net fitnesses of genotypes for any locus in any species in any environment in nature." Has the field advanced since? We believe we are heading in the right direction, albeit slowly.

The problems of accurate, or perhaps more appropriately efficient, fitness estimation are formidable. Fitness can be estimated tautologically or functionally. Both approaches are necessary because neither method alone can give a full picture. The first does not indicate what are the important aspects of the biology contributing to fitness and the second may incorrectly combine components or miss components. Our aim is to understand selection in natural populations. However, frequently there are complicating factors, and therefore it is profitable to study natural selection under laboratory conditions in which extraneous factors can be controlled. We strive for accurate and efficient estimation, and our statistical methods therefore must evolve and take advantage of new developments. Problems such as overdispersion and serial correlation, which sometimes have been glossed over, can be dealt with and may add strength to an analysis.

Frequently, field studies will provide an impetus for bringing a problem into the laboratory. Methods for detecting selection in the field, such as comparison of F_{ST} variances among different sets of loci, are strengthened by the use of DNA techniques. It is to some extent the nature of DNA variation (e.g., silent versus nonsilent sites) that allows it to be used with confidence as a null model for variation. In laboratory studies, detailed knowledge of DNA structure in and around target genes may allow a more detailed hypothesis of selection to be formulated and tested. Finally, the method of perturbation, exemplified with environmental perturbation in the application of insecticide, is a powerful approach in the dynamical analysis of selection. We believe that model systems like insecticide resistance offer the potential for reliable fitness estimation in natural populations. Bringing that problem into the laboratory and/or studying it in genetic tools like *Drosophila* may facilitate such estimation. The insecticide example also shows how we can approach analysis of selection from the level of the phenotype right through to the DNA level. However, the objection might be that application of insecticides creates artificially or unusually strong selection. Nevertheless, if such a model system works, the methods may be applied to other systems.

REFERENCES

Árnason, E. 1991. Perturbation-reperturbation test of selection *vs.* hitchhiking of the two major alleles of *Esterase-5* in *Drosophila pseudoobscura. Genetics* **129**:145–168.

Barker, J. S. F. and East, P. D. 1980. Evidence for selection following perturbation of allozyme frequencies in a natural population of *Drosophila. Nature (London)* **284**:166–168.

Bartholomew, D. J. 1987. *Latent Variable Models and Factor Analysis*. Oxford: Clarendon.

Berry, A. and Kreitman, M. 1993. Molecular analysis of an allozyme cline: alcohol dehydrogenase in *Drosophila melanogaster* on the east coast of North America. *Genetics* **134**:869–893.

Brodie, E. D., Moore, A. J., and Janzen, F. J. 1995. Visualizing and quantifying natural selection. *Trends Ecol. Evol.* **10**:313–318.

Bundgård, J. and Christiansen, F. B. 1972. Dynamics of polymorphisms: I. Selection components in an experimental population of *Drosophila melanogaster*. *Genetics* **71**:439–460.

Buroker, N. E. 1983. Population genetics of the American oyster *Crassostrea virginica* along the Atlantic coast and the Gulf of Mexico. *Mar. Biol.* **75**:99–112.

Clarke, B. 1975. The contribution of ecological genetics to evolutionary theory: detecting the direct effects of natural selection on particular polymorphic loci. *Genetics* **79**:101–113.

Clarke, G. M. and McKenzie, J. A. 1987. Developmental stability of insecticide resistant phenotypes in blowfly; a result of canalizing natural selection. *Nature (London)* **325**:345–346.

Clegg, M. T., Kahler, A. L., and Allard, R. W. 1978. Estimation of life cycle components of selection in an experimental plant population. *Genetics* **89**:765–792.

Cleveland, W. S. 1993. Visualizing Data. Summit: Hobart.

de Jong, G. 1994. The fitness of fitness concepts and the description of natural selection. *Q. Rev. Biol.* **69**:3–29.

Diggle, P. J., Liang, K.-Y., and Zeger, S. L. 1994. *Analysis of Longitudinal Data*. Oxford: Clarendon.

Draper, N. and Smith, H. 1981. *Applied Regression Analysis*. New York: Wiley.

DuMouchel, W. H. and Anderson W. W. 1968. The analysis of selection in experimental populations. *Genetics* **58**:435–449.

Endler, J. A. 1986. *Natural Selection in the Wild*. Princeton, NJ: Princeton U. Press.

ffrench-Constant, R. H., Mortlock, D. P., Shaffer, C. D., MacIntyre, R. J., and Roush, R. T. 1991. Molecular cloning and transformation of cyclodiene resistance in *Drosophila*: an invertebrate γ-aminobutyric acid subtype A receptor locus. *Proc. Natl. Acad. Sci. USA* **88**:7209–7213.

ffrench-Constant, R. H., Roush, R. T., and Cariño, F. A. 1993a. *Drosophila* as a tool for investigating the molecular genetics of insecticide resistance. In *Molecular Approaches to Fundamental and Applied Entomology*, J. Oakeshott and M. J. Whitten, eds., pp. 1–37. New York: Springer-Verlag.

ffrench-Constant, R. H., Steichen, J. C., Rocheleau, T. A., Aronstein, K., and Roush, R. T. 1993b. A single-amino acid substitution in a γ-aminobutyric acid subtype A receptor locus is associated with cyclodiene insecticide resistance in *Drosophila* populations. *Proc. Natl. Acad. Sci. USA* **90**:1957–1961.

Fisher, R. A. 1958. *The Genetical Theory of Natural Selection*. New York: Dover.

Georghiou, G. P. 1986. The magnitude of the resistance problem. In *Pesticide Resistance: Strategies and Tactics for Management*, pp. 14–43. Washington, D.C.: National Academy.

Hart, J. D. 1991. Kernel regression estimation with time series errors, *J. R. Stat. Soc. B* **53**:173–187.

Hastie, T. J. and Tibshirani, R. J. 1990. *Generalized Additive Models*. New York: Chapman & Hall.

Hutson, D. H. and Roberts, T. R. 1985. Insecticides. In *Insecticides*, D. H. Hutson and T. R. Roberts, eds., pp. 1–34. New York: Wiley.

Jones, J. S. and Yamazaki, T. 1974. Genetic background and the fitness of allozymes. *Genetics* **78**:1185–1189.

Karl, S. A. and Avise, J. C. 1992. Balancing selection at allozyme loci in oysters: implications from nuclear RFLPs. *Science* **256**:100–102.

Koehn, R. K. 1978. Physiology and biochemistry of enzyme variation: the interface of ecology and population genetics. In *Ecological Genetics: The Interface*, P. F. Brussard, ed., pp. 51–72. New York: Springer-Verlag.

Lambert, D. and Roeder, K. 1995. Overdispersion diagnostics for generalized linear models. *J. Am. Stat. Assoc.* **90**:1225–1231.

Lewontin, R. C. 1974. *The Genetic Basis of Evolutionary Change*. New York: Columbia U. Press.

Lewontin, R. C. and Krakauer, J. 1973. Distribution of gene frequency as a test of the theory of the selective neutrality of polymorphisms. *Genetics* **74**:175–195.

Long, A. D. and Singh. R. S. 1995. Molecules versus morphology: the detection of selection acting on morphological characters along a cline in *Drosophila melanogaster. Heredity* **74**:569–581.

Maynard Smith, J., Burian, R., Kauffman, S., Alberch, P., Campbell, J., Goodwin, B., Lande, R., Raup, D., and Wolpert, L. 1985. Developmental constraints and evolution. *Q. Rev. Biol.* **60**:265–287.

McCullagh, P. and Nelder, J. A. 1989. *Generalized Linear Models*. New York: Chapman & Hall.

McDonald, J. H. 1994. Detecting natural selection by comparing geographic variation in protein and DNA polymorphisms. In *Non-Neutral Evolution Theories and Molecular Data*, B. Golding, ed., pp. 88–100. New York: Chapman & Hall.

McKenzie, J. A. 1987. Insecticide resistance in the Australian sheep blowfly – messages for pesticide usage. *Chem. Ind.* **8**:266–269.

McKenzie, J. A. 1993. Measuring fitness and intergenic interactions: the evolution of resistance to diazinon in *Lucilia cuprina. Genetica* **90**:227–237.

McKenzie, J. A., Batterham, P., and Baker, L. 1990. Fitness and asymmetry modification as an evolutionary process. A study in the Australian sheep blowfly, *Lucilia cuprina* and *Drosophila melanogaster. In Ecological and Evolutionary Genetics of Drosophila*, J. S. F. Barker, W. T. Starmer, and R. J. MacIntyre, eds., pp. 57–73. New York: Plenum.

McKenzie, J. A. and Clarke, G. M. 1988. Diazinon resistance, fluctuating asymmetry and fitness in the Australian sheep blowfly, *Lucilia cuprina. Genetics* **120**:213–220.

McKenzie, J. A., McKechnie, S. W., and Batterham, P. 1994. Perturbation of gene frequencies in a natural population of *Drosophila melanogaster.* evidence for selection at the *Adh* locus. *Genetica* **92**:187–196.

McKenzie, J. A. and Whitten, M. J. 1982. Selection for insecticide resistance in the Australian sheep blowfly, *Lucilia cuprina. Experientia* **38**:84–85.

Nei, M. 1986. Definition and estimation of fixation indices. *Evolution* **40**:643–645.

Newcomb, R. D., Campbell, P. M., Russell, R. J., and Oakeshott, J. G. 1996a. cDNA cloning, baculovirus-expression and kinetic properties of the esterase, E3, involved in organophosphate resistance in *Lucilia cuprina. Insect Biochem. Mol. Biol.* **27**: 15–25.

Newcomb, R. D., East, P. D., Russell, R. J., and Oakeshott, J. G. 1996b. Isolation of α cluster esterase genes associated with organophosphate resistance in *Lucilia cuprina. Insect Mol. Biol.* **5**:211–216.

Østergård, H. and Christiansen, F. B. 1981. Selection component analysis of natural polymorphisms using population samples including mother-offspring combinations, II. *Theor. Popul. Biol.* **19**:378–419.

Palmer, A. R. and Strobeck, C. 1986. Fluctuating asymmetry: measurement, analysis, patterns. *Annu. Rev. Ecol. Syst.* **17**:391–421.

Podolsky, R. H. and Holtsford, T. P. 1995. Population structure of morphological traits in *Clarkia dudleyana* I. Comparison of F_{ST} between allozymes and morphological traits. *Genetics* 140:733–744.

Prout, T. 1965. The estimation of fitness from genotypic frequencies. *Evolution* 19:546–551.

Prout, T. 1969. The estimation of fitness from population data. *Genetics* 63:949–967.

Prout, T. 1971a. The relation between fitness components and population prediction in *Drosophila*. I: The estimation of fitness components. *Genetics* 68:127–149.

Prout, T. 1971b. The relation between fitness components and population prediction in *Drosophila*. II: Population prediction. *Genetics* 68:151–167.

Prout, T. and Barker, J. S. F. 1993. F statistic in *Drosophila buzzatii*: selection, population size and inbreeding. *Genetics* 134:369–375.

Ritland, K. 1996. Estimators for pairwise relatedness and individual inbreeding coefficients. *Genet. Res. Cambridge.* 67:175–185.

Robertson, A. 1975. Gene frequency distributions as a test of selective neutrality. *Genetics* 81:775–785.

Roush, R. T. and McKenzie, J. A. 1987. Ecological genetics of insecticide and acaricide resistance. *Annu. Rev. Entomol.* 32:361–380.

Russell, R. J., Dumancic, M. M., Foster, G. G., Weller, G. L., Healy, M. J., and Oakeshott, J. G. 1990. Insecticide resistance as a model system for studying molecular evolution. In *Ecological and Evolutionary Genetics of Drosophila*, J. S. F. Barker, W. T. Starmer, and R. J. MacIntyre, eds., pp. 293–314. New York: Plenum.

Russell, R. J., Robin, G. C., Kostakos, P., Newcomb, R. D., Boyce, T. M., Medveczky, K. M., and Oakeshott, J. G. 1996. Molecular cloning of an α-esterase gene cluster on chromosome 3R of *Drosophila melanogaster*. *Insect Biochem. Mol. Biol.* 26:235–247.

Saad, M., Game, A. Y., Healy, M. J., and Oakeshott, J. G. 1994. Associations of esterase 6 allozyme and activity variation with reproductive fitness in *Drosophila melanogaster*. *Genetica* 94:43–56.

Schaeffer, S. and Aguadé, M. 1999. Evidence for balancing, directional and background selection in molecular evolution. In *Evolutionary Genetics from Molecules to Morphology*, R. Singh and C. Krimbas, eds., Chap. 7. New York: Columbia U. Press.

Schaffer, H. E., Yardley, D., and Anderson, W. W. 1977. Drift or selection: a statistical test of gene frequency variation over generations. *Genetics* 87:371–379.

Scharloo, W. 1991. Canalization: genetic and developmental aspects. *Annu. Rev. Ecol. Syst.* 22:65–93.

Soderlund, D. M. and Bloomquist, J. R. 1990. Molecular mechanisms of insecticide resistance. In *Pesticide Resistance in Arthropods*, R. T. Roush and B. E. Tabashnik, eds., pp. 58–96. New York: Chapman & Hall.

Sokal, R. R. and Rohlf, F. J. 1995. *Biometry*. San Francisco: Freeman.

Spitze, K. 1993. Population structure in *Daphnia obtusa*: quantitative genetic and allozymic variation. *Genetics* 135:367–374.

Thompson, M., Shotkoski, F., and ffrench-Constant, R. 1993. Cloning and sequencing of the cyclodiene insecticide resistance gene from the yellow fever mosquito *Aedes aegypti*. Conservation of the gene and resistance associated mutation with *Drosophila*. *FEBS Lett.* 325:187–190.

Tsakas, S. and Krimbas, C. B. 1976. Testing the heterogeneity of F values: a suggestion and a correction. *Genetics* 84:399–401.

Tukey, J. W. 1977. *Exploratory Data Analysis*. Reading, MA: Addison-Wesley.

Watt, W. B. 1992. Eggs, enzymes and evolution: natural genetic variants change insect fecundity. *Proc. Natl. Acad. Sci. USA* 89:10608–10612.

Wedderburn, R. W. M. 1974. Quasi-likelihood functions, generalized linear models, and the Gauss-Newton method. *Biometrika* **61**:439–447.

Weir, B. S. and Cockerham, C. C. 1984. Estimating F-statistics for the analysis of population structure. *Evolution* **38**:1358–1370.

Williams, C. J., Anderson, W. W., and Arnold, J. 1990. Generalized linear modeling methods for selection component experiments. *Theor. Popul. Biol.* **37**:389–423.

Wilson, S. R., Oakeshott, J. G., Gibson, J. B., and Anderson, P. R. 1982. Measuring selection coefficients affecting the alcohol dehydrogenase polymorphism in *Drosophila melanogaster*. *Genetics* **100**:113–126.

Wilson, T. G. 1988. *Drosophila melanogaster* Diptera: Drosophilidae: A model insect for insecticide resistance studies. *J. Econ. Entomol.* **81**:22–27.

Wright, S. 1951. The genetical structure of populations. *Ann. Eugen.* **15**:323–354.

Wright, S. 1955. Classification of the factors of evolution. *Cold Spring Harbor Symp. Quant. Biol.* **20**:16–24.

Wright, S. and Dobzhansky, T. 1946. The genetics of natural populations. XII. Experimental reproduction of some of the changes caused by natural selection in certain populations of *Drosophila pseudoobscura*. *Genetics* **31**:125–156.

CHAPTER ELEVEN

Genetics of Complex Polymorphisms: Parasites and Maintenance of the Major Histocompatibility Complex Variation

PHILIP W. HEDRICK AND TIMOTHY J. KIM

11.1. Introduction

Often Dick Lewontin says or writes things with unusual clarity and prodigious insightfulness. Rather than paraphrasing his thoughts, we begin with the following timeless and enlightening quote from his presentation at the proceedings of the XI International Congress of Genetics (Lewontin, 1964a):

> In many ways the lot of the theoretical population geneticist of 1963 is a most unhappy one. For he is employed, and has been employed for the last thirty years, in polishing with finer and finer grades of jeweler's rouge those three colossal monuments of mathematical biology *The Causes of Evolution, The Genetical Theory of Natural Selection* and *Evolution in Mendelian Populations*. By the end of 1932 Haldane, Fisher, and Wright had said everything of truly fundamental importance about the theory of genetic change in populations and it is due mainly to man's infinite capacity to make more and more out of less and less, that the rest of us are not currently among the unemployed.
>
> It is true, of course, that these early formulations made very many simplifying assumptions about the nature of the genetic system, on the one hand, and the nature of the environment on the other. But the courage to deliberately treat simple systems in order to understand complex ones is the courage of genius, as Mendel's example so clearly shows. There remains for us, the *epigonai*, to reintroduce bit by bit the complexities of nature, to see to what extent these complexities really make a difference whether qualitative or quantitative in our basic formulations. We can then discard from our considerations certain complexities because they may prove to make no difference whatever and this, from the standpoint of our science, would be the happiest result possible. Other complications, however, will undoubted prove of some real significance and it will then be necessary to include them in our explanation of evolutionary change.

R. S. Singh and C. B. Krimbas, eds., *Evolutionary Genetics: From Molecules to Morphology*, vol. 1. © Cambridge University Press 2000. Printed in the United States of America. ISBN 0-521-57123-5. All rights reserved.

It seems to me that there are two major complications in populations genetic theory which are likely to be of some real importance in understanding genetical changes in populations. The first, and probably most important, is that of the variable and ever-shifting environment. We know virtually nothing about the role of spatial and temporal heterogeneity of environment in evolution and it is here that the most fruitful and most difficult area of research lies....

The second complication, about which we now know rather more, arises when we consider multi-gene systems rather than single Mendelizing factors. When we consider two or more loci, rather than one, two new factors arise which are intimately connected. These are: the interactions between loci in determining fitness (or epistasis...) and second the amount of recombination between loci.

The factors thought to influence genetic variation at the major histocompatibility complex (MHC) encompass nearly every facet of population genetics (e.g., Hedrick, 1994; Parham and Ohta, 1996; Edwards and Hedrick, 1998). In particular, the selective effects resulting from the variable environment because of the absence, presence, or abundance of different parasites (here parasite is used as a generic term for any agent that colonizes tissues or infects cells, after Parham and Ohta, 1996) appear to be the driving force in the maintenance of MHC variation. Furthermore, one could make the case that the tightly linked genes in the MHC comprise a supergene, i.e., "a group of genes acting as a mechanical unit in particular allelic combinations" (Darlington and Mather, 1949) and that it is assumed that selection is primarily responsible for the maintenance of the supergene (e.g., Hedrick et al., 1978; Bodmer, 1978; see also Trowsdale, 1993). In other words, the variable environment, consisting of parasites and their selective effects, and the tightly linked complex of related genes give the MHC the two complications that Lewontin suggested could be of real importance beyond the basic population genetics models.

There are only a few classic case studies of genetic variation that are repeatedly cited as examples of balanced polymorphisms, leading to the general impression that balanced polymorphisms are not very common among the 5000 to 100,000 genes found in most higher eukaryotes. The molecular genetic basis of the variation in some of these systems is known (e.g., in sickle-cell anemia and *Adh* in *Drosophila melanogaster*) while others are understood only from the morphological expression of the genetic types (e.g., melanism in *Biston* and band and color polymorphism in *Cepea*).

Polymorphism at the MHC provides another example of a balanced polymorphism in which many molecular details are well known. The reason for this detailed information lies in the medical importance of this region in humans for understanding the biological basis of tissue transplantation, autoimmune diseases, pathogen susceptibility, and the immune system in general. While the MHC is present in all vertebrates (e.g., Trowsdale, 1995), much of the detailed understanding of this group of genes has come from medically related research on the human MHC (called the HLA) or the mouse MHC (called H-2). However, as we will see, there are many details still not known about MHC, leading to strongly differing opinions about the most significant evolutionary factors influencing genetic variation at MHC genes.

Below, we will first briefly introduce some of the salient genetic and immune aspects of the MHC. Second, we will summarize some of the findings concerning the extent and pattern of genetic variation at the MHC, particularly in humans. Third, we will introduce the possible selective mechanisms that may be responsible for maintaining genetic variation at the MHC. Finally, we will concentrate on the potential role of the MHC in relationship to parasite resistance and evaluate the empirical evidence for resistance, discuss the models suggested for maintaining MHC variation from parasite resistance, and the estimation of selection on MHC.

11.2. Major Histocompatibility Complex

The major impetus for the early investigation of the HLA system was the need to match organ donors and recipients for the antigens important in successful tissue transplantation. We now know from intensive research by a number of laboratories that the HLA region consists of nearly 4000 kilobases (kb) of DNA ($\sim$0.001 of the human genome) covering somewhat over 2.5 map units (in the mapped region, there is $\sim$1 cM/1000 kb; Trowsdale, 1995), containing over 100 known genes, and located on the short arm of chromosome 6. It includes the class I genes (the most important of which are *A*, *B*, and *C*), the class II genes (*DR*, *DQ*, and *DP*), and the class III or complement region (Fig. 11.1) (see Trowsdale, 1995; Parham and Ohta, 1996, and references therein for more details on the MHC). A number of the other genes in this region have related immune functions, making the region rich in genes that have a part in the

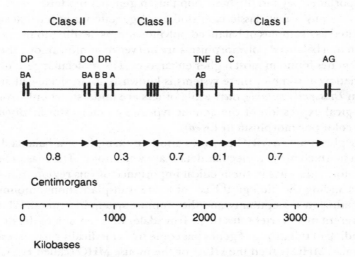

Figure 11.1. Genetic map of the human MHC region or HLA showing both the level of recombination (in centimorgans) and the physical map distance (in kilobases).

immune response. In contrast, the MHC region in chickens seems to be much reduced, with only one active class I and only one active class II gene and no apparent class III region (Kaufman et al., 1995).

Because of the role of MHC molecules in the immune response, there is an extensive amount of information about MHC molecules and their structure and function (e.g., Trowsdale, 1993, 1995). For example, the three-dimensional structure of MHC antigens from both class I and II genes are known, giving extraordinary insight into the function of the molecules coded for by these genes. The main function of many MHC molecules is to recognize short peptides and to present them to T cells to initiate the immune response. The class I molecules present peptides that are broken down from proteins within virus-infected or malignant cells, while the class II molecules present peptides that originate from foreign material (e.g., from bacteria or parasites). Most important for our consideration, these peptides are often from various organisms such as viruses, bacteria, or parasites, and their presentation may lead to an immune response that results in the destruction of the parasite. The peptide-binding properties of the MHC molecule, i.e., which peptides are likely to be associated with a given MHC molecule, are determined by the amino acids lining the peptide-binding pocket. These specific amino acids are known from the molecular structure of the class I and II molecules.

11.3. Extent of MHC Variation

11.3.1. Single-Locus Variation

A distinctive feature of the MHC is the high level of polymorphism exhibited by many of the class I and the class II loci. By use of nucleotide sequences in humans, 67 alleles have currently been described at A, 149 at B, 35 at C, 69 at DPB, 29 at DQB, and 179 at DRB (Parham and Ohta, 1996), making this region far more polymorphic than any other known human system.

In the neutral model, the expected equilibrium heterozygosity (or homozygosity) in a population is a function of the combined effects of genetic drift and mutation, thereby providing theoretical predictions against which the observed genetic variation in a population or a sample may be compared. Ewens (1972) showed that the equilibrium distribution of allele frequencies under neutrality is a function of the sample size and the number of different alleles in the sample, and Watterson (1978) showed that the homozygosity calculated from this distribution can be used to test for deviations from neutrality. Using this approach, Hedrick and Thomson (1983) compared the homozygosity for serotypic alleles at the class I loci HLA-A and HLA-B in 22 samples to neutral expectations, given the same sample size and number of alleles. In all cases, the estimated homozygosity was less (estimated heterozygosity was greater) than expected, indicating a more even allelic frequency distribution than under neutrality, and the homozygosity was significantly less ($p < 0.05$) than neutral expectations for 25 of the 44 cases. (Interestingly, Keith et al. (1985) found that the distribution of alleles at the *Xdh* locus in *D. pseudoobscura* deviated from neutrality expectations in the opposite manner, i.e., the alleles were more uneven in frequency than neutrality expectations.)

A variety of evolutionary factors, such as gene flow, population bottlenecks, unidentified alleles, and balancing selection, that potentially decrease the level of conditional homozygosity, relative to neutral expectations, were examined by Hedrick and Thomson (1983). Some form of balancing selection is the explanation most consistent with the level of conditional homozygosity observed at these loci.

Markow et al. (1993) examined the distribution of *HLA-A* and *HLA-B* alleles in the Havasupai, a small Native American tribe (<600) that inhabits an isolated side canyon of the Grand Canyon in Arizona. This distribution consisted of only three alleles at *HLA-A* and only eight alleles at *HLA-B*. These serotypic alleles have been shown to be homogeneous for nucleotide sequences within serotypes. Even in this population, which is relatively depauperate of genetic variation, the distribution was still quite even for both loci. In fact, the value of the conditional homozygosity over the two loci was only 63.4% of that expected under neutrality.

Given that there is random mating (or nearly so) in the parental generation, the genotypic frequencies in a population should be close to the Hardy–Weinberg proportions predicted by allele frequencies (e.g., Hedrick, 1985). In general, genotypic frequencies for *HLA* genes are close to these theoretical expectations; however, there are a few cases, primarily in isolated populations, in which there is a deficiency of homozygotes (excess of heterozygotes) at *HLA* loci, suggesting that there has been selection even in the current generation (Hedrick, 1990). For example, in the Havasupai, there is a statistically significant deficiency of homozygotes at both the *HLA-A* and the *HLA-B* loci, an average of 25.8% over the two loci (Markow et al., 1993). Several other nonselective factors that can reduce the proportions of homozygotes, such as avoidance of consanguineous matings, do not appear to be the cause of this deviation from Hardy–Weinberg proportions (Markow et al., 1993). A deficiency of MHC homozygotes has also been reported in natural populations of mice (Ritte et al., 1991) and pheasants (von Schantz et al., 1996). However, Boyce et al. (1997), who examined both restriction fragment length polymorphisms for MHC alleles and microsatellites in the same desert bighorn sheep, detected no homozygote deficiencies for either set of loci, and Hedrick and Parker (1998) and Kim et al. (1999) found no homozygote deficiencies in Gila Topminnow and Chinook Salmon, respectively.

11.3.2. Amino-Acid Variation

DNA sequence data and the inferred amino-acid sequence have been obtained for a number of *HLA* alleles and the amino-acid variation is concentrated in the amino-acid residues that are part of the peptide-binding region (Parham et al., 1989). To quantify the extent of this variability, Hedrick et al. (1991a)

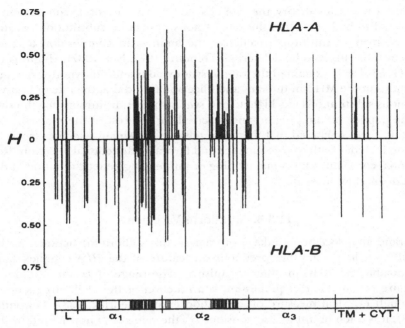

Figure 11.2. The average amino-acid heterozygosity for *HLA-A* (above the horizontal axis) and *HLA-B* (below the axis). The vertical bars along the bottom indicate the 54 amino-acid sites postulated to interact with peptides or the T-cell receptor. Also indicated are the domains of the molecule, where L is the leader domain, α_1 and α_2 are the domains that interact with the peptides and the T-cell receptor, α_3 is the domain under α_2, TM is the membrane-spanning region, and CYT is the cytoplasmic tail.

calculated the heterozygosity (H) for 366 and 363 individual amino acids at *HLA-A* and *HLA-B*, respectively. The average amino-acid heterozygosities for these sites and the amino-acid sites that interact with the peptides or the T-cell receptor are indicated in Fig. 11.2. Note the general congruence of the sites with high heterozygosity and the sites that interact with one of these other molecules.

Further, the heterozygosities among the amino-acid categories are highly significantly different, with an average heterozygosity for the peptide-binding sites of 0.264 and 0.337 for *HLA-A* and *HLA-B* to only 0.036 and 0.031 for the sites that do not interact with the peptide or the T-cell receptor. In other words, there is an order of magnitude difference in heterozygosity between the sites with the known function of presenting peptides and those sites that do not interact with other molecules. The heterozygosity at some of the peptide-binding sites is extremely high, e.g., the average heterozygosity of amino-acid sites 95, 97, 114, and 116 (four sites in the bottom of the pocket that interacts with the peptide) over the two loci is 0.599. Finally, there is a very even frequency distribution at the amino-acid level at these highly variable sites, suggesting that balancing selection may be acting on the amino-acid level.

Under neutrality theory, the rate of synonymous nucleotide substitution is predicted to be larger than the rate of nonsynonymous substitution because nonsynonymous substitutions change the amino-acid composition and are thereby more likely to be deleterious (Kimura, 1977; Nei, 1987). Hughes and Nei (1988, 1989) examined the rate of synonymous and nonsynonymous substitutions in the MHC in humans and mice and found that the rate of nonsynonymous substitutions was higher than synonymous substitutions for the sites in the peptide-binding region and reversed for the rest of the molecule. As a result, they concluded that there was positive Darwinian selection (such selection includes both overdominance and favorable directional selection; Nei, personal communication) maintaining or influencing genetic variation at the peptide-binding sites.

11.3.3. Multilocus Variation

Nonrandom association (linkage or gametic disequilibrium) between alleles at different loci is another predominant feature of the *HLA* complex (and presumably the MHC in other organisms). The range of D, the traditional measure of gametic disequilibrium, is a function of the allelic frequencies. Although no ideal measure of association exists (Hedrick, 1987; Lewontin, 1988), it is often useful to also consider D', the value of D normalized by the maximum value it could have for this combination of allelic frequencies and sign (Lewontin, 1964b).

There are characteristic combinations of alleles at different loci (haplotypes) that exhibit elevated frequencies and gametic disequilibrium (high D' values) in different ethnic and racial groups while other haplotypes do not exhibit significant gametic disequilibrium. Even under neutrality, a population at equilibrium has an expected nonrandom association between alleles at different loci because the combined effects of genetic drift and mutation, particularly when there is limited recombination. In general, there appears to be a somewhat elevated level of overall disequilibrium between *HLA* loci with known recombination levels compared with neutrality expectations (e.g., Hedrick and Thomson, 1986; Markow et al., 1993). However, it is rather difficult to determine the exact cause of this disequilibrium, i.e., whether it is due to selective forces or nonselective factors.

However, having population data on a number of highly variable, tightly linked loci allows some inference about past evolutionary events to be made if the pattern of haplotype frequencies is considered. For example, Klitz and Thomson, (1987), using the two-locus disequilibrium pattern theory they developed (Thomson and Klitz, 1987), determined that there have been approximately six *HLA-A* and *HLA-B* haplotypes that were involved in some past selective event in the sample of Europeans examined.

A number of different measures, some of which are strongly correlated, have been suggested to determine the overall gametic disequilibrium when there are multiple alleles at two loci (e.g., Hedrick and Thomson, 1986; Hedrick, 1987). The measure of Maruyama (1982) appears to decrease in a general linear fashion with increasing recombination. Using this measure, Hedrick et al. (1991b) found a general monotonic trend of decreasing disequilibrium with greater physical distance for 28 MHC locus pairs, as expected theoretically. The correlation of kilobase distance and disequilibrium was quite high, 0.84, and the correlation between the recombination estimate and disequilibrium was nearly as high, 0.81.

Recently there have been reports of important selective effects (and potentially disequilibrium and/or epistasis) for genes, other than the classic MHC loci, in the MHC. These include loci, such as the *TAP* gene in the class II region and the *TNF* in the class III region, which have both been associated with parasite resistance. These findings support the hypothesis that many of the varieties of genes involved in the immune response in the MHC are interacting and are part of a large supergene.

11.4. Microrecombination (Gene Conversion) at MHC

There is evidence that there is a high rate of recombination between alleles at some MHC genes and a lower but significant rate of exchange between alleles at different loci that are important in generating new variants. Most of these recombinants appear to exchange short segments between alleles in a nonreciprocal manner and are referred to as microrecombinants resulting from some type of gene conversion mechanism. Although the amount and the importance of microrecombination at MHC are not resolved, there are three major types of evidence that support its role for some, if not all, MHC genes.

The initial evidence for gene-conversion-like events in the MHC came from comparing DNA sequences with the finding that some alleles differed from other sequences by small segments that were present in a third allele, suggesting that there was exchange between known alleles at a gene (and between genes) to generate new sequences (e.g., Kuhner et al., 1990). The interpretation of these observations is that there is a relatively high rate of microrecombination or gene conversion (a nonreciprocal exchange of short DNA segments) between alleles for some MHC genes. Parham et al. (1995) recently examined the sequences of 64 class I alleles (18 for *HLA-A*, 40 for *HLA-B*, and 6 for *HLA-C*), which appeared to have been generated by exchange of short internal segments, and found that most of the putative exchanges involved less than 35 nucleotides (the mode was between 15 and 20 nucleotides). Parham and Ohta (1996) found evidence for only four single point mutations in *HLA* class I but at least 80 by gene conversion. These observations suggest that it is inappropriate for MHC genes to assume that each nucleotide change is independent of change at neighboring sites, an assumption fundamental to many molecular evolution calculations.

An important example of microrecombination is that the B*5301 allele, the one that confers resistance to malaria as discussed below, appears to be an allele identical to the common allele B*3501 except for codons 77 to 83 (Allsopp, et al., 1991). B*5301 differs from B*3501 by five amino acids, three of which are part of the peptide-binding region. Unlike B*5301, B*3501 does not appear to confer resistance to malaria, suggesting that the selective difference between these alleles (as well as the antigenic specificity) lies in this short segment.

Second, new alleles that appear to be the result of microrecombination between other alleles have been found in South American Amerindians and other populations (e.g., Belich et al., 1992; Watkins et al., 1992; Parham and Ohta, 1996). Because the Americas have probably been populated for only the past 10,000 to 20,000 years (~1000 human generations), the new variants, which do not appear in Asian samples, must have arisen during this period. In South Amerindians 23 out of 28 alleles found at the *HLA-B* locus are not found elsewhere and all 23 appear to be the result of microrecombination (Parham and Ohta, 1996). To have a nearly complete turnover of alleles at a locus in 1000 generations suggests that the rate of generation of new recombinant alleles must be quite high.

Third, there is direct evidence that the rate of microrecombination at some MHC loci is high. Zangenberg et al. (1995) examined the rate of interallelic gene conversion at the *HLA-DPB1* locus in sperm from males heterozygous for six regions of the highly variable exon 2. In 111,675 sperm, they observed nine interallelic conversions for a rate of 0.81×10^{-3}, nearly 1 in 10,000 gametes per generation. Hogstrand and Bohme (1994) estimated the rate of intergenic gene conversion between two loci in mice as 0.5×10^{-6}, much lower than the rate of interallelic gene conversion but high enough potentially to have an impact on the variation in the genes involved.

The rate of gene conversion appears to vary among different loci. For class I loci, *HLA-A* sequences show good allelic lineages in worldwide samples and fewer new alleles in the South American Amerindians as compared with *HLA-B*, which does not have clear allelic lineages (apparently obscured by gene conversion) and has many new alleles (e.g., Parham and Ohta, 1996). For the class II loci, the evidence for microrecombination for *HLA-DPB1* is from sequence patterns, new alleles (Moonsamy et al., 1994), and direct experimental evidence. For even *HLA-DRB1*, which is thought to have a relatively low rate of microrecombination, there is evidence of gene conversion from sequences (e.g., Gaur and Nepom, 1996) and of a novel allele in Cayapas, South American Amerindians (Titus-Trachtenberg et al., 1994).

Microrecombination cannot generate a new variation if a gene is homozygous; however, in most populations the level of heterozygosity for the polymorphic MHC loci is high. Even in populations with few alleles, such as the Havasupai, the observed heterozygosity for *HLA-A*, with only three alleles, is 69% (Markow et al., 1993). In addition, microrecombination should be intrinsically more efficient in producing new favorable alleles than point mutation because the parental sequences are already in existing alleles that have survived previous purifying selection and genetic drift. On the other extreme, the microrecombinants observed by Zangenberg et al. (1995) did not result in any frameshift mutants, i.e., recombinants with potentially low fitness. As a result, the probability of incorporation of a new variant produced by microrecombination may be higher than that of a random mutation.

11.5. Modes of Selection

Several modes of balancing selection have been proposed as important in the maintenance of variation at the MHC: parasite resistance, maternal–fetal interaction, and negative assortative mating. We will first mention here selection related to autoimmune diseases and then briefly how maternal–fetal interaction and negative assortative mating are thought to work at MHC. In the remainder of the chapter, we focus on MHC and its relationship to parasite resistance.

11.5.1. Autoimmune Diseases

The most easily identified source of selection on MHC loci is that associated with the many autoimmune diseases in humans, such as insulin-dependent diabetes, rheumatoid arthritis, Reiter's disease, ankylosing spondylitis, etc. (e.g., Tiwari and Terasaki, 1985; Thomson, 1988). For example, the frequency of particular antigens in individuals with some autoimmune diseases may be many times that of individuals in the rest of the population. However, because most of these diseases are fairly uncommon and they generally do not appear to greatly influence reproduction, their impact on genetic variation at the *HLA* loci may not be large. Further, because autoimmune diseases tend to lower fitness in the relatively few individuals with each disease, it is unlikely that they play an important role in maintaining MHC variation.

11.5.2. Maternal–Fetal Interactions

A number of studies indicate that a significant number of couples with a history of spontaneous abortions share antigens for *HLA* loci, particularly at two or three loci (e.g., Thomas et al., 1985; Alberts and Ober, 1993). The immunological explanation for these observations is that the presence of an immune response occurring when the mother and the fetus differ at *HLA* loci is necessary for proper implantation and fetal growth. This explanation has the potential to maintain many alleles at a single locus and to result in substantial gametic disequilibrium between linked loci (Hedrick and Thomson, 1988). However, the potential importance of this mechanism for maintaining polymorphism is limited to mammals because there is unlikely to be similar maternal–fetal interactions in birds, reptiles, amphibians, or most fishes. Maternal–fetal interactions could, of course, enhance the maintenance of polymorphisms in mammals or other live-bearing organisms that originally were generated as a mechanism for defense against pathogens.

11.5.3. Negative Assortative Mating

Although there is little evidence for nonrandom mating with respect to *HLA* types (Hedrick and Black, 1997; however see Ober et al., 1997), experimental studies in mice have demonstrated mating preferences with regard to H-2 types (for a review, see Brown and Eklund, 1994). There are laboratory studies both showing that males mate preferentially with females of a different MHC type and females select males of a different MHC type. From cross-fostering experiments, it appears that this preference may be primarily the result of familial imprinting (Yamazaki et al., 1988). In a seminatural mouse population, Potts et al. (1994) found a large deficiency of *H-2* homozygotes and concluded by eliminating other factors that this effect was due to female selection for males different from themselves. If there was strong negative assortative mating by females, a deficiency in homozygotes as high as that observed could be generated (Hedrick, 1992). Potts et al. (1994) suggest that such negative assortative mating based on MHC may be important in avoiding inbreeding depression. Again one could suggest that this type of negative assortative mating would enhance the maintenance of MHC polymorphism but may not be the fundamental cause of it.

11.6. Evidence for Parasite Resistance at MHC

A major function of MHC histocompatibility molecules is to present foreign antigens (peptides) to T cells to stimulate an immune response against invading parasites (e.g., Zinkernagel, 1979). Histocompatibility alleles have been shown to differ in their ability to create an immune response to a variety of infectious agents (Van Eden et al., 1983), suggesting that the epidemic diseases of the human past may have played a central role in determining the *HLA* allele and haplotype frequencies observed in human populations today (presumably MHC frequencies in other species have been influenced by similar selective factors). For example, it has been suggested that approximately 56 million Native American people died of introduced diseases as the result of settlement of the Americas by Europeans and as recently as the 1960s and 1970s mortality rates in the recently contacted Amazonian Indians populations were up to 75% (Black, 1992). Here we review the evidence for MHC resistance for most of the strong cases of association (for more details, see Plachy et al., 1992; and references therein). An additional important association, which we do not discuss below, is the correlation of the type of Epstein-Barr virus isolates and the frequency of *HLA-A11* in different human populations (de Campos-Lima et al., 1993).

11.6.1. Malaria

There has been evidence that the *HLA* variants, class I allele B*5301 and the class II haplotype associated with DRB1*1302, appear to confer resistance to malaria (Hill et al., 1991) and that susceptibility to malaria is related to variation at a class III gene, *TNF* (McGuire et al., 1994). For example, in a large Gambian sample, the frequency of allele B*5301 is 55% higher in uninfected individuals than infants infected with severe malaria. By use of the frequencies of the B*5301 allele in individuals with severe malaria and the appropriate controls, the relative risk (*RR*; see below) of severe malaria for individuals with *B*5301 can be compared with those without this allele (Hill et al., 1991). Further, allele

B*5301 is nearly absent from populations without a history of endemic malaria, such as those of northern Europeans, while nearly 40% of the individuals in a Nigerian sample were either homozygous or heterozygous for this allele (Okoye et al., 1985). Using calculations based on these findings, Hill et al. (1991) concluded that these *HLA* variants prevent as much malaria as does the sickle-cell variant in Gambia.

Molecular research has investigated the mechanism of *HLA* resistance to malaria; e.g., Davenport et al. (1995) have shown that the resistance to malaria exhibited by DRB1*1302 and not shared by the very closely related DRB1*1301 (they differ by one amino acid, Gly to Val, at position 86, a position that is part of the peptide-binding pocket) is consistent with the findings that these two alleles bind slightly different sets of peptides. They propose that the malaria parasite may contain an immunodominant epitope (peptide) that is recognized by the DRB1*1302 Gly at position 86 and not by the Val-containing DRB1*1301.

In Kenya, the frequency of B*5301 is also relatively high at approximately 16% in the general population. However, in a group of children with severe malaria, the frequency of B*5301 is quite similar to that of the population frequency (Hill et al., 1994), suggesting that the strong association of B*5301 and severe malaria in Gambia is not present in Kenya. Further, the class II haplotype important in Gambia does not appear to give protection to malaria in Kenya while there is a different haplotype that appears to confer protection (Hill et al., 1994). The basis for this difference may result from differences in the malarial parasite between west and east Africa either in surface antigens or the transmission properties (Hill et al., 1994).

11.6.2. HIV

Among individuals infected with human immunodeficiency virus type 1 (HIV-1), the course and the duration of the disease are highly variable, with some individuals developing AIDS in less than 1 year while others have remained AIDS free for over a decade (references in Kaslow et al., 1996). Some observations suggest a role of MHC in the disease progression and immune response to HIV-1 (Just, 1995). For example, Kroner et al. (1995) showed that sibling pairs of HIV-1 infected hemophiliacs that share one or two *HLA* haplotypes were significantly more concordant in their disease progression. Also, studies of a similar murine retrovirus in mice have demonstrated that different host mice strains show differential resistance to murine-acquired immune deficiency syndrome and that this resistance maps to the H-2 complex (Makino et al., 1990).

Although a number of studies suggest an association between HIV-1 and HLA type (e.g., Just, 1995), these studies suffer from several complicating issues, including the observed heterogeneity of both the HLA and the HIV and the ethnicity of the risk group. For example, HIV-1 exhibits extreme diversity and even striking intrahost variability, factors that may help the virus to evade host immune responses (Callahan et al., 1990), but make it difficult to identify HLA–HIV associations. Further, identification of control groups is also problematic in most association studies because of the mixed ethnic composition and overrepresentation of some population subgroups in many of the high-risk groups studied (Hill, 1996b).

Nonetheless, consistent associations of certain *HLA* haplotypes and HIV disease progression have been reported. For example, several studies have shown associations with the common Caucasian *HLA* haplotype *A1 B8 DR3* and rapid AIDS development (e.g., Steel et al., 1988; Kaslow et al., 1990). More recently, a study of two seropositive cohorts by Kaslow et al. (1996) used *HLA* profiles derived from serotyping the *HLA* class I and molecular typing the class II loci and the MHC region *TAP* locus that encodes proteins involved in antigen processing and transporting. They found that associations between the time to progression to AIDS and specific *HLA* combinations derived from analysis of one cohort was predictive of the progression times in the second cohort. A recent study suggests that B35 and CO4 are related to rapid onset of AIDS and that class I heterozygotes have a delayed onset of AIDS (Carrington et al., 1999).

11.6.3. Hepatitis

Infection by hepatitis B virus (HBV) results in divergent disease conditions, ranging from seemingly complete recovery to the development of chronic fulminant hepatitis and possible liver failure. The course of the disease does not appear to be determined by variations in viral virulence and most of the evidence suggests that host genetic factors play an important role in chronic carrier status (e.g., Thurz et al., 1995). Early studies, reviewed by Kaslow and Shaw (1981), generally suffered from small sample size and do not provide conclusive evidence of MHC associations. The strongest associations are with *DR* but the overall picture of resistance and susceptibility to HBV is not clear.

Among the strongest associations are those found by Almarri and Batchelor (1994) in two independent studies in Qatar, where an estimated 16% of the population is exposed to the virus, for *HLA-DR7* and carrier status or susceptibility to persistent infection and *HLA-DR2* with non-carrier status. The association of *HLA-DR7* and several other antigens with persistent infection is supported by the inability of some individuals with these genotypes to develop significant antibody responses after vaccination (Craven et al., 1986; Weissman et al., 1988). Another association was found by Thurz et al. (1995), in which DRB1*1302 and DRB1*1301 (which did not confer resistance to malaria) were less frequent among Gambians with chronic HBV, and van Hattum et al. (1987) found that an antigen type that includes DR13 was present almost twice as frequently in north European noncarriers as in carriers. Recently, Thurz et al. (1997) found an association of class II heterozygotes with apparent resistance in Gambia.

11.6.4. Marek's Disease

There are several well-supported examples of MHC–parasite resistance association in chickens (e.g., Plachy et al., 1992), and a classic example of MHC resistance to infectious disease, Marek's disease, is in chickens. In this example, animals with the haplotype B^{21} and other similar haplotypes have a much higher resistance to Marek's disease, a type of viral leukemia (Briles et al., 1977), than those with other haplotypes. In fact, individuals with haplotype B^{21} have as much as a 95% chance of survival when infected with Marek's while chickens with other haplotypes often have very high mortality from infection. The B^{21} haplotype is also found in the red jungle fowl of Borneo, the presumed ancestor of the domestic chicken, as is Marek's disease, suggesting a long association of this haplotype with Marek's resistance. If chickens in an MHC variable population are selected for resistance to Marek's, the frequency of the B^{21} haplotype increases rapidly as a consequence (Gavora et al., 1986). Kaufman et al. (1995) have found a strong negative association with the level of class I expression and resistance to Marek's, with B^{21} having very low expression, although the mechanism of resistance remains unclear. Kaufman et al. (1995) suggest that the small and simple MHC in chickens may provide an opportunity to understand the mechanisms underlying MHC–parasite resistance.

11.6.5. Low MHC Variation and Disease Susceptibility

The number of MHC genes and the amount of variation at MHC genes varies significantly among species (Trowsdale, 1995). Some researchers have suggested that low MHC variation may make a species particularly vulnerable to parasites. These correlations are tantalizing, but in published cases there do not seem to be any direct cause–effect data. In several of the organisms mentioned below, species initially thought to have low MHC variation on further investigation have been found to have significant MHC variation.

Probably the most celebrated case of low MHC variation is the cheetah, in which most reciprocal skin transplants between unrelated animals were accepted (O'Brien et al., 1985) (indicating low MHC variation), and a study of RFLPs in the MHC concluded that the cheetah had approximately one third the MHC variation of humans (Yuhki and O'Brien, 1990). For cheetahs, there is also lower variation at some other molecular markers than in other big cats (however, see Menotti-Raymond and O'Brien, 1995). As a result, the apparent high susceptibility of cheetahs to feline infectious peritonitis (FIP) (mortality in 18 out of 42 animals, O'Brien et al., 1985) in an outbreak in Oregon is also correlated with lower variation at MHC and other genes as well. However, FIP and other infectious diseases do not appear to be important immediate causes of mortality in wild cheetahs (e.g., Caro and Laurenson, 1994).

Several other studies have examined organisms that have low MHC variation and have explained these findings as the result of lifestyles in which there is less exposure to parasites. For example, the Syrian hamster, a solitary species that appears to have poor resistance to viral infections, was initially thought to lack variation in class I MHC genes (e.g., Darden and Streilein, 1984). However, a more recent study (Watkins et al., 1990) suggests that Syrian hamsters express a diverse group of class I alleles. Whales (Trowsdale et al., 1989) and southern elephant seals (Slade, 1992) also appear to have low MHC variation and, because of their aquatic environment, may be less exposed to parasites. Initially, cotton-top tamarins, which appear to be highly susceptible to a variety of parasites and are usually born as dizygotic twins and develop as stable bone marrow chimeras, were found to have low class I MHC variation (Watkins et al., 1988). However, a later study showed that there was extensive class II MHC variation in tamarins (Gyllensten et al., 1994).

11.6.6. Difficulties in Finding an MHC–Parasite Association

Even though it is generally accepted that MHC variants are important in defense against parasites, the number of documented examples demonstrating differential resistance seems relatively small to many researchers. Many of the strongest examples are in humans, because both the MHC and infectious diseases are better known in humans than for other organisms. There have been a number of suggestions as to why it has been difficult to determine a consistent relationship between MHC variants and resistance to particular parasites (most of these explanations are related to studies in humans but many are probably equally applicable to other organisms).

First, Hill (1991) suggested that many HLA–parasite studies did not have proper statistical design and, for example, did not have large enough sample sizes (enough power) to detect any but the strongest associations. Further, because there often have been a number of HLA–parasite combinations tested, associations may have been observed by chance. As a result, Hill et al. (1991), Kaslow et al. (1996), and others have used two-stage approaches in which an initial survey determines associations and the second stage examines a particular association in an independent sample from the same population. Second, until recently associations were often between serotypes and parasites and it has now been found that there is heterogeneity within serotypes in most populations, making it possible that there is association with some alleles within serotype and not others. Because HLA types are generally now determined from DNA sequences, alleles within serotypes can be examined for association, although the sample size may be reduced.

Third, from their findings in malaria, Hill et al. (1994) suggest that MHC adaptation may differ in different areas because the parasite may be geographically differentiated. Such parasite differentiation could be genetic and influence virulence, parasite transmission, or other factors important for the maintenance of the parasite. Fourth, because different loci in the MHC region may be involved in resistance and alleles at these loci may be in linkage disequilibrium, it may be difficult to determine the effect of a given allele at a locus separate from the influence of the genetic background, i.e., to obtain a large enough sample for given alleles with all background combinations. Further, alleles at different MHC loci may act epistatically either to enhance or to reduce resistance.

In addition to the above reasons, Klein and O'hUigin (1994) suggested that the present MHC alleles are ones left over from adaptation to old parasites that are not presently a problem so that testing for resistance to contemporary parasites would not give positive results. Further, other loci outside the MHC may be important in determining genetic resistance to parasites. For example, three of the 12 genes conferring resistance to malaria are in the HLA (Hill 1992) and genes other than MHC are known to influence parasite resistance in mice and other organisms (e.g., Wakelin and Blackwell, 1988). Next, some estimates of the long-term selection on MHC are of the order of 1%–2% (see discussion below). If this is so, and particularly if the extent of selection varies over time, then at any given time it may be difficult to identify strong selection. Finally, Kaufman et al. (1995) suggest that the MHC–parasite resistance association may be weaker in mammals (for which most of the association studies have been conducted) than it is in birds because the MHC region is larger and more complex in mammals than birds.

11.7. Models of Selection for Parasite Resistance

To understand how selection for parasite resistance influences genetic variation at the MHC, one should know the appropriate model to determine fitness values. As we will see below, three main types of balancing selection; heterozygote advantage, frequency-dependent selection, and variable selection in time and space, have been suggested as important for MHC, and it is not clear which of these models is most appropriate (or if all have a role at times). In fact, the three types of balancing selection may be overlapping. For example, a frequency-dependent model in which alleles, when rare, have a high fitness may also have a constant advantage for heterozygotes. Further, it is likely that the extent of selection on an allele under a heterozygote advantage or a frequency-dependent model varies over both space and time because of spatial and temporal variation in parasites.

11.7.1. Heterozygote Advantage

A number of MHC researchers have suggested that MHC heterozygotes may have a higher fitness than homozygotes (e.g., Doherty and Zingernagel, 1975; Black and Salzano, 1981), a model that in its simplest form assumes an equal and constant advantage of all heterozygotes over all homozygotes. The advantage to a constant fitness model with a given heterozygous advantage (a kind of net effect of selection) is that it depends on only one parameter, but this simplicity may ignore important biological variation in both the host and the parasite. Such heterozygote advantage (overdominant selection) is based on the suggestion that heterozygotes are able to recognize two suites of parasites, one for each allele, and are therefore likely to have a higher fitness than homozygotes that can recognize only one group of parasites. Such a model assumes that one allele is enough for recognition, implying that there is dominance for disease recognition.

If either the heterozygotes or homozygotes do not all have the same fitness inducing some asymmetry in fitnesses, then the likelihood of a stable polymorphism for multiple alleles is reduced (e.g., Kingman, 1961; Mandel, 1970). The asymmetric overdominance investigated by Takahata and Nei (1990) assumed stochastic variation in the fitness of homozygotes, different from the constant asymmetry considered in models of Kingman (1961) and others. Further, even with greater fitnesses for all heterozygotes than homozygotes, the probability of generating a stable multiallelic system of six alleles or more is very low (Lewontin et al., 1978). If new variants are generated over time, then somewhat larger stable multiple allele systems can build up (e.g., Spencer and Marks, 1988, 1992). However, even most of these examples have less than 10 alleles and generally only five or so common alleles and their distribution is not different from that expected from neutrality, unlike that observed for many MHC genes.

11.7.2. Frequency-Dependent Selection

A frequency-dependent selection model in which genotypes with a rare allele have a strong selective advantage could also maintain genetic polymorphism at MHC (Bodmer, 1972). In such models, commonly suggested for host–parasite systems in both plants and animals, a new allele would allow greater protection against parasites than more common alleles to which parasites may have evolved resistance. Takahata and Nei (1990) and Denniston and Crow (1990) showed that one model of frequency-dependent selection has the same allele frequency dynamics as the symmetric heterozygote advantage model. However, this identity is not universal, and other frequency-dependent models have very different dynamics than heterozygote advantage (e.g., Wright, 1969; Hedrick, 1985).

For example, the frequency-dependent model suggested by Huang et al. (1971) and analyzed in detail by Cockerham et al. (1972) has higher changes in allele frequency near fixation than does a comparable heterozygote advantage model, making the potential for retention of variation in a finite population higher for frequency dependence (Hedrick 1972). Specifically examining transspecies MHC polymorphism (polymorphism that was present before speciation and has been retained), Golding (1992) shows that a model in which rare alleles have a large selective advantage can maintain polymorphism longer and with less selection than the heterozygote advantage model.

After publication of the paper by Hill et al. (1991) documenting malaria resistance for HLA, there were several comments about the likelihood of heterozygote advantage versus frequency-dependent selection (Hughes and Nei, 1992a; Hill et al., 1992; Slade and McCallum, 1992; Hughes and Nei, 1992b) because Hill et al. (1991) stated that their results were more consistent with frequency dependence than heterozygosity advantage at MHC. For example, Hughes and Nei (1992a) stated that the heterozygote advantage model, as they envisaged it, depends on the simultaneous exposure of two or more parasites, suggesting that because the malaria data considered resistance to only one parasite, the relative resistance to all the parasites present in the population for the different genotypes is not known.

Table 11.1. *The Frequency of B*5301 and DRB1*1302 in Samples of size N with Malaria (Severe or Cerebral) and Controls (Mild, Severe, or Healthy Adults)[†] (Hill et al., 1991)*

	Frequency[‡] (N)	Homozygote Frequency	RR	Heterozygote Frequency	RR
B*5301					
Severe malaria	0.157(306)	0.036		0.121	
Mild control	0.243(144)	0.042	0.93	0.201	0.55
Severe control	0.216(171)	0.029	1.25	0.187	0.60
Healthy adults	0.250(112)	0.036	1.00	0.214	0.51
DRB1*1302					
Severe malaria	0.087(218)	0.0092	0.22	0.156	0.55
Cerebral malaria	0.135(413)	0.0145	0.36	0.242	0.95
Mild control	0.164(479)	0.0397		0.251	

[†] Also given are the observed homozygote and heterozygote frequencies and the RR values for these genotypes between those with malaria and a control.

[‡] For B*5301 the frequency is that of the antigen (homozygotes plus heterozygotes) while for DRB1*1302 it is that of the haplotype.

The conclusions of Hill et al. (1991) were based on the relative risk of having malaria for homozygotes versus heterozygotes. The relative risk is calculated as

$$RR = \frac{f_d(1 - f_c)}{(1 - f_d)f_c},$$ (11.1)

where f_d and f_c are the frequencies of the genotype or allele in the diseased and the control samples, respectively (e.g., Bengtsson and Thomson, 1981). For the B*5301 variant (Table 11.1, summarized from Hill et al., 1991), it appeared that the heterozygote was protected (had lower RR and therefore higher fitness) when compared with all three controls, while there was no indication of protection from malaria for the homozygotes. On the other hand, the homozygotes for the class II haplotype had higher resistance (lower RR, higher fitness) than did the heterozygotes, and both of these values for severe malaria were less than unity. If one copy of the allele confers resistance, as in the model given in Table 11.2, then it would be expected that the RR for both genotypes would be less than unity and that they would be similar in magnitude. When the allele is in low frequency (neither B*5301 nor DRB1*1302 is really rare), then most frequency-dependent models would predict that the homozygote would have a higher fitness (lower RR) than the heterozygote. Overall then, it would seem that the DRB1*1302 is potentially consistent with frequency dependence and B*5301 is not consistent with either model.

Takahata and Nei (1990) suggested that under frequency-dependent selection new alleles would have a selective advantage because parasites would not be

Table 11.2. *The Relative Survival of Three Genotypes*[†]

Parasite 1, Parasite 2	Frequency	Genotype		
		A_1A_1	A_1A_2	A_2A_2
$-,-$	$(1-e_1)(1-e_2)$	1	1	1
$+,-$	$e_1(1-e_2)$	1	1	$1-s_2$
$-,+$	$(1-e_1)e_2$	$1-s_1$	1	1
$+,+$	e_1e_2	$1-s_1$	1	$1-s_2$

[†] Parasites 1 and 2 are both absent $(-,-)$; only parasite 1 is present $(+,-)$; only parasite 2 is present $(-,+)$; or both parasites are present $(+,+)$. e_1 and e_2 are the frequencies of parasites 1 and 2 and s_1 and s_2 are the selection coefficients against A_1A_1 and A_2A_2 in the presence of parasites 2 and 1, respectively (from Hedrick et al., 1987).

adapted to them. On the other hand, Slade and McCallum (1992) hypothesized that old rare host alleles may also have a high fitness. They suggested that host alleles, common in the past, could have lost their fitness advantage because the parasite had evolved resistance to it. As a result, the host allele may decline in frequency, and subsequently the parasite adaptation may also decline or disappear. At this time, the host allele may again become adaptive and again increase in frequency. In other words, a rare old allele, as well as a rare new allele, may also have a selective advantage.

Nei and Hughes (1991) suggested that a frequency-dependent model and a heterozygous advantage model could be distinguished by looking at the fitness of a homozygote for a rare allele (of course, such a genotype would be impractically rare) and compare it with the fitness of a heterozygote for a pair of common alleles. They suggest that if the rare homozygote had a higher fitness than the heterozygote this would be support for frequency dependence, and higher fitness of the heterozygote rather than the rare homozygote would indicate heterozygote advantage (these measurements would also need to be in the presence of two or more parasites; Hughes and Nei, 1992a). If this experiment could be done, alleles could have an advantage both when rare and as heterozygotes when common so that both models may apply.

11.7.3. Variable Selection in Time and Space

Because some of the important infectious human diseases of the past are epidemic, Hedrick et al. (1987) suggested a selection model for maintaining MHC variation in which a given allele confers resistance to a given parasite and there is variable presence or absence of the parasite (selection) over time. This is different than the gene-for-gene model visualized in plant- pathogen systems (e.g., Burdon, 1987) because here the parasite variation is assumed to be between different species and the resistance due to multiple alleles at a given gene. Hill (1991) expanded this concept and suggested that such models need to incorporate spatial variation and multiple loci. Such a model can be made more biologically realistic, including factors such as linked and epistatic genes; spatial, temporal, and abundance variation in the presence of the parasite; evolution of the parasite; and mutation and microrecombination in the MHC.

In the simplest form (Table 11.2), there are only two alleles at the MHC locus with the genotypes with A_1, the homozygote A_1A_1, and the heterozygote, having resistance to parasite 1, and the genotypes with A_2, the homozygote A_2A_2, and the heterozygote, having resistance to parasite 2. The presence of the parasite may vary over time such that the probability of the presence of parasites 1 and 2 are e_1 and e_2 (and the probability of absence is one minus these values). Note that when both parasites are present in a particular generation, then there is heterozygous advantage, like that envisaged by Hughes and Nei (1992a).

To illustrate how this type of balancing selection can maintain heterozygosity to the extent observed in South American Indian populations by Black and Salzano (1981), Hedrick (unpublished) carried out simulations with different levels of selection, dominance, numbers of alleles, numbers of parasites, presence of parasites, and autocorrelation of the presence of parasites between sequential generations. An autocorrelation of zero indicates independence of the presence of the parasite over generations, while a positive autocorrelation indicates a higher likelihood of the presence of a given parasite given presence in the previous generation (e.g., Hedrick et al., 1976).

In the observed column of Table 11.3, the average heterozygosity for *HLA-A* and *HLA-B* in the three South American groups studied by Black and Salzano (1981) is given (three combinations have four alleles and three have five alleles). When selection against the homozygotes is 0.2 ($s_1 = s_1 = 0.2$) and the parasites are present in half the generations ($e_1 = e_2 = 0.5$), then the equilibrium heterozygosity is given in Table 11.3 for four autocorrelation levels. For both four and five alleles, the observed heterozygosity is bracketed by the equilibrium heterozygosity for autocorrelations of 0.5 and 0.75, showing that

Table 11.3. *Observed Hardy–Weinberg Heterozygosity in Three South American Amerindian Groups (data from Black and Salzano, 1981) and the Theoretical Expected Heterozygosity when $e_1 = e_2 = 0.5$ and $s_1 = s_1 = 0.2$ for Four Different Levels of Autocorrelation for the Presence of Parasites (Hedrick, unpublished)*

Number of alleles	Observed	Autocorrelation			
		0.0	0.25	0.5	0.75
4	0.677	0.728	0.717	0.692	0.616
5	0.713	0.774	0.762	0.737	0.671

these parameters give theoretical heterozygosity values consistent with those observed. While the model as given here does not have either finite-population size or mutation (or microrecombination), the level of selection is similar to that observed in these tribes by Black and Hedrick (1997; see below).

11.8. Estimation of Selection

Estimation of the extent of selection is a complicated and assumption-ridden process (e.g., Hedrick and Murray, 1983; Hedrick, 1985, and references therein). Even when selection is fairly substantial, it may be difficult to determine whether selection is operating. For example, the mainly aboriginal populations that show a deficiency of *HLA* homozygotes, apparently indicating a net heterozygote advantage, are ones that have a limited number of alleles (for MHC), are fairly isolated, and have little or no variation within serotype (Markow et al., 1993). On the other hand, the level of selection in many nonaboriginal populations may be hard to estimate because nearly all individuals are heterozygotes for the variable *HLA* loci, past selection pressures are not now present because of better health care and sanitation, and there is extensive variation within serotype so that many serotypic homozygotes are actually heterozygous.

There are three basic ways in which the extent of selection at MHC has been estimated. First, as just discussed, there are several studies in which there is a deficiency of homozygotes compared with Hardy–Weinberg expectations, particularly in the Havasupai Indians of Arizona (Markow et al., 1993) and several South American Amerindian groups (Black and Hedrick, 1997). In both groups, there are relatively few alleles at the *HLA* genes examined, the populations have been and still are quite isolated, and there does not appear to be variation within serotype. For example, using the difference between the expected and the observed levels of heterozygotes or homozygotes, we obtain

$$s = \frac{\sum p_i^2 - P}{\sum p_i^2(1 - P)}, \tag{11.2}$$

where p_i is the frequency of the ith allele and P is the observed frequency of homozygotes (Hedrick, 1990; see also Satta et al., 1994), the selection against *HLA-A* and *HLA-B* homozygotes in the Havasupai are 0.314 and 0.387, respectively, and for the South American tribes averages for *HLA-A* and *HLA-B* are 0.261 and 0.192, respectively, both quite high levels in the generation examined (such deviations would disappear in one generation without selection).

Second, Hill (1991) suggested that the extent of differential selection due to a particular parasite between a given genotype (or allele) and other genotypes in the population can be estimated as

$$s = m(1 - RR), \tag{11.3}$$

where m is the proportion of mortality caused by the parasite and RR is the relative risk for the genotype for the disease (e.g., Bengtsson and Thomson, 1981).

From this expression, the amount of selection increases as the proportion of mortality from the parasite increases and as RR approaches 0 (complete protection from the parasite) while the protection approaches 0 as RR approaches unity. (For RR greater than unity, higher susceptibility to infection than other genotypes, some other measure might be more appropriate.) For example, RR for individuals with B*5301 compared with non-B*5301 individuals to get severe malaria is 0.59 and for DRB1*1302 compared with those without DRB1*1302 is 0.45 (Hill et al., 1991). Assuming then that the proportion of mortality in Gambia from malaria is ~0.07 (Hill, 1991), the levels of selection favoring B*5301 and DRB1*1302 are estimated to be 0.029 and 0.038, respectively.

Third, indirect methods of estimating the amount of selection have been developed that use DNA sequence data. For example, using the unusual pattern of gene substitution for MHC, Takahata et al. (1992) and Satta et al. (1994) developed indirect methods to measure the selective difference between heterozygotes and homozygotes based on the rate of nonsynonymous and synonymous substitution observed for different loci. Satta et al. (1994) estimated two parameters, the level of selection S ($= 2N_e s$, where N_e is the effective population size and s is the selective difference between heterozygotes and homozygotes) and M based on the mean number of nonsynonymous substitutions and the relative nonsynonymous substitution rate from published sequences for seven HLA genes. The level of S varies from high values of 8200 for HLA-B and 3900 for DRB1 to a low value of 140 for DPB1, and the level of M shows a more or less inverse pattern of S (Table 11.4). Satta et al. (1994) averaged the M values (even though they are quite heterogeneous) for each class of loci and then estimated N_e, by using the relationship $N_e = M/u$, where u is the mutation rate per peptide-binding region per generation and obtained N_e estimates of ~100,000.

On the other hand, N_e can be estimated for each locus separately, as given in column 4 of Table 11.4. In theory, the estimated N_e values for different loci should be similar because N_e is a constant for the population and not a locus-specific parameter. However, the estimates of N_e here for HLA-B and HLA-DRB1 are an order of magnitude smaller, slightly over 10,000, than the value of Satta et al. (1994) and similar to the value estimated for allozyme loci by Nei (1987). The magnitude of s (estimated from $S/2N_e$) is given in the fifth column, and the value given by Satta et al. (1994) by using $N_e = 100,000$ is given in the last column for comparison.

When locus-specific estimates of N_e are used, the estimated level of selection varies greatly over loci. For example, for HLA-B there is an estimated 33.4% difference in fitness between heterozygotes and homozygotes and for HLA-DRB1 there is a 14.2% difference. On the other hand, the estimates for HLA-C and HLA-DPB1 are only 0.15% and 0.0064%, respectively. The range is several orders of magnitude larger than that estimated by Satta et al. (1994) (4.2% for HLA-B to 0.07% for DPB1). As stated above, the effective population size is not a locus-specific parameter, and finding such divergent values for different loci should cause concern for this estimation approach. Takahata (1991) suggested that balancing selection may increase the effective population size for a given

Table 11.4. *Estimates of the Amount of Selection between Heterozygotes and Homozygotes (s) for Seven HLA Loci Based on the Number of Nonsynonymous Substitutions and the Relative Nonsynonymous Substitution Rate in the Peptide Binding Region*[†]

Locus	S	M	N_e	s	s^*
A	3000	0.09	55,200	0.027	0.015
B	8200	0.02	12,300	0.334	0.042
C	530	0.29	177,900	0.0015	0.0026
DRB1	3900	0.01	13,800	0.142	0.019
DQB1	1700	0.08	110,000	0.0077	0.0085
DBB1	140	0.08	110,000	0.000064	0.0007
DQB1	550	0.14	192,600	0.0014	0.0028

[†] S, M, and s^* are from Method II of Satta et al. (1994), while N_e and s are calculated for specific loci as shown in the text.

locus but the estimates for locus-specific N_e values in Table 11.4 are smallest for the loci that are thought to be most selected, i.e., *HLA-B* and *HLA-DRB1*, the opposite of those suggested by Takahata (1991). Balancing selection can of course increase S ($= 2N_e s$) but it should not influence N_e, which is defined as a function of population-wide parameters (Kimura and Crow, 1963).

This numerical demonstration shows how the estimation of s with the technique outlined by Satta et al. (1994) is highly dependent on the estimated effective population size. The best course would seem to have a completely independent estimate of effective population size that is not a function of the rates of nonsynonymous and synonymous substitutions like that from allozymes [variation in allozymes appears consistent with neutrality expectations (Nei, 1987)], other neutral molecular markers, or even other biological information (e.g., Nunney and Elam, 1994).

In addition to problems similar to those pointed out here, estimates of selection from sequence data generally assume that single-base mutation is the mechanism for the formation of new alleles. If microrecombination is important, as it appears to be for at least several MHC loci, then changes producing new alleles are not independent at different adjacent nucleotides. Further, heterozygotes that differ at a number of amino-acid sites may have a higher fitness than heterozygotes that differ at a few sites, as suggested by Ohta (1991), so that a single measure of selection against homozygotes may not be appropriate.

11.9. Conclusions

We would like to present a clean and complete story to explain the genetic variation at the MHC. However, and not unexpectedly, it is a very complex and incomplete tale. Even as we know more about molecular variation at the MHC and more about the potential connection of MHC to parasite resistance, in some respects as many questions arise as are answered.

Theoretical studies have examined how balancing selection with or without microrecombination can explain the observed patterns of variation at MHC genes (e.g., Takahata and Nei, 1990; Ohta, 1991). In particular, the very high number of alleles (over 100 at *DRB*), the relatively even distribution of alleles, the higher rate of nonsynonymous than synonymous substitution, and disequilibrium between loci are observations that need to be explained. The factors important in determining genetic variation include the extent and the type of balancing selection, the extent and the pattern of microrecombination, and the effective population size. Because MHC has multiple alleles at multiple loci and loci may very well differ in the evolutionary factors that influence them, designing an all-encompassing model is not a simple task.

Parham and Ohta (1996) suggest that the combination of fairly weak selection favoring heterozygotes, microrecombination, and tight linkage is effective in maintaining genetic variation as observed at the MHC. They further state that the result is reminiscent of that observed by Franklin and Lewontin (1970) for linked overdominant loci in a finite populations which produced complimentary gene blocks.

In light of the discussion above about selection at MHC and the potentially high rate of microrecombination, some of the conclusions about the age of MHC alleles (50 million years) and the estimation of human effective population size based on MHC sequences (100,000 for 50 million years) (e.g., Takahata, 1991; Ayala, 1995) should be viewed with great caution. For example, how are these numbers consistent with the large number of new alleles observed (both class I and II) and the loss of the putative class I founder alleles in South American Amerindians in less than 20,000 years? It seems to us that because many things are disputed about MHC, e.g., the kind of balancing selection, the level of selection, and the extent of microrecombination and variation for all of these over loci, so that making conclusions about events that have occurred over tens of millions of years ago from MHC data is, at least, premature and, at most, untenable.

It is not clear that Dick Lewontin imagined anything as complicated as the MHC when he wrote the *Genetic Basis of Evolutionary Change* (Lewontin, 1974). However, he ended that landmark book by stating that

"The fitness at a single locus ripped from its interactive context is about as relevant to real problems of evolutionary genetics as the study of psychology of individuals isolated from their social context is to an understanding of man's sociopolitical evolution. In both cases context and interaction are not simply second-order effects to be superimposed on a primary monadic analysis. Context and interaction are of the essence."

MHC is a prime genetic example supporting these thoughts.

11.10. Acknowledgments

We appreciate the comments of Henry Erlich, Tomoko Ohta, Peter Parham, Jim Kaufman, and Yoko Satta on the manuscript, although they may not agree completely with our perceptions. This work was funded in part by the U.S. National Science Foundation.

REFERENCES

Alberts, S. C. and Ober, C. 1993. Genetic variability in the major histocompatibility complex: a review of non-pathogen-mediated selective mechanisms. *Yearb. Phys. Anthropol.* **36**:71–89.

Almarri, A. and Batchelor, J. R. 1994. HLA and hepatitis B infection. *Lancet* **344**:1194–1195.

Allsopp, C. E. M., Hill, A. V. S., Kwiakowski, D., Hughes, A., Bunce, M., Taylor, C. J., Pazmany, L., Brewster, D., McMichael, A. J., and Greenwood, B. M. 1991. Sequence analysis of HLA-Bw53, a common West African allele, suggests an origin by gene conversion of HLA-B35. *Hum. Immunol.* **30**:104–109.

Ayala, F. J. 1995. The myth of Eve: molecular biology and human origins. *Science* **270**:1930–1936.

Belich, M. P., Madrigal, J. A., Hildebrand, W. H., Zemmour, J., Williams, R. C., Luz, R., Petzl- Erler, M. L., and Parham, P. 1992. Unusual *HLA-B* alleles in two tribes of Brazilian Indians. *Nature (London)* **357**:326–329.

Bengtsson, B. O. and Thomson, G. 1981. Measuring the strength of association between HLA antigens and diseases. *Tissue Antigens* **18**:356–363.

Black F. J. 1992. Why did they die? *Science* **258**:1739–1740.

Black, F. L. and Salzano, F. M. 1981. Evidence for heterosis in the HLA system. *Am. J. Hum. Genet.* **33**:894–899.

Black, F. L. and Hedrick, P. W. 1997. Strong balancing selection at HLA loci: evidence from segregation in South Amerindian families: *Proc. Natl. Acad. Sci: USA* **94**:12452–12456.

Bodmer, W. 1972. Evolutionary significance of the HL-A system. *Nature (London)* **237**:139–145.

Bodmer, W. F. 1978. *HLA: A Super Supergene*, pp. 91–138. Vol. 72 of Harvey Lectures Series. New York: Academic.

Boyce, W. M., Hedrick, P. W., Muggli-Crockett, N. E., Kalinowski, S., Penedo, M. C. T., and Ramey, R. R. 1997. Genetic variation of major histocompatibility complex and microsatellite loci: a comparison in bighorn sheep. *Genetics* **145**:421–33.

Briles, W. E., Stone, H. A., and Cole, R. K. 1977. Marek's disease: effects of B histocompatibility alloalleles in resistant and susceptible chicken lines. *Science* **195**:193–195.

Brown, J. L. and Eklund, A. 1994. Kin recognition and the major histocompatibility complex: an integrative review. *Am. Nat.* **143**:435–461.

Burdon, J. J. 1987. *Diseases and Plant Population Biology*. Cambridge: Cambridge U. Press.

Callahan, K., Fort, M., Obah, E., Reinherz, E., and Siliciano, R. 1990. Genetic variability in HIV-1 gp 120 affects interactions with HLA molecules and T cell receptor. *J. Immunol.* **144**:3341.

Caro, T. M. and Laurenson, M. K. 1994. Ecological and genetic factors in conservation: a cautionary tale. *Science* **263**:485–496.

Carrington, M., Nelson, G. W., Martin, M. P., Kissner, T., Vlahov, D., Goedert, J. J., Kaslow, R., Buchbinder, S., Hoots, K., and O'Brien, S. J. 1999. *HLA* and HIV-1: heterozygote advantage and *B*35-Cw*04* disadvantage. *Science* **238**:1748–1752.

Cockerham, C. C., Burrows, P. M., Young, S. S., and Prout, T. 1972. Frequency-dependent selection in randomly mating populations. *Am. Nat.* **106**:493–515.

Craven, D. E., Awdeh, Z. L., Kunches, L. M., Yunis, E. J., Dienstag, J. L., Werner, B. G., Polk, B. F., Snydman, D. R., Platt, R., Crumpacker, C. S., Grady, G. F., and Alper, C. A. 1986. Nonresponsiveness to hepatitis B vaccine in health care workers. *Ann. Intern. Med.* **105**:356–360.

Darden, A. G. and Streilein, J. W. 1984. Syrian hamsters express two monomorphic class I major histocompatibility complex molecules. *Immunogenetics* **20**:603–622.

Darlington, C. D. and Mather, K. 1949. *The Elements of Genetics.* London: Allen and Unwin.

Davenport, M. P., Quinn, C. L., Chicz, R. M., Green, B. N., Willis, A. C., Lane, W. S., Bell, J. I., and Hill, A. V. S. 1995. Naturally processed peptides from two disease-resistance associated HLA-DR13 alleles show related sequence motifs and the effects of the dimorphism at position 86 of the HLA-DRb chain. *Proc. Natl. Acad. Sci. USA* **92**:6567–6571.

de Campos-Lima, Gaviolli, R., Zhang, Q.-J., Wallace, L. E., Dolcetti, R., Rowe., M., Rickinson, A. B., and Masucci, M. G. 1993. HLA-A11 epitope loss isolates of Epstein-Barr virus from a highly A11+ population. *Science* **260**:98–100.

Denniston, C. and Crow, J. F. 1990. Alternative fitness models with the same allele frequency dynamics. *Genetics* **125**:201–205.

Doherty, P. C. and Zingernagel, R. 1975. Enhanced immunologic surveillance in mice heterozygous at the H2 complex. *Nature (London)* **256**:50–52.

Edwards, S. V. and Hedrick, P. W. 1998. Evolution and ecology of MHC molecules: from genomics to sexual selection. *Trends Ecol. Evol.* **13**:305–311.

Ewens, W. J. 1972. The sampling theory of selectively neutral alleles. *Theor. Popul. Biol.* **3**:87–112.

Franklin, I. R. and Lewontin, R. C. 1970. Is the gene the unit of selection? *Genetics* **65**:707–734.

Gaur, L. K. and Nepom, G. T. 1996. Ancestral major histocompatibility complex DRB genes beget conserved patterns of localized polymorphisms. *Proc. Natl. Acad. Sci. USA* **93**:5380–5383.

Gavora, J. S., Simenson, M., Spencer, J. L., Fairfull, R. W., and Gowe, R. S. 1986. Changes in the frequency of major histocompatibility haplotypes in chickens under selection for both high egg production and resistance to Marek's disease. *J. Anim. Breed.* **103**:218–226.

Golding, B. 1992. The prospects for polymorphism shared between species. *Heredity* **68**:263–276.

Gyllensten, U., Bergstrom, T., Josefsson, A., Sundvall, M., Savage, A., Blumer, E. S., Giraldo, L. H., Soto, L. H., and Watkins, D. I. 1994. The cotton-top tamarin revisited: Mhc class I polymorphism of wild tamarins, and polymorphism and allelic diversity of the class II DQA1, DQB1, and DRB loci. *Immunogenetics* **40**:167–176.

Hedrick, P. W. 1972. Maintenance of genetic variation with a frequency-dependent model as compared to the overdominant model. *Genetics* **72**:771–775

Hedrick, P. W. 1985. *Genetics of Populations.* Boston: Jones and Bartlett.

Hedrick, P. W. 1987. Gametic disequilibrium measures: proceed with caution. *Genetics* **117**: 331–341.

Hedrick, P. W. 1990. Selection at HLA: possible explanations for deficiency of homozygotes. *Hum. Hered.* **40**:213–220.

Hedrick, P. W. 1992. Female choice and variation in the major histocompatibility complex. *Genetics* **132**:575–581.

Hedrick, P. W. 1994. Evolutionary genetics of the major histocompatibility complex. *Am. Nat.* **143**:945–964.

Hedrick, P. W. and Black, F. L. 1997. HLA and mate selection: no evidence in South Amerindians. *Amer. J. Hum. Genet.* **61**:505–511.

Hedrick, P. W., Ginevan, M., and Ewing, E. 1976. Genetic polymorphism in heterogeneous environments. *Annu. Rev. Ecol. Syst.* **7**:1–32.

Hedrick, P. W., Jain, S., and Holden, L. 1978. Multilocus systems in evolution. *Evol. Biol.* **11**:101–184.

Hedrick, P. W., Whittam, T. S., and Parham, P. 1991a. Heterozygosity at individual amino sites: extremely high levels for HLA-A and -B genes. *Proc. Natl. Acad. Sci. USA* **88**:5897–5901.

Hedrick, P. W., Klitz, W., Robinson, W. P., Kuhner, M. K., and Thomson, G. 1991b. Evolutionary genetics of HLA. In *Evolution at the Molecular Level*, R. Selander, A. Clark, and T. Whittam, eds., pp. 248–271. Sunderland, MA: Sinauer.

Hedrick, P. W. and Murray, E. 1983. Selection and measures of fitness. In *The Genetics and Biology of Drosophila*, M. Ashburner, H. Carson, and J. Thompson, eds., Vol. 3, pp. 61–104. New York: Academic.

Hedrick, P. W. and Thomson, G. 1983. Evidence for balancing selection at HLA. *Genetics* **104**:449–456.

Hedrick, P. W. and Thomson, G. 1986. A two locus neutrality test: applications to humans, *E. coli*, and lodgepole pine. *Genetics* **112**:135–156.

Hedrick, P. W. and Thomson, G. 1988. Maternal-fetal interactions and the maintenance of HLA polymorphism. *Genetics* **119**:205–212.

Hedrick, P. W., Thomson, G., and Klitz, W. 1987. Evolutionary genetics and HLA: another classic example. *Biol. J. Linn. Soc.* **31**:311–331.

Hill, A. V. S. 1991. HLA associations with malaria in Africa: some implications for MHC evolution. In *Molecular Evolution of the Major Histocompatibility Complex*, J. Klein and D. Klein, eds., pp. 403–434. Berlin: Springer-Verlag.

Hill, A. V. S. 1992. Malaria resistance genes: a natural selection. *Trans. R. Soc. Trop. Med. Hyg.* **86**:225–226.

Hill, A. V. S. 1996. HIV and HLA: confusion or complexity? *Nat. Med.* **2**: 395–396.

Hill, A. V. S., Allsop, C. E. M., Kwiatdowski, D., Anstey, N. M., Twumasi, P., Rowe, P. A., Bennett, S., Brewster, D., McMichael, A. J., and Greenwood, B. M. 1991. Common West African HLA antigens are associated with protection from severe malaria. *Nature (London)* **352**:595–600.

Hill, A. V. S., Kwiatkowski, D., McMichael, A. J., Greenwood, B. M., and Bennet, S. 1992. Maintenance of MHC polymorphism: reply. *Nature (London)* **355**:403.

Hill, A. V. S., Yates, S. N. R., Allsopp, C. E. M., Gupta, S., Gilbert, S. C., Lalvani, A., Aidoo, M., Davenport, M., and Plebanske, M., 1994. Human leukocyte antigens and natural selection by malaria. *Philos. Trans. R. Soc. London Ser. B* **346**:379–385.

Hogstrand, K. and Bohme, J. 1994. A determination of the frequency of gene conversion in unmanipulated mouse sperm. *Proc. Natl. Acad. Sci. USA* **91**:9921–9925.

Huang, S. L., Singh, M., and Kojima, K. 1971. A study of frequency-dependent selection observed in the esterase-6 locus of *Drosophila melanogaster* using a conditioned media method. *Genetics* **68**:97–104.

Hughes, A. L. and Nei, M. 1988. Pattern of nucleotide substitution at major histocompatibility complex class I loci reveals overdominant selection. *Nature (London)* **335**:167–179.

Hughes, A. L. and Nei, M. 1989. Nucleotide substitution at major histocompatibility complex class II loci: evidence for overdominant selection. *Proc. Natl. Acad. Sci. USA* **86**:958–962.

Hughes, A. L. and Nei, M. 1992a. Maintenance of MHC polymorphism. *Nature (London)* **355**:402–403.

Hughes, A. L. and Nei, M. 1992b. Models of host-parasite interaction and MHC polymorphism. *Genetics* **132**:863–864.

Just, J. J. 1995. Genetic predisposition to HIV-1 infection and acquired immune deficiency virus syndrome. A review of the literature examining association with HLA. *Hum. Immunol.* **44**:156–169.

Kaslow, R. A., Duquesnoy, R., Van Raden, L., Marrari, M., Friedman, H., Su, S., Saah, A. J., Detels, R., Phair, J., and Rinaldo, C. 1990. A1, Cw7, B8, DR3 HLA antigen combination associated with rapid decline of T-helper lymphocytes in HIV-1 infection. A report from the Muliticenter AIDS Cohort Study. *Lancet* **335**:927–930.

Kaslow, R., Carrington, M., Apple, R., Park, L., Munoz, A., Saah, A. J., Goedert, J. J., Winkler, C., O'Brien, S. J., Rinaldo, C., Detels, R., Blattner, W., Phair, J., Erlich, H., and Mann, D. L. 1996. Influence of combinations of human major histocompatibility complex genes on the course of HIV-1 infection. *Nat. Med.* **2**:405–411.

Kaslow, R. and Shaw, S. 1981. The role of histocompatibility antigens (HLA) in infection. *Epidemiol. Rev.* **3**:90–114.

Kaufman, J., Volk, H., and Wallny, H.-J. 1995. A minimal essential Mhc and an unrecognized Mhc: two extremes in selection for polymorphism. *Immunol. Rev.* **143**:63–88.

Keith, T. P., Brooks, L. T., Lewontin, R. C., Martinez-Cruado, J. C., and Rigby, D. L. 1985. Nearly identical distributions of xanthine dehydrogenase in two populations of *Drosophila pseudoobscura*. *Mol. Biol. Evol.* **2**:206–216.

Kimura, M. 1977. Preponderance of synonymous changes as evidence for the neutral theory of molecular evolution. *Nature (London)* **267**:275–276.

Kimura, M. and Crow, J. F. 1963. The measurement of effective population numbers. *Evolution* **17**:279–288.

Kingman, J. F. C. 1961. A mathematical problem in population genetics. *Proc. Cambridge. Philos. Soc.* **57**:574–582.

Klein, J. and O'hUigin, C. 1994 MHC polymorphism and parasites. *Philos. Trans. R. Soc. London Ser. B* **346**:351–358.

Klitz, W. G. and Thomson, G. 1987. Disequilibrium pattern analysis. II. Application to Danish HLA-A and -B locus data. *Genetics* **116**:633–643.

Kroner, B. L., Goedert, J. J., Blattner, W. A., Wilson, S. E., Carrington, M. N., and Mann, D. L. 1995. Concordance of human leukocyte antigen haplotype-sharing, CD4 decline and AIDS in hemophilic siblings. Multicenter Hemophilia Cohort and Hemophilia Growth and Development Studies. *AIDS* **9**:275–280.

Kuhner, M., Watts, S., Klitz, W., Thomson, G., and Goodenow, R. S. 1990. Gene conversion in the evolution of both the *H-2* and *Qa* class I genes of the murine major histocompatibility complex. *Genetics* **126**:1115–1126.

Lewontin, R. C. 1964a. The role of linkage in natural selection. *Genetics Today. Proceedings of the XI International Congress of Genetics*, pp. 517–525.

Lewontin, R. C. 1964b. The interaction of selection and linkage. I. General considerations; heterotic models. *Genetics* **49**:49–56.

Lewontin, R. C. 1974. *The Genetic Basis of Evolutionary Change.* New York: Columbia U. Press.

Lewontin, R. C. 1988. On measures of gametic disequilibrium. *Genetics* **120**:849–852.

Lewontin, R. C., Ginsburg, L. R., and Tuljapurkar, S. D. 1978. Heterosis as an explanation for large amounts of genic polymorphism. *Genetics* **88**:149–170.

Makino M., Morse, H., Fredrickson, T., and Hartley, J. 1990. H-2-associated and background genes influence the development of a murine retrovirus-induced immune syndrome. *J. Immunol.* **144**:4347

Mandel, S. P. H. 1970. The equivalence of different sets of stability conditions for multiple allelic systems. *Biometrics* **26**:840–845.

Markow, T., Hedrick, P. W., Zuerlein, K., Danilovs, J., Martin, J., Vyvial, T., and Armstrong, C. 1993. HLA polymorphism in the Havasupai: evidence for balancing selection. *Am. J. Hum. Genet.* **53**:943–952.

Maruyama, T. 1982. Stochastic integrals and their application to population genetics. In *Molecular Evolution, Protein Polymorphism, and the Neutral Theory*, M. Kimura, ed, pp. 151–166. Tokyo: Japan Scientific Societies.

McGuire, W., Hill, A. V. S., Allsopp, C. E., Greenwood, B. M., and Kwiatkowski, D. 1994. Variation in TNF-alpha promoter region associated with susceptibility to cerebral malaria. *Nature (London)* **371**:508–510.

Menotti-Raymond, M. and O'Brien, S. J. 1995. Evolutionary conservation of ten microsatellite loci four species of felidae. *J. Hered.* **86**:319–322.

Moonsamy, P. V., Aldrich, C. L., Petersdorf, E. W., Hill, A. V. S., and Begovich, A. B. 1994. Seven new DPB1 alleles and their population distribution. *Tissue Antigens* **43**:249–252.

Nei, M. 1987. *Molecular Evolutionary Genetics.* New York: Columbia U. Press.

Nei, M. and Hughes, A. 1991. Polymorphism and evolution of the major histocompatibility complex loci in mammal. In *Evolution at the Molecular Level*, R. K. Selander, A. G. Clark, and T. S. Whittam, eds., pp. 222–247. Sunderland, MA: Sinauer.

Nunney, L. and Elam, D. R. 1994. Estimating the effective population size of conserved populations. *Conserv. Biol.* **8**:175–184.

Ober, C., Weitkamp, L. R., Cox, N., Dytch, G., Kostu, E., and Elias, S. 1997. HLA and mate choice in humans. *Amer. J. Hum. Genet.* **61**:497–504.

O'Brien, S. J., Roelke, M. E., Marker, L., Newman, A., Winkler, C. A., Meltzer, D., Colly, L., Everman, J. F., Bush, M., and Wildt, D. E. 1985. Genetic basis for species vulnerability in the cheetah. *Science* **227**:1428–1434.

Ohta, T. 1991. Role of diversifying selection and gene conversion in evolution of major histocompatibility complex loci. *Proc. Natl. Acad. Sci. USA* **88**:6716–6720.

Parham, P., Lawlor, D. A., Lomen, C. E., and Ennis, P. D. 1989. Diversity and diversification of HLA-A, B, C, alleles. *J. Immunol.* **142**:3937–3950.

Parham, P., Adams, E. J., and Arnett, K. L. 1995. The origins of HLA-A, B, C polymorphism. *Immunol. Rev.* **143**:141–181.

Parham, P. and Ohta, T. 1996. Population biology of antigen presentation by MHC class I molecules. *Science* **272**:67–74.

Plachy, J., Chen, C.-L., and Hala, K. 1992. Biology of the chicken MHC (B complex). *Crit. Rev. Immunol.* **12**:47–79.

Potts, W. K., Manning, C. J., and Wakeland, E. K. 1994. The role of infectious disease, inbreeding and mating preferences in maintaining MHC genetic diversity: an experimental test. *Philos. Trans. R. Soc. London Ser. B* **346**:369–378.

Ritte, U., Neufeld, E., O'hUigin, C., Figeroa, F., and Klein, J. 1991. Origins of *H-2* polymorphism in the house mouse. II. Characterization of a model population and evidence for heterozygous advantage. *Immunogenetics* **34**:164–173.

Satta, Y., O'hUigin, C., Takahata, N., and Klein, J. 1994. Intensity of natural selection at the major histocompatibility complex loci. *Proc. Natl. Acad. Sci. USA* **91**:7184–7188.

Slade, R. W. 1992. Limited MHC polymorphism in the southern elephant seal: implications for MHC evolution and marine mammal population biology. *Proc. R. Soc. London Ser. B* **249**:163–171.

Slade, R. W. and McCallum, H. I. 1992. Overdominant vs. frequency-dependent selection at MHC loci. *Genetics* **132**:861–862.

Spencer, H. G. and Marks, R. W. 1988. The maintenance of single-locus polymorphism. I. Numerical studies of a viability selection model. *Genetics* **120**:605–613.

Spencer, H. G. and Marks, R. W. 1992. The maintenance of single-locus polymorphism. IV. Models with mutation from existing alleles. *Genetics* **130**:211–221.

Steel, C. M., Beatson, D., Cuthbert, R. J. G., Morrison, H., Ludlam, C. A., Peutherer, J. F., Simmonds, P., and Jones, M. 1988. HLA haplotype A1 B8 DR3 as a risk factor for HIV-related disease. *Lancet* **333**:1185–1188.

Takahata, N. 1991. Trans-species polymorphism of HLA molecules, founder principle, and human evolution. In *Molecular Evolution of the Major Histocompatibility Complex*, J. Klein and D. Klein, eds., pp. 29–49. Berlin: Springer-Verlag.

Takahata, N. and Nei, M. 1990. Allelic genealogy under overdominant and frequency-dependent selection and polymorphism of major histocompatibility complex loci. *Genetics* **124**:967–978.

Takahata, N., Satta, Y., and Klein, J. 1992. Polymorphism and balancing selection at major histocompatibility complex loci. *Genetics* **130**:925–938.

Thomas, M. L., Harger, J. H., Wagener, D. K., Rabin, B. S., and Gill, T. J. 1985. HLA sharing and spontaneous abortion in humans. *Am. J. Obstet. Gynecol.* **151**:1053–1058.

Thomson, G. 1988. HLA disease associations: models for insulin dependent diabetes mellitus and the study of complex human genetic disorders. *Annu. Rev. Genet.* **22**:31–50.

Thomson, G. and Klitz, W. G. 1987. Disequilibrium pattern analysis. I. Theory. *Genetics* **116**:623–632.

Thurz, M. R., Thomas, H. C., Greenwood, B. M., and Hill, A. V. S. 1997. Heterozygote advantage for HLA class-II type in hepatitis B virus infection. *Nat. Genet.* **17**:11–12.

Thurz, M. R., Kwiatkowski, D., Allsopp, C. E. M., Greenwood, B. M., Thomas, H. C., and Hill, A. V. S. 1995. Association between and MHC class II allele and clearance of hepatitis B virus in the Gambia. *N. Engl. J. Med.* **7**:11–14.

Titus-Trachtenberg, E. A., Rickards, O., De Stafano, G. F., and Erlich, H. A. 1994. Analysis of HLA class II haplotypes in the Cayapa Indians of Ecuador: a novel DRB1 allele reveals evidence for convergent evolution and balancing selection at position 86. *Am. J. Hum. Genet.* **55**:1609–167.

Tiwari, J. L. and Terasaki, P. I. 1985. *HLA and Disease Associations*. New York: Springer-Verlag.

Trowsdale, J. 1993. Genomic structure and function in the MHC. *Immunol. Today* **9**:117–122.

Trowsdale, J. 1995. "Both man & bird & beast": comparative organization of MHC genes. *Immunogenetics* **41**:1–17.

Trowsdale, J., Groves, V., and Arnason, A. 1989. Limited MHC polymorphism in whales. *Immunogenetics* **29**:19–24.

Van Eden, W., Devries, R. P., and Van Rood, J. J. 1983. The genetic approach to infectious disease with special emphasis on the MHC. *Dis. Markers* 1:221–242.

van Hattum, J., Schreuder, G. M., and Schalm, S. W. 1987. HLA antigens in patients with various courses after hepatitis B virus infection. *Hepatology* 7:11–14.

von Schantz, T., Wittzell, H., Goransson, G., Grahn, M., and Persson, K. 1996. MHC genotype and male ornamentation: genetic evidence for the Hamilton-Zuk model. *Proc. R. Soc. London Ser. B* 263:265–271.

Wakelin, D. and Blackwell, J. M. 1988. *Genetics of Resistance to Bacterial and Parasitic Infection.* London: Taylor & Francis.

Watkins, D. I., Chen, Z. W., Hughes, A. L., Lagos, A., Lewis, A. M., Shadduck, J. A., and Letvin, N. L. 1990. Syrian hamsters express diverse MHC class I gene products. *J. Immunol.* 145:3483–3490.

Watkins, D. I., Hodi, F. S., and Letvin, N. L. 1988. A primate species with limited major histocompatibility complex class I polymorphism. *Proc. Natl. Acad. Sci. USA* 85:7714–7718.

Watkins, D. I., McAdam, S. N., Liu, X., Strang, C. R., Milford, E. L., Levine, C. G., Garber, T. L., Dogon, A. L., Lord, C. I., Ghim, S. H., Troup, G. M., Hughes, A. L., and Letvin, N. L. 1992. New recombinant *HLA-B* alleles in a tribe of South American Amerindians indicate rapid evolution of MHC class I loci. *Nature (London)* 357:329–333.

Watterson, G. A. 1978. The homozygosity test of neutrality. *Genetics* 88:405–417.

Weissman, J. Tschiyose, M., Tong, M., Co, R., Chin, K., and Ettenger, R. 1988. Lack of response to recombinant hepatitis B vaccine in nonresponders to the plasma vaccine. *J. Am. Med. Assoc.* 260:1734–1738.

Wright, S. 1969. *Evolution and the Genetics of Populations. Volume 2. The Theory of Gene Frequencies.* Chicago: U. Chicago Press.

Yamazaki, K., Beauchamp, G. K., Kupniewski, D., Bard, J., Thomas, L., and Boyse, E. A. 1988. Familial imprinting determines H-2 selective mating preferences. *Science* 240:1331–1332.

Yuhki, N., and O'Brien, S. J. 1990. DNA variation of the mammalian major histocompatibility complex reflects genomic diversity and population history. *Proc. Natl. Acad. Sci. USA* 87:836–840.

Zangenberg, G. and Huang, M.-M., Arnheim, N., and Erlich, H. 1995. New HLA-DPB1 alleles generated by interallelic gene conversion detected by analysis of sperm. *Nat. Genet.* 10:407–414.

Zinkernagel, R. M. 1979. Associations between major histocompatibility antigens and susceptibility to disease. *Annu. Rev. Microbiol.* 33:201–213.

CHAPTER TWELVE

The Population Biology of Antibiotic Resistance

BRUCE R. LEVIN

It is not often that we get to see evolution in action, and when we do it is rarely a good thing, at least from a human–technological perspective. Our efforts to intervene, to control nature, are almost invariably thwarted by either ecological or genetic–evolutionary processes. We are humbled, but we never seem to learn humility. The evolution of resistance to the chemicals we use to treat or prevent infections with parasites or to control agricultural pests are two dramatic examples of the power of natural selection to frustrate our efforts to control our environment. Within the lifetime of many of the contributors to this volume, and certainly within that of Dick Lewontin, the evolution of resistance has brought us from a period of arrogant euphoria about our capacity to use chemicals to control the pests and parasites that plague us, to the current state of anxiety about the future of our agriculture and health.

In this essay I consider a subset of this broader topic of the evolution of resistance to chemicals: bacterial resistance to antimicrobial agents (antibiotics, or just drugs for convenience). In part because antibiotic resistance, like most other practical subjects, is not traditional fodder for population and evolutionary biologists, and in part to proselytize (the science as well as the consequences of antibiotic resistance), I provide more background material than is usual for this kind of treatise. My emphasis is on the two problems of the population genetics of antibiotic resistance with which I am most familiar: the conditions for the evolution and maintenance of antibiotic-resistance-encoding accessory genetic elements (plasmids, phage, and transposons) and the relationship between the incidence of antibiotic use and the frequency of antibiotic resistance. The former is a subject we have studied for some time, and although I consider it in the light of the antibiotic-resistance problem, much of what I present here is a review of published work, "new titles for old papers," to quote Marc Lipsitch. The latter is a problem on which we are currently working and therefore more interesting than anything we did earlier, at least to me.

R. S. Singh and C. B. Krimbas, eds., *Evolutionary Genetics: From Molecules to Morphology*, vol. 1. © Cambridge University Press 2000. Printed in the United States of America. ISBN 0-521-57123-5. All rights reserved.

12.1. Background

12.1.1. Historical Perspective

12.1.1.1. Antibiotics Have Made a Relatively Minor Contribution to the Decline in Infectious Disease Mortality

The development of antibiotics may well have been the single most important advance of interventive medicine (treatment) of this century. Nevertheless, in developed countries, antimicrobial chemotherapy and intervention at large have made only modest contributions to the more than 25 years the average human life span has increased this century. To be sure, most of the increase in life span can be attributed to declines in the rate of mortality due to infectious diseases. At the turn of this century in the United States and most other developed countries, infectious disease was the major immediate source of mortality. It still is in many developing countries. In 1900, in the United States, the rate of mortality due to tuberculosis alone was of a magnitude similar to that of cancer today, of the order of 200/100,000 per year, and substantially higher in the crowded areas of cities. The incidence of deaths due to pneumonia was of the same magnitude as that of tuberculosis and various diarrheas caused by bacteria, and viruses accounted for the majority of deaths during the first year of life.

However, as was so well documented in the late Thomas McKeowen's treatise on the role of medicine (McKeowen, 1976), the decline in the rate of mortality due to infectious diseases started long before the development of effective treatment. The rates of mortality due to tuberculosis, pneumonia, diphtheria, typhoid fever, and most of the other major bacterial diseases were in a continuous state of decline since the middle of the nineteenth century. Until the middle 1930s, there was virtually no effective treatment for any of these diseases. Improvements in sanitation, nutrition, living conditions, food handling, personal hygiene, and a variety of other social factors made a far greater contribution to the decline in infectious disease mortality than antibiotics and vaccination. Stated another way, nineteenth-century sanitary engineers like Thomas Crapper, the inventor of the flush toilet, contributed more to the decline in infectious disease mortality than all physicians of that century and possibly this century as well.

12.1.1.2. Antimicrobial Chemotherapy

While there were a few chemotherapeutic compounds for specific bacterial diseases, like Paul Ehrlich's Salvarsan, an arsenate used to treat syphilis, the first widely effective antibacterial chemotherapeutic agents were sulfanilamide and related drugs developed by Dormagk and his colleagues in the 1930s.[1] The first true antibiotics, antibacterial chemotherapeutic agents that were natural products of living organisms rather than the flasks of chemists, were first used for treatment in the late 1930s. These were gramicidin and tyrocidine, small peptides isolated from *Bacillus brevius* by René Dubos (a microbial ecologist) and his collaborators, primarily Roland Hotchkiss (Hotchkiss, 1990). Although Alexander Fleming had recognized the action of penicillin nearly a decade earlier, it was not used for treatment until it was purified and produced in clinically useful quantities in the early 1940s (Chain et al., 1940). Using a procedure similar to that of Dubos, Selman Waxman, another microbial ecologist, and his student G. Schatz isolated and purified streptomycin in the middle 1940s. Tetracycline and chloramphenicol soon followed.

12.1.1.3. The Resistance Problem

From the media one may get the impression that resistance-confounding antibiotic treatment is a problem that has been recognized only recently. In point of fact, microbial-resistance chemotherapeutic agents, initially known as drug fastness and attributed to training, was recognized by Ehrlich, Dormagk, Dubos, and other early chemotherapists (Moberg, 1996). Although streptomycin was initially hyped as a wonder drug for the treatment of tuberculosis, shortly after its introduction it was apparent that because of resistance, streptomycin alone would not be sufficient for the successful treatment of TB (Ryan, 1993). By the mid-1950s, it was clear that the frequency of antibiotic resistance was on the rise. Shortly after each new antibiotic was introduced, bacteria resistant to it would be observed (see Anderson, 1968; Falkow, 1975; Mitsuhashi, 1977; Watanabe, 1963; Zahner and Giedler, 1995). By the latter 1950s, plasmid-borne multiple antibiotic resistance was recognized, thanks to the impressive work of Akiba and his colleagues in Japan (for reviews see Anderson, 1968; Falkow, 1975; Mitsuhashi, 1977; Watanabe, 1963). Since then it has been downhill, with almost monotonic increases in (1) the number of species of pathogenic and commensal bacterial with resistant strains, (2) the number of resistant strains (clones) within each species, and (3) the numbers of antibiotics to which bacteria are resistant individually and collectively. In the arms race between the pharmaceutical industry and bacterial evolution, the bacteria have had the edge and, unless new measures are taken (e.g., see Zahner and Giedler, 1995), they will almost certainly win. For popular and semipopular accounts of the current status of the resistance problem see Berkowitz (1995), Bloom and Murray (1992), Cohen (1992, 1994), Levy (1992), McGowan (1991), Murray (1992), Neu (1992), and Tenover and Hughes (1996).

12.1.2. Antibiotic Action and Antibiotic Resistance

There is an impressive literature on the details of antibiotic action and the mechanisms and genetics of antibiotic resistance, much of which is summarized in the better microbiology texts, e.g., Davis et al. (1990). From the perspective of reductionist population biology, only some of these details are ·critical.

12.1.2.1. Modes of Antibiotic Action

Antibiotics are generally classified as either bactericidal or bacteriostatic, i.e., those that kill bacteria and those that just prevent their replication, respectively. Unfortunately, these qualitative distinctions are less absolute than they appear to be, and there are a variety of host and bacterial factors that contribute to the efficacy of antibiotics. In particular, the magnitude of the inhibitory effect and the rate of antibiotic-mediated mortality depend on the concentration of the antibiotic (Garrett, 1971) and, in almost all cases, the physiological state of the bacteria. Few antibiotics kill bacteria that are not dividing, and antibiotics vary in their ability to inhibit the growth or kill bacteria that replicate within somatic cells, like *Mycobacterium tuberculosis* (Mitchison, 1979).

While there are many antibiotics, especially when one includes all the minor variants and the plethora of commercial names for the same or very similar compounds, there are relatively few targets of action. Some antimicrobials, like the sulfonamides, affect specific biochemical reactions, such as folic acid synthesis. A number of antibiotics inhibit protein synthesis, either at the level of transcription, like rifampicin, or translation, like streptomycin. Some chemotherapeutic agents, like naladixic acid, affect gryases or other enzymes involved in DNA replication, and some antibiotics, like the penicillins and cephalosporins inhibit cell wall synthesis.

12.1.2.2. Mechanisms of Resistance

From the perspective of population biology, two of the most critical properties of antibiotic resistance are the cost of resistance on the fitness of bacteria in the absence of antibiotics and the effect of the resistance on the effective concentration of the antibiotic in the extracellular environment. Some resistance mechanisms, like the tetracycline pumps and the ribosomal protein mutations responsible for chromosomal resistance to streptomycin and spectinomycin or the altered penicillin-binding proteins, have very little effect on extracellular concentration of these antibiotics. Other mechanisms, like the β-lactamases responsible for much of the resistance to penicillin and other β-lactam antibiotics and chloramphenicol transacetylase, denature these antibiotics in the environment at large. One consequence of the latter is that in the presence of high concentrations of resistant bacteria, bacteria sensitive to these antibiotic can replicate. If, in addition, these antibiotic-sensitive bacteria have a fitness advantage relative to the resistant, in the presence of resistant bacteria, they can increase in frequency when rare (Lenski and Hattingh, 1986). Not only is this nice example of how frequency-dependent selection can maintain stable polymorphisms in asexual populations (Levin, 1988b), this detoxification may be of considerable importance clinically. Commensal bacteria with resistance of this sort could markedly reduce the efficacy of antibiotics susceptible to these degradative enzymes.

12.1.2.3. The Genetic Bases of Resistance

Antibiotic-sensitive bacteria can become resistant in three distinct ways: mutations of chromosomal genes, horizontal transfer and recombination of chromosomal genes from resistant strains, or the receipt, by horizontal transfer, of an accessory genetic element (a plasmid, transposon, or temperate phage) that carries resistance genes. For some species, like *M. tuberculosis*, resistance seems to occur primarily, if not exclusively, by the mutation of chromosomal genes arising within the resistant lineages. While the rates of horizontal transfer of chromosomal genes may be low in bacteria, of the order of magnitude of the mutation rate or less (Levin, 1981; Mongold, 1992), in the presence of antibiotics the power of selection for resistance is great, and there is evidence for resistance's being acquired by horizontal gene transfer and recombination. Some of the most compelling of this evidence for the natural transfer of chromosomal gene-encoded resistance is for the mosaic gene coding for the penicillin-binding proteins of *Streptococcus* and *Nessieria*, studied by Brian Spratt and colleagues, including one John Maynard Smith (Bowler et al., 1994; Coffey et al., 1993; Dowson et al., 1994). Accessory-element-borne resistance is particularly interesting (or problematic, depending on your perspective) because many of these elements, like the plasmid, can be infectiously transmitted at substantial rates (Lundquist and Levin, 1986), and some of these elements have host ranges that include phylogenetically distant species of bacteria. Moreover, some of these resistance plasmids (R plasmids) carry a number (five or even more) of different resistance-encoding genes, usually on transposons (Falkow, 1975).

12.2. The Maintenance of Resistance-Encoding Accessory Elements

From the practical perspective of antibiotic resistance, plasmid-, phage-, and transposon-borne resistances impose a problem that does not obtain for chromosomal resistance. These accessory genetic elements often code for resistance to multiple antibiotics and are capable of horizontal (infectious) transmission by a replicative process; copies rather than the elements themselves are transmitted. Consequently, in theory at least, resistance-encoding plasmids, phages, and transposons could become established in bacterial populations and could be maintained as parasites, i.e., when they impose a fitness burden on their host bacteria. The general condition for this to obtain is that the rate of infectious transfer of these elements between hosts exceeds their rates of loss because of the fitness burden they impose on their hosts and their loss by vegetative segregation. Moreover, these elements may carry multiple host-expressed genes that maintain their genetic integrity for extensive periods. As a result, even when the antibiotics for which these elements encode resistance are rarely if ever used, these unselected resistance genes could persist for extensive periods by associated linkage selection. The specific conditions for these accessory elements to be maintained by infectious transfer, as parasites, vary with how they replicate and how they are infectiously transmitted.

12.2.1. Conjugative Antibiotic Resistance, R plasmids

For conjugative plasmids in bacterial populations maintained in a homogeneous, steady-state (chemostat-like) habitat, the conditions for establishment and maintenance in a bacterial population of density N are

$$\gamma > (\alpha\rho + \tau)/N,$$

where γ is the rate constant of plasmid transfer, α is the selection coefficient of plasmid-bearing bacteria relative to plasmid free, ρ is the rate of flow through the habitat, and τ is the rate at which the plasmid is lost by vegetative segregation. The implication of this is that antibiotic-resistance R plasmids can invade populations of bacteria free of that element and will be maintained in those population even if it imposes a fitness cost, $\alpha > 0$, on the bacteria that carry them and they are occasionally lost by vegetative segregation, $\tau > 0$ (Stewart and Levin, 1977).

Whether in practice these or the analogous conditions for maintaining deleterious plasmids by infectious transfer (as parasites) in other ecological situations will obtain for contemporary R plasmids is not at all clear. On the optimistic side, there is evidence that the carriage of at least some R plasmids imposes a fitness burden, $\alpha > 0$, on their host bacteria (Levin, 1980; Simonsen, 1991, 1992) and the estimated rates of conjugative plasmid transfer γ are too low to maintain these elements in bacterial populations of realistic densities N, even if they impose very modest costs on the fitness of their host bacteria (Levin et al., 1979; Simonsen, 1990, 1992; Simonsen et al., 1990). On the more pessimistic side, the fitness cost of plasmid carriage can be quite low and may well be negligible (Simonsen, 1992), and some naturally occurring conjugative R plasmids are infectiously transferred at substantial rates and may be maintained as parasites in bacterial populations of realistic densities (Lundquist and Levin, 1986). Moreover, as a consequence of (co)evolution with the host bacterium, the cost of plasmid carriage would be anticipated to decline in time (Levin and Lenski, 1983). There is, in fact, experimental evidence for this kind of compensatory coevolution (Bouma and Lenski, 1988; Modi and Adams, 1991). Moreover, in the study by Bouma and Lenski, as a consequence of this compensatory evolution at the chromosomal level, evolved cells carrying the plasmid had a selective advantage over those that did not, even in the absence of antibiotic-mediated selection. (See Lenski et al., 1994, for a consideration of the genetic basis of this adaptation to the carriage of the plasmid.)

In addition to (co)evolution's reducing the fitness costs of plasmid carriage, associated linkage selection almost certainly plays an important role in at least the short-term persistence of plasmid-borne resistance genes. As noted above, contemporary R-plasmid and particularly the larger conjugative elements carry multiple antibiotic-resistance genes as well as other genes that could be under positive selection in our bodies and/or our environment, like resistance to mercury (Summers et al., 1993). Even intermittent selection for any of these may be sufficient to give cells carrying that plasmid and the unselected resistance genes they carry a net advantage in the population at large. There is in fact evidence for the persistence of plasmid-borne antibiotic-resistance genes that are only rarely under positive selection. For more than 20 years antibiotics like streptomycin have rarely been used clinically (at least in the more developed countries). Nevertheless, plasmids that encode for resistance to streptomycin abound in bacteria isolated from humans, animals, and plants (Sundin and Bender, 1996)[2] (see also footnote 1).

12.2.2. Nonconjugative R plasmid

While in theory there are conditions under which nonconjugative plasmids could be maintained in the absence of positive selection, as parasites, through mobilization by conjugative plasmids, those conditions are substantially more restrictive than those for self-transmissible plasmids (Levin and Stewart, 1980). First, the mobilizing conjugative plasmids have to be transmitted at rates high enough to be maintained by infectious transfer alone. Second the nonconjugative plasmids have to be mobilized for infectious transfer at relatively high rates. Third, their rate of loss by vegetative segregation and the fitness burden they impose on their host in the absence of antibiotic-mediated selection have to be vanishingly small. For some nonconjugative plasmids, mobilization can occur at substantial rates (Levin and Rice, 1980), and in the absence of positive selection for genes they carry, they are extremely stable and impose negligible costs on the fitness of their host. Nevertheless, as indicated above, it seems highly unlikely that the first of these existence conditions for parasitic nonconjugative plasmids will be met in bacterial populations of realistic densities. On the other hand, the conditions for maintaining nonconjugative R plasmids by intermittent selection favoring one or more antibiotic resistance or other gene remain almost as great as their larger and more gene-laden conjugative counterparts.

12.2.3. Antibiotic-Resistance Transposons

Some transposons have mechanisms for mediating their own infectious transfer as phage or by conjugation (Shapiro, 1983), and the conditions for maintaining these elements as parasites would be virtually identical to those of temperate phage and conjugative plasmids, respectively. In theory, non-self-transmissible transposons that encode for resistance to antibiotics could also be maintained as parasites. For those that are infectiously transmitted by transposition into conjugative plasmids these conditions are, however, restrictive (Condit et al., 1988): (1) transposition has to be replicative, i.e., copies of the transposon have to move to a new site rather than the transposon itself; (2) the cost of carriage of the transposon has to be of the order of the transposition rate or less (of the order of 10^{-3} or lower); and (3) the cells have to be infected with conjugative plasmids that are infectiously transferred at rates that exceed the cost of carrying the transposon.

From these *a priori* conditions and an experimental study with laboratory populations of *Escherichia coli* (Condit, 1990), we conjectured that they are unlikely to obtain in natural populations of bacteria and concluded that transposable elements in bacteria can be maintained only by positive selection for their carriage. It should be noted, however, that Stanley Sawyer and colleagues (Sawyer et al., 1987) came to the opposite conclusion from their analysis of the distribution of insertion sequences (transposable elements not carrying known genes) in *E. coli* from natural sources. Finally, while individual transposons encode for relatively fewer genes than plasmids, they too have many opportunities to be maintained by associated linkage selection. Individual transposons may carry more than one antibiotic-resistance gene or combinations of resistance and other host-expressed genes and resistance plasmids often carry multiple transposons.

12.2.4. Temperate Phage-Bearing Antibiotic-Resistance Genes

While purely lytic (virulent) bacteriophage have to be maintained as parasites, their temperate counterparts can also be maintained as symbionts by carrying genes that augment the fitness of their lysogenic (prophage-bearing) hosts. Thus, even if the density of a bacterial population is too low to maintain temperate phage by infectious transfer alone, as long as there is positive selection for prophage-borne genes, like those encoding for antibiotic resistance, these bacterial viruses and the resistance genes they carry will be maintained. How these resistance-determining temperate phage would fair in competition with temperate phage with the same receptor and host range that do not carry resistance genes or with purely lytic phage with these same adsorption organelles and host ranges would depend on how those resistance genes affect their rates of infectious transfer (adsorption rates, latent periods, and burst sizes) and the frequency and intensity of selection for the resistance genes they carry (Stewart and Levin, 1984).

12.3. The Incidence of Antibiotic Use and The Frequency of Resistance

If in fact there are conditions for which resistance-encoding accessory elements are maintained as parasites, it is reasonable to assume that in those populations the majority of their host bacteria would carry them. Intuitively, it would seem if an element can be maintained by infectious transfer alone, it would eventually sweep through the population. In the case of conjugative plasmids that can be maintained by infectious transfer alone, we have formally demonstrated that the frequency of their carriage in the bacterial population would approach unity (Stewart and Levin, 1977). Antibiotic-mediated selection for resistance genes carried on these elements or associated linkage selection for other characters can increase the frequencies of these elements as well as increase the rate at which they spread through the host population. For resistance-encoding accessory elements that cannot be maintained as parasites and require at least intermittent selection for persistence (which I believe is generally the case), the frequency with which bacterial populations encounter this selection regime and the fitness costs of their carriage in the absence of selection would determine their frequency in the population at large.

The latter is almost certainly the case for chromosomal resistance as well. Although chromosomal genes can also be infectiously transmitted, if their transfer is by transformation (as free DNA from dead cells) or by transduction (with a phage vector), it is unlikely to be replicative. The gene will be lost from the donor, and infectious transmission alone will not increase the number of copies of that gene in the population at large. While infectious transmission of chromosomal genes by plasmid-mediated conjugation could be replicative, at this juncture there is no evidence to suggest that conjugative transfer of chromosomal genes is anything more than a laboratory artifact (Levin, 1988a). Thus it seems reasonable to conclude that the frequency of chromosomal antibiotic resistance in bacterial populations is going to depend on the intensity of selection for resistance, which in turn depends on the incidence of antibiotic use.

12.3.1. *A Priori* Relationship between Antibiotic Use and Antibiotic Resistance

In an effort to predict the relationship between the rate of antibiotic use and the frequency of resistant bacteria, we developed a mathematical model of the population dynamics of commensal bacteria in treated hosts (Levin et al., 1997). In our model, there is an array of hosts who shed and acquire bacteria from a common reservoir, at rates g and h, respectively. We assume that in the absence of antibiotics, both in the reservoir and within individual hosts, bacteria carrying antibiotic-resistance genes are at a selective disadvantage, with a fitness of $1 - s$, where s is the familiar selection coefficient of classical population genetics. Host are treated by antibiotics at a rate T per year and we assume that, on treatment, the frequency of resistance immediately ascends to unity in that host.[3]

The results of a numerical simulation of this model are presented in Figure 12.1. As anticipated, the frequency of antibiotic resistance at equilibrium increases with the incidence of treatment and decreases with the intensity of selection against the resistant bacteria. Even with relatively low rates of treatment, of the order of once every second year, if the fitness cost of resistance is low, the frequency of resistant bacteria at equilibrium can be substantial.[4]

12.3.2. The Fitness Cost of Resistance and Adaptation to Those Costs

As noted above there is evidence for measurable fitness costs, $s > 0$, associated with the carriage of some antibiotic-resistance-encoding plasmids. This is also the case for at least some of the chromosomal genes that code for resistance. For

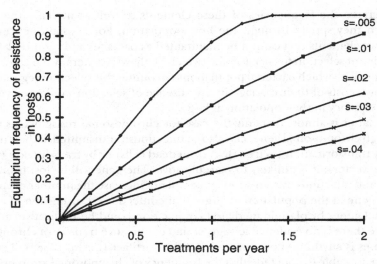

Figure 12.1. Simulation results: steady-state frequencies of antibiotic resistant commensals as a function of the annual incidence of treatment T and the fitness costs s associated with resistance. In these runs, the proportion of the bacteria in the host flowing to the common environment reservoir, f, is 0.05 per generation. The proportion of the bacteria in the treated host that are replaced each generation from the pool of bacteria in the reservoir, g, is 0.005 per generation. In the simulations, the generation times in the host and the reservoir are the same, 40 h. For more details about this model, see Levin et al. (1997).

example, isoniozaid resistance in *M. tuberculosis* is particularly costly (Meissner, 1964) as are the ribosomal protein *rpsL* mutations responsible for chromosomal resistance to streptomycin in *E. coli* and many other bacterial species (Funatsu and Wittmann, 1972; Schrag and Perrot, 1996). And, based on the above model, with high rates of selection against resistant bacteria and infrequent use of antibiotics, the equilibrium frequency of resistant bacteria could be quite low (Levin et al., 1997).

At this juncture, however, it is not at all clear how commonly chromosomal antibiotic resistance engenders a cost in fitness. Moreover, as noted above, the fitness costs associated with plasmid-encoded resistance are anticipated to decline in the course of (co)evolution between the plasmid (or accessory element) and its host bacterium. Recent evidence indicates that this is also the case for the *rpsL* chromosomal gene mutations responsible for streptomycin resistance (Schrag and Perrot, 1996). Single base substitution mutations at the 42nd codon, AAA, of *rpsL*, to ACA and AAC confer resistance to high concentrations of streptomycin and, in the absence of this antibiotic, reduce the fitness of *E. coli* by 14% and 18%, respectively. The *rpsL* gene codes for a ribosomal protein, and most if not all the cost of fitness is due to an alteration in the ribosome that diminishes the rate of protein elongation during translation. However, within 160 generations in a streptomycin-free medium, as a consequence of the substitution of second site mutation(s) – presumably at other ribosomal protein or ribosomal RNA genes – these high fitness costs are markedly reduced and rates of protein elongation restored to near-wild-type levels without significant reductions in the level of resistance to streptomycin. Moreover, in the genetic background of bacteria carrying these fitness-restoring second site mutations, wild-type streptomycin-sensitive $rpsL^+$ genes have a marked selective disadvantage relative to the resistant $rpsL$ allele (Schrag et al., 1997). Stated another way, as a consequence of evolution's reducing the fitness costs of antibiotics resistance, adaptive valleys can be established that may virtually prelude the return to the ancestral state of antibiotic sensitivity. The same, pessimistic interpretation also applies to the observation Bouma and Lenski (1988) about adaptation to the fitness cost associated with the carriage of antibiotic-resistance-encoding plasmids.

12.3.3. Retrospective Evidence

In support of the above experimental results is the observation that naturally occurring bacteria remain resistant to antibiotics that are rarely, if ever, used clinically. This may be due to associated linkage selection with other loci or positive selection for resistance due to the nonclinical application of those antibiotics, like their widespread use in animal feed and other agricultural endeavors. The results described in the preceding couple of paragraphs by Stepanie Schrag and her co-workers, along with those of Bouma and Lenski (1988) and Modi and Adams (1991), suggest a third explanation. That is, even in the absence of antibiotics, there is virtually no selection against contemporary bacteria bearing chromosomal or accessory-element-encoded resistance genes.

One prediction of the fitness-compensating mutation – adaptive valley results of the above-described $rpsL$ gene experiments – is that streptomycin-resistant $rpsL$ mutations would persist for extensive periods in the absence of streptomycin-mediated selection. To test this hypothesis, we sequenced the $rpsL$ region of the 10,000-generation-evolved strain of the E. coli B strain studied by Richard Lenski and collaborators (Lenski and Travisano, 1994) and its streptomycin-resistant $rpsL$ ancestor. The glucose-limited minimal medium that Lenski and colleagues used in their long-term evolution experiments did not contain streptomycin. Nevertheless, the 10,000-generation-evolved E. coli B, like the strain whence it was derived, was resistant to high concentrations of this antibiotic. The DNA sequence of the $rpsL$ gene of the 10,000-generation-evolved strain and its ancestor were identical and differed from the $rpsL$ DNA sequence of streptomycin-sensitive E. coli solely at codon 42, ACA instead of AAA for the streptomycin-sensitive wild type. Furthermore, just like the fitness-compensated $rpsL$ mutants of E. coli CAB281, streptomycin-sensitive $rpsL^+$ transductants of the original E. coli B strain have a disadvantage in competition with the resistant $rpsL$ transductants. One interpretation of this is that second site mutation(s) compensating for the costs of the $rpsL$ locus evolved during the more than 25 years this streptomycin-resistant strain of E. coli was maintained in laboratories.

12.4. Discussion

12.4.1. The Maintenance and Evolution
of Antibiotic-Resistance-Encoding Accessory Genetic Elements

Neglecting all of the usual hedging caveats and illusions of modesty and objectivity, I propose that the antibiotic-resistance-encoding accessory genetic elements of bacteria are not maintained as pure parasites. They have evolved and continue to persist because of antibiotic-mediated selection; also see Levin (1993). While some of these elements, like the temperate bacteriophage, may well be maintained as parasites in some bacterial populations, in the absence of at least occasional positive selection for the antibiotic resistance genes they carry those gene would eventually be lost or become nonfunctional. To be sure, part of the selection responsible for the evolution and maintenance of these antibiotic-resistance-encoding accessory elements may be due to interactions between antibiotic-producing and antibiotic-sensitive bacteria in their natural habitats (see Demain, 1995; Wiener, 1996). Nevertheless, I believe that the majority contemporary antibiotic-resistance-encoding accessory elements that currently plague us evolved in response to the human use of antibiotics. The plasmids, phage, and transposable elements carrying these genes and, in some form or other, the genes themselves almost certainly existed before the human use of antibiotics (Datta and Hughes, 1983; Hughes and Datta, 1983). I conjecture, however, that selection pressure responsible for these resistance-encoding accessory components evolving as collectives is a postantibiotic era phenomenon. For a consideration of the ecological conditions and genetics mechanisms responsible for the evolution of contemporary R plasmids, see Condit and Levin (1990), Sykora (1992), and Levin (1995). For some appealing and maybe correct ideas about the origins of the resistance genes see Davies and Gray (1984).

12.4.2. Going Back to the Antibiotic-Sensitive Days
of Yesteryear – More Opinions

It will not be easy to institute the reductions in the rates of antibiotic use that have been proposed to stem the tide of resistance (OTA, 1995). It also may no longer matter very much if we do reduce the rate and extent of antibiotic use. The evolutionary die may have already been cast. Even with the massive reductions in antibiotic use, we will not, in reasonable amounts of time, return to the carefree (hardly) days of yesteryear during which virtually all pathogenic bacteria were sensitive to the antibiotics used for treating infections with them. One can interpret the theoretical and experimental results discussed in this chapter in this pessimistic light. This is particularly so, if in general the fitness costs of resistance are compensated for in ways that establish adaptive valleys, genetic backgrounds in which sensitive alleles are less fit than resistant, as in the *E. coli rpsL* situation (Schrag et al., 1997).

On the other side of this gloomy interpretation is the observation that globally the frequency of resistant bacteria is proportional to rate of use of antibiotics (WHO, 1994). To be sure, it is possible that this is an artifact, and countries with more prudent antibiotic use policies will eventual face the same problems of antibiotic resistance as the less prudent. Nevertheless, it is also possible that the above pessimistic interpretation of the future of antibiotic resistance is in error. I hope so. Moreover, as long as the more restrained use of antibiotics does not engender clinical or public health problems, then even accepting the most pessimistic perspective of the population genetic future of antibiotic resistance, reductions in the rates of use of antibiotics, remains the rational and best policy. I cannot conceive of any way that reduced use of antibiotics will make the resistance problem any graver.

12.4.3. Does the Evolution of Antibiotic Resistance Matter?

As I indicated above, antibiotics are not the reason for most of the decline in infectious disease mortality during this century and a half. One interpretation of this is that if these chemotherapeutic agents completely cease being effective for the treatment of bacterial diseases, there would be relatively little effect on the rate of mortality in humans. This may be the case for community-acquired bacterial infections in developed countries. The majority of these bacterial infections are not lethal, even in the absence of antibiotic treatment. On the other hand, antibiotics generally reduce the morbidity of these community-acquired infections and, by accelerating the rate of clearance of the bacteria, these drugs can reduce the incidence of infection in the community (Bonhoeffer et al., 1997).

In hospital-acquired infections, antibiotics are often needed to prevent mortality as well as reduce morbidity. Almost all of the dramatic life-saving, life-extending, transplant, and other surgical procedures require the prophylactic use of antibiotics to prevent infection. In AIDS, transplant, and other patients with compromised immune defenses, antibiotic treatment can be essential to prevent the mortality of bacterial infections, including infections with normally commensal bacteria. While even in hospitals we have yet to see the complete breakdown of the efficacy of antibiotics because of resistance, in the sense of all species of potentially pathogenic bacteria being dominated by strains that are untreatable by any available antibiotic,[5] antibiotic resistance is already imposing an ever-increasing threat to the success of surgery (Low et al., 1995; Neu, 1994). Thus, even in times of political and environmental stability in the developed world, because of resistance, people are dying because of bacterial infections that would have otherwise been successfully treated with antibiotics. Finally, as noted above, the relatively low rates of symptomatic bacterial infections in the developed world can be attributed to good sanitation, nutrition, and hygiene. When, for political, social, or environmental reasons, this thin facade of cultural practices is broken or, as in the case of war, rates of trauma are increased, antibiotics treatment will become increasingly important to prevent mortality.

12.5. Conclusion

The evolution of bacteria resistant to the antibiotics used to control their populations is becoming an increasingly important clinical problem. Formerly successful antibiotic treatment regimes are failing because of resistance, and people are dying of previously curable bacterial infections. In this chapter I have tried to illustrate (proselytize) that antibiotic resistance poses problems that are intriguing, even from the precious perspective of academic population and evolutionary biology. For convenience and obvious egoism, I have restricted this consideration to the two problems of the population biology of antibiotic resistance that I know best and have worked on longest: the maintenance of antibiotic-resistance-encoding plasmids, phage, and transposons, and the relationship between the incidence of antibiotic use and the frequency of resistant bacteria. There are oodles more to do on both of these problems, at the theoretical as well as the empirical level. Antibiotic resistance (like many other aspects of disease) also poses many delicious population biological and evolutionary problems not considered here, e.g., those involved in the treatment of individual patients (see, e.g., Lipsitch and Levin, 1997a, 1997b). There are lots of wonderful things to do.[6] Moreover, in addition to the usual euphoria of doing research, while working on these types of problems, one can bask in the wonderful glow of righteousness that emanates from potentially useful endeavors. I hope I have convinced some of the readers to join in this enterprise.

12.6. Acknowledgment

I thank Marc Lipsitch and Tomoko Steen for reading this manuscript and for their many insightful and useful suggestions. This endeavor and almost all of the research reviewed here has been supported by grants from the National Institutes of Health, GM33782 and AI40662.

NOTES

1. Antiseptics were used and were effective topically for some bacteria, but not systemically (Fleming, 1995). Although inconvenient (time-consuming, labor-intensive, and requiring skill and judgement) to use, blood sera containing high concentrations of antibodies for specific pathogen-borne antigens, passive immunization, or serum therapy was used extensively during the third and fourth decades of this century and was relatively effective for treating bacterial infections like *Pneumococcus* pneumonia (Finland, 1972). Although phage therapy was used for a variety of bacterial infections, and still is in some parts of the world, its efficacy was more equivocal, but that is another story, (see Levin and Bull, 1996; Redetsky, 1996).
2. In a survey of resistance patterns of the enteric bacterial flora of children in an Atlanta day care center, an Emory undergraduate working with me, Bassen Tomeh, and my technician Nina Walker found that more than 50% of the bacteria (primarily *E. coli*) isolated from children under recent antibiotic treatment were resistant to one and usually more than one of the seven antibiotics screened (ampicillin, streptomycin, kanamycin, chloramphenicol, tetracyline, spectinomycin, and naladixic

acid). Ampicillin was the antibiotic to which there was the highest frequency of resistance, which is not unanticipated as variants of the β-lactam antibiotic are very commonly used to treat these children. More surprising is that 25% of the bacteria isolated from these fecal samples were resistant to streptomycin, which has almost neven been used cinically for more than 20 years.

3. Although the sample size is limited and the habitat may be biased, this assumption is consistent with what I found in my own enteric flora. After taking a single day's dose of tetracycline, the frequency of tetracycline resistant *E. coli* went from approximately 5×10^{-3} to effective 1.0. There was also associated linkage selection: the frequency of enteric bacteria resistant to ampicillin and kanamycin also rose in response to tetracycline treatement.

4. Currently, Frank Stewart, Marc Lipsitch, Rustom Antia, Sebastain Bonhoeffer, John Mittler, and I are developing and analyzing alternative and more general models of the rate of antibiotic treatment and the frequency of resistance bacteria transmitted through a common reservior. Rustom Antia, Sebastain Bonhoeffer, Marc Lipsitch, and I have been developing models of the population genetics (epidemiology) of antibiotic treatment of directly transmitted bacteria responsible for acute infections, In the analysis of these newer models, we are also considering the dynamics of the response to changes in the incidence of antibiotic use. Of particular concern is the rate at which resistance will decline following reductions in the rate of antibiotic use and the antibiotic use patterns that maximize the long-term efficacy antibiotics for which resistance exists or evolves.

5. Resistance to all effective antibiotics has already been observed for some strains of some bacteria, including *M. tuberculosis* and different species of *Enterococcus*.

6. As this is a rapidly moving field, a lot of good work has appeared since this chapter was submitted for publication. I offer the reader a list of the naturalized and native population and evolutionary biologists not included in the present list of references who I know are doing research in this area: Roy Anderson, Dan Andersson, Daren Austin, Jesus Balasquez, Fernando Baquero, Richard Bax, Carl Bergstrom, Johanna Björkman, Sally Blower, Jim Bull, Patrice Courvalin, Andrew Demma, Diarmaid Hughes, Karl Kristinsson, Eduardo Massad, Cristina Negri, Martin Nowak, Alan Perelson, Travis Porco, Mary Reynolds, Lone Simonsen, Jeff Smith, Ming Zhang, Renata Zappala. Needless to say, now that population biologists are on the case, the world can rest assured that the resistance problem will soon be under control.

REFERENCES

Anderson, E. S. 1968. The ecology of transferable drug resistance in the enterobacteria. *Annu. Rev. Microbiol.* **22**:131–180.

Berkowitz, F. E. 1995. Antibiotic resistance in bacteria. *South. Med. J.* **88**:797–804.

Bloom, B. R. and Murray, C. J. L. 1992. Tuberculosis – commentary on a reemergent killer. *Science* **257**:1055–1064.

Bouma, J. E. and Lenski, R. E. 1988. Evolution in bacterial – plasmids association. *Nature (London)* **335**:351–352.

Bonhoeffer, S., Lipsitch, M., and Levin, B. R. 1997. Evaluating treatment protocols to prevent resistance. *Proc. Natl. Acad. Sci. USA* **94**:12106–12111.

Bowler, L. D., Zhang, Q. Y., Riou, J. Y., and Spratt, B. G. 1994. Interspecies recombination between the penA genes of *Neisseria meningitidis* and commensal Neisseria species during the emergence of penicillin resistance in *N. meningitidis*: natural events and laboratory simulation. *J. Bacteriol.* **176**:333–337.

Chain, E., Florey, H. W., Gardner, A. D., Heatley, N. G., and Jennings, M. A., et al. 1940. Penicillin as a chemotherapeutic agent. *Lancet* **2**:226–228.

Coffey, T. J., Dowson, C. G., Daniels, M., and Spratt, B. G. 1993. Horizontal spread of an altered penicillin-binding protein 2B gene between *Streptococcus pneumoniae* and *Streptococcus oralis*. *FEMS Microbiol. Lett.* **110**:335–339.

Cohen, M. L. 1992. Epidemiology of drug resistance: implications for a post-antimicrobial era. *Science* **257**:1050–1055.

Cohen, M. L. 1994. Emerging problems of antimicrobial resistance. *Ann. Emergency Med.* **24**:454–456.

Condit, R. 1990. The evolution of transposable elements: conditions for establishment in bacterial populations. *Evolution* **44**:347–359.

Condit, R. and Levin, B. R. 1990. The evolution of plasmids carrying multiple resistance genes: the role of segregation, transposition and homologous recombination. *Am. Nat.* **135**:573–596.

Condit, R., Stewart, F. M., and Levin, B. R. 1988. The population biology of bacterial transposons: a priori conditions for maintenance as parasitic DNA. *Am. Nat.* **132**:129–147.

Datta, N. and Hughes, V. 1983. Plasmids of the same Inc groups in enterobacteria before and after the medical use of antibiotics. *Nature (London)* **306**:616–617.

Davies, J. and Gray, G. 1984. Evolutionary relationships among genes for antibiotic resistance. *Ciba Foundation Sympo.* **102**:219–232.

Davis, B. D., Dulbecco, R., Eisen, H. N., and Ginsburg, H. S. 1990. *Microbiology.* New York: Lippincott.

Demain, A. L. 1995. Why do microorganisms produce antimicrobials. In *Fifty Years of Antimicrobials: Past Perspectives and Future Prospects*, P. A. Hunter, G. K. Darbey and N. J. Russell, eds., pp. 205–228. Cambridge: Cambridge U. Press.

Dowson, C. G., Coffey, T. J., and Spratt, B. G. 1994. Origin and molecular epidemiology of penicillin-binding-protein-mediated resistance to beta-lactam antibiotics. *Trends Microbiol.* **2**:361–366.

Falkow, S. 1975. *Infectious Multiple Drug Resistance.* London: Pion.

Finland, M. 1972. Adventures with antibacterial drugs. *Clin. Phormacol. Therapeu.* **13**:469–511.

Fleming, A. 1995. Chemotherapy: yesterday, today and tomorrow. In *Fifty Years of Antimicrobials: Past Perspectives and Future Trends*, P. A. Hunter, G. K. Darby and N. J. Russell, eds., pp. 1–18. Cambridge: Cambridge U. Press.

Funatsu, G. and Wittmann, H. G. 1972. Ribosomal proteins. XXXIII. Location of amino-acid replacements in protein S12 isolated from *Escherichia coli* mutants resistant to streptomycin. *J. Mol. Biol.* **68**:547–50.

Garrett, E. R. 1971. Drug action and assay by microbial kinetics. *Prog. Drug Res.* **15**:271–352.

Hotchkiss, R. D. 1990. From microbes to medicine: Gramicidin, Rene Dubos, and the Rockefeller. In *Launching the Antibiotic Era. Personal Accounts of the Discovery and Use of Antibiotics*, C. L. Moberg and Z. A. Cohen, eds., pp. 1–18. New York: Rockefeller U. Press.

Hughes, V. and Datta, N. 1983. Conjugative plasmids in bacteria of the 'pre-antibiotic' era. *Nature (London)* **302**:725–726.

Lenski, R. E. and Hattingh, S. E. 1986. Coexistence of two competitors on one resource and one inhibitor: a chemostat model based on bacteria and antibiotics. *J. Theor. Biol.* **122**:83–93.

Lenski, R. E., Simpson, S. C., and Nguyen, T. T. 1994. Genetic analysis of a plasmid-encoded, host genotype-specific enhancement of bacterial fitness. *J. Bacteriol.* **176**:3140–3147.

Lenski, R. E. and Travisano, M. 1994. Dynamics of adaptation and diversification: a 10,000-generation experiment with bacterial populations. *Proc. Natl. Acad. Sci. USA* **91**:6808–6814.

Levin, B. R. 1980. Conditions for the existence of R-plasmids in bacterial populations. In *Proceedings of the Fourth International Symposium on Antibiotic Resistance*, pp. 197–202. Berlin: Springer-Verlag.

Levin, B. R. 1981. Periodic selection, infectious gene exchange and the genetic structure of *E. coli* populations. *Genetics* **99**:1–23.

Levin, B. R. 1988a. The evolution of sex in bacteria. In *The Evolution of Sex: A Critical Review of Current Ideas*, R. Michod and B. R. Levin, eds., Sunderland, MA: Sinauer.

Levin, B. R. 1988b. Frequency-dependent selection in bacterial populations. *Philos. Trans. R. Soc. London Ser. B* **319**:459–472.

Levin, B. R. 1993. The accessory genetic elements of bacteria: existence conditions and coevolution. *Curr. Opinion Genet. Dev.* **3**:849–854.

Levin, B. R. 1995. Conditions for the evolution of multiple antibiotic resistance plasmids: a theoretical and experimental excursion. In *The Population Genetics of Bacteria*, S. Baumberg, J. P. W. Young, S. R. Saunders, and E. M. H. Wellington, eds., pp. 175–192. Cambridge: Cambridge U. Press.

Levin, B. R. and Bull, J. J. 1996. Phage therapy revisited: the population biology of a bacterial infection and its treatment with bacteriophage and antibiotics. *Amer. Nat.* **147**:881–898.

Levin, B. R. and Lenski, R. E. 1983. Coevolution of bacteria and their viruses and plasmids. In *Coevolution*, D. J. Futuyama and M. Slatkin, eds., pp. 99–127. Sunderland, MA: Sinauer.

Levin, B. R., Lipsitch, M., Perrot, V., Schrag, S., Antia, R., Simonsen, L., Moore, N., and Stewart, F. M. 1997. The population genetics of antibiotic resistance. *Clin. Infect. Dis.* **24**:59–516.

Levin, B. R. and Rice, V. A. 1980. The kinetics of transfer of nonconjugative plasmids by mobilizing conjugative factors. *Genet. Res.* **35**:241–259.

Levin, B. R. and Stewart, F. M. 1980. The population biology of bacterial plasmids: a priori conditions for the existence of mobilizable nonconjugative factors. *Genetics* **94**:425–443.

Levin, B. R., Stewart, F. M., and Rice, V. A. 1979. The kinetics of conjugative plasmid transmission: fit of a simple mass action model. *Plasmid* **2**:247–260.

Levy, S. B. 1992. *The Antibiotic Paradox: How Miracle Drugs are Destroying the Miracle*. New York: Plenum.

Lipsitch, M. and Levin, B. R. 1997a. The within-host population dynamics of antibacterial chemotherapy: conditions for the evolution of resistance. In *Antibiotic Resistance: Origin, Evolution, Selection, and spread*, S. Levy, ed., pp. 112–130. Chichester: Wiley.

Lipsitch, M. and Levin, B. R. 1997b. The population dynamics of antimicrobial chemotherapy. *Antimicrob. Agents Chemother.* **41**:363–373.

Low, D. E., Willey, B. M., and McGeer, A. J. 1995. Multidrug-resistant enterococci: a threat to the surgical patient. *Am. J. Surg.* **169**:8S–12S.

252 *Bruce R. Levin*

Lundquist, P. D. and Levin, B. R. 1986. Transitory derepression and the maintenance of conjugative plasmids. *Genetics* 113:483–497.

McGowan Jr., J. E. 1991. Abrupt changes in antibiotic resistance. *J. Hosp. Inf.* 18:202–10.

McKeowen, T. 1976. *The Role of Medicine: Dream, Mirage or Nemesis.* Princeton, NJ: Princeton U. Press.

Meissner, G. 1964. The microbiology of the tubercule bacillus. In *Chemotherapy of Tuberculosis*, pp. 65–110. London: Butterworth.

Mitchison, D. A. 1979. Basic mechanisms of chemotherapy. *Chest* 76:771–781.

Mitsuhashi, S. 1977. *R-Factor. Drug Resistance Plasmid.* London, Tokyo: University Park.

Moberg, C. L. 1996. René Dubos, a harbinger of microbial resistance to antibiotics. *Microb. Drug Resist.* 2:287–297.

Modi, R. I. and Adams, J. 1991. Coevolution in bacterial-plasmid populations. *Evolution* 45:656–667.

Mongold, J. A. 1992. DNA repair and the evolution of transformation in *Haemophilus influenzae. Genetics* 132:893–898.

Murray, B. E. 1992. Problems and dilemmas of antimicrobial resistance. *Pharmacotherapy* 12:86S–93S.

Neu, H. C. 1992. The crisis in antibiotic-resistance. *Science* 257:1064–1073.

Neu, H. C. 1994. Emerging trends in antimicrobial resistance in surgical infections. A review. *Eur. J. Surg.* – Supplement :7–18.

OTA, 1995. *Impact of Antibiotic Resistant Bacteria.* Washington, D.C.: U.S. Congress, Office of Technology Assessment.

Radetsky, P. 1996. The good virus. *Discover Magazine.* November.

Ryan, F. 1993. *The Forgotten Plague.* Boston: Little, Brown.

Sawyer, S. A., Dykhuizen, D. E., Dubois, R. F., Green, L., Mutangadura-Mhlanaga, T., Wolczyk, D. F., and Hartl, D. L. 1987. Distribution of insertion sequences among natural isolates of *Escherichia coli. Genetics* 115:51–63.

Schrag, S. and Perrot, V. 1996. Rapid reduction in the fitness costs of antibiotic resistance by natural selection. *Nature (London)* 381:120–121.

Schrag, V., Perrot, S., and Levin, B. R. 1997. Adaptation to the fitness costs of antibiotic resistance in *Escherichia Coli. Proc. Roy. Soc. London. Ser. B.* 264:1287–1291.

Shapiro, E. J. 1983. *Mobile Genetic Elements.* New York: Academic.

Simonsen, L. 1990. Dynamics of plasmid transfer on surfaces. *J. Gen. Microbiol.* 136:1001–1007.

Simonsen, L. 1991. The existence conditions for bacterial plasmids: theory and reality. *Microb. Ecol.* 22:187–205.

Simonsen, L. 1992. *The Existence Conditions for Bacterial Plasmids.* Boston: U. Massachusetts Press.

Simonsen, L., Gordon, D. M., Stewart, F. M., and Levin, B. R. 1990. Estimating the rate of plasmid transfer: an end-point method. *J. Gen. Microbiol.* 136:2319–2325.

Stewart, F. M. and Levin, B. R. 1977. The population biology of bacterial plasmids: a priori conditions for the existence of conjugationally transmitted factors. *Genetics* 87:209–228.

Stewart, F. M. and Levin, B. R. 1984. The population biology of bacterial viruses: why be temperate. *Theor. Popul. Biol.* 26:93–117.

Summers, A. O., Wireman, J., Vimy, M. J., Lorscheider, F. L., Marshall, B. et al., 1993. Mercury released from dental "silver" fillings provokes an increase in mercury- and antibiotic-resistant bacteria in oral and intestinal floras of primates. *Antimicrob. Agents Chemother.* 37:825–834.

Sundin, G. W. and Bender, C. L. 1996. Dissemination of the strA-strB streptomycin resistance genes among commensal and pathogenic bacteria from humans, animals and plants. *Mol. Ecol.* **5**:133–143.

Sykora, P. 1992. Macroevolution of plasmids: A model for plasmid speciation. *J. Theor. Biol.* **159**:53–65

Tenover, F. C. and Hughes, J. M. 1996. The challenges of emerging infectious diseases. Development and spread of multiply-resistant bacterial pathogens. *J. Am. Med. Assoc.* **275**:300–304.

Watanabe, T. 1963. Infective heredity of multiple drug resistance in bacteria. *Bacteriol. Rev.* **27**:87–115.

WHO, 1994. Report, WHO scientific working group on monitoring and managing bacterial resistance to antimicrobial agents. *World Health Organization Rep.* BV1: 1–33.

Wiener, P. 1996. Experimental studies on the ecological role of antibiotic production in bacteria. *Evol. Ecol.* **10**:405–421.

Zahner, H. and Giedler, H.-P. 1995. The need for new antibiotics: possible ways forward. In *Fifty Years of Antimicrobials: Past Perspectives and Future Directions*, P. A. Hunter, G. K. Darby, and N. J. Russell, eds., pp. 67–84. Cambridge: Cambridge U. Press.

SECTION D

LINKAGE, BREEDING SYSTEMS, AND EVOLUTION

Introductory Remarks

IAN R. FRANKLIN

I remember Theodosius Dobzhansky presenting in 1960 a lecture on inversion polymorphisms in *Drosophila* to the Genetics Society of Australia. Sir Ronald Fisher was in the chair, and there was a feeling of unease in the audience, expecting a confrontation between the speaker, who was thought strongly aligned with Sewall Wright, and the chairman. These fears were groundless, for Fisher was at his most amiable, obviously comfortable with a message documenting strong balancing selection, coadaptation, and clinal variation. There could be no impression other than that these were the essential issues, the *Weltanschauung*, for the study of natural variation.

But cracks had already begun to appear. Genetic load arguments (Haldane, 1937; Muller, 1950) were taking hold, especially in the United States, and within a few years debates about the relative importance of substitutional loads, segregational loads, and mutational loads were at their peak. As Lewontin later explained (Lewontin, 1974), there is a long-standing tension in population genetics between, in his terminology, "classical" and "balance" theories of population structure. Flawed as these genetic load debates were, one cannot underestimate their influence in shifting allegiances away from balance theories and toward a neoclassical explanation of the nature of standing variation.

There were two important factors in this switch. First was the influence of Muller's (1950) paper on J. F. Crow and his colleagues, who did much to develop and promulgate the theory. The second was the finding and the recognition by Jack Hubby and Dick Lewontin (Hubby and Lewontin, 1966; Lewontin and Hubby, 1966) that allelic variation at a large fraction of a genome is abundant and ubiquitous. There could be only three explanations for these polymorphisms: they could be balanced, transient, or neutral. The third explanation had long been thought unlikely; Fisher (1930) had considered and rejected neutrality – see, for example, E. B. Ford's widely accepted definition and discussion of polymorphism (Ford, 1940). However, the first two explanations would seem to impose intolerable genetic loads on all biological populations. The situation worsened as molecular data began to reveal substantial gene substitutions in lineages at rates far higher than was thought possible (Haldane, 1957). The balance theory was faltering; molecular evidence suggested that near neutrality was plausible, and Kimura, with his enormous will and ability, resurrected the classical theory and drove a stake into the heart of pan-selectionism. Today, few population geneticists, observing a polymorphic gene, would assume selective balance.

Consequently this section may seem prelapsarian, harking back to those days when balance theories were in full ascendancy. All of the papers deal with topics that were considered central 40 years ago. All deal primarily with adaptive rather than purifying selection. But they are not naively pan-selectionist. For example, we *know* that self-sterility alleles, widespread through most genera of flowering plants, are highly polymorphic and maintained by balancing selection. In this case, the selective forces are mediated by frequency-dependent fertility differences, a classic example of soft selection. Similarly, the evidence for strong balancing selection as a force underlying the patterns of inversion polymorphism in *Drosophila* is overwhelming. Again, the data suggest that some form of soft selection may be involved. Finally, not even a die-hard neutralist would argue that the evolution of sex is nonadaptive, even though explanations for its origin and maintenance remain unclear.

In Chap. 13, Franklin and Feldman review some of the theories of selection at one, two, or more loci and draw attention to the analytic relationships between one-locus and multilocus theory. They confine their discussion to a restricted class of hard selection models, namely constant selective values based on only viability differences and panmictic populations. A general theoretical development of alternative selection regimes is exceedingly limited; one of the few examples is a rarely cited paper by Lewontin (1958). Multilocus theory remains one of Dick Lewontin's enduring interests, originally stimulated by a collaborative analysis with M. J. D. White (Lewontin and White, 1960) of inversion polymorphism frequencies in two chromosomes of the grasshopper *Moraba scurra*. Lewontin, his colleagues, and his students have made a major contribution to this field.

Krimbas and Powell (Chap. 14) update our understanding of one of the most intensively studied of all polymorphic systems, the inversion polymorphisms of various *Drosophila* species. This work was at its zenith in the laboratories of Dobzhansky and his colleagues in the 1950s and early 1960s, establishing beyond doubt that selective forces were at work in maintaining clinal and ecotypic variation. However, the nature of these selective forces is not fully understood. For example, in cage experiments the equilibrium values reached by inversion frequencies depend on their initial frequencies; this would not be possible if selective values were due entirely to fixed viability differences and suggests the involvement of frequency-dependent or fertility selection. Krimbas and Powell review the classical hypotheses for the origin and maintenance of inversion polymorphisms in the light of modern techniques that allow us to study more precisely the breakpoints and the genic content of these inversions.

In her paper on the evolution of breeding systems, Marcy Uyenoyama (Chap. 15) neatly blends a description of one of the best examples of balancing selection, the self-sterility alleles in plants, with coalescence theory, one of the most important developments of the current neoclassical period. The topic has a rich history in population genetics, ranging back to debates between Fisher and Wright about the role of genetic drift in maintaining large numbers of alleles. Molecular evidence has illuminated not only the precise biological mechanisms but also has shown that these polymorphisms are of great antiquity, with many alleles transcending taxonomic boundaries.

The last chapter in this section is a review by Donal Hickey of one of the most enduring topics in evolutionary biology, the origin of sexual reproduction (Chap. 16). No matter how the ground rules change for interpreting variation, the origin of sexual reproduction, the advantages of recombination, and the benefits of anisogamy remain as unresolved as ever.

REFERENCES

Fisher, R. A. 1930. *The Genetical Theory of Natural Selection*. Oxford: Clarendon.

Ford, E. B. 1940. Polymorphism and taxonomy. In *The New Systematics*, J. Huxley, ed., pp. 493–513. Oxford: Clarendon.

Haldane, J. B. S. 1937. The effect of variation on fitness. *Am. Nat.* **71**:337–349.

Haldane, J. B. S. 1957. The cost of natural selection. *J. Genet.* **54**:294–296.

Hubby, J. L. and Lewontin, R. C. 1966. A molecular approach to the study of geneic heterozygosity in natural populations. I. The number of alleles at different loci in *Drosophila pseudoobscura*. *Genetics* **54**:577–594.

Lewontin, R. C. 1958. A general method for investigating the equilibrium of gene frequency in a population. *Genetics* **43**:419–434.

Lewontin, R. C. 1974. *The genetic Basis of Evolutionary Change*. New York: Columbia U. Press.

Lewontin, R. C. and Hubby, J. L. 1966. A molecular approach to the study of geneic heterozygosity in natural populations. II. Amount of variation and degree of heterozygosity in natural populations of *Drosophila pseudoobscura*. *Genetics* **54**: 595–609.

Lewontin, R. C. and White, M. J. D. 1960. Interaction between inversion polymorphism of two chromosome pairs in a grasshopper, *Moraba Scurrva*. *Evolution* **14**:116–29.

Muller, H. J. 1950. Our load of mutations. *Am. J. Hum. Genet.* **2**:111–176.

CHAPTER THIRTEEN

The Equilibrium Theory of
One- and Two-Locus Systems

IAN R. FRANKLIN AND MARCUS W. FELDMAN

13.1. Introduction

Fisher (1922) first described the conditions for a selectively balanced equilibrium of a single gene with two alleles, and for many years heterozygote advantage was considered a primary cause for maintaining genetic variation in natural populations. Some of the commonly observed polymorphisms were clearly adaptive; sickle cell anaemia, industrial melanism, and self-incompatibility systems in plants are all examples for which selective forces could be identified. It became clear that, at many polymorphic loci, multiple alleles are common. Blood group genes, for example, are often multiallelic and, in the case of gametophytic self-sterility systems, hundreds of alleles have been identified. These observations stimulated a number of authors (Owen, 1954; Kimura, 1956a; Lewontin, 1958; Mandel, 1959) to examine the conditions for selective balance involving three or more alleles.

Also, the importance of interaction between loci in maintaining genetic variation was evident in much of the early development of population genetics theory. Fisher (1930) discussed circumstances favoring closer linkage of interacting loci, and this was later formally investigated by Kimura (1956b). Interallelic interaction was central to the development of Wright's shifting-balance theory, and experimental evidence suggested the existence of coadapted gene complexes. One of the most striking examples of balanced polymorphisms involving many individual loci, the inversion polymorphisms of *Drosophila* and other insects (such as the grasshopper *Moraba scurra*), provided the impetus for the construction and analysis of a two-locus model by Lewontin and Kojima (1960).

Interest in nonallelic interaction originated also from another quarter – quantitative genetics. Polygenic traits, such as body size or flowering time, play an important role in evolution. By definition, they are controlled by many genes,

R. S. Singh and C. B. Krimbas, eds., *Evolutionary Genetics: From Molecules to Morphology*, vol. 1. © Cambridge University Press 2000. Printed in the United States of America. ISBN 0-521-57123-5. All rights reserved.

all of which contribute to the phenotype. Even if these genes do not interact on the phenotypic scale (an implausible assumption), converting from a phenotype to a selective value generates interactions between these loci, especially if intermediate phenotypes are favored. If such selection leads to closer linkage between the loci involved or if the loci are linked to begin with, the possibility exists that, in such complexes, gametic arrays will develop in which alleles favoring an increase in the phenotype will alternate with alleles favoring a reduction. Such a complex would both act as a reservoir for genetic variability and stabilize the phenotype on intermediate values. This idea, called relational balance (Mather, 1941), motivated the two-locus studies of Bodmer and Parsons (1962). Their model was a generalization of an earlier model studied by Wright (1952), who was also prompted by considerations of optimizing selection for polygenic traits.

This chapter revisits and updates the theory of balancing selection at one or more loci.

13.1.1. The Dynamics of Gene Frequency Change

Throughout this chapter we consider only nonoverlapping generations, describing the transition from one generation to the next as if the population reproduces as a single cohort, i.e.,

$$x_{(n+1)} = g(\{x_{(n)}\}, \mathbf{W}, \mathcal{R})$$

where $x_{(n)}$ is a vector of gamete frequencies in generation n, $\mathbf{W}$ is a matrix of genotypic selective values, and $\mathcal{R}$ represents a set of recombination frequencies, if applicable.

The elements of $\mathbf{W}$, Wright's "selective values," are commonly referred to as fitnesses, a term that invokes more general notions of adaptation and evolutionary success. For the most part, we will be concerned with selective changes that are a consequence of differences in viabilities, and we shall use the symbol V rather than W to indicate this fact. Differential fertility requires that we assign selective values to mating pairs, resulting in more complicated expressions and many more parameters.

13.2. One-Locus Theory

Much of the theory of gene frequency change in populations has centered on single-locus models. A comprehensive review is beyond the scope of this chapter; we shall focus on the equilibrium behavior of one-locus multiallelic systems in which differential selection operates on zygote to adult survival.

Consider a set of alleles $A_1, A_2, \ldots, A_k$ present in the population at time t at frequencies $x_1, x_2, \ldots, x_k$ respectively. Let the selective value of an individual $A_i A_j$, receiving the allele A_i from its father and A_j from its mother, be W_{ij}. Wright's well-known equation for gene frequency change,

$$\Delta x_i = \frac{x_i(1 - x_i)}{2\bar{W}} \frac{\partial \bar{W}}{\partial x_i}, \tag{13.1}$$

indicates that, under circumscribed conditions, gene frequencies will move toward, and eventually come to rest at, local maxima for $\bar{W}$. However, it is clear that, even for a single locus, the mean selective value may not increase from generation to generation (Fisher, 1941). In particular, if selective differences in fertility are present, $\bar{W}$ does not necessarily increase (Bodmer, 1965; Pollak, 1978). Nevertheless, under viability selection, the mean viability will increase each time a population undergoes a cycle of random mating; formal proofs are found in Mandel (1959) and Kingman (1961).

We now consider the dynamics of a multiallelic system involving viability selection, in which there are no differences between sexes and $V_{ij} = V_{ji}$. From Mendel's first law and the assumption that the population mates randomly, the frequency of A_i among the offspring is

$$x_i' = x_i \frac{\sum_{j=1}^{k} x_j V_{ij}}{\bar{V}}, \tag{13.2}$$

where

$$\bar{V} = \sum_{i,j=1}^{k} x_i x_j V_{ij}, \tag{13.3}$$

the mean viability.

At equilibrium, $x_i' = x_i$, and, besides solutions with some $x_i = 0$, we have a set of nonhomogeneous linear equations,

$$\sum_j x_j V_{ij} = \bar{V}, \quad \text{all} \quad i, \tag{13.4}$$

which has a unique solution if $\det (V_{ij}) \neq 0$.

The condition that the equilibrium solution $\{x_i\}$, with $x_i > 0, i = 1, k$, be a stable point is that the $k \times k$ matrix $||V_{ij}||$ be negative definite; i.e., that the matrix $\mathbf{V}$ has only one positive eigenvalue. An equivalent, and often computationally simpler, requirement is that the principal minors of $\mathbf{V}$ alternate in sign.

The principle findings are

(1) for constant viabilities, and for any set of alleles, there can be at most one interior stationary point involving all alleles.
(2) If the matrix $\mathbf{V}$ has m positive eigenvalues, at least $(m - 1)$ alleles will be absent at a stable equilibrium.

Note that if there is an unstable equilibrium involving all alleles, the removal of one or more alleles admits the possibility of stable or unstable equilibria involving fewer alleles. For two alleles, there is only one equilibrium possible with both alleles segregating, and two equilibria with either one of the alleles fixed. The condition for stability of the polymorphic equilibrium reduces to $V_{12} > V_{11}, V_{22}$, as first shown by Fisher (1922). Mandel (1959) discussed the

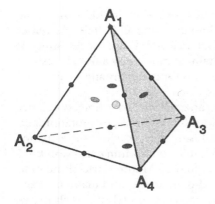

Figure 13.1. The 15 possible equilibrium states for a one-locus four-allele fixed-viability model. The equilibria are represented on the surfaces of a regular tetrahedron. Equilibria may lie on each of the four vertices (representing fixation of one of the alleles), on the six edges (two polymorphic alleles), four faces (three polymorphic equilibria), or as a single interior fully polymorphic equilibrium.

three-allele case in some detail, showing that the conditions on the viabilities cannot be written simply. Two necessary, but not sufficient, conditions are that no heterozygote can be less viable than both of the associated homozygotes and that no more than one heterozygote can be less viable than any particular homozygote.

In the three-allele case, there are seven possible stationary points; one in which all three alleles are present, three in which only two alleles are present, and three in which one of the alleles is fixed. When there are four alleles, there may exist 15 equilibria; 1 interior, 4 with three alleles, 6 with two, and 4 that are trivial (one allele fixed). In general, for k alleles, there may be up to $2^k - 1$ equilibria. It is convenient to represent the set of allele frequencies within a regular $k - 1$-dimensional simplex. For example, with four alleles, the simplex is a tetrahedron (Fig. 13.1). If the equilibria involve all k alleles, the equilibria are called interior; if one or more allele is not present, the equilibria lie on one of the boundaries of the simplex.

Kingman (1961) introduced the concept of external stability, which is crucial to our subsequent discussion of multilocus equilibria. The concept is best illustrated by considering, for example, a stable equilibrium $(\hat{x}_1, \hat{x}_2, \hat{x}_3)$ involving three alleles. This equilibrium, satisfying the above conditions, may be unstable in a wider sense if the introduction of a fourth allele results in a new equilibrium that involves that allele. The condition for the initial increase of the fourth allele is

$$\sum_{j=1,3} V_{j4}\hat{x}_j > \bar{V}, \qquad (13.5)$$

where $\bar{V}$ is the mean viability before the introduction of the new allele. This condition also ensures that the new allele will persist.

As the number of alleles increases so do the constraints on the relationships among each of the individual viabilities required for maintaining a stable, fully polymorphic equilibrium. Lewontin et al. (1978) first examined this problem by choosing entries in the viability matrix from a uniform [0, 1] distribution. They found that the likelihood of a valid, stable equilibrium involving all k alleles declines rapidly as k increases. If selective values are chosen in this manner, the probability of a stable polymorphic equilibrium involving all k alleles decreases as $\exp(-\frac{1}{2}k^2 \log k)$ (Karlin, 1981). The qualitative conclusion of the study of Lewontin et al. was that heterosis can maintain, at most, only a few alleles and therefore viability selection is unlikely to account for the many alleles observed at some enzyme loci.

However, we know that natural selection can generate, over time, outcomes that *a priori* seem highly improbable. Consider the assembly, by mutation and selection, of a complex, balanced polymorphism, and ask what is the chance that a kth allele is incorporated, given that a stable equilibrium exists with $k-1$ alleles. Spencer and Marks (1988) demonstrated by Monte Carlo simulation that viability selection can, with high probability, construct and maintain a six- or seven-allele polymorphism. Further, if we consider models in which the viabilities of new mutations are functions of the parental allele from which the mutation arose, very high levels of polymorphism may be achieved (Spencer and Marks, 1992).

The discussion above relates only to selection based on constant viabilities. More complex models of selection involving, for example, frequency-dependent selection, nonrandom mating, or differences in fertility (in which each mating type has an assigned selective value), complicate the behavior. In such cases, several equilibria may exist for any given set of alleles, and stable cycles are possible.

13.3. Multilocus Systems

Multilocus theory is crucial to a number of important concepts in modern evolutionary theory. Fisher, Wright, and Haldane, the founders of this modern theory, were overwhelmingly concerned with adaptive evolutionary change, although each also made contributions to our understanding of stochastic changes in gene frequency that arise through finite population size. Each understood that an individual's fitness could not simply be reduced to the arithmetic sum of a large number of selective values at each of the tens of thousands of individual loci. Nevertheless, most early population genetic theories were framed in terms of gene frequency changes at individual loci.

The existence of interactions between loci does not, in itself, invalidate a genetical theory of evolution based on gene frequency change at individual loci. There are several important principles. First, the rate of change in allele frequencies at any instant can be expressed in terms of the marginal selective values at the locus in question; these marginal selective values include not only differences at the locus itself but also the genetic effects at all of the other loci that contribute to survival, fecundity, or fertility. Fisher made a particular point of this in distinguishing between the average effect and average excess of a gene substitution (Fisher, 1930, 1941).

The second principle is concerned with defining an appropriate set of state variables for a kinematic description of an evolutionary process. All evolutionary change can be described by a system of equations of the form

$$x_k = \sum_{ij} x_i \mathbf{A}_{ij}^{(k)} x_j$$

or

$$x_k = x\mathbf{A}^{(k)}x,$$

where x_k are the frequencies of gametes, genotypes, or, in the case of complex nonrandom mating systems, mating types and the elements of the matrix $\mathbf{A}^{(k)}$ are functions of parameters such as viabilities, fertilities, recombination fractions, and parameters that describe the mating system. We must choose a level of description appropriate for the analysis of the process under investigation; that is, we require a set of dynamically sufficient state variables (Lewontin, 1974). We can describe the joint behavior of a number of loci in terms of gene frequencies at the individual loci only if the frequencies at different loci are uncorrelated. If such correlations exist, we must describe the transitions from one generation to the next in terms of gamete frequencies or an equivalent set of sufficient variables such as a set of gene frequencies and a set of variables $\mathcal{D}$, defined as the difference between an observed gamete frequency and its expectation, assuming independence of the constituent loci. For two loci, the members D of $\mathcal{D}$ are commonly referred to as the coefficients of linkage disequilibrium (Lewontin and Kojima, 1960). Others have called this the coefficient of gametic phase imbalance; we shall use the more common terminology.

For two loci, each with two alleles, the gametic frequencies are specified completely by two gene frequencies and a single D. If each locus has three alleles, four gene frequencies and four linkage disequilibrium parameters are required. With nonrandom mating, a more extended set of genotypic disequilibria are required (Weir and Cockerham, 1989) for specifying the genotypic arrays.

A final point, to which we return later, is that as recombination is reduced to small levels, the distinction between a single-locus system and a set of closely linked ones becomes blurred; the gametic types, or haplotypes, define a set of alleles and the dynamic behavior becomes essentially that of a one-locus multiallelic system.

13.3.1. Properties of Linked Systems in the Absence of Selection

Associations between allelic frequencies at different loci can arise in a variety of ways. As we shall see, selection can generate and maintain substantial linkage disequilibrium between interacting loci. In finite populations, however, linkage disequilibrium may be generated by sampling, both within a population and between subdivided populations. In particular, populations derived by crossing individuals from populations with different selection histories or populations that have been separated for a long time may exhibit substantial linkage disequilibrium.

In random mating populations, in the absence of selection, the dynamics of evolutionary change can always be described in terms of gamete frequencies. We cannot, however, assume linkage equilibrium unless the populations are very large and have remained so for some time. For two loci with two alleles each and a recombination fraction r, D will decline by a proportion $(1 - r)$ each generation (Jennings, 1917; Robbins, 1918). Hence, for an initial value D_0, we expect, after n generations, the disequilibrium

$$D_n = (1 - r)^n D_0.$$

Geiringer (1944) extended these results to any number of loci, and Bennett (1954) showed elegantly how functions of gene and gamete frequencies can be constructed that decline each generation as a function of a set of recombination frequencies.

In large panmictic populations, associations between alleles at different loci that arise, for example, by hybridization, will decline steadily to zero, although the rate of approach to linkage equilibrium may be slow if loci are tightly linked. However, for a population comprising several demes, all differing in gene frequencies at a set of loci, linkage disequilibrium persists overall, even if each deme is in linkage equilibrium.

13.3.1.1. Linkage Disequilibrium among Unselected Loci in Finite Populations

In the late 1960s, Sved (1968), Hill and Robertson (1968), and Ohta and Kimura (1969) demonstrated that in a random mating population, substantial linkage disequilibrium can be generated in finite populations as a result of sampling effects. The expected value of D between unselected loci is ultimately zero, but the variance in D is positive. In fact, while $E(D^2)$ will tend to zero as loci become fixed, the quantity

$$E \left[\frac{D^2}{p_A(1 - p_A)p_B(1 - p_B)} \right],$$

which is the expectation of the square of the correlation between nonallelic genes whose frequencies are p_A and p_B, will tend, approximately, to $1/4N_e r$, where N_e is the effective population size and r is the recombination fraction. This important result showed that the observation of linkage disequilibrium between two loci cannot be taken as evidence for selective interaction between these loci.

The exact form of the correlation is still a matter of debate. The conclusions drawn by various authors depend on the model (random union of gametes or random union of zygotes) and the methodology. Hill and Robertson used a moment-generating matrix to derive the expected value of D^2, as well as the quantities $E[D(1 - 2p_A)(1 - 2p_B)]$ and $E[p_A(1 - p_A)p_B(1 - p_B)]$; Ohta and Kimura used a diffusion approximation to derive the expected values of the same quantities; Watterson (1970) and Littler (1973) used a Markov chain approach, and Weir and Cockerham (1974) used descent measures. A general finding is that the ratio of the expected value of D^2 and the expected value of the product of the gene frequencies i.e., $E[D^2]/E[p_A(1 - p_A)p_B(1 - p_B)]$, which approximates the expected value of the correlation in nonallelic gene frequencies, is approximately $1/(4N_e r)$. Sved (1971), using a method that many find obscure, derived the relationship

$$E(\rho^2) = 1/(1 + 4N_e r),$$

where ρ^2 is the quantity $\frac{D^2}{p_A(1-p_A)p_B(1-p_B)}$ which seems to provide a better estimate for larger population sizes. Hill (1974) has extended the approach of Hill and Robertson (1968) to three or more loci.

13.3.2. Selectively Generated Linkage Disequilibrium

Initially we confine our discussion to a two-locus, two-allele/locus model in which selection is defined by a matrix $\mathbf{V}$ of viabilities,

$$
\begin{array}{ccccc}
 & AB & Ab & aB & ab \\
AB & V_{11} & V_{12} & V_{13} & V_{14} \\
Ab & V_{21} & V_{22} & V_{23} & V_{24}, \\
aB & V_{31} & V_{32} & V_{33} & V_{34} \\
ab & V_{41} & V_{42} & V_{43} & V_{44}
\end{array}
\tag{13.6}
$$

where V_{ij} denotes the viability of an individual that received the ith gametic type from its female parent and the jth gametic type from the male parent. We shall now make the simplifying assumption that viabilities are independent of their gametic origin, i.e., $V_{ij} = V_{ji}$ and, moreover, that the gamete frequencies in the two sexes are the same.

Let $x_i(n), i = 1, 4$, be the frequencies of the gametes AB, Ab, aB, and ab in generation n.

Then

$$\bar{V} x_i(n+1) = x_i(n) V_{i\cdot} - k(i) r D_v$$

or

$$\bar{V} \Delta x_i = x_i(n)(V_{i\cdot} - \bar{V}) - k(i) r D_v, \tag{13.7}$$

where

$$V_{i\cdot} = \sum_j x_j V_{ij},$$

$$\bar{V} = \sum_i x_i V_{i\cdot} = \sum_{ij} x_i x_j V_{ij},$$

$$D_v = x_1 x_4 V_{14} - x_2 x_3 V_{23},$$

$$k(i) = 1 \quad \text{if} \quad i = 1, 4 \quad \text{or} \quad -1 \quad \text{if} \quad i = 2, 3.$$

The above system of equations can be iterated numerically to follow the trajectory of gamete frequencies for any set of viability and recombination parameters. Note that if D_v remains zero, the equations are formally equivalent to those for a single locus with four alleles, each representing one of the four gametic types. Under these conditions, we know that the system will progress toward a local maximum for $\bar{V}$. If $D_v \neq 0$, the gamete frequencies will not necessarily move toward local maxima in $\bar{V}$ and, as Moran (1964) showed, the mean viability may, in fact, decline.

Now, consider the quantity $Z = x_1 x_4 / x_2 x_3$:

$$\Delta \log Z = \Delta \log(x_1) - \Delta \log(x_2) - \Delta \log(x_3) + \Delta \log(x_4)$$
$$\cong \Delta x_1 / x_1 - \Delta x_2 / x_2 - \Delta x_3 / x_3 + \Delta x_4 / x_4.$$

Then,

$$\bar{V} \Delta \log Z \cong (1/x_1)(x_1 V_{1.} - r D_V) - (1/x_2)(x_2 V_{2.} + r D_V)$$
$$- (1/x_3)(x_3 V_{3.} + r D_V) + (1/x_4)(x_4 V_{4.} - r D_V)$$
$$= (V_{1.} - V_{2.} - V_{3.} + V_{4.}) - r D_v \sum (1/x_i)$$
$$= \varepsilon - r D_v J \tag{13.8}$$

where

$$\varepsilon = \sum x_j v_{1j} - \sum x_j v_{2j} - \sum x_j v_{3j} + \sum x_j v_{4j}$$
$$= \sum x_j (v_{1j} - v_{2j} - v_{3j} + v_{4j})$$
$$= \sum x_j \varepsilon_j.$$

Hence, ε is an average, weighted by gamete frequencies, of four quantities ε_j, which are measures of epistasis (departures from additivity), and J is $\sum_i (1/x_i)$.

For stationary values of $\{x_i\}$, $\Delta \log Z = 0$. Hence, for any interior equilibrium of the gametic frequencies,

$$\varepsilon / J = r D_v. \tag{13.9}$$

Now, consider the behavior of Z (or $\log Z$) under evolutionary change. If initially the linkage disequilibrium is close to zero, $\Delta \log Z$ will approximate ε. For example, if ε is positive, $\log Z$ will increase. However, since $\log Z$ and D have the same sign, rD will also increase, so that $\varepsilon - r D_v J$ will decrease, reducing the change in $\log Z$ in following generations. If ε is not too large, Z will approach an equilibrium value. Similarly, again depending on constraints on ε and $r D_v J$, an approximate equilibrium for Z will be attained if D is initially large or if ε is negative.

This behavior, first formulated explicitly by Felsenstein (1965), was elaborated by Kimura (1965), as follows:

"... linked systems own a remarkable property of rapidly settling into a state which I would like to call quasi linkage equilibrium. This state is attained if gene frequencies are changing under loose linkage and relatively weak epistatic interactions. On the other hand, linkage disequilibrium may be built up indefinitely when linkage is tight, epistatic interactions are relatively strong and gene frequencies are changing towards fixation."

The significance of this concept is to resuscitate two of the most important concepts in population genetics. Kimura goes on:

"I would like to show further that for a genetic system evolving under quasi linkage equilibrium, both Wright's conception of an "adaptive topography" and Fisher's fundamental theorem of natural selection indeed hold."

Wright (1967) followed with a reformulation of the above and some numerical examples showing that a state of quasi-linkage equilibrium is reached quite rapidly, concluding with the comment that Moran's (1964) demonstration that gamete frequencies do not converge on a local maximum in $\bar{V}$ was "based on a misunderstanding of the concept."

This was unfortunate. Moran had demonstrated a fundamental fact about two-locus systems, namely that a stable polymorphic equilibrium may not lie on a maximum of $\bar{V}$, even with weak selection and loose linkage, as Wright had assumed. In fact, Wright's concept of an "adaptive topography" is more likely to be valid for tight linkage.

13.3.3. The Equilibrium Behavior of Two-Locus Systems

The equilibria of system (13.7) can be found by setting $\{\Delta x_i = 0\}_{i=1,4}$. In general, this system has proven analytically intractable for general viabilities. Much of what we know about the stationary points and their behavior as recombination or viabilities change has been gleaned from the analysis of models involving restricted parameter sets, numerical evaluation of the equilibria, or Monte Carlo simulation.

The initial work of Kimura (1956a, 1956b), Lewontin and Kojima (1960), and Bodmer and Parsons (1962) revealed three solutions to the two-locus equations. Later, Karlin and Feldman (1970) demonstrated the possible existence of an additional four equilibria. These seven potential stationary points, all of which may exist for particular fitness and recombination parameters, are interior equilibria, i.e., solutions in which all four gametic types are non-zero.

There are, in addition, eight possible solutions in which one or both loci are fixed. Therefore, for the symmetric viability two-locus two-allele models (see below), there are at least 15 different solutions to the two-locus two-allele models, seven of which are interior.

There have been numerous assertions about the number of solutions to the two-locus equations, and the subject remains unresolved. Moran (1963), using an argument that he later withdrew, claimed that there could be at most five (interior) stationary points and that at most three of these could be maxima of $\bar{V}$. Turner (1971) asserted that 21 equilibria were possible for the symmetric viability model. Arunachalum and Owen (1971) suggested, for the general two-locus model with fixed viabilities, that there may be as many as 20 solutions.

The number of stable points has been an even more contentious issue. Karlin (1975) summarized many of the numerical and analytical findings of two- and three-locus theories and presented a number of conjectures based on continuity arguments from analyses of the zero recombination case. Karlin claimed that at most two interior equilibria (and two boundary equilibria) could be stable simultaneously for $r > 0$. Feldman and Liberman (1979) showed that four boundary equilibria and two interior points could be stable for the symmetric viability model. Hastings (1985) subsequently demonstrated the stability of four interior equilibria and speculated that "it may be possible to generate an example with even eight stable equilibria, arising through a series of pitchfork bifurcations."

A number of Karlin's other 1975 conjectures have not held up. For example, the simultaneous existence of equilibria with gametic equilibrium and disequilibrium was demonstrated by Franklin and Feldman (1977), Karlin and Feldman (1978), and Hastings (1981a, 1981b). Finally, perhaps the most important principle of all, namely that stable internal equilibria show marginal overdominance, was disproved (Hastings, 1982). However, these conjectures stimulated further analytic and simulation studies on two-locus models.

13.3.3.1. Special Viability Systems

The restricted viability parameter sets fall into four main classes:

1. The additive (i.e., no epistasis) model
2. Symmetric viability models
3. Multiplicative models
4. Optimum models

As might be expected, the first model is the most tractable analytically. The dynamics and equilibrium behavior are well understood and, essentially, extend single-locus theory. The additive model is of little biological interest, except that it provides a baseline against which to judge the effect of various kinds of interaction between nonallelic genes. The symmetric viability model, while it imposes unrealistic constraints on the viabilities, allows a wide range of interactions between loci and is well known because it is the only model (aside from additivity) for which equilibria with linkage disequilibrium have been obtained explicitly. Some analytic results are also known for analogous symmetric systems involving three or more loci, and additional results have been obtained by simulation. Hence, much of what we know about the behavior of multilocus systems is derived from studies of the symmetric viability models.

The third and the fourth models (i.e., multiplicative and optimum models) have been studied for their general biological interest, but analytic results for both are limited. The multiplicative models are of interest because loci that affect viability, but are not functionally related, will interact multiplicatively. The optimum model arises naturally from the observation that, for most continuously varying traits of adaptive significance, natural selection tends to favor intermediate phenotypes and to select against extremes. While additivity on the underlying (phenotypic) scale is usually assumed, selection favoring intermediates will generate overdominance at a single locus. However, if multiple loci affect the trait (as is usually the case – a complex trait such as body size may be influenced by hundreds of loci), selection generally leads to fixation but may, if loci are linked, generate a transient associative overdominance. Despite some attention, particularly by Wright, Kojima, and Lewontin, much remains to be done for this class of model.

We present analytic results for only the symmetric viability model, as it reveals many of the equilibrium properties that we wish to discuss.

13.3.4. Symmetric Viability Models

Symmetric viability models were first proposed by Wright (1952) and Kimura (1956a, b). Both considered special cases of the more general models later introduced by Lewontin and Kojima (1960) and Bodmer and Parsons (1962). In its most general form, the viabilities can be written as

		♀ Gamete			
		AB	Ab	aB	ab
♂ Gamete	AB	v_1	v_2	v_3	1
	Ab	v_2	v_4	1	v_3
	aB	v_3	1	v_4	v_2
	ab	1	v_3	v_2	v_1

The most easily analyzed of these models, first discussed by Lewontin and Kojima (1960), has $v_4 = v_1$. This is shown below in a 3×3 table, with each axis representing the genotype at each locus. In this model there are no sex differences and the viabilities are independent of gametic origin. This model renders the equilibrium analysis tractable because the viability matrix is invariant to allelic substitution at both loci; consequently, some equilibria exhibit symmetries in gametic frequencies that allow the system of nonlinear equilibrium equations to be factored. For example, because of these symmetries, the above model has equilibria of the form $\mathbf{x} = (\frac{1}{4} + D, \frac{1}{4} - D, \frac{1}{4} - D, \frac{1}{4} + D)$. Substituting for $\mathbf{x}$ in Eq. (13.9) yields a cubic in D, indicating that there are three such symmetric solutions.

Without loss of generality, the viabilities can be rewritten in the following form:

		AA	Aa	aa
		Locus 1		
Locus 2	BB	$1 - \beta$	$1 - \alpha_1$	$1 - \beta$
	Bb	$1 - \alpha_2$	1	$1 - \alpha_2$
	bb	$1 - \beta$	$1 - \alpha_1$	$1 - \beta$

and the gametic frequencies can be transformed to a new coordinate system $\{u_i\}_{i=1,4}$, in which

$$u_1 = x_1 + x_2 - x_3 - x_4,$$
$$u_2 = x_1 - x_2 + x_3 - x_4,$$
$$u_3 = x_1 - x_2 - x_3 + x_4,$$

or

$$x_1 = \tfrac{1}{4}(1 + u_1 + u_2 + u_3),$$
$$x_2 = \tfrac{1}{4}(1 + u_1 - u_2 - u_3),$$
$$x_3 = \tfrac{1}{4}(1 - u_1 + u_2 - u_3).$$

Each u_i provides a separate measure of departure from equality of gametic frequencies. If, for example, the frequencies of the two alleles at the first locus are equal, then $u_1 = 0$. Similarly, if $p(B) = p(b) = \tfrac{1}{2}$, then $u_2 = 0$. The third variable, u_3, is a measure of the association between the frequencies at the two loci and vanishes if the sum of the *cis*-gamete frequencies equals that of the *trans*-gametes. If $u_1 = u_2 = u_3 = 0$, all four gametic frequencies are equal to $\tfrac{1}{4}$. Also, note that $D = (u_3 - u_1u_2)/4$.

Then, substituting in Eq. (13.7), the frequencies in the following generation, u_i', are given by

$$\bar{V}u_1' = \tfrac{1}{2}(1 - \beta)(u_1 + u_2u_3) + \tfrac{1}{2}(1 - \alpha_2)(u_1 - u_2u_3),$$
$$\bar{V}u_2' = \tfrac{1}{2}(1 - \beta)(u_2 + u_1u_3) + \tfrac{1}{2}(1 - \alpha_1)(u_2 - u_1u_3),$$
$$\bar{V}u_3' = \tfrac{1}{2}(1 - \beta)(u_3 + u_1u_2) + \tfrac{1}{2}(1 - 2r)(u_3 - u_1u_2), \qquad (13.10)$$

where $\bar{V} = 1 - \tfrac{1}{4}(\alpha_1 + \alpha_2 + \beta) + \tfrac{1}{4}u_1^2(\alpha_1 - \alpha_2 - \beta) + \tfrac{1}{4}u_2^2(-\alpha_1 + \alpha_2 - \beta) + \tfrac{1}{4}u_3^2(\alpha_1 + \alpha_2 - \beta)$.

Let $\kappa_1 = \frac{1}{4}(\alpha_1 - \alpha_2 - \beta)$, $\kappa_2 = \frac{1}{4}(-\alpha_1 + \alpha_2 - \beta)$, $\kappa_3 = \frac{1}{4}(\alpha_1 + \alpha_2 - \beta)$. Then, at equilibrium, Eqs. (13.10) become

$$
\begin{aligned}
u_1 \sum \kappa_i u_i^2 &= \kappa_1 u_1 + (\kappa_2 + \kappa_3) u_2 u_3, \\
u_2 \sum \kappa_i u_i^2 &= \kappa_2 u_2 + (\kappa_1 + \kappa_3) u_1 u_3, \\
u_3 \sum \kappa_i u_i^2 &= \kappa_3 u_3 + (\kappa_1 + \kappa_2) u_1 u_2 - r(u_3 - u_1 u_2).
\end{aligned} \tag{13.11}
$$

For this set of simultaneous equations there are exactly 15 solutions, as follows:

$$
\hat{u}_1 = \hat{u}_2 = \hat{u}_3 = 0, \qquad \text{or} \quad \hat{\mathbf{x}} = \left(\tfrac{1}{4}, \tfrac{1}{4}, \tfrac{1}{4}, \tfrac{1}{4}\right); \tag{13.12a}
$$

$$
\hat{u}_1 = \pm 1, \quad \hat{u}_2 = \hat{u}_3 = 0, \quad \text{or} \quad \hat{\mathbf{x}} = \left(\tfrac{1}{2}, \tfrac{1}{2}, 0, 0\right) \quad \text{and}
$$
$$
\hat{\mathbf{x}} = \left(0, 0, \tfrac{1}{2}, \tfrac{1}{2}\right); \tag{13.12b}
$$

$$
\hat{u}_2 = \pm 1, \quad \hat{u}_1 = \hat{u}_3 = 0 \quad \text{or} \quad \hat{\mathbf{x}} = \left(\tfrac{1}{2}, 0, \tfrac{1}{2}, 0\right) \quad \text{and}
$$
$$
\hat{\mathbf{x}} = \left(0, \tfrac{1}{2}, 0, \tfrac{1}{2}\right); \tag{13.12c}
$$

$$
\hat{u}_3 = \pm(1 - r/\kappa_3)^{1/2}, \quad \hat{u}_1 = \hat{u}_2 = 0, \tag{13.12d}
$$

representing the symmetric solutions $\left(\frac{1}{4} + \hat{D}, \frac{1}{4} - \hat{D}, \frac{1}{4} - \hat{D}, \frac{1}{4} + \hat{D}\right)$, where $\hat{D} = \pm \frac{1}{4}(1 - r/\kappa_3)^{1/2}$ (Lewontin and Kojima, 1960).

The remaining solutions are not as easily obtained. There are no solutions of the form $\hat{u}_i = 0$, $\hat{u}_j \neq \hat{u}_k \neq 0$ $(i \neq j \neq k)$, but 8 in which $\hat{u}_i \neq \hat{u}_j \neq \hat{u}_k \neq 0$.

Consider a new set of variables:

$$
z_1 = u_2 u_3 / u_1, \qquad z_2 = u_1 u_3 / u_2, \qquad z_3 = u_1 u_2 / u_3.
$$

Equations (13.11) become

$$
\begin{aligned}
\kappa_1 z_2 z_3 + \kappa_2 z_1 z_3 + \kappa_3 z_1 z_2 &= \kappa_1 + (\kappa_2 + \kappa_3) z_1 \\
&= \kappa_2 + (\kappa_1 + \kappa_3) z_2 \\
&= (\kappa_3 - r) + (\kappa_1 + \kappa_2 + r) z_3,
\end{aligned}
$$

which can be resolved into a quadratic in one of the variables, resulting in the solutions

$$
\hat{\mathbf{z}} = \{1, 1, 1\},
$$
$$
\hat{\mathbf{z}} = \left\{ -\frac{\kappa_1(2\kappa_1 + r)}{2L + \kappa_3 r}, -\frac{\kappa_2(2\kappa_2 + r)}{2L + \kappa_3 r}, -\frac{\kappa_3(\kappa_3 + r)(2\kappa_1 + 2\kappa_2 + r) - \kappa_1 \kappa_2 r}{(2L + \kappa_3 r)(\kappa_1 + \kappa_2 + r)} \right\},
$$
$$
\tag{13.12e}
$$

where $L = \kappa_1 \kappa_2 + \kappa_1 \kappa_3 + \kappa_2 \kappa_3$.

For each $\mathbf{z}$, there are four solutions for u, given by

$$
\hat{u}_1 = \pm\sqrt{z_2 z_3}, \qquad \hat{u}_2 = \pm\sqrt{z_1 z_3}, \qquad \hat{u}_3 = \hat{u}_1 \hat{u}_2 / z_3.
$$

Hence the solutions $z_1 = z_2 = z_3 = 1$ become

$$\hat{\mathbf{u}} = (1, 1, 1), (1, -1, -1), (-1, 1, -1), (-1, -1, 1), \qquad (13.12f)$$

representing the four vertices of the simplex

$$\hat{\mathbf{x}} = (1, 0, 0, 0), (0, 1, 0, 0), (0, 0, 1, 0), (0, 0, 0, 1).$$

The other set of solutions is

$$\hat{u}_1 = \pm \frac{1}{2L + \kappa_3 r}$$
$$\times \sqrt{-\kappa_2\kappa_3(2\kappa_2 + r)\frac{r^2 + r(2\kappa_1 + 2\kappa_2 - \kappa_3 + \kappa_1\kappa_2/\kappa_3) - 2\kappa_3(\kappa_1 + \kappa_2)}{(\kappa_1 + \kappa_2 + r)}},$$

$$\hat{u}_2 = \pm \frac{1}{2L + \kappa_3 r}$$
$$\times \sqrt{-\kappa_1\kappa_3(2\kappa_1 + r)\frac{r^2 + r(2\kappa_1 + 2\kappa_2 - \kappa_3 + \kappa_1\kappa_2/\kappa_3) - 2\kappa_3(\kappa_1 + \kappa_2)}{(\kappa_1 + \kappa_2 + r)}},$$

$$\hat{u}_3 = \frac{1}{2L + \kappa_3 r}\sqrt{\kappa_1\kappa_2(2\kappa_1 + r)(2\kappa_2 + r)}. \qquad (13.12g)$$

13.3.4.1. Behavior of the Equilibria at $r = 0$

Of the 15 equilibria of the symmetric viability model, six are functions of r and have the potential to reside in the interior of the simplex. Hence, if we include the central equilibrium, which lies within the simplex for all values of r, there are, potentially, seven interior equilibria (Fig. 13.2). The remaining eight are boundary equilibria, located at the four vertices [Eq. (13.12f)] or lying within four of the edges [Eqs. (13.12b) and (13.12c)].

At $r = 0$, the symmetric equilibria [Eqs. (13.12c)] become $(\frac{1}{2}, 0, 0, \frac{1}{2})$ and $(0, \frac{1}{2}, \frac{1}{2}, 0)$. That is, they lie on those edges (AB–ab and Ab–aB) in which both

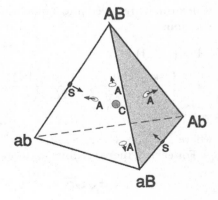

Figure 13.2. The origin of the internal equilibria in a two-locus two-allele viability model. Interior equilibria can arise from boundaries that can generate all four gametic types by recombination. These include two edges, (AB–ab) and (Ab–aB) and the four faces. Together with the fully polymorphic central equilibrium (C), there are seven in all. In the symmetric viability model discussed here, the two-edge equilibria can enter the simplex as symmetric (S) equilibria, but the equilibria on the faces always leave the simplex for small r. These may reappear as r increases as asymmetric (A) equilibria.

alleles are present at each of the two loci. Intuitively, it seems that the presence of a small amount of recombination, which will generate the remaining gametic types, may lead these equilibria into the interior of the simplex. However, if $\kappa_3 <$ 1, these equilibria move outside rather than inside the simplex as r increases.

Similarly, the four-face equilibria (three alleles present) found at $r = 0$ necessarily include both alleles at each locus; recombination will generate the fourth gametic type. Again, interior equilibria may arise from one or more of the four faces.

Setting $r = 0$ in Eqs. (13.12g) produces the four-face equilibria:

$$\mathbf{x} = \left[\frac{\kappa_3(\kappa_1 + \kappa_2)}{2L}, \frac{\kappa_2(\kappa_1 + \kappa_3)}{2L}, \frac{\kappa_1(\kappa_2 + \kappa_3)}{2L}, 0 \right], \quad \text{i.e.,} \quad ab \text{ absent;}$$

$$\mathbf{x} = \left[\frac{\kappa_2(\kappa_1 + \kappa_3)}{2L}, \frac{\kappa_3(\kappa_1 + \kappa_2)}{2L}, 0, \frac{\kappa_1(\kappa_2 + \kappa_3)}{2L} \right], \quad \text{i.e.,} \quad aB \text{ absent;}$$

$$\mathbf{x} = \left[\frac{\kappa_1(\kappa_2 + \kappa_3)}{2L}, 0, \frac{\kappa_3(\kappa_1 + \kappa_2)}{2L}, \frac{\kappa_2(\kappa_1 + \kappa_3)}{2L} \right], \quad \text{i.e.,} \quad Ab \text{ absent;}$$

$$\mathbf{x} = \left[0, \frac{\kappa_1(\kappa_2 + \kappa_3)}{2L}, \frac{\kappa_2(\kappa_1 + \kappa_3)}{2L}, \frac{\kappa_3(\kappa_1 + \kappa_2)}{2L} \right], \quad \text{i.e.,} \quad AB \text{ absent.}$$

$$(13.13)$$

Hence we have established the correspondences between the roots of the equations for the two-locus symmetric viability model and the 15 equilibria of the one-locus four-allele viability model. These relationships are summarized in Table 13.1.

13.3.4.2. Conditions for Existence of Equilibria

We must distinguish between mathematical existence (i.e., that the roots are real) and biological feasibility. For $r = 0$ and sufficiently small values of r, it appears that all equilibria are real, but they may lie outside the simplex.

Table 13.1. *Equivalence between the Equilibria of a Two-Locus Two-Allele Model and Its Corresponding One-Locus Four-Allele Model*

Eq.	Two-Locus Equilibria	Number of Solutions	One-Locus Four-allele Equilibria
13.12a	(Central)	1	Interior point (4 alleles)
13.12g	(Asymmetric)	4	Faces (3 alleles)
13.12d	(Symmetric)	2	Edges (2 alleles, neither locus fixed)
13.12b, 13.12c	(Boundary)	4	Edges (2 alleles, one locus fixed)
13.12f	(Boundary)	4	Vertices (1 allele, both loci fixed)

The central equilibrium [Eqs. (13.12a)] always exists, mathematically and biologically, for the symmetric model. The symmetric equilibria [Eqs. (13.12d)] take real values if $r/\kappa_3 < 1$. Biological existence requires that r and hence κ_3 be positive. At $r = \kappa_3$, these two equilibria meet the central equilibrium at a triple point. The asymmetric equilibria [Eqs. (13.12g)] do not exist for this model if r is near zero, but may exist for larger values of r. A succinct description of the conditions for existence of the asymmetric roots in terms of $\{\kappa_i\}$ is not possible; their behavior is illustrated in the examples below. However, we note the following:

Real solutions for **z** always exist provided that $r \neq (\kappa_1 + \kappa_3)$ or $r \neq -2(\kappa_1 + \kappa_2 + \kappa_1\kappa_2/\kappa_3)$. The existence of real roots for u requires that z_i all have the same sign. Thus critical values for the existence of u occur when the z_i change sign. From Eqs. (13.12e) it is apparent that

z_1 will change sign when $r = -2\kappa_1$ or $r = -2L/\kappa_3$,

z_2 will change sign when $r = -2\kappa_2$ or $r = -2L/\kappa_3$,

z_3 will change sign when $r = -2(\kappa_1 + \kappa_2)$, $r = -2L/\kappa_3$, or at r^*, r^{**}, defined by the roots of the quadratic equation,

$$\kappa_3 r^2 + \left(2\kappa_1\kappa_3 + 2\kappa_2\kappa_3 + \kappa_1\kappa_2 - \kappa_3^2\right)r - 2\kappa_3^2(\kappa_1 + \kappa_2) = 0. \qquad (13.14)$$

The critical points r^* and r^{**} define a region known as Ewens's gap.

Each set of $\{\kappa_i\}$ defines a set of critical values of r, values that, if they lie within the biologically acceptable range [0–0.5], mark parametric regions between the existence and the nonexistence of equilibria. For the symmetric equilibria, there is only one such value, $r = \kappa_3$, but for the asymmetric roots, there are six such r values. These are $r = -(\kappa_1 + \kappa_2)$, $r = -2L/\kappa_3$, $r = -2\kappa_1$, $r = -2\kappa_2$, r^*, and r^{**}. In addition, at $r = -2(\kappa_1 + \kappa_2)$, the asymmetric equilibria pass through the vertices, either entering or leaving the interior of the simplex.

13.3.4.3. Stability

Suppose that each u_i is perturbed from its equilibrium value $\hat{u}_i$ by a small amount δ_i, translating to an amount δ_i' in the next generation. Then, with quadratic and higher-order deviations ignored, a linear approximation to the system near an equilibrium is given, in terms of a matrix **S**, by

$$\delta' = \mathbf{S}\delta.$$

Then an equilibrium is locally stable if the absolute value of each eigenvalue of **S** is less than unity. The eigenvalues of **S** for each boundary and interior symmetric equilibrium can be written explicitly. For example, the four vertices have the same conditions for stability, namely $\kappa_1 + \kappa_3 > 0$, $\kappa_2 + \kappa_3 > 0$ and $r > -2(\kappa_1 + \kappa_2)$. In terms of the selection coefficients, these conditions become $\alpha_1, \alpha_2 < \beta$ and $r > \beta$. Similarly, the boundary edge equilibria [Eqs. (13.12b)] will be stable if $\alpha_1 - \alpha_2 > \beta$ and $\beta > \alpha_2$. The conditions on the edges [Eqs. (13.12c)] are found by interchanging α_1 and α_2.

The conditions for stability of the central equilibrium can also be written simply. These are

$$\beta > |\alpha_1 - \alpha_2|, \quad \text{and} \quad r > \tfrac{1}{4}(\alpha_1 + \alpha_2 - \beta).$$

The symmetric equilibria [Eqs. (13.12d)], if they exist, lie in the range $r = 0$ to $r = \kappa_3$. If $\beta > |\alpha_1 - \alpha_2|$ and $(\alpha_1 + \alpha_2) > \beta$, the equilibria given by $\hat{u}_1 = \hat{u}_2 = 0$, $\hat{u}_3 = \pm(1 - r/\kappa_3)^{1/2}$ are stable for r positive and near zero and for r less than and close to κ_3. However, these symmetric equilibria are not necessarily stable for all r in the interval $(0, \kappa_3)$. Ewens (1968) pointed out that if $0 < r^* < r^{**} < \kappa_3$, the symmetric equilibria may be stable for $0 < r < r^*$ and for $r^{**} < r < \kappa_3$, but unstable for $r \in (r^*, r^{**})$. A simple example is $\alpha_1 = \alpha_2 > (2 + \sqrt{2})\beta$; here a region of instability exists. This gap in stability of the symmetric roots coincides exactly with the appearance and the disappearance of the asymmetric roots. Indeed, this fact follows from the topological consideration that domains of attraction to stable equilibria within the simplex and on the boundaries should be bounded by unstable surfaces on which unstable equilibria are likely to reside.

The eigenvalues of the stability matrix for the asymmetric roots have no simple form, but these can be obtained numerically for any specific viability matrix. It is, however, unnecessary to derive them in order to obtain a complete picture of the stability of equilibria for $r \neq 0$. For the vast majority of cases, we can infer from the stability conditions on the boundaries and on the symmetric roots whether the asymmetric roots exist and are stable. For example, if there are no stable boundaries or stable interior symmetric points, it follows that the asymmetric roots must exist and be stable. Conversely, if there are stable boundaries and stable symmetric (edge) equilibria, there must be unstable asymmetric equilibria within the simplex that define the domains of attraction. These topological considerations are illustrated in the examples that follow.

13.3.4.4. Examples

As we have seen, in cases of simple heterozygous advantage (i.e., $\beta > \alpha_1, \alpha_2$), there may be either one or two stable interior equilibria. If $\kappa_3 < 0$, there is a single stable interior equilibrium with $\hat{D} = 0$. If $\kappa_3 > 0$, three equilibria exist for small r. The two symmetric equilibria from the edges (i.e., $\hat{D} \neq 0$) are stable and approach $\hat{D} = 0$ from the positive and the negative directions, merging with the central ($\hat{D} = 0$) point at $r = \kappa_3$. For $r > \kappa_3$, there is only one stable interior equilibrium, viz., that with $\hat{D} = 0$. Figure 13.3 shows an example of the equilibrium values of D plotted against increasing recombination fractions for this latter case.

Two examples of mixed underdominance and overdominance models are shown in Fig. 13.4. Here, the double homozygote has a higher viability than

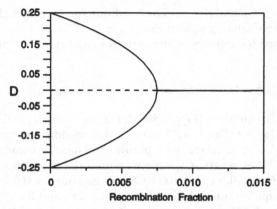

Figure 13.3. The equilibrium values of D in terms of the recombination fraction for the simple symmetric overdominance model, $\alpha_1 = \alpha_2 = 0.04$, $\beta = 0.05$. The dashed line indicates instability.

one or both single homozygotes, but both have a lower viability than the double heterozygote. Both examples illustrate Ewens's gap, in which asymmetric solutions appear within the simplex in the range (r^*, r^{**}). If these are unstable points, they separate stable boundary from stable interior solutions; if these are stable points they are separated by unstable equilibria on the boundaries and three unstable symmetric equilibria.

13.3.5. General Two-Locus Two-Allele Models: Speculation and Inference

We have shown that, for the symmetric viability model, a one-to-one correspondence can be drawn between the 15 equilibria of the one-locus four-allele model and the roots of the two-locus equations. In particular, the seven interior equilibria (Karlin and Feldman, 1970) are those that correspond, when $r = 0$, to equilibria in which neither locus is fixed. In this model, seven interior and eight boundary points are strictly the maximum number of solutions; bifurcations that generate other roots are not possible. Extensive numerical studies of more general two-locus two-allele models have confirmed that there can be at most seven interior equilibria.

This correspondence provides a general strategy for studying multilocus models. First, since the equilibria for zero recombination can be computed explicitly, these equilibria provide starting points for a numerical analysis of the system of nonlinear equations, with increasing values of r.

Second, the analysis of a one-locus multiple-allele model allows inferences to be made about the behavior of its corresponding multilocus model for tight linkage. For example, if the viability matrix allows a stable equilibrium at $r = 0$ for a set of gametic types, then any equilibrium in a subsurface that can generate all gametic types under recombination will not remain within the simplex for

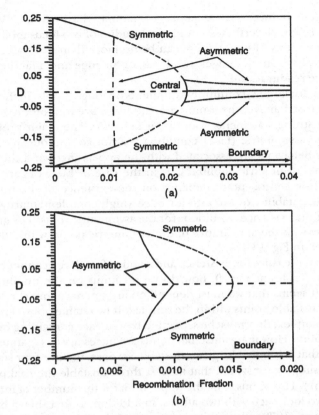

Figure 13.4. Mixed underdominance/overdominance models: (a) $\alpha_1 = \alpha_2 = 0.05$, $\beta = 0.01$. For $r < 0.01023$, there are two stable symmetric equilibria and one unstable central point. In the range $0.01023 < r < 0.02199$, there are no stable interior solutions and four stable boundaries. For $0.02199 < r < 0.225$, there are two stable interior equilibria, five unstable interior equilibria, and four stable boundaries. For $r > 0.225$, there is a single interior stable equilibrium. (b) $\alpha_1 = 0.02$, $\alpha_2 = 0.05$, $\beta = 0.05$. Four stable interior solutions (the asymmetric equilibria) appear in the range $0.007047 < r < 0.01$. If $r < 0.007047$, the symmetric equilibria are stable. If $r > 0.01$, there are no interior stable equilibria. The dashed line indicates instability.

$r \neq 0$. In particular, if the central point is stable for $r = 0$, it will be stable for $r > 0$, and this will be the only interior point.

An important issue in multilocus theory is the maximum number of stable equilibria for tight linkage. Karlin (1975), using continuity arguments from $r = 0$, speculated that a two-locus two-allele model can have at most two stable interior equilibria, based on the following theorem:

"Where there exists a stable equilibrium involving $r - k$ alleles $(1 < k < r - 1)$ then 2^k is an upper bound on the total number of possible extant stable equilibria." (Karlin, 1975, p. 385)

This theorem is incorrect, as counter examples have shown (Vickers and Cannings, 1988). Nevertheless, numerical studies of two-locus models indicate that, in the vicinity of $r = 0$, there can be no more than two stable equilibria; these may arise from two edges, two faces, or an edge and a face; examples of all three are readily found.

However, for large r, continuity arguments no longer apply. In the symmetric model discussed above, the symmetric equilibria are monotonic functions of r, allowing strong statements about their behavior. The major problem arises because the asymmetric (face) equilibria are not continuous in r. All of the anomalous behaviors are associated with the appearances and the disappearances of the asymmetric equilibria, as in the case of Ewens's gap. Just as the stability of the central point depends on the existence of the interior edge solutions, the stability of the interior edge (high complementarity) roots is a function of the existence conditions for the asymmetric roots. The gap in stability coincides with the appearance of the asymmetric points within the simplex, as illustrated in Fig. 13.4(a).

Given that the rules for existence and stability for r large cannot be deduced from a set of rules at $r = 0$, how do we set limits on the number of stable solutions? It seems that we must depend on arguments about the topology of stable and unstable points within the simplex. If two stable points are to coexist within the simplex, they must be separated by a surface containing one or more unstable points. How many unstable points are necessary to separate these? We now know that it is possible, given unstable vertices, to have four stable points within the simplex, provided that at least three unstable internal points exist [Fig. 13.4(b)]. This seems to be the upper limit for the number of interior equilibria for two loci, each with two alleles. In addition, as first shown by Feldman and Liberman (1979), there may be a total of six stable equilibria, including four stable vertices and two interior equilibria. This configuration requires four internal unstable points for separating the stable interior points from the vertices and one more (the central point) to separate the two symmetric points. Again, this seems to be a maximum, given seven internal points.

An additional complexity revealed by the numerical analysis of two-locus two-allele models is that a stable cycle can exist for a fixed viability model (Hastings, 1981b). This behavior appears for only a very restrictive set of parameters, involving a stable vertex and two closely associated unstable points within the simplex.

The analysis of symmetric viability models has overemphasized the significance of equilibria with $D = 0$. In fact, for generalized viability parameters, linkage equilibrium is never attained if there is any nonallelic interaction, although, if the interaction is weak or if recombination is loose, stable gametic frequency arrays will be in approximate linkage equilibrium. Also, in symmetric models, for tight linkage, stable equilibria arise from the central equilibria or from the fully polymorphic edges. Again, more general models do not show this behavior; stable equilibria may arise from two of the four faces or from a face and an edge. Some of these patterns are illustrated in an analysis of a symmetric model in which *cis*- and *trans*-heterozygotes have different viabilities (Nordborg et al., 1995).

13.3.6. More than Two Loci

The principles outlined above can be used to infer some of the properties of systems involving three or more loci. A partial analysis of the three-locus analog of the symmetric viability model (Feldman et al., 1974; Karlin and Liberman, 1976), confirms and extends many of the findings of two-locus models. In this case, with three loci and two alleles per locus, there are potentially 193 internal equilibrium points and 62 boundary equilibria. Indeed, examples can be constructed in which all 255 equilibria exist for tight linkage. These analyses have demonstrated that there are commonly four stable interior equilibria, but there may be as many as ten. However, three-locus models have not been explored numerically to the same degree as two-locus models, and there is a strong possibility that a larger number of stable interior equilibria are possible for intermediate recombination values. These analyses also confirmed some of the numerical findings of Franklin and Lewontin (1970), such as the coexistence of equilibria with high levels of linkage disequilibrium with equilibria in linkage equilibrium.

A few studies have investigated the properties of the central equilibrium for symmetric viability models with an arbitrary number of loci (Karlin and Avni, 1981; Christiansen, 1988). One of the main results is that if a stationary point in linkage equilibrium for all loci is stable for some R_0, this equilibrium will be stable for all larger values of R.

Numerical iteration is possible with up to six or seven loci; beyond this, the large zygotic arrays that need to be generated limit the analysis. One of the first numerical studies of multilocus equilibria was presented by Lewontin (1964a, 1964b). Lewontin obtained solutions for a number of numerical examples with five loci and selective values that are extensions of the symmetric viability and optimum models. There have been very few such studies since.

13.3.7. The Evolution of Recombination Rates

Following Fisher's early proposal that genes interacting in fitness will evolve toward closer linkage, it seemed natural to check the relationship between $\bar{V}$ and r. Kimura (1956b) showed that, for a version of the symmetric viability model, $\bar{V}$ evaluated at the stable equilibrium was a decreasing function of r, which he interpreted as favoring closer linkage between the two loci. Lewontin (1971) demonstrated that for any system of constant viabilities,

$$\left.\frac{\partial \bar{V}}{\partial r}\right|_{r=0} < 0,$$

(13.15)

which might also be viewed as a validation of Fisher's claim. This interpretation of the role of $\bar{V}$ involves a group selection argument, i.e., the population with a lower recombination between the loci would have a higher mean viability and outcompete those populations exhibiting looser recombination. No mechanism for the evolutionary modification of recombination is suggested.

Nei (1967) considered this problem by introducing a third locus, which may or may not be linked to the loci under selection and whose only function is to control the rate of recombination between the loci. The mathematical theory of the evolutionary behavior of such a gene was analyzed by Feldman (1972) in terms of the external stability of a newly arising recombination-controlling allele (M_2) in a population at a stable polymorphic equilibrium for the other two loci. Thus we consider a population of four gametic types, ABM_1, AbM_1, aBM_1 and abM_1, in equilibrium in four-chromosomal space, and ask whether M_2 introduced near this equilibrium will invade the population. This analysis showed that, for the equilibria defined by the Lewontin–Kojima model, M_2 increases if it reduces the recombination rate between the A and the B loci and is lost if M_2 increases recombination between the loci.

This result led to the search for generalizations. Feldman et al. (1980) showed that the result is true for any two-locus two-allele viability model, and Liberman and Feldman (1986) demonstrated its truth for any number of modifier alleles. This is called the reduction principle for recombination.

13.4. Discussion and Conclusions

This chapter emphasizes the strong theoretical connection between the properties of one-locus multiple-allele systems and multilocus multiallelic genetic systems. It is evident that, as linkage among a set of interacting loci becomes stronger, the equilibrium behavior will approximate that of a one-locus analog, with haplotype frequencies becoming, in effect, allele frequencies. Under these conditions, mean viabilities increase under selection and the adaptive topography principle holds. However, these principles break down as the recombination frequencies between the loci increase.

We have shown that continuity arguments allow us to deduce many of the properties of multilocus systems for tight linkage, but that these arguments become invalid as recombination increases. Nevertheless, regardless of the recombination frequencies, the zero recombination analog of the multilocus model provides a set of starting points for numerical exploration of the equilibrium points, especially if we are able to follow the trajectories of the equilibria through the complex plane.

It is generally assumed, either from genetic load arguments or from the increasing constraints on selection coefficients, that the very high degree of polymorphism observed at some loci reflect selective neutrality rather than selective balance. Spencer and Marks (1992) have shown that patterns of viabilities evolve, as do the array of gene frequencies, making it difficult to infer neutrality or selective balance from the observed standing variation. Clearly, selective balance can maintain large numbers of alleles; self-sterility alleles and major histocompatibility complex variation are examples.

The same evolutionary principles apply to multilocus systems maintained by selection. The pattern of gamete frequencies and the interactions among alleles at different loci are likely to coevolve, reflecting their selective history rather than the spectrum of mutational variation at the individual loci. In addition, the stronger the selective interaction, the greater the pressure to reduce, by some means, the recombination fraction between the interacting loci. These two factors, the evolution of close linkage and the selective sieving of new alleles as they arise, are the important elements in the development of coadapted gene complexes.

Similar principles will also operate at the intragenic level, at which almost any pair of nonneutral base substitutions will interact in their effect on gene function. Just as we can choose to consider light as a wave or a particle, a gene may be considered a single-locus or a multilocus system. We need to take the latter perspective if we wish to understand genic evolution.

REFERENCES

Arunachalum, V. and Owen, A. R. G. 1971. *Polymorphisms with Linked Loci*. London: Chapman & Hall.

Bennett, J. H. 1954. On the theory of random mating. *Ann. Eugenics* **18**:311–317.

Bodmer, W. F. 1965. Differential fertility in population genetics models. *Genetics* **51**:411–424.

Bodmer, W. F and Parsons, P. A. 1962. Linkage and recombination in evolution. *Adv. Genet.* **11**:1–100.

Christiansen, F. B. 1988. Epistasis in the multiple locus symmetric viability model. *J. Math. Biol.* **26**:595–618.

Ewens, W. J. 1968. A genetic model having complex linkage behaviour. *Theor. Appl. Genet.* **38**:140–143.

Feldman, M. 1972. Selection for linkage modification. I. Random mating populations. *Theor. Popul. Biol.* **3**:324–346.

Feldman, M. W. and Liberman, U. 1979. On the number of stable equilibria and the simultaneous stability of fixation and polymorphism in two-locus models. *Genetics* **92**:1355–1360.

Feldman, M. W., Christiansen, F. B., and Brooks, L. D. 1980. Evolution of recombination in a constant environment. *Proc. Natl. Acad. Sci. USA* **77**:4838–4841.

Feldman, M. W., Franklin, I. R., and Thomson, G. J. 1974. Selection in complex genetic systems. I. The symmetric equilibria of the three-locus symmetric viability model. *Genetics* **76**:135–162.

Felsenstein, J. 1965. The effect of linkage on directional selection. *Genetics* **52**:349–363.

Fisher, R. A. 1922. On the dominance ratio. *Proc. R. Soc. Edinburgh* **52**:321–341.

Fisher, R. A. 1930. *The Genetical Theory of Natural Selection*. Oxford: Clarendon.

Fisher, R. A. 1941. Average excess and average effect of a gene substitution. *Ann. Eugenics* **11**:53–63.

Franklin, I. R. and Feldman, M. W. 1977. Two loci with two alleles: linkage equilibrium and linkage disequilibrium can be simultaneously stable. *Theor. Popul. Biol.* **12**:95–113.

282   *Ian R. Franklin and Marcus W. Feldman*

Franklin, I. R. and Lewontin, R. C. 1970. Is the gene the unit of selection? *Genetics* **65**:701–734.

Geiringer, H. 1944. On the probability theory of linkage in Mendelian heredity. *Ann. Math. Stat.* **15**:25–57.

Hastings, A. 1981a. Simultaneous stability of $D = 0$ and $D \neq 0$ for multiplicative viabilities at two loci: an analytical study. *J. Theor. Biol.* **89**:69–81.

Hastings, A. 1981b. Stable cycling in discrete-time genetic models. *Proc. Natl. Acad. Sci. USA* **78**:7224–7225.

Hastings, A. 1982. Unexpected behavior in two locus genetic systems: an analysis of marginal underdominance at a stable equilibrium. *Genetics* **102**:129–138.

Hastings, A. 1985. Four simultaneously stable polymorphic equilibria in two locus, two allele models. *Genetics* **109**:255–261.

Hill, W. G. 1974. Disequilibrium among several linked neutral genes in finite population. II. Variances and covariances of disequilibria. *Theor. Popul. Biol.* **6**:184–198.

Hill, W. G. and Robertson, A. 1968. Linkage disequilibrium in finite populations. *Theor. Appl. Genet.* **38**:226–231.

Jennings, H. S. 1917. The numerical results of diverse systems of breeding with respect to two pairs of characters, linked or independent, with special relation to the effects of linkage. *Genetics* **2**:97–154.

Karlin, S. 1975. General two-locus selection models: some objectives, results and interpretations. *Theor. Popul. Biol.* **7**:364–398.

Karlin, S. 1981. Some natural viability systems for a multiallelic locus: a theoretical study. *Genetics* **97**:457–473.

Karlin, S. and Avni, H. 1981. Analysis of central equilibria in multilocus systems: a generalized symmetric viability regime. *Theor. Popul. Biol.* **20**:241–280.

Karlin, S. and Feldman, M. W. 1970. Linkage and selection: two locus symmetric viability model. *Theor. Popul. Biol.* **1**:39–71.

Karlin, S. and Feldman, M. W. 1978. Simultaneous stability of $D = 0$ and $D \neq 0$ for multiplicative viabilities at two loci. *Genetics* **90**:813–825.

Karlin, S. and Liberman, U. 1976. A phenotypic symmetric selection model for three loci, two alleles: the case of tight linkage. *Theoret. Pop. Biol.* **10**:334–364.

Kimura, M. 1956a. Rules for testing stability of a selective polymorphism. *Proc. Natl. Acad. Sci. USA* **42**:336–340.

Kimura, M. 1956b. A model of a genetic system which leads to closer linkage by natural selection. *Evolution* **10**:278–287.

Kimura, M. 1965. Attainment of quasi linkage when gene frequencies are changing by natural selection. *Genetics* **52**:875–890.

Kingman, J. F. C. 1961. A mathematical problem in population genetics. *Proc. Cambridge Philos. Soc.* **57**:574–582.

Lewontin, R. C. 1958. A general method for investigating the equilibrium of gene frequency in a population. *Genetics* **43**:419–434.

Lewontin, R. C. 1964a. The interaction of selection and linkage I. General considerations; heterotic models. *Genetics* **49**:49–67

Lewontin, R. C. 1964b. The interaction of selection and linkage II. Optimum models. *Genetics* **50**:757–782.

Lewontin, R. C. 1971. The effect of genetic linkage on the mean fitness of a population. *Proc. Natl. Acad. Sci. USA* **68**:984–986.

Lewontin, R. C. 1974. *The Genetic Basis of Evolutionary Change.* New York: Columbia U. Press.

Lewontin, R. C. and Kojima, K. 1960. The evolutionary dynamics of complex polymorphisms. *Evolution* 14:458–472.

Lewontin, R. C., Ginzburg, L. R., and Tuljapurkar, S. D. 1978. Heterosis as an explanation for large amounts of genic polymorphism. *Genetics* 88:149–170.

Liberman, U. and Feldman, M. W. 1986. A general reduction principle for genetic modifiers of recombination. *Theor. Popul. Biol.* 30:341–371.

Littler, R. A. 1973. Linkage disequilibrium in two-locus, finite random mating models without selection or mutation. *Theor. Popul. Biol.* 4:259–275.

Mandel, S. P. F. 1959. The stability of a multiple allelic system. *Heredity* 13:289–302.

Mather, K. 1941. Variation and selection of polygenic characters. *J. Genetics* 41:159–193.

Moran, P. A. P. 1963. Balanced polymorphisms with unlinked loci. *Aust. J. Biol. Sci.* 16:1–5.

Moran, P. A. P. 1964. On the nonexistence of adaptive topographies. *Ann. Hum. Genet.* 27:383–393.

Nei, M. 1967. Modification of linkage intensity by natural selection. *Genetics* 57:625–641.

Nordborg, M., Franklin, I. R., and Feldman, M. W. 1995. Effects of cis-trans viability selection on some two-locus models. *Theor. Popul. Biol.* 47:365–392.

Ohta, T. and Kimura, M. 1969. Linkage disequilibrium due to random genetic drift. *Genet. Res.* 13:47–55.

Owen, A. R. G. 1954. Balanced polymorphism of a multiple allelic series. *Caryologia* 6, Suppl.:1240–1241.

Pollak, E. 1978. With selection for fecundity the mean fitness does not necessarily increase. *Genetics* 90:384–389.

Robbins, R. B. 1918. Some applications of mathematics to breeding problems. III. *Genetics* 3:375–389.

Spencer, H. G. and Marks, R. W. 1988. The maintenance of a single locus polymorphism. I. Numerical studies of a viability selection model. *Genetics* 120:605–613.

Spencer, H. G. and Marks, R. W. 1992. The maintenance of a single locus polymorphism. IV. Models with mutation from existing alleles. *Genetics* 130:211–221.

Sved, J. A. 1968. The stability of linked systems with a small population size. *Genetics* 59:543–563.

Sved, J. A. 1971. Linkage disequilibrium and homozygosity of chromosome segments in finite populations. *Theor. Popul. Biol.* 2:125–141.

Turner, J. R. G. 1971. Wright's adaptive surface, and some general rules for equilibria in complex polymorphisms. *Am. Nat.* 105:267–278.

Vickers, G. J. and Cannings, C. 1988 On the number of stable equilibria in a one locus, multiallelic system. *J. Theor. Biol.* 131:273–278.

Watterson, G. A. 1970. The effect of linkage in a finite random-mating population. *Theor. Popul. Biol.* 1:72–87.

Weir, B. S. and Cockerham, C. C. 1974. Behavior of pairs of loci in finite monoecious populations. *Theor. Popul. Biol.* 6:323–354.

Weir, B. S. and Cockerham, C. C. 1989. Complete characterization of disequilibrium at two loci. In *Mathematical Evolutionary Theory*, M. W. Feldman, ed., pp. 86–110. Princeton, NJ: Princeton U. Press.

Wright, S. 1952. The genetics of quantitative variability. In *Quantitative Inheritance*, pp. 5–41. London: Her Majesty's Stationery Office.

Wright, S. 1967. "Surfaces" of selective value. *Proc. Natl. Acad. Sci. USA* 55:165–172.

CHAPTER FOURTEEN

Inversion Polymorphisms in *Drosophila*

COSTAS KRIMBAS AND JEFFREY POWELL

14.1. Introduction

Sturtevant, in the early 1920s, discovered chromosomal inversions from their inhibiting effect on genetic recombination. Indeed, nonfunctional products are produced from single recombination events occurring within the inversion's length in heterozygotes, either for pericentric inversions (including the centromere) or paracentric ones; abnormal chromatids are produced, bearing duplications, deletions, and, in some cases, two centromeres or none. Thus inversions may be considered efficient devices for maintaining intact blocks of alleles. This is true especially for the shorter inversions, which do not permit the occurrence of a second crossing over to restore the abnormal effects of the first.

Even at the time of their discovery, it was realized that inversions are not laboratory curiosities or productions but genetic devices widespread in natural populations of several species. Actually, some inversions occur over previous ones, overlapping them, thus producing complicated gene arrangements compared with what has been called the standard gene order.

14.2. Constructing Phylogenies: The Classic Version

Sturtevant and Dobzhansky (1936) were able to establish a phylogeny of gene arrangements produced by overlapping inversions based on the assumption that we can trace the origin of every copy of the same inversion encountered in nature to a single initial one, produced by a unique event in the history of the species. Thus, inversions are thought to be monophyletic. The absence of a repeated production of the same inversion was attributed to the rarity of chromosomal breaks occurring at the same specific location of the chromosome; the production of a simple inversion necessitates two simultaneous breaks at two different positions.

R. S. Singh and C. B. Krimbas, eds., *Evolutionary Genetics: From Molecules to Morphology*, vol. 1. © Cambridge University Press 2000. Printed in the United States of America. ISBN 0-521-57123-5. All rights reserved.

Phylogenetic relationships were constructed by combining elementary orderings of triads of gene arrangements that differ by simple and overlapping inversions. Let us consider such a triad:

1. A B * C D E F G H * I J K L M N O P
2. A B H G F * E D C I J K L * M N O P
3. A B H G F L K J I C D E M N O P

The sequence of capital letters indicates the different segments of the giant chromosome that are identifiable by their banding pattern. Asterisks indicate the breakage points of simple inversions. Gene arrangement 2 is produced by a simple inversion from gene arrangement 1; from 2 another simple overlapping inversion produces 3. When such triads are observed, we may ignore what is the ancestral arrangement, but we are able to order them phylogenetically as follows: 1 → 2 → 3. Thus, either gene arrangement 1 is the ancestral form from which 2 is derived, followed by 3, or the other way around, viz., 3 → 2 → 1. Finally, if gene arrangement 2 is ancestral, gene arrangements 1 and 3 are independently produced from it by single inversions: 1 ← 2 → 3. The phylogenetic derivations are symbolized by the double arrows and are articulated in the only permissible pathways: 1 ↔ 2 ↔ 3. Other pathways, such as 1 ↔ 3 ↔ 2, are unacceptable by parsimony considerations.

A number of triads sharing common members permit the construction of complex phylogenies; one such phylogeny has been completed for the gene arrangements of the third chromosome of *Drosophila pseudoobscura* and *D. persimilis* by Th. Dobzhansky (see Powell, 1992). This phylogeny also indicates the way in which two species are related, because they share a common gene arrangement, which apparently antedates species separation (Fig. 14.1). Another feature of this phylogeny is the existence of a hypothetical gene arrangement, never encountered in nature, but whose constitution is predicted precisely, differing by a simple inversion from Standard and from Santa Cruz gene arrangements. There are three instances of a predicted gene arrangement subsequently discovered in nature (two by Dobzhansky in *D. pseudoobscura* and *D. azteca* and one by Krimbas in *D. subobscura/D. madeirensis*, Fig. 14.2). The repeated confirmation of such a specific prediction provides, we believe, unusually strong support for the correctness of this method.

From gene arrangement phylogenies we may infer species phylogenies; to do so, two additional assumptions are required:

(1) that fixation of a particular gene arrangement occurs only once in a lineage. Thus the presence of sustained or creeping polymorphisms, extending beyond speciation events, are precluded. In the contrary case, that is, when repeated fixations are allowed, they may produce a pattern of inversion distributions inconsistent with this method. This assumption is similar to that made when, from a gene phylogeny, we infer the phylogeny of species bearing these genes.

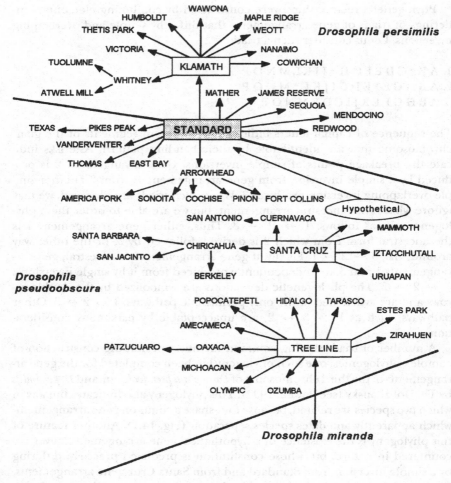

Figure 14.1. The phylogeny of *D. pseudoobscura* and *D. persimilis* gene arrangements and their relation to those of *D. miranda*. (Composite figure from various sources, but redrawn.)

(2) the absence of introgression, which would allow a gene arrangement already eliminated in an evolutionary branch to be reintroduced. Actually, this assumption is an extension of the previous one. The topology of a tree without introgression is that of a tree displaying only bifurcations, and excludes the presence of network patterns. It is an assumption analogous to the exclusion of homoplasies. These two assumptions, plus that of the unique origin of inversions (of monophyletic origin), constitute the classic version.

14.3. Repeated Production of Inversions by Transposable Elements

Some observations appear to conflict with the classic version. Breakage points are not randomly distributed along the chromosome; there is good evidence

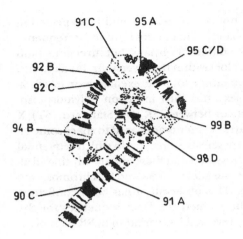

Figure 14.2. Proof for the existence of a missing link in a gene arrangement phylogeny. Two gene arrangements, O_{ST} and O_{3+4}, inversion 4 overlapping with inversion 3, which are present in natural populations of *D. subobscura*, are not linked phylogenetically in a direct way; the intermediate O_3 is missing. It was found in a close relative species, *D. madeirensis*. Here the simple loop in interspecies hybrids originated from *D. madeirensis* females homozygous for O_3 and *D. subobscura* males homozygous for O_{3+4}. (Redrawn from a published photograph by Krimbas.)

that their actual distribution reflects differences in breakage propensities in different chromosomal positions. There are hot spots, that is, points of multiple breakages. Some cases of reinversion (restoring the original gene order) or of repeated production of the same inversion have been described. Transposable elements seem to be responsible for these phenomena (Engels and Preston, 1984) and, in some cases (e.g., *hobo*), there is good evidence that transposable elements are actually responsible for the formation of naturally occurring inversions (for a review, see Krimbas and Powell, 1992; for breakage factor, not locus specific, in *D. robusta*, see Levitan, 1992; and Levitan and Verdonck, 1986).

14.4. Molecular Studies Supporting Monophyly

On the other hand, there is strong molecular evidence that each of the currently encountered inversions in natural populations is monophyletic, as postulated by the classic version. When an inversion is first formed, an array of alleles is linked together and segregates as a block. Rare recombinational events, mutations, and gene conversions may, as time goes by, introduce genic polymorphism within the inversion and decrease the extreme linkage disequilibrium among the alleles under consideration. Nevertheless, these disequilibria should be maintained for long periods.

Given the high degree of molecular variation characteristic of natural populations of *Drosophila* (e.g., Moriyama and Powell, 1996), the presence of highly polymorphic loci within the breakpoints of natural inversions is expected and could directly be used to test their monophyletic origin. Prakash and Lewontin (1968, 1971) were the first to observer a strong association between gene arrangements of the third-chromosome inversions in *D. pseudoobscura* and *D. persimilis* and allozymes at four loci within these inversions. The associations were not for specific inversions, but rather for major "phylads" of the inversion tree (phylad is a designation used by Dobzhansky to indicate an inversion residing at the root of a branch of the inversion's tree and all the inversions derived from it). These associations, however, were not complete, as would be expected in case of a complete absence of gene exchange: There were significant frequency differences between arrangements from different phylads, but inversions (and phylads) were polymorphic for alleles at loci within the inversions' breakpoints.

These polymorphisms might be due either to gene exchange or to mutation. The latter was the case for the sex ratio (SR) inversion polymorphism in *D. pseudoobscura*: Frequency differences between SR and Standard (ST) X chromosomes existed, although both arrangements appeared to share some alleles (Prakash and Merritt, 1972). A more sensitive analysis, in which sequential electrophoresis was used (Keith, 1983), revealed that these apparently shared alleles were not identical and that, in fact, the alleles on SR and ST chromosomes constituted completely separate arrays. This observation has been confirmed by Babcock and Anderson (1996), who studied the DNA sequences for the same chromosomes. Furthermore, they revealed less variation in SR than in ST chromosomes.

Several other species have been studied for the association of electrophoretic alleles and inversions. In *D. subobscura*, strong associations are often present, but there are some cases with no such association (for a review, see Krimbas, 1992, 1993a).

More revealing are some recent DNA-based studies of *Drosophila* inversions. Initially, Aquadro et al. (1991) used restriction enzymes to detect DNA variants [restriction fragment length polymorphism (RFLP)] in and around the *Amylase (Amy)* locus in *D. pseudoobscura* and *D. persimilis*; this locus shows the strongest allozyme-inversion association. Their results provided very strong evidence for monophyly of natural inversions and recombination suppression: The same inversions taken from throughout the range of the species always showed more similarity to each other than to any other gene arrangement, even to those within the same populations. Furthermore, in noninverted regions of the genome, strong linkage disequilibrium was found. More detailed sequence data for the third-chromosome polymorphisms of these species have provided strong confirmation of the RFLP study (Popadic and Anderson, 1994, 1995a). These authors, by analyzing the DNA sequence data in regard to the phylogeny, were also able to present a strong argument for the ancestral gene arrangement in the tree, thus providing unique directions of derivation of gene arrangements. Double arrows were no longer needed.

Two other species have been studied in the same way, and the data support monophyly and suppression of recombination. These species are *D. melanogaster* (Aguadé, 1988; Benassi et al., 1993) and *D. subobscura* (Rozas and Aguadé, 1993, 1994).

While suppression of recombination between chromosomes differing by paracentric inversions is supported by the data, strong evidence of another mechanism of gene exchange was detected, namely gene conversion. Embedded within a generally divergent region of the *Amy* locus of *D. pseudoobscura* is a small stretch of DNA that is identical between inversions far apart on the inversion phylogenetic tree (Popadic and Anderson, 1995b). Similarly, Rozas and Aguadé (1993, 1994) found a small region of the chromosomal protein locus they studied in *D. subobscura* to be identical or highly similar in gene arrangements that were highly dissimilar around the region. Recombination is unlikely to account for these apparent exchanges of small DNA regions because these are embedded on both sides by highly divergent DNA. Rozas and Aguadé made rough calculations that indicated that gene conversion need occur at a rate of only $\sim 10^{-7}$ per generation to account for their findings, a rate lower than that found in laboratory studies of *D. melanogaster*.

Finally, the strongest direct evidence for monophyly of naturally occurring inversions comes from DNA analysis of the breakpoints themselves. Wesley and Eanes (1994) cloned the breakpoints of the cosmopolitan *Payne* inversion in the 3L of *D. melanogaster*. They sequenced six standard arrangements and seven inverted chromosomes taken from populations from around the world. The breakpoints in all samples were identical. Furthermore, the sequence bore no similarity to any known transposable element. Likewise, the only other naturally occurring breakpoint to be cloned, in *D. subobscura*, which is a fixed difference in relation to *D. melanogaster*, was also not associated with or indicative of a transposable element (Cirera et al., 1995). This does not necessarily indicate that transposable elements are not implicated in the production of inversion breakpoints; for stable gene arrangements, these elements may have been properly excised.

14.5. A Modern Modification of the Classic Version

Thus the most recent analyses of *Drosophila* inversions by use of DNA technology provide strong support for their monophyletic origin and suppression of recombination. How, then, can we reconcile the evidence for repeated origin by transposable elements, as laboratory studies indicate, and the evidence from DNA sequence data that indicates a monophyletic origin?

While the occurrence of an inversion may be a rare but not unique event, numerous and diverse theoretical models indicate that the retention and maintenance of a newly formed inversion in a population requires several conditions that are rarely met (e.g., Nei et al., 1967; Ohta and Kojima, 1968; Fraser et al., 1966; Fraser and Burnell, 1967a, 1967b; Haldane, 1957; Deakin and Teague, 1974; Teague and Deakin, 1976; Charlesworth and Charlesworth, 1973; Charlesworth, 1974, etc. For a review see Krimbas and Powell, 1992). Therefore the entire process may be considered unique. This is the modern version – as opposed to the classic – the modern justification of the unique appearance of naturally occurring inversions.

14.6. A Modified Phylogenetic Approach

Several researchers have used a modification of the method just described to construct species phylogenies. Instead of considering the entire gene orders or gene arrangements, they have focused their attention on a single breakpoint. This is the association/dissociation method. Consider three chromosomal sequences:

$$[A–B–C]; \qquad [A–B] \cdots C; \quad \text{and, finally,} \quad A \cdots [B–C].$$

The brackets indicate, in the first case, that the three segments, A, B, and C, are located contiguously, so segment B has segment A on its left and segment C on its right. In the second instance, A is flanked by B, while C is found far apart. In the third sequence, B and C are found together while A is located apart. Associations and dissociations of segments are produced by breakages between them, constituting one of the two breakpoints of an inversion. Thus, this method is based on examining only one breakpoint at a time and not two. The most parsimonious phylogenetic ordering is the following:

$$A \cdots [B–C] \leftrightarrow [A–B–C] \leftrightarrow [A–B] \cdots C.$$

Any other ordering would require more breaks. Several authors have used the association/dissociation method to construct phylogenies of species (Carson, 1992, for the Hawaiian *Drosophila;* Brehm and Krimbas, 1993, for the *obscura* group; Wasserman, 1992, for the *repleta* group of species).

14.7. Levels of Inversion Polymorphisms

Species of *Drosophila* differ in the type and the amount of gene arrangement polymorphism. There are species with an extremely rich polymorphism, extending to all their chromosomes, including the sex chromosome. Other species have a rich polymorphism restricted to one or two chromosomes, while yet other species have a dominant gene arrangement in all their chromosomes with simple infrequent inversions. Finally, there are species without inversions.

The types of polymorphisms encountered appear to be of great interest. Some species (*D. subobscura, D. pseudoobscura, D. willistoni,* and *D. paulistorum* complex, and others) have complicated gene arrangements produced by numerous overlapping inversions. It is likely that these are very old polymorphisms, and population bottlenecks have not eliminated them. In contrast, species such as *D. melanogaster* have more restricted polymorphisms, consisting of simple inversion found in many geographically widespread natural populations (cosmopolitan inversions) with some simple inversions with very restricted geographic distributions. It is possible that these species "recently" experienced a significant population bottleneck, resulting in the fixation of one gene arrangement per chromosome, followed by an increase in population size and a subsequent spread of the species. The spread of *D. melanogaster* particularly has been facilitated by human intervention.

It is more difficult to explain the practically total absence of inversion polymorphism in some species, such as in *D. simulans.* It may be that a combination of a bottleneck and absence of a transposable element that produces breaks is responsible for this situation. We know, however, that this species harbors some transposable elements, like the *mariner.*

14.8. Natural Selection

Inversions are not selectively neutral markers. In 1943, Dobzhansky reported repeated yearly seasonal changes in gene arrangement frequencies; this convinced him that inversions are subject to selective forces.

One hypothesis, which can be traced to Mainx (1954), regards position effects, produced from the chromosome breakage, responsible for the higher fitness of the inversions and their subsequent spread. In three independent studies it was shown that the elimination or maintenance of new x-ray-produced inversions depends on the population gene pool and therefore is not an intrinsic property of these inversions. This result is contrary to the position effect hypothesis, which was therefore abandoned (Sperlich, 1963, 1966; Wallace, 1966; Vann, 1966).

Selection should be attributed to a different mechanism. Actually, it is the inhibition of recombination that is the important factor; inversions keep together blocks of alleles that segregate as a unit. One hypothesis for the selective nature of inversions is what Dobzhansky called the *classical* model: Alternative gene arrangements, differing in their allelic content, produce heterotic heterozygotes because they mutually cover recessive deleterious alleles present in alternative gene arrangements. Alternatively, according to Dobzhansky's *balanced* model, it might be that heterozygosity at some loci *per se* produces fitter genotypes than both kinds of homozygotes do, that is, overdominance. Alternative inversions should bear different alleles at a number of loci. Of course, more complicated epistatic interactions may also be postulated (see below). Whatever the mechanism, selective values are retained only when blocks of genes are kept intact. The fact that more than one gene is involved makes the study of the selection mechanism easier: the fitness effects of many genes are manifested in a single segregating unit, and therefore selective values are easily detected.

14.9. Distribution of Breakpoints

A simple test of the selective hypothesis is provided by the distribution of simple inversion lengths. A neutral distribution of breaks, the null hypothesis, may be estimated from a set of randomly positioned breakpoints. Alternatively, when recombination is selected against, we expect a paucity of large inversions that allow double crossovers to occur and, thus, permit the destruction of their gene blocks. This is what is observed in a number of species studied.

Another independent line of evidence comes from the fact that complex inversions protect the gene blocks better than simple ones do. The two extreme members of a triad of overlapping inversions, when coexisting in a population, protect their allelic contents better than when the middle member is also present. This is simply due to the fact that leakage of genetic material by a double crossover is much more likely in simple single-inversion heterozygotes than in complex heterozygotes with chromosomes differing by multiple inversions. Wallace's rule, which predicts this exclusion of the middle member in natural populations, is generally obeyed, but exceptions have also been noted.

These indications, together with others, like the nonconformity of frequency distributions to those expected under neutrality (Ewens, 1972), are indirect indications for the presence of selective processes. However, indirect evidence does not carry the conviction of a direct proof.

14.10. Laboratory Studies

In a series of classical laboratory experiments, Dobzhansky, his collaborators, and students investigated the selective regime of inversions in *D. pseudoobscura*. (Spiess, one of his students, paralleled these in *D. persimilis*). Population cage experiments showed that, under laboratory conditions, inversions are indeed selected. When two gene arrangements, differing by paracentric inversions and originating from the same natural population, were placed in a population cage, a stable equilibrium was eventually reached after several generations. Dobzhansky preferred to interpret his results by assigning constant selective values to the karyotypes, with a greater fitness of heterokaryotypes, although frequency-dependent fitnesses explained the experimental results as well or even better. Cage experiments with three different gene arrangements were also performed; the outcome could not always be predicted from two gene arrangement experiments. There is no transitivity in fitness values, and there is probably frequency dependency, as Sewall Wright argued (see Wallace in Lewontin et al., 1981, p. 807). In multiarrangement cages, the final equilibrium point often depended on the initial frequencies.

The most important outcome, however, came from experiments in cages with arrangements of mixed geographic origin. When two arrangements originated from the same natural population, a repeatable equilibrium point was reached, but when the two types of arrangements originated from distant populations, the outcome was erratic and often resulted in no stable equilibrium. This led Dobzhansky to formulate the hypothesis of coadaptation. Coadaptation in this context has two aspects. First, alleles within a gene arrangement are coadapted with one another, presumably by epistatic fitness effects. Second, the different gene arrangements are coadapted *inter se* to produce overdominance fitness in heterokaryotypes. The fact that alternative gene arrangements from the same population are coadapted, but not so when derived from different populations, indicates that, long after their origin, inversions acquire genetic variability that is subject to selective pressures. This fact was verified in laboratory experiments by Strickberger (1963). However, cage experiments are not able to provide us with a clue for what is really happening in the wild.

14.11. Natural Populations

Besides seasonal frequency changes, spatial differences are also observed. These may include altitudinal changes of frequencies as well as geographical clines of frequencies. In the past two decades, Anderson and his collaborators

(Anderson et al., 1979; Salceda and Anderson, 1988) demonstrated the existence of frequency-dependent selection in natural populations of *D. pseudoobscura*; males possessing rare arrangements were inseminating a greater number of wild females than expected from their genotype frequency. This mechanism had already been hypothesized by Spiess (1968) in the closely related species *D. persimilis*, following the studies of Petit (1958) and Petit and Ehrman (1969). However, Anderson and his collaborators provided evidence for the presence of this mechanism in nature. The above, together with habitat selection (studies of Taylor and Powell, 1977, 1978), might explain the retention of many rare variants, but they are not sufficient to explain how clines in gene arrangement frequencies are formed and maintained in nature.

In another species, *D. robusta*, there are also geographic clines of gene arrangement frequencies, mostly running from north to south. That climatic factors are likely responsible for the formation of these clines may be deduced from the existence of altitudinal clines found in conformity with the north–south ones; north gene arrangements are encountered in high altitudes but decline in frequency as altitude decreases. The evidence for seasonal changes is, in this species, equivocal. In order to obtain strong proof for natural selection's acting on gene arrangements, Levitan (1992) performed perturbation experiments of natural populations. He released great numbers of flies, raised in the laboratory and originating from a northern population, into a centrally located one and monitored the gene arrangement frequencies. Gene arrangement frequencies changed from those recorded before the release, the "natural baseline." Levitan also obtained evidence that the released flies successfully reproduced with those from the natural population. The drop of the frequencies to the natural baseline was sudden, with the greatest change occurring during the winter months following the release. The return of the frequencies to the original baseline is strong evidence for the action of natural selection on gene arrangements. One might argue that flies originating from northern populations are genetically adapted to Nordic conditions and not necessarily in the part of the genome delimited by inversions. Therefore the selection observed may not necessarily be selection on gene arrangements. The experimental results do not seem to justify this interpretation. Indeed, some genetic recombination took place in nature, since there was evidence for mating between released and wild flies. A similar perturbation experiment was performed with the release of flies carrying southern gene arrangements, with the expected results.

The situation is somewhat different for *D. subobscura*. This species is characterized by a very rich inversion polymorphism that extends to all five acrocentric chromosomes. Contrary to the above cases, in this case no altitudinal clines are observed in gene arrangement frequencies. It seems that seasonal changes do exist and have been missed in previous studies (Rodriguez-Trelles et al., 1996). Several lines of evidence seem to indicate the presence of selective mechanisms, but contrary evidence could also be cited, e.g., some exceptions to Wallace's rule and the absence of significant changes in population cage experiments, or at least not repeatable ones. On the contrary, the Ewens test indicates that selection is acting on at least two chromosomes (the same was found for the third chromosome of *D. pseudoobscura*). Finally, the frequency distribution of simple inversions compared with their length departs from the theoretical neutral distribution, indicating the influence of selection on the inversions that have been retained in natural populations.

In *D. subobscura* the presence of geographic clines running mostly north to south is extremely marked. One may visualize this by mapping the first principal component of gene arrangement frequencies in Europe (Menozzi and Krimbas, 1992). The principal components of variance are linear synthetic variables that encapsulate the information provided by numerous other variables, in our case gene arrangement frequencies, and indicate the fraction of variability derived from a common cause. The map of the first principal component of variance, which explains the greater part of the variability of inversion polymorphism (18%), seems to be extremely similar to that of the first principal component of temperature and humidity differences; its value increases when one moves from cold and humid northern locations to warm and dry southern ones. The correlation coefficient between the values of the first principal component for inversions and that of climatic variables is high (equal to 0.72, significant at $P < 0.0001$). Also, the frequencies of the ST gene arrangements, A_{ST}, J_{ST}, E_{ST}, U_{ST}, and O_{ST}, contribute the most to the first principal component of climatic values than those with latitude. It seems that yearly maximum and minimum temperatures are responsible for this high correlation.

The above data strongly indicate the presence of selection on inversion polymorphism. An additional unexpected observation reinforces this conclusion. Soon after the invasion of South America (Chile) and North America (Western USA and Canada) by *D. subobscura* in the late 1970s, (fortunately the events were closely followed by Prevosti and his collaborators) and soon after the spread of the species to new territories, clines of gene arrangements were formed as the species continued spreading to the north and to the south. The gradient of the clines increased with time and showed the same pattern as the one found in Europe. Of course, in South America the cline was reversed compared to the European one, running from the south to the north, as should be expected from the inverse direction of climatic conditions in the southern hemisphere! This is definitely a strong corroboration for the action of natural selection on gene arrangements in natural populations of this species.

How, then, can we explain the paradox of the absence of altitudinal clines in this species? It is possible that migration prevents differentiation in small distances. Probably a selective agent closely related to climate and latitude is responsible for the formation of these clines. Some ideas regarding the nature of this agent are presented below.

14.12. Is a Life-History Character the Target of Selection?

A quantitative character showing a similar cline to that of inversions is body size – larger animals are encountered in northern populations of *Drosophila*. This phenotypic cline is also a genetic one. Prevosti (1955) has shown that body size follows closely the July isotherm. In laboratory selection experiments for increased and decreased wing size (strongly correlated with body size), Prevosti (1967) found that a decrease in size was accompanied by a change of gene arrangement frequencies in favor of those found in southern populations, while in the line for an increased body size the presence of northern gene arrangements was preserved in addition to other arrangements. Measurements of body size in flies collected in nature by Krimbas (in Krimbas and Loukas, 1980) and identification of their gene arrangement partly confirmed this correlation between body size and inversions.

Partridge and Fowler (1993) recently proposed that, in *D. melanogaster*, large animals may live longer and be more fertile than smaller ones but they also have a longer period of development. In *D. subobscura*, larger males are more successful at matings. In southern climatic conditions, an extended larval period may be a liability as the larval medium deteriorates quickly. This does not seem to be the case in the north. Krimbas (1993b) suggested that the inversion polymorphism is primarily associated with the duration of the developmental period. Thus, many genes affecting the duration of larval development are linked in blocks and segregate in only five groups of units (chromosomes). This may permit an efficient and rapid change from slow to fast larval development in a few generations. Of course, secondarily, additional and different selective pressures may lead to the acquisition of other alleles related to independent aspects of the biology of the species and thus establish accessory systems that are targets to different but complementary selective agents.

We possess some evidence in support of this hypothesis. Selection experiments in *D. persimilis* for an increase and a decrease in developmental time changed the frequency of gene arrangements in the selected lines in agreement with the hypothesis presented here (Spiess and Spiess, 1964). Preliminary experiments in *D. subobscura* (Tzannidakis and Krimbas, unpublished data) show that the first pupating flies from a batch of eggs deposited the same day differ from the late pupating ones in their gene arrangements; the ST gene arrangements in all chromosomes (those predominating in northern Europe) are more frequent among the late pupating flies. Given the restricted sample size, the results were statistically significant for three out of five chromosomes (A, the sex chromosome, U, and O). Selection experiments for fast and slow lines of larval development are under way.

Thus selection for a life-history character may be able to explain the formation of geographical clines, which is the major spatial characteristic of gene arrangement polymorphism in *D. subobscura*.

14.13. Summary

Chromosomal inversions were discovered from their inhibitory effect on genetic recombination. Assuming a unique origin for every inversion, Sturtevant and Dobzhansky were able to construct their phylogeny by ordering most parsimoniously overlapping inversions. With two additional assumptions, absence of transpecific polymorphisms and of introgression (absence of homoplasy), species phylogeny is derived from inversions. DNA-based sequences near the breakpoints of inversions support the assumption of unique origin. Cases of reinversion, repeated origin of the same inversion, hot spots of breakpoints, and the generation of inversions by transposable elements, on the contrary, indicate the possibility of their rare but repeated origin. This contradiction is reconciled by the fact that maintenance of a newly originated inversion in a natural population is a very rare event, as mathematical models indicate.

Several indirect lines of evidence plead for the action of natural selection on inversions. The selective action is due to the retention of block of genes, not to position effects. In laboratory studies, Dobzhansky and collaborators shed some light on the genic content of inversions. Alternative gene arrangements from the same locality are coadapted. In laboratory experiments, larval competition is considerable, not so in nature. In nature, rare gene arrangements may be retained by sexual frequency-dependent selection. Geographic clines running mostly north to south (also seasonal cyclic changes and changes with altitude) may be explained as a product of life-history selection (for a shorter larval development), because in southern countries larval media deteriorate quickly, while in the North, on the contrary, a positive selection for greater adult size favors a longer larval period.

REFERENCES

Aguadé, M. 1988. Restriction map variation at the *Amy* locus of *Drosophila melanogaster* in inverted and noninverted chromosomes. *Genetics* 119:135–140.

Anderson, W. W., Levine, L., Olvera, O., Powell, J. R., de la Rosa, M. E., Salceda, V. M., Gaso, M. I., and Guzman, J. 1979. Evidence for selection by male mating success in natural populations of *Drosophila pseudoobscura*. *Proc. Natl. Acad. Sci. USA* 76:1519–1523.

Aquadro, C. F., Weaver, A. L., Schaeffer, S. W., and Anderson, W. W. 1991. Molecular evolution of inversions in *Drosophila pseudoobscura*: the amylase gene region. *Proc. Natl. Acad. Sci. USA* 88:305–309.

Babcock, C. S. and Anderson, W. W. 1996. Molecular evolution of sex-ratio inversion complex in *Drosophila pseudoobscura* analysis of the *esterase* gene region. *Mol. Biol. Evol.* 13:297–308.

Benassi, V., Aulard, S., Mazeau, S., and Veuille, M. 1993. Molecular variation of *Adh* and *P6* genes in an African population of *Drosophila melanogaster* and its relation to chromosomal inversions. *Genetics* 134:789–799.

Brehm, A. and Krimbas, C. B. 1993. The phylogeny of nine species of the *Drosophila obscura* group inferred by the banding homologies of the chromosomal regions. IV. Element C. *Heredity* 70:214–220.

Carson, H. L. 1992. Inversions in Hawaiian *Drosophila*. In *Drosophila Inversion Polymorphism*, C. B. Krimbas and J. R. Powell, eds., pp. 407–453. Boca Raton, FL: CRC.

Charlesworth, B. 1974. Inversion polymorphism in a two-locus genetic system. *Genet. Res.* 23:259–280 (correction, 1978, *Genet. Res.* 26:9).

Charlesworth, B. and Charlesworth, D. 1973. Selection of a new inversion in a multi-locus genetic system. *Genet. Res.* 21:167–183.

Cirera, S., Martin-Campos, J. M., Segarra, C., and Aguadé, M. 1995. Molecular characterization of the breakpoints of an inversion fixed between *Drosophila melanogaster* and *D. subobscura*. *Genetics* 139:321–326.

Deakin, M. A. B. and Teague, R. B. 1974. A generalized model of inversion polymorphism. *J. Theor. Biol.* 48:105–123.

Engels, W. R. and Preston, C. R. 1984. Formation of chromosome rearrangements by P factors in Drosophila. *Genetics* 10:657–678.

Ewens, W. 1972. The sampling theory of selectively neutral alleles. *Popul. Biol.* 3:87–112.

Fraser, A. and Burnell, D. 1967a. Simulation of genetic systems. XI. Inversion polymorphism. *Am. J. Hum. Genet.* 19:270–287.

Fraser, A. and Burnell, D. 1967b. Simulation of genetic systems. XII. Models of inversion polymorphism. *Genetics* 57:267–282.

Fraser, A., Burnell, D., and Miller, D. 1966. Simulation of genetic systems. X. Inversion polymorphism. *J. Theor. Biol.* 13:1–14.

Haldane, J. B. S. 1957. The conditions for coadaptation in polymorphism for inversions. *J. Genet.* 55:218–225.

Keith, T. P. 1983. Frequency distribution of esterase-5 alleles in two populations of *Drosophila pseudoobscura*. *Genetics* 105:135–155.

Krimbas, C. B. 1992. The inversion polymorphism of *Drosophila subobscura*. In *Drosophila Inversion Polymorphism*, C. B. Krimbas and J. R. Powell, eds., pp. 127–220. Boca Raton, FL: CRC.

Krimbas, C. B. 1993a. *Drosophila Subobscura, Biology, Genetics and Inversion Polymorphism.* Hamburg: Verlag Dr. Kovac.

Krimbas, C. B. 1993b. Natural experiments for studying selection in the wild. In *Abstracts Volume of the 17th International Congress of Genetics*, p. 61.

Krimbas, C. B. and Loukas, M. 1980. The inversion polymorphism of *Drosophila subobscura*. *Evol. Biol.* 12:163–234.

Krimbas, C. B. and Powell, J. R. 1992. Introduction. In *Drosophila Inversion Polymorphism*, C. B. Krimbas and J. R. Powell, eds., pp. 1–52. Boca Raton, FL: CRC.

Levitan, M. 1992. Chromosomal variation in *Drosophila robusta* Sturtevant. In *Drosophila Inversion Polymorphism*, C. B. Krimbas and J. R. Powell, eds., pp. 221–338. Boca Raton, FL: CRC.

Levitan, M. and Verdonck, M. 1986. 25 years of a unique chromosome-breakage system. I. Principal features and comparison to other systems. *Mutat. Res.* 161:135–142.

Lewontin, R. C., Moore, J. A., Provine, W. A., and Wallace, B. 1981. *Dobzhansky's Genetics of Natural Populations*, R. C. Lewontin, J. A. Moore, W. A. Provine, and B. Wallace, eds. New York: Columbia U. Press.

Menozzi, P. and Krimbas, C. B. 1992. The inversion polymorphism of *Drosophila subobscura* revisited: synthetic maps of gene arrangement frequencies. *J. Evol. Biol.* 5:625–641.

Moriyama, E. N. and Powell, J. R. 1996. Interspecific nuclear DNA variation in Drosophila. *Mol. Biol. Evol.* 13:261–277.

Nei, M., Kojima, K. I., and Schaffer, H. E. 1967. Frequency changes of new inversions in populations under mutation–selection equilibria. *Genetics* 57:741–750.

Ohta, T. and Kojima, K. I. 1968. Survival probabilities of new inversions in large populations. *Biometrics* 24: 501–516.

Partridge, L. and Fowler, K. 1993. Responses and correlated responses to artificial selection on thorax length in *Drosophila melanogaster. Evolution* **47**:213–226.

Petit, C. 1958. Le déterminisme génétique et psycho-physiologique de la compétition sexuelle chez *Drosophila melanogaster. Bull. Biol. France Belgique* **92**:248–329.

Petit, C. and Ehrman, L. 1969. Sexual selection in *Drosophila. Evol. Biol.* **3**:177–223.

Popadic, A. and Anderson, W. W. 1994. The history of a genetic system. *Proc. Natl. Acad. Sci. USA* **91**:6819–6823.

Popadic, A., Popadic, D., and Anderson, W. W. 1995a. Interchromosomal transfer of genetic information between gene arrangements on the third chromosome of *Drosophila pseudoobscura. Mol. Biol. Evol.* **12**:938–943.

Popadic, A. and Anderson, W. W. 1995b. Evidence for gene conversion in the amylase multigene family of *Drosophila pseudoobscura. Mol. Biol. Evol.* **12**:564–572.

Powell, J. R. 1992. Inversion polymorphism in *Drosophila pseudoobscura* and *Drosophila persimilis.* In *Drosophila Inversion Polymorphism,* C. B. Krimbas and J. R. Powell, eds., pp. 73–126. Boca Raton, FL: CRC.

Prakash, S. and Lewontin, R. C. 1968. A molecular approach to the study of genic heterozygosity. III. Direct evidence of co-adaptation in gene arrangements of *Drosophila. Proc. Natl. Acad. Sci. USA* **59**:398–405.

Prakash, S. and Lewontin, R. C. 1971. A molecular approach to the study of genic heterozygosity. V. Further evidence of co-adaptation in inversions of Drosophila. *Genetics* **69**:405–408.

Prakash, S. and Merritt, R. B. 1972. Direct evidence of genic differentiation between sex ratio and standard gene arrangements of X-chromosome in *Drosophila pseudoobscura. Genetics* **72**:169–175.

Prevosti, A. 1955. Geographical variability in quantitative traits in populations of *Drosophila subobscura. Cold Spring Harbor Symp. Quant. Biol.* **20**:294–298.

Prevosti, A. 1967. Inversion heterozygosity and selection for wing length in *Drosophila subobscura. Genet. Res.* **10**:81–93.

Rodriguez-Trelles, F., Alvarez, G., and Zapata, C. 1996. Time-series analysis of seasonal changes of the O inversion polymorphism of *Drosophila subobscura. Genetics* **142**:179–187.

Rozas, J. and Aguadé, M. 1993. Transfer of genetic information in the *rp49* region of *Drosophila subobscura* between different chromosomal gene arrangements. *Proc. Natl. Acad. Sci. USA* **90**:8083–8087.

Rozas, J. and Aguadé, M. 1994. Gene conversion is involved in the transfer of genetic information between naturally occurring inversions of *Drosophila. Proc. Natl. Acad. Sci. USA* **91**:11517–11521.

Salceda, V. M. and Anderson, W. W. 1988. Rare male mating advantage in a natural population of *Drosophila pseudoobscura. Proc. Natl. Acad. Sci. USA* **85**:9870–9874.

Sperlich, D. 1963. Experimentelle Beitrage zum Problem des positiven Heterosisefektes bei der Struktupolymorphen Art *Drosophila subobscura. Z. Vererbungs.* **90**:273–287.

Sperlich, D. 1966. Equilibria for inversions induced by x-rays strains of *Drosophila pseudoobscura. Genetics* **53**:835–842.

Spiess, E. B. 1968. Low frequency advantage in mating of *Drosophila pseudoobscura karyotypes. Am. Nat.* **102**:363–379.

Spiess, E. B. and Spiess, L. D. 1964. Selection for rate of development and gene arrangement frequencies in *Drosophila persimilis. Genetics* **50**:863–877.

Strickberger, M. W. 1963. Evolution of fitness in experimental populations of *Drosophila pseudoobscura*. *Evolution* **17**:40–55.

Sturtevant, A. H. and Dobzhansky, Th. 1936. Inversions in the third chromosome of wild race of *Drosophila pseudoobscura*, and their use in the study of the history of the species. *Proc. Natl. Acad. Sci. USA* **22**:448–450.

Taylor, C. E. and Powell, J. R. 1977. Microgeographic differentiation of chromosomal and enzyme polymorphisms in *Drosophila persimilis*. *Genetics* **85**:681–695.

Taylor, C. E. and Powell, J. R. 1978. Habitat choice in natural populations of *Drosophila*. *Oecologia* **37**:69–75.

Teague, R. B. and Deakin, M. A. 1976. Inversion polymorphism: character and stability of equilibria. *J. Theor. Biol.* **56**:75–94.

Vann, E. 1966. The fate of x-ray induced chromosomal rearrangements introduced into laboratory populations of *Drosophila melanogaster*. *Am. Nat.* **100**:425–449.

Wallace, B. 1966. Natural and radiation-induced chromosomal polymorphism in *Drosophila*. *Mutat. Res.* **3**:194–200.

Wasserman, M. 1992. Cytological evolution of the *Drosophila repleta* species group. In *Drosophila Inversion Polymorphism*, C. B. Krimbas and J. R Powell, eds., pp. 455–552. Boca Raton, FL: CRC.

Wesley, C. S. and Eanes, W. F. 1994. Isolation and analysis of the breakpoint sequences of chromosome inversion In *(3L) Payne* in *Drosophila melanogaster*. *Proc. Natl. Acad. Sci. USA* **91**:3132–3136.

CHAPTER FIFTEEN

The Evolution of Breeding Systems

MARCY K. UYENOYAMA

Few events have evolutionary consequences as pervasive as changes in the mating system. Changes in effective population size and the opportunity for recombination transform prospects for evolution across the entire genome. Such events influence not only the rate but the very nature of evolution, particularly with respect to the consequences of deleterious mutation. In this section, I address the origin and evolutionary consequences of the expression of self-incompatibility (SI) in flowering plants.

Perhaps half of all species of flowering plants express some form of genetically determined SI, which prevents fertilization by pollen produced by the same plant (de Nettancourt, 1977). In many cases, SI entirely eliminates seed set under self-pollination. A rich classical tradition has addressed the genetic basis of this unambiguous phenotype. The past decade has witnessed much progress in the characterization at the molecular level of the structure and function of genes (S-loci) that cosegregate with the expression of several major forms of SI (Nasrallah et al., 1985; Anderson et al., 1986; Foote et al., 1994; Li et al., 1994; Broothaerts et al., 1995; Sassa et al., 1996; Xue et al., 1996). Such developments have precipitated a revolution in the study of all aspects of SI, from physiological mechanism to evolutionary history (reviewed by Clark and Kao, 1994; Dodds et al., 1996). This revolution has indeed overthrown some orthodoxies, but more significantly it has fundamentally transformed the nature of the investigation, both technically, in establishing a new evidentiary basis, and conceptually, in inspiring new questions.

This discussion addresses two evolutionary issues: the origins of SI and the tempo and mode of S-locus evolution at the molecular level. At what point in the evolutionary history of plants did SI arise? Do the various extant systems of SI descend from a single origin or multiple origins? Central to questions concerning the nature of the evolutionary process is the rate of substitution at S-loci. Does hypermutability contribute to the extraordinary levels of polymorphism

R. S. Singh and C. B. Krimbas, eds., *Evolutionary Genetics: From Molecules to Morphology*, vol. 1. © Cambridge University Press 2000. Printed in the United States of America. ISBN 0-521-57123-5. All rights reserved.

characteristic of *S*-loci? How do ecological and genetic factors influence substitution rate? What consequences does the expression of SI bear for its own evolution and evolution throughout the genome? Recent developments in the molecular-level analysis of *S*-loci have begun to permit the exploration of such evolutionary questions with unprecedented levels of sensitivity and historical depth.

15.1. Origins of Homomorphic Self-Incompatibility

15.1.1. Multiple Independent Derivations

In a number of mating systems, compatibility depends on mating type, with reproduction permitted between but not within types. For example, the vertebrate sexual system comprises two mating types, distinguished by the kind of gamete produced. Like other compatibility systems, SI in angiosperms imposes restrictions on mating beyond the avoidance of self-fertilization.

Variation in flower morphology distinguishes heteromorphic from homomorphic SI, the two best-studied systems of SI in angiosperms. Under heteromorphic SI, mating types differ with respect to flower morphology, while under homomorphic SI, different mating types express and recognize different specificities through similar floral structures. Within homomorphic systems, pollen may express specificities encoded by genes carried by the pollen itself [gametophytic self-incompatibility (GSI)] or by the parent plant that produced the pollen [sporophytic self-incompatibility (SSI)]. Recent characterizations at the molecular level of loci that control the recognition of pollen specificities under both GSI and SSI have revealed a wealth of information concerning the origins of these two major forms of homomorphic SI.

Little correspondence is apparent between the incidence of homomorphic SI and phylogenetic relationships among the flowering plants (see, for example, Stebbins, 1950). This absence of phylogenetic pattern admits diametrically opposed interpretations: that SI derives from a single origin coincident with or before the origin of the flowering plants or that SI derives from multiple, independent origins. The classical view regarded the interpretation of monophyly as more parsimonious and held that all SI systems were elaborated from an ancestral GSI system (Whitehouse, 1951; de Nettancourt, 1977; Pandey, 1960).

In contrast, recent analyses of the physiological mechanisms of pollen rejection and the molecular structure of loci controlling pollen recognition suggest multiple, independent origins. No homology in sequence is apparent between the locus that controls SSI in *Brassica* (Nasrallah et al., 1985) and the loci that control GSI in the Solanaceae, Rosaceae, and Scrophulariaceae (Anderson et al., 1986; Sassa et al., 1992; Broothaerts et al., 1995; Xue et al., 1996), in field poppies (Foote et al., 1994), or in a grass (Li et al., 1994). Further, the *S*-loci in these three systems of GSI show little similarity in sequence among themselves, and GSI in lilies differs with respect to both genetic control (Lundqvist, 1991)

302 Marcy K. Uyenoyama

and physiological expression (Tezuka et al., 1993). SSI systems also appear to have derived from multiple origins. Distinct physiological mechanisms mediate pollen rejection in sunflowers and in *Brassica* (Sarker et al., 1988; Elleman et al., 1992). *Ipomoea trifida* (Convolvulaceae) expresses both SSI and stigmatic proteins encoded by loci that show high homology to the *Brassica* S-locus; however, the genes characterized to date do not cosegregate with SI (Kowyama et al., 1995). These observations suggest at least four independent derivations of GSI and two or three of SSI.

15.1.2. Times of Origin and Evolutionary Rates

In addition to providing evidence of nonhomologous origin, sequence analysis has revealed probable homology. Ribonuclease (RNase) expression directly mediates pollen rejection in the form of GSI expressed in the Solanaceae (McClure et al., 1989, 1990; Murfett et al., 1994; Lee et al., 1994; Huang et al., 1994). Transspecific evolution describes the phenomenon of higher sequence similarity in interspecific than intraspecific comparisons (Arden and Klein, 1982). Observation of this pattern among the RNase-related S-alleles (S-RNases) of the Solanaceae suggests an age for this form of GSI in excess of 28 to 30 million years (MY) (Ioerger et al., 1990).

The S-loci that control GSI in the Rosaceae (Broothaerts et al., 1995; Sassa et al., 1996) and in the Scrophulariaceae (Xue et al., 1996) also belong to this ancient superfamily of RNases. While transspecific sharing of S-RNase lineages is typical within the Solanaceae (Ioerger et al., 1990; Richman et al., 1995), transfamilial sharing across the Solanaceae, Rosaceae, and Scrophulariaceae has not been observed (Sassa et al., 1996; Xue et al., 1996). Absence of transfamilial sharing could reflect multiple independent recruitments of S-loci from the same RNase superfamily. A perhaps more probable explanation is that sufficient time has passed since divergence of the plant families to permit coalescence of S-allele lineages within families. Barring multiple recruitments of RNase genes to S-loci, the similarity in genetic structure suggests homology of GSI expression in these distantly related families. This interpretation would place the origin of RNase-mediated GSI before the divergence of subclasses Asteridae and Rosidae, which is believed to have occurred at the beginning of the Tertiary period, 65 MY ago (Cronquist, 1981, p. 854).

SSI in *Brassica* appears to have originated more recently, although transspecific sharing indicates that it arose before divergence within the genus (Dwyer et al., 1991). Joint analysis of rates of amino acid substitution in S-alleles and in homologous genes that do not participate in SI indicates that the age of this form of SSI may exceed the time since divergence between *Brassica* species by a factor of 4 to 5 (Uyenoyama, 1995). Dating the origin of this form of SSI in excess of 50 MY ago would imply that amino acid substitution occurs within the S-locus at a rate comparable with or even slower than other members of the multigene family. In contrast, assuming that this form of SSI originated within the Brassicaceae would imply hypermutability of the S-locus. (Trick and Heizmann, 1992).

15.2. Effects of Balancing Selection

By strictly enforcing heterozygosity, SI imposes an extreme form of balancing selection on the S-locus. Balancing selection of this intensity profoundly affects the tempo and mode of evolutionary change. I briefly summarize some evolutionary studies of the vertebrate major histocompatibility complex (MHC), for which the theoretical and statistical framework for the analysis of balanced polymorphisms is best developed.

15.2.1. Symmetric Overdominance in Viability

An extensive body of evidence supports the operation of balancing selection at MHC loci. Parham and Ohta (1996) and Takahata (1996) have reviewed empirical and theoretical studies of MHC variation. The extraordinarily high levels of allelic diversity (perhaps several hundred worldwide at the DRB locus in humans) reflect not hypermutability, but rather the maintenance of allelic lineages over very long periods of time (Satta et al., 1991, 1993). Class I and Class II MHC loci typically exhibit transspecific polymorphism (Figueroa et al., 1988; Lawlor et al., 1988), indicating the divergence of allelic lineages before the divergence of the taxa that carry them. Positive selection favoring allelic diversity is expected to accelerate the rate of incorporation of new mutations at specificity-determining sites (Maruyama and Nei, 1981; Takahata, 1990; Sasaki, 1992). Among the most striking empirical confirmations of a theoretical prediction is the excess in Class I and Class II loci of nonsynonymous over synonymous differences at sites that function in antigen presentation (Hughes and Nei, 1988, 1989).

15.2.2. Theoretical Framework for Balanced Polymorphisms

Perhaps the most significant development in the transformation of population genetics from the classical to the molecular levels is a change in perspective from currently segregating variation to phylogenetic history. Classical population genetics classifies genes into qualitatively defined allelic groups and assesses the number of variants and their relative frequencies. Molecular population genetics uses the extent of sequence divergence as a measure of quantitative levels of differences among genes. The theory of neutral evolution (Kimura, 1968) predicts a correlation between divergence in sequence and divergence in time. By permitting the inference of divergence times from sequence information, this theoretical framework establishes a basis for the reconstruction of evolutionary history (Nei, 1976).

Central to this phylogenetic perspective is the concept of coalescence [Tavaré (1984) and Hudson (1990) provide lucid accounts of this process]. Viewed under a reversal of time, the stochastic evolution of neutral genes is characterized by a progressive decline in the number of independent genetic lineages (Kingman, 1982). Each successive generation back in time is formed by the random assignment of genes to their ancestors, with the probability of coalescence (common ancestry) between genes determined by effective population size. This backward projection continues until the ultimate coalescence of all extant lineages into their most recent common ancestor (MRCA). For sufficiently large populations, the distribution of times between successive coalescence events is approximately exponential.

Among the most creative uses of the coalescence perspective is the reconstruction of the demographic history of populations during the period extending back to the MRCA of currently segregating allelic lineages (see Takahata, 1993). This approach introduces a third dimension to the well-established dichotomy of using genetic variation to infer the ancestry of organisms (molecular systematics) or using historical information about taxa to infer the process of genetic evolution (population genetics). Griffiths and Tavaré (1994) summarize their methods for using currently segregating neutral variation to form maximum likelihood estimates of substitution rate, time since the MRCA, and population size.

A key insight due to Takahata (1990) is that the process of coalescence of genes under symmetrically overdominant viability selection (all heterozygotes having identical viability and all homozygotes having identical reduced viability) also conforms to an exponential distribution, but on a vastly longer time scale relative to neutrality. This property permits development of a theoretical framework for the analysis of balanced polymorphism by extension of the well-developed theory of neutral evolution. Because balancing selection greatly extends the phylogenies of lineages back in time, the analysis of MHC polymorphisms permits reconstruction of demographic history over several millions of years (Takahata et al., 1992; Klein et al., 1993; Takahata, 1996).

Richman and Kohn (1997) describe an application of this approach to S-locus evolution. Vekemans and Slatkin's (1994) modification of Takahata's (1990) theoretical analysis of overdominant viability selection showed that SI generates balancing selection of an intensity beyond lethal homozygosity, the strongest form of overdominant viability selection. As a consequence, lineages of S-alleles as well as MHC alleles are expected to persist over exceedingly long periods periods of time. Relative to this time scale, the number of S-alleles converges rapidly to a level determined by the current population size and the rate of origin of new S-alleles. Because allele numbers and divergence times equilibrate on different time scales, historical fluctuations in population size can generate discrepancies between the population sizes inferred from these two aspects (see Takahata, 1993).

15.3. S-Locus Variation in Natural Populations

Richman and co-workers (Richman et al., 1995, 1996a, 1996b) surveyed the S-RNase variation in natural populations of two solanaceous species, *Solanum carolinense* and *Physalis crassifolia*. In comparison with natural populations that express other forms of GSI (for example, O'Donnell and Lawrence, 1984; Levin, 1993), the *S. carolinense* sites, in Tennessee and North Carolina, appeared to harbor unusually low numbers of S-alleles (13 to 15). Significantly more S-alleles (43 to 45) appeared to be segregating in *P. crassifolia* at a single California desert site. A phylogenetic analysis of nucleotide sequences indicated extensive transspecific sharing of S-allele lineages between *S. carolinense* and other solanaceous species; in contrast, the *P. crassifolia* S-alleles formed only two clusters (see Fig. 15.1). Generalized least-squares estimates of branch lengths within S-allele genealogies indicated that the sum of the terminal branch lengths was significantly lower for the *P. crassifolia* sequences.

Theoretical analyses of SI and overdominant viability selection (Wright, 1939; Yokoyama and Nei, 1979; Maruyama and Nei, 1981; Takahata, 1990; Takahata and Nei, 1990; Sasaki, 1992; Vekemans and Slatkin, 1994) provide a basis for the development of hypotheses to account for these strikingly different patterns. Comparison of the species surveyed suggests a negative correlation between allele number and sequence divergence: relatively few S-allele lineages of ancient divergence in *S. carolinense* and many lineages of more recent divergence in *P. crassifolia*. Higher effective population size would permit both the maintenance of more S-alleles and the persistence of lineages over longer periods of time. Consequently, differences in effective population size alone would be expected to generate a *positive* relationship between allele number and divergence time. The negative relationship observed would be more consistent with differences in the rate of origin of new S-alleles. At dynamic equilibrium, the incorporation of a new S-allele is balanced on average by the loss of an existing S-allele. As a result, higher origination rates accelerate turnover of allelic lineages in addition to promoting the segregation of more S-alleles.

Some rough calculations provide some indication of the relative contributions of differences in population size and origination rate to the contrast between the two solanaceous species. Fu and Li (1993) have shown that the expectation, over all topological relationships, of the time in the terminal branches of the phylogeny of a sample of neutral alleles is proportional to effective population size and independent of allele number. We may extend this result for neutral genealogies to S-alleles by appealing to Takahata's (1990) finding of the similarity between the processes of coalescence under neutrality and overdominant selection. This extension would suggest that the expectation of the sum of the terminal branch lengths over all possible genealogical relationships among a sample of k S-alleles is approximately $4Nf$, irrespective of the value of k, for N, the effective population size, and f, the scaling factor under SI (given by Vekemans and Slatkin, 1994).

Our surveys provide estimates for the ratios of the sums of the terminal branch lengths in *S. carolinense* and *P. crassifolia* S-allele genealogies. The empirical observations provide an indication, not of divergence time (L) itself, but of the number of substitutions in the terminal branches (L^*). Under the assumption that every S-allele generated by mutation is novel (infinite-alleles model), these two quantities are linearly related: $L^* = L\xi$, for ξ, the overall rate of substitution. Assuming further that the two species share a common value of ξ, our estimates of the sums of the terminal branch lengths in the *S. carolinense* (S) and *P. crassifolia* (P) genealogies suggest that $L_P^*/L_S^* = L_P/L_S = 0.448$. Setting

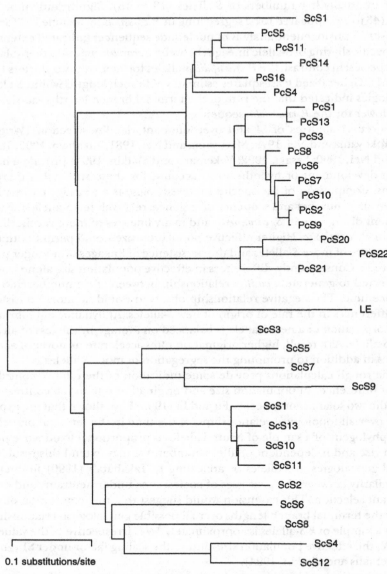

Figure 15.1. Phylogenetic relationships among *S*-alleles segregating within natural populations of *P. crassifolia* (upper figure) and *S. carolinense* (lower figure). To determine the position of the root, an *S. carolinense S*-allele (ScS1) was included in the *P. crassifolia* tree and an *S*-like RNase from *Lycopersicon esculentum* (LE) in the *S. carolinense* tree. Branch lengths represent generalized least-squares estimates of the numbers of nucleotide substitutions per site based on observed pairwise nucleotide differences corrected by the Kimura (1980) two-parameter method. The sum of the terminal branch lengths is significantly smaller in the *P. crassifolia* tree.

this value equal to $4N_P f_P/(4N_S f_S)$ gives $N_P\mu_P/(N_S\mu_S) = 96.1$ for sufficiently low rates of origination of S-alleles (μ_P and μ_S). Yokoyama and Hetherington (1982) determined the expected value of homozygosity (J) as a function of population size and origination rate. Using our estimates of the numbers of segregating S-alleles as effective allele numbers (Kimura and Crow, 1964) gives $J_P = 1/43$ and $J_S = 1/13$. These values permit separation of N and μ:

$$\frac{N_P}{N_S} = 3.8, \qquad \frac{\mu_P}{\mu_S} = 25.5.$$

These rough calculations suggest that the distinct patterns of variation observed in the two solanaceous species may primarily reflect differences in origination rate. This inference is consistent with the overall negative relationship observed between sequence divergence and allele number.

15.4. Hypotheses for Contrasting Patterns of S-Allele Variation

Results of empirical surveys indicate that the S-loci in the two solanaceous species have been subject to different evolutionary processes. With the aim of developing hypotheses to guide direct investigations, I consider some ecological and genetic factors that may have contributed to the differences observed.

15.4.1. Demographic History

Richman et al. (1996b) suggested that the *P. crassifolia* population surveyed may have experienced a bottleneck in population size. A higher S-allele number in *P. crassifolia* may reflect higher population size over the short term, the time scale over which an allele number equilibrates. In contrast, the apparently deeper divergence among *S. carolinense* S-allele lineages may reflect higher population size over the long term. This pattern is consistent with loss of S-allele variation during a bottleneck in the *P. crassifolia* population followed by regeneration during a phase of expansion.

15.4.2. Intragenic Recombination

Parham and Ohta (1996) have reviewed evidence that supports the occurrence of intragenic recombination or gene conversion within the MHC. Takahata and Satta (1998) have conducted a numerical simulation study and an analysis of nucleotide sequences to address whether recombination can account for the maintenance in humans of very large numbers of MHC alleles, particularly at the Class I B and Class II DRB loci. Ohta (personal communication) has suggested intragenic recombination as an alternative explanation for the observation of higher numbers of closely related alleles in *P. crassifolia*.

Figure 15.2 illustrates the effect of recombination on the genealogical structure among S-alleles observed in a numerical simulation study. The processes

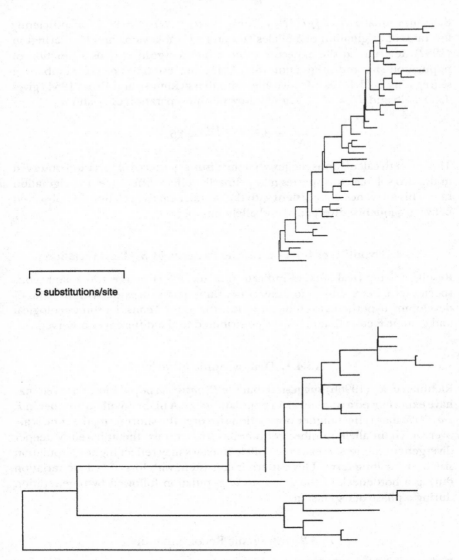

Figure 15.2. Results of a numerical simulation designed to explore the effects of intragenic recombination on the depth and topology of S-allele genealogies. Intragenic recombination occurred at the same rate (2.5×10^{-6} per gamete per generation) as point mutation in the upper panel and was absent in the lower panel. Individuals were explicitly represented as diploids in the simulation, with mating pairs formed by randomly sampling diploid genotypes. Gametes were sampled after point mutation and intragenic recombination within each parent. The generation of new S-allele specificities in maternal parents by either point mutation or intragenic recombination did not affect stigmatic rejection of pollen; however, new specificities carried by pollen were immediately expressed. Sampling of incompatible pollen resulted in the replacement of the paternal parent by another randomly sampled diploid individual.

of gametophytic SI, point mutation at specificity-determining and other sites within the S-locus, intragenic recombination, and drift in a population of 1000 diploid individuals were simulated. For a given rate of point mutation to new S-alleles, divergence times and numbers of substitutions between S-alleles were monitored in the absence of intragenic recombination and in its presence at a rate identical to the mutation rate. In the absence of recombination, all sites share a common genealogy, but in its presence, divergence times vary among sites. In order to provide a basis for comparison, the topologies and branch lengths shown in Fig. 15.2 were reconstructed from the observed numbers of substitutions by the neighbor-joining method (Saitou and Nei, 1987), even though the actual divergence times among the alleles were known for the case without recombination. In the absence of intragenic recombination, specificity-determining mutations accumulated at a rate nearly 100 times higher than the point mutation rate (2.1×10^{-4} per generation, or 75 mutations in the 359,682 generations in the entire history of all 24 segregating S-alleles). Shorter divergence times and considerable reductions in the average number of pairwise differences (from 14.3 to 4.8) were observed under intragenic recombination.

Intragenic recombination tends to elevate S-allele number (from 24 to 38 in Fig. 15.2) and to reduce substantially the number of segregating sites (from 75 to 13 in Fig. 15.2). Recombination permits the formation of more S-alleles from fewer point mutations. The most striking effect of recombination is the loss of deep divergences among allelic lineages. S-allele genealogies incorporating recombination are virtually indistinguishable from neutral genealogies with respect to Tajima's (1989) D statistic.

Whether recombination occurs within S-loci remains unresolved. Clark and Kao's (1991) analysis of S-alleles from various solanaceous species revealed no evidence of recombination, although high sequence divergence likely reduced the power of the tests. Coleman and Kao's (1992) observation in *Petunia* S-RNases of even greater sequence divergence in flanking, noncoding regions than in coding regions suggests tight linkage.

Invoking intragenic recombination to account for the contrasting patterns we observed in the two solanaceous species would entail postulating higher rates in *P. crassifolia*. Unless the rate of recombination is at least comparable with the rate of point mutation, allele number increases only moderately. Reduction in sequence divergence among lineages is the most striking qualitative consequence of intragenic recombination. While *P. crassifolia* S-alleles show lower sequence divergence on average, the two major clusters are nevertheless highly diverged (see Fig. 15.1). A significant excess of nonsynonymous over synonymous substitutions was observed only in within-cluster comparisons of *P. crassifolia* S-alleles (Richman et al., 1996a). The absence of significant excess in between-cluster comparisons of *P. crassifolia* S-alleles and among *S. carolinense* S-alleles is consistent with ancient divergence.

Whether intragenic recombination provides a plausible explanation for the contrast between the patterns of S-allele variation observed in *S. carolinense* and *P. crassifolia* remains equivocal. Frequent generation of new *S*-alleles by any mechanism would likely compromise their very function of preventing self-fertilization. I explore possible consequences of the partial breakdown of SI in the next subsection.

15.4.3. Deleterious Mutation

Foremost among the selective forces invoked to explain the origin and maintenance of SI in natural populations is inbreeding depression. Experimentally induced self-fertilization caused more severe reductions in the quality of offspring produced by plants that were derived from natural populations with a history of greater rates of outcrossing or SI (Holtsford and Ellstrand, 1990; Barrett and Kohn, 1991). While the forms of balancing selection generated by overdominance in viability and SI are similar in a number of respects, the processes by which new specificities originate may differ. Unlike overdominance in viability, the expression by a pollen grain of an *S*-allele specificity that is not recognized by the plant that produced it would compromise the very function of the system. This breach of SI may precipitate an abrupt transition from the soft selection of prezygotic rejection of incompatible gametes to the hard selection of postzygotic inbreeding depression.

A new *S*-allele that succeeds in reaching the stigmas of unrelated plants benefits from the advantage of rarity that maintains high diversity at the *S*-locus. A pollen grain that falls on a stigma of the plant that produced it would be lost whether it expresses a specificity that induces rejection or escapes recognition but forms an inviable zygote. It is the original *S*-alleles carried by the parent plant that suffer from the inbreeding depression released by the expression of a new *S*-allele specificity.

Consideration of this evolutionary cost (which may be considerable if the mechanism of pollen dispersal entails the receipt of much self-pollen) suggests that the successful incorporation of a new *S*-allele into the population may result in the replacement of the parent *S*-allele from which it descends more often than other *S*-alleles in the population. Such an event would change the specificity of an *S*-allele lineage without causing a bifurcation (see Takahata et al., 1992). This process may both promote longer persistence times of *S*-allele lineages and reduce the number of segregating *S*-alleles.

A preliminary numerical simulation study of the effects of the release of inbreeding depression on divergence times and numbers of *S*-allele lineages provides some support for this conjecture. Inbreeding depression was incorporated as a form of lineage-dependent selection, under which the viability of a zygote declines with shorter divergence time between its constituent *S*-alleles. Lineage-dependent selection tended to increase the maximum divergence time among segregating *S*-alleles. Further, the number of specificity-determining mutations accumulated within lineages increased relative to the number of segregating *S*-alleles, reflecting a higher rate of within-lineage relative to between-lineage turnovers.

Mutational load may engender its greatest evolutionary consequences in conjunction with bottlenecks in population size. Population bottlenecks reduce heterozygosity across the genome. At the S-locus, loss of variation promotes the incorporation of new S-alleles. At loci affecting viability, increased homozygosity permits the fixation of deleterious as well as benign mutations. As a result, the relative reduction in viability associated with inbreeding may decline. A reduction in lineage-dependent selection together with positive selection favoring restoration of S-allele variation may promote the incorporation of new S-alleles.

A numerical study simulating the evolution of the S-locus in absolute linkage with 100 loci affecting viability indicated that population bottlenecks can cause the number of S-alleles to rise beyond initial levels, even in the absence of differences between prebottleneck and postbottleneck population sizes. This study suggested that a negative relationship between allele number and divergence time (as observed between $S.$ $carolinense$ and $P.$ $crassifolia$) may reflect an interaction between demographic history and mutational load; in particular, bottlenecks in population size alone do not generate significant negative correlations.

15.5. Conclusions

This discussion serves to illustrate the deep restructuring of population genetics instigated by the incorporation of a phylogenetic perspective. This development mirrors the transformation of the entire discipline of evolutionary genetics, from its origins in quantitative genetics at the phenotypic level, through allozyme studies of population genetics, to the development of molecular evolutionary genetics. It is an extraordinary testament to the scientific vision of R. C. Lewontin that at each of these recent junctures in the development of the discipline he has been a major figure in the shaping of the next epoch.

In his landmark volume, Lewontin (1974, Chap. 1) described the phenotypically based biometric approach to evolution as dynamically insufficient. It is genetic lineages that maintain continuity across generations; phenotypic lineages do not in fact exist. As a consequence, purely phenotypic descriptions of evolutionary dynamics are necessarily incomplete. By virtue of its mechanistic orientation, theoretical population genetics would appear to provide a self-consistent description of evolutionary change. However, Lewontin (1974, Chap. 5) enumerated three reasons for its failure in this respect: insufficient accuracy of estimates of key parameters; intractability of dynamical descriptions at the level of entire genomes, the true unit of selection; and its ahistoricity, in fundamental conflict with the very nature of the evolutionary process.

More than 20 years later, formidable obstacles to the development of a unified theory of evolution remain; yet the transformation of population genetics into molecular evolutionary genetics permits a growing optimism. Qualitative comparisons among regions or kinds of nucleotide substitutions can obviate the need for quantitative estimates of particular parameters (for example: Hughes and Nei, 1988, 1989; McDonald and Kreitman, 1991; Ohta, 1995; Akashi, 1995).

Close linkage and disequilibrium among sites no longer serve only to obscure the evolutionary process; rather, the levels of monomorphism as well as polymorphism at selectively neutral sites can provide sensitive indicators of processes operating in neighboring regions (for example, Berry et al., 1991; Begun and Aquadro, 1992; Braverman et al., 1995). Perhaps the most significant aspect of the emergence of molecular evolutionary genetics is its phylogenetic perspective (Nei, 1976). Molecular population genetics grows increasingly triumphant in exploiting currently segregating variation to reveal the mechanistic basis of evolutionary change on an historical scale of truly evolutionary proportions.

15.6. Acknowledgments

This study was supported in part by the Japan Society for the Promotion of Science and the National Institutes of Health (PHS grant GM-37841). I thank I. Franklin for perceptive comments on the manuscript and N. Takahata for clarity of thought as well as generous financial support.

REFERENCES

Akashi, H. 1995. Inferring weak selection from patterns of polymorphism and divergence at 'silent' sites in *Drosophila melanogaster*. *Genetics* **122**:607–615.

Anderson, M. A., Cornish, E. C., Mau, S.-L., Williams, E. G., Hoggart, R., et al. 1986. Cloning of cDNA for a stylar glycoprotein associated with expression of self-incompatibility in *Nicotiana alata*. *Nature (London)* **321**:38–44.

Arden, B. and Klein, J. 1982. Biochemical comparison of major histocompatibility complex molecules from different subspecies of *Mus musculus*: evidence for trans-specific evolution of alleles. *Proc. Natl. Acad. Sci. USA* **79**:2342–2346.

Barrett, S. C. H. and Kohn, J. R. 1991. Genetic and evolutionary consequences of small population size in plants: implications for conservation. In *Conservation of Rare Plants: Biology and Genetics*, D. A. Falk and K. E. Holsinger, eds., pp. 3–30. Oxford: Oxford U. Press.

Begun, D. J. and Aquadro, C. F. 1992. Levels of naturally occurring DNA polymorphism correlate with recombination rates in *D. melanogaster*. *Nature (London)* **356**:519–520.

Berry, A. J., Ajioka, J. W., and Kreitman, M. 1991. Lack of polymorphism on the *Drosophila* fourth chromosome resulting from selection. *Genetics* **129**:1111–1117.

Braverman, J. M., Hudson, R. R., Kaplan, N. L., Langley, C. H., and Stephan, W. 1995. The hitchhiking effect on the site frequency spectrum of DNA polymorphism. *Genetics* **140**:783–795.

Broothaerts, W., Janssens, G. A., Proost, P., and Broekaert, W. F. 1995. cDNA cloning and molecular analysis of two self-incompatibility alleles from apple. *Plant Mol. Biol.* **27**:499–511.

Clark, A. G. and Kao, T.-h. 1991. Excess nonsynonymous substitution at shared polymorphic sites among self-incompatibility alleles of Solanaceae. *Proc. Natl. Acad. Sci. USA* **88**:9823–9827.

Clark, A. G. and Kao, T.-h. 1994. Self-incompatibility: theoretical concepts and evolution. In *Genetic Control of Self-Incompatibility and Reproductive Development in Flowering*

Plants, E. G. Williams, A. E. Clarke, and R. B. Knox, eds., pp. 220–241. Boston: Kluwer Academic.

Coleman, C. E. and Kao, T.-h. 1992. The flanking regions of two *Petunia inflata* S-alleles are heterogeneous and contain repetitive sequences. *Plant Mol. Biol.* 18:725–737.

Cronquist, A. 1981. *An Integrated System of Classification of Flowering Plants*. New York: Columbia U. Press.

Dodds, P. N., Clarke, A. E., and Newbigin, E. 1996. A molecular perspective on pollination in flowering plants. *Cell* 85:141–144.

Dwyer, K. G., Balent, M. A., Nasrallah, J. B., and Nasrallah, M. E. 1991. DNA sequences of self-incompatibility genes from *Brassica campestris* and *B. oleracea*: polymorphism predating speciation. *Plant Mol. Biol.* 16:481–486.

Elleman, C. J., Franklin-Tong, V., and Dickinson, H. G. 1992. Pollination in species with dry stigmas: the nature of the early stigmatic response and the pathway taken by pollen tubes. *New Phytol.* 121:413–424.

Figueroa, F., Günther, E., and Klein, J. 1988. MHC polymorphism predating speciation. *Nature (London)* 256:265–267.

Foote, H. C. C., Ride, J. P., Franklin-Tong, V. E., Walker, E. A., Lawrence, M. J., et al. 1994. Cloning and expression of a distinctive class of self-incompatibility (S) gene from *Papaver rhoeas* L. *Proc. Natl. Acad. Sci. USA* 91:2265–2269.

Fu, Y.-X. and Li, W.-H. 1993. Statistical tests of neutrality of mutations. *Genetics* 133:693–709.

Griffiths, R. C. and Tavaré, S. 1994. Ancestral inference in population genetics. *Stat. Sci.* 9:307–319.

Holtsford, T. P. and Ellstrand, N. C. 1990. Inbreeding effects in *Clarkia tembloriensis* (Onagraceae) populations with different natural outcrossing rates. *Evolution* 44:2031–2046.

Huang, S., Lee, H.-S., Karunanadaa, B., and Kao, T.-h. 1994. Ribonuclease activity of *Petunia inflata* S proteins is essential for rejection of self-pollen. *Plant Cell* 6:1021–1028.

Hudson, R. R. 1990. Gene genealogies and the coalescent process. In *Oxford Surveys in Evolutionary Biology*, D. Futuyma and J. Antonovics, eds., Vol. 7, pp. 1–44. New York: Oxford U. Press.

Hughes, A. L. and Nei, M. 1988. Pattern of nucleotide substitution at major histocompatibility complex class I loci reveals overdominant selection. *Nature (London)* 335:167–170.

Hughes, A. L. and Nei, M. 1989. Nucleotide substitution at major histocompatibility complex class II loci: evidence for overdominant selection. *Proc. Natl. Acad. Sci. USA* 86:958–962.

Ioerger, T. R., Clark, A. G., and Kao, T.-h. 1990. Polymorphism at the self-incompatibility locus in Solanaceae predates speciation. *Proc. Natl. Acad. Sci. USA* 87:9732–9735.

Kimura, M. 1968. Evolutionary rate at the molecular level. *Nature (London)* 217:624–626.

Kimura, M. 1980. A simple method for estimating evolutionary rates of base substitutions through comparative studies of nucleotide sequences. *J. Mol. Evol.* 16:111–120.

Kimura, M. and Crow, J. F. 1964. The number of alleles that can be maintained in a finite population. *Genetics* 49:725–738.

Kingman, J. F. C. 1982. On the genealogy of large populations. *J. Appl. Prob.* **19A**: 27–43.

Klein, J., Takahata, N., and Ayala, F. J. 1993. MHC polymorphism and human origins. *Sci. Am.* **269**:78–83.

Kowyama, Y., Kakeda, K., Nakano, R., and Hattori, T. 1995. SLG/SRK-like genes are expressed in the reproductive tissues of *Ipomoea trifida*. *Plant Sex. Reprod.* **8**:333–338.

Lawlor, D. A., Ward, F. E., Ennis, P. D., Jackson, A. P., and Parham, P. 1988. HLA-A and B polymorphisms predate the divergence of humans and chimpanzees. *Nature (London)* **256**:268–271.

Lee, H.-S., Huang, S., and Kao, T.-h. 1994. S proteins control rejection of incompatible pollen in *Petunia inflata*. *Nature (London)* **367**:560–563.

Levin, D. A. 1993. its-gene polymorphism in *Phlox drummondii*. *Heredity* **71**:193–198.

Lewontin, R. C. 1974. *The Genetic Basis of Evolutionary Change.* New York: Columbia U. Press.

Li, X., Nield, J., Hayman, D., and Langridge, P. 1994. Cloning a putative self-incompatibility gene from the pollen of the grass *Phalaris coerulescens*. *Plant Cell* **6**:1923–1932.

Lundqvist, A. 1991. Four-locus its-gene control of self-incompatibility made probable in *Lilium martagon (Liliaceae)*. *Hereditas* **114**:57–63.

Mainx, F. (1954). Die Bedeutung der chromosomalen Struktur für das Evolutionsgeschehen in natürlichen Populationem. *Verhanlungen der Deutsches Zoologisches Gesellschaft in Tubingen* **1954**:280–281.

Maruyama, T. and Nei, M. 1981. Genetic variability maintained by mutation and over-dominant selection in finite populations. *Genetics* **98**:441–459.

McClure, B. A., Gray, J. E., Anderson, M. A., and Clarke, A. E. 1990. Self-incompatibility in *Nicotiana alata* involves degradation of pollen rRNA. *Nature (London)* **347**:757–760.

McClure, B. A., Haring, V., Ebert, P. R., Anderson, M. A., Simpson, R. J., et al. 1989. Style self-incompatibility gene products of *Nicotiana alata* are ribonucleases. *Nature (London)* **342**:955–957.

McDonald, J. H. and Kreitman, M. 1991. Adaptive protein evolution at the *Adh* locus in *Drosophila*. *Nature (London)* **351**:652–654.

Murfett, J., Atherton, T. L., Mou, B., Gasser, C. S., and McClure, B. A. 1994. S-RNase expressed in transgenic *Nicotiana* causes S-allele-specific pollen rejection. *Nature (London)* **367**:563–566.

Nasrallah, J. B., Kao, T.-h., Goldberg, M. L., and Nasrallah, M. E. 1985. A cDNA clone encoding an S-locus-specific glycoprotein from *Brassica oleracea*. *Nature (London)* **318**:263–267.

Nei, M. 1976. The cost of natural selection and the extent of enzyme polymorphism. *Trends Biochem. Sci.* **1**:N247–N248.

de Nettancourt, D. 1977. *Incompatibility in Angiosperms.* Berlin: Springer-Verlag.

O'Donnell, S. and Lawrence, M. J. 1984. The population genetics of the self-incompatibility polymorphism in *Papaver rhoeas*. IV. The estimation of the number of alleles in a population. *Heredity* **53**:495–507.

Ohta, T. 1995. Gene conversion vs point mutation in generating variability at the antigen recognition site of major histocompatibility complex loci. *J. Mol. Evol.* **41**:115–119.

Pandey, K. K. 1960. Evolution of gametophytic and sporophytic systems of self-incompatibility in angiosperms. *Evolution* **14**:98–115.

Parham, P. and Ohta, T. 1996. Population biology of antigen presentation by MHC class I molecules. *Science* **272**:67–74.

Richman, A. D. and Kohn, J. R. 1996. Learning from rejection: the evolutionary biology of single-locus self-incompatibility. *Trends Ecol. Evol.* **11**:49–502.

Richman, A. D., Kohn, J. R., and Uyenoyama, M. K. 1996a. Its-allele diversity in a natural population of ground cherry *Physalis crassifolia* (Solanaceae) assessed by RT-PCR. *Heredity* **76**:497–505.

Richman, A. D., Uyenoyama, M. K., and Kohn, J. R. 1996b. Allelic diversity and gene genealogy at the self-incompatibility locus in the Solanaceae. *Science* **273**:1212–1216.

Richman, A. D., Kao, T.-h., Schaeffer, S. W., and Uyenoyama, M. K. 1995. Surveys of *S*-allele sequence diversity in natural populations of horsenettle, *Solanum carolinense*. *Heredity* **75**:405–415.

Saitu, N. and Nei, M. 1987. The neighbor-Joining method: a new method for reconstructing phylogenetic trees. *Mol. Biol. Evol.* **4**:406–425.

Sarker, R. H., Elleman, C. J., and Dickinson, H. G. 1988. Control of pollen hydration in Brassica requires continued protein synthesis, and glycosylation is necessary for intraspecific incompatibility. *Proc. Natl. Acad. Sci. USA* **85**:4340–4344.

Sasaki, A. 1992. The evolution of host and pathogen genes under epidemiological interaction. In *Population Paleo-Genetics*, N. Takahata, ed., pp. 247–263. Tokyo: Japan Scientific Society.

Sassa, H., Hirano, H., and Ikehashi, H. 1992. Self-incompatibility-related RNases in styles of Japanese pear (*Pyrus serotina* Rehd.). *Plant Cell Physiol.* **33**:811–814.

Sassa, H., Nishio, T., Kowyama, Y., Hirano, H., Koba, T., and Ikehashi, H. 1996. Self-incompatibility (*S*) alleles of the Rosaceae encode members of a distinct class of the T_2/S ribonuclease superfamily. *Mol. Gen. Genet.* **250**:547–557.

Satta, Y., O'hUigin, C., Takahata, N., and Klein, J. 1993. The synonymous substitution rate of the major histocompatibility complex loci in primates. *Proc. Natl. Acad. Sci. USA* **90**:7480–7484.

Satta, Y., Takahata, N., Schönbach, C., Gutknecht, J., and Klein, J. 1991. Calibrating evolutionary rates at major histocompatibility complex loci. In *Molecular Evolution of the Major Histocompatibility Complex*, J. Klein and D. Klein, eds., pp. 51–62. Heidelberg: Springer-Verlag.

Stebbins, G. L. 1950. *Variation and Evolution in Plants*. New York: Columbia U. Press.

Tajima, F. 1989. Statistical method for testing the neutral mutation hypothesis by DNA polymorphism. *Genetics* **123**:585–595.

Takahata, N. 1990. A simple genealogical structure of strongly balanced allelic lines and trans-species evolution of polymorphism. *Proc. Natl. Acad. Sci. USA* **87**:2419–2423.

Takahata, N. 1993. Evolutionary genetics of human paleo-populations. In *Mechanisms of Molecular Evolution*, N. Takahata and A. G. Clark, eds., pp. 1–21. Sunderland, MA: Sinauer.

Takahata, N. 1996. Neutral theory of molecular evolution. *Curr. Opinions Genet. Dev.* **6**:767–772.

Takahata, N. and Nei, M. 1990. Allelic genealogy under overdominant and frequency-dependent selection and polymorphism of major histocompatibility loci. *Genetics* **124**:967–978.

Takahata, N., Satta, Y., and Klein, J. 1992. Polymorphism and balancing selection at major histocompatibility loci. *Genetics* **130**:925–938.

Takahata, N. and Satta, Y. 1998. Selection, convergence, and intragenic recombination in HLA diversity. *Genetica* **102/103**:157–169.

Tavaré, S. 1984. Line-of-descent and genealogical processes, and their applications in population genetic models. *Theor. Popul. Biol.* **26**:119–164.

Tezuka, T., Hiratsuka, S., and Takahashi, S. Y. 1993. Promotion of the growth of self-incompatible pollen tubes in lily by cAMP. *Plant Cell Physiol.* **34**:955–958.

Trick, M. and Heizmann, P. 1992. Sporophytic self-incompatibility systems: Brassica *S* gene family. *Int. Rev. Cytol.* **140**:485–524.

Uyenoyama, M. K. 1995. A generalized least-squares estimate for the origin of sporophytic self-incompatibility. *Genetics* **139**:975–992.

Vekemans, X. and Slatkin, M. 1994. Gene and allelic genealogies at a gametophytic self-incompatibility locus. *Genetics* **137**:1157–1165.

Whitehouse, H. L. K. 1951. Multiple-allelomorph incompatibility of pollen and style in the evolution of the angiosperms. *Ann. Bot.* **14**:198–216.

Wright, S. 1939. The distribution of self-sterility alleles in populations. *Genetics* **24**:538–552.

Xue, Y. B., Carpenter, R., Dickinson, H. G., and Coen, E. S. 1996. Origin of allelic diversity in *Antirrhinum S*-locus RNases. *Plant Cell* **8**:805–814.

Yokoyama, S. and Hetherington, L. E. 1982. The expected number of self-incompatibility alleles in finite plant populations. *Heredity* **48**:299–303.

Yokoyama, S. and M. Nei. 1979. Population dynamics of sex-determining alleles in honey bees and self-incompatibility alleles in plants. *Genetics* **91**:601–626.

CHAPTER SIXTEEN

The Evolution of Sex and Recombination

DONAL A. HICKEY

16.1. Introduction

Sexual reproduction is ubiquitous among eukaryotic organisms and is especially common in multicellular plants and animals. The most obvious biological effect of sex is the production of genotypically variable offspring. In most discussions of the topic, this effect is assumed, quite reasonably, to also be the main biological function of sex. Consequently questions about the origin and maintenance of sex are rephrased as questions about the nature of the selective forces that act to increase rates of genetic recombination (for a recent review, see Feldman et al., 1997). In this chapter, I give a brief overview of the various theories that have been proposed to explain the evolution of meiotic recombination. Despite the extensive literature on this topic, there is still no consensus regarding the selective forces that led to the evolution of sex (Crow, 1994). Here, I will not offer yet another solution to "the problem of sex" but I do propose that the problem may be made more tractable by recognizing that the evolution of eukaryotic sex is a multistage process, and that the advantages of recombination *per se* may not have been relevant for the evolution of both the earliest and the latest stages of the process. The evolution of the earliest stages might have been the result of direct selection on DNA repair genes and mobile genetic elements, while the later stages, e.g., the evolution of males, might best be explained in terms of selection for a more efficient breeding strategy, given that sexuality already existed. These proposals are consistent with the view that recombination itself confers a major evolutionary advantage, but it is not necessary to try to explain every aspect of the evolution of sex in terms of this advantage.

Previous discussions of the evolution of sex have subdivided the topic based on the types of model under consideration (e.g., Williams, 1975; Maynard Smith, 1978; Bell, 1982; Ghiselin 1988; Felsenstein, 1988; Kondrashov, 1993). In this review, I try to arrange the various theories that have been put forward

R. S. Singh and C. B. Krimbas, eds., *Evolutionary Genetics: From Molecules to Morphology*, vol. 1. © Cambridge University Press 2000. Printed in the United States of America. ISBN 0-521-57123-5. All rights reserved.

according to a broad evolutionary chronology. For instance, we can think of the evolution of sex as consisting of three principal phases. The earliest phase encompasses the evolution of the molecular and cytological processes that constitute a prerequisite for the existence of meiotic recombination. The second phase encompasses the selective forces that must have been at work to perfect and maintain significant levels of meiotic recombination in virtually all eukaryotic lineages. The third, and most recent, phase is the elaboration of specific modes of sexual reproduction in particular eukaryotic lineages. A familiar example of this last phase is the evolution of obligate biparental reproduction in many plants and animals. I believe that this attempt to establish a broad chronology for the evolution of sex may help us to decipher what the selective forces were that shaped the various stages and, more particularly, to eliminate certain proposals as being improbable. Specifically, we can discount the possibility that the more recently evolved aspects of sex constituted a necessary prerequisite for the evolution of the earlier stages. For example, I argue that the evolution of males is too recent to be considered a significant factor in the evolution of meiosis.

16.2. Phase 1. The Premeiotic World

A complete explanation of the evolution of sex must include not only a description of the selective forces that prevent sex from being lost in extant species, but also an explanation of the selective forces that gave rise to the meiotic process in the first place. Although it would seem reasonable to assume that the forces that ensure the maintenance of sex in recent evolution are the same as those responsible for its first appearance, this is by no means certain. Indeed, many other complex adaptations are believed to have evolved in stages, and it has been pointed out that some of the selective forces that led to the origin of sex may be unrelated to the evolutionary benefits of meiotic recombination (Hickey, 1993; Kondrashov, 1993). In fact, there are now several proposed explanations for the evolutionary origin of many aspects of sex, none of which invoke the benefits of recombination. Such theories are important in that they provide a possible example for the step-by-step evolution of this complex reproductive cycle.

It has been suggested that sex originally evolved, not because of the genetic benefits of recombination, but because it provided a template to repair double-stranded DNA breaks in haploid cells (Michod, 1993; Bernstein and Bernstein, 1991; Long and Michod, 1995). According to this view, the first sexual organisms would take up exogenous DNA in a manner similar to that seen in bacteria that are naturally competent for transformation. The attraction of this theory lies in the fact that many of the molecular processes involved in DNA recombination are very similar to the processes involved in DNA repair. This proposal has been tested experimentally by use of bacterial transformation (Mongold, 1992), and the conclusion of those experiments was that the theory was not strongly supported. Recent molecular data on transgenic mammals, however, lend striking support of the DNA repair theory. Transgenic mice lacking functional DNA repair enzymes showed predicted defects in DNA replication repair, but they also showed defects in homologous chromosome pairing during meiosis (Baker et al., 1995) and a high frequency of nonhomologous pairing (de Wind et al., 1995). This suggests strongly that both DNA repair and the meiotic pairing of chromosomes share some molecular components. This, in turn, suggests that both processes have a common evolutionary origin.

An alternative view of DNA uptake during bacterial transformation is that its primary function is nutritional rather than a template for DNA repair (Redfield, 1993). Evidence in favor of this view comes from the finding that heterologous DNA can serve as a substrate in bacterial transformation experiments. Consequently, most of the DNA taken up by the cell could not serve as a template for repair. Instead, DNA uptake is advantageous to the cell because it serves a nutritional function. Once the process of DNA uptake was in place, however, it would allow for the possibility of homologous sequences entering the cell and, perhaps, recombining with the host chromosome before it was degraded and digested (Redfield, 1993).

Both the DNA repair and nutritional DNA theories attempt to explain a relatively primitive type of gene exchange, as exemplified by bacterial transformation. As such, they describe the very earliest steps in the evolution of sex. A somewhat more advanced feature, but one still common to all sexual eukaryotes, is the process of gamete fusion to form zygotes. It is difficult to imagine how efficient processes of zygote formation could be selected for before the evolution of meiosis. It is even more unlikely that meiosis could arise in the absence of zygote formation. One solution to this evolutionary chicken-and-egg paradox is the proposal that conjugation could evolve because of selection on mobile genetic elements (Hickey, 1982; Rose, 1983; Hickey, 1984; Hickey and Rose, 1988; Hurst, 1991; Hickey, 1993; Bell, 1993). There is now some molecular evidence to support this molecular symbiont theory of the evolution of sex (Hurst, 1991). One of the most direct bits of evidence for a link between mobile elements and eukaryotic sex comes from the finding that a gene that controls mating-type switching in yeast is related to mobile intron sequences (Keeling and Roger, 1995). The fact that the mating-type locus in *Chlamydomonas* also contains highly rearranged DNA sequences (Ferris and Goodenough, 1994) suggests that mobile elements also play a role in this case. Other evidence includes the demonstration that bacterial plasmids and transposons can induce conjugation (Gawron-Burke and Clewell, 1982; Levin, 1988) and that the activity of a yeast transposon is linked to the mating cycle (Kinsey and Sandmeyer, 1995).

Not only conjugation but also some of the later stages in the evolution of the sexual process may have evolved independently from the benefits of recombination. The spread of mobile genetic elements depends on repeated rounds of conjugation; thus, such elements would be selected to provoke nonpermanent associations between genomes (Hickey, 1993). This is what we see, in fact, during the spread of conjugative plasmids in bacteria, and it has been demonstrated experimentally that the sexual cycle facilitates plasmid spread in yeast populations (Futcher et al., 1988). This simple alternation between syngamy and genetic reduction might also benefit the ancestral eukaryotic host genome. The resulting ploidy cycle can lessen the mutation load, compared with permanent diploidy or polyploidy (Bengtsson, 1992; Kondrashov, 1994a). In addition, the ploidy cycle provides a means whereby favorable mutations could be made homozygous (Kirkpatrick and Jenkins, 1989), and it can provide conditions favorable to the evolution of diploidy (Otto and Goldstein, 1992). Once the ploidy cycle was established because of the selection on both mobile elements and deleterious mutations, there would be direct selection to improve the processes of homologous chromosome pairing and segregation. If these genomes were fragmented, the independent assortment of these genomic fragments (chromosomes) would automatically result in genetic recombination between genes on separate chromosomes. It has recently been shown that this independent segregation of chromosomes can go a long way toward providing some of the evolutionary benefits previously ascribed to meiotic recombination (M. Antezana, personal communication).

These theories, like those described below for the maintenance of sex, all seem plausible, and most are supported, at least indirectly, by some empirical evidence. But several different, seemingly competing theories cannot all be correct. I believe that there may be very little real contradiction among these seemingly competing theories. First, although the primary biological function of bacterial transformation may not be DNA repair, it is clear that the genes involved in the meiotic pairing of eukaryotic chromosomes are the same genes that control certain aspects of DNA replication repair. Most biologists agree that replication repair is the more fundamental and evolutionarily ancient process. Consequently we must conclude that the evolution of meiotic chromosome pairing is built on prior evolutionary foundations that, themselves, had evolved because of their function in DNA replication repair. The nutritional DNA theory accounts for selection on individual cells to ingest DNA molecules. This would, in turn, set the stage for the evolution of horizontally transmitted genes, by direct selection on ingested sequences to resist digestion within the cell. Once such elements had evolved, they would be selected further to promote their own spread, and this could lead eventually to the evolution of conjugative transposons and plasmids. These mobile elements could impose a ploidy cycle on their host genomes. I do not mean to imply that sex did, actually, evolve by this exact sequence of steps but, rather, I suggest that a plausible progression of events leads to a situation resembling a primitive form of sexual reproduction, without having to invoke either group selection or the selective benefits of genetic recombination.

16.3. Phase 2. Selection for Meiotic Recombination

The evolutionary story outlined above describes how one lineage of ancient eukaryotic ancestors could have evolved many of the key attributes of a sexual organism, including the possibility for limited genetic recombination between genes within the genome. During the subsequent evolution of eukaryotes, recombination mechanisms were perfected and maintained in virtually all of the major lineages, despite some notable exceptions (Judson and Normark, 1996). This ubiquity of sexual reproduction provides the most compelling evidence that meiotic recombination must have an essential biological function in plants, animals, and fungi. The problem has been to establish the exact nature of this function. The trivial solution to the problem is to state that the function of meiosis is to facilitate genetic recombination, but this simply redefines the problem as one of identifying the biological function of recombination. This problem has received a lot of attention from evolutionary biologists, resulting in a bewildering variety of explanations (see Kondrashov, 1993; Hurst and Peck, 1996).

The first formal proposal for a selective advantage for genetic recombination was made by Fisher (Fisher, 1930). The Fisher–Müller model (Fisher, 1930; Müller, 1932; Crow and Kimura, 1965; Eshel and Feldman, 1970; Crow, 1988) presents an intuitively appealing explanation for the advantage of allelic recombination. Essentially, the model states that recombination can facilitate adaptive evolution by allowing favorable mutations at different loci to be recombined into a single highly adapted genotype. The particular advantage of recombination comes from the fact that several rare advantageous mutations are not required to occur sequentially in a single lineage; the mutations can occur in different individuals but, through random mating, these will eventually be recombined into a single genotype. Although appealing, this theory suffers from a number of difficulties. For instance, in its simplest form, this explanation implies group selection, i.e., populations that undergo outcrossing evolve faster than those that do not. Williams (1975) described a number of situations for which recombination would provide an advantage to a breeding pair of organisms rather than merely to the entire population, but this type of individual selection is effective only when there are very large numbers of offspring produced by each parent. The second problem with this theory is that the very process, recombination, that can produce the lucky combination of several favorable mutations is equally efficient at destroying this combination once it has been produced (Lewontin, 1971; Maynard Smith, 1978). Consequently the issue becomes one of balancing the advantages of creating favorable combinations of alleles against the disadvantage of breaking up these combinations once they are formed. Models of individual selection, acting at loci that modify recombination rates, have been used to evaluate the trade-off between the constructive and the destructive effects of recombination. The outcome of selection on the modifiers of recombination depends on the nature of the epistatic interactions between loci (Eshel and Feldman, 1970; Feldman et al., 1997).

The Fisher–Müller model is usually described in terms of constant selection coefficients, based on unchanging environmental conditions. A relaxation of the assumption on environmental constancy provides new possibilities for positive selection for increased recombination rates (Charlesworth, 1993).

Environmental fluctuations can result in unpredictable changes in the relative fitnesses of given genotypes. For instance, a genotype that was at a selective advantage in the parental generation may lose this advantage if the environment changes in the following generation. It has been argued (Williams, 1975) that a successful evolutionary strategy would be for each genotype to produce a range of different offspring genotypes, thus increasing the probability that at least one of the offspring will be well adapted to the changed environmental conditions. This line of reasoning underlies many models that explain the maintenance of sex and recombination as an adaptation that provides genetic variability. Often, this genetic variability is seen to be a specific adaptation to fluctuating environmental conditions. In very large populations of asexual individuals with high initial levels of genetic variability, coupled with high reproductive and dispersion rates, migration of the offspring could provide the same advantages as genetic recombination, provided that the pattern of temporal fluctuations in the environment was not synchronized over all environmental patches. In asexual populations of moderate size, however, changing environmental conditions will result in the loss of genetic variability (Maynard Smith, 1988; Crow, 1992). This is due to the fact that, in asexual populations, selection acts on genotypes; and once some genotypes have been culled from a local population by selection, they cannot be reconstituted by recombination. A population of sexual individuals, will, in contrast, be able to reconstitute the lost genotypic variability. In this case, sex allows a much larger phenotypic space to be explored, and in an elastic fashion, i.e., phenotypic variation that was eliminated by selection can later be re-created by recombination (Haldane, 1932; Hurst and Peck, 1996). Consequently, sex can increase the adaptive range of the sexual offspring, given that the existing allelic variation is of biological importance (Lewontin, 1974).

This latter model, based on existing allelic variability, combined with fluctuating selection pressures, is more realistic than the original Fisher–Müller model. It turns out, however, that there are certain restrictions on the patterns of environmental variability. Specifically, recombination will be favored only if the temporal variation in the environment produces selection for linkage disequilibrium of varying sign (Charlesworth, 1976). This requirement is met when the environmental change is due to biotic factors, such as the increasing challenge from coevolving parasites (Jaenike, 1978; Bell and Maynard Smith, 1987; Hamilton et al., 1990; Howard and Lively, 1994). More generally, it has been shown that stabilizing selection with a moving optimum leads to selection for higher recombination (Maynard Smith, 1988; Crow, 1992).

Both the Fisher–Müller and the fluctuating selection models focus on recombination between favored mutations or allelic variants. Several other models have been developed to examine the effects of recombination on deleterious mutations (Müller, 1964; Kimura and Maruyama, 1966; Felsenstein, 1974; Kondrashov, 1982; Felsenstein, 1988). Deleterious mutations are potentially of greater importance than favorable mutations because of their expected higher rate of occurrence. Essentially, two types of models have been studied. In finite populations of asexual genotypes, slightly deleterious mutations might accumulate in every genotype, such that the nonmutant genotype would be completely lost. Subsequently, all fitnesses would be tested relative to a suboptimal genotype, and the process could repeat itself, i.e., a second deleterious mutation could be incorporated into the less mutated genotype. This process is referred to as Müller's ratchet (Müller, 1964), and it would eventually lead to the extinction of finite asexual populations because of the gradual loss of wild-type alleles at many loci. A sexual population of comparable size would not suffer the same fate because nonmutant genotypes could be reconstituted by recombination.

A related model based on deleterious mutations was developed by Kondrashov (1982). This model does not depend on small population sizes; rather, it depends on large genome sizes. This is because, in larger genomes, the total mutation rate per genome per generation will be higher. When the mutation rate per genome surpasses unity, i.e., when the number of genes approximates the reciprocal of the mutation rate, most genomes will accumulate a mutation at some locus each generation. Thus the class of nonmutant genotypes will quickly approach zero, even in the absence of genetic drift. Again, genetic recombination can solve this problem by salvaging nonmutant blocks from a variety of genomes and recombining them into a single wild-type genotype. Experimental support for this model has recently been reported (Devisser et al., 1996).

Just as in the models based on favorable mutations, both the Müller's ratchet and deleterious mutation models are also sensitive to the fitness interactions between loci. Synergy between the effects of deleterious mutations will serve to slow the ratchet (Charlesworth, et al., 1993; Kondrashov, 1994b; Peck, 1994) and will also favor recombination in large populations that are subject to a high deleterious mutation rate (Kondrashov, 1993; Butcher, 1995). Elimination of multiply mutated genotypes by truncation selection would produce this synergy in negative effects between different loci, thus making the deleterious mutation model quite realistic. In this case, the advantage of recombination lies in the break down of nonrandom associations between loci that build up as a consequence of truncation selection. In this way, it is analogous to the advantage under the fluctuating selection models that focus on advantageous mutations. Müller's ratchet depends critically on the population size, while the deleterious genomic mutation rate model depends on the number of genes per genome. It has also been suggested that the effects of limited population size and deleterious mutations may themselves act synergistically (Müller, 1964; Gabriel et al., 1993).

Which of the above explanations for the selective advantage of genetic recombination is correct? It may be that all of them are. We can look at many of these studies, not as descriptions of the benefits of genetic recombination, but as specific explorations of the various evolutionary pitfalls that may be encountered in its absence. For instance, the evolutionary pressure from parasitism may be the most important factor promoting genotypic variation in large multicellular host species (Ebert and Hamilton, 1996), whereas Müller's ratchet is more important when one is dealing with small populations and/or large genomes. Likewise, the potential for individual selection among genetically variable offspring is greatest in those species with very large brood sizes, but sex is maintained even in species with very small brood sizes. We could, of course, conclude that sex is maintained in different lineages for different reasons, but this is not intuitively satisfying. It is also not very realistic. Natural populations are subject to both favorable and unfavorable mutations simultaneously (e.g., see Peck, 1994), and real genomes are likely to contain some loci that are subject to sustained directional selection while others are responding to short-term environmental fluctuations. I propose that the general advantage of recombination is that it simply decouples the effects of selection acting at different loci in a single genome. Without recombination, the genotype would not only be the unit of selection (Lewontin, 1974) but it would also be the unit of evolution. As a result, selection at any locus would tend to erode genetic variability at all loci (e.g., see Charlesworth et al., 1995; Hudson and Kaplan, 1995). Recombination, however, by continually destroying genotypic combinations, ensures that individual genes become the units of evolution. Stated in another way, it is recombination that translates genotypic selection in any given generation into genic selection, when summed over many generations.

16.4. Phase 3. Evolution of Reproductive Patterns among Sexual Organisms

The meiotic process is highly conserved among plants, animals, and fungi. Consequently it is reasonable to presume that meiosis and recombination were well established in the ancestral eukaryotes that gave rise to these three lineages more than one billion years ago. This is important to remember when considering the possible trade-offs between the costs and benefits of sex in extant organisms. Among mammals, for instance, the perceived reproductive cost of males is often balanced against the evolutionary benefits of genetic recombination (Weismann, 1887; Maynard Smith, 1978). If we consider the chronology of the process, however, we realize that the ancient ancestors of mammals must have had perfectly well-evolved recombination mechanisms in place hundreds of millions of years before the evolution of males. This means that the reproductive cost of males cannot be balanced against the benefits of meiosis and recombination – even if those benefits could be shown to exceed the required factor of 2. Consequently, I believe that we must look for alternative explanations for the evolution of males. On the positive side, showing that the costs and/or benefits of males are not directly relevant to the evolution of meiosis and recombination, as such, would mean that our self-imposed requirement to show a twofold per generation advantage for recombination could be removed.

Essentially, because the evolution of outcrossing and recombination predated the evolution of males, the evolution of the latter cannot have been a crucial factor in the evolution of the former. Why then, should a species that already possessed a perfectly good recombination mechanism evolve two sexes? The answer may lie in the fact that the evolution of males can serve to reduce the cost of natural selection (Haldane, 1957). Haldane pointed out that the reproductive cost of substituting one allele with another requires a certain number of selective deaths and that the number of deaths is independent of the intensity of selection. The total number of deaths is surprisingly large, often an order of magnitude greater than the total number of individuals in the population. Haldane remarked that this problem is especially acute in slowly breeding animals such as cattle, for which "one cannot cull even half of the females, even though only one in a hundred of them combines the various qualities desired." He went on to say that "the situation with respect to natural selection is comparable" (Haldane, 1957). A cattle breeder would agree with Haldane's calculation but would not be unduly worried about its implications. This is because the breeder would not normally cull the females, but would choose the "one-in-a-hundred" bull that "combines the various qualities desired" to breed with all the females in the herd. In this way, the frequency of the selected genes can rise very rapidly, and the cost of the artificial selection is absorbed by the variance in breeding success between bulls, rather than by selective deaths among the offspring. This is why we can change the genetic composition of cattle populations quite rapidly despite the fact that cows produce only a single offspring per year. If this principle also applies to natural populations, the evolution of males serves to retain the capacity for relatively rapid adaptive evolution, even "in slowly breeding animals such as cattle." In other words, I propose that the well-recognized reproductive cost of males (Maynard Smith, 1978) is offset by a very large reduction in the cost of natural selection, as defined by Haldane (1957).

In general, there is an evolutionary trade-off between the benefits of producing few, high-quality offspring and maintaining the reproductive excess that is a prerequisite for being able to adapt quickly to changing environmental conditions. This reproductive excess is necessary to allow rapid selection, while maintaining the population size. Could it be that this dilemma has been solved in many sexual species by adoption of a twofold reproductive strategy? Rather than evolving toward a compromise brood size that would both maximize investment per individual offspring and, at the same time, allow a sufficient level of reproductive excess to facilitate adaptive evolution, in these species female reproduction is specialized to ensure the transmission of genes to the next generation, while male reproduction facilitates the rapid adjustment of allelic frequencies in response to selection.

This proposal that males are an adaptation that prevents a drop in evolutionary rates in those species that have reduced rates of intrinsic increase is consistent with the observed phylogenetic patterns of occurrence of males. It can also explain the fact that, although a moderate level of sexual reproduction might be enough to reap the benefits of recombination (Green and Noakes, 1995), many lineages, such as mammals, are exclusively sexual in their reproduction. It might also provide an alternative explanation for the observation of male-driven evolution (Chang and Li, 1995). Finally, the idea that the evolution of males may be distinct from the evolution of sex is not entirely new. Geodakyan (1965; cited in Kondrashov, 1993) stated that the evolution of two sexes may provide an advantage other than genetic heterozygosity of the offspring. Specifically, he suggested that it allows more stringent selection not involving an additional reproductive cost.

To test the validity of this proposal for natural populations, we would need to make direct measures of the variance in offspring number of individual wild males and females. The development of reliable molecular diagnostics should make this task possible. Recent studies of DNA sequence variation in the Y chromosomes of primates indicate that there is very little polymorphism within species, but significant divergence between species (Hammer, 1995; Dorit et al., 1995; Burrows and Ryder, 1997). These patterns of nucleotide variation indicate that there is a high rate of turnover of Y chromosomes within natural populations. Various biological explanations have been offered to explain this fact, such as selective sweeps or small effective male population sizes but, regardless of the underlying cause, the finding of rapid evolutionary turnover of Y-chromosome sequences provides direct evidence in support of higher variances in the reproductive success of males.

Many other features of sexual reproduction can be viewed as being distinct from and more recent than the evolution of genetic recombination per se. Among them are the evolution of mating types (Hurst, 1995) and mating-type switching mechanisms (Birdsell and Wills, 1996). Finally, perhaps we should consider the possibility that sex in our own species has acquired an important, but recent, evolutionary role in maintaining social cohesion, in addition to its genetic and reproductive functions.

16.5. Summary

Sex probably first originated as a hybrid of several preexisting molecular processes that, in combination, produced a significant new evolutionary advantage for organisms with large genomes. As pointed out by Williams (1988), recombination may have started as a spandrel, in the sense of Gould and Lewontin (1979). By placing the evolution of sex in a chronological context, we are better able to understand the changing evolutionary pressures that have shaped this complex process.

The evolutionary success of sex stems from the ability of recombination to alleviate the negative effects of the complication of linkage in a variety of different ways. Many specific models have been proposed to explain the evolutionary maintenance of sex and recombination, and all these models are valid within the limits of their assumptions. It would be useful, however, to have a more general theory within which the existing models could constitute special cases. Almost three decades ago, Franklin and Lewontin (1970) asked "Is the gene the unit of selection?" The answer, I believe, is no, the unit of selection is the entire genotype. The unit of evolution, however, is not the genotype because genotypes are subject to repeated randomization by recombination. Thus the power of recombination lies in its ability to translate genotypic selection into genic evolution.

Finally, I propose that the evolution of males does not represent an evolutionary adaptation that facilitates recombination. Rather, it serves to reduce the dependence of evolutionary rates on the intrinsic rate of increase of the population. Thus the reproductive cost of males does not have to be paid in terms of the advantages of genetic recombination. The cost is offset more directly by a reduction in the cost of natural selection.

REFERENCES

Baker, SM., Browner, CE, Zhang, L., Plug, A. W., Robatzek, M., Warren, G., Elliott, E. A., Yu, J., Ashley, T., Arnheim, N., Flavell, R. A., and Liskay, R.M. 1995. Male mice defective in the mismatch repair gene PMS2 exhibit abnormal chromosome synapsis in meiosis. *Cell* **82**:309–319.

Bell, G. 1982. *The Masterpiece of Nature: The Evolution and Genetics of Sexuality.* Berkeley, CA: U. California Press.

Bell, G. 1993. The sexual nature of the eukaryotic genome. *J. Hered.* **84**:351–359.

Bell, G. and Maynard Smith, J. 1987. Short-term selection for recombination among mutually antagonistic species. *Nature (London)* **328**:66–68.

Bengtsson, B. O. 1992. Deleterious mutations and the origin of the meiotic ploidy cycle. *Genetics* **131**:741–744.

Bernstein, C. and Bernstein, H. 1991. *Aging, Sex and DNA Repair.* San Diego, CA: Academic.

Birdsell, J. and Wills, C. 1996. Significant competitive advantage conferred by meiosis and syngamy in the yeast Saccharomyces cerevisiae. *Proc. Natl. Acad. Sci. USA* **93**:908–912.

Burrows, W. and Ryder, O. A. 1997. Y-chromosome variation in great apes. *Nature (London)* **385**:125–126.

Butcher, D. 1995. Müller's ratchet, epistasis and mutation effects. *Genetics* **141**:431–437.

Chang, B. H. J. and Li, W. H. 1995. Estimating the intensity of male-driven evolution in rodents by using X-linked and Y-linked Ube-1 genes and pseudogenes. *J. Mol. Evol.* **40**:70–77.

Charlesworth, B. 1976. Recombination modification in a fluctuating environment. *Genetics* **83**:181–195.

Charlesworth, B. 1993. The evolution of sex and recombination in a varying environment. *J. Hered.* **84**:345–350.

Charlesworth, D., Morgan, M. T., and Charlesworth, B. 1993. Mutation accumulation in finite outbreeding and inbreeding populations. *Genet. Res.* **61**:39–56.

Charlesworth, D., Charlesworth, B. and Morgan, M. T. 1995. The pattern of neutral molecular variation under the background selection model. *Genetics* **141**:1619–1632.

Crow, J. F. 1988, The importance of recombination. In *The Evolution of Sex: An Examination of Current Ideas*, R. E. Michod and B. R. Levin, eds., pp. 56–73. Sunderland, MA: Sinauer.

Crow, J. F. 1992. An advantage of sexual reproduction in a rapidly changing environment. *J. Hered.* **83**:169–173.

Crow, J. F. 1994. Advantages of sexual reproduction. *Dev. Genet.* **15**:205–213.

328   *Donal A. Hickey*

Crow, J. F. and Kimura, M. 1965. Evolution in sexual and asexual populations. *Am. Nat.* **99**:439–450.

Devisser, J. A. G. M., Hoekstar, R. F., and Vandenende, H. 1996. The effects of sex and deleterious mutations on fitness in Chlamydomonas. *Proc. R. Soc. London Ser. B* **263**:193–200.

de Wind, N., Dekker, M., Berns, A., Radman, M., and te Riele, H. 1995. Inactivation of mouse Msh2 gene results in mismatch repair deficiency, methylation tolerance, hyperrecombination, and predisposition to cancer. *Gene* **82**:321–330.

Dorit, R. L., Akashi, H., and Gilbert, W. 1995. Absence of polymorphism at the ZFY locus on the human Y chromosome. *Science* **268**:1183–1185.

Ebert, D. and Hamilton, W. D. 1996. Sex against virulence: the coevolution of parasitic diseases. *Trends Ecol. Evol.* **11**:79–82.

Eshel, I. and Feldman, M. W. 1970. On the evolutionary effect of recombination. *Theor. Popul. Biol.* **1**:88–100.

Feldman, M. W., Otto, S. P., and Christiansen, F. B. 1997. Population genetic perspectives on the evolution of recombination. *Annu. Rev. Genet.* **30**:261–295.

Felsenstein, J. 1974. The evolutionary advantage of recombination. *Genetics* **78**:737–756.

Felsentstein, J. 1988. Sex and the evolution of recombination. In *The Evolution of Sex: An Examination of Current Ideas*, R. E. Michod and B. R. Levin, eds., pp. 74–86. Sunderland, MA: Sinauer.

Ferris, P. J. and Goodenough, U. W. 1994. The mating-type locus of *Chlamydomonas reinhardtii* contains highly rearranged DNA sequences. *Cell* **76**:1135–1145.

Fisher, R. A. 1930. *The Genetical Theory of Natural Selection*. Oxford: Clarendon.

Franklin, I. and Lewontin, R. C. 1970. Is the gene the unit of selection? *Genetics* **65**:707–734.

Futcher, B., Reid, E., and Hickey, D. A. 1988. Maintenance of the 2μm circle plasmid of *Saccharomyces cerevisiae* by sexual transmission: an example of selfish DNA. *Genetics* **118**:411–415.

Gabriel, W., Lynch, M., and Bürger, R. 1993. Müller's ratchet and mutational meltdowns. *Evolution* **47**:1744–1757.

Gawron-Burke, C. and Clewell, D. B. 1982. A transposon in *Streptomyces faecalis* with fertility properties. *Nature (London)* **300**:1–3.

Geodakyan, V. A. 1965. The role of sexes in transmission and transformation of genetic information. *Probl. Peredachi Inf.* **1**:105–122.

Ghiselin, M. T. 1988. The evolution of sex: a history of competing points of view. In *The Evolution of Sex: An Examination of Current Ideas*, R. E. Michod and B. R. Levin, eds., pp. 7–23. Sunderland, MA: Sinauer.

Gould, S. J., and Lewontin, R. C. 1979. The spandrels of San Marco and the Panglossian paradigm: a critique of the adaptationist program. *Proc. R. Soc. London Ser. B* **205**:581–598.

Green, R. F. and Noakes, D. L. G. 1995. Is a little bit of sex as good as a lot? *J. Theor. Biol.* **174**:87–96.

Haldane, J. B. S. 1932. *The Causes of Evolution*. New York: Harper.

Haldane, J. B. S. 1957. The cost of natural selection. *J. Genet.* **55**:511–524.

Hamilton, W. D., Axelrod, R., and Tanese, R. 1990. Sexual reproduction as an adaptation to resist parasites: a review. *Proc. Natl. Acad. Sci. USA* **87**:3566–3573.

Hammer, M. F. 1995. A recent common ancestry for human Y chromosomes. *Nature (London)* **378**:376–378.

Hickey, D. A. 1982. Selfish DNA: a sexually transmitted nuclear parasite. *Genetics* **101**:519–531.

Hickey, D. A. 1984. DNA can be a selfish parasite. *Nature* **311**:417–418.

Hickey, D. A. 1993. Molecular symbionts and the evolution of sex. *J. Hered.* **84**:410–414.

Hickey, D. A. and Rose, M. R. 1988. The role of gene transfer in the evolution of eukaryotic sex. In *The Evolution of Sex: An Examination of Current Ideas*, R. E. Michod and B. R. Levin, eds., pp. 161–175. Sunderland, MA: Sinauer.

Howard, R. S. and Lively, C. M. 1994. Parasitism, mutation accumulation and the maintenance of sex. *Nature (London)* **367**:554–557.

Hudson, R. R. and Kaplan, N. L. 1995. Deleterious background selection with recombination. *Genetics* **141**:1605–1617.

Hurst, L. D. 1991. Sex, slime and selfish genes. *Nature (London)* **354**:23–24.

Hurst, L. D. 1995. Selfish genetic elements and their role in evolution: the evolution of sex and some of what it entails. *Philos. Trans. R. Soc. London Ser. B* **249**:321–332.

Hurst, L. D. and Peck, J. R. 1996. Recent advances in understanding of the evolution and maintenance of sex. *Trends Ecol. Evol.* **11**:46–52.

Jaenike, J. 1978. A hypothesis to account for the maintenance of sex within populations. *Evol. Theory* **3**:191–194.

Judson, O. P. and Normark, B. B. 1996. Ancient asexual scandals. *Trends Ecol. Evol.* **11**:41–46.

Keeling, P. J. and Roger, A. J. 1995. A selfish pursuit of sex. *Nature (London)* **375**:283.

Kimura, M. and Maruyama, T. 1966. The mutation load with epistatic gene interactions in fitness. *Genetics* **54**:1337–1351.

Kinsey, P. T. and Sandmeyer, S. B. 1995. Ty3 transposes in mating populations of yeast: a novel transposition assay for Ty3. *Genetics* **139**:81–94.

Kirkpatrick, M. and Jenkins, C. D. 1989. Genetic segregation and the maintenance of sexual reproduction. *Nature (London)* **339**:300–301.

Kondrashov, A. S. 1982. Selection against harmful mutations in large sexual and asexual populations. *Genet. Res.* **40**:325–332.

Kondrashov, A. S. 1993. Classification of hypotheses on the advantage of amphimixis. *J. Hered.* **84**:372–387.

Kondrashov, A. S. 1994a. The asexual ploidy cycle and the origin of sex. *Nature (London)* **370**:213–216.

Kondrashov, A. S. 1994b. Müller's ratchet under epistatic selection. *Genetics* **136**:1469–1473.

Levin, B. R. 1988. The evolution of sex in bacteria. In *The Evolution of Sex: An Examination of Current Ideas*, R. E. Michod and B. R. Levin, eds., pp. 194–211. Sunderland., MA: Sinauer.

Lewontin, R. C. 1971. The effect of genetic linkage on the mean fitness of a population. *Proc. Natl. Acad. Sci. USA* **68**:984–986.

Lewontin, R. C. 1974. *The Genetic Basis of Evolutionary Change*. New York: Columbia U. Press.

Long, A. and Michod, R. E. 1995. Origin of sex for error repair. 1. Sex, diploidy and haploidy. *Theor. Popul. Biol.* **47**:18–55.

Maynard Smith, J. 1978. *The Evolution of Sex*. Cambridge: Cambridge U. Press.

Maynard Smith, J. 1988. Selection for recombination in a polygenic model – the mechanism. *Genet. Res.* **51**:59–63.

Michod, R. E. 1993. Genetic error, sex and diploidy. *J. Hered.* **84**:360–371.

Mongold, J. A. 1992. DNA-repair and the evolution of transformation in *Haemophilus influenzae. Genetics* **132**:893–898.

Müller, H. J. 1932. Some genetic aspects of sex. *Am. Nat.* **66**:118–138.

Müller, H. J. 1964. The relation of recombination to mutational advance. *Mutat. Res.* **1**:2–9.

Otto, S. P. and Goldstein, D. B. 1992. Recombination and the evolution of diploidy. *Genetics* **131**:745–751.

Peck, J. R. 1994. A ruby in the rubbish – beneficial mutations, deleterious mutations and the evolution of sex. *Genetics* **137**:597–606.

Redfield, R. J. 1993. Genes for breakfast: the have-your-cake-and-eat-it-too of bacterial transformation. *J. Hered.* **84**:400–404.

Rose, M. R. 1983. The contagion mechanism for the origin of sex. *J. Theor. Biol.* **101**:137–146.

Weismann, A. 1887. On the signification of the polar globules. *Nature (London)* **36**:607–609.

Williams, G. C. 1975. *Sex and Evolution*. Princeton, NJ: Princeton U. Press.

Williams, G. C. 1988. Retrospect on sex and kindred topics. In *The Evolution of Sex: An Examination of Current Ideas*, R. E. Michod and B. R. Levin, eds., pp. 287–298. Sunderland, MA: Sinauer.

QUANTITATIVE GENETICS AND PHENOTYPIC EVOLUTION

Introductory Remarks

RICHARD FRANKHAM

Knowledge of quantitative genetics is essential to an understanding of phenotypic evolution. Adaptive evolutionary change requires phenotypic variation, differential fitness of phenotypes, and heritability of fitness (Lewontin, 1970). The critica. ...ues in evolutionary quantitative genetics are

1. Is there genetic variation for quantitative characters? How much?
2. How is the genetic variation organized?
3. How is the genetic variation maintained?
4. Are quantitative characters subject to genetic drift?
5. How does natural selection work? What forms of selection are there, and how important are the different forms?

This section of the book addresses these questions. The first of these questions can clearly be answered in the affirmative; almost all quantitative characters show genetic variation (Lewontin, 1974). However, it is still not possible to quantify the variation in terms of polymorphism and heterozygosity at underlying loci. This led Lewontin to study variation at the level of allozymes and DNA, where variation can be quantified, even though we know little about its evolutionary significance.

Chapter 17 by Lande addresses a range of theoretical issues in evolutionary quantitative genetics, mainly directed toward questions 4 and 5. These issues include genetic drift, phenotypic trajectories, measuring the intensity of selection, and aspects of sexual selection. Much of what was done in quantitative genetics up to the later 1970s was done in the context of optimizing animal and plant breeding plans. Lande's contributions have led to a resurgence of interest in wild populations.

Chapter 18 by Frankham and Weber addresses questions 2, 3, and 4. They review empirical evidence on the nature of quantitative genetic variation and how it is maintained for peripheral characters. Genetic variation for quantitative characters is controlled by Mendelian genes with a range of effects, typically involving a small number with large effects and increasing numbers with smaller and smaller effects. There is controversy concerning the mechanism responsible for the maintenance of genetic variation for characters not closely related to reproductive fitness. The mechanism currently favored is a balance among mutation, genetic drift, and natural selection operating by means of deleterious pleiotropic effects of mutations on reproductive fitness.

The critical issue of the maintenance of genetic variation for life-history traits (and fitness itself) is addressed by Charlesworth and Hughes (Chap. 19). They derive theoretical predictions for models based on balancing selection, mutation–selection balance, and directional selection (including antagonistic pleiotropy and differential selection in different environments) and evaluate these against empirical evidence. They conclude that mutation–selection equilibrium is an important factor, but that a component of the genetic variation is maintained by some form of balancing selection. Chapter 20 by Demetrius reviews his consideration of entropy as an approach to life-history evolution.

The question of how selection operates remains a major issue in evolutionary biology. Darwin recognized that altruism posed a challenge to his theory of evolution. Lewontin (1970) pointed out that selection may operate at the level of the gene, chromosome, individual, kin, group, species, and ecosystem. Kin selection and reciprocal altruism are recognized as plausible solutions to the evolution of altruism (see Maynard Smith, Chap. 30). The theory of life-history evolution has been reviewed by Charlesworth (1994). Most of his contributions to this subject were made while he was working in Lewontin's laboratory. Following Williams (1966), the issue of group selection has been dismissed by most theoreticians, but controversy continues. In fact, empirical studies in a range of species (*Tribolium*, plants, and chickens) show that group selection is effective (see Goodnight and Stevens, 1997) and typically more effective than individual selection in those experiments. These results are probably due to interactions among individuals (cannibalism and competition), and they are predicted by the rather neglected theory of Griffing (1967). The critical unanswered question is, how important is group selection in evolution?

In spite of its fundamental importance, quantitative genetics was considered rather moribund and unfashionable by the late 1960s. The reemergence of quantitative genetics was mainly sparked by four events, Lande's seminal 1976 paper (Lande, 1976), by the discovery of selection response for a quantitative character due to changes in rRNA gene copy number (Frankham et al., 1978), by the discovery of P-element-induced quantitative genetic variation (Mackay, 1988; Torkamanzehi et al., 1992), and the development of more powerful methods for mapping QTL (Lander and Botstein, 1989; see Falconer and Mackay, 1996). The application of molecular methods to quantitative genetics, such as recent QTL mapping work, particularly by Trudy Mackay and Chuck Langley, now offers real prospects of obtaining convincing answers to many of the fundamental questions in the field (see Frankham and Weber, Chap. 18).

While Lewontin is better known for his contributions in other topics in this book, he has made substantial contribution to evolutionary quantitative genetics. His first academic appointment was at North Carolina State University, where he interacted with the strong group in quantitative genetics, especially

Ralph Comstock. That influence is evident in his papers dealing with the interaction between selection and linkage (Lewontin, 1964a, 1964b; Lewontin and Hull, 1967; Franklin and Lewontin, 1970; Lewontin, 1971; see Franklin and Feldman, Chap. 13). He is widely quoted for his review (Lewontin, 1974) indicating that most quantitative characters in most populations of outbreeding species show genetic variation. In addition, Lewontin made important contributions concerning the units of selection (Lewontin, 1970), frequency-dependent selection (Lewontin, 1955; Lewontin and Matsuo, 1963), adaptation of populations to varying environments (Lewontin, 1957; Lewontin and Birch, 1966), selection for colonizing ability (Lewontin, 1965), heterozygosity and homeostasis (Lewontin, 1956), reaction norms, and genotype – environment interactions (Gupta and Lewontin, 1982). In Lewontin (1984) he drew together his early interests with later work on allozymes, pointing out the hazards involved in drawing inferences from comparisons of phenotypic differences among populations with gene frequency differences at allozyme loci.

REFERENCES

Charlesworth, B. 1994. *Evolution in Age-Structured Populations*, 2nd ed. Cambridge: Cambridge U. Press.

Falconer, D. S., Mackay, T. F. C. 1996. *Introduction to quantitative genetics*, 4th ed. London: Longman.

Frankham, R., Briscoe, D. A., and Nurthen, R. K. 1978. Unequal crossing over at the rRNA locus as a source of quantitative genetic variation. *Nature (London)* **272**:80–81.

Franklin, I. and Lewontin, R. C. 1970. Is the gene the unit of selection? *Genetics* **65**:707–734.

Goodnight, C. J. and Stevens, L. 1997. Experimental studies on group selection: what they tell us about group selection in nature. *Am. Nat.* **150**(Suppl.):59–79.

Griffing, B. 1967. Selection with reference to biological groups. I. Individual and group selection applied to populations of unordered groups. *Aust. J. Biol. Sci.* **20**:127–139.

Gupta, A. P. and Lewontin, R. C. 1982. A study of reaction norms in natural populations of *Drosophila pseudoobscura. Evolution* **36**:934–948.

Lande, R. 1976. The maintenance of genetic variability by mutation in a polygenic character with linked loci. *Genet. Res.* **26**:221–235.

Lander, E. S. and Botstein, D. 1989. Mapping Mendelian factors underlying quantitative traits using RFLP linkage maps. *Genetics* **121**:185–199.

Lewontin, R. C. 1955. The effect of population density and competition on viability in *Drosophila melanogaster. Evolution* **9**:27–41.

Lewontin, R. C. 1956. Studies on homeostasis and heterozygosity. I. General consideration. Abdominal bristle number in second chromosome homozygotes of *Drosophila melanogaster. Am. Nat.* **90**:237–255.

Lewontin, R. C. 1957. The adaptations of populations to varying environments. *Cold Spring Harbor Symp. Quant. Biol.* **22**:395–408.

Lewontin, R. C. 1964a. The interaction of selection and linkage. I. General considerations; heterotic models. *Genetics* **48**:49–67.

Lewontin, R. C. 1964b. The interaction of selection and linkage. II. Optimum models. *Genetics* **50**:757–782.

Lewontin, R. C. 1965. Selection for colonizing ability. In *The Genetics of Colonizing Species*, H. Baker, ed., pp. 77–94. New York: Academic Press.

Lewontin, R. C. 1970. The units of selection. *Annu. Rev. Ecol. Syst.* 1:1–18.

Lewontin, R. C. 1971. The effect of genetic linkage on the mean fitness of a population. *Proc. Natl. Acad. Sci. USA* 68:984–986.

Lewontin, R. C. 1974. *The Genetic Basis of Evolutionary Change*. New York: Columbia U. Press.

Lewontin, R. C. 1984. Detecting population differentiation in quantitative characters as opposed to gene frequencies. *Am. Nat.* 123:115–124.

Lewontin, R. C. and Birch, L. C. 1966. Hybridization as a source of variation for adaptation to new environments. *Evolution* 20:315–336.

Lewontin, R. C. and Hull, P. 1967. The interaction of selection and linkage. III. Synergistic effects of blocks of genes. *Der Züchter* 37:92–98.

Lewontin, R. C. and Matsuo, Y. 1963. Interaction of genotypes determining viability in *Drosophila busckii*. *Proc. Natl. Acad. Sci. USA* 49:270–278.

Mackay, T. F. C. 1988. Transposable element-induced quantitative genetic variation in *Drosophila*. In *Proceedings of the Second International Conference on Quantitative Genetics*, B. S. Weir, E. J. Eisen, M. M. Goodman, and G. Namkoong, eds., pp. 219–235. Sunderland, MA: Sinauer.

Torkamanzehi, A., Moran, C., and Nicholas, F. W. 1992. *P* element transposition contributes substantial new variation for a quantitative trait in *Drosophila melanogaster*. *Genetics* 131:73–78.

Williams, G. C. 1966 *Adaptation and Natural Selection: A Critique of Some Current Evolutionary Thought*. Princeton, NJ: Princeton U. Press.

CHAPTER SEVENTEEN

Quantitative Genetics and Phenotypic Evolution

RUSSELL LANDE

17.1. Introduction

The major gap in Darwin's (1859) theory of evolution concerned mechanisms of inheritance. Between the time of Mendel's experiments in the 1860s and their rediscovery in 1900, biometricians led by Galton and Pearson developed statistical methods for studying phenotypic resemblance among relatives in quantitative or metrical characters commonly believed to obey blending inheritance. Because blending inheritance appears to entail rapid loss of hereditary variation, Darwin (1872) accepted environmental variation and Lamarckian inheritance as mechanisms for generating abundant heritable variation in quantitative characters, without knowing that particulate genes preserve their integrity and variation in Mendelian segregation, as later described by the Hardy–Weinberg law. In the first few decades of this century, the theories of Fisher and Wright and the experiments of Johannsen, Nilsson-Ehle, East, and Castle, reconciled Mendelism with biometry by establishing that quantitative characters are influenced by multiple Mendelian genes and environmental effects (Provine, 1971).

Subsequently, quantitative genetics became the province of applied animal and plant breeders, who used the theories of Wright and Fisher to measure correlations between relatives and to predict the results of artificial selection for improvement of economically important quantitative characters such as milk yield in cattle, egg production in chickens, and grain yield in corn (Smith, 1936; Lush, 1937; Falconer, 1960, 1981). In academia, genetic studies turned to genes with major effects and characters with a simple genetic basis rather than complex polygenic phenotypes. During the second half of the century, the ascendency of molecular biology was followed by the advent of molecular population genetics and evolution (Lewontin, 1974; Kimura, 1983). In the last quarter of this century evolutionary biologists have revitalized their interest in

R. S. Singh and C. B. Krimbas, eds., *Evolutionary Genetics: From Molecules to Morphology*, vol. 1. © Cambridge University Press 2000. Printed in the United States of America. ISBN 0-521-57123-5.

quantitative genetics to provide tools for studying mechanisms of evolution of complex phenotypes that originally concerned Darwin.

Applied plant and animal breeding and classical population genetics theory provide the necessary empirical and theoretical background for the development of a quantitative genetic theory of phenotypic evolution. This chapter outlines some aspects of this theory. After reviewing the genetic basis of quantitative inheritance, I describe an adaptive topography for phenotypic evolution analogous to Wright's adaptive topography for gene frequencies, which then suggests a logical method for measuring phenotypic natural selection. Quantitative genetic models of sexual dimorphism and sexual selection confirm ideas suggested by Darwin and Fisher. Random genetic drift in quantitative characters has been analyzed by derivation of a theory of neutral phenotypic evolution analogous to Kimura's neutral theory of molecular evolution and by modeling of the interaction of selection and random genetic drift in phenotypic evolution, as done by Wright for gene frequencies. Stimulated in large part by these theories, quantitative inheritance, the measurement of selection, and the evolutionary dynamics of quantitative characters in natural populations, are again subjects of active study by evolutionary biologists.

17.2. Quantitative Inheritance

Fisher (1918) invented the analysis of variance to partition phenotypic variance of a quantitative character into a sum of variance components caused by additive, dominance and epistatic genetic effects, and environmental effects. He showed that correlations between different types of relatives, such as parents and offspring, full siblings, and half siblings can be used to estimate different components of genetic variance. Wright (1921) invented the simpler and less general method of path analysis for a similar purpose. Along with extensive experimental data from a variety of plants and animals, these theories were instrumental in establishing the Mendelian basis of quantitative genetic inheritance and revealed that the resemblance between parents and offspring is determined almost exclusively by the statistically additive component of genetic variance. From this fundamental result it is straightforward to derive how the response to selection on a quantitative character depends on the amount of additive genetic variance and the intensity of selection (Lush, 1937). The formula most often used to predict the results of artificial selection on a single character or a linear combination of characters called a selection index, denoted as z, is known as the breeder's equation, $R_z = (G_{zz}/P_{zz}) S_z$. The selection response per generation in character z, R_z, is the difference between the mean phenotype of parents before selection and the mean phenotype of their offspring before selection in the offspring generation. The proportion of phenotypic variance P_{zz} caused by additive genetic effects G_{zz} is known as the heritability h^2. The selection differential on character z, S_z, is the difference between the mean phenotypes of selected and unselected parents within a generation. A fraction of the (additive × additive) epistatic genetic variance also contributes to parent–offspring resemblance (Falconer, 1960; Falconer and Mackay, 1996), but for characters other than major components of fitness, its magnitude is usually small (Crow and Kimura, 1970) and its contribution to selection response is transient because the linkage disequilibria generated by epistatic selection decays with recombination (Griffing, 1960).

Pleiotropy and linkage disequilibria produce correlated responses to selection in unselected characters that are genetically correlated with selected characters. The correlated response per generation in an unselected character y caused by selection on character z is $CR_y = (G_{yz}/P_{zz})S_z$, where G_{yz} is the additive genetic correlation between characters y and z, which can be estimated from the phenotypic resemblance between parental character z and offspring character y and vice versa (Falconer, 1960; Falconer and Mackay, 1996).

Artificial selection experiments demonstrate the polygenic basis of quantitative genetic inheritance through the common observation that lines selected either to increase or decrease almost any trait in any species change their mean phenotypes by 3 to 5 phenotypic standard deviations within a few dozen generations, widely exceeding the range of variation in the original unselected or base population, which could not occur with only one or a few genes that affect the character segregating in the base population (Robertson, 1980; Falconer and Mackay, 1996).

Genetic differences between natural varieties and species tend to be of the same kind as genetic polymorphism within populations (Dobzhansky, 1951). Biometrical analysis of experimental crosses between varieties or species that differ greatly in a quantitative character corroborate the neo-Darwinian view that major phenotypic changes occur by the accumulation of multiple genetic steps (Wright, 1968, Chap. 15; Lande, 1981a), as is also confirmed by careful mapping studies with visible or molecular markers linked to quantitative trait loci (Coyne, 1985; Beavis, 1994). Because of the extra effort needed to score molecular markers, mapping studies that use molecular markers frequently have small sample sizes (<100 F_2 or backcross individuals). This produces underestimates of the actual and effective numbers of loci contributing to quantitative variation because small samples are inadequate to detect genes of small effect and sampling errors tend to magnify the estimated effects of genes of large effect. Major phenotypic changes can sometimes be produced by one or a few genes of large effect, the most likely situation being when a large population sustains strong directional selection, as in the evolution of pesticide resistance (Lande, 1983; Roush and McKenzie, 1987) and some cases of artificial selection on commercial species (Doebley and Stec, 1993).

Phenotypic distributions for quantitative characters in a population often are approximately normal or can be transformed to approximate normality by a simple change in the scale of measurement, for example, to a logarithmic scale (Wright, 1968, Chaps. 10, 11; Falconer, 1960; Falconer and Mackay, 1996). The breeding value of an individual for a particular trait is the sum of additive effects of all alleles carried by the individual that influence the trait. When several or many genes with subequal effects influence a character, the distribution of breeding values in a population also tends to approximate normality by the central limit theorem of statistics (see also Turelli and Barton, 1994).

Plateaus or limits in response to high-speed artificial selection on populations of moderate or large size usually are not caused by lack of additive genetic variance in the selected character, as evidenced by rapid responses to reversed selection; rather, artificial selection often is impeded by countervailing natural selection directly against extreme expression of the character or indirectly because of detrimental pleiotropic effects of loci that affect the selected trait. Another common impediment to artificial selection is that in eukaryotic genomes there are thousands of loci producing deleterious mutations, some of which inevitably will be in linkage disequilibrium with loci affecting the character (Falconer, 1960; Falconer and Mackay, 1996). The large sizes of most natural populations and the enormously long times available for evolution permit recombination to erode unfavorable linkage disequilibria and allow natural selection to overcome the constraints of genetic correlations (if these are not complete), producing the astonishing range of phenotypic evolution revealed in the modern diversity of life and its fossil record.

17.3. Phenotypic Evolution by Natural Selection

Wright (1931, 1932, 1969) derived an adaptive topography for gene frequencies in a population, which is the basis for his shifting-balance theory of evolution. Assuming that genotypic fitnesses are constant in time, the population mates randomly, and different loci are in approximate linkage equilibrium, Wright showed that gene frequency changes increase the mean fitness in the population. Thus a population can be represented as a point on a multidimensional surface with height equal to the mean fitness and other dimensions equal to gene frequencies at different loci. Natural selection moves the population uphill on the adaptive topography until the population reaches an adaptive peak or a local maximum of mean fitness.

To interpret morphological evolution in the mammalian fossil record, Simpson (1953) made extensive use of an analogy to Wright's adaptive topography in which phenotypic characters replaced gene frequencies and phenotypic evolution increased fitness. Simpson's phenotypic analogy to Wright's adaptive topography for gene frequencies remained qualitative, without specifying the measures of either fitness or phenotypes in the adaptive topography.

A general model of adaptive phenotypic evolution should incorporate multivariate patterns of phenotypic and genetic variation and natural selection. A vector of phenotypic character values of an individual is denoted as $z = (z_1, z_2, \ldots, z_n)^T$, where the superscript T denotes vector–matrix transposition. The distribution of phenotypes in population $p(z)$ is assumed to be multivariate normal, as is the distribution of breeding values. For a population with discrete nonoverlapping generations, $W(z)$ signifies the expected Wrightian fitness (viability times fecundity) of individuals with phenotype z. Genetic constraints on phenotypic evolution are represented by a matrix G in which diagonal elements G_{ii} are additive genetic variances of the characters and off-diagonal elements G_{ij} are additive genetic covariances between pairs of characters.

The change in the mean phenotype vector per generation is (Lande, 1979)

$$\Delta \bar{z} = \mathbf{G} \nabla \ln \bar{W}. \tag{17.1a}$$

The surface of mean fitness, $\bar{W} = \int p(\mathbf{z}) W(\mathbf{z}) d\mathbf{z}$, is a function of the mean phenotype vector $\bar{z}$, constructed by averaging individual fitnesses in the population for every possible $\bar{z}$, holding constant the shape of the phenotype distribution represented by the phenotypic variance–covariance matrix $\mathbf{P}$. The gradient vector of the logarithm of mean fitness or selection gradient is $\beta = \nabla \ln \bar{W}$. Phenotypic evolution is not in the steepest uphill direction on the mean fitness surface, which is given by the selection gradient, but rather is modified by the constraints of inheritance contained in $\mathbf{G}$.

The ith element of the selection gradient, $\beta_i = \partial \ln \bar{W}/\partial \bar{z}_i$, gives the proportional change in mean fitness with respect to a small change in $\bar{z}_i$, holding constant all other aspects of the phenotypic distribution (sliding the phenotype along the $\bar{z}_i$ axis). The coefficient β_i therefore can be interpreted as the intensity of selection acting directly on character z_i to change its mean phenotype. Adaptive evolution of a particular character z_i is determined by its additive genetic variance and the intensity of direct selection on it, $G_{ii}\beta_i$, as well as indirect selection on every character with which it is genetically correlated:

$$\Delta \bar{z}_i = \sum_{j=1}^{n} G_{ij} \beta_j. \tag{17.1b}$$

Hence a character may evolve in a direction opposite to the selection acting directly on it because of indirect selection on correlated characters.

Although evolution is not in the steepest uphill direction on the phenotypic adaptive topography, if phenotypic fitnesses are constant in time and $\mathbf{G}$ and $\mathbf{P}$ do not change during evolution, the mean phenotype always evolves uphill, increasing the mean fitness in the population (Lande, 1979), so that $\Delta \bar{W} \geq 0$. Equation (17.1a) therefore shows that Simpson's phenotypic adaptive topography for phenotypes has height $\bar{W}$, with other dimensions being the mean phenotypes of different characters (see Fig. 17.1).

If the environment changes, so that phenotypic fitnesses are not constant, then phenotypic evolution may decrease the mean fitness of the population. Maladaptive phenotypic evolution also can occur in a constant environment if individual fitnesses are frequency dependent (Wright 1969), as commonly occurs with intraspecific resource competition (Lewontin 1955) and sexual selection (Lande 1980b, 1981b).

17.4. Measurement of Selection

A central problem in the analysis of adaptive phenotypic evolution is to separate direct and indirect selection, that is, to estimate the intensity of selection acting directly on each of a set of characters separated from the indirect selection acting through correlated characters. This can be done either by a retrospective

340 *Russell Lande*

ADAPTIVE TOPOGRAPHY FOR PHENOTYPES

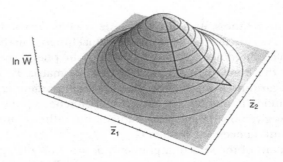

Figure 17.1. A phenotypic adaptive topography illustrating the evolutionary dynamics of two quantitative characters in response to natural selection toward a single adaptive peak. The two characters have equal phenotypic variance and heritability with a high positive additive genetic correlation. Initially the joint mean phenotype is far from the adaptive peak. The mean phenotypes do not evolve in the steepest uphill direction, shown by the selection gradient (heavy line) orthogonal to the contours of mean fitness. The actual evolutionary trajectory is a curved path going uphill on the adaptive landscape (heavy curve).

estimate of the net selection gradient responsible for an observed phenotypic change across a span of generations or by measurement of ongoing phenotypic selection within a generation. The net selection gradient producing an observed phenotypic change across generations can be reconstructed by summing Eq. (17.1a) through time to obtain (Lande, 1979)

$$\sum_{\tau=0}^{t-1} \beta(\tau) = \mathbf{G}^{-1}[\bar{\mathbf{z}}(t) - \bar{\mathbf{z}}(0)], \qquad (17.2)$$

assuming that $\mathbf{G}$ is constant, the phenotypic change has an entirely genetic basis and was caused by selection, and all selected characters are included in the analysis. $\mathbf{G}$ can be expected to change in response to changing selection pressures and because of random genetic drift in small populations, but it could also remain nearly constant for long periods if phenotypic evolution occurs by a gradual shift in the optimum phenotype while the shape of the adaptive peak is maintained by functional constraints (Lande, 1979, 1980a). Empirical tests for homogeneity of $\mathbf{G}$ among populations at different taxonomic levels in natural and experimental populations have yielded various results (see Brodie, 1993; Shaw et al., 1995, and references therein).

Ongoing phenotypic selection can be measured in living populations without any genetic assumptions by observation of phenotypic changes within a generation. The classical breeder's equation and formula (17.1a) reveal that adaptive evolution is a product of distinct processes of phenotypic selection (within generations) and inheritance (between generations). Phenotypic natural selection therefore can be estimated from measurements of individual fitness (or components of fitness) and individual phenotypes within a generation, regardless of inheritance. The appropriate methodology follows from another expression for the selection gradient (Lande and Arnold, 1983),

$$\beta = \mathbf{P}^{-1}\text{cov}[w, \mathbf{z}], \qquad (17.3a)$$

in which w represents the relative fitness of an individual; $w = W/\bar{W}$. This has the standard form of a vector of partial regression coefficients in a linear multiple regression of relative individual fitness on individual phenotypes,

$$w = \text{const} + \sum_{i=1}^{n} \beta_i z_i + \varepsilon \qquad (17.3b)$$

where ε is a residual deviation from the best-fitting linear regression. The partial regression coefficient β_i gives the expected change in relative individual fitness produced by a small change in character z_i, holding all other characters constant. This confirms the interpretation of the selection gradient as the intensity of directional selection acting directly on each of the characters separated from effects of correlated characters.

The appearance of the selection gradient in dynamic equations for phenotypic evolution recommends it above other possible approaches to the measurement of selection on phenotypic characters. As with all multiple regression analyses, this method for measuring phenotypic selection cannot distinguish correlation from causation. The estimated selection could be caused by selection on unmeasured characters correlated with those included in the study. The method does ensure, however, that the selection measured on a particular character is not caused by indirect selection on another measured character. Whenever possible, the measurement of phenotypic selection should be supplemented by observations and experiments on the functional, behavioral, and ecological context of selection to help establish its causal basis.

Higher-order regressions permit the estimation of stabilizing (or disruptive) and correlational aspects of natural selection related to the geometries of the individual and mean fitness surfaces (Lande and Arnold, 1983; Phillips and Arnold, 1989). Curve-fitting techniques are also available for more detailed graphical representation of one- or two-dimensional surfaces of individual fitness (Schluter, 1988, Schluter and Nychka, 1994; Brodie et al., 1995).

Analogous formulas for the dynamics of phenotypic evolution and the measurement of natural selection can be derived for species with overlapping generations (Charlesworth, 1994; Lande, 1982). The definition of fitness appropriate for a species can be deduced from its life history, in terms of the probability of survival to age x and fecundity at age x, denoted respectively as l_x and m_x. Fisher (1930) defined the absolute Malthusian fitness of a genotype as the exponential rate of increase of a pure population of this type that would be achieved after it approaches a stable age distribution (see also Charlesworth, 1994). For a population near a stable age distribution, an approximate expression for the expected relative fitness of individuals with phenotype $\mathbf{z}$ is

$$w(\mathbf{z}) = T^{-1} \int_0^{\infty} e^{-rx} l_x(\mathbf{z}) m_x(\mathbf{z}) \, \mathrm{d}x, \qquad (17.4)$$

in which T is the generation time and r is the instantaneous rate of population growth (Lande, 1982).

Combined studies of both selection and inheritance in single populations and among closely related species, can provide further insight into the mechanisms of phenotypic evolution (e.g., Grant, 1986).

17.5. Sexual Dimorphism and Sexual Selection

Males of many species have conspicuous or exaggerated secondary sexual characters. Darwin (1874) distinguished sexual selection from natural selection to explain the common occurrence of exaggerated secondary sexual characters that appear to be nonadaptive or maladaptive with respect to survival and reproduction of a species. Sexual selection by means of classical Darwinian mechanisms of intermale competition for mates and female choice of mates gives some males a relative (frequency-dependent) advantage over conspecific males by acquiring either more mates (in polygamous species) or mates capable of producing more offspring (in monogamous species). Two primary mechanisms of sexual selection are competition between males for access to mates and female mate choice. It is easy to see that competition among males can cause the evolution of male characters used as weapons of combat or in bluffing opponents, and it is equally obvious that female mate choice can cause the evolution of sexual advertisement traits in males. Darwin failed to understand why females of many species should evolve preferences for male characters that either are not used as weapons or are exaggerated beyond the optimal expression for intraspecific combat. His belief that females of many species possessed an esthetic sense drove many of his contemporaries, notably Wallace (1889), to reject Darwin's theory of sexual selection.

Darwin documented two major generalizations about sexual selection. The first is that closely related species often differ most in secondary sexual characters of males, which are therefore among the most rapidly evolving phenotypic traits, explaining their importance in the systematics of many taxa. The second is that in species with males possessing exaggerated secondary sexual characters, females often show a rudimentary expression of the same traits, which Darwin called the transference of characters between the sexes.

The transference of characters between the sexes and the evolution of female mating preference remained mysterious to Darwin because he lacked a mechanistic understanding of inheritance. Genetic mechanisms for the evolution of sexual dimorphism and female mating preference were first described by Fisher (1930, 1958). A common misconception (e.g., Bonner and May, 1981) is that sex-limited characters generally are sex linked, because in species with heterogametic males, X-linked recessive mutations are expressed in males but not in females unless they are homozygous. This explanation fails for birds and butterflies, in which males often have conspicuous secondary sexual characters, yet females are the heterogametic sex. Experimental evidence indicates that most sexually dimorphic quantitative characters have polygenic, primarily autosomal inheritance, with the sex chromosome contributing roughly in proportion to its relative size in the genome (Lande, 1980b; Carson and Lande, 1984; Shaw, 1996). Homologous male and female characters usually have a high genetic correlation because of similar (pleiotropic) expression in the sexes. Autosomal genes can nevertheless be differentially expressed in males and females because of interaction with sex-determining genes and their products, e.g., circulating sex hormones, and even autosomal genes can show sex-limited expression (Lande, 1980b; Carson and Lande, 1984; Shaw, 1996).

Quantitative genetic models of the evolution of sexual dimorphism, including high genetic correlations between the sexes, confirm Fisher's description of an initial relatively fast elaboration in both sexes of characters under sexual selection only in males, followed by a relatively slow evolution of sexual dimorphism in which male secondary sexual characters become more exaggerated while the homologous female characters diminish in expression. The genetic correlation of homologous characters between the sexes and the final slow phase in the evolution of sexual dimorphism help to explain Darwin's observation on the transference of characters between the sexes and why females of a species often retain in rudimentary form characters that are greatly exaggerated in males (Lande, 1980b).

Fisher's mechanism for the evolution of female mating preferences in polygamous species involves the joint evolution of two sex-limited traits: a male secondary sexual character and a female mating preference. Fisher described how an initial advantage for a male character, such as longer tails in a bird, would lead to adaptive evolution of longer tails and would also confer an indirect advantage to females that preferred males with longer tails. Thus female mating preferences can originate as a correlated response to natural selection on male characters, the genetic correlation between male morphological trait and female perceptual/psychological trait arising because of assortative mating caused by genetic variation in female mating preferences. Once established, female mating preference selects not only for further increases in the male character, but also indirectly selects for further increase in female mating preference as a correlated response. This positive feedback is capable of producing an unstable runaway process, in which the male character and the female mating preference for it both increase exponentially with time, as the male character evolves past the point at which it has ceased to have any advantage under natural selection alone. According to Fisher, runaway sexual selection may be checked by severe natural selection against the most extreme males or against females who become so choosy as to substantially delay or avoid mating.

Quantitative genetic models confirm Fisher's mechanism for the evolution of female mating preferences in polygamous species (Lande, 1981b). For polygamous species, in which males do not protect or provision their mates or assist in raising offspring, the models reveal that female mating preferences are selectively neutral. This follows from noting that the number of offspring produced by a female is independent of her mate choice. In classical population genetics theory for a stable population, lifetime fitness is measured as the number of zygotes produced per zygote. The practice of accounting for fitness that starts and stops in the middle of a generation, e.g., counting the number of successful offspring produced by an adult, confounds selection and heredity, leading to spurious measures of fitness from which evolution cannot be correctly predicted (Prout, 1965).

Selective neutrality of female mating preferences in polygamous species implies indeterminate evolution. Even with stabilizing natural selection on the male character, regardless of how exaggerated or maladaptive it becomes, there is a degree of female mating preference that will exactly counterbalance natural selection on the male trait. There thus emerges a line of possible equilibria for the male secondary sexual character and the female mating preference, which may be either unstable (as in Fisher's runaway process) or stable, depending on the genetic parameters. Even if the line of equilibria is stable, the interaction of natural and sexual selection with random genetic drift can produce rapid diversification among populations. The indeterminate and potentially rapid evolution of sexual dimorphisms helps to explain Darwin's observation that closely related species often differ most in male secondary sexual characters in ways that appear to be nonadaptive or maladaptive (Lande, 1981b).

Many students of sexual selection still adhere to erroneous definitions of fitness and the panselectionist belief that all phenotypic evolution is adaptive. One popular version of the perennial good genes hypothesis of sexual selection is that of a facultative handicap through which relatively fit males are able to develop more extreme characters to advertise their fitness, including resistance to parasites. Although there is empirical support for facultative expression of secondary sexual characters, this mechanism fails to explain Darwin's observation that closely related species often differ most in male characters. If the primary function of secondary sexual characters were to indicate fitness, it would be sufficient for all species in a taxonomic group to have identical, or very similar, secondary sexual characters. Mechanisms for nonadaptive and maladaptive aspects of sexual selection suggested by Darwin and Fisher are now well established, but subject to continued controversy (Andersson, 1994). It also has been proposed that inherent sensory bias may be an important factor in sexual selection through female mating preference (Kirkpatrick and Ryan, 1991).

17.6. Phenotypic Evolution by Random Genetic Drift

The neutral theory of phenotypic evolution, describing stochastic evolution of the mean phenotype by random genetic drift in a finite population, is important in generating null hypotheses with which to test for the operation of selection. Random genetic drift in a population with discrete nonoverlapping generations can be analyzed by considering that in each generation the mean phenotype is measured among an infinite number of progeny, from which an effective number N_e of adults are sampled at random to produce the next generation of progeny. The vector of mean phenotypes then has a sampling variance–covariance matrix between generations equal to that for the vector of mean breeding values within generations, G/N_e (Lande, 1976a, 1979).

If $\mathbf{G}$ fluctuates around a constant average value $\bar{\mathbf{G}}$, then after t generations the probability distribution of the mean phenotype will be approximately multivariate normal with expected value at the initial point and variance–covariance matrix $t\bar{\mathbf{G}}/N_e$. Since a proportion $1/(2N_e)$ of the heterozygosity or (purely) additive genetic variance is expected to be lost per generation within populations (Wright, 1931; Latter and Novitski, 1969), $\bar{\mathbf{G}}$ should remain nearly constant on a time scale up to at least $N_e/5$ generations.

The predicted rate of diversification among populations by random genetic drift was used to analyze differentiation in wing length among replicate population cages of *Drosophila pseudoobscura* kept at different temperatures for 12 years, after which the mean wing lengths in the populations were measured in a common environment (Anderson, 1973). The null hypothesis that the populations diverged solely by random genetic drift was strongly rejected; the observed variance in mean phenotypes among populations was much larger than that predicted by random genetic drift alone, implying the operation of diversifying selection among environments (Lande, 1977).

Over long time scales, $\bar{\mathbf{G}}$ should be maintained in a balance between mutation and random genetic drift. Assuming purely additive genetic variance and covariance among characters and a mutation model with a constant input of additive genetic variance and covariance from pleiotropic mutations, $\Delta\bar{\mathbf{G}} = -\bar{\mathbf{G}}/(2N_e) + \mathbf{U}$. This yields the equilibrium value of $\bar{\mathbf{G}} = 2N_e\mathbf{U}$ (Clayton and Robertson, 1955; Lande, 1979), in which $\mathbf{U}$ is the mutational variance–covariance matrix that specifies the amount of new additive genetic variability created each generation by spontaneous mutation. Using the equilibrium value of $\bar{\mathbf{G}}$ the probability distribution of the mean phenotype vector after t generations of random genetic drift is multivariate normal with variance–covariance matrix $2t\mathbf{U}$ (Lande, 1979). Hence on a long time scale the diversification in the mean phenotype vector among populations is proportional to the mutational variance–covariance matrix and independent of N_e. This is analogous to the well-known result of the neutral theory of molecular evolution, that the substitution rate (of amino acids or DNA bases) equals the mutation rate (Kimura, 1968). The factor of 2 in the formula for the neutral theory of phenotypic evolution occurs because new additive mutations that arise as heterozygotes within populations double in phenotypic effect when eventually fixed as homozygotes and converted to variation among populations (Wright, 1969).

This theory, in conjunction with data on the mutability of quantitative characters, can be used to analyze long-term phenotypic evolution as revealed in the fossil record, assuming that the observed phenotypic changes have a predominantly genetic basis. Mutation accumulation experiments that measure the rate of phenotypic diversification among replicate highly inbred lines can be used to estimate the mutational variance (and covariance) of quantitative characters (Lande, 1976b; Lynch, 1988). Artificial selection on initially highly inbred lines also can be used to accumulate new mutations and to estimate the mutational variance of the selected character (Clayton and Robertson, 1955; Lopéz and Lopéz-Fanjul, 1993). The mutational variance in quantitative morphological characters is typically on the order of 10^{-3} times the environmental variance in the character.

Lynch (1990) analyzed the diversity of skeletal measurements of several mammalian taxa and showed that the rate of long-term phenotypic evolution generally is too slow to be explained solely by random genetic drift. An analysis of morphological traits in *Drosophila* by Spicer (1993) reached a similar conclusion. Thus, although natural selection can act as an important diversifying factor for populations in different environments, in the long run it is also conservative, preventing the evolution of extremely exaggerated phenotypes.

17.7. Interaction of Selection and Random Genetic Drift

A general understanding of phenotypic evolution in finite populations requires analyzing the interaction of selection and random genetic drift. This can be most easily accomplished with diffusion equations, which provide an accurate approximation at least when selection is weak and N_e is not very small (Lande, 1976a). The simplest case of a single character with no selection and constant G is formally equivalent to classical Brownian motion with diffusion coefficient $G/(2N_e)$. In this case, the average time taken for the mean phenotype to evolve by random genetic drift a distance of (plus or minus) z units is $(N_e/G)z^2$ generations, which at the mutation–drift equilibrium for G becomes $z^2/(2U)$.

More generally, a diffusion process is characterized by its first two infinitesimal moments and its behavior at defined boundaries. The multivariate diffusion process describing the interaction of natural selection and random genetic drift has an infinitesimal mean vector equal to the deterministic change caused by selection [Eq. (17.1)] and the infinitesimal variance–covariance matrix $G/(2N_e)$. The stationary probability distribution of the mean phenotype vector, achieved after a long time, is $c\bar{W}^{2N_e}$, where c is a normalization constant. The probability of a particular mean phenotype depends on only the mean fitness and the effective population size and is essentially independent of the genetic constraints in G, provided that there is some additive genetic variance in all directions (G is not singular). An intuitive explanation for this remarkably simple result is that both the response to selection and the rate of random genetic drift are proportional to G and in the long run these effects exactly cancel. More rapid random genetic drift away from adaptive peaks in the directions of largest additive genetic variance is precisely countered by more rapid response to selection back toward the adaptive peaks in the same directions. The form of the stationary distribution indicates that in populations of moderate or large size the mean phenotype spends very little time at appreciably suboptimal states below the highest adaptive peak(s).

This raises the question of how long it takes for the mean phenotype of a population to move by random genetic drift from one adaptive peak to another, possibly higher adaptive peak, which is a crucial process in Wright's shifting-balance theory of evolution. The expected waiting time until a shift between phenotypic adaptive peaks is approximately $k(\bar{W}_p/\bar{W}_v)^{2N_e}$ generations, where $\bar{W}_p$ is the height of the first adaptive peak, $\bar{W}_v$ is the height of the adaptive valley or saddle that the population must cross to reach the new peak, and k is a function of the curvature of the adaptive landscape around the first peak and around the valley or saddle (Barton and Charlesworth, 1984; Lande, 1985, 1986). The expected time until a transition between adaptive peaks increases exponentially with the product of $2N_e$ and the logarithm of $\bar{W}_p/\bar{W}_v$. A shift between adaptive peaks is therefore unlikely to occur, even on a geological time scale, unless the effective population size is small and the depth of the adaptive valley or saddle relative to the first peak is shallow. For consideration of geographic aspects of peak shifts, see Rouhani and Barton (1987) and Lande (1989).

In contrast, the expected duration of the actual transition between adaptive peaks, from when the population finally passes the first peak until it first reaches the new adaptive peak, is nearly independent of N_e and extremely short compared with the waiting time until a shift occurs between adaptive peaks (Lande, 1985, 1986). Because of the small population sizes and short times involved, transitional stages in shifts between phenotypic adaptive peaks generally will be below the resolution of the fossil record.

This pattern of prolonged duration around an initial phenotype before a relatively rapid transition to a new stable phenotype is reminiscent of punctuated equilibrium, but produced by classical evolutionary mechanisms rather than some novel and mysterious mechanism to which Eldredge and Gould (1972) allude. Because only shallow adaptive valleys have an appreciable chance of being crossed by random genetic drift, the associated postzygotic reproductive isolation created by peak shifts does not approach the high degree of reproductive isolation typically observed between closely related species (Barton and Charlesworth, 1984), although such an event may later be reinforced by further postzygotic or prezygotic isolation (Coyne and Orr, 1989).

17.8. Conclusions

Quantitative genetic models of phenotypic evolution have contributed to a resurgence of interest in several areas of evolution, including quantitative inheritance, estimation of spontaneous mutation in quantitative traits, the measurement of natural selection, analysis of sexual selection, life-history evolution, and random genetic drift and its interaction with selection. Quantitative genetic methods for analyzing phenotypic inheritance, selection, and evolution should continue to play a central role in evolutionary studies of the objects of primary interest to Darwin and many modern evolutionists and ecologists: the complex phenotypes of whole organisms.

REFERENCES

Anderson, W. W. 1973. Genetic divergence in body size among experimental populations of *Drosophila pseudoobscura* kept at different temperatures. *Evolution* **27**:278–284.

Andersson, M. 1994. *Sexual Selection*. Princeton, NJ: Princeton U. Press.

Barton, N. H. and Charlesworth, B. 1984. Genetic revolutions, founder effects, and speciation. *Annu. Rev. Ecol. Syst.* **15**:133–164.

Beavis, W. D. 1994. The power and deceit of QTL experiments: lessons from comparative QTL studies. In *Proceedings of the 49th Annual Corn & Sorghum Industry Research Conference*, pp. 250–266.

Bonner, J. T. and May, R. M. 1981. Introduction, In C. Darwin. 1871. *The Descent of Man, and Selection in Relation to Sex*, pp. vii–xli. London: Murray. (Reprinted by Princeton U. Press.)

Brodie III, E. G. 1993. Homogeneity of the genetic variance-covariance matrix for antipredator traits in two natural populations of the garter snake, *Thamnophis ordiniodes. Evolution* **47**:844–854.

Brodie III, E. G., Moore, A. J., and Janzen, F. J. 1995. Visualizing and quantifying natural selection. *Trends Ecol. Evol.* **10**:313–318.

Carson, H. L. and Lande, R. 1984. The inheritance of a secondary sexual character in *Drosophila silvestris. Proc. Natl. Acad. Sci. USA* **81**:6904–6907.

Charlesworth, B. 1994. *Evolution in Age-Structured Populations*, 2nd ed. New York: Cambridge U. Press.

Clayton, G. A. and Robertson, A. 1955. Mutation and quantitative variation. *Am. Nat.* **89**:151–158.

Coyne, J. A. 1985. Genetic studies of three sibling species of *Drosophila* with relationship to theories of speciation. *Genet. Res.* **46**:169–192.

Coyne, J. A. and Orr, H. A. 1989. Patterns of speciation in *Drosophila. Evolution* **43**:362–381.

Crow, J. F. and Kimura, M. 1970. *An Introduction to Population Genetics Theory*. New York: Harper & Row.

Darwin, C. 1859. *On the Origin of Species*, 1st ed. London: Murray. (Reprinted 1964, Cambridge, MA: Harvard U. Press.)

Darwin, C. 1872. *The Origin of Species*, 6th ed. London: Murray. (Reprinted 1936, New York: Random House.)

Darwin, C. 1874. *The Descent of Man, and Selection in Relation to Sex*, 2nd ed. London: Murray.

Dobzhansky, Th. 1951. *Genetics and the Origin of Species*, 3rd ed. New York: Columbia U. Press.

Doebley, J. and Stec, A. 1993. Inheritance of the morphological differences between maize and teosinte: comparison of results for two F_2 populations. *Genetics* **134**:559–570.

Eldredge, N. and Gould, S. J. 1972. Punctuated equilibria: an alternative to phyletic gradualism. In *Models in Paleobiology*, T. J. M. Schopf, ed., pp. 82–115. San Francisco: Freeman, Cooper.

Falconer, D. S. 1960. *Introduction to Quantitative Genetics*, 1st ed. London: Oliver and Boyd.

Falconer, D. S. 1981. *Introduction to Quantitative Genetics*, 2nd ed. London: Longman.

Falconer, D. S. and Mackay, T. F. C. 1996. *Introduction to Quantitative Genetics*, 4th ed. London: Longman.

Fisher, R. A. 1918. The correlations between relatives on the supposition of Mendelian inheritance. *Trans. R. Soc. Edinburgh.* **52**:399–433.

Fisher, R. A. 1930. *The Genetical Theory of Natural Selection*, 1st ed. Oxford: Oxford U. Press.

Fisher, R. A. 1958. *The Genetical Theory of Natural Selection*, 2nd ed. New York: Dover.

Grant, P. R. 1986. *Ecology and Evolution of Darwin's Finches*. Princeton, NJ: Princeton U. Press.

Griffing, B. 1960. Theoretical consequences of truncation selection based on the individual phenotype. *Aust. J. Biol. Sci.* **13**:307–343.

Kimura, M. 1968. Evolutionary rate at the molecular level. *Nature (London)* **217**:624–626.

Kimura, M. 1983. *The Neutral Theory of Molecular Evolution.* Cambridge: Cambridge U. Press.

Kirkpatrick, M. and Ryan, M. J. 1991. The evolution of mating preferences and the paradox of the lek. *Nature (London)* **350**:33–38.

Lande, R. 1976a. Natural selection and random genetic drift in phenotypic evolution. *Evolution* **30**:314–334.

Lande, R. 1976b. The maintenance of genetic variability by mutation in a polygenic character with linked loci. *Genet. Res.* **26**:221–235.

Lande, R. 1977. Statistical tests for natural selection on quantitative characters. *Evolution* **31**:442–444.

Lande, R. 1979. Quantitative genetic analysis of multivariate evolution applied to brain: body size allometry. *Evolution* **33**:402–416.

Lande, R. 1980a. The genetic covariance between characters maintained by pleiotropic mutations. *Genetics* **94**:203–215.

Lande, R. 1980b. Sexual dimorphism, sexual selection, and adaptation in polygenic characters. *Evolution* **34**:292–305.

Lande, R. 1981a. The minimum number of genes contributing to quantitative variation between and within populations. *Genetics* **99**:541–553.

Lande, R. 1981b. Models of speciation by sexual selection on polygenic traits. *Proc. Natl. Acad. Sci. USA* **78**:3721–3725.

Lande, R. 1982. A quantitative genetic theory of life history evolution. *Ecology* **63**:607–615.

Lande, R. 1983. The response to selection on major and minor mutations affecting a metrical trait. *Heredity* **50**:47–65.

Lande, R. 1985. Expected time for random genetic drift of a population between stable phenotypic states. *Proc. Natl. Acad. Sci. USA* **82**:7641–7645.

Lande, R. 1986. The dynamics of peak shifts and the pattern of morphological evolution. *Paleobiology* **12**:343–354.

Lande, R. 1989. Fisherian and Wrightian theories of speciation. *Genome* **31**: 221–227.

Lande, R. and Arnold, S. J. 1983. The measurement of selection on correlated characters. *Evolution* **37**:1210–1226.

Latter, B. H. D. and Novitski, C. E. 1969. Selection in finite populations with multiple alleles. I. Limits to directional selection. *Genetics* **62**:859–876.

Lewontin, R. C. 1955. The effects of population density and competition on viability in *Drosophila melanogaster. Evolution* **9**:27–41.

Lewontin, R. C. 1974. *The Genetic Basis of Evolutionary Change.* New York: Columbia U. Press.

Lopéz, M. A. and Lopéz-Fanjul, C. 1993. Spontaneous mutation for a quantitative trait in *Drosophila melanogaster.* I. Response to artificial selection. *Genet. Res.* **61**:107–116.

Lush, J. L. 1937. *Animal Breeding Plans.* Ames, IA: Iowa State U. Press.

Lynch, M. 1988. The rate of polygenic mutation. *Genet. Res.* **51**:137–148.

Lynch, M. 1990. The rate of morphological evolution in mammals from the standpoint of the neutral expectation. *Am. Nat.* **136**:727–741.

Phillips, P. C. and Arnold, S. J. 1989. Visualizing multivariate selection. *Evolution* **43**:1209–1222.

Prout, T. 1965. The estimation of fitnesses from genotypic frequencies. *Evolution* 19:546–551.

Provine, W. B. 1971. *The Origins of Theoretical Population Genetics.* Chicago: U. Chicago Press.

Robertson, A. 1980. *Selection Experiments in Laboratory and Domestic Animals.* Farnham Royal: Commonwealth Agricultural Bureaux.

Rouhani, S. and Barton, N. H. 1987. Speciation and the shifting balance in a continuous population. *Theor. Popul. Biol.* 31:465–492.

Roush, R. T. and McKenzie, J. A. 1987. Ecological genetics of insecticide and acaricide resistance. *Annu. Rev. Entomol.* 32:361–380.

Schluter, D. 1988. Estimating the form of natural selection on a quantitative trait. *Evolution* 42:849–861.

Schluter, D. and Nychka, D. 1994. Exploring fitness surfaces. *Am. Nat.* 143:597–616.

Shaw, F. H., Shaw, R. G., Wilkinson, G. S., and M. Turelli. 1995. Changes in genetic variances and covariances: G whiz! *Evolution* 49: 1260–1267.

Shaw, K. L. 1996. Polygenic inheritance of a behavioral phenotype: interspecific genetics of song in the Hawaiian cricket genus *Laupala. Evolution* 50:256–266.

Simpson, G. G. 1953. *The Major Features of Evolution.* New York: Columbia U. Press.

Smith, H. F. 1936. A discriminant function for plant selection. *Ann. Eugen.* 7:240–250.

Spicer, G. S. 1993. Morpological evolution of the *Drosophila virilis* species group as assessed by rate tests for natural selection on quantitative characters. *Evolution* 47:1240–1254.

Turelli, M. and Barton, N. H. 1994. Genetic and statistical analysis of strong selection on polygenic traits: what, me normal? *Genetics* 138:913–941.

Wallace, A. R. 1889. *Darwinism; an Exposition of the Theory of Natural Selection, with Some of Its Applications.* London: Macmillan.

Wright, S. 1921. Systems of mating. *Genetics* 6:111–173.

Wright, S. 1931. Evolution in Mendelian populations. *Genetics* 16:97–159.

Wright, S. 1932. The roles of mutation, inbreeding, crossbreeding and selection in evolution. In *Proceedings of the Sixth International Congress on Genetics*, Vol. 1, pp. 356–366.

Wright, S. 1968. *Evolution and the Genetics of Populations, Vol. 1. Genetic and Biometric Foundations.* Chicago: U. Chicago Press.

Wright, S. 1969. *Evolution and the Genetics of Populations, Vol. 2. The Theory of Gene Frequencies.* Chicago: U. Chicago Press.

CHAPTER EIGHTEEN

Nature of Quantitative Genetic Variation

RICHARD FRANKHAM AND KENNETH WEBER

18.1. Introduction

An understanding of the inheritance of continuously varying characters is essential in the study of evolution and in the application of genetics to animal and plant breeding and conservation biology. The genetic basis of quantitative characters is still poorly understood and difficult to study. Of all quantitative characters, those related to reproductive fitness are the least understood. This chapter focuses on evolutionary issues and on experimental evidence rather than on theoretical developments.

Quantitative characters are determined by genotypes that interact with environmental conditions. Genetic variation for quantitative characters is due to multiple polymorphic Mendelian genes that show segregation, chromosomal location, linkage, and differing degrees of dominance (see East, 1916; Sax, 1923; Wright, 1968; Falconer and Mackay, 1996). When compared with simple Mendelian traits, quantitative characters involve the additional complications of major environmental effects, multiple genes, linkage, linkage disequilibrium, and gene interactions.

The classical infinitesimal model assumes that quantitative genetic variation is due to the action of many loci with small, approximately equal and additive effects (Fisher, 1918). Alleles at different loci are usually assumed to be in linkage equilibrium, and are usually assumed to be unlinked. These assumptions are based on mathematical convenience. As detailed below, they are not supported by the empirical evidence, but they form the basis of most quantitative genetic theory. This deviation between reality and the assumptions of theory does not matter in some circumstances (e.g., predicting short-term response to selection). However, it is important in other circumstances (especially long-term selection response), and the sensitivity of conclusions to violations of assumptions is rarely investigated.

R. S. Singh and C. B. Krimbas, eds., *Evolutionary Genetics: From Molecules to Morphology*, vol. 1.
© Cambridge University Press 2000. Printed in the United States of America. ISBN 0-521-57123-5. All rights reserved.

18.2. Genetic Variation

Quantitative genetic variation represents the raw material for evolutionary change. Virtually all quantitative characters in outbred populations show genetic variation (Lewontin, 1974; Wright, 1977, 1978; Falconer and Mackay, 1996), including reproductive fitness, size, shape, chemical composition, physiology, behavior, recombination, and disease resistance. Genetic variation is even detectable in the shape of small subregions of the wing of *Drosophila melanogaster* fewer than 100 cells across (Weber, 1992). The few cases in which genetic variation has not been detected generally involve bilateral symmetry (Maynard Smith and Sondhi, 1960; Coyne, 1987). Yet even bilateral symmetry must sometimes show genetic variation, as pronounced left–right asymmetry has evolved in many characters.

18.3. Effects of Natural Selection

There are fundamental differences among quantitative characters that reflect the closeness of their relationship to reproductive fitness and the forces of natural selection they have experienced (Fig. 18.1). Reproductive fitness itself is

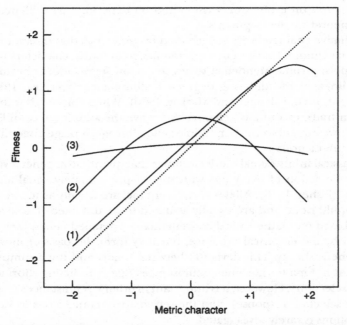

Figure 18.1. Relationships between phenotype and reproductive fitness. The dotted line is for fitness itself, (1) represents a character subject to strong directional selection (a major component of fitness), (2) represents a character experiencing stabilizing selection, and (3) represents a trait under very weak selection that is near neutral with respect to fitness. (2) and (3) are referred to as peripheral characters in the text. Disruptive selection would result in a pattern that was the inverse of (2). (Reproduced with permission of Addison-Wesley Longman from Falconer and Mackay, 1996, p. 338.)

subject to directional natural selection. Consequently, it is expected to have lower genetic variation than peripheral characters [curves (2) and (3) in Fig. 18.1] (Fisher, 1930) and to exhibit directional dominance, gene interactions (Mather, 1973), extreme gene frequencies (see Falconer and Mackay, 1996), and nonlinear offspring–parent regressions (Frankham, 1991). Conversely, peripheral characters are typically subject to weaker stabilizing selection that favors intermediate phenotypes. These are predicted to show more additive genetic variation, little directional dominance or epistasis, more intermediate gene frequencies, and generally linear offspring–parent regressions. New mutations are expected to have primarily deleterious effects on reproductive fitness, but relatively symmetrical effects on peripheral characters.

Experimental evidence generally supports these predictions. Lower heritabilities for fitness than peripheral characters are clearly found for all well-studied groups (Mousseau and Roff, 1987; Roff and Mousseau, 1987; Falconer and Mackay, 1996). However, this may be due to greater environmental variances, rather than lower additive genetic variances (Houle, 1992). Directional dominance is found for fitness characters as indicated by inbreeding depression and heterosis in outbreeding populations of all well-studied species (see Wright, 1977; Ralls and Ballou, 1983; Charlesworth and Charlesworth, 1987; Frankham, 1995; Falconer and Mackay, 1996). Chromosome substitution experiments in *Drosophila* have revealed directional dominance and gene interactions for fitness characters, but more additivity for peripheral characters (Kearsey and Kojima, 1967). Asymmetrical gene frequencies (favorable alleles at high frequencies and deleterious alleles at low frequencies) for fitness characters are indicated by the significant asymmetry in response to selection for fitness characters (Frankham, 1991). Peripheral characters appear to exhibit relatively symmetrical responses to selection (Wright, 1977; Falconer and Mackay, 1996). Spontaneous and induced mutations have predominantly deleterious effects on reproductive fitness, but little net directional effect on peripheral characters (Mackay, 1990), as predicted.

18.4. Organization of Quantitative Genetic Variation

The distributions of number of loci, alleles per locus, and effects of alleles are the basic parameters required for accounting for quantitative genetic variation. Mapping of quantitative trait loci (QTL) is beginning to yield convincing estimates of these parameters (see Mackay, 1995). Considerable uncertainties still surround the results from QTL mapping: The minimum detectable QTL effect is dependent on the size of experiments, the methods leave some uncertainty whether located effects are due to single genes, and there is a random chance of detecting nonexistent QTL or missing real ones (see Mackay, 1995; Frankel, 1995; Stuber, 1995; Falconer and Mackay, 1996). The number of QTL detected will always underestimate the true number affecting the character as QTL of small effect will be missed. Although some repeat studies on the same material have mapped somewhat different QTL (Stuber, 1995), there is a clear overall trend for estimated QTL map position and effects to cluster in the same genomic regions (Falconer and Mackay, 1996). The studies by Mackay and colleagues (see Mackay, 1995), who used a variety of different approaches, have yielded concordant conclusions.

18.4.1. Genes of Large Effect

Genes of large effect have now been found for a large number of quantitative characters in laboratory and domestic animals and plants (see Tanksley, 1993; Haley, 1995; Falconer and Mackay, 1996). While early evidence of this from Karp (1936), Thoday (1961), and others was subject to many methodological problems and potential artifacts (see McMillan and Robertson, 1974; Falconer and Mackay, 1996), recent evidence is increasingly convincing. Detailed genetic analyses of QTL of large effect in *Drosophila*, by use of deletion mapping, complementation analyses, and/or molecular analyses, have confirmed the presence of a single gene in all cases we know (Frankham, 1980b; Moran, 1990; Mackay, 1995). Large effects on abdominal bristle number due to molecular variation at both the *scute* (Mackay and Langley, 1990) and the *scabrous* loci (Lai et al., 1994) have been found.

18.4.2. Number of QTL

How many QTL contribute to typical quantitative traits? This can best be formulated as the question of how many QTL explain the majority of the genetic variation. It involves four questions: (1) How many QTL can mutate following saturation mutagenesis? (2) How many polymorphic QTL account for the standing genetic variation in outbred populations? (3) How many QTL account for response to artificial selection? (4) How many QTL differentiate two homozygous populations? The theoretical maximum number of QTL that could be detected is expected to be greatest following saturation mutagenesis, less for outbred populations, and least for crosses between two inbred lines (these have the equivalent of only two gametes). For selection lines, the number of QTL should increase with effective size of the line, the intensity of selection, and the duration of selection, since these increase both the probability of fixing alleles from the base population and new mutations. In general, the number of QTL detected depends on the size of QTL location experiments and the number of markers. The number of QTL should be greatest for fitness characters, but these will be hardest to detect as they show the highest environmental variances.

The limited available data are in line with these predictions. From 1 to 22 QTL have been identified for characters in animals and plants (Fig. 18.2). Susceptibility to diabetes in the outbred human population showed the second highest number of QTL (Davies et al., 1994, Chap. 20); the inbred line crossed the least. For inbred line crosses, fitness showed a trend for QTL number to be greater than for other characters. Further, Stuber et al. (1992) detected more QTL for grain yield than for other characters in the same inbred line

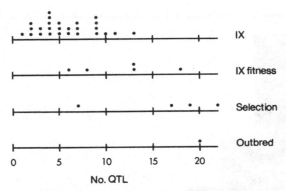

Figure 18.2. Distribution of number of QTL detected in different species from crosses between inbred lines (IX), for grain yield in inbred line crosses (IX fitness), from selection lines (Selection), and from an outbred populations (Outbred). (Data from Tanksley, 1993; Goldman et al., 1993; Davies et al., 1994; Falconer and Mackay, 1996).

cross. High numbers of QTL were detected in long-term selection lines; 17 for bristle number in *Drosophila* for chromosome 3 (Shrimpton and Robertson, 1988), 22 and 19 for protein and starch content in maize (Goldman et al., 1993), while only 7 QTL were detected for chromosomes 1 and 3 only in a line selected for a shorter duration for bristle number in *Drosophila* (Long et al., 1995).

While all estimates of the number of QTL are minimum estimates, there can be little doubt that substantial proportions of quantitative genetic variation for individual characters are due to less than 20 loci. Current evidence indicates that variation is due to a small proportion of QTL loci with large effects and an increasing proportion with smaller and smaller effects, as found in tomatoes (Fig. 18.3) and for bristle number in *Drosophila* (Shrimpton and Robertson, 1988). The infinitesimal model of a nearly infinite number of loci, each with a small effect, is clearly disproven.

18.4.3. Allele Frequencies

We know little about gene frequencies for QTL, apart from the above-mentioned asymmetry for fitness characters. Robertson (1966) inferred that rare alleles of large effect did not make a major contribution to selection response for sternopleural bristle number in *Drosophila* on the basis of the effects of population-size bottlenecks on selection response. However, this information came from an old cage population whose effective size was low, such that it had lost much of its original genetic variation (Lopez-Fanjul and Hill, 1973b; Malpica and Briscoe, 1981; Briscoe et al., 1992).

Alleles of large effect are expected to be at low frequencies and alleles of small effect at more intermediate frequencies, based on models of either mutation and stabilizing selection or mutations having deleterious pleiotropic effects

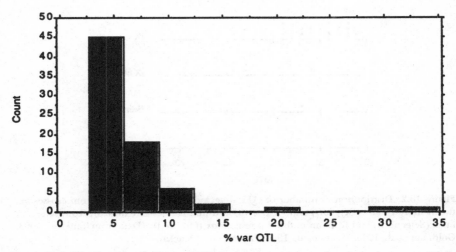

Figure 18.3. Distribution of effects of QTL in tomatoes, expressed as the percentage of the variation the QTL accounts for. (Reproduced with permission of Annual Reviews Inc. from Tanksley, 1993.)

on fitness that are correlated with their phenotypic effects. However, Lai et al. (1994) reported alleles with large effects on bristle number in *Drosophila* at intermediate frequencies. A similar conclusion is suggested by the findings of Stuber et al. (1992) in maize. QTL were found in similar locations on chromosomes 5, 1, and 9 in 16/20, 13/18, and 13/17 crosses, respectively. More information on QTL frequencies is required before the significance of these data can be assessed.

Do populations differ in allele frequencies for QTL? Lopez-Fanjul and Hill (1973a, 1973b) inferred that three old laboratory cage populations contained similar alleles at slightly different frequencies. They all contained less genetic variation than a new wild population. Contamination could not explain the similarity of the laboratory populations as one was maintained in a different laboratory from the other two.

18.5. Gene Action

18.5.1. Degree of Dominance

QTL exhibit the full range of dominance from recessive, to dominance, and to overdominance, with the majority being in the additive range (Fig. 18.4). The average degree of dominance for peripheral and fitness characters is clearly different, as described above. Most QTL affecting grain yield in maize exhibited overdominance (Stuber et al., 1992), but later analyses indicated that this could be due to linked loci (associative overdominance) rather than true single-locus

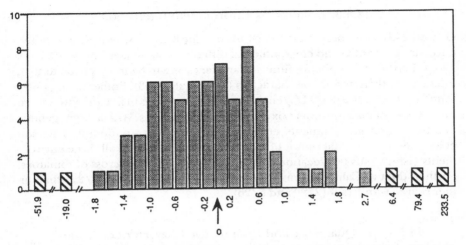

Figure 18.4. Dominance of QTL alleles in tomatoes (reproduced with permission of Annual Reviews Inc. from Tanksley, 1993). Values of −1 are recessive, 0 additive, 2 dominant, and greater than 2 overdominant. Striped bars are significantly outside the recessive–dominant range.

overdominance, although overdominance could not be ruled out (Cockerham and Zeng, 1996). Comprehensive evidence on the dominance of QTL for fitness characters is awaited with interest.

18.5.2. Interactions among Loci

Nonadditive interactions among loci are to be expected from the biochemistry of gene action. However, this may not translate into substantial proportions of nonadditive genetic variation (Lush, 1945). Tests for nonadditive genetic variation frequently have little power (Barker, 1970; Cheverud and Routman, 1995). Critical evidence comes from chromosome substitution experiments in *Drosophila* and recent QTL mapping data. These indicate that nonadditive gene action is common for reproductive fitness characters, while peripheral characters show more additive genetic variation (Kearsey and Kojima, 1967). QTL data for bristle number in *Drosophila* (Mackay, 1995) indicates that the primary gene action may show more nonadditivity than previously indicated, even for characters showing primarily additive genetic variation. Similarly, Cheverud and Routman (1995) reported that 15% of two-locus QTL combinations for body weight in mice showed significant physiological epistasis. Conversely, Tanksley (1993) concluded that the evidence for gene interactions in plants is weak. However, Lark et al. (1995) found substantial evidence for gene interactions in soybeans in which the identification of an effect at one locus was conditional on a specific allele at another locus. Such conditional interactions would have been missed in most other QTL mapping experiments.

18.5.3. Genotype × Environment Interaction

Genotypes by environment interactions are most likely when genetic differences are substantial and environmental differences large (see Hull and Gowe, 1962). Further, reproductive fitness characters seem to be more prone to such interactions than peripheral characters (Hull et al., 1963). Stuber et al. (1992) found little evidence of QTL by environment interactions in a cross between two inbred lines of maize grown at six different U.S. locations. With a larger genetic difference and environments encompassing different continents, Paterson et al. (1991) found that only 4 of 29 QTL were identified in all three environments (Israel, plus two locations in California) in a species cross of tomatoes. Further, the two California environments shared more QTL (11/25) than either shared with Israel (7/20 and 5/26).

18.5.4. Do Qualitative and Quantitative Character Loci Differ?

Mather (1944) suggested that qualitative and quantitative loci were qualitatively different and located primarily in euchromatin and heterochromatin, respectively. However, the best current information is that there are no clear differences in location and that they are often different alleles at the same loci (Mackay and Langley, 1990; Beavis et al., 1991; Lai et al., 1994; Haley, 1995; Mackay, 1995; Stuber, 1995; Touzet et al., 1995; Gibson and Hogness, 1996; Huang et al., 1996).

18.6. Maintenance of Quantitative Genetic Variation

The ultimate source of genetic variation is mutation. QTL clearly mutate, the cumulative rate being approximately 10^{-3} times the environmental variation for a range of characters (Lande, 1976; Lynch, 1988; Falconer and Mackay, 1996). QTL appear to mutate by the same mechanisms shown by qualitative genes (Frankham, 1990). Spontaneous mutations (see Lynch, 1988; Keightley et al., 1993), mutations induced by radiation (Hill and Caballero, 1992), and transposable elements have been documented (see Mackay, 1988; Moran, 1990; Frankham et al., 1991). Unequal crossing over in the rRNA tandon produced mutations with different gene copy numbers affecting abdominal bristle number, and it involved an X–Y translocation (Frankham, 1988).

Genetic variation for reproductive fitness could be maintained by mutation–selection equilibrium, heterozygote advantage, frequency-dependent selection, mutation–drift–antagonistic pleiotropic effects of genes on different fitness components or selection of varying direction in temporally or spatially varying environments. Genetic variation for peripheral quantitative characters could be maintained by the mechanisms described above for fitness characters, and in addition by neutral mutation–random genetic drift, mutation–drift–stabilizing selection, and mutation–drift–natural selection operating by means of deleterious pleiotropic effects of mutations on fitness related to their phenotypic effects (deviation from the mean) on the peripheral character (Barton and Turelli, 1989; Falconer and Mackay, 1996). Critical evidence is limited for both fitness and peripheral characters, and we concur with Barker's (1995) conclusion that there is no clear consensus.

Overdominance is out of favor as a means for maintaining quantitative genetic variation, based largely on work on heterosis in maize. However, this conclusion is less secure than usually assumed. The average dominance of alleles controlling yield in maize declines from overdominance in the F_2 of crosses between inbred lines to dominance in advanced generations (see Moll et al., 1964; Eberhart, 1977). If all alleles at all loci show the same dominance, then this rules out overdominance. However, alleles show a range of dominance from overdominant, to recessive (Fig. 18.4). Further, deleterious alleles affecting viability in *Drosophila* vary in dominance, with lethals nearly recessive and mildly deleterious alleles showing effects nearer additivity (Falconer and Mackay, 1996). Consequently the maize data are compatible with a mixture of dominant, overdominant, and additive loci. While overdominant QTL have been found for grain yield in maize and for viability in *Arabidopsis* (Stuber et al., 1992; Mitchell-Olds, 1995), these could be due to associative overdominance, as may be the case in maize (Stuber, 1995). Recent molecular evidence indicates that at least some loci are subject to balancing selection (Brookfield and Sharp, 1994; Kreitman and Ashaki, 1995).

Charlesworth and Hughes provide a detailed review of evidence relating to the maintenance of genetic variation for life history traits in Chap. 19. Mutation–selection balance is widely acknowledged as an important factor maintaining genetic variation for reproductive fitness (Rose et al., 1987; Mukai, 1988; Falconer and Mackay, 1966; Charlesworth and Hughes, 1996), although most authors conclude that it is not a complete explanation. Charlesworth and Hughes (1996) conclude that some type of balancing selection is required for some of the genetic variation. Mukai (1988) concluded that mutation–selection balance was responsible for maintaining genetic variability for larval viability in *D. melanogaster*, with selection in heterogeneous environments being an additional factor in some populations, but not in others. Some allozyme loci appear to be maintained by differential selection in different environments (Powers et al., 1991; Watt, 1994). Dolan and Robertson (1975) and Mukai (1988) concluded that frequency-dependent selection was of little importance for viability in *Drosophila*. Rose et al. (1987) concluded that antagonistic pleiotropy and mutation–selection balance were both involved in maintaining genetic variation for fitness components. A proportion of loci affecting fitness appears to be maintained by some form of balancing selection.

Lande (1976) proposed that mutation, random genetic drift, and stabilizing selection maintained quantitative genetic variation for peripheral characters. This model has been challenged by Turelli (1984) and others (see Bulmer, 1989; Tachida and Cockerham, 1988). Charlesworth and Charlesworth (1995) found that quantitative genetic variation was lower in selfing than outbreeding species of plants and concluded that this favored the mutation–drift–pleiotropy or perhaps the mutation–drift–stabilizing-selection hypotheses and was least compatible with the heterozygote advantage and neutral models. Conversely, Lopez-Fanjul and Hill (1973a, 1973b) inferred that polymorphic QTL frequencies for bristle number in *Drosophila* were similar in three unrelated old laboratory populations, results that are not compatible with mutation–selection–drift models. They suggested that similar forces of balancing selection generated the similarity among the three laboratory populations. Mackay (1981) found that heterogeneous environments had different effects on different characters; sternopleural bristle number and body weight showed evidence for greater genetic variation in heterogeneous environments, while abdominal bristle number did not. In reviews of the evidence, both Barton and Turelli (1989) and Falconer and Mackay (1996) tentatively favored a mutation–drift–pleiotropic model, as it takes into account observed properties of mutations, but Falconer and Mackay (1996) were concerned about its compatibility with estimates of mutational variance and the strength of stabilizing selection.

It is most improbable that a single mechanism is maintaining genetic variation at all QTL; some loci are likely to be maintained by each of the above-mentioned mechanisms. The question at issue is the relative importance of the different mechanisms. There is clearly a need for further incisive experimental work on this issue.

18.7. Predictions

18.7.1. Do QTL Behave as Predicted from Population Genetics Theory?

It is widely assumed that QTL behave as predicted by population genetics theory. The limited experimental evidence is consistent with this. Changes in the frequency of major genes resulted in the predicted changes in quantitative genetic variation (Frankham and Nurthen, 1981). QTL clearly show genetic drift as indicated by the effects of population size on selection response (Jones et al., 1968; Eisen, 1975; Weber, 1990; Weber and Diggins, 1990) and from the behavior of a major gene (Frankham and Nurthen, 1981). Further, population-size bottlenecks have resulted in the predicted reductions in subsequent responses to selection for peripheral characters (Robertson, 1966; James, 1971; Franklin, 1980; Frankham, 1980a; Brakefield and Saccheri, 1994). Highly inbred populations show limited quantitative genetic variation, as expected (Wright, 1968; Falconer and Mackay, 1996).

Selection is inferred to change allelic frequencies for QTL, and this has been explicitly demonstrated in some cases (Frankham and Nurthen, 1981). Linkage disequilibrium for QTL due to genetic drift, and its decay has been documented (Eberhart, 1977; Dudley, 1994; Frankham, 1999).

18.7.2. Is Selection Response Predictable?

In the short term, selection response is usually predicted satisfactorily by the equation $R = Sh^2$ (Sheridan, 1988; Hill and Caballero, 1992; Falconer and

Mackay, 1996). The most notable disagreement with predictions is for lines that have previously been subjected to artificial selection (Frankham et al., 1968; Sheridan, 1988). Further, response to selection for reproductive fitness characters is significantly asymmetrical (Frankham, 1991).

By contrast, there are no simple equations to predict long-term response to selection because it depends on gene effects and frequencies. Robertson's (1960) theory of limits in artificial selection predicts that the limit to selection due to base population genetic variation will depend on effective population size (N_e) and the selection differential. These qualitative predictions have been verified (Jones et al., 1968; Eisen, 1975; Weber, 1990; Weber and Diggins, 1990). However, Robertson's prediction of the selection limit ($2N_eR_1$, where R_1 is the first-generation response) has no correspondence to reality above as low as $N_e = 20$ (Weber and Diggins, 1990). The long-term asymptotic response to selection from new mutations was initially thought to be reasonably predictable (Hill, 1982; Frankham, 1983) with Hill's (1982) equation, but recent evidence indicates that selection against new mutations constrains selection response (Mackay et al., 1994).

18.8. The Future

The major questions about the nature of number of polymorphic genes, number of alleles per locus, frequencies, effects, and modes of gene action responsible for quantitative genetic variation will shortly be answered for peripheral characters by use of molecular approaches and QTL mapping. It is critical that such information be obtained for reproductive fitness characters as they are the most important in understanding evolution. For many issues, this will confirm our present wisdom, but, given our ignorance, surprises can be expected in the same manner as has occurred in molecular genetics with the discoveries of introns and transposons.

The question of how quantitative genetic variation is maintained is of fundamental importance. QTL mapping combined with fine structure mapping, complementation, deletion mapping, and molecular analyses should allow questions of the importance of overdominance and the genetical basis of heterosis to be answered in an unequivocal fashion. Information on the joint distribution of QTL allele frequencies and effects will help resolve how variation is maintained. The distribution of mutational effects is of major importance in conservation biology (Lande, 1995). What proportion of mutations is favorable versus deleterious? What is the distribution of effects (proportions with large versus small effects)? Comparisons of QTL locations and allele frequencies among species (e.g., Paterson et al., 1991; Stuber, 1995; Teutonico and Osborn, 1995) are likely to be highly informative.

A major change in quantitative genetics will be an increasing emphasis on biologically realistic models and less emphasis on models that reflect statistical convenience. Information on the biochemistry of gene action has been sadly neglected in quantitative genetics, with the notable exception of the works of Wright (1977), Kacser (Kacser and Burns, 1981; Keightley and Kacser, 1987;

Kacser, 1989), and the Dykhuizen–Hartl group (see Dean et al., 1988). The assumption of additive interlocus gene action is most improbable based on what we know about the biochemistry of gene action, and QTL studies are pointing in this direction. The biological reality of quantitative genetic models is often of minor importance in predicting short-term consequences, but in the long term it is critical.

Future research will be more concerned with gene function: Do allelic differences arise from changes in coding sequence, or regulatory sequences in the promoter, or introns? Do homologous loci produce quantitative genetic variation in related species? There is some evidence that they may in plants (see Paterson et al., 1991; Stuber, 1995; Teutonico and Osborn, 1995). Insecticide resistance typically involves homologous genes in different species; in some cases it is known to involve the same amino acid substitution and in one case the same base substitution (McKenzie and Batterham, 1994).

The major impediments to progress in quantitative genetic research have been lack of suitable markers, the cost and labor involved in raising and scoring individuals, and lack of knowledge of the biochemistry of the characters. All of these are now solvable. Ample molecular and cytogenetic markers now exist or can be developed. In *Drosophila*, the labor involved in scoring can be greatly reduced by using automated characters, such as those developed by Weber and colleagues (Weber, 1988; Huey et al., 1992; Frankham et al., 1992). A number of characters exist with reasonable clues as to their biochemistry, such as eye pigments and ethanol tolerance in *D. melanogaster.* Major advances in quantitative genetics can be expected if these three approaches are combined, along with the use of transposable elements to generate tagged mutations.

18.9. Summary

1. Virtually all quantitative characters in outbred populations exhibit genetic variation.
2. There are fundamental differences between reproductive fitness and peripheral characters in average dominance of alleles, in symmetry of allelic frequencies, in the importance of gene interactions, and the average effect of new mutations.
3. The infinitesimal model that shows that variation for quantitative characters is due to many genes of large and approximately equal effect with additive interactions must be rejected. There is now convincing evidence for genes of large effect, for a distribution of gene effects with few loci of large effect, and for increasing numbers with smaller and smaller effects. Alleles show the full range of dominance, and interactions among loci are not uncommon.
4. QTL appear to have chromosomal locations, methods of mutation, and molecular characteristics similar to those of major genes, and in a growing number of cases appear to be different alleles at the same loci.
5. QTL appear to behave as predicted from population genetics theory.

6. Biologically realistic models of quantitative genetic variation that reflect knowledge of the biochemistry of gene action and empirical information on the
characteristics of QTL are likely to become an increasingly important part of
evolutionary biology.

18.10. Acknowledgments

We thank Dick Lewontin for his encouragement and support at critical stages
of our careers, Heidi Manning for assistance with the bibliography, and Barry
Brook, Brian Charlesworth, Lloyd Demetrius, Dean Gilligan, Bill Hill, Russell
Lande, Trudy Mackay, and Derek Spielman for comments on a draft manuscript.
Research by RF is supported by the Australian Research Council and Macquarie
University research grants and that by KEW by U.S. National Science Founda-
tion research grants.

REFERENCES

Barker, J. S. F. 1970. Interlocus interactions: a review of experimental evidence. *Theor.
Popul. Biol.* **16**:323–346.

Barker, J. S. F. 1995. Quantitative genetic models: past, present and future challenges.
SABRAO J. **27**:1–15.

Barton, N. H. and Turelli, M. 1989. Evolutionary quantitative genetics: how little do
we know? *Annu. Rev. Genet.* **23**:337–370.

Beavis, W. D., Grant, D., Albertsen., M. and Fincher, R. 1991. Quantitative trait loci
for plant height in four maize populations and their associations with qualitative
genetic loci. *Theor. Appl. Genet.* **83**:141–145.

Brakefield, P. M. and Saccheri, I. J. 1994. Guidelines in conservation genetics and
the use of population cage experiments with butterflies to investigate the effects of
genetic drift and inbreeding. In *Conservation Genetics*, V. Loeschke, J. Tomiuk, and
S. K. Jain, eds., pp. 165–179. Basel, Switzerland: Birkhauser-Verlag.

Briscoe, D. A., Malpica, J. M., Robertson, A., Smith, G. J., Frankham, R., Banks, R. G.,
and Barker, J. S. F. 1992. Rapid loss of genetic variation in large captive populations
of *Drosophila* flies: implications for the genetic management of captive populations.
Conserv. Biol. **6**:416–425.

Brookfield, J. F. Y. and Sharp, P. M. 1994. Neutralism and selectionism face up to DNA
data. *Trends Genet.* **10**:109–111.

Bulmer, M. G. 1989. Maintenance of genetic variability by mutation-selection balance:
a child's guide through the jungle. *Genome* **31**:761–767.

Charlesworth, B. and Hughes, K. 1996. Age-specific inbreeding depression and components of genetic variance in relation to the evolution of genescence. *Proc. Natl.
Acad. Science USA* **93**:6140–6145.

Charlesworth, D. and Charlesworth, B. 1987. Inbreeding depression and its evolutionary consequences. *Annu. Rev. Ecol. Syst.* **18**:237–268.

Charlesworth, D. and Charlesworth, B. 1995. Quantitative genetics in plants: the effect
of the breeding system on genetic variability. *Evolution* **49**:911–920.

Cheverud, J. M. and Routman, E. J. 1995. Epistasis and its contribution to genetic
variance components. *Genetics* **139**:1455–1461.

Cockerham, C. C. and Zeng, Z.-B. 1996. Design III with marker loci. *Genetics* **143**:1437–
1456.

Coyne, J. A. 1987. Lack of response to selection for directional asymmetry in *Drosophila melanogaster. J. Hered.* **78**:119.

Davies, J. L., Kawaguchi, Y., Bennent, S. T., Copeman, J. B., Cordell, H. J., Pritchard, L. E., Reed, P. W., Gough, S. C. L., Jenmins, S. C., Palmer, S. M., Balfour, K. M., Rowe, B. R., Farrall, M., Barnett, A. H., Bain, S. C., and Todd, J. A. 1994. A genome-wide search for human type 1 diabetes susceptibility genes. *Nature (London)* **371**:130–136.

Dean, A. M., Dykhuizen, D. E., and Hartl, D. L. 1988. Theories of metabolic control in quantitative genetics. In *Proceedings of the Second International Conference on Quantitative Genetics*, B. S. Weir, E. J. Eisen, M. M. Goodman, and G. Namkoong, eds., pp. 536–548. Sunderland, MA: Sinauer.

Dolan, R. and Robertson, A. 1975. The effect of conditioning the medium in *Drosophila* in relation to frequency-dependent selection. *Heredity* **35**:311–316.

Dudley, J. W. 1994. Linkage disequilibrium in crosses between Illinois maize strains divergently selected for protein percentage. *Theor. Appl. Genet.* **87**:1016–1020.

East, E. M. 1916. Studies on size inheritace in *Nicotiana. Genetics* **1**:164–176.

Eberhart, S. A. 1977. Quantitative genetics and practical corn breeding. In *Proceedings of the International Conference on Quantitative Genetics*, E. Pollak, O. Kempthorne, and T. B. Bailey Jr., eds., pp. 491–502. Ames, IA: Iowa State U. Press.

Eisen, E. J. 1975. Population size and selection intensity effects on long-term selection response in mice. *Genetics* **79**:305–323.

Falconer, D. S. and Mackay, T. F. C. 1996. *Introduction to Quantitative Genetics*, 4th ed. Harlow: Longman.

Fisher, R. A. 1918. The correlation between relatives on the supposition of Mendelian inheritance. *Trans. R. Soc. Edinburgh* **52**:399–433.

Fisher, R. A. 1930. *The Genetical Theory of Natural Selection*. Oxford: Clarendon.

Frankel, W. N. 1995. Taking stock of complex genetics in mice. *Trends Genet.* **11**:471–477.

Frankham, R. 1980a. The founder effect and response to artificial selection in *Drosophila*. In *Selection Experiments in Laboratory and Domestic Animals*, A. Robertson, ed., pp. 87–90. Farnham Royal: Commonwealth Agricultural Bureaux.

Frankham, R. 1980b. Origin of genetic variation in selection lines. In *Selection Experiments in Laboratory and Domestic Animals*, A. Robertson, ed., pp. 56–68. Farnham Royal: Commonwealth Agricultural Bureaux.

Frankham, R. 1983. Origin of genetic variation in selection lines. In *Proceedings of the Thirty-Second Annual Breeders' Roundtable*. pp. 1–18.

Frankham, R. 1988. Exchanges in the rRNA multigene family as a source of genetic variation. In *Proceedings of the Second International Conference on Quantitative Genetics*, B. S. Weir, E. J. Eisen, M. M. Goodman, and G. Namkoong, eds., pp. 236–242. Sunderland, MA: Sinauer,.

Frankham, R. 1990. Contribution of novel sources of genetic variation to selection response. In *Proceedings of the fourth World Congress on Genetics and Applied Livestock Production*, pp. 185–194.

Frankham, R. 1991. Are responses to artificial selection for reproductive fitness traits consistently asymmetrical? *Genet. Res.* **56**:35–42.

Frankham, R. 1995. Conservation genetics. *Annu. Rev. Genet.* **29**:305–327.

Frankham, R. 1999. Modeling problems in conservation genetics using laboratory animals. In *Quantitative Methods in Conservation Biology*, S. Ferson, ed. New York: Springer-Verlag.

Frankham, R. and Nurthen, R. K. 1981. Forging links between population and quantitative genetics. *Theor. Appl. Genet.* **59**:251–263.

Frankham, R., Jones, L. P., and Barker, J. S. F. 1968. The effects of population size and selection intensity in selection for a quantitative character in *Drosophila*. III. Analyses of the lines. *Genet. Res.* **12**:267–283.

Frankham, R., Torkamanzehi, A., and Moran, C. 1991. P element transposon-induced quantitative genetic variation for inebriation time in *Drosophila melanogaster*. *Theor. Appl. Genet.* **78**:869–886.

Frankham, R., Weber, K. E., Rousseau, S., Davidson, T., Fanning, P., and Clisby, B. 1992. An apparatus for measurement and selection on larval-pupal development time. *Dros. Inf. Serv.* **71**:175–176.

Franklin, I. R. 1980. Evolutionary change in small populations. In *Conservation Biology: An Evolutionary–Ecological Perspective*, M. E. Soulé and B. A. Wilcox, eds., pp. 135–149. Sunderland, MA: Sinauer.

Gibson, G. and Hogness, D. S. 1996. Effect of polymorphism in the *Drosophila* regulatory gene *Ultrabithorax* on homeotic stability. *Science* **271**:200–203.

Goldman, I. L., Rocheford, T. R., and Dudley, J. W. 1993. Quantitative trait loci influencing protein and starch concentration in the Illinois long term selection maize strains. *Theor. Appl. Genet.* **87**:217–224.

Haley, C. 1995. Livestock QTL – bringing home the bacon? *Trends. Genet.* **11**:488–492.

Hill, W. G. 1982. Predictions of response to artificial selection from new mutations. *Genet. Res.* **40**:255–278.

Hill, W. G. and Caballero, A. 1992. Artificial selection experiments. *Annu. Rev. Ecol. Syst.* **23**:287–310.

Houle, D. 1992. Comparing evolveability and variability of quantitative traits. *Genetics* **130**:195–204.

Huang, N., Courtois, B., Khush, G. S., Lin, H., Wang, G., Wu, P., and Zheng, K. 1996. Association of quantitative trait loci for plant height with major dwarfing genes in rice. *Heredity* **77**:130–137.

Huey, R. B., Crill, W. D., Kingsolver, J. G., and Weber, K. E. 1992. A method for rapid measurement of heat or cold resistance of small insects. *Funct. Ecol.* **6**:489–494.

Hull, P. and Gowe, R. S. 1962. The importance of interaction detected between genotypic and environmental factors for characters of economic significance in poultry. *Genetics* **47**:143–159.

Hull, P., Gowe, R. S., Slen, S. B., and Crawford, R. D. 1963. A comparison of the interaction, with two types of environment, of pure strain or strain crosses of poultry. *Genetics* **4**:370–381.

James, J. W. 1971. The founder effect and response to artificial selection. *Genet. Res.* **16**:241–250.

Jones, L. P., Frankham, R., and Barker, J. S. F. 1968. The effects of population size and selection intensity in selection for a quantitative character in *Drosophila*. II. Long-term response to selection. *Genet. Res.* **12**:249–266.

Kacser, H. 1989. Quantitative variation and the control analysis of enzyme systems. In *Evolution and Animal Breeding: Reviews on Molecular and Quantitative Approaches in Honour of Alan Robertson*, W. G. Hill and T. F. C. Mackay, eds., pp. 219–226. Wallingford: C.A.B. International.

Kacser, H. and Burns, J. A. 1981. The molecular basis of dominance. *Genetics* **97**:639–666.

Karp, M. L. 1936. The number and distribution of genes in the third chromosome of *D. melanogaster*, affecting the number of sternital bristles. *C.R. (Dokl.) Acad. Sci. URSS* **110**:43–46.

Kearsey, M. J. and Kojima, K. 1967. The genetic architecture of body weight and egg hatchability in *Drosophila melanogaster*. *Genetics* **56**:23–37.

Keightley, P. D. and Kacser, H. 1987. Dominance, pleiotropy and metabolic structure. *Genetics* **117**:319–329.

Keightley, P. D., Mackay, T. F. C., and Caballero, A. 1993. Accounting for bias in estimates of the rate of polygenic mutation. *Proc. R. Soc. London Ser. B* **253**:291–296.

Kreitman, M. and Akashi, A. 1995. Molecular evidence for natural selection. *Annu. Rev. Ecol. Syst.* **26**:403–422.

Lai, C., Lyman, R. F., Long, A. D., Langley, C. H., and Mackay, T. F. C. 1994. Naturally occurring variation in bristle number and DNA polymorphisms at the *scabrous* locus of *Drosophila melanogaster*. *Science* **266**:1697–1702.

Lande, R. 1976. The maintenance of genetic variability by mutation in a polygenic character with linked loci. *Genet. Res.* **26**:221–235.

Lande, R. 1995. Mutation and conservation. *Conserv. Biol.* **9**:782–791.

Lark, K. G., Chase, K., Adler, F., Manusur, L. M., and Orf, J. H. 1995. Interactions between quantitative trait loci in soybean in which trait variation at one locus is conditional upon a specific allele at another. *Proc. Natl. Acad. Sci. USA* **92**:4656–4660.

Lewontin, R. C. 1974. *The Genetic Basis of Evolutionary Change*. New York: Columbia U. Press.

Long, A. D., Mullaney, S. L., Reid, L. A., Fry, J. D., Langley, C. H., and Mackay, T. F. C. 1995. High resolution mapping of genetic factors affecting abdominal bristle number in *Drosophila melanogaster*. *Genetics* **139**:1273–1291.

Lopez-Fanjul, C. and Hill, W. G. 1973a. Genetic differences between populations of *Drosophila melanogaster* for a quantitative trait. I. Laboratory populations. *Genet. Res.* **22**:51–68.

Lopez-Fanjul, C. and Hill, W. G. 1973b. Genetic differences between populations of *Drosophila melanogaster* for a quantitative trait. II. Wild and laboratory populations. *Genet. Res.* **22**:69–78.

Lush, J. L. 1945. *Animal Breeding Plans*, 3rd ed. Ames, IA: Iowa State U. Press.

Lynch, M. 1988. The rate of polygenic mutation. *Genet. Res.* **51**:127–148.

Mackay, T. F. C. 1981. Genetic variation in varying environments. *Genet. Res.* **37**:79–93.

Mackay, T. F. C. 1988. Transposable element-induced quantitative genetic variation in *Drosophila*. In *Proceedings of the Second International Conference on Quantitative Genetics*, B. S. Weir, E. J. Eisen, M. M. Goodman, and G. Namkoong, eds., pp. 219–235. Sunderland, MA: Sinauer.

Mackay, T. F. C. 1990. Distribution of effects of new mutations affecting quantitative traits in *Drosophila melanogaster*. *In Proceedings of the 4th World Congress on Genetics and Applied Livestock Production*, pp. 219–228.

Mackay, T. F. C. 1995. The genetic basis of quantitative variation: number of sensory bristles of *Drosophila melanogaster* as a model system. *Trends Genet.* **11**:464–470.

Mackay, T. F. C., Fry, J. D., Lyman, R. F., and Nuzhdin, S. V. 1994. Polygenic mutation in *Drosophila melanogaster*: estimates from response to selection of inbred strains. *Genetics* **136**:937–951.

Mackay, T. F. C. and Langley, C. H. 1990. Molecular and phenotypic variation in the achaete- scute region of *Drosophila melanogaster*. *Nature (London)* **348**:64–66.

Malpica, J. M. and Briscoe, D. A. 1981. Effective population number estimates of laboratory populations of *Drosophila melanogaster. Experientia* **37**:947–948.

Mather, K. 1944. The genetical activity of heterochromatin. *Proc. R. Soc. London Ser. B* **132**:308–332.

Mather, K. 1973. *Genetical Structure of Populations.* London: Chapman & Hall.

Maynard Smith, J. and Sondhi, K. C. 1960. The genetics of a pattern. *Genetics* **45**:1039–1050.

McKenzie, J. A. and Batterham, P. 1994. The genetic, molecular and phenotypic consequences of selection for insecticide resistance. *Trends Ecol. Evol.* **9**:166–169.

McMillan, I. and Robertson, A. 1974. The power of methods for detecting major genes affecting quantitative characters. *Heredity* **32**:349–356.

Mitchell-Olds, T. 1995. Interval mapping of viability loci causing heterosis in *Arabidopsis. Genetics* **140**:1105–1109.

Moll, R. H., Lindsey, M. F., and Robinson, H. F. 1964. Estimates of genetic variances and level of dominance in maize. *Genetics* **49**:411–423.

Moran, C. 1990. The role of transposable elements in generating quantitative genetic variation. In *Proceedings of the 4th World Congress on Genetics and Applied Livestock Production*, pp. 229–237.

Mousseau, T. A. and Roff, D. A. 1987. Natural selection and the heritability of fitness components. *Heredity* **59**:181–197.

Mukai, T. 1988. Genotype-environment interaction in relation to the maintenance of genetic variability in populations of *Drosophila melanogaster.* In *Proceedings of the Second International Conference on Quantitative Genetics*, B. S. Weir, E. J. Eisen, M. M. Goodman, and G. Namkoong, eds., pp. 21–31. Sunderland, MA: Sinauer.

Paterson, A. H., Damon, S., Hewitt, J. D., Zamir, D., Rabinowitch, H. D., Lincoln, S. E., Lander, E. S., and Tanksley, S. D. 1991. Mendelian factors underlying quantitative traits in tomatoes: comparison across species, generations and environments. *Genetics* **127**:181–197.

Powers, D. A., Lauerman, T., Crawford, D., and DiMichele, L. 1991. Genetic mechanisms for adapting to a changing environment. *Annu. Rev. Genet.* **25**:629–659.

Ralls, K. and Ballou, J. 1983. Extinction: lessons from zoos. In *Genetics and Conservation: A Reference for Managing Wild Animal and Plant Populations*, C. M. Schonewald-Cox, S. M. Chambers, B. MacBryde, and L. Thomas, eds., pp. 164–184. Menlo Park, CA: Benjamin/Cummings.

Robertson, A. 1960. A theory of limits in artificial selection. *Proc. R. Soc. London Ser. B* **153**:234–249.

Robertson, A. 1966. Artificial selection in plants and animals. *Proc. R. Soc. London Ser. B* **164**:341–349.

Roff, D. A. and Mousseau, T. A. 1987. Quantitative genetics and fitness: lessons from *Drosophila. Heredity* **58**:103–118.

Rose, M. R., Service, P. M., and Hutchinson, E. W. 1987. Three approaches to trade-offs in life-history evolution. In *Genetic Constraints on Adaptive Evolution*, V. Loeschcke, ed., pp. 91–105. Berlin: Springer-Verlag.

Sax, K. 1923. The association of size differences with seed coat pattern and pigmentation in *Phaseolus vulgaris. Genetics* **8**:552–560.

Sheridan, A. K. 1988. Agreement between estimated and realised genetic parameters. *Anim. Breed. Abst.* **56**:877–889.

Shrimpton, A. and Robertson, A. 1988. The isolation of polygenic factors controlling bristle score in *Drosophila melanogaster.* II. Distribution of third chromosome bristle effects within chromosome sections. *Genetics* **118**:445–459.

Stuber, C. W. 1995. Mapping and manipulating quantitative traits in maize. *Trends. Genet.* **11**:477–481.

Stuber, C. W., Lincoln, S. E., Wolff, D. W., Helentjaris, T., and Lander, E. S. 1992. Identification of genetic factors contributing to heterosis in a hybrid from two elite maize inbred lines using molecular markers. *Genetics* **132**:823–839.

Tachida, H. and Cockerham, C. C. 1988. Variance components and fitness under stabilising selection. *Genet. Res.* **51**:47–53.

Tanksley, S. D. 1993. Mapping polygenes. *Annu. Rev. Genet.* **27**:205–233.

Teutonico, R. A. and Osborn, T. C. 1995. Mapping loci controlling vernalization requirement in *Brassica rapa. Theor. Appl. Genet.* **91**:1279–1283.

Thoday, J. M. 1961. Location of polygenes. *Nature (London)* **191**:368–370.

Touzet, P., Winkler, R. G., and Helentjaris, T. 1995. Combined genetic and physiological analysis of a locus contributing to quantitative variation. *Theor. Appl. Genet.* **91**:200–205.

Turelli, M. 1984. Heritable variation via mutation-selection balance: Lerch's zeta meets the abdominal bristle. *Theor. Popul. Biol.* **25**:138–193.

Watt, W. B. 1994. Allozymes in evolutionary genetics: self-imposed burden or extraordinary tool? *Genetics* **136**:11–16.

Weber, K. E. 1988. An apparatus for measurement of resistance to gas-phase agents. *Dros. Inf. Serv.* **67**:91–93.

Weber, K. E. 1990. Increased selection response in large populations. I. Selection for wing-tip height in *Drosophila melanogaster* at three population sizes. *Genetics* **125**:579–584.

Weber, K. E. 1992. How small are the smallest selectable domains of form? *Genetics* **130**:345–353.

Weber, K. E. and Diggins, L. T. 1990. Increased selection response in larger populations. II. Selection for ethanol vapor resistance in *Drosophila melanogaster*, at two population sizes. *Genetics* **125**:585–597.

Wright, S. 1968. *Evolution and the Genetics of Populations. Vol. 1. Genetic and Biometric Foundations.* Chicago: U. Chicago Press.

Wright, S. 1977. *Evolution and the Genetics of Populations. Vol. 3. Experimental Results and Evolutionary Deductions.* Chicago: U. Chicago Press.

Wright, S. 1978. *Evolution and the Genetics of Populations. Vol. 4. Variability Within and Among Natural Populations.* Chicago: U. Chicago Press.

CHAPTER NINETEEN

The Maintenance of Genetic Variation in Life-History Traits

BRIAN CHARLESWORTH AND KIMBERLY A. HUGHES

19.1. Abstract

We present three models of quantitative genetic variation in life-history traits, two of which involve the maintenance by selection of alleles at intermediate frequencies. The other model assumes that alleles with deleterious effects on life-history traits are present at low frequencies and are maintained by the balance between mutation and selection. We compare the predictions of these models with quantitative genetic data on *Drosophila melanogaster*.

The results imply that a detrimental mutation isolated from an equilibrium population typically causes heterozygous and homozygous reductions in net fitness of at least 1% and 5%, respectively. A newly arisen detrimental mutation often seems to affect several life-history traits simultaneously, with an average effect on a given trait of ~40% of its overall effect on fitness under laboratory conditions.

It seems likely that mutation–selection balance can account for only part of the observed genetic variance in most traits, although a balancing-selection model in which heterozygotes are superior to homozygotes for all traits controlled by a given locus can usually be ruled out. Comparisons of measurements of dominance variance with the inbreeding decline suggest that most life-history traits are influenced by many genes with small effects. Virgin male longevity and sperm precedence may be influenced by a small number of polymorphic loci with relatively large effects, suggesting that more detailed genetic investigations of these traits would be rewarding.

19.2. Introduction

Richard Lewontin has long argued that a crucial task for evolutionary geneticists is to understand how genetic variation is maintained in natural populations.

R. S. Singh and C. B. Krimbas, eds., *Evolutionary Genetics: From Molecules to Morphology*, vol. 1. © Cambridge University Press 2000. Printed in the United States of America. ISBN 0-521-57123-5. All rights reserved.

He has also emphasized both the difficulty and the importance of relating variation at the level of interesting phenotypes to variation at the loci that control them (Lewontin, 1974, Chaps. 1 and 2). This difficulty is particularly acute when the phenotype is fitness itself. The long-debated question of whether or not genetic variation in fitness primarily reflects contributions of low-frequency deleterious alleles maintained by the balance between selection and mutation, or has a substantial contribution from variants maintained at intermediate frequencies by selection, is still unanswered.

While increasingly sophisticated methods are being developed to test for the action of natural selection on nucleotide site variants (Kreitman and Akashi, 1995), the very small selection intensities that seem usually to be involved in molecular variation (Gillespie, 1991) mean that any one instance of a DNA variant that is found to be under selection will probably contribute only a minute fraction of the total genetic variation in fitness in a population. A complete understanding of the causes of within-population variation in fitness is thus unlikely to be achieved by the scrutiny of examples of selection on DNA variants at single loci.

There is still scope, therefore, for approaches that attempt to relate estimates of quantitative genetic parameters to the population genetic mechanisms that may affect variation in fitness. The expected lifetime reproductive success of a genotype usually provides an adequate approximation to its net fitness (Charlesworth, 1994, Chap. 4). This is difficult to measure, especially when male mating success and fertility need to be considered. In practice, therefore, students of natural populations are forced to study components of the life history rather than net fitness (Clutton-Brock, 1988; Burt, 1995). Specialized techniques have, however, been developed in *Drosophila* and bacteria that allow variables that are very close to representing net fitness to be measured under controlled laboratory conditions (Latter and Sved, 1994; Lenski and Travisano, 1994).

We define a life-history trait as one for which an increase in its value results in an increase in net fitness w, holding all other traits constant. In other words, for trait k with current value z_k we have

$$\frac{\partial w}{\partial z_k} > 0. \tag{19.1}$$

Traits that satisfy this criterion include developmental rate, age-specific survival probabilities, female fecundity, and male mating and fertilization success. The life history of a genotype can thus in principle be represented by a vector z whose components satisfy this inequality and completely determine fitness (Lande, 1982; Charlesworth, 1990a, 1993).

Fisher's fundamental theorem of natural selection (Fisher, 1930, Chap. 2) implies that additive genetic variance in fitness is exhausted by selection, so that no heritability for fitness is expected in a population that is at equilibrium under selection alone. This result holds under rather general conditions (Kimura, 1958; Charlesworth, 1987). It has led to the often-repeated statement that life-history traits, with their close connection to fitness, should exhibit little or no additive genetic variance, at least in large, randomly mating populations (Haldane, 1949; Robertson, 1955; Roff and Mousseau, 1987; Mousseau and Roff, 1987). But this expectation is contradicted by the fact that life-history traits often display substantial amounts of additive genetic variation (Charlesworth, 1987; Mukai, 1988). In fact, when trait measurements are standardized by dividing by the population mean, life-history traits seem to show significantly more additive genetic variance on average than morphometric traits (Houle, 1992).

Several explanations for the existence of relatively large amounts of additive variation in life-history traits can be imagined (Charlesworth, 1987; Mukai, 1988). Our purpose here is to explore the extent to which simple population genetic models can explain data on easily measured quantitative genetic parameters, such as additive and dominance components of variance and inbreeding depression. We shall discuss data exclusively on *D. melanogaster*, partly because we are most familiar with work on this species, and partly because the sophistication of its genetic technology has allowed more detailed investigations of these questions than are possible in other organisms. As a result, a large body of information on genetic variation in life-history traits of *D. melanogaster* is available.

19.3. Models of the Maintenance of Genetic Variation

Here we summarize the results of single-locus models on the assumption that the contributions from many loci can be adequately approximated by summing over loci, as is conventionally done in quantitative genetics (Falconer and Mackay, 1996). Two alleles per locus will usually be assumed. If individual locus effects are small, variances of trait values expressed as proportions of the mean can be be treated as approximately additive, even if there are multiplicative effects across loci (Wright, 1968, Chap. 10). As recommended by Charlesworth (1987) and Houle (1992), we present variances for traits scaled in this way, which has the advantage of facilitating comparisons among traits with different dimensions. There are three main possibilities with respect to the effects of a locus on the life-history trait under consideration, whose value will usually be denoted by z, dropping the subscript k for convenience (details of the parameters of the models are given in Table 19.1).

19.3.1. Pure Balancing Selection

Alleles are maintained at intermediate frequencies by balancing selection. This could involve overdominance (heterozygote advantage) with constant fitnesses (Fisher, 1922) or frequency dependence of genotypic fitnesses [implying either underdominance or overdominance at equilibrium, when the three genotypes at a locus have different fitnesses (Lewontin, 1958)]. In addition, genotypic differences in z among genotypes are assumed to be directly proportional to the corresponding differences in fitness, with a constant of proportionality α_i for the ith locus. This type of assumption has been used in several studies of

Table 19.1. *Three Models of Genetic Variation*

Genotypes at locus i	$A_i A_i$	$A_i a_i$	$a_i a_i$
Frequencies	p_i^2	$2 p_i q_i$	q_i^2
Model 1 (Balancing Selection)			
Relative fitnesses	$1 - s_i$	1	$1 - t_i$
Relative trait values	$1 - \alpha_i s_i$	1	$1 - \alpha_i t_i$

α_i measures the genotypic effect on trait value z relative to the corresponding effect on fitness. The locus may affect fitness through several different life-history traits, each of which has its own value of α_i [see Eqs. (19.2)].

At equilibrium, $p_i = \frac{t_i}{(s_i + t_i)}$, $q_i = \frac{s_i}{(s_i + t_i)}$ (Fisher, 1922).

Model 2 (Directional Selection)			
Genotypes at locus i	$A_i A_i$	$A_i a_i$	$a_i a_i$
Relative fitnesses		Unspecified	
Relative trait values	1	$1 - h_i \delta z_i$	$1 - \delta z_i$

The effect of homozygosity for allele a_i, relative to the trait value of homozygotes for A_i, is measured by δz_i, which may be positive or negative. The equilibrium allele frequencies cannot be determined without specification of the relationship between fitness and trait value, e.g., through a model of antagonistic pleiotropic effects of different life-history traits on fitness (Rose, 1982; Charlesworth and Hughes, 1996).

Model 3 (Mutation–Selection Balance)			
Genotypes at locus i	$A_i A_i$	$A_i a_i$	$a_i a_i$
Relative fitnesses	1	$1 - h_i s_i$	$1 - s_i$
Relative trait values	1	$1 - h_i \delta z_i$	$1 - \delta z_i$
		$= 1 - \alpha_i h_i s_i$	$= 1 - \alpha_i s_i$

For partially recessive mutations ($h_i > 0$), $q_i = \frac{u_i}{h_i s_i} \ll 1$, where u_i is the rate of mutation from wild-type to mutant alleles at locus i. For fully recessive mutations ($h_i = 0$), $q_i = \sqrt{\frac{u_i}{s_i}}$ (Haldane, 1927).

variation in viability in *Drosophila* (Mukai and Yamaguchi, 1974; Mukai et al., 1974), and is a good approximation if the effects of each locus are small [see Eqs. (19.2) below].

19.3.2. Directional Selection

Alleles are again assumed to be segregating at intermediate frequencies, but one of the two homozygotes takes the highest value of z of the three genotypes. This does not necessarily correspond to the ordering of genotypes with respect to their fitnesses. This uncoupling of the effects of a locus on z and on fitness can occur if there is antagonistic pleiotropy between the homozygous effects of alleles on different life-history traits, i.e., the two homozygotes at a locus differ in opposite directions for different traits. If deleterious allelic effects on life-history traits are sufficiently recessive, such antagonism can lead to heterozygote advantage with respect to net fitness and the maintenance of a stable polymorphism (Rose, 1982, 1985; Curtsinger et al., 1994; Charlesworth and Hughes, 1996). Negative additive genetic correlations between pairs of life-history traits may be established under this model (Dickerson, 1955; Robertson, 1955; Rose, 1982; 1985; Charnov, 1989; Charlesworth, 1990a; Houle, 1991), although this is not inevitable for all pairs of traits if patterns of trade-offs are sufficiently complex (Pease and Bull, 1988; Charlesworth, 1990a).

A similar uncoupling may occur if variation is maintained by temporal or spatial variation in fitnesses, such that there are genotype–environment interactions with respect to fitness and trait values (Gillespie, 1991, Chap. 4; Mukai, 1988) or if there is frequency-dependent selection (Mukai et al., 1974). These processes can lead to a situation in which the relative values of the genotypes at a locus in the genotypic or environmental context in which z is measured differ considerably from those in nature or from their means over their temporal or spatial distributions. Selection acting to preserve genetic variation in nature is then quite consistent with directional selection on fitness-related traits measured in the laboratory, and additive genetic variance in fitness may be observed when there is none in nature (Comstock and Moll, 1963; Mukai, 1988). There is no necessity for negative genetic correlations between antagonistic pairs of traits in this case (Service and Rose, 1985).

19.3.3. Mutation–Selection Balance

Variation is maintained by the balance between the input of deleterious alleles by mutation and their elimination by selection. Genotypic differences in z are assumed to be directly proportional to the corresponding differences in fitness, and mutant alleles always reduce trait values. The mutant alleles would normally be expected to be held at low frequencies. This model is thus mathematically a special case of model 2 (see Table 19.1), with the frequencies of the mutant alleles being determined by the standard formulas for mutation–selection equilibrium (Haldane, 1927). This model has been extensively analyzed in relation to the *Drosophila* data on variation in egg-to-adult viability (Mukai et al., 1972, 1974; Mukai and Yamaguchi, 1974; Crow and Simmons, 1983; Tachida et al., 1983; Mukai, 1988; Crow, 1993).

Fitness may be affected by several life-history traits controlled by the same locus. Evidence for such multiple effects of mutant alleles on life-history traits has been discussed by Simmons and Crow (1977) and Houle et al. (1994). If homozygosity for a given allele at the ith locus causes a small reduction of δz_{ik} in the kth trait, below the maximal value for genotypes at this locus, the net reduction in fitness is approximated by

$$\delta w_i = \sum_j \left(\frac{\partial w}{\partial z_{ij}} \right) \delta z_{ij}. \tag{19.2a}$$

α_i for the kth trait is then given by

$$\alpha_{ik} = \frac{\delta z_{ik}}{\sum_j \left(\frac{\partial w}{\partial z_j}\right)\delta z_{ij}}. \tag{19.2b}$$

Broadly speaking, models 1 and 2 correspond to the balanced hypothesis for the maintenance of natural variation, and model 3 corresponds to the classical hypothesis (Dobzhansky, 1955; Lewontin, 1974). While models 1 and 2 share the (somewhat ill-defined) property of allele frequencies that are not close to zero or one, they make very different predictions about observable statistical properties of variation, as will be discussed below.

19.4. Expected and Observed Population Parameters

We now examine the predictions of these models with respect to various population parameters and compare them with parameter estimates from data on *D. melanogaster*. These predictions will be developed in terms of the standard variables of quantitative genetic variation and genetic load theory, such as genetic variance components and inbreeding load. Detailed descriptions of the underlying theory can be found in standard treatments, such as Crow (1993) and Falconer and Mackay (1996).

19.4.1. Additive and Dominance Variances

In random-mating populations, it is straightforward to estimate additive and dominance components of the genetic variance (V_A and V_D, respectively). This can be done for the whole genome by use of breeding designs that provide estimates of the degree of resemblance between full-sibs and half-sibs, if complications such as epistasis and linkage disequilibrium are ignored (Falconer and Mackay, 1996). (The consequences of relaxing these assumptions are examined in Section 19.4.) In *D. melanogaster*, multiply inverted balancer chromosomes allow the isolation of sets of wild-type chromosomes from natural or laboratory populations (Lewontin, 1974, p. 39). Intercrosses between such sets create diallel crosses, which enable measurement of the effects of just one of the three major chromosomes (X, second, or third). Partitioning of the components of variance in such diallels enables V_A and V_D to be estimated (Mukai et al., 1974; Hughes, 1995a).

Predictions from the three models are summarized in Table 19.2. These are simple extensions of standard results (Mukai et al., 1974; Tachida et al., 1983; Hughes, 1995b; Charlesworth and Hughes, 1996). Under model 1, V_D is the only component of genetic variance, since no additive genetic variance for fitness exists at equilibrium and genotypic differences in z are directly proportional to differences in fitness (Haldane, 1949; Robertson, 1955; Mukai et al., 1974). Under model 2, both V_A and V_D are expected at equilibrium, although the ratio V_A/V_D is expected to be higher for late-life traits, which are more weakly

Table 19.2. *Genetic Variance Components for a Life-History Trait under the Three Models*

Model	Additive Variance (V_A)	Dominance Variance (V_D)	Homozygous Variance (V_G)
1	0	$\displaystyle\sum_i \frac{(s_i t_i \alpha_i)^2}{(s_i + t_i)^2}$	$\displaystyle\sum_i \frac{s_i t_i [(s_i - t_i)\alpha_i]^2}{(s_i + t_i)^2}$
2	$\displaystyle\frac{1}{2}\sum_i p_i q_i [1 + (1 - 2h_i) \times (q_i - p_i)]^2 (\delta z_i)^2$	$\displaystyle\sum_i [p_i q_i (1 - 2h_i)\delta z_i]^2$	$\displaystyle\sum_i p_i q_i (\delta z_i)^2$
3			
$h_i > 0$:	$\displaystyle 2\sum_i u_i h_i s_i \alpha_i^2$	$\displaystyle\sum_i \frac{[u_i (1 - 2h_i)\alpha_i]^2}{h_i^2}$	$\displaystyle\sum_i \frac{u_i s_i \alpha_i^2}{h_i}$
$h_i = 0$:	$\displaystyle 2\sum_i \left(\frac{u_i}{s_i}\right)\sqrt{u_i s_i}\,\alpha_i^2$	$\displaystyle\sum_i u_i s_i \alpha_i^2$	$\displaystyle\sum_i (\sqrt{u_i s_i})\alpha_i^2$

selected than traits expressed early in life (Charlesworth and Hughes, 1996). With intermediate allele frequencies, V_A will generally be larger than V_D, even with a high degree of recessivity of the allele that causes a reduced value of z. For example, if allele frequencies are equal to one-half at all loci, we have $V_A/V_D = 8$ when the dominance coefficient h is 0.25 at all loci, and $V_A/V_D = 2$ with complete recessivity. Model 2 cannot therefore be ruled out by the observation of substantial additive variance relative to dominance variance (Rose, 1985).

On model 3, a high V_A/V_D ratio is expected for an early-life-history trait, unless mutant effects on z are completely recessive, which seems unlikely to be true for all loci (see below). Traits that are expressed very late in life, for which deleterious alleles can rise to high frequencies, may have a relatively high value of V_D (Charlesworth and Hughes, 1996). These results for model 3 follow from the fact that the contribution to V_D from a locus segregating for a rare, partially recessive allele is proportional to the square of the frequency of the rare allele, whereas V_A is proportional to its frequency (Mukai et al., 1974).

Table 19.3 gives some examples of estimates of variance components for *D. melanogaster* life-history traits, expressed in terms of the corresponding standard deviations measured as percentages of the population trait means (additive and dominance coefficients of variation, CV_A and CV_D, respectively). In most cases, there is statistically significant V_A, but significant V_D is found only occasionally, notably for longevity and sperm precedence. V_A is usually much greater than V_D, with the exception of female longevity and sperm precedence. This seems to refute model 1 as a major mechanism for maintaining variation in such life-history traits as viability, female fecundity, and male mating success (see Mukai et al., 1974), but does not clearly distinguish between models 2 and 3, especially as the variance components are subject to large errors of estimation, so that the

Table 19.3. *Coefficients of Additive and Dominance Variation for some Life-History Traits*

Character	CV_A	CV_D
Female longevity	5.3	17.5
Early female fecundity	12.5	5.0
(Whole genome, *IV* laboratory population; Rose and Charlesworth, 1981)		
Egg-to-adult viability	9.5	3.5
(Second chromosome, North Carolina natural population; Mukai et al., 1974)		
Early male mating success	7.5	0
Late male mating success	13.9	0
Virgin male longevity	8.3	3.6
(Third chromosome, *IV* laboratory population; Hughes, 1995a)		
Sperm precedence of first-mated male		
	0	40.1
Sperm precedence of second-mated male		
	0	25.7
(Third chromosome, *IV* laboratory population; Hughes, 1997)		
Zero entries denote cases in which the relevant variance component estimates were negative.		

ratios CV_A/CV_D mostly have wide confidence intervals. Sperm precedence, with its unusually high level of dominance variance and apparent absence of additive variance, is the only candidate for a trait in which variation is controlled by model 1.

19.4.2. Homozygous Genetic Variances

Lines that are homozygous for whole chromosomes can be constructed by use of balancer chromosomes. Provided that the genetic background has been homogenized across lines, the between-line component of variance provides an estimate of the variance contributed by the homozygous effects of alleles that are segregating in the population, V_G (this should not be confused with the total genetic variance in an outbred population, which is equal to the sum of V_A and V_D). Expressions for V_G under the different models are given in Table 19.2. Under model 1, the ratio V_A/V_G should be very small, since V_A is expected to be zero, unless allele frequencies at all loci affecting z are equal to one-half, in which case V_G is also zero. In contrast, V_D/V_G should substantially exceed one, unless selection against one homozygote is much stronger than against the other.

No simple and general prediction about the magnitude of V_A/V_G is provided by model 2. If there is no dominance or if allele frequencies are equal to one-half at all loci, it is easily seen that $V_A/V_G = 0.5$. Values of V_A/V_G that are close to 0.5 therefore suggest that allele frequencies or dominance coefficients are generally intermediate, under model 2. A value of V_A/V_G that is close to zero suggests that the rare alleles at each locus are mostly recessive or nearly recessive.

Under model 3, with nonrecessive mutant effects the ratio $V_A/2V_G$ can be written as

$$\frac{V_A}{2V_G} = \frac{\sum_i u_i h_i s_i \alpha_i^2}{\sum_i u_i s_i \alpha_i^2} \bigg/ \frac{\sum_i u_i s_i \alpha_i^2}{\sum_i \dfrac{u_i s_i \alpha_i^2}{h_i}}. \tag{19.3}$$

This ratio is thus equal to the product of the weighted harmonic mean and arithmetic means of h_i (with weights of $u_i s_i \alpha_i^2$) and is therefore likely to be smaller than 0.25 for traits that show inbreeding depression, for which a mean dominance coefficient of less than 0.5 is required (see below). With completely recessive mutant effects at all loci under model 3, the ratio $V_A/2V_G$ is of the order of the ratio of the mutation rate to the selection coefficient at each locus and is thus likely to be very small.

The values of $\sqrt{V_G}$ (expressed as a coefficient of variation) and $\sqrt{V_A/(2V_G)}$ (which provides a measure of mean dominance under model 3) are shown in Table 19.4 for the data of Takano et al. (1987) on egg-to-adult viability. These are for sets of nonlethal chromosomes from which chromosomes with viabilities less than 50% of the balancer heterozygotes (which presumably carry genes with major deleterious effects) have been excluded. Chromosomes with such high viabilities when homozygous are often referred to as mildly detrimentals, whereas all chromosomes that are nonlethal when homozygous are called detrimentals (Mukai et al., 1972). We follow this terminology here. The results of Hughes (1995b, 1997) for male mating success, sperm precedence, and longevity are also shown in Table 19.4.

Table 19.4. *Coefficients of Homozygous Genetic Variation for Some Life-History Traits*

Character	CV_G	$CV_A/(\sqrt{2}CV_G)$
Egg-to-adult viability	10.9	0.17
(Second chromosome, Northern Japan population; Takano et al., 1987)		
Egg-to-adult viability	20.2	0.17
(Second chromosome, Southern Japan population; Takano et al., 1987)		
Early male mating success	20.1	0.26
Late male mating success	19.6	0.50
Virgin male longevity	15.1	0.39
(Third chromosome, *IV* laboratory population; Hughes, 1995b)		
Sperm precedence of first-mated male		
	53.9	0
Sperm precedence of second-mated male		
	52.9	0
(Third chromosome, *IV* laboratory population; Hughes, 1997)		

The estimates of $\sqrt{V_A/(2V_G)}$ for viability are both ~0.17, whereas the estimates for the other life-history variables (except sperm precedence) are higher (0.26 or more) and are consistently larger than the mean h_i estimated by the regression method described in subsection 19.3.3. (but see Caballero et al., 1997) This may reflect a contribution of model 2 variability to these traits. The sampling errors of these estimates are high, however, so this conclusion needs to be viewed with caution. For male longevity, there is is a clear rejection of the hypothesis of predominantly recessive mutant effects under model 3 (Hughes, 1995b). For viability, the dominance coefficient estimated by the regression method described below agrees quite well with $\sqrt{V_A/(2V_G)}$, so that there is no reason to reject model 3 on this basis alone. Apart from the two measures of sperm precedence, which have extraordinarily high levels of homozygous genetic variation, the values of CV_D in Table 19.3 are consistently very much lower than the corresponding CV_G values in Table 19.4. This is inconsistent with model 1 unless allele frequencies are very extreme at all loci, which seems improbable.

19.4.3. Relations between Heterozygous and Homozygous Line Means

Another method for discriminating between mutational variation and selectively maintained variation involves estimation of the regression of the trait values for the heterozygotes formed by crossing pairs of lines (each of which is homozygous for a single independent chromosome) on the sums of the trait values for the pairs of homozygous lines that contribute to the heterozygotes (Mukai and Yamaguchi, 1974; Caballero et al., 1997). The expected value of the regression coefficient b for chromosomes extracted from a natural population was derived by Mukai et al. (1972) and Mukai and Yamaguchi (1974).

Under model 1, no relation between homozgygous and heterozygous trait values is expected (Mukai and Yamaguchi, 1974). The general formula under models 2 and 3 is

$$b = \frac{\sum_i p_i q_i (\delta z_i)^2 [h_i + q_i(1 - 2h_i)]}{\sum_i p_i q_i (\delta z_i)^2 (1 + 2q_i)}. \tag{19.4}$$

For model 3, second-order terms in the q_i can be neglected, so that b provides a weighted estimate of the mean of the h_i [the weights are $p_i q_i (\delta z_i)^2 = p_i q_i (\alpha_i s_i)^2$]. Under model 2, b is expected to be positive if the h_i are mostly <0.5, as seems likely from the data on inbreeding effects (see below) and as is required by the antagonistic pleiotropy model (Rose, 1982), but its quantitative relation to the mean of h_i is unclear in general. For example, if allele frequencies at all loci are equal to one-half, it is easy to see that $b = 0.25$, independent of the distribution of h_i.

In practice, it is hard to measure b accurately. Several large-scale experiments on the viability effects of detrimental second chromosomes have yielded values between 0.20 and 0.30 (Mukai et al., 1972, 1974; Takano et al., 1987), and data on the viability effects of chromosome 3 gave $b = 0.40$ (Watanabe et al., 1976). A measure of female fertility gave $b = 0.07$ (Watanabe and Ohnishi, 1975).

The data of Hughes (1995b) for the third chromosome gave b values of 0.07, 0.27, and 0.30 for early male mating success, late male mating success, and male longevity, respectively. The high sampling errors of these estimates means that it is hard to evaluate the significance of these differences among estimates, but they are consistently less than 0.5. Their overall mean is approximately 0.23, but a value of 0.32 is obtained after correcting for bias due to sampling error (Caballero et al., 1997). This seems to rule out model 1 once again, but is consistent with either model 2 or 3.

The mean value of the dominance coefficient for newly arisen mutations can similarly be estimated from chromosomes that have accumulated spontaneous mutations (see Subsection 19.3.4 below), but has an even higher sampling error than that for mutations sampled from nature. Mukai and Yamazaki (1968) obtained an estimate of ~0.40 for detrimental viability mutations on the second chromosome. This is much larger than the mean estimated for chromosomes from nature, as would be expected on model 3 if mutations differ in their degree of dominance, such that more dominant alleles have shorter persistence times before elimination by selection (Mukai, 1969b; Mukai and Yamaguchi, 1974; Watanabe et al., 1976). Other data suggesting nearly intermediate dominance for detrimental viability mutations are reviewed by Caballero and Keightley (1994).

Houle et al. (1997) obtained estimates of −0.03, 0.12, 0.37, 0.26, and −0.07 for second-chromosome detrimental mutations affecting early female fecundity, late female fecundity, male longevity, female longevity, and early male mating success, respectively. Apart from the value for early female fecundity, which differs significantly from 0.5, these all have very high sampling errors and cannot be taken very seriously. Nevertheless, the fact that the mean dominance coefficient is apparently only 0.13 for these traits suggests that their spectrum of heterozygous mutational effects may be different from those affecting viability. Further investigation of this point is important, since model 3 can clearly be rejected as a complete explanation of genetic variability if estimates of mean dominance coefficients from nature generally turn out to be larger than for new mutations.

19.4.4. Mutational Parameters and Homozygous Variances

The mutational variance V_M is defined as the amount of new variance that arises per generation as a result of spontaneous mutation (Simmons and Crow, 1977). We can estimate V_M for homozygous mutations for a single chromosome of D. melanogaster by accumulating mutations independently on wild-type chromosomes that are all derived from the same extracted chromosome and by assaying chromosomal homozygotes periodically. Such accumulation can be done by means of backcrosses of single males heterozygous for the balancer and wild-type to balancer stock females, thereby sheltering the wild-type chromosome from selection against all mutations except for those with strongly deleterious effects when heterozygous (Mukai, 1964, 1969a; Mukai et al., 1972; Ohnishi, 1977; Houle et al., 1994).

Under models 1 and 2, no strong relation between mutational and populational variance is expected. Under model 3, it is intuitively obvious that there
must be a relation between mutational parameters and the properties of an
equilibrium population, and this intuition is confirmed by detailed analysis
(see Crow, 1993; Houle et al., 1996). With the parameters defined in Table
19.1, V_M is given by

$$V_M = \sum_i u_i \alpha_i^2 s_i^2. \tag{19.5}$$

From Table 19.2, it is easily seen that under model 3 the ratio V_G/V_M provides
an estimate of the weighted harmonic mean of $h_i s_i$, H, with individual locus
contributions weighted by $u_i \alpha_i^2 s_i^2$.

Empirical estimates of mutational and homozygous genetic variances can be
compared with these equations. Mukai et al. (1972) obtained an estimate of V_M
of approximately 9.4×10^{-5} for mildly detrimental viability mutations on the
second chromosome, while Ohnishi (1977) obtained a value of approximately
5.0×10^{-5}. With the lower estimate of V_G for mildly detrimental chromosomes
from Table 19.4, on the grounds that the higher estimate from the southern
Japanese population may well reflect a contribution to genetic variance from
balancing selection (Takano et al., 1987), the two estimates of V_M give $H =$
0.008 and 0.004, respectively.

Houle et al. (1994) obtained estimates of V_M for six fitness components (early
female fecundity, later female fecundity, male longevity, female longevity, early
male mating success, and late male mating success). These give a mean value
of $V_M = 6.8 \times 10^{-4}$, rather higher than the value obtained for detrimental
viability mutations by Mukai et al. (1972) and Ohnishi (1977), 3.3×10^{-4} and
2.1×10^{-4}, respectively. If the values for the two estimates of male mating success
(which have very high sampling variances and are not statistically significant)
are omitted, the mean V_M is 2.3×10^{-4}. From Table 19.4, we have a mean V_G
value of 0.033 for three male fitness components (early male mating success,
late male mating success, and longevity) for third chromosomes sampled from
a laboratory population. The relative size of the euchromatic portion of the
third chromosome is ~1.2 times that of the second (Charlesworth et al., 1992),
so that V_M for the third chromosome should probably be adjusted by the same
factor. Overall, the average estimate of H for these traits is 0.008, if the lower
estimate of mean V_M is used. This is identical to the value for mildly detrimental
viability mutations, obtained with the value of V_M of Mukai et al. (1972).

Crow and Simmons (1983) and Crow (1993) have suggested another estimator for the harmonic mean of $h_i s_i$. This is obtained by comparison of the
mutational decline in mean viability of chromosomal homozygotes associated
with mildly detrimental viability mutations, $\sum u_i \alpha_i s_i$, with the reduction below
zero in log mean viability associated with homozygosity for mildly detrimental
chromosomes extracted from a natural population, $\sum u_i \alpha_i s_i / (h_i s_i)$. The ratio
of these two quantities estimates the harmonic mean of $h_i s_i$, weighted by $u_i \alpha_i s_i$.

Table 19.5. *Inbreeding Load (B)*
under the Three Models of Variation

Model	B
1	$\sum_i \dfrac{s_i t_i \alpha_i}{(s_i + t_i)}$
2	$\sum_i p_i q_i (1 - 2h_i)\delta z_i$
3	
$h_i > 0$:	$\sum_i u_i \left(\dfrac{1}{h_i} - 2\right)\alpha_i$
$h_i = 0$:	$\sum_i (\sqrt{u_i s_i})\alpha_i$

A problem with this approach is that the balancer method of measuring viability is based on the ratio of balancer heterozygotes to wild-type flies in a culture, so that the viabilities of chromosomal homozygotes are usually measured relative to those of chromosomal heterozygotes, not to the viability of mutation-free individuals. The logarithm of this relative measure is equivalent to the inbred load B, defined in Subsection 19.3.6 below. The formula for B under model 3 (Table 19.5) implies that it is an underestimate of $\sum u_i \alpha_i s_i / (h_i s_i)$, by $2 \sum u_i \alpha_i$. The use of B to estimate the homozygous reduction in fitness thus biases the estimate of mean heterozygous effect on fitness.

This bias can be corrected as follows. There is evidence from mutation accumulation experiments that the diploid mutation rate $U = 2\sum u_i$ for the second chromosome may be at least 0.4 (Mukai et al., 1974; Ohnishi, 1977; Keightley, 1994). Using a value of $U = 0.4$ in conjunction with the estimate of the mean value of α_i of 0.37 derived in Subsection 19.3.5 gives a value of 0.010 for the harmonic mean of $h_i s_i$, instead of 0.025, as estimated by Crow (1993), with Crow's values of 0.15 for B for mildly detrimental mutations and 0.003 for the rate of mutational decline in viability.

The relatively good agreement between the different estimates of the harmonic mean of $h_i s_i$ is consistent with the hypothesis that mutation–selection balance may contribute a large portion of the standing variance in fitness and its components in *Drosophila* and suggests that, on average, detrimental mutations have substantial heterozygous effects on fitness. But in themselves, the results do not prove the validity of model 3, since variation maintained by selection under models 1 and 2 can contribute to both the inbreeding decline and to the genetic variance. It is useful to note that contributions from variation maintained by selection cause both types of estimate to be downwardly biased, so that the conclusion that deleterious mutations have a mean heterozygous effect on fitness of ~1% is, if anything, a conservative one (see Subsection 19.3.5). Given the evidence for dominance coefficients of the order of 0.2 for detrimental mutations in an equilibrium population (section 19.3.3), these results suggest that a typical deleterious mutation found in a natural population has a homozygous selection coefficent of at least 0.05.

19.4.5. Mutational Parameters and Additive Genetic Variances

The question of whether the amount of genetic variance is consistent with mutation–selection balance alone can also be approached as follows. Assuming the same $\alpha_i = \alpha$ for all loci, the expression in Table 19.2 for the equilibrium additive variance under model 3 for nonrecessive mutations becomes

$$V_A = U\alpha^2 \overline{hs}, \tag{19.6}$$

where $\overline{hs}$ is the arithmetic mean of $h_i s_i$, with weights of u_i (Mukai et al., 1974; Tachida et al., 1983; Mukai, 1988).

A prediction of V_A can be obtained by use of an estimate of 0.01 for $\overline{hs}$, which is expected to be larger than the harmonic mean H discussed in Subsection 19.3.4 (see below). The ratio of the estimate of V_M for fitness of 17×10^{-4} (Houle et al., 1992) to the mean for the two estimates of V_M for nonlethal viability mutations and for accurately measured fitness components (Houle et al., 1994) suggests an average value of α_i^2 (weighted by s_i^2) of approximately $2.4/17.0 = 0.14$. With $U = 0.4$, these parameter values give a predicted V_A of 0.0006. This is very low compared with the value of 0.005 suggested as representative for the contribution from a major autosome to variation in traits other than sperm precedence (Table 19.3). It is even inconsistent with the lowest reported values of V_A for viability effects of the second chromosome [0.0023–0.0027 for northern Japanese populations (Mukai, 1988)]. Even if $\alpha = 0.5$, as suggested by Mukai (1988), the predicted V_A is only 0.001.

But the use of H to estimate $\overline{hs}$ gives excessive weight to smaller $h_i s_i$ values; if there is a very wide distribution of $h_i s_i$, such as an exponential distribution, its arithmetic mean could be as large as double the harmonic mean (see Mukai et al., 1972). This would bring the estimate of V_A up to 0.0012 with $\alpha^2 = 0.14$ and to 0.0025 with $\alpha^2 = 0.25$. In addition, from the argument in Subsection 19.3.4, the presence of nonmutational variation means that the true value of H is approximately equal to the estimated value, divided by the proportion of V_G from mutational sources, θ. Assume as a first approximation that the proportion of V_A due to mutational causes is also equal to θ, so that the expected V_A is $1/\theta$ times the mutational component. Since $\overline{hs} = H$, we have

$$\theta \geq \sqrt{\frac{U\alpha^2 H}{V_A}}. \tag{19.7}$$

Equating observed and expected V_A and substituting into (19.7) $U = 0.4$, $\alpha^2 = 0.14$, $H = 0.01$, we find $\theta \geq 0.33$ when $V_A = 0.005$ and $\theta \geq 0.47$ when $V_A = 0.0025$. If $\overline{hs} = 2H$, then θ is $\sqrt{2} = 1.4$ times these values. This suggests that, even for populations with the lowest levels of V_A for viability, only 67% of the variance at most is likely to be due to mutation–selection balance; the fraction is much smaller with the more typical value, $V_A = 0.005$, or with a much smaller variance of the distribution of mutational effects.

It should be noted that this conclusion disagrees with those of Kusakabe and Mukai (1984) and Houle et al. (1996). Kusakabe and Mukai relied on a comparison of V_A with inbreeding load to test model 3. As shown in Section 19.3.6 below, the effect of inbreeding seems to be too large to be entirely accounted for by model 3 if epistasis is ignored, so that the inflation of both V_A and inbreeding load by a similar degree would give a false picture of agreement with model 3. With synergistic epistasis among deleterious mutations (Mukai, 1969a; Kondrashov, 1988; Crow, 1993), the load will be increased over the value expected without epistasis (Charlesworth, 1998), but there will be little effect on V_A (Charlesworth and Barton, 1996). This could mask a contribution from nonmutational sources to V_A and thereby generate a spurious agreement with model 3. Houle et al. (1996) use a method similar to that of Subsection 19.3.4, but applied to the ratio V_A/V_M rather than V_G/V_M. This means that the effect of dominance of deleterious mutation is not taken into account, and the resulting estimate of the harmonic mean of the $h_i s_i$ is seriously downwardly biased.

Unless $U = 0.4$ is a gross underestimate of the deleterious mutation rate for the second chromosome, these results imply that recurrent mutation might contribute between 33% and 67% of the variation in a typical average life-history trait, suggesting that there is often likely to be a significant contribution to variability from sources other than deleterious mutations. The very much higher additive variances for viability (of the order of 0.02) found in southern Japanese populations, compared with the values in northern Japanese or U.S. populations, strongly suggest that there is a contribution from balancing selection to variation in viability in these populations (Mukai and Nagano, 1983; Tachida et al., 1983; Mukai, 1988; Takano et al., 1987).

19.4.6. Evidence from Inbreeding Effects

The dominance coefficient estimates discussed above can be compared with the results of studies of the inbreeding load B, measured as the difference between chromosomal heterozygotes and homozygotes in log mean values for fitness components (Greenberg and Crow, 1960; Simmons and Crow, 1977; Charlesworth and Charlesworth, 1987; Mukai, 1988; Charlesworth and Hughes, 1996). The predictions of the three models about the magnitude of the inbreeding load on the three models are summarized in Table 19.5. Multiplicative fitness effects across loci are assumed.

For model 3, Table 19.5 implies that the inbreeding load can be predicted from the previously estimated parameters, if α_i, u_i, and h_i are distributed independently. With $U = 0.48$ for the third chromosome and a mean h_i of 0.25, the expected B for fitness is 0.48, assuming a lack of covariance between h_i

and u_i. With mean $\alpha_i \approx 0.37$ (Subsection 19.3.5), the expected B for a fitness component is 0.18.

For the third chromosome of *D. melanogaster*, we have a mean B estimate of 0.28 for detrimental viability mutations (Simmons and Crow, 1977; Table 19.2) and estimates of 0.19, 0.50, 0.09 for early and late male mating success and male longevity, respectively (Hughes, 1995b; Table 19.1). For first male sperm precedence and second male sperm precedence, B values of 0.29 and 0.40 were obtained by Hughes (1996). The mean B over all these traits is 0.30 (which is 67% larger than the predicted value) or 0.27 if sperm precedence is excluded, as may be warranted because of its unusual features (see Subsections 19.3.1 and 19.3.2). Two estimates of the inbreeding effects on net fitness for chromosome 3 give a mean value for B of 1.72 (Tracey and Ayala, 1974; Sved, 1975), which is far greater than expected. Mutational variation thus seems to be capable of explaining only ~30% of the observed inbred load for net fitness and ~60% of the average load for the fitness components considered here.

This suggests that one or more life-history traits, which are not included among those measured, might contribute substantially to variation in net fitness, through alleles maintained by selection. Alternatively, some traits may tend to be affected by much more recessive alleles than is indicated by the measurements of mean dominance coefficients, or the total mutation rate for fitness could be much higher than we have assumed. Watanabe and Ohnishi (1975) reported a B value of 0.54 for a measure of the female fertility of nonsterile second-chromosome homozygotes, which is much greater than the theoretical expectation of 0.15. This suggests that female fertility may be an important contributor to the load for net fitness. It is interesting that their estimate of B for fertility was only 0.075, with an upper confidence limit of 0.17, which suggests a greater degree of recessivity for alleles affecting female fertility than for the other life-history components. This is consistent with the relatively high load of 0.48 for net fitness effects of the X chromosome (Wilton and Sved, 1979), which contrasts with the near-zero load for viability for this chromosome (Eanes et al., 1985). Such a difference would be expected if there is a substantial contribution to net fitness from fertility genes with female-limited effects.

It is harder to make numerical predictions about the contributions to inbred load from loci maintained at intermediate frequencies, but there are some constraints on the relation between dominance variance and B, which can be explored as follows. For all three models, it is easily seen from Tables 19.2 and 19.5 that, if we write B_i for the inbred load associated with locus i, we have

$$B = \sum_i B_i = m\bar{B}, \qquad (19.8a)$$

$$V_D = \sum_i B_i^2 = m(\bar{B}^2 + V_B), \qquad (19.8b)$$

where m is the number of loci affecting the trait, $\bar{B}$ is the mean inbred load per locus, and V_B is the variance in inbred load among loci.

Formulas similar to these were previously derived by Charlesworth (1969) and Mukai et al. (1974) for model 1, but they clearly apply much more generally. They can be used to place lower and upper bounds on m and $\bar{B}$, since they imply that

$$\frac{B^2}{V_D} \leq m, \qquad (19.9a)$$

$$\frac{V_D}{B} \geq \bar{B}. \qquad (19.9b)$$

On the assumption that most variation is maintained by mutation–selection balance, we would expect the bound on m to be large and the bound on $\bar{B}$ to be small, since a large number of loci must contribute to inbred load on this model, and the load per locus is of the order of the mutation rate for partially recessive mutations. If a significant fraction of variation involves models 1 or 2, then the converse might be true.

This suggests that the use of these inequalities may provide a useful, and fairly assumption-free, way of detecting variability maintained by selection. The chief practical difficulty in doing this is that few accurate estimates are available of dominance variance for life-history traits for which we also have estimates of inbred load. The most extensive data are for second chromosome effects on viability. The tabulation of V_D estimates for several populations by Mukai (1988) give a mean estimate of V_D of 0.0008 for the second chromosome and a mean B of 0.27 for nonlethal chromosomes. Simmons and Crow (1977; Table 19.1) give a mean estimate of B of 0.13 for mildly detrimental chromosomes and a mean of 0.24 for all detrimentals. The lower bounds to m from the lower and the higher values of B are 22 and 91, respectively. The corresponding upper bounds to $\bar{B}$ are 0.0006 and 0.0003. For chromosome 3, Hughes (1995a, 1995b) reported a fairly accurate estimate of V_D for virgin male longevity of 0.0013 and a B of 0.081. These give $m \geq 5$ and $\bar{B} \leq 0.02$. This suggests that a relatively small number of polymorphic loci may influence male longevity. Female longevity also has a relatively high dominance variance (Table 19.3), suggesting that it may behave in a similar way. There is supporting evidence from the existence of negative additive correlations between female longevity and early fecundity (Rose and Charlesworth, 1981; Service et al., 1988) and from effects of polymorphic superoxide dismutase alleles on longevity (Tyler et al., 1994).

For first male sperm precedence, we have $m \geq 0.52$ and $\bar{B} \leq 0.56$ and for second male sperm precedence we have $m \geq 2.4$ and $\bar{B} \leq 0.165$ (Hughes, 1996). These two traits are apparently quite highly correlated genetically, so that the estimates are probably not independent. The results indicate that sperm precedence may be influenced by a relatively small number of polymorphic loci with rather large effects, consistent with the findings of Clark et al. (1995) of associations between parameters of sperm displacement for chromosomal homozygotes and molecular variation at candidate loci, notably the male accessory gland protein genes on chromosome 2.

19.5. Discussion

We have reviewed the results of over 25 years of investigations of the quantitative genetics of life-history traits of *D. melanogaster* and have reached the following conclusions.

(1) There is probably a substantial contribution from deleterious alleles maintained by mutation to the standing genetic variance in fitness-related traits and to the genetic load revealed by inbreeding. Deleterious mutations must usually be partially recessive, rather than fully recessive or additive in their effects, in order for this to be true. This is consistent with measurements of the average level of dominance of newly arisen deleterious mutations, although more extensive data on mutational parameters for traits other than viability are badly needed.

(2) Nevertheless, mutational contributions are unlikely to be sufficient to explain all of the observed genetic variation and inbreeding depression. Current estimates of mutation and selection parameters suggest that there is too much additive genetic variance and too high a reduction in life-history traits under inbreeding to be accounted for by a purely mutational model. In some instances, such as male longevity and sperm precedence, relatively high dominance variance compared with additive genetic variance and inbreeding depression strongly indicates the existence of polymorphic loci with alleles of relatively large effect.

(3) A model of selection in which heterozygotes are superior to homozygotes with respect to all life-history traits controlled by the loci in question (and in all environments) is decisively rejected by the data, with the possible exception of sperm precedence. There is nearly always too little dominance variance relative to additive genetic variance, and the genetic variance of chromosomal homozygotes is too high relative to that of chromosomal heterozygotes, for this model to be accepted.

(4) Variation that is not mutational in origin must therefore usually reflect directional selection at the level of individual traits in the environment and genetic background in which they are measured. This may arise from antagonistic pleiotropic deleterious effects of partially recessive alleles on different life-history traits (leading to net heterozygote advantage), from temporally and spatially varying selection pressures, or from frequency-dependent selection.

These conclusions have been derived by use of a number of simplifying assumptions, and it is clearly important to examine how sensitive they are to these assumptions. They are also subject to the caveat that the data are all derived from laboratory experiments, so that there may be only a loose relation between measured trait values of genotypes and their values in nature. If the model enshrined in Eqs. (2) approximates reality, this should not be a serious source of error, since we need assume only that the effect of a genotype on fitness in nature is an increasing function of its effect on the trait in the laboratory, which is not a particularly strong assumption for a life-history trait.

The first important simplification is that additive or multiplicative models have been used to predict variance components and inbreeding effects. The conclusion that there is usually low dominance variance relative to additive variance might be thought to be an artifact of this assumption, since epistatic variance components have been neglected in the partitionings of variance reviewed here. Tachida and Cockerham (1988) have discussed this issue for a specific model of variance in fitness caused by stabilizing selection on a quantitative trait, but similar considerations apply more generally (Griffing, 1956; Kempthorne, 1957, Chaps. 19 and 20). If additive × additive epistasis exists, it will give similar contributions to the estimates of V_A and V_D derived from diallel analyses and a smaller contribution to the estimate of V_A than to V_D from sib analyses. Epistasis involving dominance effects clearly contributes to the estimate of only V_D. Hence epistasis will tend to reduce the estimate of V_A/V_D, so that conclusions based on observing large values of this ratio remain robust. As noted by Tachida and Cockerham (1988), cases such as viability, female fecundity, and male mating success, for which the estimates of V_A/V_D are large, do not support the notion that epistasis contributes much to the genetic variance in fitness-related characters. This seems to rule out some models of stabilizing selection, such as the Gaussian model of allelic effects (Kimura, 1965; Lande, 1975), which predict such epistatic variance in fitness. In contrast, the apparent absence of epistatic variance is consistent with mutation–selection balance with synergistic epistasis, which generates only a small amount of epistatic variance (Charlesworth and Barton, 1996). But a substantial amount of epistatic variance due to dominance terms could enter into the genetic variance of chromosomal homozygotes, V_G, so that some caution should be exercised with regard to conclusions based on high values of this parameter.

We have also neglected linkage disequilibrium in the analysis of variance components. This is more difficult to deal with theoretically. There are two models of mutational variation for which linkage disequilibrium contributions to variance in fitness have been computed, stabilizing selection (Tachida and Cockerham, 1988) and synergistic selection (Charlesworth and Barton, 1996). The results for both of these cases show that the linkage disequilibrium contribution to the estimated additive variance is very small compared with the true V_A. It is therefore unlikely that there is a serious error in our conclusion that V_A is generally too large to be explained solely by mutation–selection balance.

In contrast, synergistic epistasis between the effects of mutations could considerably increase the predicted inbreeding load B. This question has been examined by Charlesworth (1998), who showed that a moderate degree of synergism is capable of explaining observed values of B. While this result cannot be taken too seriously, given the uncertainties in the parameters involved, it is clear that synergistic epistasis may greatly enhance the expected inbred load in *Drosophila*, so that the conclusion, that mutational load is inadequate to account for observed values of B, may be premature. Unfortunately, experiments to detect epistatic effects on inbred load have yielded conflicting results (Simmons and Crow, 1977; Seager and Ayala, 1982; Seager et al., 1982; Clark, 1987). This may reflect methodological problems, rather than truly weak effects of synergism (Charlesworth, 1990b). In view of the important implications of synergistic interactions between deleterious mutations for a variety of evolutionary questions (Kondrashov, 1988), more research is badly needed.

If synergism really does play an important role, we then face the problem of reconciling the excess additive variance for many life-history traits with the lack of an apparent contribution of variation maintained under model 2 to inbreeding effects. One possibility is that the loci involve little directionality of their dominance coefficients (see Table 19.5). This is consistent with the existence of heterozygote advantage as a result of antagonistic pleiotropy, if the direction of dominance varies across loci, but there is a high degree of dominance at each locus (Rose, 1982). Negative correlations between the traits involved with respect to dominance deviations would be expected in this case, but these have usually not been looked for with sufficient power. The other mechanisms that could act to maintain variation under model 2 have no particular requirements with respect to dominance, nor do they have strong implications for genetic correlations among traits.

As described above, traits with significant dominance variance relative to additive genetic variance or inbred load, such as longevity and sperm precedence (Table 19.3), are candidates for traits influenced by a relatively small number of loci with alleles maintained polymorphic by selection. The criteria enshrined in expressions (9) offer the hope that life-history traits that are under this type of genetic control can be identified and then be subjected to more detailed analyses by the methods of quantitative trait loci mapping (Long et al., 1995) or by identifying effects of candidate genes (Lai et al., 1995). This might pave the way to uniting the study of traits that are the direct target of natural selection with analysis of the underlying molecular causes.

19.6. Acknowledgments

The research described here was supported by grants from the U.S. National Science Foundation and the National Institutes of Health. We thank Deborah Charlesworth and an anonymous reviewer for their comments on the manuscript. We are especially grateful to James Crow for his detailed critique of this paper, which corrected several errors and ambiguities in the original version. B. C. is supported by the Royal Society.

REFERENCES

Burt, A. 1995. The evolution of fitness. *Evolution* 49:1–8.

Caballero, A. and Keightley, P. D. 1994. A pleiotropic nonadditive model of variation in quantitative traits. *Genetics* 138:883–900.

Caballero, A., Keightley, P. D., and Turelli, M. 1997. Average dominance for polygenes: drawback of regression estimates. *Genetics* 147:1487–1490.

Charlesworth, B. 1969. *Genetic Variation in Viability in Drosophila Melanogaster.* Ph.D. Dissertation, University of Cambridge, Cambridge, England.

Charlesworth, B. 1987. The heritability of fitness. In *Sexual Selection: Testing the Alternatives*, J. W. Bradbury and M. B. Andersson, eds. pp. 21–40. Chichester, U.K.: Wiley.

Charlesworth, B. 1990a. Optimization models, quantitative genetics, and mutation. *Evolution* 44:520–538.

Charlesworth, B. 1990b. Mutation–selection balance and the evolutionary advantage of sex and recombination. *Genet. Res.* 55:199–221.

Charlesworth, B. 1993. Natural selection on multivariate traits in age-structured populations. *Proc. R. Soc. London Ser. B* 251:47–52.

Charlesworth, B. 1994. *Evolution in Age-Structured Populations.* 2nd ed. Cambridge: Cambridge U. Press.

Charlesworth, B. 1998. The effect of synergistic epistasis on the inbreeding load. *Genet. Res.* 71:85–89.

Charlesworth, B. and Barton, N. H. 1996. Recombination load associated with selection for increased recombination. *Genet. Res.* 67:27–41.

Charlesworth, B. and Hughes, K. A. 1996. Age-specific inbreeding depression and components of genetic variance in relation to the evolution of senescence. *Proc. Natl. Acad. Sci. USA* 93:6140–6145.

Charlesworth, B., Lapid, A., and Canada, D. 1992. The distribution of transposable elements within and between chromosomes in a population of *Drosophila melanogaster.* II. Inferences on the nature of selection against elements. *Genet. Res.* 60:115–130.

Charlesworth, D. and Charlesworth, B. 1987. Inbreeding depression and its evolutionary consequences. *Annu. Rev. Ecol. Syst.* 18:237–268.

Charnov, E. L. 1989. Phenotypic evolution under Fisher's Fundamental Theorem of natural selection. *Heredity* 62:97–106.

Clark, A. G. 1987. A test of multilocus interaction in *Drosophila melanogaster. Am. Nat.* 130:283–299.

Clark, A. G., Aguadé, M., Prout, T., Harshman, L. G., and Langley, C. H. 1995. Variation in sperm displacement and its association with accessory gland protein loci in *Drosophila melanogaster. Genetics* 139:189–201.

Clutton-Brock, T. H., ed. 1988. *Reproductive Success.* Chicago: U. Chicago Press.

Comstock, R. E. and Moll, R. H. 1963. Genotype-environment interactions. In *Statistical Genetics and Plant Breeding*, W. D. Hanson and H. F. Robinson, eds., pp. 164–194. Washington, D.C.: National Academy of Sciences – National Research Council.

Crow, J. F. 1970. Genetic loads and the cost of natural selection. In *Mathematical Topics in Population Genetics*, K. Kojima, ed., pp. 128–177. Berlin: Springer-Verlag.

Crow, J. F. 1993. Mutation, mean fitness, and genetic load. *Oxford Surv. Evol. Biol.* 9:3–42.

Crow, J. F. and Simmons, M. J. 1983. The mutation load in Drosophila. In *The Genetics and Biology of Drosophila*, Vol. 3c, M. Ashburner, H. L. Carson, and J. N. Thompson, eds., pp. 1–35. London: Academic.

Curtsinger, J. W., Service, P. M., and Prout, T. 1994. Antagonistic pleiotropy, reversal of dominance, and genetic polymorphism. *Am. Nat.* 144:210–228.

Dickerson, G. E. 1955. Genetic slippage in response to selection. *Cold Spring Harbor Symp. Quant. Biol.* 20:213–224.

Dobzhansky, T. 1955. A review of some fundamental concepts and problems of population genetics. *Cold Spring Spring Harbor Symp. Quant. Biol.* 20:1–15.

Eanes, W. F., Hey, J., and Houle, D. 1985. Homozygous and hemizygous viability variation on the X chromosome of *Drosophila melanogaster. Genetics* 111:831–844.

Falconer, D. S. and Mackay, T. F. C. 1996. *Introduction to Quantitative Genetics*, 4th ed. London: Longman.

Fisher, R. A. 1922. On the dominance ratio. *Proc. R. Soc. Edinburgh* 52:312–341.

Fisher, R. A. 1930. *The Genetical Theory of Natural Selection*. Oxford: Oxford U. Press.

Gillespie, J. H. 1991. *The Causes of Molecular Evolution*. Oxford: Oxford U. Press.

Greenberg, R. and Crow, J. F. 1960. A comparison of the effect of lethal and detrimental chromosomes from Drosophila populations. *Genetics* 45:1153–1168.

Griffing, B. 1956. A generalised treatment of the use of diallel crosses in quantitative inheritance. *Heredity* 10:31–50.

Haldane, J. B. S. 1927. A mathematical theory of natural and artificial selection. Part V. Selection and mutation. *Proc. Cambridge Philos. Soc.* 23:838–844.

Haldane, J. B. S. 1949. Parental and fraternal correlations in fitness. *Ann. Eugen.* 14:288–292.

Houle, D. 1991. Genetic covariance of fitness correlates: what genetic correlations are made of and why it matters. *Evolution* 45:630–648.

Houle, D. 1992. Comparing evolvability and variability of quantitative traits. *Genetics* 130:195–204.

Houle, D., Hoffmaster, D. K., Assimacopoulous, S., and Charlesworth, B. 1992. The genomic mutation rate for fitness in *Drosophila*. *Nature (London)* 359:58–60.

Houle, D., Hughes, K. A., Hoffmaster, D. K., Ihara, J. T., Assimacopoulos, S., and Charlesworth, B. 1994. The effect of spontaneous mutation on quantitative traits. I. Variances and covariances of life history traits. *Genetics* 138:773–785.

Houle, D., Morikawa, B., and Lynch, M. 1996. Comparing mutational variabilities. *Genetics* 143:1467–1483.

Houle, D., Hughes, K. A., and Assimacopoulos, S., and Charlesworth, B. 1997. The effects of spontaneous mutation on quantitative traits. II. Dominance of mutations with effects on life-history traits. *Genet. Res.* 70:27–34.

Hughes, K. A. 1995a. Evolutionary genetics of male life history characters in *Drosophila melanogaster*. *Evolution* 49:521–537.

Hughes, K. A. 1995b. The inbreeding decline and average dominance of genes affecting male life-history characters in *Drosophila melanogaster*. *Genet. Res.* 65:41–52.

Hughes, K. A. 1997. Quantitative genetics of sperm competition in *Drosophila melanogaster*. *Genetics* 145:139–151.

Keightley, P. D. 1994. The distribution of mutation effects on viability in *Drosophila melanogaster*. *Genetics* 138:1–8.

Kempthorne, O. 1957. *An Introduction to Genetic Statistics*. New York: Wiley.

Kimura, M. 1958. On the change of population mean fitness by natural selection. *Heredity* 12:145–167.

Kimura, M. 1965. A stochastic model concerning the maintenance of genetic variability in quantitative characters. *Proc. Natl. Acad. Sci. USA* 54:731–736.

Kondrashov, A. S. 1988. Deleterious mutations and the evolution of sexual reproduction. *Nature (London)* 336:435–440.

Kreitman, M. and Akashi, H. 1995. Molecular evidence for natural selection. *Annu. Rev. Ecol. Syst.* 26:403–422.

Kusakabe, S. and Mukai, T. 1984. The genetic structure of natural populations of *Drosophila melanogaster*. XVII. A population carrying genetic variability explicable by the classical hypothesis. *Genetics* 108:393–408.

Lai, C., Lyman, R. F., Langley, C. H., and Mackay, T. F. C. 1995. Naturally occurring variation in bristle number and DNA polymorphisms at the *scabrous* locus of *Drosophila melanogaster. Science* **266**:1697–1702.

Lande, R. 1975. The maintenance of genetic variability by mutation in a polygenic character. *Genet. Res.* **26**:221–235.

Lande, R. 1982. A quantitative genetic theory of life history evolution. *Ecology* **63**:607–615.

Latter, B. D. H. and Sved, J. A. 1994. A reevaluation of data from competitive tests shows high levels of heterosis in *Drosophila melanogaster. Genetics* **137**:509–511.

Lenski, R. E. and Travisano, M. 1994. Dynamics of adaptation and diversification: a 10,000-generation experiment with bacterial populations. *Proc. Natl. Acad. Sci. USA* **91**:6608–6818.

Lewontin, R. C. 1958. A general method for investigating the equilibrium of gene frequency in a population. *Genetics* **43**:421–433.

Lewontin, R. C. 1974. *The Genetic Basis of Evolutionary Change.* New York: Columbia U. Press.

Long, A. D., Mullaney, S. L., Reid, L. A., Fry, J. D., Langley, C. H., and Mackay, T. F. C. 1995. High resolution mapping of genetic factors affecting abdominal bristle number in *Drosophila melanogaster. Genetics* **139**:1273–1291.

Mousseau, T. A. and Roff, D. A. 1987. Natural selection and the heritability of fitness components. *Heredity* **59**:181–197.

Mukai, T. 1964. The genetic structure of natural populations of *Drosophila melanogaster.* I. Spontaneous mutation rate of polygenes controlling viability. *Genetics* **50**:1–19.

Mukai, T. 1969a. The genetic structure of natural populations of *Drosophila melanogaster.* VII. Synergistic interactions of spontaneous mutant polygenes affecting viability. *Genetics* **61**:749–761.

Mukai, T. 1969b. The genetic structure of natural populations of *Drosophila melanogaster.* VIII. Natural selection on the degree of dominance of viability polygenes. *Genetics* **63**:476–478.

Mukai, T. 1988. Genotype-environment interaction in relation to the maintenance of genetic variability in populations of *Drosophila melanogaster.* In *Proceedings of the 2nd International Conference on Quantitative Genetics,* B. S. Weir, E. J. Eisen, M. M. Goodman, and G. Namkoong, eds., pp. 21–31. Sunderland, MA: Sinauer.

Mukai, T., Cardellino, R. A., Watanabe, T. K., and Crow, J. F. 1974. The genetic variance for viability and its components in a local population of *Drosophila melanogaster. Genetics* **78**:1195–1208.

Mukai, T., Chigusa, S. I., Mettler, L. E., and Crow, J. F. 1972. Mutation rate and dominance of genes affecting viability in *Drosophila melanogaster. Genetics* **72**:335–355.

Mukai, M. and Nagano, S. 1983. The genetic structure of natural populations of *Drosophila melanogaster.* XVI. Excess of additive genetic variance of viability. *Genetics* **105**:115–134.

Mukai, T. and Yamaguchi, O. 1974. The genetic structure of natural populations of *Drosophila melanogaster.* XI. Genetic variability in a local population. *Genetics* **76**:339–366.

Mukai, T. and Yamazaki, T. 1968. The genetic structure of natural populations of *Drosophila melanogaster.* V. Coupling-repulsion effect of spontaneous mutant polygenes controlling viability. *Genetics* **59**:513–535.

Ohnishi, O. 1977. Spontaneous and ethyl methanesulfonate-induced mutations controlling viability in *Drosophila melanogaster.* II. Homozygous effects of polygenic mutations. *Genetics* **87**:529–545.

Pease, C. M. and Bull, J. J. 1988. A critique of methods for measuring life-history trade-offs. *J. Evol. Biol.* 1:293–303.

Robertson, A. 1955. Selection in animals: synthesis. *Cold Spring Harbor Symp. Quant. Biol.* 20:225–229.

Roff, D. A. and Mousseau, T. A. 1987. Quantitative genetics and fitness: lessons from *Drosophila. Heredity* 58:103–118.

Rose, M. R. 1982. Antagonistic pleiotropy, dominance, and genetic variation. *Heredity* 48:63–78.

Rose, M. R. 1985. Life history evolution with antagonistic pleiotropy and overlapping generations. *Theor. Popul. Biol.* 28:342–358.

Rose, M. R. and Charlesworth, B. 1981. Genetics of life history in *Drosophila melanogaster.* I. Sib analysis of adult females. *Genetics* 97:173–186.

Seager, R. D. and Ayala, F. J. 1982. Chromosome interactions in *Drosophila melanogaster.* I. Viability studies. *Genetics* 102:467–483.

Seager, R. D., Ayala, F. J., and Marks, R. W. 1982. Chromosome interactions in *Drosophila melanogaster.* II. Total fitness. *Genetics* 102:485–502.

Service, P. M., Hutchinson, E. W., and Rose, M. R. 1988. Multiple genetic mechanisms for the evolution of senescence in *Drosophila melanogaster. Evolution* 42:708–716.

Service, P. M. and Rose, M. R. 1985. Genetic covariation among life-history components: the effect of novel environments. *Evolution* 39:943–945.

Simmons, M. J. and Crow, J. F. 1977. Mutations affecting fitness in *Drosophila* populations. *Annu. Rev. Genet.* 11:49–78.

Sved, J. A. 1975. Fitness of third chromosome homozygotes in *Drosophila melanogaster. Genet. Res.* 25:197–200.

Tachida, H. and Cockerham, C. C. 1988. Variance components of fitness under stabilizing selection. *Genet. Res.* 51:47–53.

Tachida, H., Matsuda, M., Kusakabe, S., and Mukai, M. 1983. Variance component analysis for viability in an isolated population of *Drosophila melanogaster. Genet. Res.* 42:207–217.

Takano, T., Kusakabe, S., and Mukai, T. 1987. The genetic structure of natural populations of *Drosophila melanogaster.* XX. Comparisons of genotype-environment interaction in viability between a northern and a southern population. *Genetics* 117:245–254.

Tracey, M. L. and Ayala, F. J. 1974. Genetic load in natural populations: is it compatible with the hypothesis that many polymorphisms are maintained by natural selection? *Genetics* 77:569–589.

Tyler, R. H., Brar, H., Singh, M., Latorre, A., Graves, J. L., Mueller, L. D., Rose, M. R., and Ayala, F. J. 1994. The effect of superoxide dismutase alleles on aging in *Drosophila.* In *Genetics and Evolution of Aging,* M. R. Rose and C. E. Finch, eds., pp. 161–167. Dordrecht, The Netherlands: Kluwer.

Watanabe, T. K. and Ohnishi, S. 1975. Genes affecting productivity in natural populations of *Drosophila melanogaster. Genetics* 80:807–819.

Watanabe, T. K., Yamaguchi, O., and Mukai, T. 1976. The genetic variability of third chromosomes in a local population of *Drosophila melanogaster. Genetics* 82:63–82.

Wilton, A. N. and Sved, J. A. 1979. X-chromosomal heterosis in *Drosophila melanogaster. Genet. Res.* 34:303–315.

Wright, S. 1968. *Evolution and the Genetics of Populations. Vol. 1. Genetic and Biometric Foundations.* Chicago: U. Chicago Press.

CHAPTER TWENTY

Population Genetics and Life-History Evolution

LLOYD DEMETRIUS

20.1. Introduction

The Darwinian theory of evolution by natural selection revolves around two main principles. The first pertains to the production of genetic variability. The second refers to the sorting of this variability by natural selection acting at the phenotypic level. The selective component of the evolutionary process, which leads to the replacement of one constellation of types by another, suggests a directional aspect to this dynamical system, an idea that has had a powerful impact on theoretical studies to explain the diversity of organisms and their adaptation.

The theoretical issue that the evolutionary phenomenon generates may be expressed as follows: What kind of mathematical abstraction is best suited to represent the biological reality of adaptive change by mutation and selection?

In 1930, Fisher, who was very much influenced by the achievement of Boltzmann in developing a statistical mechanics formalism to represent macroscopic changes in physical systems, proposed a new mathematical abstraction for evolutionary processes based on the concept of mean fitness. This new notion was exploited to establish a directionality theorem – the increase in mean fitness under natural selection – that was claimed to be an analog of the second law of thermodynamics (Fisher, 1930). The fundamental theorem of natural selection, as the theorem was christened, asserts that the rate of increase in the mean fitness at any time of any organism is equal to its genetic variance in fitness at that time. Fitness in this context is measured in terms of viability. The theorem therefore implies that as the gene frequencies in a population change because of viability differences between genotypes, the mean fitness will increase monotonically until a state of equilibrium is attained. Thus the theorem is simply a statement about the relative viability of individuals within a

R. S. Singh and C. B. Krimbas, eds., *Evolutionary Genetics: From Molecules to Morphology*, vol. 1. © Cambridge University Press 2000. Printed in the United States of America. ISBN 0-521-57123-5. All rights reserved.

population and conveys no consequence regarding the absolute reproduction and survival of the population. As Richard Lewontin has remarked in one of his many essays on the concept of adaptation, natural selection does not necessarily result in increased population numbers and higher reproductive rate; populations subject to natural selection may even be smaller and have inferior growth rates (Levins and Lewontin, 1985, Chap. 1).

Fisher's theorem has nevertheless exercised a strong influence in theoretical population genetics; witness the extensive literature (see, for example, Kingman, 1961; Edwards, 1990; Karlin, 1992) that the theorem has spawned. The mean fitness paradigm has also had a dominant effect on the work of Fisher's contemporaries. The Wrightian synthesis of evolution by drift and natural selection was formulated within the context of this paradigm (Wright, 1931, 1942). The cornerstone of Wright's evolutionary work, the shifting-balance theory, invoked the concept of mean fitness to parameterize the adaptive topography of the evolutionary process. Peaks of the adaptive surface corresponded to local and global maxima of the mean fitness function. Evolutionary change was analytically described as a stochastic process involving shifts from one adaptive peak to another through the combined action of random drift, selection, and migration.

The question that these observations raise is this: To what extent does the concept, mean fitness, embody the biological reality expressed by the evolutionary process?

Mean fitness is a property that pertains to the average reproduction and survivorship of individuals in the population. Such averages are within-population properties and need not be related to population attributes such as growth rate and density. Hence theories that revolve around the notion of mean fitness may not pertain to the life-history characteristics of a population or, more generally, to macroevolutionary phenomena such as adaptation, extinction, and speciation. The Wright–Fisher paradigm embodies a kinematics of the evolution of genotypes with little direct contact to the dynamics of populations and species in ecological communities. The limitation of the mean fitness concept to address problems that arise in the evolution of phenotypic properties was recognized by several theoretical biologists in the 1960s. This acknowledgement gave rise to two new classes of analytical theories: quantitative genetics and evolutionary ecology. Quantitative genetics (see Falconer, 1960) deals with the effect of selection on quantitative characters. The theory recasts Mendelian inheritance in phenotypic terms and analyzes the dynamics of phenotypic traits under directional and stabilizing-selection regimes. Ecological genetics (see MacArthur, 1962) considers the effect of selection on population parameters, such as growth rate, denoted by r, and carrying capacity, denoted by K. This theory, in contrast to that of quantitative genetics, considers Mendelian inheritance in genetic terms and studies the dynamical changes in the population variables under density-independent and density-dependent ecological constraints. The MacArthur theory embodied two tenets: (1) Under density-independent conditions, natural selection increases population growth rate, and (b) under density-dependent conditions, natural selection increases the carrying capacity. These two principles, which form the basis of what is now known as the $r-K$ theory, exerted a strong influence on evolutionary ecology, partly on account of the simplicity of its tenets and also on account of the ecological elements that were introduced in the models. Studies by Clarke (1972), Roughgarden (1971), Kimura (1978), Naglyaki (1979), Ginzburg (1983) on selection in density-dependent populations have considerably extended certain aspects of the theory. The $r-K$ models thus constitute a basis for a theory of ecological genetics – the first system to integrate in genetic terms the process of Mendelian inheritance with the dynamics of selection on ecological variables. The MacArthur models and their extensions ignored demographic heterogeneity.

All natural populations, however, are characterized by a variability in the age at which individuals reproduce and die. This variability has its origins in the developmental process: Demographic variability is also observed in a population of genetically identical cells, descendants from a single mother cell. Demographic heterogeneity thus constitutes an intrinsic feature of all populations of replicating organisms. The fundamental nature of this property entails that any model of phenotypic evolution that purports to explain the adaptation of populations must involve age, or some proxy such as size, as an independent variable. The absence of demographic structure in the framework of the $r-K$ theory vitiates any attempt of this model to explain evolutionary change in natural populations.

The significance of demographic structure in mathematical models of evolutionary change seems to have been first articulated by Lewontin (1965) in a numerical study of the effect of small changes in the age-specific fecundity and mortality variables on the evolution of colonizing ability. The extension of Lewontin's work given by Demetrius (1969) in terms of the Leslie model and the elaborations by Caswell (1978), among others, underscore the pertinence of age structure in understanding evolutionary dynamics.

Attempts to introduce demographic structure and to create a new evolutionary synthesis have been made by several authors (Pollak and Kempthorne, 1970, 1971; Charlesworth and Giesel, 1972; Naglyaki, 1976; Crow, 1978; Lande, 1982). One of the more concerted programs to integrate demography in population genetics models is organized in Charlesworth (1994). This effort revolved around two tenets: (1) The condition for invasion of a rare mutant is determined by its effect on the growth rate r; and (2) genotypic differences in r control the outcome of selection with density independence, and genotypic differences in the carrying capacity K act similarly with density dependence. These two tenets were derived under certain stringent conditions. First, the assertion that the selective dynamics of rare mutants is predicted by r assumes that the invasion process is deterministic. This property, however, holds only when population size is sufficiently large that fluctuations due to demographic stochasticity can be neglected. Second, the assertion that differences in K determine the outcome of selection under density-dependent conditions rests on the assumption that the vital rates are functions of the total number of individuals in a so-called critical age group. However, individuals of different ages may exert different constraints on the age-specific fecundity and mortality; consequently, vital rates will now be functions of a generalized population size, a weighted sum of the individuals in the different age classes. The nature of these assumptions indicates that the parameters r and K will not in general constitute adequate descriptors of evolutionary dynamics in age-structured populations.

The problem that now emerges can be formulated as follows: What kind of mathematical abstraction is best applicable in representing evolutionary change in populations defined by demographic heterogeneity and finite size?

The theory introduced by Demetrius (1974) devised a new formalism to resolve this problem. This body of work, which we now call directionality theory (see Arnold et al., 1994, for extensions to general deterministic and random dynamical systems), was developed in a series of papers (see Demetrius, 1974, 1975, 1977, 1983, 1992a, 1992b, 1997, for the principal trends). The directionality theory drew extensively from studies in statistical mechanics and the ergodic theory of dynamical systems to introduce a new family of demographic concepts to describe the evolutionary dynamics of populations. A critical element in this study is the notion that heterogeneity in age-specific birth and death rates is a property that is intrinsic to all populations of replicating organisms. The concept of population entropy was introduced to characterize this heterogeneity: A population has positive entropy if replication occurs at several distinct stages in the life cycle and zero entropy if replication is concentrated at a single stage.

Other critical parameters that appear in this theory are the population growth rate r, a demographic variable that goes back to Lotka (1925), and the demographic variance σ^2, a measure of the variance in the ages of mothers at the birth of their offspring (Arnold et al., 1994). The importance of these two parameters resides in the fact that they provide criteria for the invasion and extinction of mutant genes (Demetrius and Gundlach, 1999). These criteria entail that the selective advantage of a mutant with respect to the wild type, denoted s, is given by

$$s = \Delta r - \frac{1}{N}\Delta\sigma^2, \tag{20.1}$$

where N denotes the total population size and the parameters r and σ^2 refer to the Malthusian parameter and demographic variance, respectively, of a genotype.

Expression (20.1) should be contrasted with

$$s = \Delta r, \tag{20.2}$$

which defines selective advantage in classical population models.

Directionality theory is the study of the kinematics of the evolutionary process when selective advantage is described by Eq. (20.1). Our analysis (Arnold et al., 1994) of evolutionary change under mutation and selection showed that the patterns of change in the demographic variables, as one constellation of types replaces another, are critically determined by the ecological conditions and the demographic state of the population. Our study of ecological constraints distinguished between density-dependent and density-independent effects. In our analysis of demographic constraints, we differentiated between slow and rapid exponential growth and between weak and strong iteroparity. By exploiting these distinctions, which have a precise analytical representation, we derived the following relations between ecological states and evolutionary changes in entropy and growth rate. (Demetrius 1992a, 1997; Arnold et al., 1994).

(I) *Density-dependent control.* A unidirectional increase in entropy, the growth rate at equilibrium remains zero.

(II) *Density-independent control.* (a) Slow exponential population growth: an increase in entropy and growth rate; (b) rapid exponential growth, weak iteroparity: a decrease in entropy, and an increase in growth rate; (c) rapid exponential growth, strong iteroparity: random, nondirectional change in entropy, and growth rate.

I should emphasize at this point that the directionality theorems expressed in (I) and (II) are different in character from their analogs in classical population genetics. Fisher's theorem, for example, considers dynamical changes in populations to be the outcome of a single process, natural selection. The theorem is a statement about the monotonic increase in mean viability as gene frequencies in the population change because of differential viability of the genotypes. Directionality theorems (I) and (II), by contrast, consider evolutionary dynamics to be the outcome of a dual process: mutation and natural selection.

The model thus considers a population at equilibrium, defined by a class of demographic parameters, growth rate, and entropy. Mutation introduces new types in the population and thus perturbs the equilibrium state. The selective interaction between the ancestral and the mutant types drives the combined population to some new equilibrium state defined by a new set of values for the demographic parameters. The directionality theorems are statements regarding the global change in entropy and growth rate as the mutation–selection process drives the population from one equilibrium state to the next.

It is of some interest to underscore certain implications of principles (I) and (II) and to relate them to earlier efforts to derive general evolutionary tenets.

As regards (I), we note that the density-dependent condition entails stationary growth constraints, a property that is the ecological analog of the irreversible adiabatic constraints that characterize physical systems. Assertion (I), describing an increase in population entropy under density-dependent conditions, thus constitutes an analog of the second law of thermodynamics (Demetrius, 1992b, 1997). We also note that the mean fitness theorem, contrary to Fisher's claim, has no connection with the second law. Mean fitness is simply a Liapunov function of the dynamical system that describes changes in gene frequency within a population. It is not an aggregate population property, and hence it ascribes no ordering to populations within an evolutionary lineage.

As regards (II), we observe that the density-independent condition entails exponential population growth. Assertion (II) implies that when density-independent conditions obtain, evolutionary change may result in a decrease in population growth rate. Moreover, this decrease is characteristically nondirectional: It may be followed by an increase, as new alleles are introduced in the population by mutation and then ordered by selection. Our theory thus points to two critical biological realities that are largely absent from the repertoire of classical age-structured models, namely, (1) the nonadaptive property of some evolutionary changes in life-history variables – when certain ecological conditions prevail, evolution can lead to a decreased growth rate; (2) an inherent stochasticity in the evolutionary process – fluctuations in age-specific birth and death rates will induce variations in population numbers, with the result that the invasion dynamics of new mutants now becomes a stochastic process.

The notion that natural selection can result in nonadaptive states and that historical contingency imposes constraints on predicting evolutionary trends has been articulated by Lewontin (1978, 1987) in several critiques of the optimality paradigm. These critiques have exploited the technique of gedanken experiments to elucidate an inherent stochasticity in the evolutionary process that runs counter to adaptive criteria and optimality arguments. The statements embodied in the evolutionary principle (II) (c), which describes random, nondirectional changes in entropy and growth rate, can be construed as analytic expressions of the observations gleaned from these gedanken models.

In this chapter I provide an account of the main concepts that underlie this new synthesis of demography and genetics. I also discuss the implications of the new theory toward the understanding of variation in life-history patterns and trends in morphological traits such as body size.

20.2. Heterogeneity in Life History

Heterogeneity in birth and death rates is a fundamental property in any population of replicating organisms. This condition has its origin in processes that underlie the ontogeny of the individual. In cellular systems, it derives from the random inequalities between cells such as the unequal distribution of metabolic components that occurs at cell division. In multicellular and higher organisms, the heterogeneity is a consequence of the small variations in timing and in the sequence of developmental events that translate the genetic program into adult morphology. Accordingly, any genetically homogeneous population of organisms will be characterized by a variability in their phenotypic states – size, shape, physiology – and a concomitant variability in their birth and death rates.

The problem that now emerges is this: What kind of mathematical object best characterizes this property of demographic variability?

The individuals in a population of replicating organisms can be parameterized in terms of variables such as age, size, developmental stage, etc. Since heterogeneity in birth and death rates is a result of random chance events during organismal development, any analytical descriptor of this heterogeneity should be independent of the parameterization invoked. There exists a unique descriptor that satisfies this independence condition: It is the dynamical entropy of the birth–death process that defines the replicating organisms (Demetrius, 1974, 1975). This quantity, denoted by H, is called population entropy.

In systems parameterized by the variable age, the analytical expression for entropy is given by

$$H = -\frac{\sum p_j \log p_j}{\sum j p_j},\qquad (20.3)$$

where p_j represents the probability that the ancestor of a randomly chosen newborn is in age class j. The numerator $S = -\sum p_j \log p_j$ describes the variability in the age of reproducing individuals. The quantity $T = \sum j p_j$ describes the generation time, the mean age of mothers at the birth of their offspring. Population entropy, as observed from Eq. (20.3) has units of inverse time and hence has the same dimension as growth rate. Entropy, however, unlike growth rate, is generally independent of the ecological constraints that impinge on the population. This important characteristic can be shown by a comparison of the dynamics of two genetically identical populations, one subject to a density-independent ecological regime and the other to a density-dependent regime; the effect of density on vital rates and fecundity are assumed to be independent of age. At demographic equilibrium, the first population will be described by a positive growth rate, the second by zero growth rate. However, the entropy of both populations will be identical. This follows from the fact, which can easily be shown, that p_j, the probability that the ancestor of a randomly chosen newborn has age j, will be identical in both the exponentially growing and the stationary population.

Entropy is also a measure of population stability: It describes the rate of decrease of fluctuations in population numbers due to chance realizations in the age-specific birth and death rates (Demetrius, 1977; Arnold et al., 1994). The intensity of fluctuations in population numbers constitutes an index of the adaptation of the population to a given environmental state. Hence evolutionary trends in entropy reflect the adaptive response of populations to different environmental regimes.

20.3. The Population Dynamics

The dynamical system that describes changes in age structure can be represented either in terms of the Leslie model, in which time is a discrete variable, or the von Forster equation, in which time is continuous. The continuous model leads to a simpler representation, and for this reason we will adopt this model in this exposition. We consider an age-dependent model of population growth described as follows. Let $u(x, t)$ denote the number of individuals of age x and time t. Population size $N(t)$ is given by

$$N(t) = \int_0^\infty u(x, t)\, dx.$$

The changes in $u(x, t)$ are determined by $\mu[x, N(t)]$ and $m[x, N(t)]$, which represent the age-specific death rates and birth rates, respectively. The dynamics of $u(x, t)$ are

$$\frac{\partial u}{\partial x} + \frac{\partial u}{\partial t} = -\mu[x, N(t)]u(x, t), \tag{20.4a}$$

$$u(0, t) = \int_0^\infty m[x, N(t)]u(x, t)\, dx. \tag{20.4b}$$

Steady-state solutions of Eqs. (20.4) are known to exist under certain constraints on the variables μ and m.

Linear models correspond to the case in which μ and m are independent of the population size $N(t)$. The steady state in this case is described by a stable age distribution and a population size $N(t)$ that increases exponentially at a rate r, which is the unique real root of the equation

$$1 = \int_0^\infty \exp(-rx)V(x)\, dx, \tag{20.5}$$

where $V(x) = \ell(x)m(x)$ and

$$\ell(x) = \exp\left[-\int_0^x \mu(y)\, dy\right].$$

The function $p(x)$ defined by

$$p(x) = \exp(-rx)V(x)$$

satisfies

$$\int_0^\infty p(x)dx = 1$$

and represents a probability density function.

In nonlinear models, μ and m are assumed to be increasing and decreasing functions, respectively, of $N(t)$. The steady state in these models will also be described by a stable age distribution, but the population size at steady state will be constant in time. Hence at steady state, μ and m will be functions of age only. Population growth rate will satisfy an equation of the form of Eq. (20.5) with $r = 0$.

Population entropy H can also be defined for both classes of models. The expression for H, the continuous analog of Eq. (20.3), is given by

$$H = -\frac{\int_0^\infty p(x) \log p(x)}{\int_0^\infty x p(x)}.$$

20.3.1. Demographic Parameters

Population growth rate r represents a macroscopic parameter, defined as in Eq. (20.5), by averaging over the age-specific variables. A new class of macroscopic variables can be obtained by a perturbation expansion of the function $r(\delta)$, the growth rate that corresponds to the net reproductive function $V(x)^{1+\delta}$ (Demetrius, 1996).

We have

$$1 = \int_0^\infty \exp[-r(\delta)] V(x)^{1+\delta} \, dx.$$

The Taylor expansion of $r(\delta)$ yields

$$r(\delta) = r(0) + \delta r'(0) + \frac{\delta^2}{2!} r''(0) + \frac{\delta^3}{3!} r'''(0) + \cdots,$$

where $r'(0) \equiv \Phi, r''(0) \equiv \sigma^2, r'''(0) \equiv \kappa$, and

$$\Phi = \frac{\int_0^\infty p(x) \log V(x) \, dx}{\int_0^\infty x p(x) \, dx}, \qquad \sigma^2 = \frac{\int_0^\infty [W(x)]^2 p(x) \, dx}{\int_0^\infty x p(x) \, dx},$$

$$\kappa = \int_0^\infty [W(x)]^3 p(x) \, dx - 3 \frac{\int_0^\infty [W(x)]^3 p(x) \, dx \int_0^\infty x W(x) p(x) \, dx}{\int_0^\infty x p(x) \, dx}$$

$$+ 2 \int_0^\infty [W(x)]^2 p(x) \, dx. \tag{20.6}$$

The function $W(x) = xH + \log p(x)$.

The quantity Φ is called the reproductive potential, σ^2 is the demographic variance, and the function $\gamma = \kappa + 2\sigma^2$ is the reproductive index. The functions Φ and γ describe different aspects of the demographic structure of the population, which are now illustrated.

Since $r = H + \Phi$, a property that is easy to verify, we have the following implications:

$$\Phi < 0 \Rightarrow r < H, \qquad \Phi > 0 \Rightarrow r > H. \tag{20.7}$$

The condition $\Phi < 0$ thus describes a slowly growing population and $\Phi > 0$ a rapidly increasing population. The parameter γ is a measure of the skewness of the net fecundity distribution; $\gamma < 0$ describes a population whose reproductive activity is concentrated during the earlier or the later stages of the life cycle; $\gamma > 0$ represents a population whose reproductive activity is spread over several age classes.

The signs assumed by the parameters Φ and γ thus provide a basis for classifying a population into different groups according to their demographic state and the ecological condition it experiences.

20.4. The Evolutionary Dynamics

Evolutionary change is determined by the concatenation of two processes. The first, which is nonadaptive, consists of the production of genetic variability through mutation. The second, which is adaptive, is the ordering of the genetic variability by natural selection.

20.4.1. Mutation

The mathematical model that describes the dynamical consequences of mutation assumes that the population is described by a genotype $A_1 A_1$ at demographic equilibrium. The population is described by a net reproductive function $V_1(x)$ with demographic parameters r_1, Φ_1, H_1, and γ_1, as defined in Section 20.3. The introduction of a mutant into the population is represented initially by $A_1 A_2$ heterozygotes with a net reproductive function $V_2(x)$ defined by

$$V_2(x) = V_1(x)W(x)^\delta,$$

where

$$\int \log V_1(x)\, \mathrm{d}x = \int \log W(x)\, \mathrm{d}x.$$

The parameter δ denotes the magnitude of the mutation.

Let Δr, ΔH, and $\Delta \sigma^2$ denote the differences in Malthusian parameter, entropy, and demographic variance, respectively, between the mutant heterozygote $A_1 A_2$ and the homozygote $A_1 A_1$. These quantities are known to satisfy, for δ of small absolute value, the relations

$$\Delta r = \Phi \delta, \qquad \Delta H = -\sigma^2 \delta, \qquad \Delta \sigma^2 = \gamma \delta. \tag{20.8}$$

Since $\sigma^2 > 0$, Eq. (20.8) leads to the following mutation relations:

$$\Phi < 0 \Rightarrow \Delta r \Delta H > 0, \qquad \Phi > 0 \Rightarrow \Delta r \Delta H < 0; \tag{20.9a}$$
$$\gamma < 0 \Rightarrow \Delta H \Delta \sigma^2 > 0, \qquad \gamma > 0 \Rightarrow \Delta H \Delta \sigma^2 < 0. \tag{20.9b}$$

20.4.2. Invasion–Extinction

The invasion–extinction dynamics of the mutant gene, that is, its ultimate establishment in the population, is analyzed in terms of a stochastic model. The ideas invoked in this study go back to Feller (1951), who provided a general review of diffusion processes in genetics. Subsequent applications of the results and techniques of diffusion processes to the Wright–Fisher and related models have appeared in Crow and Kimura (1970), Ludwig (1974), Gillespie (1974), Karlin and Levikson (1974), and Ewens (1979). The development given in Demetrius (1997) and Demetrius and Gundlach (1999) has analogs to these studies, in particular, to the models analyzed in Gillespie (1974). However, the models I developed differ in certain fundamental respects from the systems studied in these earlier works. The Wright–Fisher models describe populations without age structure, and the statistical variables, mean and variance, which appear in the diffusion equations, refer to the mean and variance of allele frequencies. The genetic models I analyze are characterized by demographic structure, and the statistical variables, mean and variance, now pertain to the Malthusian parameter and the demographic variance, respectively. In my analysis, the derivation of the diffusion equations that describe the population process draws extensively from the ergodic theory of age-structured populations (Demetrius, 1983). This theory states that the response of total population numbers to statistical fluctuations in age-specific birth and death rates can be characterized in terms of the macroscopic variables, population growth rate r, as given by Eq. (20.5), and the demographic variance σ^2, as given by Eqs. (20.6).

Our analysis of the invasion condition for a mutant gene derives from a statistical model that considers fluctuations in the numbers $N_1(t)$ and $N_2(t)$ of the ancestral and the mutant types, respectively – these fluctuations are generated by random perturbations in the age-specific fecundity and mortality. The density functions $f_1(N_1, t)$ and $f_2(N_2, t)$ of the ancestral and the mutant populations are solutions of the corresponding Fokker–Planck equation described in terms of the demographic parameters (r_1, σ_1^2) and (r_2, σ_2^2), respectively.

The analysis given in Demetrius (1997) and Demetrius and Gundlach (1999) considered the invasion dynamics of mutants as a two-stage process. The first stage pertains to the invasion of single mutants – a process that can be treated in terms of the Wright–Fisher model. As time proceeds and new mutants enter the population, the individual mutants will ultimately become structured by age. The second stage involves the invasion dynamics of this structured population of mutants. My study focuses on this stage and analyzes the process on the assumption that the mutant population size N_2 is small compared with the total population size N.

The stochasticity in the invasion process derives from chance fluctuations, which are modelled by a white-noise process, in the age-specific birth and death rates. The probability density $\psi(p, t)$ of the stochastic process that describes the change in the frequency of the mutant p as a function of time t was shown to satisfy

$$\frac{\partial \psi}{\partial t} = -\alpha(p)\frac{\partial \psi}{\partial p} + \frac{1}{2}\beta(p)\frac{\partial^2 \psi}{\partial p^2}, \tag{20.10}$$

where

$$\alpha(p) = \beta(1-p)\left(\Delta r - \frac{1}{N}\Delta\sigma^2\right), \tag{20.11a}$$

$$\beta(p) = \frac{p(1-p)}{N}\left[\sigma_1^2 p + \sigma_2^2(1-p)\right]. \tag{20.11b}$$

The analysis of Eq. (20.10) shows that the selective advantage s, the parameter that describes the invasion–extinction process, is given by

$$s = \Delta r - \frac{1}{N}\Delta\sigma^2. \tag{20.12}$$

This expression for the selective advantage permits the characterization of invasion–extinction criteria in terms of the quantities Δr and $\Delta\sigma^2$. We obtain the following:

A(i) $\Delta r > 0$, $\Delta\sigma^2 < 0$: Invasion occurs with a probability that is independent of population size.

A(ii) $\Delta r < 0$, $\Delta\sigma^2 > 0$: Extinction almost always occurs.

A(iii) $\Delta r > 0$, $\Delta\sigma^2 > 0$: Invasion occurs with a probability that is an increasing function of population size.

A(iv) $\Delta r < 0$, $\Delta\sigma^2 < 0$: Invasion occurs with a probability that is a decreasing function of population size.

20.4.3. The Dynamics of Selection

The analysis of the invasion process assumes that the new mutants that arise will be in very low frequency. This property implies that the ancestral and the mutant types during this initial phase of the evolutionary process can be considered as separately regulated. The assumption of separate regulation is no longer valid when mutants have attained a high frequency. At this stage, genetic and ecological interactions between the ancestral and the mutant types will occur. Given that the ancestral type X_1 is homozygous (A_1A_1) and the mutant X_2 is heterozygous (A_1A_2), new types X_3, which are homozygous (A_2A_2), will be generated by mating between the heterozygotes. The evolutionary trajectory is now determined by the selective dynamics of the three genotypes. This can be described, by using Mendel's laws, in terms of a set of coupled partial differential equations representing the changes in the age distribution of the types as functions of the age-specific fecundity and mortality of each type

(Demetrius, 1992a). Analogous formulations of selection equations are given in Charlesworth (1994).

The dynamical equations for the changes in numbers $u_i(x, t)$ of the genotype X_i are

$$\frac{\partial u_i}{\partial x} + \frac{\partial u_i}{\partial t} = -\mu_i(x, t)u_i(x, t), \qquad i = 1, 2, 3, \ldots, \qquad (20.13)$$

where $\mu_i(x, t)$ represents the age-specific survivorship of the genotype X_i.

The birth rate $\mu_i(o, t)$ of the genotypes can be derived from the Mendelian laws. These are given by the set of equations

$$u_1(0, t) = \int_0^\infty (m_1 u_1 + m_2 u_2) F_2 \, dx, \qquad (20.14a)$$

$$2u_2(0, t) = \int_0^\infty (m_1 u_1 F_1 + m_2 u_2 + m_3 u_3 F_2) \, dx, \qquad (20.14b)$$

$$u_3(0, t) = \int_0^\infty (m_2 u_2 + m_3 u_3) F_1 \, dx, \qquad (20.14c)$$

where

$$F_1 \equiv \frac{u_2(x, t) + u_3(x, t)}{U(x, t)}, \qquad F_2 \equiv \frac{u_1(x, t) + u_2(x, t)}{U(x, t)},$$

$$U(x, t) \equiv u_1(x, t) + 2u_2(x, t) + u_3(x, t).$$

The derivation of Eqs. (20.14) requires that (a) the fecundity of a mating between a given pair of individuals be determined uniquely by the age and genotypes of the female, and (b) mating is random with respect to age and genotype.

The equilibrium solutions of Eqs. (20.13) and (20.14) describe the ultimate effect of the selective process. Fixation of the mutant corresponds to the elimination of the types X_1 and X_2. A polymorphism corresponds to the presence of all three types.

20.4.4. Directional Changes under Mutation and Selection

The directional changes in macroscopic parameters pertain to changes as the population evolves from an equilibrium state represented by the ancestral genotype X_1 with, say, entropy H_1 to a new equilibrium state described by a new population [which may be polymorphic (X_1, X_2, X_3) or monomorphic (X_3) with entropy $\hat{H}$].

The expressions $\tilde{\Delta} H$, $\tilde{\Delta} r$, which denote the change in entropy and growth rate, respectively, as the population evolves from one equilibrium state to the next, and the expressions ΔH, Δr, which denote the changes that characterize the invading mutant, can be shown to satisfy, Demetrius, 1992,

$$\Delta H \tilde{\Delta} H > 0; \quad \Delta r \tilde{\Delta} r > 0. \tag{20.15}$$

These relations assert that the *global* directional change in macroscopic variables as one population replaces another under the dual processes of mutation and selection is positively correlated with the *local* directional changes induced by the invading mutant itself.

Conditions (20.15), together with mutation relations (20.9) and the invasion–extinction criteria A(i)–A(iv), can be exploited to relate the demographic state (weak or strong iteroparity) and the ecological constraints (density-dependent, density-independent conditions) with global directional trends in the demographic parameters. The correspondence between demographic–ecological states and global trends in entropy, and growth rate are summarized in terms of the directionality theorems stated in the introduction.

The theorems fall into two patterns:

(I) A unidirectional change in life-history variables under (a) stationary growth constraints ($r = 0$), (b) slow exponential growth ($\Phi < 0$), (c) rapid exponential growth ($\Phi > 0$) and weak iteroparity ($\gamma < 0$).

(II) Random, nondirectional changes in life-history variables under conditions of rapid exponential growth ($\Phi > 0$) and strong iteroparity ($\gamma > 0$).

There exist intrinsic limits to the directional changes expressed in (I), owing to constraints on the ability of new mutants to become established in the population. The degree of genetic polymorphism at a given locus can be shown to increase for the mutation–selection process when ecological conditions that define (I) obtain. However, a limit will ultimately be attained, described by the state in which the genome becomes invulnerable to the invasion of new alleles. This limiting condition derives from a result due to Kingman (1980), who showed that the expectation ρ that a new mutant takes its place in a new equilibrium population scales according to the relation $\rho \sim \exp(-\mu k)$, where μ is a function that depends on the fitness of the different alleles and k denotes the number of alleles at the locus. The expression for ρ implies that large polymorphisms, once established, are highly resistant to invasion by a new mutant; moreover, this resistance increases exponentially with the number of alleles.

We can therefore assert that, under ecological conditions that induce stationary or slow population growth, the life-history variable, entropy, for example, will increase to some upper limit that may be inferior to the mathematically defined state of maximal entropy. Analogously, in weakly iteroparous populations under conditions of rapid exponential growth, entropy will decrease to some lower limit that will in general be distinct from the extremal case $H = 0$.

I should emphasize at this juncture that the theorems described here are based on single-locus models. However, life-history traits such as growth rate, entropy, and demographic variance are composite characters that are invariably under the control of genes at several loci. The question that now emerges is this: To what extent are the results pertaining to invasion–extinction properties and directional trends in demographic parameters valid for multilocus systems? Theoretical studies suggest (see for example Crow and Kimura, 1970) that, provided gene effects are approximately additive across loci and that linkage is loose, the results for multilocus systems can be derived by integration of the contributions to phenotypic change from individual loci. This implies that the general results based on a single locus reviewed in this article provide a good approximation to the evolutionary dynamics of life-history variables.

20.5. Body Size and Evolutionary Trends

Body size is a central property of any organism. Adult size is a multivariate character that is the end result of the physiological processes that determine the ontogeny of the organism. Size imposes constraints on the rate of metabolic events and thus regulates the relationship between the organism and the external environment.

Body size, or some measurable proxy of size such as molar area, is readily preserved in fossils. As a result, paleontologists have used size to investigate patterns of evolutionary change within phyletic lineages. One of the most significant trends, codified as Cope's rule, describes the tendency of certain animal groups to increase in body size within single unbranching lineages (see for example McKinney, 1990).

The theoretical problem that this series of empirical observations generates can be expressed as follows: Can the trend toward increased size within phyletic lineages be explained in terms of the synthesis of population genetics and life-history evolution that have been outlined? We now explore this problem by appealing to an analytic relation between population entropy and body size.

20.5.1. Entropy and Body Size

Population entropy is a measure of demographic heterogeneity – a property that is inherent in any population of replicating entities, cells, or higher organisms. This heterogeneity has its origins in the physiological processes that determine the ontogeny of the individual. It derives from the fact that genetically identical individuals in a population of replicating organisms will be characterized by small variations in the sequence of developmental events that translate the genetic program into adult morphology.

The morphometric variable, adult size, is the end result of this sequence of developmental events. We now address the following problem: Does there exist an allometric relation between entropy and body size? Suggestions that such a relation exists derives from the fact that within a taxon, such as mammals, physiological, life-history, and morphological variables (Y) are power functions of body mass (W):

$$Y = aW^b. \tag{20.16}$$

The parameter a represents a proportionality coefficient. The exponent b is known to fall into certain patterns according to the dimension of the variable Y: capacities of transport organs ($b \simeq 1$); volume rates, such as metabolic rates ($b \simeq 3/4$); cycle time, such as generation time, ($b \simeq 1/4$) (Calder, 1984).

Population entropy is a life-history property. As observed from Eq. (20.3), in the case of models in which individuals are parameterized by variable age, entropy H is given by the ratio S/T. We now present an analytic argument to show that for populations subject to stationary growth constraints, the entropy S is isometric to body size.

Living organisms are maintained by the continuous expenditure of energy. They exist so long as the energy from resources in the environment can be obtained to maintain the metabolic process required for reproduction and survival. Our model assumes that the energy acquired by an individual from resources throughout life (metabolic energy) is converted uniquely to energy, contributing to reproduction and survivorship (demographic energy).

We now consider the adult organism as a metabolic system, converting resources into metabolic energy, and describe it in terms of a series of coupled reactions between chemical components X_i. This analysis is based on thermodynamic principles applied to chemical reactions, as discussed for example in de Groot and Mazur (1984, Chap. 3).

The rate of production of metabolic energy dQ/dt is given by

$$\frac{dQ}{dt} = T \int^V \sigma \, dV,$$

where σ denotes the entropy production per unit volume and unit time, T is the temperature, and V is the volume of the organism. Now $T\sigma = \rho \sum_i \mu_i (\partial c_i/\partial t)$, where μ_i and c_i denote the chemical potential and the concentration of the molecular species X_i, respectively, and ρ is the density (mass per unit volume) of the organism. We therefore obtain

$$\frac{dQ}{dt} = kW, \tag{20.17}$$

where $W = \int^V \rho \, dV$ represents the total body mass, and $k = \sum_i \mu_i (\partial c_i/\partial t)$. Hence the metabolic energy acquired from resources by an individual throughout life will be proportional to body size.

To relate the entropy S with body size, we now appeal to certain facts in demographic theory (Demetrius, 1983). We note that the function p_j defined in Eq. (20.3) can be expressed by $p_j = \exp(-\omega_j)/Z$, where $Z = \sum_j \exp(-\omega_j)$, and $\omega_j = -\log V_j$, where V_j denotes the net reproductive function of individuals in age class (j). The quantity Z is the net reproductive rate, the number of offspring per individual who from birth is subject to the reproductive schedule described by the function V_j. The parameter $\log Z$ can be expressed as the difference between the entropy S and an energy function E:

$$\log Z = -\sum_j p_j \log p_j - \sum_j p_j \omega_j. \qquad (20.18)$$

The quantity $E \equiv \sum p_j \omega_j$, which we call demographic energy, describes the expected net offspring production over the course of an individual life.

Now, under stationary growth constraints, we have $Z = 1$. In this case, Eq. (20.18) implies that the entropy S coincides with E, the demographic energy. By invoking the assumption that the energy acquired from resources by an individual throughout life is converted uniquely into net offspring production and appealing to Eq. (20.17), we infer that when stationary growth conditions obtain, the entropy S is isometric with body size, that is, expression (20.16) holds for the entropy function S with the exponent $b = 1$.

Now, as we observe from Eq. (20.16), the generation time T scales with body size with exponent $b = 1/4$. The entropy H can be expressed by the ratio S/T; hence H scales with body size with exponent $b = 3/4$.

The energetic variable, metabolic rate, is also known to scale with body size with exponent $b = 3/4$. Hence the function H, which has the dimension of inverse time, can be considered a measure of the metabolic rate of the organism, whereas the function S, a pure number, can be considered as a measure of its metabolic capacity.

20.5.2. Trends in Body Size

Trends in body size are codified as Cope's rule, which is defined as a widespread tendency of animal groups to evolve toward larger sizes. There are many exceptions to this rule, and a large body of literature has emerged in efforts to explain evolutionary trends in size (see for example Stanley, 1973; Peters, 1983; Brown and Maurer, 1986; Brown et al., 1993).

The scaling relations between the entropy functions S and H with body size, together with the evolutionary trends in entropy under different ecological conditions, as expressed by the directionality theorems, now yield a new set of predictions that provide a perspective on the ecological forces that determine trends in body size. We can now delineate a set of relations between ecological states and evolutionary trends in size as follows:

(I) Slow or stationary growth: increase in body size.

(II) Rapid exponential growth, weak iteroparity: decrease in body size.

(III) Rapid exponential growth, strong iteroparity: random, nondirectional changes in body size.

These predictions imply that trends toward an increase in body size are not universal but highly constrained by the ecological conditions that the species experience. When density-dependent conditions operate to impose constraints on population growth, mutations that result in an increased body size will be favored, and the individuals will evolve from smaller to larger size. When density-independent conditions prevail and growth is rapid, the evolutionary trends now depend on the degree of iteroparity of the population. In the case of weak iteroparity, mutants that are characterized by a decreased body size will enjoy a selective advantage and the individuals will evolve from larger to smaller body size. In the event of strong iteroparity, the stochastic elements of the invasion process of new mutants will dominate and random increases or decreases in body size will occur.

There are constraints on directional changes (increase or decrease) in body size. This is a consequence of the constraints on the changes in entropy we have already discussed. Accordingly, under stationary growth conditions, for example, the mutation–selection process will ultimately result in a state of evolutionary stability where no further increase in body size occurs.

There now is a large set of experimental studies concerning trends in body size in bacterial populations (see Lenski and Travisano, 1994, and references therein). These authors have studied the dynamics of adaptation and diversification in bacterial populations over periods as long as 10,000 generations. This body of work provides data to assess the validity of several predictions advanced in the theory that was outlined.

20.6. Conclusion

An important aspect of the history of theoretical population genetics can be expressed in terms of efforts to provide mechanistic explanations of trends in phenotypic traits over evolutionary time. Mean fitness and population growth rate are two concepts that have exercised a significant influence in attempts to address this problem.

Mean fitness and the theorem that asserts its increase under natural selection form the basis of a large body of work in population genetics. However, the evolutionary dynamics of mean fitness involves an assertion regarding the relative fitnesses of individuals within the population and hence has little consequence for the understanding of adaptation and extinction, properties that are determined by absolute population properties.

The claim that selection maximizes population growth rate and that the invasion of new mutants is completely determined by the growth rate parameter are notions that drive a large class of studies in ecological genetics. However, under density-dependent constraints, the typical condition for many natural populations, growth rate is zero. Also, in populations with finite size, mutants with inferior growth rate may become fixed in the population. The population growth rate thus constitutes an inadequate descriptor of evolutionary dynamics:

Principles based uniquely on the growth rate parameter provide very little perspective on trends in morphological and phenotypic traits of organisms over evolutionary time.

The understanding of adaptation and extinction and an explanation of evolutionary trends in terms of the dynamics of gene frequency change are now known to involve three parameters: population growth rate and demographic variance, quantities that jointly determine the invasion–extinction dynamics of new mutants, and population entropy, a demographic invariant that is a measure of population stability. Entropy is isometric to the morphometric variable, body size; it describes the metabolic capacity of the organism and hence represents a significant indicator of the physiological condition of individuals in the population.

The main tenet of what was described as directionality theory is characterized by the following principle: Under conditions of stationary or slow population growth, evolution by mutation and natural selection results in a unidirectional increase in entropy, whereas when growth is rapid, entropy may decrease or its trajectory may be random and nondirectional. This general property underscores the effect ecological factors exert on population dynamics and provides new perspectives on ecological determinants of evolutionary trends in morphological and phenotypic traits.

20.7. Acknowledgment

I would like to thank Brian Charlesworth, James Crow, and Warren Ewens for their comments on an earlier draft of the manuscript.

REFERENCES

Arnold, L., Demetrius, L., and Gundlach, M. 1994. Evolutionary formalism for products of positive random matrices. *Ann. Appl. Probabi.* **4**:859–901.

Bonner, J. T. 1988. *The Evolution of Complexity.* Princeton, NJ: Princeton U. Press.

Brown, J. H. and Maurer, B. A. 1986. Body size, ecological dominance and Cope's rule. *Nature (London)* **324**:248–250.

Brown, J. H., Marquet, P. A., and Taper, M. L. 1993. Evolution of body size: consequences of an energetic definition of fitness. *Am. Nat.* **142**:573–584.

Calder, W. 1984. *Size, Function and Life History.* Cambridge, MA: Harvard U. Press.

Caswell, H. 1978. A general formulation for the sensitivity of population growth rate to changes in the life-history parameters. *Theor. Popul. Biol.* **24**:215–230.

Charlesworth, B. 1994. *Evolution in Age-Structured Populations.* Cambridge: Cambridge U. Press.

Charlesworth, B. and Giesel, J. T. 1972. Selection in populations with overlapping generations. II. Relations between gene frequency and demographic variables. *Am. Nat.* **106**:388–401.

Clarke, B. 1972. Density dependent selection. *Am. Nat.* **106**:1–13.

Crow, J. 1978. Gene frequency and fitness change in an age-structured population. *Ann. Hum. Genet.* **42**:335–370.

Crow, J. and Kimura, M. 1970. *An Introduction to Population Genetics Theory.* New York: Harper & Row.

de Groot, S. R. and Mazur, P. 1984. *Non-Equilibrium Thermodynamics.* New York: Dover.

Demetrius, L. 1969. The sensitivity of population growth rate to perturbations in the life-cycle components. *Math. Biosci.* 4:129–136.

Demetrius, L. 1974. Demographic parameters and natural selection. *Proc. Natl. Acad. Sci. USA* 21:4645–4647.

Demetrius, L. 1975. Natural selection and age-structured populations. *Genetics* 79:535–544.

Demetrius, L. 1977. Measures of fitness and demographic stability. *Proc. Natl. Acad. Sci. USA* 74:384–386.

Demetrius, L. 1983. Statistical mechanics and population biology. *J. Stat. Phys.* 30:709–753.

Demetrius, L. 1992a. Growth rate, entropy and evolutionary dynamics. *Theor. Popul. Biol.* 41:208–220.

Demetrius, L. 1992b. Thermodynamics of evolution. *Physica A* 189:417–438.

Demetrius, L. 1997. Directionality principles in thermodynamics and evolution. *Proc. Natl. Acad. Sci. USA* 94:3491–3498.

Demetrius, L. and Gundlach, M. 1999. Evolutionary dynamics in random environments. In *Stochastic Dynamics,* H. Crauel and M. Gundlach, eds., pp. 371–392. New York: Springer-Verlag.

Edwards, A. W. F. 1990. Fisher, $\bar{W}$ and the fundamental theorem. *Theor. Popul. Biol.* 38:276–284.

Ewens, W. J. 1979. *Mathematical Population Genetics.* New York: Springer.

Falconer, D. S. 1960. *Introduction to Quantitative Genetics.* New York: Ronald.

Feller, W. 1951. Diffusion processes in genetics. In *Proceedings of the 2nd Berkeley Symposium of Math Statistics and Probability,* pp. 227–248.

Fisher, R. A. 1930. *The Genetical Theory of Natural Selection.* Oxford: Claredon.

Gillespie, J. 1974. Natural selection for within variance in offspring number. *Genetics* 76:601–606.

Ginzburg, L. 1983. *Mathematical Evolutionary Theory.* New York: Benjamin.

Karlin, S. 1992. R. A. Fisher and evolutionary theory. *Stat. Sci.* 7:13–33.

Karlin, S. and Levikson, B. 1974. Temporal fluctuations in selection intensities: case of small population size. *Theor. Popul. Biol.* 6:383–412.

Kimura, M. 1978. Change of gene frequency by natural selection under population number regulation. *Proc. Natl. Acad. Sci. USA* 75:1934–1937.

Kingman, J. F. C. 1961. A mathematical problem in population genetics. *Proc. Cambridge Philos. Soc.* 57:574–582.

Kingman, J. F. C. 1980. *The Mathematics of Genetic Diversity.* Philadelphia: Society for Industrial and Applied Mathematics.

Lande, R. 1982. A quantitative genetic theory of life-history evolution. *Ecology* 63:607–615.

Lenski, R. and Travisano, M. 1994. Dynamics of adaptation and diversification: a 10,000-generation experiment with bacterial populations. *Proc. Natl. Acad. Sci. USA* 91:6808–6814.

Levins, R. and Lewontin, R. 1985. *The Dialectical Biologist.* Cambridge, MA: Harvard U. Press.

Lewontin, R. 1965. Selection for colonizing ability. In *The Genetics of Colonizing Species,* H. G. Baker and G. L. Stebbins, eds., pp. 77–94. New York and London: Academic.

Lewontin, R. 1978. Adaptation. *Sci. Am.* XX:114–128.

Lewontin, R. 1987. The shape of optimality. In *The Latest and the Best*, by J. Dupré, ed. Cambridge, MA: MIT.

Lotka, A. J. 1925. *Elements of Physical Biology.* Baltimore: Williams and Watkins (reprinted and revised as *Elements of Mathematical Biology*, 1956. New York: Dover).

Ludwig, D. 1974. *Stochastic Population Theories. Lecture Notes in Biomathematics*, Vol. 3. Berlin, Heidelberg, New York: Springer-Verlag.

MacArthur, R. H. 1962. Some generalized theorems of natural selection. *Proc. Natl. Acad. Sci. USA* 59:123–138.

McKinney, M. L. 1990. Trends in body size evolution. In *Evolutionary Trends*, K. J. McNamara, eds., pp. 75–120. Tucson, AZ: U. Arizona Press.

Naglyaki, T. 1976. The evolution of one and two locus systems. *Genetics* 83:583–600.

Naglyaki, T. 1979. The dynamics of density- and frequency-dependent selection. *Proc. Natl. Acad. Sci. USA* 76:483–451.

Peters, R. H. 1983. *The Ecological Implications of Body Size.* Cambridge: Cambridge U. Press.

Pollack, E. and Kempthorne, O. 1970. Malthusian parameters in genetic populations. Part I. Haploid and selfing models. *Theor. Popul. Biol.* 2:315–345.

Pollack, E. and Kempthorne, O. 1971. Malthusian parameters in genetic populations. Part II. Random mating populations in infinite habitats. *Theor. Popul. Biol.* 2:351–390.

Roughgarden, J. 1971. Density-dependent natural selection. *Ecology* 52:453–468.

Stanley, S. M. 1973. An explanation of Cope's rule evolution. *Evolution* 27:1–26.

Wright, S. 1931. Evolution in Mendelian populations. *Genetics* 16:97–159.

Wright, S. 1942. Statistical genetics and evolution. *Bull. Am. Math. Soc.* 48:223–246.

SECTION F

GENE FLOW, POPULATION STRUCTURE, AND EVOLUTION

Introductory Remarks

MONTGOMERY SLATKIN

Gene flow and population structure have never occupied a central place in evolutionary biology, although studies of geographic variation and biogeography have always been an integral part of the subject. If the most important question in evolutionary genetics is what underlies evolutionary change (Lewontin, 1974), then variation among populations or other subgroups is not likely to provide that answer. The working of natural selection on existing variation within each subpopulation is a more likely starting point.

The study of geographic variation does play an important part in tests of evolutionary hypotheses. Populations, subspecies, and species that live in different places can be seen as representing independent outcomes of evolutionary experiments. Evidence for divergence in different environments and convergence in similar environments has been important for demonstrating the power of natural selection, and evidence for genetic differentiation for no apparent reason demonstrates the power of genetic drift. As Lewontin (1974, p. 212) has emphasized, however, observed patterns of variation are difficult to interpret in the absence of knowledge about migration. Populations in different areas are independent only to the extent that ongoing gene flow and recent historical association allow them to be. Many field studies are directed to finding how independent different populations actually are and some effort has gone into developing theory to help interpret data generated from those field studies.

Wright emphasized the importance of population structure in his shifting-balance theory, introduced in 1931 and 1932, and discussed by him many times since. In his view, selection among populations and among species is important for evolutionary progress. He emphasized that genetic drift and possibly group selection in a highly subdivided population could allow a population to make an evolutionary transition that would not be possible in a large randomly mating population. Wright's shifting-balance theory has been subject to extensive debate. The consensus seems to be that, while the shifting-balance process is theoretically possible, it is doubtful whether it has occurred commonly in nature or whether it has led to important evolutionary transitions (Coyne et al., 1997)

Haldane and Wright initiated the theoretical analysis of population structure. Haldane (1930) showed that sufficiently strong immigration could prevent adaptation to local conditions. Wright (1931) showed that relatively weak immigration could prevent population divergence at neutral loci. Later Wright (1943) and Malécot (1948) examined the problem of how local genetic differentiation at neutral loci could evolve in a continuously distributed population. Although Wright's term, isolation by distance, came to apply to this situation, it is Malécot's method of analysis that proved to be more useful.

Wright (1951) also initiated the use of geographic variation in allele frequencies as a method for estimating average levels of genetic exchange in a subdivided population. He showed how to partition the overall inbreeding coefficient of an individual into components describing the extent of inbreeding within each population, F_{IS}, and the extent of geographic variation in allele frequencies among populations, F_{ST}. He also showed that in an island model at equilibrium, values of F_{ST} for neutral alleles were related to the extent of gene flow, $F_{ST} = 1/(1 + 4Nm)$, where N is the effective size of a local population and m is the immigration rate. This theoretical development led Wright and many others to indirectly estimate Nm from values of F_{ST} estimated from electrophoretic data in many species (Wright, 1978).

Lewontin's own research has not often focused on population structure, although he did develop one of the first models of group selection (Lewontin and Dunn, 1960; Lewontin, 1962). This model is still one of the very few models of group selection to be applied to any data. Lewontin and Krakauer (1973) made a somewhat less successful attempt to use geographic variation in inbreeding coefficients to test the neutral mutation theory. As Nei and Maruyama (1975) and Robertson (1975) pointed out, this approach ignored the correlations among populations, making the method statistically unsound. Lewontin (1972) emphasized that geographic variation in humans, specifically variation among races, accounted for only a small fraction of the total variation. That result is widely used to refute claims of extensive genetic differences among human races.

In this section, there are four papers, all dealing with the general topic of gene flow and population structure. In Chap. 21 I review some of the recent developments that show how population structure affects gene genealogies. That approach provides a relatively simple way to characterize a wide variety of possible demographic processes. Zouros and Rand (Chap. 22) review the extensive studies of geographic variation in mitochondrial DNA. Because of the greater level of variability and the relative ease of analysis, mitochondrial DNA has provided the bridge from more traditional studies of geographic variation in allozymes to the studies of microsatellites and sequence variation in nuclear DNA. Parker and Niklasson (Chap. 23) and Cohan (Chap. 24) review important topics that are often ignored in most discussions of gene flow and population structure. Asexual organisms are similar in many ways to mitochondrial DNA, in that the lack of genetic exchange between individuals makes a phylogenetic or genealogical approach most natural. The maintenance and reestablishment of asexuality are of particular importance to the understanding of the evolution of sexual reproduction. The study of genetic variation in bacterial populations lies in between that of mitochondrial DNA and of nuclear DNA in higher organisms.

Genetic exchange between individuals occurs in several ways but at rates that vary widely among species. And bacterial populations are ideal for many kinds of laboratory studies of evolution and adaptation.

REFERENCES

Coyne, J. C., Barton, N. H., and Turelli, M. 1997. A critique of Sewall Wright's shifting balance theory of evolution. *Evolution* 51:643–671.

Haldane, J. B. S. 1930. A mathematical theory of natural and artificial selection. Part VI. Isolation. *Proc. Cambridge Philos. Soc.* 26:220–230.

Lewontin, R. C. 1962. Interdeme selection controlling a polymorphism in the house mouse. *Am. Nat.* 96:65–78.

Lewontin, R. C. 1972. The apportionment of human diversity. *Evol. Biol.* 6:381–398.

Lewontin, R. C. 1974. *Genetic Basis of Evolutionary Change.* New York: Columbia U. Press.

Lewontin, R. C. and Dunn, L. C. 1960. The evolutionary dynamics of a polymorphism in the house mouse. *Genetics* 45:702–722.

Lewontin, R. C. and Krakauer, J. 1973. Distribution of gene frequency as a test of the theory of the selective neutrality of polymorphisms. *Genetics* 74:175–195.

Lewontin, R. C. and Krakauer, J. 1975. Testing the heterogeneity of F values. *Genetics* 80:397–398.

Malécot, G. 1948. *Les Mathématiques de l'Hérédité.* Paris: Masson et Cie.

Nei, M. and Maruyama, T. 1975. Lewontin-Krakauer test for neutral genes. *Genetics* 80:395.

Robertson, A. 1975. Gene frequency distributions as a test of selective neutrality. *Genetics* 81:775–785.

Wright, S. 1931. Evolution in mendelian populations. *Genetics* 16:97–159.

Wright, S. 1932. The roles of mutation, inbreeding, crossbreeding and selection in evolution. In *Proceedings of the VI International Congress on Genetics*, pp. 356–366.

Wright, S. 1943. Isolation by distance. *Genetics* 28:1114–1138.

Wright, S. 1951. The genetical structure of populations. *Ann. Eugen.* 15:323–354.

Wright, S. 1978. *Evolution and the Genetics of Populations. Vol. 4. Variability within and among Natural Populations.* Chicago: U. Chicago Press.

CHAPTER TWENTY ONE

A Coalescent View of Population Structure

MONTGOMERY SLATKIN

Population structure, as the term is currently used in evolutionary biology, includes anything that makes a species not conform to the simplest assumptions made in population genetic models, namely random mating and constant population size. Those assumptions are so restrictive that all populations, including those used in laboratory experiments, probably have some structure. Nevertheless, most of population genetics theory and most interpretations of population genetics data ignore population structure. One reason is that there is an infinitude of structures that can be envisioned and no obvious classification of them. So any population genetics theory that wants to include population structure is forced to focus on special cases with little hope of achieving generality. When population structure does play a role, as it does in Wright's shifting balance theory and in some theories of speciation, it often enters as a *deus ex machina* and has attributed to it just the right properties to make an otherwise unachievable transition possible.

In this chapter, I will attempt to clarify the subject of population structure by reviewing the subject from a genealogical perspective. There are indeed many kinds of population structures that can be imagined but there are relatively few effects that population structure can have on gene genealogies. From those effects and from assumptions about mutation, it is possible to infer some general effects of population structure on patterns of genetic variation. I will not attempt an exhaustive classification but instead try to outline broad patterns. I will start by reviewing briefly coalescent theory in a randomly mating population in order to have a frame of reference for later discussions. Then I will discuss the effects of changes in population size and in particular the effects of population growth. Then, I will discuss the effects of population subdivision, both in populations at equilibrium and not at equilibrium. I will concentrate on the effect of population structure on patterns of genetic variation at neutral

R. S. Singh and C. B. Krimbas, eds., *Evolutionary Genetics: From Molecules to Morphology*, vol. 1.
© Cambridge University Press 2000. Printed in the United States of America. ISBN 0-521-57123-5. All rights reserved.

loci, possibly as influenced by linked selected loci. Finally, I will discuss briefly how natural selection can mimic effects of different population structures.

21.1. Gene Genealogies under Random Mating

To understand the effects of population structure on gene genealogies, we need to consider first what happens in the absence of structure. The analysis of gene genealogies is based on coalescent theory introduced by Kingman (1982), but it has its roots in earlier theories of Malécot, Felsenstein, and Griffiths, among others. Tavaré (1984) and Hudson (1990) provide convenient summaries of the subject. A sample of neutral alleles has a genealogical history that can be described by a Markov process introduced by Kingman (1982). As we go backwards in time from the present, the n copies of the locus sampled remain distinct until two or more of them are descended from a single ancestral copy. This event is called a *common ancestry event* or a *coalescent event*. The number of lineages is reduced, and the remaining lineages are then distinct until the next coalescent event. The coalescent events continue until only one lineage, which represents the common ancestor of all copies in the sample, remains. A typical gene genealogy is shown in Fig. 21.1.

The genealogical history of a sample does not tell us anything about the genetic state of each copy in the sample. But differences among the different copies can have arisen by mutation only since the earliest coalescent event (the root of the gene genealogy). Furthermore, if the distribution of mutations on the genealogy is known, the genetic states in the sample are easily determined. This distinction between the genealogy and the mutation is of fundamental importance. The genealogy is determined by demographic processes such as mating system, the history of population growth, and population subdivision, in the present and the past. Mutation is usually independent of demographic processes and depends instead on what kind of locus is being examined.

Under the assumption of neutrality and when the sample size n is small relative to the total population size N, Kingman (1982) showed that, to a first approximation, only two lineages coalesce at each event and that the distribution

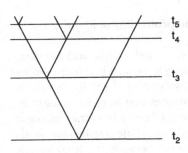

Figure 21.1. An illustration of a gene genealogy representing the history of a sample of five copies of a locus. Each line represents a lineage of ancestry, and points where lineages join represent coalescent events that indicate that two lineages were descended from a single earlier lineage. The times, $t_5, \ldots, t_2$, are the times of the coalescent events. In a randomly mating population of constant size, the intervals between successive coalescence times, $t_i - t_{i-1}$ have an exponential distribution given by Eq. (21.1) in the text, and for such a population, those intervals will become longer at earlier times in the past, as shown.

of times between successive coalescent events has a geometric distribution:

$$\Pr(t \mid i) = \frac{i(i-1)}{4N}\left[1 - \frac{i(i-1)}{4N}\right]^{t-1} \approx \frac{i(i-1)}{4N}e^{-i(i-1)t/(4N)}, \qquad (21.1)$$

where i is the number of lineages remaining and t is the number of generations until the next coalescent event. The second approximation is useful in many applications. Furthermore, Kingman (1982) showed that the probability of any topology of the gene genealogy can be generated from the assumption that when i lineages remain, any of the $i(i-1)/2$ possible coalescent events is equally likely to occur. The implication of Eq. (21.1) is that, in a population of constant size, the times between successive coalescent events become longer in the more distant past and that the age of the root is of the order of $4N$ generations in the past.

21.2. Adding Mutations to Gene Genealogies

The separation of coalescence from mutation can be illustrated with the following example. A common question in population genetics is whether two copies in a locus are in the same genetic state or not. We can calculate this probability by considering a sample of size $n = 2$, for which the distribution of times until they coalesce is a geometric distribution with mean $2N$, given by Eq. (21.1) with $i = 2$. For two alleles to be in the same state, either there is no mutation before the coalescence or mutations have occurred in such a way that the two copies are in the same state. To illustrate, consider first the infinite alleles model in which every mutation creates a new allele. In that case, the probability of identity is just the probability that no mutations occur before the coalescence, because every mutation destroys genetic identity and no mutation can recover it. The probability of no mutation occuring before coalescence at t is $e^{-2\mu t}$, where the 2 is present because there are two lineages on which mutation can occur. In that case, the probability of identity P is found by averaging over the distribution of t to obtain

$$P = \frac{1}{2N}\int e^{-t/(2N)-2\mu t}\,dt = \frac{1}{1+4N\mu}, \qquad (21.2)$$

(Hudson, 1990), which is the classic result for the probability of identity under the infinite alleles model.

If instead we consider the stepwise mutation model (Ohta and Kimura, 1973), which is thought to be appropriate for many microsatellite loci (Valdes et al., 1993), we can proceed in the same way. Two copies of the locus will be in the same state if the mutations on the two branches that join them result in no net change in allele size. Let j_+ be the number of mutations that increase allele size by one and j_- be the number that decrease allele size by one. Both j_+ and j_- have a Poisson distribution with mean μt, because the total length of the genealogy is $2t$, but only half of the mutations will increase or decrease allele size. Given t, the probability that $j_+ = j_-$ is the probability of getting the same result from two draws from a Poisson distribution with mean μt:

$$\Pr(j_+ = j_- \mid t) = \sum_{j=0}^{\infty} \frac{(\mu t)^{2j}}{(j!)^2} e^{-2\mu t} = I_0(2\sqrt{\mu t}), \qquad (21.3)$$

where $I_0(.)$ is a modified Bessel function (Abramowitz and Stegun 1965, p. 375). Averaging over the distribution of t yields the results of Ohta and Kimura (1973), $P = 1/\sqrt{1 + 8\mu t}$.

The derivation of these results with the coalescent approach has the advantage of showing the relationship between seemingly different models and it also provides us with a guide to predicting the effects of anything that changes the distribution of coalescence times. If population subdivision resulted in a different distribution of coalescence times, that distribution could be combined with Eq. (21.2) or Eq. (21.3) to calculate the new values of P for each mutation model. Furthermore, even when the entire distribution of t cannot be derived, approximate results can sometimes be obtained.

21.3. Population Growth

Variation in population size will change the distribution of coalescence times but in a way that is particularly simple to understand. As noted by Kingman (1982) and implemented by Slatkin and Hudson (1991) and Griffiths and Tavaré (1994), changes in population size can be absorbed into a transformation of the time scale in the coalescent process for a population of constant size. That is, we can define a transformed time scale

$$\tau(t) = \int_0^t \frac{dt'}{2N(t')} \qquad (21.4)$$

so that the distribution of times between successive coalescent events measured in terms of τ is given by Eq. (21.1) with $2N$ set to 1. What is happening is that time is being stretched or compressed on the new scale, depending on whether $N(t)$, the population size at time t in the past, is increasing or decreasing.

A model in which the population has grown in the past is of particular relevance for human populations because historical records indicate that many and probably most human populations have grown substantially in the past 50,000 to 100,000 years. Such rapid population growth has a predictable effect on gene genealogies. In a population of constant size, Eq. (21.1) predicts that branch lengths tend to be longer further in the past. The shortest branches tend to be the terminal branches. In contrast, a gene genealogy from a rapidly growing population tends to have shorter internal branches, with the terminal branches being the longest (Slatkin and Hudson, 1991). In the extreme case, with very rapid population growth, the gene genealogy becomes starlike, with all the internal branches being much shorter than the external branches.

This difference in gene genealogy has some predictable effects on genetic variability. If we again consider the infinite alleles model, the number of allelic classes represented by a single copy depends on the number of mutations on the terminal branches on the gene genealogy. Mutations that occur on internal branches will result in alleles represented in two or more copies. Clearly, rapid population growth will result in more singletons that would be found in a population of constant size, which is what Nei et al. (1975) found in their simulations.

For the infinite sites model, Slatkin and Hudson (1991) showed that the distribution of pairwise differences in nucleotide sequence would be similar to a Poisson distribution, which is the distribution of pairwise differences that would result if the gene genealogy were a star genealogy, with no internal branches. Such distributions have been found in the distribution of pairwise differences in mtDNA sequences in several human populations (Di Rienzo and Wilson 1991), suggesting that those populations have undergone rapid growth in the past. Griffiths and Tavaré (1994) have recently shown that it is possible to estimate the growth rate of a population from the configuration of sequences of a sample of mtDNA.

21.4. Population Subdivision

Population subdivision results in different individuals that have restricted and nonoverlapping sets of possible mates. We can imagine many causes of population subdivision but there are two broad categories that are worth distinguishing, depending on whether there is currently some gene flow or not. If there is no ongoing gene flow among subpopulations, then any genetic similarity is the result of historical association. If there is ongoing gene flow, then genetic similarity of different subpopulations is continually being renewed by the movement of individuals. In the latter case, historical association can also be important, although there is the possibility that an equilibrium can be reached. The differences between the two cases are illustrated in Fig. 21.2. There are many biological situations in which there is no obvious partitioning of individuals into discrete subpopulations, but in which there is restricted dispersal – Wright's (1943) model of isolation by distance. The dispersal that does occur results in ongoing gene flow.

In general, population subdivision makes gene genealogies longer than they would be in comparable randomly mating populations. The reason is that it will take time for different ancestral lineages to find themselves in the same subpopulation or in the same neighborhood so that they can coalesce. The full theory of the coalescent process in a subdivided population seems difficult to solve in general, and there seem to be no results comparable with Eq. (21.1) for the distribution of times between successive coalescent events. Notohara (1988), Takahata (1988), and Takahata and Slatkin (1990) present the general theory for models of gene flow at equilibrium, and Wakeley (1996) obtained

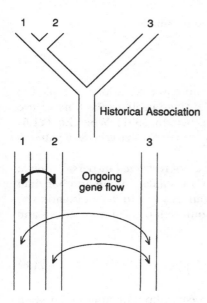

Figure 21.2. An illustration of the differences between two possible explanations for why there would be greater genetic similarity between populations 1 and 2 than between either of them and population 3. If there is no ongoing gene flow, then the pattern of historical association of these populations would determine which populations were more similar, as in the upper figure. If there is ongoing gene flow, as in the lower figure, then the exchange of more migrants between populations 1 and 2 than between 2 and 3 or 1 and 3, as indicated by the relative weights of the lines, would result in the same pattern of genetic similarity. Ongoing gene flow could mask any prior historical relationships among the three populations.

some results that contrast the effects of ongoing gene flow and of historical association.

For models of gene flow in populations at equilibrium, Strobeck (1987), Hey (1991), and I (Slatkin, 1991, 1993) obtained rather general results for samples of size 2. A surprising result is that the average coalescence time of two copies of a locus drawn from the same subpopulation is just twice the total number of copies of the locus in the entire population, independently of the pattern of gene flow among subpopulations (Strobeck, 1987). For example, if there are n subpopulations, each containing N individuals, then the average coalescence time is $4Nn$, provided that all the subpopulations are connected by gene flow. That implies that the average number of differences in DNA sequence between two mtDNAs drawn from the same population is $4Nn\mu$, where μ is the mutation rate (Strobeck, 1987). The average number of differences between copies from different populations does depend on the level and pattern of gene flow, suggesting that the difference within and between populations could be used to estimate levels of gene flow.

The complete analysis of genetic data from subdivided populations will probably have to rely on simulation methods of the kind being developed by Griffiths and Tavaré (1994) and by Felsenstein and his co-workers (Kuhner et al., 1995). The more traditional analysis of data from subdivided populations has relied on summary statistics such as Wright's (1951) F_{ST} and various genetic distances. I showed that many of these statistics were closely related to ratios of coalescence times when mutation rates are very small (Slatkin, 1991). For example, Wright (1951) defined F_{ST} to be the excess in correlation between alleles within the same subpopulation, but when mutation rates are small:

$$F_{ST} \approx \frac{\bar{t} - \bar{t}_0}{\bar{t}}, \tag{21.5}$$

where $\bar{t}$ is the average coalescence time of two copies of a locus sampled at random from the population and $\bar{t}_0$ is the average coalescence time of two copies sampled from the same subpopulation (Slatkin, 1991). When Eq. (21.5) is used, it is relatively easy to derive expected values of F_{ST} in a variety of patterns of migration (Slatkin, 1991).

When there is isolation by distance as a result of restricted dispersal, then at equilibrium nearby subpopulations will be more similar to one another than to more distant population. I suggested that an easy way to detect isolation by distance is to use values of F_{ST} computed for different pairs of populations and then to use the quantity

$$\hat{M} = \frac{1}{4}\left(\frac{1}{F_{ST}} - 1\right) \tag{21.6}$$

to describe the extent of genetic similarity between subpopulations. The reason for using $\hat{M}$ is that it has a relatively simple dependence on geographic distance in populations at equilibrium. For example, $\hat{M}$ decreases with the inverse of distance in a one-dimensional habitat and roughly with the inverse of the square root of distance in a two-dimensional habitat (Slatkin, 1991).

21.5. Nonequilibrium Subdivided Populations

We have already seen that rapid population growth results in a gene genealogy quite different from what is expected in a population of constant size. Statistical tests for differences can reveal some past history of population growth. The analysis of subdivided populations has a similar goal, namely distinguishing between populations that are at an equilibrium under their current patterns and levels of gene flow from populations that are not at equilibrium. At present, there is no good test based on only the distribution of allele frequencies or DNA sequences among populations. For quantities like F_{ST} the same patterns can result either from gene flow in equilibrium population or from historical association. There are a sufficient number of parameters in both classes of models that there is no obvious way to distinguish conclusively between them.

I have shown that patterns of apparent isolation by distance might in principle reveal a recent colonization of a large geographic area provided that, other than during the colonization process, dispersal was restricted (Slatkin, 1993). A recently expanded population would not have had time to establish a detectable pattern of isolation by distance among all populations. Isolation by distance would become established first between nearby populations and then would spread more slowly to more distant populations. The problem is that dispersal patterns would have to change abruptly and be consistent over large geographic distances for this theory to be applied and numerous loci would have to be examined before the overall pattern becomes evident. Nevertheless, Hellberg (1995) has found some evidence for this type of nonequilibrium population structure in corals.

I think that it will always be difficult for strong conclusions about subdivided populations to be drawn from the analysis of genetic data alone, at least until very large data sets are available. There are too many kinds of gene flow and too many historical events imaginable for genetic data to point conclusively to a single explanation. I think that much stronger conclusions can be drawn when there is additional information available about the species being studied. The situation I have focused most attention on is one in which F_{ST} indicates relatively high levels of gene flow in species for which direct studies indicate very limited dispersal. I have argued that such species are not at a genetic equilibrium and that historical association accounts for the relatively low F_{ST} values. I have suggested that several species, including *Drosophila pseudoobscura*, are in this category (Slatkin, 1987). We currently lack robust statistical methods that would provide more rigorous tests of population equilibrium.

21.6. Natural Selection

Our understanding of gene genealogies under selection is much more limited, although there are some results available and our intuition can guide us in cases in which the formal theory has not yet been developed. In many cases, the effects of natural selection are similar to the effects of some change in population structure. For example, if all alleles are overdominant, Takahata (1990) showed that, with an appropriate rescaling of time, the coalescent process for the different alleles is equivalent to one for neutral alleles. The difference is that overdominance causes the intervals between successive coalescent events to be much longer than in the absence of selection, so different alleles have a gene genealogy that is the same as one for neutral alleles drawn from a much larger population. There is an important difference, however, because different copies of the same allele will have an underlying gene genealogy that is quite different. Vekemans and Slatkin (1994) used Takahata's method to obtain a similar result for alleles at a gametic self-incompatibility locus, and they showed that gene genealogies within allelic classes tend to be rather short. The overall gene genealogy, then, resembles one from a subdivided population, with each allelic class representing a distinct subpopulation.

Selection that has resulted in the substitution of an allele can be regarded as equivalent to past population growth. The potential ancestors of a sample taken today can come only from those lineages that had the advantageous allele in the past. If a locus is polymorphic because one allele is in the process of being substituted, then different allelic classes would each have their own genealogies. The advantageous allele would appear to be from a growing population because its numbers had increased recently, and the disadvantageous allele would appear to be from a declining population because its numbers would be decreasing.

21.7. Recombination

What I have said so far treats each locus as if there is no recombination. Only in that case is it correct to talk about the genealogy of a sample. People who work on coalescent theory do their best to conceal the fact that recombination makes a mess of gene genealogies. The general problem seems intractably hard to analyze, except by simulation, but we can understand two extremes that are useful. If recombination occurs but is extremely rare compared with mutation, the genome can be regarded as comprising unrecombined blocks, each of which has its own gene genealogy. Adjacent blocks differ only because of a single recombination event, so the gene genealogies for adjacent blocks can be obtained from each other by a single branch swap (Hudson, 1990; Hein, 1990). In the other extreme, recombination is so common relative to mutation that each polymorphic site or locus will occupy its own unrecombined block, and gene genealogies of adjacent polymorphic sites or loci will differ by several recombination events (Hudson, 1990). In this case, a genealogical approach is probably not helpful, and it is just as well to proceed by using average quantities such as heterozygosities or the numbers of segregating sites. With or without recombination, the extent of linkage disequilibrium between sites is sensitive to the demographic history of a population (Slatkin, 1994).

21.8. Conclusion

Understanding the effect of different kinds of population structure on gene genealogies is an important step toward making robust predictions about patterns of genetic variation in natural populations. Although a great many population structures can be imagined and only partial analysis has been achieved for some of them, the theory is sufficiently well developed that useful generalizations can be made. In this chapter, I have emphasized the effect of different kinds of population structure on gene genealogies. With a gene genealogy in a population of constant size as the standard of comparison, there are two major effects of population structure. The first is a change in the relative intervals between coalescent events. In a neutral gene genealogy, the intervals become successively longer in the more distant past. Rapid population growth and natural selection in favor of advantageous mutations both can result in the opposite pattern, with the intervals becoming shorter further in the past. Population subdivision results in a more complex pattern, with numerous coalescent events occurring in the recent past, reflecting the recent common ancestry of different copies in the same population; but then much longer intervals occur between coalescent events in the more distant past, reflecting the much longer times needed for copies initially in different subpopulations to finally be in the same subpopulation.

Changes in the relative times between successive coalescent events change the distribution of configurations of genetic types in a sample from what would be expected in a population of constant size.

The second effect is to change the time until the most recent common ancestor of the gene genealogy. In general, rapid population growth and natural selection reduce the time to common ancestry because at more distant times in the past, there are fewer potential ancestors. In contrast, population subdivision always results in longer and possibly much longer times to common ancestry. Overdominant selection has the same effect. Longer times to common ancestry result in more extensive genetic polymorphism than in a comparable population of constant size. How much more variation and the character of that variation depend on the exact model of population subdivision, but under a variety of assumptions, the relative coalescence times within and between populations provide a guide to the general patterns of population subdivision.

The theory of population subdivision is far from complete, and much of it will be sufficiently complicated that simple general solutions may never be obtained. I have tried to show that it is considerably easier to consider the effects of population structure on gene genealogies rather than on genetic variation itself. The relationship between gene genealogies and patterns of genetic variation is well understood and is independent of the demographic factors that produced the gene genealogy, so there is no need to reconsider that relationship in trying to understand the genetic consequences of different demographic models.

21.9. Summary

I have reviewed the different effects of population structure on gene genealogies, using the gene genealogy of a neutral locus in a randomly mating population of constant size as a reference. Rapid population growth and natural selection in favor of one allele will result in gene genealogies that have longer terminal branches and successively shorter internal branches than in a population of constant size, and the time of the most recent common ancestor is generally shorter. Population subdivision results in gene genealogies that have predictable patterns of coalescent events within and between populations and longer overall gene genealogies. Overdominant selection has the same effect but on gene genealogies within and between allelic classes. Changes in gene genealogies can often be related to patterns of variation at genetic loci subject to different kinds of mutation pressure.

REFERENCES

Abramowitz, M. and Stegun, I. A. 1965. *Handbook of Mathematical Functions*. New York: Dover.

Di Rienzo, A. and Wilson, A. C. 1991. Branching patterns in the evolutionary tree for human mitochondrial DNA. *Proc. Natl. Acad. Sci. USA* **88**:1597–1601.

Griffiths, R. C. and Tavaré, S. 1994. Sampling theory for neutral alleles in a varying environment. *Philos. Trans. R. Soc. London Ser. B* **344**:403–410.

Hein, J. 1990. Reconstructing evolution of sequences subject to recombination using parsimony. *Math. Biosci.* **98**:185–200.

Hellberg, M. E. 1995. Stepping-stone gene flow in the solitary coral, *Balanophyllia elegans. Mar. Biol.* **123**:573–581.

Hey, J. 1991. A multi-dimensional coalescent process applied to multi-allelic selection models and migration models. *Theor. Popul. Biol.* **39**:30–48.

Hudson, R. R. 1990. Gene genealogies and the coalescent process. *Oxford Surv. Evol. Biol.* **7**:1–44.

Kingman, J. F. C. 1982. The coalescent. *Stochast. Process. Appl.* **13**:235–248.

Kuhner, M. K., Yamato, J., and Felsenstein, J. 1995. Estimating effective population size and mutation rate from sequence data using Metropolis-Hastings sampling. *Genetics* **140**:1421–1430.

Nei, M., Maruyama, T., and Chakraborty, R. 1975. The bottleneck effect and genetic variability in populations. *Evolution* **29**:1–10.

Notohara, M. 1988. The coalescent and the genealogical process in a geographically structured population. *J. Math. Biol.* **29**:59–75.

Ohta, T. and Kimura, M. 1973. A model of mutation appropriate to estimate the number of electrophoretically detectable alleles in a finite population. *Genet. Res.* **22**:201–204.

Slatkin, M. 1987. Gene flow and the geographic structure of natural populations. *Science* **236**:787–792.

Slatkin, M. 1991. Inbreeding coefficients and coalescence times. *Genet. Res.* **58**:167–175.

Slatkin, M. 1993. Isolation by distance in equilibrium and non-equilibrium populations. *Evolution* **47**:264–279.

Slatkin, M. 1994. Linkage disequilibrium in growing and stable populations. *Genetics* **137**:331–336.

Slatkin, M. and Hudson, R. R. 1991. Pairwise comparisons of mitochondrial DNA sequences in stable and exponentially growing populations. *Genetics* **129**:555–562.

Strobeck, C. 1987. Average number of nucleotide differences in a sample from a single subpopulation: a test for population subdivision. *Genetics* **117**:149–153.

Takahata, N. 1988. The coalescent in two partially isolated diffusion populations. *Genet. Res.* **52**:213–222.

Takahata, N. 1990. A simple genealogical structure of strongly balanced allelic lines and trans-species evolution of polymorphism. *Proc. Natl. Acad. Sci. USA* **87**:2419–2423.

Takahata, N. and Slatkin, M. 1990. Genealogy of neutral genes in two partially isolated populations. *Theor. Popul. Biol.* **38**:331–350.

Tavaré, S. 1984. Line-of-descent and genealogical processes, and their application in population genetic models. *Theor. Popul. Biol.* **26**:119–164.

Valdes, A. M., Slatkin, M., and Freimer, N. B. 1993. Allele frequencies at microsatellite loci: the stepwise mutation model revisited. *Genetics* **133**:737–749.

Vekemans, X. and Slatkin, M. 1994. Gene and allelic genealogies at a gametophytic self-incompatibility locus. *Genetics* **137**:1157–1165.

Wakeley, J. 1996. The variance of pairwise nucleotide differences in two populations with migration. *Theor. Popul. Biol.* **49**:39–57.

Watterson, G. A. 1975. On the number of segregating sites in genetical models without recombination. *Theor. Popul. Biol.* **7**:256–276.

Wright, S. 1943. Isolation by distance. *Genetics* **28**:114–138.

Wright, S. 1951. The genetical structure of populations. *Ann. Eugen.* **15**:323–354.

CHAPTER TWENTY TWO

Population Genetics and Evolution of Animal Mitochondrial DNA

ELEFTHERIOS ZOUROS AND DAVID M. RAND

22.1. Introduction

Mitochondrial DNA (mt DNA) has played an extremely important role in the development of evolutionary genetics. This contribution is due to the nature of data that can be obtained (homologous nucleotide sequences within and between populations and species) and due to the unique genetics of the molecule (uniparental asexual transmission of a nonrecombining set of genes). As such, some workers have used mtDNA to infer the evolutionary history of organisms while others have used organisms as distinct sources of mtDNA to study the evolution of genomes. We review some aspects of the population genetics and evolution of mtDNA that are often not emphasized: the population genetics of the cytoplasm, the consequences of paternal leakage, the contrasts among mtDNA, sex chromosomes and autosomes in population structure, and the role of mildly deleterious mutations in population genetic and macroevolutionary patterns. There are many new theoretical and empirical questions to be answered about the evolutionary genetics of mtDNA that will inform us about organismal and molecular evolution alike.

22.2. The Cytoplasm as a Population

The unique genetics of mtDNA (e.g., lack of recombination, maternal inheritance) have been central to the great contribution this molecule has made to virtually all fields of population and evolutionary biology. But the contribution of mtDNA goes beyond the interesting patterns in nature that the marker has helped unveil: Its unique genetics has forced theoreticians and experimentalists alike to see the world through the narrow but focused lenses of uniparental, cytoplasmic descent. This has opened up new terrain in population genetics. While there is no doubt that the exponential growth of allozyme studies through

R. S. Singh and C. B. Krimbas, eds., *Evolutionary Genetics: From Molecules to Morphology*, vol. 1. © Cambridge University Press 2000. Printed in the United States of America. ISBN 0-521-57123-5. All rights reserved.

the 1960s and 1970s stimulated tremendous advances in population genetics, most of this work involved population genetics with the markers. The genetics of mtDNA forced researchers to think about the population genetics of the marker as well as the population genetics that could be done with the marker. This additional level of variation stems, in part, from the cytoplasmic, hierarchical nature of mtDNA transmission.

In most metazoans, mtDNA is both polyploid and haploid (i.e., multiple genomes exist within a cell and they are transmitted asexually). The mitochondrial genome exists as a population of 2–10 copies per organelle; there are hundreds to thousands of organelles per cell; there are multiple cells that make up each tissue (clearly only the germ line tissue is directly relevant to the transmission of variation); there are multiple individuals per deme; demes are nested within populations; populations are nested within species, and so on up the hierarchy. Taking an average of 1000 mitochondria per cell, an average of 10 mitochondrial DNA (mtDNA) molecules per mitochondrion, and an average of 20,000 nucleotide pairs for a mtDNA molecule leads to 2×10^8 nucleotide pairs in a cell, which is 2% of a human cell's DNA content. The dynamics of this hierarchical structure of mtDNA variation, imposed by the cellular and subcellular packages in which mtDNA is inherited, have important implications for mtDNA as both the subject and the object of population genetic study.

The population genetics of the cytoplasm has been a fundamental factor in the evolution of organelle genomes. The obligatory coexistence of an alpha protobacterium with an amoeboid cell has been in existence for some 2000 million years (MY). Very few researchers today question Margulis' (1981) endosymbiotic theory for the origin of the eukaryotic cell. This symbiosis is undoubtedly among the most dramatic events in the history of life on Earth, secondary in significance only to the origin of life itself. This point was emphasized recently by Margulis (1996), who argued that any classification of life has to be based on the origins of symbiosis. It appears now that there may never have been a eukaryotic organism that contained no mitochondria at some point in its evolutionary history. Extant species with no mitochondria (trichomonads, archezoan diplomonads, microspiridia) most likely carried mitochondria at some point in their phylogenetic history and lost them secondarily (Clark and Roger, 1995). Hence, the cytoplasm has been a population, indeed a community, for the majority of time that life has existed on earth. The population genetics (or community ecology) of this cytoplasm may have played a central role in the evolution of multicellularity through the competition among organelle genomes. This dynamic is ongoing, but despite considerable progress on how organelles are inherited (Gillham, 1994), much remains to be learned about the microevolutionary dynamics of the cytoplasm (e.g., Mikelsaar, 1987).

The observed divergence of a specific nucleotide site in mtDNA between populations or species can be traced, ultimately, to a mutation in a germ line cell of a single female (see below for exceptions to maternal inheritance). The establishment of a mutant haplotype in a population thus requires drift (or selection) to fixation in the cytoplasm and the subsequent drift (or selection) to fixation in the organismal population. While this pattern has long been recognized (e.g., Lewontin, 1970; Upholt and Dawid, 1977), there have been very few studies that have empirically examined the consequences of this "subindividual" variation (heteroplasmy) for interindividual population genetics of mtDNA (Birky, 1983; Rand and Harrison, 1989b; Arnason and Rand, 1992; Brown et al., 1992). The few cases that have studied this problem have generally addressed variation in the size of the mtDNA molecule stemming from insertion–deletion mutations (indels), and not point mutations (but see Monnat and Loeb, 1985). This is due to the fact that heteroplasmy for size is easily detected, in part resulting from a mutation rate for indels that is several orders of magnitude higher than that for point mutations. Accordingly there are relatively few reported cases of nucleotide site heteroplasmy, but they are accumulating (*Drosophila:* Hale and Singh, 1986; Kondo et al., 1990. Fish: Bentzen et al., 1988; Magoulas and Zouros, 1993. Cows: Koehler et al., 1991. Humans: Gill et al., 1994; Ivanov et al., 1996). The increasing number of examples of nucleotide site heteroplasmy should provide useful material for cytoplasmic-through-species analyses of variation that focus on the incipient stages of mtDNA divergence.

The hierarchical analysis is essentially an F-statistical approach that apportions variation to specific levels in a hierarchy (e.g., Lewontin, 1972). In a typical study of population structure, F_{st} quantifies the proportion of variation that lies among subpopulations of the species relative to the total variation. This value sheds light on the balance of mutation, drift, and migration in determining the levels of differentiation among populations (e.g., Slatkin, 1987). F-statistical analyses through a cytoplasmic hierarchy can shed light on the balance of mutation and drift that determines the fate of novel mtDNA haplotypes in germ cells (Rand and Harrison, 1989b). The available data show that the lowest-level F-statistic in cytoplasmic studies (variation within individual organisms, relative to the total variation) can be substantial for mtDNA size variation (Rand and Harrison, 1989b; Arnason and Rand, 1992; Brown et al. 1992). Qualitative assessments of the comparable F-statistics for nucleotide site variation show that this intraindividual component is trivial, yet basic biology requires that it be nonzero (see previous paragraph).

The relevant population genetic factors that can affect which hierarchical level carries the greatest proportion of genetic variation depends primarily on the mutation rate, the number of segregating units in the cytoplasm, and the number of germ cell generations per animal generation (plus the higher-level factors influencing typical organismal populations such as breeding effective size, sex ratios, migration patterns, etc.). Solignac et al. (1984) first applied Wright's formula for the evolution of the variance in allele frequency among populations to the problem of mtDNA transmission. This is described by

$$V_n = p(1 - p)(1 - \{1 - 1/N_{eo}\}^{g_n}), \tag{22.1}$$

where V_n is the variance among populations (i.e., cytoplasms) at the nth generation, p is the frequency of one allele in the population at the start (or the frequency of an mtDNA variant in the heteroplasmic mother), N_{eo} is the effective size of the population (i.e., the effective number of segregating organelles in the cytoplasm), and g is the number of germ cell generations per animal generation. Since the mutation rate, N_{eo}, and g may differ among organisms with different cellular and developmental properties, there could be important consequences for mtDNA sequence evolution. As illustrated in Fig. 22.1, species with many ($g = 50$) germ cell generations per animal generation should exhibit high variance in mtDNA frequencies among offspring of heteroplasmic females across a range of N_{eo} because of rapid sorting out to homoplasmy during multiple cell divisions. Early studies examining the inheritance of mtDNA variants in heteroplasmic lines identified differences among yeasts (Birky, 1983), insects (Solignac et al., 1984; Rand and Harrison, 1986), and mammals (Hauswirth and Laipis, 1982). As expected, yeasts exhibited a faster sorting out than did insects because of lower values of N_{eo} (Birky, 1983; Solignac et al., 1984; Rand and Harrison, 1986). Cows appear to exhibit very rapid sorting out (one to four animal generations; Koehler et al., 1991), and recent data from the remains of Czar Nicholas II of Russia and his relatives show that heteroplasmy can drift to homoplasmy in four generations (Ivanov et al., 1996).

The dynamics illustrated in Figure 22.1 appears to be consistent with the empirical results that mammals, which have approximately 20–50 germ cell generations per animal generation, should show rapid sorting out of heteroplasmy. However, the observations that a cytoplasmic population of mtDNA haplotypes can shift from fixation for one type to fixation for a distinct, previously undetected mtDNA in one or two animal generations (Koehler et al., 1991) underscores the importance of the cytoplasmic population genetic view of mtDNA variation and evolution in animals.

From Eq. (20.1) it would seem that an organism with many germ cell generations per animal generation would be predisposed to retard mtDNA evolution, since new mutations that enter the cytoplasmic population would have such an unlikely chance of drifting to fixation. Since this very pattern has been documented, the effective number of mtDNAs per germ line cell may be considerably lower than basic cell biology might predict. Perhaps single mtDNA templates are chosen at random for replication and serve at the mother template from which the majority of new mtDNA molecules are derived. This would be analogous to a randomly generated selective sweep of the cytoplasmic population. Coupled with a high point mutation rate, such a cellular dynamic could promote quite a high rate of mtDNA evolution. The fact that Monnat and Loeb (1985) observed almost no intraindividual mtDNA variation in humans when the interindividual variation was one in every 200 nucleotides suggests that the cytoplasmic population genetics of somatic tissue is distinct from that of the germ line. Either the mtDNA mutation rate in somatic tissues is very low or there is a strong mechanism that homogenizes the nucleotide sequences within the mitochondrion or within the cell.

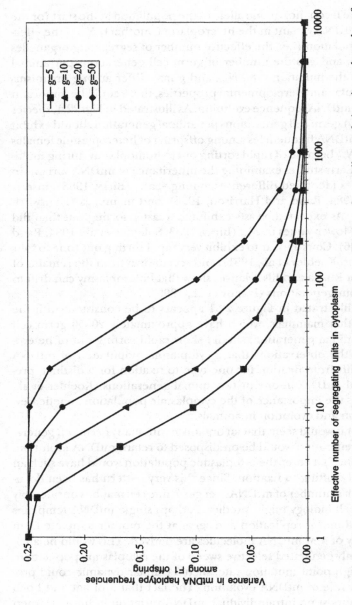

Figure 22.1. Variation among offspring of a heteroplasmic mother as a function of organelle population size and number of germ cell generations. On the vertical axis is the variance in mtDNA haplotype frequency among an F1 sample of offspring from a heteroplasmic mother assumed to carry two haplotypes in equal frequency [see Eq. (20.1)]. The horizontal axis shows the effective population size of the mtDNAs subjected to the genetic drift process imposed by cell division. The value of N_{eo} really refers to the number of segregating units (mtDNA, organelles, or cluster of organelles) that generates the observed variance in haplotype frequency. The four curves show the effect that differing numbers of germ cell generations per animal generation have on the observed variation in haplotype frequencies after one generation of heteroplasmic transmission.

434

These factors may bear on the data showing different rates of mtDNA evolution between endotherms and ectotherms (Martin and Palumbi, 1993; Rand, 1994). Explanations for these rate differences have focused on differences in mutation rates, but the population genetics of mtDNA haplotypes through the cytoplasmic hierarchy of germ line development may contribute to differences in the probability of fixation of mutant haplotypes (Rand and Harrison, 1986; Beckenbach, 1994).

22.3. Maternal Transmission

The maternal inheritance of mtDNA in animals was established as soon as genetic markers for its detection became available (Hutchison et al., 1974). Very soon it became obvious that uniparental transmission is a general property not only of the mitochondrial genome but of all extranuclear genomes. As Table 22.1 shows, uniparental inheritance seems to be the common feature of organelle transmission, not whether the transmission occurs through the female or the male parent. The chlamydomonas and redwood cases make this point quite clearly. In the first species the two organelles are transmitted through different mating types. In redwoods the mtDNA became pollen transmitted in violation of the conifer rule of mtDNA transmission through the ovule. Clearly, the transmitting sex is irrelevant as long as the uniparental inheritance is preserved.

Because of its generality, uniparental transmission of extranuclear DNA became the interest of extensive theoretical work. The dominant theory (Cosmides and Tooby, 1981; Hoekstra, 1990; Hurst and Hamilton, 1992; Hastings, 1992) sees the evolution of uniparental transmission as the nuclear genome's way of preventing the costly war between different extranuclear genomes for higher representation in the cytoplasm of daughter cells (in cytokinensis there is no mechanism to play the role of the fair divisor that mitosis plays in karyokinesis). Godelle and Reboud (1995) have suggested a new explanation, one that

Table 22.1. *Uniparentality of Organelle DNA Transmission*

	Transmission	
	mtDNA	*cpDNA*
Chlamydomonas reinhardtii	−type	+type
Yeast	+type	n.a.
Conifers (redwood)	maternal	paternal
Redwood	paternal	paternal
Angiosperms	maternal	maternal
Animals (-mussels)	maternal	n.a.
Mussels Type F	maternal	n.a.
Mussels Type M	paternal	n.a.

replaces the interorganelle conflict with organelle adaptation. The cytoplasm can be seen as the organelle's exclusive habitat. The organelle genome that is best adapted to the germ line that donates the bulk of the zygote's cytoplasm (usually the female germ line) will eventually establish itself. Uniparental inheritance will then evolve as the result of two-level selection, one at the haploid level (gametes) and the other at the diploid level (zygote). Both models involve interactions between nuclear and organelle genomes and, because of this, it is not clear what sort of empirical observation could distinguish between the two.

The empirical evidence on the mechanism of mtDNA transmission is limited, but it strongly points to nuclear control. There are many ways for preventing paternal transmission, and several of these are known to operate in different species (Birky, 1995). In gray fish the sperm lacks mitochondria, in tunicates the mitochondria do not enter the egg, and in mammals the mitochondria enter the egg but are destroyed. The experimental approach to the study of maternal inheritance is complicated by the fact that the ratio of sperm-to-egg mitochondria in the one-cell zygote is very small (perhaps less than 10^{-4}), which means that if during the first cell divisions the rate of organelle division is smaller than the rate of cell division, sperm mtDNA will be eliminated from the organism by random drift. One way to counter this is to backcross the F_1 female repeatedly from two strains differing in mtDNA to males from the same strain. If there is a contribution by the sperm and if there is a small probability that the sperm's mtDNA will be retained in the female and find its way into the egg, a detectable amount of paternal mtDNA may be found in the line after many generations of backcrossing. This protocol was tried in moths (Lansman et al., 1983) and in mice (Gyllensten et al., 1985). No paternal mtDNA was observed in the moths, but a positive result was seen in mice when a sensitive polymerase chain reaction (PCR)-based assay was used (Gyllensten et al., 1991).

The study by Kaneda et al. (1995) is at present the most informative regarding the control of mtDNA inheritance in animals. Working also with mice, they have shown that the oocyte is armed with the ability to recognize and destroy the sperm's mitochondria. The recognition mechanism resides in the outer surface of sperm mitochondria. Sperm mitochondria from a foreign species may escape the egg's recognition mechanism, which explains why most examples of paternal mtDNA inheritance come from interspecific cross. It also suggests that even though hybrid crosses provide a way of detecting male transmission, repeating backcrossing may in fact reduce the probability of detecting paternal inheritance: As the backcross proceeds, the female's nuclear genome becomes more and more similar to the male's, and therefore the chances are increasing that her eggs will carry the factors that would recognize and destroy the sperm mitochondria. Examples of paternal mtDNA inheritance are now known in *Drosophila* (Satta et al., 1988; Kondo et al., 1990), anchovy (Magoulas and Zouros, 1993), and mice (Gyllensten et al., 1991), so that it may be assumed that a leaky contribution from the father (accounting for $\sim 10^{-4}$ of an individual's DNA) may be the rule rather than the exception in animals.

There are a number of consequences of paternal leakage for population genetics and systematics. As illustrated in Fig. 22.2, paternal leakage increases

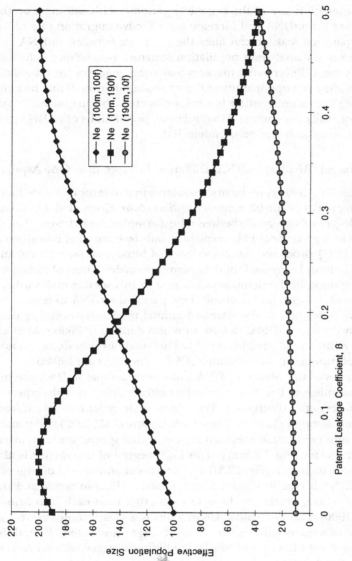

Figure 22.2. The effect of paternal leakage on the effective population size of mtDNA. Paternal leakage is defined by β, which ranges from zero (strict maternal inheritance) to 1.0 (strict paternal inheritance). Only the range from 0.0 to 0.5 is shown since the curves are symmetric. The three curves indicate different sex ratios, and the effective population size is determined by the relation $N_e = N_m \cdot N_f / [(1-\beta)^2 \cdot N_m + \beta^2 \cdot N_f]$ (Birky et al., 1989) where N_m and N_f are the effective population size of males and females and β is the paternal leakage coefficient.

the effective population size of the organelle genome. This will increase the effect of selection on mtDNA and increase the effective migration rate of the marker. Thus paternal leakage can alter the contrasts between mtDNA and nuclear markers in the analysis of population structure (see below). A further consequence is that differentiated markers may come together into the same cytoplasm, providing the opportunity to detect recombination. While no compelling evidence for recombination has been detected, this may stem in part from the lack of available markers. Homoplasmic populations of mtDNA may undergo recombination but remain undetected.

22.4. The Mussel (*Mytilus*) mtDNA: Different in More than One Aspect

Fisher and Skibinski (1990) were the first to observe that heteroplasmy was more common among males of the blue mussel *Mytilus edulis*. Hoeh et al. (1991) argued that the degree of divergence between the two molecules in heteroplasmic individuals was so high that could be explained only by biparental inheritance. Zouros et al. (1992) produced direct evidence of biparental inheritance and also noted that heteroplasmy was found exclusively in males. These observations found an explanation in the demonstration that in this species males inherit mtDNA from both parents and transmit their paternal mtDNA to their sons. Females behave according to the standard animal model, transmitting their one, maternally derived mtDNA, to both sons and daughters (Skibinski et al., 1994a, 1994b; Zouros et al., 1994a, 1994b). This mode of inheritance is now known as doubly uniparental inheritance (DUI) (Zouros et al., 1994a).

The differences of the *Mytilus* mtDNA from other animal mtDNA are not restricted to transmission. The molecule is also radically different with respect to gene order and sequence divergence. The *Mytilus* gene order is different from any other molluscan order currently known (Hoffmann et al., 1992) and is more different from other molluscan species than these other species are from insect or mammalian species [Fig. 22.3(a)]. This high degree of divergence is also found at the DNA sequence [Fig. 22.3(b)]. In terms of amino acid divergence the *Mytilus* mtDNA is equally distant from the snail *Albinaria turrita* and the mammal *Homo sapiens*, which are closer to each other than each is to *Mytilus* (Hoeh et al., 1996) The nematode *Ascaris suum* is a close runner-up in this atypical pattern of sequence divergence. Both the egg-transmitted (F) and the sperm-transmitted (M) lineages of *Mytilus* are highly diverged with relation to other animal mtDNA, but there is evidence that the M lineage is faster evolving than the F (Stewart et al., 1995; Rawson and Hilbish, 1995). The most likely explanation for the faster evolution of the M molecule is relaxed selection (Stewart et al., 1996), which in turn is explained by the fact that the role of this molecule is basically restricted in the germ line, whereas the F molecule occurs exclusively in all female tissues and is the sole or main molecule in male somatic tissues.

The introgression of mitochondrial DNA among related species was one of the most impressive early findings in mtDNA population genetics (Ferris et al.,

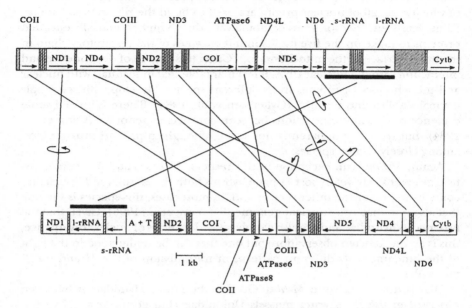

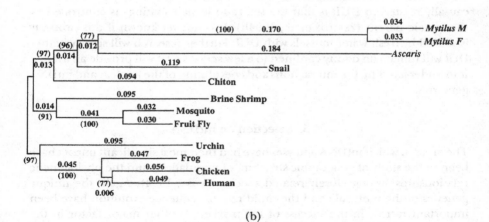

Figure 22.3. The highly divergent mtDNA molecule of *Mytilus*. (a) Differences in gene arrangement from that of *Drosophila* (from Hoffmann et al., 1992), (b) Nucleotide divergence from typical representatives of invertebrate and vertebrate species (from Hoeh et al., 1996).

1983; Powell, 1983). The phenomenon has been observed in most major divisions of the animal kingdom. There are over 300 proteins in the mitochondrion, of which ~90, all of nuclear origin, are used to build the ribosomes. Another 50 nuclearly encoded proteins combine with the 13 mitochondrially encoded proteins to make up the five multisubunit enzymes of the respiratory chain in the inner surface of the mitochondrion. Given this degree of dependence and interaction with the nuclear DNA, it is remarkable that individuals with nuclear and mitochondrial genomes from different species "develop with seemingly normal viability and fertility" (Gyllensten et al., 1985). There is indeed some evidence of epistatic interactions between conspecific genomes (Cann et al., 1984), but these are apparently not strong enough to prevent introgression among closely related species.

Again, *Mytilus* is an exception in this respect. *M. edulis* and *M. trossulus* hybridize extensively, yet in post F_1 male hybrids both the M and the F molecules come from one or the other species and, in most cases, this species is the one that also dominates the nuclear genome (Saavedra et al., 1996). Given that the two species hybridize and are also closely related in terms of nuclear divergence, this is an unexpected observation, but one that can be understood in the light of the fundamentally different patterns of transmission of the M and the F molecules.

DUI is not exclusive to *Mytilus*. Outside the family Mytilidae, it has been observed in the fresh water mussels Unionidae (Liu et al., 1996). The two families are separated by ~400 MY. Thus the origin of DUI can be either very old (with secondary losses) or it occurred independently in separate lineages. There is not at present enough evidence to distinguish between these two alternatives. One of the most intriguing observations in mussels, which may or may not be causally related to DUI, is that the sex ratio in pair matings is controlled by the female parent (Zouros et al., 1994b). It is not yet known if this property is shared by fresh water mussels with DUI. Further research will show whether DUI will remain an oddity confined to a few species or it will provide an avenue into understanding the interactions and coevolution of the nuclear and mtDNA genomes.

22.5. Selection on mtDNA

The areas in which mtDNA analyses have had their most significant impact have been in the study of geographic structure of populations and the phylogenetic relationships among closely related species (Avise, 1994). Again the unique genetics of the molecule and the rapid rate of sequence evolution have been important factors in the success of the marker. Another major factor in the wide use of mtDNA has been the general belief that variation in mtDNA is neutral with respect to fitness (see e.g., Ballard and Kreitman, 1995). However, it is at these levels of the hierarchy (species level and below) that selection on any marker can be the most misleading in attempting to translate molecular variation into organismal history (Rand et al., 1994; Rand, 1996).

Because mtDNA does not recombine, genetic hitchhiking (Maynard Smith and Haigh, 1974) will affect all sites on the molecule. Advantageous mutations that increase in frequency to fixation will sweep all linked polymorphisms out of the population (e.g., Berry et al., 1991; Begun and Aquadro, 1992, 1993). Similarly, background selection against deleterious mutations can reduce variation (Charlesworth et al., 1993). Thus, selection acting on mtDNA variation is likely to reduce levels of polymorphism over those expected under a strict neutral model. Evidence for this pattern has been provided by sequence surveys of mtDNA in *D. melanogaster* and *D. simulans* (Ballard and Kreitman, 1994; Rand et al., 1994). While nuclear DNA variation in *D. simulans* is generally much higher than in *D. melanogaster* (Aquadro, 1992), mtDNA polymorphism in *D. simulans* is much lower in the former than in the latter. Using an HKA approach, Ballard and Kreitman (1994) and Rand et al. (1994) showed that mtDNA polymorphism in *D. simulans* is reduced significantly relative to nuclear polymorphism, even after correcting for the lower effective population size of mtDNA stemming from a maternal haploid transmission.

While these data suggest that selection may be acting on mtDNA variation, the reduced polymorphism may be due to the presence of cytoplasmic incompatibilities due to *Wolbachia* endosymbionts (e.g., Turelli and Hoffmann, 1991; Werren et al., 1995). It has been shown that the *Wolbachia* infection can spread rapidly through natural populations of *D. simulans* (Turelli and Hoffmann, 1991), which, from the perspective of mtDNA polymorphism, could have the same effect as the selective sweep of a favored mtDNA mutation. While there is some evidence that mtDNA variation is reduced in North American samples of *D. melanogaster* (Ballard and Kreitman, 1994; Rand et al., 1994), this pattern is not as strong as reduced variation in *D. simulans*. Recent reports have documented endosymbiont-induced cytoplasmic incompatibilities in *D. melanogaster* (Solignac et al., 1994), but these appear to be weaker than in *D. simulans*. It may be that the greater level of mtDNA polymorphism in *D. melanogaster* than in *D. simulans* reflects a more ancient relationship between *Wolbachia* and *D. melanogaster* than with *D. simulans*. If the coevolutionary arms race between host and parasite has lead to a (temporary) condition in which *Wolbachia* is less virulent in *D. melanogaster*, the greater mtDNA polymorphism in this species may reflect the accumulation of new mutations since a previous reduction of mtDNA polymorphism induced by cytoplasmic incompatibilities.

Despite the wide use of mtDNA as a marker, explicit tests of neutrality in mtDNA have been used only in *Drosophila*, mice, and humans (reviewed in Ballard and Kreitman, 1995; Rand and Kann, 1996). Significant reductions in mtDNA polymorphism have been detected only in *D. simulans*. It may be that selective sweeps driven by nonneutral mutations in the mtDNA itself are sufficiently rare or that the relaxation time to normal levels of polymorphism is sufficiently short that it is very difficult to detect mitochondrial selective sweeps. Alternatively, selection on cotransmitted factors such as *Wolbachia* may be a more common source of departures from neutrality than mitochondrial mutations, at least in arthropods (see, e.g., Werren et al., 1995). However, our understanding of the generality of the neutrality test results published so far will remain limited until species other than model organisms are subjected to the neutrality tests.

22.6. Mildly Deleterious Evolution?

Departures from neutral evolution in mtDNA have also been detected from comparisons between silent (synonymous) and replacement (nonsynonymous) sites. McDonald and Kreitman (1991) showed for the *Adh* locus that an excess of replacement substitutions was observed as fixed between *Drosophila* species, relative to the silent:replacement ratio observed for polymorphisms. Examples of this pattern have been detected at other nuclear loci (Eanes et al., 1993) and have been interpreted as evidence for adaptive fixation of amino acid variants (but see Brookfield and Sharp, 1994). A very different pattern of polymorphism and divergence has been detected for several mitochondrial genes: an excess of replacement *polymorphisms* relative to the silent:replacement ratio observed as fixed between species (Ballard and Kreitman, 1994; Nachman et al., 1994, 1996; Rand et al., 1994; Rand and Kann, 1996). The most striking example of this is at the *ND3* gene in mice where the silent:replacement ratio is 23:2 between *Mus domesticus* and *M. spretus*, but is 13:11 in a sample of 56 *M. domesticus* (i.e., almost 10 times more replacement polymorphism than that expected under neutrality; Nachman et al., 1994). Such a pattern is suggestive of some form of balancing selection. However, all cases of excess amino acid polymorphism in mtDNA to date show greater levels of silent heterozygosity than replacement heterozygosity, and there is no clear evidence that the replacement polymorphisms fall on deep branches of the haplotype phylogenies (Nachman et al., 1994, 1996; Rand and Kann, 1996). Hence balancing selection does not appear to fit the data.

One explanation that is consistent with these nonneutral patterns is that amino acid variants are mildly deleterious and are observed as polymorphisms that do not persist in populations long enough to contribute to fixed differences (Ohta, 1992; Nachman et al., 1994, 1996; Rand and Kann, 1996). A prediction of this nearly neutral hypothesis is that polymorphic silent sites should exhibit a wider range of frequencies than replacement sites and there should be a greater skew toward rare frequencies at replacement sites. These predictions are upheld (Nachman et al., 1994; 1996; Rand and Kann, 1996), but unfortunately these patterns could just as likely be produced by relaxed selection. Since the only species in which excess amino acid variation has been well documented are those that are commensal with humans (*Drosophila* and mice), or humans themselves, the relaxed selection hypothesis seems plausible.

On the empirical front, the next step is to obtain additional data from species for which relaxed selection (such as permitted by a recent population expansion) is not likely and from related species with a wide range of effective population sizes. These sorts of data could be used to test the mildly deleterious hypothesis from a different perspective. The pattern Ohta (1995) commonly presents is a negative relationship between rate of amino acid substitution and effective population size, arguing that selection is more effective at eliminating deleterious mutations in large populations (see e.g., DeSalle and Templeton, 1988; Lynch, 1996). The mildly deleterious hypothesis would also predict that levels of amino acid polymorphism should be negatively correlated with effective population size, assuming comparable organisms with similar mutation rates. Suggestive of this pattern are some preliminary data from *D. pseudoobscura* mtDNA. For 1500 base pairs of the mitochondrial *ND5* gene in 20 wild lines, a deficiency of amino acid polymorphisms has been detected in *D. pseudoobscura* in which a slight excess of replacement polymorphism is evident in a comparable sample of *D. melanogaster. D. pseudoobscura* is known to have a larger effective population size than *D. melanogaster* (Schaeffer and Miller, 1992; Aquadro, 1992), although relaxed selection in the latter cannot be ruled out as yet.

Gillespie (1995, and references therein) has argued that the restrictive conditions under which the mildly deleterious model can apply make it unlikely as a model sufficient to account for patterns of molecular variation. Other models may suffice. However, comparisons of rates of amino acid substitution and levels of amino acid polymorphism between independent pairs of sister taxa with very different effective population sizes might provide the phylogenetic context in which to put the mildly deleterious hypothesis to a stringent test.

Resolving the issue of the generality of selection on mtDNA has important implications for basic problems in molecular evolution and more applied aspects of evolutionary genetics in which mtDNA is used as a marker. A common use of mtDNA is to infer the age of the most recent common ancestor (MRCA) of a sample of sequences based on divergence to a sister species (e.g., Cann et al., 1987) or to estimate effective population size (e.g., Fu, 1994). In the former case, a selective sweep will bring the MRCA closer to the recent and the pattern of excess amino acid polymorphism will push the MRCA back in time. Clearly these patterns are mutually exclusive, but we have no estimate of how common or general either of the cases is. More data will address this gap in the literature and, if collected with the competing mechanistic hypotheses in mind, may help distinguish alternative models of molecular evolution.

22.7. Population Structure and Phylogeography

The insight that mtDNA variation has provided for population structure, biogeography, and phylogenetic relationships of closely related species is so great that it is now difficult to find examples of these evolutionary problems that have not used mtDNA. This literature has been reviewed thoroughly by Avise (1994). We suggest that the power of mtDNA analyses has, until relatively recently, allowed researchers to get away with using only a single marker in addressing these complex questions. In the allozyme days, a dozen independent loci comprised a modest sample of markers with which to infer evolutionary patterns. Through the 1980s and early 1990, countless papers used mtDNA alone to study the evolutionary history of a species or group of species. While there are many exceptions to this pattern (e.g., Martin and Simon, 1988; Baker et al., 1989; Rand and Harrison, 1989a), cases in which different markers have generated contrasting results have actually become noteworthy in recent years (Degnan, 1993; Karl and Avise, 1992; Pogson et al., 1995). In some respects it is disconcerting to note that over 20 years ago, a discordant pattern of variation among loci was a controversial issue in population genetics and tests of the neutral theory (Lewontin and Krakauer, 1975; Nei and Chakravarti, 1977).

Clearly, the progress that has been made lies in our ability to infer that one marker (or set of markers) is likely to be neutral, such as sequence variation in noncoding DNA (e.g., Berry and Kreitman, 1993; McDonald, 1994). Such markers can provide an informative standard against which other possibly selected markers such as some allozymes can be judged. In some cases, mtDNA variation has been used as the neutral standard (e.g., Karl and Avise, 1992). In this light, one area of study that has been underexploited is contrasts between markers with different effective populations sizes, e.g., mtDNA, sex chromosomes, and autosomes. Under neutrality and equal migration rates between the sexes, mtDNA and Y chromosomes should show the greatest differentiation among populations, X chromosomes somewhat less differentiation (N_e = three times that of mtDNA), and autosomes the least population structure (N_e = four times that of mtDNA).

Birky et al. (1989) worked out the conditions under which an autosomal or an organellar marker would show greater population subdivision for a range of breeding and migrating sex ratios and a range of paternal leakage values. Avise (1995) has provided a graphical approach to inferring differential migration patterns between the sexes. (e.g., panmixia for Y chromosomes and subdivision for mtDNA suggests wide male dispersal and limited female dispersal). There are some nice examples that have taken advantage of nuclear versus mtDNA differences in a qualitative way (DeSalle et al., 1987; Hansen and Loeschcke, 1996). What has not been worked out is a formal statistical test for patterns of population subdivision that use loci with different effective populations sizes. While Tajima's (1989) and Fu's (1996) tests can detect departures from neutral equilibrium or evidence of population subdivision, these are based on single loci, and the interlocus contrasts should provide some additional power of inference.

Figure 22.4 is a plot of the ratios of F_{st} for a mitochondrial marker to the F_{st} for autosomal, X-chromosome, and Y-chromosome markers for different values of Nm. F_{st} is defined as

$$F_{st} \approx 1/(Nm + 1), \qquad (22.2)$$

where N is the effective population size and m is the migration rate. This plot is based on the assumptions stated above (relative values of N for mtDNA, Y-linked, X-linked, and autosomal loci are 1:1:3:4, respectively, equal migration for the sexes, and neutral markers). Note that the ratio of F_{st} does not approach

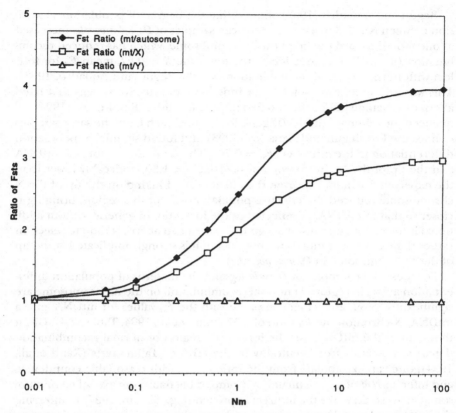

Figure 22.4. Ratios of F_{st} values for mtDNA relative to F_{st} values for nuclear markers for a range of effective migration rates (Nm). F_{st}s are calculated as $F_{st} \approx 1/(Nm + 1)$, where N is the relative effective population size for mtDNA, Y-linked, X-linked, or autosomal markers (1:1:3:4, respectively). The curves indicating the F_{st} ratios assume equal sex ratios, equal migration rates for the sexes, and neutrality of each marker.

the ratio of effective population sizes except for large values of Nm, which follows trivially from approximation (22.2). Taken together, markers from each of these chromosomal locations could provide ratios of F_{st} interpretable as independent estimates of Nm. Departures from the expected patterns would provide strong inference about the population biology of the organism, to the extent that each locus were neutral (the contrasts between Y-linked and mtDNAs would provide a test of the equality of migration rates for the sexes). To be an effective test, the sampling error on these ratios needs to be worked out (e.g., Ford and Aquadro, 1996). Moreover, relatively few systems are amenable to molecular analyses in which nuclear markers can clearly be mapped to Y, X, and autosomal loci. Nevertheless some data from *D. melanogaster* can be applied in this manner.

Begun and Aquadro (1993) showed that the degree of population subdivision between North American and African samples of *D. melanogaster* increased as one moved toward the tip of the X chromosome, which has reduced recombination (mean F_{st} for three loci at the tip of the $X = 0.57$; mean F_{st} for four loci with normal rates of recombination $= 0.29$; Begun and Aquadro, 1993). This was interpreted as evidence for independent selective sweeps that fixed alternative haplotypes in the two geographic localities. Rand et al. (1994) sequenced the mitochondrial *ND5* gene for 10 lines each from the same samples of lines used by Begun and Aquadro (1993) and found significant population differentiation in the mtDNAs ($F_{st} = 0.74$). The ratio of F_{st} values for mtDNA and the tip of the X chromosome is $0.74/0.57 = 1.30$, noticeably lower than the expected 3:1 ratio. However, the effect of hitchhiking on the tip of the X chromosome reduced the effective population size of this region, bringing it closer to that of mtDNA, as reflected in the low ratio. In general without additional information, we cannot assign the cause of an aberrant ratio to selection rather than a low migration rate (but selection is strongly implicated at the tip of the X chromosome in *D. melanogaster*).

For these same samples of *D. melanogaster*, the patterns of population differentiation at loci in regions of normal recombination on the X chromosome are significantly lower, and when compared with the F_{st} values for mtDNA, give a mtDNA/X-chromosome F_{st} ratio of 2.57 (Rand et al., 1994, Table 22.4). Given that neither the mtDNA nor the loci in the region of normal recombination depart significantly from neutrality by the HKA or Tajima tests (Rand et al., 1994), one can extrapolate from the ratio of F_{st}, with reasonable confidence, and infer an *Nm* of approximately 5. Comparable data from a Y-linked or autosomal locus could make the indirect approach (e.g., Slatkin, 1987) to inferring the patterns of population structure and gene flow very informative. However, the confidence limits on these sorts of comparisons need to be worked out.

In light of population cage data suggesting that mtDNAs may not behave as neutral markers (Kilpatrick and Rand, 1995, and references therein), cytonuclear interactions could alter the realized migration rates of mtDNA markers. This could increase or decrease the geographic range of various mtDNA haplotypes, which could be subject to corroboration by use of contrasting patterns of F_{st} for other nuclear markers.

22.8. Macroevolution of mtDNA

The presymbiosis history of present-day mitochondrial genomes can be traced in their ribosomal RNA sequences (e.g., Denovan-Wright et al., 1996). Yet, the differences in size, content, and organization among mitochondrial genomes and the variety of mtDNA replication and transcription mechanisms that can be found among and within the basic divisions of eukaryotes (protoctista, plants, fungi, and animals) are of such a magnitude that has forced some researchers to question the hypothesis of a unique symbiotic event (Mikelsaar, 1987) or to propose multiple origins for parts of the molecule (Gray et al., 1989). Whether these differences are, indeed, the relics of multiple and independent origins or the results of coevolutionary forces acting in concert on nuclear and mtDNA genomes, or, simply, the outcomes of historical accidents after the symbiotic event remains the domain of fertile speculation.

Against this background of radical differences within protists, plants, and fungi, the animal mtDNA appears remarkably conservative (Attardi, 1985). Its similarity in size, genetic content, and gene order and its uniformity in the mechanisms of replication and processing of the genetic information are the more remarkable when we consider that this kingdom is as old as plants or fungi. It is this structural and functional uniformity (together with the maternal inheritance and the absence of recombination) that has made mtDNA a powerful tool in evolutionary studies in animals, much more so than it has been the case for other eukaryotes.

The uniformity of the animal mtDNA and the evolutionary conservatism it implies are not, however, as pervasive and general as it was originally believed. The length of the molecule may vary by a factor of 3 [from less than 14 kb in the nematode *Caenorhabditis elegans* (Okimoto et al., 1992) to more than 40 kb in the scallop *Placopecten magellanicus* (Snyder et al., 1987)], the gene order along the circular molecule can be rather different from the one commonly found in metazoans [examples are the nematodes *C. elegans* and *A. suum* (Okimoto et al., 1992), the sea mussel *M. edulis* (Hoffmann et al., 1992), and the land snail *A. coerulea* (Hatzoglou et al., 1995)], heteroplasmy can vary from virtually zero to virtually 100% [e.g. in weevils of the genus *Pissodes* (Boyce et al., 1989) and the scallop *P. magellanicus* (Gjetvaj et al., 1992)], uniparental inheritance may be replaced by other types of transmission (Skibinski et al., 1994a, 1994b; Zouros et al., 1994a, 1994b), and some molecules may evolve at a much faster rate than others (Hoeh et al., 1996). Worse, these differences do not seem to obey taxonomic borders, except that size variation is considerably more constrained in endotherms (Rand, 1993).

22.9. Size Variation and Gene Order

The issue of mtDNA size variation has been reviewed (Harrison, 1989; Moritz, 1991; Rand, 1993). While there are a number of interesting points that mtDNA size variation can tell us about the evolution of repetitive DNA, mtDNA size variation caused by duplication events may provide hints about the mechanism through which novel gene arrangements appear in the evolution of the animal mtDNA. In lizards, the boundaries of duplication events seem to be associated with tRNA sequences (Moritz, 1991, and references therein). The secondary structure potential of tRNAs may be causally related to the introduction of size variation. Further evidence for this model has been provided by Stanton et al. (1994), which is consistent with earlier suggestions that tRNA mobility has been a major theme in the evolution of gene order in mtDNA (Jacobs et al., 1988).

With the exception of the mussel (see above) the most remarkable gene order differences are found in the nematodes *C. elegans* and *A. suum* (Okimoto et al., 1992) and in the land snail *A. coerulea* (Hatzoglou et al., 1995) where it appears to have been associated with reduction of the size of the molecule. In *A. coerulea* there is hardly a region of the molecule that does not code for a gene (the largest sequence with unknown function is only 42 bp), posing questions about the position and the role of the control region. Also the existence of overlapping tRNAs raises the possibility for posttranscription editing. Certain patterns of conservation of gene order can be seen across the animal kingdom, but it is not clear whether they are selectively conserved or they represent accidental conservations. At present, there is no evidence to suggest that there can be gene arrangements intrinsically incompatible with mitochondrial function. The accidental events leading to new gene orders must, however, be very rare, and this can be exploited to search for deep splits in the animal phylogenetic tree (Boore et al., 1995).

22.10. Rates of Evolution

One of the earliest observations from animal mtDNA sequence comparisons was that the molecule evolves at a high rate. More specifically, rRNA and tRNA genes evolve faster than their nuclear counterparts and third codon positions of the protein-coding genes evolve faster than corresponding third positions of nuclear genes or even pseudogenes. A number of explanations were suggested for this apparent high rate of evolution. Some compelling explanations are the lack of a replication error repair mechanism and the higher chance of mutations to survive and accumulate because of the large number of mtDNA copies within the organelle and within the cell (Wilson et al., 1985).

However, additional explanations are needed to account for variation among taxa in the rates of mtDNA evolution. The slower rate of mtDNA evolution in ectotherms relative to endotherms has been attributed to mutation rate differences relating to body size, metabolic rate, and a longer nucleotide generation time (Martin and Palumbi, 1993). Alternatively, a relaxed functional constraint in the homogeneous cellular environment of endotherms may permit faster mtDNA evolution (Thomas and Beckenbach, 1989; reviewed in Rand, 1994). Differences between silent and replacement substitution rates in different groups provide conflicting data that support either the mutation or the relaxed functional constrain hypotheses (Adachi et al., 1993; Martin, 1995). While these hypotheses and the approaches that use the comparative method with silent and replacement sites are appealing, a number of the issues covered in this chapter may be equally important.

The patterns of development and germline sequestration differ between fish and mammals. Differences in the number of germ cell generations per animal generation (see above, and Beckenbach, 1994) or modes of organelle segregation may covary with the endothermic and ectothermic habits. Alternatively, the relative importance of mildly deleterious mutations may be distinct in the taxa for which mtDNA has been shown to evolve at different rates. Some mammals may have had relatively small effective population sizes because of demic population structure in historical times. Sharks and salmon, on the other hand, may have had historically very large populations. These conditions could allow faster mtDNA evolution in mammals that may not be related directly to the metabolic rate issue.

These competing explanations for the patterns of mtDNA variation and evolution are good problems for comparative biology: they are intriguing questions that can be tested and falsified, but require a broad spectrum biological insight. What better way to get a cell biologist interested in population genetics (or a population geneticist interested in cell biology) than to consider the roles that mutation, selection, and drift might play in the transmission of mtDNA? The cytoplasmic, hierarchical thinking that mtDNA imposes accounts for why this molecule will continue to be such an important subject and object of evolutionary inquiry.

REFERENCES

Adachi, J., Cao, Y., and Hasegawa, M. 1993. Tempo and mode of mitochondrial DNA evolution in vertebrates at the amino acid level: rapid evolution in warm blooded vertebrates. *J. Mol. Evol.* **36**:270–281.

Aquadro, C. F. 1992. Why is the genome variable: insights from *Drosophila. Trends Genet.* **8**:355–362.

Arnason, E. and Rand, D. M. 1992. Heteroplasmy of short tandem repeats in Atlantic cod (*Gadus morhua*). *Genetics* **132**:211–220.

Attardi, G. 1985. Animal mitochondrial DNA: an extreme example of genetic economy. *Int. Rev. Cytol.* **93**:93–145.

Avise, J. C. 1994. *Molecular Markers, Natural History, and Evolution.* New York: Chapman & Hall.

Avise, J. C. 1995. Mitochondrial DNA polymorphism and a connection between genetics and demography of relevance to conservation. *Conserv. Biol.* **9**:686–690.

Ballard, J. W. O. and Kreitman, M. 1994. Unraveling selection in the mitochondrial genome of Drosophila. *Genetics* **138**:757–772.

Ballard, J. W. O. and Kreitman, M. 1995. Is mitochondrial DNA a strictly neutral marker? *Trends Ecol. Evol.* **10**:485–488.

Baker, R. J., David, S. K., Bradley, R. D., Hamilton, M. J., and Van Den Bussche, R. A., 1989. Ribosomal DNA, mitochondrial DNA, chromosomal and allozymic studies of a contact zone in the pocket gopher. *Geomys. Evol.* **43**:63–75.

Beckenbach, A. T. 1994. Mitochondrial haplotype frequencies in oysters: neutral alternatives to selection models. In *Non-Neutral Evolution: Theories and Molecular Data,* B. Golding, ed., pp. 188–198. New York: Chapman & Hall.

Begun, D. J. and Aquadro, C. F. 1992. Levels of naturally occurring DNA polymorphism correlate with recombination rates in *D. melanogaster. Nature (London)* **356**:519–520.

Begun, D. J. and Aquadro, C. F. 1993. African and North American populations of *Drosophila melanogaster* are very different at the DNA level. *Nature (London)* **356**:519–520.

Bentzen, P., Legget, W. C., and Brown, G. C. 1988. Length and restriction site heteroplasmy in the mitochondrial DNA of American shad (*Alosa sapidissima*). *Genetics* **118**:509–518.

Berry, A. J., Ajioka, J. W., and M. Kreitman. 1991. Lack of polymorphism on the *Drosophila* fourth chromosome resulting from selection. *Genetics* **129**:1111–1117.

Berry, A. J. and Kreitman, M. 1993. Molecular analysis of an allozyme cline:alcohol dehydrogenase in *Drosophila melanogaster* on the east coast of North America. *Genetics* **134**:869–893.

Birky Jr., C. W. 1983. Relaxed cellular controls and organelle heredity. *Science* **222**:468–475.

Birky Jr., C. W., Fuerst, P., and Maruyama, T. 1989. Organelle gene diversity under migration, mutation, and drift: equilibrium expectations, approach to equilibrium, effects of heteroplasmic cells, and comparison to nuclear genes. *Genetics* **121**:613–627.

Birky Jr., C. W. 1995. Uniparental inheritance of mitochondrial and chloroplast genes: mechanisms and evolution. *Proc. Natl. Acad. Sci. USA* **92**:11331–11338.

Boore, J. L., Collins, T. M., Stanton, D., Daehler, L. L., and Brown, W. M. 1995. Deducing the pattern of arthropod phylogeny from mitochondrial DNA rearrangements. *Nature (London)* **376**:163–165.

Boyce, T. M., Zwick, M. E., and Aquadro, C. F. 1989. Mitochondrial DNA in the bark weevils: size, structure and heteroplasmy. *Genetics* **123**:825–836.

Brookfield, J. F. Y. and Sharp, P. M. 1994. Neutralism and selectionism face up to DNA data. *Trends Genet.* **10**:109–111.

Brown, J. R., Beckenbach, A. T., and Smith, M. J. 1992. Mitochondrial length variation and heteroplasmy in populations of white sturgeon (*Acipenser transmontanus*). *Genetics* **132**:221–228.

Cann, R. L., Brown, W. M., and Wilson, A. C. 1984. Polymorphic sites and mechanism of evolution in human mitochondrial DNA. *Genetics* **106**:479–499.

Cann, R. L., Stoneking, M., and Wilson, A. C. 1987. Mitochondrial DNA and human evolution. *Nature (London)* **325**:31–36.

Charlesworth, B., Morgan, M. T., and Charlesworth, D. 1993. The effect of deleterious mutations on neutral molecular variation. *Genetics* **134**:1289–1303.

Clark, C. G. and Roger, A. J. 1995. Direct evidence for secondary loss of mitochondria in *Entamoeba histolytica*. *Proc. Natl. Acad. Sci. USA* **92**:6518–6521.

Cosmides, L. M. and Tooby, J. 1981. Cytoplasmic inheritance and intragenomic conflict. *J. Theor. Biol.* **89**:83–129.

Degnan, S. M. 1993. The perils of single gene trees – mitochondrial versus single-copy nuclear DNA variation in white-eyes (Aves: Zosteropidae). *Mol. Ecol.* **2**:219–225.

Denovan-Wright, E. M., Sankoff, D., Spencer, D., and Lee, R. W. 1996. Evolution of fragmented mitochondrial ribosomal RNA genes in *Chlamydomonas*. *J. Mol. Evol.* **42**(4): 382–391.

DeSalle, R. and Templeton, A. 1988. Founder effects and the rate of mitochondrial DNA evolution in Hawaiian Drosophila. *Evolution* **42**:1076–1084.

DeSalle, R., Templeton, A., Mori, I., Pletscher, S., and Johnson, J. S. 1987. Temporal and spatial heterogeneity of mtDNA polymorphisms in natural populations of *Drosophila mercatorum*. *Genetics* **116**:215–223.

Eanes, W. F., Kirchner, M., and Yoon, J. 1993. Evidence for adaptive evolution of the *G6pd* gene in the *Drosophila melanogaster* and *Drosophila simulans* lineages. *Proc. Nat. Acad. Sci. USA* **90**:7475–7479.

Ferris, S. D., Sage, R. D., Huang, S-M., Neilson, J. T., Ritte, U., and Wilson, A. C. 1983. Flow of mitochondrial DNA across a species boundary. *Proc. Natl. Acad. Sci. USA* **80**:2290–2294.

Fisher, C. and Skibinski, D. O. F. 1990. Sex-biased mitochondrial DNA heteroplasmy in the marine mussel *Mytilus. Proc. R. Soc. London Ser. B* **242**:149–156.

Ford, M. J. and Aquadro, C. F. 1996. Selection on X-linked genes during speciation in the *Drosophila athabasca* complex. *Genetics* **144**:689–703.

Fu, Y. X. 1994. A phylogenetic estimator of effective population size or mutation rate. *Genetics* **136**:685–692.

Fu, Y. X. 1996. New statistical tests of neutrality for DNA samples from a population. *Genetics* **143**:557–570.

Gill, P., Ivanov, P. L., Kimpton, C., Piercy, R., Benson, N., Tully, G., Evett, I., Hagelberg, E., Sullivan, K. 1994. Identification of the remains of the Romanov family by DNA analysis. *Nat. Genet.* **6**:130–135.

Gillham, N. 1994. *Organelle Genes and Genomes.* New York: Oxford U. Press.

Gillespie, J. 1995. On Ohta's hypothesis: most amino acid substitutions are deleterious. *J. Mol. Evol.* **40**:64–69.

Gray, M. W., Cedergren, R., Abel, Y., and Sankoff, D. 1989. On the evolutionary origin of the plant mitochondrion and its genome. *Proc. Natl. Acad. Sci. USA* **86**:2267–2271.

Gjetvaj, B., Cook, D. I., and Zouros, E. 1992. Repeated sequences and large-scale size variation of mitochondrial DNA: a common feature among scallops (Bivalvia: Pectinidae). *Mol. Biol. Evol.* **9**:106–124.

Godelle, B. and Reboud, X. 1995. Why are organelles uniparentally inherited? *Proc. R. Soc. London Ser. B* **259**:27–35.

Gyllensten, U., Wharton, D., and Wilson, A. C. 1985. Maternal inheritance of mitochondrial DNA during backcrossing of two species of mice. *J. Hered.* **76**:321–324.

Gyllensten, U., Wharton, D., Joseffson, A., and Wilson, A. C. 1991. Paternal inheritance of mitochondrial DNA in mice. *Nature (London)* **352**:255–257.

Hale, L. R. and Singh, R. S. 1986. Extensive variation and heteroplasmy in size of mitochondrial DNA among geographic populations of *Drosophila melanogaster. Proc. Natl. Acad. Sci. USA* **78**:8813–8817.

Hansen, M. M. and Loeschcke, V. 1996. Temporal variation in mitochondrial DNA haplotype frequencies in a brown trout (*Salmo trutta* L.) population that shows stability in nuclear allele frequencies. *Evolution* **50**:454–457.

Harrison, R. G. 1989. Animal mitochondrial DNA as a marker in population and evolutionary biology. *Trends. Ecol. Evol.* **4**:6–11.

Hastings, I. M. 1992. Population genetics aspects of deleterious cytoplasmic genomes and their effect on the evolution of sexual reproduction. *Genet. Res.* **59**:215–225.

Hatzoglou, E., Rodakis, G. C., and Lecanidou, R. 1995. Complete sequence and gene organization of the mitochondrial genome of the land snail *Albinaria coerulea. Genetics* **140**:1353–1366.

Hauswirth, W. W. and Laipis, P. J. 1982. Mitochondrial DNA polymorphism in a maternal lineage of Holstein cows. *Proc. Natl. Acad. Sci. USA* **79**:4686–4690.

Hoeh, W. R., Blackley, K. H., and Brown, W. M. 1991. Heteroplasmy suggests limited biparental inheritance of *Mytilus* mitochondrial DNA. *Science* **251**:1488–1490.

Hoeh, W. R., Stewart, D. T., Sutherland, B. W., and Zouros, E. 1996. Cytochrome c oxidase sequence comparisons suggest an unusually high rate of mitochondrial DNA evolution in *Mytilus* (Mollusca:Bivalvia). *Mol. Biol. Evol.* **13**:418–421.

Hoekstra, R. F. 1990. Evolution of uniparental inheritance of cytoplasmic DNA. In *Organizational Constraints on the Dynamics of Evolution,* J. Maynard Smith and G. Vida, eds., pp. 269–278. Manchester: Manchester U. Press.

Hoffmann, R. J., Boore, G. L., and Brown, W. M. 1992. A novel mitochondrial genome organization for the blue mussel *Mytilus edulis. Genetics* **131**:397–412.

Hurst, L. D. and Hamilton, W. D. 1992. Cytoplasmic fusion and the nature of sexes. *Proc. R. Soc. London Ser. B* **258**:287–298.

Hutchison III, C. A., Newbold, J. E., Potter, S. S., and Edgell, M. H. 1974. Maternal inheritance of mammalian mitochondrial DNA. *Nature (London)* **251**:536–538.

Ivanov, P. L., Wadhams, M. J., Roby, R. K., Holland, M. M., Weedn, V. W., and Parsons, T. J. 1996. Mitochondrial DNA sequence heteroplasmy in the Grand Duke of Russia Georgij Romanov establishes the authenticity of the remains of Tsar Nicholas II. *Nat. Genet.* **12**:417–420.

Jacobs, H., Elliot, D. J., Math, V. B., and Farquharson, A. 1988. Nucleotide sequence and gene organization of sea urchin mitochondrial DNA. *J. Mol. Biol.* **202**:185–217.

Kaneda, H., Hayashi, J.-I., Takahama, S., Taya, C., Lindahl, K. F., and Yonekawa, H. 1995. Elimination of paternal mitochondrial DNA in intraspecific crosses during early mouse embryogenesis. *Proc. Natl. Acad. Sci. USA* **92**:4542–4546.

Karl, S. A. and Avise, J. C. 1992. Balancing selection at allozyme loci in oysters: implications from nuclear RFLPs. *Science* **256**:100–102.

Kilpatrick, S. R. and Rand, D. M. 1995. Conditional hitchhiking of mitochondrial DNA: frequency shifts of *Drosophila melanogaster* mtDNA variants depend on nuclear genetic background. *Genetics* **141**:1113–1124.

Koehler, C. M., Lindberg, G. L., Brown, D. R., Beitz, D. C., Freeman, A. E., Mayfield, J. E., and Myers, A. M. 1991. Replacement of bovine mitochondrial DNA by a sequence variant within one generation. *Genetics* **129**:247–255.

Kondo, R., Matsuura, E. T., Ishima, H., Takahata, N., and Chigusa, S. I. 1990. Incomplete maternal transmission of mitochondrial DNA in Drosophila. *Genetics* **126**:657–663.

Lansman, R. A., Avise, J. C., and Huettel, M. D. 1983. Critical experimental test of the possibility of paternal leakage of mitochondrial DNA. *Proc. Natl. Acad. Sci. USA* **80**:1969–1971.

Lewontin, R. C. 1970. The units of selection. *Annu. Rev. Ecol. Syst.* **1**:1–15.

Lewontin, R. C. 1972. The apportionment of human diversity. *Evol. Biol.* **6**:381–398.

Lewontin, R. C. and Krakauer, J. 1975. Testing the heterogeneity of *F* values. *Genetics* **80**:397–398.

Liu, H-P., Mitton, J. B., and Wu, S.-K. 1996. Gender-specific mitochondrial DNA within and among populations of the freshwater mussel, *Anodonta grandis grandis. Evolution* **50**:952–957.

Lynch, M. 1996. Mutation accumulation in transfer RNAs: molecular evidence for Muller's ratchet in mitochondrial genomes. *Mol. Biol. Evol.* **13**:209–220.

Magoulas, A. and Zouros, E. 1993. Restriction-site heteroplasmy in Anchovy (*Engraulis encrasicolus*) indicates incidental biparental inheritance of mitochondrial DNA. *Mol. Biol. Evol.* **10**:319–325.

Margulis, L. 1981. *Symbiosis in Cell Evolution: Life and its Environment in the Early Earth.* San Francisco: Freeman.

Margulis, L. 1996. Archaeal-eubacterial mergers in the origin of eukarya: phylogenetic classification of life. *Proc. Natl. Acad. Sci. USA* **93**:1071–1076.

Martin, A. P. 1995. Metabolic rate and directional nucleotide substitution in animal mitochondrial DNA. *Mol. Biol. Evol.* **12**:1124–1131.

Martin, A. P. and Palumbi, S. R. 1993. Body size, metabolic rate, generation time and the molecular clock. *Proc. Natl. Acad. Sci. USA* **90**:4087–4091.

Martin, A. P. and Simon, C. 1988. Anomolous distribution of nuclear and mitochondrial DNA markers in periodical cicadas. *Nature (London)* **336**:237–239.

Maynard Smith, J. and Haigh, J. 1974. The hitchhiking effect of a favorable gene. *Genet. Res.* **23**:23–35.

McDonald, J. H. 1994. Detecting natural selection by comparing geographic variation in protein and DNA polymorphisms. In *Non-Neutral Evolution*, B. Golding, ed., pp. 88–100. New York: Chapman & Hall.

McDonald, J. H. and Kreitman, M. 1991. Adaptive protein evolution at the *Adh* locus in *Drosophila*. *Nature (London)* **351**:652–654.

Mikelsaar, R. 1987. A view of early cellular evolution. *J. Mol. Evol.* **25**:168–183.

Monnat Jr., R. J. and Loeb, L. A. 1985. Nucleotide sequence preservation of human mitochondrial DNA. *Proc. Natl. Acad. Sci. USA* **82**:2895–2899.

Moritz, C. 1991. Evolutionary dynamics of mitochondrial DNA duplications in parthenogenetic Geckos, *Heteronotia binoei*. *Genetics* **129**:221–230.

Nachman, M. W., Boyer, S. N., and Aquadro, C. F. 1994. Non-neutral evolution at the mitochondrial NADH dehydrogenase subunit 3 gene in mice. *Proc. Natl. Acad. Sci. USA* **91**:6364–6368.

Nachman, M. W., Brown, W. M., Stoneking, M., and Aquadro, C. F. 1996. Non-neutral mitochondrial DNA variation in humans and chimpanzees. *Genetics* **142**:953–963.

Nei, M. and Chakravarti, A. 1977. Drift variances of Fst and Gst statistics obtained from a finite number of isolated populations. *Theor. Popul. Biol.* **11**:307–325.

Ohta, T. 1992. The nearly neutral theory of molecular evolution. *Ann. Rev. Ecol. Syst.* **23**:263–286.

Ohta, T. 1995. Synonymous and nonsynonymous substitutions in mammalian genes and the nearly neutral theory. *J. Mol. Evol.* **40**:56–63.

Okimoto, R., MacFarlane, J. L., Clary, D. O., and Wolstenholme, D. R. 1992. The mitochondrial genomes of two nematodes *Caenorhabditis elegans* and *Ascaris suum*. *Genetics* **130**:471–498.

Pogson, G. H., Mesa, K. A., and Boutilier, R. G. 1995. Genetic population structure and gene flow in the Atlantic cod *Gadus morhua*: a comparison of allozyme and nuclear RFLP loci. *Genetics* **139**:375–385.

Powell, J. R. 1983. Interspecific cytoplasmic gene flow in the absence of nuclear gene flow: evidence from Drosophila. *Proc. Natl. Acad. Sci. USA* **80**:492–495.

Rand, D. M. 1993. Endotherms, ectotherms and mitochondrial genome-size variation. *J. Mol. Evol.* **37**:281–295.

Rand, D. M. 1994. Thermal habit, metabolic rate and the evolution of mitochondrial DNA. *Trends Ecol. Evol.* **9**:125–131.

Rand, D. M. 1996. Neutrality tests of molecular markers and the connection between DNA polymosphism, demography, and conservation biology. *Conserv. Biol.* **10**:665–671.

Rand, D. M. and Harrison, R. G. 1986. Mitochondrial DNA transmission genetics in crickets. *Genetics* **114**:955–970.

Rand, D. M. and Harrison, R. G. 1989a. Ecological genetics of a mosaic hybrid zone: mitochondrial, nuclear and reproductive differentiation of crickets by soil type. *Evolution* **43**:432–449.

Rand, D. M. and Harrison, R. G. 1989b. Molecular population genetics of mitochondrial DNA size variation in crickets. *Genetics* **121**:551–569.

Rand, D. M., Dorfsman, M., and Kann, L. M. 1994. Neutral and non-neutral evolution of Drosophila mitochondrial DNA. *Genetics* **138**:741–756.

Rand, D. M. and Kann, L. M. 1996. Excess amino acid polymorphism in mitochondrial DNA: contrasts among genes from *Drosophila*, mice, and humans. *Mol. Biol. Evol.* 13:735–748.

Rawson, P. D. and Hilbish, T. J. 1995. Evolutionary relationships among the male and female mitochondrial DNA lineages in the *Mytilus edulis* species complex. *Mol. Biol. Evol.* 12:893–901.

Saavedra, C., Stewart, D. T., Stanwood, R. R., and Zouros, E. 1996. Species-specific segregation of gender-associated mitochondrial DNA types in an area where two mussel species (*Mytilus edulis* and *Mytilus trossulus*) hybridize. *Genetics* 143:1359–1367.

Satta, Y., Toyohara, N., Ohtaka, C., Tatsuno, Y., Watanabe, T. K., Matsura, E. T., Chigusa, S. I., and Takahata, N. 1988. Dubious maternal inheritance of mitochondrial DNA in *D. simulans* and evolution of *D. mauritiana*. *Genet. Res. Cambridge* 52:1–6.

Schaeffer, S. W. and Miller, E. L. 1992. Molecular population genetics of an electrophoretically monomorphic protein in the alcohol dehydrogenase region of *Drosophila melanogaster*. *Genetics* 132:163–178.

Skibinski, D. O. F., Gallagher, C., and Beynon, C. M. 1994a. Mitochondrial DNA inheritance. *Nature (London)* 368:817–818.

Skibinski, D. O. F., Gallagher, C., and Beynon, C. M. 1994b. Sex-limited mitochondrial DNA transmission in the marine mussel *Mytilus edulis*. *Genetics* 138:801–810.

Slatkin, M. 1987. Gene flow and the geographic structure of populations. *Science* 236:787–792.

Smith, D. R. and Brown, W. M. 1990. Restriction endonuclease cleavage size and length polymorphism in mitochondrial DNA of *Apis mellifera mellifera* and *A. m. carnina* (Hymenoptera, Apidae). *Ann. Entomol. Soc. Am.* 83:81–88.

Snyder, M., Fraser, A. R., LaRoche, J., Gardner-Kepkay, K. E., and Zouros, E. 1987. Atypical mitochondrial DNA from the deep-sea scallop *Placopecten magellanicus*. *Proc. Natl. Acad. Sci. USA* 84:7595–7599.

Solignac, M., Genermont, J., Monnerot, M., and Mounolou, J.- C. 1984. Genetics of mitochondria in Drosophila: inheritance in heteroplasmic strains of *D. mauritiana*. *Mol. Gen. Genet.* 197:183–188.

Solignac, M., Vautrin, D., and Rousset, F. 1994. Widespread occurrence of the proteobacteria *Wolbachia* and partial cytoplasmic incompatibility in *Drosophila melanogaster*. *C. R. Acad. Sci. Paris* 317:461–470.

Stanton, D. J., Daehler, L. L., Moritz, C. C., and Brown, W. M. 1994. Sequences with the potential to form stem-and-loop structures are associated with coding-region duplications in animal mitochondrial DNA. *Genetics* 137:233–241.

Stewart, D. T., Saavedra, C., Stanwood, R. R., Ball, A. O., and Zouros, E. 1995. Male and female mitochondrial DNA lineages in the blue mussel (*Mytilus edulis*) species group. *Mol. Biol. Evol.* 12:735–747.

Stewart, D. T., Kenchington, E., Singh, R. K., and Zouros, E. 1996. Degree of selective constraint as an explanation of the different rates of evolution of gender-specific mitochondrial DNA lineages in the mussel *Mytilus*. *Genetics* 143:1349–1357.

Tajima, F. 1989. Statistical method for testing the neutral mutation hypothesis by DNA polymorphism. *Genetics* 123:585–595.

Thomas, W. K. and Beckenbach, A. T. 1989. Variation in salmonid mitochondrial DNA: evolutionary constraints and mechanisms of substitution. *J. Mol. Evol.* 29:233–245.

Turelli, M. and Hoffmann, A. A. 1991. Rapid spread of an inherited incompatibility factor in California *Drosophila*. *Nature (London)* 353:440–442.

Upholt, W. I. and Dawid, I. B. 1977. Mapping mitochondrial DNA of sheep and goats: rapid evolution in the D-loop region. *Cell* 11:571–583.

Werren, J. H., Zhayng, W., and Guo, L. R. 1995. Evolution and phylogeny of *Wolbachia*: reproductive parasites of arthropods. *Proc. R. Soc. London Ser. B* 261:55–63.

Wilkinson, G. S. and Chapman, A. M. 1991. Length and sequence variation in evening bat D-loop mtDNA. *Genetics* 128:607–617.

Wilson, A. C., Cann, R. L., Carr, S. M., George, M., Gyllensten, U. B., Helm-Bychowski, K. M., Higuchi, R. G., Palumbi, S. R., Prager, E. M., Dage, R. D., and Stoneking, M. 1985. Mitochondrial DNA and two perspectives on evolutionary genetics. *Biol. J. Linn. Soc.* 26:375–400.

Zouros, E., Freeman, K. R., Ball, A. O., and Pogson, G. H. 1992. Direct evidence for extensive paternal mitochondrial DNA inheritance in the marine mussel *Mytilus*. *Nature (London)* 359:412–414.

Zouros, E., Ball, A. O, Saavedra, C., and Freeman, K. R. 1994a. Mitochondrial DNA inheritance. *Nature (London)* 368:818.

Zouros, E., Ball, A. O, Saavedra, C., and Freeman, K. R. 1994b. An unusual type of mitochondrial DNA inheritance in the blue mussel *Mytilus*. *Proc. Natl. Acad. Sci. USA* 91:7463–7467.

CHAPTER TWENTY THREE

Genetic Structure and Evolution
in Parthenogenetic Animals

E. DAVIS PARKER JR. AND MONICA NIKLASSON

23.1. Introduction

Research on the evolution and the ecology of parthenogenesis in animals has been a major theme in European evolutionary biology since the pioneering work of Seiler (1923), Vandel (1928), and Suomalainen (1950). In contrast, before 1970, research in North America on parthenogenetic animals was mainly restricted to their aberrant cytogenetics or to systematic treatments, with the several notable exceptions discussed below. Parthenogenetic taxa were relegated to the status of interesting deviations from the eukaryotic sexual cycle of meiosis and fertilization but that had little evolutionary potential (Mayr, 1963; Dobzhansky, 1970).

In the past 25 years, renewed interest in the problem of sex as an individual adaptation has stimulated comparisons of related parthenogenetic and sexual taxa to identify the distributional and ecological correlates of breeding system variation (Levin, 1975; Glesener and Tilman, 1978; Lynch, 1984) and to test models for the maintenance of sex (Lively et al., 1990; Moritz et al., 1991; Lively, 1992; Brown et al., 1995).

Additionally, the adoption of protein electrophoresis for population genetic problems (Frydenberg et al., 1965; Hubby and Lewontin, 1966) provided a method to study ecogenetic problems in parthenogenesis research such as (1) the phylogeny of parthenogenesis (Uzzell and Goldblatt, 1967; Neaves, 1969; Vrijenhoek, 1972; Uzzell and Darevsky, 1975; Johnson, 1992), (2) the amount and origin of clonal diversity within parthenogenetic taxa (Suomalainen and Saura, 1973; Parker and Selander, 1976; Parker et al., 1977; Hebert and Crease, 1983; Harshman and Futuyma, 1985; Menken and Wiebosch-Steeman, 1988; Dessauer and Cole, 1989; Honeycutt and Wilkinson, 1989; Dybdahl and Lively, 1995), and the ecological genetics of clonal variation (Vrijenhoek, 1978; Christensen, 1979; Mitter et al., 1979; Jaenike et al., 1980; Ochman et al., 1980; Weider and Hebert, 1987; Christensen et al., 1988).

R. S. Singh and C. B. Krimbas, eds., *Evolutionary Genetics: From Molecules to Morphology*, vol. 1. © Cambridge University Press 2000. Printed in the United States of America. ISBN 0-521-57123-5. All rights reserved.

More recently, data on nucleotide sequence variation have begun to refine our understanding of both the origins and amounts of clonal diversity (Moritz et al., 1989; Turner et al., 1990; Browne and Hoopes, 1990; Elder and Schlosser, 1995; Theisen et al., 1995), and the phylogeny of parthenogenesis (Judson and Normark, 1996; Normark, 1996).

In this chapter, we review the impact of the mechanisms of clonal origin (or phylogeny) on the genetic structure of parthenogenetic animals as a background for a discussion of selection in parthenogenetic taxa as it relates to the evolution of phenotypic adaptability and the phenomenon of geographical parthenogenesis. Besides providing an introduction to the literature, our major goal is to stress that comparisons between related sexual and parthenogenetic taxa are an underutilized general tool in population biology. However, such comparisons can be misleading without knowledge of levels of clonal diversity and of the phenotypic properties of sexual and parthenogenetic relatives.

Our focus is on facultative and obligate female parthenogenesis (thelytoky). Thus the evolutionary sequence under study is from ancestral sexuality to descendent parthenogenesis. Unlike other forms of asexual reproduction, the offspring produced by thelytokous females are ecologically comparable with those produced by sexual females. We discuss cyclically parthenogenetic taxa only as they relate to the origins of obligate thelytoky and clonal diversity. Moreover, we exclude discussion of microbe-induced parthenogenesis (Stouthamer et al., 1990) and hybridogenetic taxa (Schultz, 1971). General reviews of parthenogenesis include Bell (1982) for a list of clonal animals, Lynch (1984) for geographical parthenogenesis, and Suomalainen et al. (1987) for cytology and genetics. Jackson et al. (1985) and Hughes (1989) discuss ecological and physiological aspects of clonal reproduction in general. Useful collections of papers on parthenogenesis in animals are those edited by Oliver (1971), Grassle and Shick (1979), Dawley and Bogart (1989), and Lourenço (1994).

23.2. The Genetic Structure of Parthenogenetic Animal Species

23.2.1. The Origin of Parthenogenesis

Categorization of modes of origin of parthenogenesis must take four processes into consideration: (1) egg maturation, (2) polyploidization, (3) spontaneous versus hybrid origin, and (4) monophyletic versus polyphyletic origins of clonal diversity.

(1) The first challenge for an incipient parthenogen is egg maturation, which determines much of the initial pattern of clonal variation exhibited (Asher, 1970; Uzzell, 1970; Asher and Nace, 1971). While remarkably diverse, these mechanisms can be divided into two categories that reflect whether inheritance is recombinational (automixis) or clonal (apomixis) (White, 1973). In automictic (meiotic) thelytoky, somatic ploidy is restored through fusion of polar bodies, suppression of one of the meiotic divisions, or premeiotic or postmeiotic chromosome duplication. The extreme form of automixis, duplication

of a haploid egg nucleus, generates completely homozygous offspring in a single generation of parthenogenetic reproduction (e. g., Stille, 1985). Other forms of automixis are similar to selfing in the rate of increase in the inbreeding coefficient (Asher, 1970). In automictic species, chromosomal rearrangements and lack of chromosome homology (for example, in hybrid parthenogens) can suppress chiasma formation and recombination, leading to a condition known as functional apomixis (Stalker, 1956; Uzzell, 1970; Asher and Nace, 1971; Scali et al., 1995). In apomictic (mitotic) thelytoky, there are one or two maturation divisions that do not involve synapsis and chiasma formation. True and functionally apomictic mechanisms are usually thought of as enforcing clonal reproduction. However, aberrations typical of all cell divisions, such as quadrivalent formation (Asher and Nace, 1971) or simple nondisjunction at anaphase, can theoretically lead to recombined offspring (White, 1973).

Variations on the basic parthenogenetic theme are characteristic of certain all-female groups. Gynogenesis or pseudogamy is the penetration of the egg by a sperm, followed by clonal development of the egg, with no genetic contribution of the male parent (Uzzell, 1964; Schultz, 1971). It is particularly interesting because it requires close ecological and behavioral interaction between the parthenogenetic taxon and a sexual host species on which it depends for sperm (Moore and McKay, 1971; Schenck and Vrijenhoek, 1989).

(2) Many parthenogenetic animal taxa are polyploid. During the early stages of parthenogenesis, polyploidy can be generated by backcrosses between newly arisen parthenogens and their sexual ancestor (Parker et al., 1977). Likewise, in rare parthenogenesis in normally sexual species, different aberrant automictic egg maturation mechanisms can lead to polyploidy (Stalker, 1954).

(3) Many parthenogenetic animals, especially vertebrates, are hybrids, reproducing as perpetual F_1's (Dawley and Bogart, 1989). Additionally, hybrid parthenogenesis has been documented in *Warramaba* grasshoppers (Honeycutt and Wilkinson, 1989), *Campeloma* snails (Johnson, 1992), *Bacillus* stick insects (Scali et al., 1995) and cladocerans (Hebert et al., 1993; Müller and Seitz, 1993). In the Cladocera, *Bacillus* and *Campeloma*, both hybrid and non-hybrid parthenogens are known to involve the same ancestral sexual species. Polyploidy is common in hybrid parthenogenetic lineages of both invertebrates and vertebrates, with up to three different sexual species involved in the parentage of some triploid and tetraploid clonal complexes (Dawley and Bogart, 1989; Manaresi et al., 1993).

(4) Parthenogenesis can arise either by a single origin of a parthenogenetic clone from a sexual ancestor (monophyletic origin) or by the acquisition of parthenogenesis by several different females of the sexual ancestor (polyphyletics origin).

Templeton (1982) pointed out that it has often been assumed wrongly that obligate secondary parthenogenesis arises suddenly or instantaneously. Although data on the genetics and initial stages of parthenogenesis are sparse, several workers have demonstrated a capacity for rare parthenogenesis in normally sexual species (tychoparthenogenesis). Stalker (1954) found 3 of 28 normally sexual species of Drosophilidae produced a low frequency (10^{-4}–10^{-5}) of viable offspring from unfertilized eggs. He was further able to increase these frequencies ~100-fold through selection. Carson (1967) was likewise able to increase the frequency of viable diploid progeny from unfertilized eggs from ~0.1% to ~6% in some strains of *Drosophila mercatorum*, which also showed geographic variation in the ability to reproduce parthenogenetically.

In other insects, Hamilton (1953) induced four generations of thelytokous reproduction in the locust, *Schistocerca gregaria*, with an increase in survival from 32.7% to 41.7%, suggesting viability selection. Likewise, Roth and Willis (1956) found evidence for an increase in tychoparthenogenesis in the cockroach *Periplaneta americana*. Two of four other normally sexual species of cockroach also produced rare viable offspring from unfertilized eggs.

These experiments on tychoparthenogenesis in normally sexual species indicate genetic variation for parthenogenesis in normally sexual species. Stalker (1954) listed a series of "rather rare events" that must take place if a tychothelytokous form is to establish itself as an obligate parthenogen: (1) the impaternate progeny must themselves be fertile, (2) selection must improve the fitness of parthenogenetically produced offspring, and (3) populations of the strains developing functional thelytoky would have to be isolated from bisexual forms to prevent backcrosses from disrupting the parthenogenetic reproduction. For this third criterion, Stalker hypothesized that small local populations with little migration would be the kind of population structure in which parthenogenesis is likely to evolve (see also Templeton, 1982). Similarly, Hamilton (1967), noting the high frequency of thelytokous species and races among parasitoid Hymenopterans and mites, suggested that thelytoky was an end point on the evolutionary trajectory toward fewer males in highly inbred haplo–diploid species of arthropods. Likewise, Nur (1971) noted that most species of parthenogenetic coccids (Homoptera) that have homozygosity-reinforcing automixis are derived from haplo–diploid sexual species in which the males are fragile and short lived.

There is evidence for this kind of population structure in Italian populations of the stick insect *B. rossius*, which are capable of either sexual or parthenogenetic reproduction. In this species, sexual populations are small and highly inbred, with little or no long-distance migration. Mated females produce normal male and female progeny while eggs in virgin females undergo gamete duplication. Females of *B. rossius* from all-female populations can mate and reproduce sexually in the lab, suggesting that this species is in the earliest stages of the evolution of parthenogenesis (Scali et al., 1995). Thus the attainment of facultative parthenogenesis by gamete duplication in this species may have been facilitated by the prior selection for a highly inbred population structure.

Of special interest are cases in which specific genes have been predicted or identified that induce parthenogenesis. Jaenike and Selander (1979) modeled the spread of a dominant gene in simultaneous hermaphrodites that causes parthenogenetic development of eggs while allowing normal spermatogenesis. They concluded that such a gene will spread to fixation, while there would be concomitant selection for the reduction of male reproductive effort, which is verified by the reduction or loss of male-related functions in parthenogenetic earthworms. Evidence for such genes has been found in aphids (Blackman, 1972) and cladocerans (Innes and Hebert, 1988) in which both obligate and cyclically thelytokous populations occur.

23.2.2. The Origin of Clonal Diversity

Although clonal diversity in parthenogenetic taxa has been detected in a number of different characters, including morphology (Zweifel, 1965), cytology (Stalker, 1956; Basrur and Rothfels, 1959), and histocompatibility (Kallman, 1962; Maslin, 1967; Moore and Eisenberry, 1979), the richest source of information on the origins of diversity comes from allozyme studies. An early review of the data on clonal diversity suggested that most, if not all, parthenogenetic taxa were highly variable and that polyphyletic origin, recombination at egg maturation, and mutation all contributed to genetic diversity (Parker, 1979), although there is evidence for a strictly monophyletic origin of parthenogenesis in some widespread thelytokous taxa (Stoddard, 1983; Menken and Wiebosch-Steeman, 1988; review in Suomalainen et al., 1987).

Polyphyletically derived lineages initially differ at a large number of loci, which make them morphologically (Zweifel, 1965), physiologically (Roth, 1974; Bulger and Schultz, 1979; Parker, 1984; Niklasson and Parker, 1994), and ecologically (Scudday, 1973; Vrijenhoek, 1978; Mitter et al., 1979) distinct. With time, divergence by recombination and mutation (Lynch, 1985) will blur the distinction between monophyly and polyphyly in the origins of specific clones due to convergent mutation at marker loci. This will be particularly true in cases in which the sexual ancestor to a clonal complex either is extinct or has not been identified, leading to uncertainty about the origin of clonal diversity (Tomiuk and Loeschcke, 1992).

23.3. Evolution in Parthenogenetic Species

Given clonal diversity in a parthenogenetic taxon, we now need to consider the further evolution of the complex. One model that has received attention is the evolution of general- purpose genotypes (GPGs) in connection with the phenomenon of geographic parthenogenesis. We also discuss currently used definitions of GPGs and methods used for estimating general purposeness, using examples and data from our own work on the parthenogenetic cockroach *Pycnoscelus surinamensis* (Niklasson and Parker, 1994; Parker and Niklasson, 1995).

23.3.1. Geographic Parthenogenesis

Most parthenogens display a geographical and/or ecological distribution that is different from that of their sexual relatives. This phenomenon of geographical parthenogenesis was generalized by Vandel (1928), who first noted that parthenogenetic arthropods tend to have a wider and more northern distribution than their closest sexual relatives. Other general patterns in the distribution of parthenogens and their sexual ancestors have since been shown. These include association of parthenogens with glacial regression and their sexual ancestors with unglaciated refugia (Suomalainen et al., 1987; Hughes, 1989), association of parthenogens with peripheral habitats (Enghoff, 1976) or with island habitats (Cuellar, 1977; Glesener and Tilman, 1978), and the spread of parthenogenetic taxa during human-assisted colonizations (Roth, 1974; Hughes, 1996).

In cases of microgeographic distributions of parthenogenetic and sexual relatives in areas of sympatry, the parthenogens are found mainly in more disturbed or ecotonal habitats (Wright and Lowe, 1968; Roth, 1974; Enghoff, 1976). However, exceptions to this rule occur, as in several Danish populations of the collembolan *Mesaphorura macrochaeta*, in which sexuals are found associated with isolated clumps of grasses on the highly disturbed outer beaches, but are replaced a few meters inland with more dispersed parthenogenetic individuals in the more continuously distributed and stable grasslands (Petersen, 1978).

The hypotheses proposed to explain geographical parthenogenesis can be categorized in a number of ways. One such classification distinguishes models that differ in their assumptions about differences between ancestral sexual and descendent parthenogenetic genotypes in their phenotypic responses to environmental conditions (Table 23.1). The first set of hypotheses focuses on the shift in genetic system from sexuality to parthenogenesis and makes no explicit predictions concerning adaptive differences between sexual and asexual genotypes other than in their reproductive capacity and in the level of genetic variation present. The *"Parthenogenesis"* hypothesis should be considered the null model for geographical parthenogenesis since parthenogens are viewed as being successful as colonizers only because of their uniparental reproduction. The other hypotheses in the first category are not really exclusive of the parthenogenesis hypothesis, but merely propose more specific reasons for geographic patterns. All work from the assumption that genotypes in parthenogens and sexuals is directly comparable, that is, there has not been any selection in the parthenogenetic lineages (see also Lively and Howard, 1994).

It is the second set of hypotheses that interest us here. These models assume that the origin of parthenogenesis or selection subsequent to origin has led to qualitative differences between parthenogenetic and sexual genotypes. The most inclusive model is the "General-purpose-genotype" hypothesis. The concept of a GPG was developed by Baker (1965) to describe the life history syndrome associated with weedy species of plants. He defined it as a genotype characterized by the ability to grow in a multitude of climates and edaphic situations, i.e., a genotype with a wide environmental tolerance. Baker's original description was explicitly comparative: Colonizing species of weeds were more tolerant of physical stresses, more plastic in flowering phenology, and/or more likely to be self-compatible or apomictic than their closest noncolonizing relatives. In parthenogenetic animals, circumstantial evidence for a wide ecological and physiological tolerance is provided by the broad geographical distribution

Table 23.1. *Classification of Hypotheses for Geographical Parthenogenesis based on Assumptions about the Evolved Differences between the Phenotypic Characteristics of Genotypes Selected under Sexual and Parthenogenetic Reproduction*

Assumptions/Hypotheses	Reason for Different Distribution
I. No qualitative differences between sexual and parthenogenetic genotypes	
A. *"Parthenogenesis"* (White, 1948; Williams, 1975; Maynard Smith, 1978)	Rapid reproduction of parthenogens in open habitats, single parthenogens able to found new populations.
B. *"Biotic uncertainty"* (Levin, 1975; Glesener and Tilman, 1978)	Failure of parthenogens to compete with sexual ancestors where the biotic component of environment is important because of their (parthenogens') inability to coevolve with predators, parasites, and competitors.
C. *"Tangled bank"* (Bell, 1982)	Genetically variable sexuals better able to match spatial heterogeneity of environment, and more complex, tropical habitats are more spatially complex.
D. *"Destabilizing hybridization"* (Lynch, 1984)	Lower fitness of parthenogens in contact zone because of hybridization with sexual ancestor.
II. Parthenogenetic and sexual genotypes are qualitatively different in phenotypic responses	
A. *"General-purpose genotype"* (Baker, 1965; Parker et al., 1977; Templeton, 1982)	Selection among clones allows those with wide tolerance limits to survive over many generations; selection in sexuals cannot produce genotypes adapted to all possible alternative environments.
B. *"Elevated ploidy"* (Vandel, 1940; Suomalainen, 1962)	Polyploid genotypes are better able to withstand physical environmental stress.
C. *"Hybridization"* (Stebbins, 1950; Wright and Lowe, 1968; Schultz, 1971)	Hybrid parthenogens are better able to withstand physical environmental stress and/or can exploit more/different environments than sexual ancestors.
D. *"Frozen niche variation"* (Vrijenhoek, 1978, 1979)	Multiclonal parthenogenetic populations can more effectively utilize available niche space than broadly adapted sexuals, making them more resistant to competitive exclusion by sexual ancestors.

displayed by a number of parthenogens relative to their sexual ancestors. However, the GPG model cannot be resolved from the others (for example, the frozen niche variation model, see below) without knowledge of clonal diversity and the niche characteristics of individual genotypes (Scudday, 1973; Jaenike et al., 1980; Hanley et al., 1994).

The GPG hypothesis to explain geographic parthenogenesis in animals was first outlined by Parker et al. (1977) as an explanation for the colonizing success of the parthenogenetic cockroach *P. surinamensis*. They argued that selection among an array of continuously generated clones from a sexual population of *Pycnoscelus* in a temporally varying environment would result in the persistence of only the clones with broad ecological tolerance. Templeton (1982) emphasized that the reproductive transition from sexual to parthenogenetic reproduction is accompanied by a radical shift in the unit of selection that could cause major differences in the evolutionary trajectories of related parthenogenetic and sexual populations even under identical environmental conditions.

The "Hybridization" hypothesis is applicable to parthenogens of hybrid origin and is essentially a more restricted version of the GPG hypothesis, with the hybrid nature of the parthenogenetic genotypes being the reason for their enhanced or altered tolerance. This hypothesis is difficult to test independently of other models for the success of parthenogens since extant hybrid parthenogens also show extensive clonal diversity and have been subject to interclonal selection (Bulger and Schultz, 1979; Wetherington et al., 1987).

The "Elevated ploidy" hypothesis argues that polyploid races are more tolerant to extreme physical stresses and that selection for parthenogenesis in seasonal environments is indirect, being a consequence of the higher frequency of polyploidy among parthenogenetic animal and apomictic plants (Vandel, 1940). Lindroth (1954) found triploid parthenogenetic *Otiorrynchus* weevils to be more cold tolerant than their diploid sexual ancestors, although selection in the parthenogens, rather than their polyploid nature, could have been the cause of the increased tolerance. In nonhybrid parthenogenetic animals, polyploidy is usually associated with backcrosses between initially diploid parthenogens and their sexual ancestor, with all three kinds of genotypes showing similar heterozygosities (e.g., Parker et al., 1977) and a test of the polyploidy model independently of the GPG hypothesis should involve model system with sexuals and both diploid and triploid clones. In *Daphnia pulex*, sexual (cyclical) and obligate clonal diploids occur at a temperate site in Canada (77 pond populations), obligate diploid and polyploid parthenogens are approximately equally frequent at a low arctic site (364 populations) and almost all clones are polyploid at a high arctic site (164 populations), suggesting increased tolerance of polyploids at sites with more severe winters (Beaton and Hebert, 1988). In experimental comparisons of colonizing diploid and triploid clones of *P. surinamensis*, triploid clones generally have lower reproductive rates and desiccation resistance than diploid clones (Roth, 1974; Niklasson and Parker, 1994; Parker and Niklasson, 1995).

The "Frozen niche variation" model of Vrijenhoek (1978, 1979) was developed to explain coexistance between sexual and parthenogenetic relatives, but can be extended to predict that it is clonal diversity that enables a parthenogenetic taxon to occupy wider geographic areas than sexual ancestors (Parker, 1979). Thus this hypothesis stands as a logical alternative to the GPG model and there is some good evidence for clonal niche or major habitat separation in parthenogenetic *Cnemidophorus* lizards (review in Parker, Walker and Paulisson, 1989) and in the colonizing parthenogenetic snail, *Potamopyrgus antipodarum*, in Europe (Hughes, 1996; Jacobsen and Forbes, 1997).

23.3.2. The Evolution of General-Purpose Genotypes in Clonal Complexes

The GPG model as an explanation of geographic parthenogenesis is simple and testable. However, as with any model, its assumptions and predictions must be critically evaluated (Fig. 23.1). The first assumption is that there is a sexual

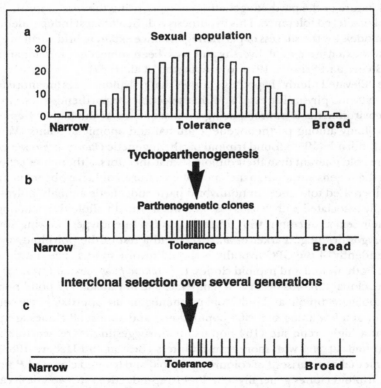

Figure 23.1. Hypothesis for the origin of GPGs in colonizing parthenogenetic taxa. See text for explanation of various stages.

species with tychoparthenogenetic ability. Second, that there is a distribution of breadth of tolerance to abiotic extremes in the sense of Lewontin (1956) in this species [Fig. 23.1(a)]. This phenotypic distribution is given as a bell-shaped curve to emphasize that in sexual species, tolerance to environmental extremes has been shown to be a polygenic trait, subject to artificial selection (Hoffman and Parsons, 1991). The curve does not represent the tolerance of individuals with different phenotypes over a single environmental gradient nor does the breadth of the curve reflect the proportion of individuals that can survive in suboptimal conditions, going in either direction away from the mean distribution, as has been modeled by Lynch and Gabriel (1987). Their model of the evolution of environmental tolerance concerned the effect of environmental heterogeneity on the shape (mean and variance) of the tolerance curve in an asexual population. Here, the phenotypic distribution reflects variation among individuals in their breadth of tolerance, with the optimum (mean) breadth of the sexual population determined by the temporal and spatial pattern of variation in the abiotic variable. The exact shape, position, and width of this distribution is not important to the GPG model, only that individual genotypes differ in their breadth of tolerance.

The next assumption is that the parthenogenetic mode of reproduction will be favored in the marginal areas where the population is sparse and it is hard to find a mate (Stalker, 1954; Gerritsen, 1980; Templeton, 1982). We further assume that there is no genetic correlation between the tendency to produce successful clones and the breadth of tolerance of the cloned genotypes. Thus the tolerance to abiotic extremes among the continuously produced parthenogenetic clones will reflect the variation possessed by the sexual population. Since the unit of selection under parthenogenetic reproduction is the entire genotype, long-term selection among the parthenogenetic clones in a temporally varying environment will favor the clones with a wide tolerance to abiotic extremes [Fig. 23.1(b)]. Hence, clones with the highest geometric mean fitness (i.e., GPGs) over time are more likely to persist in the population, while clones with a more narrow fitness range (i.e., specialists) are going to flourish during the time the environment is favorable but go extinct as soon as the abiotic conditions change.

Likewise, the shift in the unit of selection from single-gene (allele) to multilocus genotype may qualitatively affect the phenotypic characteristics that are selected in sexuals versus parthenogens. For example, it may be possible to select directly for genotypes that can survive over a long-term sequence of environments, while in sexuals, genes are selected more for their additive properties (Templeton et al., 1976), and some optimal level of tolerance will be determined by selection responses to shorter term environmental sequences and the nature of recombination of polygenes for tolerance breadth.

Evidence for differing patterns of selection in parthenogenetic and sexual relatives is sparse but suggestive. Wöhrmann and Tomiuk (1988) found that cyclically parthenogenetic (holocyclic) populations of the rose aphid *Macrosiphium rosae* maintained much more electrophoretic variability than populations characterized by permanent thelytoky (anholocyclic), which were almost monoclonal but that also had extensive between population diversity. Their results suggest that genotypic selection may be more intense than genic selection in reducing genetic variance. Multiple colonizations are necessary to allow scope for selection among clones. For example, as many as 14 clones of 4 morphotypes of the parthenogenetic snail Potamopyrgus antipodarum, at least two of which are ecologically distinct, have colonized Europe from New Zealand (Foltz et al., 1984, Hauser et al., 1992, Jacobsen, Forbes and Skovgaard, 1996).

With such strong selection acting on parthenogenetic clones, the prediction of the GPG model is that generalist genotypes persist over time while specialists go extinct. This will result in a continuing increase of generalist genotypes in the clonal population. Therefore, these kinds of clones will then be more likely to be included in a dispersal event [Fig. 23.1(c)]. The probability of surviving dispersal and of establishing populations in the new habitat will also depend on the degree of tolerance of the individuals since both events are likely to include abiotic extremes.

Thus, this model, the GPG hypothesis, to explain geographical parthenogenesis, has few assumptions, all of which are probably true:

1. Genetic differences among individuals in their breadth of tolerance (Parker et al., 1977; Hoffman and Parsons, 1991)
2. The disparity in units of selection between parthenogenetic and sexually reproducing individuals (Wright, 1977; Templeton, 1982)
3. Selection for parthenogenesis on the edge of a species range (Stalker, 1954; Templeton, 1982)
4. Selectively important temporal variation in the abiotic environment.

23.3.3. Current Definitions of General-Purpose Genotypes

Having the model, we must now ask how we can adequately test it. In the search for GPGs in parthenogens, the definition is usually based on (1) the range of ecological conditions the genotype is able to exploit or (2) the extent of the geographical area the genotype occupies. Both criteria are in concordance with the original definition by Baker, emphasizing a wide ecological tolerance.

Experimental tests of the hypothesis have been scarce, especially for fitness-related characters. In these tests, the definitions of GPGs have been based on different criteria: (1) the genotype's mean performance across environments (Niklasson and Parker, 1994), (2) the genotype's degree of sensitivity to environmental changes (Bierzychudek, 1989; Weider, 1993), (3) the degree of survival tolerance to environmental stress (Parker and Niklasson, 1995; Gade and Parker, 1997), and (4) a combination of mean performance and sensitivity (Michaels and Bazzaz, 1989).

The definitions based on adaptability, the ability of an individual or population to adapt to a wide variety of environments (Lewontin, 1956), as an indicator

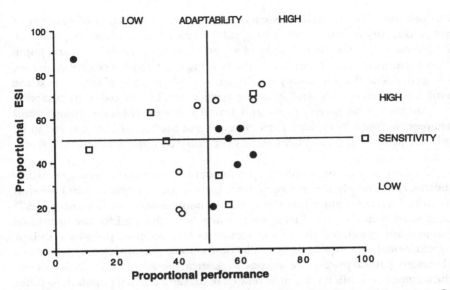

Figure 23.2. Proportional distribution pattern of adaptability and sensitivity to temperature effects for life-history characters (●, adult age at first brood; □, number of offspring; ○, longevity) among seven clones of the parthenogenetic cockroach *P. surinamensis*. (Data from Niklasson and Parker, 1994).

of general purposeness, are in accordance with Baker's original definition of a GPG and both mean performance across environments and stress tolerance are examples of adaptability. Conversely, the use of a genotype's sensitivity as the sole indicator of a GPG is not consistent with the original definition since there is no *a priori* reason to expect a GPG to be insensitive to environmental changes. Baker (1965, p. 155) stressed clearly that plasticity in some characters was necessary to maintain homeostasis (or adaptability) in survival and reproduction. To illustrate the lack of correlation between adaptability and environmental sensitivity (Fig. 23.2) we plotted sensitivity against adaptability (see Michaels and Bazzaz, 1989) among seven different clones of the parthenogenetic cockroach *P. surinamensis* data (from Niklasson and Parker, 1994). Sensitivity is measured by the environmental sensitivity index ESI (Niklasson and Parker, 1994), and adaptability is measured as the arithmetic mean for lifetime survival and reproduction of replicate groups of each of the clones measured over four different temperatures. Thus low sensitivity to environmental changes is in itself not a good indicator of GPGs.

23.4. Conclusions

A science of population biology that can provide insight into the survival and reproduction of organisms in changing environments has been an elusive goal (Lewontin, 1957, 1968, 1974). The evolution of female parthenogenesis can provide valuable model systems to extend the integration of patterns of genotypic, phenotypic, and ecological variations in natural populations. Problems such as the maintenance of sex and fitness variation in natural populations, the genetics of niche width, the limiting similarity in coexisting species, the genetics and evolutionary significance of phenotypic plasticity, and the evolution of colonizing and invading species can be amenable to comparative analyses of parthenogenetic and sexual relatives. However, comparisons between secondarily evolved parthenogens and their sexual ancestors can be misleading if the peculiar characteristics of clonal complexes are not taken into consideration.

Contrary to prior expectations, parthenogenetic taxa are highly genetically diverse. Additionally, the nature of the origins of parthenogenesis and the shift in unit of selection imply that the collective homeostasis (see Lewontin, 1957) utilized by sexually reproducing species may be exchanged for the individual homeostasis generated by such processes as formation of polyploid and/or hybrid clonal genotypes and the selection among polyphyletically generated clones for general-purpose genotypes. In classical models of the evolution of sex, the assumption is often made that related sexual and asexual populations differ in amounts of genetic variation and in frequency of genetic recombination and that sexual and asexual genotypes are similar in their phenotypic responses to environmental heterogeneity (e.g., Williams, 1975; Maynard Smith, 1978; but see Case and Taper, 1986; Lynch and Gabriel, 1987; Lively and Howard, 1994, for more realistic models). What is clear from study of real parthenogenetic populations is that we can not assume that "all other things are equal" in comparisons of parthenogens and their sexual ancestors. Only detailed experimental analyses of parthenogenetic taxa and their sexual ancestors can provide not only an explanation for the evolution of parthenogenesis in a particular case but also a justification for their use as models in addressing general population biological questions.

23.5. Acknowledgments

We thank Valery Forbes, Roger Hughes, Valerio Scali, Jürgen Tomiuk, and an unnamed reviewer for valuable suggestions. This work was supported in part by grants to EDP from the Danish National Research Council and to EMN from the Nordica Foundation.

REFERENCES

Asher, J. H. 1970. Parthenogenesis and genetic variability. II. One-locus models for various diploid populations. *Genetics* **66**:369–391.

Asher, J. H. and Nace, G. W. 1971. The genetic structure and evolutionary fate of parthenogenetic amphibian populations as determined by markovian analysis. *Am. Zool.* **11**:381–398.

Baker, H. G. 1965. Characteristics and modes of origin of weeds. In *Genetics of Colonizing Species*, H. G. Baker and G. L. Stebbins, eds., pp.137–172. New York: Academic.

Basrur, V. R. and Rothfels, K. H. 1959. Triploidy in natural populations of the black fly *Cnephia mutata* (Malloch). *Can. J. Zool.* 37:571–589.

Beaton, M. J. and Hebert, P. D. N. 1988. Geographic parthenogenesis and polyploidy in *Daphnia pulex*. *Am. Nat.* 132:837–845.

Bell, G. 1982. *The Masterpiece of Nature. The Evolution and Genetics of Sexuality.* Berkeley, CA: U. California Press.

Bierzychudek, P. 1989. Environmental sensitivity of sexual and apomictic *Antennaria*: do apomicts have general-purpose genotypes? *Evolution* 43:1456–1466.

Blackman, R. L. 1972. The inheritance of life-cycle differences in *Myzus persicae* (Sulz.) (Hem., Aphididae). *Bull. Entomol. Res.* 62:281–294.

Brown, S. G., Kwan, S., and Shero, S. 1995. The parasitic theory of sexual reproduction: parasitism in unisexual and bisexual geckos. *Proc. R. Soc. London Ser. B* 260:317–320.

Browne, R. A. and Hoopes, C. W. 1990. Genotype diversity and selection in asexual brine shrimp. *Evolution* 44:1035–1051.

Bulger. A. J. and Schultz, R. J. 1979. Heterosis and interclonal variation in thermal tolerance in unisexual fishes. *Evolution* 33:848–859.

Carson, H. L. 1967. Selection for parthenogenesis in *Drosophila mercatorum*. *Genetics* 55:157–171.

Case, T. J. and Taper, M. L. 1986. On the coexistence and coevolution of asexual and sexual competitors. *Evolution* 40:366–387.

Christensen, B. 1979. Differential distribution of genetic variants in triploid parthenogenetic *Trichoniscus pusillus* (Isopoda, Crustacea) in a heterogeneous environment. *Hereditas* 91:179–182.

Christensen, B., Noer, H., and Theisen, B. F. 1988. Differential response to humidity and soil type among clones of triploid parthenogenetic *Trichoniscus pusillus* (Isopoda, Crustacea). *Hereditas* 108:213–217.

Cuellar, O. 1977. Animal parthenogenesis. *Science* 197:837–843.

Dawley, R. M. and Bogart, J. P., eds. 1989. *Evolution and Ecology of Unisexual Vertebrates.* Albany, NY: Bulletin 466, New York State Museum.

Dessauer, H. C. and Cole, C. J. 1989. Diversity between and within nominal forms of unisexual teiid lizards. In *Ecology and Evolution of Unisexual Vertebrates*, R. Dawley and J. P. Bogart, eds., pp. 49–71. Albany, NY: Bulletin 466, New York State Museum.

Dobzhansky, T. 1970. *Genetics of the Evolutionary Process.* New York: Columbia U. Press.

Dybdahl, M. F. and Lively, C. M. 1995. Diverse, endemic and polyphyletic clones in mixed populations of a freshwater snail (*Potamopyrgus antipodarum*). *J. Evol. Biol.* 8:385–398.

Elder Jr., J. F. and Schlosser, I. J. 1995. Extreme clonal uniformity of *Phoxinus eos/neogaeus* gynogens (Pisces: Cyprinidae) among variable habitats in northern Minnesota beaver ponds. *Proc. Natl. Acad. Sci. USA* 92:5001–5005.

Enghoff, H. 1976. Parthenogenesis and bisexuality in the millipede, *Nemasoma varicorne* C. L. Koch, 1847 (Diplopoda: Blaniulidae). Morphological, ecological and biogeographical aspects. *Vidensk. Medd. Dansk Naturhist. Foren.* Khobenhavn 139:21–59.

Foltz, D. W., Ochman, H., Jones, J. S., and Selander, R. K. 1984. Genetic heterogeneity within and among morphological types of the parthenogenetic snail *Potamopyrgus Jenkinsi* (Smith, 1889). Journal of Molluscan Studies. 50:242–245.

Frydenberg, O., Miller, D., Naendal, G., and Sick, K. 1965. Haemoglobin polymorphism in Norwegian cod populations. *Hereditas* 53:257–271.

Gade, B. and Parker Jr., E. D. 1997. The effect of life cycle stage and genotype on dessication tolerance in the colonizing parthenogenetic cockroach *Pycnoscelus surinamensis* and its sexual ancestor *P. indicus. J. Evol. Biol.* 10:479–493.

Gerritsen, J. 1980. Sex and parthenogenesis in sparse populations. *Am. Nat.* 115:718–742.

Glesener, R. R. and Tilman, D. 1978. Sexuality and the components of environmental uncertainty: Clues from geographic parthenogenesis in terrestrial animals. *Am. Nat.* 112:659–673.

Grassle, J. F. and Shick, J. M. 1979. Introduction to the symposium: ecology of asexual reproduction in animals. *Am. Zool.* 19:667–668.

Hamilton, A. G. 1953. Thelytokous parthenogenesis for four generations in the desert locust (*Schistocerca gregaria* Forsk.). *Nature (London)* 172:1153–1154.

Hamilton, W. D. 1967. Extraordinary sex ratios. *Science* 156:477–488.

Hanley, K. A., Bolger, D. T., and Case, T. J. 1994. Comparative ecology of sexual and asexual gecko species (*Lepidodactylus*) in French Polynesia. *Evol. Ecol.* 8:438–454.

Harshman, L. G. and Futuyma, D. J. 1985. The origin and distribution of clonal diversity in *Alsophila pometaria* (Lepidoptera: Geometridae). *Evolution* 39:315–324.

Hauser, L., Carvalho, G. R., Hughes, R. N., and Carter, R. E. 1992. Clonal structure of the introduced freshwater snail *Potamopyrgus antipodarum* (Prosobranchia: Hydrobiidae), as revealed by DNA fingerprinting analysis. *Proc. R. Soc. London B* 249:19–25.

Hebert, P. D. N. and Crease, T. J. 1983. Clonal diversity in populations of *Daphnia pulex* reproducing by obligate parthenogenesis. *Heredity* 51:353–369.

Hebert, P. D. N., Schwartz, S. S., Ward, R. D., and Finston, T. L. 1993. Macrogeographic patterns of breeding system diversity in the *Daphnia pulex* group. I. Breeding systems of Canadian populations. *Heredity* 70:148–161.

Hoffman, A. A. and Parsons, P. A. 1991. *Evolutionary Genetics and Environmental Stress.* Oxford: Oxford U. Press.

Honeycutt, R. L. and Wilkinson, P. 1989. Electrophoretic variation in the parthenogenetic grasshopper *Warramaba virgo* and its sexual relatives. *Evolution* 43:1027–1044.

Hubby, J. L. and Lewontin, R. C. 1966. A molecular approach to the study of genic heterozygosity in natural populations. I. The number of alleles at different loci in *Drosophila pseudoobscura. Genetics* 54:577–594.

Hughes, R. N. 1989. *A Functional Biology of Clonal Animals.* London: Chapman & Hall.

Hughes, R. N. 1996. Evolutionary ecology of parthenogenetic strains of the prosobranch snail, *Potamopyrgus antipodarum* (Gray) (=*P. jenkinsi* (Smith)). In *Molluscan Reproduction*, pp. 101–113. Malacological Review, Suppl. 6. Ann Arbor, MI: U. Michigan Press.

Innes, D. J. and Hebert, P. D. N. 1988. The origin and genetic basis of obligate parthenogenesis in *Daphnia pulex. Evolution* 42:1024–1035.

Jackson, J. B. C., Buss, L. W., and Cook, R. E., eds. 1985. *Population Biology and Evolution of Clonal Organisms.* New Haven, CT: Yale U. Press.

Jacobsen, R., Forbes, V. E., and Skovgaard, O. 1996. Genetic population structure of the prosobranch snail *Potamopyrgus antipodarum* (Gray) in Denmark using PCR-RAPD fingerprints. *Proc. R. Soc. London B* 263:1065–1070.

Jacobsen, R. and Forbes, V. E. 1997. Clonal variation in life history traits and feeding rates in the gastropod, *Potamopyrgus antipodarum*: performance across a salinity gradient. *Function. Ecol.* 11:260–267.

Jaenike, J. and Selander, R. K. 1979. Evolution and ecology of parthenogenesis in earthworms. *Am. Zool.* **19**:729–737.

Jaenike, J., Parker Jr., E. D., and Selander, R. K. 1980. Clonal niche structure in the parthenogenetic earthworm *Octolasion tyrtaeum. Am. Nat.* **116**:196–205.

Johnson, S. G. 1992. Spontaneous and hybrid origins of parthenogenesis in *Campeloma decisum* (freshwater prosobranch snail). *Heredity* **68**:253–261.

Judson, O. P. and Normark, B. B. 1996. Ancient asexual scandals. *Trends Ecol. Evol.* **11**:41–46.

Kallman, K. D. 1962. Population genetics of the gynogenetic teleost *Molliensia formosa* (Girard). *Evolution* **16**:497–504.

Levin, D. A. 1975. Pest pressure and recombination systems in plants. *Am. Nat.* **109**:437–451.

Lewontin, R. C. 1956. Studies on homeostasis and heterozygosity. I. General considerations. Abdominal bristle number in second chromosome homozygotes of *Drosophila melanogaster. Am. Nat.* **90**:237–255.

Lewontin, R. C. 1957. The adaptations of populations to varying environments. *Cold Spring Harbor Symp. Quant. Biol.* **22**:395–408.

Lewontin, R. C., ed. 1968. *Population Biology and Evolution.* Syracuse, NY: Syracuse U. Press.

Lewontin, R. C. 1974. *The Genetic Basis of Evolutionary Change.* New York: Columbia U. Press.

Lindroth, C. H. 1954. Experimentelle Beobachtungen an parthenogenetischem und bisexuellem *Otiorrhynchus dubius* Stroem (Col., Curculionidae). *Entomol. Tidskr.* **75**:111–116.

Lively, C. M. 1992. Parthenogenesis in a freshwater snail: reproductive assurance versus parasitic release. *Evolution* **46**:907–913.

Lively, C. M. and Howard, R. S. 1994. Selection by parasites for clonal diversity and mixed mating. *Philos. Trans. R. Soc. London Ser. B* **346**:271–281.

Lively, C. M., Craddock, C., and Vrijenhoek, R. C. 1990. Red queen hypothesis supported by parasitism in sexual and clonal fish. *Nature (London)* **344**:864–866.

Lourenço, W. R., ed. 1994. Colloque parthénogenèse géographique. *Biogeographica* **70**:1–48.

Lynch, M. 1984. Destabilizing hybridization, general-purpose genotypes and geographic parthenogenesis. *Q. Rev. Biol.* **59**:257–290.

Lynch, M. 1985. Spontaneous mutations for life history characters in an obligate parthenogen. *Evolution* **39**:804–818.

Lynch, M. and Gabriel, W. 1987. Environmental tolerance. *Am. Nat.* **129**:283–303.

Manaresi, S., Marescalchi, D., and Scali, V. 1993. The trihybrid genome constitution of *Bacillus linceorum* (Insecta, Phasmatodea) and its karyotypic variations. *Genome* **36**:317–326.

Maslin, T. P. 1967. Skin grafting in the bisexual teiid lizard *Cnemidophorus sexlineatus* and in the unisexual *C. tesselatus. J. Exp. Zool.* **166**:137–150.

Maynard Smith, J. 1978. *The Evolution of Sex.* Cambridge: Cambridge U. Press.

Mayr, E. 1963. *Animal Species and Evolution.* Cambridge, MA: Harvard U. Press.

Menken, S. B. J. and Wiebosch-Steeman, M. 1988. Clonal diversity, population structure, and dispersal in the parthenogenetic moth *Ectoedemia argyropeza. Entomol. Exp. Appl.* **49**:141–152.

Michaels, H. J. and Bazzaz, F. A. 1989. Individual and population responses of sexual and apomictic plants to environmental gradients. *Am. Nat.* **134**:190–207.

Mitter, C., Futuyma, D. J., Schneider, J. C., and Hare, J. D. 1979. Genetic variation and host plant relations in a parthenogenetic moth. *Evolution* **33**:777–790.

Moore, W. S. and Eisenberry, A. B. 1979. The population structure of an asexual vertebrate, *Poeciliopsis 2 monacha-lucida* (Pisces: Poeciliidae). *Evolution* **33**:563–578.

Moore, W. S. and McKay, F. E. 1971. Coexistence in unisexual–bisexual species complexes of *Poeciliopsis* (Pisces: Poeciliidae). *Ecology* **52**:791–799.

Moritz, C., Brown, W. M., Densmore, L. D., Wright, J. W., Vyas, D., Donnellan, S., Adams, M., and Baverstock, P. 1989. Genetic diversity and the dynamics of hybrid parthenogenesis in *Cnemidophorus* (Teiidae) and *Heteronotia* (Gekkonidae). In *Ecology and Evolution of Unisexual Vertebrates*, R. Dawley and J. P. Bogart, eds., pp. 87–112. Albany, NY: Bulletin 466, New York State Museum.

Moritz, C., McCallum, H., Donnellan, S., and Roberts, J. D. 1991. Parasite loads in parthenogenetic and sexual lizards (*Heteronotia binoei*): support for the red queen hypothesis. *Proc. R. Soc. London Ser. B* **244**:145–149.

Müller, J. and Seitz, A. 1993. Habitat partitioning and differential vertical migration of some *Daphnia* genotypes in a lake. *Arch. Hydrobiol. Beih.* **39**:167–174.

Neaves, W. B. 1969. Adenosine deaminase phenotypes among sexual and parthenogenetic lizards in the genus *Cnemidophorus* (Teiidae). *J. Exp. Zool.* **171**:175–184.

Niklasson, M. and Parker Jr., E. D. 1994. Fitness variation in an invading parthenogenetic cockroach. *Oikos* **71**:47–54.

Normark, B. 1996. Phylogeny and evolution of parthenogenetic weevils of the *Arimagustessellatus* species complex (Coleoptera: Curculionidae:Naupactini): evidence from mitochondrial DNA sequences. *Evolution* **50**:734–745.

Nur, U. 1971. Parthenogenesis in coccids (Homoptera). *Am. Zool.* **11**:301–308.

Ochman, H., Stille, B., Niklasson, M., Selander, R. K., and Templeton, A. R. 1980. Evolution of clonal diversity in the parthenogenetic fly *Lonchoptera dubia*. *Evolution* **34**:539–547.

Oliver Jr., J. H. 1971. Introduction to the symposium on parthenogenesis. *Am. Zool.* **11**:241–243.

Parker Jr., E. D. 1979. Ecological implications of clonal diversity in parthenogenetic morphospecies. *Am. Zool.* **19**:753–762.

Parker Jr., E. D. 1984. Reaction norms of development rate among diploid clones of the parthenogenetic cockroach *Pycnoscelus surinamensis*. *Evolution* **38**:1186–1193.

Parker Jr., E. D. and Selander, R. K. 1976. The organization of genetic diversity in the parthenogenetic lizard *Cnemidophorus tesselatus*. *Genetics* **84**:791–805.

Parker Jr., E. D. and Niklasson, M. 1995. Desiccation resistance in invading parthenogenetic cockroaches: a search for the general purpose genotype. *J. Evol. Biol.* **8**:331–337.

Parker Jr., E. D., Selander, R. K., Hudson, R. O., and Lester, L. J. 1977. Genetic diversity in colonizing parthenogenetic cockroaches. *Evolution* **31**:836–842.

Parker Jr., E. D., Walker, J. M., and Paulisson, M. A. 1989. Clonal diversity in *Cnemidophorus*: ecological and morphological consquences. In *Ecology and Evolution of Unisexual Vertebrates*, R. Dawley and J. P. Bogart, eds., pp. 72–86. Albany, NY: Bulletin 466, New York State Museum.

Petersen, H. 1978. Sex-ratios and the extent of parthenogenetic reproduction in some collembolan populations. In *Proceeding of the First International Seminary on Apterygota*. R. Dallai, ed., pp. 19–35. Siena, Italy: Accademia delle Scienze di Siena dette dèFisiocritici.

Roth, L. M. 1974. Reproductive potential of bisexual *Pycnoscelus indicus* and clones of its parthenogenetic relative, *Pycnoscelus surinamensis. Ann. Entomol. Soc. Am.* **67**:215–223.

Roth, L. M. and Willis, E. R. 1956. Parthenogenesis in cockroaches. *Ann. Entomol. Soc. Am.* **49**:195–204.

Scali, V., Tinti, F., Mantovani, B., and Marescalchi, D. 1995. Mate recognition and gamete cytology features allow hybrid species production and evolution in *Bacillus* stick insects. *Boll. Zool.* **62**:59–70.

Schenck, R. A. and Vrijenhoek, R. C. 1989. Coexistence among sexual and asexual *Poeciliopsis*: foraging behavior and microhabitat selection. In *Ecology and Evolution of Unisexual Vertebrates*, R. Dawley and J. P. Bogart, eds., pp. 39–48. Albany, NY: Bulletin 466, New York State Museum.

Schultz, R. J. 1971. Special adaptive problems associated with unisexual fish. *Am. Zool.* **11**:351–360.

Scudday, J. F. 1973. A new species of the *Cnemidophorus tesselatus* group from Texas. *J. Herpetol.* **7**:363–371.

Seiler, J. 1923. Geschlechtschromosomenuntersuchungen an Psychiden. IV. Die Parthenogenese der Psychiden. *Z. indukt.Abstamm.-u. Vererb.Lehre* **31**:1–99.

Stalker, H. D. 1954. Parthenogenesis in *Drosphila. Genetics* **39**:4–34.

Stalker, H. D. 1956. On the evolution of parthenogenesis in *Lonchoptera* (Diptera). *Evolution* **10**:345–359.

Stebbins, G. L. 1950. *Variation and Evolution in Plants.* New York: Columbia U. Press.

Stille, B. 1985. Population genetics of the parthenogenetic gall wasp *Diplolepis rosae* (Hymenoptera, Cynipidae). *Genetica* **67**:145–151.

Stoddard, J. A. 1983. The accumulation of genetic variation in a parthenogenetic snail. *Evolution* **37**:546–554.

Stouthamer, R., Luck, R. F., and Hamilton, W. D. 1990. Antibiotics cause parthenogenetic *Trichogramma* (Hymenoptera/Trichogrammatidae) to revert to sex. *Proc. Natl. Acad. Sci. USA* **87**:2424–2427.

Suomalainen, E. 1950. Parthenogenesis in animals. *Adv. Genet.* **3**:193–253.

Suomalainen, E. and Saura, A. 1973. Genetic polymorphism and evolution in parthenogenetic animals. I. Polyploid Curculionidae. *Genetics* **74**:489–508.

Suomalainen, E., Saura, A., and Lokki, J. 1987. *Cytology and Evolution in Parthenogenesis.* Boca Raton, FL: CRC.

Templeton, A. R. 1982. The prophecies of parthenogenesis. In *Evolution and Genetics of Life Histories*, H. Dingle and J. P. Hegmann, eds., pp. 75–101. New York: Springer-Verlag.

Templeton, A. R., Sing, C. F., and Brokaw, B. 1976. The unit of selection in *Drosophila mercatorum* I. The interaction of selection and meiosis in parthenogenetic strains. *Genetics* **82**:349–376.

Theisen, B. F., Christensen, B., and Arctander, P. 1995. Origin of clonal diversity in triploid parthenogenetic *Trichoniscus pusillus pusillus* (Isopoda, Crustacea) based on allozyme and nucleotide sequence data. *J. Evol. Biol.* **8**:71–80.

Tomiuk, J. and Loeschcke, V. 1992. Evolution of parthenogenesis in the *Otiorhynchus scaber* complex. *Heredity* **68**:391–397.

Turner, B. J., Elder Jr., J. F., Laughlin, T. F., and Davis, W. P. 1990. Genetic variation in clonal vertebrates detected by simple sequence DNA fingerprinting. *Proc. Natl. Acad. Sci. USA* **87**:5653–5657.

Uzzell Jr., T. M. 1964. Relations of the diploid and triploid species of the *Ambystoma jeffersonianum* complex (Amphibia, Caudata). *Copeia* **1964**:257–300.

Uzzell Jr., T. M. and Goldblatt, S. M. 1967. Serum proteins of salamanders of the *Ambystoma jeffersonianum* complex, and the origin of the triploid species of this group. *Evolution* **21**:345–354.

Uzzell, T. 1970. Meiotic mechanisms of naturally occurring unisexual vertebrates. *Am. Nat.* **104**:433–445.

Uzzell, T. and Darevsky, I. S. 1975. Biochemical evidence for the hybrid origin of the parthenogenetic species of the *Lacerta saxicola* complex (Sauria: Lacertidae), with a discussion of some ecological and evolutionary implications. *Copeia* **1975**:204–222.

Vandel, A. 1928. La parthénogenèse géographique. Contribution à l'étude biologique et cytologique de la parthénogenèse naturelle. I. *Bull. Biol. France-Belgique* **62**:164–281.

Vandel, A. 1940. La parthénogenèse géographique. IV. Polyploïdie et distribution géographique. *Bull. Biol. France-Belgique* **74**:94–100.

Vrijenhoek, R. C. 1972. Genetic relationships of unisexual-hybrid fishes to their progenitors using lactate dehydrogenase isozymes as gene markers (*Poeciliopsis*, Poeciliidae). *Am. Nat.* **106**:754–766.

Vrijenhoek, R. C. 1978. Coexistance of clones in a heterogeneous environment. *Science* **199**:549–552.

Vrijenhoek, R. C. 1979. Factors effecting clonal diversity and coexistance. *Am. Zool.* **19**:787–797.

Wetherington, J. D., Kotora, K. E., and Vrijenhoek, R. C. 1987. A test of the spontaneous heterosis hypothesis for unisexual vertebrates. *Evolution* **41**:721–731.

Weider, L. J. 1993. A test of the "general purpose" genotype hypothesis: differential tolerance to thermal and salinity stress among *Daphnia* clones. *Evolution* **47**:965–969.

Weider, L. J. and Hebert, P. D. N. 1987. Ecological and physiological differentiation among low-arctic clones of *Daphnia pulex*. *Ecology* **68**:188–198.

White, M. J. D. 1973. *Animal Cytology and Evolution*, 3rd ed. Cambridge: Cambridge U. Press.

Williams, G. C. 1975. *Sex and Evolution*. Princeton, NJ: Princeton U. Press.

Wöhrmann, K. and Tomiuk, J. 1988. Life history strategies and genotypic variability in populations of aphids. *J. Genet.* **67**:43–52.

Wright, J. W. and Lowe, C. H. 1968. Weeds, polyploids, parthenogenesis, and the geographical and ecological distribution of all-female species of *Cnemidophorus*. *Copeia* **1968**:128–138.

Wright, S. 1977. *Evolution and the Genetics of Populations, Vol. 3. Experimental Results and Evolutionary Deductions*. Chicago: U. Chicago Press.

Zweifel, R. G. 1965. Variation in and distribution of the unisexual lizard, *Cnemidophorus tesselatus*. *Am. Mus. Novit.* **2235**:1–49.

CHAPTER TWENTY FOUR

Genetic Structure of Prokaryotic Populations

FREDERICK M. COHAN

Until the mid-1970s, evolutionary genetics was almost exclusively the domain of zoologists and botanists (e.g., Lewontin, 1974). It should therefore come as no surprise that microbiologists have borrowed their intuition about evolutionary processes from studies of plants and animals. For example, zoologists and botanists have long recognized that genetic exchange is a powerful force of evolutionary cohesion among populations: Populations with the potential to exchange genes are prevented from diverging, while populations that have lost the potential to exchange genes are free to diverge without bound (Stebbins, 1950; Dobzhansky, 1951; Mayr, 1963). Many microbiologists have assumed that the potential to exchange genes is equally important in determining patterns of divergence in bacteria (e.g., Matic et al., 1996). However, the character of genetic exchange in bacteria is very different from that in the higher eukaryotes, and, as I will show, genetic exchange plays very different roles in the eukaryotic and the prokaryotic worlds. Several of the important evolutionary consequences of genetic exchange that are taken for granted by zoologists and botanists do not apply to bacteria; nevertheless, genetic exchange can foster adaptation in bacteria in ways that are impossible in the eukaryotes.

24.1. The Character of Genetic Exchange in the Bacterial World

24.1.1. The Rarity of Bacterial Genetic Exchange

Whereas genetic exchange in most animals and plants is obligately tied to reproduction, genetic exchange is extremely rare for bacteria in nature. This was shown first in allozyme surveys of bacterial populations (reviewed by Selander and Musser, 1990). Nearly every species investigated has shown high levels of linkage disequilibrium that are consistent with very little or no recombination among strains. Even those occasional allozyme surveys yielding low linkage

R. S. Singh and C. B. Krimbas, eds., *Evolutionary Genetics: From Molecules to Morphology*, vol. 1. © Cambridge University Press 2000. Printed in the United States of America. ISBN 0-521-57123-5. All rights reserved.

disequilibrium levels are not strong evidence against rare recombination, since very low rates of recombination, only 10–20 times that of mutation, are sufficient to break up linkage disequilibrium (Maynard Smith et al., 1993).

Surveys of DNA sequence diversity have confirmed that homologous recombination does occur at least occasionally in nature. In these studies, recombination has been implicated when the DNA sequences of different gene sequences yield different phylogenetic relationships among strains (DuBose et al., 1988; Dykhuizen and Green, 1991; Maynard Smith et al., 1991; Milkman and Bridges, 1993; Guttman and Dykhuizen, 1994a). The variance in sequence divergence among strains can be used to infer the frequency of homologous recombination within natural populations of bacteria (Hudson, 1987). The rate of recombination within *Escherichia coli* has thus been estimated at 10^{-8} per gene segment per genome per generation (Whittam and Ake, 1993), and the recombination rates within *Bacillus subtilis* and within *B. mojavensis* have been estimated at 10^{-7} (Roberts and Cohan, 1995).

Guttman and Dykhuizen (1994a) have recently estimated the rate of homologous recombination between species of bacteria. A survey of four genes in 12 *E. coli* strains was used to reconstruct the history of recombination and mutation in these strains since their last common ancestor. Since the common ancestor, several lineages incorporated sequence homologs from other species, but not one mutation appeared in any of the lineages. They estimated that between-species recombination occurs at a rate of 5×10^{-9} per nucleotide site per genome per generation. While this rate is not high, the authors argued persuasively that recombination with other species is a more significant source of neutral sequence variation than mutation.

24.1.2. The Promiscuity of Bacterial Genetic Exchange

Genetic exchange is much more promiscuous in bacteria than is the case for animals and plants. Eukaryotic species that are as little as 2% divergent in DNA sequence are usually unable to exchange genes (DeSalle and Hunt, 1987). Bacteria, on the other hand, can easily undergo homologous recombination with organisms as divergent as 25% in DNA sequence (and possibly more) (Shen and Huang, 1986; Duncan et al., 1989; Rayssiguier et al., 1989; Maynard Smith et al., 1991; Roberts and Cohan, 1993; Guttman and Dykhuizen, 1994a; Zawadzki et al., 1995).

Bacteria can also capture new gene loci from other organisms (Lan and Reeves, 1996). This may occur in nature as a side effect of homologous recombination, such that a heterologous gene from the donor is integrated along with closely flanking homologous donor DNA (a possibility demonstrated by Hamilton et al., 1989). Alternatively, heterologous genes may be integrated along with a transposable element, or heterologous genes from a lysogenic phage may become a permanent part of the host chromosome if the phage loses its ability to be induced (Campbell, 1981).

The past two decades of systematic bacteriology have suggested a vast potential for gene capture in the bacterial world: hundreds of whole-genome hybridization experiments have shown that most pairs of closely related strains do not contain exactly the same set of genes (Johnson, 1986); this conclusion has recently been supported by a more precise genomic subtraction technique (Lan and Reeves, 1996). Strains that are almost identical in the sequences of genes they share can diverge up to 10%–15% in the fraction of their genomes that are not homologous (e.g., Roberts et al., 1996; Lan and Reeves, 1996). This suggests a high rate of incorporation of genes from other species.

In addition to integrating new genes and new alleles into their chromosomes, bacteria can also accept and express new genes on plasmids from extremely divergent species (Young and Levin, 1992).

Given the highly promiscuous nature of bacterial genetic exchange, one might conclude that there are no bounds to bacterial genetic exchange, that each bacterium can exchange genes equally well with all other bacteria. There are, however, several important constraints on bacterial genetic exchange. Recombination that depends on vectors (phage or plasmids) is limited by the host ranges of the respective vectors. Restriction endonuclease activity of the recipient greatly reduces the rate of recombination by transduction and conjugation (e.g., McKane and Milkman, 1995), although restriction has only a modest effect on recombination by transformation (Trautner et al., 1974; Cohan et al., 1991). Finally, homologous recombination is limited by the resistance to integration of divergent DNA sequences (te Riele and Venema, 1982; Shen and Huang, 1986; Rayssiguier et al., 1989). The rate of recombination decays exponentially with donor–recipient sequence divergence (Roberts and Cohan, 1993; Zawadzki et al., 1995; Vulie et al., 1997; Majewski and Cohan, 1998).

While there is some degree of sexual isolation between closely related bacterial taxa, I shall soon show that the integrity of bacterial populations' adaptations does not rely at all on sexual isolation.

24.1.3. The Small Size of Recombined Segments

The amount of genetic material exchanged in prokaryotes is much smaller than that in eukaryotes. Whereas meiosis and fertilization in eukaryotes yield recombinant individuals that are a 1 : 1 mix of DNA from each parent, in bacteria the DNA exchanged is a small fraction of the genome, and the DNA is transferred in one direction (Smith, 1989). The sizes of segments transduced or transformed in the laboratory are frequently less than several kilobases (McKane and Milkman, 1995; Zawadzki and Cohan, 1995). Surveys of sequences in nature have supported the conclusion that each recombination event is highly localized within the chromosome (Maynard Smith et al., 1991).

Let us next consider the consequences of rare, promiscuous, and unidirectional transfer of small DNA segments for bacterial evolution.

24.2. The Evolutionary Consequences of Genetic Exchange in Bacteria

24.2.1. The Integrity of Population-Specific Adaptations

Microbiologists have inherited from zoologists and botanists the notion that populations cannot evolve ecological differences without being sexually isolated from one another. That is, the rate of recombination between ecologically distinct populations must be greatly reduced compared with that within populations. This notion is correct for animals and plants because the rate of recombination within each population is so high. If recombination between animal populations were to occur at this same high rate, divergence between these populations would be impossible (Antonovics et al., 1971).

In contrast, the evolution of adaptive divergence in bacteria does not require sexual isolation (Cohan, 1994a, 1994b). Even if recombination between populations occurred at the same rate as recombination within populations, recombination would not be sufficient to reverse adaptive divergence. This is because the frequency of maladaptive foreign alleles (coding for another population's adaptations) is determined as a balance between the intensity of selection (s) against those alleles and the rate of between-population recombination (c_b): c_b/s. Given the low rates of recombination observed in bacteria, the frequency of maladaptive foreign alleles will always be extremely low even in the absence of sexual isolation. In contrast to the case for animals and plants, the evolution of sexual isolation is not a necessary step in adaptive divergence between bacterial populations.

24.2.2. Transfer of Adaptations across Taxa

While genetic exchange is too rare to constrain adaptive divergence between bacterial populations, the promiscuous nature of bacterial genetic exchange allows adaptations to be transferred between very divergent groups (Young and Levin, 1992). This gives bacteria an advantage unknown in the eukaryotic world, since each eukaryotic species must evolve any new adaptation based on its own mutations.

Homologous recombination is one mechanism for the transfer of adaptations across taxa. For example, some strains of *Neisseria gonorrhoeae* have acquired penicillin resistance by homologous recombination with *N. flavescens* (Maynard Smith et al., 1991). Alternatively, some adaptations have been transferred by heterologous recombination, in which a gene locus from the donor has been added to the recipient chromosome (gene capture). For example, some virulence genes of *Salmonella* appear to have been added from distantly related species (Ochman and Groisman, 1994). Finally, adaptations have been transferred and expressed by plasmids; dozens of antibiotic-resistance operons have been transferred across species in this manner (Young and Levin, 1992).

The promiscuous nature of bacterial genetic exchange is not the only factor facilitating the transfer of adaptations across bacterial taxa (Cohan, 1994b).

Also important is the modular nature of bacterial adaptations. An entirely new operon may be expressed as a stand-alone module, without any need to adapt to the cell's physiology. For instance, Isberg and Falkow (1985) have shown that introducing a single virulence gene from *Yersinia* into *E. coli* immediately transforms the recipient into a weak pathogen.

The transfer of adaptations is also facilitated by the ability of bacteria to limit genetic exchange to a very small number of genes (Cohan, 1994b; Zawadzki and Cohan, 1995). This is because genes that can be adaptively transferred across taxa are a very small set that confer general adaptations (e.g., antibiotic-resistance genes), whose benefits are not restricted to the ecological and genetic context of a single taxon. The intertaxon transfer of such a gene is more likely to be successful if it can be transferred alone, without the cotransfer of more narrowly adaptive alleles whose benefits are limited to a single taxon.

The rare but promiscuous nature of genetic exchange in bacteria provides the advantage of sex with distant relatives, without any disadvantage. A single genetic exchange event in which an adaptation is transferred across taxa can profoundly alter the course of adaptive evolution in the recipient species. Nevertheless, owing to the rarity of bacterial recombination, the introduction of maladaptive alleles from other taxa cannot threaten the integrity of a bacterial population's unique adaptation.

That bacteria can take up modules of adaptation from other taxa raises an interesting evolutionary issue unique to microbial systems. How much of adaptive evolution in the bacterial world is due to changes in existing genes and how much is due to gene capture? Groisman and Ochman (1994) have recently addressed this issue with regard to the evolution of virulence in *S. enterica*. They found that the majority of gene loci involved in virulence were also found in closely related nonpathogenic species, suggesting a minor role for the capture of new genes in the evolution of virulence. This result stands in contrast to the great potential for evolution by gene capture demonstrated by genomic hybridization (Johnson, 1986) and confirmed by genomic substraction experiments (Lan and Reeves, 1996). It will be interesting to assess more generally the relative importance of gene capture versus changes in existing genes in bacterial evolution.

As important as the capture of new genes is for bacteria, this mode of adaptive evolution may be even more important for the plasmids and bacteriophage that infect bacteria. Plasmids and phage have both acquired adaptations by taking up genetic modules that constitute a large fraction of their genomes (Hughes and Datta, 1983; Inamine and Burdett, 1985; Campbell, 1988; Willson et al., 1989; Highton et al., 1990; Oberto et al., 1994): plasmids and phage that are otherwise unrelated may share as little as a single gene locus.

We have recently developed an approach for exploring the relative importance of gene capture in the adaptive evolution of plasmids. The approach is based on a coalescence model for predicting patterns of neutral sequence divergence within and between ecologically distinct populations of plasmids, for any gene that they share. The model predicts substantial neutral sequence divergence between plasmid populations under a broad diversity of conditions: Ecological populations of plasmids should frequently be identified as separate sequence-similarity clusters for any gene shared among populations.

This result allows us to investigate the genetic basis of adaptation in the plasmid world by surveying the sequences of genes shared among many plasmids. If allelic differences are usually responsible for population differences, then we should expect to find multiple sequence-similarity clusters (or ecological populations) within a group of plasmids known to contain exactly the same complement of genes. If gene capture is largely responsible for population differences, then all of the plasmids bearing a particular complement of genes should appear together as a separate sequence-similarity cluster for any gene segment shared among plasmids.

24.2.3. The Purging of Diversity under Periodic Natural Selection

Zoologists and botanists have taken for granted that natural selection has only a limited effect on the genetic diversity within an animal or plant species, but this is not the case for bacteria. Consider the consequences of an asexual (or rarely sexual) bacterial cell's acquiring an adaptive mutation, which allows the cell to outcompete other members of its ecological population. In the absence of recombination, the original adaptive mutant and its clonal descendants will replace all the other cells in the population. Because the entire genome originally associated with the adaptive mutation remains intact as it sweeps through the population, the population loses its genetic diversity at all loci (Atwood et al., 1951; Koch, 1974; Levin, 1981). Every adaptive mutation occurring in an asexual (or rarely sexual) population has the potential to set in motion this purging of genetic diversity (an event known as periodic selection). Given the low rates of recombination observed for bacteria, each periodic selection event is expected to purge 99% or more of the population's variation at all loci (Cohan, 1994b).

Natural selection is much less effective in purging genetic diversity from an animal or plant population. Owing to the much higher rate of recombination, the adaptive mutation is quickly transferred into many genetic backgrounds. The adaptive mutation can then increase in frequency without bringing with it the entire genome of the original mutant.

24.2.4. Multilocus Associations of Alleles

One hypothesized advantage of sex lies in its ability to create new associations of alleles at different loci (Williams, 1975). It appears that bacterial genetic exchange, as rare as it is, may be sufficient to create new multilocus associations of alleles. Maynard Smith et al. (1993) have shown that the rate of recombination need be only 10 times greater than the rate of mutation to effectively randomize multilocus associations. Apparently recombination has been sufficient in at least a few bacterial species to randomize associations of alleles (as evidenced by zero linkage disequilibrium; Maynard Smith et al., 1993). Bacterial genetic exchange is thus frequent enough to foster adaptations by bringing together new adaptive combinations of alleles.

24.2.5. Neutral Sequence Diversity

It is a universal feature of life that organisms fall into clusters of similar organisms, with very few individuals falling outside these clusters. Organisms fall into clusters on the basis of their phenotypes and on the basis of their molecular sequence characteristics; clusters appear in both prokaryotes and eukaryotes, and in asexual, rarely sexual, and highly sexual organisms (Jones and Sneath, 1970; Raven, 1980; Mayr, 1982). Let us next consider the forces that cause bacteria to form discrete clusters based on DNA sequence similarity, as has been observed in many recent systematic studies (e.g., Normand et al., 1996; Roberts et al., 1996).

A model of sequence divergence must take into account that the genes typically sequenced by bacterial systematists (e.g., 16S rRNA, *rpoB*, *gndA*) are not likely to be involved in ecological differences between bacterial populations. The genes studied are thought to perform the same "housekeeping" duties in every species: Alleles of these genes may be assumed to be functionally interchangeable across taxa (Osuna et al., 1991; Sedgwick et al., 1991; Mikulskis and Cornelis, 1994; Yu et al., 1995). Also, a model of sequence diversity must take into account that nearly all nucleotide substitutions detected in surveys are likely to be neutral in their fitness effects (Kimura, 1983). So the patterns of sequence divergence observed by systematists involve substitutions that are of no fitness consequence in genes that are not involved in population-specific adaptations.

Cohan (1994a, 1995) has developed a coalescence model for predicting the sequence divergence observed in systematic studies of bacteria. The model requires a somewhat unorthodox concept of an ecological population of bacteria, a concept based on the fate of an adaptive mutation. Because of the competitive advantage conferred by an adaptive mutation, the original mutant cell and its clonal (or nearly clonal) descendants will replace all competing cells of the same population. However, because different populations use different resources, an adaptive mutant from one population is not expected to outcompete members of other populations. An ecological population is thus defined as the domain of competitive superiority of an adaptive mutant.

This definition yields an interesting prediction about sequence divergence within and between such ecological populations (Cohan, 1994a, 1994b). Each periodic selection event will purge the diversity within the population of the adaptive mutant, but it will have very little effect on the divergence between populations. Each round of periodic selection thereby enhances the distinctiveness of ecological populations, and fosters the divergence of different ecological populations into separate sequence-similarity clusters.

On the other hand, the tendency for bacterial populations to form separate sequence clusters is opposed by recombination between populations. It is a

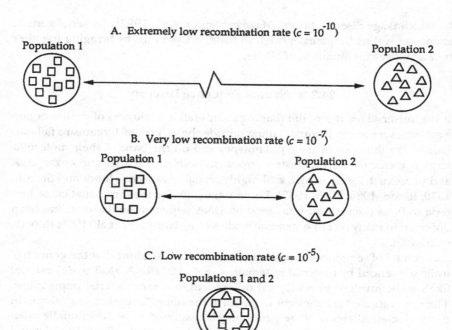

Figure 24.1. Patterns of sequence divergence within and between ecological populations, based on the coalescence model of Cohan (1994a). The model assumes no sexual isolation between populations, such that the rates of recombination within and between populations are equal. The divergence within populations is represented by the diameter of each population's circle, and the divergence between populations is represented by the length of the line connecting the populations. The individual organisms of populations 1 and 2 are represented by squares and triangles, respectively. A. With extremely low recombination rates (e.g., $c = 10^{-10}$), there is unbounded neutral sequence divergence between populations, such that every nucleotide site that can mutate without a fitness cost will be substituted at least once. B. With very low recombination rates (e.g., $c = 10^{-7}$), the populations are still divergent, but interpopulation divergence is bounded at an intermediate level. C. With low recombination rates (e.g., $c = 10^{-5}$), populations that are ecologically distinct are not distinguishable by neutral sequence variation, so that the average divergence within populations equals the average divergence between populations. The ranges of recombination rates that yield each of these three classes of outcomes depend on the values of other parameters, such as the frequency and the intensity of periodic selection events.

quantitative question whether the diversifying effect of periodic selection or the homogenizing effect of interpopulation recombination will dominate the pattern of sequence divergence. The coalescence model shows how the distinctiveness of ecological populations is determined by the interplay of these factors.

The coalescence model has shown that three classes of outcomes are possible for sequence divergence within and between populations (Figs. 24.1; Cohan, 1994a). The results in Fig. 24.1 are based on the extreme case of no sexual isolation, such that the recombination rates within and between populations are equal. The first class of outcomes occurs when the rates of recombination (both within and between populations) are extremely low (e.g., 10^{-10} in Fig. 24.1A) or when periodic selection is extremely frequent. Here populations diverge without bound in neutral sequence characters, so that every nucleotide site that can mutate harmlessly will become substituted at least once. Thus unbounded neutral divergence is possible in bacteria even when recombination rates within and between populations are equal (provided that the rate of recombination within populations is sufficiently low). This is in contrast to the case for animal (and many plant) populations. Sequence divergence between animal populations is impossible unless the rate of recombination between populations is sharply reduced from the high rate occurring within populations.

The second class of outcomes occurs under higher rates of recombination (e.g., 10^{-7} in Fig. 24.1B) or lower rates of periodic selection. Here different ecological populations are still distinctive in DNA sequence, but a bound is placed on population divergence. Instead of eventually becoming substituted at every site that can mutate harmlessly, the populations reach an equilibrium level of sequence divergence, where the diversifying effect of periodic selection is balanced by the homogenizing effect of recombination between populations. The outcomes in Figs. 24.1A and 24.1B represent the two possible fates of divergence for closely related sequence-similarity clusters observed in nature: They may continue to diverge without bound in DNA sequence or they may reach an equilibrium level of divergence. While we cannot yet predict the ultimate fate of sequence divergence for most pairs of clusters, it appears that recombination is too rare, at least in *Bacillus*, to constrain the sequence divergence between taxa: *B. subtilis* and all its closest relatives are destined to diverge without bound in neutral sequence characters (Cohan, 1995).

Finally, with still higher rates of recombination (e.g., 10^{-5} in Fig. 24.1C), different ecological populations become indistinguishable on the basis of DNA sequence data (for genes that are functionally interchangeable across populations). Here the average sequence divergence within populations equals that between populations. Recombination is low enough not to obstruct adaptive divergence between populations, but is high enough to prevent neutral sequence divergence.

Given the variety of possible outcomes for sequence divergence between bacterial populations, how should we interpret sequence-similarity clusters observed among strains in nature? One possibility is that each cluster represents a different ecological population (the ecological model, shown in Figs. 24.1A and 24.1B). This would be the case under frequent and intense periodic selection (which makes populations more distinctive) and/or low rates of recombination between populations (so that recombination fails to homogenize the populations). Alternatively, Fig. 24.1C shows that under strong recombination, several ecological populations may be indistinguishable from one another, all falling within a single sequence cluster. In this case, each sequence-similarity cluster would include a set of ecological populations. This will occur when there is a high rate of recombination among the populations of a set, but much lower

rates of recombination between populations of different sets (the sexual isolation model of clustering). Here adaptive and neutral divergence become
uncoupled (Cohan, 1994a, 1994b).

It is not clear whether bacterial recombination rates in nature are ever high
enough to uncouple neutral and adaptive divergence between populations, as
occurs in the sexual isolation model of clustering. The limited data on recombination rates (Roberts and Cohan, 1995; Whittam and Ake, 1993; Selander and
Musser, 1990) indicate much smaller recombination rates than are required for
obstructing neutral divergence between populations (Palys et al., 1997). Nevertheless, some recent studies in bacterial systematics suggest (at least weakly) a
possible decoupling of neutral and adaptive divergence. Several bacterial taxa
that have long been distinguishable as separate ecological populations (e.g., on
the basis of differences in host ranges) have recently been found to be indistinguishable by DNA sequence data (Cohan, 1994a). For example, there is no
more difference in 16S rRNA sequence between strains of different *Xanthomonas
campestris* pathovars than between strains of the same pathovar (Vauterin et al.,
1990). Also, *B. psychrophilus* and *B. globisporus* fail to be distinguished by 16S
rRNA sequence (Fox et al., 1992).

It is tempting to interpret these results (and others) as evidence that between-
population recombination is allowing adaptive but not neutral divergence. However, the apparent homogeneity among populations could be an artifact of the
extremely low rate of 16S rRNA evolution. Indeed, there is essentially no variation in 16S rRNA either within or between the pairs of populations cited.

It will be interesting to investigate these apparent cases of decoupling of
adaptive and neutral divergence more closely by surveying sequence variation
at more variable gene loci (e.g., most protein-coding genes). If the levels of divergence within and between populations are identical at protein-coding genes
as well as 16S rRNA, we may conclude that recombination is indeed preventing
neutral divergence between populations. To this end, we recently surveyed *B.
psychrophilus* and *B. globisporus* (previously shown indistinguishable by 16S rRNA
sequence; Fox et al., 1992) for sequence variation at the pyruvate kinase gene,
and found these species to be easily distinguished. While more data are needed,
we may tentatively conclude that bacterial recombination is too rare to prevent
neutral divergence between ecological populations: Each DNA sequence cluster appears to represent a separate ecological population, subject to its own
periodic selection events (Palys et al., 1997).

Just as one sequence cluster is unlikely to contain two or more ecological
populations, it is also unlikely for one ecological population to contain cells
from two or more separate sequence clusters. This is because such divergence
within a population would be unstable with respect to the diversity-purging
effect of periodic selection. Each adaptive mutant occurring within the population would drive to extinction cells from all the clusters of the population;
the cluster bearing the original adaptive mutant would be all that survives this
purging of diversity. It appears, then, that two long-standing, highly divergent
clusters cannot each contain cells from the same population.

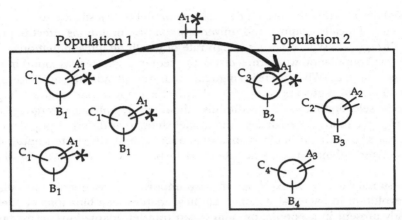

Figure 24.2. Transfer of a generally adaptive mutation from one ecological population (population 1) to another (population 2) results in a local periodic selection event in the second population. The figure represents a moment soon after the adaptive mutation (shown as an asterisk) has swept through population 1, so that little variation remains in that population. The arrow shows the segment containing the adaptive mutation and the closely linked A_1 being transferred into population 2 by homologous recombination. This recombinant cell is now able to outcompete other members of population 2, and so a periodic selection event in population 2 is begun. Population 2 then becomes fixed (or nearly fixed) for the entire genome of the original adaptive recombinant, including allele A_1 from population 1. The two populations thus become homogenized for the short segment that cotransferred with the adaptive mutation, but not for other portions of the genome.

This conclusion yields the following prediction: A single periodic selection event cannot purge genetic diversity from more than one cluster. This is because the competitive superiority of an adaptive mutant is confined to a single ecological population, and each population is limited to a single cluster. Surprisingly, this prediction is contradicted by the only documented example of periodic selection in nature. A periodic selection event purged sequence diversity within a small chromosomal region from all the various sequence clusters of *E. coli* (Guttman and Dykhuizen, 1994b), while the clusters have retained their distinctiveness at other loci (Milkman and Bridges, 1993).

This paradoxical result may be explained by an alternative "adapt globally, act locally" model of population structure (Fig. 24.2). In this model, the domain of competitive superiority of an adaptive mutant (i.e., the cell) is still its ecological population, but the adaptive mutation (i.e., the allele) can be recombined into another population and then confer higher fitness in its new population. This causes a local periodic selection event within each population receiving the adaptive mutation. Here the adaptive recombinant and its clonal descendants will replace other cells of the recombinant's population. This process will tend to homogenize different populations (and different sequence clusters) for any small segment that is cotransferred between populations along with the adaptive mutation (Fig. 24.2). This model appears to resolve the paradox that periodic selection can purge diversity from several ecological populations.

We have recently explored a coalescence model to investigate how the frequencies of locally adaptive and universally adaptive mutations affect population divergence. When there is a high rate of locally adaptive mutation, each ecological population will form a distinctive sequence cluster (as found in the previous models). When there is a high rate of universally adaptive mutation and when the DNA segments cotransferred with the adaptive mutation are large, periodic selection may strongly diminish the neutral sequence divergence between populations. Nevertheless, the model shows that different populations will always be distinguishable as separate sequence-similarity clusters under the rates of recombination thus far observed in bacteria (Majewski and Cohan 1999).

In summary, genetic exchange has two important consequences for adaptive evolution in bacteria: It can create new adaptive combinations of alleles already present in a population, and it can transfer adaptations across taxa. However, genetic exchange is too rare to protect genetic variation from the diversity-purging effect of natural selection. Genetic exchange is also too rare to prevent adaptive divergence between bacterial populations. In *Bacillus* and perhaps other taxa, genetic exchange is not even sufficient to constrain neutral divergence among taxa. As far as these processes are concerned, the rate of recombination might as well be zero (Roberts and Cohan, 1995).

REFERENCES

Antonovics, J., Bradshaw, A. D., and Turner, R. G. 1971. Heavy metal tolerance in plants. *Adv. Ecol. Res.* **7**:1–85.

Atwood, K. C., Schneider, L. K., and Ryan, F. J. 1951. Periodic selection in *Escherichia coli. Proc. Natl. Acad. Sci. USA* **37**:146–155.

Campbell, A. 1981. Evolutionary significance of accessory DNA elements in bacteria. *Ann. Rev. Microbiol.* **35**:55–83.

Campbell, A. 1988. Phage evolution and speciation. In *The Bacteriophages*, R. Calendar, ed., Vol. 1, pp. 1–14. New York: Plenum.

Cohan, F. M. 1994a. The effect of rare but promiscuous genetic exchange on evolutionary divergence in prokaryotes. *Am. Nat.* **143**:965–986.

Cohan, F. M. 1994b. Genetic exchange and evolutionary divergence in prokaryotes. *Trends Ecol. Evol.* **9**:175–180.

Cohan, F. M. 1995. Does recombination constrain neutral divergence among bacterial taxa? *Evolution* **49**:164–175.

Cohan, F. M., Roberts, M. S., and King, E. C. 1991. The potential for genetic exchange by transformation within a natural population of *Bacillus subtilis. Evolution* **45**:1393–1421.

DeSalle, R. and Hunt, J. 1987. Molecular evolution in Hawaiian drosophilids. *Trends Ecol. Evol.* **2**:212–216.

Dobzhansky, R. 1951. *Genetics and the Origin of Species*, 3rd ed. New York: Columbia U. Press.

DuBose, R. F., Dykhuizen, D. E., and Hartl, D. L.. 1988. Genetic exchange among natural isolates of bacteria: recombination within the *phoA* gene of *Escherichia coli. Proc. Natl. Acad. Sci. USA* **85**:7036–7040.

Duncan, K. E., Istock, C. A., Graham, J. B., and Ferguson, N. 1989. Genetic exchange between *Bacillus subtilis* and *Bacillus licheniformis*: variable hybrid stability and the nature of bacterial species. *Evolution* **43**:1585–1609.

Dykhuizen, D. E. and Green, L. 1991. Recombination in *Escherichia coli* and the definition of biological species. *J. Bacteriol.* **173**:7257–7268.

Fox, G. E., Wisotzkey, J. D., and Jurtshuk Jr., P. 1992. How close is close: 16S rRNA sequence identity may not be sufficient to guarantee species identity. *Int. J. Syst. Bacteriol.* **42**:166–170.

Groisman, E. A. and Ochman, H. 1994. How to become a pathogen. *Trends Microbiol.* **2**(8): 289–294.

Guttman, D. S. and Dykhuizen, D. E. 1994a. Clonal divergence in *Escherichia coli* as a result of recombination, not mutation. *Science* **266**:1380–1383.

Guttman, D. S. and Dykhuizen, D. E. 1994b. Detecting selective sweeps in naturally occurring *Escherichia coli*. *Genetics* **138**:993–1003.

Hamilton, C. M., Aldea, M., Washburn, B. K., Babitzke, P., and Kushner, S. R. 1989. New method for generating deletions and gene replacements in *Escherichia coli*. *J. Bacteriol.* **171**:4617–4622.

Highton, P. J., Chang, Y., and Myers, R. J. 1990. Evidence for the exchange of segments between genomes during the evolution of lambdoid bacteriophages. *Mol. Microbiol.* **4**:1329–1340.

Hudson, R. R. 1987. Estimating the recombination parameter of a finite population model without selection. *Genet. Res.* **50**:242–250.

Hughes, V. M. and Datta, N. 1983. Conjugative plasmids in bacteria of the 'pre-antibiotic' era. *Nature (London)* **302**:725–726.

Inamine, J. M. and Burdett, V. 1985. Structural organization of a 67-kilobase streptococcal conjugative element mediating multiple antibiotic resistance. *J. Bacteriol.* **161**:620–626.

Isberg, R. R. and Falkow, S. 1985. A single genetic locus encoded by *Yersinia pseudotuberculosis* permits invasion of cultured animal cells by *Escherichia coli* K-12. *Nature (London)* **317**:262–264.

Johnson, J. L. 1986. Nucleic acids in bacterial classification. In *Bergey's Manual of Systematic Bacteriology*, P. H. A. Sneath, N. S. Mair, M. E. Sharpe, and J. G. Holt, eds., Vol. 2, pp. 972–975. Baltimore: Williams & Wilkins.

Jones, D. and Sneath, P. H. A. 1970. Genetic transfer and bacterial taxonomy. *Bacteriol. Rev.* **34**:40–81.

Kimura, M. 1983. *The Neutral Theory of Molecular Evolution*. Cambridge: Cambridge U. Press.

Koch, A. L. 1974. The pertinence of the periodic selection phenomenon to prokaryotic evolution. *Genetics* **77**:127–142.

Lan, R. and Reeves, P. R. 1996. Gene transfer is a major factor in bacterial evolution. *Mol. Biol. Evol.* **13**:47–55.

Levin, B. R. 1981. Periodic selection, infectious gene exchange and the genetic structure of *E. coli* populations. *Genetics* **99**:1–23.

Lewontin, R. C. 1974. *The Genetic Basis of Evolutionary Change*. New York: Columbia U. Press.

Majewski, M. and Cohan, F. M. 1998. The effect of mismatch repair and heteroduplex formation on sexual isolation in Bacillus. *Genetics* **148**:13–18.

Majewski, J. and Cohan, F. M. 1999. Adapt globally, act locally: a model of periodic selection in bacteria. *Genetics* **152**:1459–1474.

488 *Frederick M. Cohan*

Matic, I., Taddei, F., and Radman, M. 1996. Genetic barriers among bacteria. *Trends Microbiol.* **4**:69.

Maynard Smith, J., Dowson, C. B., and Spratt, B. G. 1991. Localized sex in bacteria. *Nature (London)* **349**:29–31.

Maynard Smith, J., Smith, N. H., O'Rourke, M., and Spratt, B. G. 1993. How clonal are bacteria? *Proc. Natl Acad. Sci. USA* **90**:4384–4388.

Mayr, E. 1963. *Animal Species and Evolution.* Cambridge, MA: Harvard U. Press.

Mayr, E. 1982. *The Growth of Biological Thought: Diversity, Evolution, and Inheritance,* Chap. 6. Cambridge, MA: Harvard U. Press.

McKane, M. and Milkman, R. 1995. Transduction, restriction and recombination patterns in *Escherichia coli. Genetics* **139**:35–43.

Mikulskis, A. V. and Cornelis, G. R. 1994. A new class of proteins regulating gene expression in enterobacteria. *Mol. Microbiol.* **11**:77–86.

Milkman, R. and Bridges, M. M. 1993. Molecular evolution of the *Escherichia coli* chromosome. *Genetics* **133**:455–468.

Normand, P., Orso, S., Cournoyer, B., Jeannin, P., Chapelon, C., Dawson, J., Evtushenko, L., and Misra, A. K. 1996. Molecular phylogeny of the genus *Frankia* and related genera and emendation of the family Frankiaceae. *Int. J. Syst. Bacteriol.* **46**:1–9.

Oberto, J., Sloan, S. B., and Weisberg, R. A. 1994. A segment of the phage HK022 chromosome is a mosaic of other lambdoid chromosomes. *Nucleic Acids Res.* **22**:354–356.

Ochman, H. and Groisman, E. A. 1994. The origin and evolution of species differences in *Escherichia coli* and *Salmonella typhimurium. Experientia* **69**:479.

Osuna, R., Boylan, S. A., and Bender, R. A.. 1991. *In vitro* transcription of the histidine utilization (*hutUH*) operon from *Klebsiella aerogenes. J. Bacteriol.* **173**:116–123.

Palys, T., Nakamura, L. K., and Cohan, F. M. 1997. Discovery and classification of ecological diversity in the bacterial world: the role of DNA sequence data. *Int. J. Syst. Bacteriol.* **47**:1145–1156.

Raven, P. H. 1980. Hybridization and the nature of species in higher plants. *Can. Bot. Assoc. Bull. Suppl.* **13**(1):3–10.

Rayssiguier, C., Thaler, D. S., and Radman, M. 1989. The barrier to recombination between *Escherichia coli* and *Salmonella typhimurium* is disrupted in mismatch-repair mutants. *Nature (London)* **342**:396–401.

Roberts, M. S. and Cohan, F. M. 1993. The effect of DNA sequence divergence on sexual isolation in Bacillus. *Genetics* **134**:401–408.

Roberts, M. S. and Cohan, F. M. 1995. Recombination and migration rates in natural populations of *Bacillus subtilis* and *Bacillus mojavensis. Evolution* **49**:1081–1094.

Roberts, M. S., Nakamura, L. K., and Cohan, F. M. 1996. *Bacillus vallismortis* sp. nov., a close relative of *Bacillus subtilis,* isolated from soil in Death Valley, California. *Int. J. Syst. Bacteriol.* **46**:470–475.

Sedgwick, S. G., Lodwick, D., Doyle, N., Crowne, H., and Strike, P. 1991. Functional complementation between chromosomal and plasmid mutagenic DNA repair genes in bacteria. *Mole. Gen. Genet.* **229**:428–436.

Selander, R. K. and Musser, J. M. 1990. Population genetics of bacterial pathogenesis. In *Molecular Basis of Bacterial Pathogenesis,* B. H. Iglewski and V. L. Clark, eds., pp. 11–36. San Diego, CA: Academic.

Shen, P. and Huang, H. V. 1986. Homologous recombination in *Escherichia coli*: dependence on substrate length and homology. *Genetics* **112**:441–457.

Smith, G. R. 1989. Homologous recombination in *E. coli*: multiple pathways for multiple reasons. *Cell* **58**:807–809.

Stebbins, G. L. 1950. *Variation and Evolution in Plants*. New York: Columbia U. Press.

te Riele, H. P. J. and Venema, G. 1982. Molecular fate of heterologous bacterial DNA in competent *Bacillus subtilis*. II. Unstable association of heterologous DNA with the recipient chromosome. *Genetics* **102**:329–340.

Trautner, T. A., Pawlek, B., Bron, S., and Anagnostopoulos, C. 1974. Restriction and modification in *B. subtilis*. *Mole. Gen. Genet.* **131**:181–191.

Vauterin, L., Swings, J., Kersters, K., Gillis, M., Mew, T. W., Schroth, M. N., Palleroni, N. J., Hildebrand, D. C., Stead, D. E., Civerolo, E. L., Hayward, A. C., Maraite, H., Stall, R. E., Vidaver, A. K., and Bradbury, J. F. 1990. Towards an improved taxonomy of *Xanthomonas*. *Int. J. Syst. Bacteriol.* **40**:312–316.

Vulic, M., Dionisio, F., Taddei, F., and Radman, F. 1997. Molecular keys to speciation: DNA polymorphism and the control of genetic exchange in enterobacteria. *Proc. Natl. Acad. Sci. USA* **94**:9763–9767.

Whittam, T. S. and Ake, S. E. 1993. Genetic polymorphisms and recombination in natural populations of *Escherichia coli*. In *Molecular Paleopopulation Biology*, N. Takahata and A. G. Clark, eds., pp. 223–245. Tokyo: Japan Scientific Society.

Williams, G. C. 1975. *Sex and Evolution*. Princeton, NJ: Princeton U. Press.

Willson, P. J., Albritton, W. L., Slaney, L., and Setlow, J. K. 1989. Characterization of a multiple antibiotic resistance plasmid from *Haemophilus ducreyi*. *Antimicrob. Agents Chemother.* **33**:1627–1630.

Young, J. P. W. and Levin, B. R. 1992. Adaptation in bacteria: unanswered ecological and evolutionary questions about well-studied molecules. In *Genes in Ecology*. R. J. Berry, T. J. Crawford, and G. M. Hewitt, eds., pp. 169–191. Oxford: Blackwell.

Yu, H., Schurr, M. J., and Deretic, V. 1995. Functional equivalence of *Escherichia coli* sigma E and *Pseudomonas aeruginosa AlgU*: *E. coli rpoE* restores mucoidy and reduces sensitivity to reactive oxygen intermediates in *algU* mutants of *P. aeruginosa*. *J. Bacteriol.* **177**:3259–3268.

Zawadzki, P. and Cohan, F. M. 1995. The size and continuity of DNA segments integrated in Bacillus transformation. *Genetics* **141**:1231–1243.

Zawadzki, P., Roberts, M. S., and Cohan, F. M. 1995. The log-linear relationship between sexual isolation and sequence divergence in Bacillus transformation is robust. *Genetics* **140**:917–932.

SECTION G

POPULATION GENETICS AND SPECIATION

Introductory Remarks

RAMA S. SINGH

The species concepts, theories, and patterns of speciation, together, are supposed to help us understand the origin and the diversification of taxa around us. The general methodological revolution in science and technology, i.e., describing a completed structure versus asking how that structure came to be (i.e., the developmental process), applies to the species problem as well, and we are always faced with the dual questions of What are species? and How do they originate? A species concept is supposed to tell us about the nature of the units whose origin we want to understand, but how can we formulate a species concept without having some idea of what species are? The two most distinguishing features of species are resemblance and reproductive cohesion within and differences and reproductive discontinuities between groups. The two features together define Mayr's (1942) biological species concept (BSC) as "groups of actually or potentially interbreeding natural populations which are reproductively isolated from other such groups." All species concepts focusing on reproduction agree on the first part of the definition but not on the second. Thus Dobzhansky (1951, p. 262) defined the species as "groups of populations the gene exchange between which is limited or prevented in nature by one, or a combination of several, reproductive isolating mechanisms," and stated that "species is the most inclusive Mendelian population." His definition of species implied a direct role for selection in the development of reproductive isolation that he hypothesized earlier (Dobzhansky, 1940), based on the suggestion of Fisher (1930). Later on he limited the role of selection to premating isolation through selective reinforcement in sympatric situations (Dobzhansky, 1972). Whether or not reproductive isolation should be part of the species definition and whether or not speciation occurs by selective reinforcement are the main features of the contributions by Hampton Carson and Judith Masters in this section.

Drawing from his vast experience with the Hawaiian *Drosophila* fauna, Carson discusses the importance of sexual selection in speciation and links it to his organizational theory of speciation (OTS), which calls for a buildup of balanced polymorphisms in sexual traits as a result of the repeated cycles of male–female mate choice encounters. Carson's treatment has two main points. First, he uses the dynamics of the interbreeding populations, the recombining gene pool, to revive Dobzhansky's (1937) genetic species concept (GSC) and dismisses the between-group "reproductive isolation" aspect of the BSC as unnecessary. Carson considers his view in line with that of Darwin (1859) and states that reproductive isolation between species is not directly selected and is an incidental by-product of natural and sexual selection in adaptation. Carson criticizes the two most commonly used experimental methods of measuring premating isolation between species, i.e., the use of isofemale lines and the marker genes, as unnatural and he argues that the very process of making isofemale lines or of introducing a marker gene in a different genetic background disturbs the species' balanced genetic system, and as a result all measurements of premating isolation in these (laboratory) systems are suspect. Carson's call for a revision of the BSC is similar to those of Paterson's (1985) recognition species concept (RSC) except that the traits involved in the latter are a subset of the traits involved in mate choice and sexual selection, whose only role is to bring males and females together in the mating arena and are less important in the final mate choice. Second, Carson supports Kaneshiro's (1989) sexual selection theory, which assumes the presence of genetic variation in sexual traits affecting male and female mate choice and predicts a loss in key male sexual recognition traits and a decrease in female discrimination as a result of founder effects in recently derived populations. The differences in the levels of female discrimination between ancestral and derived populations have been used by Kaneshiro to predict the direction of the evolution of populations (Kaneshiro, 1989). In view of these results, Carson cautions against using the presence of partial premating isolation between allopatric populations as evidence of incipient speciation; he thinks of them simply as the consequence of ancestral - derived (populations) mating relationship, as predicted by Kaneshiro's theory.

Judith Masters has summarized the role of selection in various, competing theories of speciation. These theories fall into three categories: speciation by direct selection of isolating mechanisms through genetic reinforcement, as envisaged by Wallace (1889) and Dobzhansky (1951), speciation through incidental effect of natural selection during adaptation, as proposed in the biological (Mayr, 1963), the phylogenetic (Cracraft, 1987; Mishler and Brandon, 1987), and the recognition (Paterson, 1985) species concepts, and speciation by natural selection in combination with founder effects (Carson, 1989). Masters first argues against species concepts that separate patterns from processes, as is the case in the phylogenetic species concept, and focuses on the species concepts (i.e., BSC, GSC, and RSC) that are capable of telling us about the processes of speciation. She follows this with arguments against speciation by genetic reinforcement and summarizes recent theoretical works that bear on this point. Finally, she compares Carson's theory of speciation focusing on sexual selection with that of Paterson's which is based on sexual recognition, and sets out specific predictions that can help us choose between them. She argues against speciation by sexual selection, specially as specified by Kaneshiro (1989) and favors speciation by natural selection involving mate recognition systems in ecologically shifted habitats.

Coyne and Orr have summarized the recent progress on the genetics of speciation, pointing out that study of speciation has become respectable and has grown rapidly, attracting both theoreticians and experimentalists. Their focus is the genetic studies of reproductive isolation and they have treated all types of isolation, premating and postmating in both plants and animals. In this respect they have made Dobzhansky's (1951, 1970) treatment of this subject up to date by focusing on the best-studied cases of genetic analysis. They have treated sexual selection, Haldane's rule, and reinforcement in detail as these topics have featured prominently in the literature and have provided a major contribution to our understanding of the genetic basis of speciation. More importantly, they have put together important questions that need to be answered if we are to make further progress in this field.

In spite of the spectacular successes of the current genetic research programs on speciation, in the broader context there continues to be serious debates with respect to species concepts, geographic isolation, and the role of natural selection, the importance of sexual and nonsexual traits, and the relevance of current speciation theories to plants and asexual organisms. In my contribution, I have made an attempt to take a longer, historical look at the problem of speciation, pointing out how construction of speciation theories have changed since Darwin (1859) with our changing view of the species, from a loosely organized genetic system to a developmentally complex and highly integrated system, making species as an individual. I have focused on the types of traits involved in reproductive isolation and have proposed a unified theory of speciation based on the dichotomy of the species gene pool into weakly coupled sexual and nonsexual components. The theory is meant to be general and applicable to both plants and animals and sexual and asexual organisms, with varying degrees of emphasis on the organisms' breeding systems and the types of traits and the population processes.

It has been pointed out by Lewontin (1997) that by equating the origin of species with the origin of reproductive isolation, Dobzhansky had removed "the problem of the actual speciation from the concern of most population geneticists." Acquisition of new adaptations were relegated to the problem of adaptive population divergence and species were treated as an incidental outcome of genetic divergence in geographic isolation, with selective reinforcement of premating isolation in sympatry. While attainment of reproductive isolation is an important aspect of speciation, reproductive isolation alone does not capture the meaning of speciation: Acquisition of new adaptations is also an essential, and in some way more important, aspect of speciation. The genetic studies of differences between incipient or newly arising species (Lewontin, 1974) have told us a great deal about the nature and the amount of genetic changes that occur during speciation. While the genetic studies of postzygotic reproductive isolation and Haldane's rule naturally have mostly treated genes as if their involvement is only incidental, it is hoped that the increasing knowledge of species-specific adaptations and their genetic analysis would bring the real problem of speciation, i.e., the process of speciation, back in population genetics. All contributors to this section agree on this point.

REFERENCES

Carson, H. L. 1989. Genetic imbalance, realigned selection, and the origin of species. In *Genetics, Speciation, and the Founder Principle*, L. V. Giddings, K. Y. Kaneshiro, and W. W. Anderson, eds., pp. 345–362. New York: Oxford U. Press.

Cracraft, J. 1987. Species concepts and the ontology of evolution. *Biol. Philos.* 2:329–346.

Darwin, C. 1859. *On the Origin of Species*. London: Murray.

Dobzhansky, Th. 1937, 1951. *Genetics and the Origin of Species*. New York: Columbia U. Press.

Dobzhansky, Th. 1940. Speciation as a stage in evolutionary divergence. *Amer. Nat.* 74:312–321.

Dobzhansky, Th. 1970. *Genetics of the Evolutionary Process*. New York: Columbia U. Press.

Dobzhansky, Th. 1972. Species of Drosophila. *Science* 177:664–669.

Fisher, R. A. 1930. *The Genetical Theory of Natural Selection*. Oxford: Clarendon.

Kaneshiro, K. Y. 1989. The dynamics of sexual selection and founder effects in species formation. In *Genetics, Speciation, and the Founder Principle*, L. V. Giddings, K. Y. Kaneshiro, and W. W. Anderson, eds., pp. 279–296. New York: Oxford U. Press.

Lewontin, R. C. 1974. *The Genetic Basis of Evolutionary Change*. New York: Columbia U. Press.

Lewontin, R. C. 1997. Dobzhansky's *Genetics and the Origin of Species*: is it still relevant? *Genetics* 147: 351–355.

Mayr, E. 1942. *Systematics and the Origin of Species*. New York: Columbia U. Press.

Mayr, E. 1963. *Animal Species and Evolution*. Cambridge, MA: Harvard U. Press.

Mishler, B. D. And Brandon, R. N. 1987. Individuality, pluralism, and the phylogenetic species concept. *Biol. Philos.* 2:397–414.

Paterson, H. E. H. 1985. The recognition concept of species. In *Species and Speciation*, E. S. Vrba, ed., pp. 21–29. Pretoria: Transval Museum Monograph No. 4.

Wallace, A. R. 1889. *Darwinism*. London: Macmillan.

CHAPTER TWENTY FIVE

Sexual Selection in Populations: The Facts Require a Change in the Genetic Definition of the Species

HAMPTON L. CARSON

25.1. Introductory Statement

As a naturalist with a particular interest in Darwinism and populations under selection, I was strongly drawn, more than 50 years ago, to the genetic definition of the species proposed by Dobzhansky (1937) and Mayr (1942). As stated by Mayr, species are groups of actually or potentially interbreeding natural populations that are reproductively isolated from other such groups. I adopted this idea and for years did all my research with its assumptions in mind. I now advocate stopping the definition with a period after the word "populations."

The first part of the definition had and still has great advantages: It suggests that the species should be viewed as a series of gene pools and thus as a dynamic entity capable of genetic change through time, especially as a powerful field for gene recombination (Carson, 1957). As Mayr emphasized, it led to populational rather than typological thinking about the sexual species and it allowed the easy application of mathematical population genetics.

The second part of the definition (isolation), first seriously criticized by Paterson (1978), has become less useful for me over the years, although such dissatisfaction is not widely shared (see Coyne and Orr, 1989). Experience has shown that allopatric differentiation prevails in nature and the long lists of interspecific "reproductive isolating mechanisms" (premating and postmating) may not have evolved as isolating functions but may be merely incidental, secondary accompaniments of divergence, the side effects of other genetic alterations affecting adaptations. The gene pools of plants tend to be open to hybridization to an considerable degree, leaving little room for the *ad hoc* evolution of genetic isolation. The new data on sexual selection in animals in general and in *Drosophila* in particular suggest that premating courtship characters arise in local populations not by virtue of possible effects on reproductive isolation but simply as fitness enhancers within the local population.

R. S. Singh and C. B. Krimbas, eds., *Evolutionary Genetics: From Molecules to Morphology*, vol. 1. © Cambridge University Press 2000. Printed in the United States of America. ISBN 0-521-57123-5. All rights reserved.

In this chapter I urge a return to the original Darwinian view that designates selection, both natural and sexual, as the major force that drives descent with genetic change, relegating reproductive isolation to an incidental outcome of other selective processes. Particular attention is given to what I believe are flaws in the interpretation of some of the behavioral work in *Drosophila.* Following the suggestion of Ehrlich and Raven (1969), I conclude that selection operating within the deme is the central process in evolution, leading to genetically based adaptation, involving both the ambient and sexual environments.

25.2. Selection and the Genetic System

Selection that affects the differential reproduction of genetic material in populations of a sexual species is the major driver of evolution. Natural and sexual selection are similar forces that are often difficult to distinguish; both operate most efficiently in local, natural demes of interbreeding individuals. All incorporated changes occur strictly at the level of the local interbreeding population. Such populations are composed of genetically related individuals of both sexes. Under sexual reproduction, each individual varies from the next in its total DNA composition. It thus carries a unique but temporary encapsulation of a specific combination of DNA and thus represents a particular individual variant of the genetic code. Relating a precise phenotype to the many genes of minor individual effect, as in quantitative characters in diploid organisms, has been difficult.

Individual gene combinations have only a brief life. The minority of individuals that successfully reproduce must collapse their sexually chosen genetic endowment down into haploid cells, the gametes. In turn, only an exceedingly small proportion of these gametes participate in the formation of the novel zygotic combinations from which a novel but equally temporary array of combinations will come in the ensuing generation. Selection acts throughout the life of the individual, rendering the DNA passage between generations a process from which many individuals are wholly or partially excluded.

25.3. Natural Selection

Each individual variant of the DNA code is monitored by selection from the beginning to the end of its life. In any generation, the great majority of these recombinational variants, including all of their DNA, are lost forever to the Earth. Under conditions of stable population size, screening due to natural selection entrusts the biological future to a minority of individuals. These carry sets of codes that have proved adequate for survival under the conditions imposed by the ambient environment, through which natural selection operates. This severe and pervasive aspect of natural selection was thoroughly explored by Darwin and continues to be pivotal in view of the role now known to be played by the genetic code in adaptational response. Darwin, however, already saw that natural selection, with its strong emphasis on forces in the ambient environment, was insufficient to describe those crucial aspects of selection that directly impinge on individuals as they arrive at sexual maturity and are faced with the final challenge: reproduction.

Participation in successful sexual reproduction represents a final rigorous test of overall Darwinian fitness within a genetically variable population. The demands made on the individual are greatly intensified at this late stage of the life cycle. Again it was Darwin who perceived that the selection that occurs during the mate-finding stage of the sexual life cycle was of such importance that it required special types of observational and theoretical treatment. This was developed by him in detail under the heading of sexual selection, culminating in his work *The Descent of Man* (1871). Nevertheless, sexual selection, as a fully recognized sister process of natural selection, has not been accorded the theoretical prominence in modern biology that it now appears to deserve.

25.4. Sexual Selection

25.4.1. General Attributes

Descriptive accounts of sexual selection in animals have been extensive, culminating especially in the reviews of Bateson (1983), Eberhard (1986), and Andersson (1994). Unfortunately, theoretical and empirical analysis of the population genetics underlying the process has not been equally thorough or extensive, despite the influential theoretical work of Fisher (1930) and O'Donald (1980).

Sexual selection is a process that starts operating at full sexual maturity and exists as an extension of natural selection. Sexual maturity imposes on the individual a new set of crucial, and often final, tests of relative reproductive capacity or what is generally called Darwinian fitness.

The particular genotypic combination carried by the individual finally gets its opportunity to make a direct genetic contribution to the next generation. The competition among male animals involves such attributes as variable secondary sexual characters that promote individual success in fighting with other males for access to females. Separate physiological or morphological characters of males are directed towards the choice mechanisms manifested by females (Ryan, 1990). These may be in the nature of advertisements that reveal, to female perception, the relative fitness of the individual male that is courting her (see Williams, 1966). Indeed, the females within a population show significant variation in their ability to discriminate among a field of genetically variable males (Kirkpatrick, 1982, 1987).

The geneticist is thus presented with complex suites of crucial characters unique to each sex. Continuous genetic variability appears to exist in these characters, compounding the difficulty of empirical genetic analysis (Lande, 1980). Despite this, most geneticists have relied on unrealistically simplified genetic theoretical models (see O'Donald, 1980); the experimentalist who uses empirical methods has tended not to enter the field.

25.4.2. The Reproductive Elite

Descriptive accounts and marking experiments, however, make it clear that sexual selection closely monitors the fitness of the individuals that attempt reproduction. The data emphasize that this system operates demographically within the smallest population unit, the local interbreeding population or deme. Since sexual selection results in differential reproduction of some individuals at the expense of others, the process creates a reproductive elite. Thus a small array of individuals that outbreed others is established within the population. Such an array or class, however, is destroyed in each ensuing generation and must be reformulated. This class is often very small compared with the census size of adult, healthy, courting individuals. Sexual selection thus exerts an additional major influence, supplementing that of natural selection, that functions to partially regulate the important population parameter N_e, the effective population size.

Local populations may harbor many individuals that are to some degree incompetent in sexual behavior. For example, approximately one third of the males in laboratory lines of *D. silvestris* fall into this category (Carson, 1986, 1987). Mating failure implies that their reproductive fitness is less, although empirical data are required for establishing this point. Nevertheless, some of these apparently vigorous and healthy individuals are entirely excluded from reproduction. As Darwin rightly saw, this situation leads to strong positive evolution of secondary sexual characters in a relevant population. Sexual selection thus superimposes on natural selection a new system that is instituted following the attainment of sexual maturity, in this sense becoming a powerful supplement to natural selection.

25.4.3. Isolation Based on Mate Choice

Accordingly, a type of sexual isolation occurs within the deme, mediated by sexual selection. This is imposed on the deme by mate choice. Indeed, by its very nature, sexual selection affects the deme in this divisive manner, separating the reproductively fitter from those that are less so. This phenomenon amounts to a degree of premating isolation of the unfit with regards to reproduction. The term sexual isolation, on the other hand, has been widely used to describe what is basically a different phenomenon, namely interdemic premating isolation.

25.4.4. Limitations on Studies of Sexual Isolation

Interdemic mating behavior related to isolation is virtually impossible to study in a state of nature. The behavioral isolation concept has arisen from theoretical consideration of what mating events might take place if the two sexes from separate demes were artificially assembled and tested for compatibility. Empirically, this is possible only with organisms that may be handled in the laboratory or garden plot.

A very large literature has developed in which laboratory material is used to study mating behavior in *Drosophila* (Ehrman and Parsons, 1976). Such studies normally make use of laboratory strains that are established from specimens collected, for example, in two geographically separate localities that might be called A and B. Mating tests are then made in various combinations. Conclusions are drawn concerning the degree of mating compatibility, or lack of it, between specimens originating from the two areas. In the hope of standardizing the results, most investigators make use of strains that have been established by the isofemale method (Parsons, 1977). An isofemale line, or laboratory stock, is originally derived from a single, wild female specimen collected in the designated locality. The method makes use of the fact that the female has already been inseminated in nature by one or more wild males.

Procedures for comparing the behavioral characteristics of localities A and B ideally call for making a substantial number of isofemale lines (perhaps 10) from A and an equal number from B. This permits the comparison of results obtained within a locality with those between localities. Basically what is done is to isolate virgin males and females from each isoline and then perform a series of control crosses within A lines and within B lines. Finally, various interlocality tests follow. The latter have included female-choice, male-choice, or multiple-choice designs.

This sort of procedure is sometimes faulted by making an unjustified typological assumption, namely that all isoline males and females from A, for example, have identical behavioral syndromes. This makes the assumption that there is no differential sexual selection within a single isoline. There is increasing evidence that this assumption is not justified. As mentioned above, the use of marked males in *D. silvestris* isolines shows that ~30% of the males from an isoline are not accepted by females from the same line (Carson, 1986). This means that when sexual selection is operating, any attempt to compare the sexual behavior from locality A with locality B may yield ambiguous results if multiple choice systems are used. Differences in sexual selection between strains, even between those from the same locality, may be based on a number of intrastrain coadaptations, involving chemical, auditory, and tactile signals, for example. Accordingly, I feel that most of the tests purporting to be showing positive sexual isolation between localities or between laboratory populations in *Drosophila* species are seriously flawed. Unfortunately, the extent and the strength of sexual selection within a strain are largely undocumented phenomena in most of the populations that have been studied. This results in bringing the observation of premating sexual isolation between strains present under considerable suspicion.

There are, of course, advantages and disadvantages to the isofemale method. The development of laboratory cultures through isolation of single wild females was originally developed for dealing with mixtures of morphologically identical sibling species occurring sympatrically. In this case, the isofemale procedure provides a population sample that is both workable and reliable for making a preliminary systematic identification. When the method is used to establish stocks representative of a locality, however, a drawback is that a single wild female will not be a fully informative sample of the genetic variability in the population. This is true, although such a single individual may carry as much as 65% of the total variability of the population from which it is drawn (Nei et al., 1975). Empirical studies, however, have shown that laboratory isofemale lines can carry extensive polymorphism in balanced state over many generations (Dobzhansky and Spassky, 1947; Carson, 1961, 1987).

25.4.5. The Genetics of Sexual Selection

The characters involved in final mate selection appear to consist of a complex and integrated set of genetically influenced behavioral and morphological characters. The basis for these crucial differences between individuals appears to be multigenic, including systems that may involve a major locus and an accompanying set of modifying genes, making possible a cascade of regulatory gene effects. In natural populations, this variability appears, like most genetic variability relevant to fitness, to be held in a balanced state of polymorphism. Out of these polymorphisms, recombination results in a high probability that each sexually produced individual is genetically unique.

This concept, which considers the characters involved in sexual selection in the context of the genetics of quantitative characters, was developed by Lande (1980) and Lande and Arnold (1983). This work has done much to show that the classical models of runaway selection and fixation and a perceived antagonism between sexual and natural selection are only a very small part of the story.

The major result of such a case appears to be that which is suggested by the behavior of most other genetic markers, namely, a balanced polymorphism wherein sexually competing fields of males and females reappear each generation in genetically recombined arrays. As deme succeeds deme in time, there is likely to be retained, or indeed built up, a relatively stable equilibrium of those genes that are exposed to selection. More simplistic views of the collective genotype of a deme are best avoided in view of the extensive variability revealed by modern work on quantitative genetics and nuclear DNA variability.

Less important in this formulation are those characters of the species that merely aid in the accumulation of males or females, or both, in a general mating arena. These characters do not appear to be directly involved in the final choice of an individual mate. Accordingly, many species-specific recognition signals (see Paterson, 1985) have this general quality and are not crucial monitors of the ultimate fitness of the breeding individual.

A well-developed sexual selection system in a deme appears to contribute to the persistence and integrity of a local population and may be a major factor in the resistance to gene flow manifested by many demes in a state of nature (see Ehrlich and Raven, 1969). As will be argued later in this discussion, demographic insults to the deme (e.g., hybridization or reduction in population size, or indeed, the establishment and experimental use of an isofemale line) may perturb the system of sexual selection within the deme. This may result in a reorganization of the genetic basis of the sexual selection system, driving the deme away from the equilibrium state in the manner comparable with a shift in adaptive peak.

25.4.6. Perturbation of Wild Stocks by Inserted Marker Genes

This same perturbation difficulty also arises from the common practice of engineering genetic combinations of genes for use in laboratory behavioral tests. Thus the use of single-gene markers such as *white* or *yellow* in *D. melanogaster* cannot be introduced into a wild strain by crossing without perturbing the natural sexual selection inherent in the strain (Bösiger, 1974). Not only might such a gene have pleiotropic behavioral effects on the normal functioning of sexual selection in the strain, but also the succession of laboratory crosses needed to insert such a gene marker into a test stock would be expected to break up the polygenic sexual selection system characteristic of the line.

25.5. Age of Populations and the Strength of Sexual Selection

25.5.1. Sexual Behavior and the Hawaiian *Drosophila* Affair

Some populations, either at the species level or within the species, clearly have been in existence for long periods of time whereas others must be derived more recently. The biota of the geologically dynamic Hawaiian islands, however, provides a unique opportunity to age existing populations from geological data on the ages of volcanoes and lava flows on which these populations occur. From Necker in the northwest to Hawaii Island in the southeast, each Hawaiian island is successively younger (McDougall, 1969). Hawaii Island, with its five volcanoes, is both very large and very young, having reached its present size by continuous and, indeed continuing, volcanic action over the past 0.43 million years (Moore and Clague, 1992). The phases of its growth have been monitored and dated by K–Ar conversion, magnetic declination shifts, reconstruction of ancient shorelines, and carbon dating of the more recent lava flows. The four major islands immediately to the northwest of Hawaii Island also represent an age succession; the oldest high island in the chain is Kauai, at 5.1 million years.

These crucial geological data have only recently become available to the evolutionary biologist as an aid in interpreting the history of the diverse endemic biota of the islands (see reviews in Wagner and Funk, 1995). *Drosophila* is well represented in the fauna of the islands. There are hundreds of phylogenetically closely related species, most of which are endemic to single islands. Genetic markers confirm a geographical succession of increasingly younger endemic populations, leading from older faunas in the northern islands through intermediates to newer ones in the south (see Carson, 1990, for a recent review). The data support the idea of a pattern of colonization from north to south that occurs in succession as each island has emerged above the ocean surface, slightly southeast of the previous one (see Carson and Clague, 1995). Most of these successive populations on the newer islands appear to be at the specific level of differentiation. In some cases, species endemic to single islands show intraspecific differentiation from older to younger lava flows.

The extraordinary sexual dimorphism and behavior of the Hawaiian *Drosophila* were first described by Spieth (1966). The theory that sexual selection was a prominent factor in their evolution was later proposed by Ringo (1977) and Carson (1978) but little empirical detail were available at the time. The first behavioral studies were started during a visit by Dobzhansky to the islands (see Ahearn et al., 1974). Kaneshiro (1976) and Ohta (1978) carried out the first crucial quantitative behavioral experiments on groups of related species and species populations of Hawaiian *Drosophila*. When these authors made mating tests, comparing populations from older islands with very similar populations endemic with newer islands, a striking correlation was discovered. Females from the older populations (or species) universally show very strong mating discrimination. They not only reject certain males from their own populations, as expected in sexual selection by means of female choice, but they strongly reject males from younger populations. An added and very important fact is that females from the younger populations show considerably less discrimination, both against their own males and against males from the older populations (see Kaneshiro and Boake, 1987). The data from these experiments indicate that each newly derived population appears to have lost some degree of female discrimination relative to the older population. From these observations, repeated in several species groups, came the hypothesis that sexual selection, based primarily on female choice, increases in intensity under sexual selection over time as a local population ages. This intensity is markedly lessened in daughter populations derived from them. The differentiation of these broad lineages of species therefore appears to result from periodic episodes that are marked by continuing evolution of a sexual selection that is strongly based on female choice.

The study of behavioral change with age of population has been carried several steps further in the Hawaiian *Drosophila*. These data are of special interest since they involve populations of a single species (*D. silvestris*). Collections were made from recent, geologically dated lava flows on the new island of Hawaii (Kaneshiro and Kurihara, 1981). In this instance, experiments were carried out with isofemale lines and, as in the studies involving different species, the populations from the older lava flows on the island generally have the more discriminatory females. Successively younger populations have females that show an increase in promiscuous female mating behavior. The *silvestris* behavioral series has been paralleled by a study of morphological changes in a novel secondary sexual leg character of males, which appears to play a role in the sexual selection process by female choice (Carson, 1982).

25.5.2. Kaneshiro and the Direction of Evolution

From the data on these mating asymmetries, Kaneshiro (1983) proposed the theory that the direction of evolution from older to newer populations can be read from the behavioral changes. Although most of the colonizations of

SEXUAL SELECTION IN POPULATIONS 503

D. silvestris are from older to newer lava flows, one crucial event in the series involves the colonization of Kohala or Mauna Kea volcano from Hualalai, which is a geologically newer mountain (see Carson, 1997). This finding injects a note of caution, although in this case the behavioral data are supported by data from counts of the tibial cilia (Carson, 1982) and from mitochondrial DNA (DeSalle, 1995).

Further evidence of shifts in the intensity of female choice comes from populations of several species that have been established in the laboratory for some years. These newer experimental populations show female behavior that is significantly less discriminatory than wild females from the known ancestral forms in nature or from the original isofemale lines established earlier in the laboratory (Arita and Kaneshiro, 1979, for *D. adiastola* from Maui; Ahearn, 1980, for *D. silvestris*).

Buildup of high female discrimination (female choice: Kirkpatrick, 1982) in *Drosophila* mating systems under sexual selection and its decline with the stepwise founding of new populations is thus well documented by direct observational and empirical data with populations of known age. The facts support the view of Kaneshiro (1980) that these behavioral relationships may be used as a diagnostic tool to distinguish older populations from younger ones. In particular, such behavior may be used in situations and species for which independent geological evidence on age of populations is not available.

25.5.3. Comparable Studies of Behavior in Non-Hawaiian Species

Using three species of the *melanogaster* subgroup, Watanabe and Kawanishi (1979) performed a small series of behavior experiments, the design of which provides data comparable with the interspecific studies of Kaneshiro (1976) and Ohta (1978). Strains were used of each of the species *mauritiana*, *simulans*, and *melanogaster*. The latter two have worldwide distributions, whereas the *mauritiana* is endemic to the island of Mauritius in the Indian Ocean. Asymmetrical mating relationships were demonstrated in experiments involving the three experimental pairings of the sexes of these species. The asymmetries are very similar to those found in Hawaiian *Drosophila*. The greatest amount of female discrimination was found in *D. mauritiana*, an intermediate amount in *simulans*, and the least amount in *melanogaster*. The relationships may be represented thus:

mauritiana → *simulans* → *melanogaster.*

Molecular, morphological, chromosomal, and geographical evidence fail to provide a decisive indication of the direction of evolution for these species and there are no data comparable with those from Hawaii, for which unique biogeographical evidence is provided by a geologically dated set of substrates harboring populations of successively younger populations. These, in turn, can provide a novel clue to the understanding of the observed differences in mating behavior, as argued above. Watanabe and Kawanishi, however, chose not to invoke the Kaneshiro theory but rather speculated that *D. melanogaster* represents an ancestral form, whereas the exact opposite series, given in the arrow diagram above, would be expected from the Hawaii data. Kaneshiro (1980) and Giddings and Templeton (1983) have criticized the Watanabe and Kawanishi view. The latter has had some argumentative support (e.g., Wasserman and Koepfer, 1980; Markow, 1981; Charlesworth et al., 1982).

Kaneshiro (1980) has also pointed out that the remarkable asymmetries in the data obtained by Dobzhansky and Streisinger (1944) on *D. prosaltans* can be geographically understood under his hypothesis. This has permitted a new and provocative theory on the origin and spread of this species. Another case involves crosses between *D. pseudoobscura* from Bogota, Colombia, and strains from the mainland of North America (Singh, 1983). Females from the mainland discriminate against Bogota males more than vice versa. This fits the Kaneshiro hypothesis in accordance with the widely held view that Bogota is a recently isolated population. Unfortunately, few other new empirical data on intraspecific populations, with careful intrademic controls for sexual selection, have appeared recently in the literature.

25.6. Intraspecific Populations of *D. melanogaster*

25.6.1. The Case of the Zimbabwe Strains

D. melanogaster is a worldwide species of unknown origin, although Africa has been suspected (Lachaise et al., 1988). New data have been presented recording novel mating behavior of certain strains from Zimbabwe (Wu et al., 1995). The original collection (50 isofemale lines) was made in 1990 from Sengwa Wildlife Preserve (Begun and Aquadro, 1993). This refers to the Sengwa Wildlife Research Area (SWRA, located at 18 10 S, 28 13 E) in the Chirisa Safari Area centered at the junction of the Sengwa and the Lutope Rivers (Peter Frost, personal communication).

Using homozygous X-chromosome lines prepared by a balancer technique, restriction site analysis by Begun and Aquadro (1993, 1995), indicated that some of the Zimbabwe material and one strain from Kenya have exceptionally high genetic variability, approximately twice as much as samples of this species from North and South America and China. The data analysis, however, does not indicate how many of the 50 isolines from Zimbabwe showed this unusual property.

I concentrate here on a reexamination of the behavioral data of Wu et al., as I consider the case instructive on the application of sexual behavior in evolutionary studies. The authors used nine of the original isofemale lines from Zimbabwe, concentrating on three of them. Mating behavior tests were made with various strains of what the authors term common *melanogaster* from Africa (Botswana, Malawi, Ivory Coast, and Northern Zambia) as well as strains from France, California, Australia, and Canada. Preliminary no-choice experiments (five virgin females and males of a designated origin placed together in a clean vial) were carried out. Such a design, with intrastrain controls, is reasonably favorable for detecting sexual selection, although the results from separate isolines have been pooled in the reporting of results. The most striking result is the strong disinclination of females from seven of the Zimbabwe isolines to mate with common *melanogaster* males from other localities. These strains, stated to be mostly isofemale lines collected in the 1980s, are far from ideal material against which to test the new strains from Zimbabwe. For example, two of the stocks used, from Malawi and Ivory Coast, are stated to be isochromosome 2, having been crossed and backcrossed into Curly balancer stocks; these would be unlikely to retain the original behavioral syndrome of the respective geographical sites of origin.

The no-choice experiments of Wu et al. were unfortunately discontinued in favor of a multiple-choice design in which 50–60 virgin females and males from each of two compared lines were marked and released into a population cage and watched for one hour. Copulating pairs were aspirated from the cage, the strain of origin of each fly was determined, and the aspirated flies returned to the cage. Results of these tests have generally confirmed the unique behavioral attributes of some of the Zimbabwe strains, especially those labeled Z53 and Z30.

The authors analyze the data from the point of view of sexual isolation rather than sexual selection. For example, they have presented a single joint measure of sexual isolation for each experiment, given as a joint discrimination index (DI) between the two strains tested. Since the behavior of the two kinds of females in each experiment is often significantly different, the DI has limited interpretive value. In calculating the DI, data that are not homogenous have thus been pooled. In order to apply the female-choice/sexual selection hypothesis, I have regrouped the data from the 16 experiments from Table 2 of Wu et al. that involved Zimbabwe strain females. In order to emphasize sexual selection, I have calculated a separate C value for these results, as was done for comparable experiments by Kaneshiro (1976) and Ohta (1978). This statistic tests for any deviation from random mating by each category of female in each experiment (Table 25.1).

Z53 strain has females that show exclusively homogamic pairings in experiments with males of five strains of common *melanogaster* (experiments 3-7). Discrimination is slightly less against males from Z29 (experiment 10A). Furthermore, some males from Zambia (north and east of adjacent Zimbabwe) are accepted by Z53 females (experiment 14). Although not included in Table 25.1, Zambia females also show strong discrimination against common *melanogaster* males (Wu et al., 1995, experiments 12 and 13 of Table 2). On the other hand, there is no evidence of discrimination by Z53 females against Z30 or Z56 males (experiments 15A and 16), giving rise to a hypothesis that they are alike; indeed their performance resembles that of Zambia. In two tests (experiments 1 and 2) Z30 females also discriminate strongly against common *melanogaster*. Z30 females show minor acceptance of Z29 males (experiment 11) but, in accordance with earlier experiments, the Z29 females strongly accept, in fact, significantly prefer, Z53 males (experiment 15B). Finally, experiments 8, 9, and

Table 25.1. *Discriminatory Behavior of Females from Four Zimbabwe (Z) Strains (Data and Experimental Results from Table 2 of Wu et al., 1995)*

Experiment Number	Homogamic Matings (No.)	Heterogamic Matings (No.)	C*
Z53 females with common *melanogaster* males			
3	27	0	5.20[†]
4	21	0	4.58[†]
5	41	0	6.40[†]
6	49	0	7.00[†]
7	46	0	6.78[†]
Z53 females with Z29 males (10A) and with Zambia males (14)			
10A	25	8	2.96[†]
14	27	9	3.00[†]
Z53 females with Z30 males (15A) and Z56 males (16)			
15A	26	26	0.0
16	24	23	0.15
Z30 females with common *melanogaster* males			
1	38	0	6.16[†]
2	26	2	4.54[†]
Z30 females with Z29 males (11) and Z53 males (15B)			
11	22	4	3.53[†]
15B	9	25	−2.70[†]
Z29 females with common *melanogaster* males: California (8); Ivory Coast (9)			
8	11	7	0.94
9	38	12	3.68[†]
Z29 females with Z53 males			
10B	24	22	0.33

* $C = 2\sqrt{n}(p - 0.5)$: p is the observed frequency of homogamic matings and n is the total number of matings by a given type of female. At the 5% confidence interval, the null hypothesis that mating is random is accepted when $-1.96 < C < +1.96$ (Kaneshiro and Kurihara, 1978).
[†] Significant deviation from random mating at or below the 5% confidence level.

10B document the semiconformance of the Z29 strain to the pattern of Z53 and Z30. These three Zimbabwe isolines, along with the single experiment with Z56 and the Zambia stock (LA69 of Wu et al.) thus may represent an unique Zimbabwe entity; all show a high level of female choice relative to the males of a number of other *melanogaster* strains. The Kenya stock studied by Begun and Aquadro (1993) may belong in this grouping.

Despite the uncertainty about the geographical integrity of some of the stocks used in these tests, the implications of some of the experiments are nevertheless geographically interesting and suggestive. Both Z30 and Z53 females, for example, discriminate strongly against males from France (experiments 1 and 3), suggesting that France may represent a newer, derived population. Cohet and David (1980) reported a very similar result, pairing a strain from the Congo with one from France. Although they ascribed the result to a weak performance by French males, an alternative interpretation is that the African strain is older and has females that show high discrimination.

Although female choice may play a leading role in sexual selection in *Drosophila*, other courtship elements are surely also important in the outcome of both intrastrain and interstrain behavior in the laboratory. Examples are mating speed and fighting between males. Female pheromones influence wing display in males and show a worldwide polymorphism (Ferveur et al., 1995). In Hawaiian *Drosophila*, male struggles, lek behavior, and acoustical displays are integral parts of courtship. Experiments involving different strains should be designed with these complications in mind; indeed, the complexity of the sexual environment within the deme appears to have been greatly underestimated both for *melanogaster* and other species.

On the other hand, multiple-choice experiments, such as those of Wu et al., in which 220–240 specimens of each of two strains are unnaturally mixed in a population cage, are vulnerable to the problem of mixed signals. Because the genetic basis of sexual selection appears to be a genetically complex multigenic intrademe process, mixing hybrid specimens between two behaviorally different strains may result in confusion, especially within a confined space.

In view of these complicating possibilities, it is remarkable that female discrimination as a mating syndrome is strongly manifested even in the face of unfavorable or mixed conditions in the mating arena. Indeed, this robustness may be a rough measure of its importance in the mating systems as they exist in natural demes.

25.6.2. Conclusion on this *melanogaster* Case

Wu et al. state: "These observations suggest that we are seeing the early stages of speciation in this group and that it is driven by sexual selection." I surely agree that sexual selection may serve as a driver that results in novel evolutionary divergence between populations; in fact, that is the theme of this chapter. Under a system of sexual selection in Hawaii, however, *Drosophila* data show that females from phylogenetically older populations are strongly discriminatory and that the opposite is true of females from younger populations. Accordingly, I suggest that, rather than an early stage of speciation *in statu nascendi*, Wu et al. may have discovered a previously unrecognized older endemic population or sibling species of *D. melanogaster* from East Africa.

25.7. Genetic Definition of Species

The complexity of sexual selection strongly suggests that reproductive isolation between groups may result as an incidental outcome of natural and sexual selection within the interbreeding group. Such a view calls for revision of some prominent concepts of species. For example, I suggest that genetic isolation as a necessary marker of the completion of speciation should be abandoned. Genetic isolation exists between many full species but the evidence that this arises *ad hoc* from selection is, in my view, not well supported by data. On this point, I agree with Paterson (1978) who, expanding on an oral presentation given in 1976, was the first to boldly make this proposal.

The extraordinary morphological and physiological diversity of the Hawaiian drosophilids has added important perspectives in evolutionary biology in general. Genetic differentiation in many different lineages of these flies strongly favors morphological and behavioral differentiation of males through sexual selection. This has occurred allopatrically and, because of founder effects, without any intergroup selective reinforcement, as the flies have allopatrically colonized the newer volcanoes and lava flows arising in succession towards the southeast. Nevertheless, as shown by the laboratory data of Yang and Wheeler (1969), hybridizations between widely different species are possible. In the absence of choice, sexual selection does not in itself automatically confer strong sexual isolation. A further point can be made with regard to the species pair *D. silvestris* and *D. heteroneura*. These show geologically very recent morphological differentiation but nevertheless are still compatible biologically, hybridizing both in nature and in the laboratory and showing little postmating incompatibility (Ahearn and Templeton, 1989; Carson et al., 1994).

The hypothesis is proposed that the genetic parameters of sexual selection may eventually become widened over time so as to eventually produce, as side effects, the reproductive incompatibilities that are often found between many older related pairs of species. Accordingly, I call for an abandonment of the view that ultimate reproductive isolation is ascribable to the primary evolutionary thrust of *ad hoc* selection, either natural or sexual.

Darwinian fitness, engineered by selection in a slightly different form in each local population, is the major process that imposes evolutionary change on genetically variable sexual popualtions. Unless destroyed by loss of environment or by demographic insults, the local population may continue to be driven by processes that maximize the Darwinian fitness of the individual. This is the biological law that dominates local populations in both survival and reproductive aspects.

25.8. Summary

Recent premating behavioral research has confirmed Darwin's view that sexual selection parallels natural selection in importance for the understanding of the evolutionary process as it occurs in sexual populations. As defined by Darwin, sexual selection "... depends on the success of certain individuals over others of the same sex, in relation to the propagation of the species...." Data defining the genetic aspects of this theory indicate that sexual selection operates best in a small geographically localized population (deme); this is the site for the formation of the crucial high-fitness sexual pairings. Genetic variability within such demes is high; the members of the population appear to represent genetic recombinants. Each individual thus has an unique genetic construction that encodes the underlying quantitative characters relating to mate selection. Frequently, there appears to be a strong component of female choice that discriminates against less fit males. Male-to-male interactions also act to exclude less fit males from the mating arena but the process is rarely complete and generally precedes exposure to female choice. At the time of reproduction, the genetic uniqueness of each individual, whether female or male, is thus put to the final test of Darwinian fitness. Each deme thus has its own genetic elite, made up of those individuals that leave more offspring than others. Full appreciation of the fitness-maximizing role of sexual selection in local populations has been held back by theories and experimental designs that ascribe many of these same behavioral attributes to a wholly different function, namely, premating, intergroup sexual isolation. I propose a return to the original Darwinian view. Thus the genetically based reproductive isolation often observed between species is simply an incidental outcome stemming ultimately from the process of selection, natural or sexual or both, that occurs within the freely interbreeding group. Accordingly, concepts that define species strictly in terms of the genetics of reproductive isolating mechanisms need revision.

25.9. Acknowledgments

This paper is affectionately dedicated to Dick Lewontin, whose bold approach to science and life has been a source of inspiration for me. I thank my long-term colleague and friend Kenneth Y. Kaneshiro for many profitable discussions.

REFERENCES

Ahearn, J. N. 1980. Evolution of behavioral reproductive isolation in a laboratory stock of *Drosophila silvestris*. *Experentia* **36**:63–64.

Ahearn, J. N., Carson, H. L., Dobzhansky, Th., and Kaneshiro, K. Y. 1974. Ethological isolation among three species of the *planitibia* subgroup of Hawaiian *Drosophila*. *Proc. Natl. Acad. Sci. USA* **71**:901–903.

Ahearn, J. N. and Templeton, A.R. 1989. Interspecific hybrids of *Drosophila heteroneura* and *Drosophila silvestris*. I. Courtship success. *Evolution* **43**:347–361.

Andersson, M. 1994. *Sexual Selection*. Princeton, NJ: Princeton U. Press.

Arita, L. H. and Kaneshiro, K. Y. 1979. Ethological isolation between two stocks of *Drosophila adiastola* Hardy. *Proc. Hawaii Entomol. Soc.* **13**:31–34.

Bateson, P. 1983. *Mate Choice*. Cambridge: Cambridge U. Press.

Begun, D. J. and Aquadro, C. F. 1993. African and North American populations of *Drosophila melanogaster* are very different at the DNA level. *Nature (London)* **365**:548–550.

Begun, D. J. and Aquadro, C. F. 1995. Molecular variation at the *vermilion* locus in geographically diverse populations of *Drosophila melanogaster* and *D. simulans*. *Genetics* **140**:1019–1032.

Bösiger, E. 1974. The role of sexual selection in the maintenance of the genetical heterogeneity of *Drosophila* populations and its genetic basis. *Front. Biol.* **38**:176–184.

Carson, H. L. 1957. The species as a field for gene recombination. In *The Species Problem*. *Am. Assoc. Adv. Sci. Pub.* **50**:23–38.

Carson, H. L. 1961. Relative fitness of genetically open and closed experimental populations of *Drosophila melanogaster*. *Genetics* **46**:553–567.

Carson, H. L. 1978. Speciation and sexual selection in Hawaiian Drosophila. In *Ecological Genetics: The Interface*, P. F. Brussard, ed., pp. 93–107. New York: Springer-Verlag.

Carson, H. L. 1982. Evolution of Drosophila on the newer Hawaiian volcanoes. *Heredity* **48**:3–25.

Carson, H. L. 1986. Sexual selection and speciation. In *Evolutionary Processes and Theory*. S. Karlin and E. Nevo, eds., pp. 391–409. London: Academic.

Carson, H. L. 1987. High fitness of heterokaryotypic individuals segregating naturally within a long-standing laboratory population of *Drosophila silvestris*. *Genetics* **116**:415–422.

Carson, H. L. 1990. Evolutionary process as studied in population genetics: clues from phylogeny. *Oxford Surv. Evol. Biol.* **7**:129–156.

Carson, H. L. 1997. Sexual selection: A driver of genetic change in Hawaiian Drosophila. *Jour. Hered.* **88**:343–352.

Carson, H. L. and Clague, D. A. 1995. Geology and biogeography of the Hawaiian Islands. In *Hawaiian Biogeography*, W. L. Wagner and V. A. Funk, eds., pp. 14–29. Washington, D.C.: Smithsonian Institution.

Carson, H. L., Val, F. C., and Templeton, A. R. 1994. Change in male secondary sexual characters in artificial interspecific hybrid populations. *Proc. Natl. Acad. Sci. USA* **91**:6315–6318.

Charlesworth, B., Lande, R., and Slatkin, M. 1982. A neo-Darwinian commentary on macroevolution. *Evolution* **36**:474–498.

Cohet, Y. and David, J. R. 1980. Geographical divergence and sexual behavior: comparisons of mating systems in French and Afrotropical populations of *Drosophila melanogaster*. *Genetica* **54**:161–165.

Coyne, J. A. and Orr, H. A. 1989. Two rules of speciation. In *Speciation and its Consequences*, D. Otte and J. A. Endler, eds., pp.180–207. Sunderland, MA: Sinauer.

Darwin, C. 1871. *The Descent of Man and Selection in Relation to Sex*. London: Murray.

DeSalle, R. 1995. Molecular approaches to biogeographic analysis of Hawaiian Drosophilidae. In *Hawaiian Biogeography*, W. L. Wagner and V. A. Funk, eds., pp. 72–89. Washington, D.C.: Smithsonian Institution.

Dobzhansky, Th. 1937. *Genetics and the Origin of Species*. New York: Columbia U. Press.

Dobzhansky, Th. and Spassky, B. 1947. Evolutionary changes in laboratory cultures of *Drosophila pseudoobscura*. *Evolution* **1**:191–216.

Dobzhansky, Th. and Streisinger, G. 1944. Experiments on sexual isolation in Drosophila. II. Geographic strains of *Drosophila prosaltans*. *Proc. Natl. Acad. Sci. USA* **30**:340–345.

Eberhard, W. G. 1986. *Sexual Selection and Animal Genitalia*. Cambridge, MA: Harvard U. Press.

Ehrlich, P. R. and Raven, P. H. 1969. Differentiation of populations. *Science* **165**:1228–1232.

Ehrman, L. and Parsons, P. A. 1976. *The Genetics of Behavior*. Sunderland, MA: Sinauer.

Ferveur, J.-F., Cobb, M., Boukella, H., and Jallon, J.-M. 1996. World-wide variation in *Drosophila melanogaster* sex pheromone: behavioural effects, genetic bases and potential evolutionary consequences. *Genetica* **97**:73–80.

Fisher, R. A. 1930. *The Genetical Theory of Natural Selection*. Oxford: Clarendon.

Frost, Peter (personal communication).

Giddings, L. V. and Templeton, A. R. 1983. Behavioral phylogenies and the direction of evolution. *Science* **220**:372–377.

Kaneshiro, K. Y. 1976. Ethological isolation and phylogeny in the *planitibia* subgroup of Hawaiian *Drosophila*. *Evolution* **30**:740–745.

Kaneshiro, K. Y. 1980. Sexual isolation, speciation and the direction of evolution. *Evolution* **34**:437–444.

Kaneshiro, K. Y. 1983. Sexual selection and direction of evolution in the biosystematics of Hawaiian Drosophilidae. *Annu. Rev. Entomol.* **28**:161–178.

Kaneshiro, K. Y. and Boake, C. R. B. 1987. Sexual selection and speciation: issues raised by Hawaiian *Drosophila*. *Trends Ecol. Evol.* **2**:207–212.

Kaneshiro, K. Y. and Kurihara, J. S. 1981. Sequential differentiation of sexual isolation in populations of *Drosophila silvestris*. *Pac. Sci.* **35**:177–183.

Kirkpatrick, M. 1982. Sexual selection and the evolution of female choice. *Evolution* **36**:1–12.

Kirkpatrick, M. 1987. Sexual selection by female choice in polygynous animals. *Annu. Rev. Ecol. Syst.* **18**:43–70.

Lachaise, D. L., Cariou, M.-L., David, J.R., Lemeunier, F., Tsacas, L., and Ashburner, M. 1988. Historical biogeography of the *Drosophila melanogaster* species subgroup. *Evol. Biol.* **22**:159–225.

Lande, R. 1980. Sexual dimorphism, sexual selection, and adaptation in polygenic characters. *Evolution* **34**:292–305.

Lande, R. and Arnold, S. J. 1983. The measurement of selection on correlated characters. *Evolution* **37**:1210–1226.

Markow, T. M. 1981. Mating preference is not predictive of the direction of evolution in experimental populations of *Drosophila*. *Science* **213**:1405–1407.

Mayr, E. 1942. *Systematics and the Origin of Species*. New York: Columbia U. Press.

McDougall, I. 1969. Potassium–argon ages from lavas of the Hawaiian Islands. *Bull. Geol. Soc. Am.* **80**:2597–2600.

Moore, J. G. and Clague, D. A. 1992. Volcano growth and evolution of the island of Hawaii. *Bull. Geol. Soc. Am.* **104**:1471–1484.

Nei, M., Maruyama, T., and Chakraborty, R. 1975. The bottleneck effect and genetic variability in populations. *Evolution* **29**:1–10.

O'Donald, P. 1980. *Genetic Models of Sexual Selection*. Cambridge: Cambridge U. Press.

Ohta, A. T. 1978. Ethological isolation and phylogeny in the *grimshawi* species complex of Hawaiian *Drosophila*. *Evolution* **32**:485–492.

Parsons, P. A. 1977. Isofemale strains and quantitative traits in natural populations of *Drosophila*. *Am. Nat.* **111**:613–621.

Paterson, H. E. H. 1978. More evidence against speciation by reinforcement. *S. Afr. J. Sci.* **74**:369–371.

Paterson, H. E. H. 1985. The recognition concept of species. In *Species and Speciation*, E. S. Virba, ed., pp. 21–34. Pretoria: Transvaal Museum Monograph No 4.

Ringo, J. M. 1977. Why 300 species of Hawaiian *Drosophila*? The sexual selection hypothesis. *Evolution* **31**:695–696.

Ryan, M. J. 1990. Sexual selection, sensory systems and sensory exploitation. *Oxford Surv. Evol. Biol.* **7**:157–195.

Singh, R.S. 1983. Genetic differentiation for allozymes and fitness characters between mainland and Bogota populations of *Drosophila pseudoobscura*. *Can. J. Genet. Cytol.* **25**:590–604.

Spieth, H. T. 1966. Courtship behavior of Hawaiian Drosophilidae. *Univ. Texas Publ.* **6615**:245–313.

512 Hampton L. Carson

Wagner, W. L. and Funk, V. A., eds. 1995. *Hawaiian Biogeography.* Washington, D.C.: Smithsonian Institution.

Wasserman, M. and Koepfer, H. R. 1980. Does asymmetrical mating preference show the direction of evolution? *Evolution* **34**:1116–1124.

Watanabe, T. K. and Kawanishi, M. 1979. Mating preference and the direction of evolution in *Drosophila. Science* **205**:906–907.

Williams, G. C. 1966. *Adaptation and Natural Selection.* Princeton, NJ: Princeton U. Press.

Wu, C.-I., Hollocher, H., Begun, D. J., Aquadro, C. F., Xu, Y., and Wu, M.-L. 1995. Sexual isolation in *Drosophila melanogaster*, a possible case of incipient speciation. *Proc. Natl. Acad. Sci. USA* **92**:2519–2523.

Yang, H. and Wheeler, M. R. 1969. Studies on interspecific hybridization within the picture-winged group of endemic Hawaiian *Drosophila. Univ. Texas Publ.* 6919:133–170.

The Role of Selection in Speciation

JUDITH MASTERS

"It is part of the dialectic of science that the apparent solution of a problem usually reveals that we have not asked the right question in the first place, or that a much more difficult and intractable problem lies just below the surface that has been so triumphantly cleared away. And in the process of redefinition of the issues, the old parties remain, sometimes under new rubrics, but always with old points of view."   (Lewontin, 1974, p. 29)

26.1. Introduction: The Problem of Linking Process and Pattern

The title of Darwin's great work, *On the Origin of Species by means of Natural Selection, or the Preservation of Favoured Races in the Struggle for Life*, describes unambiguously the problem to which he claimed to have found a solution: He believed the pattern of biodiversity to be explicable by a simple extrapolation of his process of natural selection. Considerable quantities of time and paper have been expended since November, 1859, debating the success of Darwin's project. By the close of the nineteenth century, three distinct opinions regarding the role of selection[1] in speciation had emerged. One hundred years later, these three approaches still represent the major contenders in the controversy, and can be summarized as follows:

1. Species are products of direct selection for discontinuity (Wallace).
2. Speciation occurs as a byproduct of adaptation to local environmental conditions (Darwin).
3. Adaptation by natural selection can account for only part of the process of speciation (Romanes).

This contribution offers a summary of the manner in which these ideas were expressed by Victorian scientists, followed by a critical evaluation of their application in contemporary speciation studies. The distinctive modes of operation

R. S. Singh and C. B. Krimbas, eds., *Evolutionary Genetics: From Molecules to Morphology*, vol. 1. © Cambridge University Press 2000. Printed in the United States of America. ISBN 0-521-57123-5. All rights reserved.

of natural and sexual selection are also discussed. Although this chapter is more concerned with process than with the pattern of species that are recognized, the two aspects of the problem are so intimately connected that it is impossible to discuss one without the other. The three models for the action of selection in speciation are therefore discussed with respect to theories regarding the kinds of species purported to be generated by these processes.

The fact that the above three routes of explanation have been in existence, and fiercely contested, for such a long time deserves further comment. At least part of the reason for this century-old stalemate must be attributable to the difficulty of subjecting evolutionary theories to rigorous testing, and choosing between alternative explanations. As Stamos (1996) has pointed out, species concepts are not falsifiable hypotheses; they are not refuted by the discovery of counterexamples and therefore must be classified as metaphysical statements. But metaphysical statements can generate falsifiable predictions, and therein lies their usefulness to science. In my assessment of the role of selection in currently popular theories of species and speciation, I shall pay particular attention to the testable predictions that are generated.

26.2. Victorian Arguments Concerning Selection and Speciation

Darwin's attitude toward speciation derives from his epistemological view of species, which is exemplified by his use of the analogy of artificial selection. The logic underlying this analogy runs as follows: If we compare the range of morphologies that have been induced by breeders within species of domesticated plants and animals – e.g., between a cabbage rose and a wild rose, or between a toy terrier and a wild dog – we find differences of a degree that, had they appeared in nature, would have required the organisms' presenting them to be allocated to distinct species or even distinct genera. Artificial selection is therefore capable of generating what are virtually new species. The struggle for existence in nature is both more severe and more ubiquitous than is artificial selection; hence natural selection must be viewed as more than adequate to the task of generating the characters indicative of new specific taxa.

The artificial selection analogy, however, has substantial drawbacks for demonstrating a role for selection in speciation. First, despite the extensive powers envisioned for natural selection, it is surely no accident that the level of variation seen in domesticated taxa is not mirrored in natural populations, in which the struggle for existence appears to have played a far more conservative role. Second, and more crucially, while centuries of selective breeding by agriculturalists have indeed brought about dramatic changes in whole suites of organismal characters, these alterations are ephemeral. Left to their own devices, domesticated breeds would revert rapidly to their original stocks. Inadvertently, what the artificial selection analogy highlights is the inadequacy of a purely epistemological species concept: For most evolutionary biologists, the term species implies more than the fact that they are sufficiently distinct to be recognized as such (but see the phylogenetic species concept, below).

Soon after entering the public domain, Darwin's view of the origin of species encountered criticism, which gave rise to two alternative responses. The first stemmed from the difficulty many workers experienced in postulating a selective origin for species-defining characters. One of the most articulate defenders of this position was Romanes (1897), a committed Darwinian who saw natural selection as "the keynote of organic nature, ... one of the principal chords of the universe." Nevertheless, he felt that this mechanism could accommodate neither the large number of specific and higher taxonomic characters that apparently had no adaptive value, nor the fact that hybrid sterility could not have arisen through natural selection. In Romanes' view, factors in addition to natural selection were required for explaining speciation.

The above critique caused concern to Darwin's closest colleagues, notably Wallace and Huxley (Kottler, 1985). In their respective opinions, if natural selection could not be demonstrated to be involved directly in the generation of species-defining criteria, it could not rightly be held as the major force behind the origin of species. Wallace (1889, p. 174) was outspokenly defensive of the view that selection should be capable of accumulating and increasing "infertility variations of incipient species." Darwin vacillated on this issue through the various editions of *On the Origin of Species*, and he and Wallace debated it vehemently by correspondence in early 1868 (Darwin and Seward, 1903; Kottler, 1985). It was an argument that caused Darwin no small measure of distress. But ultimately, in the sixth edition of *On the Origin of Species* (1872, p. 261), he expressed the opinion to which he adhered for the remainder of his life: "... the sterility of species when first crossed, and that of their hybrid offspring, cannot have been acquired ... by the preservation of successive profitable degrees of sterility. It is an incidental result of differences in the reproductive systems of the parent-species." He terminated his correspondence with Wallace on this topic with the following statement: "Natural selection cannot effect what is not good for the individual. . . . Lessened fertility is equivalent to a new source of destruction" (Darwin and Seward, 1903).

Despite Darwin's resistance, Wallace took his idea further: "We may fairly suppose, also, that as soon as any sterility appears some disinclination to *cross unions* will appear, and this will further tend to the diminution of the production of hybrids" (Wallace, 1889, italics original). This hypothesis was adopted as the rationale for Dobzhansky's model of speciation by reinforcement, discussed below.

26.3. Contemporary Views on Selection and Speciation

26.3.1. Species are Products of Direct Selection for Discontinuity

Only one species concept has been identified consistently with such a process, i.e., Dobzhansky's version of the biological species concept (BSC). The BSC has been defined by Mayr (1963, p. 663) as "[a] concept of the category species based on the reproductive isolation of the constituent populations from other species." Reproductive isolation is ensured by the development of isolating mechanisms, which are "perhaps the most important set of attributes a species has, because they are, by definition, the species criteria" (p. 89). Similar sentiments are to be found in Dobzhansky (1951).

It follows that an investigation of the role of selection in the origin of biological species is equivalent to a study of the role of selection in the development of reproductive isolating mechanisms. Here, Dobzhansky and Mayr differed considerably. Dobzhansky (1951, pp. 207–208) was at pains to clarify an active role for natural selection in the generation of the species-defining criteria:

". . . not all genetic differences produce even partial isolation, and there is no reason to believe that isolation results automatically when a certain number of genetic differences have accumulated. . . . Isolating mechanisms encountered in nature appear to be *ad hoc* contrivances which prevent the exchange of genes between nascent species, rather than incongruities originating in accidental changes in the gene functions."

Dobzhansky's BSC – particularly in its earlier incarnations (1940, 1951) – incorporated several of Wallace's ideas concerning speciation. He implicated selection directly in the development of both premating isolation (differences in courtship signals or behaviors; divergences in the timing of mating, or the ecological conditions necessary for it; mechanical differences in copulatory organs) and postmating isolation (gamete and zygote mortality; hybrid inviability and sterility). However, in later writings (e.g., Dobzhansky, 1972) only premating mechanisms received the distinction of a selective origin [Fisher (1930) took a similar view]. According to Dobzhansky's model of speciation by reinforcement, when two populations come into secondary recontact after a period of geographic isolation, postmating isolation provides the selective pressure to drive the evolution of premating isolating mechanisms by reinforcing any propensities to assortative mating already present in the populations.

Reinforcement is one of the most fiercely contested areas of speciation theory. A practical objection that has been raised against it is the fact that no support for the model can be derived from the several attempts to replicate the appropriate circumstances in the laboratory, even when selection was maintained for up to 20 generations (Rice and Hostert, 1993). Other objections have a theoretical basis.

In order for reinforcement to occur, a significant degree of hybridization must take place between incipient species. Paterson (1978, 1982, 1993b) has argued that, given such circumstances in nature, reinforcement is not the most likely outcome. Rather, the cause of the hybrid disadvantage stands to be eliminated by selection. Where selection against hybrids is absolute, this means the extinction of the smaller population; under circumstances for which selection against hybrids is less severe, the populations will fuse, with elimination only of the genes or chromosomal rearrangements responsible for the disadvantage.

Mathematical investigations of the conditions necessary for speciation by reinforcement by Lambert et al. (1984), Spencer et al. (1986), and Liou and Price

(1994), and a laboratory study by Harper and Lambert (1983) have tended to support Paterson's predictions. Spencer et al. investigated a variety of conditions influencing levels of gene flow and concluded that reinforcement could occur only when hybrid fitness was zero, or close to it, for an extended period. This situation would be difficult to maintain in the face of strong selection to increase hybrid fitness, and, in their model, higher fitness values for hybrids made extinction a more likely outcome than reinforcement. Furthermore, the selective pressure for the evolution of characters to facilitate assortative mating would decrease in intensity once these traits became established, indicating that forces other than selection would be required to complete the process. The model derived by Liou and Price (1994) emulated that of Spencer et al. in many features, but enhanced the probability of a reinforcement outcome by making hybrids inviable instead of sterile (hence eliminating them from the pool of prospective mates) and by including sexual selection with its consequent genetic linkage of male traits and female preferences. Nevertheless, they still found reinforcement to be unfeasible unless substantial divergence with regard to "both their premating and postzygotic isolating mechanisms" had taken place in the populations before secondary recontact. Thus both models predict that reinforcement is likely to occur only under a restricted set of conditions that are unlikely to be met with frequently in nature; and when these conditions are met, the role of reinforcement is necessarily secondary.

Kelly and Noor (1996) recently presented a model that focuses attention on selection for increased female mating discrimination rather than altered female preference, as was the case in previous models. Their results indicated that reinforcement would be rendered more likely under such a selection regime. However, like its predecessors, this model also requires that significant degrees of both premating and postmating divergence occur before recontact in order for it to be workable. Supporters of the view that crucial features of the fertilization system are subjected to stabilizing selection (see below) would argue that the allopatric divergence of such traits would be constrained.

There is an eminently testable prediction deriving from the model of speciation by reinforcement: Where closely related taxa occur in sympatry, they should show higher levels of prezygotic isolation than do allopatric populations of the same taxa (Dobzhansky et al., 1968). Dobzhansky et al. (1968) tested this prediction using data from five species of *Drosophila* and obtained results that contradicted their expectation. However, Coyne and his co-workers have published a series of papers indicating that the expectation is met for a wide range of *Drosophila* species (Coyne and Orr, 1989; Coyne, 1992; Noor, 1995). Kelly and Noor (1996) suggested that the genetic architecture of *Drosophila* makes reinforcement particularly likely in this group.

Studies of reproductive isolation in allopatric and sympatric populations of salamanders by Verrell and his colleagues (Verrell and Arnold, 1989; Tilley et al., 1990; Verrell and Tilley, 1992) turned up no evidence that reinforcement had been involved in the evolution of sexual incompatibility in the genus *Desmog-nathus*. Sanderson et al. (1992) discovered similarly negative evidence across a hybrid zone between two taxa of fire-bellied toads, *Bombina bombina* and *B. variegata*.

There is a caveat that needs to be borne in mind in all laboratory estimates of reproductive isolation. Difficulties in the characterization of isolating mechanisms (see Section 26.2 below) mean that any discrepancy in insemination frequency or time to copulation between allopatric and sympatric populations is regarded as evidence of premating isolation. This assumption is not always justified. For example, Etges(1992) and Brazner and Etges (1993) recently re-examined a purported case of incipient speciation by reinforcement in geographically isolated populations of *D. mojavensis*, one of which is sympatric with *D. arizonae*. Their studies showed that the degree of behavioral isolation observed was strongly dependent on the nature of the food type on which the larvae had been reared. Laboratory food engendered strong isolation between flies derived from populations collected on the Baja peninsula and on the mainland, whereas fermenting cactus, which forms the natural food source of the species, showed no such effect. Larval substrates affect the composition of cuticular hydrocarbons in these species.

Reinforcement is a coherent theory with testable predictions. Its weakness lies in the difficulty of characterizing premating isolating mechanisms, an issue discussed in Subsection 26.3.2.

26.3.2. Speciation Occurs as a By-Product of Adaptation to Local Environmental Conditions

Three contemporary species concepts fall within this purview.

(1) Mayr's version of the BSC: Perceived difficulties with Dobzhansky's BSC have led most modern authors to follow the BSC version championed by Mayr (1963) as well as by Muller (1942). In this view, isolating mechanisms are seen as accidental by-products of allopatric divergence (Coyne, 1992; King, 1993) or pleiotropic by-products of other selected loci (Coyne et al., 1988). In their review of the experimental literature, Rice and Hostert (1993) showed that drift alone was insufficient to generate reproductive isolation, which required consistent regimes of divergent selection among the target populations. Under these circumstances, reproductive isolation evolved as a correlated response by means of incidental pleiotropy or genetic hitchhiking. The isolating mechanisms thus evolved constitute incompatibilities that arise at various levels of organization: genetic, morphological, behavioral, and ecological. The process of allopatric divergence has been modeled by Nei et al. (1983).

Coyne et al. (1988) and Chandler and Gromko (1989) have argued that the force behind the evolution of isolating mechanisms is irrelevant; what counts is their presence. But this approach introduces severe problems of definition: Without a coherent scenario for their origin, isolating mechanisms must be characterized purely by their effects on gene flow. Does any feature that impedes gene exchange between populations qualify as an isolating mechanism? If this is the case, then geographic isolation should logically be included (Masters and Spencer, 1989). How distantly related do species have to be before the concept of an isolating mechanism becomes irrelevant? What if reproductive isolation is shown to be largely an artifact of laboratory rearing conditions (Etges, 1992; Brazner and Etges, 1993) or infection by endosymbionts (Masters and Spencer, 1989)?

Apart from the confusion that this lack of definition engenders, it also denies the possibility of formulating falsifiable predictions for studies of speciation. If the majority of isolating mechanisms that have a genetic basis can be explained in terms of pleiotropy or hitchhiking, then attempts to explain the speciation process by focusing on such diverse effects are doomed to failure. Logically, such an exercise is equivalent to trying to explain the structure of the arches of the basilica of San Marco in terms of the triangular spandrels required to accommodate them, to use the terminology of Gould and Lewontin's (1979) paradigmatic example of misconstrued adaptationism.

(2) The phylogenetic species concept (PSC): A similar form of argument follows from an adherence to some formulations of the PSC (Cracraft, 1983, 1987; Donoghue, 1985; Mishler and Brandon, 1987; Nixon and Wheeler, 1990), although most of these authors eschew the concept of reproductive isolation. Phylogenetic species are recognized on the basis of diagnosability; fixation of a character or suite of characters to enable such diagnosis is agnostic as to the action of selection, and changes due to speciation are not viewed as distinct from adaptation to local conditions. Once again, there is no mechanism to differentiate between changes that are ephemeral and those that signify a speciation event (Masters and Rayner, 1998).

The PSC is largely supported by proponents of the view that concepts of pattern and process should be divorced from one another in the reconstruction of evolutionary history (see also McKitrick and Zink, 1988; Patterson, 1988; Chandler and Gromko, 1989; Mallet, 1995). This action, however, generates a significant philosophical problem, since it treats species as simple aggregates of individuals, recognizable only by their shared possession of essential traits. Lewontin (1983) has argued against this kind of static reductionism in evolutionary theory, whereby organisms are regarded as passive bearers of traits [actually termed *semaphoronts* by Nixon and Wheeler (1990)] that are then acted on by the environment. A collection of stuffed or pinned specimens in a museum drawer may well bring such a static picture to mind, but the traits or the bodies bearing them are not what constitute a species. Rather, the species exists because of interactions among the organisms, and it is the manner in which the traits function during such interactions that makes them significant. Process theories are crucial to estimations of the value of traits in species identification.

(3) The recognition concept (RC): Paterson (1985, 1993b) has argued that the fundamental force behind the generation of diversity stems not from gamete wastage through hybridization, but from the need for coadaptation between

mating partners. Species, in this view, are groups of organisms that share system components essential to successful fertilization and syngamy (the fusion of gametes to form a zygote). The fertilization system includes social, behavioral, physiological, and biochemical characters. An important subset of fertilization mechanisms is the specific-mate recognition system (SMRS), i.e., the male–female signal-response chain that allows prospective mating partners to identify each other.

There has been a tendency in the literature to view Paterson's RC as simply the flip side of the BSC (presumably Dobzhansky's version): Surely positive selection for the recognition of appropriate mates is equivalent to selection for failure to recognize inappropriate ones? The reason these two approaches are not equivalent can be seen in the diverse selection regimes they imply. In Paterson's formulation, the requirement for coadaptation of elements of the fertilization system between males and females induces stabilizing selection: Any organism that is unable to produce or respond to components of the system will be removed from the gene pool and not contribute to the next generation. This ensures that characters crucial to mating and fertilization "are expected to be largely stabilized throughout the range of the species" (Paterson, 1985, p. 28). Selection for failure of recognition implies no such vector for stabilization, and Mayr (1963) predicted that prezygotic isolating mechanisms should show equivalent degrees of variation to any other characteristics.

It should be noted that the prediction that aspects of the fertilization system should be stabilized and less variable than other characters does not equate with their being invariant (Paterson 1993c), although several authors have made this equation (Coyne, et al., 1988; Dagley et al., 1994; Bakker and Pomiankowski, 1995). It is obvious that a necessarily complex biological system could never be phenotypically or genetically invariant.

A second postulate of the RC is the close link between the fertilization system and the species' preferred habitat (i.e., the habitat at speciation). This is especially true of the SMRS components that must be transmitted within a particular set of environmental conditions. For example, different auditory frequencies attenuate at different rates and are variously affected by scatter when they encounter objects in the environment (e.g., Michelsen, 1978); hence, frequency structures of long-distance vocalizations must be adapted to the transmission characteristics of the habitat. The fact that members of sexual species tend to seek out and remain within particular preferred habitats generates a second source of stabilizing selection on the fertilization system. Vrba (1995) has championed this aspect of the RC, describing species as "habitat-specific complex systems."

There is a strong prediction from this postulate: Because of the close link between habitat and fertilization system, speciation occurs only when the habitat deteriorates to the point at which the fertilization system no longer functions effectively. That is, speciation must be accompanied by significant niche shifts. As a corollary, sister species should not have identical resource requirements.

Paterson's model has been challenged both on the basis of the patterns of discontinuity it predicts and the selective processes it highlights. The most useful of these challenges has come from Carson and is discussed in the next section.

26.3.3. Natural Selection Can Account for Only Part of the Process of Speciation

That Carson is the modern intellectual heir to Romanes' approach to speciation, is evident from the following quotation (Carson 1985, p. 381, my italics):

"Evolution produces both adaptations and species. The mechanisms and modes whereby adaptations arise in populations are reasonably well understood both in nature and in experimental populations. . . . Speciation, on the other hand, involves events whereby the more or less permanent subdivision of an older population occurs. *Adaptation is only part of what happens.*"

Like the RC, Carson's (1985) organization theory of speciation (OTS) views species as stabilized systems. In this case, however, stability is conferred by genome dynamics rather than by interactions between potential mating partners. In Carson's view, a species is characterized by a complex and coadapted genomic system, the integrity of which is not easily perturbed. The breakdown of the integrated, polygenic system that must accompany speciation takes part in two phases:

Phase 1: Gene pool disorganization by chance factors
Phase 2: Gene pool reorganization by selection
    Mode A: Adaptations leading to altered sexual reproduction
    Mode B: Adaptations to the ambient (nonsexual) environment

The initial disorganization of the genome is seen as occurring in a subpopulation of the parent species for which selection pressures have been relaxed. Founder events or *in situ* reduction of population size, hybridization, and shifting balance are all potential factors that could contribute to this breakdown. Selection plays a role only in the second phase, in which it is seen as affecting primarily one of two discrete character sets. If selection follows the mode A regime, it affects those characters Paterson would include within the fertilization system, and neospecies (or newly formed species) will be characterized by differences in adaptations associated directly with sexual reproduction. If it follows the mode B regime, it affects those features necessary for survival in the ambient (or nonsexual) environment, and neospecies will be adapted to different environmental conditions.

Carson predicts that the character set most likely to form the focus of selective reorganization is that which evolved most recently within the genomic system. This is because older genetic pathways have become more canalized and less open to perturbation. In animals, the traits most likely to be affected are those associated with sexual reproduction (mode A). In plants, nonsexual adaptations (mode B) predominate as the primary selective responses of emerging species. Of course, with the passage of time both types of adaptations will become apparent; once one form of adaptive response becomes fixed and stabilized, the genomic system is free to respond to the other form.

While aspects of the OTS sound superficially similar to those of the RC, the predictions deriving from the two models are very different. The first of these relates to the relative stability of characters involved in specific-mate recognition and fertilization. Whereas the RC predicts that such traits should be strongly canalized as a result of generations of stabilizing selection, the OTS predicts that they will often be less canalized because of recent genomic reorganization and hence may show greater ranges of variation than characters not related to sexual reproduction. This receives more discussion in the following section.

A second prediction from Carson's model is that "...the novel selective regime that follows a genetic disorganization, as an incipient genetic change leading in the direction of species formation, will tend to mount *either* selective response A *or* B ... but not both simultaneously" (Carson 1985, p. 383, italics original). In animal neospecies in particular, speciation should often not be accompanied by significant shifts in preferred habitats that would require environmental readjustments. These predictions run counter to those of the RC, in which speciation events require a degree of habitat change which in turn drives the change in the fertilization system. Hence, for the RC, neospecies should show selective responses A and B simultaneously.

Alternative testable predictions are rare and welcome occurrences in a field as rampant with contingency as speciation. I have conducted a preliminary test of the latter prediction using data from a clade of primate species that have, on molecular evidence, radiated within the last 1 million years (Masters, 1998). The taxa are lesser galagos (Primates, Prosimii) – nocturnal primates that communicate chiefly by loud calls and scent marks. In accordance with the predictions of both the RC and the OTS, the taxa show marked and consistent differences in the vocalizations they use in long-distance communication and mate attraction, indicating that mode A adaptation has occurred. When the distributions of these taxa were examined, there were also indications that a degree of habitat shifting had occurred during the speciation events. Figure 26.1 shows the capture localities of the lesser galagos comprising the collection of the Natural History Museum, London, overlain on a map of White's (1983) African phytochoria. Phytochoria are vegetation zones that reflect levels of plant endemism and fall into two categories: Regional centers of endemism (RCEs) are areas that have at least 1000 endemic plant species that constitute at least 50% of the total flora; between the RCEs are areas of very low plant endemism, known as transition mosaics (TMs), which are characterized by mixed floras spilling over from the neighboring RCEs (White, 1983). The cores of the distributions of three of the four galago species investigated were anchored in RCEs, with the margins of their ranges extending into TMs. The fourth taxon, which

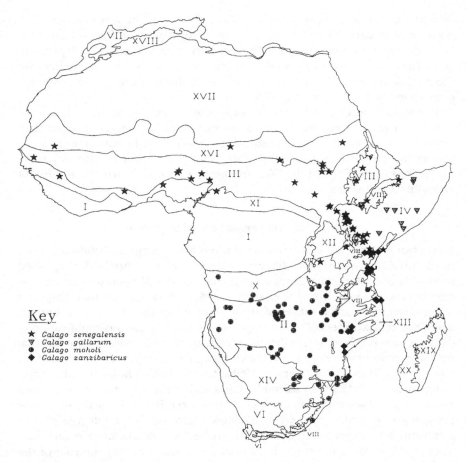

Figure 26.1. Map showing the capture localities of the lesser galago specimens comprising the collection in the Natural History Museum, London, superimposed on the main phytochoria of Africa, after White (1983). The African phytochoria are identified as follows: Regional Centers of Endemism: I, Guineo-Congolian; II, Zambezian; III, Sudanian; IV, Somalia-Masai; V, Cape; VI, Karoo-Namib; VII, Mediterranean; VIII, Afromontane; IX, Afroalpine (not shown). Transition Mosaics: X, Guinea-Congolia/Zambezia; XI, Guinea-Congolia/Sudania; XII, Lake Victoria; XIII, Zanzibar-Inhambane; XIV, Kalahari-Highveld; XV, Tongaland-Pondoland; XVI, Sahel; XVII, Sahara; XVIII, Mediterranean/Sahara. See text for discussion.

is morphologically the most primitive of the clade, had the major part of its range in a TM. The habitat within this transition zone is similar to what has been postulated as the ancestral habitat type for the galago clade as a whole (McCrossin, 1992).

The coincidence of these plant and animal distributions suggests that the forces driving speciation among the plant taxa were also responsible for divergence among the primates. Indeed, it is one of the assumptions of vicariance biogeography that similar patterns of distribution reflect similar influences during evolutionary history. These results imply that mode B adaptations, i.e., adaptations to the nonsexual environment, occurred in conjunction with the modifications to the systems of mate recognition, thereby supporting the predictions of the RC over the OTS.

In contrast, other primate species of similarly recent divergence, i.e., some of the African guenons, do not appear to show shifts in preferred habitat characteristics (Gautier-Hion et al., 1988), although a detailed environmental study has yet to be carried out. Estimating the relative frequencies with which the predictions of the RC and the OTS are met is a potentially fruitful and highly accessible area for future research.

26.4. Natural versus Sexual Selection

These two regimes are often conflated when authors fail to distinguish mate recognition from mate choice (Paterson, 1993a); yet they must be distinguished in models of evolutionary change, since their modes of operation and consequences are not identical. Andersson's (1994, p. 207) opinion that "[s]pecies recognition traits are a subset of those involved in the choice of a suitable mate, one crucial aspect of which is species identity," is an oversimplification of the processes involved in specific-mate recognition and overlooks the fact that sexual selection can only operate once specific-mate recognition has already taken place [Ryan, 1990; Paterson, 1993a; Carson (1995) holds that the major function of the SMRS is to assemble males and females so that the important events of sexual selection can take place]. It also renders invisible the dissimilar predictions generated by the two schemes regarding (a) the degree of conservatism to be expected in characters selected for specific-mate recognition as opposed to those evolving under sexual selection and (b) the role of the environment.

Various models have been proposed to implicate sexual selection in speciation by elaborating on Fisher's (1930) runaway process (e.g., Lande, 1981; Pomiankowski and Iwasa, 1995; Iwasa and Pomiankowski, 1995). In this scenario a female preference for a male epigamic trait may arise if the preferred character is also favored by natural selection. Once established, the intensity of the preference will be increased, because the sons of the females exhibiting the preference (also the sons of the males exhibiting the trait) will be preferred in the following generation. The preference and the trait thus evolve together, even beyond the point at which the trait is favored by natural selection.

Kaneshiro (1989) has devised a somewhat different "intuitive" model to support his view that "sexual selection is *the most important factor in the initial stages* of species formation" (p. 285, my italics). His definition of sexual selection is important to his argument; it is "the differential mating ability of males regardless of their contribution to the overall genetic fitness of the species" (p. 283). The model assumes that, within an interbreeding population, there is variation in mating ability among the males ranging from unsuccessful to highly successful, and in discrimination among the females from nondiscriminant to highly discriminant.

The selective pressures impinging on males and females will be very different. The most successful males will obviously be most strongly selected for. But in contrast to more traditional models of sexual selection, Kaneshiro hypothesized that highly discriminant females will be selected against, because the frequency of highly successful males will be low. Hence females will be under selection to broaden their range of acceptable males, to ensure some reproductive success. Because there will be a genetic correlation between the male and female phenotypes, the differential selective regimes operating on the two sexes will act as a stabilizing agent to prevent runaway sexual selection.

If a population passes through a bottleneck, such as a founder event, the females will be under still more pressure to broaden their range of acceptable males. Additionally, because of the loss of variability consequent on the bottleneck, elements of species-specific courtship behavior are likely to be lost (Kaneshiro, 1976, 1980, 1989). Hence, more recently evolved SMRSs should have fewer components than more ancestral ones. As population size increases, sexual selection becomes "a dominant force in mate selection" (Kaneshiro, 1989, p. 293). Carson was thinking along these lines for his source of mode A selection in the reorganization phase of the OTS (see Carson, 1986).

It is unclear to me why a decrease in female discriminant ability, and hence a narrowing of the differential in mating success among males, during the founder event and in subsequent generations, should be considered evidence that "sexual selection is the most important factor in the initial stages of species formation." In my understanding of past and current usages of the term sexual selection, the above scenario exhibits precisely the opposite result; the role of sexual selection is shown to be rather minimal, and secondary at that. What is also not clear is the source of novelty in emerging SMRSs. There is a limit to how many times elements can be lost before the entire system breaks down. But terminological and other quibbles aside, this is a logically coherent model with readily testable predictions.

Both the Fisherian models of speciation by sexual selection and Kaneshiro's intuitive model require a high degree of flexibility in mating systems for their operation. Very little attention is paid to the need for coadaptation between prospective mating partners. When coadaptation between mating partners was incorporated into one such model, Liou and Price (1994) concluded that sexual selection could influence the divergence of mating systems only if a substantial degree of assortative mating was present to begin with.

As a general rule, models of sexual selection generate the prediction that characters used in courtship and mating should show significantly higher levels of intraspecific variation than features not associated with these activities (see West-Eberhard, 1983, for review). By contrast, the RC predicts that aspects of the fertilization system crucial to specific-mate recognition will show less variation than nonsexual characters (see above). The predictions are clear and mutually exclusive, but adequate data to choose between these alternatives are rare and conflicting [compare, e.g., the conclusions of Henderson and Lambert (1982) with those of Pomiankowski and Møller (1995)].

It is quite possible that different components of mating systems serve different functions and are therefore subject to different kinds of forces. For example, the advertisement calls of galagos show strong intraspecific consistency in the most energetic frequency bands used and in the durations of individual units and interunit intervals, but a great deal of individual variation in call durations and the use of higher-level harmonics (Masters, 1991). This combination of stabilized and idiosyncratic variation allows for the encoding of different types of information. Any sophisticated system of communication can be expected to incorporate diverse sorts of components, and it would be an error to view them as monolithic structures subjected to monotonic selection regimes. Ryan (1990) and Ryan and Keddy-Hector (1992) cite further examples of this phenomenon. An understanding of the relative significance of natural and sexual selection in the evolution of courtship signals will require controlled experiments that allow dissection of the respective functions of their components.

The second area in which theories of mate choice and mate recognition show divergence relates to the role of the environment. According to the RC, speciation occurs because habitat changes drive adaptive responses in the system of mate recognition and fertilization. In most descriptions of sexual selection, the broader environmental context appears, at best, as the backdrop against which the selective processes are held to take place.

For example, Carson (1987) has described how *D. silvestris* males raised from isofemale lines and given access to virgin females in Plexiglas cages show significant differentials in mating success. Fully one third of the males produced no offspring at all, while a second third was responsible for two thirds of the matings. The successful males included a disproportionately high frequency of heterokaryotypes, indicating some form of heterosis, which Carson believed could be responsible for the maintenance of stable chromosome polymorphisms both in laboratory stocks and in nature. But are we justified in extrapolating directly from an artificial laboratory system to nature? Is it because of bad genes that a substantial percentage of the males attempted to mate but were refused, or is it because some crucial environmental cue was missing? If the experiment had been performed in a natural environment, are we sure that the same males would account for most of the matings? Only if the environment has no influence on fitness, which is clearly wrong.

Similarly, when Ryan and Keddy-Hector (1992) sought an explanation for the observation that female anurans often show a preference for conspecific male calls of lower frequency, they considered only two alternatives: (a) the good genes hypothesis, i.e., that lower calls are emitted by larger bodied, and therefore older, males that have demonstrated their superior survival abilities; or (b) that the preference is related to the tuning properties of the female's peripheral auditory system. The latter hypothesis was supported by experimental data, demonstrating most elegantly the physiological coadaptation between spectral qualities of calls and tuning properties of the receivers that are crucial to the efficient functioning of a communication system. However, they overlooked an equally important component – the medium through which signals must be transmitted, the properties of which are highly complex and must affect signal structures. To go back to a fairly simple example mentioned earlier, there is a tendency for higher frequencies to be more subject to attenuation in cluttered environments, so that lower frequencies travel more efficiently. In all studies of auditory communication systems in insects and frogs reviewed by Ryan and Keddy-Hector, more intense calls were preferred over less intense ones, which suggests immediately the significance of sound transmission properties. In fact, in some studies, it was observed that "intensity can cancel or even reverse a preference that is based on other call characters" (Ryan and Keddy-Hector, 1992, p. S10), including frequency preference. An examination of such a communication system that ignores the properties of the transmission environment is necessarily incomplete.

26.5. Conclusions

I have presented evidence to show that the arguments concerning the role of selection in speciation are more than a century old and are just as vehemently debated today as they were in the halls of the Victorian scientific societies. The venerable age of these recurrent arguments is a sure indicator that we have not yet asked all the right questions, or perhaps not asked them in the right way. Many of our most firmly held tenets are not falsifiable hypotheses but metaphysical statements. Nevertheless these metaphysical statements often yield falsifiable predictions that provide the basis for rigorous application of the scientific method.

I have adopted the stand that, when processes of origin and maintenance are highlighted, the BSC actually comprises two distinct concepts. Only Dobzhansky's BSC generates readily falsifiable predictions, because the role of selection in the evolution of reproductive isolation is a coherent one. In the more commonly held version of the BSC, which I have called Mayr's version for the sake of convenience, reproductive isolation generally arises as a correlated response by means of pleiotropy or genetic hitchhiking under divergent selection regimes. Such isolation effects are likely to be so diverse in their nature and modes of origin that consistent predictions regarding the speciation process are denied. A modified form of this argument can also be applied to the PSC.

The RC and the OTS generate contradictory predictions that are readily accessible to investigation. The RC predicts that characters crucial to mate recognition and fertilization should be more strongly stabilized than other traits, while the OTS predicts that such traits should often be more labile and respond more rapidly to selection. The latter prediction also derives from theories of sexual selection. This is a potentially fruitful area for future research.

The RC and OTS also differ with regard to the significance attributed to environmental change. According to the RC, the environment is the primary motor for change in the fertilization system. For the OTS, sexual adaptation can and often does take place in the absence of environmental change.

The role of the environment is one of the great neglected areas of speciation biology. For example, studies of reproductive isolation are generally conducted under controlled laboratory conditions that ignore the environmental context within which mate attraction, courtship, and mating take place. Investigations of sexual selection generally suffer from the same omission. The preferred habitats of species have been ignored by species and speciation biologists to such an extent, that we simply do not have the fine-grained data to assess the role of environmental change in speciation. This is an area in which more research is desperately needed.

It is my view that the best means we have of breaking the 100-year deadlock regarding the link between neo-Darwinian theories of pattern and process is to insist that our metaphysical constructs generate falsifiable predictions and to subject them to rigorous testing.

26.6. Acknowledgments

I thank Rama Singh for inviting me to contribute to this volume and for his patience and encouragement while I drafted this essay. Several people made the time and effort to help me clarify my ideas and make them intelligible. These include Hampton Carson, Laurie Godfrey, Chris Green, Dick Lewontin, Adrian Lister, Jim Mallet, Des Maxwell, Hugh Paterson, Dick Rayner, Rama Singh, and Hamish Spencer. Finally, I thank Dick Lewontin for his support and enthusiasm, and for never allowing me to get away with saying something I didn't mean.

NOTE

1. I include both natural and sexual selection under this rubric, and although I agree with Paterson (1993a) that both terms are vague in definition and often difficult to distinguish in practice, I shall not attempt to redefine them here.

REFERENCES

Andersson, M. 1994. *Sexual Selection*. Princeton, NJ: Princeton U. Press.

Bakker, T. C. M. and Pomiankowski, A. 1995. The genetic basis of female mate preferences. *J. Evol. Biol.* 8:129–171.

Brazner, J. C. and Etges, W. J. 1993. Pre-mating isolation is determined by larval rearing substrates in cactophilic *Drosophila mojavensis*. II. Effects of larval substrates on time to copulation, mate choice and mating propensity. *Evol. Ecol.* 7:605–624.

Carson, H. L. 1985. Unification of speciation theory in plants and animals. *Syst. Bot.* 10:380–390.

Carson, H. L.1986. Sexual selection and speciation. In *Evolutionary Processes and Theory*, S. Karlin and E. Nevo, eds., pp. 391–409. New York: Academic.

Carson, H. L. 1987. High fitness of heterokaryotypic individuals segregating naturally within a long-standing laboratory population of *Drosophila silvestris*. *Genetics* 116:415–422.

Carson, H. L.1995. Fitness and the sexual environment. In *Speciation and the Recognition Concept: Theory and Application*, D. M. Lambert and H. G. Spencer, eds., pp. 123–137. Baltimore: Johns Hopkins U. Press.

Chandler, C. R. and Gromko, M. H. 1989. On the relationship between species concepts and speciation processes. *Syst. Zool.* 38:116–125.

Coyne, J. A. 1992. Genetics and speciation. *Nature (London)* 355:511–515.

Coyne, J. A. and Orr, H. A. 1989. Patterns of speciation in *Drosophila*. *Evolution* 43:362–381.

Coyne, J. A., Orr, H. A., and Futuyma, D. J. 1988. Do we need a new species concept? *Syst. Zool.* 37:190–200.

Cracraft, J. 1983. Species concepts and speciation analysis. *Curr. Ornithol.* 1:159–187.

Cracraft, J. 1987. Species concepts and the ontology of evolution. *Biol. Philos.* 2:329–346.

Dagley, J. R., Butlin, R. K., and Hewitt, G. M. 1994. Divergence in morphology and mating signals, and assortative mating among populations of *Chorthippus parallelus*. (Orthoptera: Acrididae). *Evolution* 48:1202–1210.

Darwin, C. R. 1872. *On the Origin of Species*, 6th ed. London: Murray.

Darwin, F. and Seward, A. C. 1903. *More Letters of Charles Darwin*, Vol. I. London: Murray.

Dobzhansky, T. 1940. Speciation as a stage in evolutionary divergence. *Am. Nat.* 74:312–321.

Dobzhansky, T. 1951. *Genetics and the Origin of Species*, 3rd ed. New York: Columbia U. Press.

Dobzhansky, T. 1972. Species of *Drosophila*. *Science* 177:664–669.

Dobzhansky, T., Ehrman, L., and Kastritsis, P. A. 1968. Ethological isolation between sympatric and allopatric species of the *obscura* group of *Drosophila*. *Anim. Behav.* 16:79–87.

Donoghue, M. J. 1985. A critique of the Biological Species Concept and recommendations for a phylogenetic alternative. *Bryologist* 88:172–181.

Etges, W. J. 1992. Premating isolation is determined by larval substrates in cactophilic *Drosophila mojavensis*. *Evolution* 46:1945–1950.

Fisher, R. A.1930. *The Genetical Theory of Natural Selection*. Oxford: Oxford U. Press.

Gautier-Hion, A., Bourlière, F., Gautier, J.-P., and Kingdon, J. 1988. *A Primate Radiation. Evolutionary Biology of the African Guenons*. Cambridge: Cambridge U. Press.

Gould, S. J. and Lewontin, R. C. 1979. The spandrels of San Marco and the Panglossian paradigm: A critique of the adaptationist programme. *Proc. R. Soc. London Ser. B* 205:581–598.

Harper, A. A. and Lambert, D. M. 1983. The population genetics of reinforcing selection. *Genetica* 62:15–23.

Henderson, N. R. and Lambert, D. M. 1982. No significant deviation from random mating of worldwide populations of *Drosophila melanogaster*. *Nature (London)* 300:437–440.

Iwasa, Y. and Pomiankowski, A. 1995. Continual change in mate preferences. *Nature (London)* 377:420–422.

Kaneshiro, K. Y. 1976. Ethological isolation and phylogeny in the *planitibia* subgroup of Hawaiian *Drosophila*. *Evolution* 30:740–745.

530   *Judith Masters*

Kaneshiro, K. Y. 1980. Sexual isolation, speciation and the direction of evolution. *Evolution* 34:437–444.

Kaneshiro, K. Y. 1989. The dynamics of sexual selection and founder effects in species formation. In *Genetics, Speciation, and the Founder Principle*, L. V. Giddings, K. Y. Kaneshiro, and W. W. Anderson, eds., pp. 279–296. New York: Oxford U. Press.

Kelly, J. K. and Noor, M. A. F. 1996. Speciation by reinforcement: a model derived from studies of drosophila. *Genetics* 143:1485–1497.

King, M. 1993. *Species Evolution: The Role of Chromosome Change.* Cambridge: Cambridge U. Press.

Kottler, M. J. 1985. Charles Darwin and Alfred Russel Wallace: Two decades of debate over natural selection. In *The Darwinian Heritage*, D. Kohn, ed., pp. 367–432. Princeton, NJ: Princeton U. Press.

Lambert, D. M., Centner, M. R., and Paterson, H. E. H. 1984. Simulation of the conditions necessary for the evolution of species by reinforcement. *S. Afr. J. Sci.* 80:308–311.

Lande, R. 1981. Models of speciation by sexual selection on polygenic traits. *Proc. Natl. Acad. Sci.* 78:3721–3725.

Lewontin, R. C. 1974. *The Genetic Basis of Evolutionary Change.* New York: Columbia U. Press.

Lewontin, R. C. 1983. Gene, organism and environment. In *Evolution from Molecules to Men*, D. S. Bendall, ed., pp. 273–285. Cambridge: Cambridge U. Press.

Liou, L. W. and Price, T. D. 1994. Speciation by reinforcement of premating isolation. *Evolution* 48:1451–1459.

Mallet, J. 1995. A species definition for the Modern Synthesis. *Trends Evol. Ecol.* 10:294–299.

Masters, J. C. 1991. Loud calls of *Galago crassicaudatus* and *G. garnettii* and their relation to habitat structure. *Primates* 32:153–167.

Masters, J. C. 1998. Speciation in the lesser galagos. *Folia Primatol.* 69 (Suppl. 1):357–370.

Masters, J. C. and Rayner, R. J. 1996. The recognition concept and the fossil record: putting the genetics back into phylogenetic species. *S. Afr. J. Sci.* 92:225–231.

Masters, J. C. and Spencer, H. G. 1989. Why we need a new genetic species concept. *Syst. Zool.* 38:270–279.

Mayr, E. 1963. *Animal Species and Evolution.* Cambridge, MA: Harvard U. Press.

McCrossin, M. L. 1992. New species of bushbaby from the middle Miocene of Maboko Island, Kenya. *Am. J. Phys. Anthropol.* 89:215–233.

McKitrick, M. C. and Zink, R. M. 1988. Species concepts in ornithology. *Condor* 90:1–14.

Michelsen, A. 1978. Sound reception in different environments. In *Sensory Ecology: Review and Perspectives*, M. A. Ali, ed., pp. 345–373. New York: Plenum.

Mishler, B. D. and Brandon, R. N. 1987. Individuality, pluralism, and the phylogenetic species concept. *Biol. Philos.* 2:397–414.

Muller, H. J. 1942. Isolating mechanisms, evolution and temperature. *Biol. Symp.* 6:71–125.

Nei, M., Maruyama, T., and Wu, C.-I. 1983. Models of the evolution of reproductive isolation. *Genetics* 103:557–579.

Nixon, K. C. and Wheeler, Q. D. 1990. An amplification of the phylogenetic species concept. *Cladistics* 6:211–223.

Noor, M. A. 1995. Speciation driven by natural selection. *Nature (London)* 375:674–675.

Paterson, H. E. H. 1978. More evidence against speciation by reinforcement. *S. Afr. J. Sci.* 74:369–371.

Paterson, H. E. H. 1982. Perspective on speciation by reinforcement. *S. Afr. J. Sci.* **78**:53–57.

Paterson, H. E. H. 1985. The recognition concept of species. In *Species and Speciation*, E. S. Vrba, ed., pp. 21–29. Pretoria: Transvaal Museum Monograph No. 4.

Paterson, H. E. H. 1993a. Animal species and sexual selection. In *Evolutionary Patterns and Processes*, D. Edwards and D. Lees, eds., pp. 209–228. London: Linnean Society of London.

Paterson, H. E. H. 1993b. *Evolution and the Recognition Concept of Species.* Baltimore: Johns Hopkins U. Press.

Paterson, H. E. H. 1993c. Variation and the specific-mate recognition system. In *Perspectives in Ethology, 10: Behavior and Evolution*, P. P. G. Bateson et al. eds., pp. 209–227. New York: Plenum.

Patterson, C. 1988. The impact of evolutionary theories on systematics. In *Prospects in Systematics*, D. Hawksworth, ed., pp. 59–91. Oxford: Clarendon.

Pomiankowski, A. and Iwasa, Y. 1995. What causes diversity in male sexual characters? *Rev. Suisse Zool.* **102**:883–894.

Pomiankowski, A. and Møller, A. P. 1995. A resolution of the lek paradox. *Proc. R. Soc. London Ser. B* **260**:21–29.

Rice, W. R. and Hostert, E. E. 1993. Laboratory experiments on speciation: What have we learned in 40 years? *Evolution* 47:1637–1653.

Romanes, G. J. 1897. *Darwin, and After Darwin. I. The Darwinian Theory.* London: Longmans, Green.

Ryan, M. J. 1990. Signals, species, and sexual selection. *Am. Sci.* **78**:46–52.

Ryan, M. J. and Keddy-Hector, A. 1992. Directional patterns of female mate choice and the role of sensory bias. *Am. Nat.* **139** (Suppl.):S4–S35.

Sanderson, N., Szymura, J. M., and Barton, N. H. 1992. Variation in mating call across the hybrid zone between the fire-bellied toads *Bombina bombina* and *B. variegata.* *Evolution* **46**:595–607.

Spencer, H. G., McArdle, B. H., and Lambert, D. M. 1986. A theoretical investigation of speciation by reinforcement. *Am. Nat.* **128**:241–262.

Stamos, D. N. 1996. Popper, falsifiability, and evolutionary biology. *Biol. Philos.* **11**:161–191.

Tilley, S. G., Verrell, P. A., and Arnold, S. J. 1990. Correspondence between sexual isolation and allozyme differentiation: a test in the salamander *Desmognathus ochrophaeus.* *Proc. Natl. Acad. Sci. USA* **87**:2715–2719.

Verrell, P. A. and Arnold, S. J. 1989. Behavioral observations of sexual isolation among allopatric populations of the mountain dusky salamander, *Desmognathus ochrophaeus.* *Evolution* **43**:745–755.

Verrell, P. A. and Tilley, S. G. 1992. Population differentiation in plethodontid salamanders: divergence of allozymes and sexual compatibility among populations of *Desmognathus imitator* and *D. ochrophaeus* (Caudata: Plethodontidae). *Zool. J. Linn. Soc.* **104**:67–80.

Vrba, E. S. 1995. Species as habitat-specific, complex systems. In *Speciation and the Recognition Concept: Theory and Application*, D. M. Lambert, and H. G. Spencer, eds., pp. 3–44. Baltimore: Johns Hopkins U. Press.

Wallace, A. R. 1889. *Darwinism.* London: Macmillan.

West-Eberhard, M. J. 1983. Sexual selection, social competition, and speciation. *Q. Rev. Biol.* **58**:155–183.

White, F. 1983. *The Vegetation of Africa.* Paris: UNESCO.

CHAPTER TWENTY SEVEN

The Evolutionary Genetics of Speciation

JERRY A. COYNE AND H. ALLEN ORR

27.1. Summary

Over the past decade there has been renewed interest in the genetics of speciation, which has yielded a number of new models and empirical results. Defining speciation as "the origin of reproductive isolation between two taxa," we review recent theoretical studies and all relevant data, emphasizing the regular patterns seen among genetic analyses. Finally, we point out some important and tractable questions about speciation that have been neglected.

When we last reviewed the evolution of reproductive isolation (Coyne and Orr, 1989a), we complained that workers on speciation were considered poor cousins in the family of evolutionists, mired in endless and untestable speculations about a process that no one could witness. Since then, the study of speciation has grown increasingly respectable, recruiting ever more experimentalists and theorists. A number of new phenomena have been uncovered and new theories offered to explain them. Here we summarize recent progress on the genetics of speciation, highlighting areas in which important and tractable questions remain unanswered.

27.2. What are Species?

Any discussion of the genetics of speciation must begin with the observation that species are real entities in nature, not subjective human divisions of what is really a continuum among organisms. We have previously summarized the evidence for this view and counterarguments by dissenters (Coyne, 1994). The strongest evidence for the reality of species is the existence of distinct groups living in sympatry (separated by genetic and phenotypic gaps) that are recognized consistently by independent observers. To a geneticist, these disjunct

R. S. Singh and C. B. Krimbas, eds., *Evolutionary Genetics: From Molecules to Morphology*, vol. 1. © Cambridge University Press 2000. Printed in the United States of America. ISBN 0-521-57123-5. All rights reserved.

groups suggest a species concept based on gene flow. As Dobzhansky (1935, p. 281) noted: "Any discussion of these problems should have as its logical starting point a consideration of the fact that no discrete groups of organisms differing in more than a single gene can maintain their identity unless they are prevented from interbreeding with other groups.... Hence, the existence of discrete groups of any size constitutes evidence that some mechanisms prevent their interbreeding, and thus isolate them." This conclusion inspired Dobzhansky (1935) and Mayr's (1942) biological species concept (BSC), which considered species to be groups of populations reproductively isolated from other such groups by isolating mechanisms – genetically based traits that prevent gene exchange. The list of such mechanisms is familiar to all evolutionists, and includes those acting before fertilization (prezygotic mechanisms, such as mate discrimination and gametic incompatibility) and after fertilization (postzygotic mechanisms, including hybrid inviability and sterility).

Like most evolutionists, we adopt the BSC as the most useful species concept, and our discussion of the genetics of speciation will accordingly be limited to the genetics of reproductive isolation. We recognize that this view of speciation is not universal: Systematists in particular often reject the BSC in favor of concepts involving diagnostic characters (Cracraft, 1989; Baum and Shaw, 1995; Zink and McKitrick, 1995). We have argued against these concepts elsewhere (Coyne et al., 1988; Coyne, 1992a, 1993a, 1994) and will not repeat our contentions here. We note only that the recent burst of work on speciation reflects almost entirely the efforts of those adhering to the BSC. In fact, every recent study on the genetics of speciation is an analysis of reproductive isolation.

27.3. Why Are There Species?

One of the most important but neglected questions about speciation is why organisms are divided up into many discrete groups instead of constituting a few extremely variable types. The answer to the question of why species exist may not be the same as the answer to how do species arise? The only coherent discussion of this problem is that of Maynard Smith and Szathmáry (1995, pp. 163–167), who give three possible reasons for the existence of discrete species: (1) species might represent stable, discontinuous states of matter; (2) species might be adapted to discontinuous ecological niches; or (3) reproductive isolation (which can arise only in sexually reproducing taxa) might create gaps between taxa by allowing them to evolve independently.

The second and the third hypotheses seem most plausible. There are several ways to distinguish between them. One is to determine if asexually reproducing groups form taxa just as distinct in sympatry as do sexually reproducing groups. Although such comparisons are hampered by the rarity of large asexual groups, bacteria seem reasonable candidates. While little work has been published, two studies (Roberts and Cohan, 1995; Roberts et al., 1996) show that forms of *Bacillus subtilis* from the American desert fall into discrete clusters in sympatry. Moreover, coalescence models (Cohan, 1998) show that a combination of new mutations conferring ecological difference, periodic selection, and limited gene flow can create distinct taxa of bacteria living in a single habitat.

The only empirical work on clustering in asexual eukaryotes is that of Holman (1987), who determined that the nomenclature of bdelloid rotifers (not known to have a sexual phase) was more stable than that of sexually reproducing relatives through successive revisions of taxonomic monographs. From this he concluded that because they are recognized more consistently, asexual rotifers are actually more distinct than their sexual counterparts. While intriguing, this result is hardly conclusive, and we badly need similar studies based not on nomenclature but on genetic and phenotypic cluster analyses. We hasten to add that while studies of clustering in asexuals are worthwhile, they must not be overinterpreted. While such studies might show that ecological specialization can produce discrete asexual forms, it does not follow that such specialization explains clustering in sexuals: The maintenance of discrete forms by selection in sexuals is far more difficult and could be far rarer.

We suggest two other approaches not discussed by Maynard Smith and Szathmáry. First, if distinct niches alone (and not reproductive isolation) can explain the existence of species, then sympatric speciation should be common. The whole premise that geographic isolation is essential for speciation rests on Dobzhansky and Mayr's idea that the swamping effect of gene flow prevents the evolution of reproductive isolation. But if reproductive isolation is not important in explaining the existence of species, adaptation to distinct niches could occur in sympatry, and phylogenetic analyses should often reveal that the most recently evolved pairs of species are sympatric. Although there is some evidence for sympatric speciation based on niche use in fish (e.g., Schliewen et al., 1994), we know of no other studies featuring similarly rigorous phylogenetic analyses. Perturbation experiments could also address this problem. For example, sexual isolation between sympatric species could be overcome by hybridization in the laboratory and the resulting (fertile) hybrids reintroduced into their original habitat. If the ecological-niche explanation is correct, the hybrids should revert to the parental types (or to similar but distinct types). But if reproductive isolation helps maintain species distinctness, the hybrids should either revert to a single parental type or remain a hybrid swarm. Such a study would, however, require many years of observation in most organisms.

Of course, the existence of species in sexual taxa could well depend on both distinct adaptive peaks and reproductive isolation. But it is hard to believe that ecological niches alone can explain distinct species, if for no other reason than such species, lacking reproductive isolating mechanisms, would hybridize. If they were then to remain distinct, hybrids would have to suffer a fitness disadvantage because of their inability to find a suitable niche, and this disadvantage is a form of postzygotic isolation.

27.4. Studying Reproductive Isolation

27.4.1. What Is Novel about Speciation?

Some of our colleagues have suggested that speciation is not a distinct field of study because – as a by-product of conventional evolutionary forces like selection and drift – the origin of species is simply an epiphenomenon of normal population–genetic processes. But even if speciation is an epiphenomenon, it does not follow that the mathematics or genetics of speciation can be inferred from traditional models of evolution in single lineages. Under the BSC, the origin of species involves reproductive isolation, a character that is unique because it requires the joint consideration of two species and usually an interaction between the genomes of two species (Coyne, 1994). The distinctive feature of the genetics of speciation is therefore epistasis. This is necessarily true for all forms of postzygotic isolation, in which an allele that yields a normal phenotype in its own species causes hybrid inviability or sterility on the genetic background of another (see below). Epistasis also occurs in many forms of prezygotic isolation. Sexual isolation, for example, usually requires the coevolution of male traits and female preferences, so that the fitness of a male trait depends on whether the choosing female is conspecific or heterospecific.

These ubiquitous (and complex) interactions between the genomes of two species guarantee that the mathematics of speciation will differ from that describing evolutionary change within species and that speciation may well show emergent properties not seen in traditional models. Indeed, such properties have already been seen for postzygotic isolation (e.g., the snowball effect; see below).

Two motives usually underlie genetic analyses of speciation. First, just as with quantitative-trait-locus (QTL) analyses of ordinary characters, we would like to understand the genetic basis of cladogenesis. That is, we would like to know the number of genes involved in reproductive isolation, the distribution of their phenotypic effects, and their location in the genome. Second, we expect genetic analyses of reproductive isolation to shed light on the process of speciation, as different evolutionary processes should leave different genetic signatures. The observation of more genes causing hybrid male than hybrid female sterility has suggested, for example, that these critical substitutions were driven by sexual selection (Wu and Davis, 1993; True et al., 1996).

27.4.2. What Traits Should We Study?

Because speciation is complete when reproductive isolation stops gene flow in sympatry, the genetics of speciation properly involves the study of only those isolating mechanisms evolving up to that moment. The further evolution of reproductive isolation, while interesting, is irrelevant to speciation. Although widely recognized, this point is understandably often ignored in practice. If speciation is allopatric and several isolating mechanisms evolve simultaneously, it is hard to know which will be important in preventing gene flow when the taxa become sympatric. *Drosophila simulans* and *D. mauritiana*, for example, are allopatric, and in the laboratory show sexual isolation, sterility of F_1 hybrid males, and inviability of both male and female backcross hybrids. We have no idea which of these factors would be most important in preventing gene exchange in sympatry, or if other unstudied factors – like ecological differences – would play a role.

It seems likely, in fact, that several isolating mechanisms evolve simultaneously in allopatry and act together to both prevent gene flow in sympatry and allow coexistence. (Although reproductive isolation is sufficient for speciation, different species must coexist in sympatry in order to be seen.) There are two reasons why multiple isolating mechanisms seem likely. In theory, no single isolating mechanism except for distinct ecological niches or some types of temporal divergence can at the same time completely prevent gene flow and allow coexistence in sympatry. Two species solely isolated by hybrid sterility, for example, cannot coexist: One will become extinct through excessive hybridization or ecological competition. Species subject only to sexual isolation are ecologically unstable because they occupy identical niches. Second, direct observation often shows that complete reproductive isolation in nature often involves several isolating mechanisms. Schluter (1997), for example, describes several species pairs having incomplete prezygotic isolation. When hybrids are formed, however, they are ecologically unsuited for the parental habitats, and do not thrive.

We know little about the temporal order in which reproductive isolating mechanisms appear. The only study comparing the rates of evolution of different forms of reproductive isolation is Coyne and Orr's (1989b; 1997) analysis of prezygotic versus postzygotic isolation in *Drosophila*. In these studies, prezygotic isolation (mate discrimination) between species is on average a stronger barrier to gene flow than is postzygotic isolation (Fig. 27.1). This disparity, however, is due entirely to much faster evolution of sexual isolation in sympatric than allopatric species pairs, suggesting – as we discuss below – the possibility of direct selection for sexual isolation in sympatry. Among allopatric taxa, prezygotic and postzygotic isolation arise at similar rates. When both prezygotic and postzygotic forms of reproductive isolation are considered simultaneously, the total strength of reproductive isolation increases quickly (Fig. 27.2). We are unaware of any analogous data on the rate at which total reproductive isolation increases in other taxa. It would certainly be worth obtaining such information, as it would allow one to see if the speciation clock ticks at the same rate in different taxa. In other taxa, the data are far more impressionistic. In many groups of plants, such as orchids and the genus *Mimulus*, temporal or pollinator isolation sometimes evolves faster than postzygotic isolation, because related species produce fertile offspring when forcibly crossed in the greenhouse but fail to hybridize in nature (Grant, 1981). Students of bird evolution have noted that prezygotic isolation often seems to evolve well before hybrid sterility and inviability (Prager and Wilson, 1975; Grant and Grant, 1996).

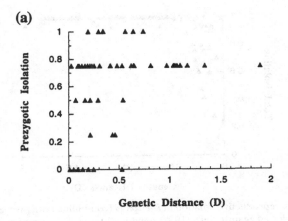

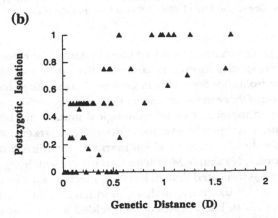

Figure 27.1. Strength of reproductive isolation in *Drosophila* plotted against Nei's (1972) electrophoretic genetic distance (an index of divergence time). Each point represents the average among pairs within a species group, so points are evolutionarily independent. (a) Prezygotic (sexual) isolation, (b) postzygotic isolation. See Coyne and Orr (1989b, 1997) for further details.

But these conclusions must be seen as preliminary. We require additional and more systematic studies in which different forms of reproductive isolation are assessed among pairs of species that diverged at about the same time. Such work is especially practicable in plants, as ecological differences can be studied in the greenhouse, pollinator isolation can be studied *in situ*, and postzygotic isolation can be studied through forced crossing.

27.4.3. A Summary of Genetic Studies

Because there are relatively few studies of the genetics of speciation, we have summarized them all in Table 27.1. We used two criteria for including a study

538   *Jerry A. Coyne and H. Allen Orr*

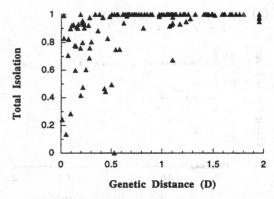

Figure 27.2. Total reproductive isolation in *Drosophila* (combining both prezygotic and postzygotic isolation) plotted against Nei's (1972) genetic distance. Each point represents a single pair of species. See Coyne and Orr (1989b, 1997) for further details.

in this table. First, the character studied must be known to cause reproductive isolation between species in either nature or the laboratory or be plausibly involved in such isolation. Second, the genetic analysis must have been fairly rigorous, with one of three methods used: (1) classical genetic analyses, in which species differing in molecular or morphological mutant markers are crossed and the segregation of reproductive isolation with the markers examined. Here we included only those studies in which markers were distributed among all major chromosomes; (2) simple Mendelian analyses in which segregation ratios in backcrosses or F_2s indicated that an isolating mechanism was due to changes at a single locus; or (3) biometric analyses, in which measurement of character means and variances in backcrosses or F_2s yielded a rough estimate of gene number. Table 27.1 also gives the actual or the minimum number of genes involved for each isolating mechanism. The footnotes give more detail about the type of genetic analysis, whether each pair of species was sympatric of allopatric, and a brief summary and critique of the results.

Several caveats are necessary. First, despite our attempts to comprehensively comb the literature, we have surely missed some studies. Second, the quality of the analyses is uneven: Some classical genetic studies involved detailed mapping experiments with many markers, while others relied on only one marker per chromosome. Third, in most cases the number of genes causing a single reproductive isolating mechanism is an underestimate. This may reflect a limited number of markers, failure to test all possible interactions between chromosomes, or the use of biometric approaches, which nearly always underestimate true gene number. Finally, data are given for single isolating mechanisms, but several mechanisms may often operate together to impede gene flow in nature.

The most striking feature of Table 27.1 is the imbalance of both species and isolating mechanisms. Roughly 75% of all the studies involve *Drosophila*, with only seven other pairs of taxa, mostly plants from the genus *Mimulus*.

Table 27.1. *Summary of Existing Genetic Analyses of Reproductive Isolation between Closely Related Species*

Species Pair	Trait	Number of genes	Footnote*
D. heteroneura/D. silvestris	Head shape	9	1
D. melanogaster/D. simulans	Hybrid inviability	≥9	2
	Female pheromones	≥5	3
D. mauritiana/D. simulans	Hybrid male sterility	≥15	4
	Hybrid female sterility	≥4	5
	Hybrid inviability	≥5	5
	Male sexual isolation	≥2	6
	Female sexual isolation	≥3	6
	Genital morphology	≥9	7
	Shortened copulation	≥3	8
D. mauritiana/D. sechellia	Female pheromones	≥6	9
D. simulans/D. sechellia	Hybrid male sterility	≥6	10
	Hybrid inviability	≥2	11
	Female sexual isolation	≥2	12
D. mojavensis/D. arizonae	Hybrid male sterility	≥3	13
	Male sexual isolation	≥2	14
	Female sexual isolation	≥2	14
D. pseudoobscura/D. persimilis	Hybrid male sterility	≥9	15
	Hybrid female sterility	≥3	15
	Sexual isolation	≥3	16
D. pseudoobscura USA/Bogota	Hybrid male sterility	≥5	17
D. buzatti/D. koepferae	Hybrid male inviability	≥4	18
	Hybrid male sterility	≥7	19
D. subobscura/D. madeirensis	Hybrid male sterility	≥6	20
D. virilis/D. littoralis	Hybrid female viability	≥5	21
D. virilis/D. lummei	Male courtship song	≥4	22
	Hybrid male sterility	≥6	23
D. hydei/D. neohydei	Hybrid male sterility	≥5	24
	Hybrid female sterility	≥2	24
	Hybrid inviability	≥4	25
D. montana/D. texana	Hybrid female inviability	≥2	26
D. virilis/D. texana	Hybrid male sterility	≥3	27
Ostrina nubialis, Z and E races	Female pheromones	1	28
	Male perception of pheromones	2	28
Laupala paranigra/L. kohalensis	Song pulse rate	≥8	29
Spodoptera latifascia/S. descoinsi	Pheromone blend	1	30
Xiphophorus helleri/X. maculatus	Hybrid inviability	2	31
M. lewisii/M. cardinalis	8 floral traits	(see note)	32
M. guttatus/M. micranthus	Bud growth rate	8	33
	Duration of bud development	10	33

(Continued)

Table 27.1. (*Continued*)

Species pair	Trait	Number of genes	Footnote*
Mimulus, four taxa	(see note)	(see note)	34
M. guttatus pops.	Hybrid inviability	2 (system 1)	35
		≥2 (system 2)	35
M. guttatus/ M. cupriphilus	Flower size	3–7	36
Helianthus annuus/ H. petiolarus	Pollen viability	≥14	37

* The footnote to each study includes a summary of the results, emphasizing any interesting features. The text gives our criteria for including studies in this table.

1. (Templeton, 1977; Val, 1977). Sympatric species, biometric analysis. Head shape difference conjectured (but not known) to be involved in sexual isolation between the two species. Templeton (1977, p. 636) concludes that shape difference "is a polygenic trait determined by alleles with predominantly additive effects." No confidence limits are given for either of the two estimates of gene number (9 in one study and 10 in the other).

2. (Pontecorvo, 1943). Sympatric species, genetic analysis. (Both species are cosmopolitan human commensals.) Gene number is a minimum estimate and includes factors from both species. Results based on analysis of pseudobackcross hybrids between marked triploid *D. melanogaster* females and irradiated *D. simulans* males.

3. (Coyne, 1996a). Sympatric species, genetic analysis. (Both species are cosmopolitan human commensals.) Genes affect ratio of two female cuticular hydrocarbons that appear to be involved in sexual isolation.

4. (True et al., 1996; Wu et al., 1996). Allopatric species, genetic analysis. The analysis of True et al. implicates at least 14 regions of the *D. mauritiana* genome that cause male sterility when introgressed into a *D. simulans* background. Wu et al. (1996) report 15 genes on the *D. simulans* X chromosome that cause hybrid sterility; by extrapolating to the entire genome, they estimate that at least 120 loci cause hybrid male sterility. This latter figure may, however, be an overestimate, as some of the X-linked regions studied were known in advance to have large effects on sterility. See also Coyne (1989) and Davis and Wu (1996).

5. (True et al., 1996). Allopatric species, genetic analysis. Estimate based on homozygous introgression of *D. mauritiana* segments into *D. simulans*; we have given our minimum estimate of gene number based on the size of introgressions. Regions causing inviability invariably affect both males and females. No large effect of the X chromosome for either character. There could be many more factors affecting both female sterility and inviability, as 19 sites out of 87 cytological positions produced hybrid inviability and 12 sites caused female sterility. Density of male steriles is probably higher than that of female steriles: 65 out of the 185 regions caused male sterility. See also Davis et al. (1994) and Hollocher and Wu (1996).

6. (Coyne, 1989, 1992b, 1996b). Allopatric species, genetic analysis. Character studied was sexual isolation between *D. mauritiana* females and *D. simulans* males. The major chromosomes had different effects in the two sexes, implying that different genes are involved in female discrimination versus the male character that females discriminate against.

7. (True et al., 1997). Allopatric species, genetic analysis. Character measured was shape of the posterior lobe of the male genitalia. The interspecific difference in this character is thought, but not proven, to be involved in shortened copulation time between *D. mauritiana* males and *D. simulans* females, which itself causes reduced progeny number in hybrid matings (see note 8 below and Coyne, 1983).

8. (Coyne, 1993b). Allopatric species, genetic analysis. Shortened copulation time between *D. simulans* females and *D. mauritiana* males contributes to the reduced number of progeny in interspecific crosses.

9. (Coyne and Charlesworth, 1997). Allopatric species, genetic analysis. Ratio of two female cuticular hydrocarbons that appears to affect sexual isolation between these species.

10. (Coyne and Kreitman, 1986; Coyne and Charlesworth, 1989; Cabot et al., 1994; Hollocher and Wu, 1996). Allopatric species, genetic analysis. At least three genes on the X, two on the second, and one on the third chromosome cause sterility in hybrid males.

11. (Hollocher and Wu, 1996). Allopatric species, genetic analysis. Two out of three large nonoverlapping regions on the *D. sechellia* second chromosome (the only chromosome analyzed) cause inviability when introgressed as homozygotes into *D. simulans*.

12. (Coyne, 1992b). Allopatric species, genetic analysis. Only two of three major chromosomes are involved in female traits leading to reduced mating propensity of *D. sechellia* females with *D. simulans* males.

13. (Vigneault and Zouros, 1986; Pantazidis et al., 1993). Sympatric species, genetic analysis. Trait studied was sperm motility in hybrid males. The Y chromosome and two autosomes affect the character, while two other chromosomes do not. The X chromosome was not studied.

14. (Zouros, 1973, 1981). Sympatric species, genetic analysis. Different pairs of chromosomes (and hence genes) are involved in sexual isolation among males versus females (see note 6 above).

15. (Orr, 1987, 1989b). Sympatric species, genetic analysis. Large X effect was observed in both male and female hybrid sterility. Genes causing male and female sterility are probably different, given the different locations and effects. See also Wu and Beckenbach (1983).

16. (Noor, 1997). Sympatric species, genetic analysis. In each of the two backcrosses, one X-linked and one autosomal gene affect the male traits discriminated against by heterospecific females. Because the two X-linked genes map to different locations, the minimum estimate of genetic divergence between the species is three loci. See also Tan (1946).

17. (Orr 1989a, 1989b). Genetic analysis between U.S. populations and an allopatric isolate from Bogota, Colombia. Hybrid male sterility is due largely to X-autosomal incompatibilities. Recent work shows that at least three genes are involved on the Bogota X chromosome and two on the U.S. autosomes. Nonetheless, large regions of genome have no discernible effect on hybrid sterility (e.g., chromosome 4, half of chromosome 2, Y chromosome, etc.).

18. (Carvajal et al., 1996). Sympatric species, genetic analysis. Male inviability in backcrosses is caused by at least two factors on the X chromosome of *D. koepferae*. Two X-linked loci from *D. koepferae* cause male lethality on a *D. buzzatii* genetic background; this inviability can be rescued by cointrogression of two autosomal segments from *D. koepferae*. Therefore at least four loci are involved.

19. (Naveira and Fontdevila, 1986, 1991). Sympatric species, genetic analysis. This is probably a gross underestimate of gene number, particularly on the X, as heterospecific X-chromosomal (but not autosomal) segments of any size cause complete male sterility. Marin (1996), however, argues that Naveira and Fontdevila overestimated the number of autosomal factors that cause hybrid male sterility.

20. (Khadem and Krimbas, 1991). Sympatric species, genetic analysis. Male sterility quantified as testis size. Both X-linked and autosomal factors affect testis size in hybrids; the X chromosome has the largest effect.

21. (Mitrofanov and Sidorova, 1981). Sympatric species (*D. virilis* is a cosmopolitan species associated with humans), genetic analysis. Character measured was reduced viability of offspring from backcross females.

22. (Hoikkala and Lumme, 1984) Sympatric species, genetic analysis. Four factors (at least one on each autosome) affect the interspecific difference in number of pulses in pulse train of male courtship song. No apparent effect of X. Character is possibly but not yet definitely known to be involved in sexual isolation between these species.

23. (Heikkinen and Lumme, 1991). Sympatric species, genetic analysis. F_1 males are weakly sterile. Backcross analysis uncovered X-2, X-4, X-5 and Y-3, Y-4, Y-5 incompatibilities. Slight female sterility was also detected in backcross.

24. (Schäfer, 1978). Sympatric species (*D. hydei* is a cosmopolitan species associated with humans), genetic analysis. F_1 hybrids are fertile, but backcross hybrids show severe sterility. Female sterility involves a 3–4 incompatibility; some hint of X-A incompatibilities were also suggested, but not proven. Male sterility involves 3-4, X-3, X-4, Y-3, Y-4, and Y-5 incompatibilities.

25. (Schäfer, 1979). Genetic analysis, see note 24 for geographic distribution. Backcross hybrid inviability involves chromosomes X, 2, 3, and 4. Chromosomes 5 and 6 play no role. Thus, while a modest number of genes are involved, lethality is not truly polygenic.

26. (Patterson and Stone, 1952). Sympatric species, genetic analysis. F_1 females having *D. montana* mothers are inviable. Analysis showed that inviability results from an incompatibility between a dominant X-linked factor from *D. texana* and recessive maternally acting factor(s) from *D. montana*. While the *D. texana* factor appears to be a single gene (mapped near the *echinus* locus), nothing is known about the number or location of the maternally acting factor(s).

27. (Lamnissou et al., 1996). Sympatric species (*D. virilis* is a cosmopolitan species associated with humans), genetic analysis. Male sterility appears only in backcrosses. Sterility is primarily caused by a Y_{tex}–5_{vir} incompatibility, although milder Y_{tex}–2-3vir incompatibility also occurs. (2-3 is fused in *D. texana*; thus 2-3 behaves as single linkage group in backcrosses.)

28. (Lofstedt et al., 1989; Roelofs et al., 1987). Genetic analysis, sympatric races of European corn borers in New York State. Characters studied were female pheromone blend (ratios of two long-chain acetates, whose interspecific difference is apparently due to a single autosomal locus). The electrophysiological response to these pheromones by male antennae sensilla is due to another unlinked autosomal locus, and the behavioral response of males to female pheromones is due to a single X-linked factor.

29. (Shaw, 1996). Allopatric species, biometric analysis. Trait studied was male song pulse rate of two species of Hawaiian crickets, which may be involved in sexual isolation. Reciprocal F_1 crosses indicate no disproportionate effect of the X chromosome.

30. (Monti et al., 1997). Sympatric species, genetic analysis. Trait studied was ratio of two pheromones, apparently due to a single factor. The timing of female emission of the pheromones, which differs between the species, appears to be polygenic, and the authors give no estimate of number of factors. Male perception of traits must, of course, also differ if there is to be reproductive isolation, so that, as in corn borers (see note 28), sexual isolation must be due to changes in at least two genes.

31. (Wittbrodt et al., 1989). Sympatric species, genetic analysis. Hybrid inviability between *X. helleri* (swordtail) and *X. maculatus* (platyfish) is caused by appearance of malignant melanomas, which are often fatal. These cancers are caused by the interaction between a dominant, X-linked oncogene encoding a receptor tyrosinane kinase and an autosomal suppressor that is either missing in the swordtail or dominant in the platyfish. Some backcross hybrids inherit the oncogene without the suppressor, yielding melanomas.

32. (Bradshaw et al., 1995). Sympatric species, genetic analysis. Traits studied by QTL analysis include flower color, corolla and petal width, nectar volume and concentration, and stamen and pistil length. All of these traits affect whether a flower is pollinated by bumblebees (*M. lewisii*) or hummingbirds (*M. cardinalis*). Species are at least partially reproductively isolated by pollinator difference. For most traits, the species difference involved at least one chromosome region of large effect. The difference in carotenoids in petal lobes was governed by a single gene.

33. (Fenster et al., 1995) Sympatric species, biometric analysis. Bud growth rate and duration reflect difference in flower size between these species: *M. micranthus* is largely selfing and apparently derived from the outcrossing *M. guttatus*. It is not known whether this difference causes reproductive isolation in nature, but this seems likely, given the decrease in gene flow caused by selfing. No evidence for factors of large effect. Lower bound of 95% confidence interval for gene number is 3.2 for bud growth rate and 4.6 for duration of bud development.

THE EVOLUTIONARY GENETICS OF SPECIATION 543

34. (Fenster and Ritland, 1994). Biometric analysis. Four taxa of controversial status, named *M. guttatus, M. nasutus,* (both outcrossing), and *M. micranthus* and *M. laciniatus* (predominantly selfing). The latter three taxa are sometimes.classified as subspecies of *M. guttatus.* Traits studied included differences in flowering time, corolla length, corolla width, stamen level, pistil length, stigma – anther separation. Selfing species have shorter and narrower corollas, shorter stamens and pistils, and less stigma – anther separation than outcrossers. Flower characters are therefore associated with breeding systems. Minimum number of genes for character differences averaged across all taxa varied between 5 and 13 per trait. Standard errors are large. Our impression is that these phenotypic differences involve several to many genes, with no single locus causing most of the difference in any character between any two taxa. Within a cross, positive genetic correlations were often seen between many traits, so that genes causing these character differences are not necessarily independent.
35. (Macnair and Christie, 1983; Christie and Macnair, 1984, 1987). Allopatric populations, genetic analysis. Complementary lethal loci are polymorphic (one in each population) in two North American populations of *M. guttatus.* There are two separate systems of inviability. The first involves only two genes, both polymorphic for complementary lethals. The other involves one locus (possibly the gene involved in copper tolerance) that interacts with an unknown number of genes in a nontolerant population.
36. (Macnair and Cumbes, 1989). Sympatric species, biometric analysis. Seven flower-size characters studied, some of which (e.g, height, width, pistil length) are probably related to difference in breeding systems between these species *M. cupriphilus* selfs much more often than does *M. guttatus,* and this difference in breeding systems may contribute to reproductive isolation. Stamen/pistil length ratio is also important in reproductive isolation, but it was not studied genetically. Each character difference was due to between three and seven genes. High genetic correlations were observed between many characters, so that traits are not genetically independent.
37. (Rieseberg, 1997). Sympatric species, genetic analysis. Within the colinear portions of the genome of these two sunflower species, Rieseberg estimates that at least 14 chromosomal segments are responsible for inviability of hybrid pollen. Thus approximately 40 genes are involved if one assumes a similar density of factors in the rearranged portions of the genome.

Moreover, nearly two thirds of the *Drosophila* work is on hybrid sterility and inviability. There is no published genetic study of ecological isolation (but see below). We obviously must extend such studies to other groups and other forms of reproductive isolation.

For convenience, we discuss prezygotic and postzygotic isolation separately.

27.5. Prezygotic Isolation

Although there are many forms of prezygotic isolation, including differences in ecology, behavior, time of reproduction, gametic compatibility, and (in plants) pollinators, only one form – sexual isolation – has been the subject of much theory and experiment.

27.5.1. Sexual Isolation

Many evolutionists have noted that closely related animal species, particularly those involved in adaptive radiations, seem to differ most obviously in sexually dimorphic traits. This impression has been confirmed by two phylogenetic studies of birds. Barraclough et al. (1995) found a positive correlation between the speciosity of groups and their degree of sexual dimorphism, suggesting a link between sexual selection and speciation. (It is not likely that species are simply recognized more easily in strongly dimorphic groups, for the authors note that females of such species can also be distinguished easily.) Mitra et al. (1996) found that taxa with promiscuous mating systems contain more species than their nonpromiscuous sister taxa. Promiscuously mating species are, of course, more likely to experience strong sexual selection.

The connection between sexual selection and sexual isolation may seem obvious. After all, it is a tenet of neo-Darwinism that reproductive isolation is a by-product of evolutionary change occurring within populations. Two populations undergoing sexual selection may readily diverge in both male traits and female preferences, and the natural outcome of this would be sexual isolation between the populations. This idea is much easier to grasp than, say, the notion that adaptation among isolated populations would cause sterility or inviability of their hybrids; under sexual selection, the pleiotropic effect of the diverging genes (sexual isolation) is obviously connected to their primary effect (exaggeration of male traits or female preferences).

Nonetheless, consideration of the role of sexual selection in speciation was remarkably late in coming. The earliest paper even mentioning this topic appears to be that of Nei (1976), who made a model of sexual selection in which one gene controlled the male trait and another the female preference. Exploring the conditions that could lead to the joint evolution of these traits, Nei noted that such a process could cause reproductive isolation. On the empirical side, Ringo (1977) proposed that sexual selection might explain both the adaptive radiation and the strong sexual dimorphism of Hawaiian *Drosophila*.

The most extensive theoretical work on this problem is that of Lande (1981, 1982). In 1981, he showed that sexual isolation could be the by-product of sexual selection if random genetic drift in small populations triggered the runaway process suggested by Fisher (1930). Later workers (e.g., Kirkpatrick, 1996) have shown, however, that such instability is unlikely if natural selection acts on female preference. Lande's 1982 model, an explicit quantitative-genetic treatment of clinal speciation by means of sexual selection, is more realistic. Here he showed that adaptive geographic differentiation in male traits could be amplified by the evolution of female preferences, resulting in reproductively isolated populations along a cline. Data that may support this model come from the guppy *Poecilia reticulata*, in which local populations have differentiated so that females prefer to mate with local rather than foreign males (Endler and Houde, 1995).

There is only one theoretical analysis of sympatric speciation resulting from sexual selection. Turner and Burrows (1995) constructed a genetic model of a male trait affected by four loci and a female preference for that trait affected by a single locus with two alleles. Under some conditions, this model produced sympatric taxa showing complete sexual isolation; but these results may be highly dependent on their assumptions. (One such assumption was that the preference locus showed complete dominance, but more realistic assumptions of intermediate dominance and additional loci would almost certainly reduce the probability of speciation.)

The only theoretical study of allopatric speciation resulting from sexual isolation is that of Iwasa and Pomiankowski (1995). Their quantitative-genetic model of sexual selection on a male trait and on female preference assumed that there was a fitness cost to increased female preference. This cost results in a cyclical fluctuation of both trait and preference; different populations undergoing the same selection regime could fall out of synchrony, causing their sexual isolation. Kirkpatrick and Barton (1995), however, note that this model neglects the possibility of stabilizing selection's acting on the male trait around its optimum value for survival and that such selection (even if very weak) could eliminate the cycling. Moreover, related species do not usually show different stages of elaboration and diminution of single male-limited traits, but instead the exaggeration of completely different traits. [Some species of bowerbirds, for example, have different forms of elaborate male plumage, while others lack such plumage but have males who build elaborate bowers (Gilliard, 1956).] A useful extension of Iwasa and Pomiankowski's model might incorporate either novel environments or genetic drift that could launch populations on different trajectories of sexual selection. Schluter and Price (1993) also suggest that, if the secondary sexual traits of males reflect their fitness or physiological condition, differences among habitats that change the female's ability to detect different traits could lead to sexual isolation among populations.

Sexual isolation is often asymmetric, i.e., species or populations show strong isolation in only one direction of the hybridization. This is a common phenomenon in *Drosophila* (Watanabe and Kawanishi, 1979; Kaneshiro, 1980), is also seen in salamanders (Arnold et al., 1996), and may be ubiquitous in other animals. Kaneshiro (1980) offered an explanation of this pattern based on biogeography, but newer data contradict his theory (Cobb et al., 1989; David et al., 1974). Arnold et al. (1996) proposed that mating asymmetry is a transitory phenomenon that decays rapidly as populations diverge, but such asymmetry is often seen among even distantly related species of *Drosophila* (Coyne and Orr, 1989b). (Eventually, however, asymmetry must disappear as sexual isolation becomes complete in both directions.) There are likely to be other explanations of asymmetry, such as the combination of open-ended female preferences and sexual selection operating in only one of two populations. If asymmetry does prove ubiquitous, it may provide important clues about how sexual selection causes speciation.

Unfortunately, there are too few data about the genetics of sexual isolation to confirm or motivate any theory. The few studies listed in Table 27.1 show only that sexual isolation may sometimes have a simple basis, as in races of *Ostrina nubialis*, in which complete sexual isolation is apparently based on changes at only three loci (one each for differences in female pheromones, male perception, and male attraction), or can be more complicated, as in *D. mauritiana/D. sechellia*, in which differences in female pheromones alone involve at least five loci. One conclusion that seems reasonable, based as it is on three independent studies (Table 27.1), is that for a given pair of species, the genes causing sexual isolation of males differ from those causing sexual isolation in females. This is not surprising, as there is no obvious reason why genes affecting male traits should be identical to those affecting female perception. (It is possible that sexual selection could cause male and female genes to be closely linked by selecting for genetic correlations between trait and preference, but there is no theory addressing this possibility.) This lack of correlation is worth verifying, however, as its absence is assumed in many models of sexual isolation (e.g., Spencer et al., 1986).

Given the likely importance of sexual selection in animal speciation, more genetic analyses are clearly needed. Fortunately, the advent of QTL analysis has made such studies feasible in any pair of species that produces fertile hybrids.

27.5.2. Reinforcement

One of the greatest controversies in speciation concerns reinforcement: The process whereby two allopatric populations that have evolved some postzygotic isolation in allopatry undergo selection for increased sexual isolation when they later become sympatric. Reinforcement was introduced and popularized by Dobzhansky (1937), who apparently considered it the necessary last step of speciation. Its wide popularity may have reflected its seductive assumption of a creative (and not an incidental) role for natural selection in speciation (Coyne, 1994). While theoretical studies of reinforcement appeared only recently, two analyses of *Drosophila* (Ehrman, 1965; Wasserman and Koepfer, 1977)·supported the idea: In species pairs with overlapping ranges, sexual isolation was stronger when populations derived from areas of sympatry than from allopatry.

In the 1980s, however, several critiques eroded the popularity of reinforcement. First, Templeton (1981) pointed out that the pattern of stronger sexual isolation among sympatric than allopatric populations could be caused not by reinforcement but by differential fusion, in which species could persist in sympatry only if they had evolved sufficiently strong sexual isolation in allopatry. Thus stronger isolation in sympatry might not reflect direct selection, but the loss by fusion of weakly isolated populations. Moreover, it became clear that some of the data offered in support of reinforcement were flawed (Butlin, 1987). Finally, the first serious theoretical treatment of reinforcement (Spencer et al., 1986) showed that even under favorable conditions (e.g., complete sterility of hybrids), extinction of populations occurred more often than reinforcement.

Recently, however, a combination of empirical and theoretical work has resurrected the popularity of reinforcement. In an analysis of 171 pairs of *Drosophila* species, Coyne and Orr (1989b, 1997) found that recently diverged pairs show far more sexual isolation when sympatric than allopatric (Fig. 27.3). [An independent analysis of these data by Noor (1997), making less restrictive assumptions, arrived at similar conclusions.] Although in 1989 there were no theoretical studies showing that reinforcement was feasible, two such investigations have appeared recently. Liou and Price (1994) and Kelly and Noor

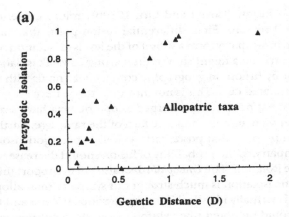

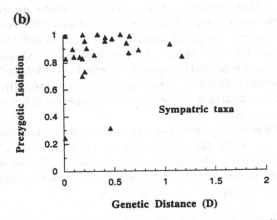

Figure 27.3. Prezygotic isolation in *Drosophila* plotted against Nei's (1972) electrophoretic genetic distance. Each point represents the average among pairs within a species group. (a) Allopatric taxa, (b) Sympatric taxa. See Coyne and Orr (1989b, 1997) for further details.

(1996) showed that reinforcement can occur frequently even if hybrids have only moderate postzygotic isolation. The important difference between these models and that of Spencer et al. (1986) is that the former explicitly allowed for sexual selection, which greatly enhanced the evolution of sexual isolation.

Newer data also support reinforcement. These include a reanalysis of the earlier literature, finding many more possible examples (Howard, 1993), and two new studies of species pairs with partially overlapping ranges (Noor, 1995; Saetre et al., 1997). The work of Saetre et al. on two species of European fly-catchers is especially interesting, as the reduced hybridization in sympatry is caused by the divergence of male plumage occurring in that area, a difference that was presumably caused by sexual selection.

Although the pattern of stronger isolation in sympatry shown in Fig. 27.3 might in principle be explained by several processes, including reinforcement and differential fusion (Coyne and Orr, 1989b), reinforcement seems most likely for several reasons. First, differential fusion posits that cases of strong sexual isolation in sympatry form a subset of the levels of isolation seen in allopatry: while allopatric taxa might show either strong or weak isolation, the latter cases disappear by fusion on geographic contact, leaving us with a preexisting set of strongly isolated taxa. This is not, however, the pattern seen in *Drosophila*: Fig. 27.3 shows that no recently diverged allopatric taxa have sexual isolation as strong as that seen among sympatric taxa of the same age. Furthermore, differential fusion predicts that postzygotic as well as prezygotic isolation will be stronger in sympatry, as the probability of fusion should decrease with any form of reproductive isolation. The *Drosophila* data also fail to support this prediction: While prezygotic isolation is much stronger in sympatry than allopatry, postzygotic isolation is virtually identical in the two groups (Coyne and Orr, 1997).

Finally, it should be noted that reinforcement does not necessarily require the preexistence of hybrid sterility or inviability, but might also result when allopatric populations evolve some behavioral or ecological difference that leaves hybrids behaviorally or ecologically maladapted. Stratton and Uetz (1986), for example, observed that hybrids between two species of wolf spiders were behaviorally sterile, so that hybrid males were rejected by females of both species, while hybrid females refused to mate with any male. Similarly, Davies et al. (1997) observed that hybrid females between two species of butterflies suffer a reduced tendency to mate.

The data and theory reviewed above strongly suggest that reinforcement of sexual isolation can occur. This conclusion represents one of the most radical changes of views about speciation over the past decade. Future work must determine whether reinforcement is rare or ubiquitous across animal taxa and whether it exists in plants.

27.5.3. Ecological Isolation

Ecological isolation must play a role in maintaining biological diversity since – even if it is not a primary isolating mechanism – ecological differences are required for the sympatric coexistence of taxa. Consider, for instance, the fate of a newly arisen polyploid plant species. Although all autopolyploids or allopolyploids are automatically postzygotically isolated from their parental species (hybridization produces mostly sterile triploid hybrids), any polyploid that does not differ ecologically from its ancestral species will be quickly driven extinct through competition, rendering it unavailable for study.

Ecological isolation actually subsumes three phenomena:

(1) Individuals of different species live in the same region and may encounter each other, but confine mating and/or reproduction to different habitats so that hybrids are not formed [e.g., the sympatric, host-specific *Drosophila* that breed on different cacti (Ruiz and Heed, 1988)]. This is a form of prezygotic reproductive isolation and the only type of reproductive isolation that can by itself both cause complete speciation and allow persistence of species in sympatry.

This form of isolation may be the most common result of sympatric speciation (Rice and Hostert, 1993).

(2) Species live in different subniches of the same area and rarely, if ever, come into contact (e.g., the spadefoot toads *Scaphiopus holbrooki hurteri* and *S. couchi*). Although their ranges overlap extensively, heterospecific toads almost never meet because they are restricted to different soil types (Wasserman, 1957). Although this situation corresponds to Dobzhansky's (1937, p. 231) definition of ecological isolation, such species are effectively allopatric.

(3) Species live in different subhabitats of the same area and sometimes come into contact with one another and mate, forming hybrids that are not well adapted to available habitat. Such cases may be common and have been studied extensively in stickleback fish (Schluter, 1996). They constitute examples of postzygotic isolation that, while technically a form of hybrid inviability, depend on ecological details of the environment and not on inherent problems of development. Like any other form of postzygotic isolation, ecological inviability could trigger reinforcement between sympatric species (Coyne and Orr, 1989b). Unlike other forms of postzygotic isolation, however, it may – depending on how population size is regulated in the different habitats – allow two species to coexist without prezygotic isolation. In addition, postzygotic isolation based on ecological divergence need not involve complementary gene interactions (epistasis) between alleles of two species. Instead, the genetics of this type of ecological isolation would presumably resemble the genetics of ordinary adaptation [about which we unfortunately know little (Orr and Coyne, 1992)].

Schluter (1996, 1997) makes a strong case for the importance of ecological isolation (especially type 3) in the origin and persistence of species. Unfortunately, there has been only one study of the genetics of ecological isolation. The island endemic *D. sechellia* breeds exclusively on the normally toxic fruit of *Morinda citrifolia*. All of *D. sechellia*'s relatives, including its presumed ancestor *D. simulans*, succumb to *Morinda*'s primary toxin, octanoic acid. Although *D. sechellia* and *D. simulans* were originally allopatric, they have recently become sympatric and breed on separate hosts. Recent work (Jones, 1998) shows that *D. sechellia*'s resistance to octanoic acid involves at least five semidominant alleles, distributed over all three of the major chromosomes, with the largest effect mapping to chromosome three.

27.5.4. Pollinator Isolation

Pollinator isolation probably represents a common form of reproductive isolation in plants (Grant, 1981). A variant of this is the isolation of insect-pollinated plants from self-compatible species, a mechanism described in *Mimulus*. The few data at hand (Table 27.1) show that the difference between outcrossing and inbreeding is due to several genes, but that differences in flower shape, color, or nectar reward that attract different pollinators may be due to one or a few major genes (Prazmo, 1965; Bradshaw et al., 1995). This latter observation, if common, might suggest a rapid form of speciation.

27.5.5. Postmating, Prezygotic Isolation

Biologists have begun to appreciate that sexual selection is not limited to obvious behavioral and morphological traits that act before copulation, but can
include cryptic characters that act between copulation and fertilization. Such
selection [as well as sexually antagonistic selection (Rice, 1996)] can lead to
female control of sperm usage, male – male sperm competition within multiply
inseminated females, and the mediation of such competition by the female.
Selection acting between copulation and fertilization has also been invoked to
explain the striking diversity of male genitalia among animal species, bizarre
conformations of female reproductive tracts, and postcopulation courtship behavior by males (Eberhard, 1985, 1996). Moreover, the relatively rapid evolution of proteins involved in reproduction (Coulthart and Singh, 1988; Metz and
Palumbi, 1996; Tsaur and Wu, 1997) is also consistent with sexual selection.

Just as sexual selection acting on male plumage or courtship behavior can
pleiotropically produce sexual isolation, so postcopulation, prezygotic sexual
selection can produce cryptic sexual isolation detectable only after fertilization. Such isolation may take the form of either blocked heterospecific fertilization, such as the insemination reaction of *Drosophila* (Patterson, 1946), or
the preferential use of conspecific sperm when a female is sequentially inseminated by heterospecific and conspecific males. This latter phenomenon has
recently been described in grasshoppers (Bella et al., 1992), crickets (Gregory
and Howard, 1994), flour beetles (Wade et al., 1994) and *Drosophila* (Price,
1997). In three species of *Drosophila*, single heterospecific inseminations produce large numbers of hybrid offspring, but females doubly inseminated by
both conspecific and heterospecific males produce very few hybrids. Such reproductive isolation thus depends on competition between conspecific and
heterospecific sperm, and resembles interspecific pollen competition in some
plants, in which heterospecific pollen tubes grow more slowly than conspecific
tubes, yielding reproductive isolation detectable only after double pollination
(Rieseberg et al., 1995). The only genetic analysis among these studies is that of
Price (1997), who showed that, among females, conspecific sperm precedence
was dominant, so that F_1 female hybrids between *D. simulans* and *D. mauritiana*
induce the same sperm preference as do *D. simulans* females. Gametic incompatibilities may be among the earliest-evolving forms of reproductive isolation,
and are worthy of more attention.

27.6. Postzygotic Isolation

Postzygotic isolation occurs when hybrids are unfit. Evolutionists have, historically, pointed to three types of genetic differences as causes of these fitness problems: species may have different chromosome arrangements, different ploidy
levels, or different alleles that do not function properly when brought together
in hybrids. Although each of these modes of speciation has enjoyed its advo-
cates, it is now clear that the latter two are by far the most important.

Speciation by autopolyploidy and allopolyploidy is clearly common in plants,
as about ~60% of angiosperms are of polyploid origin (Masterson, 1994). We

THE EVOLUTIONARY GENETICS OF SPECIATION

will not, however, discuss polyploid speciation here as it has been thoroughly reviewed elsewhere (e.g., Grant, 1981; Ramsey and Schemske, 1998). Instead, we concentrate on the question of whether postzygotic isolation in animals is based on chromosomal or genic differences. We will conclude that genes play a far more important role in hybrid sterility and inviability than do structural differences in chromosomes.

27.6.1. Chromosomal Speciation

The notion that chromosome rearrangements are a major cause of hybrid sterility was once very popular (White, 1969, 1978) and still enjoys a few adherents (e.g., King, 1993). The idea of chromosomal speciation, however, suffers from both theoretical and empirical difficulties. The theoretical problems have been discussed elsewhere (Lande, 1979; Walsh, 1982; Barton and Charlesworth, 1984), but the empirical problems have not been widely recognized.

The first such difficulty is that many species producing sterile hybrids are homosequential, that is, they do not differ in chromosome arrangement. White (1969, p. 77), the greatest champion of chromosomal speciation, argued that such cases "represent only a small fraction of the total number of species complexes that have been extensively studied." But this argument is potentially misleading: Many of the taxa showing fixed chromosomal differences are old and may have accumulated many – if not all – of their chromosomal differences after the actual speciation event. Resolving this issue requires systematic comparisons of the extent of chromosomal divergence with both the age of taxa (as determined by molecular data) and the strength of postzygotic isolation. In addition, although sterile hybrids often suffer meiotic pairing problems, this sterility is often limited to the heterogametic sex (see below). This is not expected if sterility results from the disruption of chromosome pairing, which in most cases should afflict both males and females. Indeed, in some species such as Drosophila, intraspecific chromosomal rearrangements typically sterilize females only (male Drosophila have no recombination), while hybrid sterility is far more severe in males (Ashburner, 1989).

Third, even in species hybrids whose chromosomes fail to pair during meiosis, we do not know whether this failure is caused by differences in chromosome arrangement or differences in genes. As Dobzhansky (1937) emphasized, both rearrangements and gene mutations can disrupt meiosis within species and so, presumably, within species hybrids. The best attempt to disentangle these causes remains the first: D. pseudoobscura and D. persimilis, which differ by at least four inversions, produce sterile hybrid males (Dobzhansky, 1937). Meiotic pairing in hybrids is abnormal and univalents are common. Dobzhansky (1933) showed that islands of tetraploid spermatocytes are often found in hybrid testes. All chromosomes in these $4N$ cells have a homologous pairing partner and hence should show improved pairing if pairing problems in hybrid cells reflect structural differences between chromosomes. However, univalents are just as common in the hybrid tetraploid as in diploid spermatocytes. The hybrid meiotic problems must therefore have a genetic and not a chromosomal basis. This test has apparently not been repeated in any other hybridization.

A further problem with chromosomal speciation is that it depends critically on the semisterility of hybrids who are heterozygous for chromosome rearrangements. It is not widely appreciated, however, that rearrangements theoretically expected to be deleterious as heterozygotes (e.g., fusions and pericentric inversions) in reality often enjoy normal fitness, probably because recombination or missegregation simply does not occur (see discussion in Coyne et al., 1987). Any putative case of chromosomal speciation requires proof that different rearrangements actually cause semisterility in heterozygotes, and almost no studies have met this standard.

Finally, direct genetic analyses over the past decade have shown conclusively that postzygotic isolation in animals is typically caused by genes, not by large chromosome rearrangements. Many of these genes have been well mapped, and several have been genetically characterized or even cloned (Wittbrodt et al., 1989; Orr, 1992; Perez et al., 1993). Our main task, therefore, is to understand how the evolution of genic differences can produce hybrid sterility and inviability.

27.6.2. The Dobzhansky–Muller Model

Understanding the evolution of postzygotic isolation is difficult, because the phenotypes we are hoping to explain – the inviability and sterility of hybrids – seem maladaptive. The difficulty is best seen by considering the simplest possible model for the evolution of postzygotic isolation: change at a single gene. One species has genotype *AA*, the other *aa*, while *Aa* hybrids are completely sterile. Regardless of whether the common ancestor was *AA* or *aa*, fixation of the alternative allele cannot occur because the first mutant individual has genotype *Aa* and so is sterile. Using the metaphor of adaptive landscapes, it is hard to see how two related species can come to reside on different adaptive peaks unless one lineage passed through an adaptive valley.

This problem was finally solved by Bateson (1909), Dobzhansky (1937), and Muller (1942), who noted that, if postzygotic isolation is based on incompatibilities between two or more genes, hybrid sterility and inviability can evolve unimpeded by natural selection. If, for example, the ancestral species had genotype *aabb*, a new mutation at one locus (allele *A*) could be fixed by selection or drift in one isolated population as the *Aabb* and the *AAbb* genotypes are perfectly fit. Similarly, a new allele (*B*) at the other locus could be fixed in a different population as the *aaBb* and *aaBB* genotypes are perfectly fit. But while each population is fit, it is entirely possible that when these populations come into contact, the resulting *AaBb* hybrids would be sterile or inviable. The *A* and *B* alleles have never been tested together within a genome, and so may not function properly when brought together in hybrids.

Alleles showing this pattern of epistasis are called complementary genes. Such genes need not, of course, have drastic effects on hybrid fitness; any particular incompatibility might lower hybrid fitness by only a small amount. It should also be noted that the Dobzhansky–Muller model is agnostic about the evolutionary causes of substitutions that ultimately produce hybrid sterility or inviability: Purely adaptive or purely neutral evolution within populations can give rise to complementary genes and thus to postzygotic isolation.

The Dobzhansky–Muller model is the basis for almost all modern work in the genetics of postzygotic isolation. There is now overwhelming evidence that hybrid sterility and inviability do indeed result from such between-locus incompatibilities (reviewed in Orr, 1997). Curiously, theoretical studies of the Dobzhansky–Muller model have been slow in coming. Recent analyses, however, predict that the evolution of postzygotic isolation should show several regularities.

27.6.3. Patterns in the Genetics of Postzygotic Isolation

Long before any formal studies of the Dobzhansky–Muller model, Muller (1942) predicted that the alleles causing postzygotic isolation must act asymmetrically. To see this, consider the two-locus case sketched above. Although the A and B alleles might be incompatible in hybrids, their allelomorphs a and b must be compatible. This is because the $aabb$ genotype must represent an ancestral state in the evolution of the two species.

There is now good evidence that genic incompatibilities do in fact act asymmetrically. The best data come from Wu and Beckenbach's (1983) study of male sterility in *D. pseudoobscura–D. persimilis* hybrids. When an X-linked region from one species caused sterility on introgression into the other species' genome, they found that the reciprocal introgression had no such effect. This observation has now been confirmed in many additional *Drosophila* hybridizations (e.g., Orr and Coyne, 1989).

The second pattern expected under the Dobzhansky–Muller model was pointed out more recently. If hybrid sterility and inviability are caused by the accumulation of complementary genes, the severity of postzygotic isolation, as well as the number of genes involved, should snowball much faster than linearly with time (Orr, 1995; see also Menotti-Raymond et al., 1997). This follows from the fact that any new substitution in one species is potentially incompatible with the alleles from the other species at all of those genes that have previously diverged. (In our discussion above, the B substitution is potentially incompatible with the previous A substitution in the other species.) Later substitutions are therefore more likely to cause hybrid incompatibilities than earlier ones. Consequently, the cumulative number of hybrid incompatibilities increases much faster than linearly with the number of substitutions K. If all incompatibilities involve pairs of loci, the expected number of Dobzhansky–Muller incompatibilities increases as K^2 or (assuming a rough molecular clock) as the square of the time since species diverged (Orr, 1995). If incompatibilities sometimes involve interactions between more than two loci (see below), the number of hybrid incompatibilities will rise even faster. This snowballing effect requires that we interpret genetic studies of postzygotic isolation with caution. Because the genetics of hybrid sterility and inviability will quickly grow complicated as species diverge, it is easy to overestimate the number of genes required to cause strong reproductive isolation (see Orr, 1995, and below).

The limited data we possess are consistent with a snowballing effect, although they do not prove it (see Orr, 1995, for a discussion). The simplest prediction of the snowballing hypothesis is that the number of mapped genes causing hybrid sterility or inviability should increase quickly with molecular genetic distance between species. But given the enormous difficulties inherent in accurately mapping and counting speciation genes, it may be some time before such direct contrasts are possible.

Third, while we have assumed that hybrid incompatibilities involve pairs of genes, analysis of the Dobzhansky–Muller model shows that more complex hybrid incompatibilities, involving interactions among three or more of genes, should be common. The reason is not intuitively obvious but is easily demonstrated mathematically (Cabot et al., 1994; Orr, 1995). Certain paths to the evolution of new species are barred because they would require passing through intermediate genotypes that are sterile or inviable. It is easy to show, however, that the proportion of all imaginable paths to speciation allowed by selection increases with the complexity of hybrid incompatibilities (Orr, 1995). Thus, for the same reason that two-gene speciation is easier than single-gene speciation so three-gene is easier than two-gene speciations, and so on.

The evidence for complex incompatibilities is now overwhelming. They have been described in the *D. obscura* group (Muller, 1942), the *D. virilis* group (Orr and Coyne, 1989), the *D. repleta* group (Carvajal et al., 1996), and the *D. melanogaster* group (Cabot et al., 1994; Davis et al., 1994). Indeed, such interactions could prove more common than the two-locus interactions discussed at length by Bateson, Dobzhansky, and Muller.

Theoretical analysis of the Dobzhansky–Muller model also has yielded results that contradict popular ideas about the effect of population subdivision on speciation. Many evolutionists have maintained, for example, that speciation is most likely in taxa subdivided into small populations. This is, however, demonstrably untrue, at least for postzygotic isolation. Orr and Orr (1996) showed that if the substitutions ultimately causing Dobzhansky–Muller incompatibilities are driven by natural selection – as seems likely (Christie and Macnair, 1984) – the waiting time to speciation grows longer as a species of a given size is splintered into ever smaller populations. If, on the other hand, the substitutions causing hybrid problems are originally neutral, population subdivision has little effect on the time to speciation. In no case is the accumulation of hybrid incompatibilities greatly accelerated by small population size. Unfortunately, we have little empirical data bearing on this issue. Although it might seem that the effect of population size on speciation rates could be estimated by comparing the rate of evolution of postzygotic isolation on islands versus continents, this comparison is confounded by the likelihood of stronger selection in novel island habitats, which itself might drive rapid speciation.

Finally, much of the theoretical work on the evolution of postzygotic isolation has been devoted to explaining one of the most striking patterns characterizing the evolution of hybrid sterility and inviability – Haldane's rule. Because this large and confusing literature has recently been reviewed elsewhere (e.g., Wu et al., 1996; Orr, 1997), we will not attempt a thorough discussion here. Instead, we briefly consider the data from nature and sketch the leading hypotheses offered to explain them.

27.6.4. Haldane's Rule

In 1922, Haldane noted that, if only one hybrid sex is sterile or inviable, it is nearly always the heterogametic (XY) sex. More recent and far more extensive reviews (Coyne, 1992c) show that Haldane's rule is obeyed in all animal groups that have been surveyed, e.g., Drosophila, mammals, Orthoptera, birds, and Lepidoptera (the latter two groups have heterogametic females). Indeed, it is likely that Haldane's rule characterizes postzygotic isolation in all animals having chromosomal sex determination. Moreover, these surveys show that Haldane's rule is consistently obeyed. In Drosophila, for instance, out of 114 species crosses producing sterile hybrids of one sex only, in 112 it is the males (Coyne, 1992c). Comparative work in Drosophila also shows that Haldane's rule represents an early stage in the evolution of postzygotic isolation: Hybrid male sterility or inviability arises quite quickly, while female effects appear only much later (Coyne and Orr, 1989b, 1997).

For obvious reasons, Haldane's rule has received a great deal of attention. It represents one of the strongest patterns in evolutionary biology and perhaps the only pattern characterizing speciation. In addition, the rule implies that there is some fundamental similarity in the genetic events causing speciation in all animals.

Although many hypotheses have been offered to explain Haldane's rule, most have been falsified. We will not consider these failed explanations here (see Orr, 1997; Wu et al., 1996; Coyne et al., 1991; and Coyne, 1992c). Instead, we briefly review the three explanations of Haldane's rule that remain feasible. There is strong evidence for two of these and suggestive evidence for the third. In a field that has historically been rife with disagreement, a surprisingly good consensus has emerged that some combination of these hypotheses explains Haldane's rule (Orr, 1997; Wu et al., 1996; True et al., 1996).

The first hypothesis, the dominance theory, posits that Haldane's rule reflects that the recessivity of X-linked genes causes hybrid problems. This idea was first suggested by Muller (1942), and his verbal theory was later formalized by Orr (1993b) and Turelli and Orr (1995). The mathematical work shows that heterogametic hybrids suffer greater sterility and inviability than homogametic hybrids whenever the alleles causing hybrid incompatibilities are, on average, partially recessive ($\bar{d} < 1/2$; this parameter incorporates both the effects of dominance per se and any correlation between dominance and severity of hybrid effects; see Turelli and Orr, 1995). The reason is straightforward.

Although XY hybrids suffer the full hemizygous effect of all X-linked alleles causing hybrid problems (dominant and recessive), XX hybrids suffer twice as many X-linked incompatibilities (as they carry twice as many Xs). These two forces balance when $\bar{d} = 1/2$. But if $\bar{d} < 1/2$, the expression of recessives in XY hybrids outweighs the greater number of incompatibilities in XX hybrids, and Haldane's rule results. Obviously, the dominance theory can account not only for Haldane's rule, but also for the well-known large effect of the X chromosome on hybrid sterility and inviability (Wu and Davis, 1993; Turelli and Orr, 1995).

There is now strong evidence that dominance explains Haldane's rule for hybrid inviability. In particular, the dominance theory predicts that, in *Drosophila* hybridizations obeying Haldane's rule for inviability, hybrid females who are forced to be homozygous for their X chromosome should be as inviable as F_1 hybrid males. (Such unbalanced females possess an F_1 malelike genotype in which all recessive X-linked genes are fully expressed.) In both of the species crosses in which this test has been performed, unbalanced females are, as expected, completely inviable (Orr, 1993a; Wu and Davis, 1993). Similarly, there is evidence from haplodiploid species that hybrid backcross males (who are haploid) suffer more severe inviability than their diploid sisters (Breeuwer and Werren, 1995).

There is also weaker indirect evidence that the alleles causing hybrid sterility act as partial recessives: Hollocher and Wu (1996) and True et al., (1996) found that while most heterozygous introgressions from one *Drosophila* species into another are reasonably fertile, many homozygous introgressions are sterile. It therefore seems likely that dominance contributes to Haldane's rule for both hybrid inviability and hybrid sterility. Last, it is worth noting that the dominance theory – unlike several alternatives – should hold in all animal taxa, regardless of which sex is heterogametic (Orr and Turelli, 1996).

The second hypothesis posits that Haldane's rule reflects the faster evolution of genes ultimately causing hybrid male than female sterility (Wu and Davis, 1993; Wu et al., 1996). Wu and his colleagues offer two explanations for this faster-male evolution: (1) In hybrids, spermatogenesis may be disrupted far more easily than oogenesis, and (2) sexual selection may cause genes expressed in males to evolve faster than those expressed in females. While there is now good evidence for faster-male evolution (see below), this theory cannot be the sole explanation of Haldane's rule. First, it cannot explain Haldane's rule for sterility in those taxa having heterogametic females, e.g., birds and butterflies. After all, spermatogenesis and sexual selection involve males per se, while Haldane's rule extends to all heterogametic hybrids, male or not. Second, the faster-male theory cannot account for Haldane's rule for inviability in any taxa. Because there is strong evidence that genes causing lethality are almost always expressed in both sexes (reviewed in Orr, 1997), it seems unlikely that hybrid male lethals can evolve faster than female lethals. Finally, it is not obvious that sexual selection would inevitably lead to faster substitution of male than female alleles. One can easily imagine, for instance, forms of sexual selection in which each substitution affecting female preference is matched by a substitution affecting expression of a male character. If sexual selection causes faster evolution of male sterility, one may need to consider processes like male-male competition in addition to male–female coevolution.

Despite these caveats, there is now good evidence – at least in *Drosophila* – that alleles causing sterility of hybrid males accumulate much faster than those affecting females (True et al., 1996; Hollocher and Wu, 1996). (Unfortunately, both of these studies analyzed the same species pair; analogous data from other species are badly needed.) While we cannot be sure of the mechanism involved, it certainly appears that faster-male evolution plays an important role in Haldane's rule for sterility in taxa with heterogametic males.

The last hypothesis, the faster-X theory, posits that Haldane's rule reflects the more rapid divergence of X-linked than autosomal loci (Charlesworth et al., 1987; Coyne and Orr, 1989b). Charlesworth et al. (1987) showed that, if the alleles ultimately causing postzygotic isolation were originally fixed by natural selection, X-linked genes will evolve faster than autosomal genes if favorable mutations are on average partially recessive ($\overline{h} < 1/2$). (It must be emphasized that this theory requires only that the favorable effects of mutations on their normal conspecific genetic background be partially recessive; nothing is assumed about the dominance of alleles in hybrids. Conversely, the dominance theory requires only that the alleles causing hybrid problems act as partial recessives in hybrids; nothing is assumed about the dominance of these alleles on their normal conspecific genetic background.) Under various cases, this faster evolution of X-linked genes can indirectly give rise to Haldane's rule (Orr, 1997).

There is some evidence that X-linked hybrid steriles and lethals do in fact evolve faster than their autosomal analogs. In their genome-wide survey of speciation genes in the *D. simulans–D. mauritiana* hybridization, True et al. (1996) found a significantly higher density of hybrid male steriles on the X chromosome than on the autosomes. Hollocher and Wu (1996), however, found no such difference in a much smaller experiment. Thus, while faster-X evolution may contribute to Haldane's rule, the evidence for it is considerably weaker than that for both the dominance and faster-male theories. [The faster-X theory also suffers several other shortcomings described elsewhere (Orr, 1997).] Definitive tests of the faster-X theory must await new experiments that, following True et al., allow direct comparison of the number of hybrid steriles and lethals on the X versus autosomes.

In sum, there is now strong evidence for both the dominance and faster-male theories of Haldane's rule. Future work must include better estimates of the dominance of hybrid steriles, better tests of the faster-X theory, and, most important, genetic analyses of Haldane's rule in taxa having heterogametic females. Although both the dominance and faster-male theories make clear predictions about the genetics of postzygotic isolation in these groups (Orr, 1997), we have virtually no direct genetic data from these critical taxa. Last, it is important to determine if Haldane's rule extends beyond the animal kingdom, particularly to those species of plants having heteromorphic sex chromosomes.

27.6.5. Hybrid Rescue Mutations

No discussion of postzygotic isolation would be complete without mentioning a recent and remarkable discovery – hybrid rescue mutations. These are alleles that, when introduced singly into *Drosophila* hybrids, rescue the viability or fertility of normally inviable or sterile individuals (Watanabe, 1979; Takamura and Watanabe, 1980; Hutter and Ashburner, 1987; Hutter et al., 1990; Sawamura et al., 1993a, 1993b, 1993c; Davis et al., 1996). All of the rescue mutations studied to date involve the *D. melanogaster–D. simulans* hybridization. In one direction of this cross, hybrid males die as late larvae; in the other, hybrid females die as embryos. All surviving hybrids are completely sterile (Sturtevant, 1920). Several mutations are known that rescue the larval inviability: lethal hybrid rescue (*Lhr*) from *D. simulans* (Watanabe, 1979), and hybrid male rescue (*Hmr*) and *In(1)AB* from *D. melanogaster* (Hutter and Ashburner, 1987; Hutter et al., 1990). Despite some early confusion, it now appears that these rescue mutations have no effect on hybrid embryonic inviability, which is instead rescued by a different set of mutations: zygotic hybrid rescue (*Zhr*) from *D. melanogaster* (Sawamura et al., 1993c), and maternal hybrid rescue (*mhr*) from *D. simulans* (Sawamura et al., 1993a). The fact that larval and embryonic lethality are rescued by nonoverlapping sets of mutations strongly suggests that these forms of isolation have different developmental bases (Sawamura et al., 1993).

Recently, Davis et al. (1996) described a mutation that rescues, albeit weakly, the fertility of *D. melanogaster–D. simulans* hybrid females. Although little is known about this allele, its discovery suggests that it may be possible to bring all of the genetic and molecular technology available in *D. melanogaster* to bear on speciation. (This species has previously been of limited use in the genetics of speciation as it produces no fertile progeny when crossed to any other species.)

The discovery of hybrid rescue genes has several important implications. First, it suggests that postzygotic isolation may have a simple genetic basis. It seems quite unlikely that a single mutation could restore the viability of hybrids if lethality were caused by many different developmental problems. But a simple developmental basis implies in turn a simple genetic basis: If many genes were involved, it seems unlikely that they all would affect the same developmental pathway. These inferences have been roughly confirmed by more recent work on the developmental basis of larval inviability in *D. melanogaster–D. simulans* group hybrids, which shows that lethal hybrids suffer a profound mitotic defect that can be reversed by introduction of hybrid rescue mutations (Orr et al., 1997). This suggests (but does not prove) that hybrid inviability in this case results from a single developmental defect – failure to condense chromosomes during mitosis. [Such a suggestion may seem contradicted by recent introgression experiments showing that many different chromosome regions cause postzygotic isolation when introgressed from one *Drosophila* species into another [(True et al., 1996; Hollocher and Wu, 1996). But the overwhelming majority of these introgressions have discernible effects only when homozygous and, so, would not play any role in F_1 hybrid inviability or sterility (Hollocher and Wu, 1996).]

The discovery of hybrid rescue genes may also provide an important shortcut to the cloning and characterization of speciation genes. It is possible that rescue mutations are alleles of the genes that normally cause hybrid inviability or sterility (Hutter and Ashburner, 1987). If so, characterization of these mutations might quickly lead to the molecular isolation of speciation genes. Alternatively, rescue mutations might be second-site suppressors, i.e., mutations at some second set of loci that suppress the problems caused by a different, primary set of genes.

27.6.6. The Role of Endosymbionts in Speciation

Recent work on postzygotic isolation has pointed to a novel cause of hybrid incompatibilities – cytoplasmic incompatibility (CI) resulting from infection by cellular endosymbionts. Although CI has now been found in at least five orders of insects (Stevens and Wade, 1990; Hoffmann and Turelli, 1997), most work has focused on two model systems, the fly *D. simulans* (Hoffmann et al., 1986; Turelli and Hoffmann, 1995) and the small parasitic wasp *Nasonia* (Perrot-Minnot et al., 1996; Werren, 1997).

In both systems, hybrid embryonic lethality results when males from infected populations or species are crossed to females from uninfected populations or species. Moreover, in both cases the infective agent is the *rickettsia*-like endosymbiont *Wolbachia* (Turelli and Hoffmann, 1995; Werren, 1997). Remarkably, antibiotic treatment of infected lines cures *Wolbachia* infections, allowing fully compatible crosses. Because infected-by-infected crosses are compatible – while the same crosses using genetically identical but cured females are incompatible – the presence of *Wolbachia* in females must confer immunity to the effects of fertilization by sperm from infected males. The mechanism underlying this immunity is unclear.

Because CI only occurs when naive uninfected cytoplasm is fertilized by sperm from infected males, CI is typically unidirectional. Recently, however, cases have been found in both *Drosophila* (O'Neill and Karr, 1990) and *Nasonia* (Breeuwer and Werren, 1995; Perrot-Minnot et al., 1996) in which two different populations or species are infected by different varieties of *Wolbachia*. Crosses between individuals carrying these different types of *Wolbachia* are bidirectionally incompatible, so that postzygotic isolation is complete. This represents a remarkable mode of speciation in which no changes are required in an organism's genome.

Although *Wolbachia* infections will surely prove common – and speciation workers must begin routine testing for them – there are reasons for questioning whether CI will prove an important cause of speciation. First, *Wolbachia* cannot explain Haldane's rule, which is ubiquitous among animals and characterizes (at least in *Drosophila*), an early and nearly obligate step in the evolution of

560   *Jerry A. Coyne and H. Allen Orr*

postzygotic isolation (Coyne and Orr, 1989b, 1997). Although CI results in dead embryos, we know of no cases in which *Wolbachia* causes lethality of one sex only. Similarly, *Wolbachia* seems unlikely to be a common cause of hybrid sterility: Among the many genetic analyses of hybrid sterility, a single case alone involves an endosymbiont (Somerson et al., 1984).

27.7. Conclusions

The study of speciation has grown increasingly respectable over the past decade. The reasons are obvious: A number of important questions have been resolved by experiment and a number of patterns explained by theory. This progress reflects several fundamental but rarely recognized changes in our approach to speciation. First, the field has grown increasingly genetical. As a consequence, a large body of grand but notoriously slippery questions (How important are peak shifts in speciation? Is sympatric speciation common?) have been replaced with a collection of simpler questions (Is the Dobzhansky-Muller model correct? What is the cause of Haldane's rule?). While it would be fatuous to claim that these new questions are more important than the old, there is no doubt that they are more tractable. Second, the connection between theory and experiment has grown increasingly close. While speciation once seemed riddled with amorphous and untestable verbal theories, the past decade of work has produced a body of mathematical theory, yielding clear and testable predictions about the basis of reproductive isolation. Last, but most important, many of these predictions have been tested.

Despite this progress, many questions about speciation remain unanswered. Throughout this chapter we have tried to highlight those questions that seem to us both important and tractable. Most fall into two broad sets. The first concerns speciation in taxa that have been relatively ignored: Does reinforcement occur in plants? Do prezygotic and postzygotic isolation evolve at approximately the same rate in most taxa as in *Drosophila*? Do plants with heteromorphic sex chromosomes obey Haldane's rule? Do hybrid male steriles still evolve faster than female steriles in taxa having heterogametic females? How distinct are sympatric asexual taxa?

The second set of questions concerns the evolution of prezygotic isolation, which has received less attention than postzygotic isolation. How complex is the genetic basis of sexual isolation? How common is reinforcement? Why is sexual isolation so often asymmetric? What is the connection between adaptive radiation and sexual isolation?

It may seem that trading yesterday's grand verbal speculations for today's smaller, more technical studies risks a permanent neglect of the larger questions about speciation. We believe, however, that more focused pursuits of tractable questions will ultimately produce better answers to these bigger questions. Just as no mature theory of population genetics was possible until we understood the facts of inheritance, so no mature view of speciation seems possible until we understand the origins and mechanics of reproductive isolation.

27.8. Acknowledgments

As Lewontin's student (JAC) and grandstudent (HAO), we are most grateful to Dick for his constant scientific and personal inspiration. Our work is supported by grants from the U.S. National Institutes of Health to JAC and to HAO and from the David and Lucile Packard Foundation to HAO. We thank M. Turelli for helpful comments and discussion of the snowball effect. An earlier version of this paper was published under the same title in the Philosophical Transactions of the Royal Society Series B (1998), Vol. 353, pp. 287–305.

The study of Jones (1998) described on page 549, should be included in Table 27.1; it is the only existing genetic analysis of ecological isolation.

REFERENCES

Arnold, S. J., Verrell, P. A., and Tilley, S. G. 1996. The evolution of asymmetry in sexual isolation: a model and a test case. *Evolution* 50:1024–1033.

Ashburner, M. 1989. *Drosophila: A Laboratory Handbook.* New York: Cold Spring Harbor Laboratory.

Barraclough, T. G., Harvey, P. H., and Nee, S. 1995. Sexual selection and taxonomic diversity in passerine birds. *Proc. Roy. Soc. London Ser. B* 259:211–215.

Barton, N. H. and Charlesworth, B. 1984. Genetic revolutions, founder effects, and speciation. *Annu. Rev. Ecol. Syst.* 15:133–164.

Bateson, W. 1909. Heredity and variation in modern lights. In *Darwin and Modern Science*, A. C. Seward, ed., pp. 85–101. Cambridge: Cambridge U. Press.

Baum, D. and Shaw, K. L. 1995. Genealogical perspectives on the species problem. *Monogr. Syst. Bot.* 53:289–303.

Bella, J. L., Butlin, R. K., Ferris, C., and Hewitt, G. M. 1992 Asymmetrical homogamy and unequal sex ratio from reciprocal mating-order crosses between *Chorthippus parallelus* subspecies. *Heredity* 68:345-352.

Bradshaw, H. D., Wilbert, S. M., Otto, K. G., and Schemske, D. W. 1995. Genetic mapping of floral traits associated with reproductive isolation in monkey flowers (*Mimulus*). *Nature (London)* 376:762–765.

Breeuwer, J. A. J. and Werren, J. H. 1995. Hybrid breakdown between two haplodiploid species: the role of nuclear and cytoplasmic genes. *Evolution* 49:705–717.

Butlin, R. 1987 Speciation by reinforcement. *Trends Ecol. Evol.* 2:8–13.

Cabot, E. L., Davis, A. W., Johnson, N. A., and Wu, C.-I. 1994. Genetics of reproductive isolation in the *Drosophila simulans* clade: complex epistasis underlying hybrid male sterility. *Genetics* 137:175–189.

Carvajal, A. R., Gandarela, M. R., and Naveira, H. F. 1996. A three-locus system of interspecific incompatibility underlies male inviability in hybrids betwen *Drosophila buzzattii* and *D. koepferi. Genetica* 98:1–19.

Charlesworth, B., Coyne, J. A., and Barton, N. 1987. The relative rates of evolution of sex chromosomes and autosomes. *Am. Nat.* 130:113-146.

Christie, P. and Macnair, M. R. 1984. Complementary lethal factors in two North American populations of the yellow monkey flower. *J. Hered.* 75:510-511.

Christie, P. and Macnair, M. R. 1987. The distribution of postmating reproductive isolating genes in populations of the yellow monkey flower, *Mimulus guttatus. Evolution* 41:571–578.

Cobb, M., Burnet, B., Blizard, R., and Jallon, J.-M. 1989. Courtship in *Drosophila sechellia*: its structure, functional aspects, and relationship to those of other members of the *Drosophila melanogaster* species subgroup. *J. Insect Behav.* **2**:63–89.

Cohan, F. M. 1998. Genetic structure of bacterial populations. In *Evolutionary Genetics from Molecules to Morphology*, R. Singh and C. Krimbas, eds. Cambridge: Cambridge U. Press.

Coulthart, M. D. and Singh, R. S. 1988. High level of divergence of male reproductive tract proteins between *Drosophila melanogaster* and its sibling species, *D. simulans*. *Mol. Biol. Evol.* **5**:182–191.

Coyne, J. A. 1983. Genetic basis of differences in genital morphology among three sibling species of Drosophila. *Evolution* **37**:1101–1118.

Coyne, J. A. 1989. The genetics of sexual isolation between two sibling species, *Drosophila simulans* and *Drosophila mauritiana*. *Proc. Natl. Acad. Sci. USA* **86**:5464–5468.

Coyne, J. A. 1992a. Much ado about species. *Nature (London)* **357**:289–290.

Coyne, J. A. 1992b. Genetics of sexual isolation in females of the *Drosophila simulans* species complex. *Genet. Res. Cambridge* **60**:25–31.

Coyne, J. A. 1992c. Genetics and speciation. *Nature (London)* **355**:511–515.

Coyne, J. A. 1993a. Recognizing species. *Nature (London)* **364**:298.

Coyne, J. A. 1993b. The genetics of an isolating mechanism between two sibling species of Drosophila. *Evolution* **47**:778–788.

Coyne, J. A. 1994. Ernst Mayr and the origin of species. *Evolution* **48**:19–30.

Coyne, J. A. 1996a. Genetics of differences in pheromonal hydrocarbons between *Drosophila melanogaster* and *D. simulans*. *Genetics* **143**:353-364.

Coyne, J. A. 1996b. Genetics of sexual isolation in male hybrids of *Drosophila simulans* and *D. mauritiana*. *Genet. Res.* **68**:211–220.

Coyne, J. A. and Charlesworth, B. 1989. Genetic analysis of X-linked sterility in hybrids between three sibling species of Drosophila. *Heredity* **62**:97–106.

Coyne, J. and Charlesworth, B. 1997. Genetics of a pheromonal difference affecting sexual isolation between *Drosophila mauritiana* and *D. sechellia*. *Genetics* **145**:1015–1030.

Coyne, J. A., Charlesworth, B., and Orr, H. A. 1991. Haldane's rule revisited. *Evolution* **45**:1710–1713.

Coyne, J. A. and Kreitman, M. 1986. Evolutionary genetics of two sibling species, *Drosophila simulans* and *D. sechellia*. *Evolution* **40**:673–691.

Coyne, J. A. and Orr, H. A. 1989a. Two rules of speciation. In *Speciation and its Consequences*, D. Otte and J. Endler, eds., pp. 180–207. Sunderland, MA: Sinauer.

Coyne, J. A. and Orr, H. A. 1989b. Patterns of speciation in *Drosophila*. *Evolution* **43**:362–381.

Coyne, J. A. and Orr, H. A. 1997. Patterns of speciation in Drosophila revisited. *Evolution* **51**:295–303.

Coyne, J. A., Orr, H. A., and Futuyma, D. J. 1988. Do we need a new species concept? *Syst. Zool.* **37**:190–200.

Cracraft, J. 1989. Speciation and its ontology: the empirical consequences of alternative species concepts for understanding patterns and processes of differentiation. In *Speciation and its Consequences*, D. Otte and J. A. Endler, eds., pp. 28–59. Sunderland, MA: Sinauer.

David, J., Bocquet, C., Lemeunier, F., and Tsacas, L. 1974. Hybridation d'une nouvelle espéce *Drosophila mauritiana* avec *D. melanogaster* et *D. simulans*. *Ann. Genét.* **17**:235–241.

Davies, N., Aiello, A., Mallet, J., Pomiankowski, A., and Silberglied, R. E. 1997. Speciation in two neotropical butterflies: extending Haldane's rule. *Proc. R. Soc. London Ser. B* **264**:845–851.

Davis, A. W., Noonburg, E. G., and Wu, C. I. 1994. Evidence for complex genetic interactions between conspecific chromosomes underlying hybrid female sterility in the *Drosophila simulans* clade. *Genetics* **137**:191–199.

Davis, A. W., Roote, J., Morley, T., Sawamura, K., Herrmann, S., and Ashburner, M. 1996. Rescue of hybrid sterility in crosses between *D. melanogaster* and *D. simulans*. *Nature (London)* **380**:157–159.

Davis, A. W. and Wu, C.-I. 1996. The broom of the sorcerer's apprentice: the fine structure of a chromosomal region causing reproductive isolation between two sibling species of Drosophila. *Genetics* **143**:1287–1298.

Dobzhansky, T. 1933. On the sterility of the interracial hybrids in *Drosophila pseudoobscura*. *Proc. Natl. Acad. Sci. USA* **19**:397–403.

Dobzhansky, T. 1935. A critique of the species concept in biology. *Philos. Sci.* **2**:344–345.

Dobzhansky, T. 1937. *Genetics and the Origin of Species*. New York: Columbia U. Press.

Eberhard, W. G. 1985. *Sexual Selection and Animal Genitalia*. Cambridge, MA: Harvard U. Press.

Eberhard, W. G. 1996. *Female Control: Sexual Selection by Cryptic Female Choice*. Princeton, NJ: Princeton U. Press.

Ehrman, L. 1965. Direct observation of sexual isolation between allopatric and between sympatric strains of the different *Drosophila paulistorum* races. *Evolution* **19**:459–464.

Endler, J. A. and Houde, A. E. 1995. Geographic variation in female preferences for male traits in *Poecilia reticulata*. *Evolution* **49**:456–468.

Fenster, C. B., Diggle, P. K., Barrett, S. C. H., and Ritland, K. 1995. The genetics of floral development differentiating two species of *Mimulus* (Scrophulariaceae). *Heredity* **74**:258–266.

Fenster, C. B. and Ritland, K. 1994. Quantitative genetics of mating system divergence in the yellow monkey flower species complex. *Heredity* **73**:422–435.

Fisher, R. A. 1930. *The Genetical Theory of Natural Selection*. Oxford: Oxford U. Press.

Gilliard, E. T. 1956. Bower ornamentation versus plumage characters in bower-birds. *Auk* **73**:450–451.

Grant, B. R. and Grant, P. R. 1996. Cultural inheritance of song and its role in the evolution of Darwin's finches. *Evolution* **50**:2471–2487.

Grant, V. 1981. *Plant Speciation*. 2nd ed. New York: Columbia U. Press.

Gregory, P. G. and Howard, D. J. 1994. A postinsemination barrier to fertilization isolates two closely related ground crickets. *Evolution* **48**:705–710.

Haldane, J. B. S. 1922. Sex-ratio and unisexual sterility in hybrid animals. *J. Genet.* **12**:101–109.

Heikkinen, E. and Lumme, J. 1991. Sterility of male and female hybrids of *Drosophila virilis* and *Drosophila lummei*. *Heredity* **67**:1–11.

Hoffmann, A. A. and Turelli, M. 1997. Cytoplasmic incompatibility in insects. In *Influential Passengers*, S. L. O'Neill, J. H. Werren, and A. A. Hoffmann, eds., Oxford: Oxford U. Press.

Hoffmann, A. A., Turelli, M., and Simmons, G. M. 1986. Unidirectional incompatibility between populations of *Drosophila simulans*. *Evolution* **40**:692–701.

Hoikkala, A. and Lumme, J. 1984. Genetic control of the difference in male courtship sound between *D. virilis* and *D. lummei*. *Behav. Genet.* **14**:827–845.

Hollocher, H. and Wu, C.-I. 1996. The genetics of reproductive isolation in the *Drosophila simulans* clade: X versus autosomal effects and male vs. female effects. *Genetics* **143**:1243–1255.

Holman, E. W. 1987. Recognizability of sexual and asexual species of rotifers. *Syst. Zool.* **36**:381–386.

Howard, D. J. 1993. Reinforcement: origin, dynamics, and fate of an evolutionary hypothesis. In *Hybrid Zones and the Evolutionary Process*. R. G. Harrison, ed., pp. 46–69. Oxford: Oxford U. Press.

Hutter, P. and Ashburner, M. 1987. Genetic rescue of inviable hybrids between *Drosophila melanogaster* and its sibling species. *Nature (London)* **327**:331–333.

Hutter, P., Roote, J., and Ashburner, M. 1990. A genetic basis for the inviability of hybrids between sibling species of Drosophila. *Genetics* **124**:909–920.

Iwasa, Y. and Pomiankowski, A. 1995. Continual change in mate preferences. *Nature (London)* **377**:420–422.

Jones, C. D. 1998. The genetic basis of *Drosophila sechellia*'s resistance to a host plant toxin. *Genetics* **149**:1899–1908.

Kaneshiro, K. Y. 1980. Sexual isolation, speciation, and the direction of evolution. *Evolution* **34**:437–444.

Kelly, J. K. and Noor, M. A. F. 1996. Speciation by reinforcement: a model derived from studies of Drosophila. *Genetics* **143**:1485–1497.

Khadem, M. and Krimbas, C. B. 1991. Studies of the species barrier between *Drosophila subobscura* and *D. madeirensis*. 1. The genetics of male hybrid sterility. *Heredity* **67**:157–165.

King, M. 1993. *Species Evolution*. Cambridge: Cambridge U. Press.

Kirkpatrick, M. 1996. Good genes and direct selection in the evolution of mating preferences. *Evolution* **50**:2125–2140.

Kirkpatrick, M. and Barton, N. 1995. Déjà vu all over again. *Nature (London)* **377**:388–389.

Lamnissou, K., Loukas, M., and Zouros, E. 1996. Incompatibilities between Y chromosome and autosomes are responsible for male hybrid sterility in crosses between *Drosophila virilis* and *Drosophila texana*. *Heredity* **76**:603–609.

Lande, R. 1979. Effective deme sizes during long-term evolution estimated from rates of chromosomal rearrangement. *Evolution* **33**:234–251.

Lande, R. 1981. Models of speciation by sexual selection on polygenic traits. *Proc. Natl. Acad. Sci. USA* **78**:3721–3725.

Lande, R. 1982. Rapid origin of sexual isolation and character divergence in a cline. *Evolution* **36**:213–223.

Liou, L. W. and Price, T. D. 1994. Speciation by reinforcement of prezygotic isolation. *Evolution* **48**:1451–1459.

Lofstedt, C., Hansson, B. S., Roelofs, W., and Bengtsson, B. O. 1989. No linkage between genes controlling female pheromone production and male pheromone response in the European corn borer, *Ostrinia nubilalis* Hübner (Lepidoptera; Pyralideae). *Genetics* **123**:553–556.

Macnair, M. R. and Christie, P. 1983. Reproductive isolation as a pleiotropic effect of copper tolerance in *Mimulus guttatus*. *Heredity* **50**:295–302.

Macnair, M. R. and Cumbes, Q. J. 1989. The genetic architecture of interspecific variation in Mimulus. *Genetics* **122**:211–222.

Marin, I. 1996. Genetic architecture of autosome-mediated hybrid male sterility in Drosophila. *Genetics* **142**:1169–1180.

Masterson, J. 1994. Stomatal size in fossil plants: evidence for polyploidy in majority of angiosperms. *Science* **264**:321-424.

Maynard-Smith, J. and Szathmáry, E. 1995. *The Major Transitions in Evolution*. Oxford: Freeman/Spektrum.

Mayr, E. 1942. *Systematics and the Origin of Species*. New York: Columbia U. Press.

Menotti-Raymond, M., David, V. A., and O'Brien, S. J. 1997. Pet cat hair implicates murder suspect. *Nature (London)* **386**:774.

Metz, E. C. and Palumbi, S. R. 1996. Positive selection and sequence rearrangements generate extensive polymorphism in the gamete recognition protein binding. *Mol. Biol. Evol.* **13**:397–406.

Mitra, S., Landel, H., and Pruett-Jones, S. 1996. Species richness covaries with mating system in birds. *Auk* **113**:544–551.

Mitrofanov, V. G. and Sidorova, N. V. 1981. Genetics of the sex ratio anomaly in *Drosophila* hybrids of the *virilis* group. *Theor. Appl. Gene.* **59**:17–22.

Monti, L., Génermont, J., Malosse, C., and Lalanne-Cassou, B. 1997. A genetic analysis of some components of reproductive isolation between two closely related species, *Spodoptera latifascia* (Walker) and *S. descoinsi* (Lalanne-Cassou and Silvain) (Lepidoptera: Noctuidae). *J. Evol. Biol.* **10**:121–143.

Muller, H. J. 1942. Isolating mechanisms, evolution, and temperature. *Biol. Symp.* **6**:71–125.

Naveira, H. and Fontdevila, A. 1986. The evolutionary history of *Drosophila buzzatii*. The genetic basis of sterility in hybrids between *D. buzzattii* and its sibling *D. serido* from Argentina. *Genetics* **114**:841–857.

Naveira, H. and Fontdevila, A. 1991. The evolutionary history of *Drosophila buzzatii*. XXI. Cumulative action of multiple sterility factors on spermatogenesis in hybrids of *D. buzzatii* and *D. koepferae*. *Heredity* **67**:57–72.

Nei, M. 1972. Genetic distances between populations. *Am. Nat.* **106**:282–292.

Nei, M. 1976. Mathematical models of speciation and genetic distance. In *Population Genetics and Ecology*, S. Karlin and E. Nevo, eds., pp. 723–766. New York: Academic.

Noor, M. 1995. Speciation driven by natural selection in *Drosophila*. *Nature (London)* **375**:674–675.

Noor, M. A. F. 1997. How often does sympatry affect sexual isolation in Drosophila? *Am. Nat.* **149**:1156–1163.

Noor, M. A. F. 1997. Genetics of sexual isolation and courtship dysfunction in male hybrids of *Drosophila pseudoobscura* and *D. persimilis*. *Evolution* **51**:809–815.

O'Neill, S. L. and Karr, T. L. 1990. Bidirectional incompatibility between conspecific populations of *Drosophila simulans*. *Nature (London)* **348**:178–180.

Orr, H. A. 1987. Genetics of male and female sterility in hybrids of *Drosophila pseudoobscura* and *D. persimilis*. *Genetics* **116**:555–563.

Orr, H. A. 1989a. Genetics of sterility in hybrids between two subspecies of *Drosophila*. *Evolution* **43**:180–189.

Orr, H. A. 1989b. Localization of genes causing postzygotic isolation in two hybridizations involving *Drosophila pseudoobscura*. *Heredity* **63**:231–237.

Orr, H. A. 1992. Mapping and characterization of a "speciation gene" in Drosophila. *Genet. Res.* **59**:73–80.

Orr, H. A. 1993a. Haldane's rule has multiple genetic causes. *Nature (London)* **361**:532–533.

Orr, H. A. 1993b. A mathematical model of Haldane's rule. *Evolution* **47**:1606–1611.

Orr, H. A. 1995. The population genetics of speciation: the evolution of hybrid incompatibilities. *Genetics* **139**:1805–1813.

Orr, H. A. 1997. Haldane's rule. *Annu. Rev. Ecol. Syst.* **28**, 195–218.

Orr, H. A. and Coyne, J. A. 1989. The genetics of postzygotic isolation in the *Drosophila virilis* group. *Genetics* **121**:527–537.

Orr, H. A. and. Coyne, J. A. 1992. The genetics of adaptation: a reassessment. *Am. Nat.* **140**:725–742.

Orr, H. A. and Orr, L. H. 1996. Waiting for speciation: the effect of population subdivision on the time to speciation. *Evolution* **50**:1742–1749.

Orr, H. A. and Turelli, M. 1996. Dominance and Haldane's rule. *Genetics* **143**:613–616.

Orr, H. A., Madden, L. D., Coyne, J. A., Goodwin, R., and Hawley, R. S. 1997. The developmental genetics of hybrid inviability: a mitotic defect in *Drosophila* hybrids. *Genetics* **145**:1031–1040.

Pantazidis, A. C., Galanopoulos, V. K., and Zouros, E. 1993. An autosomal factor from *Drosophila arizonae* restores normal spermatogenesis in *Drosophila mojavensis* males carrying the *D. arizonae* Y chromosome. *Genetics* **134**:309–318.

Patterson, J. T. 1946. A new type of isolating mechanism in Drosophila. *Proc. Natl. Acad. Sci. USA* **32**:202–208.

Patterson, J. T. and Stone, W. S. 1952. Evolution in the genus *Drosophila*. New York: MacMillan.

Perez, D. E., Wu, C.-I., Johnson, N. A., and Wu, M.-L. 1993. Genetics of reproductive isolation in the *Drosophila simulans* clade: DNA-marker assisted mapping and characterization of a hybrid-male sterility gene, *Odysseus* (*Ods*). *Genetics* **134**:261–275.

Perrot-Minnot, M.-J., Guo, L. R., and Werren, J. H. 1996. Single and double infections with Wolbachia in the parasitic wasp *Nasonia vitripennis*: effects on compatibility. *Genetics* **143**:961–972.

Pontecorvo, G. 1943. Viability interactions between chromosomes of *Drosophila melanogaster* and *Drosophila simulans*. *J. Genet.* **45**:51–66.

Prager, E. R. and Wilson, A. C. 1975. Slow evolutionary loss of the potential for interspecific hybridization in birds: a manifestation of slow regulatory evolution. *Proc. Natl. Acad. Sci. USA* **72**:200–204.

Prazmo, W. 1965. Cytogenetic studies on the genus *Aquilegia*. III. Inheritance of the traits distinguishing different complexes in the genus *Aquilegia*. *Acta Soc. Bot. Pol.* **34**:303–437.

Price, C. S. C. 1997. Conspecific sperm precedence in Drosophila. *Nature (London)* **388**:363–366.

Ramsey, J. M. and Schemske, D. W. 1998. The dynamics of polyploid formation and establishment in flowering plants. *Annu. Rev. Ecol. Syst.*, **29**:467–501.

Rice, W. R. 1996. Sexually antagonistic male adaptation triggered by experimental arrest of female evolution. *Nature (London)* **381**:232–234.

Rice, W. R. and Hostert, E. E. 1993. Laboratory experiments on speciation: what have we learned in 40 years? *Evolution* **47**:1637–1653.

Rieseberg, L. H., Desrochers, A. M., and Youn, S. J. 1995. Interspecific pollen competition as a reproductive barrier between sympatric species of *Helianthus* (Asteraceae). *Am. J. Bot.* **82**:515–519.

Rieseberg, L. H. 1997. Genetic mapping as a tool for studying speciation. In *Molecular Systematics of Plants*, 2nd ed., D. E. Soltis, P. S. Soltis, and J. J. Doyle, eds. New York: Chapman & Hall.

Ringo, J. M. 1977. Why 300 species of Hawaiian Drosophila? The sexual selection hypothesis. *Evolution* **31**:694–696.

Roberts, M. S. and Cohan, F. M. 1995. Recombination and migration rates in natural populations of *Bacillus subtilis* and *Bacillus mojavensis*. *Evolution* **49**:1081–1094.

Roberts, M. S., Nakamura, K. L., and Cohan, F. M. 1996. *Bacillus vallismortis* sp. nov., a close relative of *Bacillus subtilis*, isolated from soil in Death Valley, California. *Int. J. Syst. Bacteriol.* **46**:470–475.

Roelofs, W., Glover, T., Tang, X.-H, Sreng, I., Robbins, P., Eckenrode, C., Löfstedt, C., Hansson, B. S., and Bengtson, B. 1987. Sex pheromone production and perception in European corn borer moths is determined by both autosomal and sex-linked genes. *Proc. Natl. Acad. Sci. USA* **84**:7585–7589.

Ruiz, A. and Heed, W. B. 1988. Host-plant specificity in the cactophilic *Drosophila mulleri* species complex. *J. Anim. Ecol.* **57**:237–249.

Saetre, G.-P., Moum, T., Bures, S., Král, M., Adamjan, M., and Moreno, J. 1997. A sexually selected character displacement in flycatchers reinforces premating isolation. *Nature (London)* **387**:589–592.

Sawamura, K., Taira, T., and Watanabe, T. K. 1993a. Hybrid lethal systems in the *Drosophila melanogaster* species complex. I. The *maternal hybrid rescue* (*mhr*) gene of *Drosophila simulans*. *Genetics* **133**:299–305.

Sawamura, K., Watanabe, T. K., and Yamamoto, M.-T. 1993b. Hybrid lethal systems in the *Drosophila melanogaster* species complex. *Genetica* **88**:175–185.

Sawamura, K., Yamamoto, M.-T., and Watanabe, T. K. 1993c. Hybrid lethal systems in the *Drosophila melanogaster* species complex. II. The zygotic hybrid rescue (*Zhr*) gene of *D. melanogaster*. *Genetics* **133**:307–313.

Schäfer, U. 1978. Sterility in *Drosophila hydei* X *D. neohydei* hybrids. *Genetica* **49**:205–214.

Schäfer, U. 1979. Viability in *Drosophila hydei* X *D. neohydei* hybrids and its regulation by genes located in the sex heterochromatin. *Biol. Zentralbl.* **98**:153–161.

Schliewen, U. K., Tautz, D., and Pääbo, S. 1994. Sympatric speciation suggested by monophyly of crater lake cichlids. *Nature (London)* **368**:629–632.

Schluter, D. 1996. Ecological causes of adaptive radiation. *Am. Nat.* **148**(Suppl.):S40–S64.

Schluter, D. 1997. Ecological causes of speciation. In *Endless Forms: Species and Speciation*. D. J. Howard and S. H. Berlocher, eds. Oxford: Oxford U. Press.

Schluter, D. and Price, T. 1993. Honesty, perception, and population divergence in sexually selected traits. *Proc. R. Soc. London Ser. B* **253**:117–122.

Shaw, K. L. 1996. Polygenic inheritance of a behavioral phenotype: interspecific genetics of song in the Hawaiian cricket genus *Laupala*. *Evolution* **50**:256–266.

Somerson, N. L., Ehrman, L., Kocka, J. P., and Gottlieb, F. J. 1984. Streptococcal L-forms isolated from *Drosophila paulistorum* semispecies cause sterility in male progeny. *Proc. Natl. Acad. Sci. USA* **81**:282–285.

Spencer, H. G., McArdle, B. H., and Lambert, D. M. 1986. A theoretical investigation of speciation by reinforcement. *Am. Nat.* **128**:241–262.

Stratton, G. E. and Uetz, G. W. 1986. The inheritance of courtship behavior and its role as a reproductive isolating mechanism in two species of *Schizocosa* wolf spiders (Araneae: Lycosidea). *Evolution* **40**:129–141.

Stevens, L. and Wade, M. J. 1990. Cytoplasmically inherited reproductive incompatibility in *Tribolium* flour beetles: the rate of spread and effect on population size. *Genetics* **124**:367–372.

Sturtevant, A. H. 1920. Genetic studies on *Drosophila simulans*. I. Introduction. Hybrids with *Drosophila melanogaster*. *Genetics* **5**:388–500.

Takamura, T. and Watanabe, T. K. 1980. Further studies on the lethal hybrid rescue (Lhr) gene of *Drosophila simulans*. *Jpn. J. Genet.* **55**:305–408.

Tan, C. C. 1946. Genetics of sexual isolation between *Drosophila pseudoobscura* and *Drosophila persimilis. Genetics* 31:558–573.

Templeton, A. R. 1977. Analysis of head shape differences between two interfertile species of Hawaiian *Drosophila. Evolution* 31:630–641.

Templeton, A. R. 1981. Mechanisms of speciation – a population genetics approach. *Annu. Rev. Ecol. Syst.* 12:23–48.

True, J. R., Weir, B. S, and Laurie, C. C. 1996. A genome-wide survey of hybrid incompatibility factors by the introgression of marked segments of *Drosophila mauritiana* chromosomes into *Drosophila simulans. Genetics* 144:819–837.

True, J. R., Liu, J., Stam, L. F., Zeng, Z.-B., and Laurie, C. C. 1997. Quantitative genetic analysis of divergence in male secondary sexual traits between *Drosophila simulans* and *D. mauritiana.* Evolution 51:816–832.

Tsaur, S.-C. and Wu, C.-I. 1997. Positive selection and the molecular evolution of a gene of male reproduction, *Acp26Aa* of *Drosophila. Mol. Biol. Evol.* 14, 544–549.

Turelli, M. and Hoffmann, A. A. 1995. Cytoplasmic incompatibility in *Drosophila simulans*: dynamics and parameter estimates from natural populations. *Genetics* 140:1319–1338.

Turelli, M. and Orr, H. A. 1995. The dominance theory of Haldane's rule. *Genetics* 140:389–402.

Turner, G. F. and Burrows, M. T. 1995. A model of sympatric speciation by sexual selection. *Proc. R. Soc. London Ser. B* 260:287–292.

Val, F. C. 1977. Genetic analysis of the morphological differences between two interfertile species of Hawaiian *Drosophila. Evolution* 31:611–629.

Vigneault, G. and Zouros, E. 1986. The genetics of asymmetrical male sterility in *Drosophila mojavensis* and *Drosophila arizonensis* hybrids: interaction between the Y chromosome and autosomes. *Evolution* 40:1160–1170.

Wade, M. J., Patterson, H., Chang, N. W., and Johnson, N. 1994. Postcopulatory, prezygotic isolation in flour beetles. *Heredity* 72:163–167.

Walsh, J. B. 1982. Rate of accumulation of reproductive isolation by chromosome rearrangements. *Am. Nat.* 120:510–532.

Wasserman, A. O. 1957. Factors affecting interbreeding in sympatric species of spadefoots (genus *Scaphiopus*). *Evolution* 11:320–338.

Wasserman, M. and Koepfer, H. R. 1977. Character displacement for sexual isolation between *Drosophila mojavensis* and *Drosophila arizonensis. Evolution* 31:812–823.

Watanabe, T. K. 1979. A gene that rescues the lethal hybrids between *Drosophila melanogaster* and *D. simulans. Jpn. J. Genet.* 54:325–331.

Watanabe, T. K. and Kawanishi, M. 1979. Mating preference and the direction of evolution in *Drosophila. Science* 205:906–907.

Werren, J. H. 1997. *Wolbachia* and speciation. In *Endless Forms: Species and Speciation*. D. Howard and S. Berlocher, eds. Oxford: Oxford U. Press.

White, M. J. D. 1969. Chromosomal rearrangements and speciation in animals. *Annu. Rev. Genet.* 3:75–98.

White, M. J. D. 1978. *Modes of Speciation*. San Francisco: Freeman.

Wittbrodt, J., Adam, D., Malitschek, B., Maueler, W., Raulf, F., Telling, A., Robertson, S. M., and Schartl, M. 1989. Novel putative receptor tyrosine kinase encoded by the melanoma-inducing *Tu* locus in *Xiphophorus. Nature (London)* 341:315–421.

Wu, C.-I. and Beckenbach, A. T. 1983. Evidence for extensive genetic differentiation between the sex-ratio and the standard arrangement of *Drosophila pseudoobscura* and *D. persimilis* and identification of hybrid sterility factors. *Genetics* 105:71–86.

Wu, C.-I. and Davis, A. W. 1993. Evolution of postmating reproductive isolation: the composite nature of Haldane's rule and its genetic bases. *Am. Nat.* **142**:187–212.

Wu, C.-I., Johnson, N., and Palopali, M. F. 1996. Haldane's rule and its legacy: why are there so many sterile males? *Trends. Ecol. Evol.* **11**:281–284.

Zink, R. M. and McKitrick, M. C. 1995. The debate over species concepts and its implications for ornithology. *Auk* **112**:701–719.

Zouros, E. 1973. Genic differentiation associated with the early stages of speciation in the mulleri subgroup of *Drosophila*. *Evolution* **27**:601–621.

Zouros, E. 1981. The chromosomal basis of sexual isolation in two sibling species of Drosophila: *D. arizonensis* and *D. mojavensis*. *Genetics* **97**:703–718.

CHAPTER TWENTY EIGHT

Toward a Unified Theory of Speciation

RAMA S. SINGH

28.1. Introduction

The most visible aspect of the diversity of life, besides the sheer variety of organisms and the differences between plants and animals, is its clumpiness, i.e., organisms appear strikingly similar within groups but different between groups. The neo-Darwinian theory of evolution, with its emphasis on gradual evolution, requires a mechanism for producing this clumpiness in the diversity of life, and this is the problem of speciation. Speciation still remains one of the most fundamental unsolved problems in evolutionary biology. The objective of this chapter is fourfold. First, I will provide a brief historical overview of speciation, concepts, and theories to show that the reality or nonreality of the species problem, beyond that of evolution or change as shown by Darwin (1859), is directly related to our perception of continuity or discontinuity in nature. I will argue that while the overwhelming discontinuity of organisms in nature is clearly the result of accumulated changes occurring both within and between lineages over time, nevertheless there is a widespread tendency to mistakenly equate the problem of speciation to the problem of the origin of this total accumulated discontinuity. Second, a brief review of the recent progress in the area of the genetic studies of species differences and reproductive isolation will be followed by a description of a new emerging sex gene pool theory of speciation that presupposes a dichotomy of the species gene pool – sex gene pool and nonsex gene pool (Singh, 1990; Singh and Zeng, 1994). Third, I will try to answer the following question: Is a unified theory of speciation possible? The question is an important one if we are to aim for generality. Currently we are living in an era of plurality, which may be the correct approach for studying speciation, and the prevailing general notion is that each group of organisms has a unique mechanism of speciation. My own view is that a unified theory of speciation is indeed possible provided it allows

R. S. Singh and C. B. Krimbas, eds., *Evolutionary Genetics: From Molecules to Morphology*, vol. 1. © Cambridge University Press 2000. Printed in the United States of America. ISBN 0-521-57123-5. All rights reserved.

for a balanced combination of general and organism-specific principles of speciation.

28.2. The Nature of the Species Problem

It is commonly stated that Darwin did not solve the problem of speciation but rather the problem of evolution or change within species. This is of course not what Darwin himself thought. Darwin provided a solution to the species problem, and to him the problem of change (evolution) and the problem of diversity (speciation) were not as decoupled as they are now viewed. While Darwin has been blamed for a variety of reasons for his emphasis on gradualism, ranging from sociopolitical to biological, it is quite clear that his emphasis on the arbitrary nature of species was a logical necessity for him as no theory of evolution that relies on a material basis of (mechanistic) change would have succeeded unless the barriers or gaps between species were considered less formidable than they appear in general perception.

The alternative competing theory, of course, is a theory of punctuated evolution that proposes to consolidate most of the significant, if not all, evolutionary changes to brief periods or bursts at the time of lineage diversifications (Gould, and Eldredge, 1977; Gould, 1982). We are of course still arguing about these two models of evolution and speciation but the theory of punctuated evolution is a reformulation of a long tradition that goes back to the works of early geneticists (Bateson, 1894; De Vries, 1906), embryologists (Goldschmidt, 1940), and systematists and paleontologists (see Mayr, 1963, for detailed references) who argued for a decoupling of the processes of within and between lineage change and sought for factors of speciation besides those considered under population genetics theory (Fisher, 1930; Wright, 1931; Haldane, 1932). The neo-Darwinian revolution had a remarkable success – an agreement on the population genetic mechanisms of change (Provine, 1971), an evolutionary synthesis and unification of biology (Mayr and Provine, 1980), Dobzhansky's (1937, 1951) and Mayr's (1963) polygenic/allopatric theory of speciation, and Simpson's (1944, 1953) enlightening interpretation of paleontological evolution in the light of population genetics theory and Wright's adaptive landscape analogy (Mayr 1982a, Wright 1982). But in spite of this success the initial position of the proponents of nongradual evolution has not changed. And while the continued success of molecular biology and new knowledge about development and differentiation may lessen or even remove the objections leveled by traditional embryologists and developmental biologists, there may not be a dramatic change in the area of systematics and paleontology. If anything, the new discoveries of molecular variation, with a preponderance of neutral variation on one hand and modular aspects of gene regulation and developmental programs on the other (King and Wilson, 1975; Kauffman, 1993), make the position of punctuated equilibrium even more determined.

This should not be surprising. Our world views are shaped by the nature of the material we deal with and scientists are no exception. While population genetics deals with processes of populational change and uses forward arguments for patterns of population differentiation and speciation, systematics and paleontology deal with specimens that are widely separated in morphology, or in space and time, and these fields necessarily make use of backward arguments to infer the processes of evolutionary change. But even the best series of fossils will leave a lot to be desired in terms of providing adequate material for inferring successive, early stages of speciation. Thus, while modern population genetics theory may claim to be sufficient to explain all evolutionary changes, palaeontology and population genetics will not meet as their domains lie on different and nonoverlapping time scales. The microevolutionary–macroevolutionary debate has become embedded in evolutionary biology and will remain so as its resolution has less to do with facts and more to do with ideologies.

However, molecular methods of evolutionary genetics provide new hope; while both population genetics and palaeontology have sufficient explanatory power to account for both gradual and nongradual evolution, the availability of molecular methodologies makes the speciation problem very much solvable.

28.3. Geographic Theories of Speciation – A Brief Overview

All geographic theories of speciation can be classified with respect to one or more of the following four criteria: (1) geographic isolation (allopatric versus sympatric speciation), (2) population demography (geographic versus founder-effect speciation), (3) presence or absence of secondary contact coupled with selective reinforcement, and (4) gradual versus cladistic speciation.

The geographical theory of speciation developed by Jordan, Rensch, Mayr, and others holds that biological species are the end products of geographical differentiation and race formation (Mayr, 1942, 1963; Bush, 1975; Grant, 1985). Gene flow between populations is assumed to be a powerful force of genetic homogenization, and hence spatial isolation is a basic requirement of this theory. The sympatric model of speciation, particularly in its application to outcrossing organisms with complete overlapping ranges, stands at the other extreme on the axis of geographic isolation. The model requires disruptive selection and ecological divergence strong enough to overpower gene flow between diverging populations (Thoday and Gibson, 1962; Maynard Smith, 1966; Bush, 1969). While strong selection at times may occur in nature (Ford, 1964; Endler, 1986), it is more likely to be an exception rather than the rule (the extensive amount of genetic polymorphisms in nature has taught us this much!). Sympatric models of speciation are therefore being developed that require weaker or no selection. These models involve either organisms that are less mobile and have narrow ecological zones, such as many invertebrates and plants (Grant, 1971) or variables such as behavioral isolation (Andersson, 1994) that produce genetic isolation without geographic separation. According to these models, sympatric speciation in plants with self-fertilization, hybridization, or allopolyploidy and in animals with ecological differentiation and assortative mating or sexual selection is a possibility; sympatric speciation in the original sense, i.e., in outcrossing organisms with complete overlapping ranges, is not likely. An allopatric model of speciation should always be considered the *null hypothesis* for speciation.

A second factor that has influenced the theories of speciation is variation in population size. Wright's (1931) concept of the selection–drift interaction in small populations and Simpson's (1944) concept of quantum evolution and adaptive peak shifts in small populations fueled the imagination of many evolutionary biologists and produced a number of theories of speciation, essentially identical in nature but under different names. These include Mayr's (1954, 1963, 1982b) founder effect and genetic revolution or peripatric theory, Lewis' (1962) catastrophic selection theory, Carson's (1975, 1989) founder-flush or organization theory, Templeton's (1980) genetic transilience theory, Gould and Eldredge's (1977) punctuated equilibrium, White's (1978) stasipatric speciation, and Wilson's (King and Wilson, 1975; Wilson et al., 1975) theory of speciation by regulatory genes and social structuring of (small) mammalian populations. A common assumption in all these models of speciation by founder effect is that speciation occurs in small, isolated populations (also see Stanley, 1979; Vrba, 1980; Paterson, 1985). A demographic shake-up of the genetic system under inbreeding is supposed to produce novel adaptive genotypic complexes, new mate-recognition systems, rapid morphological change, rapid reproductive isolation, and finally rapid speciation. The role of small, isolated populations in speciation ranges from being very specific and causal (such as in the punctuated equilibrium and the peripatric model) to being more general, promoting factors (Levin, 1993).

A third factor in shaping the theories of speciation has been the role of selective reinforcement after previously geographically isolated populations make recontact and become sympatric. The role of reinforcement in completing reproductive isolation in sympatric species goes back to A. R. Wallace (1889), with the so-called Wallace effect, and the theory was later developed and elaborated by Fisher (1930), Huxley (1942), and Dobzhansky (1951). More recently the theory of speciation by reinforcement has come under heavy criticism (Paterson, 1978, 1985), leading to a great deal of new work (see Coyne and Orr, 1997). The theory is in principle testable as the recontact of geographically isolated (and presumably genetically differentiated) populations would require certain levels of postzygotic isolation for making reinforcement possible, and the hybridization would have an effect on the level of genetic and ecological differentiation between the sympatric taxa.

The last factor that has become part of the ongoing debate on the mechanisms of speciation is the phylogenetic or cladistic concept of speciation, which treats species as unitary evolutionary lineages that evolve by splitting. Its emphasis is on the classification of past and present evolutionary lineages and has less to do with species gene pools (it calls for the abandonment of the biological species concept) or with the mechanisms of change. The concepts of stasis and punctuation have exerted a powerful influence on our view of the dynamics of genetic and phenotypic change. This influence will remain regardless of evidence as these concepts paint a dichotomous theoretical picture of the organisms' developmental program that gives us the two contrasting views of macroevolution – stasis versus change.

28.4. Testing Geographic Theories of Speciation

Since speciation cannot be studied in action and since sympatric speciation remains but a theoretical possibility, is it possible to tell which mode of speciation has been most common in the past? The answer is yes. There are three features of allopatric, sympatric, and founder-effect speciation that can be used to infer mechanisms of past speciation in a group of taxa. These are effective size of the species, species distribution pattern, and genetic evidence for the pattern of past gene flow. One approach has been to use phylogenetic evidence and distributional data on different major taxonomic groups to make inferences about the allopatric versus sympatric modes of speciation. Using this approach with three groups of fishes, three groups of frogs, and one group of birds, Lynch (1989) reported that over 70% speciation events involved the divergence of major geographically isolated population systems (vicariant speciation), a result in agreement with those of other workers (Wiley, 1981; Cracraft, 1982; Wiley and Mayden, 1985). Speciation in peripheral populations accounted for 15% and sympatric speciation for 6%. Lynch defines vicariant versus peripheral speciation arbitrarily on the basis of the size of the daughter species (the smaller daughter species has an area of no more than 5% of the larger daughter species). Population growth immediately following speciation is an unknown that will affect the proportion of vicariant and peripheral speciation but it will not affect the proportion of sympatric speciation. I echo Lynch's (1989, p. 529) sentiment that in spite of the lack of incontestable evidence, "there has been a persistent argument that sympatric speciation is not only possible . . . but is possibly common." Fortunately now there is a shift in the arguments for sympatric speciation in that its role is being more clearly defined and preferentially applied to organisms such as insects, mites, and nematodes, which would provide the best opportunity for sympatric speciation to occur (Futuyma and Mayer, 1980; Bush, 1994).

A second approach is to make use of the DNA sequence variation and structure of gene trees to test whether or not the patterns of sequence divergence are consistent among all loci during speciation. Under the null model the relationship between intraspecific and interspecific variation is the same for all loci, and any deviation can be used to make inference about founder effect, species effective size, intraspecific and interspecific gene flow, natural selection, and population structure. This approach is ideal with recently evolved species groups containing members with varying degrees of evolutionary divergence. Jody Hey and his colleagues (Kliman and Hey, 1993; Hilton and Hey, 1996; Wang and Hey, 1996) have used this approach with several groups of *Drosophila* species and have reported results that support previous findings but also some new and interesting results. In the *D. pseudoobscura* group they found evidence for large effective size in *D. pseudoobscura* and *D. persimilis*, extensive intraspecific gene flow, and small population size for *D. p. bogotana* (Wang and Hey, 1996). In the *D. simulans* complex, their data support large population size for *D. simulans* and the independent speciation of *D. sechellia* and *D. mauritiana*, with the former having a much smaller population size than the latter (Kliman and Hey, 1993). In the *virilis* group, they found a lack of divergence between *D.a. americana* and *D.a. texana*, and evidence for two distinct groups in *D. novamexicana* (Hilton and Hey, 1996). The comparison of multiple gene trees among related species is a powerful approach to infer past patterns of population differentiation and gene flow during speciation.

28.5. The Genetic Basis of Speciation and the Significance of Haldane's Rule

There are basically two approaches to the genetic basis of speciation studies. The first, applied extensively during 1970s and 1980s by use of protein electrophoretic techniques, is the genetic characterization of differences between closely related species and evaluation of its significance, with respect to the amount and the type of genetic divergence, for species formation (see Chap. 5 in this volume). The most universal generalizations that emerged from these studies were summarized by Lewontin as follows:

1. "The overwhelming preponderance of genetic differences between closely related species is latent in the polymorphisms existing within species." (Lewontin, 1974, p. 179)
2. "The first stage of speciation, the acquisition of primary reproductive isolation in geographical solitude, does not require major overhaul of the genotype and may result from chance change in a few loci." (Lewontin, 1974, p. 186).

Continuation of this method of analysis by the sequencing of selected genes chosen with respect to their relevance in speciation is a promising approach to pursue.

The second, more successful, approach has been the genetic analysis of premating and postmating reproductive isolation between closely related species, which was introduced by Dobzhansky (1951) and which has been widely used with *Drosophila* (Coyne, 1992; Orr, 1995). A major aspect of this approach, centering on postmating isolation (hybrid inviability and sterility) is what is known as Haldane's rule. In a survey of four diverse animal taxa (including a single species pair sample from *Drosophila*), Haldane (1922) revealed that, "when in the F_1 offspring of two different animal races one sex is absent, rare, or sterile, that sex is the heterozygous sex."

Intrinsic in its definition is the fact that Haldane's rule is not the consequence of the hybrid's sex (i.e., male versus female), but rather may be the result of the hybrid's sex chromosomal makeup (i.e., heterogametic versus homogametic). For example, in genera such as birds and Lepidoptera, the female is the heterogametic sex (WZ) but in *Drosophila* and mammalian species, males are the heterogametic sex (XY). Naturally, studies looking into the genetic basis of Haldane's rule have placed a large emphasis on the interaction of loci from the heterospecific chromosomes (i.e., X–Y, X autosome, and Y autosome) within the less fit interspecific hybrid. In *Drosophila*, hypotheses that involve genic interactions between the Y chromosome and the X chromosome have been proposed (Haldane, 1932; Muller, 1942; Coyne, 1985) but lack experimental evidence (Zeng and Singh, 1993). However, a number of studies have shown that hybrid male sterility can be induced by the Y chromosome of one species in certain interspecific hybrids (Zouros, 1981; Vigneault and Zouros, 1986; Lamnissou et al., 1996).

Most explanations of Haldane's rule revolve around heterospecific interactions, in the hybrid, between loci found on the X chromosome of one species and loci found on one of the autosomal chromosomes of the other species. Such an interaction had first been suggested by Haldane (1922) and Muller (1940). In one version, termed as the X-autosomal imbalance hypothesis, Dobzhansky (1937) proposed that each species has a particular balance of genes on each chromosome that has become different from other species through events such as translocations. Unlike female hybrids, male hybrids have deficiencies (or redundancies) on the X chromosome relative to the paternal set of autosomes. This genic imbalance renders the male hybrid sterile. In another version, the loss of complementary interactions between X chromosome and autosomal loci, caused by the gradual divergence of loci in each species, may result in sterility (Muller, 1940).

While Haldane's rule focused our attention onto heterospecific chromosomal interactions in the hybrid, another pattern from interspecific hybridization became apparent. Evidence has accumulated that the X chromosome has a disproportionate effect on hybrid incompatibilities compared with other chromosomes (Coyne and Orr, 1989). This observation was noted over 50 years ago by Dobzhansky (1936) in his backcross analyses of species crosses with the *D. pseudoobscura* group. Various theories have been proposed that try to explain this supposed large X-chromosome effect and Haldane's rule (Orr, 1997). For example, Charlesworth et al. (1987) argued that both the large X effect and Haldane's rule could be the result of a faster rate of evolution on the X chromosome compared with the autosomes. This pattern results because X-linked advantageous recessive alleles would have a higher rate of fixation compared with autosomal alleles.

Turelli and Orr (1995) have tried to explain Haldane's rule and the large X effect through a modified version of Muller's dominance theory. They state that the preponderance of incompatibilities in the heterogametic interspecific hybrid is due to recessive alleles being expressed in the heterogametic sex. Of course, two of these alleles will be expressed in the homogametic sex (see Zeng, 1996, for a simple model based on additivity effects and no assumption of correlation between the degree of dominance and amount of fitness change). It could then be shown that when the alleles involved in such incompatibilities as hybrid sterility and inviability are partially recessive (i.e.,

$d < 0.5$), Haldane's rule would be the inevitable result (Orr, 1993; Turelli and Orr, 1995).

What is the biological significance of Haldane's rule? In other words, what does Haldane's rule tells us about the genetic mechanisms of speciation or about how characters change during speciation? So far, research on Haldane's rule has almost exclusively focused on providing a unitary genetic basis for Haldane's rule as well as for the presumed large X effect. The assertion that Haldane's rule can be explained by the dominance theory (hemizygous expression of recessive alleles) or by the rapid evolution of the X chromosome (hemizygous expression of advantageous alleles) says less about the biology of speciation and more about the minimum population genetic assumptions needed to explain the rule. The dominance theory reveals nothing about the role of the recessive genes within species, and the theory of rapid X chromosome evolution says nothing about why otherwise advantageous genes should have deleterious pleiotropic effects on reproduction. And questions remain unanswered on the nature, number, and kinds of genes involved in the incompatibility of hybrids from closely related species. Thus both theories are formulated with minimum and necessary assumptions in the light of the hemizygosity of the X chromosome to provide a general explanation for Haldane's rule. In addition, we must also think about mechanisms of speciation in organisms that do not have well differentiated sex chromosomes (e.g., flowering plants).

Recently, a number of studies utilizing *Drosophila* have deviated from the traditional inquiries of chromosomal interactions and have re-examined the large effect of the X chromosome on hybrid incompatibilities. One major criticism is that in males, the X chromosome is hemizygous and cannot be fairly compared with heterozygous autosomal loci in females. This problem could be effectively solved if hemizygous X-chromosomal introgressions from one species onto another species' background are compared with similarly sized homozygous autosomal introgressions. This has in fact been done, and most striking is the observation that loci effecting hybrid male sterility evolve much more rapidly than loci controlling hybrid male inviability or any female incompatibility (True et al., 1996; Hollocher and Wu, 1996). Such observations support a faster-male hypothesis, which has been proposed as an alternative explanation to Haldane's rule in taxa in which the heterogametic sex is male (Wu and Davis, 1993). Although genotypic explanations of Haldane's rule and the large effect of the X chromosome are appealing since they explain such effects in genera in which males are the heterogametic sex, the hope for a unitary explanation of Haldane's rule has distracted us from a true resolution. Instead, patterns within Haldane's rule that pertain to a certain taxon (i.e., faster evolution of hybrid male sterility in *Drosophila*) need to be investigated.

Investigations of Haldane's rule may have steered attention away from the ultimate goal of understanding the genetic basis of speciation. However, two general and relevant conclusions can be drawn from the genetic studies on Haldane's rule that point toward a mechanistic explanation of incipient speciation: (1) The majority of the epistatic interactions affecting hybrid sterility appear to involve one sex chromosome (X or Y) and an autosome, and often

578   *Rama S. Singh*

the sex autosomal interactions are asymmetric, giving rise to unidirectional hybrid sterility (Coyne, 1992; Zeng and Singh, 1993). (2) Hybrid (male) sterility evolves faster than hybrid inviability (Wu and Davis, 1993). This implies a postzygotic dichotomy between fertility (sex-related) and inviablility (nonsex-related) genes. In Section 28.6 we provide arguments for a genetic theory of speciation that centers on the rapid evolution of sex-related genes and traits within species.

28.6. High Divergence of Sexual Traits and Speciation

28.6.1. Importance of Sexual Selection in Higher Organisms

Following Darwin's use of the term sexual selection to denote the selection process of mate choice, the term has been used in a restrictive sense to study the problem of female choice and sexual dimorphism (Bateson, 1983; Eberhard, 1985; Andersson, 1994). The large body of literature on sexual selection has proliferated into three different areas, all of which have direct or indirect relevance to mechanisms of speciation: (1) ecological or functional basis of female choice, (2) mate choice differences and premating isolation, and (3) rapid divergence of sexual traits. The role of mating system variation and sexual selection in speciation has been emphasized but it has been treated, more or less, as an additional factor to the traditional geographic theories of speciation, with little or no contact between the two approaches. With increasing documentation of the divergence of sexual traits between species, sexual traits are becoming the focus of attention in many studies of speciation.

28.6.2. Morphological and Behavioral Traits

Several studies on the analysis of genetic divergence between species, phenotypic breakdown of hybrids between closely related species, and species-specific behavioral and morphological characters have provided a framework to the question of the nature of genetic changes during speciation, and they all suggest a possible link between sex-related traits and speciation.

The early studies of Dobzhansky and Pontecorvo describing the breakdown of phenotypic characteristics in hybrids of closely related species provide a hint to the possible role of genes affecting genitalia during the early stages of species formation. The progeny resulting from the cross *D. pseudoobscura* male × *D. persimilis* female possessed atrophied testes (Dobzhansky, 1934). Pontecorvo (1943) showed that, regardless of the direction of the cross, testes were shrunk in the sterile male progeny obtained from crossing *D. melanogaster* and *D. simulans*. In both the Dobzhansky and Pontecorvo studies, interspecific hybrids were phenotypically normal except for their sterility and atrophied testes.

When premating barriers are the key components of interspecific isolation, male secondary sexual traits or the mate-recognition system are the most diverged between related species. A thoroughly studied example is the Hawaiian *Drosophila* group. *D. heteroneura* and *D. sylvestris* are capable of producing viable and fertile F1 hybrids; however, the males of the species differ drastically in the shape of their heads, which is used in male–male competition by *D. heteroneura* males (Spieth, 1981). The courtship songs in Hawaiian *Drosophila* species show evolutionary innovations, such as high-frequency sounds and abdominal vibrations, with respect to continental *Drosophila* species (Hoy et al., 1988). Differences in the last steps of the courtship sequence between two closely related species of the planitibia subgroup of the Hawaiian *Drosophila* groups, *D. sylvestris* and *D. planitibia*, seem to complicate females acceptance of heterospecific males (Hoikkala and Kaneshiro, 1993).

There are two possible explanations for the role played by sexual selection in the divergence of mating systems and preferred male traits. One relies on the perfection of species recognition systems by stabilizing selection's favoring a tightly coupled system of signals and responses. For example, Kaneshiro and Boake (1987) have suggested that the loss, through founder events, of behavioral components of a stabilized signal-response chain system in the ancestral species may have promoted speciation. The existence of intraspecific constraints in mate-recognition signals has been used as an indication of the possible role of stabilizing selection in the establishment of species isolation (Gerhardt, 1982; Butlin et al., 1985). A mate-recognition system under stabilizing selection should be relatively invariant within species.

Alternatively, the mating-recognition signals have been proposed to be a flexible system in which males may evolve rapidly to match female preferences outside their original species range. Some evidence on the flexibility of females preferences come from studies in which females preferred heterospecific signals added to homospecific ones (Ryan and Rand, 1993) or females recognize traits from ancestral or extant species (Ryan and Rand, 1995). The existence of such flexibility in female preferences is not in agreement with previous models of speciation proposing that mate-recognition systems should be well buffered within species but disrupt easily during speciation events (Carson, 1985; Paterson, 1985; Kaneshiro and Boake, 1987).

Sexual selection also provides a basis for the common pattern of rapid and divergent evolution in the morphology of genitalia structures among organisms that reproduce through internal fertilization (Eberhard, 1985). Eberhard suggested that these structures are not simply required for sperm transfer and that their complexity is an indication of their function as internal courtship devices under sexual selection by female choice. Eberhard and Cordero (1995) have extended this proposal to seminal substances in the ejaculate (chemical genitalia) that influence the mating behavior and reproductive physiology of the females (Eberhard and Cordero, 1995). We have come a long way from the classical examples of sexual dimorphisms such as plumage color and tail length in birds, and a rapidly increasing array of external and internal sexual traits is being discovered to have variation and sexual dimorphism.

28.6.3. Molecular Traits

An increasing array of available gene sequences provide an opportunity to study the molecular nature of sex-related gene evolution (i.e., genes involved in mating recognition, fertilization, expressed in the genitalia, or involved in sex determination). These genes have been independently sequenced and analyzed by different authors in different organisms, and they show an interesting pattern of rapid divergence between related species.

For example, a gene involved in mate recognition (*fus-1*) and a sex-determination gene (*mid*) analyzed in the green algae *Chlamydomonas reinhardtii* have shown extremely low codon bias compared with other genes of *Chlamydomonas reinhardtii* (Ferris et al., 1996; Ferris and Goodenough, 1997). These results have lead these authors to suggest a strong mutational pressure as responsible for the rapid evolution of sex genes in this group (Ferris et al., 1997).

Marine invertebrates have shown patterns of high interspecific sequence divergence for sex-related genes that support intense positive selection affecting the evolution of these genes. Among abalone species, the sperm gene *lysin* is known to be involved in fertilization by creating a hole in the egg vitelline envelope. This gene and an 18-kD protein of a yet unknown function that coats the acrosomal process of the abalone sperm have shown a higher proportion of nonsynonymous (Ka) than synonymous (Ks) changes among closely related species (Lee et al., 1995; Swanson and Vacquier, 1995). Lysin's corresponding egg receptor has also recently been implicated in a coevolutionary process (Swanson and Vacquier, 1998). In a sequence analysis study of a gamete recognition gene (*bindin*) among three species of sea urchins, Metz and Palumbi (1996) found an unusually high ratio of polymorphic and fixed replacement to silent substitutions.

Male accessory gland secretions in *Drosophila* affect females' receptivity to remating and increases egg laying (Chen et al., 1988). Some of the accessory gland genes (*Acp26A, Acp29B, Acp36DE,* and *Acp53E*) have also been linked to sperm displacement (Clark et al., 1995). Only *Acp26A* gene sequence divergence has been analyzed between closely related species of the *melanogaster* complex. The gene showed high levels of nonsynonymous substitutions among the four species of the *melanogaster* complex and divergence of cleavage sites in polymorphic alleles of *D. melanogaster* and its sibling species (Aguade et al., 1992). Esterase-6 is mainly expressed in the ejaculatory duct of *D. melanogaster, D. simulans,* and *D. mauritiana* (Stein et al., 1984), it is transferred during copulation to the females, affecting their reproductive behavior (Scott, 1986), and it has shown a higher replacement to synonymous site divergence between *D. melanogaster* and *D. simulans* (Karotam et al., 1993). A recent survey of coding sequence divergence between species of *Drosophila* have shown an elevated proportion of nonsynonymous changes for sex-related genes between two closely related species (*D. melanogaster* and *D. simulans*). The high proportion of nonsynonymous changes reached a plateau when more distantly related species (*D. melanogaster* and *D. pseudoobscura*) were compared. This result suggests the possibility that strong directional selection shapes the evolution of sex-related genes at the time of species formation (Civetta and Singh, 1998a). We had reached the same conclusion on the basis of rapid divergence of male reproductive tract protein divergence between *D. melanogaster* and *D. simulans* a decade earlier (Coulthart and Singh, 1988a,b).

Among mammals, the sequence of the homeobox gene *Pem* proved to be the one showing the highest nonsynonymous divergence between rat and mouse in a homeodomain region (Maiti et al., 1996). Although Maiti et al., (1996) found no direct evidence that *Pem* regulates reproductive physiology, the gene was selectively expressed in reproductive tissues such as testis, epididymis, and ovary.

Gene sequences involved in sex determination have also shown to be rapidly evolving between related species. This pattern is common to such diverse groups as nematodes (de Bono and Hodgkin, 1996; Kuwabara and Hodgkin, 1996), *Drosophila* (O'Neil and Belote, 1992), and mammals (Whitfield et al., 1993; Tucker and Lundrigan, 1993; Pamilo and O'Neil, 1997).

Finally, at least one example from plants seems to suggest that the pattern of rapid sex-related gene evolution may extend beyond the animal kingdom. The example comes from a sequence analysis of the self-incompatibility locus that functions in the recognition of pollen. A comparison among alleles from four species of *Solanaceae* revealed high sequence diversity with a homogeneous synonymous rate but a heterogeneous proportion of nonsynonymous changes among different regions of the gene. An excess of nonsynonymous changes were found in sites that are assumed to be linked to protein function (Clark and Kao, 1991, also see Chapter 15).

28.6.4. Extending the Role of Sexual Selection

In the preceding paragraphs, no distinction was made between mating (pre-copulatory) and copulatory devices, as copulation itself is part of the mating courtship required for siring offspring. Eberhard (1996) provides several examples in diverse groups of animals in which copulatory and postcopulatory behaviors such as biting, rubbing, vocalizations, postcopulatory displays, postcopulatory feeding, genitalic movements, etc., still seem to be still part of the courtship repertoire to increase a female's receptivity.

Sexual selection is the result of the differential mating ability among males, and such differences are based on courtship. However, studies of sexual selection have been traditionally limited to secondary sexual traits and behavioral components of the mating (precopulatory) system. If receptivity by the female is not guaranteed by the time copulation has started, both copulatory and postcopulatory devices are still important in order to ensure a female's receptivity of the sperm and fertilization. Therefore all components of the reproductive strategy (chemical, behavior, morphology) are expected to be influenced by sexual selection. Studies in beetles (Bella et al., 1992), grasshoppers (Wade et al., 1994), crickets (Gregory and Howard, 1994), *Drosophila* (Clark et al., 1995), birds (Birkhead and Muller, 1992), and rodents (Dewsbury, 1984) show that female choice or male–male competition is still part of the reproductive strategy, even after sperm has been transferred to the female. These few examples in a diverse group of organisms suggest that sexual selection is not to be restricted to the precopulatory stages, leading to mating. We have suggested (Civetta and Singh, 1998b) that separate terms be adopted to denote the narrow and the broad meanings of the term sexual selection: sexual selection in the narrow sense to denote the traditional meaning limited to secondary sexual traits and premating behaviors, and sexual selection in the broad sense to denote all aspects of sexuality, including both premating and postmating traits.

28.7. A Unified Theory of Speciation

In 1994 in a brief outline of what I have called the unified theory of speciation, we stated that "All currently available models of speciation... feature some combination of genetic, populational, and ecological factors, and the models differ depending upon which factors they emphasize the most. The geographic models of speciation emphasize isolation, the founder effect models population bottlenecks, and so forth. Since each of these models was developed with a particular group of organisms in mind, each can be said to contain some element of truth but not the whole truth. No single theory of speciation can be both general enough to apply to all organisms and specific enough to provide a detailed picture of speciation in a given group of organisms" (Singh and Zeng, 1994).

The point is that the various mechanisms of speciation are not equally likely to apply to all organisms because of differences in their breeding system, life history, and ecology. Here, focusing on the rapid evolution of sexual traits, I first outline a general genetic theory of speciation applicable to all higher sexual organisms and then provide a brief description of how this general theory can be linked to speciation in plants and asexual organisms, taking into account the major reproductive and life-history features of these organisms.

28.7.1. The Sex Gene Pool Theory of Speciation

As noted above, theories of speciation by sexual selection have been with us for a long time. However, they have been restrictive in their application to certain groups of organisms and either directly implicate variation and evolution of mating behavior traits involved in male–female recognition (Kaneshiro, 1989; Paterson, 1985) or emphasize pleiotropic effects on fitness or mating behavior traits (Carson, 1985, 1989; Templeton, 1981, 1989). What I propose here is not so much new but a change of paradigm – to make all sex-related genes and traits the focus of attention in genetic studies of speciation.

From our point of view an organism can be thought of, broadly speaking, as the outcome of three types of genetic programs – those determining development and differentiation, brain and behavior, and gonads and genitalia. From the point of view of adaptation and speciation these three categories can be reduced to two categories affecting primarily survival or reproduction. Accordingly, the species gene pool can be thought of as consisting of two components – the sex gene pool, involving all genes affecting mating behavior, germline development and gametogenesis, and fertilization, and the nonsex gene pool, involving all other genes affecting development and differentiation, metabolic

Table 28.1. *The Two-Component Gene Pool Theory of Adaptation and Speciation*

Gene Pool/Traits	Adaptation/Survival	Speciation/ Reproductive Isolation
Sex gene pool Mating behavior, gametogenesis, and fertilization system	Minor effect	Major effect
Nonsex gene pool Development, metabolic functions, and survival behavior	Major effect	Minor effect

functions, and survival behavior (Table 28.1). While there may be a large number of genes affecting both types of traits, i.e., survival and reproduction, there must be a group of genes whose primary role is successful transfer of genetic material to the next generation. I consider the sex gene pool to have a direct and major role in the evolution of reproductive isolation, by affecting mating behavior and gamete formation, and comparatively a minor role in survival. In contrast, the nonsex gene pool would have a major role in the survival of organisms and a minor role in the evolution of reproductive isolation. This dichotomy of the species gene pool is not intended to be too strongly decoupled as, in individual situations, genes from both categories can be shown to have a major role of the opposite type. For example, in species with strong predators, mating behavior can expose individuals to death! Similarly many genes involved in development and differentiation will also be involved in mating behavior and/or genetogenesis. The fact that hybrid sterility evolves faster than hybrid inviability (Wu, 1992) lends support to the above two-component model of the gene pool. The consequence of this observation for designing genetic theories of speciation has not been fully appreciated.

The two-component gene pool system is assumed to have a direct bearing on the nature of the genetic changes that occur during the early stages of speciation. Current theories of speciation have advocated the substitution of neutral mutations, with fitness change following a peak shift (Wagner et al., 1994), of advantageous mutations with negative epistatic interactions (Dobzhansky, 1951, 1970), or of advantageous mutations with a pleiotropic effect on fertility and fertility modifiers (Charlesworth et al., 1987). Under the present sex gene pool theory, I propose that fixation of suboptimal or mildly deleterious mutations between geographically isolated populations affecting any aspect of mating and reproduction will have a predominant effect on reproduction and fertility, which are likely to be under stronger selection than viability in the new environment. The deleterious effects on reproduction and fertility will force the evolution of independent fitness modifiers that will be incompatible between species. In addition, because of sex-chromosome hemizygosity, the heterogametic sex will be more likely to experience fitness-related changes than the homogametic sex. Any change in one sex will lead to an immediate (in terms of generations) compensatory genetic change in the other sex, resulting in a sexual coevolution (and not an arms race; Rice, 1996). Thus the evolution of independent fertility modifiers and sexual coevolution will drive the species divergence of sex-related traits between species in the early generation of geographic isolation until the populations reach a new equilibrium in the new environment. Substitution of advantageous mutations with pleiotropic effect on fertility will have a similar effect, but I believe such genes will have a minor role in the beginning. The theory as presented here is meant to encompass all aspects of sexuality, including but not limited to female preference, and it is for this reason that we have used a new and more broader term, the sex gene pool theory, to denote its inclusive nature. Theories of sexual selection, in the sense of Carson (1989), Kaneshiro (1989), or Lande (1981, 1982), are special mechanisms of speciation that can apply to both sympatric and allopatric speciation and vary in their relevance to different groups of organisms. Our presentation here implies variation and evolution of sex-related genes as the major mode of change in speciation in all sexual organisms involving natural or sexual selection.

The above case of accelerated genetic change in sex-related genes and traits during the early stages of speciation has consequences, many of which are observed in different organisms: (1) The strength of decoupling between the sex and the nonsex gene pools can allow change in one component without effecting a change in the other. Species with partial or complete reproductive isolation without significant morphological differences (the so-called sibling species) and species with morphological differences without reproductive isolation (such as species of Oaks; Grant, 1971) are expected under this theory. (2) The animal and plant breeding literature have many examples that show that morphology and size traits can be altered by the magnitude of several standard deviations without affecting fertility. (3) The heightened expression of incompatibility for sexual traits and lack thereof for nonsexual traits in species crosses (Civetta and Singh, 1998b). (4) Premating and postmating isolation may evolve independently and remain decoupled in the early stages of speciation, and an early completion of premating isolation may lead to an incomplete development (or even a lack of) postmating isolation. This is often the case in many organisms with strong premating behavioral components such as birds (Grant and Grant, 1992, 1996) and the Hawaiian *Drosophila* (Carson, 1989). (5) Hybrid sterility evolves faster than hybrid inviability (Wu, 1992).

The differential sensitivity of the two types of gene pools will lead to different criteria for speciation in evolutionary genetics (genetic incompatibility) and palaeontology (morphological differences). A recognition of two types of gene pools with their built in propensity to affect sexual and nonsexual traits independently, and the differential time frame for the evolution of fertility and survival traits can go a long way toward resolving the speciation debate between the two fields.

28.7.2. Sympatric or Microallopatric Speciation?

Speciation, or more precisely the development of reproductive isolation, cannot take place without some physical or behavioral isolation of populations in space or time. The theory of allopatric speciation was developed with highly mobile organisms in mind, and thus most textbooks, when describing the theory, overemphasize the necessity of physical barriers between populations. But the theory requires only effective genetic isolation between populations and this can be achieved in a variety of ways – some with and some without major physical barriers. A great many species of plants have restricted mobility of seeds or pollens, and so physical barriers can be achieved more readily in plants than animals (Ehrlich and Raven, 1969). Many species of insects and invertebrates also have narrow ecological zones or elaborate premating behavioral traits that require less stringent physical barriers in order to become genetically isolated. I believe the role of behavioral isolation in sympatric situation has been underestimated, as it is commonly assumed that the individuals will mate within their cruising range. Many probable cases of sympatric speciation may indeed turn out to be the product of microallopatric speciation. If we focus on genetic isolation, then the problem of sympatric speciation is not whether it can occur without isolation (as it cannot without some form of isolation) but whether diversifying natural selection is strong enough to create substantial genetic divergence in the face of repeated gene flow (Schluter, 1996). It is not surprising that all sympatric models of speciation use some form of assortative mating linked to the trait under selection to achieve reproductive isolation. Thus there is an increasing emphasis on the role of behavioral isolation to achieve reproductive isolation without ecological differentiation. It is not surprising then that there is a shift in the models of sympatric speciation from being based on resource competition and character displacement (Maynard Smith, 1966; Tauber and Tauber, 1977; Kondrashov, 1986; Bush, 1994) to models based on sexual selection and geographic differentiation (Lande, 1981, 1982), to models involving selective reinforcement (Endler, 1977; Lande, 1982; Butlin, 1989; Liou and Price, 1994), to models that are not based on ecological differentiation and involve behavioral isolation only through a variety of mechanisms, such as social competition (West-Eberhard, 1983), female-choice-based sexual selection (Turner and Burrows, 1995), or mating-success-dependent diffusion (Payne and Krakauer, 1997). Even the proposal that mammals have speciated faster because of the social structuring of populations (Bush et al., 1977) can be thought of as allopatry in sympatry (or microallopatry) resulting from behavioral isolation. A similar sentiment was expressed by Goldschmidt (1940) when he asserted that the various Hindu hereditary castes of India have achieved behavioral isolation without being segregated in space (as they all live side by side) but they have not lost their genetic potential to interbreed. Assortative mating within alternative habitats is a built in mechanism for physically splitting the population; habitat – directed selection would rarely be strong enough in nature to overcome gene flow without having some physical isolation in space or time first.

Some of the most interesting model systems such as the three-spined stickle-backs (Schluter and McPhail, 1993; Schluter and Nagel, 1995; Schluter, 1996; Grundle and Schluter, 1998), the lake Whitefish (Bernatchez et al., 1996; Pigeon et al., 1997), and the Cichlid fishes from Lakes Malawi and Victoria (Meyer, 1993; Schliewen et al., 1994; Turner, 1994) that provide the most prob-able examples for sympatric speciation also need to be looked at from the point of the behavioral segregation of habitats. For example, the possibility of size-based mating preference in the sticklebacks and the whitefish and lim-ited dispersal and mouth brooding in the Cichlids may be important factors in addition to the resource-based ecological differences to which these species may be exposed. Resource-based divergence, parallel speciation, and recency of the species are the main arguments for proposing selection-driven sympatric speciation in these organisms. Mating behavior or microspatial segregation of habitats may turn out to be an important factor in all these cases, and the ar-gument for sympatric speciation in the traditional sense (strong local natural selection overriding gene flow) is not likely in outcrossing organisms.

28.7.3. Unification of Speciation Theories in Plants and Animals

The characteristics of open growth habit, lack of separate germline tissues, and vegetative reproduction make plant genetic systems developmentally more open and tolerant to genetic changes, interspecific hybridization, and intro-gression without severe effect on reproductive isolation and species integrity (Stebbins, 1950, 1982; Grant, 1971). This is reflected in the fact that over 70% of plants are polyploids, and plants, particularly trees, shrubs, and perennial herbs, generally show incomplete and quite often a complete absence of repro-ductive isolation between species. So while the allopatric model of speciation applies to plants, as "some degree of hybrid inviability, hybrid semisterility, and hybrid breakdown is in fact found between races in many plant species" (Grant, 1971), there is something different about the plant genetic system because in plants, as Grant puts it,

"Related species are interfertile, and their isolation in nature depends largely on ecological and other external factors. . . . Species belonging to different circles of affinity – sections, subgenera, or genera – are isolated internally by various blocks to crossing or to successful growth of the hybrid embryos. . . . These gene-controlled incompatibility barriers appear suddenly in the taxonomic structure of the plant group. . . . The internal channels of gene exchange between species are open within relatively wide taxonomic limits, then beyond these limits become closed (Grant, 1971, p. 103).

In addition, plants are also characterized, because of their sessile nature, as being more likely to experience strong selection pressure (Jain and Bradshaw, 1966) and reduced gene flow (Ehrlich and Raven, 1969; Levin 1979). These characteristics and the open genetic system make plants more likely than animals to experience quantum (by inbreeding, fixation of chromosomal rearrangements, allopolyploidy, and hybrid speciation) and sympatric speciation (Grant, 1971).

In view of the above, it would seem that there is hardly any basis for considering a unified theory of speciation that can be applied to both plants and animals. Actually there is a basis, and it is reflected in the fact that plant species do not show strong reproductive isolation in spite of morphological, physiological, and ecological divergence. A unified theory of speciation in plants and animals can be based on the nature of their sexual and nonsexual genetic systems and on the nature of the response to changes during speciation. First, as discussed above, the sexual and nonsexual genetic systems are in general more decoupled in plants than animals. Thus, depending on the breeding systems, some plants, such as insect-pollinated species (Janzen, 1977), may behave very much like sexual animal species (insect–plant coevolution is equivalent to intersex coevolution in animals), while others, such as open-pollinated and self-pollinated species, may show much less rapid changes in their breeding systems. The coevolution of pollen and stigma can function as a type of sexual selection (Charnov and Bull, 1977; Bawa, 1980). Animal pollination, mainly by insects, is associated with high rates of diversification in flowering plants (Stanley, 1979; Stebbins, 1981; Eriksson and Bremer, 1992). Second, because of their sessile nature, their lack of behavioral avoidance, and reduced gene flow, plants are, *ceteris paribus*, more likely to experience stronger (environmental) selection pressure than animals (Jain and Bradshaw, 1966). Plants are more likely than animals to show nonsexual trait differences as a result of adaptation and speciation in the early stages of speciation.

A strikingly similar view has been presented in a remarkable paper (that has escaped my attention until recently) by Carson (1985) on the unification of speciation theories in plants and animals. Carson applies his organization theory of speciation, based on coadapted gene complexes, open and closed genetic systems, and differences in the sexual and nonsexual genetic systems between plants and animals, and comes to the conclusion that (1) the adaptive changes in the sexual and nonsexual traits can be seen as two alternative mechanisms of divergence during speciation, (2) that the sequence of change in animals is generally first in the sexual traits followed by the nonsexual traits, whereas in plants it is the other way around, and (3) that the genetic changes during speciation tend to occur in the recently evolved phenotypes as the fundamental or ancient genetic pathways tend to be fixed or deleterious when perturbed.

I agree with Carson about the sexual and nonsexual genetic system as alternative mechanisms of genetic change during speciation and about the nature of differences in the degree of coupling between the sexual and nonsexual genetic systems and its effect on varying degree of reproductive isolation in plants versus animals. The focus here is on the nature of traits that are more or less likely to change during speciation and this is not limited to founder-effect speciation and would equally apply to allopatric speciation. Therefore, while I agree with Carson on the nature of differences in the degree of openness of plant and animal genetic systems, I do not share his views of animal species having "highly coadapted gene complexes" that need to be "disorganized/reorganized" by inbreeding during founder-effect speciation. The idea of a strict open and closed genetic system among animals also does not hold, as sexual traits, which are part of the closed genetic system and that are intimately involved in speciation, do not seem to be any less genetically variably than nonsexual traits (Civetta and Singh, 1998a, 1998b). This is not to say that all traits are equally variable or that traits from all developmental stages are equally variable. It is true that all members of a species share some common genetic features and that some aspects of development would appear invariant at higher levels of taxonomic organizations. But there is absolutely no evidence to say that the genetic variations within and between species are of different kinds. I believe the most important and useful criterion for the division of the species gene pool is the functional criterion along the line of sexual versus nonsexual systems and this would capture the essential feature of Carson's view of speciation, including the observation that secondary sexual traits are more likely to be involved during speciation as observed in the Hawaiian *Drosophila* (Carson, 1985). I also think that just as premating and postmating isolation in the early stages of speciation can be thought of as alternative mechanisms of developing genetic barriers between species, changes in sexual and nonsexual gene pools can be thought of as two separate modes of genetic change during speciation.

28.7.4. Unification of Speciation Theories in Sexual and Asexual Organisms

The concept of the Mendelian population, which is the core of the biological species concept, obviously does not apply to organisms with uniparental reproduction (self-fertilization, parthenogenesis, pseudogamy, and vegetative reproduction). As a result, the dominant view in evolutionary biology has been that the species concept, or at least the biological species concept, does not apply to uniparental organisms (Simpson, 1961; Mayr, 1963, 1987) and that these organisms do not form species (Eldredge, 1985; Hull, 1987; Ghiselin, 1987). The problem of using the biological species concept, of course, has not stopped biologists from identifying and naming species of asexual organisms. In fact, identification of phenotypic clusters or discontinuities in some asexual organisms, such as the bdelloid rotifers, appears to be as good, if not better than, sexual organisms (Holman, 1987). Occupation of specialized ecological niches in general, and mutational adaptation in microorganisms in particular, is held responsible for the observed phenotypic discontinuities in the asexuals. The real question is therefore not whether good species can be identified in asexual organisms, as they certainly can be, but whether they are the same kind of species as in the sexual organisms. For the proponents of the biological species concept the answer is obviously no. To them, the species concept requires more than phenotypic discontinuities, and asexual organisms for a variety of reasons may maintain phenotypic cohesion but they are merely a collection of living objects with no sharing of genes among individuals. Furthermore, the potential to generate future species is obviously much lower among asexuals than sexuals. Sexual reproduction is more than just a device to maintain cohesion and sexual species are real evolutionary units (with a shared gene pool among individuals) and the asexual species are not.

The proponents of the cohesive species concept (see Templeton, 1989) focus on the within-species cohesion, regardless of the mechanisms, as the key feature of the species concept and treat sexual and asexual organisms on an equal footing in the application of this species concept. They maintain that reproductive isolation is an unnecessary aspect of the biological species concept, that many good species are able to maintain their cohesion without substantial reproductive isolation, and that interbreeding is just one of the many devices that produce and maintain species cohesion. Random genetic drift, ecological specialization, periodic selection, stabilizing selection, and common ecological factors are some of the processes that can create and maintain cohesion within species and discontinuities between species.

The cohesion species concept is the primary and logical way to define an asexual species as genetic, developmental, and ecological cohesion can be maintained by demographic processes and interbreeding or sexual reproduction is not necessary. Thus, by focusing on the outcome (i.e., cohesion) rather than the mechanisms, the cohesion species concept provides a basis for unifying the speciation theories in sexual and asexual organisms. Where it fails, though, is that in an intention to apply to all organisms equally, the cohesion concept loses the significance of the most important aspect of sexual species, i.e., that only sexual reproduction is capable of generating the kind of multicellular diversity that we see in the biosphere. Microorganisms, through mutation, selection, and genetic differentiation, can maintain long-term taxonomic diversity (Rosenzweig et al., 1994) but truly asexual lineages cannot.

A second route to unify the theories of speciation between sexuals and asexuals would be to look for a species concept (and a theory of speciation) that uses the same criteria and that applies to both groups equally. Recent work with genetic variation and population structure of bacteria show that the early verdict on the inapplicability of the biological species concept to bacterial species may have been premature. Genetic studies for which gene enzymes, restriction fragment length polymorphisms, and DNA sequence variation are used with a number of bacterial species have shown that bacterial species are not all alike and show a full range of genetic structures from being very clonal, as in *Escherichia coli* (Selander and Levin, 1980; Caugant et al., 1981; Milkman and Bridges, 1993), to being very panmictic as in *Neisseria gonorrhoeae* (Maynard Smith et al., 1991, 1993). Localized sex within species and its absence between species appear to be the rule in bacteria. The level of recombination appears to be high enough to genetically link bacterial strains within species (Dykhuizen and Green, 1991) and low enough to maintain genetic distinctness between species (Sharp and Li, 1987). Dykhuizen and Green (1991) have argued that in bacterial species different gene trees should be similar in the absence of recombination but different in the presence of recombination. Thus clustering of strains based on DNA sequences of different genes can be used to differentiate species in bacteria. Since the argument is based on the expectation from the biological species concept, which would produce similar gene trees from individuals from different species but different gene trees from individuals from the same species, the application of the biological species concept in bacterial species is justified.

Another line of work that draws similarity between bacteria and higher organisms is that of Radman and his colleagues. They have shown that bacteria make use of their sexual systems to control the levels of DNA exchange and recombination between species (Rayssiguier et al., 1989; Radman and Wagner, 1993; Matic et al., 1995). Working with mutant strains of *E. coli* and *Salmonella typhimurium*, they have shown that while these species are genetically isolated, with appropriate mutations at loci controlling the SOS and the mismatch repair systems, they are able to exchange DNA in spite of the 20% sequence divergence in their DNA. These authors maintain that DNA mismatch repair systems also operate in higher organisms, which, because of lack of proper pairing and disjunction, can lead to hybrid sterility and speciation. Hunter et al. (1996) have shown the existence of a similar mismatch repair system in interspecific yeast hybrids, but the effect of the mismatch repair genes (increasing meiotic recombination, decreasing chromosome nondisjunction, and improving spore viability) was small. Whether a system like that in bacteria also works in higher organisms is a separate issue. The point of interest to us here is the fact that bacteria, just like higher organisms, can exchange DNA and produce recombinants by controlling sexual genes. In some sense, bacteria are like plants in that they can engage in wide crosses through controlling their sexual systems, and at the same time they are ecologically specialized through the use of their metabolic genetic systems.

28.8. A New Concept of Species

The unified theory of speciation proposed above centers on the dichotomy of the species gene pool, sexual versus nonsexual, and its differential applicability to different groups of organisms with varying degrees of sexuality in animals, plants, and microorganisms. The emphasis on sexual traits (all sexual traits, not just those involved in sexual selection involving female preference) as the targets of genetic change during the speciation of sexual animals and plants differentiates this theory from others. Organisms with less predominant sexual traits are more likely to differentiate as a result of genetic changes in the non-sexual gene pool system. It follows then that, if species concepts and species definitions are meant to be relevant to our understanding of the mechanisms of speciation, the dichotomy of trait suites and species gene pool should be taken into account in designing the species concept(s).

An investigation into the mechanisms of speciation requires that we can define the unit that changes or evolves during speciation. In sexual organisms this evolutionary unit can be anything above the level of individuals: a local population (Ehrlich and Raven, 1969; Levin, 1979, 1993; Carson, 1989), a set of populations (Mayr, 1963), or the entire species (Gould and Eldredge, 1977). Some hold that the unit of differentiation can be any of the various levels of taxonomic organization, not only species, in which case there is no such thing as speciation as we know it, and hence we get rid of the problem all together (Nelson, 1989)!

There are of course many concepts of species, actually more than theories of speciation. Sometimes there is more than one theory of speciation associated with the same species concept, as is the case with the biological species concept, which applies to allopatric, paraptric, and sympatric speciation. In several cases an old species concept is redefined and its application is modified, as is the case with the isolation concept (a renaming of the biological species concept – see Templeton, 1989) and the recognition concept (Paterson, 1985). In one case, an existing species concept is discarded and a new one along with a new mechanism of speciation is suggested, as is the case with the phylogenetic species concept (Cracraft, 1982; there are several versions of the phylogenetic species concept) and the cohesive species concept (Templeton, 1989). Then there are species concepts that provide different, but complementary, perspectives on the nature of species as units of evolutionary change without necessarily specifying a different mechanism of speciation. This is the case with the evolutionary concept (Simpson, 1944) and the ecological concept (Schluter, 1996). And then there is, of course, the taxonomic concept of species which is nonevolutionary (Sneath and Sokal, 1973). Each concept is presented as the only satisfactory alternative to the others, which are all deemed unsatisfactory! Indeed it would be hard to find one person, other than the proponents of the various concepts themselves, who is completely satisfied with any of the numerous species concepts.

What is wrong here actually can be said to be symptomatic of all evolutionary biology. Evolutionary biologists habitually pursue their science, like physicists and chemists, following the general scientific method of searching for the truth by looking at the cause and effect of the phenomenon being investigated. The very essence of evolutionary biology – that natural selection is blind, that it destroys history and that it usually comes up with different solutions to the same

592 Rama S. Singh

problem – is often forgotten in our search for generality. This is literally true with the problem of species concept as all species concepts have something true in them and yet none of them satisfactorily explain all aspects of speciation. Thus the biological species concept is meant to emphasize the fundamental nature of species as an evolutionary unit, and it is a relational concept that recognizes both cohesion within a group and distinctness among groups. The recognition species concept focuses only on the sexual recognition system that ensures cohesiveness within group (by inbreeding) but ignores isolation between groups. Templeton's (1989) cohesive species concept is meant to be inclusive, to be applicable to both sexual and asexual organisms. But this concept treats all evolutionary factors producing cohesion as equally important, and so it diminishes the significance of what many would call the core of speciation concept – sexual reproduction. In addition, it is not operational, certainly less so than the biological species concept, which it is meant to replace. All phylogenetic species concepts are pattern driven and independent of evolutionary processes, and the ecological species concept is heuristic but the most nonoperational of them all. Thus clearly each species concept captures some key features of the species characteristics but none of them capture it all.

So what is the solution? One solution, of course, is that we can expect from the current debate on species concepts a more general, if not universal, species concept (a new or a modified version of the existing ones) to evolve that will capture all the important natures of species as evolutionary units and all the important biological processes that give rise to speciation. This is particularly relevant to all biological species concepts that focus on the processes rather than the product of speciation. This does not mean that other species concepts are not relevant, only that they are less relevant to how genetic differentiation and speciation occur.

A second solution is to adopt a plural concept of species (Cain, 1954; Kitcher, 1984a, 1984b; Mishler and Donoghue, 1982; Mishler and Brandon, 1987). One brand of pluralism (Kitcher, 1984a, 1984b) implies that there are many possible species classification systems that can apply in a given situation, depending on the needs and the interest of the investigators. A second brand of pluralism (Mishler and Donoghue, 1982) implies a single general-purpose classification system for all situations, with possibly different criteria in each particular case. This second brand of pluralism is considered transitional and applies only "until a prevailing monistic concept is broken up" (Mishler and Brandon, 1987, p. 403). The first brand of pluralism has been criticized (Sober, 1984; Hull, 1987; Ghiselin, 1987) for the reason, among others, that it violates the concept of parsimony. Qualitatively the second brand of pluralism (Mishler and Donoghue, 1982) is different from the first brand only to the extent that a general-purpose classification system can be satisfactorily applied to all organisms.

A third solution, considered here, would be to treat the functional relationship among the three aspects, or dimensions, of the organism–environment relationship – i.e. genotype (gene pool compatibility), phenotype (morphology and behavior), and ecology – as indeterministic in the sense that in the early stages of speciation (and in many cases for a very long time), there is no exact predictable, one-to-one relationship between the levels of divergence in these three domains of population evolution. The indeterminacy is not because of bad biology but because the very nature of genetical, phenotypic, and ecological differences between organisms are such that an incremental change in one domain cannot be said to arise from a similar incremental change in the other domain except in the sense of average. We all know about sibling species, with little phenotypic divergence but complete reproductive isolation on one hand, and many plant species with substantial morphological divergence but little reproductive isolation on the other. This functional indeterminacy among the three domains of population evolution is at the very root of our problem of coming up with a satisfactory universal species concept. In other words, it is not possible to construct a universal species concept that will apply in all situations. This problem is implied in the plural species concept proposed by Mishler and Donoghue (1982), but they do not advocate an indeterminacy of outcome and try to solve the problem by proposing adoption of different criteria in different situations. In other words, they think that the problem of relation among the three domains of change (genetics, phenotype, and ecology) is a transitory one that will ultimately resolve itself.

The concept of functional indeterminacy also applies to the differential role of sexual versus nonsexual traits in the speciation and reproductive isolation of plants versus animals. If we accept this indeterminacy, then we need to differentiate between theoretical and practical species concepts. The theoretical species concept can be of a general-purpose type, as advocated by Mishler and Donoghue (1982), that specifies the overall relationship between individuals within and between species. The practical species concept can be tailored to the needs of particular situation in population genetics, ecology, or systematics. The theoretical definition will capture the universal nature of species as evolutionary units without being specific about the kinds of change or levels of reproductive isolation between taxa. A theoretical definition of species focusing on the gene pool is offered here:

A species is a group of actually or potentially interbreeding natural populations with a sufficient degree of genetical, ecological, or spatiotemporal isolation that would maintain the developmental or reproductive compatibility (or cohesiveness) of the gene pool within groups and incompatibility (or distinctness) between groups.

In the above definition, monophyly is implied but not specified. It would apply to both sexual and asexual organisms as potential interbreeding is meant to cover both geographically separated sexual populations as well as asexual species. The definition leaves the nature of isolation between groups open (genetical, ecological, or spatial) in the case of sexuals and implies phenotypic or gene pool distinctness in the case of asexuals. Because of the indeterministic nature of the outcome of population differentiation in the early stages of speciation, as well as to differentiate it from the plural species definition of Mishler and Donoghue (1982), I call this species concept the multidimensional gene pool concept. In practice, the plural and the multidimensional species concepts are similar, but in theory the former is transitional and a lack of species resolution in particular cases implies a temporary lack of sufficient knowledge, whereas the latter presumes indeterminacy as an inherent functional property of the gene pool and as a result the species relationships can be resolved only approximately (or even arbitrarily) in the early stages but quite naturally in the later stages of speciation. The two principles, functional indeterminacy and gene pool multidimensionality, emphasize the reality of species relationship and provide a way to save the universality of the theoretical species concept without becoming bogged down with its application in each and every case.

28.9. What Limits Speciation?

That speciation is a slow process, beyond the reach of normal experimental observation, is a truism in evolutionary biology. Even the most rapidly speciating groups of organisms, such as the cichlids and the Hawaiian *Drosophila*, are expected to take from tens to hundreds of thousands of years to produce species. It is true, however, that the documentations of rapidly evolving groups coupled with the general finding of small genetic differences between closely related species has changed our perception of speciation from being inherently a rare event to one that has genetic and/or ecological limitations. Mayr's (1963) theory of speciation by genetic revolution assumed a genetic limit to speciation; the amount of differences between species were perceived as being large and speciation required large-scale genetic changes for accomplishing reproductive isolation. The three decades of evolutionary genetics of species have taught us that it is not so. The genetic changes between closely related species are small and the number of genes involved in speciation, i.e., reproductive isolation, still smaller. These results are universal and speak against a genetic limit to speciation.

It would appear then that the limit to speciation is ecological, and it probably results from a high probability of extinction in populations that are most likely to go under speciation. While the vicariant speciation, with large populations, would more or less ensure the survival of the newly separated populations, all other models of speciation (including instant speciation such as by polyploidy) invoking speciation in small populations face the problem of high rate of extinction of the newly isolated small populations. This is regardless of whether the newly isolated population is in the middle of the parental species or at the periphery. In the former case the incipient populations are likely to be demographically run over by the parental species and in the latter by the vagaries of the environment that is likely to prevail at the edge of the species range. This is by assuming, of course, that the new population has not acquired new adaptations, a new ecological niche, that will minimize the chances of extinction. As Lewontin (1997) puts it, "Unless the populations have come to occupy different peaks in the adaptive landscape, the local populations or geographical races are not likely to be in the preliminary stages in species formation." We can conclude that speciation is a slow process, not because the required genetic changes are large and take too long to accumulate, but because most populations that find themselves in the preliminary stages of speciation are likely to go extinct.

28.10. Conclusions

We can summarize the main points as follows:

(1) All through this century there has been a steady shift in our view of species as evolutionary units, from a loosely organized genetic and population system (Darwin, 1859) to a very organized, developmentally complex genetic system. Wright's (1931) shifting-balance theory, with its emphasis on the role of random genetic drift and inbreeding in evolution, and the more recent treatment of species as individuals by the cladists have had major roles in this shift. This has led to a view of speciation that requires substantial genetic changes between populations. This view is not supported by the nature of genetic differences, from allozymes to DNA sequence, between closely related species.

(2) The allopatric and sympatric theories of speciation are extreme examples from a continuum involving geographical isolation and natural selection. The continued insistence on the importance of sympatric speciation, without much clear evidence, has produced much debate and little light. An allopatric model of speciation should always be considered the null hypothesis for speciation. Current emphasis on models of sympatric speciation involving sexual–behavioral isolation with or without ecological differentiation is useful approach to pursue.

(3) The general observations (rules?) of speciation, derived from the genetic studies of species differences, Haldane's rule, and reproductive isolation, and from ecological consideration can be summarized as follows: (i) The genetic differences between closely related species are small, and the major proportion of these are latent in the polymorphisms existing within species, (ii) the first stage of speciation, the acquisition of primary reproductive isolation in geographical populations, does not require a major overhaul of the genotype and may result from change in a few loci, (iii) the hemizygosity of the sex chromosomes and epistatic interactions between the sex chromosomes and autosomes provides a general basis for Haldane's rule, as well as for the appearance of asymmetry, giving rise to unidirectional sterility in species hybrids, (iv) hybrid sterility evolves faster than hybrid inviability, and in sympatric situations premating isolation evolves faster than postmating isolation, (v) the limiting factor in speciation is ecological: unless the geographical populations or races have achieved new adaptations and new niches, they are not likely to complete speciation.

(4) In the past, theories of speciation have emphasized genetic, geographic, and demographic processes and rather little attention has been paid, with the exception of secondary sexual traits, to the types of traits that are likely to change during speciation. In this chapter a proposal is made to view the species gene

pool as made of two separate but coupled systems, sexual and nonsexual, and it is proposed that speciation and reproductive isolation require changes in the former system and survival and adaptations in the latter. A unified sexual gene pool theory of speciation is proposed and it is shown that with varying emphasis the theory can apply to both plants and animals and to sexuals and asexuals.

(5) A generalized, multidimensional gene pool concept of species is proposed that recognizes that the evolutionary relationship among genetical, morphological, and ecological changes during the early stages of speciation is indeterministic and unresolvable, and that in spite of its plural nature, reproductive compatibility (actual or potential) between taxa will remain a fundamental criteria of species concept and speciation.

28.11. Acknowledgments

I would like to thank Alberto Civetta, Rob Kulathinal, Subodh Jain, and Dick Morton for their critical comments on the numerous versions of this manuscript.

REFERENCES

Aguade, M., Miyashita, N., and Langley, C. H. 1992. Polymorphism and divergence in the Mst26A male accessory gland gene region in Drosophila. *Genetics* **132**:755–770.

Andersson, M. 1994. *Sexual Selection*. Princeton, NJ: Princeton U. Press.

Bateson, P. 1983. *Mate Choice*. Cambridge: Cambridge U. Press.

Bateson, W. 1894. *Materials for the Study of Variation*. London: Macmillan.

Bawa, K. 1980. Evolution of diocey in flowering plants. *Annu. Rev. Ecol. Syst.* **11**:15–39.

Bella, J. L., Butlin, R. K., Ferris, C., and Hewitt, G. M. 1992. Asymmetrical homogamy and unequal sex ratio from reciprocal mating-order crosses between *Chorthippus parallelus* subspecies. *Heredity* **68**:345–352.

Bernatchez, L., Vuorinen, J. A., Bodaly, R. A., and Dodson, J. J. 1996. Genetic evidence for reproductive isolation and multiple origins of sympatric trophic ecotypes of Whitefish. *Evolution* **50**:624–635.

Birkhead, T. R. and Muller, A. P. 1992. *Sperm Competition in Birds: Evolutionary Causes and Consequences*. London: Academic.

Bush, G. L. 1969. Sympatric host race formation and speciation in frugivorous flies of the genus *Rhagoletis* (Diptera, Tephritidae). *Evolution* **23**:237–251.

Bush, G. L. 1975. Modes of animal speciation. *Annu. Rev. Ecol. Syst.* **6**:339–364.

Bush, G. L. 1994. Sympatric speciation in animals: new wine in old bottles. *Trends Ecol. Evol.* **9**:285–288.

Bush, G. L., Case, S. M., Wilson, A. C., and Patton, J. L. 1977. Rapid speciation and chromosomal evolution in mammals. *Proc. Natl. Acad. Sci. USA* **74**:3942–3946.

Butlin, R. K. 1989. Reinforcement of premating isolation. In *Speciation and its Consequences*, D. Otte and J. A. Endler, eds., pp. 158–179. Sunderland, MA: Sinauer.

Butlin, R. K., Hewitt, G. M., and Webb, S. F. 1985. Sexual selection for intermediate optimum in *Chorthippus brunneus* (Orthoptera: Acrididae). *Anim. Behav.* **33**:1281–1292.

Cain, A. J. 1954. *Animal Species and their Evolution*. London: Hutchinson.

Carson, H. L. 1975. Genetics of speciation at the diploid level. *Am. Nat.* **109**:73–92.

Carson, H. L. 1985. Unification of speciation theory in plants and animals. *Syst. Bot.* **10**:380–390.

Carson, H. L. 1989. Genetic imbalance, realigned selection, and the origin of species. In *Genetics, Speciation, and the Founder Principle*, L. V. Giddings, K. Y. Kaneshiro, and W. W. Anderson, eds., pp. 345–362. New York: Oxford U. Press.

Caugant, D. A., Levin, B. R., and Selander, R. K. 1981. Genetic diversity and temporal variation in the *E. coli* population of a human host. *Genetics* **98**:467–490.

Charlesworth, B., Coyne, J. A., and Barton, N. H. 1987. Relative rates of evolution of sex chromosomes and autosomes. *Am. Nat.* **130**:113–146.

Charnov, E. L. and Bull, J. 1977. When is sex environmentally determined? *Nature (London)* **266**:828–830.

Chen, P. S., Stumm-Zollinger, E., Aigaki, T., Balmer, J., Bienz, M., and Bohlen, P. 1988. A male accessory gland peptide that regulates reproductive behavior of female *D. melanogaster*. *Cell* **54**:291–298.

Civetta, A. and Singh, R. S. 1998a. Sex related genes, directional sexual selection and speciation. *Mol. Biol. Evol.* **15**:901–909.

Civetta, A. and Singh, R. S. 1998b. Sex and speciation: genetic architecture and evolutionary potential of sexual versus non-sexual traits in the sibling species of the *Drosophila melanogaster* complex. *Evolution* **52**:1080–1092.

Clark, A. G., Aguade, M., Prout, T., Harshman, L. G., and Langley, C. H. 1995. Variation in sperm displacement and its association with accessory gland protein loci in *Drosophila melanogaster*. *Genetics* **139**:189–201.

Clark, A. G. and Kao, T.-H. 1991. Excess nonsynonymous substitution at shared polymorphic sites among self-incompatibility alleles of Solanaceae. *Proc. Natl. Acad. Sci. USA* **88**:9823–9827.

Coulthart, M. B. and Singh, R. S. 1988a. Low genic variation in male-reproductive-tract proteins of *Drosophila melanogaster* and *D. simulans. Mol. Biol. Evol.* **5**:167–181.

Coulthart, M. B. and Singh, R. S. 1988b. High levels of divergence of male reproductive tract proteins between *Drosophila melanogaster* and its sibling species, *D. simulans. Mol. Biol. Evol.* **5**:182–191.

Coyne, J. A. 1985. The genetic basis of Haldane's rule. *Nature (London)* **314**:736–738.

Coyne, J. A. 1992. Genetics and speciation. *Nature (London)* **355**:511–515.

Coyne, J. A. and Orr, A. 1989. Two rules of speciation. In *Speciation and its Consequences*. D. Otte and J. A. Endler, eds., pp. 189–211. Sunderland, MA: Sinauer.

Coyne, J. A. and Orr, H. A. 1997. "Patterns of speciation in Drosophila" revisited. *Evolution* **51**:295–303.

Cracraft, J. 1982. Geographic differentiation, cladistics and biogeography: reconstructing the tempo and mode of evolution. *Am. Zool.* **22**:411–424.

Darwin, C. (1859). *On the Origin of Species*. London: Murray.

de Bono, M. and Hodgkin, J. 1996. Evolution of sex determination in Caenorhabditis: unusually high divergence of tra-1 and its functional consequences. *Genetics* **144**:587–595.

De Vries, H. 1906. *Species and Varieties, Their Origin by Mutation* (lectures delivered at the University of California, 2nd ed). Chicago: Open Court.

Dewsbury, D. A. 1984. Sperm competition in muroid rodents. In *Sperm Competition and the Evolution of Animal Mating Systems*, R. L. Smith, ed., pp. 547–571. New York: Academic.

Dobzhansky, Th. 1934. Studies on hybrid sterility. I. Spermatogenesis in pure and hybrid *Drosophila pseudoobscura*. *Z. Zellforsch. Mikrosk. Anat.* 21:169–223.

Dobzhansky, Th. 1936. Studies on hybrid sterility. II. Localization of sterility factors in *Drosophila pseudoobscura* hybrids. *Genetics* 21:113–135.

Dobzhansky, Th. 1937, 1951. *Genetics and the Origin of Species.* New York: Columbia U. Press.

Dobzhansky, Th. 1970. *Genetics of Evolutionary Process.* New York: Columbia U. Press.

Dykhuizen, D. E. and Green, L. 1991. Recombination in *Escherichia coli* and the definition of biological species. *J. Bacteriol.* 173:7257–7268.

Eberhard, W. G. 1985. *Sexual Selection and Animal Genitalia.* Cambridge, MA: Harvard U. Press.

Eberhard, W. G. 1996. *Female Control: Sexual Selection by Cryptic Female Choice.* Princeton, NJ: Princeton U. Press.

Eberhard, W. G. and Cordero, C. 1995. Sexual selection by cryptic female choice on male seminal products – a new bridge between sexual selection and reproductive physiology. *TREE* 10:493–496.

Ehrlich, P. R. and Raven, P. H. 1969. Differentiations of populations. *Science* 165:1228–1232.

Eldrege, N. 1985. *Unfinished Synthesis: Biological Hierarchies and Modern Evolutionary Thought.* New York: Oxford U. Press.

Endler, J. A. 1986. Natural selection in the wild. *Monogr. Popul. Biol. 21.* Princeton, NJ: Princeton Univ. Press.

Endler, J. A. 1977. Geographic variation, speciation, and clines. *Monogr. Popul. Biol.* (10). Princeton, NJ: Princeton U. Press.

Eriksson, O. and Bremer, B. 1992. Pollination systems, dispersal modes, life forms, and diversification rates in Angiosperm families. *Evolution* 46:258–266.

Ferris, P. J. and Goodenough, U. W. 1997. Mating type in *Chlamydomonas* is specified by *mid*, the minus-dominance gene. *Genetics* 146:859–869.

Ferris, P. J., Woessner, J. P., and Goodenough, U. W. 1996. A sex recognition glycoprotein is encoded by the plus mating-type gene fus1 of *Chlamydomona reinhardtii. Mol. Biol. Cell* 7:1235–1248.

Ferris, P. J., Pavlovic, C., Fabry, S., and Goodenough, U. W. 1997. Rapid evolution of sex-related genes in Chlamydomonas. *Proc. Natl. Acad. Sci. USA* 94:8634–8639.

Fisher, R. A. 1930. *The Genetical Theory of Natural Selection.* Oxford: Clarendon.

Ford, E. B. 1964. *Ecological Genetics.* London: Methuen.

Futuyma, D. J. and Mayer, G. C. 1980. Non-allopatric speciation. *Syst. Zool.* 29:254–271.

Gerhardt, H. C. 1982. Sound pattern recognition in some North American tree frogs (Anura: Hylidae): implications for mate choice. *Am. Zool.* 22:581–595.

Ghiselin, M. J. 1987. Species concepts, individuality, and objectivity. *Biol. Philos.* 2:127–143.

Goldschmidt, R. B. 1940. *The Material Basis of Evolution.* New Haven, CT: Yale U. Press.

Gould, S. J. 1982. Darwinism and the expansion of evolutionary theory. *Science* 216: 380–387.

Gould, S. J. and Eldredge, N. 1977. Punctuated equilibria: the tempo and mode of evolution reconsidered. *Paleobiology* 3:115–151.

Grant, P. R. and Grant, B. R. 1992. Hybridization of bird species. *Science* 256:193–197.

Grant, P. R. and Grant, B. R. 1996. Speciation and hybridization in island birds. *Philos. Trans. R. Soc. London Ser. B* 351:765–772.

Grant, V. 1971. *Plant Speciation.* New York: Columbia U. Press.

Grant, V. 1985. *The Evolutionary Process. A Critical Review of Evolutionary Theory.* New York: Columbia U. Press.

Gregory, P. G. and Howard, D. J. 1994. A postinsemination barrier to fertilization isolates two closely related ground crickets. *Evolution* 48:705–710.

Grundle, H. and Schluter, D. 1998. Reinforcement of Stichlebach mate preference: sympatry breeds contempt. *Evolution* 52:200–208.

Haldane, J. B. S. 1922. Sex ratio and unisexual sterility in hybrid animals. *J. Genet.* 12:101–109.

Haldane, J. B. S. 1932. *The Causes of Evolution.* Ithaca, NY: Cornell U. Press.

Hilton, H. and Hey, J. 1996. DNA sequence variation at the *Period* locus reveals the history of speciation and speciation events in the *Drosophila virilis* group. *Genetics* 144:1015–1025.

Hoikkala, A. and Kaneshiro, K. 1993. Change in the signal-response sequence responsible for asymmetric isolation between *Drosophila planitibia* and *Drosophila sylvestris. Proc. Natl. Acad. Sci. USA* 90:5813–5817.

Hollocher, H. and Wu, C.-I. 1996. The genetics of reproductive isolation in the *Drosophila simulans* clade: X *vs.* autosomal effects and male *vs.* female effects. *Genetics* 143:1243–1255.

Holman, E. W. 1987. Recognizability of sexual and asexual species of rotifers. *Syst. Zool.* 36:381–386.

Hoy, R. R., Hoikkala, A., and Kaneshiro, K. 1988. Hawaiian courtship songs: evolutionary innovation in communication signals of *Drosophila. Science* 240:217–219.

Hull, D. L. 1987. Genealogical actors in ecological role. *Biol. Philos.* 2:168–184.

Hunter, N., Chambers, S. R., Louis, E. J., and Borts, R. H. 1996. The mismatch repair system contributes to meiotic sterility in an interspecific yeast hybrid. *EMBO J.* 15:1726–1733.

Huxley, J. 1942. *Evolution: The Modern Synthesis.* New York: Allen and Unwin.

Jain, S. K. and Bradshaw, A. D. 1966. Evolution in closely adjacent plant populations. I. The evidence and its theoretical analysis. *Heredity* 21:407–441.

Janzen, D. H. 1977. A note on optimal mate selection by plants. *Am. Nat.* 111:365–371.

Kaneshiro, K. Y. 1989. The dynamics of sexual selection and founder effects in species formation. In *Genetics, Speciation and the Founder Principle,* L. V. Giddings, K. Y. Kaneshiro, and W. W. Anderson, eds., pp. 279-296. New York: Oxford U. Press.

Kaneshiro, K. Y. and Boake, C. R. B. 1987. Sexual selection and speciation: issues raised by Hawaiian Drosophila. *TREE* 2:207–212.

Karotam, J., Delves, A. C., and Oakeshott, J. G. 1993. Conservation and change in structural and 5' flanking sequences of esterase 6 in sibling Drosophila species. *Genetica* 88:11–28.

Kauffman, S. A. 1993. *The Origins of Order.* New York: Oxford U. Press.

Kondrashov, A. S. 1986. Multilocus models of sympatric speciation. III. Computer simulations. *Theor. Popul. Biol.* 29:1–15.

King, M. C. and Wilson, A. C. 1975. Evolution at two levels in humans and chimpanzees. *Science* 188:107–116.

Kitcher, P. 1984a. Species. *Philos. Sci.* 51:308–333.

Kitcher, P. P. 1984b. Against the monism of the moment. Philos. Sci. 51:606–630.

Kliman, R. M. and Hey, J. 1993. DNA sequence variation at the *Period* locus within and among species of the *Drosophila melanogaster* complex. *Genetics* 133:375–387.

Kuwabara, P. E. and Hodgkin, J. 1996. Interspecies comparisons reveals evolution of control regions in the nematode sex-determining gene tra-2. *Genetics* **144**:597–607.

Lamnissou, K., Loukas, M., and Zouros, E. 1996. Incompatibilities between Y chromosome and autosomes are responsible for male hybrid sterility in crosses between *Drosophila virilis* and *D. texana. Heredity* **76**:603–609.

Lande, R. 1981. Models of speciation by sexual selection on polygenic traits. *Proc. Natl. Acad. Sci. USA* **78**:3721–3725.

Lande, R. 1982. Rapid origin of sexual isolation and character divergence in a cline. *Evolution* **36**:213–233.

Lee, Y.-H., Ota, T., and Vacquier, V. D. 1995. Positive selection is a general phenomenon in the evolution of abalone sperm lysin. *Mol. Biol. Evol.* **12**:231–238.

Levin, D. A. 1979. The nature of plant species. *Science* **204**:381–384.

Levin, D. A. 1993. Local speciation in plants. *Syst. Bot.* **18**:197–208.

Lewis, H. 1962. Catastrophic selection as a factor in speciation. *Evolution* **16**:257–271.

Lewontin, R. C. 1974. *The Genetic Basis of Evolutionary Change.* New York: Columbia U. Press.

Lewontin, R. C. 1997. Dobzhansky's *Genetics and the Origin of Species*: is it still relevant? *Genetics* **147**:351–355.

Liou, L. W. and Price, T. D. 1994. Speciation by reinforcement of premating isolation. *Evolution* **48**:1451–1459.

Lynch, J. D. 1989. The gauge of speciation: on the frequencies of modes of speciation. In *Speciation and Its Consequences*, D. Otte and J. A. Endler, eds., pp. 527–553. Sunderland, MA: Sinauer.

Maiti, S., Doskow, J., Sutton, K., Nhim, R. P., Lawlor, D. A., Levan, K., Lindsey, J. S., and Wilkinson, M. F. 1996. The Pem homeobox gene: rapid evolution of the homeodomain, X chromosomal localization, and expression in reproductive tissue. *Genomics* **34**:304–316.

Matic, I., Rayssiguier, C., and Radman, M. 1995. Interspecies gene exchange in bacteria: the role of SOS and mismatch repair systems in evolution of species. *Cell* **80**:507–515.

Maynard Smith, J. 1966. Sympatric speciation. *Am. Nat.* **100**:637–650.

Maynard Smith, J., Dowson, C. G., and Spratt, B. G. 1991. Localized sex in bacteria. *Nature (London)* **349**:29–31.

Maynard Smith, J., Smith, N. H., O'Rourke, M., and Spratt, B. G. 1993. How clonal are bacteria? *Proc. Natl. Acad. Sci. USA* **90**:4384–4388.

Mayr, E. 1942. *Systematics and the Origin of Species.* New York: Columbia U. Press.

Mayr, E. 1954. Change of genetic environment and evolution. In *Evolution as a Process*, J. Huxley, A. C. Hardy, and E. B. Ford, eds., London: Allen and Unwin.

Mayr, E. 1963. *Animal Species and Evolution.* Cambridge, MA: Harvard U. Press.

Mayr, E. 1982a. Speciation and macroevolution. *Evolution* **36**:1119–1132.

Mayr, E. 1982b. Processes of speciation in animals. In *Mechanisms of Speciation*, C. Barigozzi, ed., pp. 1–19. New York: Liss.

Mayr, E. 1987. The ontological status of species. *Biol. Philos.* **2**:145–166.

Mayr, E. and Provine, W. G. 1980. *The Evolutionary Synthesis: Perspectives on the Unification of Biology.* Cambridge, MA: Harvard U. Press.

Metz, E. C. and Palumbi, S. R. 1996. Positive selection and sequence rearrangements generate extensive polymorphism in the gamete recognition protein binding. *Mol. Biol. Evol.* **13**:397–406.

Meyer, A. 1993. Phylogenetic relationships and evolutionary processes in east African Cichlids fishes. *Trends Ecol. Evol.* **8**:279–284.

Milkman, R. and Bridges, M. M. 1993. Molecular evolution of the *Escherichia coli* chromosome. IV. Sequence comparisons. *Genetics* **133**:455–468.

Mishler, B. D. and Brandon, R. N. 1987. Individuality, pluralism, and the phylogenetic species concept. *Biol. Philos.* **2**:397–414.

Mishler, B. D. and Donoghue, M. J. 1982. 'Species concept: a case for pluralism.' *System. Zool.* **31**:491–503.

Muller, H. J. 1940. Bearing of the *Drosophila* work on systematics. In *The New Systematics*, J. Huxley, ed., pp. 185–268. Oxford: Clarendon.

Muller, H. J. 1942. Isolating mechanisms, evolution and temperature. *Biol. Symp.* **6**:71–125.

Nelson, G. 1989. Species and taxa: systematics and evolution. In *Speciation and Its Consequences*, D. Otte and J. A. Endler, eds., pp. 60–81. Sunderland, MA: Sinauer.

O'Neil, M. T. and Belote, J. M. 1992. Interspecific comparison of the transformer gene of Drosophila reveals an unusually high degree of evolutionary divergence. *Genetics* **131**:113–128.

Orr, H. A. 1993. A mathematical model of Haldane's rule. *Evolution* **47**:1606–1611.

Orr, H. A. 1995. The population genetics of speciation: the evolution of hybrid incompatibility. *Genetics* **139**:1805–1813.

Orr, H. A. 1997. Haldane's rule. *Annu. Rev. Ecol. Syst.* **28**:195–218.

Pamilo, P. and O'Neill, R. J. W. 1997. Evolution of the Sry genes. *Mol. Biol. Evol.* **14**:49–55.

Paterson, H. E. H. 1978. More evidence against speciation by reinforcement. *S. Afri. J. Sci.* **74**:369–371.

Paterson, H. E. H. 1985. The recognition concept of species. In *Species and Speciation*, E. S. Vrba, ed., pp. 21–29. Pretoria: Transvaal Museum Monograph No. 4.

Payne, R. J. H. and Krakauer, D. C. 1997. Sexual selection, space, and speciation. *Evolution* **51**:1–9.

Pigeon, D., Chouinard, A., and Bernatchez, L. 1997. Multiple modes of speciation involved in the parallel evolution of sympatric morphotypes of Lake Whitefish (*Coregonus clupeaformis*, Salmonide). *Evolution* **51**:196–205.

Pontecorvo, G. 1943. Hybrid sterility in artificially produced recombinants between *Drosophila melanogaster* and *D. simulans*. *Proc. R. Soc. Edinburgh Ser. B* **41**:385–397.

Provine, W. B. 1971. *The Origins of Theoretical Population Genetics*. Chicago: U. Chicago Press.

Radman, M. and Wagner, R. 1993. Mismatch recognition in chromosomal interactions and speciation. *Chromosoma* **102**:369–373.

Rayssiguier, C., Thaler, D. S., and Radman, M. 1989. The barrier to recombination between *Escherichia coli* and *Salmonella typhimurium* is disrupted in mismatch-repair mutants. *Nature (London)* **342**:396–401.

Rice, W. R. 1996. Sexually antagonistic male adaptation triggered by experimental arrest of female evolution. *Nature (London)* **381**:232–234.

Rosenzweig, R. F., Sharp, R. R., Treves, D. S., and Adams, J. 1994. Microbial evolution in a simple unstructured environment:genetic differentiation in *Escherichia coli*. *Genetics* **137**:903–917.

Ryan, M. J. and Rand, A. S. 1993. Species recognition and sexual selection as a unitary problem in animal communication. *Evolution* **47**:647–657.

Ryan, M. J. and Rand, A. S. 1995. Female responses to ancestral advertisement calls in Tngara frogs. *Science* **269**:390–392.

Schliewen, U. K., Tautz, D., and Paabo, S. 1994. Sympatric speciation suggested by monophyly of crater lake cichlids. *Nature (London)* **368**:629–632.

Schluter, D. 1996. Ecological causes of adaptive radiation. *Am. Nat.* **148**:s40–s64.

Schluter, D. and McPhail, J. D. 1993. Character displacement and replicate adaptive radiation. *Trends Ecol. Evol.* **8**:197–200.

Schluter, D. and Nagel, L. M. 1995. Parallel speciation by natural selection. *Am. Nat.* **146**:292–301.

Scott, D. 1986. Inhibition of female Drosophila melanogaster remating by a seminal fluid protein (esterase 6). *Evolution* **40**:1084–1091.

Selander, R. K. and Levin, B. R. 1980. Genetic diversity and structure in populations of *Escherichia coli. Science* **210**:545–547.

Sharp, P. M. and Li, W. H. 1987. The rate of synonymous substitution in enterobacterial genes is inversely related to codon usage bias. *Mol. Biol. Evol.* **4**:222–230.

Simpson, G. G. 1944. *Tempo and Mode in Evolution.* New York: Columbia U. Press.

Simpson, G. G. 1953. *The Major Features of Evolution.* New York: Columbia U. Press.

Simpson, G. G. 1961. *Principles of Animal Taxonomy.* New York: Columbia U. Press.

Singh, R. S. 1990. Pattern of species divergence and genetic theories of speciation. In *Population Biology,* K. Wohrmann and S. K. Jain, eds., pp. 231–265. Berlin: Springer-Verlag.

Singh, R. S. and Zeng, L.-W. 1994. Genetic divergence, reproductive isolation and speciation. In *Non-Neutral Evolution,* B. Golding, ed., pp. 217–232. Toronto: Chapman & Hall.

Sneath, P. H. A. and Sokal, R. R. 1973. *Numerical Taxonomy.* 2nd ed. San Francisco: Freeman.

Sober, E. 1984. Discussion: sets, species, and evolution: Comments on Philip Kitcher's "Species." *Philos. Sci.* **51**:334–341.

Spieth, H. T. 1981. *Drosophila heteroneura* and *Drosophila silvestris:* head shapes, behavior and evolution. *Evolution* **35**:921–930.

Stanley, S. M. 1979. *Macroevolution – Pattern and Process.* San Francisco: Freeman.

Stebbins, G. L. 1950. *Variation and Evolution in Plants.* New York: Columbia U. Press.

Stebbins, G. L. 1981. Why are there so many species of flowering plants? *Bioscience* **31**:573–577.

Stebbins, G. L. 1982. Plant speciation. In *Mechanisms of Speciation,* C. Barigozzi, ed., pp. 21–39. New York: Liss.

Stein, S. P., Tepper, C. S., Able, N. D., and Richmond, R. C. 1984. Studies of esterase 6 in *Drosophila melanogaster.* XVI synthesis occurs in the male reproductive tract (anterior ejaculatory duct) and is modulated by juvenile hormone. *Insect Biochem.* **14**:527–532.

Swanson, W. J. and Vacquier, V. D. 1995. Extraordinary divergence and positive Darwinian selection in a fusagenic protein coating the acrosomal process of abalone spermatozo. *Proc. Natl. Acad. Sci. USA* **92**:4957–4961.

Swanson, W. J. and Vacquier, V. D. 1998. Correlated evolution in a egg receptor from a rapidly evolving Abalone sperm protein. *Science* **281**:710–712.

Tauber, C. A. and Tauber, M. J. 1977. A genetic model of sympatric speciation through habitat diversification and seasonal isolation. *Nature (London)* **268**:702–705.

Templeton, A. R. 1980. Modes of speciation and inferences based on genetic distances. *Evolution* **34**:719–729.

Templeton, A. R. 1981. Mechanisms of speciation – a population genetic approach. *Annu. Rev. Ecol. Syst.* **12**:23–48.

Templeton, A. R. 1989. The meaning of species and speciation: a genetic perspective. In *Speciation and Its Consequences*, D. Otte and J. A. Endler, eds., Sunderland, MA: Sinauer.

Thoday, J. M. and Gibson, J. B. 1962. Isolation by disruptive selection. *Nature (London)* **193**:1164–1166.

True, J. R., Weir, B. S., and Laurie, C. C. 1996. A genome-wide survey of hybrid incompatibility factors by the introgression of marked segments of *Drosophila mauritiana* chromosomes into *Drosophila simulans*. *Genetics* **112**:819–837.

Tucker, P. K. and Lundrigan, B. L. 1993. Rapid evolution of the sex determining locus in old world mice and rats. *Nature (London)* **364**:715–717.

Turelli, M. and Orr, H. A. 1995. The dominace theory of Haldane's rule. *Genetics* **140**:389–402.

Turner, G. F. 1994. Speciation mechanisms in Lake Malawi cichlids: a critical review. *Arch. Hydrobiol. Beith. Limnol.* **44**:139–160.

Turner, G. F. and Burrows, M. T. 1995. A model of sympatric speciation by sexual selection. *Proc. R. Soc. London Ser. B* **260**:287–292.

Vigneault, G. and Zouros, E. 1986. The genetics of asymmetrical male sterility in *Drosophila movanensis* and *Drosophila arizonensis* hybrids: interactions between the Y chromosome and autosomes. *Evolution* **40**:1160–1170.

Vrba, E. S. 1980. Evolution, species and fossils: How does life evolve? *S. Afr. J. Sci.* **76**:61–84.

Wade, M. J., Patterson, H., Chang, N. W., and Johnson, N. 1994. Postcopulatory, prezygotic isolation in flour beetles. *Heredity* **72**:163–167.

Wagner, A., Wagner, G. P., and Similion, P. 1994. Epistasis can facilitate the evolution of reproductive isolation by peak shifts: a two-locus two-allele model. *Genetics* **138**:533–545.

Wallace, A. R. 1889. *Darwinism.* London: Macmillan.

Wang, R.-L. and Hey, J. 1996. The speciation history of *Drosophila pseudoobscura* and close relatives: inferences from DNA sequence variation at the Period locus. *Genetics* **144**:1113–1126.

West-Eberhard, M. J. 1983. Sexual selection, social competition, and speciation. *Q. Rev. Biol.* **58**:155–183.

White, M. J. D. 1978. *Modes of Speciation.* San Francisco: Freeman.

Whitfield, L. S., Lovell-Badge, R., and Goodfellow, P. N. 1993. Rapid sequence evolution of the mammalian sex-determining gene SRY. *Nature (London)* **364**:713–715.

Wiley, E. O. 1981. *Phylogenetics: The Theory and Practice of Phylogenetic Systematics.* New York: Wiley.

Wiley, E. O. and Mayden, R. L. 1985. Species and speciation in phylogenetic systematics, with examples from the North American fish fauna. *Ann. Mo. Bot. Gard.* **72**:596–635.

Wilson, A. C., Bush, G. L., Case, S. M., and King, M. C. 1975. Social structuring of mammalian popualtions and the rate of chromosomal evolution. *Proc. Natl. Acad. Sci. USA* **72**:5061–5065.

Wright, S. 1931. Evolution in Mendelian populations. *Genetics* **16**:97–159.

Wright, S. 1982. Character change, speciation and higher taxa. *Evolution* **36**:427–434.

Wu, C.-I. 1992. A note on Haldane's rule: hybrid inviability *versus* hybrid sterility. *Evolution* **46**:1584–1587.

Wu, C.-I. and Davis, A. W. 1993. Evolution of postmating reproductive isolation: the composite nature of Haldane's rule and its genetic basis. *Am. Nat.* **142**:187–212.

Zeng, L.-W. 1996. Resurrecting Muller's theory of Haldane's rule. *Genetics* **143**:603–607.

Zeng, L.-W. and Singh, R. S. 1993. The genetic basis of Haldane's rule and the nature of asymmetric hybrid male sterility between *Drosophila simulans, D. mauritiana,* and *D. sechellia. Genetics* **134**:251–260.

Zouros, E. 1981. An autosome-Y chromosome combination that causes sterility in *D. momavensis* and *D. arizonensis* hybrids. *Drosophila Inf. Serv.* **56**:167–168.

SECTION H

BEHAVIOR, ECOLOGY, AND EVOLUTION

Introductory Remarks

SUBODH JAIN

> Most of what is unusual about man can be summed up in one word: culture.
> —(Barlow, 1991)

Just as no two snowflakes are identical, no two evolutionary biologists would agree on the full scope and relative rankings of currently developing topics in evolution, especially when it encompasses philosophy and social sciences. Many authors are speaking about the so-called Modern Synthesis and neo-Darwinism as if the triumphant new knowledge gained by them might be fading away in relation to the new challenges in morphometrics, phylogeny, ecology, and sociobiology. Even genetic foundations as laid down in early population genetics are considered inadequate; Provine (1986) found that synthesis was more of an evolutionary constriction, that is, accepting rather few variables as a crucial start. However, the edifice would largely stand even if it had to invent new ways to deal with the dynamic sufficiency and capricious historicity problems. So much is in the air about complexity, holism, uncertainty, chaos, and self-organization's yielding order out of chaos, both within biology and outside it. But one need not be surprised or perturbed; no one in our graduate school years felt that we had more than good working hypotheses in the so-called synthetic theory of evolution; after all, an evolutionist must certainly accept change and openendedness.

Evolution textbooks referred to Darwin's puzzling problems such as sexual selection, asexuality, social cooperative behavior, altruism, instinct, speciation, and macroevolution as special topics. I believe those appended topics have now moved to the centerfold pages (new book titles such as Depew and Weber's *Darwinism Evolving* or Wesson's *Beyond Natural Selection* make it clear, but is it necessary?). Futuyma (1986) concluded that "entire fields such as behavior and ecology have become incorporated into evolutionary biology." In any event, with population genetic and ecological theories always seeking to meet new challenges, we can truly expect another exciting era of postmodern resynthesis! This section underscores such developments toward a "new age" evolutionary biology. Thus the four chapters in this section cover quite divergent areas such as

605

phylogeny, comparative analyses in adaptation biology, social behavior, gender-related sociobiology, population ecology, and more.

Joe Felsenstein first provides a gentle historical reading about variation, selection, and an overall meaning of fitness parameters, noting how morphologists failed to find use for population genetics. Molecular evolution with all its power and glamor apparently did even less for the ecologists and morphometric evolutionists (yes, Martha, there are many kinds of evolutionists today). I personally found the useful paradigm of developing rules, establishing certain derived features, and then studying instances very valuable, even for following his chapter. Here, certain rules are developed for studying coalescence and cladism, which nicely allow statistics to be derived for specific empirical tests of the inferred phylogenies. Thus, going from molecular evolution to phylogenies by using certain developmental, genomic, ecological, and statistical rules is nicely illustrated here. And what we may label as the Felsenstein rule, well developed in his masterpiece paper (Felsenstein, 1982), advised us thus: acknowledge uncertainties in both theory and data, and then develop scientific ways around them. This in my view would help avoid pitfalls of many pseodocontroversies, or most claims about proofs and disproofs. You may quote me. Also note Felsenstein's comments: developmental biology has had no impact on morphometrics, whether constraints are important in evolution is an old controversy, ecological theory has made no notable advances for evolutionists; the change from population genetics to evolutionary genetics has been due to the shift from theory driven to more data driven approaches. This is enough in my view to demand a serious reading of this section.

Maynard Smith's chapter is equally elegant and delightful. With his inimitable prose he reviews the germane issues of female choice in the sexual selection theory and various alternative ideas that might have delayed a proper scientific approach until the gender revolution changed things. The key issues are whether choices are based largely on adaptive processes or on certain internal or structural characteristics. Many philosophical commentaries on Darwinism for humans are raised in the literature on the evolution of behavior, and Maynard Smith emphatically points out that even the most mechanistic biologists need not reject the notion of choice. Physiology and ethology need not be antagonistic to evolutionary thinking. For many elaborate discussions of this issue, Bendall (1993) is an excellent source. Maynard Smith also questions the idea of modules with somehow preprogrammed human behavior controls, but predicts that some genetic support for a modular theory should be forthcoming. The old debate between the early Mendelians and biometricians is interestingly found to be resolved if, as Maynard Smith claims, most writers on the subject have reached a mathematical understanding of bifurcation in which continuous variation could lead to discontinuities in much of population biology literature I found the major genes versus polygenes issue rather sketchily handled. Thus there is a lot here for the future behavioral evolutionists to think about seriously.

Deborah Gordon's chapter deals with the evolution of social behavior, treating it as an optimality problem, with emphasis on interindividual variation due to individual's interactions as members of a group, colony, or social unit as much as their own genetic machinery. Her central point, that without the interactive context individual fitness arithmetic would not accurately account for social behavior, is well taken. Thus Lewontin's ideas about the dialectical nature of biology are reinforced here. For good experimental support, for example, Gordon's own work on harvester ants is reviewed in Real (1994) in which we find numerous other evolutionary models for studying behavior.

My own chapter attempts to review how much or how well have population genetics and ecology made their marriage work (sure, all marriages have domestic problems). It appeared that the literature has grown abundantly on certain topics such as population structure, metapopulation dynamics, evolution of life histories, breeding systems, and evolution under stress, all of which warranted ecological and genetic treatments. The older problems of population regulation, evolution of niche diversity in coexisting species, and even the discovery of various meanings of population fitness or adaptive landscapes are not reviewed, but overall it is interesting to find an almost chorus call for more field research, more pluralistic explanations, and simply more conscious effort to work together.

Here, we should note the large amount of unfinished business just in dealing with the measurements of fitnesses and rates of evolutionary change, using simple genetic polymorphisms, metapopulations, species assemblages on the one hand and the ideas of interactions between genes, social members of a group, and species on the other. With all our nice holistic expectations about coadapted systems of genes or species, much of evolutionary theory is still far behind. On the other hand, information age writers are predicting that new computer designs and software developments will learn from evolution how to develop more thinking–adapting (almost self-organizing) algorithms! Neurobiologists are also keen about the evolutionary origins of neural networking and our brain functions. The old question whether or how humans are unique is now acutely raised by the animal behavior researchers as well as by the cyberworld thinkers. Kelley (1994) wrote that "the shock is not that evolution has been transported from carbon to silicon; silicon and carbon are actually very similar elements." However, evolutionary biologists will surely find that our uniqueness will depend on some threshold multivariate differences in choice-based behavior, learning and adapting ingeniously, and of course other cultural evolution characteristics. In a thoughtful prologue, Maynard Smith and Holiday (1979) had predicted that molecular evolution and use of prokaryotes in evolutionary research would grow rapidly, and so they did, but their comment that the "units of selection muck" was by no means resolved, is in my view superfluous today.

It is also nice to remind ourselves of the oft-cited dictum from Dobzhansky that nothing in biology makes sense without evolutionary biology, but also we discover that vice versa is equally true, as noted in this section repeatedly. Furthermore, as noted by Lewontin on several occasions, we must think along the lines of teaching undergraduate-level population genetics and evolution with the most diverse and exciting problems of biology in the first place and theory only when or where it is truly insightful.

REFERENCES

Barlow, C., ed. 1991. *From Gaia to Selfish Genes*. Cambridge, MA: Massachusetts Institute of Technology Press.

Bendall, D. S., ed. 1993. *Evolution from Molecules to Men*. Cambridge: Cambridge U. Press.

Felsenstein, J. 1982. Numerical methods for inferring evolutionary tress. *Q. Rev. Biol.* 57:379–404.

Futuyma, D. J. 1986. *Evolutionary Biology*. Sunderland, MA: Sinauer.

Kelley, K. 1994. *Out of Control: The New Biology of Machines*. Reading, MA: Addison-Wesley.

Maynard Smith, J. and Holiday, R., eds. 1979. The evolution of adaptation by natural selection. *Proc. R. Soc. London Ser.* B 205:ii.

Provine, W.B. 1986. *Sewall Wright and Evolutionary Biology*. Chicago: U. Chicago Press.

Real, L., ed. 1994. *Behavioral Mechanisms of Evolution*. Chicago: U. Chicago Press.

CHAPTER TWENTY NINE

From Population Genetics to Evolutionary Genetics: A View through the Trees

JOSEPH FELSENSTEIN

This is a personal and impressionistic account of the situation in evolutionary genetics and molecular evolution as it seemed to one participant in theoretical and methodological developments in those fields. It is based in part on a Presidential Address delivered to the Society for the Study of Evolution at its 1993 annual meeting in Snowbird, Utah.

29.1. Theory Reaches its Zenith

During the years I spent in the Lewontin lab (1964–1967), population genetics was a quite different field than evolutionary genetics is now. Theoretical population genetics had a tradition dating back before Hardy and Weinberg and was a reasonably mature field. We spent our time making various evolutionary forces collide with each other, as experimental physicists do with particles. The intention was to try to discover general laws about which force would prevail. There was a great need for total generality, for a simple reason there were almost no data to be had, so that we could not constrain our conclusions to particular parts of the parameter space. As we could not know how large selection coefficients were or what were the patterns of gene interaction, we needed results that would be general across all of them. The theory was beautiful indeed: the Kolmogorov equations, as masterfully used by Kimura, and the variance components machinery of Fisher and Wright, burnished to a high polish by Kempthorne and Cockerham.

Outside of this beautiful but small world was the frightening reality of interspecies differences, about which we could say almost nothing. In about 1970, evolutionary biology looked to us as it does in Fig. 29.1. Our central obsession was finding out what function evolution would try to maximize. Population geneticists used to think, following Sewall Wright, that mean relative fitness $\bar{w}$ would be maximized by natural selection. This is true for one locus, as long as

R. S. Singh and C. B. Krimbas, eds., *Evolutionary Genetics: From Molecules to Morphology*, vol. 1. © Cambridge University Press 2000. Printed in the United States of America. ISBN 0-521-57123-5.

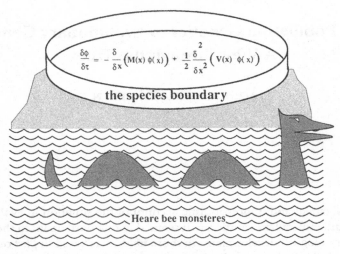

Figure 29.1. A cartoon of the state of population genetics in 1970.

the relative fitnesses remain constant. But by 1966 work on multilocus systems had advanced from its inception by Kimura (1956) and Lewontin and Kojima (1960). We now knew (Moran, 1966) that $\bar{w}$ could decline as a result of natural selection and recombination, even decline steadily. It had already been known to Fisher and Wright that if the fitnesses changed secularly or as a result of gene frequency dependence, the result would usually be a failure to maximize mean relative fitness. But Moran's result was shocking – the failure to maximize $\bar{w}$ had invaded even the case of constant relative fitnesses.

If only we could find out what function other than $\bar{w}$ was maximized by natural selection in the presence of recombination, we would have some insight into what compromises occur between the genetic machinery and natural selection. We might then also be able to see what selection would do to the genetic system itself, and to what extent mean fitness, and hence adaptation, would be expected to be increased in evolution.

Two technological revolutions came to the field. One was the availability of computers. They were first used in biology in solving the selection–migration cline (Fisher, 1950; see Wilkes, 1975). In 1954, the original and eccentric Nils Aall Barricelli was the first to use them for genetic simulation (Barricelli, 1954), partly because he was one of the first to have access to a computer. Barricelli's work was not published until later in English, and so it fell to Fraser (1957a, 1957b) to do the first widely noticed genetic simulations. By the early 1960s genetic simulation, numerical iteration of deterministic equations, numerical solution of equilibria in deterministic cases, and numerical evaluation of equilibrium distributions in stochastic ones were all standard tools. They made the investigation of multilocus natural selection possible.

Joseph Felsenstein

The result has been that the models used in molecular evolution are particularly simple, but in some respects disappointing. They are primarily models that specify what does *not* happen. Since we do not know whether the changes that do happen are the result of positive natural selection or neutral mutation, there is no way to predict how much natural selection is expected to improve molecular function. I often find that when people outside the field ask what theory predicts about molecular evolution, they are disappointed to find that its predictions are so negative.

29.6. Sequence Samples from Populations

While molecular evolution was maturing as a field, the same molecular technology was slowly invading evolutionary genetics. The invasion has been slow because so much evolutionary genetics was supported by the aptly named NSF.[1] It is still much cheaper to do electrophoretic surveys at many loci than to get one population sample of molecular sequences. It is noticeable how long it has taken, since Kreitman's pioneering studies (1983), for population samples of nuclear molecular sequences to become common.

A major exception must be made for one tiny piece of eukaryotic DNA. Mitochondrial population samples have been widely used. They got much attention with the famous "mitochondrial Eve" study of Cann et al. (1987). At first, mitochondria were used because their DNA was easy to extract. That ceased to be an advantage with the development of the polymerase chain reaction, but other advantages (maternal inheritance, haploidy, and rapid rate of evolution) have been cited as additional reasons to concentrate on the mitochondrial genome.

Initially, few of the researchers using mitochondrial data related their trees to population processes. The presence of an "Eve" seemed to be an unusual consequence of maternal inheritance. But every nucleotide in the genome should in principle have its own "Eve" or "Adam," as a result of the coalescent process at that locus. How far along the genome one can go and still find that same coalescent tree, so that every gene copy comes from that same ancestor, depends on the rate of recombination and the effective population size. The techniques of molecular evolution have invaded evolutionary genetics as people began to infer these gene trees. At the same time, evolutionary genetic theory turned out to provide the theoretical framework for analyzing these gene trees. The basic theoretical work was done by J. F. C. Kingman (1982a, 1982b). His work generalized Sewall Wright's result (1930) for two copies. Wright pointed out that two copies of a gene in a closed randomly mating population would have a common ancestor an average of $2N_e$ generations ago, with the distribution being approximately exponential. Kingman's "n-coalescent" process (Fig. 29.2) is the generalization of this to n copies. The time during which all n copies have had distinct ancestral lineages is exponentially distributed, with mean $4N_e/n(n-1)$ generations. Before that, there were $n-1$ copies, to which the same rule applies, and so on.

These undoubted successes have, however, affected only life-history traits. They leave us without any better understanding of the elephant's trunk or the Panda's thumb. Some of the "monsters" who lurked outside the species barrier in the 1970s were actually morphological evolutionists, who roamed the hallways of our departments, desperately seeking someone to help them analyze their data. Among ourselves, population geneticists used to boast about how little help we had given. "Fred Bloggs came to see me with his meaningless snail data. I threw him out" (laughter).

Under these circumstances, morphologists could be forgiven for concluding that population genetics theory was useless to them and for trying to find a different evolutionary theory that spoke to their concerns. One of the attractions of species-selection forms of punctuationism, for example, was that in one stroke they demoted all within-population genetic phenomena to the mere generation of variability, with the interesting natural selection occurring above the species level. This was a satisfying response to unhelpful population geneticists.

And yet, population genetics had already delivered insights into morphological evolution. The theory has succeeded in giving us a quantitative feel for the relative strength of various evolutionary forces. We can use it to show how slow change would be if driven only by mutation, how small selection coefficients must be to have their effect swamped by random genetic drift, and how much migration would be needed to overwhelm local adaptation. All of these might also have been attempted with another theory of inheritance, such as the blending inheritance theory of Fleeming Jenkin (1867), but the results would have been very different. Our current understanding of when migration, mutation, or genetic drift can be invoked to explain an unusual pattern derives directly from the mathematical implications of Mendelian inheritance, as assimilated into evolutionary theory during the period of the Modern Synthesis. The fact that we could not be of much practical help to morphological evolutionists has tended to obscure this fact.

29.2. Enter the Data

What was needed was data, and here a second technological revolution had a more dramatic effect. When I was a graduate student, Lewontin and Hubby (1966) were working on some obscure project involving electrophoretic gels. It seemed a bit odd, so we other members of the Lewontin lab felt fortunate in not being made to work on it. As soon as their paper had appeared, it was immediately obvious that the field was transformed: where there had been little data, there was now a lot. A wave of excitement swept population genetics, and extravagant promises were made. If we could only compute the right statistic from them, electrophoretic gene frequencies would solve the problem of discriminating between neutrality and selection. But this was not to be, and as the years wore by, both lecture audiences and granting agencies became wary of the claim that we would solve this problem Real Soon Now. The reason for the failure was simple – natural selection, with millions of generations and vast population sizes, can detect far smaller selection coefficients than we can hope to in any experiment. It should therefore have been anticipated that we were not going to be able to discriminate between neutrality and weak selection as well as evolution could.

Nevertheless, the revolution wrought by the arrival of molecular biology has been vast. It was now relatively easy to generate multiple-locus gene frequency data within species by electrophoretic techniques.

The need for methods to analyze these data was great, and population geneticists responded. One landmark was Ewens's (1972) derivation of the likelihood for a single-locus sample from an infinite-alleles model, with the surprising conclusion that the number of alleles was the sufficient statistic for the parameter $\theta = 4N_e\mu$, the scaled product of effective population size and neutral mutation rate. As real data are likely to contain some deleterious alleles as well, we cannot actually draw from this the conclusion that gene frequencies do not matter in this estimation. Watterson (1975) put forward the infinite-sites model and explored the estimation problems that it raises, suggesting the number of segregating sites statistic as useful for estimating θ.

Another landmark was Nei's (1972) genetic distance, which became an important tool near the species level. Even if we could not resolve the neutrality-selection controversy, we could use tools based on neutrality to ask other questions about population structure and history. In fact, one might say that the failure to resolve the neutrality-selection controversy actually strengthened the use of these tools. For if there was no easy way to discriminate between the gene frequency patterns expected under neutrality and under selection, one could simply use the neutrality theory to generate expectations, secure in the conviction that the presence of selection would change these patterns very little.

Twenty years on, we can see that these methods of data analysis had a greater effect than the pure theory done during the same period. To a considerable extent the change from population genetics to evolutionary genetics has been a change from a field that was theory-driven to one that is driven by data analysis.

29.3. Into the Trees – The Rise of Molecular Evolution

While population genetics was turning into evolutionary genetics, other kinds of molecular data and other uses of computers were also revolutionizing the analysis of between-species data. With molecular sequences available, precisely comparable data at the genetic level became available across multiple species. Previously, there had been some chromosome banding data, but only in certain groups. Morphological data suffered from having an unknown genetic basis, so that it could not give information at the genetic level.

It took only a few years from the first protein sequences to the first sequence comparisons between species. Zuckerkandl and Pauling (1962) were the pioneers. They foresaw a "chemical paleogenetics" in which ancestral protein sequences could be reconstructed from contemporary comparisons, and in their work the molecular clock was first postulated. As the 1960s progressed, protein sequences for multiple species accumulated. The publication of the *Atlas of Protein Sequences* by the late Margaret Dayhoff (Eck and Dayhoff, 1966) was important in bringing evolution to the attention of molecular biologists and molecular biology to the attention of evolutionists. Dayhoff took an un-compromisingly evolutionary approach to the data she compiled; the *Atlas* was, from the start, full of phylogenies and discussions of phylogenetic me-thods.

Suddenly, molecular biology started answering open questions about the evolutionary history of organisms. In the hands of Morris Goodman (1963a, 1963b), Vincent Sarich, and Allan Wilson (Sarich and Wilson, 1967), humans were relocated within the great apes (rather than being a sister group to them), and within the African apes, and finally made the nearest relative of the chim-panzees by Sibley and Ahlquist (1984). The date of human–chimp divergence was reduced to 5 million years by Wilson and Sarich (1969). Similar revolutions affected other groups. Woese and Fox (1977) separated the Archebacteria from the Eubacteria, alough the Archaebacteria subsequently proved to be a bit less archaic than their name implied. In many other groups, molecular evidence had a less disruptive effect, confirming the historical reconstructions of mor-phologists more than it contradicted them. But especially where morphology faltered, molecular evolution was magically effective.

29.4. Numerical Methods for Molecular Evolution

As the number of species in molecular data sets increased beyond a few, it be-came necessary to come up with a numerical methology for reconstructing phy-logenies. This literature had started with Edwards and Cavalli-Sforza's paper of 1964, which is one of the true landmarks in the phylogenetic literature. Sadly, it is very little known among people making phylogenies, because of the misappre-hension that this literature sprang from the work of Willi Hennig (1950, 1966). The availability of computers in the 1960s led to the development of methods for inferring phylogenies. The parsimony and likelihood methods were introduced by Edwards and Cavalli-Sforza, for blood group polymorphism gene frequen-cies. Camin and Sokal (1965) described the first discrete-characters parsimony method, and Cavalli-Sforza and Edwards (1967) one of the first distance-matrix methods. Molecular evolutionists were involved in this process very early: Eck and Dayhoff (1966) carried out the first molecular parsimony analysis, and Walter Fitch (Fitch and Margoliash, 1967) used cytochrome sequences to pio-neer distance matrix methods. Sanger had produced the first protein sequence in almost the same year that computers became widely available to scientists; 10 years later, all of the major methods of analysis of molecular sequence data had been introduced, and most had already been applied to molecular data.

At first, these methods did not seem to be making any assumptions about evo-lution. But as distance-matrix methods, and later likelihood methods, became more popular, it was apparent that they necessarily involved models of evolution. This point has not been easily appreciated by systematists; it has taken outsiders to bring models to their attention. It is noticeable that, of the originators of the major methods of inferring phylogenies, only a few were trained as systematists. The others were population geneticists (Cavalli-Sforza and Edwards), biochemists (Walter Fitch and Margaret Dayhoff), medical microbiologists (Peter Sneath), or statisticians (Jerzy Neyman). The mathematical perspective did come easily to systematists.

29.5. Unilluminating Models

The simplest models in molecular evolution (e.g, Jukes and Cantor, 1969) were models of random hits, in which each site was equally likely to change, with alternative nucleotides equiprobable when it did change. These models were based on the observation of an approximate molecular clock, but without any specification of what evolutionary forces were creating these changes.

Lewontin and Hubby (1966) had already suggested neutral mutation as a possible mechanism for electrophoretic polymorphism. It was Kimura (1968; Kimura and Ohta, 1971) who first integrated within- and between-species molecular observations by suggesting that both electrophoretic polymorphism and sequence evolution were the consequence of the same process, neutral mutation. This had the effect on evolutionary genetics of bringing more data to bear, but an even greater effect on molecular evolution. It supplied it with a theory, ready-made.

It was immediately apparent that natural selection was acting as well. Some proteins were evolving rapidly (fibrinopeptides, for example) and others slowly (histones). These differences could easily be accomodated in the neutral mutation theory, which did not postulate that all mutations were neutral, only that all that caused polymorphism were. There could be a fraction of all mutants that did not cause polymorphism or sequence change because they were deleterious. It could vary from site to site and from locus to locus. It was to be expected that critically important regions of sequence would have most mutants be deleterious and that these would be eliminated by purifying selection. Models of molecular evolution have undergone modification in recent years to take these heterogeneities of rate of evolution into account.

Most evolutionary geneticists and molecular evolutionists have agreed that most noticeable heterogeneities of rate of evolution were caused by the presence of this purifying selection. Both committed neutralists and zealous selectionists have been in agreement on this. Where they would disagree was whether the polymorphisms that were seen, and the sequence changes that were seen, resulted from natural selection or genetic drift. So far, there has been little progress in telling these forces apart, from either within- or between-species data. This has been frustrating, but, as we have noted, it has had the happy effect that data could be analyzed by modified neutral mutation models, without much threat from the inadequacy of the model.

The result has been that the models used in molecular evolution are particularly simple, but in some respects disappointing. They are primarily models that specify what does *not* happen. Since we do not know whether the changes that do happen are the result of positive natural selection or neutral mutation, there is no way to predict how much natural selection is expected to improve molecular function. I often find that when people outside the field ask what theory predicts about molecular evolution, they are disappointed to find that its predictions are so negative.

29.6. Sequence Samples from Populations

While molecular evolution was maturing as a field, the same molecular technology was slowly invading evolutionary genetics. The invasion has been slow because so much evolutionary genetics was supported by the aptly named NSF.[1] It is still much cheaper to do electrophoretic surveys at many loci than to get one population sample of molecular sequences. It is noticeable how long it has taken, since Kreitman's pioneering studies (1983), for population samples of nuclear molecular sequences to become common.

A major exception must be made for one tiny piece of eukaryotic DNA. Mitochondrial population samples have been widely used. They got much attention with the famous "mitochondrial Eve" study of Cann et al. (1987). At first, mitochondria were used because their DNA was easy to extract. That ceased to be an advantage with the development of the polymerase chain reaction, but other advantages (maternal inheritance, haploidy, and rapid rate of evolution) have been cited as additional reasons to concentrate on the mitochondrial genome.

Initially, few of the researchers using mitochondrial data related their trees to population processes. The presence of an "Eve" seemed to be an unusual consequence of maternal inheritance. But every nucleotide in the genome should in principle have its own "Eve" or "Adam," as a result of the coalescent process at that locus. How far along the genome one can go and still find that same coalescent tree, so that every gene copy comes from that same ancestor, depends on the rate of recombination and the effective population size. The techniques of molecular evolution have invaded evolutionary genetics as people began to infer these gene trees. At the same time, evolutionary genetic theory turned out to provide the theoretical framework for analyzing these gene trees. The basic theoretical work was done by J. F. C. Kingman (1982a, 1982b). His work generalized Sewall Wright's result (1930) for two copies. Wright pointed out that two copies of a gene in a closed randomly mating population would have a common ancestor an average of $2N_e$ generations ago, with the distribution being approximately exponential. Kingman's "n-coalescent" process (Fig. 29.2) is the generalization of this to n copies. The time during which all n copies have had distinct ancestral lineages is exponentially distributed, with mean $4N_e/n(n-1)$ generations. Before that, there were $n-1$ copies, to which the same rule applies, and so on.

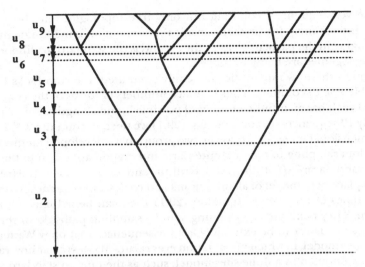

Figure 29.2. Kingman's coalescent process.

Kingman's work is mathematically elegant, but easily applied in practice. We do not need to use Kingman's technical machinery to understand and use the distribution of trees that his work predicts. The name coalescent has come to be applied to any gene tree. Kingman's prediction allows us to make a statistical analysis of sequence samples from populations. In principle, when developed enough, this will allow a statistical analysis of the evidence which discriminates between the "Out of Africa" case for human evolution and the "Multiregional Hypothesis." At the moment, there is not a truly statistical methodology for analyzing the data in this controversy.

The key to analyzing population samples of sequences is not to focus too obsessively on finding the true gene tree. One will usually have too few varying sites in the molecule to make an accurate estimate of the gene tree. It therefore becomes essential to take into account the noise in our estimate of the gene tree. There are two main sources of statistical error in these inferences. One is the randomness of the coalescent tree itself, which is drawn from Kingman's distribution. The other is the error in our estimate of the coalescent tree, which depends on the randomness of the mutational processes at molecular sites.

The fundamental equation for likelihood inference in coalescent trees is (Felsenstein, 1988)

$$L = \sum_{G} \text{prob}(G \,|\, \alpha)\text{prob}(D \,|\, G, \mu), \qquad (29.1)$$

where L is the likelihood (the probability of the data given the parameters), α is the collection of parameters for the population processes (effective population sizes, migration rates, etc.), and μ is the neutral mutation rate per site. The summation over G sums over all possible coalescent trees that could connect the observed sequences, counting not only tree topologies but also branch lengths. Interestingly, the first term inside the summation is the Kingman prior (or its analog in the particular case); the second is the standard likelihood used when phylogenies are being evaluated.

Although these two quantities are easily computed, the sum looks like an impossible one to compute. There are vast numbers of forms of the coalescent tree, and each has an infinite number of possible branch lengths. For example, with only 10 sequences, to compute Eq. (29.1) we need to compute 2.571×10^9 nine-dimensional integrals! However, two groups have developed methods that use random sampling to take a Monte Carlo integration approach to this sum. Griffiths and Tavaré (Griffiths, 1989; Griffiths and Tavaré, 1994a, 1994b) have taken in place of G the set of all mutational and coalescent sequences of events, without times of the events. Equation (29.1) then can be written in a recursive form. They compute the resulting sum by sampling paths down through the recursion. This can be extremely fast for sequences that obey Watterson's infinite-sites model, in which no mutation ever recurs. It is less clear how rapidly it can be computed in a finite-sites model, such as the ones in standard use in studies of molecular evolution.

Our own lab (Kuhner et al., 1995) has used a Metropolis–Hastings sampler to wander through the space of all possible coalescent trees (Fig. 29.3). We do

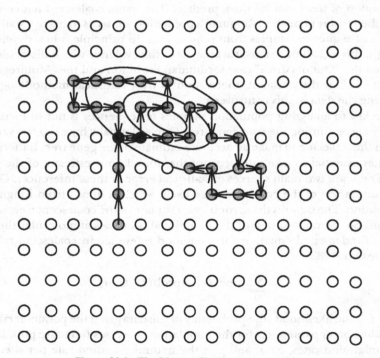

Figure 29.3. The Metropolis–Hastings sampler.

this by starting with a reasonable tree and gradually modifying it. This is an example of a Markov Chain Monte Carlo method. Such methods are gaining wide acceptance for solving complex statistical problems. The Metropolis–Hastings sampler works by acceptance or rejection of the resulting tree, depending on the value for it of the quantity being summed in Eq. (29.1). After a large sample of possible trees has been built up, an estimate of the likelihood surface can be made. Mary Kuhner and Jon Yamato's COALESCE program, available in our LAMARC package of coalescent likelihood programs,[2] seems to compute and maximize L effectively on workstations and fast microcomputers with large enough memory.

In effect, the mathematics of likelihoods on phylogenies, and that of coalescent priors, has come together to provide a set of tools for likelihood analysis of population samples of sequences. I suspect that these will become the standard methods for analyzing these data.

The Metropolis–Hastings sampler averages over our uncertainty about the coalescent tree, taking our knowledge of that tree into account properly. One might imagine that this could be made unnecessary. Suppose that there was little uncertainty about the coalescent tree. For example, if we sequenced entire mitochondria, we could be very sure of its tree topology and fairly sure of its branch lengths. In such a case, that would eliminate the need for the Markov Chain Monte Carlo sampling. Yet it would not solve the problem. Looking at a large number of sites does eliminate the uncertainty that comes from the mutational variability. But it looks at only one coalescent tree, since the whole mitochondrion shares one coalescent. In addition, there is the risk that hitchhiking natural selection could affect that coalescent, biasing our estimate of the effective population size.

To eliminate the noise due to the variability of the coalescent itself, we must look at many regions of the genome, each of which would have its own coalescent. That implies that we need population studies of nuclear DNA. And that raises the issue of recombination. It is not hard to generalize Kingman's formulas to allow for recombination (see Hudson and Kaplan, 1985). In the presence of recombination, the coalescent tree becomes a collection of trees (see Hudson, 1990) or a network with loops in it. However, both Griffiths and Marjoram (1996) and my own lab have developed random-sampling methods for trees with recombination. When these methods are available, they will make it possible to analyze population samples of nuclear DNA. That will in turn allow evolutionary geneticists to subdue the variability of the coalescent tree itself, not just the uncertainty of our estimate of it.

29.7. Enter Genomics

These same molecular and computer technologies have brought modern genomics into existence. As it is a high-budget technology, it has been even slower to show up in evolutionary biology, but its presence is now noticeable. Although many of the current genomics projects in evolutionary biology are for organisms that have economic importance, it is becoming possible for evolutionary biologists to search for major genes affecting traits that they study.

One does not need to be very daring to predict that this trend will continue. Of course, if traits are controlled by large numbers of loci of small effect, it will fail to explain them. To the extent that it succeeds, the classical variance components techniques will recede from view. In some cases the identification of quantitative trait loci can resolve old dilemmas. For example, suppose that we see a trait differing between two populations. Has this difference arisen because of genetic drift or selection? And if selection, was it acting on this trait or on some correlated trait?

If genetic drift acted, we would expect all the loci involved to have differentiated in gene frequency to approximately equal extents, though in varying directions. If natural selection acted, we would expect the loci of largest effect to show proportionately more differentiation than those of smaller effect, but all loci to show differentiation in similar directions. If natural selection acted on a correlated trait, we would expect differentiation only in those loci that affected the correlated trait.

Figures 29.4 and 29.5 show the two natural selection cases, in an imaginary example. The first one is the result of natural selection on that character, the second one the result of natural selection on another character, which is affected

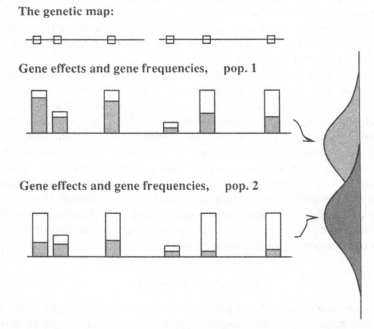

Figure 29.4. A genetic map showing differentiation of gene frequencies of loci drift.

The genetic map:

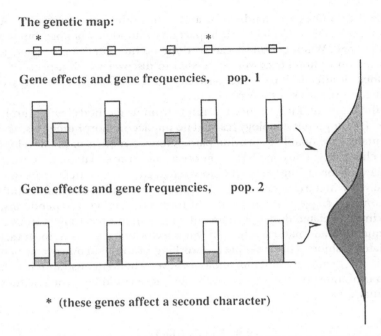

Gene effects and gene frequencies,   pop. 1

Gene effects and gene frequencies,   pop. 2

* (these genes affect a second character)

Figure 29.5. A genetic map showing differentiation of gene frequencies of loci affecting a quantitative character when population divergence is by natural selection on a correlated character.

only by the loci that are shown with asterisks. Of course, in reality the analysis is more complicated than this: there is expected to be a certain amount of heterogeneity in all these patterns as a result of genetic drift, and, given that, we have to be able to do the appropriate statistical analysis. But the point is that without the genomics, we could not hope to untangle whether the trait in question was the actual target of natural selection. With the genomics, it is at least possible to ask the question. Other confounded evolutionary forces may be separated for the first time as well.

There has also been an explosive growth of comparative genomics. The existence of genetic maps for multiple species gives us data to discuss rates and processes of genome rearrangement. Sankoff and Goldstein (1989) have begun the process of developing probabilistic models of genome rearrangement. We will surely see many more uses of these in evolutionary biology.

29.8. Enter Morphometrics?

Computer technology has led to the development of methods of image capture and analysis, and those to the development of morphometrics. Under the leadership of Fred Bookstein and Jim Rohlf, methods are being developed for analyses of outlines and landmarks, and even both at the same time. At the moment, this activity has had little impact on evolutionary biology, but that will surely change. When genomics can be done on quantitative characters affecting shape, morphometrics will be needed to discover which aspects of shape each locus is affecting. In between-species comparisons, there would seem to be much room for use of morphometrics.

At the moment, morphometrics suffers from its methods being purely geometric. Developmental biology has had no impact on morphometrics, as developmentalists are unable to provide parameterized models of the development of the characters. Thus one has to choose as the basic variables of interest ones that may not correspond to the developmental parameters. In fact, the information may flow in the other direction. Morphometric studies of within-population variation may suggest which aspects of form are varying independently, and which in a correlated direction. The difficulty with this program of study is that both mutation and natural selection will affect which aspects of form vary in a correlated fashion. This is simply a reworking of the old controversy between quantitative geneticists and morphologists over whether constraints are important in evolution. With genomics available, there could be some resolution in particular cases.

29.9. Really Big Trees

As phylogenies grow in size, further questions open up. It will become possible to assess statistically the role of species selection, for example. Kirkpatrick and Slatkin (1993) have studied statistics for testing imbalance in phylogenies. With such statistics, we could see whether there was significant evidence that a trait had been affected by species selection. If those clades that had higher values of the trait also seemed to be more speciose, this would be evidence for a role of species selection. We need to have large phylogenies to even attempt this, so it will not be happening very soon. It is possible to take a large phylogeny and search for that linear (or even nonlinear) combination of characters that shows the greatest correlation with clade size.

If the data of paleontologists could be recast from survival of taxa to phylogenies that have quantitative characters on them, many questions about long-term trends in evolution as well as natural selection above the species level could be addressed.

Evolutionary biology in the 1990s therefore no longer looks like Fig 29.1. The species boundary holds few terrors, and within- and between-species variation can be studied together (Fig. 29.6).

29.10. Continuing Frustrations

In some areas there does not seem to be the progress that we need. We need a developmental biology that finds repeatable patterns, and hence canonical models, among different developmental processes. But if development, like

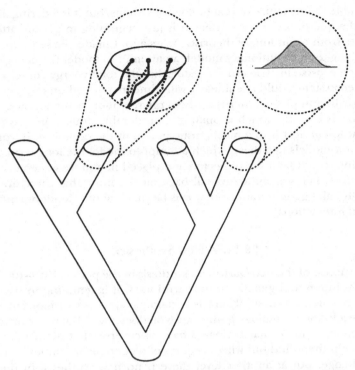

Figure 29.6. A cartoon of the state of evolutionary biology in the 1990s.

life, "is just one damned thing after another," then general models of developmental biology will lend little aid to evolutionary biology. The same is true of ecology. Although it is intrinsically a more important field than evolution, it is noticeable how little successful theory is available in ecology. When I was a graduate student, Robert MacArthur's work seemed to promise that strong ecological models of the Lotka–Volterra type would be available. Those models have succumbed to skepticism, leaving ecology as one of the few biological fields in which postmodernism has made any inroads.[3] We are thus unable to connect our molecular and morphological inferences with strong and stable developmental or ecological models.

29.11. Emergent Properties?

In the absence of developmental generalities, one interesting approach has been to seek generalizations that will apply to large classes of developmental systems. Stuart Kauffman has for many years been investigating the evolutionary implications of randomly connected developmental systems (summarized in Kauffman, 1993). His *NK* model has two parameters that allow the complexity of the gene interaction to be varied. There is certainly reason to believe that it does not resemble the structure of real biological systems, but it is a daring attempt to pose the question more generally. It is just beginning to get the attention it deserves from evolutionary theorists. Years ago, I made (Felsenstein, 1978) another long-term evolutionary model, an attempt to model the energetics of an evolving ecosystem. The crucial result, that the total energy content of the evolving ecosystem would rise linearly with time, depended on one tail of the initial distribution of the efficiency of energy retention being a power curve. Even if this is realistic, which is open to considerable doubt, there is no way at present to connect it to a mechanistic model of the organism. It would be wonderful if models like these, which attempt to model the long-term course of evolution, could be interconnected and placed in a more general context. If such efforts fail, we can always fall back on the hope that there are some empirically valid generalizations that can be made about developmental and ecological interactions.

29.12. A New Synthesis?

After the success of the neo-Darwinian synthesis in the period 1920–1950, connecting evolution and genetics in powerful ways, it is tempting to try to discern another new synthesis. There is certainly a synthesis of molecular biology and evolutionary biology going on. Nor does the influence run in only one direction, as molecular biologists have discovered that population variation can help them find out which regions of a genome are under constraint against change. But at another level there is no new synthesis. In the original synthesis, the formal structure of genetic systems came to evolutionary biology and resulted in the mathematical theory of population genetics, easily the most elaborate body of theory in biology. In the present syntheses, no new theory has yet arisen. The forces invoked are those that we already knew about within populations. We have a post-neo-Darwinian synthesis without a post-neo-Darwinism or, at least, without a new theory. We have many new statistical and computational methods, but all use the preexisting theory as their base.

Making such a theory is a major challenge. It probably involves asking a new set of questions and seeking generalizations at a higher level. It may need to wait for developmental biology and ecology to come up with some new generalizations. Or it may simply need a new generation of evolutionary biologists who can ask a new set of questions.

29.13. Acknowledgments

This chapter is a revised version of a Presidential Address delivered at the annual meeting of the Society for the Study of Evolution in June, 1993 in Snowbird, Utah. I am grateful to David Fogel for pointing out the priority of the late Nils Aall Barricelli in the invention of genetic simulation. Some of the research reported was supported by grants 2 R55 GM41716-04 and 1 R01

GM51929-01 from the National Institutes of Health and grants DEB-9207558, and BIR-9527687 from the National Science Foundation.

NOTES

1. In the U.S., NSF means National Science Foundation and is also the abbreviation on your bank statement when you have Not Sufficient Funds.
2. Available on the World Wide Web at http://evolution.genetics.washington.edu/lamarc.html.
3. (joke)

REFERENCES

Altenberg, L. and Feldman, M. W. 1987. Selection, generalized transmission and the evolution of modifier genes. I. The reduction principle. *Genetics* **117**:559–572.

Barricelli, N. A. 1954. Esempi numerici di processi di evoluzione. *Methodos*, 45–68.

Camin, J. H. and Sokal, R. R. 1965. A method for deducing branching sequences in phylogeny. *Evolution* **19**:311–326.

Cann, R. L., Stoneking, M., and Wilson, A. C. 1987. Mitochondrial DNA and human evolution. *Nature (London)* **325**:31–36.

Cavalli-Sforza, L. L. and Edwards, A. W. F. 1967. Phylogenetic analysis: models and estimation procedures. *Evolution* **32**:550–570 (also published in *Am. J. Hum. Genet.* **19**:233–257).

Charlesworth, B. 1978. Model for the evolution of Y chromosomes and dosage compensation. *Proc. Natl. Acad. Sci., USA* **75**:5618–5622.

Eck, R. V. and Dayhoff, M. O. 1966. *Atlas of Protein Sequence and Structure 1966*. Silver Spring, MD: National Biomedical Research Foundation.

Edwards, A. W. F. and Cavalli-Sforza, L. L. 1964. Reconstruction of evolutionary trees. In *Phenetic and Phylogenetic Classification*, V. H. Heywood and J. McNeill, eds., pp. 67–76. London: Systematics Association Publ. No. 6.

Ewens, W. J. 1972. The sampling theory of selectively neutral alleles. *Theor. Popul. Biol.* **3**:87–112, 240, 376.

Felsenstein, J. 1978. Macroevolution in a model ecosystem. *Am. Natur.* **112**:177–195.

Felsenstein, J. 1988. Phylogenies from molecular sequences: inference and reliability. *Annu. Rev. Genet.* **22**:521–565.

Fisher, R. A. 1950. Gene frequencies in a cline determined by selection and diffusion. *Biometrics* **6**:353–361.

Fitch, W. M. and Margoliash, E. 1967. Construction of phylogenetic trees. *Science* **155**:279–284.

Fraser, A. S. 1957a. Simulation of genetic systems by automatic digital computers. I. Introduction. *Aust. J. Biol. Sci.* **10**:484–491.

Fraser, A. S. 1957b. The simulation of genetic systems by automatic digital computers. II. The effects of linkage on the rates of advance under selection. *Aust. J. Biol. Sci.* **10**:492–499.

Goodman, M. 1963a. Man's place in the phylogeny of the primates as reflected in serum proteins. In *Classification and Human Evolution*, S. L. Washburn, ed., pp. 204–234. Chicago: Aldine.

Goodman, M. 1963b. Serological analysis of the systematics of recent hominoids. *Hum. Biol.* **35**:377–436.

Griffiths, R. C. 1989. Genealogical tree probabilities in the infinitely-many-site model. *J. Math. Biol.* **27**: 667–680.

Griffiths, R. C. and Tavaré, S. 1994a. Sampling theory for neutral alleles in a varying environment. *Philos. Trans. R. Soc. London Ser. B* **344**:403–410.

Griffiths, R. C. and Tavaré, S. 1994b. Ancestral inference in population genetics. *Stat. Sci.* **9**:307–319.

Griffiths, R. C. and Marjoram, P. 1996. Ancestral inferences from samples of DNA sequences with recombination. *J. Comp. Biol.* **3**:479–502.

Hamilton, W. D. 1967. Extraordinary sex ratios. *Science* **156**:477–488.

Hennig, W. 1950. *Grundzüge einer Theorie der Phylogenetischen Systematik*. Berlin: Deutscher Zentralverlag.

Hennig, W. 1966. *Phylogenetic Systematics* (translated by D. D. Davis and R. Zangerl). Urbana, IL: U. of Illinois Press.

Hudson, R. R. 1990. Gene genealogies and the coalescent process. *Oxford Surv. Evol. Biol.* **7**:1–44.

Hudson, R. R. and Kaplan, N. L. 1985. Statistical properties of the number of recombination events in the history of a sample of DNA sequences. *Genetics* **111**:147–164.

Jenkin, F. 1867. Origin of Species. *N. Br. Rev.* **46**:277–318.

Jukes, T. H. and Cantor, C. 1969. Evolution of protein molecules. In *Mammalian Protein Metabolism*, M. N. Munro, ed., pp. 21–132. New York: Academic.

Kauffman, S. A. 1993. *The Origins of Order*. New York and London: Oxford U. Press.

Kimura, M. 1956. A model of a genetic system which leads to closer linkage by natural selection. *Evolution* **10**:278–287.

Kimura, M. 1968. Evolutionary rate at the molecular level. *Nature (London)* **217**:624–626.

Kimura, M. and Ohta, T. 1971. Protein polymorphism as a phase of molecular evolution. *Nature (London)* **229**:467–469.

Kingman, J. F. C. 1982a. The coalescent. *Stoch. Process. Appl.* **13**:235–248.

Kingman, J. F. C. 1982b. On the genealogy of large populations. *J. Appl. Probabil.* **19A**:27–43.

Kirkpatrick, M. and Slatkin, M. 1993. Searching for evolutionary patterns in the shape of a phylogenetic tree. *Evolution* **47**:1171–1181.

Kreitman, M. 1983. Nucleotide polymorphism at the alcohol dehydrogenase locus of *Drosophila melanogaster. Nature (London)* **304**:412–417.

Kuhner, M. K., Yamato, J., and Felsenstein, J. 1995. Estimating effective population size and mutation rate from sequence data using Metropolis-Hastings sampling. *Genetics* **140**:1421–1430.

Lewontin, R. C. and Kojima, K. 1960. The evolutionary dynamics of complex polymorphisms. *Evolution* **14**: 458–472.

Lewontin, R. C. and Hubby, J. L. 1966. A molecular approach to the study of genic heterozygosity in natural populations. II. Amount of variation and degree of heterozygosity in natural populations of *Drosophila pseudoobscura. Genetics* **54**:595–609.

Maynard Smith, J. 1971. The origin and maintenance of sex. In *Group Selection*, G. C. Williams, ed., pp. 163–171. Chicago: Aldine-Atherton.

Moran, P. A. P. 1966. On the nonexistence of adaptive topographies. *Ann. Hum. Genet.* **27**:383–393.

Nei, M. 1972. Genetic distance between populations. *Am. Nat.* **106**:283–292.

Sankoff, D. and Goldstein, M. 1989. Probabilistic models of genome shuffling. *Bull. Math. Biol.* **51**:117–124.

Sarich, V. M. and Wilson, A. C. 1967. Immunological time scale for hominid evolution. *Science* **158**:1200–1203.

Sibley, C. G. and Ahlquist, J. E. 1984. The phylogeny of the hominoid primates, as indicated by DNA–DNA hybridization. *J. Mol. Evol.* **20**:2–15.

Wagner, G. P. and Altenberg, L. 1996. Perspective: complex adaptations and the evolution of adaptability. *Evolution* **50**:967–976.

Watterson, G. A. 1975. On the number of segregating sites in genetical models without recombination. *Theor. Popul. Biol.* **7**:256–276.

Wilkes, M. V. 1975. How Babbage's dream came true. *Nature (London)* **257**:541–544.

Wilson, A. C. and Sarich, V. M. 1969. A molecular time scale for human evolution. *Proc. Natl. Acad. Sci. USA* **63**:1088–1093.

Woese, C. R. and Fox, G. E. 1977. Phylogenetic structure of the prokaryotic domain: the primary kingdoms. *Proc. Natl. Acad. Sci. USA* **74**:5088–5090.

Wright, S. 1930. Evolution in Mendelian populations. *Genetics* **16**:97–159.

Zuckerkandl, E. and Pauling, L. 1962. Molecular disease, evolution, and genetic heterogeneity. In *Horizons in Biochemistry*, M. Marsha and B. Pullman, eds., pp. 189–225. New York: Academic.

CHAPTER THIRTY

Attitudes to Animal Behavior

JOHN MAYNARD SMITH

30.1. Introduction

Richard Lewontin has contributed to science not only by his own work on evolution theory and molecular variation and by his influence on the many young scientists who have worked with him, but also by asking us to think about the relationships between the science we do and the world we do it in. Although I have not always shared his views on the latter topic, I do share his conviction that scientists should take it seriously. It seemed natural, therefore, when I was asked to write an essay on the evolution of animal behavior for this volume, that I should write about the history of ideas about behavior and the interaction between studies of animals and humans.

It is inevitable that our theories about animal behavior should be influenced by our knowledge of humans and vice versa. This seems to me not only inevitable but entirely proper, provided that we are aware of what we are doing. But it has helped to give rise to an astonishing range of views about animal behavior – animals as machines, as blank slates upon which experience can write, as lumbering robots programmed by their genes, as creatures endowed with emotions and aesthetic tastes. These different attitudes arise in part from a wish either to emphasize the difference between animals and humans or to emphasize their similarities. Descartes' mechanistic view of animals reflected his theological need to separate them from humans, whereas Darwin's ascription to them of an aesthetic sense arose from his wish to narrow the gap and so make the evolution of humans from animals more plausible.

These differences of attitude, between historical periods and between contemporaries, are the subject of this essay. A question that at once arises is how far the explanations lie within science itself and how far in factors external to science. Faced with this question, working scientists tend to be internalists and historians of science externalists. This is natural enough. For a scientist to take

R. S. Singh and C. B. Krimbas, eds., *Evolutionary Genetics: From Molecules to Morphology*, vol. 1. © Cambridge University Press 2000. Printed in the United States of America. ISBN 0-521-57123-5. All rights reserved.

an externalist view is to assume that one's own scientific views are influenced by social factors that are largely irrelevant to their truth or falsehood. It is equally natural that historians of science should often favor externalist explanations: It saves them from the need to understand the scientific issues. Lewontin, of course, is an exception to this oversimple dichotomy, but I am not. The explanations I come up with end to be internal ones: External influences seem to lack explanatory power and to be impossible to test. I accept, however, that like most scientists I am ill equipped to write history. I would like to be regarded as a honey guide for historians: I may be unable to dig up the nest to get the honey, but I can indicate to the real diggers where the honey may lie buried.

30.2. Darwin, Wallace, and the Problem of Sexual Selection

Since I cannot, in a short essay, tackle the whole of behavior, I have chosen one topic, sexual selection, but I shall not hesitate to digress from time to time. The problem is simply formulated. Darwin proposed the theory of sexual selection in general, and female choice in particular, in *The Descent of Man and Selection in Relation to Sex*, in 1871. The topic is today one of the most investigated and debated in evolutionary biology. Yet, for almost 100 years after the publication of the *Descent*, virtually no serious research was done. (One important exception, R. A. Fisher's 1915 paper, is discussed below: It does not remove the problem, because it was largely ignored at the time). Why this gap? The question is particularly attractive because there is an obvious sociological explanation. Female choice and the notion that women might be in charge were unlikely to be popular in 1870, but by 1970 the feminist movement had made it plausible. I cannot point to any evidence to support this explanation, but I find it hard to believe that the timing of the current interest in sexual selection is entirely coincidental.

As a start, it is curious that the views of the Darwin and Wallace on sexual selection diverged sharply (my account of this divergence is based largely on Cronin, 1991). Wallace came to reject the idea of female choice, except to the limited degree that females might choose the most healthy and vigorous males. He did not think that they could have a preference for gaudy ornamentation or display. Oddly, he argued that the elaborate ornaments of the males of some species were an unselected consequence of superabundant vigor: In modern terms, he thought there was a pleiotropism, or developmental constraint, linking ornament and vigor. In contrast, he explained the dull coloration of females as protective coloration, supporting his argument with comparative data (for example, the absence of color dimorphism in many hole-nesting birds). Darwin, on the other hand, explained male ornament as the selective consequence of an aesthetic taste in females, but, again uncharacteristically, he did not accept Wallace's selective explanation for the dull coloration in females. Instead, he appealed to the constraint of heredity. The genetic factors causing bright coloration in males, he suggested, did not affect females: In modern terms, they were sex limited.

We are faced with a curious situation. Both Darwin and Wallace explained the characteristics of one sex, females for Darwin and males for Wallace, as the unselected consequence of a developmental process that could not readily be altered. If either view had been proposed by Lewontin, it would be less puzzling, because he is distrustful of the idea that selection can mold every aspect of the phenotype, but for Darwin and Wallace, who in other contexts were eager to point to the power of selection, it is unexpected. Cronin, who gives an account of their views, can offer no explanation, and neither can I.

A serious weakness of Darwin's theory of female choice was his failure to offer an explanation of why females should choose. Instead, he left female choice as a matter of aesthetic taste. There is a reason why he should have been satisfied with such an explanation. The existence of an aesthetic sense in humans had been quoted as the qualitative difference between humans and other animals, and hence as an argument against the descent of humans from apes. If Darwin could show that animals, too, have an aesthetic sense, this objection would be answered.

30.3. The Rejection of Choice

Darwin insisted to the end of his life on the importance of sexual selection. After his death, the role of male–male competition was widely accepted, but the notion of female choice was almost universally rejected. Why? Despite the possibility that it was female choice that was rejected, what evidence there is suggests that it was choice itself. Cronin (1991) gives a number of quotations objecting to the idea of choice: typical is the following, from Groos in 1898, "It would be absurd to affirm that all bird songs originate in a conscious and critical act of judgement on the part of the female.... The Darwinian principle is materially strengthened by [eliminating the idea of]... conscious aesthetic choice on the part of the female." I think this rejection arises from an increasingly experimental and mechanistic approach to biology in the period 1870–1900, stimulated by the work of Bernard and Pasteur in France and by the growth of biological research in German universities during that period. This professionalization of biology was accompanied by increased opportunities for experimental research and a desire by the researchers to make biology more like other branches of science. The result was a flowering of physiology and cell biology. The distinction between physiological and evolutionary explanation dates from this time, although the terms causal and functional for the two types of explanation and the recognition that biological structures require both types of explanation came much later. The professionalization of biological research and the growth of an experimental philosophy were necessary steps forward, but the antagonism between physiological and evolutionary approaches was an unhappy side effect that still hampers behavioral research.

The hostility to the notion of choice was, I think, based on a misunderstanding, although in the absence of a selective explanation for the choice it was perhaps an excusable one. Choice was taken to mean a process of conscious deliberation on the part of the animal, analogous to the process whereby a woman might decided to elope with her lover rather than marry the wealthy prospective husband favored by her family. But an animal can possess a structure or behavior that makes it more likely to mate with one kind of male than another. That is all that is required to cause selection on male traits: Whether the animal is aware of what it is doing is a separate and much harder question, which we do not have to answer. But is the choice behavior of accidental or the result of selection, and if the latter, what kind of selection? These are the main issues in current debates about female choice. But my immediate point is that there is no need for the most mechanistic of biologists to reject the notion of choice.

This misunderstanding is of a kind not uncommon in biology. A word with a well-understood meaning in human affairs is used by biologists in what Darwin might have called a "loose metaphorical sense." A classic example is the phrase selfish gene. People complain that genes cannot be selfish, because to be selfish one must be aware of what one is doing, and genes are not aware. Of course genes are not aware, but I find it useful, for example when analyzing cases of intragenomic conflict, to think "if I were a gene in that situation, I would do so-and-so." I find biology hard enough without having to deny myself that helpful way of thinking: I know that I must also be able to formulate the argument without recourse to human cognition. What is important is that we should not mistake such semantic misunderstandings for disagreements of substance. The commonest example of such a mistake is made by those who reject natural selection on the grounds that only people, and not nature, can select. The rejection of choice was, I think, a similar mistake.

30.4. Genetics and the Modern Synthesis

The rise of mechanistic and physiological biology led to an eclipse of Darwinism in general, and not just of ideas about sexual selection. It did not regain its ascendancy until after the development of a science of genetics in the period 1900–1920. The immediate effect on evolutionary biology of the rediscovery of Mendel's laws was the emergence of two camps; the Mendelians, who saw mutation and not selection as the cause of the appearance of new species, and the biometricians, who denied the existence of genes and saw selection as primary. For many years I found this debate incomprehensible, but a recent reading of Galton brought a glimmer of light (Maynard Smith, 1991). The leaders of the two sides in the biometricians versus Mendelians debate, Karl Pearson and William Bateson, were both followers of Galton. The latter had seen two kinds of variation as qualitatively different: on the one hand continuous variation, for example in seed size, which had led him to the concept of the regression towards the mean, and on the other discontinuous variation, as exemplified by sports. This sharp distinction between two kinds of variation, continuous and discontinuous, was a problem for nineteenth-century scientists: It is felt less acutely today, in part because of a mathematical understanding of bifurcation, whereby continuous change in parameters can result in sudden, qualitative change in outcome. To digress, I think that Marx and Engels may have retained Hegel's dialectic, while rejecting his idealism, for a similar reason: Hegel's "change of quantity into quality" is a bifurcation by another name. Bateson and Pearson, although both followers of Galton, championed different aspect of his thought, the discontinuous and the continuous.

It was not until this debate had been resolved that a revised Darwinism, based on Mendelian genetics, could be established. The outlines of a resolution had been achieved by 1930, mainly by Fisher, Haldane, and Wright. But there was no immediate return to the problem of sexual selection (again, except for Fisher, 1930). The reason is interesting, but essentially internal to biology. After 1930, the crucial issue for evolutionary biologists was seen to be the origin and nature of species. Despite the title of his greatest work, the nature of species was not central for Darwin. In the period 1930–1950 it became so. As a result, male ornament and display were seen as signals of species identity, and female choice as a choice of a male of the right species, and not as choice between conspecifics. Choice was back, but in a new role. In the period of the Modern Synthesis, sexual selection in Darwin's sense was almost totally ignored. There is no reference to the phrase in Huxley (1942) or Dobzhansky (1951). The only reference in Mayr (1942) says "many phenomena that have been recorded in the past as furthering intraspecific sexual selection are actually specific recognition marks."

30.5. Reflex Behavior versus Ethology: Skinner versus Chomsky

The rise of ethology in the years after 1945, stemming from the work of Lorenz, Tinbergen, and Von Frisch, might have been expected to bring sexual selection to center stage, particularly because of their willingness to look at what animals do in the wild, but, curiously, it failed to do so. I can best illustrate the impact of ethology in Britain (it arrived later in the U.S.) by describing my own experiences as a zoology undergraduate at University College London in the years 1947–1950. The standard textbook on behavior was by Frankel and Gunn. Essentially, it was an account of how flatworms, maggots, and other invertebrates orient to light and other stimuli. As an ex-engineer, I found it a pleasure: the animals were behaving like robots that I could have designed myself. The lectures were given by G. P. Wells (son of H. G.). His favorite animal was the lugworm, *Arenicola*, whose complex 20-min cycle of feeding and defecating he had worked out: *Arenicola* was a somewhat more complex robot, like the clock in Berne. The ethologists were not mentioned (except by Helen Spurway in her final-year genetics lectures: she had become fascinated by their work and had interested Haldane also). However, I was stimulated by my fellow undergraduates, Aubrey Manning and David Blest, to read Lorenz and Tinbergen. After graduating, they both went to Oxford, where Tinbergen and Lack were active. I remained in London to work with Haldane, judging that my combination of myopia and mathematics would make it more fruitful, but I kept contact with the ethologists in Oxford.

Through this experience, I became familiar with the reflex view of animals and the ethologists' instinct view, according to which animals could without training generate complex behaviors in response to simple stimuli (fixed action patterns) and inherit the ability to respond, on its first appearance, to a complex signal (innate releasing mechanism). For the ethologists, animals were born with something in their heads. The contrast between the two views was made more vivid by Wells' brilliant lectures, since he saw himself in the reflex tradition, but had been taught by his animal to recognize a fixed action pattern. Before discussing the attitude of the ethologists to sexual selection, I want to digress to trace the connection between modern behaviorism and earlier views about the behavior of simple invertebrates and even of plants (for a fuller account, see Boakes, 1984) and then to describe how the debate between the reflex and instinct views was later replayed in the debate about human language between B. F. Skinner and Noam Chomsky.

The view of behavior summarized in Frankel and Gunn's textbook had its origin in physiological studies carried out, mainly on invertebrates, in Germany in the latter part of the last century. A central figure was Jacques Loeb. Working at the University of Wurzburg, he was influenced by colleagues in the department of botany, who interpreted the reactions, or tropisms, of the growing tips of roots and shoots to light and gravity in terms of local chemical and physical forces. Adopting the same word, tropism, Loeb interpreted the behavior of maggots, cockroaches, and caterpillars along similar lines. In 1891, Loeb moved to the new University of Chicago, where he was later to influence J. B. Watson, the founder of behaviorism, who did his graduate work at Chicago.

What is common to the views of the two men, apart from their commitment to an experimental approach, is the insistence that behavior is to be explained without recourse either to consciousness as a cause of behavior or to internal states or structures in the mind. For example, in 1913 Watson wrote "One can assume either the presence or absence of consciousness anywhere in the phylogenetic scale without affecting the problems of behavior one jot or one tittle, and without influencing in any way the mode of experimental attack upon them." In hindsight, it is easy to see that it was indeed helpful to banish the Cartesian notion of consciousness as a cause of behavior, separate from physiology, but that the insistence that internal states and structures in the brain be likewise banished has had less happy results.

Students of the behavior of animals in the wild, working mainly in departments of zoology, were liberated by the ethologists from a behaviorist straightjacket, although behaviorism continued to dominate studies of laboratory rats and pigeons in departments of psychology. In human psychology, the inadequacy of the behaviorist approach emerged most clearly in debates about language. In accordance with behaviorist doctrine, Skinner saw learning to talk as essentially similar to learning any other task, and dependent on reinforcement— that is, on reward and punishment. In contrast Chomsky and his supporters have emphasized the special nature of our ability to talk – special both in being unique to humans and unique to language, rather than just a result of our general capacity to learn. The ability to learn to talk depends on innate structures in the mind. I will return below to their reasons for holding this view. For the present, I want only to emphasize the parallels between the Chomsky–Skinner debate and the earlier debate between the ethologists and their critics.

30.6. Why the Etholoigsts Ignored Sexual Selection

It is time to return to sexual selection. The outlook of the ethologists seemed ideally suited to a study of female choice, both because of their interest in what animals do in the wild and because their belief that animals can inherit the ability to respond to a complex stimulus meant that the phenomenon of choice presented no difficulty. Both Tinbergen and Lack were ingenious in using experimental and comparative data to demonstrate selection. Yet they did little work on sexual selection. As it happens, both Margaret Bastock, then a graduate student in Oxford, and I (Maynard Smith, 1956) watched the courtship behavior of *Drosophila*, although in different species. The contrast between our conclusions is instructive. Bastock (1956) investigated the curious fact that wild-type *D. melanogaster* females do not mate with males expressing the mutant yellow. She concluded that yellow males lack motivation. I stumbled over the fact that outbred *D. subobscura* females reject inbred males. I had been working on the effects of inbreeding on fitness. Since eggs supposedly fertilized by an inbred male often failed to hatch, I started watching pairs to make sure that mating had in fact happened, and found that often it had not; I had not planned to study sexual selection. I concluded that inbred males are rejected because they fail to keep up with the female during a rapid side-stepping dance. The female's dance could therefore be seen as a mechanism enabling her to choose between males. I could not accept Bastock's interpretation of low motivation because an inbred male, after courting a female for up to an hour, would sometimes approach the female directly and attempt to mount, always unsuccessfully.

At the time, I discussed my results and interpretation with Bastock and other students in the ethology group at Oxford, but failed to convince them of my interpretation. They had come to think of instincts as universal, common to all members of a species: It was this universality that made an instinct recognizable. Differences of behavior between individuals – for example, between an animal feeding, or courting, or preening – were explained as caused by differences in motivation, as were differences in the intensity with which a particular activity was carried out. For me, trained in genetics by Haldane and Spurway, evidence for genetic causation came from analyzing the differences between members of a species: For the ethologists, it came from recognizing characteristics shared by all members of a species (or a sex within a species). The distinction is important: We will meet it again at the end of this chapter.

If this view is right, the ethologists prepared the way for an investigation of sexual selection, but their lack of interest in individual differences inhibited them from carrying it out. I am interested that Peter Marler should have reached the same conclusion: He wrote (1985, quoted in Cronin, 1991) "We are still in debt to our ethological progenitors for the insight that what appears to the uninitiated observer as a series of continuously varying movements, too chaotic to be scientifically manageable, typically proves to have at its core actions that are stereotyped and species-specific.... [But sexual selection is] about the extent to which behaviour varies between members of the same species, and even within the same population. This variation...is the raw material on which the forces of sexual selection can operate."

30.7. Genetics and Sexual Selection

The next step came with the introduction of specifically genetical thinking into behavior, particularly by Hamilton (1964) and Trivers (1972, 1974), and the introduction of game-theoretic methods (Hamilton, 1967; Trivers, 1971; Maynard Smith and Price, 1973; Maynard Smith, 1976). The replacement of ethology by behavioral ecology as the dominant approach to behavior during that period is curious. Again, I think the explanation is internal to science. Many ethologists around 1960 saw the next step as lying in neurobiology. If animals had something in their heads, that something had to be neurological structure. There is no question that they were right, but unfortunately, at that time, neurological techniques were inadequate, and progress was slow. What Hamilton, Trivers, and the game-theoretic methods proposed by Price and myself offered were soluble problems. What is more, they were problems that could be solved by going into the field, which is what most ethologists wanted to do anyway, and not by sitting in the laboratory.

The study of sexual selection, however, required not only a general ambition to explain behavior in terms of selection: It also required appropriate genetical models. Here the pioneer was O'Donald (1962, 1967). As a student of Fisher's, O'Donald was aware both of Darwin's views and of Fisher's suggested explanation of female choice. He tells me, however, that it was unplanned observations of the Arctic Skua that led him to choose sexual selection as the focus of his study: The existence of two color morphs in the skua gave him a chance to test Darwin's explanation for female choice in monogamous species. What Fisher and O'Donald offered was an explanation for the evolution of female choice. The particular explanation they offered is not the only possible one. Since 1980, there has been a flood of models and of tests of models. These have been admirably reviewed by Andersson (1994): All I wish to do here is to emphasize the range of ideas now in the field. First, there is a distinction between female choice of elaborate ornaments and of male traits, like the dancing ability of male *Drosophila* or the ability of males in courtship feeding, that are direct indices of fitness. Hamilton and Zuk (1982) have made the promising suggestion that the bright colors and displays of males indicate their resistance to parasites. Considering only ornaments, which was after all the problem that concerned Darwin, there are three main categories of explanation. Fisher argued (and it was this idea that was formalized by O'Donald) that it pays a female to mate with a male with a generally preferred type of ornament because, if she does so, her sons will possess the ornament, and so be preferred by females. Zahavi (1975) argued that ornaments are costly signals whose intensity is correlated with fitness, because only fit males can afford to make costly signals. Finally, Ryan (1990) argues that males are exploiting a sensory bias already present in females, either accidentally or for selective reasons unconnected with mating: This is a modern version of the idea that Darwin himself held.

This chapter should perhaps end here, but I want to comment briefly on two recent developments not specifically connected with sexual selection. First, there are signs of a return to the ambition of the ethologists and to seek neurological explanations for some of the classical behavioral phenomena of ethology: Examples are bird song (Nottebohm, 1989), food storage (Sherry et al., 1992) and imprinting (Horn, 1985). The explanation is surely internal to science. It depends on technical progress in neurobiology: The ambition was always there.

30.8. The Evolution of Human Behavior

The other recent development is the attempt, following Wilson (1975), to apply sociobiological ideas to humans. Although my views on this attempt are different from Lewontin's, I think it would be wrong not to discuss them. I was critical of Wilson, but not because I saw him, or others following his lead, as politically damaging. I did not see human sociobiology as intended to lend support to politically reactionary policies. I do not doubt that genetic arguments have been used to support policies that I deeply dislike, from eugenics to racial discrimination. To argue on the basis of unreliable statistics for the existence of genetic differences between races is indeed politically damaging, not least because it invites us to think of people, not as individuals, as we should, but as representatives of ill-defined races. Genetic studies of individual differences are of greater potential value, particularly in medicine, although we should be aware that they can be misused. In most cases, however, the sociobiologists were not concerned with genetic differences, either between races or between individuals. Like the ethologists before them, they were looking for genetically based human universals. The one kind of difference that they did discuss was that between the sexes: I will return to this topic at the end of this chapter.

My disagreements with human sociobiology arose because I saw it, not as politically dangerous, but as scientifically naive. It seemed implausible, in a species in which so much information is transmitted culturally rather than genetically, that much of behavior could be explained, as I think it can in other animals, as a direct means of maximizing inclusive fitness. I also found the attempt by Lumsden and Wilson (1981) to analyze the coevolution of genes and culture to be mathematically flawed (Maynard Smith and Warren, 1982). In retrospect, I think I may have been too critical, rather than too little critical.

Some investigations (for example, Dickemann, 1979; Irons, 1979) have been surprisingly (to me) successful in explaining along sociobiological lines particular patterns of human behavior in specific circumstances.

At present we are seeing a new wave of human sociobiology under the title of evolutionary psychology (see, for example, Barkow et al., 1992; Jackendoff, 1993). This shares with the earlier sociobiology the conviction that the human mind is not a *tabula rasa* upon which experience writes, but has structures best understood as evolutionary adaptations. It differs in recognizing that the environments in which humans now live are very different from those in which they evolved, and, more importantly, in knowing more about the psychological and sociological establishment that it is criticizing. The main theoretical tenet is the existence of modules in the mind: that is, of separate but intercommunicating structures, evolved to perform different functions (Sperber, 1994). The strongest reason for holding this belief lies in the work of Chomsky on linguistics. If, as he argues, there is a specific structure or organ in the human mind that makes it possible for children to learn to talk, then that is a module in precisely the sense intended by the evolutionary psychologists. And if a module for language, why not for other functions? This is a conclusion well appreciated by Chomsky himself. In what could be a manifesto for the evolutionary psychology movement, he wrote in 1976 "We may think of human nature as a system of a sort familiar in the biological world: a system of 'mental organs' based on physical mechanisms.... [The system] manifests itself in our unique capacity to develop a concept of number and abstract space; to construct scientific theories in certain domains; to create certain systems of art, myth and ritual, to interpret human actions, to develop and comprehend certain systems of social institutions, and so on" (reprinted in Chomsky, 1987, p. 197).

Two points require comment. Are there really modules? Is the whole program politically dangerous? First, modules: As in the case of the ethologists' instincts, they are to be recognized by their universality and adaptive complexity. Such properties are certainly suggestive, but I do not think they are sufficient. Consider an example. Atran (1990) has argued that there is a module for the identification and classification of living kinds. That is, it is a cultural universal that organisms are identified as belonging to natural kinds (an animal may be a dog, or a cat, but not both, and must belong to some kind), and that these kinds are then grouped hierarchically. I would like to think that this reflects a genetically programmed module, if only because it would explain the delusion suffered by my microbiological colleagues that bacteria must belong to species, as well as the cladistic passion for constructing phylogenetic trees. But could not the beliefs in natural kinds, and in hierarchy, be culturally universal because they are true? Universality and adaptedness can be learned as well as genetically programmed.

How, then, are genetically specified modules to be recognized? The case for a language module depends not only on universality, but on rapid learning not dependent on rewards and punishment and on the ability to learn and apply rules of which the learner is not conscious. Additional support comes from genetics. If a structure is genetically specified, then there should be mutations that affect it (although pleiotropism may make them hard to spot). Strong evidence for one such gene exists (Gopnik, 1990). Although the data will not be easy to collect, genetic support for a modular theory should be forthcoming in time. At present, the evidence for most of the proposed modules seems to me suggestive rather than decisive.

As far as the political implications are concerned, I am in general unworried. Why should it be a threat to human freedom to think that our linguistic ability depends on a genetically programmed language organ? There is, however, one topic, that of human sexual differences, that is more contentious, because evidence of a genetic contribution to the behavioral differences between men and women might seem to threaten sexual equality. Wilson and Daly (1992) argue that male violence toward women is an extreme (and nonadaptive) expression of a genetically influenced male proprietoriness – that is, a desire by a man to acquire sole sexual access to a woman. They support this conclusion by analyzing police records and by a cross-cultural analysis of legal and social systems.

Should feminists be particularly critical of these ideas? Only, I think, if one holds one of two beliefs. If it were the case that what is natural is right, or that any behavior for which we have a genetic predisposition is unalterable, then feminists would rightly regard the findings of Daly and Wilson as threatening. But what is natural is not necessarily right: Infanticide is not uncommon among animals, but in humans it is wrong. And genetic predispositions are not unalterable: Humans are so constituted that they can become addicted to cigarettes, but it is possible to give up smoking. It is the fallacious notion of genetic determinism that is at fault. I find the conclusions of Daly and Wilson persuasive. If they are right, it seems better that we should know it.

30.9. Conclusions

Two themes recur in the history I have reviewed. The first is the distinction between those who see behavior as depending on complex structures in the mind and those who see it as a direct response to environmental input. Examples are the debate between the ethologists and proponents of tropisms, and between Chomsky and Skinner over language. At root, this is a debate about method. Is the proper method of science to describe what happens, eschewing needless hypotheses, or to suggest theories that could account for what is observed and test them? It is interesting that precisely this issue underlaid the debate about genetics. What separated the biometricians and the Mendelians was the refusal of the former, and particularly Karl Pearson, to admit hypotheses into science. On this issue, I have no doubt where my sympathies lie. If scientists had consistently followed Pearson's, or Skinner's, methodology, there would have been no atomic theory, and so no modern chemistry, and no genetics, and so little twentieth-century biology.

The second distinction is between a tradition originating with Darwin and Wallace, interested in evolution, explaining behavior in terms of its adaptive function and stressing diversity rather than universal mechanisms, and a tradition originating in the German universities in the latter part of the nineteenth century, committed to experiment, seeking physiological mechanisms, and universal, or at least general, explanations. In this debate, I think it is foolish to favor one side or the other, except in the choice of one's own research projects. The two traditions should be complementary, not antagonistic. Any specific behavior requires explanation in terms of physiological mechanism, developmental process, and evolutionary history. Sadly, proponents of causal and of functional explanations often see one another as enemies. In a world with limited resources for research, this may be hard to avoid, but we should at least try.

REFERENCES

Andersson, M. 1994. *Sexual Selection.* Princeton, NJ: Princeton U. Press.

Atran, S. 1990. *Cognitive Foundations of Natural History.* Cambridge: Cambridge U. Press.

Barkow, J. H., Cosmides, L. and Tooby, J., eds. 1992. *The Adapted Mind: Evolutionary Psychology and the Generation of Culture.* New York: Oxford U. Press.

Bastock, M. 1956. A gene mutation that changes a behaviour pattern. *Evolution* 10:421–439.

Boakes, R. 1984. *From Darwin to Behaviourism.* Cambridge: Cambridge U. Press.

Chomsky, N. 1987. *The Chomsky Reader,* J. Peck, ed. New York: Random House.

Cronin, H. 1991. *The Ant and the Peacock.* Cambridge: Cambridge U. Press.

Dickemann, M. 1979. Female infanticide, reproductive strategies and social stratification: a preliminary model. In *Evolutionary Biology and Human Social Behaviour,* N.A. Chagnon and W. Irons, eds., pp. 321–367. Duxbury, MA: Duxbury.

Dobzhansky, Th. 1951. *Genetics and the Origin of Species,* 3rd ed. New York: Columbia U. Press.

Fisher, R. A. 1915. The evolution of sexual preference. *Eugen. Rev.* 7:84–92.

Fisher, R. A. 1930. *The Genetical Theory of Natural Selection.* Oxford: Clarendon.

Gopnik, M. 1990. Feature-blind grammar and dysphasia. *Nature (London)* 344:715.

Groos, K. 1898. *The Play of Animals: A Study of Animal Life and Instinct.* London: Chapman & Hall.

Hamilton, W. D. 1964. The genetical evolution of social behaviour. *J. Theor. Biol.* 7:1–52.

Hamilton, W. D. 1967. Extraordinary sex ratios. *Science* 156:477–499.

Hamilton, W. D. and Zuk, M. 1982. Heritable true fitness and bright birds: a role for parasites? *Science* 218:384–387.

Horn, G. 1985. *Memory, Imprinting and the Brain.* Oxford: Clarendon.

Huxley, J. S. 1942. *Evolution, the Modern Synthesis.* London: Allen and Unwin.

Irons, W. 1979. Natural selection, adaptation and human social behaviour. In *Evolutionary Biology and Human Social Behaviour,* N.A. Chagnon and W. Irons, eds., pp. 4–39. Duxbury, MA: Duxbury.

Jackendoff, R. S. 1993. *Patterns in the Mind: Language and Human Nature.* New York: Harvester Wheatsheaf.

Lumsden, C. J. and Wilson, E.O. 1981. *Genes, Mind and Culture: The Coevolutionary Process.* Cambridge, MA: Harvard U. Press.

Marler, P. 1985. Foreword in Ryan, M. J. *The Tungar Frog: A Study in Sexual Selection and Communication.* Chicago: Chicago U. Press.

Maynard Smith, J. 1956. Fertility, mating behaviour and sexual selection in *Drosophila subobscura. J. Genet.* **54**:261–279.

Maynard Smith, J. 1976. *Evolution and the Theory of Games.* Cambridge: Cambridge U. Press.

Maynard Smith, J. 1991. Galton and Evolutionary Theory. In *Sir Francis Galton, FRS: The Legacy of His Ideas.* London: Macmillan.

Maynard Smith, J. and Price, G.R. 1973. The logic of animal conflict. *Nature (London)* **246**:15–18.

Maynard Smith, J. and Warren, N. 1982. Models of cultural and genetic change. *Evolution* **36**:620–627.

Mayr, E. 1942. *Systematics and the Origin of Species.* New York: Columbia U. Press.

Nottebohm, F. 1989. From bird song to neurogenesis. *Sci Am.* **260**:74–79.

O'Donald, P. 1962. The theory of sexual selection. *Heredity* **17**:541–552.

O'Donald, P. 1967. A general model of sexual and natural selection. *Heredity* **22**:499–518.

Ryan, M. J. 1990. Sexual selection, sensory systems and sensory exploitation. *Oxford Surv. Evol. Biol.* **7**:157–195.

Sherry, D. F., Jacobs, L.F., and Gaulin, S.J.C. 1992. Spatial memory and adaptive specialisation of the hippocampus. *TINS* **15**:298–303.

Sperber, D. 1994. The modularity of thought and the epidemiology of representations, In *Mapping the Mind,* Hirschfeld and Lebman, eds., pp. 39–67. Cambridge: Cambridge U. Press.

Trivers, R. L. 1971. The evolution of reciprocal altruism. *Q. Rev. Biol.* **46**:35–57.

Trivers, R. L. 1972. Parental investment and sexual selection. In *Sexual Selection and the Descent of Man,* B. Campbell, ed., pp. 136–179. London: Heinemann.

Trivers, R. L. 1974. Parent-offspring conflict. *Am. Zool.* **14**:249–264.

Wilson, E. O. 1975. *Sociobiology: The New Synthesis.* Cambridge, MA: Harvard U. Press.

Wilson, M. and Daly, M. 1992. The man who mistook his wife for a chattel. In *The Adapted Mind: Evolutionary Psychology and the Generation of Culture,* J. H. Barkow, L. Cosmides, and J. Tooby, eds., pp. 289–322. New York: Oxford U. Press.

Zahavi, A. 1975. Mate selection – a selection for a handicap. *J. Theor. Biol.* **53**:205–214.

CHAPTER THIRTY ONE

The Evolution of Social Behavior

DEBORAH M. GORDON

The past 50 years of thinking about the evolution of social behavior can be divided into two phases. The earlier one, which ended ~20 years ago, used a comparative approach with the goal of finding a gradient across taxa in the extent of sociality, using some measure of the degree of interdependence among individuals, such as the extent of cooperation or of aggregation. This approach sought to trace the trajectory along which social behavior evolved and then to examine how natural selection might have intervened at each step. When closely related species differ in social behavior, perhaps investigation of the ecology of each species can provide an explanation for why natural selection would favor each variant in its environment.

For approximately the past 20 years, most discussion of the evolution of social behavior has been about the evolution of individual traits. In this second, more recent phase, the view of what social behavior is has changed. Social behavior is viewed as a collection of individual optimization problems: How does an individual get the best mate, the most copulations, the biggest or best territory, the most food for its offspring, and so on.

Often the transition to this second phase is located at the publication of Williams' (1966) rejection of Wynne-Edwards' notion of population- or species-level selection (e.g., by Alcock, 1993). It is argued that since selection cannot have acted on societies, groups, or any unit above the individual, then an explanation of the evolution of social behavior must account for why an individual's behavior promotes its own fitness relative to that of other individuals.

In this chapter I will suggest that this latter view of social behavior is inadequate for the kinds of evolutionary explanation people are currently seeking. A restricted view of social behavior leads to inaccurate counting of the effects of social behavior on individual fitness.

R. S. Singh and C. B. Krimbas, eds., *Evolutionary Genetics: From Molecules to Morphology*, vol. 1. © Cambridge University Press 2000. Printed in the United States of America. ISBN 0-521-57123-5. All rights reserved.

31.1. How Does Social Behavior Affect Fitness?

Defining social behavior as an individual's effort to maximize its gains blurs the distinction between social behavior and other types of behavior. The field of behavioral ecology currently divides behavior into a set of problems that an individual must solve: obtain food, obtain a mate, do whatever is necessary for its offspring to survive to reproduce, and so on. The optimal-foraging problem a solitary individual faces in deciding how hard to work for a morsel of food yielding a certain number of calories, is seen as more or less equivalent to the problem an individual in a territorial social group has in negotiating which space it uses during the mating season.

But social behavior is something other than individual problem solving, because an animal's social environment and its interactions with that environment constantly redefine the problems the animal has to solve. The social layer of behavior poses a particular set of difficult evolutionary questions.

What distinguishes individual and social behavior is that social behavior arises out of, and makes sense only in the context of, interaction. An individual-based explanation of natural selection on the mating displays of sage grouse at a lek might take into account how much energy a male has to spend strutting around at a lek and the increment in reproductive success accruing to him from each copulation. But the display is a display only for the females that see it; A male sage grouse that displays alone in the forest is not engaged in mating behavior. Moreover the consequences of the display for the males' reproductive success depend on what the females do: what other males those females mate with; if, where, and when the females lay eggs; how they feed the nestlings; and so on. Some of this, in turn, depends on characteristics of each female and may vary measurably from one female to the next. Some of the females' behavior depends on chance events to which different females would react differently, some on chance events that would affect any female the same way, some on the ways that females interact with each other and with other males; all of which have their own norms of reaction.

An individual's social behavior does not exist without the social behavior of the others with which it interacts, and its social behavior generates that of others. This makes it impossible to think of social behavior as a trait that an individual simply carries around, ready to deploy when the occasion arises.

Imagine a group of people all playing catch. Each player can throw a ball to someone else, who might then throw it to someone else, and so on. Each player also catches balls thrown to them by others, and throws them on. The balls come in two colors, red or green. The score depends on the color of the balls. Red balls do not affect the score – only green balls are important. Every time a person catches a green ball, they get a dollar. The more dollars, the better. To calculate who is likely to get rich at this game, one might first look for those attributes of individuals that make them likely to catch more balls. For example, one might estimate that people with the largest hands are likely to do better. On top of that, the people most likely to be thrown green balls instead of red ones are likely to do better.

This sort of thinking corresponds to a simple individual-based view of the evolution of social behavior. Dollars correspond to fitness, greater catching ability corresponds to some genetic or other factor that predetermines fitness, and interaction (being thrown a ball) affects the score only when it confers fitness, e.g., being chosen as a mate by an individual with high fitness, or being in a group of foragers with a skilled hunter, and so on.

Suppose the game is more complicated, because the main factor that determines the score, the color of the balls, is subject to change. The balls change color, from red to green or from green to red, when people catch or throw them. People vary in their effects on the color of the balls: Some are more likely than others to turn the balls green. Moreover, the way balls change color can depend on current conditions, which vary in time and space. There are times, for example, when all the balls caught in one corner of the field tend to turn into green ones. People move around the field, so particular individuals move in and out of this corner.

Now it is impossible to predict how rich people will become playing this game merely by measuring the size of their hands, and it is impossible to characterize a person in advance as someone likely to be thrown a green ball. To make such a prediction one would need to know how each person changes the color of the ball when he or she catches it and throws it, and how each person's effect on ball color depends on the local situation.

The same is true of social behavior. How an individual's behavior affects its own fitness and that of others depends on dynamics that vary with time, space, individuals, and type of interaction.

Lewontin's essay on the interaction of organism and environment (Lewontin, 1983) raises this as a general problem in evolutionary biology. He contrasts a theory of evolution in which fitness is predetermined with one in which fitness is a consequence of the relation of organism, environment, and chance, a relation that is constantly in flux. At the ecological level, this means recognizing that species do not adapt to preexisting niches, but rather create and modify their niches by living there.

Social behavior offers an even more obvious example of this. Lewontin points out that stones at the base of a tree are not part of a bird's niche unless the bird uses them somehow, e.g., to smash nuts. But the stones are there, even if they are not there ecologically for a bird that does not come into contact with them. The social environment of an animal, however, is not even physically there, although it is crucial to the animal's ecology. It is generated by the animal's own social behavior in relation to that of others. A territory, for example, is often defined as a defended space. What makes it a territory is the interactions an individual has with others; its location and boundaries are in social space. On a map, without the animals that interact in relation to it, the territory does not exist. If the animals disappeared, or stopped engaging in territorial behavior, the territory would disappear.

All of this makes the evolution of social behavior very difficult to think about. Individual-based theory leaves open some important questions. The first is whether we understand the rules of the game well enough to estimate fitness.

We need to be sure that it is really catching a green ball that earns a participant a dollar. The second question is about the dynamics relating interactions
and fitness. The way a particular behavior affects individual fitness varies, as
individuals develop and in response to changing conditions. If balls change
color in a predictable way, so that individuals at a particular developmental
stage or at certain times or places are likely to catch green balls, understanding
those dynamics will be essential.

31.2. Evolution of Social Behavior in Social Insects

The study of the evolution of social behavior in social insects provides a good
example of the difficulty of measuring the effects of social behavior on fitness.
The first studies of the evolution of social behavior in social Hymenoptera (ants,
bees, and wasps) were based on the comparative approach, attempting to rank
all the extant species according to some scheme of less to more social. There
are many tens of thousands of species of social insects, and both the bees and
the wasps contain many solitary species. The species that live in colonies are
clearly social. It is less clear that the species in which individuals live alone are
not social, since we are beginning to learn more about some of solitary species
and discover that individuals interact with each other (Schmid-Hempel and
Schimd-Hempel, 1993). The approach taken was to try somehow to arrange
all the species in between the solitary and the social. Wheeler (1930) ranked
animal societies according to the permanence of association among individuals
and the number of activities, such as feeding, migration, hibernation, and,
most important, reproduction, that depend on interactions among individuals.
Michener (1974) ranked bee species according to several criteria, including
cooperative work in nest construction, contact between generations, feeding
interactions, and the existence of division of labor.

An individual-based approach to the evolution of social insect behavior was
inspired by Hamilton's (1964) notion of inclusive fitnes. The focus shifted
to worker sterility and the way reproduction is partitioned among nestmates.
Wilson (1971) redefined levels of sociality in social insects in terms of who lays
the eggs and who feeds the brood.

In most current work on the evolution of sociality in social insects, social
behavior is seen to consist either of reproducing or helping a nestmate to reproduce (e.g., Sherman et al., 1995). These are very different kinds of behavior.
Reproducing means laying eggs. Helping a nestmate to reproduce could mean
all kinds of things, of which one is not laying eggs, i.e., the absence of egg-
laying behavior. The rest of the behavior that could be considered to help the
reproduction of nestmates includes not only the entire range of behavior that
individual social insects perform, but also the organization of that behavior in
a colony-wide system of task allocation.

At any moment in an ant colony, workers are performing a variety of tasks,
and some are inactive. Which ones are helping the queen to lay eggs and
which are not? It is not clear that we understand the rules of this game well

enough to estimate how an individual's behavior contributes to fitness, its own, its nestmates or that of the colony. Almost any act by a social insect worker could be seen to contribute to the probability that the current batch of eggs will continue to develop: collecting food, keeping the nest intact and clean, feeding the larvae that might one day tend the current eggs when those eggs become larvae. Even doing nothing could be seen as a way to hoard valuable resources, saving calories until they are really needed. By contrast, many of the same acts could be seen as promoting the fitness of one nestmate at the expense of another. If there is more than one queen or one queen that mates more than once, both of which are very common in social insects, then the colony consists of different subfamilies. Workers might preferentially feed, or care for, or otherwise help close relatives and ignore or injure others.

The problem of estimating how an individual's behavior affects the fitness of others and the fitness of the colony is further complicated because some things that workers do are necessary to maintain the whole colony. If the nest collapses, all the workers, of every subfamily, might perish. Do we count some of a worker's behavior as contributing fractionally to every individual's fitness and to colony fitness, while some of its behavior contributes more to the fitness of one subfamily than another? How exactly do we divide this up? A worker that preferentially cares for the eggs of her own patriline is helping to produce workers that might act to maintain the fitness of the colony as a whole, including other patrilines, if those new workers devote themselves to maintaining the whole nest. Do we subtract the future contributions of a larva to the well-being of other patrilines, when calculating the increment to her own patriline conferred by a worker's preferential treatment of a larva? To get the right answers to these questions we would need to know the function of each act of a worker and how that act affects the rest of the colony.

A simple contrast between laying eggs or helping someone else to lay eggs is obviously inadequate. The same individual can do both. Workers (who are not mated) often produce haploid male eggs in many ant species and in honey bees. In some ant species, workers can produce diploid females parthenogenetically. Among all of the behavior that is not the act of laying eggs, many actions could be interpreted as either promoting or injuring the fitness of another egg-laying individual. Which is really taking place at any instant and what determines how the fitness consequences of an act may vary? Suppose we knew which kinds of behavior promote the fitness of an egg layer. Going back to the green-ball-catching game, suppose we knew that catching a green ball earns the catcher a dollar. We still need to know how the relation between a worker's behavior and the fitness of the egg layer changes in different circumstances. How and when do the balls change color?

It may be that sometimes a worker behaves in ways that benefit the colony as a whole, sometimes in ways that benefit one subfamily only. (And we should keep in mind that this is a simplification; a worker's behavior could benefit or harm parts of various subfamilies.) The same behavior could have different consequences. This depends in part on the way the colony interacts with its environment. For example, collecting food might promote the fitness of the larvae of one patriline if the food cannot be stored and the queen is producing eggs fertilized by one male's sperm that week. But the same food might promote the fitness of the larvae of another patriline as well if the food can be stored 6 weeks, and 3 weeks from now the queen starts producing eggs by using another male's sperm. The fitness consequences of behavior further depend, in part, on chance. For example, suppose some workers happen to be standing near a cluster of eggs when an animal steps on the nest, crushing part of it, and the workers' bodies protect the eggs from the impact of falling soil. Standing near the eggs would have a strong effect on fitness only when an accident happens to occur.

To count up the effects of behavior on reproductive success, we need to know the basics of how individuals interact. In a relatively small number of well-studied species of social insects, we have learned to identify what individuals are doing. But we are far from understanding the impact of every individual's behavior on the rest of the colony. Moreover we would need to know the dynamics of how this impact changes as the colony's environment and colony requirements fluctuate. A certain kind of behavior by a worker may be more likely to promote the fitness of a batch of eggs in certain conditions, when a particular food is available, or when the colony is producing sexual forms for the annual mating flight, or when the onset of the rainy season means new workers will be needed to repair the nest. These dynamics must enter into the measurement of the effects of behavior on fitness.

31.3. What Next?

A first step toward progress in understanding the evolution of social behavior in social insects would be to admit how little we know about it and to recognize the extremely primitive nature of current theory. The theory is primitive because it seeks to balance the act of reproduction, laying eggs, against all the acts that constitute participation in the social organization that supports the laying of eggs and the rest of the life cycle of the egg. Balancing these two kinds of behavior is misleading because we do not know enough about the fitness consequences of all of the behavior that goes in to colony organization. We are not ready to apply the same currency to egg laying and the rest of a social insect's behavior.

The next step is further empirical study of social insect behavior, what individuals do and how this affects others. This would probably enable us to make better guesses about the dynamics that relate behavior and fitness, and this might lead to better theory.

Another step is the empirical study of evolution itself: how natural selection is currently acting on social behavior. This consists of looking for a correlation between variation in social behavior and variation in reproductive success.

Many studies in behavioral ecology investigate the relation of behavior and fitness. For the most part these are based on a view of social behavior in which the fitness consequences of behavior depend on fixed, heritable attributes of the individual. For example, there are studies of the relation between variation in plumage coloration in male birds and variation in reproductive success. The premise is that the probability a male will be chosen as a mate by a female, and thus the probability he will reproduce, is related to his plumage color.

What would it take to study variation in a system of interactions and the correlation between such variation and variation in reproductive success, recognizing that the relation between interactions and fitness is dynamic? The premise is that what is heritable is some tendency to enter into a particular set of interactions and that what varies among individuals is the sort of interactions they might join and produce. The relation between those interactions and fitness is not simple and would have to be investigated in a variety of conditions.

Section 31.4 outlines some possibilities for an empirical study of the evolution of one aspect of social behavior in seed-eating ants.

31.4. Is Natural Selection Acting on Foraging Behavior in Harvester Ants?

The red seed-eating ant *Pogonomyrmex barbatus* is part of a guild of seed-eating species, including many ants, rodents, and birds, that compete for food in the desert of the southwestern U.S. Colonies compete intraspecifically for foraging area (Gordon, 1992), and more crowded colonies produce fewer sexuals (Gordon and Wagner, 1997). These results indicate that there is some link between the amount of food a colony obtains and its reproductive success. Here, since colonies are the reproductive units, natural selection would act on colonies as individuals. Selection may currently be acting on the social behavior that determines the amount of foraging that a colony accomplishes.

In this species, ants perform four tasks outside the nest: foraging for seeds, at distances up to about 20 m from the nest; patrolling, which includes searching for places to collect seeds and investigating disturbances around the nest; midden work, sorting and piling the refuse, or midden; and nest maintenance, the construction and maintenance of the chambers and tunnels inside the nest. The numbers of workers engaged in each task vary over time in response to changing colony needs and changing environmental conditions. Task allocation is the process that leads to the shifting assignment of individuals to various tasks (Gordon, 1996).

It is clear that a colony's foraging behavior is a consequence of interactions among workers. An alternative might be that the amount of foraging a colony accomplishes depends in a simple way on the number of individuals in the colony that are specialized or predetermined as foragers. This proves not to be the case. Workers switch from one task to another, and the workers currently allocated to a task may decide to remain inactive in a particular situation. A worker's decisions, whether to switch tasks and whether to perform a task actively or instead to remain inside the nest, are both influenced by the behavior of other workers (Gordon, 1987, 1989).

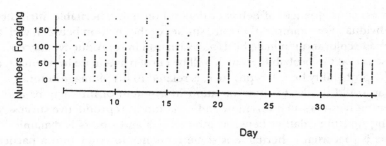

Figure 31.1. Variation among colonies in foraging intensity. Each dot shows the numbers foraging, within 1.3 m of the nest entrance, in one colony. Within each day, there is considerable variation among colonies in foraging intensity. From day to day, there are trends in the intensity of foraging common to all colonies. These trends are probably due to temperature and humidity conditions.

Variation among colonies in the amount of food collected is a consequence of variation among colonies in how many ants forage and where they forage. Foragers rarely return to the nest until they find food (Gordon, 1991), so the number of active foragers on a given day is a reasonable measure of the amount of food the colony collects that day. How many ants forage depends on the process of task allocation, which is based on individual decisions about whether to forage in particular situations. Where a colony forages also depends in part on its interactions with neighboring colonies (Gordon, 1992).

Figure 31.1 shows the numbers of ants foraging at 8:30 a.m. in 37 colonies in a 12-hectare site (Gordon, 1991). The vertical range of points at each day shows the extent of variation among colonies in numbers foraging. Some of this within-day variation is simply measurement error, and more measurements of the same colonies, averaged, might show a smaller range. Some of the variation is due to differences among colonies in the environmental conditions that affect numbers foraging. Such conditions include recent interactions with neighboring colonies (Gordon, 1992), amount of food available (Gordon, 1991), and so on. In addition, some of the variation is due to differences among colonies in size, which would lead to differences in numbers of ants available to forage.

The important point here, however, is that some of the within-day variation among colonies shown in Fig. 31.1 may be due to differences among colonies in task allocation. Colonies may differ slightly in the processes that determine how many workers forage actively. Empirical work shows that some colonies forage more intensively, staying active on more days, than others, and such colony-specific differences persist from year to year (Gordon, 1991). This may reflect variation among colonies in the rules that lead ants to forage actively. If we knew these rules, we could measure variation in their application. Suppose we already knew the rule was something simple such as "Forage when the temperature at the nest entrance reaches 41degrees + x." Then one could simply measure temperatures at many nest entrances, count the foragers, and solve for x. But we know the rules are not so simple. In particular, we know that an individual's decision whether to forage depends in part on the numbers of ants engaged in other tasks such as nest maintenance work (Gordon, 1987, 1989).

To determine whether natural selection is currently acting on foraging behavior in this species, several steps are needed. The first is to discover the rules individuals use in deciding whether to forage. This work is in progress; it seems likely that individuals use the rate or number of interactions with other workers as a cue in such decisions (Gordon, 1996).

The second step will be to measure variation among colonies in these rules and to consider whether such variation is correlated with variation in reproductive success. More crowded colonies produce fewer reproductives (Gordon and Wagner, 1997), suggesting that competition for food affects reproductive success. While some of the variation in numbers of reproductives is due to crowding, there are clearly other sources of variation as well. One factor may be a colony-specific tendency to forage, and work in progress will measure this.

Finally, it would be necessary to measure the heritability of task allocation. Colonies reproduce in an annual mating aggregation that contains the virgin queens and males produced each year by all the mature colonies in the region. The offspring of a colony are the new colonies founded by the parent colony's virgin queens or founded by a queen that mated with one of the parent colony's males. Queens often mate with more than one male. Task allocation could be considered heritable if the offspring of a colony resemble their parents in the ways that individuals make task decisions. In practice, heritability of task allocation is still impossible to study in harvester ants because we cannot yet identify the offspring colonies of a parent colony. However, genetic studies of task differences in honeybee patrilines (Robinson and Page, 1989) suggest that some aspects of task allocation can be heritable: Bees of different patrilines vary, for example, in tendency to forage actively.

The project outlined here considers whether variation in a system of interactions, those that determine the numbers of active foragers, is correlated with variation in reproductive success. If such a study found a correlation between foraging behavior and reproductive success and it could be established that foraging behavior is heritable, then this would demonstrate that natural selection is acting on foraging behavior.

The project outlined here does not specify the dynamics that relates social behavior and fitness. Instead, the processes that determine fitness are being treated as a black box. The study would measure the outcome, in reproductive success, of the interactions among ants that produce foraging behavior. How each of the interactions contributes to colony reproductive success and how the fitness consequences of each interaction depend on current conditions are not measured. These dynamics have an outcome, for example, in numbers of sexuals that a colony produces, but the processes that lead to this outcome remain mysterious. If the colony with the highest total number of active foragers (over the course of a foraging season) also has the highest reproductive success, then we know that interactions that promote active foraging also promote colony fitness. We still do not know how such interactions promote fitness. That would require an understanding of the details of the interactions and their impact on the rest of the colony and the link between the task decisions of foragers and the production of workers and sexuals.

31.5. Can We Explain the Evolution of Social Behavior?

I have argued that accurate tabulation of the effect of social behavior on individual fitness will require different measures than are currently in use, because the dynamics that relates interactions among individuals and fitness varies among individuals and depends on environmental conditions. When one is discussing social insects, individual can mean insects within colonies or colonies. A colony is a reproductive unit, and selection can act on colonies as individuals, but selection may act on the ways that the behavior of individual insects affects colony fitness. Extending the discussion more generally, individual can refer to any reproducing organism. The general point here is that an individual's fitness depends not only on how the individual acts, but on how the individual affects others, how those others respond to it, how their interactions modify the environment they all share, and how all of these relations are affected by changes in the environment. The theory we would need does not yet exist, except perhaps, as Lewontin's (1983) essay mentions, in the evolutionary theory of species interactions such as the coevolution of mutualistic or host–parasite species. In the absence of theory, we could attempt empirical studies that examine how natural selection is operating, without attempting to tabulate, one by one, the ways that individuals affect each others' fitness. The results of such empirical studies might show the way to future theory, by showing some of the details of how social interactions affect reproductive success.

These amount to suggestions for how to modify the standard approach to evolutionary questions. If we want to explain the evolution of a trait, we ask what about that trait makes it likely to increase the reproductive success of those individuals that carry it. New theory is required that models individual fitness as local and dynamic, prone to change in space and time, and as a function of interaction among individuals. But to me it seems unlikely that we will ever generate a tidy list of how natural selection produced all the social behavior we now observe. It is unrealistic to expect that every trait will have an adaptive explanation, however we count up the costs and benefits for fitness. Some traits may not have one – they may not promote reproductive success. For some traits, there may be such an explanation but it is inaccessible to us now because of historical change in the forces that affected natural selection on that trait.

The biology of the future probably holds new ways to think about the evolution of social behavior. Perhaps we will come to think of evolution of social behavior as a maintaining rather than a favoring force: The evolution of social behavior has led individuals to behave in ways that allow the group that supports them and their reproduction to keep going. This involves the notion that an individual's fitness depends on the social organization it lives within, and that the heritability of an individual's behavior involves the continuation from one generation to the next of the group's social organization. An evolutionary explanation for some aspect of social behavior would do its accounting in terms of individual fitness, but would consist of asking how this system of interactions allows individuals to continue reproducing themselves and the system of interactions. This might be a different question from why this set of interactions allows some individuals to get more food, energy, copulations, or territory than other individuals.

Darwin's problem in *On the Origin of Species* was to explain the diversity of life, past and present. The problem of understanding the evolution of social behavior is a very different one. We know very little about social behavior in the evolutionary past. Animal behavior textbooks tend to emphasize, not diversity, but similarity in the social behavior of many species of animals with diverse anatomies, senses, and means of interaction. Animal behavior is a young science, and it may be that apparent similarity in the social behavior of different taxa is an artifact of widespread application to animals of individual-based, economic views of human behavior. When we know more about how animal societies work, they may come to appear more diverse. It may be, however, that the evolutionary history of social behavior is mainly one of convergence, not differentiation.

We cannot yet generalize about the evolution of social behavior because the relation of behavior and fitness is still an open empirical question about most social behavior in most animals. This relation is complex because an individual's social behavior develops within and helps to generate a web of reciprocal interactions, and the function of this web changes with development and with a changing environment. As we work to discover the details of the relation of social behavior and fitness in particular species, we may develop new ways to ask evolutionary questions.

31.6. Acknowledgments

Many thanks to M. Brown, B. Crow, P. Godfrey-Smith, K. Human, S. Oyama, P. Taylor, and D. Wagner for comments on the manuscript.

REFERENCES

Alcock, J. 1993. *Animal Behavior*. Sunderland, MA: Sinauer.

Gordon, D. M. 1987. Group-level dynamics in harvester ants: young colonies and the role of patrolling. *Anim. Behav.* **35**:833–843.

Gordon, D. M. 1989. Dynamics of task switching in harvester ants. *Anim. Behav.* **38**:194–204.

Gordon, D. M. 1991. Behavioral flexibility and the foraging ecology of seed-eating ants. *Am. Nat.* **138**:379–411.

Gordon, D. M. 1992. How colony growth affects forager intrusion in neighboring harvester ant colonies. *Behav. Ecol. Sociobiol.* **31**:417–427.

Gordon, D. M. 1996. The organization of work in social insect colonies. *Nature (London)* **380**:121–124.

Gordon, D. M. and Wagner, D. 1997. Neighborhood density and reproductive potential in harvester ants. *Oecologia,* **109**:556–560.

Hamilton, W. D. 1964. The evolution of social behavior. *J. Theor. Biol.* **31**:295–311.

Lewontin, R. C. 1983. The organism as subject and object of evolution. *Scientia* **118**:63–82.

Michener, C. D. 1974. *The Social Behavior of the Bees.* Cambridge, MA: Belknap Press of Harvard U. Press.

Robinson, G. E. and Page Jr., R. E. 1989. Genetic determination of nectar foraging, pollen foraging, and nest-site scouting in honey bee colonies. *Behav. Ecol. Sociobiol.* **24**:317–323.

Schmid-Hempel, P. and Schmid-Hempel, R. 1993. Transmission of a pathogen in *Bombus terrestris,* with a note on division of labour in social insects. *Behav. Ecol. Sociobiol.* **33**:319–327.

Sherman, P. W., Lacey, E. A., Reeve, H. K., and Keller, L. 1995. The eusociality continuum. *Behav. Ecol.* **6**:102–108.

Wheeler, W. M. 1930. Societal Evolution. In *Human Biology and Racial Welfare,* E. V. Cowdry, ed., pp. 139–155. New York: Hoeber.

Williams, G. C. 1966. *Adaptation and Natural Selection.* Princeton: Princeton U. Press.

Wilson, E. O. 1971. *The Insect Societies.* Cambridge: Belknap.

CHAPTER THIRTY TWO

Population Genetics and Evolutionary Ecology: A Progress Report

SUBODH JAIN

"From principles is derived probability, but truth or certainty is obtained only from facts"   – Nathaniel Hawthorne

"The mystic leans towards celebration, the mathematician to cerebration"
                                        – John Barrow, Pi in the Sky

In this chapter we review some topics in evolution with an emphasis on the historical beginnings and the ensuing progress in the field of ecological genetics. Populations and species vary in terms of genotypic composition and adapt to changing environments; this was primarily the domain of remarkable early foundations of population genetics. Ecologists, on the other hand, study populations of a species in terms of demography, size, and geographical distribution, which are determined by its niche axes, competition, and other environmental factors. Thus various measures of geographical diversity and fitness and theories for predicting population size and variation at equilibria were developed independently by the genetic and ecological approaches. However, it became quite clear in the 1950s and 1960s that their joining together was imminent. My favorite classic, Julian Huxley's (1943) *Evolution: The Modern Synthesis*, had already established the need for such a multidisciplinary synthesis; to quote from the preface, "The time is ripe for a rapid advance in our understanding of evolution. Genetics, developmental physiology, ecology, systematics, paleontology, cytology, and mathematical analysis, have all provided new facts or new tools of research. The need today is for concerted attack and synthesis." It was noted that every biological fact has three distinct aspects: adaptive–functional, mechanistic–physiological; historical (in today's terms, optimality, form–function, and contingencies); and that fitness is not just the net rate of differential propagation or persistence but we one must also consider survivorship, mate choice, and reproduction along with the complexities of allometry,

R. S. Singh and C. B. Krimbas, eds., *Evolutionary Genetics: From Molecules to Morphology*, vol. 1. © Cambridge University Press 2000. Printed in the United States of America. ISBN 0-521-57123-5. All rights reserved.

genome organization, and speciation models! Truly speaking, even going back
to Charles Darwin and to the works of J. B. S. Haldane, R. A. Fisher, and
Sewall Wright, numerous ecological aspects of evolution were already recognized. So I found this challenge of historical review (progress report by a 1960s
baby-boomer biologist) doubly difficult: tracing the roots of our separate child-
hood, on one hand, and documenting joint progress, on the other, from a
massive literature in evolutionary ecology. Therefore this chapter is admittedly
a sketchy attempt, with a survey of selected few examples from evolutionary
studies. My primary goal is to draw attention to the numerous impressive developments and the ample scope they offer for both theoretical and applied
evolutionary research. Population genetics has to embrace this broader arena
beyond the molecular, phylogenetic, and strictly genetic framework.

32.1. Some Early Beginnings

Already in the 1940s we had learned about variation in terms of both Mendelian
gene polymorphisms and ecotype-specific quantitative traits. Although the paradox of variation (Lewontin, 1974) has found a growing number of explanations
(both monistic and pluralistic), the kinds and detailed analyses of selection in
natural populations and of their adaptive features are still ranked highly in the
evolutionary research goals. Ford (1964) emphasized the need for field studies, and as noted by Cain and Provine (1992), Charles Elton or Jens Clausen
could be nominated as the founders of ecological genetics. The Cold Spring
Harbor Symposia (1955, 1957, 1959) clearly recorded a consensus view that
population genetics needed to move beyond the single locus and constant selection models into the ecological contexts of complex genotype–phenotype
relationships, frequency-dependent and density-dependent selection, varying
population sizes, evolving species ranges, and such. Soon after came several
highly influential symposium volumes. For example, *The Genetics of Colonizing
Species* (Baker and Stebbins, 1965) remains a classic to this day for raising numerous questions, e.g., are colonizing species successful because of certain
genetic system features or life-history attributes, or both? How important is a
historical accident or a random event? Can we even characterize a species or
species assemblages as colonizing and noncolonizing? (Incidentally, the subject
of invasions in community context has loomed large only in the past decade
under global conservation and pest management projects.) A short but equally
remarkable volume edited by Lewontin (1968) further offered many original
ideas; to wit, Hutchinson's treatment of species as ecological units in classifying
niches; Dobzhansky's discussion of fitness versus adaptation; Wallace's model of
hard versus soft selection (in search for combining genetic variation and population size dynamics); Carson's founder-flush model (based on dramatically
varying population sizes) for interdeme selection and speciation; Robertson's
secondary theorem of natural selection based on covariation of a selected trait
and some direct fitness measure; Levins' short article on fitness sets for changing environments; and much more! Each one of these seeds in the evolutionary

biology forest became a tree and now their progenies form a patchwork of many research theme parks. Lewontin (1968) wrote a wonderful preface here but, as noted below, the warranted synthesis of these themes remains unfinished to date.

A few more observations may be added from the 1960s. At the Tenth International Genetics Congress (1963), Harper, Pimentel, and Levins, among others, discussed ecological genetics in terms of coevolution in two-species systems, evolution under competition in plants, and the evolution of adaptive strategies. MacArthur's influential writings on niche evolution and ecological consequences of natural selection joined those of Lack, Birch, Ehrlich, Chitty, and others. In an elegant essay, Lack (1965) outlined the goals of evolutionary ecology to seek explanations of population numbers, distribution, and geographical range of a species in relation to its evolutionary history, but he considered population cycles to be simply a consequence of population interactions. He distinguished complex, multilocus adaptations from the ecological genetic studies of simple polymorphisms and recognized the role of phenotypic plasticity as well as the need for comparative studies! His statement, "overenthusiasm for the theory of natural selection led to fantastic explanations" of ecological findings, preceded a key message in Gould and Lewontin (1979). Instead of inspired guesses that tend to bias one's theories or belief webs, field experiments are warranted. Over three decades later, many of these ideas are widely appearing as pseudoexplanations in our conclusion sections.

Developing a course in population biology during the 1960s looked very attractive to me, as we initiated research on colonizing annuals in plant genera *Avena, Bromus,* and *Trifolium*; many botanists and agronomists were clearly drawn to new prospects in ecological genetics. However, a population ecologists' meeting in Oosterbeek (1968) showed rather minor influence of evolutionary thinking on population dynamics. Botanists largely stayed on the sidelines, with the typological jump from individual level adaptation and niche/habitat descriptors in physiological ecology to the patterns of vegetational and life-form diversity at the community level. Their population biology thinking was often limited to an interest in the diversity of breeding systems and ecotypic variation. The symposium on colonizing species organized by Baker and Stebbins, both botanists, did provide a fair stake for initiating plant population studies; and in fact, Allard and his associates began work on some basic genetic features of evolving inbreeding populations. I must acknowledge strong influence of many early writings of Lewontin (e.g., on multilocus models, plasticity, interdeme selection–drift model for *t* alleles in mice, life-history change in colonizing species, estimation of Malthusian parameter, and nature's capriciousness) for stimuli in my own work on grasslands and also for relief from Mayr's (1963) difficult commentaries on population genetics. Most certainly, the potential pay-offs from the various new attempts for evolutionary neosynthesis as well as new applications appeared enchanting.

Dobzhansky's (1970) book provided a gem for us: Ecological ideas were clearly made relevant to the population genetic research as he wrote about melanism, host–parasite coevolution, insecticide resistance, herbivory in plants, biotic interactions, nonconstant selective values, and substitution load (why did ecologists fault population genetics for ignoring phenotypic evolution?) (see also Creed, 1971, for nice reviews of research on polymorphisms and new ideas in quantitative genetics). Later, however, Spiess (1977) narrowed the scope a bit by emphasizing neutrality versus selectionist debate, gene interaction, and genomic/karyotypic evolution. He did note that highly theoretical population genetics serves those students rather little who need it most and that "too few cases of comprehensive estimates (overall fitnesses) or pinpointing stages in the life cycle" for selection had been worked out. The outstanding books of Levins (1968), Giesel (1974), and Ricklefs (1979) outlined a lot of interesting prospects and early joint findings that truly excited my population biology class in the 1970s. This second phase of our field needs to be revisited as we document progress below. Christiansen and Fenchel (1977) discussed elegantly some examples of selection from the ecological perspective. Ewens (1979) succinctly reviewed many similar ideas for the ecological genetic paradigm. It is therefore puzzling that Real (1994) stated: "*now* we should emphasize the importance of population size and structure, the interaction among populations through migration and dispersal. . . ."

32.2. Ecologists' Interest in Evolution

We may ask whether there is a consensus now about the ecological research agenda for evolution and how ecologists view the ecological genetics' contributions (see Table 32.1 for some cross-disciplinary parallels) to the topics of local adaptation in races or ecotypes, interpopulation variation due to selection and drift, niche evolution, and host–pathogen coevolution. Genetic and species diversity are two complementary ways to look at many common interests in theory. Examples of melanism, mimicry, flowering time, and sex-ratio variation offered highly rewarding population studies in contrast to such topics as the genetic aspects of population regulation, the ecological factors of speciation, and the genetic basis of niche evolution. Microevolution, adaptations in terms of life histories, and certain population dynamic processes are commonly inferred from the patterns of distribution of species in relation to the patterns of environments, habitats, and ecosystems. Lack (1965) credits Gordon Orians for defining evolutionary ecology that seeks "to explain numbers, distribution, and other ecology in relation to the evolutionary history of a species." Lack emphasized that we need to know which features of a species of population are evolutionary adaptations and which are merely consequences of population dynamics, and he noted the need to study changes in the human-impacted environments.

Here are some notable quotes to establish primarily a diversity of views. Endler (1992, p. 329) concluded "there is every reason to regard ecology and evolution as two aspects of the same subject." To quote Pimm (1991), p. xi,

Table 32.1. *Cross Links between Population Genetics and Ecology through Parallel and Joint Interests in Biodiversity and Evolution*

Population Genetics	Cross Links	Ecology
Genetic variation (major genes, polygenes)	Diversity descriptors; null/neutral models	Variation in species distributions and abundance
Multilocus, systems, epistasis, coadaptation Historicity (mutations, unique events)	Interactive systems with coevolving species	Succession; patch dynamics
Equilibrium p, q; clines; races	Spatial diversity patterns, disequilibrium/ flux cycles	Species diversity and community dynamics
Genetic systems (recombination, matings, longevity)	Life cycles, life forms, and generation turnover	Kinds of species in differing habitats, niche evolution
Population substructure; gene flow; Variation and evolution at different spatiotemporal scales	History and evolution in populations studied at several hierarchical scales; measures of population fitness and ecological well-being	Populations interconnected by dispersal; Ecologists' interest in variation and adaptive changes within and between species
Metapopulation/ interdeme selection, genetics of traits, genotypic frequency analyses	Genetic variation and population dynamics treated jointly	Ecological details of life table, λ, reproduction, local extinction or range expansion
Adaptation to stress – metal tolerance, herbicide resistance variation/plasticity	Need details from long-term studies of genetic variation; incomplete without ecology	Generalists and specialists; rates of response to changing environment
Frequency-dependent and density-dependent selection; details of selection in ecological contexts	All these topics require ecology; incomplete without genetics, e.g., mimicry, host–pathogen	Population regulation; predator–prey or other multispecies dynamics
Extinction of small population; inbreeding/ loss of variation	Extinction risk assessment with genetic, demographic, and environmental factors	Extinction of small populations; demographic stochasticity of population size, matings, migration

"What ecologists do, what they know is important, and what they speculate about are often different. Very little ecology deals with any processes that last more than a few years, involve more than a handful of species, and cover an area of more than a few hectares." Strong words! Do the ecological and evolutionary time scales differ? Endler (1992) again: "The dichotomy between ecological time and evolutionary time makes no sense at all" (p. 322). These fields should not be even separated! Endler gives a list of processes in which genetic variation possibly affects ecological phenomena and another list of major unsolved questions in ecology with possible effects of natural selection (an impressive treatment indeed except that we have lots of maybes or possibles here).

In a very succinct discussion, Feldman (1989) noted several developments in evolutionary ecology: Ecological genetic theory requires evolutionary forces to be functions of population numbers (therefore density, spatial dispersion, etc.); phenotypic fitness to be determined by both life cycle components of selection and heritability; viability and fertility studies to require extraordinary experimental efforts in fitness studies; and the demography of age-structured migration to be dealt with in behavioral ecology and genetic analyses. Clearly, this defines some new dimensions of ecological genetics; it is hardly dead! Real (1994), in his preface, further extends this agenda, p. xi. "The great excitement that now flourishes in evolution and ecological genetics (besides methodological advances of molecular and quantitative genetic studies) is linked to several important conceptual advances: emergence of a rich theory for the evolution of neutral characters, the operation of ecological processes coupled to evolutionary processes at the metapopulation level, and a focusing of attention on life history phenomena relating genetic processes and population dynamics." In contrast, Harvey et al. (1992) stated that "when history is written, the fading of ecological genetics from the scientific firmament, after a promising flurry in the 1950's and 1960's, may become generally interpreted in terms of the failure of its practitioners to broaden their hierarchical perspectives" (which essentially means several nested scales of time and space).

We shall now review briefly a few selected topics such as adaptive significance of variation, life-history evolution, breeding system dynamics, and applied population biology.

32.3. Variation Analyses, Ecology, and Evolutionary Insights

Genetic assays of variation in natural populations have progressed rapidly from the few Mendelian and karyotypic polymorphisms to the polygenic variation in the 1940s, to allozymes in the 1970s, to the DNA markers and genomic variation. The use of molecular markers and numerous theoretical models have greatly improved the database on breeding systems, gene flow, interspecies gene exchanges, and phylogenetic trees. Earliest uses of single population models of gene frequency changes under viability selection have been accordingly replaced by elaborate theories of subdivided populations evolving under varied combinations of mutation, migration, selection, and drift (see Barton and Clark, 1990, for an elegant treatment); ecological inputs are increasingly sought in describing environmental factors and the different kinds of selective forces. Many ecologists have shown little interest in the evolution of many morphological traits (see Endler, 1986, for numerous examples) because of some preemptive ideas about the eminent relevance of life-history traits (growth, reproduction, and mortality, and some include dispersal); accordingly, quantitative genetics seemed to replace the old ecological genetic interests in polymorphisms.

In his classic book published a quarter of century ago, Lewontin (1974) thoroughly reviewed the questions of evolutionary importance of variation and the classical versus balanced models for maintaining variation. The answers were quite equivocal about the nature of available data as well as the population genetic theory being insufficient. Endler (1986), on the other hand, developed surveys of selection studies that in fact delighted most readers in evolutionary ecology looking for the evidence for strong selection. But more recently, many chapters in two books (Berry et al., 1992; Rose and Lauder, 1996) note the lack of long-term empirical studies required for detecting or measuring selection in nature. It is not obvious how much detail each one of us wishes to see in going from a few good examples to generalities. There are clearly many leaps of faith in going from variation surveys to their evolutionary explanations. For example, even for a simple story of cyanogenesis in *T. repens*, which was explained as a two-locus system evolving under selective predation and latitudinal (clinal) pattern, Hughes (1991) recently summarized evidence for much more complex genetics, selection, population dynamics, and even stochastic elements. Another long-standing example of variation, *t* haplotypes in house mouse populations, analyzed theoretically in terms of segregation–distortion and interdeme selection (Lewontin, 1962), has now yielded many new results; the original claim for low-level, widespread polymorphisms has to be revised and numerous details of migration and selective mating are to be considered (Durand et al., 1997). Similar developments toward pluralistic, complex, and more uncertain hypotheses are found in the studies of melanism, gynodioecy, and many other polymorphisms.

Even simple theoretical ideas are not to be trusted. Just consider two sets of examples. First, can we predict loss of genetic variation in small populations? Young and Merriam (1994) found in a study of *Acer saccharum* with its increasingly fragmented habitat that variation in small patches had increased presumably because of higher interpatch outcrossing rates. An important issue not discussed by the authors should be whether or why evolution occurred toward increased gene flow (the implication being that variation has to be maintained). We have not yet established any theorem prescribing any general conditions for adaptive mechanisms that would regulate the levels of variation without a specific character→performance→fitness logic (e.g., see below for herbicide and metal tolerance). Likewise, Young et al. (1996) concluded from their studies of several rare plant species that "not all fragmentation events lead to genetic losses" and different types of variation (allozyme, quantitative) in fact gave different results. Even more basic genetic issues in variation analyses are recognized as one finds that, because of mutational and transmission genetic features, allozyme and DNA markers would be expected to yield different variation patterns (Mitton, 1994). Second, can we predict losses in population fitness and greater extinction risks from certain measures of inbreeding depression (Frankham, 1995; Pimm, 1991)? In natural populations there are serious difficulties in establishing this relationship since the ecological features of life table, environmental context of growth rate and regulation, and certain adaptive adjustments might help avoid the inbreeding effects (Thornhill, 1993).

Quantitative genetic analysis of variation and evolutionary changes enjoy the advantage of phenotypic descriptors from the ecologists' viewpoint. However, for developing a genotype–phenotype–fitness framework of selection responses within and across generations, several constraints are to be recognized (see Fig. 32.1). Clearly, one could begin research with certain known genotypic arrays [e.g., mapped quantitative trait loci (QTL) or simply distinct parents with known heritable divergence] and proceed to analyze the role of various constraints (earlier literature referred to them as added complexities or simply departures from the simplest theory). Progress in dissecting many of these constraints by the use of new experimental approaches needs to be reviewed. In fact, even the simplest properties of genes controlling quantitative traits are under scrutiny not only in terms of the number but also their relative effects, nonadditivity, and nature of phenotypic expression. Echoing Lewontin's (1974) paradox of "more description, less adaptive relevance," Turrelli (1988) concluded his review in quantitative genetics as follows: "An outstanding problem is to delimit class of questions for which the details of genetic variation are relatively unimportant. This class may well exclude many interesting questions concerning the maintenance and dynamics of genetic variance." Ecologists emphasize phenotypic evolution for which three different kinds of analyses are useful, namely those describing optimality, evolutionary stable strategies, and response to selection. Their comparisons (Table 32.2) emphasize differences in both theory and empirical expectations, and clearly their joint uses are highly likely to succeed. (For a lucid discussion of optimality, see Schmid-Hempel, 1990.) Olivieri and Gouyon (1996) compared the ESS and OS (optimality) criteria for the evolution of migration rates in a metapopulation. Only partial adaptation is found as one defines various hierarchical units of selection.

Two basic ideas emerge from all this: We have certainly very nice and novel ways of studying genetic variation in space and time as well as different features of hereditary machinery, but the ecological and evolutionary interpretations of adaptive outcomes would require rather extensive and integrative research involving long-term field studies (a novel result? hardly so).

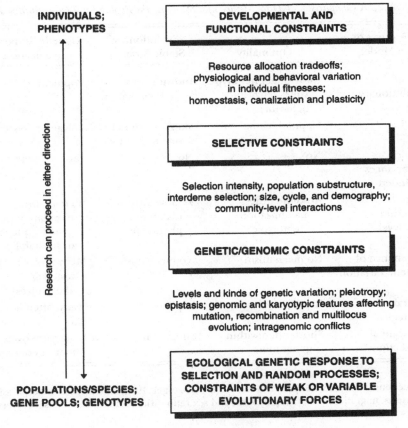

Figure 32.1. Constraints affecting rates and limits of evolutionary change.

32.4. Life-History Evolution

Several chapters in this volume have dealt with selection measurements based on the life-cycle components (viability, fecundity, and mating systems). Density dependence or frequency dependence is a part of many such studies, as is adaptation to varying environments. As noted above, ecologists emphasize that selection primarily acts on a phenotype made up of many life-history descriptors (e.g., body size, first age of reproduction, seed number per fruit) that are under polygenic control with varying heritability and provide for adaptive changes in a strategic suite of traits. Thus the evolution of life histories requires special treatment. Plant life histories have been studied as a part of describing community dynamics, niche partitioning by species, life forms adapting to major climatic–physiographic features of abiotic environments, and just plain reproductive patterns. Response to grazing, heavy metals, drought, domestication, pollution, etc., entails some specific trait(s) but rarely have these been examined in terms

Table 32.2. *A Comparison of Three Methods for Studying the Evolution of Quantitative Traits*

Features of a Model	Optimality	Evolutionary Stable Strategy	Genetic Response to Selection
Level of evolutionary response	Individual phenotype (physiology and behavior)	Individualization and population	Population
Genetics	Haploid/clonal one locus or polygenic	Haploid/clonal one or two loci	Polygenic; sexual
Genetic covariances included	No	No	Yes
Selection criterion and level	Absolute fitness maximization; phenotypic	Relative fitness maximum; phenotypic	Increase in population fitness; gene frequencies or heritability
Maintenance of genetic variation	No mechanism	No mechanism	Many mechanisms tested as alternatives
Constraints on strategies	Explicit	Explicit	Many, often implicit
Time scale of predictions*	Short, often within generation	Short term	Long term over many generations

* Predictions have been compared (modified from Mangel, 1992) in studies of superparasitism in insects, migration in metapopulations, and sex ratio variation in birds and insects (detailed review in Antolin, 1993).

of entire life-history shifts. Colonizing species and weeds described in terms of *r–K* selection dichotomy clearly brought out this new focus on the evolution of life histories, as did also Cole's paradox on semelparity–iteroparity, Janzen's question as to why bamboos wait so long to flower, or Robin's query about why so few plants are biennial.

Answers to such questions have spurred a lot of modeling and search for optimality or evolutionary stable strategies. Optimal life history, of course, varies with the environmental patterns (predictability, effects on juvenile versus adult mortality), basic modes of propagation, (seed, vegetative, stolons), metapopulation structure, and so on (see Smith, 1991, for a lucid review). For example, iteroparity is favored if juvenile mortality is high but the cost of reproduction is smaller, but this prediction changes when both means and variances of life table parameters are considered. In a desert annual, theory predicts high seed dormancy if chances of dispersal and recolonizing are small and if subpopulations live under autocorrelated environments. Facultative biennials are not really all that rare if one looks in the right places. Assumptions are often made, without many case-specific details, about the size–fecundity correlation, trade-offs with or without resource limitation, or density-dependent changes (see Begon et al., 1986, for further discussion of basic habitat descriptions and life history classification). Optimal life histories evolve only when genetic assumptions ensure ubiquitous additive genetic variation; this rosy picture soon fades as we introduce genetic correlations based on pleiotropy or linkage disequilibrium and other constraints in evolutionary response. Very little is known, even in theory, about the added uncertainties due to drift, hitchhiking, epistatic gene sets, and various segregation or transmission genetic irregularities.

Even within this basic framework note that (1) life-history studies have rarely dealt with changes in population numbers, i.e., the basic population regulation problem; (2) genotypic changes are not a part of most predictions; and (3) therefore, maintenance of genetic variation (and perhaps, polymorphic life-history traits) are underemphasized, except in the explicit treatments of mutation–selection balance or antagonistic pleiotropy. Arguments about whether or how well life-history is predictable without genetics are emerging; however, combining genetic and ecological measures of selective forces and definitions of various fitness measures and units of evolution will take serious effort, not self-glorifying camp building. Several nice theoretical examples are found in Antonovics (1994), Frank (1989), Olivieri and Gouyon (1997), and Godfray et al. (1992). Here, natural selection and evolutionary changes are modeled explicitly by use of the gene frequency and population size variables. One may ask whether in any well-studied species, the principal life-cycle components of selection identified in demographic genetic studies are the same as the evolving key life-history features? Are potential selection stages as revealed by path analysis or another multivariate method truly indicating when and how much selection is modifying the genotypic makeup of a population? Management of gene pools and metapopulations in conservation biology will also be asking where do the genetic and ecological predictors of success or failure meet.

Life-history evolution also emphasizes phenotypic plasticity since many environmental challenges are met with by nongenetic variation in phenotype. This is not new, but several new approaches to the description of phenotype–environment matching have come forth. Heritability studies by use of an overall norm-of-reaction measure of genotype–environment interactions requires special experimental design. Fitness consequences are another matter, depending on the choice of test environments and the extent of phenotypic trait description. Is plasticity under the control of specific genes? This has caused rather unneeded polemics about the nature of plasticity genes and gene regulation, something we know so little about (Schlichting, 1986). Constraints are in the popular jargon but not much is known; moreover, developmental and ecological limits to evolutionary changes (rates, directions) were always recognized. In summary, genetic and phenotypic understanding of variance components are serious matters of dispute as one sees so much guessing of gene

number, gene action, and genome integration features in all of evolutionary
biology.

32.5. Evolution of Mixed Selfing and Random Mating Systems

By use of relative rates of selfing or outcrossing, a single breeding system parameter, a great deal has been written on the evolutionary features of inbreeding
and outbreeding plants. Much of the early literature dealt with just two classes.
Lande and Schemske (1985) theoretically predicted this bimodality based on
how inbreeding might or might not purge deleterious recessives (inbreeding
depression balanced against certain benefits of inbreeding). The few examples
they chose were a biased sample and not sufficiently helpful to test their model.
In recent years, much empirical information on breeding system parameters
has been gathered and various new models allowing for more complex mating
patterns, resource allocation choices, cost-benefit considerations, and, most notably, spatial subdivision/gene flow aspects have added up to a far richer array
of models that in fact predict mixed mating systems.

This is too large a topic to discuss many aspects here, but we can see how
genetics and ecology students have come to ask similar questions: (1) Why
do many plant species maintain mixed selfing and random mating systems?
(2) What sorts of genetic and ecological features play an important role in the
evolution of breeding systems? Clearly, these questions relate to the following
oft-cited theses: Selfing species originate from outbreeders; many species show
bimodality (predominant outbreeding versus predominating selfing) but retain variability; many genetic variation studies show a general pattern (such as
less within- and more between-population variation in selfers; many selfers are
highly successful colonizers and even endemics; selfers show lower resource allocation to the male gender; selfers might often show an r-selected life history.
Many of these ideas have good empirical support.

Mather, Baker, Stebbins and others had basically summarized the field in the
1950s as a primarily genetic solution: Homozygosity allows local adaptedness
by inbreeders, but occasional outcrossing would be needed for some long-
term evolutionary change; the solution is long term, and involves gene flow
among populations. A simple model of heterozygote advantage and balanced
polymorphism showed how predominant inbreeders can maintain variation; a
theme was initially put forth from studies in barley, wild oats, and fescues. Soon
one had new insights about the joint role of linkage and epistasis, inspired by
the work of Kojima, Lewontin, and Franklin, among others. Models quickly
demanded evidence, which the data readily obliged to yield (life was simpler in
those years)! Evolution of selfing species was not covered by such studies, but
several parallel models readily treated random fixation of translocations, and
origin of polyploids, allowing sympatric speciation.

Reproductive assurance under selfing and Fisher's model of runaway selfing advantage remained undeniable until a new flurry of models with new
ideas on the kinds and rates of mutations changing dominance and inbreeding

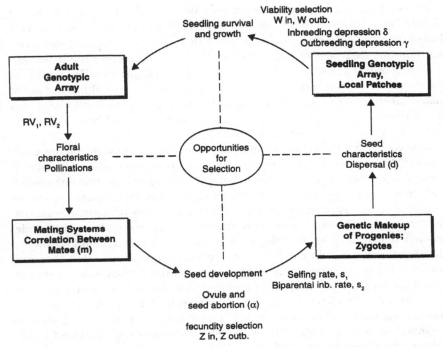

Figure 32.2. Life-cycle components of selection affecting the evolution of mating systems in plants. Parameters RV_1, RV_2 (relative success as an outcrossing parent), w_{in}, w_{out} (relative survival rates), m (genetic correlation between mates), α (paternity correlation with seed abortion), etc., define stages of selection. Note that we have already listed 13 or more variables in modelling selection.

depression, role of drift and linkage, and even the potential role of recombination in adapting to varying environments. Ecologists, on the other hand, emphasized that simplest panmixia assumed for the outcrossing component should be modified to invoke outcrossing measures for male and female separately, pollen carryover, geitonogamy, pollinator behavior, resource allocation constraints (gender allocation, sexual selection), along with the life-history-related measures of reproductive fitness (see Fig. 32.2, based on Waller, 1993, and Uyenoyama et al., 1993). Population structure becomes relevant and mixed mating can now result in theory from many different combinations of genetic and ecological factors (see Lyons and Antonovics, 1991, for an elegant study in *Leavenworthia*). But how much do we really know in nature about the genetic control of variation in selfing rates, seed dispersal, or the heritability of associated resource allocation features? We need much more fieldwork on the measures of fitness as we speak of local or global fitness, individual and group selection, and different time scales. Now we have many more ways to look at the evolutionary importance of inbreeding depression parameters (e.g., kinds of loci, linkage disequilibrium, association between selfing locus and viability loci, actual impact on net reproductive fitness). We need also to verify the role of outbreeding depression based on the local variation patterns in space. A great deal of reproductive ecology writings deal with the gender allocation resources, net reproductive fitness of male and female parents, and cost-benefit analyses of inbred–outbred outputs. Pollen discounting (i.e., limits on selfed flowers in pollen transfer into outcrosses) has been identified as a critical factor. Finally, spatial substructure of populations and various units of selection along with the dispersal pattern appear critical in any future theory of breeding system evolution (see *Thymus* gynodioecy below).

Gynodioecy is an excellent example of breeding system with numerous attempts to bring genetic and ecological components together, whereas most writers tend to separate the genetic issues (heterozygote advantage, transmission advantage of females, gene flow) from the ecological aspects of optimal sex ratios and reproductive patterns. Koelewijn (1993) thoroughly reviewed the gynodioecy research in *Plantago coronopus*. A single-locus, nuclear male sterility model requires females to have twofold selective advantage to maintain gynodioecy. Most known examples in nature are nucleocytoplasmic, and models have considered numerous fitness components, based on (a) overdominance/hybridity advantage (or its converse δ), (b) progeny of females with increased variance, or (c) pleiotropic effects of male sterility factors. An experiment on seven populations of *T. hirtum* found rather weakly frequency-dependent seed set advantage of females (Jain, 1994). Inbreeding depression measured in two generations of selfing matched the expectation from heterozygosity levels in gynodioecious versus hermaphrodite progenies. Overall, the range expansion and colonization patterns in this species strongly suggest a genetic advantage of gynodioecious populations. For gynodioecy in *Thymus*, in which populations vary in female frequencies under fire-impacted changes in vegetation, Olivieri et al. (1990) proposed a model for the observed patterns of restorer allele dynamics (Fig. 32.3). Gynodioecy shows cycles of female frequency in different patches undergoing successional vegetation dynamics. Different cytoplasmic male sterility alleles and their specific restorers can result in dynamics based on frequency-dependent selection's favoring females in a metapopulation. Similar coevolutionary oscillation of host and pathogen genes have been extensively modeled by Frank (1989), as noted below.

32.6. Plant–Pathogen Coevolution

Crop breeders have continually been challenged to breed new varieties as newly evolved pathogen races begin to attack resistant varieties. A gene-for-gene model has been fairly successful for such a dynamic coevolutionary system. However, generalities were soon doomed as new knowledge was sought on mutation rates in host and pathogen; mode of inheritance (allelism, number of loci, dominance), sexuality (versus asexual reproduction), cost of virulence (in the parasite and of resistance in the host), epidemiological factors, and much more (Frank, 1995). Burdon (1987) noted that nonagricultural communities had not

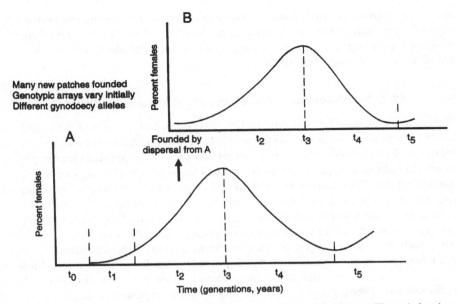

Figure 32.3. Evolutionary dynamics of nuclear cytoplasmic gynodioecy (e.g., *Thymus*) showing cycles due to selection, irrigation, and successional patch dynamics. Older (Source) population (A) and a newly established subpopulation (B) show changes in the frequency of females; at time t_0, gynodioecy arises because of mutation or hybridization, t_2 = gynodioecy favored, t_3 = a restorer allele arrives, t_4 = gynodioecy reduced, t_5, new cytoplasmic male sterility helps repeat the cycle.

been seriously studied for host–pathogen interactions, and his book forcefully initiated a new era. He discussed population genetic aspects in terms of the role of inbreeding and linked gene systems as well as the role of environment, so that simple theory would not serve well. For a good review of genetics, trade-offs and race specific interactions, Simms (1996) provides a pithy review of many new developments. Certain examples show high costs of virulence and a false dichotomy of models based on major genes versus polygenes. Mode of selection and its interplay with numerous variables of population subdivision, migration and random events have been discussed by Antonovics (1994) in relation to an elegant set of studies in *Silene* where gynodioecy is found, and transmission of pathogen depends on male function (life history, gynodioecy, host–pathogen interactions and metapopulation models intersect here). Burdon and his colleagues have tested many of these theoretical ideas. However, as noted succinctly by Levin (1992), "the influences of the historical record and stochastic events predominate. Hence, whereas explanation is an achievable goal, prediction is often impossible." Such difficulties (uncertainties) are inherent in any large nonlinear system with multiple time–space scales in hierarchical linkages. Numerous protocol requirements in selection and coevolution studies were discussed by Rausher (1996). It is interesting to note in contrast how successful the molecular approaches to disease resistance genetics are for finding few general patterns of resistance genetics in *Arabidopsis* (Kunkel, 1996). Will the paradigms of simple and complex findings by the molecular and evolutionary researchers meet? We need a crystal ball here.

32.7. Metal Tolerance and Herbicide Resistance in Plants

Population biologists have widely started to recognize many uses of their findings in the applied fields such as agriculture, bioremediation, and nature conservation. For example, colonizing species can provide important clues to the potential risks of new pests or weeds evolving from genetically engineered organism releases. They are also crucial in protecting nature reserves from invading aliens. Bradshaw and his collaborators have contributed significantly to the habitat restoration work on derelict mining lands through a long-term research program on metal tolerance genetics and ecology. Evolution of herbicide and insecticide resistance in agricultural ecosystems is an important economic concern. Next, we briefly review some developments in our knowledge of metal tolerance, herbicide resistance, and host–pathogen coevolution, all, in the words of Slobodkin (1988), great intellectual challenges for applied ecologists.

Evolution of metal tolerance in grasses has been cited as an example of clinal variation in many writings; Bradshaw's work provided the selection–migration background, although genetic basis of metal tolerance was rather polygenic or complex. A key result of screening results was highly variable levels of variability for this trait in different species, such that Bradshaw (1984) referred to genostasis, a caveat against assuming ubiquitous variation and evolutionary capabilities. A few cases of Mendelian inheritance have been reported for copper tolerance (Shaw, 1990; Baker et al., 1992). Reciprocal transplant experiments have established the role of selection. Several mechanisms of tolerance (e.g., hyperaccumulation, lower uptake, or general physiological tolerance) have been studied in some detail, but the dominant theme for evolutionary ecologists is the wide-ranging success of habitat restoration studies. More basic research is warranted here.

Like metal tolerance, evolution of herbicide resistance in agricultural weedy species also involves strong selection and numerous dramatic microevolutionary examples (Caseley et al., 1991; Powles and Holtum, 1994). Here, genetics and biochemistry have been very well studied by the use of laboratory-screened mutants as well as naturally occurring variants. The most common model assumes spread of a rare resistant allele at one or very few additive loci, and through allogamous interpopulation gene flow, resistance becomes widespread within 10 to 30 generations. Thus a weedy species might soon exhibit multilocus resistance to a series of herbicides (although such a cross-resistance mechanism is still poorly understood). Simple population genetic theory is used without much attention to such complexities such as the lack of good selective value estimates, frequency- and density-dependent components of viability selection, and cost of resistance. Furthermore, natural populations are reported to carry lower resistance, often polygenic, and here fecundity selection is invoked heavily. Allogamy, large seed bank, and other life-history features of different species are greater issues here so that genetic models have not been put forth. For both metal tolerance and herbicide resistance, clearly more basic research in evolutionary ecology is warranted if we wish to understand the interactive genetic and ecological processes.

32.8. Conservation Biology

Here, many ideas from population genetics and ecology are forcefully brought together. Increased fragmentation and the fate of small populations presumably lead to higher probability of extinction based on loss of variation (and therefore loss of evolutionary flexibility), random walk to extinction (demographic and stochastic elements together), breakdown of social and reproductive deme structure, etc. With pollination and other serious environmental degradation processes, challenges to a species maintaining minimum viable sizes (locally and globally) are analyzed by use of various risk assessment models in population terms (Burgman et al., 1993). Several articles have attempted to compare the effectiveness of genetic and ecological (variation versus population size) management choices in conservation programs, and clearly both must be considered jointly (Burgman and Lamont, 1992). Frankel et al. (1995) refer to the need for an "extensive population structure, open recombination system and measures to weaken selection pressures" (see Fig. 32.4). Many genetic and ecological ideas for conservation in practice are covered in the minimum viable population approaches of risk assessment. Theoretical treatment of metapopulation dynamics provided a big circus tent for many topics such as evolution of population structure, dispersal, connectivity, nonequilibrium dynamics, and hierarchical units of selection/evolutionary change (Hanski and Gilpin, 1997). Both population numbers (or patch sizes and density) and genetic variation for any trait affecting local or global fitness measures, matings, migration, and even competitive ability or life-history features affecting the effective population size (N_e/N) restrictions are relevant to the fate of metapopulations. Here, as noted succinctly by Hastings and Harrison (1994), metapopulation is too loosely applied to any subdivided structure involving discrete or continuous patches, and whether patches are unequal in size, habitability or migration rates. They questioned theoretical results suggesting interdeme selection to be important without a closer look at empirical results; often we find a conflict between the predicted most likely scenarios and the rarer, more episodic consequences.

Evolution of dispersal has been treated by several authors. Cohen and Levin (1991) analyzed the role of dormancy versus dispersal as ESS, although Levin and Castillo-Chavez (1990) found attempts to treat such problems from the optimization theory to yield incorrect answers. McCauley (1993) and Wade (1996) have contributed impressively to the theory and empirical work on interdeme selection, emphasizing founder effects and the details of dispersing events, and found that varying life-history characteristics make answers very

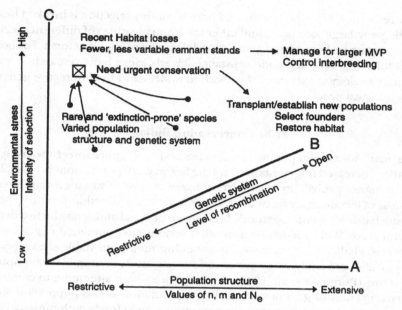

Figure 32.4. Major axes (A,B,C) of concern in conservation biology. Population structure is extensive when n = number of populations, m = migration rate, and N_e = effective population size are all large. Genetic systems with high recombination have higher outbreeding, shorter generation time, karyotype that allows extensive recombination, etc. Axis C represents measures of environmental deterioration and increased selective pressures. Conservation strategies involve both genetic and ecological ways to reverse negative trends in n, m, and N_e, and to facilitate evolutionary response. The MVP (minimum viable population) concept is based on risk analyses.

elusive. That dispersal generally helps global persistence of metapopulations is widely accepted, but a chaotic model of population size variation locally could lead to the evolution of dispersal, even without externally imposed temporal heterogeneity. Wiener and Feldman (1993) dealt with the evolution of dispersal in the context of inbreeding depression due to small local neighborhood sizes, and here their elegant analysis showed how critical various details are about the genetic component of dispersibility and of breeding system, population history, etc. Effective population size (N_e) in a metapopulation is an important issue since the ratio of N_e to N_a (actual size) is often reduced under subdivided structure. Frankham (1995) and others have discussed various protocols and empirical problems in estimating N_e/N_a. A molecular phylogenetic study of self-incompatibility alleles found N_e to be large, primarily because of the greater success of newly arising alleles in establishing new populations!

Undoubtedly, we connect this knowledge to the island biogeography and the design of nature conservation programs. Whether islands are comparable with recently fragmented habitat patches is not certain, nor can we treat genetic diversity and species diversity conservation with the same strategies. In fact, Pimm (1991) has critically examined these topics; we are again reminded that predicting the extinction of estimating minimum viable population sizes is not an easy matter; population histories and actual inbreeding depression amounts matter, as also do the various demographic and community structure details for each of the taxonomic/life form sets of examples (see Stacey and Taper, 1992, for environmental stochasticity and small population persistence). One clearly needs caveats for counterintuition. Are small fragments worth preserving? Why do certain small, genetically depauperate populations persist for a long time? How small is too small and when could we treat a series of remnant stands as a metapopulation? The SLOSS (single large reserve or many small reserves) issue in conservation biology has both ardent supporters and opponents, and not surprisingly, because of many variables, different examples do not help one to choose optimally (Holsinger, 1993). In addition, the relative impacts of demographic, environmental, and genetic stochasticity are not independently measurable such that ecologists could successfully argue under certain assumptions that adequate demographic protection overrides any genetic considerations (Halley and Manasse, 1993). Genetics may point to the loss of variation and evolutionary potential as imminent risks toward extinction, but quite candidly, very few long-term studies are at hand to support this admittedly highly plausible idea (Gray, 1996).

32.9. Concluding Remarks

Starting with solid foundations of population biology and enthusiastic research agenda in the 1950s and 1960s, we have certainly traveled a long way, jumping among different bandwagons (fine if they are all with us), with many younger biologists aboard, and discovering many useful new molecular, theoretical, and natural history ideas. We can strongly disagree with the comment of Harvey et al. (1992) on the fading of ecological genetics from the scientific firmament, although one might be impressed by so many calls for more attention to long-term field studies with multidisciplinary goals. Clearly, this chapter has only cursorily touched on many topics in evolutionary ecology. For population dynamics, excellent updates are given by Rhodes et al. (1996), Cappucino and Price (1996), and Hanski and Gilpin (1997); however, the evolutionary discussions are only scantily treated, which most likely reflects the paucity of information on the genetic aspects of population regulation. Hard and soft selection models need a critical review, and we need to follow up on Prout's (1980) classic work on density-dependent selection components in order to make progress here. Better accounts are found in the coevolutionary literature (e.g., chapters by Futuyma and May and by Ennos in Berry et al., 1992). We can assess progress in the following few examples. Three decades of research on colonizing species have produced several impressive data sets and models for range expansion (biogeographic theory of invasions; Williamson, 1996), numerous genetic studies on tracing the founders by use of molecular markers, and of course, documentation of correlated life-history features (several variations on the *r*-selection theme). However, genetic factors in helping colonizing success have not been often identified, and we have rather little demographic genetic data on the local colonizing events. Predicted associations with inbreeding and polyploidy are also open to further research. What do we know about the various theories on the evolution of inbreeding in plants? As noted above, we have a much larger choice of models, and getting empirical support would require the estimation of a horde of parameters over several space–time scales. Model organisms such as *Plantago* (Kuiper and Bos, 1992) and *Clarkia* (Raven, 1979) have been proposed for large collaborative projects, but one soon finds that, unlike the geneticists' advantage of largely representative genetic properties of a model, ecologists seek models for studying some unique, uncommon biological processes. Evolution of life histories has probably the best prospects for uniting genetic and ecological research efforts, with new ideas on population substructure, dispersal, stages of selection, and genuine interest in frequency- and density-dependent population dynamics. Finally, in applied fields such as conservation biology we see progress in moving from early naive adoptions of island biogeography theory and of genetic drift/random-walk-to-extinction models to more integration and realism in developing strategies.

In all of the topics reviewed in this chapter, several common themes are discussed: (1) the genetic basis of various traits and fitness measures (major genes, polygenes, additive or more complex); (2) relative fitnesses or selection coefficients in genetic models versus ecological fitness measures derived from population dynamics; (3) units of selection and evolutionary change as defined in relation to breeding, social, and spatial substructure of population(s); and (4) intensity of selection to be directly (causally) related to environmental factors. Long polemic exchanges on these and related personalized views seem unproductive in my view if we adopt certain consensus definitions and operational research methods. A few remarkably lucid writings might explain this position. Antolin (1993) elegantly reviewed the genetics and ecology of variable sex ratios with a nice comparative survey of three main theoretical explanations, listing models, explicit assumptions, and main theoretical as well as empirical findings. The need for more solid genetic information is well argued, and one can find clear ways through the jungle of models. In searching for a comprehensive treatment of evolutionary studies in a select group of organisms, Beaumont (1994) provides an exemplary volume; the rich details of variation, population structure and comparative ecology are well treated. Another nice example of evolutionary ecology research dealing with life-history evolution is found in Reznick and Travis (1996), who candidly discussed the findings and their current limitations with a systematic review of diverse research methods. The elegant work of Clark (1993) provides another example of impressive theoretical solution to a long-standing problem: how small populations can retain a large number of self-incompatibility alleles. Such examples could be multiplied readily.

Thus progress is undeniably impressive in the availability of new tools for studying genetic variation, genome-level properties, controlled gene transfers, marking–release–recapture techniques, detailed environmental measurements, etc., and in the computer-aided theoretical and data-processing domains. Certainly we have even more statistics to get around less desired outcomes, and reading research grants or papers for methods is not child's play. Just mastering multivariate analyses and various cladistic acronyms should earn kudos from our peers. And we can also note strides toward new ecological genetics by comparing earlier books (e.g., Baker and Stebbins, 1965; Lewontin, 1968) with those of Ricklefs (1979, 2nd edition), Berry et al. (1992), and Real (1994) and asking the obvious questions: Do colonizing species have unique genetic or ecological features? How often do niches evolve for two competing species to coexist? Do organisms respond to varying environments with polymorphism and plasticity as alternatives? Are widespread species more variable genetically? Do we find central–marginal comparisons useful for interpreting adaptive roles of variation and recombination? Such questions remind us of the past, whereas new questions refer to metapopulation dynamics, chaotic population dynamics, and sexual selection. Ecological interest in evolution is far more interwoven with the questions on coevolution, life-history strategies, the connections among various fitness measures and population well being, and even species-level selection within a community framework (Sober, 1984). Comparative studies have promised us that phylogenies matter to the ecologists and perhaps speciation models do too. Use of quantitative genetic thinking in evolutionary ecology would be helped by the use of mapped QTLs, if we can see through the genetic makeup of QTLs beyond statistical end products. To cash on current advances in evolutionary ecology fully, geneticists must also combine interests in molecular and high levels of biological form and function. Richard Lewontin (1979) has often reminded us of dialectical nature of biological questions, one of which is the genotype–environment interactions; the issues of Mendelian and quantitative variation have been overplayed as a dichotomy in my view, and removing it would be very useful.

In early days of ecological genetics, we asked for a theory of evolutionary change that incorporates density-dependent and frequent-dependent selection, environmental heterogeneity, regulation of population numbers, or coevolution of two interacting species. We had rather few empirical data sets on migration, mating system, selection or drift. Most assuredly, our research toolbox has become far more sophisticated in both genetic and outdoor ecological research as well as in the use of analytical and statistical analyses. Each of the evolutionary factors has been proposed in numerous component subfactors for further reductionist details and search for ultimate causes of evolution; reading Haldane's (1932) *Causes of Evolution*, which had so many ideas without experiments, gives us one yardstick of progress. Perhaps it is the nature of evolutionary biology that we find delightfully new details as if our puzzle is always gaining more pieces with even shapes of them evolving in a bigger context (Frank, 1997). Applied population biology, on the other hand, should

find increasingly greater uses in the restoration of derelict mine lands, weed
management in agricultural lands, and in the management of nature reserves,
and therefore a firmer place in our educational programs (Botsford and Jain,
1992).

32.10. Acknowledgments

I sincerely thank Costas Krimbas and Rama Singh for their invitation to contribute to this volume and for their patience and editorial advice. Many discussions during our field trips with students contributed ideas for which I feel
grateful and fortunate.

REFERENCES

Antolin, M. F. 1993. Genetics of biased sex ratios in subdivided populations: models,
assumptions, and evidence. *Oxford Surv. Evol. Biol.* 9:238–270.
Antonovics, J. 1994. The interplay of numerical and gene-frequency dynamics in
host-pathogen systems. In *Ecological Genetics*, L. Real, ed., pp. 129–170. Chicago:
U. Chicago Press.
Baker, A. J., Proctor, J., and Reeves, R. D., eds. 1992. *The Vegetation of Ultrafamic (Ser-
pentine) Soils.* Andover, MA: Intercept.
Baker, H. G. and Stebbins, G. L., eds. 1965. *The Genetics of Colonizing Species.* New York:
Academic.
Barton, H. N. and Clark, A. 1990. Population structure and processes in evolution.
In *Population Biology*, K. Wöhrmann and S. K. Jain, eds., pp. 115–173. New York:
Springer-Verlag.
Beaumont, A. R., ed. 1994. *Genetics and Evolution of Aquatic Organisms.* London:
Chapman & Hall.
Begon, M., Harper, J. L., and Townsend, C. R. 1986. *Ecology: Individuals, Populations,
and Communities.* Sunderland, MA: Sinauer.
Botsford, L. W. and S. K. Jain. 1992. Applying the principles of population biol-
ogy: assessment and recommendations. In *Applied Population Biology*, pp. 263–285.
Dordrecht, The Netherlands: Kluwer.
Berry, R. J., Crawford, T. C., and Hewitt, G. M., eds. 1992. *Genes in Ecology.* London:
Blackwell.
Bradshaw, A. D. 1984. The importance of evolutionary ideas in ecology – and vice-versa.
In *Evolutionary Ecology*, B. Shorrocks, ed., pp. 1–25. Oxford: Blackwell.
Burdon, J. J. 1987. *Disease and Plant Population Biology.* Cambridge: Cambridge U.
Press.
Burgman, M. A. and Lamont, B. B. 1992. A stochastic model for the viability of *Banksia
cuneata* populations: environmental, demographic and genetic effects. *J. Appl. Ecol.*
29:719–727.
Burgman, M. A., Ferson, S., and Akcakaya, H. R. 1993. *Risk Assessment in Conservation
Biology.* London: Chapman & Hall.
Cain, A. J. and Provine, W. B. 1992. Genes and ecology in history. In *Genes in
Ecology*, R. J. Berry, T. C. Crawford, and G. M. Hewitt, eds., pp. 3–28. London:
Black-well.
Capuccino, N. and Price, P., eds. 1996. *Population Dynamics.* New York: Academic.

Caseley, J. C., Cussons, G. W., and Aitkin, R. K., eds. 1991. *Herbicide Resistance in Weeds and Crops*. Oxford: Butterworth–Henemann.

Christiansen, F. B. and Fenchel, T. M., eds. 1977. *Measuring Selection in Natural Populations*. New York: Springer-Verlag.

Clark, A. G. 1993. Evolutionary inferences from molecular characterization of self-incompatibility alleles. In *Mechanisms of Molecular Evolution*. N. Takahata and A. G. Clark, eds., pp. 79–107. Tokyo: Japan Scientific Societies.

Cohen, D. and Levin, S. A. 1991. Dispersal in patchy environments: the effects of temporal and spatial structure. *Theor. Popul. Biol.* **39**:63–99.

Creed, R., ed. 1971. *Ecological Genetics and Evolution*. Oxford: Blackwell.

Dobzhansky, Th. 1970. *Genetics of the Evolutionary Process*. New York: Columbia U. Press.

Durand, D., Ardlie, K., Buttel, L., Levin, S. A., and Silver, L. M. 1997. Impact of migration and fitness on the stability of lethal t-haplotype polymorphisms in *Mus musculus*: a computer study. *Genetics* **145**:1093–1108.

Endler, J. A. 1986. *Natural Selection in the Wild*. Princeton, NJ: Princeton U. Press.

Endler, J. A. 1992. Genetic heterogeneity and ecology. In *Genes in Ecology*, R. J. Berry, T. J. Crawford, and G. M. Hewitt, eds., pp. 315–334. Oxford: Blackwell.

Ewens, W. J. 1979. *Mathematical Population Genetics*. London: Methuen.

Feldman, M. W. 1989. Discussion: ecology and evolution. In *Perspectives in Ecological Theory*, J. Roughgarden, R. M. May, and S. A. Levin, eds., pp. 135–138. Princeton, NJ: Princeton U. Press.

Ford, E. B. 1964. *Ecological Genetics*. London: Methuen.

Frank, S. A. 1989. The evolutionary dynamics of cytoplasmic male sterility. *Am. Nat.* **133**:345–376.

Frank, S. A. 1995. Recognition and polymorphism in host-parasite genetics. *Philos. Trans. R. Soc. London Ser. B* **246**:283–293.

Frank, S. A. 1997. The design of natural and artificial adaptive systems. In *Adaptation*, M. R. Rose and G. V. Lauder, eds., pp. 451–506. San Diego, CA: Academic.

Frankel, O. H., Brown, A. H. D., and Burdon, J. J. 1995. *The Conservation of Plant Biodiversity*. Cambridge: Cambridge U. Press.

Frankham, R. 1995. Effective population size/adult population size ratios in wildlife: a review. *Genet. Res.* **66**:95–107.

Giesel, J. T. 1974. *The Biology and Adaptability of Natural Populations*. St. Louis, MO: Mosby.

Godfray, H. C. J., Cook, L. M., and Hassell, M. P. 1992. Population dynamics, natural selection and chaos. In *Genes in Ecology*, R. J. Berry, T. C. Crawford, and G. M. Hewitt, eds., pp. 55–85. Oxford: Blackwell.

Gould, S. J. and Lewontin, R. C. 1979. The spandrels of San Marco and the Panglossian paradigm: a critique of the adaptationist programs. *Proc. R. Soc. London Ser. B.* **205**:581–598.

Grant, P. R. 1986. *Ecology and Evolution of Darwin's Finches*. Princeton, NJ: Princeton U. Press.

Gray, A. J. 1996. The genetic basis of conservation biology. In *Conservation Biology*, I. Spellenberg, ed., pp. 107–121. Essex: Longman.

Halley, J. M. and Manasse, R. S. 1993. A population dynamics model for annual plants subject to inbreeding depression. *Evol. Ecol.* **7**:15–24.

Hanski, I. A. and M. E. Gilpin, eds. 1997. *Metapopulation Biology: Ecology, Genetics and Evolution*. San Diego, CA: Academic.

Harvey, P. H., Nee, S., Moores, A. O., and Partridge, L. 1992. Three hierarchical views of life: phylogenies and metapopulations. In *Genes in Ecology*, R. J. Berry, T. C. Crawford, and G. M. Hewitt, eds., pp. 123–138. Oxford: Blackwell.

Hastings, A. and Harrison, S. 1994. Metapopulation dynamics and genetics. *Annu. Rev. Ecol. Syst.* 25:167–188.

Holsinger, K. E. 1993. The evolutionary dynamics of fragmented plant populations. In *Biotic Interactions and Global Change*, P. M. Kareiva, E. G. Kingsolver, and R. B. Huey, eds., pp. 198–216. Sunderland, MA: Sinauer.

Hughes, M. A. 1991. The cyanogenic polymorphism in *Trifolium repens* L. (white clover). *Heredity* 66:105–115.

Jain, S. K. 1994. Genetics and demography of rare plants and patchily distributed colonizing species. In *Conservation Genetics*, V. Loescheke, J. Tomiuk, and S. K. Jain, eds., pp. 291–307. Basel, Switzerland: Birkhauser.

Koelewijin, H. P. 1993. On the genetics and ecology of sexual reproduction in *Plantago coronopus*. Ph.D. dissertation, Utrecht University.

Kuiper, P. J. C., and Bos., H., eds. 1992. *Plantago: A Multidisciplinary Study*. Berlin: Springer-Verlag.

Kunkel, B. 1996. A useful weed put to work: genetic analysis of disease resistance in *Arabidopsis thaliana*. *Trends Genet.* 12:63–68.

Lack, D. 1965. Evolutionary ecology. *J. Appl. Ecol.* 2:247–255.

Lande, R. and Schemske, D. W. 1985. The evolution of self-fertilization and inbreeding depression in plants. I. Genetic models. *Evolution* 39:24–40.

Levin, S. 1992. The problem of pattern and scale in ecology. *Ecology* 73:1943–1967.

Levin, S. and C. Castillo-Chavez. 1990. Topics in evolutionary ecology. In *Mathematical and Statistical Developments of Evolutionary Theory*, S. Lessard, ed., pp. 327–358. Dordrecht, The Netherlands: Kluwer.

Levins, R. 1968. *Evolution in Changing Environments*. Princeton, NJ: Princeton U. Press.

Lewontin, R. C. 1962. Interdeme selection controlling a polymorphism in the house mouse. *Am. Nat.* 96:65–78.

Lewontin, R. C., ed. 1968. *Population Biology and Evolution*. Syracuse, NY: Syracuse U. Press.

Lewontin, R. C. 1974. *The Genetic Basis of Evolutionary Change*. New York: Columbia U. Press.

Lewontin, R. C. 1979. Fitness survival and optimality. In *Analysis of Ecological Systems*, D. J. Horn, G. R. Stairs, and R. D. Mitchell, eds., pp. 3–21. Columbus, OH: Ohio U. Press.

Lyons, E. E. and Antonovics, J. 1991. Breeding system evolution in *Leavenworthia*: breeding system variation and reproductive success in natural populations of *Leavenworthia crassia* (Cruciferae). *Am. J. Bot.* 78:270–287.

Mangel, M. 1992. Descriptions of superparasitism by optimal foraging theory, evolutionarily stable strategies and quantitative genetics. *Evol. Ecol.* 6:152–169.

Mayr, E. 1963. *Animal Species and Evolution*. Cambridge: Belknap.

McCauley, D. E. 1993. Genetic consequences of extinction and recolonization in fragmented habitats. In *Biotic Interactions and Global Change*, P. M. Kareiva, J. G. Kingsolver, and R. B. Huey, eds., pp. 143–151. Sunderland, MA: Sinauer.

Mitton, J. B. 1994. Molecular approaches to population biology. *Annu. Rev. Ecol. Syst.* 25:45–69.

Olivieri, I. and Gouyon, P. H. 1997. Evolution of migration rate and other traits:

the metapopulation effect. In *Metapopulation Biology: Ecology, Genetics and Evolution.* I. A. Hanski and M. E. Gilpin, eds., pp. 293–323. San Diego, CA: Academic.

Olivieri, I., Couvet, D., and Gouyon, P. H. 1990. The genetics of transplant populations: research at the metapopulation level. *Trends Ecol. Evol.* **5**:207–210.

Pimm, S. L. 1991. *The Balance of Nature.* New York: Chapman & Hall.

Powles, S. B. and Holtum, J. A. M., eds. 1994. *Herbicide Resistance in Plants.* Boca Raton, FL: CRC.

Prout, T. 1980. Some relationship between density-independent selection and density-dependent population growth. *Evol. Biol.* **13**:1–68.

Rausher, M. D. 1996. Genetic analysis of coevolution between plants and their natural enemies. *Trends Genet.* **12**:212–217.

Raven, P. H. 1979. Future directions in plant population biology. In *Topics in Plant Population Biology,* O. T. Solbrig, S. Jain, G. B. Johnson, and P. H. Raven, eds., pp. 461–481. New York: Columbia U. Press.

Real, L. 1994. *Ecological Genetics.* Chicago: U. Chicago Press.

Reznick, D. and Travis, J. 1996. The empirical study of adaptation in natural populations. In *Adaptation,* M. R. Rose and G. V. Lauder, eds., pp. 243–289. San Diego, CA: Academic.

Rhodes Jr., O. E., Chesser, R. K., and Smith, M. H., eds. 1996. *Population Dynamics in Ecological Space and Time.* Chicago: U. Chicago Press.

Ricklefs, R. E. 1979. *Ecology,* 2nd ed. New York: Chiron.

Rose, M. R. and Lauder, G. V., eds. 1996. *Adaptation.* San Diego, CA: Academic.

Schlichting, C. D. 1986. The evolution of phenotypic plasticity in plants. *Annu. Rev. Ecol. Syst.* **17**:667–693.

Schmid-Hempel, P. 1990. In search of optima: equilibrium models of phenotypic evolution. In *Population Biology: Ecological and Evolutionary Implications,* K. Wohrmann and S. K. Jain, eds., pp. 321–347. Berlin: Springer-Verlag.

Shaw, A. J., ed. 1990. *Heavy metal tolerance in plants: Evolutionary aspects.* Boca Raton, FL: CRC.

Simms, E. L. 1996. The evolutionary genetics of plant-pathogen systems. *BioScience* **46**:136–145.

Slobodkin, L. B. 1988. Intellectual problems of applied ecology. *BioScience* **38**:337–342.

Smith, R. H. 1991. Genetic and phenotypic aspects of life history evolution in animals. *Adv. Ecol. Res.* **21**:63–119.

Sober, E. 1984. *The Nature of Selection: Evolutionary Theory in Philosophical Focus.* Cambridge, MA: MIT.

Spiess, E. B. 1977. *Genes in Populations.* New York: Wiley.

Stacey, P. B. and Taper, M. 1992. Environmental variation and the persistence of small populations. *Ecol. Appl.* **2**:18–29.

Thornhill, N. W., ed. 1993. *The Natural History of Inbreeding and Outbreeding.* Chicago: U. Chicago Press.

Turrelli, M. 1988. Population genetic models for polygenic variation and evolution. In *Quantitative Genetics,* B. S. Weir, E. J. Eisen, M. M. Goodman, and G. Namkoong, eds., pp. 601–618. Sunderland, MA: Sinauer.

Uyenoyama, M. K., Holsinger, K. E., and Waller, D. M. 1993. Ecological and genetic factors directing the evolution of self-fertilization. *Oxford Surv. Evol. Biol.* **9**:327–380.

Wade, M. J. 1996. Adaptation in subdivided populations: kin selection and interdemic selection. In *Adaptation,* M. R. Rose and G. V. Lauder, eds., pp. 381–405. San Diego, CA: Academic.

Waller, D. M. 1993. The statistics and dynamics of mating systems evolution. In *The Natural History of Inbreeding and Outbreeding*, N. W. Thornhill, ed., pp. 97–116. Chicago: U. Chicago Press.

Wiener, P. and Feldman, M. W. 1993. The effects of the mating system on the evolution of migration in a spatially heterogeneous population. *Evol. Ecol.* 7:251–269.

Williamson, M. 1996. *Biological Invasions.* London: Chapman & Hall.

Young, A., Boyle, T., and Brown, T. 1996. The population genetic consequences of habitat fragmentation for plants. *Trends Ecol. Evol.* 11:413–418.

Young, A. G. and Merriam, H. G. 1994. Effects of forest fragmentation on the spatial genetic structure of *Acer saccharum* Marsh (Sugar maple) populations. *Heredity* 72:201–208.

Publications of R. C. Lewontin

1952. An elementary text on evolution. Review of *A Textbook of Evolution* by E. O. Dodson. *Evolution* **6(2)**:247–248.

1953. The effect of compensation on populations subject to natural selection. *Am. Nat.* **87**:375–381.

1954. Review of *Problems of Life. Am. J. Sci.* **252**:123–124.

1954. Review of *Biology and Language. Am. J. Sci.* **252**:124–126.

1954. Familial occurrence of migraine headache: A study of heredity. *A.M.A. Arch. Neurol. Psych.* **72**:325–334 (with H. Goodell and H. G. Wolff).

1955. The effects of population density and composition on viability in *Drosophila melanogaster. Evolution* **9**:27–41.

1956. Estimation of the number of different classes in a population. *Biometrics* **12**:211–223 (with T. Prout).

1956. Studies on homeostasis and heterozygosity. I. General considerations. Abdominal bristle number in second chromosome homozygotes of *Drosophila melanogaster. Am. Nat.* **90**:237–255.

1956. A reply to Professor Dempster's comments on homeostasis (Letters to the Editors). *Am. Nat.* **90**:386–388.

1957. The adaptations of populations to varying environments. *Cold Spring Harbor Symp. Quant. Biol.* **22**:395–408.

1958. A general method for investigating the equilibrium of gene frequency in a population. *Genetics* **43**:419–434.

1948. Studies of heterozygosity and homeostasis. II. Loss of heterosis in a constant environment. *Evolution* **12**:494–503.

1959. On the anomalous response of *Drosophila pseudoobscura* to light. *Am. Nat.* **93**:321–328.

1959. The goodness-of-fit test for detecting natural selection. *Evolution* **13**:561–564 (with C. C. Cockerham).

1960. *Quantitative Zoology*, 2nd ed. New York: Harcourt Brace (with G. G. Simpson and A. Roe).

1960. Interaction between inversion polymorphism of two chromosome pairs in the grasshopper *Moraba scurra. Evolution* **14**:116–129.

1960. The evolutionary dynamics of a polymorphism in the house mouse. *Genetics* **45**:705–722 (with L. C. Dunn).

1960. The evolutionary dynamics of complex polymorphisms. *Evolution* **14**:458–472 (with K. Kojima).

1960. Review of *Introduction to Quantitative Genetics* by D. S. Falconer. *Am. Sci.* **48**:274A–276A.

1961. Evolution and the theory of games. *J. Theor. Biol.* **1**:382–403.

1961. Review of *Biochemical Genetics. Am. Sci.* **49**:190A.

1962. Review of *Introduction to the Mathematical Theory of Genetic Linkage* by N. T. J. Bailey. *Am. Sci.* **50**:320A–322A.

1962. Review of *An Outline of Chemical Genetics* by B. Strauss. *Hum. Biol.* 235–236.

1962. Interdeme selection controlling a polymorphism in the house mouse. *Am. Nat.* **96**:65–78.

1963. Interaction of genotypes determining viability in *Drosophila busckii. Proc. Natl. Acad. Sci.* **49**:270–278 (with Y. Matsuo).

1963. Relative fitness of geographic races of *Drosophila serrata. Evolution* **17**:72–83 (with L. C. Birch, T. Dobzhansky, and P. O. Elliott).

1963. Models, mathematics, and metaphors. *Synthese* **15**:222–244.

1963. Cytogenetics of the grasshopper *Moraba scurra*. VII. Geographic variation of adaptive properties of inversions. *Evolution* **17**:147–162 (with M. J. D. White and L. E. Andrew).

1964. The interaction of selection and linkage. I. General considerations: Heterotic models. *Genetics* **49**:49–67.

1964. A molecular messiah: The new gospel of genetics? Essay review of *The Mechanics of Inheritance* by F. Stahl. *Science* **145**:525.

1964. The role of linkage in natural selection. Proceedings of the XI International Congress of Genetics, The Hague, September 1964. *Genet. Today* 517–525.

1964. The interaction of selection and linkage. II. Optimum models. *Genetics* **50**:757–782.

1964. Review of *Elizabethan Acting. The Seventeenth Century News.*

1964. The capacity for increase in chromosomally polymorphic and monomorphic populations of *Drosophila pseudoobscura. Heredity* **19**:597–614 (with T. Dobzhansky and O. Pavlovsky).

1965. Selection in and of populations. In *Ideas in Modern Biology*, (Proceedings of the XVI International Congress of Zoology, vol. 6) J. A. Moore, ed., pp. 299–311. Garden City, NY: National History Press.

1965. The robustness of homogeneity tests in 2 X N tables. *Biometrics* **21**:19–33 (with J. Felsenstein).

1965. Selection for colonizing ability. In *The Genetics of Colonizing Species*, Herbert Baker, ed., pp. 77–94 New York: Academic Press.

1965. Review of *The Effects of Inbreeding on Japanese Children. Science* **150**:332–333.

1965. Review of *Stochastic Models in Medicine and Biology. Am. Sci.* **53**:254A–255A.

1966. Adaptation and natural selection (essay review). *Science* **152**:338–339.

1966. Is nature probable or capricious? *Bio Science* **16**:25–27.

1966. Differences in bristle-making abilities in scute and wild-type *Drosophila melanogaster. Genet. Res.* **7**:295–301 (with S. S. Y. Young).

1966. On the measurement of relative variability. *Syst. Zool.* **15**:141–142.

1966. Stable equilibria under optimizing selection. *Proc. Natl. Acad. Sci.* **56**:1345–1348 (with M. Singh).

1966. A molecular approach to the study of genic heterozygosity in natural populations. I. The number of alleles at different loci in *Drosophila pseudoobscura. Genetics* **54**:577–594 (with J. L. Hubby).

1966. A molecular approach to the study of genic heterozygosity in natural populations. II. Amount of variation and degree of heterozygosity in the natural populations of *Drosophila pseudoobscura. Genetics* 54:595–609 (with J. L. Hubby).

1966. Hybridization as a source of variation for adaptation to new environments. *Evolution* 20:315–336 (with L. C. Birch).

1966. Review of *The Theory of Inbreeding. Science* 150:1800–1801.

1967. The genetics of complex systems. *Proceedings of the 5th Berkeley Symposium on Mathematical Statistics and Probability,* Vol. IV. Berkeley, CA: U. of Calif. Press, pp. 439–455.

1967. The interaction of selection and linkage. III. Synergistic effect of blocks of genes. *Der Zuchter* 37:93–98 (with P. Hull).

1967. The principle of historicity in evolution. In P. S. Moorhead and M. M. Kaplan, *Mathematical Challenges of the Neo-Darwinian Theory of Evolution.* Wistar Symposium Monograph No. 5, pp. 81–94.

1967. An estimate of average heterozygosity in man. *Am. J. Human. Genet.* 19:681–685.

1967. Population genetics. *Ann. Rev. Genet.* 1:37–70.

1968. A molecular approach to the study of genic heterozygosity in natural populations. III. Direct evidence of coadaptation in gene arrangements of Drosophila. *Proc. Natl. Acad. Sci.* 59:398–405 (with S. Prakash).

1968. A note on evolution and changes in the quantity of genetic information. In *Towards a Theoretical Biology,* I, pp. 109–110. Aldin (with C. H. Waddington).

1968. Essay review of *Phage and the Origins of Molecular Biology,* J. Cairns et al., eds., (Cold Spring Harbor, New York: Cold Spring Harbor Laboratory of Quantative Biology, 1966, xii + 340). *J. Hist. Biol.* 1(1):155–161.

1968. The concept of evolution. In "Evolution," *International Encyclopedia of the Social Sciences,* pp. 202–210. Macmillan and The Free Press.

1968. The effect of differential viability on the population dynamics of *t* alleles in the house mouse. *Evolution* 22:262–273.

1968. Selective mating, assortative mating, and inbreeding: Definitions and implications. *Eugen. Quar.* 15:141–143 (with D. Kirk and J. Crow).

1969. The bases of conflict in biological explanation. *J. Hist. Biol.* 2(1):35–45.

1969. On population growth in a randomly varying environment. *Proc. Natl. Acad. Sci.* 62(4):1056–1060 (with D. Cohen).

1969. A molecular approach to the study of genic heterozygosity in natural populations. IV. Patterns of genic variation in central, marginal and isolated populations of *Drosophila pseudoobscura. Genetics* 61:841–858 (with S. Prakash and J. L. Hubby).

1969. The meaning of stability. (Reprinted from *Diversity and Stability in Ecological Systems*) *Brookhave Symp. Biol.* 22:13–24.

1970. On the irrelevance of genes. In *Towards a Theoretical Biology,* 3: Drafts. Edinburgh, UK: Edinburgh U. Press, pp. 63–72.

1970. Race and intelligence. *Bull. Atom. Sci.* 26(Mar):2–8.

1970. Further remarks on race and the genetics of intelligence. *Bull. Atom. Sci.* 26(May):23–25.

1970. Genetic variation in the horseshoe crab (*Limulus polyphemus*), a phylogenetic "relic." *Evolution* 24:402–414 (with R. K. Selander, S. Y. Yang, and W. E. Johnson).

1970. Is the gene the unit of selection? *Genetics* 65:707–734 (with I. Franklin).

1970. The units of selection. *Ann. Rev. Ecol. Syst.* 1:1–18.

1971. Genes in populations – end of the beginning. Review of *An Introduction to Population Genetics Theory* by J. F. Crow and M. Kimura. *Quart. Rev. Biol.* 46:66–67.

1971. Evolutionary significance of linkage and epistasis. In *Biomathematics Vol. I: Mathematical Topics in Population Genetics*, pp. 367–388 (with K. Kojima).

1971. The effect of genetic linkage on the mean fitness of a population. *Proc. Natl. Acad. Sci. USA* **68**:984–986.

1971. The Yahoos ride again. *Evolution* **25**:442.

1971. Science and ethics. *BioScience* **21**:799.

1971. A molecular approach to the study of genic heterozygosity in natural populations. V. Further direct evidence of coadaptation in inversions of *Drosophila*. *Genetics* **69**:405–408 (with S. Prakash).

1972. Testing the theory of natural selection. *Nature* **236**:181–182.

1972. The apportionment of human diversity. *Evol. Biol.* **6**:381–398.

1972. Comparative evolution at the levels of molecules, organisms and populations. *Proceedings of the VI Berkeley Symposium on Mathematical Statistics and Probability* **5**:23–42 (with G. L. Stebbins).

1973. Distribution of gene frequency as a test of the theory of the selective neutrality of polymorphism. *Genetics* **74**:175–195 (with J. Krakauer).

1974. Population genetics. *Ann. Rev. Gen.* **7**:1–17.

1974. Molecular heterosis for heat-sensitive enzyme alleles. *Proc. Natl. Acad. Sci. USA* **71**:1808–1810 (with R. S. Singh and J. L. Hubby).

1974. *The Genetic Basis of Evolutionary Change*. New York: Columbia U. Press.

1974. Annotation: The analysis of variance and the analysis of causes. *Am. J. Hum. Gen.* **26**:400–411.

1974. Darwin and Mendel – the materialist revolution. In *The Heritage of Copernicus: Theories "More Pleasing to the Mind,"* J. Neyman, ed., pp. 166–183. Cambridge, MA: MIT Press.

1975. The problem of genetic diversity. *Harvey Lec. Ser.* **70**:1–20.

1975. Selection in complex genetic systems. III. An effect of allele multiplicity with two loci. *Genetics* **79**:333–347 (with M. W. Feldman, I. R. Franklin, and F. B. Christiansen).

1975. Review of *The Modern Concept of Nature* by H. J. Muller. *Soc. Biol.* **22**:96–98.

1975. The heritability hang-up. *Science* **190**:1163–1168 (with M. W. Feldman).

1975. Genetic aspects of intelligence. *Ann. Rev. Genet.* **9**:382–405.

1976. Review of *Race Differences in Intelligence* by J. C. Loehlin, G. Lindzey, and J. N. Spuhler. *Am. J. Hum. Genet.* **28**:92–97.

1976. Adattamento genetico. *Enciclopedia del Novecento* **1**:61–68.

1976. Genetic heterogeneity within electrophoretic "alleles" of xanthine dehydrogenase in *Drosophila pseudoobscura*. *Genetics* **84**:609–629 (with R. S. Singh and A. A. Felton).

1976. The problem of Lysenkoism. In *The Radicalisation of Science* by H. Rose and S. Rose, eds., pp. 32–64. London: MacMillan (with R. Levins).

1976. The fallacy of biological determinism. *The Sciences* **16**:6–10.

1976. Sociobiology – a caricature of Darwinism. *Phil. Sci. Assoc.* **2**:22–31.

1977. The relevance of molecular biology to plant and animal breeding. *Proceedings of the International Conference on Quantitative Genetics*, Ames, IA: Iowa State U. Press, pp. 55–62.

1977. Population genetics. *Proceedings of the 5th International Congress of Human Genetics, Excerpta Medica International Congress* Series No. 441, pp. 13–18.

1977. Adattamento. *Enciclopedia Einaudi* **1**:198–214.

1977. Biological determinism as a social weapon. In *Biology as a Social Weapon*, pp. 6–18. Minneapolis, MN: Burgess.

PUBLICATIONS OF R. C. LEWONTIN   683

1977. Caricature of Darwinism. Book review of *The Selfish Gene* by R. Dawkins. *Nature* **266**:283–284.

1978. Heterosis as an explanation for large amounts of genic polymorphism. *Genetics* **88**:149–170 (with L. R. Ginzburg and S. D. Tuljapurkar).

1978. Fitness, survival and optimality. In *Analysis of Ecological Systems* by D. J. Horn et al., eds., Columbia, OH: Ohio State U. Press.

1978. The extent of genetic variation at a highly polymorphic esterase locus in *Drosophila pseudoobscura*. *Proc. Natl. Acad. Sci. USA* **75**:5090 (with J. Coyne and A. A. Felton).

1978. Evoluzione. *Enciclopedia Einaudi* **5**:995–1051.

1978. Adaptation. *Sci. Am.* **239**(3):212–228.

1979. Sociobiology as an adaptationist program. *Behav. Sci.* **24**:5–14.

1979. Theodosius Dobzhansky. (Biographical article) In *International Encyclopedia of the Social Sciences*. New York: The Free Press, MacMillan.

1979. Single- and multiple-locus measures of genetic distance between groups. *Am. Nat.* **112**(988):1138–1139.

1979. The genetics of electrophoretic variation. *Genetics* **92**:353–361 (with J. A. Coyne and W. F. Eanes).

1979. The spandrels of San Marco and the Panglossian paradigm: a critique of the adaptationist programme. *Proc. Royal Soc. Lond. B.* **205**:581–598 (with S. J. Gould).

1979. The sensitivity of gel electrophoresis as a detector of genetic variation. *Genetics* **93**:1019–1037 (with J. A. M. Ramshaw and J. A. Coyne).

1979. *Mutazione/selezione. Enciclopedia Einaudi* **9**:647–695.

1980. Dialectics and reductionism in ecology. *Synthese* **43**:47–78 (with R. Levins).

1980. Sociobiology: Another biological determinism. *Int. J. Health Sci.* **10**(3):347–363.

1980. Economics down on the farm. Review of *Farm and Food Policy: Issues of the 1980's* by D. Paarlberg. *Nature* **287**:661–662.

1980. The political economy of food and agriculture (World Agricultural Research Project). *Int. J. Health Sci.* **10**:161–170.

1981. *An Introduction to Genetic Analysis*, 2nd ed., San Francisco: W. H. Freeman (with D. T. Suzuki and A. J. F. Griffiths).

1981. Evolution/Creation debate: A time for truth. *BioScience* **31**:559.

1981. Sleight of hand. Review of *Genes, Mind and Culture* by C. J. Lumsden and E. O. Wilson. *The Sciences* **21**:23–26.

1981. Review of *The Mismeasure of Man* by S. J. Gould. *NY Rev. Books.*

1981. Gene flow and the geographical distribution of a molecular polymorphism in *Drosophila pseudoobscura*. *Genetics* **98**:157–178 (with J. S. Jones, S. H. Bryant, J. A. Moore, and T. Prout).

1981. Theoretical population genetics in the evolutionary synthesis. In *The Evolutionary Synthesis*, E. Mayr and W. Province, eds. Cambridge, MA: Harvard U. Press.

1981. *Dobzhansky's Genetics of Natural Populations I-XLIII*. New York: Columbia U. Press (with J. A. Moore, W. B. Provine, and B. Wallace, ed.).

1981. L'Evolution. *La Pensee* **223**:16–24.

1982. Artifact, cause and genic selection. *Phil. Sci.* **49**:157–180 (with E. Sober).

1982. A study of reaction norms in natural populations of *Drosophila pseudoobscura*. *Evolution* **36**:934–938 (with A. Gupta).

1982. Review of *Matter, Life and Generation* by S. A. Roe. *The Sciences*, December.

1982. Prospectives, perspectives, and retrospectives. Review of *Perspectives on Evolution* by R. Milkman. *Paleobiology* **8**(3):309–313.

1982. Organism and environment. In *Learning, Development and Culture: Essays in Evolutionary Epistemology*, H. Plotkin, ed., Chichester, England, UK: Wiley.

1982. Review of *Evolution and the Theory of Games* by J. M. Smith, Cambridge U. Press. *Nature* **300**:113–114.

1982. *Human Diversity*. Sci. Am., Redding, CT: W. H. Freeman.

1982. Elementary errors about evolution. Review of *Intentional systems in cognitive ethology:* The '*Panglossian paradigm*' defined by D. Dennett. *Behav. Brain Sci.* **6**:367.

1983. The corpse in the elevator. Reviews of *Against Biological Determinism and Toward a Liberating Biology* by The Dialectics of Biology Group. *NY Rev. Books*.

1983. Biological determinism. In *The Tanner Lectures on Human Values* Vol. IV, p. 147–183. Salt Lake City, UT: U. Utah Press.

1983. Science as a social weapon. In *Occasional Papers I*, pp. 13–29. Amherst, MA: Inst. for Advanced Study in the Humanities, U. Massachusetts.

1983. Review of books by J. Miller & B. Van Loon, J. M. Smith, B. G. Gale, J. Gribben & J. Cherfas, N. Eldrige, I. Tattersall, D. Futuyma, P. Kitcher, and M. Ruse. *NY Rev. Books*.

1983. The organism as the subject and object of evolution. *Scientia* **188**:65–82.

1983. Gene, organism and environment. In *Evolution from Molecules to Men*, D. S. Bendall, ed., pp. 273–285. Cambridge: Cambridge U. Press.

1983. Introduction. In *Scientists Confront Creationism*, L. R. Godfrey and J. R. Coles, eds., W. W. Norton.

1983. Discussion: Reply to Rosenberg on genic selectionism. *Phil. Sci.* **50**:648–650 (with E. Sober).

1984. Detecting population differences in quantitative characters as opposed to gene frequencies. *Am. Nat.* **123**(1):115–124.

1984. *Not In Our Genes: Biology, Ideology and Human Nature*. New York: Pantheon (with S. Rose and L. Kamin).

1984. Review of *Women in Science: Portraits of a World in Transition* by V. Gornick. New York: *NY Rev. Books*.

1984. Le determinisme biologique comme arme social. In *Les Enjeux du Progress*, A. Cambrosio and R. Duchesne, eds., pp. 233–251. Québec, PQ, Canada: Presses de l'U. du Québec.

1985. Nearly identical allelic distributions of xanthine dehydrogenase in two populations of *Drosophila pseudoobscura*. *Mol. Biol. Evol.* **2**:206–216 (with T. P. Keith, L. D. Brooks, J. C. Martinez-Cruzado, and D. L. Rigbny).

1985. Population genetics. *Ann. Rev. Genet.* **19**:81–102.

1985. Population genetics. In *Evolution, Essays in Honor of John Maynard Smith*, J. Greenwood and M. Slatkins, eds., pp. 3–18. Cambridge, England, UK: Cambridge U. Press.

1985. *The Dialectical Biologist*. Cambridge, MA: Harvard University Press (with R. Levins).

1985. Review of books by R. W. Clark, C. Darwin, V. Orel, L. J. Jordanova, and J. P. Changeeuex. *NY Rev. Books* **32**(15).

1985. In *Current Contents*. This week's citation classic. October (with J. L. Hubby).

1986. Technology, research, and the penetration of capital: The case of agriculture. *Monthly Rev.* **38**:21–34 (with J. P. Berlan).

1986. The political economy of hybrid corn. *Monthly Rev.* **38**:35–47 (with J. P. Berlan).

1986. Review of *In the Name of Eugenics* by D. J. Kevles. *Rev. Symp. Isis* **77**(2):314–317.

1986. *Education and Class*. Oxford England, UK: Oxford U. Press (with M. Schiff).

1986. A comment on the comments of Rogers and Felsenstein. *Am. Nat.* **127**(5):733–734.

1986. An Introduction to *Genetic Analysis*, 3rd. ed., by D. T. Suzuki, A. J. F. Griffiths, J. H. Miller and R. C. Lewontin. New York: W. H. Freeman and Co.

1986. Breeder's rights and patenting of life forms. *Nature* **322**:785–788 (with J. P. Berlan).

1986. How important is genetics for an understanding of evolution? In *Science as a Way of Knowing*, Vol III, p. 811–820. American Society of Zoologists.

1987. The shape of optimality. In *The Latest on the Best: Essays on Evolution and Optimality*, J. Dupre, ed., pp. 151–159. Cambridge, MA: MIT Press.

1987. Sequence of the structural gene for xanthine dehydrogenase (*rosy* locus) in *Drosophila melanogaster. Genetics* **116**:64–73 (with T. P. Keith, M. A. Riley, M. Kreitman, D. Curtis, and G. Chambers).

1987. Polymorphism and heterosis: Old wine in new bottles and *vice versa. J. Hist. Biol.* **20**:337–349.

1988. A general asymptotic property of two-locus selection models. *Theor. Popul. Biol.* **34**(2):177–193 (with M. W. Feldman).

1988. On measures of gametic disequilibrium. *Genetics* **120**:849–852.

1988. Aspects of wholes and parts in population biology. In *Evolution of Social Behavior and Integrative Levels*, G. Greenberg and E. Tobach, eds., pp. 31–52. NJ: Erlbaum (with R. Levins).

1988. La paradoja de la adaptacion biologica. In *Polemicas contemporáneas en Evolución*, A. O. Franco, ed. pp. 57–65. SA, Mexico: AGT Editor.

1989. Inferring the number of evolutionary events from DNA coding sequence differences. *Mol. Biol. Evol.* **6**(1):15–32.

1989. Review of *Controlling Life: Jacques Loeb and the Engineering Ideal in Biology* by P. Pauly, and *Topobiology: An Introduction to Molecular Embryology* by G. Edelman. *NY Rev. Books.*

1989. DNA sequence polymorphism. In *Essays in Honor of Alan Robertson*, W. G. Hill, and T. F. C. Mackey eds., pp. 33–37. Edinburgh, Scotland, UK: U. Edinburgh Press.

1989. On the characterization of density and resource availability, *Am. Nat.* **134**:513–524 (with R. Levins).

1989. Distinguishing the forces controlling genetic variation at the *Xdh* locus in *Drosophila pseudoobscura, Genetics* **123**:359–369 (with M. A. Riley and M. E. Hallas).

1989. Review of *Evolutionary Genetics* by J. Maynard Smith. *Nature* **339**:107 (May 11).

1990. The political economy of agricultural research: The case of hybrid corn. In *Agroecology*, C. R. Carroll, J. H. Vandermeer, and P. Rosset, eds., pp. 613–626. New York: McGraw-Hill.

1990. Review of *Wonderful Life: The Burgess Shale and the Nature of History* by S. J. Gould, Norton. *NY Rev. Books.*

1990. The evolution of cognition. In *Thinking, an Invitation to Cognitive Sciences*, Vol 3., D. N. Osherson and E. E. Smith, eds., pp. 229–240. Cambridge, MA: MIT Press.

1991. *How much did the brain have to change for speech?* (Pinker and Bloom Commentary) San Diego, CA: Academic Press.

1991. Review of *The Structure and Confirmation of Evolution Theory* by E. A. Lloyd, Greenwood Press, New York, 1988. *Biol. and Phil.* **6**:461–466.

1991. Foreword. In *Organism and the Origins of Self*, A. I. Tauber, ed., pp. xiii–xix. Boston: Kluwer Academic.

1991. Facts and the factitious in natural science. *Crit. Inq.* **18**:140–153.

1991. *Biology as Ideology: The Doctrine of DNA.* Ontario, Canada: Stoddart.

1991. Population genetics. *Encyl. Hum. Biol.* **6**:107–115.

1991. Perspectives: 25 years ago in Genetics: Electrophoresis in the development of evolutionary genetics: Milestone or Millstone? *Genetics* **128**:657–662.

1991. Population genetic problems in forensic DNA typing. *Science* **254**:1745–1750 (with D. L. Hartl).

1992. Genotype and phenotype. In *Keywords in Evolutionary Biology*, E. F. Keller and E. A. Lloyd, eds. pp. 137–144. Cambridge, MA: Harvard U. Press.

1992. Biology. In *Academic Press Dictionary of Science and Technology*, C. Morris, ed., p. 261. San Diego, CA: Academic Press.

1992. The dream of the human genome. *NY Rev. Books* **39**(10):31–40.

1992. Polemiche sul genoma umano, I. *La Riv. dei Libri*, Oct. 7–10.

1992. Polemiche sul genoma umano, II. *La Riv. dei Libri*, Nov. 6–9.

1992. The dimensions of selection. *J. Phil. Sci.* **60**:373–395 (with P. Godfrey-Smith).

1992. Open peer commentary: Gene talk on target. *Soc. Epist.* **6**(2):179–181.

1992. Letter to the Editor: Which population? *Am. J. Hum. Genet.* **52**:205.

1992. Inside and Outside: Genetics, environment and organism. *Heinz Werner Lecture Series*, Vol. XX. Worcester, MA: Clark U. Press.

1992. DNA data banking and the public interest. In *DNA on Trial*, P. R. Billings, ed., pp. 141–149. Cold Spring Harbor Press. (with N. L. Wilker, S. Stawksi, and P. R. Billings).

1993. *Biology as Ideology: The Doctrine of DNA.* New York: Harper Collins.

1993. *The Doctrine of DNA. Biology as Ideology.* Penguin Books.

1993. Correlation between relatives for colorectal cancer mortality in familial adenomatous polyposis. *Ann. Hum. Genet.* **57**:105–115 (with S. Presciuttini, L. Bertario, P. Saia, and C. Rossetti).

1993. Letters *Science* **260**:473–474 (with D. L. Hartl).

1993. Risposta a Sgaramella (Reply to Sgaramella). *La Riv. dei Libri.*

1993. Biology as Ideology. In *The Dancer and the Dance*, E. Sagarra and M. Sagarra, eds., pp. 54–64. Trinity Jameson Quartercentenary Symposium.

1994. Comment: The use of DNA profiles in forensic context. In *DNA Fingerprinting: A Review of the Controversy*, by K. Roeder. *Stat. Sci.* **9**(2):259–262.

1994. *Biologia come Ideologia: La Dottrina del DNA.* Italy: Bollati Boringhieri.

1994. Women *versus* the biologists. *NY Rev. Books* **41**(7):31–35.

1994. Forensic DNA typing dispute. *Nature* (Correspondence) **372**:398.

1994. DNA fingerprinting. *Science* (Letters) **266**:201 (with D. L. Hartl).

1994. Here we go again. Neri e bianchi per mi pari sono. *La Republica* (Milan), Oct. 25, pp. 24–25.

1994. Response to Goldberg and Hardy. *NY Rev. Books.*

1994. Holism and reductionism in ecology. *CNS* **5**(4):33–40 (with R. Levins).

1994. Facts and the factitious in natural sciences. In *Questions of Evidence*, J. Chandler, A. I. Davidson, and H. Harootunian, eds., pp. 478–491. Chicago: U. Chicago Press.

1994. A rejoinder to William Wimsatt. In *Questions of Evidence*, J. Chandler, A. I. Davidson, and H. Harootunian, eds., pp. 504–509. Chicago: U. Chicago Press.

1994. Response to Lander and Budowle. *Nature* (Letter to the Editor) **372**:398.

1994. *De DNA Doctrine.* Amsterdam, The Netherlands: Uitgevery Bert Bakker.

1995. *Human Diversity.* New York: Scientific American Library.

1995. Promesses, promesses. *Genetique et Evolution* **11**:II–V.

1995. Genes, environment and organisms. In *Hidden Histories of Science*, R. B. Silvers, ed., pp. 115–139. *NY Rev. Books.*

1995. The detection of linkage disequilibrium in molecular sequence data. *Genetics* **140**:377–388.

1995. A la recherche du temps perdu. *Configurations* **2**:257–265.

1995. Sex, lies and social science. *NY Rev. Books* **42**(7):24–29.

1995. Il potere del progetto (The power of the project). *SFERA Magazine* **43**:11–34. Rome and Milan.

1995. Theodosius Dobzhansky – A theoretician without tools. In *Genetics of Natural Populations: The Continuing Importance of Theodocius Dobzhansky*, L. Levine, ed., pp. 87–101. New York: Columbia U. Press.

1995. The dream of the human genome. In *Politics and the Human Body*, J. B. Elshtain and J. T. Lloyd, eds., pp. 41–66. Nashville, TN: Vanderbilt U. Press.

1995. IV. Primate models of human traits. In *Aping Science* by B. P. Reines, pp. 17–35.

1996. What does electrophoretic variation tell us about protein variation? *Mol. Biol. Evol.* **13**(2):427–432. (with A. Barbadilla and L. M. King).

1996. La recherche du temps perdue. In *Science Wars*, ed., Durham, NC: Duke U. Press.

1996. *Of genes and genitals.* Cambridge, MA: Transition Publishers.

1996. The end of natural history? *CNS* **7**(1):1–4 (with R. Levins).

1996. Pitfalls of genetic testing. *NEJM* (Sounding Board) **334**(18):1192–1194 (with R. Hubbard).

1996. Authors' Reply. *NEJM* (Correspondence) **335**:1236–1237 (with R. Hubbard).

1996. In defense of science. *Society* **33**(4):30–31.

1996. Evolution as engineering. In *Integrative Approaches to Molecular Biology.* J. Collado, T. Smith, and B. Magasnik, eds., pp. 1–10. Cambridge, MA: MIT Press.

1996. Primate models of human traits. In *Monkeying with Public Health.* R. Reines and S. Kaufman, eds., pp. 17–35. WI: One Voice.

1996. Indiana Jones meets King Kong. Review of three books by M. Crichton. *NY Rev. Books* (February 29).

1996. Letter. (Zola biography review) *NY Rev. Books* (May '96).

1996. *Evolution and Religion.* Am. Jewish Congress.

1996. Letter. (On Horton's essay on genetics and homosexuality.) *NY Rev. Books.*

1996. Population genetic issues in the forensic use of DNA. In *The West Companion to Scientific Evidence.* D. Faigman, ed., pp. 673–696.

1996. The evolution of cognition: questions we will never answer. In *An Invitation to Cognitive Science—Methods, Models, and Conceptual Issues*, Vol. 4, D. Scarborough and S. Sternberg, eds., pp. 132–197, 2nd ed. Cambridge, MA: MIT Press.

1996. Detecting heterogeneity of substitutions along DNA and protein sequences. P. J. E. Goss and R. C. Lewontin, eds., *Genetics* **143**:589–602.

1996. False dichotomies. *CNS* **7**(3):27–30 (with R. Levins).

1996. The return of old diseases and the appearance of new ones. *CNS* **7**(2):103–107 (with R. Levins).

1997. Billions and billions of demons. Review of *The Demon Haunted World: Science as a Candle in the Dark* by C. Sagan. *NY Rev. Books.*

1997. Nucleotide variation and conservation at the dpp locus, a gene controlling early development. *Genetics* **145**(2):311–323 (with B. Richter, M. Long, E. Nitasaka).

1997. A scientist meets a visionary. In *Bucky's 100.* (In press.)

1997. Population genetics. In *Encyclopedia of Human Biology*, 2nd ed. New York: Academic Press.

1997. Genetics, plant breeding and patents: Conceptual contradictions and practical problems in protecting biological innovation. *Plant Genet. Res. Newsletter* **112**:1–8 (with M. de Miranda Santos).

1997. The evolution of cognition: Questions we will never answer. In *Invitation to Cognitive Science*, Vol. 4., D. N. Osherson, D. Scarborough, and S. Sternberg, eds., pp. 107–132. Cambridge, MA: MIT Press.

1997. Dobzhansky's "Genetics and the Origin of Species": Is it still relevant? *Genetics* **147**:351–355.

1997. The Cold War and the transformation of the Academy. In *The Cold War and the University*, Vol. 1., pp. 1–34. New York: The New Press.

1997. The biological and the social. *CNS* **8**(3):89–92 (with R. Levins).

1997. Chance and Necessity. *CNS* **8**(2):95–98 (with R. Levins).

1997. Confusing over cloning. Review of: *Cloning Human Beings: Report and Recommendations of the National Bioethics Advisory Commission* (June, 1997). *NY Rev. Books* **44**(16).

1997. Biologie as Ideologie: Ursache und Wirkung bei der Tuberkulose, den menschlichen Genen und in der Landwirtschaft. *Streitbarer Materialismus* **21**:111–128.

1997. A question of biology: Are the races different? In *Beyond Heroes and Holidays*, E. Lee, D. Menkart, and M. Okazawa-Rey, eds., Network of Educators on the Americas.

1998. What do population geneticists know and how do they know it? In *Biology and Epistemology*, R. Creath and J. Maienschein, eds., Cambridge, UK: Cambridge U. Press. (In press.)

1998. *Gene, organismo, ambiente.* Rome: Laterza. (In press.)

1998. The maturing of capitalist agriculture: The farmer as proletarian. *Monthly Rev. Press.* (In press.)

1998. The economics of "hybrid" corn. *Am. Econom. Rev.* (with J. P. Berlan).

1998. Forward. In *The Ontogeny of Information*, S. Oyama, R. Gray, and P. Griffiths, eds., Duke U. Press. (In press.)

1998. Review of *"Unto Others": The Evolution and Psychology of Unselfish Behavior.* by E. Sober and D. S. Wilson. Cambridge, MA: Harvard U. Press. *NY Rev. Books.*

1998. *Gene, Organismo, e Ambiente.* Rome: Guis. Laterza and Figli Spa.

1998. The maturing of capitalist agriculture: Farmer as Proletarian. *Monthly Rev.* **50**(3):72–85.

1998. Survival of the Nicest? *NY Rev. Books* **45**(16):59–63.

1998. Foreword. In *Building a New Biocultural Synthesis*, A. H. Goodman and T. L. Leatherman, eds., pp. xi–xv (with R. Levins).

1998. The confusion over cloning. In *Flesh of My Flesh. The Ethics of Cloning Humans*, G. E. Pense, ed., pp. 129–139. Lanham, MD: Rowman and Littlefield.

1998. How different are natural and social science? *CNS* **9**(1):85–89 (with R. Levins).

1998. Does anything new ever happen? *CNS* **9**(2):53–56 (with R. Levins).

1998. Life on other worlds. *CNS* **9**(4):39–42 (with R. Levins).

1999. Menace of the genetic-industrial complex. *The Guardian Weekly.*

1999. *Modern Genetic Analysis.* New York: W. H. Freeman (with A. J. F. Griffiths, W. M. Gelbart, and J. H. Miller).

1999. Foreword. In *The Ontogeny of Information*, S. Oyama, R. Gray, and P. Griffiths, eds., Duke U. Press. (In press.)

1999. Locating regions of differential variability in DNA and protein sequences. *Genetics.* (with H. Tang). (In press.)

1999. *Dream of the Human Genome – and Other Confusions.* R. Silvers, ed. *NY Rev. Books.* (In press.)

1999. *Gene, Organism and Environment.* Cambridge, MA: Harvard U. Press. (In press.)

1999. Genotype and phenotype. In *International Encyclopedia of the Social and Behavioral,* N. J. Smelser and P. D. Baltes, eds. Elsevier Science, Pergamon. (In preparation.)

1999. The problems of population genetics. In *Evolutionary Genetics from Molecules to Morphology,* Vol. 1., R. S. Singh and C. Krimbas, eds. Cambridge, England, UK: Cambridge U. Press.

1999. Are we programmed? *CNS* 10(2):71–75 (with R. Levins).

1999. *Does Culture Evolve?* (with J. Fracchia) (In preparation.)

Index

692   SUBJECT INDEX

Balance
  mutation-selection, 136, 197, 359, 369,
    370, 373, 381, 382, 383, 385,
    387, 663
  recombination-selection, 478, 483
  selective, 258, 280, 281
Balanced
  equilibrium, 16, 258
Balancer chromosome, 374, 376, 377, 379,
    381, 504, 505
Behavioural ecology, 642, 646, 658
bent, 91
bindin, 580
Biogeographic theory, 671
Biological species concept, 515, 527, 588
Biston, 11, 205
Blood group, 52, 258, 614
Bombina bombina, 518
Bombina variegata, 518
boss, 89
Bottleneck, 136, 185, 208, 290, 307, 311,
    355, 360, 525, 582
Brassica, 301, 302

Caenorhabditis elegans, 447
CaMKII, 91
Canalization, 9, 521, 522
Candidate gene, 385, 388
Catastrophic selection theory, 573
Cepea, 205
Character displacement, 585
Charge state model, 36, 61, 62, 65, 66, 104
Chlamydomonas reinhardtii, 580
Chromosome
  arrangement, 551
  banding, 613
  breakage, 291
  heterospecific, 576
  homologous, 120
  rearrangements, 552
  sex, 295, 343, 577, 584
  wild-type, 45, 374, 379
  X, 91, 132, 135, 136, 288, 384, 444, 446,
    504, 556, 557, 576, 577
  Y, 326, 444, 576, 611
ci^D, 91, 92, 135
Cladogenesis, 535
Coadaptation, 36, 44, 46, 47, 55, 258, 281,
    292, 499, 519, 520, 521, 525, 527
Coadapted, 46, 292, 296, 587, 588
Coalescent, 16, 32, 38, 39, 105, 111, 126,
    127, 131, 132, 137, 303–305,
    418–427, 616–619
Coalescent model, 105, 126, 482, 486, 534

Coalescent theory, 38, 105, 418, 419, 426
Codon bias, 17, 82, 83, 86–98, 580
Coevolution, 96, 281, 322, 440, 441, 535,
    557, 584, 587, 636, 650, 655, 656,
    666–668, 671, 673
Colias butterflies, 11, 58
Compensatory evolution, 240
Covariance, 8, 186, 339, 344, 345, 346, 383
Crassostrea virginica, 193

Daphnia pulex, 463
Darwinian fitness. See Fitness
Deficiency, 16, 129, 208, 214, 215, 224,
    443
Deleterious mutation model, 323
Deletion, 284
  analysis, 61
  mapping, 354, 361
Desmognathus, 518
Differential equations, 32, 404
Differential mortality, 11, 58
Diploidy, 320
Dispersal, 310, 422, 424, 425, 444, 466,
    586, 656, 659, 662, 665, 669, 672
Disruptive selection. See Selection
Distance matrix, 614
Diversity
  clonal, 456, 457, 460, 463, 464, 466
  genetic. See Genetic
  haplotypic, 18, 104, 111, 114, 121
  intra-genic, 114, 116, 120
  nucleotide, 86, 88, 126, 127, 476, 481,
    485, 581
DNA
  diversity, 102
  exogenous, 318
  extranuclear, 435
  haplotype, 18
  heterologous, 319
  homologous, 476
  mitochondrial, 422, 423, 430–433, 435,
    436, 438, 440–449, 503
  noncoding, 95, 444
  nuclear, 105, 193, 440, 441, 500, 619
  polymorphism, 102, 104, 111, 136, 190,
    193, 199, 288, 370
  recombinant, 60
  recombination, 90, 110, 318
  repair, 317–320, 590
  repetitive, 447
  replication, 238, 319, 320
  strand bias, 97
  transformation, 477
DNA repair theory, 319

Printed in the United States
By Bookmasters